Systems and Control

THE OXFORD SERIES IN ELECTRICAL AND COMPUTER ENGINEERING

ADEL S. SEDRA, Series Editor

Allen and Holberg, *CMOS Analog Circuit Design, 2nd Edition*
Bobrow, *Elementary Linear Circuit Analysis, 2nd Edition*
Bobrow, *Fundamentals of Electrical Engineering, 2nd Edition*
Burns and Roberts, *An Introduction to Mixed-Signal IC Test and Measurement*
Campbell, *The Science and Engineering of Microelectronic Fabrication, 2nd Edition*
Chen, *Analog & Digital Control System Design*
Chen, *Linear System Theory and Design, 3rd Edition*
Chen, *System and Signal Analysis, 2nd Edition*
Chen, *Digital Signal Processing*
Comer, *Digital Logic and State Machine Design, 3rd Edition*
Cooper and McGillem, *Probabilistic Methods of Signal and System Analysis, 3rd Edition*
DeCarlo and Lin, *Linear Circuit Analysis, 2nd Edition*
Dimitrijev, *Understanding Semiconductor Devices*
Franco, *Electric Circuits Fundamentals*
Granzow, *Digital Transmission Lines*
Guru and Hiziroğlu, *Electric Machinery and Transformers, 3rd Edition*
Hoole and Hoole, *A Modern Short Course in Engineering Electromagnetics*
Jones, *Introduction to Optical Fiber Communication Systems*
Krein, *Elements of Power Electronics*
Kuo, *Digital Control Systems, 3rd Edition*
Lathi, *Modern Digital and Analog Communications Systems, 3rd Edition*
Lathi, *Signal Processing and Linear Systems*
Lathi, *Linear Systems and Signals*
Martin, *Digital Integrated Circuit Design*
McGillem and Cooper, *Continuous and Discrete Signal and System Analysis, 3rd Edition*
Miner, *Lines and Electromagnetic Fields for Engineers*
Parhami, *Computer Arithmetic*
Roberts and Sedra, *SPICE, 2nd Edition*
Roulston, *An Introduction to the Physics of Semiconductor Devices*
Sadiku, *Elements of Electromagnetics, 3rd Edition*
Sarma, *Introduction to Electrical Engineering*
Schaumann and Van Valkenburg, *Design of Analog Filters*
Schwarz, *Electromagnetics for Engineers*
Schwarz and Oldham, *Electrical Engineering: An Introduction, 2nd Edition*
Sedra and Smith, *Microelectronic Circuits, 4th Edition*
Stefani, Savant, Shahian, and Hostetter, *Design of Feedback Control Systems, 4th Edition*
Van Valkenburg, *Analog Filter Design*
Warner and Grung, *Semiconductor Device Electronics*
Warner and Grung, *MOSFET Theory and Design*
Wolovich, *Automatic Control Systems*
Yariv, *Optical Electronics in Modern Communications, 5th Edition*
Żak, *Systems and Control*

Systems and Control

Stanislaw H. Żak

School of Electrical and Computer Engineering
Purdue University

New York Oxford
OXFORD UNIVERSITY PRESS
2003

Oxford University Press

Oxford New York
Auckland Bangkok Buenos Aires Cape Town Chennai
Dar es Salaam Delhi Hong Kong Istanbul Karachi Kolkata
Kuala Lumpur Madrid Melbourne Mexico City Mumbai
Nairobi São Paulo Shanghai Taipei Tokyo Toronto

Published by Oxford University Press, Inc.
198 Madison Avenue, New York, New York, 10016
http://www.oup-usa.org

Library of Congress Cataloging-in-Publication Data

Żak, Stanislaw H.
Systems and Control / Stanislaw H. Żak.
p. cm.
Includes bibliographical references and index.
ISBN 0-19-515011-2
1. Linear control systems. I. Title.
TJ220 .Z35 2002
629.8'32—dc21 2002066792

Printing number: 9 8 7 6 5 4 3 2 1

Printed in the United States of America
on acid-free paper

To

J. M. J.

CONTENTS

PREFACE

This book is about modeling, analysis, and control of dynamical systems. Its objective is to familiarize the reader with the basics of dynamical system theory while, at the same time, equipping him or her with the tools necessary for control system design. The emphasis is on the design in order to show how dynamical system theory fits into practical applications. Various methods of a controller design are discussed.

This text can be used in a variety of ways. It is suitable for a senior- or graduate-level course on systems and control. The text can also be used in an introductory course on nonlinear systems. Finally, it can be used in an introductory course to modern automatic control. No special background is necessary to read this text beyond basic differential equations and elements of linear algebra. The book is also suitable for self-study. To accomplish this goal, the required mathematical techniques and terminology are reviewed in the Appendix. Depending on the background, the reader may refer to the material in the Appendix as needed.

Whenever one wishes to analyze a problem arising in the "real" world, the first order of business is the construction of a mathematical model of the problem. In this book we discuss mathematical modeling of problems from mechanical and electrical engineering, as well as from physics and biology. Constructing models of dynamical systems using a linguistic description that employs IF–THEN rules is also considered. Two main types of dynamical systems are common in applications: those for which the time variable is discrete and those for which the time variable is continuous. When the time variable is discrete, the dynamics of such systems are often modeled using difference equations. In the case when the time is continuous, ordinary differential equations are frequently chosen for modeling purposes. Both types of models are considered in the book. Nonlinear systems are emphasized to acknowledge the critical role that nonlinear phenomena play in science and technology. The models presented are the ones that can be used to design controllers. These models are constructed from the control engineering point of view. Most of the time, engineers develop models of man-made objects. This is in contrast with physicists, who are more interested in modeling nature. Models in the last chapter of this book, however, are of interest to engineers, physicists, or biologists. This chapter is an invitation to control engineering of the future where control designers will be challenged to tackle highly complex, nonlinear, dynamical systems using multidisciplinary tools.

As can be concluded from the above remarks, the scope of the book is quite broad. This is in order to show the multidisciplinary role of nonlinear dynamics and control. In particular, neural networks, fuzzy systems, and genetic algorithms are presented and a self-contained introduction to chaotic systems is provided. Lyapunov's stability theory and its extensions are used in the analysis of dynamical system models. The objective of the stability analysis is to determine the system behavior without solving the differential, or difference, equations modeling the system. In fact, the Lyapunov theory is used as a unifying medium for different types of dynamical systems analyzed. In particular, optimal, fuzzy, sliding mode, and chaotic controllers are all constructed with the aid of the Lyapunov method and its variants. Furthermore, a class of neural networks is analyzed using Lyapunov's method. Classical methods and the techniques of post-modern control engineering, such as fuzzy logic, neural networks, and genetic algorithms, are presented in a unified fashion. It is shown how these new tools of a control engineer supplement the classical ones.

In addition to worked examples, this book also contains exercises at the end of each chapter. A number of solved problems, as well as the exercises, require the use of a computer. Although there are now many quality software packages available that could be used with this book, MATLAB* was chosen because it appears to be the most widely used math-tools program in engineering circles, and it is also popular among science students. Student editions of MATLAB and SIMULINK are sufficient for the worked examples and exercises in this book. The solutions manual, with complete solutions to all the exercises in the book, is available from the publisher to the instructors who adopt this book for their courses.

There is a companion web site for this book, where the current errata are available as well as the MATLAB m-files of examples solved in the text. The URL of the site is http://dynamo.ecn.purdue.edu/~zak/systems/Index.html.

Acknowledgments

I am indebted to my wife, Mary Ann, and to my children, Michaela, George, Elliot, and Nick, for their support during the writing of this book.

I express my gratitude to numerous students on whom this text was tested. I have benefited tremendously while working with Professor Stefen Hui. His insights and discussions helped clarify many concepts discussed in this book. Dr. Yonggon Lee contributed many simulations included in the book and in the solutions manual. It was a pleasant experience working with Peter Gordon, Justin Collins, and the staff of Oxford University Press during the production of the book. Finally, I wish to thank the National Science Foundation for the much appreciated support during the last stages of this project.

It is most unfortunate that I could not find a co-author, because now I must bear the final responsibility for all the errors and omissions. So dear Reader if you find an error, no matter how minor, please let me know about it. Errors found by the author or by readers are corrected every printing.

Stanislaw H. Żak

---

*MATLAB is a registered trademark of the MathWorks, Inc., 24 Prime Park Way, Natick, MA 01760-1500. Phone: 508-647-7000, fax: 508-647-7001, e-mail: info@mathworks.com, http://www.mathworks.com.

Dynamical Systems and Modeling

To catch a mouse alive, try this trick: When you see the beast running around a room, open a black umbrella and hold it on the floor at an angle that allows the mouse to run in (it helps if someone else can chase the mouse). Your prey, drawn to dark places, will probably head straight for the umbrella. Then quickly close it, go outside, and set your prisoner free.

—*Reader's Digest Practical Problem Solver* [241]

1.1 What Is a System?

A *system* is a collection of interacting components. An electric motor, an airplane, and a biological unit such as the human arm are examples of systems. A system is characterized by two properties, which are as follows:

1. The interrelations between the components that are contained within the system
2. The system boundaries that separate the components within the system from the components outside

The system boundaries can be real or imagined. They are elastic in the sense that we may choose, at any stage of the system analysis, to consider only a part of the original system as a system on its own. We call it a *subsystem* of the original system. On the other hand, we may decide to expand the boundaries of the original system to include new components. In Figure 1.1, we represent a system's boundaries using a box. The interactions between the system components may be governed, for example, by physical, biological, or economical laws. In dealing with systems, we are interested in the effects of external quantities upon the behavior of the system quantities. We refer to the external quantities acting on the system as the *inputs* to the system. The condition or the state of the system is described by the *state variables*. The state variables provide the information that, together with the knowledge of the system inputs, enables us to determine the future state of the system. "A *dynamical system* consists of a set of possible states, together with a rule that determines the present state in terms of past states" [7, p. 1]. An axiomatic description of a dynamical system is presented in Section 1.3. In practice it is often not possible or too expensive to measure or determine the values of all of the state variables. Instead, only their subset or combination can be measured. The system quantities whose behavior can be measured or observed are referred to as the system *outputs*.

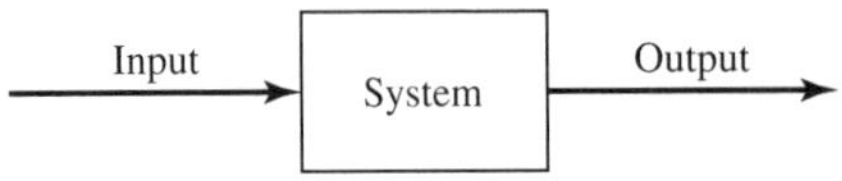

Figure 1.1 Representation of a system.

In engineering applications, when dealing with dynamical systems, we are interested in specifying the system inputs that force the system states or outputs to behave with time in some prespecified manner. That is, we are interested in *controlling* the system states or outputs. This is accomplished by means of a *controller* whose task is to produce the required system's inputs that in turn result in the desired system's outputs. An interconnection of the system and a controller is called a *control system*. In Figures 1.2 and 1.3, we show two different types of control systems that we discuss in the next section. Constructing a controller is a part of the *control problem*. The essential elements of the control problem, as described by Owens [218, p. 181] are as follows:

1. A specified objective for the system
2. A model of a dynamical system to be controlled
3. A set of admissible controllers
4. A means of measuring the performance of any given control strategy to evaluate its effectiveness

We now examine these elements one by one. The objective of a control system is to complete some specified task. This can be expressed as a combination of (a) constraints on the output or state variables and (b) limits on the time available to complete the control objective. For example, the objective of a controller might be to force the system output to settle within a certain percentage of the given value after prespecified time.

We now discuss the modeling part of the control problem. The first step in the controller design procedure is the construction of a *truth model* of the dynamics of the process to be controlled. The truth model is a *simulation model* that includes all the relevant characteristics of the process. The truth model is too complicated for use in the controller design. Thus, we need to develop a simplified model that can be used to design a controller. Such a simplified model is labeled by Friedland [91] as the *design model*. The design model should capture the essential features of the process. A common model of a dynamical system is the finite set of ordinary differential equations of the form

$$
\begin{aligned}
\dot{\boldsymbol{x}}(t) &= \boldsymbol{f}(t, \boldsymbol{x}(t), \boldsymbol{u}(t)), \qquad \boldsymbol{x}(t_0) = \boldsymbol{x}_0,\\
\boldsymbol{y}(t) &= \boldsymbol{h}(t, \boldsymbol{x}(t), \boldsymbol{u}(t)),
\end{aligned}
$$

where the state $\boldsymbol{x} \in \mathbb{R}^n$, the input $\boldsymbol{u} \in \mathbb{R}^m$, the output $\boldsymbol{y} \in \mathbb{R}^p$, and $\boldsymbol{f}$ and $\boldsymbol{h}$ are vector-valued functions with $\boldsymbol{f} : \mathbb{R} \times \mathbb{R}^n \times \mathbb{R}^m \to \mathbb{R}^n$ and $\boldsymbol{h} : \mathbb{R} \times \mathbb{R}^n \times \mathbb{R}^m \to \mathbb{R}^p$. Another common model of a dynamical system is the finite set of difference equations,

$$
\begin{aligned}
\boldsymbol{x}(k+1) &= \boldsymbol{f}(k, \boldsymbol{x}(k), \boldsymbol{u}(k)), \qquad \boldsymbol{x}(k_0) = \boldsymbol{x}_0,\\
\boldsymbol{y}(k) &= \boldsymbol{h}(k, \boldsymbol{x}(k), \boldsymbol{u}(k)),
\end{aligned}
$$

where $\boldsymbol{x}(k) = \boldsymbol{x}(kh)$, $\boldsymbol{u}(k) = \boldsymbol{u}(kh)$, h is the sampling interval, and $k \geq 0$ is an integer.

In our subsequent discussion, we regard the dynamical system and its model as equivalent. During the modeling process, properties of the system, physical constraints, and technical requirements should be taken into account. By itself the pair $(\boldsymbol{f}(t, \boldsymbol{x}, \boldsymbol{u}), \boldsymbol{h}(t, \boldsymbol{x}, \boldsymbol{u}))$ has little engineering significance until constraints on the state $\boldsymbol{x}$, the control $\boldsymbol{u}$, and the output $\boldsymbol{y}$ are also given. Control input signals are obtained from physical devices capable of providing only a limited amount of energy. We refer to the class of controllers that can be considered for the given control design problem as the set of admissible controllers.

We can evaluate the performance of any given control law by visual inspection of the transient characteristics of the controlled system after the design is complete. But this method of assessment of the effectiveness of the given control strategy is inaccurate. We therefore try to quantify the means of evaluating performance in the form of a performance index or cost functional at the beginning of the design process. We construct a numerical index of performance whose value reflects the quality of any admissible controller in accomplishing the system objective. This performance index is assumed to have the property that its numerical value decreases as the quality of the controller increases. This is the reason why we sometimes refer to the performance index as the penalty functional. The "best" controller then is the one that produces the smallest value of the performance index. We call an admissible controller that simultaneously ensures the completion of the system objective and the minimization of the performance index an *optimal controller*.

1.2 Open-Loop Versus Closed-Loop

We distinguish between two types of control systems. They are:

- Open-loop control systems
- Closed-loop control systems

An open-loop control system usually contains the following:

1. A process to be controlled, labeled *plant*
2. The controlling variable of the plant, called the *plant input*, or just *input* for short
3. The controlled variable of the plant, called the *plant output*, or just *output* for short
4. A *reference input*, which dictates the desired value of the output
5. A *controller* that acts upon the reference input in order to form the system input forcing the output behavior in accordance with the reference signal

Note that the plant and controller themselves can be considered as systems on their own. A schematic representation of an open-loop system is depicted in Figure 1.2. In an open-loop control system the output has no influence on the input or reference signal. The controller operates without taking into account the output. Thus, the plant input is formed with no influence of the output. A household appliance such as an iron is a simple example of an open-loop control system. In this example, we consider the iron itself as the plant or process to be controlled. Here we set the temperature control mechanism for the desired temperature. This setting—for example, very hot—is the reference input. We can view the potentiometer, inside the iron, as a controller, or a control actuator. It regulates the temperature of the iron by allowing the necessary amount of resistance to produce the desired temperature. The plant's output is the temperature of the heated iron. Not every control system is an open-loop system. Another type of a control system is a closed-loop system. We can convert an open-loop system into a closed-loop system by adding the following components:

6. The *feedback loop* where the output signal is measured with a sensor and then the measured signal is fed back to the *summing junction*
7. The summing junction, where the measured output signal is subtracted from the reference (command) input signal in order to generate an *error signal,* also labeled as an *actuating signal*

A schematic of a closed-loop control system is shown in Figure 1.3. In a closed-loop system the error signal causes an appropriate action of the controller, which in turn "instructs" the plant to behave in a certain way in order to approach the desired output, as specified by the reference input signal. Thus, in the closed-loop system, the plant output information is fed back to the controller, and the controller then appropriately modifies the plant output behavior. A controller, also called a *compensator,* can be placed either in the *forward loop,* as in Figure 1.3, or in the *feedback loop.*

If the goal of the controller is to maintain the output at a constant value, then we have a *regulator* control system. A "centrifugal speed regulator," or the "flyball governor," commonly known as Watt's governor, is an example of a regulator in a closed-loop system. A schematic of Watt's governor is shown in Figure 1.4. The task of the centrifugal speed regulator is to control the setting of a throttle valve to maintain the desired engine speed. The nominal speed is set by adjusting the valve in the throttle. As the engine speeds up, the weights are thrown outwards, the throttle partially closes, and the engine slows down. As the engine slows down below its nominal speed, the throttle is opened up and the engine gains its speed. The operation of an engine equipped with a centrifugal speed regulator can be represented using the block diagram of Figure 1.3. The reference input to the closed-loop system is the desired engine speed. The desired engine speed is obtained by appropriately setting the valve. The system output is the actual engine speed. The flyball governor is our controller, while the engine represents the plant. The control signal generated by the controller is the setting of the throttle valve.

We mention at this juncture that Watt's application of the centrifugal speed governor in 1788, to regulate the speed of the steam engine, marks the starting point for the development of automatic control—and, in particular, feedback control—as a science [192]. Before 1788,

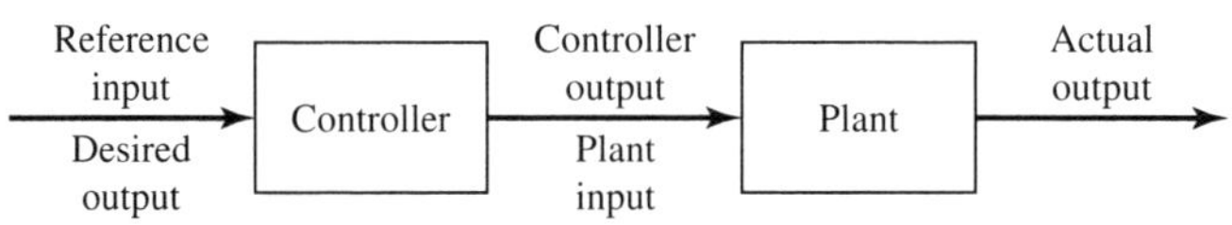

Figure 1.2 Open-loop control system.

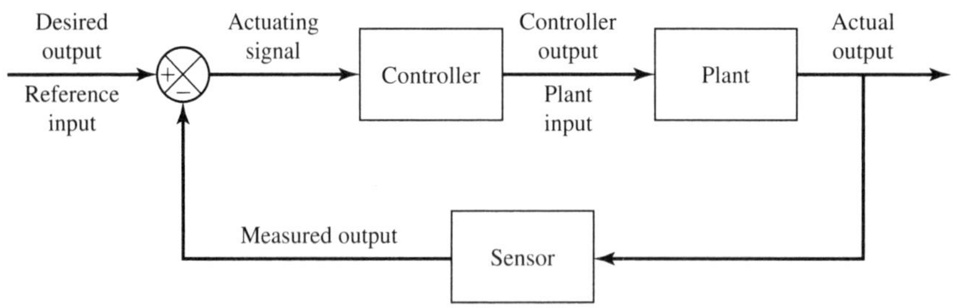

Figure 1.3 Closed-loop control system.

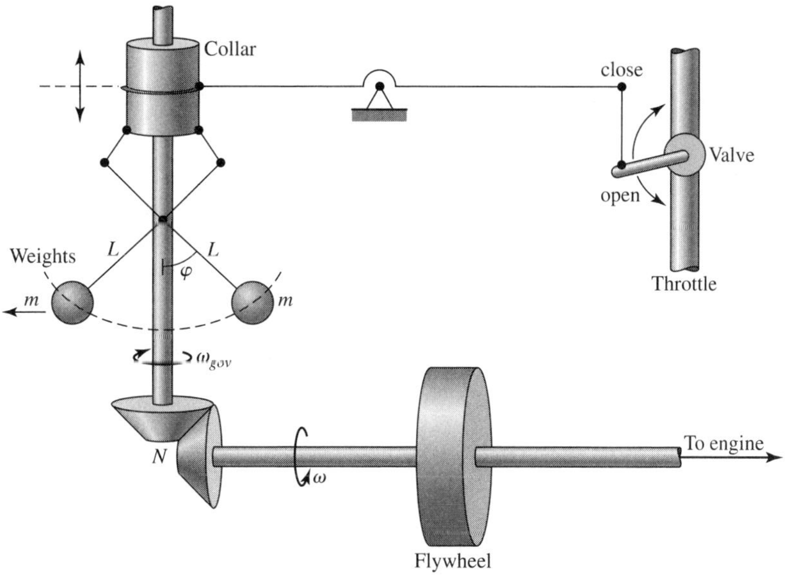

Figure 1.4 A simplified schematic of the flyball governor.

steam engines were controlled manually using the throttle valve. The flyball governor spread widely and quickly. It has been estimated that by 1868 there were about 75,000 flyball governors working in England alone [192]. The flyball governor is credited for the hatching of the industrial revolution. We devise a mathematical model of this revolutionary invention in Subsection 1.7.1, and in Section 4.7 we perform a quantitative analysis of its dynamical behavior.

If the goal of the controller is to force the output to follow a desired trajectory, then we have a *servomechanism* or *tracking* control system. As we mentioned before, to design a controller, we need first to analyze the plant quantitatively. The analysis requires a mathematical, or linguistic, description of the interrelations between the system quantities themselves as well as the interrelations between system quantities and system inputs. The mathematical description of these interrelations is called *mathematical modeling*. We devote Section 1.4 to mathematical modeling. Constructing models using linguistic description of the process to be controlled will be discussed in Subsection 8.5.1.

1.3 Axiomatic Definition of a Dynamical System

Following Kalman [147], we can define a dynamical system formally using the following axioms:

1. There is given a state space $\mathcal{X}$; there is also T, an interval in the real line representing time.
2. There is given a space $\mathcal{U}$ of functions on T that represent the inputs to the system.
3. For any initial time t_0 in T, any initial state $\boldsymbol{x}_0$ in $\mathcal{X}$, and any input $\boldsymbol{u}$ in $\mathcal{U}$ defined for $t \geq t_0$, the future states of the system are determined by the *transition mapping*

$$\boldsymbol{\Phi} : T \times \mathcal{X} \times \mathcal{U} \rightarrow \mathcal{X}$$

written as

$$\boldsymbol{\Phi}(t_1; t_0, \boldsymbol{x}(t_0), \boldsymbol{u}(t)) = \boldsymbol{x}(t_1).$$

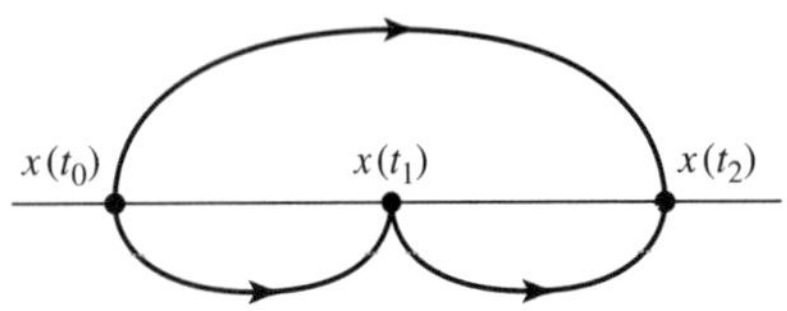

Figure 1.5 Illustration of the semigroup axiom.

Moreover, the following axioms are satisfied:

4. The identity property of the transition mapping, that is,
$$\boldsymbol{\Phi}(t_0; t_0, \boldsymbol{x}(t_0), \boldsymbol{u}(t_0)) = \boldsymbol{x}(t_0).$$
5. The semigroup property of the transition mapping, that is,
$$\boldsymbol{\Phi}(t_2; t_0, \boldsymbol{x}(t_0), \boldsymbol{u}(t)) = \boldsymbol{\Phi}(t_2; t_1, \boldsymbol{\Phi}(t_1; t_0, \boldsymbol{x}(t_0), \boldsymbol{u}(t)), \boldsymbol{u}(t)).$$
The semigroup property axiom states that it is irrelevant whether the system arrives at the state at time t_2 by a direct transition from the state at time t_0, or by first going to an intermediate state at time t_1, and then having been restarted from the state at time t_1 and moving to the state at time t_2. In either case the system, satisfying the semigroup axiom, will arrive at the same state at time t_2. This property is illustrated in Figure 1.5.
6. The causality property, that is,
$$\boldsymbol{\Phi}(t; t_0, \boldsymbol{x}(t_0), \boldsymbol{u}_1(t)) = \boldsymbol{\Phi}(t; t_0, \boldsymbol{x}(t_0), \boldsymbol{u}_2(t)) \qquad \text{for } t_0, t \in T$$
$$\text{if and only if } \boldsymbol{u}_1(t) = \boldsymbol{u}_2(t) \text{ for all } t \in T.$$
7. Every output of the system is a function of the form
$$\boldsymbol{h} : T \times \mathcal{X} \times \mathcal{U} \to \mathcal{Y},$$
where $\mathcal{Y}$ is the output space.
8. The transition mapping $\boldsymbol{\Phi}$ and the output mapping $\boldsymbol{h}$ are continuous functions.
9. If in addition the system satisfies the following axiom, then it is said to be *time-invariant:*
$$\boldsymbol{\Phi}(t_1; t_0, \boldsymbol{x}(t_0), \boldsymbol{u}(t)) = \boldsymbol{\Phi}(t_1 + \tau; t_0 + \tau, \boldsymbol{x}(t_0), \boldsymbol{u}(t)),$$
where $t_0, t_1 \in T$.

Thus, a dynamical system can be defined formally as a quintuple
$$\{T, \mathcal{X}, \mathcal{U}, \boldsymbol{\Phi}, \mathcal{Y}\}$$
satisfying the above axioms.

Our attention will be focused on dynamical systems modeled by a set of ordinary differential equations
$$\dot{x}_i = f_i(t, x_1, x_2, \ldots, x_n, u_1, u_2, \ldots, u_m), \qquad x_i(t_0) = x_{i0}, \qquad i = 1, 2, \ldots, n,$$
together with p functions
$$y_j = h_j(t, x_1, x_2, \ldots, x_n, u_1, u_2, \ldots, u_m), \qquad j = 1, 2, \ldots, p.$$
The system model state is $\boldsymbol{x} = [x_1\ x_2\ \cdots\ x_n]^T \in \mathbb{R}^n$. The system input is $\boldsymbol{u} = [u_1\ u_2\ \cdots\ u_m]^T \in \mathbb{R}^m$, and the system output is $\boldsymbol{y} = [y_1\ y_2\ \cdots\ y_p]^T \in \mathbb{R}^p$. In vector notation the above system

model has the form

$$\dot{\boldsymbol{x}} = \boldsymbol{f}(t, \boldsymbol{x}, \boldsymbol{u}), \qquad \boldsymbol{x}(t_0) = \boldsymbol{x}_0,$$
$$\boldsymbol{y} = \boldsymbol{h}(t, \boldsymbol{x}, \boldsymbol{u}),$$

where $\boldsymbol{f} : \mathbb{R} \times \mathbb{R}^n \times \mathbb{R}^m \to \mathbb{R}^n$ and $\boldsymbol{h} : \mathbb{R} \times \mathbb{R}^n \times \mathbb{R}^m \to \mathbb{R}^p$ are vector-valued functions. In further considerations, we regard a dynamical system under considerations and its model represented by the above equations as equivalent. We now illustrate dynamical system axioms on a simple example.

◆ Example 1.1

We consider a dynamical system with no external inputs modeled by the first-order linear differential equation

$$\dot{x}(t) = ax(t), \qquad x_0 = x(t_0), \tag{1.1}$$

where a is a constant, $x \in \mathcal{X} = \mathbb{R}$, and $T = \mathbb{R}$. The solution to the above differential equation has the form

$$x(t) = e^{a(t-t_0)} x(t_0).$$

The above system model can be characterized by the triple

$$\{T, \mathcal{X}, \Phi\},$$

where

$$\Phi : T \times \mathcal{X} \to \mathcal{X}$$

is the transition mapping. In this example,

$$\Phi(t; t_0, x(t_0)) = e^{a(t-t_0)} x(t_0).$$

It is clear that Φ is a continuous function. The transition function $\Phi = e^{a(t-t_0)} x(t_0)$ satisfies the semigroup property. Indeed,

$$\begin{aligned}\Phi(t_2; t_1, \Phi(t_1; t_0, x(t_0))) &= e^{a(t_2-t_1)} \left(e^{a(t_1-t_0)} x(t_0)\right) \\ &= e^{a(t_2-t_0)} x(t_0) \\ &= \Phi(t_2; t_0, x(t_0)).\end{aligned}$$

The identity property axiom is also satisfied:

$$\begin{aligned}\Phi(t_0; t_0, x(t_0)) &= e^{a(t_0-t_0)} x(t_0) \\ &= x(t_0).\end{aligned}$$

Finally, observe that the system model given by (1.1) is time invariant because

$$\begin{aligned}\Phi(t_1; t_0, x(t_0)) &= e^{a(t_1-t_0)} x(t_0) \\ &= e^{a(t_1+\tau-(t_0+\tau))} x(t_0) \\ &= \Phi(t_1 + \tau; t_0 + \tau, x(t_0)).\end{aligned}$$

1.4 Mathematical Modeling

The goal of mathematical modeling is to provide a mathematical description of the interrelations between the system quantities as well as the relations between these quantities and external inputs. A mathematical model can be viewed as "a mathematical representation of the significant, relevant aspects of an existing or proposed physical system. (Significance and relevance being in relation to an application where the model is to be used)" [181, p. xi]. Another definition of mathematical model that we found in reference 76, p. 1, is: "A mathematical model is a system of equations that purports to represent some phenomenon in a way that gives insight into the origins and consequences of its behavior." (For the controller design purposes, we first construct a truth model of the given plant and then its design model.) In general, the more accurate the mathematical model is, the more complex the modeling equations are. This means that during the mathematical modeling process we may be forced to find a middle course between accuracy of the model and its complexity. Our goal is to generate the most accurate possible description of the given system while still being able to carry out the model analysis and computations in a reasonable amount of time. At first, we tend to make oversimplifying assumptions in order to obtain a very simple model. However, we should follow the principle of maximum simplicity, which states that "everything should be as simple as possible—but no simpler." After extracting some information about the system behavior based on this first, possibly crude model, it may be desirable to construct a more accurate model. We try to create a mathematical description of a given physical process that incorporates realistic assumptions and physical constraints. We then extract information from this model and compare it with the experimental data. We perform this comparative study to determine the value of the obtained model. If the model appears to be inadequate for our needs, then it must be adjusted. The process of mathematical modeling is depicted in Figure 1.6. In the process of mathematical modeling, one can use different types of equations and their combinations like algebraic, ordinary and partial differential equations,

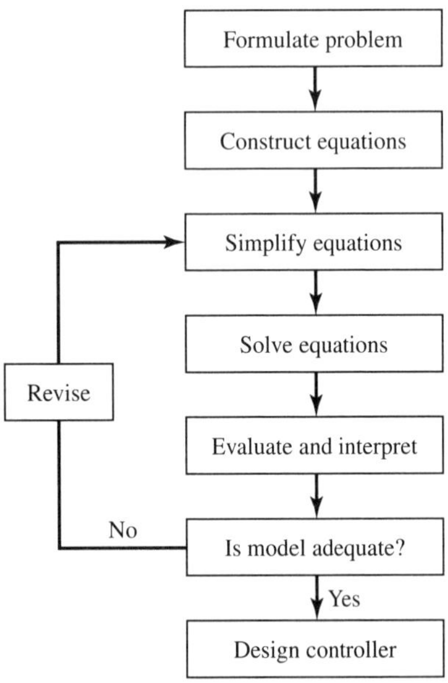

Figure 1.6 Diagram of mathematical modeling process and its use in the controller design process.

difference, integral, or functional equations. In this book, only difference and ordinary differential equations are used to model dynamical systems.

In the following, we discuss the Lagrange equations of motion that are useful in mathematical modeling of dynamical systems. We first review relevant concepts that are needed to derive the Lagrange equations. We then construct mathematical models of a few dynamical systems.

1.5 Review of Work and Energy Concepts

The purpose of this section is to review and reacquaint ourselves with some basic notions from physics: work and energy. After a brief review of work and energy concepts, we derive the Lagrange equations of motion and use them to devise mathematical models of dynamical systems.

Suppose we are given a particle of a constant mass m subjected to a force $\boldsymbol{F}$. Then, by Newton's second law we have

$$\boldsymbol{F} = m\frac{d\boldsymbol{v}}{dt} = m\boldsymbol{a}. \tag{1.2}$$

Here the force $\boldsymbol{F}$ and the velocity $\boldsymbol{v}$ are vectors. Therefore, they can be represented as

$$\boldsymbol{F} = \begin{bmatrix} F_{x_1} & F_{x_2} & F_{x_3} \end{bmatrix}^T \quad \text{and} \quad \boldsymbol{v} = \begin{bmatrix} v_{x_1} & v_{x_2} & v_{x_3} \end{bmatrix}^T = [\dot{x}_1 \quad \dot{x}_2 \quad \dot{x}_3]^T.$$

Using the above notation, we represent equation (1.2) as

$$\begin{bmatrix} F_{x_1} \\ F_{x_2} \\ F_{x_3} \end{bmatrix} = m \begin{bmatrix} \dfrac{dv_{x_1}}{dt} \\ \dfrac{dv_{x_2}}{dt} \\ \dfrac{dv_{x_3}}{dt} \end{bmatrix}. \tag{1.3}$$

Suppose now that a force $\boldsymbol{F}$ is acting on a particle located at a point A and the particle moves to a point B. The work δW done by $\boldsymbol{F}$ along infinitesimally small distance $\delta \boldsymbol{s}$ is

$$\begin{aligned} \delta W &= \boldsymbol{F}^T \delta \boldsymbol{s} \\ &= \begin{bmatrix} F_{x_1} & F_{x_2} & F_{x_3} \end{bmatrix} \begin{bmatrix} \delta x_1 \\ \delta x_2 \\ \delta x_3 \end{bmatrix} \\ &= F_{x_1}\,\delta x_1 + F_{x_2}\,\delta x_2 + F_{x_3}\,\delta x_3. \end{aligned}$$

where $\delta \boldsymbol{s} = [\delta x_1 \; \delta x_2 \; \delta x_3]^T$. The work W_{AB} done on the path from A to B is obtained by integrating the above equation:

$$W_{AB} = \int_A^B \left(F_{x_1}\,dx_1 + F_{x_2}\,dx_2 + F_{x_3}\,dx_3\right).$$

We now would like to establish a relation between work and kinetic energy. For this observe that

$$\ddot{x}\,dx = \frac{d\dot{x}}{dt}\,dx = \dot{x}\,d\dot{x}.$$

Using the above relation and Newton's second law, we express W_{AB} in a different way as

$$\begin{aligned} W_{AB} &= \int_A^B (m\ddot{x}_1\,dx_1 + m\ddot{x}_2\,dx_2 + m\ddot{x}_3\,dx_3) \\ &= \int_A^B m(\dot{x}_1\,d\dot{x}_1 + \dot{x}_2\,d\dot{x}_2 + \dot{x}_3\,d\dot{x}_3) \\ &= m\left(\frac{\dot{x}_1^2}{2} + \frac{\dot{x}_2^2}{2} + \frac{\dot{x}_3^2}{2}\right)\Bigg|_A^B \\ &= m\left(\frac{\|\boldsymbol{v}_B\|^2}{2} - \frac{\|\boldsymbol{v}_A\|^2}{2}\right) \\ &= \frac{m\|\boldsymbol{v}_B\|^2}{2} - \frac{m\|\boldsymbol{v}_A\|^2}{2}, \end{aligned} \tag{1.4}$$

where $\boldsymbol{v}_A = [\dot{x}_{1A}\ \ \dot{x}_{2A}\ \ \dot{x}_{3A}]^T$, $\boldsymbol{v}_B$ is defined in an analogous manner, and $\|\boldsymbol{v}\| = \boldsymbol{v}^T\boldsymbol{v}$. Observing that $m\|\boldsymbol{v}_B\|^2/2$ is the kinetic energy of the particle at the point B and $m\|\boldsymbol{v}_A\|^2/2$ is its kinetic energy at the point A, we obtain

$$W_{AB} = K_B - K_A.$$

The above relation can be used to define the kinetic energy of a particle as the work required to change its velocity from some value $\boldsymbol{v}_A$ to a final value $\boldsymbol{v}_B$. The relation $W_{AB} = K_B - K_A = \Delta K$ is also known as the *work–energy theorem for a particle*.

A force is conservative if the work done by the force on a particle that moves through any round trip is zero. In other words, a force is conservative if the work done by it on a particle that moves between two points depends only on these points and not on the path followed.

We now review the notion of *potential energy*, or the *energy of configuration*. Recall that if the kinetic energy K of a particle changes by ΔK, then the potential energy U must change by an equal but opposite amount so that the sum of the two changes is zero; that is,

$$\Delta K + \Delta U = 0. \tag{1.5}$$

This is equivalent to saying that any change in the kinetic energy of the particle is compensated for by an equal but opposite change in the potential energy U of the particle so that their sum remains constant; that is,

$$K + U = \text{constant}.$$

See Exercise 1.1 for a rigorous proof of the above relation. The potential energy of a particle represents a form of stored energy that can be recovered and converted into the kinetic energy. If we now use the work–energy theorem and equation (1.5), then we obtain

$$W = \Delta K = -\Delta U.$$

The work done by a conservative force depends only on the starting and the end points of motion and not on the path followed between them. Therefore, for motion in one dimension, we obtain

$$\Delta U = -W = -\int_{x_0}^{x} F(s)\,ds. \tag{1.6}$$

We thus can write

$$F(x) = -\frac{dU(x)}{dx}. \tag{1.7}$$

A generalization of equation (1.6) to motion in three dimensions yields

$$\Delta U = -\int_{x_{10}}^{x_1} F_{x_1}\, ds - \int_{x_{20}}^{x_2} F_{x_2}\, ds - \int_{x_{30}}^{x_3} F_{x_3}\, ds,$$

and a generalization of equation (1.7) to motion in three dimensions is

$$\boldsymbol{F}(\boldsymbol{x}) = -\begin{bmatrix} \frac{\partial U}{\partial x_1} & \frac{\partial U}{\partial x_2} & \frac{\partial U}{\partial x_3} \end{bmatrix}^T = -\nabla U(\boldsymbol{x}), \tag{1.8}$$

where $\boldsymbol{x} = [x_1\ x_2\ x_3]^T$.

For a formal proof of the property of a conservative vector field that the work done by it on a particle that moves between two points depends only on these points and not on the path followed, see Exercise 1.2.

We now represent Newton's equation in an equivalent format that establishes a connection with the Lagrange equations of motion, which are discussed in the following section. To proceed, note that

$$K = \frac{m\|\dot{\boldsymbol{x}}\|^2}{2} = \frac{m\dot{\boldsymbol{x}}^T\dot{\boldsymbol{x}}}{2}. \tag{1.9}$$

Hence,

$$\frac{\partial K}{\partial \dot{x}_i} = m\dot{x}_i, \qquad i = 1, 2, 3. \tag{1.10}$$

Combining (1.8) and (1.10), we represent Newton's equations as

$$\frac{d}{dt}\left(\frac{\partial K}{\partial \dot{x}_i}\right) + (\nabla U)_i = 0, \qquad i = 1, 2, 3. \tag{1.11}$$

Let

$$L = K - U. \tag{1.12}$$

The function L defined above is called the *Lagrangian function* or just the Lagrangian. Note that

$$\frac{\partial L}{\partial \dot{x}_i} = \frac{\partial K}{\partial \dot{x}_i}, \tag{1.13}$$

and

$$\frac{\partial L}{\partial x_i} = -\frac{\partial U}{\partial x_i}. \tag{1.14}$$

Substituting (1.13) and (1.14) into (1.11), we obtain

$$\frac{d}{dt}\left(\frac{\partial L}{\partial \dot{x}_i}\right) - \frac{\partial L}{\partial x_i} = 0, \qquad i = 1, 2, 3. \tag{1.15}$$

Equations (1.15) are called the Lagrange equations of motion, in Cartesian coordinates, for a single particle. They are just an equivalent representation of Newton's equations given by (1.3).

1.6 The Lagrange Equations of Motion

A single particle with no constraints is free to take up any position in the three-dimensional space. This is the reason we need three coordinates to describe its position. In the case of a system of p particles, there are $3p$ degrees of freedom. In general, if there are constraints imposed on the particles, then the number of the degrees of freedom is

$$n = 3p - \text{number of constraints}.$$

For example, if the system consists of only one particle in space and if the particle is restricted to move along a line, then $n = 3-2 = 1$. We can use many different coordinate systems to describe a location of a particle in space. We will use the letter q as a symbol for a coordinate regardless of its nature. A generalized coordinate may represent, for example, angular displacement or electric charge. This indicates that the results we are going to obtain will be of general nature and will not be restricted to mechanical systems. We refer to q as a generalized coordinate. Thus, if

$$\begin{aligned} x &= x(q_1, q_2, q_3), \\ y &= y(q_1, q_2, q_3), \\ z &= z(q_1, q_2, q_3), \end{aligned}$$

then

$$\dot{x} = \frac{\partial x}{\partial q_1}\dot{q}_1 + \frac{\partial x}{\partial q_2}\dot{q}_2 + \frac{\partial x}{\partial q_3}\dot{q}_3,$$

and so on. We assume that the functions $x, y, z \in C^2$; that is, they are twice continuously differentiable, which implies that their second partial derivatives are equal. Suppose now that a particle is confined to move on a surface. Therefore, to describe the location of the particle, we need only two generalized coordinates. The position of the particle on a surface in a three-dimensional space is then described as

$$\begin{aligned} x &= x(q_1, q_2), \\ y &= y(q_1, q_2), \\ z &= z(q_1, q_2), \end{aligned}$$

and hence

$$\begin{aligned} \delta x &= \frac{\partial x}{\partial q_1}\delta q_1 + \frac{\partial x}{\partial q_2}\delta q_2, \\ \delta y &= \frac{\partial y}{\partial q_1}\delta q_1 + \frac{\partial y}{\partial q_2}\delta q_2, \\ \delta z &= \frac{\partial z}{\partial q_1}\delta q_1 + \frac{\partial z}{\partial q_2}\delta q_2. \end{aligned} \tag{1.16}$$

We know that the work δW done by the force $\boldsymbol{F}$ is $\delta W = F_x\delta x + F_y\delta y + F_z\delta z$. On the other hand, this work is equal to the change in the kinetic energy. Therefore,

$$\boxed{m(\ddot{x}\,\delta x + \ddot{y}\,\delta y + \ddot{z}\,\delta z) = F_x\,\delta x + F_y\,\delta y + F_z\,\delta z} \tag{1.17}$$

The above equation is known as *D'Alembert's* equation.

If we have a system of particles acted upon by forces $\boldsymbol{F}_1, \ldots, \boldsymbol{F}_p$ and if each of the particles undergoes displacements $\delta \boldsymbol{s}_1, \ldots, \delta \boldsymbol{s}_p$, then the total work done by the forces is

$$\delta W = \sum_{i=1}^{p} \left(F_{x_i}\, \delta x_i + F_{y_i}\, \delta y_i + F_{z_i}\, \delta z_i \right). \tag{1.18}$$

Substituting (1.16) into D'Alembert's equation and rearranging, we obtain

$$\begin{aligned} \delta W &= m \left(\ddot{x} \frac{\partial x}{\partial q_1} + \ddot{y} \frac{\partial y}{\partial q_1} + \ddot{z} \frac{\partial z}{\partial q_1} \right) \delta q_1 + m \left(\ddot{x} \frac{\partial x}{\partial q_2} + \ddot{y} \frac{\partial y}{\partial q_2} + \ddot{z} \frac{\partial z}{\partial q_2} \right) \delta q_2 \\ &= \left(F_x \frac{\partial x}{\partial q_1} + F_y \frac{\partial y}{\partial q_1} + F_z \frac{\partial z}{\partial q_1} \right) \delta q_1 + \left(F_x \frac{\partial x}{\partial q_2} + F_y \frac{\partial y}{\partial q_2} + F_z \frac{\partial z}{\partial q_2} \right) \delta q_2. \end{aligned} \tag{1.19}$$

Let

$$F_{q_1} = F_x \frac{\partial x}{\partial q_1} + F_y \frac{\partial y}{\partial q_1} + F_z \frac{\partial z}{\partial q_1} \quad \text{and} \quad F_{q_2} = F_x \frac{\partial x}{\partial q_2} + F_y \frac{\partial y}{\partial q_2} + F_z \frac{\partial z}{\partial q_2}.$$

Then, (1.19) can be represented as

$$\delta W = F_{q_1}\, \delta q_1 + F_{q_2}\, \delta q_2.$$

To proceed further, note that

$$\frac{d}{dt} \left(\dot{x} \frac{\partial x}{\partial q_1} \right) = \ddot{x} \frac{\partial x}{\partial q_1} + \dot{x} \frac{d}{dt} \left(\frac{\partial x}{\partial q_1} \right).$$

Hence

$$\boxed{\ddot{x} \frac{\partial x}{\partial q_1} = \frac{d}{dt} \left(\dot{x} \frac{\partial x}{\partial q_1} \right) - \dot{x} \frac{d}{dt} \left(\frac{\partial x}{\partial q_1} \right)} \tag{1.20}$$

The time derivative of $x = x(q_1, q_2)$ is

$$\dot{x} = \frac{\partial x}{\partial q_1} \dot{q}_1 + \frac{\partial x}{\partial q_2} \dot{q}_2. \tag{1.21}$$

Differentiating the above partially with respect to $\dot{q}_1$ yields

$$\boxed{\frac{\partial \dot{x}}{\partial \dot{q}_1} = \frac{\partial x}{\partial q_1}} \tag{1.22}$$

Finally, we will show that

$$\boxed{\frac{d}{dt} \left(\frac{\partial x}{\partial q_1} \right) = \frac{\partial \dot{x}}{\partial q_1}} \tag{1.23}$$

Indeed, because $x = x(q_1, q_2)$ and the partial derivative $\partial x/\partial q_1$ is in general a function of q_1 and q_2, we have

$$\frac{d}{dt}\left(\frac{\partial x}{\partial q_1}\right) = \frac{\partial}{\partial q_1}\left(\frac{\partial x}{\partial q_1}\right)\dot{q}_1 + \frac{\partial}{\partial q_2}\left(\frac{\partial x}{\partial q_1}\right)\dot{q}_2. \tag{1.24}$$

Taking the partial derivative of (1.21) with respect to q_1 yields

$$\begin{aligned}\frac{\partial \dot{x}}{\partial q_1} &= \frac{\partial}{\partial q_1}\left(\frac{\partial x}{\partial q_1}\right)\dot{q}_1 + \frac{\partial}{\partial q_1}\left(\frac{\partial x}{\partial q_2}\right)\dot{q}_2 \\ &= \frac{\partial}{\partial q_1}\left(\frac{\partial x}{\partial q_1}\right)\dot{q}_1 + \frac{\partial}{\partial q_2}\left(\frac{\partial x}{\partial q_1}\right)\dot{q}_2,\end{aligned} \tag{1.25}$$

because by assumption $x, y, z \in \mathcal{C}^2$ and hence

$$\frac{\partial}{\partial q_1}\left(\frac{\partial x}{\partial q_2}\right) = \frac{\partial}{\partial q_2}\left(\frac{\partial x}{\partial q_1}\right).$$

With the above in mind, comparing the right-hand sides of (1.24) and (1.25), we obtain (1.23). We now use (1.20), (1.22), and (1.23) to arrive at the Lagrange equation of motion for a single particle. We first substitute (1.22) and (1.23) into (1.20) to get

$$\ddot{x}\frac{\partial x}{\partial q_1} = \frac{d}{dt}\left(\dot{x}\frac{\partial \dot{x}}{\partial \dot{q}_1}\right) - \dot{x}\frac{\partial \dot{x}}{\partial q_1}. \tag{1.26}$$

Note that

$$\frac{\partial}{\partial \dot{q}_1}\left(\frac{\dot{x}^2}{2}\right) = \dot{x}\frac{\partial \dot{x}}{\partial \dot{q}_1} \tag{1.27}$$

and

$$\frac{\partial}{\partial q_1}\left(\frac{\dot{x}^2}{2}\right) = \dot{x}\frac{\partial \dot{x}}{\partial q_1}. \tag{1.28}$$

Substituting (1.27) and (1.28) into (1.26) gives

$$\ddot{x}\frac{\partial x}{\partial q_1} = \frac{d}{dt}\left(\frac{\partial(\dot{x}^2/2)}{\partial \dot{q}_1}\right) - \frac{\partial(\dot{x}^2/2)}{\partial q_1}.$$

We can obtain similar expressions for y and z. Taking this into account, when $\delta q_2 = 0$, equation (1.19) becomes

$$\begin{aligned}\delta W &= \left(\frac{d}{dt}\left(\frac{\partial}{\partial \dot{q}_1}\frac{m(\dot{x}^2+\dot{y}^2+\dot{z}^2)}{2}\right) - \frac{\partial}{\partial q_1}\frac{m(\dot{x}^2+\dot{y}^2+\dot{z}^2)}{2}\right)\delta q_1 \\ &= \left(F_x\frac{\partial x}{\partial q_1} + F_y\frac{\partial y}{\partial q_1} + F_z\frac{\partial z}{\partial q_1}\right)\delta q_1 \\ &= F_{q_1}\,\delta q_1.\end{aligned} \tag{1.29}$$

Let $K = \frac{1}{2}m(\dot{x}^2 + \dot{y}^2 + \dot{z}^2)$ denote the kinetic energy of the particle. Then, we can represent (1.29) as

$$\boxed{\frac{d}{dt}\left(\frac{\partial K}{\partial \dot{q}_1}\right) - \frac{\partial K}{\partial q_1} = F_{q_1}} \tag{1.30}$$

The equation above is called the *Lagrange equation of motion* for the q_1 coordinate. Using the same arguments as above, we can derive the Lagrange equation of motion for the q_2 coordinate. In general there are as many the Lagrange equations of motion as there are degrees of freedom of the particle.

Recall that the expression $\delta W = F_x \delta x + F_y \delta y + F_z \delta z$ is a work done by a force $\boldsymbol{F} = [F_x \;\; F_y \;\; F_z]^T$ for a general displacement $\delta \boldsymbol{s} = [\delta x \;\; \delta y \;\; \delta z]^T$. In terms of generalized coordinates and when, say, $\delta q_2 = 0$, we have

$$\begin{aligned}\delta W &= \left(F_x \frac{\partial x}{\partial q_1} + F_y \frac{\partial y}{\partial q_1} + F_z \frac{\partial z}{\partial q_1}\right)\delta q_1 \\ &= F_{q_1}\delta q_1 \\ &= \delta W_{q_1}.\end{aligned}$$

A generalized force F_{q_r} is of such nature that the product $F_{q_r}\delta q_r$ is the work done by driving forces when q_r alone is changed by δq_r. We mention that a generalized force does not have to be a force in the usual sense. For example, if q_r is an angle, then F_{q_r} must be a torque in order that $F_{q_r}\delta q_r$ be work.

We now derive a more general version of the Lagrange equations of motion for a single particle. In our derivation we will need the notion of *nonconservative* forces, which we discuss next. First recall that we interpret the kinetic energy of a particle as its ability to do work by virtue of its motion. This motion is the result of conservative and nonconservative forces acting upon the particle. A force is nonconservative if the work done by the force on a particle that moves through any round trip is not zero. Thus a force is nonconservative if the work done by that force on a particle that moves between two points depends on the path taken between these points.

Let us now suppose that in addition to the conservative forces, a single nonconservative force due to friction acts on the particle. Denote by $\sum W_c$ the sum of the work done by the conservative forces, and denote by W_f the work done by friction. We then have

$$W_f + \sum W_c = \Delta K. \tag{1.31}$$

Thus, if a nonconservative force, like friction, acts on a particle, then the total mechanical energy is not constant, but changes by the amount of work done by a nonconservative force according to the relation (1.31). Let us now denote by E the final mechanical energy of the particle and denote by E_0 its initial mechanical energy. Then, we can write

$$E - E_0 = \Delta E = W_f.$$

The work done by friction, the nonconservative force, is always negative. Hence, the final mechanical energy E is less than the initial mechanical energy. The "lost" mechanical energy is transformed into *internal energy* U_{int}. In other words, the loss of mechanical energy equals the gain in internal energy and we write

$$\Delta E + U_{int} = 0.$$

In general, we have

$$\Delta K + \sum \Delta U + U_{int} + \text{(change in other forms of energy)} = 0.$$

This means that the total energy, kinetic plus potential plus internal plus all other forms of energy, does not change. This is the *principle of the conservation of energy*. To proceed further, recall that any generalized force F_{q_i} acting on our particle can be expressed as

$$F_{q_i} = F_x \frac{\partial x}{\partial q_i} + F_y \frac{\partial y}{\partial q_i} + F_z \frac{\partial z}{\partial q_i}.$$

Assuming that the forces are conservative, we get

$$F_{q_i} = -\left(\frac{\partial U}{\partial x} \frac{\partial x}{\partial q_i} + \frac{\partial U}{\partial y} \frac{\partial y}{\partial q_i} + \frac{\partial U}{\partial z} \frac{\partial z}{\partial q_i} \right).$$

We thus have

$$F_{q_i} = -\frac{\partial U}{\partial q_i}.$$

Therefore, the Lagrange equations of motion for a single particle for conservative forces can be represented as

$$\frac{d}{dt}\left(\frac{\partial K}{\partial \dot{q}_i} \right) - \frac{\partial K}{\partial q_i} = -\frac{\partial U}{\partial q_i}.$$

We can rearrange the above equations as

$$\frac{d}{dt}\left(\frac{\partial K}{\partial \dot{q}_i} \right) - \frac{\partial}{\partial q_i}(K - U) = 0.$$

We now introduce the *Lagrangian function* L, defined as

$$L = K - U.$$

Note that because U is not a function of $\dot{q}_i$, we have

$$\frac{\partial K}{\partial \dot{q}_i} = \frac{\partial L}{\partial \dot{q}_i}.$$

Taking the above into account, we can represent the Lagrange equations of motion for conservative forces as

$$\frac{d}{dt}\left(\frac{\partial L}{\partial \dot{q}_i} \right) - \frac{\partial L}{\partial q_i} = 0.$$

If some of the forces acting on the particle are nonconservative, the Lagrange equations of motion for our particle will take the form

$$\boxed{\frac{d}{dt}\left(\frac{\partial L}{\partial \dot{q}_i}\right) - \frac{\partial L}{\partial q_i} = \tilde{F}_{q_i}} \tag{1.32}$$

where $\tilde{F}_{q_i}$ are nonconservative forces acting on the particle.

◆ Example 1.2

Consider the simple pendulum shown in Figure 1.7. The simple pendulum is an idealized body consisting of a point mass M, suspended by a weightless inextensible cord of length l. The simple pendulum is an example of a one-degree-of-freedom system with a generalized coordinate being the angular displacement θ. The pendulum kinetic energy is

$$K = \tfrac{1}{2} M l^2 \dot{\theta}^2.$$

We assume that the potential energy of the pendulum is zero when the pendulum is at rest. Hence, its potential energy is

$$U = Mgl(1 - \cos\theta),$$

where $g = 10$ m/sec^2 is the acceleration due to gravity. The Lagrangian function is

$$\begin{aligned} L &- K - U \\ &= \tfrac{1}{2} M l^2 \dot{\theta}^2 - Mgl(1 - \cos\theta) \\ &= 6\dot{\theta}^2 - 60(1 - \cos\theta). \end{aligned}$$

The Lagrange equation describing the pendulum motion is

$$\frac{d}{dt}\left(\frac{\partial L}{\partial \dot{\theta}}\right) - \frac{\partial L}{\partial \theta} = 0.$$

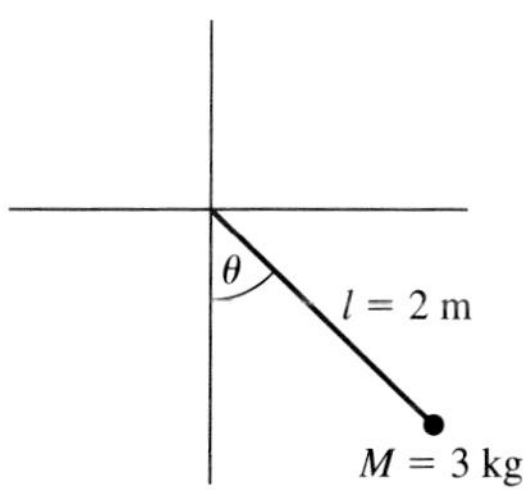

Figure 1.7 A simple pendulum.

In our example the above equation evaluates to

$$12\ddot{\theta} + 60 \sin\theta = 0,$$

or, equivalently,

$$\ddot{\theta} = -5 \sin\theta.$$

◆ Example 1.3

In Figure 1.8, the θ–r robot manipulator is shown. This example was adapted from Snyder [266, Chapter 10]. A lumped mass representation of the θ–r robot manipulator is shown in Figure 1.9. The mass $m_1 = 10$ kg represents the mass of the outer cylinder positioned at its center of mass. The constant distance $r_1 = 1$ m designates the fixed distance between the center of the mass of the outer cylinder and the center of rotation. The mass of the load is represented by $m_2 = 3$ kg and is assumed to be located at the end of a piston of a telescoping arm that is a variable radial distance r from the hub or center of rotation. The angle of rotation of the manipulator arm is θ. The inputs to the system are assumed to be (a) a torque T_θ

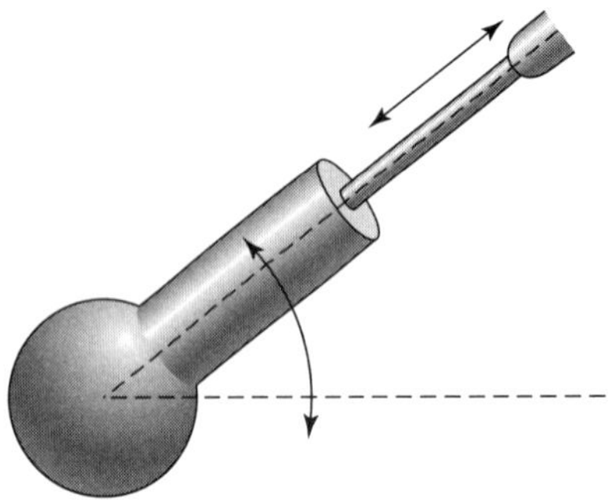

Figure 1.8 The θ–r robot manipulator.

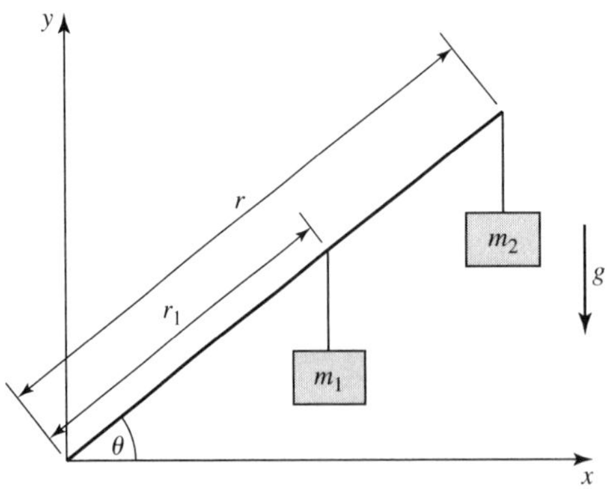

Figure 1.9 A lumped mass representation of the θ–r robot manipulator.

applied at the hub in the direction of θ and (b) a translational force F_r applied in the direction r. Finally, $g = 10\ \text{m/sec}^2$ is the gravitational constant. We construct the Lagrange equations of motion for the system, and then we represent the obtained model in a state-space format. We first find the total kinetic energy of the system. The position of the mass m_1 is

$$\begin{aligned} x_1 &= r_1 \cos\theta, \\ y_1 &= r_1 \sin\theta. \end{aligned}$$

Note that r_1 is a constant. Differentiating x_1 and y_1 with respect to time yields

$$\begin{aligned} \dot{x}_1 &= -r_1\dot{\theta}\sin\theta, \\ \dot{y}_1 &= r_1\dot{\theta}\cos\theta. \end{aligned}$$

The magnitude squared of the velocity vector of the mass m_1 is

$$\begin{aligned} v_1^2 &= \dot{x}_1^2 + \dot{y}_1^2 \\ &= r_1^2\dot{\theta}^2\sin^2\theta + r_1^2\dot{\theta}^2\cos^2\theta \\ &= r_1^2\dot{\theta}^2. \end{aligned}$$

Therefore, the kinetic energy of the mass m_1 is

$$K_1 = \tfrac{1}{2}m_1 v_1^2 = \tfrac{1}{2}m_1 r_1^2\dot{\theta}^2.$$

We now derive an expression for the kinetic energy of the second mass. The position of the mass m_2 is

$$\begin{aligned} x_2 &= r\cos\theta, \\ y_2 &= r\sin\theta. \end{aligned}$$

Note that the distance r is not a constant. Differentiating x_2 and y_2 with respect to time yields

$$\begin{aligned} \dot{x}_2 &= \dot{r}\cos\theta - r\dot{\theta}\sin\theta, \\ \dot{y}_2 &= \dot{r}\sin\theta + r\dot{\theta}\cos\theta. \end{aligned}$$

The magnitude squared of the velocity vector of the mass m_2 is

$$\begin{aligned} v_2^2 &= (\dot{x}_2)^2 + (\dot{y}_2)^2 \\ &= (\dot{r}\cos\theta - r\dot{\theta}\sin\theta)^2 + (\dot{r}\sin\theta + r\dot{\theta}\cos\theta)^2 \\ &= \dot{r}^2 + r^2\dot{\theta}^2. \end{aligned}$$

Hence, the kinetic energy of the mass m_2 is

$$K_2 = \tfrac{1}{2}m_2 v_2^2 = \tfrac{1}{2}m_2(\dot{r}^2 + r^2\dot{\theta}^2).$$

The total kinetic energy of the system is

$$K - K_1 + K_2 = \tfrac{1}{2}m_1 r_1^2\dot{\theta}^2 + \tfrac{1}{2}m_2(\dot{r}^2 + r^2\dot{\theta}^2).$$

We now find the potential energy of the system. The potential energy of the mass m_1 is

$$U_1 = m_1 g r_1 \sin\theta.$$

The potential energy of the mass m_2 is

$$U_2 = m_2 g r \sin\theta.$$

The total potential energy of the system is

$$U = U_1 + U_2 = m_1 g r_1 \sin\theta + m_2 g r \sin\theta.$$

The Lagrangian function for the system is

$$L = K - U = \tfrac{1}{2} m_1 r_1^2 \dot\theta^2 + \tfrac{1}{2} m_2 \dot r^2 + \tfrac{1}{2} m_2 r^2 \dot\theta^2 - m_1 g r_1 \sin\theta - m_2 g r \sin\theta.$$

The manipulator has two degrees of freedom. Hence, we have two Lagrange equations of motion for this system:

$$\frac{d}{dt}\left(\frac{\partial L}{\partial \dot\theta}\right) - \frac{\partial L}{\partial \theta} = T_\theta,$$

$$\frac{d}{dt}\left(\frac{\partial L}{\partial \dot r}\right) - \frac{\partial L}{\partial r} = F_r.$$

The first Lagrange equation yields

$$m_1 r_1^2 \ddot\theta + m_2 r^2 \ddot\theta + 2 m_2 r \dot r \dot\theta + g\cos\theta (m_1 r_1 + m_2 r) = T_\theta.$$

The second Lagrange equation gives

$$m_2 \ddot r - m_2 r \dot\theta^2 + m_2 g \sin\theta = F_r.$$

We next represent the above modeling equations in state-space format. (One can use, for example, MATLAB's Symbolic Math Toolbox to perform needed calculations.) Let the state variables be defined as

$$z_1 = \theta, \qquad z_2 = \dot\theta, \qquad z_3 = r, \qquad z_4 = \dot r.$$

Then, we can write

$$\begin{bmatrix} \dot z_1 \\ \dot z_2 \\ \dot z_3 \\ \dot z_4 \end{bmatrix} = \begin{bmatrix} z_2 \\ \dfrac{-2 m_2 z_2 z_3 z_4 - g\cos z_1 (m_1 r_1 + m_2 z_3) + u_1}{m_1 r_1^2 + m_2 z_3^2} \\ z_4 \\ z_2^2 z_3 - g \sin z_1 + u_2/m_2 \end{bmatrix}, \tag{1.33}$$

where $u_1 = T_\theta$ and $u_2 = F_r$.

In the following section, we develop models of a few more dynamical systems.

1.7 Modeling Examples

1.7.1 Centrifugal Governor

In this subsection, we model the centrifugal governor. In our modeling, we follow Pontryagin [235, pp. 213–220]. For an alternative approach to modeling the governor see Maxwell [197] and Fuller [92]. Some—for example, MacFarlane [192, p. 250]—claim that Maxwell founded the theory of automatic control systems in his 1868 paper "On governors." However, this landmark paper is written in a terse and cryptic style. Fuller [92] does a wonderful job elucidating, correcting, and expanding Maxwell's paper. Mayr [199, Chapter X] gives a fascinating historical account of the application of the centrifugal governor to the speed regulation of the steam engine and tracks the origins of the flyball governor.

A schematic of the governor is depicted in Figure 1.4, while forces acting on the governor are shown in Figure 1.10. For simplicity of notation, we assume that the length of the arms, on which the weights are fastened, are of unity length. If the weights rotate with angular velocity ω_{gov}, then the arms form an angle φ with the vertical position. Let v_{weight} denote the linear velocity of the weights, and let r be the radius of the circle along which the weights rotate. Then, $v_{weight} = \omega_{gov} r$. Note that $r = \sin\varphi$. The resulting centrifugal force acting on each weight is

$$\begin{aligned} m\frac{v_{weight}^2}{r} &= m\omega_{gov}^2 r \\ &= m\omega_{gov}^2 \sin\varphi. \end{aligned}$$

Simultaneously, a gravitational force mg acts on each weight. In addition, a frictional force in the hinge joints acts on each weight. We assume that the frictional force is proportional to $\dot{\varphi}$

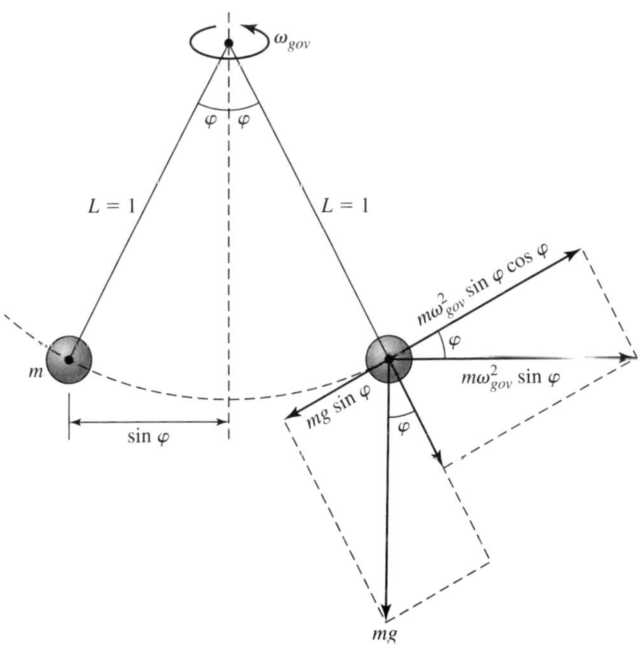

Figure 1.10 Forces acting on the weights in the governor.

with the coefficient of friction b. Summing forces along the axis perpendicular to an arm yields an equation of motion of the weight:

$$\begin{aligned} m\ddot{\varphi}L &= m\omega_{gov}^2 \sin\varphi\cos\varphi - mg\sin\varphi - b\dot{\varphi} \\ &= \tfrac{1}{2}m\omega_{gov}^2\sin 2\varphi - mg\sin\varphi - b\dot{\varphi}. \end{aligned} \tag{1.34}$$

Note that $m\ddot{\varphi}L = m\ddot{\varphi}$ because by assumption, $L = 1$. The governor is connected to the engine via a gear with the gear ratio N such that

$$\omega_{gov} = N\omega, \tag{1.35}$$

where ω is the angular velocity of the flywheel. Substituting (1.35) into (1.34) gives

$$m\ddot{\varphi} = \tfrac{1}{2}mN^2\omega^2\sin 2\varphi - mg\sin\varphi - b\dot{\varphi}.$$

Let I be the moment of inertia of the flywheel, τ_1 the torque delivered by the engine, and τ_l the load torque. Then, the equation modeling the motion of the flywheel is

$$I\dot{\omega} = \tau_1 - \tau_l. \tag{1.36}$$

The collar of the governor is connected with the valve so that at the nominal value of $\varphi = \varphi_0$ the engine delivers the nominal torque τ_d that results in the desired angular velocity $\omega = \omega_d$ of the flywheel. We model the interaction of the collar and the valve using the equation

$$\tau_1 = \tau_d + \kappa(\cos\varphi - \cos\varphi_0), \tag{1.37}$$

where $\kappa > 0$ is a proportionality constant that depends on the type of the governor. Substituting (1.37) into (1.36) yields

$$\begin{aligned} I\dot{\omega} &= \tau_1 - \tau_l \\ &= \tau_d + \kappa(\cos\varphi - \cos\varphi_0) - \tau_l \\ &= \kappa\cos\varphi - \tau, \end{aligned}$$

where $\tau = \tau_l - \tau_d + \kappa\cos\varphi_0$. We thus have two equations that model the governor:

$$\begin{aligned} m\ddot{\varphi} &= \tfrac{1}{2}mN^2\omega^2\sin 2\varphi - mg\sin\varphi - b\dot{\varphi}, \\ I\dot{\omega} &= \kappa\cos\varphi - \tau. \end{aligned}$$

We use the state variables

$$x_1 = \varphi, \quad x_2 = \dot{\varphi}, \quad \text{and} \quad x_3 = \omega$$

to represent the above model in state-space format, i.e., as a system of first-order differential equations:

$$\begin{aligned} \dot{x}_1 &= x_2, \\ \dot{x}_2 &= \frac{1}{2}N^2x_3^2\sin 2x_1 - g\sin x_1 - \frac{b}{m}x_2, \\ \dot{x}_3 &= \frac{\kappa}{I}\cos x_1 - \frac{\tau}{I}. \end{aligned} \tag{1.38}$$

1.7.2 *Ground Vehicle*

In this subsection, we model a ground vehicle in a turning maneuver, such as a step lane-change maneuver. Our modeling process follows that of Wong [303] and of Will and Żak [299]. During such a step lane-change maneuver, the vehicle undergoes translational as well as rotational motion. When modeling a moving vehicle, it is convenient to use a reference coordinate frame attached to and moving with the vehicle. This is because with respect to the vehicle coordinate frame, moments of inertia of the vehicle are constant. We use the Society of Automotive Engineers (SAE) standard coordinate system as shown in Figure 1.11. We use lowercase letters to label axes of the vehicle coordinate system. Following the SAE convention, we define the coordinates as follows:

- The x coordinate is directed forward and on the longitudinal plane of symmetry of the vehicle.
- The y coordinate has a lateral direction out the right-hand side of the vehicle.
- The z coordinate is directed downward with respect to the vehicle.
- The yaw velocity about the z axis is denoted $\dot{\theta}$.

In developing our simple model, we neglect the effects of roll and pitch. The origin of the vehicle coordinate system is located at the vehicle center of gravity (CG). We also use a global, or fixed, coordinate system. Its axes are labeled using uppercase letters as shown in Figure 1.12. To form

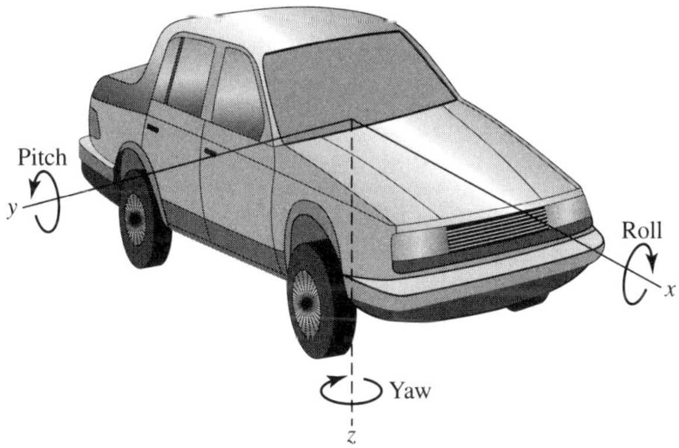

Figure 1.11 The Society of Automotive Engineers (SAE) coordinate system attached to the vehicle.

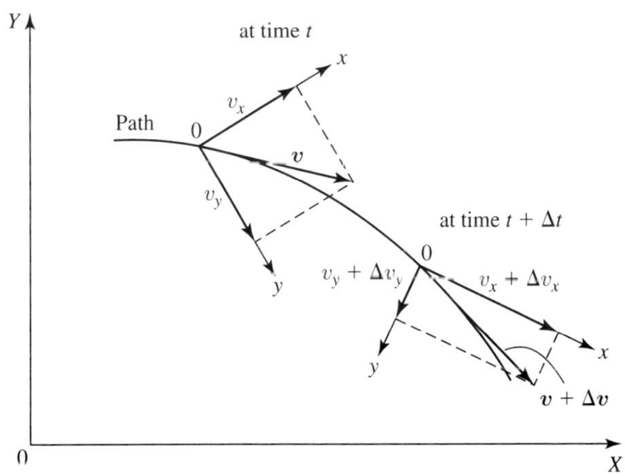

Figure 1.12 Analyzing vehicle motion using coordinate frame attached to the vehicle; snapshots of the vehicle velocity at times t and $t + \Delta t$.

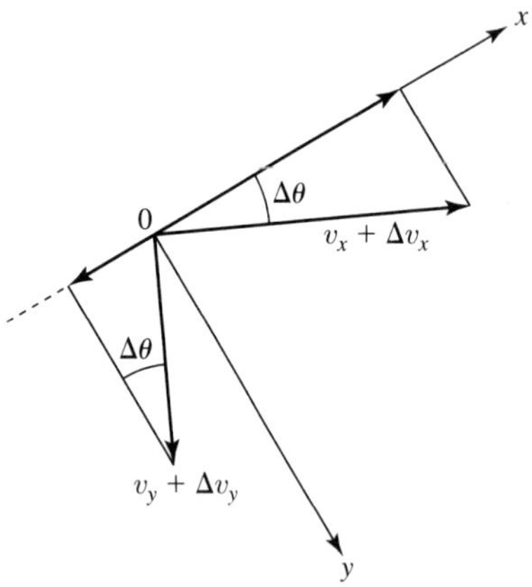

Figure 1.13 Computing the change in the longitudinal and lateral components of the vehicle velocity.

the equations of motion of the vehicle, we need an expression for the vehicle CG acceleration. We first compute the change in the longitudinal component of the velocity. Using Figures 1.12 and 1.13 and assuming small angles, we obtain

$$\begin{aligned}(v_x + \Delta v_x)\cos\Delta\theta - (v_y + \Delta v_y)\sin\Delta\theta - v_x &\approx \Delta v_x - v_y\,\Delta\theta \\ &= \Delta\dot{x} - \dot{y}\,\Delta\theta.\end{aligned}$$

We form the Newton quotient using the above. Then, taking the limit, we obtain the total longitudinal acceleration component a_x:

$$\begin{aligned}a_x &= \lim_{\Delta t\to 0}\frac{\Delta v_x - v_y\,\Delta\theta}{\Delta t} \\ &= \dot{v}_x - v_y\dot{\theta} \\ &= \ddot{x} - \dot{y}\omega_z,\end{aligned} \tag{1.39}$$

where $\omega_z = \dot{\theta}$. Using an argument similar to the one above, we obtain the lateral acceleration component a_y to be

$$\begin{aligned}a_y &= \dot{v}_y + v_x\dot{\theta} \\ &= \ddot{y} + \dot{x}\omega_z.\end{aligned} \tag{1.40}$$

We assume that the vehicle is symmetric with respect to the x–z plane. We use the so-called bicycle model that is shown in Figure 1.14, where δ_f denotes the front wheels' steering angle and δ_r is the rear wheels' steering angle. We denote by $2F_{xf}$ lumped longitudinal force at the front axle and denote by $2F_{xr}$ lumped longitudinal force at the rear axle. The lumped lateral force at the front axle is denoted $2F_{yf}$, while lumped lateral force at the rear axle is denoted $2F_{yr}$. Summing the forces along the x axis and assuming small angles, we obtain

$$\begin{aligned}m(\ddot{x} - \dot{y}\omega_z) &= 2F_{xf}\cos\delta_f - 2F_{yf}\sin\delta_f + 2F_{xr}\cos\delta_r + 2F_{yr}\sin\delta_r \\ &\approx 2F_{xf} + 2F_{xr}.\end{aligned} \tag{1.41}$$

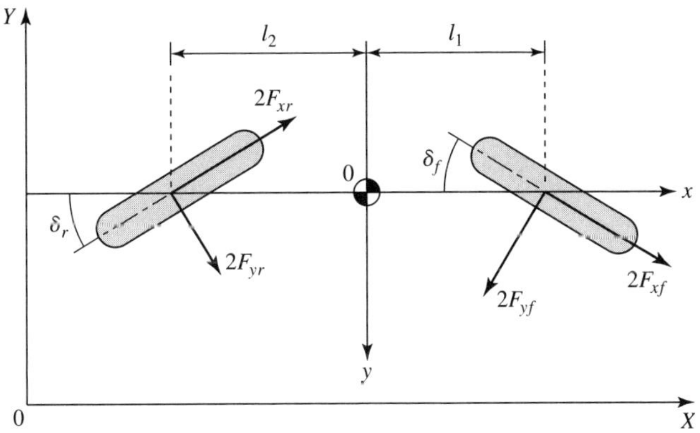

Figure 1.14 Bicycle model of a vehicle.

Next, summing the forces along the y axis and assuming small angles yields

$$\begin{aligned} m(\dot{y} + \dot{x}\omega_z) &= 2F_{xf}\sin\delta_f + 2F_{yf}\cos\delta_f + 2F_{yr}\cos\delta_r - 2F_{xr}\sin\delta_r \\ &\approx 2F_{yf} + 2F_{yr}. \end{aligned} \tag{1.42}$$

The third equation of motion is obtained by summing the torques with respect to the z axis. Let I_z denote the moment of inertia of the vehicle about the z axis. Then, referring again to Figure 1.14 and assuming small angles, we obtain

$$\begin{aligned} I_z\dot{\omega}_z &= 2l_1F_{yf}\cos\delta_f + 2l_1F_{xf}\sin\delta_f - 2l_2F_{yr}\cos\delta_r + 2l_2F_{xr}\sin\delta_r \\ &\approx 2l_1F_{yf} - 2l_2F_{yr}. \end{aligned} \tag{1.43}$$

In our further analysis, we assume $v_x = \dot{x} =$ constant; that is, the driver is neither accelerating nor braking the vehicle. This means that we need to use only the last two equations of motion in our model. To proceed, we need to define tire slip angles and cornering stiffness coefficients. The tire slip angle is the difference between the tire's heading direction, which is the steer angle, and the tire's actual travel direction. In other words, the tire slip angle is the angle between the desired direction of motion of the tire specified by the steer angle and the actual direction of travel of the center of the tire contact. We denote by $C_{\alpha f}$ the cornering stiffness coefficient of the front tires and denote by $C_{\alpha r}$ the cornering stiffness coefficient of the rear tires. The coefficients $C_{\alpha f}$ and $C_{\alpha r}$ are measured experimentally. A typical relation between a tire slip angle and a cornering force is shown in Figure 1.15. It is a nonlinear relation, where the cornering coefficient is

$$C_{\alpha w} = \frac{dF_{yw}}{d\alpha_w},$$

where $w = f$ or $w = r$. The lateral forces F_{yf} and F_{yr} that appear in (1.41)–(1.43) are functions of the corresponding tire slip angles α_f and α_r and cornering stiffness coefficients $C_{\alpha f}$ and $C_{\alpha r}$. In our simple linearized model, we assume that $C_{\alpha f} = C_{\alpha r} =$ constant, and model dependencies between the lateral forces and tire slip angles are expressed as

$$F_{yf} = C_{\alpha f}\alpha_f \tag{1.44}$$

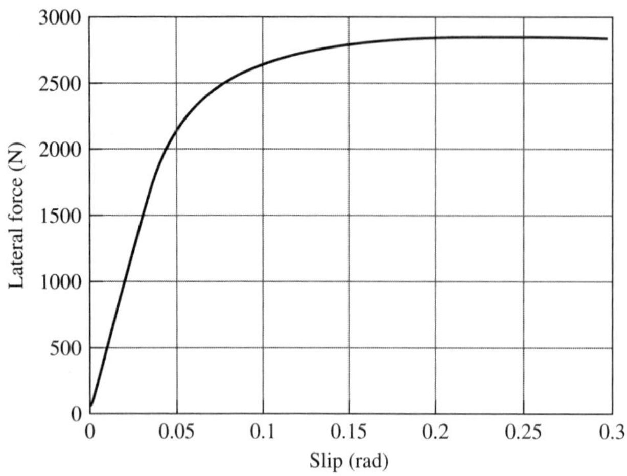

Figure 1.15 Typical cornering characteristic.

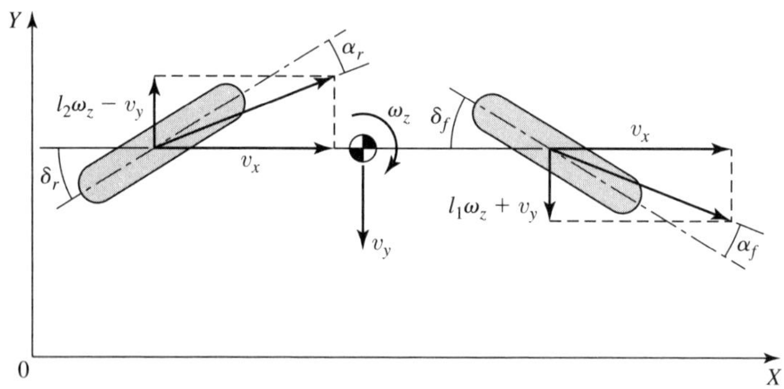

Figure 1.16 Computing tire slip angles.

and

$$F_{yr} = C_{\alpha r}\alpha_r. \tag{1.45}$$

We formulate expressions for the tire slip angles using Figure 1.16. Assuming small angles, we form an expression for the front tire slip angle:

$$\begin{aligned}\alpha_f &= \delta_f - \tan^{-1}\left(\frac{l_1\omega_z + v_y}{v_x}\right)\\ &\approx \delta_f - \frac{l_1\omega_z + v_y}{v_x}\\ &= \delta_f - \frac{l_1\omega_z + \dot{y}}{\dot{x}}.\end{aligned} \tag{1.46}$$

The rear tire slip angle is

$$\begin{aligned}
\alpha_r &= \delta_r - \tan^{-1}\left(-\frac{l_2\omega_z - v_y}{v_x}\right) \\
&= \delta_r + \tan^{-1}\left(\frac{l_2\omega_z - v_y}{v_x}\right) \\
&\approx \delta_r + \frac{l_2\omega_z - v_y}{v_x} \\
&= \delta_r + \frac{l_2\omega_z - \dot{y}}{\dot{x}}.
\end{aligned} \tag{1.47}$$

Hence

$$F_{yf} = C_{\alpha f}\left(\delta_f - \frac{l_1\omega_z + \dot{y}}{\dot{x}}\right), \tag{1.48}$$

and

$$F_{yr} = C_{\alpha r}\left(\delta_r + \frac{l_2\omega_z - \dot{y}}{\dot{x}}\right). \tag{1.49}$$

Substituting the above two equations into (1.42) and assuming small angles, we obtain

$$m\ddot{y} + m\dot{x}\omega_z = 2C_{\alpha f}\delta_f - 2C_{\alpha f}\frac{l_1\omega_z + \dot{y}}{v_x} + 2C_{\alpha r}\delta_r + 2C_{\alpha r}\frac{l_2\omega_z - \dot{y}}{v_x}.$$

Rearranging terms and performing some manipulations, we get

$$\ddot{y} = -\left(\frac{2C_{\alpha f} + 2C_{\alpha r}}{mv_x}\right)\dot{y} + \left(-v_x - \frac{2C_{\alpha f}l_1 - 2C_{\alpha r}l_2}{mv_x}\right)\omega_z + \frac{2C_{\alpha f}}{m}\delta_f + \frac{2C_{\alpha r}}{m}\delta_r. \tag{1.50}$$

Substituting the expressions for F_{yf} and F_{yr} into (1.43) and assuming small angles, we obtain

$$I_z\dot{\omega}_z = 2l_1C_{\alpha f}\left(\delta_f - \frac{l_1\omega_z + \dot{y}}{v_x}\right) - 2l_2C_{\alpha r}\left(\delta_r + \frac{l_2\omega_z - \dot{y}}{v_x}\right).$$

Rearranging and performing manipulations yields

$$\dot{\omega}_z = -\left(\frac{2l_1C_{\alpha f} - 2l_2C_{\alpha r}}{I_zv_x}\right)\dot{y} - \left(\frac{2l_1^2C_{\alpha f} + 2l_2^2C_{\alpha r}}{I_zv_x}\right)\omega_z + \frac{2l_1C_{\alpha f}}{I_z}\delta_f - \frac{2l_2C_{\alpha r}}{I_z}\delta_r. \tag{1.51}$$

Later, we will analyze a vehicle performance in the fixed coordinates. In particular, we will be interested in an emergency lane-change maneuver. This requires that we transform the vehicle lateral variable into the fixed coordinates. Assuming small angles and referring to Figure 1.17, we obtain

$$\begin{aligned}
\dot{Y} &= -\dot{x}\sin\psi - \dot{y}\cos\psi \\
&\approx -v_x\psi - \dot{y}.
\end{aligned} \tag{1.52}$$

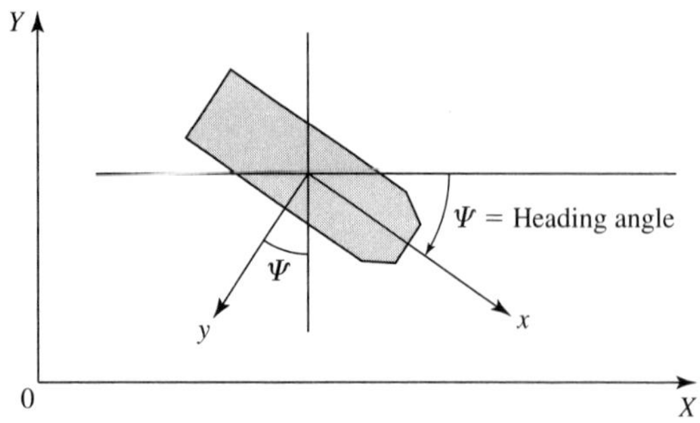

Figure 1.17 Computing the velocity of a vehicle with respect to the fixed coordinates.

Note that

$$\dot{X} = \dot{x}\cos\psi - \dot{y}\sin\psi.$$

For the step lane change, the yaw rate of the vehicle is equal to the yaw rate error; that is,

$$\omega_z = \dot{\psi}.$$

We further assume that

$$\theta = \psi. \tag{1.53}$$

We can now represent the vehicle modeling equations in a state-space format. Let

$$x_1 = \dot{y}, \qquad x_2 = \theta, \qquad x_3 = \omega_z, \qquad x_4 = Y.$$

Then, using (1.50), (1.51), (1.52), and (1.53), we obtain

$$\begin{bmatrix}\dot{x}_1\\ \dot{x}_2\\ \dot{x}_3\\ \dot{x}_4\end{bmatrix} = \begin{bmatrix} -\dfrac{2C_{\alpha f}+2C_{\alpha r}}{mv_x} & 0 & -v_x - \dfrac{2C_{\alpha f}l_1 - 2C_{\alpha r}l_2}{mv_x} & 0\\ 0 & 0 & 1 & 0\\ -\dfrac{2l_1C_{\alpha f} - 2l_2C_{\alpha r}}{I_z v_x} & 0 & -\dfrac{2l_1^2C_{\alpha f} + 2l_2^2C_{\alpha r}}{I_z v_x} & 0\\ -1 & -v_x & 0 & 0\end{bmatrix}\begin{bmatrix}x_1\\ x_2\\ x_3\\ x_4\end{bmatrix}$$
$$+\begin{bmatrix}\dfrac{2C_{\alpha f}}{m} & \dfrac{2C_{\alpha r}}{m}\\ 0 & 0\\ \dfrac{2l_1C_{\alpha f}}{I_z} & -\dfrac{2l_2C_{\alpha r}}{I_z}\\ 0 & 0\end{bmatrix}\begin{bmatrix}\delta_f\\ \delta_r\end{bmatrix}. \tag{1.54}$$

A possible vehicle parameters are given in Table 1.1.

1.7.3 Permanent Magnet Stepper Motor

In this subsection, we develop a model of a permanent magnet (PM) stepper motor. Stepper motors can be viewed as digital positioning devices. Specifically, stepper motors are electromechanical motion devices that convert input information in digital form into output that has

Table 1.1 Typical Vehicle Data

$C_{\alpha f} = C_{\alpha r}$	30,000.00 N/rad
l_1	1.20 m
l_2	1.22 m
m	1280 kg
I_z	2500 kg·m^2
v_x	18.3 m/sec

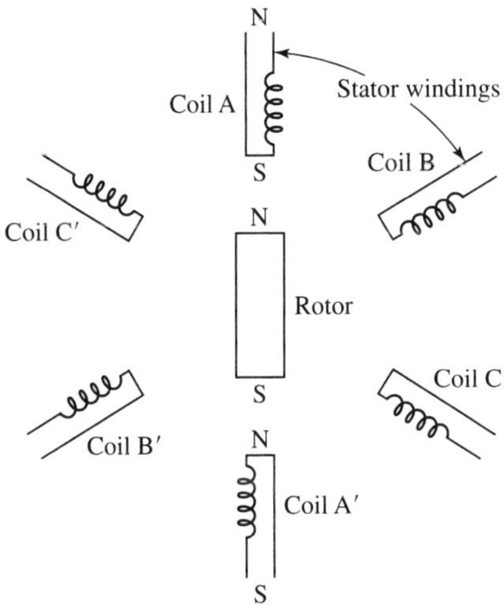

Figure 1.18 A simplified schematic of an operation of a stepper motor.

mechanical form. The stepper motor can be driven by digital pulses of the following form: move clockwise or move counterclockwise. We now describe an idea of an operation of a PM stepper motor and then derive its modeling equations.

A simplified organization of a stepper motor is shown in Figure 1.18. The armature, or rotor, is a magnet. At any instant of the operation, exactly one pair of windings of the stator is energized. A pair of windings represents a phase. Each pair of the stator windings is wound in such a way that their magnetic fields are collinear—for example, windings A and A′. The rotor will then rotate until its magnetic field is aligned with the fields of the energized pair of the stator windings. Figure 1.18 shows the situation where the pair of windings A and A′ is energized. If the rotor is at rest, as shown in Figure 1.18, and the pair A, A′ is de-energized while the pair B, B′ is energized, the rotor will rotate to the new position so that its magnetic field is aligned with the magnetic field of the pair B, B′. In this example of the three-phase motor, a single step results in the rotor motion of $360°/6 = 60°$. We next give a more detailed description of the electrical organization of the stepper motor and construct its model.

The basic PM stepper motor consists of a slotted stator and a rotor. There can be two sets of teeth on the rotor. These two sets of teeth are out of alignment with each other by a tooth width, as shown in Figure 1.19. Specifically, the rotor teeth of the left end-cap are displaced one-half pitch from the teeth on the right end-cap. A magnet, or an electromagnet, that is mounted on

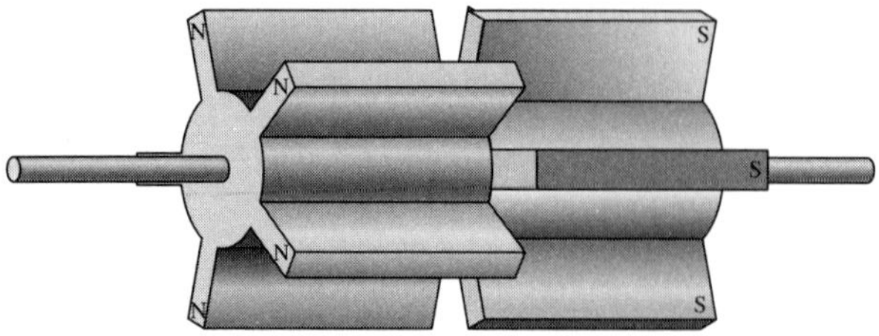

Figure 1.19 A possible rotor structure of a PM stepper motor.

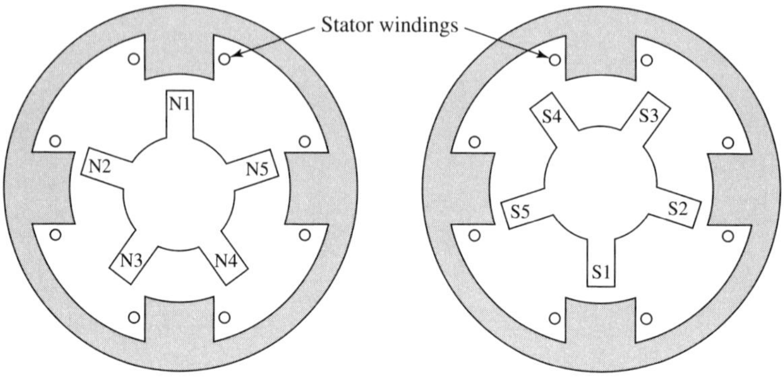

(a) Axial view at a

(b) Axial view at b

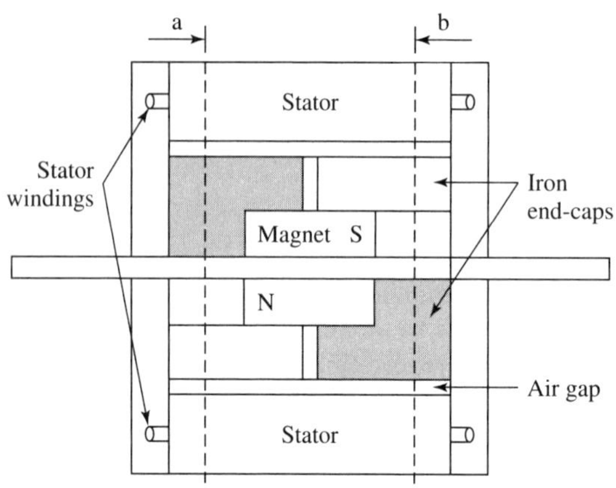

(c) Side, cross-sectional view

Figure 1.20 Construction features of a four-pole stator and a five-pole rotor PM stepper motor.

the rotor magnetizes the iron end-caps. The left end-cap is magnetized as a north pole, while the right end-cap is magnetized as a south pole. The slotted stator is equipped with two or more individual coils. Figure 1.20 shows a four-pole stator and a five-pole rotor structure.

We now discuss the way in which the PM stepper motor can be used to perform precise positioning. Assume that the starting point is as represented in Figure 1.21. Here, coil A–A′ is energized so that the upper stator pole is given an S-pole polarity and the corresponding

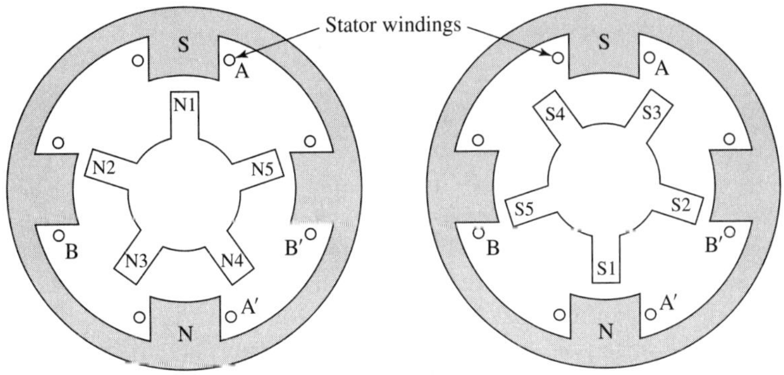

Figure 1.21 Coil A–A′ energized for an S–N orientation, while coil B–B′ de-energized.

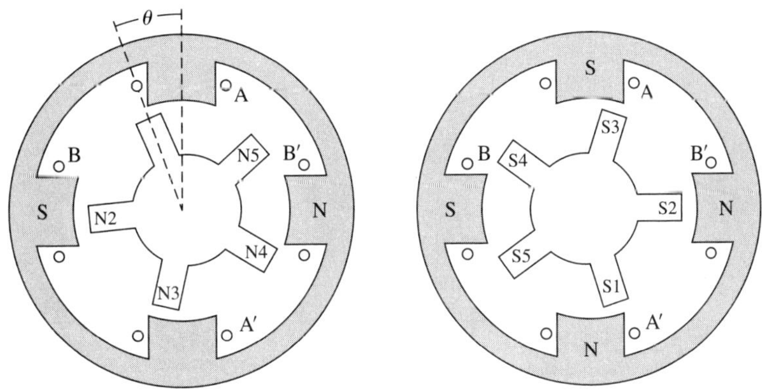

Figure 1.22 Coil A–A′ de-energized, while coil B–B′ energized for an S–N orientation.

lower stator is given an N-pole polarity. The rotor remains in this position as long as stator coil voltage—and hence stator polarities—remains unchanged. This is because the net torque acting on the rotor is zero. If we now de-energize coil A–A′ and energize coil B–B′ so that the left pole becomes an S-pole and the right pole assumes an N-polarity, then this results in the new steady-state position shown in Figure 1.22. We can proceed by alternatively energizing the two stator coils. The PM stepper motor can thus be controlled by digital pulses applied to its two inputs: move clockwise or move counterclockwise.

Our next objective is to devise a mathematical model of the PM stepper motor. In our discussion we use the following notation. We denote by R_t the number of rotor teeth. The angular displacement between rotor teeth is called the tooth pitch and denoted T_p. The angular displacement of the rotor as we change the input voltage from one pair of windings to the other is called the step length and denoted S_l. The number of phases—that is, the number of pairs of stator windings—is denoted N_p. In our example, we have $N_p = 2$, $R_t = 5$, and $T_p = 2\pi/R_t = 72^\circ$. In this example, it takes four input voltage switchings, and hence steps, to advance the rotor one tooth pitch. Indeed, because

$$T_p = 2N_pS_l = \frac{2\pi}{R_t}, \tag{1.55}$$

we obtain

$$S_l = \frac{\pi}{R_t N_p} = 18^\circ. \tag{1.56}$$

We denote the rotor angular displacement by θ.

Following Krause and Wasynczuk [169, pp. 350–351], we now trace the main path of the flux linking the A–A′ windings for the rotor position shown in Figure 1.23. The flux leaves the left end-cap through the rotor tooth at the top that is aligned with the stator tooth having the A part of the A–A′ winding. The flux then travels up through the stator tooth in the stator iron. Next, the flux splits and travels around the circumference of the stator, as shown in Figure 1.23(a). The flux returns to the south pole of the rotor through the stator tooth positioned at the bottom of the A′ part of the A–A′ winding, as shown in Figure 1.23(b). We denote the peak flux linkage produced by the rotor permanent magnet in the A–A′ stator winding by $n\phi_m$, where n is the number of turns of the winding. For $\theta = 0^\circ$, corresponding to the rotor position in Figure 1.23, the flux linkage in the A–A′ winding of the stator is maximal and is approximately equal to $n\phi_m$.

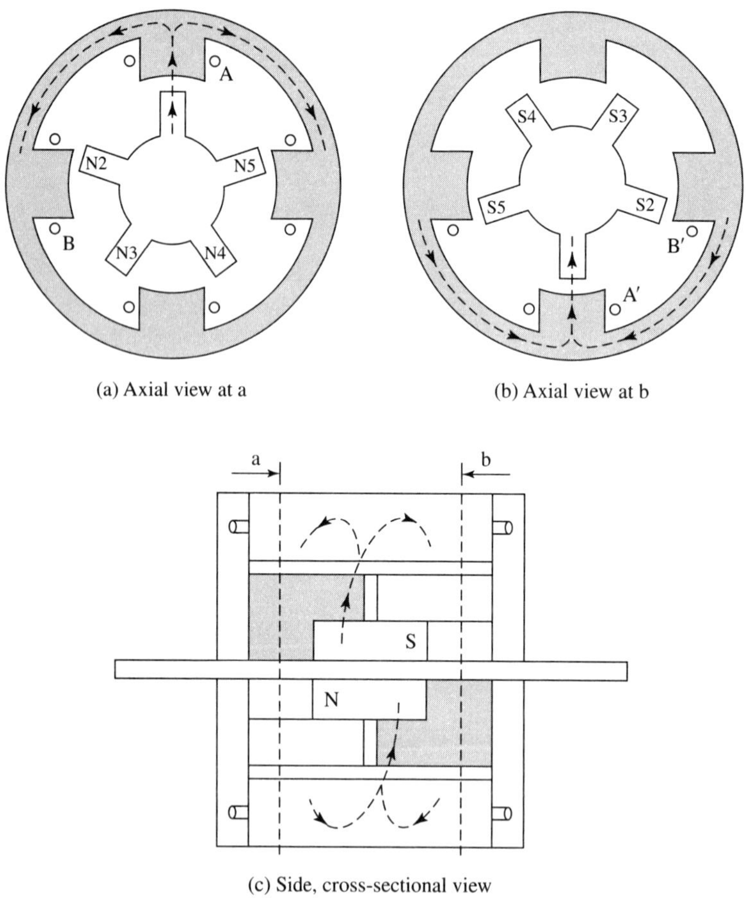

Figure 1.23 The main path of the flux linking the A–A′ winding.

Table 1.2 Symbols Used in Modeling Equations of the PM Stepper Motor

T_p	Tooth pitch
S_l	Step length
R_t	Number of teeth on rotor
N_p	Number of phases, that is, stator winding pairs
θ	Angular displacement of the rotor
ω	Angular velocity of the rotor
$v_a(t)$	Voltage applied to phase a, that is, A–A′ stator winding
$v_b(t)$	Voltage applied to phase b, that is, B–B′ stator winding
$i_a(t)$	Current in phase a
$i_b(t)$	Current in phase b
R	Stator winding resistance per phase
L	Stator winding inductance per phase
B	Viscous friction coefficient
I	Rotor inertia
K_b	Back emf constant

For $\theta = 90°/R_t$, corresponding to the rotor position in Figure 1.22, the flux linkage in the A–A′ winding of the stator is minimal, approximately equal to zero. We model the flux linkage, ψ_a, produced by the permanent magnet in the A–A′ stator winding as

$$\psi_a = n\phi_m \cos(R_t\theta). \tag{1.57}$$

The (main) flux, linking the B–B′ winding for the rotor position shown in Figure 1.22, would enter the stator tooth on which the B part of the B–B′ winding is located from the rotor tooth on the left. The flux then would travel around the circumference of the stator and return to the rotor through the stator pole on which the B′ part of the B–B′ winding is wound. For $\theta = 90°/R_t$, corresponding to the rotor position in Figure 1.22, the flux linkage in the B–B′ winding of the stator is maximal and is approximately equal to $n\phi_m$. For $\theta = 0°$, corresponding to the rotor position in Figure 1.23, the flux linkage in the B–B′ winding is minimal, approximately equal to zero. We model the flux linkage in the B–B′ phase of the stator by

$$\psi_b = n\phi_m \sin(R_t\theta). \tag{1.58}$$

In the following development, we use symbols that are described in Table 1.2. Applying Kirchhoff's voltage law to the circuit of phase a yields

$$v_a(t) = Ri_a(t) + L\frac{di_a(t)}{dt} + e_a(t),$$

where e_a is the back emf induced in phase a; that is,

$$\begin{aligned} e_a &= \frac{d\psi_a}{dt} \\ &= \frac{d\psi_a}{d\theta}\frac{d\theta}{dt} \\ &= -nR_t\phi_m \sin(R_t\theta)\frac{d\theta}{dt} \\ &= -K_b\,\omega \sin(R_t\theta), \end{aligned}$$

where $K_b = nR_t\phi_m$. Hence, the voltage equation for phase a can be represented as

$$v_a = Ri_a + L\frac{di_a}{dt} - K_b\,\omega\sin(R_t\theta).$$

Applying now Kirchhoff's voltage law to the circuit of phase b yields

$$v_b(t) = Ri_b(t) + L\frac{di_b(t)}{dt} + e_b(t),$$

where e_b is the back emf induced in phase b; that is,

$$\begin{aligned} e_b &= \frac{d\psi_b}{dt} \\ &= \frac{d\psi_b}{d\theta}\frac{d\theta}{dt} \\ &= nR_t\phi_m\cos(R_t\theta)\frac{d\theta}{dt} \\ &= K_b\,\omega\cos(R_t\theta), \end{aligned}$$

where, as before, $K_b = nR_t\phi_m$. Hence, the voltage equation for phase b can be represented as

$$v_b = Ri_b + L\frac{di_b}{dt} + K_b\,\omega\cos(R_t\theta).$$

Let T_a and T_b denote torques produced by the currents i_a and i_b, respectively. Then, applying Newton's second law to the rotor yields

$$I\frac{d^2\theta}{dt^2} + B\frac{d\theta}{dt} = T_a + T_b.$$

The PM stepper motor is an electromechanical device in which electric power is transformed into mechanical power. The mechanical power, P_a, due to the torque T_a is

$$P_a = T_a\frac{d\theta}{dt}. \tag{1.59}$$

Assuming that there are no losses, this mechanical power is equal to the power developed in phase a:

$$P_a = i_a\frac{d\psi_a}{dt} = i_a\frac{d\psi_a}{d\theta}\frac{d\theta}{dt}. \tag{1.60}$$

Equating the right-hand sides of (1.59) and (1.60) and performing simple manipulations, we obtain

$$T_a = i_a\frac{d\psi_a}{d\theta} = -i_a K_b\sin(R_t\theta).$$

Applying similar arguments to phase b; we get

$$T_b = i_b\frac{d\psi_b}{d\theta} = i_b K_b\cos(R_t\theta).$$

Taking the above equations into account, we represent the equation of motion of the rotor as

$$I\frac{d^2\theta}{dt^2} + B\frac{d\theta}{dt} = T_a + T_b = -i_a K_b\sin(R_t\theta) + i_b K_b\cos(R_t\theta).$$

Combining the above equations, we obtain a model of a two-phase (i.e., a two-input) PM stepper motor. We have

$$
\begin{aligned}
\frac{d\theta}{dt} &= \omega, \\
\frac{d\omega}{dt} &= -\frac{K_b}{I} i_a \sin(R_t\theta) + \frac{K_b}{I} i_b \cos(R_t\theta) - \frac{B}{I}\omega, \\
\frac{di_a}{dt} &= -\frac{R}{L} i_a + \frac{K_b}{L}\omega \sin(R_t\theta) + \frac{1}{L} v_a, \\
\frac{di_b}{dt} &= -\frac{R}{L} i_b - \frac{K_b}{L}\omega \cos(R_t\theta) + \frac{1}{L} v_b.
\end{aligned}
$$

We now define the state variables: $x_1 = \theta$, $x_2 = \omega$, $x_3 = i_a$, and $x_4 = i_b$. We let $K_1 = B/I$, $K_2 = K_b/I$, $K_3 = R_t$, $K_4 = K_b/L$, $K_5 = R/L$, $u_1 = \frac{1}{L} v_a$, and $u_2 = \frac{1}{L} v_b$. Using the above notation, we represent the PM stepper motor model in state space format:

$$
\begin{bmatrix} \dot{x}_1 \\ \dot{x}_2 \\ \dot{x}_3 \\ \dot{x}_4 \end{bmatrix} = \begin{bmatrix} x_2 \\ -K_1x_2 - K_2x_3\sin(K_3x_1) + K_2x_4\cos(K_3x_1) \\ K_4x_2\sin(K_3x_1) - K_5x_3 \\ -K_4x_2\cos(K_3x_1) - K_5x_4 \end{bmatrix} + \begin{bmatrix} 0 & 0 \\ 0 & 0 \\ 1 & 0 \\ 0 & 1 \end{bmatrix} \begin{bmatrix} u_1 \\ u_2 \end{bmatrix}. \tag{1.61}
$$

1.7.4 Stick Balancer

In this subsection we develop a nonlinear model of the stick balancer, also known as the cart with an inverted pendulum. In our derivation of the model, we use Newton's laws, in a fashion similar to that of Kwakernaak and Sivan [173] and Ogata [216], where linearized models of the stick balancer were developed. An alternative derivation, using the method of D'Alembert, can be found in Cannon [41, Section 22.4].

A free-body diagram of a cart on which a stick (i.e., an inverted pendulum) is mounted is shown in Figure 1.24. Let $H = H(t)$ and $V = V(t)$ be horizontal and vertical reaction forces, respectively, in the pivot. That is, the force H is the horizontal reaction force that the cart exerts upon the pendulum, whereas $-H$ is the force exerted by the pendulum on the cart. Similar convention applies to the forces V and $-V$. The x and y are the coordinates of the fixed, nonrotating coordinate frame x–y. The angular displacement of the stick from the vertical position is $\theta = \theta(t)$. The mass of the cart is M, while the mass of the stick is m. The length of the stick is $2l$, and its center of gravity is at its geometric center. The control force applied to the cart is labeled u. We assume that the wheels of the cart do not slip. We model the frictional force of the cart wheels on the track by

$$
f_c = \mu_c \operatorname{sign}(\dot{x}), \tag{1.62}
$$

where μ_c is the cart friction coefficient. Friedland [91, p. 201] refers to this model of friction as the classical Coulomb friction model. Let (x_G, y_G) be the coordinates of the center of gravity of the stick. Then,

$$
\begin{aligned}
x_G &= x + l\sin(\theta), \\
y_G &= l\cos(\theta).
\end{aligned} \tag{1.63}
$$

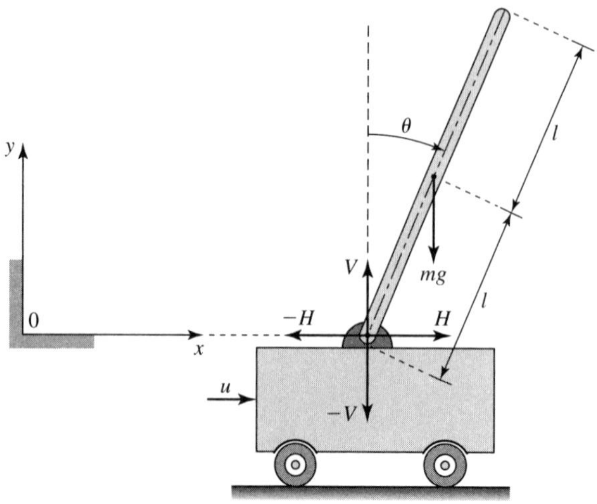

Figure 1.24 Free-body diagram of a cart with an inverted pendulum.

We are now ready to write down equations modeling the system. The equation that describes the rotational motion of the stick about its center of gravity is obtained by applying the rotational version of Newton's second law. Summing the moments about the center of gravity of the stick, we obtain

$$I\frac{d^2\theta}{dt^2} = Vl\sin(\theta) - Hl\cos(\theta), \tag{1.64}$$

where

$$I = \int_{-l}^{l} r^2\,dm = \frac{ml^2}{3} \tag{1.65}$$

is the moment of inertia of the stick with respect to its center of gravity.

We next write the equation that describes the horizontal motion of the center of gravity of the stick. Applying Newton's second law along the x axis yields

$$m\frac{d^2}{dt^2}(x + l\sin(\theta)) = H. \tag{1.66}$$

Performing the required differentiation, we get

$$m(\ddot{x} + l[-\dot{\theta}^2\sin(\theta) + \ddot{\theta}\cos(\theta)]) = H. \tag{1.67}$$

The equation that describes the vertical motion of the center of gravity of the stick is obtained by applying Newton's second law along the y axis. We have

$$m\frac{d^2}{dt^2}(l\cos(\theta)) = V - mg. \tag{1.68}$$

Performing the needed differentiation yields

$$ml(-\dot{\theta}^2\cos(\theta) - \ddot{\theta}\sin(\theta)) = V - mg. \tag{1.69}$$

Finally, we apply Newton's second law to the cart to get

$$M\frac{d^2x}{dt^2} = u - H - f_c. \tag{1.70}$$

Substituting (1.70) into (1.67) gives

$$m\ddot{x} + ml\ddot{\theta}\cos(\theta) - ml\dot{\theta}^2\sin(\theta) + f_c = u - M\ddot{x}. \tag{1.71}$$

Now substituting equations (1.69) and (1.70) into (1.64) yields

$$I\ddot{\theta} = (mg - ml\dot{\theta}^2\cos(\theta) - ml\ddot{\theta}\sin(\theta))l\sin(\theta) + f_c l\cos(\theta) - (u - M\ddot{x})l\cos(\theta). \tag{1.72}$$

We next substitute $u - M\ddot{x}$ from (1.71) into (1.72) and perform manipulations to get

$$I\ddot{\theta} = mgl\sin(\theta) - ml^2\ddot{\theta} - m\ddot{x}l\cos(\theta). \tag{1.73}$$

Let

$$a = \frac{1}{m+M}.$$

Then, we can represent (1.71) as

$$\ddot{x} = -mal\ddot{\theta}\cos(\theta) + mal\dot{\theta}^2\sin(\theta) - af_c + au. \tag{1.74}$$

We substitute (1.74) into (1.73) to obtain

$$\ddot{\theta} = \frac{mgl\sin(\theta) - m^2l^2a\dot{\theta}^2\sin(2\theta)/2 + mal\cos(\theta)f_c - mal\cos(\theta)u}{I - m^2l^2a\cos^2(\theta) + ml^2}. \tag{1.75}$$

Let $x_1 = \theta$ and $x_2 = \dot{\theta}$. Using the expression for I, given by (1.65), we represent (1.75) in state-space format as follows:

$$\begin{aligned}
\dot{x}_1 &= x_2, \\
\dot{x}_2 &= \frac{g\sin(x_1) - mlax_2^2\sin(2x_1)/2 + a\cos(x_1)f_c}{4l/3 - mla\cos^2(x_1)} - \frac{mal\cos(x_1)u}{4ml^2/3 - m^2l^2a\cos^2(x_1)}.
\end{aligned} \tag{1.76}$$

We next substitute $\ddot{\theta}$, obtained from (1.73), into (1.74) to get

$$\ddot{x} = \frac{-mag\sin(2x_1)/2 + ax_2^2\sin(x_1)4ml/3 + (u - f_c)4a/3}{4/3 - ma\cos^2(x_1)}. \tag{1.77}$$

Let $x_3 = x$ and $x_4 = \dot{x}$. Then, by combining (1.76) and (1.77) we obtain a state-space model of the inverted pendulum on a cart of the form

$$\begin{bmatrix}\dot{x}_1\\ \dot{x}_2\\ \dot{x}_3\\ \dot{x}_4\end{bmatrix} = \begin{bmatrix} x_2 \\ \dfrac{g\sin(x_1) - mlax_2^2\sin(2x_1)/2}{4l/3 - mla\cos^2(x_1)} \\ x_4 \\ \dfrac{-mag\sin(2x_1)/2 + alx_2^2\sin(x_1)4m/3}{4/3 - ma\cos^2(x_1)} \end{bmatrix} + \begin{bmatrix} 0 \\ \dfrac{-a\cos(x_1)}{4l/3 - mla\cos^2(x_1)} \\ 0 \\ \dfrac{4a/3}{4/3 - ma\cos^2(x_1)} \end{bmatrix}(u - f_c). \tag{1.78}$$

While designing a controller for the stick balancer system, we will use an equivalent representation of (1.78):

$$\begin{bmatrix}\dot{x}_1\\ \dot{x}_2\\ \dot{x}_3\\ \dot{x}_4\end{bmatrix} = \begin{bmatrix} x_2 \\ \dfrac{g\sin(x_1)}{4l/3 - mla\cos^2(x_1)} \\ x_4 \\ \dfrac{-mag\sin(2x_1)/2}{4/3 - ma\cos^2(x_1)}\end{bmatrix} + \begin{bmatrix} 0 \\ \dfrac{-a\cos(x_1)}{4l/3 - mla\cos^2(x_1)} \\ 0 \\ \dfrac{4a/3}{4/3 - ma\cos^2(x_1)}\end{bmatrix}\left(u - f_c + mlx_2^2\sin(x_1)\right). \quad (1.79)$$

1.7.5 Population Dynamics

We first model dynamics of a single population. Then, we use the obtained model to investigate two interacting populations.

Let $x = \phi(t)$ be the population of the given species at a time t. The simplest model of the population dynamics of the given species is the one resulting from the hypothesis that the net rate of change of x is proportional to the current value of x. Let r be the rate of growth of the population. Then, the simple model of the population dynamics is

$$\dot{x} = rx. \quad (1.80)$$

Solving (1.80) subject to the initial condition, $x(0) = x_0$, yields

$$x = e^{rt}x_0. \quad (1.81)$$

In the above linear model, the growth rate, r, is constant. It does not account for the limitation of space, food supply, and so on. Therefore, we modify the unrestricted growth rate r to take into account the fact that the environment of the population can only support a certain number, say K, of the species. The constant K is called the *carrying capacity* of the environment or the *saturation level*. Specifically, if $x > K$, then because there is not enough food or space to support x, more species will die than are born, which results in a negative growth rate. On the other hand, if $x < K$, then there is an extra food and space and the population growth is positive. One function that satisfies the above requirements has the form

$$r\left(1 - \frac{x}{K}\right).$$

Using the above model of the growth rate, we obtain the following model of the population growth:

$$\dot{x} = r\left(1 - \frac{x}{K}\right)x. \quad (1.82)$$

Let $c = r/K$. Then, we represent (1.82) as

$$\boxed{\dot{x} = (r - cx)x} \quad (1.83)$$

Equation (1.83) is known as the *Verhulst equation* or the *logistic equation*. According to Boyce and DiPrima [32, p. 60], Verhulst, a Belgian mathematician, introduced equation (1.83) in 1838 to model human population growth. For reasons somewhat unclear, Verhulst referred to it as the logistic growth; hence (1.83) is often called the logistic equation.

We will use the logistic equation to analyze interacting species. We first model two interacting species that do not prey upon each other, but compete for the same resources—for example, for a common food supply. Let x_1 and x_2 be the populations, at time t, of the two competing species. We assume that the population of each of the species in the absence of the other is governed by a logistic equation. Thus, a model of two noninteracting species has the form

$$\begin{aligned} \dot{x}_1 &= (r_1 - c_1 x_1)x_1, \\ \dot{x}_2 &= (r_2 - c_2 x_2)x_2, \end{aligned} \tag{1.84}$$

where r_1 and r_2 are the unrestricted growth rates of the two populations, while r_1/c_1 and r_2/c_2 are their saturation levels. When both species are present, each will affect the growth rate of the other. Because x_1 and x_2 use the same resources, the growth rate of x_1 will be lowered by an amount proportional to the size of x_2. We modify the growth equation for species x_1 as

$$\dot{x}_1 = (r_1 - c_1 x_1 - a_1 x_2)x_1, \tag{1.85}$$

where a_1 is a measure by which species x_2 lowers the growth rate of x_1. Similarly, the growth rate of x_2 is reduced by $a_2 x_1$, where a_2 is a measure of the degree by which species x_1 lowers the growth rate of x_2. Therefore, the growth equation for species x_2 becomes

$$\dot{x}_2 = (r_2 - c_2 x_2 - a_2 x_1)x_2. \tag{1.86}$$

In summary, equations (1.85) and (1.86) model two species that interact by competing for a common resource like food or other supply.

We will now study the situation in which one species, the predator, preys upon the other species, the prey, and there is no competition between the predator and prey for a common resource. Let x_1 and x_2 denote the populations of the prey and predator, respectively. We will make the following modeling assumptions:

1. When there is no predator—that is, $x_2 = 0$—the prey enjoys an exponential growth governed by

$$\dot{x}_1 = r_1 x_1, \qquad r_1 > 0. \tag{1.87}$$

2. When there is no prey—that is, $x_1 = 0$—the predator perishes. We model this scenario using the equation

$$\dot{x}_2 = -r_2 x_2, \qquad r_2 > 0. \tag{1.88}$$

3. The larger the population of the predator, the more prey is eaten. The larger the population of the prey, the easier hunting is for the predator. We model the growth rate decrease of the prey by including in (1.87) the term $-dx_1x_2$, $d > 0$. The growth rate increase of the predator is modeled by adding the term ex_1x_2 in (1.88), where $e > 0$.

The resulting model of the predator–prey relationship is

$$\boxed{\begin{aligned} \dot{x}_1 &= (r_1 - dx_2)x_1 \\ \dot{x}_2 &= (-r_2 + ex_1)x_2 \end{aligned}} \tag{1.89}$$

Equations (1.89) are known as the *Lotka–Volterra equations*. They were published by Lotka in 1925 and by Volterra in 1926.

Notes

A landmark paper on mathematical description of linear dynamical systems from a controls point of view is by Kalman [147]. The first chapter of Sontag's book [269] is a nice, easy to read, and comprehensive introduction to the subject of mathematical control theory. Sontag's chapter is easily available on the web. For more examples of dynamical system models from mechanical and electrical engineering, the reader is referred to Truxal [283], Banks [19], Raven [240], Mohler [207], and D'Azzo and Houpis [57]. Wertz [298] has many excellent real-life models from space science. A well-written paper on vehicle modeling and control for broad audience is by Ackermann [2]. In addition to references 303 and 299, vehicle and tire dynamics modeling are investigated in detail by Nalecz [211], Nalecz and Bindemann [212, 213], and Smith and Starkey [263, 264]. For more on stepper motors, the reader is referred to Acarnley [1], Kenjo [153], Kenjo and Nagamori [154], Del Toro [61, Chapter 9], Krause and Wasynczuk [169, Chapter 8], and Zribi and Chiasson [319]. For further analysis of the equations modeling competing species and the predator–prey equations, we recommend Boyce and DiPrima [32, Sections 9.4 and 9.5]. Our derivation of the logistic equation follows that of Sandefur [253]. The logistic equation as well as equations modeling interacting species can be used for mathematical modeling of tumor dynamics and interactions between tumor and immune system. A survey of models for tumor–immune system dynamics can be found in the well-edited book by Adam and Bellomo [4].

Mayr writes on page 109 in reference 199, "It is still widely believed that the steam-engine governor is the oldest feedback device, and that James Watt had not only invented but also patented it. While both errors are easily refuted, we are still not able to reconstruct the history of this invention in all desired completeness." Watt did not patent the governor. He did not invent it either. On page 112 of his book [199], Mayr adds the following: "But the application of the centrifugal pendulum in a system of speed regulation of steam engines was a new breakthrough for which the firm of Boulton & Watt, if not James Watt himself, clearly deserves the credit."

EXERCISES

1.1 Prove the law of conservation of energy as stated in the following theorem:

> **Theorem 1.1** Given a conservative vector field $\boldsymbol{F}$—that is, $\boldsymbol{F} = -\nabla U$, where U is the potential energy—assume that a particle of mass m, satisfying Newton's law, moves on a differentiable curve $\boldsymbol{C}$. Then, the sum of the potential energy and kinetic energy is constant.

Hint Denote by $\boldsymbol{C}(t)$ a differentiable curve along which a particle of mass m moves. Hence,

$$\boldsymbol{F}(\boldsymbol{C}(t)) = m\ddot{\boldsymbol{C}}(t).$$

Prove that

$$U(\boldsymbol{C}(t)) + \tfrac{1}{2}m\|\dot{\boldsymbol{C}}(t)\|^2 = \text{constant}.$$

To show that the above relation holds, differentiate $U(\boldsymbol{C}(t)) + \frac{1}{2}m\|\dot{\boldsymbol{C}}(t)\|^2$ with respect to time and show that the time derivative is zero; hence the sum of the energies is constant.

1.2 Show that in a conservative vector field the work done by the field on a particle that moves between two points depends only on these points and not on the path followed.

Hint Note that the work done against the conservative force field $\boldsymbol{F}$ along a curve $\boldsymbol{C}$ between points $\boldsymbol{C}(t_1)$ and $\boldsymbol{C}(t_2)$ is

$$\int_{t_1}^{t_2} \boldsymbol{F}(\boldsymbol{C}(t))^T \dot{\boldsymbol{C}}(t)\,dt.$$

By assumption, $\boldsymbol{F}$ is a conservative vector field. Hence,

$$\boldsymbol{F} = -\nabla U. \tag{1.90}$$

Combining the above yields

$$\int_{t_1}^{t_2} \boldsymbol{F}(\boldsymbol{C}(t))^T \dot{\boldsymbol{C}}(t)\,dt = -\int_{t_1}^{t_2} \nabla U(\boldsymbol{C}(t))^T \dot{\boldsymbol{C}}(t)\,dt.$$

Use the chain rule to represent the expression inside the integral on the right-hand side as

$$\nabla U(\boldsymbol{C}(t))^T \dot{\boldsymbol{C}}(t) = \frac{dU(\boldsymbol{C}(t))}{dt}.$$

1.3 An example of a closed-loop feedback system is a toilet flushing device connected to a water tank. A toilet system is shown in Figure 1.25. The control objective is to maintain water in the tank at a constant level. Draw a block diagram of the system.

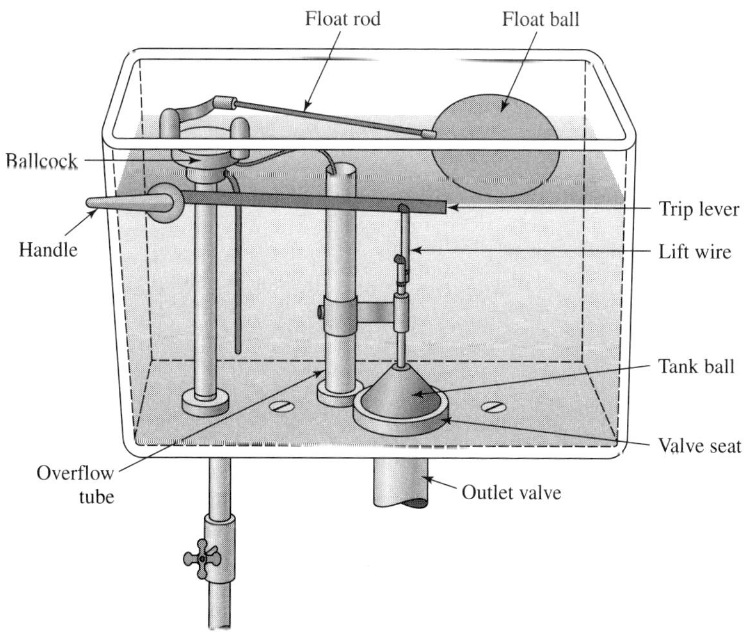

Figure 1.25 Toilet-flushing system of Exercise 1.3. (Adapted from reference 241, p. 211.)

Table 1.3 Numerical Data for the Two-Tank Pumped Storage System Model for Exercise 1.4

Tank 1 diameter	2 m
Tank 2 diameter	3 m
R_1	10 sec/m^2
R_2	20 sec/m^2

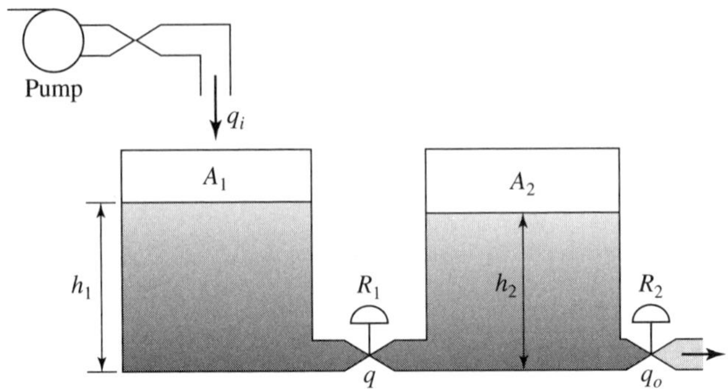

Figure 1.26 A two-tank pumped storage system for Exercise 1.4.

1.4 The goal of this exercise is to develop a model of a two-tank pumped storage system. Figure 1.26 illustrates such a system. A pumped storage system can be used to supply water to a turbine producing electric power. The two storage tanks are connected together through a valve that may be modeled as a linear resistance R_1. The resistance of the turbine supply pump is modeled by a linear resistance R_2. Suppose that we wish to design a controller that would control the inflow rate q_i in order to maintain the output flow rate q_o at some desired value q_d. We assume that the linearized flow rates $q(t)$ and $q_o(t)$ are proportional to the corresponding water heights; that is,

$$q(t) = \frac{h_1(t) - h_2(t)}{R_1},$$

$$q_o(t) = \frac{h_2(t)}{R_2}.$$

Let A_1 and A_2 represent the surface areas in the two tanks. Then, the principle of conservation of mass for an incompressible fluid yields

$$q_i(t) - q(t) = A_1 \dot{h}_1(t),$$

$$q(t) - q_o(t) = A_2 \dot{h}_2(t).$$

Numerical data of a two-tank pumped storage system is given in Table 1.3. Find the transfer function $\mathcal{L}(q_o(t))/\mathcal{L}(q_i(t))$.

1.5 The objective of this exercise is to derive a model of an automotive suspension. The role of an automotive suspension is to support the vehicle body on the axles. A simple linear model of the suspension assembly associated with a single wheel, called the quarter-car model, is shown in Figure 1.27. The vehicle body is represented by the "sprung" mass, m_s, while the tire and axle are represented by the "unsprung" mass, m_u. Model the tire

Table 1.4 Parameters of the Vehicle Suspension of Exercise 1.5

m_s	355 kg
m_u	46 kg
k_s	14384 N/m
b_s	1860 N/m/sec
k_t	300,000 N/m

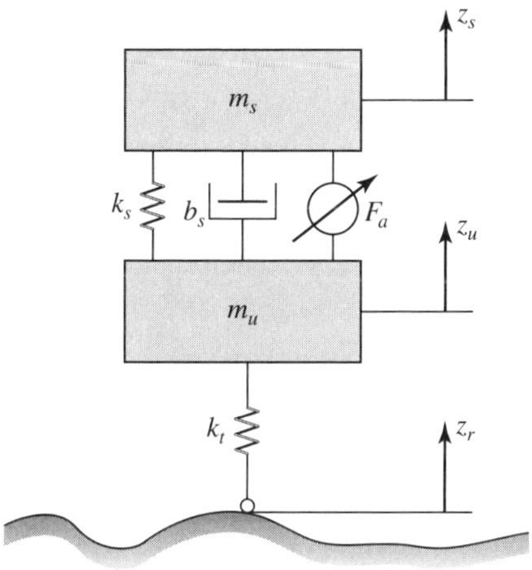

Figure 1.27 Linear quarter-car model for Exercise 1.5.

as a spring with spring coefficient k_t, and assume its damping to be negligible. The suspension consists of the spring, the shock absorber, and the variable force element. The role of the variable force element is to generate a force that compensates for the uneven road, which is the source of the system disturbances. Assume that the variable force element F_a can instantaneously provide any desired force. The above described model of an automotive suspension is referred to as a linear active suspension. Write down equations modeling the dynamics of the two-degree-of-freedom system depicted in Figure 1.27 by applying Newton's law to each mass of the system and taking into account the positive direction of the z coordinates. Define the state variables to be

$$x_1 = z_s - z_u = \text{suspension stroke},$$
$$x_2 = \dot{z}_s = \text{sprung mass velocity},$$
$$x_3 = z_u - z_r = \text{tire deflection},$$
$$x_4 = \dot{z}_u = \text{unsprung mass velocity}.$$

Let $u = F_a$ be the system input, and let $d(t) = \dot{z}_r(t)$ be the disturbance input due to the road roughness. Represent the obtained linear model of active suspension in state-space format. Use data of a commercial vehicle given in Table 1.4. The data come from Sunwoo and Cheok [272]. Substitute the vehicle data into the modeling equations. A design goal of an automotive engineer may be to design a control law to isolate the sprung mass from road disturbances. In other words, he or she may wish to design a control law such that the vertical velocity of the sprung mass x_2 is as close to 0 as possible in spite of road disturbances. Assume that we can measure the vertical velocity of the sprung mass. Write the output equation.

To investigate the effect of road disturbances on the vertical velocity of the sprung mass, set $u = 0$ and $\delta(t) = d\sin(6t)$. Generate a plot of y versus t for t in the range [0 sec, 5 sec] and $d = 0.5$.

1.6 The objective of this exercise is to develop a model of a single-axis machine tool drive in which a motor drives a lead screw through a gear. Figure 1.28 illustrates such a system. This model comes from Driels [68, pp. 613–619]. The system operates as follows. The lead screw moves the table with a workpiece on it under the cutter to perform the desired operation. The table is positioned by a control system. We wish to place the table in a desirable position as rapidly and accurately as possible. This means that while modeling the machine tool drive system we should, in particular, take into account the flexibility of the lead screw. The length of the lead screw implies that its rotation at the gear box will not be the same as its rotation at the table. Thus, the angular twist along the length of the lead screw cannot be neglected. Model the lead screw flexibility using a torsional spring as illustrated in Figure 1.29. The system input is the motor torque τ_m, and the system output is the rotation of the end of the lead screw labeled θ_l. Assume that the damping torque, due to viscous friction, acting on the motor is proportional to its angular velocity $\dot{\theta}_m = \omega_m$ with the coefficient of viscous friction denoted c_m. Denote the motor moment of inertia by I_m, and denote by τ_1 the load torque on the motor gear due to the rest of the gear train. Write the equation modeling the dynamics of the motor shaft.

Let τ_2 be the torque transmitted to the load shaft, and let $N = N_1/N_2$ be the gear ratio, where N_1 denotes the number of teeth of the first gear, and N_2 is the number of teeth of the second gear. The work done by the first gear is equal to that of the second.

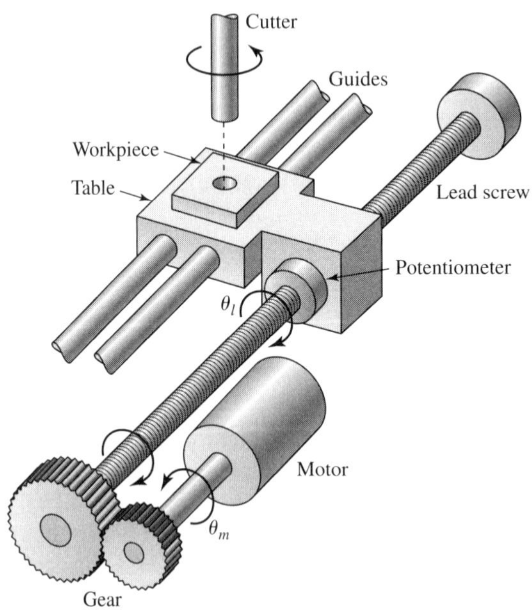

Figure 1.28 A schematic of a machine tool drive system of Exercise 1.6. (Adapted from Driels [68].)

Table 1.5 Data for the Machine Tool Drive System Model for Exercise 1.6

I_m	0.0001 Nm·sec^2/rad
I_l	0.001 Nm·sec^2/rad
$c_m = c_l$	0.01 Nm·sec/rad
k	1 Nm/rad
N	1/5

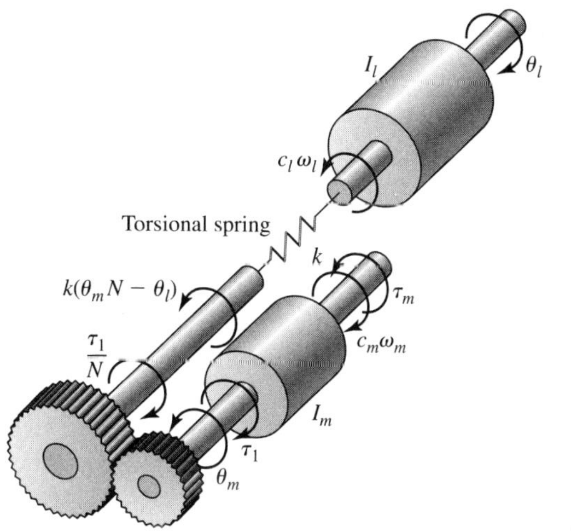

Figure 1.29 A mechanical model of the machine tool drive system for Exercise 1.6.

Model the load shaft flexibility with a torsional spring. The torsional spring coefficient is k. Assume that the moment of inertia of the lead screw is negligible compared to that of the load. Denote the moment of inertia of the load by I_l. The torque due to viscous friction of the load is $c_l\omega_l$, where c_l is the coefficient of the viscous friction of the load. The angular displacement of the end of the lead screw is θ_l. Write the equation modeling the dynamics of the load. Define the state variables to be

$$x_1 = \theta_m, \qquad x_2 = \theta, \qquad x_3 = \theta_l, \qquad x_4 = \dot{\theta}_l.$$

Represent the modeling equations in state-space format. Typical numerical data for a machine tool drive system are shown in Table 1.5.

1.7 Consider a bicycle model of the front-wheel-steering vehicle shown in Figure 1.30. Construct a two-dimensional state-space model of the vehicle, where $x_1 = \dot{y}$ and $x_2 = \dot{\theta}$. The input $u = \delta_f$ and the output $y = x_2$. In your derivation assume small angles. Vehicle data are given in Table 1.6.

Table 1.6 Vehicle Data for Exercise 1.7

$l_1 = l_2$	1 m
v_x	10 m/sec
m	1000 kg
I_z	1000 kg·m^2
$C_f = C_r$	10,000 N/rad

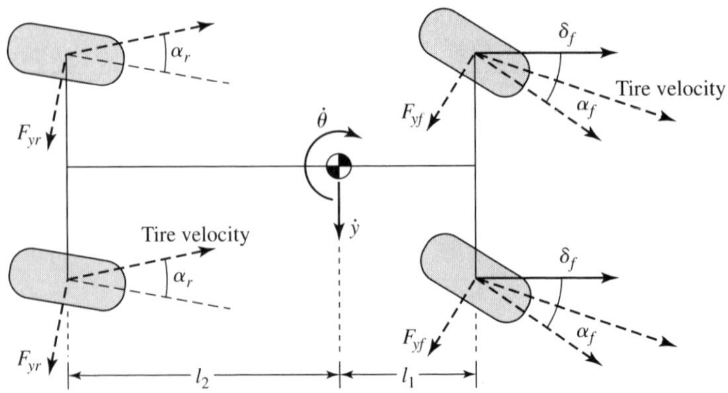

Figure 1.30 Bicycle model of a vehicle for Exercise 1.7.

1.8 Consider the two-mass–spring system depicted in Figure 1.31.

(a) Form the Lagrangian of the system.

(b) Write the Lagrange equations of motion for the system.

(c) Represent the obtained Lagrange equations of motion in state-space format.

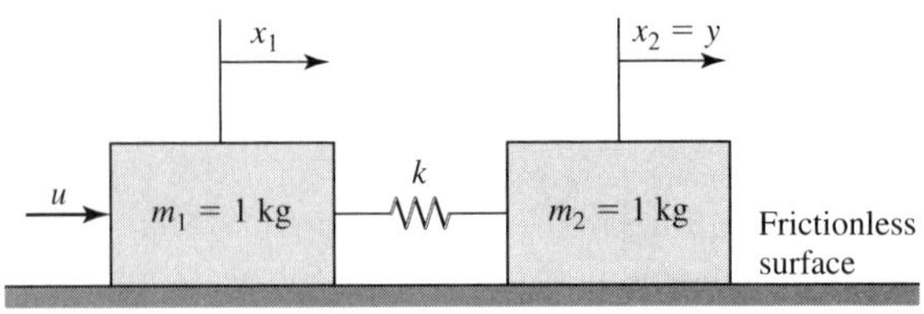

Figure 1.31 The two-mass–spring system for Exercise 1.8.

1.9 Consider the mechanical system shown in Figure 1.32.

(a) Form the Lagrangian for the system.

(b) Write the Lagrange equation of motion for the system.

(c) Represent the system model in state space format.

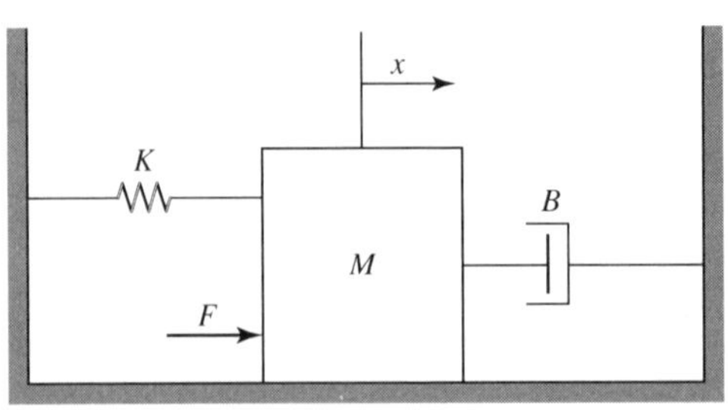

Figure 1.32 Mechanical system for Exercise 1.9.

1.10 Consider a dynamical system that consists of a cart with an inverted pendulum (point mass on a massless shaft) attached to it as depicted in Figure 1.33.

(a) Write the Lagrange equations of motion for the system.

(b) Represent the obtained model in state-space format using the state variables

$$x_1 = x, \qquad x_2 = \dot{x}, \qquad x_3 = \theta, \qquad x_4 = \dot{\theta}.$$

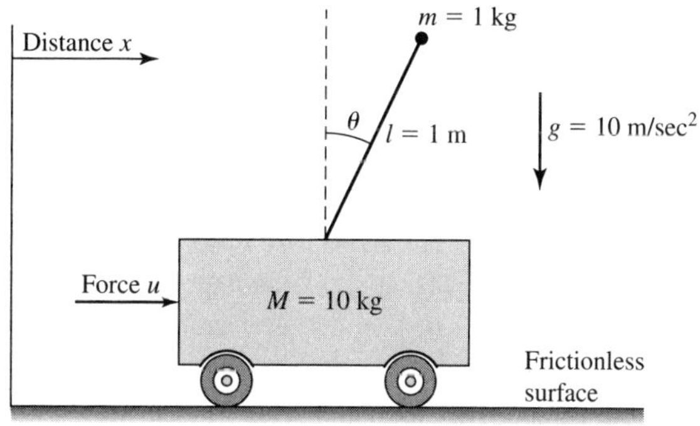

Figure 1.33 An inverted pendulum moving on a cart for Exercise 1.10.

1.11 Consider a schematic of a field-controlled DC motor shown in Figure 1.34. Let $x_1 = \theta$, $x_2 = \dot{\theta}$, $x_3 = i_f$, and $y = x_1$. Treat the field inductance L_f as a parasitic element and denote it as μ.

(a) Form the actual plant model P_μ in state-space format.

(b) Find the reduced-order design model P_0.

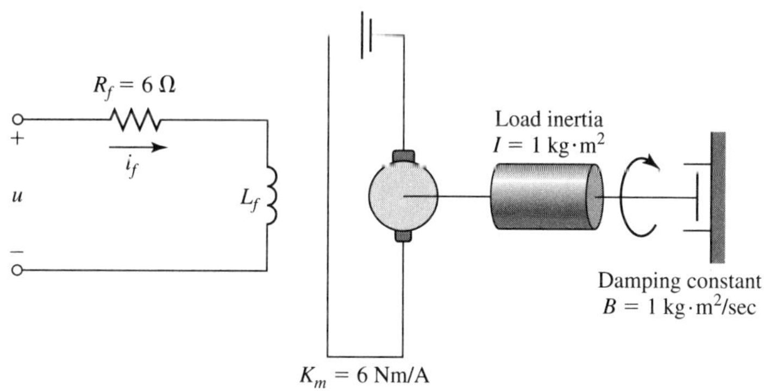

Figure 1.34 A diagram of a field-controlled DC motor for Exercise 1.11.

CHAPTER 2 Analysis of Modeling Equations

> To be sure, mathematics can be extended to any branch of knowledge, including economics, provided the concepts are so clearly defined as to permit accurate symbolic representation. That is only another way of saying that in some branches of discourse it is desirable to know what you are talking about.
>
> —James R. Newman

Most of dynamical systems analyzed in this book are modeled by ordinary differential equations. A main use of a mathematical model is to predict the system transient behavior. Unlike linear systems, where closed-form solutions can be written in terms of the system's eigenvalues and eigenvectors, finding analytical, exact, solutions to nonlinear differential equations can be very difficult or impossible. However, we can approximate the solutions of differential equations with difference equations whose solutions can be obtained easily using a computer. In this chapter we discuss a number of methods for solving differential equations. The first class of methods allows us to graphically determine solutions to second-order differential equations. Then, we discuss numerical techniques for solving differential equations. After that, we present two methods of linear approximation of nonlinear systems.

2.1 State-Plane Analysis

The *state plane,* which is a state space of two-dimensional systems, is also called the *phase plane*. Analysis in the state plane is applicable to linear and nonlinear systems modeled by second-order ordinary differential equations. The state-plane methods are graphical procedures for solving such equations. Using state-plane methods, one can graphically determine the transient response of a second-order dynamical system. We now introduce relevant terms. Consider a class of second-order systems whose dynamics are modeled by the differential equation

$$\ddot{x} = f(x, \dot{x}). \tag{2.1}$$

Define the state variables

$$x_1 = x \quad \text{and} \quad x_2 = \dot{x}.$$

Then, equation (2.1) can be represented as

$$\begin{aligned}\dot{x}_1 &= x_2,\\ \dot{x}_2 &= f(x_1, x_2).\end{aligned} \tag{2.2}$$

The plane with coordinates x_1 and x_2 is labeled as the state plane or phase plane. A solution of (2.1) can be illustrated by plotting x versus t, or by plotting x_2 versus x_1 using t as a parameter. To each value of the state $\boldsymbol{x}(t) = [x_1(t)\ \ x_2(t)]^T$ there corresponds a point in the (x_1, x_2) plane called the *representative point* (RP). As t varies, the RP describes a curve in the state plane called a *trajectory*. A family of trajectories is called a *phase portrait*.

2.1.1 Examples of Phase Portraits

◆ Example 2.1

Consider the spring–mass system shown in Figure 2.1, where x is the displacement of the mass M from equilibrium, and k is the linear spring constant. The spring is assumed to be linear; that is, it obeys Hooke's law. Applying Newton's law, we obtain the equation of motion of the mass:

$$M\ddot{x} + kx = 0.$$

Let $\omega^2 = k/M$. Then, the above equation can be represented in standard form:

$$\ddot{x} + \omega^2 x = 0. \tag{2.3}$$

Observe that

$$\ddot{x} = \dot{x}\frac{\ddot{x}}{\dot{x}} = \dot{x}\frac{d\dot{x}/dt}{dx/dt} = \dot{x}\frac{d\dot{x}}{dx}.$$

Using the above, we represent equation (2.3) as

$$\dot{x}\frac{d\dot{x}}{dx} + \omega^2 x = 0.$$

Separating the variables yields

$$\frac{\dot{x}}{\omega^2}\,d\dot{x} + x\,dx = 0.$$

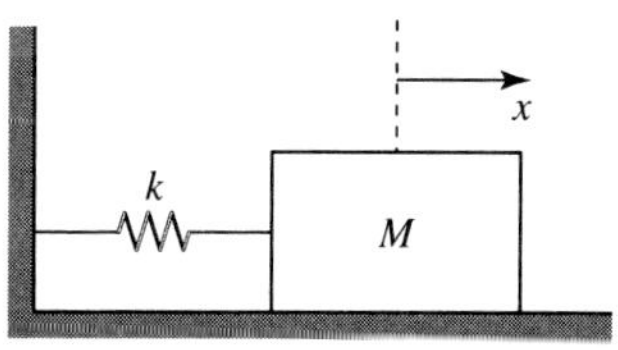

Figure 2.1 A spring–mass system of Example 2.1.

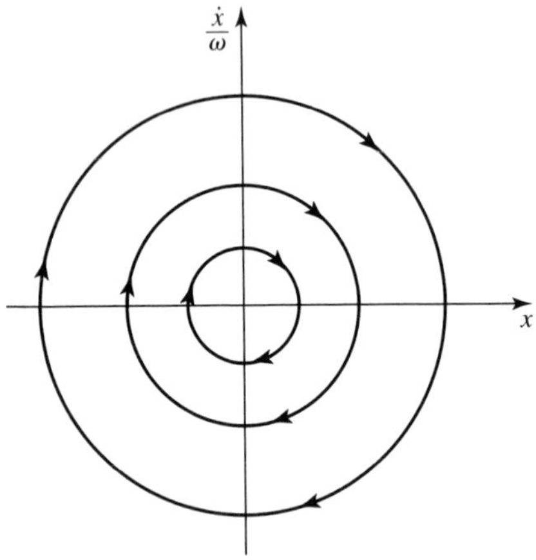

Figure 2.2 A phase portrait for the spring–mass system of Example 2.1.

Integrating the above gives

$$\left(\frac{\dot{x}}{\omega}\right)^2 + x^2 = A^2,$$

where A is the integration constant determined by the initial conditions. We see that a phase portrait of equation (2.3) consists of a family of circles centered at the origin of the $(x, \dot{x}/\omega)$ plane as illustrated in Figure 2.2. Different circular trajectories, in this plane, correspond to different values of the constant A representing initial conditions. Note that

$$x = \omega \int \frac{\dot{x}}{\omega}\, dt,$$

where $\omega > 0$. Hence, for positive values of $\dot{x}/\omega$ the displacement x increases. This means that for the values $\dot{x}/\omega > 0$ the arrows on the trajectories indicate x moving to the right. For negative values of $\dot{x}/\omega$—that is, for the values of $\dot{x}/\omega$ below the horizontal axis—the arrows on the trajectories indicate x decreasing by going to the left. Therefore, as time increases, the RP moves in a clockwise direction along the trajectory.

◆ Example 2.2

In this example, we consider a model of a satellite shown in Figure 2.3. We assume that the satellite is rigid and is in a frictionless environment. It can rotate about an axis perpendicular to the page as a result of torque τ applied to the satellite by firing the thrusters. The system input is the applied torque, and the system output is the attitude angle θ. The satellite's moment of inertia is I. Applying the rotational

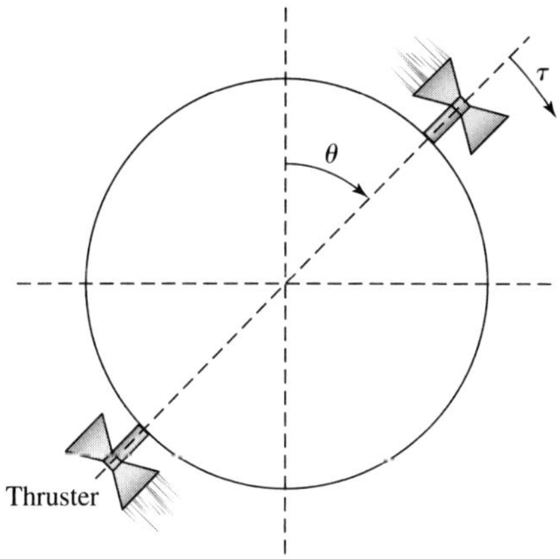

Figure 2.3 A model of a rigid satellite of Example 2.2.

version of Newton's law, we obtain the satellite's equation of motion:

$$I\frac{d^2\theta}{dt^2} = \tau,$$

which can be represented in the following equivalent form:

$$\frac{d^2\theta}{dt^2} = \frac{1}{I}\tau.$$

Let

$$\frac{1}{I}\tau = u,$$

then the equation describing the satellite's motion becomes

$$\frac{d^2\theta}{dt^2} = u. \tag{2.4}$$

We assume that when the thrusters fire, the thrust is constant; that is, $u = \pm U$. For simplicity, let $U = 1$. Using the state variables $x_1 = \theta$ and $x_2 = \dot{\theta}$, we obtain the satellite's state-space model:

$$\begin{bmatrix} \dot{x}_1 \\ \dot{x}_2 \end{bmatrix} = \begin{bmatrix} 0 & 1 \\ 0 & 0 \end{bmatrix} \begin{bmatrix} x_1 \\ x_2 \end{bmatrix} + \begin{bmatrix} 0 \\ 1 \end{bmatrix} u, \tag{2.5}$$

where $u = \pm 1$. We first consider the case when $u = 1$. With this control, the system equations are

$$\dot{x}_1 = x_2,$$
$$\dot{x}_2 = 1.$$

Solving the second equation yields

$$x_2 = t + c_1,$$

where c_1 is an integration constant. We substitute $x_2 = t + c_1$ into $\dot{x}_1 = x_2$ to get

$$\dot{x}_1 = t + c_1.$$

Solving the above equation, we obtain

$$\begin{aligned} x_1 &= \int (t + c_1)\,dt \\ &= \tfrac{1}{2}(t + c_1)^2 + c_2 \\ &= \tfrac{1}{2}x_2^2 + c_2, \end{aligned}$$

where c_2 is an integration constant. In the state plane the above equation represents a family of parabolas open toward the positive x_1 axis. The representative point moves upward along these parabolas because $\dot{x}_2 = 1$; that is, $\dot{x}_2 > 0$. Typical system trajectories for this case are shown in Figure 2.4.

Similarly, when $u = -1$, we have

$$\dot{x}_2 = -1.$$

Solving the above differential equation yields

$$x_2 = -t + \tilde{c}_1,$$

where $\tilde{c}_1$ is an integration constant. Substituting $x_2 = -t + \tilde{c}_1$ into $\dot{x}_1 = x_2$ and then solving the resulting equation yields

$$x_1 = -\tfrac{1}{2}x_2^2 + \tilde{c}_2,$$

where $\tilde{c}_2$ is an integration constant. In the state plane, this equation represents

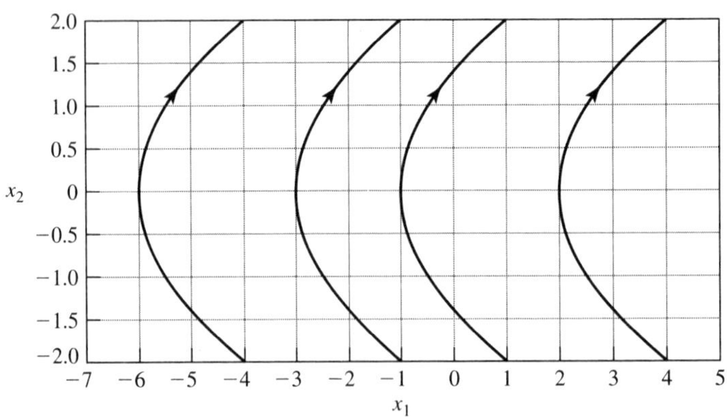

Figure 2.4 Trajectories of the system (2.5) for $u = 1$.

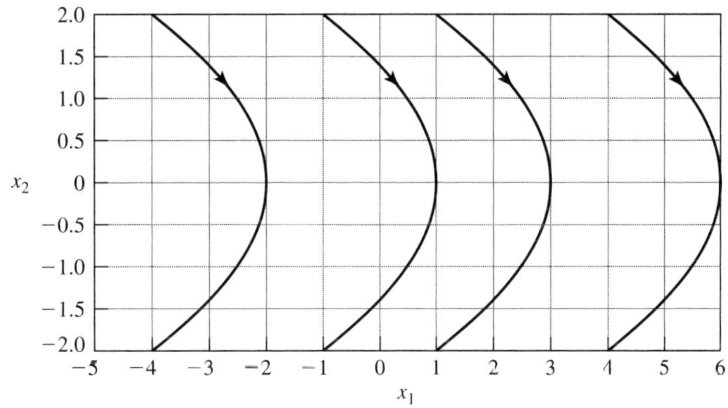

Figure 2.5 Trajectories of the system (2.5) for $u = -1$.

a family of parabolas open toward the negative x_1 axis. The representative point moves downward along these parabolas because $\dot{x}_2 = -1$; that is, $\dot{x} < 0$. Typical system trajectories for this case are shown in Figure 2.5.

In the above examples, systems' trajectories were constructed analytically. Analytical methods are useful for systems modeled by differential equations that can be easily solved. If the system of differential equations cannot be easily solved analytically, we can use graphical or numerical methods. We now describe a graphical method for solving second-order differential equations or a system of two first-order differential equations.

2.1.2 The Method of Isoclines

An *isocline* is a curve along which the trajectories' slope is constant. In other words, an isocline is a locus (a set of points), of the trajectories' constant slope. To obtain equations that describe isoclines, we consider a system of two first-order differential equations of the form

$$\begin{aligned} \dot{x}_1 &= \frac{dx_1}{dt} = f_1(x_1, x_2), \\ \dot{x}_2 &= \frac{dx_2}{dt} = f_2(x_1, x_2). \end{aligned} \tag{2.6}$$

Dividing the second of the above equations by the first one, we obtain

$$\frac{dx_2}{dx_1} = \frac{f_2(x_1, x_2)}{f_1(x_1, x_2)}. \tag{2.7}$$

Thus, we eliminated the independent variable t from the set of the first-order differential equations given by (2.6). In equation (2.7), we consider x_1 and x_2 as the independent and dependent variables, respectively. Let

$$m = m(x_1, x_2) = \frac{dx_2}{dx_1}.$$

Note that m is just the slope of the tangent to the trajectory passing through the point $[x_1 \;\; x_2]^T$. The locus of the trajectory constant slope,

$$\frac{dx_2}{dx_1} = m(x_1, x_2) = \text{constant},$$

can be obtained from the equation

$$f_2(x_1, x_2) = m f_1(x_1, x_2).$$

The curve that satisfies the above equation is an isocline corresponding to the trajectories' slope m because a trajectory crossing the isocline will have its slope equal to m. The idea of the method of isoclines is to construct several isoclines in the state plane. This then allows one to construct a field of local tangents m. Then, the trajectory passing through any given point in the state plane is obtained by drawing a continuous curve following the directions of the field.

◆ Example 2.3

Consider the spring–mass system from Example 2.1. Defining the state variables $x_1 = x$ and $x_2 = \dot{x}$, we obtain the state equations

$$\begin{aligned}\dot{x}_1 &= x_2,\\ \dot{x}_2 &= -\omega^2 x_1.\end{aligned}$$

Let, for simplicity, $\omega = 1$. Then,

$$\frac{dx_2}{dx_1} = m = -\frac{x_1}{x_2}.$$

From the above we obtain the equation of isoclines:

$$x_2 = -\frac{1}{m} x_1.$$

This equation describes a family of straight lines through the origin. In Figure 2.6, we show some of the isoclines and a typical trajectory for the above spring–mass system. Note that in this example the trajectory slope is m, while the isocline slope is $-1/m$. This means that trajectories of the spring–mass system are perpendicular to the isoclines; that is, the trajectories are circles centered at the origin of the state plane.

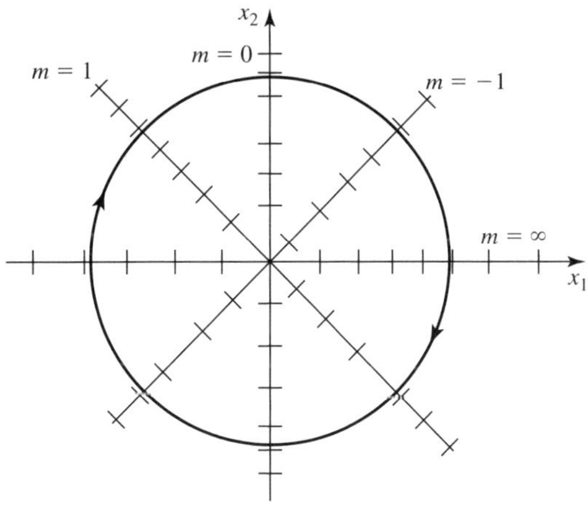

Figure 2.6 Isoclines and a typical trajectory for the spring–mass system of Example 2.3.

◆ Example 2.4

Given a second-order differential equation

$$\ddot{x} = -4\eta\dot{x} - 4x, \tag{2.8}$$

where η is a parameter. We will find values of η for which there are isoclines along which the trajectory slope and the isocline slope are equal. Such isoclines, if they exist, are called the *asymptotes*.

Let $x_1 = x$ and $x_2 = \dot{x}$. Then, the second-order differential equation given by (2.8) can be represented as a system of first-order differential equations:

$$\begin{aligned}\dot{x}_1 &= x_2,\\ \dot{x}_2 &= -4\eta x_2 - 4x_1.\end{aligned}$$

The isocline equation is

$$\frac{dx_2}{dx_1} = \frac{-4\eta x_2 - 4x_1}{x_2} = m,$$

or, equivalently,

$$x_2 = -\left(\frac{4}{m+4\eta}\right)x_1. \tag{2.9}$$

A trajectory slope is equal to an isocline slope if and only if

$$-\frac{4}{m+4\eta} = m;$$

that is,

$$m^2 + 4\eta m + 4 = 0. \tag{2.10}$$

The discriminant of (2.10) is

$$\Delta = 4\sqrt{\eta^2 - 1}.$$

Hence, the condition for the existence of asymptotes is

$$|\eta| \geq 1. \tag{2.11}$$

◆ Example 2.5

Consider the system of differential equations

$$\dot{x}_1 = x_1 + x_2 + 1,$$
$$\dot{x}_2 = -x_2 + 2.$$

(a) Find the equation of the isocline corresponding to the trajectory slope $m = 1$.

(b) Does the given system have asymptotes? If yes, then write the equations of the asymptotes.

The trajectory slope is

$$\frac{dx_2}{dx_1} = m = \frac{-x_2 + 2}{x_1 + x_2 + 1}.$$

Hence, the equation of the isocline corresponding to the trajectory slope $m=1$ is obtained by solving

$$\frac{-x_2 + 2}{x_1 + x_2 + 1} = 1.$$

We obtain

$$x_2 = -\tfrac{1}{2}x_1 + \tfrac{1}{2}.$$

The trajectory slope is

$$m = \frac{-x_2 + 2}{x_1 + x_2 + 1}. \tag{2.12}$$

From (2.12), we obtain an equation of isoclines of the form

$$x_2 = -\frac{m}{m+1}x_1 - \frac{m-2}{m+1}.$$

An asymptote is the isocline for which its slope and the trajectory slope are equal.

Therefore, we can obtain the equations of asymptotes by solving

$$m = -\frac{m}{m+1}.$$

We obtain

$$m^2 + 2m = m(m+2) = 0.$$

Hence, the equations of asymptotes are

$$x_2 = 2 \quad \text{and} \quad x_2 = -2x_1 - 4.$$

We can use a computer to draw phase portraits interactively. Below is an example of MATLAB's script that was used to sketch a phase portrait of the van der Pol equation shown in Figure 2.7.

```
t0=0; % initial time
tf=20; % final time
tspan=tf-t0;
x0=[-4 -4]'; % initial condition
button=1;
% we will now draw internal axes
p=4*[-1 0;1 0];
clf;plot(p(:,1),p(:,2))
hold on
plot(p(:,2),p(:,1))
axis(4*[-1 1 -1 1])
    while(button==1)
% select a new initial condition and press the left button. Press
% the right button if you want to stop.
      [t,x]=ode45(@my_xdot,tspan,x0); % invoke your own m.file
      plot(x(:,1),x(:,2))
      [x1,x2,button]=ginput(1);
      x0=[x1 x2]';
    end
```

The van der Pol equation has the form

$$\ddot{x} - \mu(1 - x^2)\dot{x} + x = 0, \tag{2.13}$$

where μ is a parameter. We can equivalently represent the van der Pol equation as

$$\begin{aligned} \dot{x}_1 &= x_2, \\ \dot{x}_2 &= \mu\left(1 - x_1^2\right)x_2 - x_1. \end{aligned} \tag{2.14}$$

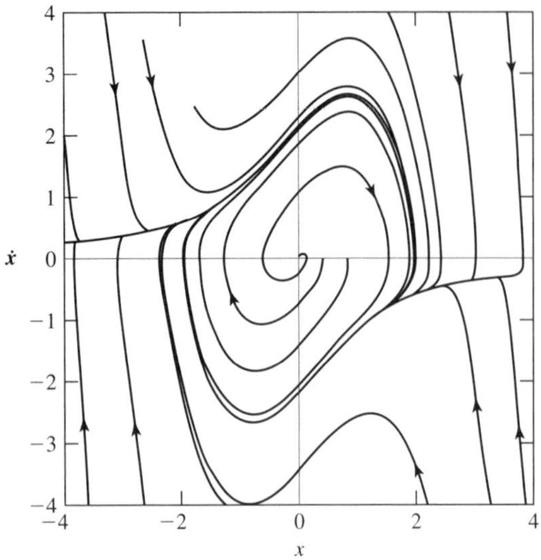

Figure 2.7 A phase-plane portrait of the van der Pol equation for $\mu = 1$.

An inspection of the phase portrait of the van der Pol equation reveals the existence of a closed trajectory in the state plane. This solution is an example of a limit cycle.

Definition 2.1 A limit cycle is a closed trajectory in the state plane, such that no other closed trajectory can be found arbitrarily close to it.

Note that the differential equation modeling the spring–mass system does not have a limit cycle because the solutions are not isolated. Limit cycles can only appear in nonlinear systems; they do not exist in linear systems.

Theorem 2.1 (Bendixson) If f_1 and f_2 in (2.7) have continuous partial derivatives in a domain $\mathcal{A}$ bounded by a closed path $\boldsymbol{C}$ and if $\partial f_1/\partial x_1 + \partial f_2/\partial x_2$ does not vanish and is of constant sign in that domain, then (2.7) can have no limit cycle in that domain.

Proof We prove the theorem by contradiction. Equation (2.7) is satisfied on any of its trajectory. Therefore, if (2.7) has a limit cycle, then clearly

$$f_1\,dx_2 - f_2\,dx_1 = 0. \tag{2.15}$$

We now apply Green's theorem, stated on page 674, to the vector field

$$[-f_2 \quad f_1]^T$$

to obtain

$$\int_C f_1\,dx_2 - f_2\,dx_1 = \iint_A \left(\frac{\partial f_1}{\partial x_1} + \frac{\partial f_2}{\partial x_2}\right) dx_1\,dx_2. \tag{2.16}$$

By (2.15), the right-hand side of (2.16) must vanish, which contradicts the hypothesis. Thus, if $\partial f_1/\partial x_1 + \partial f_2/\partial x_2$ does not vanish and is of constant sign in a domain $\mathcal{A}$ bounded by a closed path $\boldsymbol{C}$, then no limit cycle can exist in $\mathcal{A}$. The proof of the theorem is complete.

◆ Example 2.6

We apply Bendixson's theorem to the van der Pol equation represented by (2.14). We have

$$f_1 = x_2 \quad \text{and} \quad f_2 = \mu(1 - x_1^2)x_2 - x_1.$$

We calculate $\partial f_1/\partial x_1 + \partial f_2/\partial x_2$ to get

$$\partial f_1/\partial x_1 + \partial f_2/\partial x_2 = \mu(1 - x_1^2).$$

It follows from the Bendixson theorem that there are no limit cycles in the region $|x_1| < 1$ because $\partial f_1/\partial x_1 + \partial f_2/\partial x_2 > 0$ in that region.

Bendixson's theorem gives us a necessary condition for a closed curve to be a limit cycle. It is useful for the purpose of establishing the nonexistence of a limit cycle. In general, it is very difficult to establish the existence of a limit cycle. Sufficient conditions for the existence of a limit cycle are discussed, for example, by Hochstadt [123, Chapter 7]. A method for predicting limit cycles, called the describing function method, is presented in Section 2.5.

2.2 Numerical Techniques

In this section we discuss numerical techniques for solving differential equations. We begin our presentation of numerical techniques that allow us to determine approximate solutions to the given system of differential equations with a method based on a Taylor series expansion. This section is based on the Class Notes of Prof. V. B. Haas [109].

2.2.1 The Method of Taylor Series

The basis for many numerical methods for solving differential equations is the Taylor formula. Suppose we are given a state-space model of the form

$$\dot{\boldsymbol{x}}(t) = \boldsymbol{f}(t, \boldsymbol{x}(t), \boldsymbol{u}(t)), \qquad \boldsymbol{x}(t_0) = \boldsymbol{x}_0. \tag{2.17}$$

The method we will discuss depends on the fact that if $\boldsymbol{f}(t, \boldsymbol{x}, \boldsymbol{u})$ is differentiable with respect to its $n + m + 1$ variables $x_1, \ldots, x_n, u_1, \ldots, u_m, t$ in some $(n + m + 1)$-dimensional region $\mathcal{D}$ and if $\boldsymbol{x}(t)$ is a solution of (2.17), where $(t_0, \boldsymbol{x}_0, \boldsymbol{u}(t_0))$ lies in $\mathcal{D}$, then we can expand $\boldsymbol{x}(t)$ into a Taylor series to obtain

$$\begin{aligned}
\boldsymbol{x}(t) &= \boldsymbol{x}_0 + \dot{\boldsymbol{x}}(t_0)(t - t_0) + \ddot{\boldsymbol{x}}(t_0)\frac{(t - t_0)^2}{2} + \cdots \\
&= \boldsymbol{x}_0 + \boldsymbol{f}(t_0, \boldsymbol{x}_0, \boldsymbol{u}(t_0))(t - t_0) + \left[\frac{\partial \boldsymbol{f}(t_0, \boldsymbol{x}_0, \boldsymbol{u}(t_0))}{\partial t} + \sum_{i=1}^{n} \frac{\partial \boldsymbol{f}(t_0, \boldsymbol{x}_0, \boldsymbol{u}(t_0))}{\partial x_i}\dot{x}_i \right. \\
&\quad \left. + \sum_{j=1}^{m} \frac{\partial \boldsymbol{f}(t_0, \boldsymbol{x}_0, \boldsymbol{u}(t_0))}{\partial u_j}\dot{u}_j\right]\frac{(t - t_0)^2}{2} + \cdots.
\end{aligned}$$

If we stop after the qth term of the Taylor series solution, then the remainder after q terms of a Taylor series gives us an estimate of the error. We can expect the Taylor series solution to give a fairly accurate approximation of the solution of the state-space equations only in a small interval about t_0. Thus, if $\Delta t = t - t_0$ is some small time interval such that the Taylor series converges for $t \in [t_0,\ t_0 + \Delta t]$, we can use the expansion

$$\boldsymbol{x}(t) = \boldsymbol{x}_0 + \frac{d\boldsymbol{x}(t_0)}{dt}(t - t_0) + \frac{d^2\boldsymbol{x}(t_0)}{dt^2}\frac{(t-t_0)^2}{2} + \cdots$$

to evaluate $\boldsymbol{x}(t_1) = \boldsymbol{x}(t_0 + \Delta t)$. Then, we can start over again solving $\dot{\boldsymbol{x}}(t) = \boldsymbol{f}(t, \boldsymbol{x}(t), \boldsymbol{u}(t))$ with the new initial condition $\boldsymbol{x}(t_1)$. We illustrate the method with simple examples.

◆ Example 2.7

Consider the first-order differential equation

$$\frac{dx}{dt} = t^2 - x^2, \qquad x(1) = 1.$$

Expanding $x = x(t)$ into a Taylor series about $x(1) = 1$ yields

$$x(t) = x(1) + \frac{dx(1)}{dt}(t-1) + \frac{d^2x(1)}{dt^2}\frac{(t-1)^2}{2} + \frac{d^3x(1)}{dt^3}\frac{(t-1)^3}{3!} + \frac{d^4x(1)}{dt^4}\frac{(t-1)^4}{4!}\cdots.$$

Because $dx/dt = t^2 - x^2$, we obtain by successive differentiations

$$\begin{aligned}
\frac{d^2x}{dt^2} &= 2t - 2x\frac{dx}{dt},\\
\frac{d^3x}{dt^3} &= 2 - 2\left(\frac{dx}{dt}\right)^2 - 2x\frac{d^2x}{dt^2},\\
\frac{d^4x}{dt^4} &= -6\frac{dx}{dt}\frac{d^2x}{dt^2} - 2x\frac{d^3x}{dt^3},\\
&\vdots
\end{aligned}$$

Evaluating the above at $t = 1$ yields

$$\begin{aligned}
\frac{dx(1)}{dt} &= 0,\\
\frac{d^2x(1)}{dt^2} &= 2,\\
\frac{d^3x(1)}{dt^3} &= -2,\\
\frac{d^4x(1)}{dt^4} &= 4,\\
&\vdots
\end{aligned}$$

Hence,

$$x(t) = 1 + (t-1)^2 - \frac{(t-1)^3}{3} + \frac{(t-1)^4}{6} + \cdots.$$

◆ Example 2.8

In this example, we consider the second-order differential equation

$$\frac{d^2x}{dt^2} + 3\frac{dx}{dt} + 6tx = 0,$$

subject to the initial conditions $x(0)=1$ and $dx(0)/dt=1$. Using $d^2x/dt^2 = -3dx/dt - 6tx$, we obtain by successive differentiation

$$\begin{aligned}
\frac{d^3x}{dt^3} &= -3\frac{d^2x}{dt^2} - 6t\frac{dx}{dt} - 6x,\\
\frac{d^4x}{dt^4} &= -3\frac{d^3x}{dt^3} - 6t\frac{d^2x}{dt^2} - 12\frac{dx}{dt},\\
\frac{d^5x}{dt^5} &= -3\frac{d^4x}{dt^4} - 6t\frac{d^3x}{dt^3} - 18\frac{d^2x}{dt^2},\\
&\vdots
\end{aligned}$$

Taking into account the initial conditions, we evaluate all coefficients at $t = 0$ to get

$$\begin{aligned}
\frac{d^2x(0)}{dt^2} &= -3,\\
\frac{d^3x(0)}{dt^3} &= 3,\\
\frac{d^4x(0)}{dt^4} &= -21,\\
\frac{d^5x(0)}{dt^5} &= 117,\\
&\vdots
\end{aligned}$$

Hence,

$$\begin{aligned}
x(t) &= x(0) + \frac{dx(0)}{dt}t + \frac{d^2x(0)}{dt^2}\frac{t^2}{2} + \frac{d^3x(0)}{dt^3}\frac{t^3}{3!} + \frac{d^4x(0)}{dt^4}\frac{t^4}{4!} + \frac{d^5x(0)}{dt^5}\frac{t^5}{5!} + \cdots\\
&= 1 + t - \frac{3}{2}t^2 + \frac{1}{2}t^3 - \frac{7}{8}t^4 + \frac{117}{120}t^5 + \cdots.
\end{aligned}$$

We now illustrate the method of Taylor series when applied to a system of first-order differential equations.

◆ Example 2.9

We consider the following system of two first-order equations:

$$dx_1/dt = f_1(t, x_1, x_2),$$
$$dx_2/dt = f_2(t, x_1, x_2).$$

The initial conditions at $t = t_0$ are $x_1(t_0) = x_{10}$, $x_2(t_0) = x_{20}$. Having the initial conditions, we can determine the values $dx_1(t_0)/dt$ and $dx_2(t_0)/dt$ from the original equations. We then differentiate dx_1/dt and dx_2/dt to obtain

$$\frac{d^2x_1}{dt^2} = \frac{d}{dt} f_1\left(t, x_1, x_2, \frac{dx_1}{dt}, \frac{dx_2}{dt}\right),$$
$$\frac{d^2x_2}{dt^2} = \frac{d}{dt} f_2\left(t, x_1, x_2, \frac{dx_1}{dt}, \frac{dx_2}{dt}\right).$$

Now, d^2x_1/dt^2 and d^2x_2/dt^2 can be calculated by substituting known quantities on the right-hand sides of the above equations. By successive differentiation and substitution, we can determine d^3x_1/dt^3 and d^3x_2/dt^3 as well as higher derivatives at $t = t_0$. The solution to the given system is then computed as

$$x_1(t) = x_{10} + \frac{dx_1(t_0)}{dt}(t - t_0) + \frac{d^2x_1(t_0)}{dt^2}\frac{(t - t_0)^2}{2} + \frac{d^3x_1(t_0)}{dt^3}\frac{(t - t_0)^3}{3!} + \cdots,$$
$$x_2(t) = x_{20} + \frac{dx_2(t_0)}{dt}(t - t_0) + \frac{d^2x_2(t_0)}{dt^2}\frac{(t - t_0)^2}{2} + \frac{d^3x_2(t_0)}{dt^3}\frac{(t - t_0)^3}{3!} + \cdots.$$

The Taylor series method can be programmed on a computer. However, this is not a very efficient method from a numerical point of view. In the following, we discuss other methods for numerical solution of the state equations that are related to the method of Taylor series. We first present two simple numerical techniques known as the forward and backward Euler methods.

2.2.2 Euler's Methods

Suppose that we wish to obtain the solution of the modeling equation

$$\dot{\boldsymbol{x}}(t) = \boldsymbol{f}(t, \boldsymbol{x}(t), \boldsymbol{u}(t))$$

in the time interval $[t_0,\ t_f]$, subject to the initial condition $\boldsymbol{x}(t_0) = \boldsymbol{x}_0$ and the input vector $\boldsymbol{u}(t)$. We divide the interval $[t_0,\ t_f]$ into N equal subintervals of width

$$h = \frac{t_f - t_0}{N}.$$

We call h the *step length*. We set

$$t_k = t_0 + kh.$$

To proceed further, we approximate the derivative $d\boldsymbol{x}(t)/dt$ at time t_k by

$$\frac{d\boldsymbol{x}(t_k)}{dt} \approx \frac{\boldsymbol{x}(t_{k+1}) - \boldsymbol{x}(t_k)}{h}$$

and write

$$\frac{\boldsymbol{x}(t_{k+1}) - \boldsymbol{x}(t_k)}{h} = \boldsymbol{f}(t_k, \boldsymbol{x}(t_k), \boldsymbol{u}(t_k)).$$

Hence, for $k = 1, 2, \ldots, N$, we have

$$\boxed{\boldsymbol{x}(t_{k+1}) = \boldsymbol{x}(t_k) + h\boldsymbol{f}(t_k, \boldsymbol{x}(t_k), \boldsymbol{u}(t_k))}$$

The above is known as the *forward Euler* algorithm. We see that if h is sufficiently small, we can approximately determine the state at time $t_1 = t_0 + h$ from the initial condition $\boldsymbol{x}_0$ to get

$$\boldsymbol{x}(t_1) = \boldsymbol{x}_0 + h\boldsymbol{f}(t_0, \boldsymbol{x}_0, \boldsymbol{u}(t_0)).$$

Once we have determined the approximate solution at time t_1, we can determine the approximate solution at time $t_2 = t_1 + h$, and so on. The forward Euler method can also be arrived at when instead of considering the differential equation $\dot{\boldsymbol{x}}(t) = \boldsymbol{f}(t, \boldsymbol{x}(t), \boldsymbol{u}(t))$, we start with its equivalent integral representation:

$$\boldsymbol{x}(t) = \boldsymbol{x}_0 + \int_{t_0}^{t} \boldsymbol{f}(\tau, \boldsymbol{x}(\tau), \boldsymbol{u}(\tau))\, d\tau. \tag{2.18}$$

We then use the *rectangular rule* for numerical integration that can be stated as follows. The area under the curve $z = g(t)$ between $t = t_k$ and $t = t_k + h$ is approximately equal to the area of the rectangle $ABCD$ as shown in Figure 2.8; that is,

$$\int_{t_k}^{t_k+h} g(t)\, dt \approx hg(t_k). \tag{2.19}$$

Applying the rectangular rule of integration to (2.18), we obtain the forward Euler method.

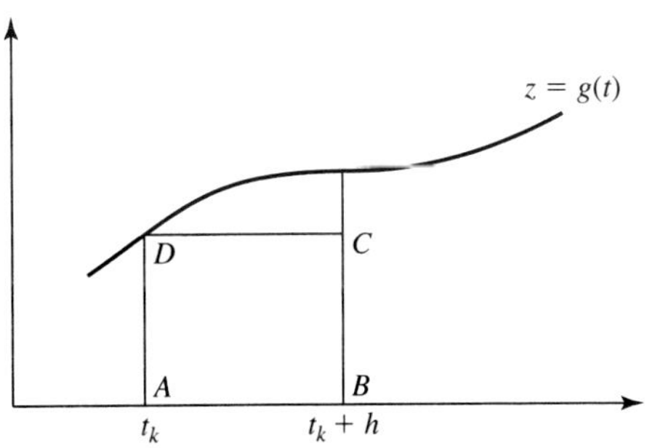

Figure 2.8 Illustration of the rectangular rule of integration.

The *backward Euler* method differs from the forward Euler method in the way we approximate the derivative:

$$\frac{d\boldsymbol{x}(t_k)}{dt} \approx \frac{\boldsymbol{x}(t_k) - \boldsymbol{x}(t_{k-1})}{h}.$$

This approximation gives the backward Euler formula:

$$\boxed{\boldsymbol{x}(t_{k+1}) = \boldsymbol{x}(t_k) + h\boldsymbol{f}(t_{k+1}, \boldsymbol{x}(t_{k+1}), \boldsymbol{u}(t_{k+1}))}$$

In the above formula, $\boldsymbol{x}(t_{k+1})$ is a function of itself. For this reason, we say that the backward Euler method is an implicit integration algorithm. Implicit integration algorithms require additional computation to solve for $\boldsymbol{x}(t_{k+1})$. For further discussion of this issue, we refer to Parker and Chua [222].

Euler's methods are rather elementary, and thus they are not as accurate as some of the more sophisticated techniques that we are going to discuss next.

2.2.3 Predictor–Corrector Method

The predictor–corrector method, also known as Heun's method, is based on the trapezoidal rule for numerical integration. The trapezoidal rule states that the area under the curve $z = g(t)$ between $t = t_k$ and $t = t_k + h$ is approximately equal to the area of the trapezoid $ABCD$ as depicted in Figure 2.9; that is,

$$\int_{t_k}^{t_k+h} g(t)\,dt \approx h\frac{g(t_k) + g(t_k+h)}{2}. \tag{2.20}$$

Suppose we wish to find $\boldsymbol{x}(t)$ at $t = t_1 > t_0$. Applying the trapezoid rule of integration to (2.17), we obtain

$$\boldsymbol{x}(t_1) = \boldsymbol{x}_0 + \frac{h}{2}(\boldsymbol{f}(t_0, \boldsymbol{x}_0, \boldsymbol{u}(t_0)) + \boldsymbol{f}(t_1, \boldsymbol{x}(t_1), \boldsymbol{u}(t_1))).$$

The above is an implicit algorithm, and it may be viewed as a merging of the forward and backward Euler algorithms. If the function $\boldsymbol{f}$ is nonlinear, then we will, in general, not be able

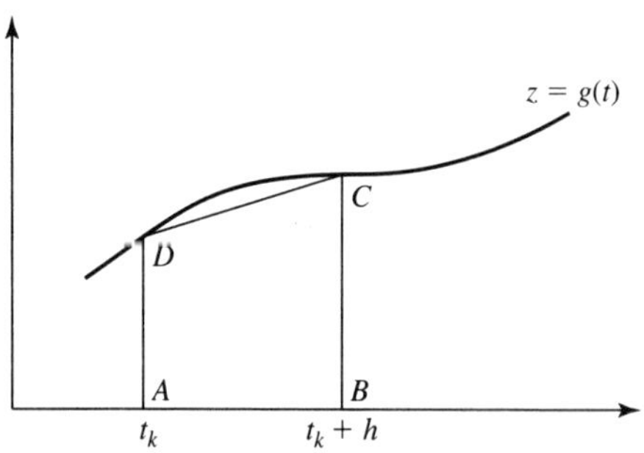

Figure 2.9 The trapezoidal rule of integration.

to solve the above equation for $\boldsymbol{x}(t_1)$ exactly. We attempt to obtain $\boldsymbol{x}(t_1)$ iteratively. Let

$$\begin{aligned} h &= t_{k+1} - t_k, \\ \boldsymbol{x}_1^* &= \boldsymbol{x}_0 + h\boldsymbol{f}(t_0, \boldsymbol{x}_0, \boldsymbol{u}(t_0)), \\ \boldsymbol{m}_0 &= \boldsymbol{f}(t_0, \boldsymbol{x}_0, \boldsymbol{u}(t_0)), \\ \boldsymbol{m}_1 &= \boldsymbol{f}(t_1, \boldsymbol{x}_1^*, \boldsymbol{u}(t_1)). \end{aligned}$$

A first approximation to $\boldsymbol{x}(t_1)$ is

$$\boxed{\boldsymbol{x}_1^* = \boldsymbol{x}_0 + h\boldsymbol{f}(t_0, \boldsymbol{x}_0, \boldsymbol{u}(t_0))}$$

and is called a *predictor*. The next approximation is

$$\boxed{\boldsymbol{x}(t_1) = \boldsymbol{x}_0 + \frac{h}{2}(\boldsymbol{m}_0 + \boldsymbol{m}_1)}$$

and is called a *corrector*.

Suppose now that we wish to find $\boldsymbol{x}(t)$ at $t = t_f > t_0$. We first divide the interval $[t_0,\ t_f]$ into N equal subintervals using evenly spaced subdivision points to obtain

$$t_0 < t_1 < t_2 < \cdots < t_N = t_f.$$

Once we have found $\boldsymbol{x}(t_1)$, we can apply the same procedure to find $\boldsymbol{x}(t_2)$, and so on.

◆ Example 2.10

We use the predictor–corrector method to find $x(1)$, where the step length h equals 0.5, for the following differential equation:

$$\frac{dx}{dt} = tx^2 - x, \qquad x(0) = 1. \tag{2.21}$$

The calculations are summarized in Table 2.1.

Table 2.1 Solving the Differential Equation (2.21) Using the Predictor–Corrector Method

t_0	x_0	t_1	m_0	x_1^*	m_1	$m = (m_0 + m_1)/2$	Δx
0	1	0.5	−1.0	0.5	−0.375	−0.688	−0.344
0.5	0.656	1.0	−0.441	0.435	−0.246	−0.344	−0.172
1.0	0.484						

We could use $\boldsymbol{x}(t_1)$ obtained from the corrector formula to construct a next approximation to $\boldsymbol{x}(t_1)$. In general, we can use the formula

$$\boldsymbol{x}^{(k)}(t_1) = \boldsymbol{x}_0 + \frac{h}{2}\left(\boldsymbol{f}(t_0, \boldsymbol{x}_0, \boldsymbol{u}(t_0)) + \boldsymbol{f}\left(t_1, \boldsymbol{x}^{(k-1)}(t_1), \boldsymbol{u}(t_1)\right)\right)$$

to evaluate $\boldsymbol{x}(t_1)$ until two successive iterations agree to the desired accuracy. This general algorithm is referred to as a second-order predictor–corrector method.

An n-dimensional system of first-order differential equations can be worked analogously. We illustrate the procedure for $n = 2$, where

$$\frac{dx}{dt} = f(t, x, y), \qquad x(t_0) = x_0,$$
$$\frac{dy}{dt} = g(t, x, y), \qquad y(t_0) = y_0.$$

Let

$$m_0 = f(t_0, x_0, y_0),$$
$$n_0 = g(t_0, x_0, y_0).$$

Then, the predictor equations are

$$x_1^* = x_0 + m_0\, h,$$
$$y_1^* = y_0 + n_0\, h.$$

Let

$$m_1 = f(t_1, x_1^*, y_1^*),$$
$$n_1 = g(t_1, x_1^*, y_1^*).$$

Then, the corrector equations are

$$x(t_1) = x_0 + \frac{m_0 + m_1}{2} h,$$
$$y(t_1) = y_0 + \frac{n_0 + n_1}{2} h.$$

Using the above, we can generate the corresponding formulas for t_k.

We will now show that the predictor–corrector method yields a solution that agrees with the Taylor series solution through terms of degree two. We consider the scalar case, where f is assumed to be differentiable with respect to its arguments and $\partial^2 f/\partial t \partial x = \partial^2 f/\partial x \partial t$. First observe that $m_1 = f(t_1, x_1^*)$ can be written as

$$\begin{aligned} f(t_1, x_1^*) &= f(t_0 + h, x_0 + m_0 h) \\ &\approx f(t_0, x_0) + [f_t \quad f_x]\begin{bmatrix} h \\ m_0 h \end{bmatrix} + \frac{1}{2}[h \quad m_0 h]\begin{bmatrix} f_{tt} & f_{tx} \\ f_{xt} & f_{xx} \end{bmatrix}\begin{bmatrix} h \\ m_0 h \end{bmatrix}, \end{aligned}$$

where $f_t = \partial f/\partial t$, $f_x = \partial f/\partial x$, and so on. Substituting the above into the corrector equation yields

$$\begin{aligned} x(t_1) &= x_0 + \frac{h}{2}(m_0 + m_1) \\ &= x_0 + \frac{h}{2}(m_0 + m_0) + \frac{h}{2}\left([f_t \quad f_x] \begin{bmatrix} h \\ m_0 h \end{bmatrix} \right. \\ &\quad \left. + \frac{1}{2}[h \quad m_0 h] \begin{bmatrix} f_{tt} & f_{tx} \\ f_{xt} & f_{xx} \end{bmatrix} \begin{bmatrix} h \\ m_0 h \end{bmatrix} \right) \\ &= x_0 + m_0 h + (f_t + f_x m_0)\frac{h^2}{2} + \left(f_{tt} + 2 f_{tx} m_0 + f_{xx} m_0^2\right)\frac{h^3}{4}. \end{aligned}$$

The Taylor series expansion, on the other hand, gives

$$\begin{aligned} x(t_1) &= x_0 + \dot{x}\, h + \ddot{x}\, \frac{h^2}{2} + x^{(3)} \frac{h^3}{6} + \cdots \\ &= x_0 + f(t_0, x_0)\, h + (f_t + f_x \dot{x})\frac{h^2}{2} \\ &\quad + (f_{tt} + 2 f_{tx} \dot{x} + f_{xx} (\dot{x})^2 + f_x \ddot{x})\frac{h^3}{6} + \cdots, \end{aligned}$$

where f_t, f_x, $\dot{x}$, and so on, are all evaluated at $t = t_0$. Comparing the right-hand sides of the last two equations, we see that the solution obtained using the predictor–corrector method agrees with the Taylor series expansion of the true solution through terms of degree two.

2.2.4 Runge's Method

Runge's method is an improvement over the predictor–corrector method in the sense that it agrees with the Taylor series solution through terms of degree three. The method depends upon integration by means of Simpson's rule, which says that the area under the curve $x = g(t)$ between t_0 and $t_0 + h$ is given approximately by

$$\int_{t_0}^{t_0+h} g(t)\, dt \approx \frac{h}{6}(x_0 + 4x_1 + x_2), \tag{2.22}$$

where $x_0 = g(t_0)$, $x_1 = g(t_0 + h/2)$, and $x_2 = g(t_0 + h)$. The above formula is derived by passing a parabola through the three points (t_0, x_0), $(t_0 + h/2, x_1)$, and $(t_0 + h, x_2)$ and assuming that the area under the given curve is approximately equal to the area under the parabola, as illustrated in Figure 2.10. To verify Simpson's formula, we translate the axes so that in the new coordinates $(\tilde{t}, \tilde{x})$ we have $t = t_0 + h/2 + \tilde{t}$, $x = \tilde{x}$. In the new coordinates the three points become

$$(-h/2, x_0), \qquad (0, x_1), \qquad (h/2, x_2).$$

The parabola through the above points is described by the equation

$$\tilde{x} = a\tilde{t}^2 + b\tilde{t} + c.$$

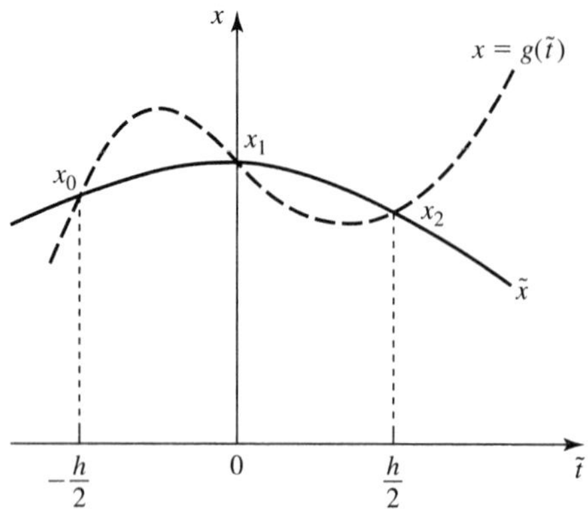

Figure 2.10 Integration using Simpson's rule.

Hence

$$
\begin{aligned}
x_0 &= a\frac{h^2}{4} - b\frac{h}{2} + c,\\
x_1 &= c,\\
x_2 &= a\frac{h^2}{4} + b\frac{h}{2} + c.
\end{aligned}
$$

Using the above three equations, we calculate

$$
\begin{aligned}
c &= x_1,\\
b &= \frac{x_2 - x_0}{h},\\
a &= \frac{2}{h^2}(x_0 - 2x_1 + x_2).
\end{aligned}
$$

Then, the area under the parabola is

$$
\begin{aligned}
\int_{-h/2}^{h/2} (a\tilde{t}^2 + b\tilde{t} + c)\, d\tilde{t} &= \int_{-h/2}^{h/2} \left(\frac{2}{h^2}(x_0 - 2x_1 + x_2)\tilde{t}^2 + \frac{x_2 - x_0}{h}\tilde{t} + x_1 \right) d\tilde{t}\\
&= \frac{h}{6}(x_0 + 4x_1 + x_2).
\end{aligned}
$$

If we are given the differential equation

$$
\frac{dx}{dt} = f(t, x), \qquad x(t_0) = x_0,
$$

we integrate it to obtain

$$
x(t_1) = x(t_0) + \int_{t_0}^{t_1} f(s, x(s))\, ds.
$$

Then, we approximate the integral

$$\int_{t_0}^{t_1} f(s, x(s))\, ds$$

by

$$\frac{h}{6}(m_0 + 4m_1 + m_3),$$

where

$$\begin{aligned} h &= t_1 - t_0, \\ m_0 &= f(t_0, x_0), \\ m_1 &= f\left(t_0 + \frac{h}{2}, x_0 + m_0\frac{h}{2}\right), \\ m_2 &= f(t_0 + h, x_0 + m_0 h), \\ m_3 &= f(t_0 + h, x_0 + m_2 h). \end{aligned}$$

Thus, we may write

$$x(t_1) = x_0 + \Delta x,$$

where $\Delta x = mh$, and $m = \frac{1}{6}(m_0 + 4m_1 + m_3)$.

◆ **Example 2.11**

We use Runge's method to find $x(1)$, where the step length h equals 0.5, for the differential equation (2.21). We found $x(1)$ using the predictor–corrector method in Example 2.10. The calculations for Runge's method are summarized in Table 2.2.

Table 2.2 Solving the Differential Equation (2.21) Using Runge's Method

t_0	x_0	m_0	m_1	m_2	m_3	m	Δx
0	1	−1	−0.610	−0.375	−0.482	−0.654	−0.327
0.5	0.673	−0.446	−0.325	−0.248	−0.248	−0.332	−0.166
1.0	0.507						

Runge's method can, of course, be used for systems of first order differential equations. If, for example,

$$\begin{aligned} \frac{dx}{dt} &= f(t, x, y), \qquad x(t_0) = x_0, \\ \frac{dy}{dt} &= g(t, x, y), \qquad y(t_0) = y_0, \end{aligned}$$

then

$$x(t_1) = x_0 + \Delta x,$$
$$y(t_1) = y_0 + \Delta y,$$

where

$$\Delta x = \frac{h}{6}(m_0 + 4m_1 + m_3),$$
$$\Delta y = \frac{h}{6}(n_0 + 4n_1 + n_3).$$

In the above expressions,

$$\begin{aligned}
m_0 &= f(t_0, x_0, y_0),\\
m_1 &= f(t_0 + h/2, x_0 + m_0\,h/2, y_0 + n_0\,h/2),\\
m_2 &= f(t_0 + h, x_0 + m_0 h, y_0 + n_0 h),\\
m_3 &= f(t_0 + h, x_0 + m_2 h, y_0 + n_2 h),\\
n_0 &= g(t_0, x_0, y_0),\\
n_1 &= g(t_0 + h/2, x_0 + m_0\,h/2, y_0 + n_0\,h/2),\\
n_2 &= g(t_0 + h, x_0 + m_0 h, y_0 + n_0 h),\\
n_3 &= g(t_0 + h, x_0 + m_2 h, y_0 + n_2 h).
\end{aligned}$$

2.2.5 Runge–Kutta Method

This method is an improvement over the Runge method, and it agrees with the Taylor series solution through terms of degree four; for this reason it is sometimes called the fourth-order Runge–Kutta algorithm. It is based on the use of the formula

$$x(t_1) = x(t_0) + \frac{h}{6}(m_0 + 2m_1 + 2m_2 + m_3), \tag{2.23}$$

where

$$\begin{aligned}
m_0 &= f(t_0, x_0),\\
m_1 &= f(t_0 + h/2, x_0 + m_0\,h/2),\\
m_2 &= f(t_0 + h/2, x_0 + m_1\,h/2),\\
m_3 &= f(t_0 + h, x_0 + h\,m_2).
\end{aligned}$$

Thus, we may write

$$x(t_1) = x_0 + \Delta x,$$

where $\Delta x = mh$, and $m = \frac{1}{6}(m_0 + 2m_1 + 2m_2 + m_3)$.

◆ Example 2.12

We apply the Runge–Kutta method to find $x(1)$, where the step length h equals 0.5, for the differential equation (2.21) that we used Examples 2.10 and 2.11. Our calculations for the Runge–Kutta method are summarized in Table 2.3.

Table 2.3 Solving the Differential Equation (2.21) Using the Runge–Kutta Method

t_0	x_0	m_0	m_1	m_2	m_3	m	Δx
0	1	−1.0	−0.609	−0.668	−0.444	−0.666	−0.333
0.5	0.667	−0.444	−0.324	−0.328	−0.250	−0.333	−0.167
1.0	0.500						

The above problem has been worked by the predictor–corrector, Runge's, and the Runge–Kutta methods. This problem can be solved exactly by means of the Taylor series. The Taylor series solution is

$$x(t) = 1 - t + t^2 - t^3 + \cdots.$$

The series converges to $1/(1+t)$ for $-1 < t < 1$. A check on the differential equation shows that

$$x(t) = \frac{1}{1+t}$$

is indeed a solution for all $t > -1$. Thus, $x(1) = 0.500$.

The algorithms presented above require only one input point, $x(t_k)$, at each step to compute $x(t_{k+1})$. Such algorithms are referred to as single-step algorithms. In the multistep algorithms, a number of previous points are used, along with the corresponding values of f, to evaluate $x(t_{k+1})$; that is, the multistep algorithms reuse past information about the trajectory when evaluating its new point. In general, it may be difficult to decide which type of algorithm is more efficient because the performance of a particular algorithm is problem dependent. Well-known multistep algorithms are: Adams–Bashforth, Adams–Moulton, and Gear. More information about the above-mentioned algorithms can be found in Parker and Chua [222, Section 4.1].

2.3 Principles of Linearization

The replacement of a nonlinear system model by its linear approximation is called *linearization*. The motivation for linearization is that the dynamical behavior of many nonlinear system models can be well approximated within some range of variables by linear system models. Then, we can use well-developed techniques for analysis and synthesis of linear systems to analyze a nonlinear system at hand. However, the results of analysis of nonlinear systems using their linearized models should be carefully interpreted in order to avoid unacceptably large errors due to the approximation in the process of linearization.

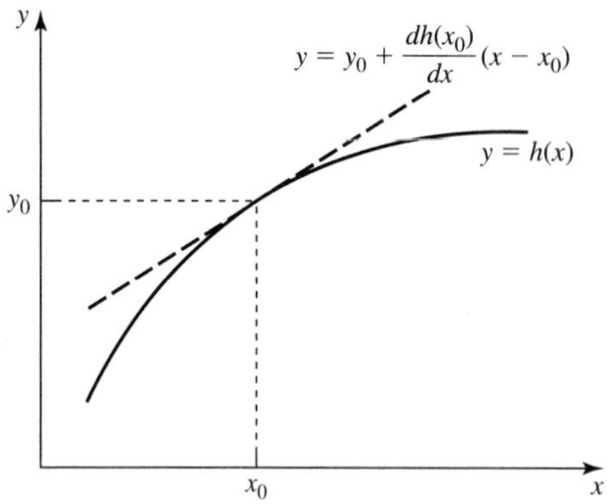

Figure 2.11 Linear approximation to a function $y = h(x)$.

We begin our discussion of linear approximations of nonlinear systems by considering a nonlinear element with a state variable x and an output variable y that are related by the equation

$$y = h(x),$$

where the function $h : \mathbb{R} \to \mathbb{R}$ is continuously differentiable; that is, $h \in C^1$. Let x_0 be an operating point. We expand h into a Taylor series about the operating point x_0 to obtain

$$\begin{aligned} y &= h(x) \\ &= h(x_0) + \frac{dh(x_0)}{dx}(x - x_0) + \text{higher-order terms.} \end{aligned}$$

Linearization of $h(x)$ about the point x_0 consists of replacing h by the linear approximation

$$\begin{aligned} y &= h(x_0) + \frac{dh(x_0)}{dx}(x - x_0) \\ &= y_0 + \frac{dh(x_0)}{dx}(x - x_0), \end{aligned} \tag{2.24}$$

where $y_0 = h(x_0)$. Let $\Delta y = y - y_0$ and $\Delta x = x - x_0$. Then, we can represent (2.24) as

$$\Delta y = \frac{dh(x_0)}{dx}\Delta x.$$

Over a small range of Δx, the line (2.24) is a good approximation to the curve $y = h(x)$ in a neighborhood of the operating point x_0; see Figure 2.11 for an illustration of the above statement. We now illustrate the process of linearization on the simple pendulum.

◆ Example 2.13

Consider the simple pendulum shown in Figure 2.12. Let g be the gravity constant. The tangential force component, $mg \sin\theta$, acting on the mass m returns the pendulum to its equilibrium position. By Newton's second law we obtain

$$F = -mg \sin\theta.$$

The equilibrium point for the pendulum is $\theta = 0°$. Note that $F(0°) = 0$. Therefore,

$$\Delta F = F \quad \text{and} \quad \Delta\theta = \theta.$$

We also have

$$\frac{dF}{d\theta}(0°) = -mg\cos 0° = -mg.$$

Hence, the linearized model about the equilibrium position $\theta = 0°$ has the form

$$F = -mg\theta.$$

Thus, for small angular displacements θ, the force F is proportional to the displacements. This approximation is quite accurate for $-\pi/4 \leq \theta \leq \pi/4$, as can be seen in Figure 2.13.

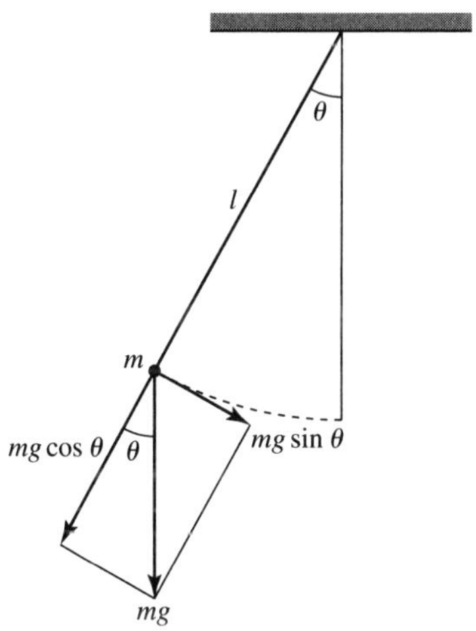

Figure 2.12 Forces acting on the simple pendulum.

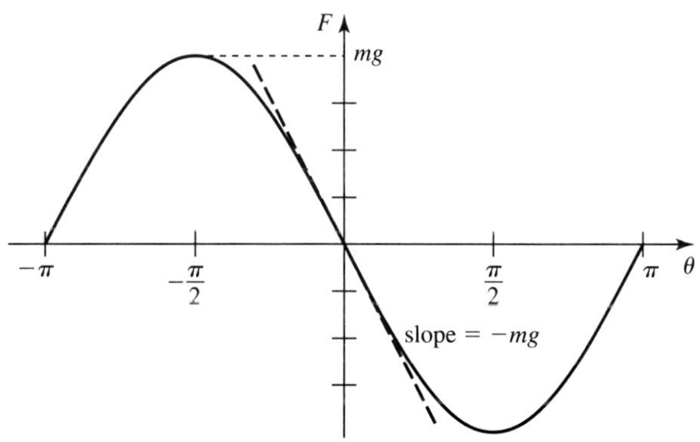

Figure 2.13 Linearizing the equation modeling the simple pendulum.

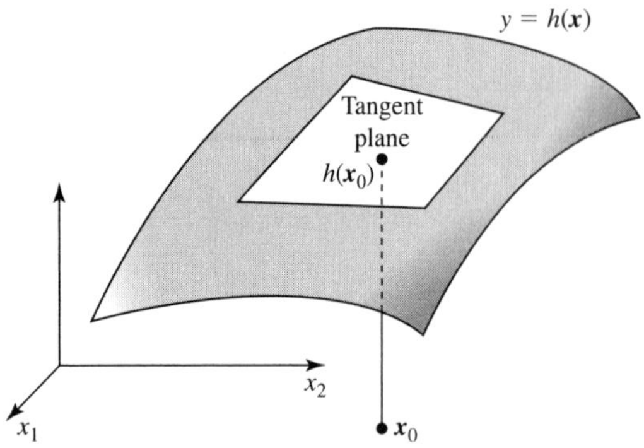

Figure 2.14 Tangent plane as a linear approximation for a function of two variables.

If $h : \mathbb{R}^n \to \mathbb{R}$—that is, $y = h(x_1, x_2, \ldots, x_n)$, which means that the dependent variable depends upon several variables—the same principle applies. Let

$$\boldsymbol{x}_0 = [x_{10} \quad x_{20} \quad \cdots \quad x_{n0}]^T$$

be the operating point. The Taylor series expansion of h about the operating point $\boldsymbol{x}_0$ yields

$$y - h(\boldsymbol{x}_0) = \nabla h(\boldsymbol{x}_0)^T (\boldsymbol{x} - \boldsymbol{x}_0) + \text{higher-order terms},$$

where

$$\nabla h(\boldsymbol{x}_0)^T = \left[\left.\frac{\partial h}{\partial x_1}\right|_{\boldsymbol{x}=\boldsymbol{x}_0} \quad \left.\frac{\partial h}{\partial x_2}\right|_{\boldsymbol{x}=\boldsymbol{x}_0} \quad \cdots \quad \left.\frac{\partial h}{\partial x_n}\right|_{\boldsymbol{x}=\boldsymbol{x}_0} \right].$$

Geometrically, the linearization of h about $\boldsymbol{x}_0$ can be thought of as placing a tangent plane onto the nonlinear surface at the operating point $\boldsymbol{x}_0$ as illustrated in Figure 2.14.

2.4 Linearizing Differential Equations

We now consider a dynamical system modeled by a set of nonlinear differential equations:

$$\begin{aligned}
\frac{dx_1}{dt} &= f_1(x_1, x_2, \ldots, x_n, u_1, u_2, \ldots, u_m),\\
\frac{dx_2}{dt} &= f_2(x_1, x_2, \ldots, x_n, u_1, u_2, \ldots, u_m),\\
&\vdots\\
\frac{dx_n}{dt} &= f_n(x_1, x_2, \ldots, x_n, u_1, u_2, \ldots, u_m).
\end{aligned}$$

We assume that the functions f_i, $i = 1, 2, \ldots, n$, are continuously differentiable. The above set of differential equations can be represented in vector form as

$$\dot{\boldsymbol{x}} = \boldsymbol{f}(\boldsymbol{x}, \boldsymbol{u}). \tag{2.25}$$

Let $\boldsymbol{u}_e = [u_{1e} \; u_{2e} \; \cdots \; u_{me}]^T$ be a constant input that forces the system (2.25) to settle into a constant equilibrium state $\boldsymbol{x}_e = [x_{1e} \; x_{2e} \; \cdots \; x_{ne}]^T$; that is, $\boldsymbol{u}_e$ and $\boldsymbol{x}_e$ satisfy

$$\boldsymbol{f}(\boldsymbol{x}_e, \boldsymbol{u}_e) = \boldsymbol{0}.$$

We now perturb the equilibrium state by allowing

$$\boldsymbol{x} = \boldsymbol{x}_e + \Delta\boldsymbol{x}, \qquad \boldsymbol{u} = \boldsymbol{u}_e + \Delta\boldsymbol{u}.$$

Taylor's expansion yields

$$\begin{aligned}\frac{d}{dt}\boldsymbol{x} &= \boldsymbol{f}(\boldsymbol{x}_e + \Delta\boldsymbol{x}, \boldsymbol{u}_e + \Delta\boldsymbol{u}) \\ &= \boldsymbol{f}(\boldsymbol{x}_e, \boldsymbol{u}_e) + \frac{\partial \boldsymbol{f}}{\partial \boldsymbol{x}}(\boldsymbol{x}_e, \boldsymbol{u}_e)\,\Delta\boldsymbol{x} + \frac{\partial \boldsymbol{f}}{\partial \boldsymbol{u}}(\boldsymbol{x}_e, \boldsymbol{u}_e)\,\Delta\boldsymbol{u} + \text{higher-order terms},\end{aligned}$$

where

$$\frac{\partial \boldsymbol{f}}{\partial \boldsymbol{x}}(\boldsymbol{x}_e, \boldsymbol{u}_e) = \left.\begin{bmatrix} \dfrac{\partial f_1}{\partial x_1} & \cdots & \dfrac{\partial f_1}{\partial x_n} \\ \vdots & & \vdots \\ \dfrac{\partial f_n}{\partial x_1} & \cdots & \dfrac{\partial f_n}{\partial x_n} \end{bmatrix}\right|_{\substack{\boldsymbol{x}=\boldsymbol{x}_e \\ \boldsymbol{u}=\boldsymbol{u}_e}} \quad \text{and} \quad \frac{\partial \boldsymbol{f}}{\partial \boldsymbol{u}}(\boldsymbol{x}_e, \boldsymbol{u}_e) = \left.\begin{bmatrix} \dfrac{\partial f_1}{\partial u_1} & \cdots & \dfrac{\partial f_1}{\partial u_m} \\ \vdots & & \vdots \\ \dfrac{\partial f_n}{\partial u_1} & \cdots & \dfrac{\partial f_n}{\partial u_m} \end{bmatrix}\right|_{\substack{\boldsymbol{x}=\boldsymbol{x}_e \\ \boldsymbol{u}=\boldsymbol{u}_e}}$$

are the Jacobian matrices of $\boldsymbol{f}$ with respect to $\boldsymbol{x}$ and $\boldsymbol{u}$, evaluated at the equilibrium point, $[\boldsymbol{x}_e^T \ \ \boldsymbol{u}_e^T]^T$. Note that

$$\frac{d}{dt}\boldsymbol{x} = \frac{d}{dt}\boldsymbol{x}_e + \frac{d}{dt}\Delta\boldsymbol{x} = \frac{d}{dt}\Delta\boldsymbol{x}$$

because $\boldsymbol{x}_e$ is constant. Furthermore, $\boldsymbol{f}(\boldsymbol{x}_e, \boldsymbol{u}_e) = \boldsymbol{0}$. Let

$$\boldsymbol{A} = \frac{\partial \boldsymbol{f}}{\partial \boldsymbol{x}}(\boldsymbol{x}_e, \boldsymbol{u}_e) \quad \text{and} \quad \boldsymbol{B} = \frac{\partial \boldsymbol{f}}{\partial \boldsymbol{u}}(\boldsymbol{x}_e, \boldsymbol{u}_e).$$

Neglecting higher-order terms, we arrive at the linear approximation

$$\frac{d}{dt}\Delta\boldsymbol{x} = \boldsymbol{A}\Delta\boldsymbol{x} + \boldsymbol{B}\Delta\boldsymbol{u}.$$

Similarly, if the outputs of the nonlinear system model are of the form

$$\begin{aligned} y_1 &= h_1(x_1, x_2, \ldots, x_n, u_1, u_2, \ldots, u_m), \\ y_2 &= h_2(x_1, x_2, \ldots, x_n, u_1, u_2, \ldots, u_m), \\ &\ \ \vdots \\ y_p &= h_p(x_1, x_2, \ldots, x_n, u_1, u_2, \ldots, u_m) \end{aligned}$$

or in vector notation

$$\boldsymbol{y} = \boldsymbol{h}(\boldsymbol{x}, \boldsymbol{u}),$$

then Taylor's series expansion can again be used to yield the linear approximation of the above output equations. Indeed, if we let

$$\boldsymbol{y} = \boldsymbol{y}_e + \Delta\boldsymbol{y},$$

then we obtain

$$\Delta\boldsymbol{y} = \boldsymbol{C}\Delta\boldsymbol{x} + \boldsymbol{D}\Delta\boldsymbol{u},$$

where

$$C = \frac{\partial \boldsymbol{h}}{\partial \boldsymbol{x}}(\boldsymbol{x}_e, \boldsymbol{u}_e) = \left.\begin{bmatrix} \frac{\partial h_1}{\partial x_1} & \cdots & \frac{\partial h_1}{\partial x_n} \\ \vdots & & \vdots \\ \frac{\partial h_p}{\partial x_1} & \cdots & \frac{\partial h_p}{\partial x_n} \end{bmatrix}\right|_{\substack{\boldsymbol{x}=\boldsymbol{x}_e \\ \boldsymbol{u}=\boldsymbol{u}_e}} \in \mathbb{R}^{p\times n}$$

and

$$D = \frac{\partial \boldsymbol{h}}{\partial \boldsymbol{u}}(\boldsymbol{x}_e, \boldsymbol{u}_e) = \left.\begin{bmatrix} \frac{\partial h_1}{\partial u_1} & \cdots & \frac{\partial h_1}{\partial u_m} \\ \vdots & & \vdots \\ \frac{\partial h_p}{\partial u_1} & \cdots & \frac{\partial h_p}{\partial u_m} \end{bmatrix}\right|_{\substack{\boldsymbol{x}=\boldsymbol{x}_e \\ \boldsymbol{u}=\boldsymbol{u}_e}} \in \mathbb{R}^{p\times m}$$

are the Jacobian matrices of $\boldsymbol{h}$ with respect to $\boldsymbol{x}$ and $\boldsymbol{u}$, evaluated at the equilibrium point $[\boldsymbol{x}_e^T \ \ \boldsymbol{u}_e^T]^T$.

◆ Example 2.14

In this example, we linearize the model of Watt's governor that we derived in Section 1.7.1. Recall that the equations modeling the governor have the form

$$\begin{aligned} \dot{x}_1 &= x_2, \\ \dot{x}_2 &= \frac{1}{2} N^2 x_3^2 \sin 2x_1 - g \sin x_1 - \frac{b}{m} x_2, \\ \dot{x}_3 &= \frac{\kappa}{I} \cos x_1 - \frac{\tau}{I}. \end{aligned}$$

We linearize the above model of an equilibrium state of the form

$$\boldsymbol{x}_e = [x_{1e} \quad 0 \quad x_{3e}]^T.$$

Equating the right-hand sides of the above engine–governor equations to zero yields

$$\begin{aligned} x_{2e} &= 0 \\ N^2 x_{3e}^2 &= \frac{g}{\cos x_{1e}} \\ \cos x_{1e} &= \frac{\tau}{\kappa} \end{aligned}$$

The Jacobian matrix of the engine–governor model evaluated at the equilibrium state is

$$\left.\frac{\partial \boldsymbol{f}}{\partial \boldsymbol{x}}\right|_{\boldsymbol{x}=\boldsymbol{x}_e} = \begin{bmatrix} 0 & 1 & 0 \\ N^2 x_{3e}^2 \cos 2x_{1e} - g \cos x_{1e} & -\frac{b}{m} & N^2 x_{3e} \sin 2x_{1e} \\ -\frac{\kappa}{I} \sin x_{1e} & 0 & 0 \end{bmatrix}.$$

We use the following trigonometric identities

$$\cos 2\alpha = \cos^2\alpha - \sin^2\alpha, \qquad \sin 2\alpha = 2\sin\alpha\,\cos\alpha$$

and the previously obtained relation, $N^2 x_{3e}^2 = g/\cos x_{1e}$, to eliminate $N^2 x_{3e}^2$ from the Jacobian matrix. Performing needed manipulations, we obtain

$$\left.\frac{\partial \boldsymbol{f}}{\partial \boldsymbol{x}}\right|_{\boldsymbol{x}=\boldsymbol{x}_e} = \begin{bmatrix} 0 & 1 & 0 \\ -\dfrac{g\sin^2 x_{1e}}{\cos x_{1e}} & -\dfrac{b}{m} & \dfrac{2g\sin x_{1e}}{x_{3e}} \\ -\dfrac{\kappa}{I}\sin x_{1e} & 0 & 0 \end{bmatrix}.$$

Hence, the linearized model of the engine–governor system can be represented as

$$\begin{bmatrix} \Delta\dot{x}_1 \\ \Delta\dot{x}_2 \\ \Delta\dot{x}_3 \end{bmatrix} = \begin{bmatrix} 0 & 1 & 0 \\ -\dfrac{g\sin^2 x_{1e}}{\cos x_{1e}} & -\dfrac{b}{m} & \dfrac{2g\sin x_{1e}}{x_{3e}} \\ -\dfrac{\kappa}{I}\sin x_{1e} & 0 & 0 \end{bmatrix} \begin{bmatrix} \Delta x_1 \\ \Delta x_2 \\ \Delta x_3 \end{bmatrix}. \tag{2.26}$$

2.5 Describing Function Method

The describing function method allows us to apply familiar frequency domain techniques, used in the linear system analysis, to the analysis of a class of nonlinear dynamical systems. The method can be used to predict limit cycles in nonlinear systems. The describing function method can be viewed as a "harmonic linearization" of a nonlinear element. The method provides a "linear approximation" to the nonlinear element based on the assumption that the input to the nonlinear element is a sinusoid of known, constant amplitude. The fundamental harmonic of the element's output is compared with the input sinusoid to determine the steady-state amplitude and phase relation. This relation is the describing function for the nonlinear element. The describing function method is based on the Fourier series. In our review of Fourier series, we use the notion of a scalar product over a function space, which we discuss next. Our presentation of the scalar product in the context of spaces of functions follows that of Lang [175]. The reader familiar with the Fourier series may go directly to Subsection 2.5.3, where the describing function method analysis is presented.

2.5.1 Scalar Product of Functions

Let $\mathcal{C}([a,b])$ be the set of continuous functions on the interval $[a,b] \subset \mathbb{R}$.

Definition 2.2 The scalar product of two real-valued functions $u, v \in \mathcal{C}([a,b])$ is

$$\langle u, v\rangle = \int_a^b u(t)v(t)\,dt.$$

Using simple properties of the integral, we can verify that the above scalar product satisfies the following conditions:

1. For all $u, v \in C([a, b])$, we have $\langle u, v\rangle = \langle v, u\rangle$.
2. If u, v, w are elements of $C([a, b])$, then $\langle u, v + w\rangle = \langle u, v\rangle + \langle u, w\rangle$.
3. If x is a number, then $\langle xu, v\rangle = x\langle u, v\rangle = \langle u, xv\rangle$.
4. For all $v \in C([a, b])$, we have $\langle v, v\rangle \geq 0$, and $\langle v, v\rangle > 0$ if $v \neq 0$.

The functions $u, v \in C([a, b])$ are said to be *orthogonal* if $\langle u, v\rangle = 0$. A set of functions from $C([a, b])$ is said to be *mutually orthogonal* if each distinct pair of functions in the set is orthogonal.

◆ Example 2.15

The functions $\psi_n(t) = \sin n\omega t$, $\phi_n(t) = \cos n\omega t$, $n = 1, 2, \ldots$, and $\phi_0(t) = 1$ form a mutually orthogonal set of functions on the interval $[t_0, t_0 + 2\pi/\omega]$. This can be verified by direct integration. For example, for positive m and n such that $m \neq n$, we have

$$\begin{aligned}
\langle \psi_m, \psi_n\rangle &= \int_{t_0}^{t_0+2\pi/\omega} \sin m\omega t \sin n\omega t \, dt \\
&= \frac{1}{2}\int_{t_0}^{t_0+2\pi/\omega} (\cos(m-n)\omega t - \cos(m+n)\omega t)\, dt \\
&= \frac{1}{2\omega}\left(\frac{\sin(m-n)\omega t}{m-n} - \frac{\sin(m+n)\omega t}{m+n}\right)\Bigg|_{t_0}^{t_0+2\pi\omega} \\
&= 0. \qquad (2.27)
\end{aligned}$$

Note that $m + n \neq 0$ because m and n are positive. Using similar computations, we can show that for $m \neq n$ we obtain

$$\langle \phi_m, \phi_n\rangle = 0 \qquad (2.28)$$

and that for all non-negative m and positive n we have

$$\langle \phi_m, \psi_n\rangle = 0. \qquad (2.29)$$

We define the *norm* of $v \in C([a, b])$ to be

$$\|v\| = \sqrt{\langle v, v\rangle}.$$

◆ Example 2.16

Consider $\psi_n(t) = \sin n\omega t$ on the interval $[t_0, t_0 + 2\pi/\omega]$. Then,

$$
\begin{aligned}
\|\psi_n\|^2 &= \int_{t_0}^{t_0+2\pi/\omega} \sin n\omega t \sin n\omega t \, dt \\
&= \int_{t_0}^{t_0+2\pi/\omega} \sin^2 n\omega t \, dt \\
&= \frac{1}{2}\int_{t_0}^{t_0+2\pi/\omega} (1 - \cos 2n\omega t) \, dt \\
&= \frac{1}{2}\left(t - \frac{\sin 2n\omega t}{2n\omega}\right)\Bigg|_{t_0}^{t_0+2\pi/\omega} \\
&= \frac{\pi}{\omega}.
\end{aligned}
\tag{2.30}
$$

Similarly,

$$\|\phi_n\|^2 = \frac{\pi}{\omega}, \qquad n = 1, 2 \ldots, \tag{2.31}$$

and

$$\|\phi_0\|^2 = \frac{2\pi}{\omega}. \tag{2.32}$$

So far we discussed only continuous functions. In many applications we have to work with more general functions, specifically with piecewise continuous functions.

Definition 2.3 A function f is said to be *piecewise continuous* on an interval $[a, b]$ if the interval can be partitioned by a finite number of points, say t_i, so that:

1. $a = t_0 < t_1 < \cdots < t_n = b$.
2. The function f is continuous on each open subinterval (t_{i-1}, t_i).
3. The function f approaches a finite limit as the end points of each subinterval are approached from within the subinterval; that is, the limits

$$\lim_{\substack{h\to 0\\ h>0}} f(t_i - h) \quad \text{and} \quad \lim_{\substack{h\to 0\\ h>0}} f(t_i + h)$$

both exist.

The graph of a piecewise continuous function is shown in Figure 2.15. In further discussion, we assume that two piecewise functions f and g are equal if they have the same values at the points where they are continuous. Thus, the functions shown in Figure 2.16 are considered to be equal, though they differ at the points of discontinuity.

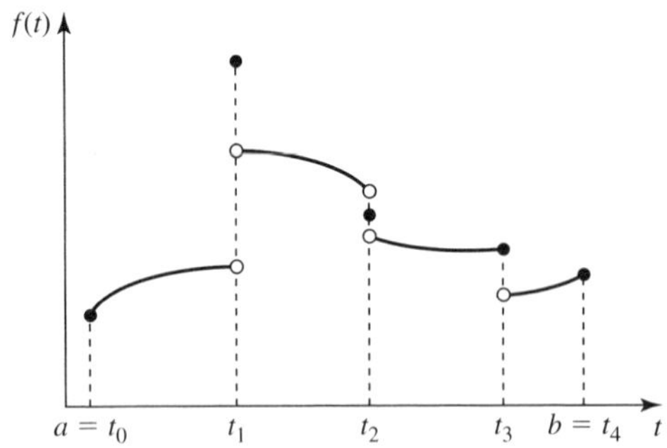

Figure 2.15 A graph of a piecewise continuous function.

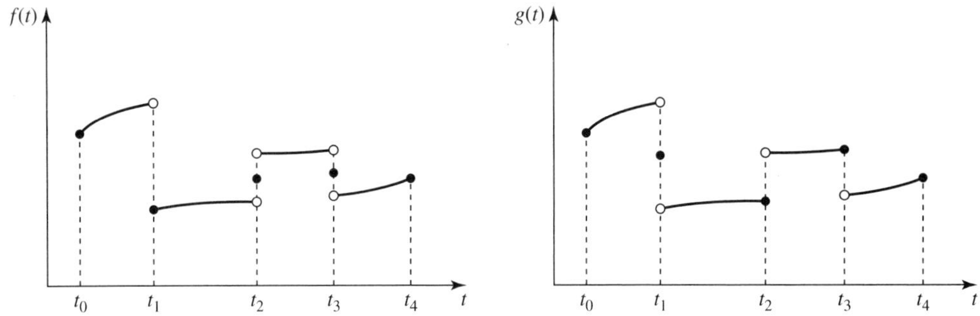

Figure 2.16 Two piecewise continuous functions that we consider to be equal.

Theorem 2.2 Let V be the space of functions that are piecewise continuous on the interval $[a, b]$. Let $f \in V$. Then, $\| f\| = 0$ if and only if $f(x) = 0$ for all but finite number of points x in the interval $[a, b]$.

Proof ($\Leftarrow$) It is clear that if $f(x) = 0$ except for a finite number of $x \in [a, b]$, then

$$\| f\|^2 = \int_a^b f^2(x)\,dx = 0.$$

($\Rightarrow$) Suppose f is piecewise continuous on $[a, b]$, and let $a = t_0 < t_1 < \cdots < t_n = b$ be a partition of the interval $[a, b]$ such that f is continuous on each subinterval $[t_{i-1}, t_i]$ except possibly at the end points. Suppose that $\| f\| = 0$. Then, also $\| f\|^2 = \langle f, f\rangle = 0$. The above means that

$$\int_a^b f^2(x)\,dx = 0,$$

where the integral is the sum of the integrals over the intervals $[t_{i-1}, t_i]$; that is,

$$\int_a^b f^2(x)\,dx = \sum_{i=1}^{n} \int_{t_{i-1}}^{t_i} f^2(x)\,dx = 0.$$

Each integral satisfies

$$\int_{t_{i-1}}^{t_i} f^2(x)\,dx \geq 0.$$

Hence, each such integral must be equal to 0. The function f is continuous on (t_{i-1}, t_i). Therefore,

$$f^2(x) = 0 \qquad \text{for} \quad t_{i-1} < x < t_i,$$

which means that

$$f(x) = 0 \qquad \text{for} \quad t_{i-1} < x < t_i.$$

Hence, $f(x) = 0$ except at a finite number of points. The proof is complete.

The above theorem implies that the scalar product on the space V of piecewise continuous functions is positive definite. In other words, for a piecewise continuous function $f \in V$, we have $\langle f, f \rangle \geq 0$, and $\langle f, f \rangle = 0$ if and only if $f = 0$ for all but a finite number of points in the interval $[a, b]$.

2.5.2 *Fourier Series*

We begin with the definition of a periodic function.

Definition 2.4 A function $f : \mathbb{R} \to \mathbb{R}$ is said to be periodic with the period T if the domain of f contains $t + T$ whenever it contains t and if

$$f(t + T) = f(t)$$

for every value of t. The smallest positive value of T for which $f(t + T) = f(t)$ is called the fundamental period of f.

It follows from the above definition that if T is a period of f, then $2T$ is also a period, and so is any integral multiple of T

A periodic function, satisfying certain assumptions spelled out later, may be represented by the series

$$f(t) = A_0 + \sum_{k=1}^{\infty} (A_k \cos k\omega t + B_k \sin k\omega t), \tag{2.33}$$

where $\omega = 2\pi/T$, with T being the fundamental period of f. On the set of points where the series (2.33) converges, it defines f, whose value at each point is the sum of the series for that value of t, and we say that the series (2.33) is the Fourier series for f.

Suppose that the series (2.33) converges. We use the results of the previous subsection to find expressions for the coefficients A_k and B_k. We first compute A_k coefficients. Multiplying (2.33) by $\phi_m = \cos m\omega t$, where $m > 0$ is a fixed positive integer, and then integrating both sides with respect to t from t_0 to $t_0 + 2\pi/\omega$ yields

$$\langle \phi_m, f \rangle = A_0 \langle \phi_m, \phi_0 \rangle + \sum_{k=1}^{\infty} A_k \langle \phi_m, \phi_k \rangle + \sum_{k=1}^{\infty} B_k \langle \phi_m, \psi_k \rangle, \tag{2.34}$$

where $\phi_0 = 1$. It follows from (2.27), (2.28), and (2.29) that the only nonvanishing term on the right-hand side of (2.34) is the one for which $k = m$ in the first summation. Using (2.31), we obtain

$$
\begin{aligned}
A_m &= \frac{\langle \phi_m, f\rangle}{\langle \phi_m, \phi_m\rangle} \\
&= \frac{\omega}{\pi}\int_{t_0}^{t_0+2\pi/\omega} f(t)\cos m\omega t\,dt, \qquad m = 1, 2, \ldots. \qquad (2.35)
\end{aligned}
$$

To obtain A_0, we integrate (2.33). Using (2.32), we get

$$
A_0 = \frac{\langle \phi_0, f\rangle}{\langle \phi_0, \phi_0\rangle} = \frac{\omega}{2\pi}\int_{t_0}^{t_0+2\pi/\omega} f(t)\,dt. \qquad (2.36)
$$

An expression for B_m may be obtained by multiplying (2.33) by $\sin m\omega t$ and then integrating term by term from t_0 to $t_0 + 2\pi/\omega$. Using the orthogonality relations (2.27)–(2.29) yields

$$
B_m = \frac{\langle \psi_m, f\rangle}{\langle \psi_m, \psi_m\rangle} = \frac{\omega}{\pi}\int_{t_0}^{t_0+2\pi/\omega} f(t)\sin m\omega t\,dt, \qquad m = 1, 2, \ldots. \qquad (2.37)
$$

Suppose now that a function f is given that is periodic with period $2\pi/\omega$ and integrable on the interval $[t_0, t_0+2\pi/\omega]$. Then, coefficients A_k and B_k can be computed using (2.35)–(2.37), and a series of the form (2.33) can be formally constructed. However, we do not know whether this series converges for each value of t and, if so, whether its sum is $f(t)$. The answer is given by the theorem that we state after explaining the notation used. We denote the limit of $f(t)$ as t approaches c from the right by $f(c+)$; that is,

$$
f(c+) = \lim_{t\to c+} f(t).
$$

Similarly, $f(c-) = \lim_{t\to c-} f(t)$ denotes the limit of $f(t)$ as $t \to c$ from the left. The mean value of the right- and left-hand limits at the point c is

$$
(f(c+) + f(c-))/2.
$$

Note that at any point c where f is continuous, we have

$$
f(c+) = f(c-) = f(c).
$$

Theorem 2.3 Assume that f and its derivative are piecewise continuous on the interval $[t_0, t_0 + 2\pi/\omega]$. Furthermore, suppose that f is defined outside the interval $[t_0, t_0 + 2\pi/\omega]$ so that it is periodic with period $2\pi/\omega$. Then, f has a Fourier series of the form (2.33) whose coefficients are given by (2.35)–(2.37). The Fourier series converges to $(f(t+) + f(t-))/2$ at all points of the interval $[t_0, t_0 + 2\pi/\omega]$.

With this knowledge of the Fourier series, it is now possible to discuss the describing function method for analyzing a class of nonlinear control systems.

2.5.3 *Describing Function in the Analysis of Nonlinear Systems*

The describing function method can be considered as an extension of the Nyquist stability criterion to nonlinear systems. In our further discussion, we sometimes use a double box to represent a nonlinear element as illustrated in Figure 2.17.

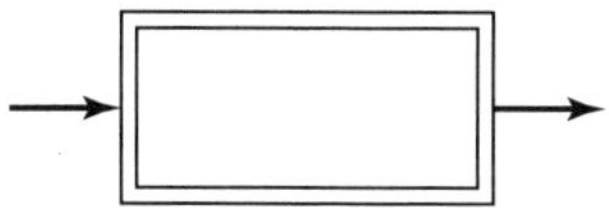

Figure 2.17 A symbol for a nonlinear element.

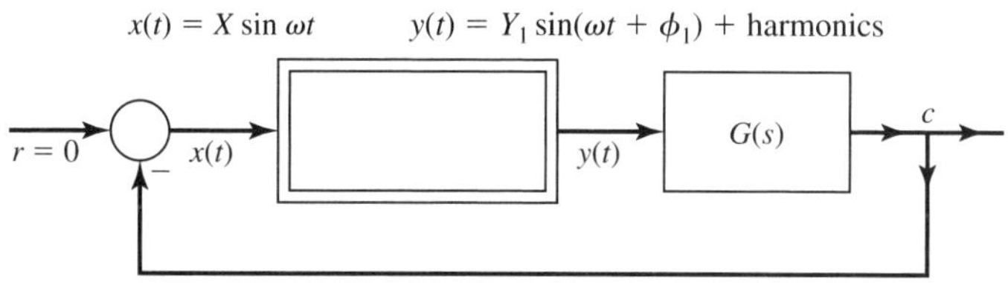

Figure 2.18 A nonlinear feedback system with a nonlinear element in the forward path.

Consider a feedback system that contains a nonlinear element as shown in Figure 2.18. Assume that the input to the nonlinear element is a sinusoidal signal $x(t) = X \sin \omega t$, where X is the amplitude of the input sinusoid. The output of the nonlinear element is, in general, not sinusoidal in response to a sinusoidal input. Suppose that the nonlinear element output is periodic with the same period as its input and that it may be expanded in the Fourier series,

$$\begin{aligned} y(t) &= A_0 + \sum_{k=1}^{\infty} (A_k \cos k\omega t + B_k \sin k\omega t) \\ &= A_0 + \sum_{k=1}^{\infty} Y_k \sin(k\omega t + \phi_k). \end{aligned} \tag{2.38}$$

Thus $y(t)$ consists of a fundamental component

$$Y_1 \sin(\omega t + \phi_1),$$

together with a mean level A_0 and harmonic components at frequencies $2\omega, 3\omega, \ldots$. The describing function method relies on the assumption that only the fundamental harmonic is significant. This assumption is called the *filtering hypothesis*. The filtering hypothesis can be expressed as

$$|G(j\omega)| \gg |G(kj\omega)| \qquad \text{for} \quad k = 2, 3, \ldots,$$

which requires that the linear element, modeled by the transfer function $G(s)$, be a low-pass filter. The filtering hypothesis is often valid because many, if not most, control systems are low-pass filters. This results in higher harmonics being more attenuated compared with the fundamental harmonic component. The nonlinear element is then approximately modeled by its describing function,

$$N(X) = \frac{Y_1}{X} \angle \phi_1 = \frac{\text{fundamental component of output}}{\text{input}}. \tag{2.39}$$

The describing function is, in general, a complex-valued function of the amplitude X of the input sinusoid.

The stability of the nonlinear closed-loop system depicted in Figure 2.18 can be analyzed using Nyquist's criterion applied to the associated linear system shown in Figure 2.19. The

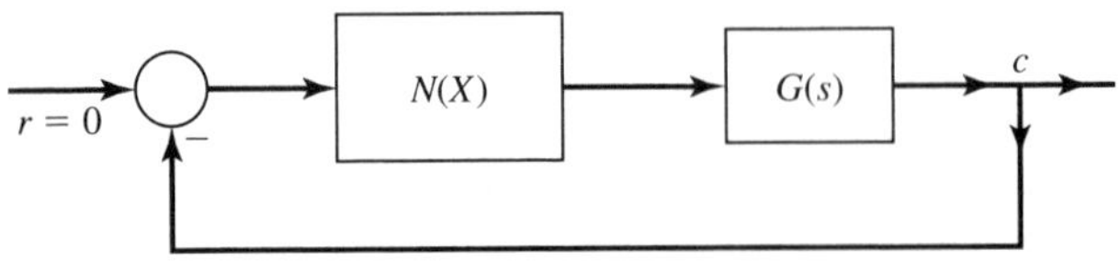

Figure 2.19 Associated linear system used in the describing function analysis of the nonlinear system shown in Figure 2.18.

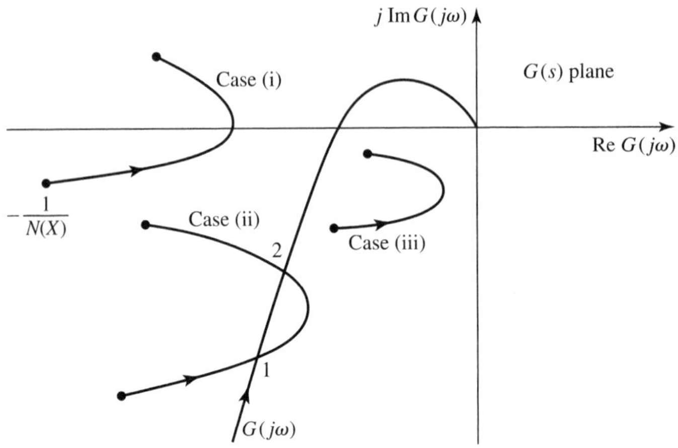

Figure 2.20 Stability analysis using describing function.

associated linear closed-loop system characteristic equation is

$$1 + N(X)G(s) = 0. \tag{2.40}$$

In the $G(j\omega)$ complex plane, the locus traced out by $-1/N$ as X varies is considered a generalization of the $(-1, j0)$ point in the linear case. The stability of the nonlinear system is determined by computing the encirclements of the $-1/N$ locus by the $G(j\omega)$ plot. We might have the situations shown in Figure 2.20. In case (i), the $-1/N$ locus is not encircled by the $G(j\omega)$ plot. Assuming that the poles of $G(s)$ are in the left-half complex plane, the closed-loop system is asymptotically stable. For a limit cycle to exist, we must have an intersection between the polar plot of $G(j\omega)$ and the $-1/N$ locus. We have two intersections between $-1/N$ and $G(j\omega)$ in case (ii). At intersection point 1, any increase in X will cause the operating point to be encircled and will consequently cause X to grow. If there is a decrease in X, then the oscillations will decay. Thus, point 1 represents an unstable limit cycle. At intersection point 2, an increase in X would shift the operating point outside the $G(j\omega)$ plot resulting in the decaying oscillations. Point 2 represents a stable limit cycle. In case (iii), any oscillation will grow without limit.

◆ Example 2.17

Consider the nonlinear control system shown in Figure 2.21 that was analyzed in Power and Simpson [237, pp. 339–341]. We first derive the describing function of the ideal relay. When the input of the ideal relay is greater than zero, its output is M. When the relay's input is less than zero, its output is $-M$. Thus, the ideal relay

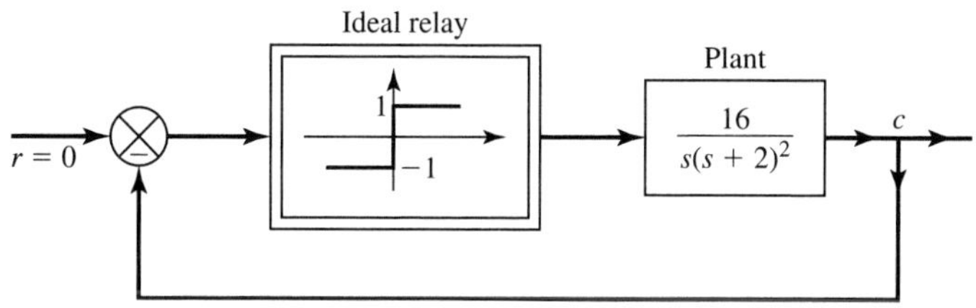

Figure 2.21 Nonlinear feedback control system analyzed in Example 2.17.

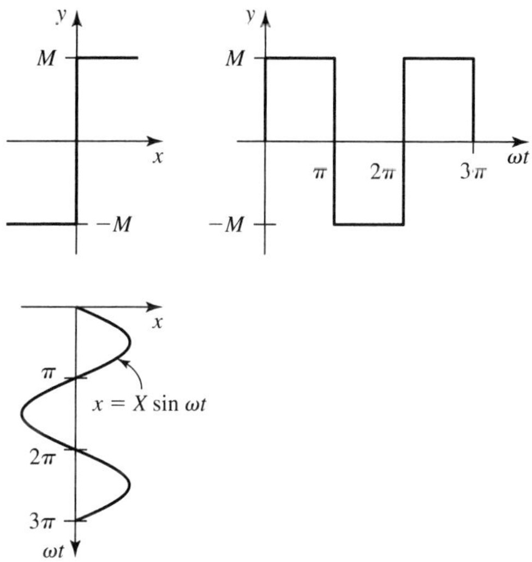

Figure 2.22 Input and output waveforms of an ideal relay.

can be described as

$$y = \begin{cases} M & \text{for } x > 0 \\ -M & \text{for } x < 0. \end{cases} \tag{2.41}$$

We assume that the input to the relay is a sinusoid, $x(t) = X \sin \omega t$, drawn vertically down from the relay characteristic in Figure 2.22. The output $y(t)$ of the relay is a rectangular wave. It is drawn horizontally to the right of the relay characteristic in Figure 2.22. This rectangular wave is an odd function. For an odd function, we have $A_k = 0$, $k = 0, 1, \ldots,$ in its Fourier series. Thus, the Fourier series of the relay's output has the form

$$y(t) = \sum_{k=1}^{\infty} B_k \sin k\omega t.$$

The fundamental harmonic of $y(t)$ is

$$B_1 \sin \omega t = Y_1 \sin \omega t,$$

where

$$
\begin{aligned}
B_1 &= \frac{\omega}{\pi}\int_0^{2\pi/\omega} y(t)\sin\omega t\,dt \\
&= \frac{2\omega}{\pi}\int_0^{\pi/\omega} M\sin\omega t\,dt \\
&= \frac{2M\omega}{\pi}\int_0^{\pi/\omega} \sin\omega t\,dt \\
&= -\frac{2M}{\pi}\cos(\omega t)\Big|_0^{\pi/\omega} \\
&= \frac{4M}{\pi}.
\end{aligned} \tag{2.42}
$$

Hence, the describing function of the ideal relay is

$$N(X) = \frac{Y_1}{X}\angle 0^\circ = \frac{4M}{\pi X}.$$

A normalized plot of the describing function of the ideal relay nonlinearity is depicted in Figure 2.23. In our example, $M = 1$ and thus the describing function of the relay is

$$N(X) = \frac{4}{\pi X}.$$

The sinusoidal transfer function of the linear plant model is

$$G(j\omega) = \frac{16}{j\omega(j\omega+2)^2} = \frac{-64\omega + j16(\omega^2-4)}{\omega(4+\omega^2)^2}.$$

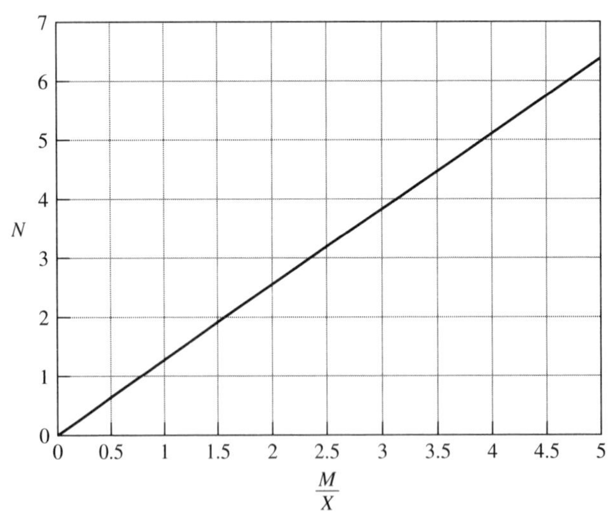

Figure 2.23 A plot of the describing function for the ideal relay.

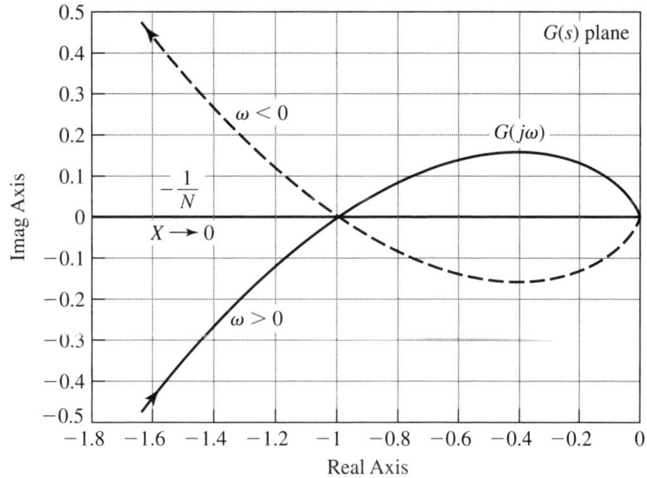

Figure 2.24 Plots of $G(j\omega)$ and $-1/N$ of Example 2.17.

At $\omega = 2$ rad/sec the polar plot of $G(j\omega)$ intersects the plot of $-1/N$ (see Figure 2.24), to form a stable limit cycle. The amplitude of the limit cycle is obtained from the relation $G(j2) = -1/N$; that is,

$$\frac{-64}{(4+4)^2} = -\frac{\pi X}{4}.$$

Hence, the amplitude of the limit cycle is

$$X = \frac{4}{\pi},$$

while the amplitude of the fundamental harmonic of the plant output, c, is

$$\frac{4}{\pi}|G(j2)| = \frac{4}{\pi}.$$

Note that the filtering hypothesis is satisfied in this example. Indeed, the Fourier series of the output of the relay is

$$\frac{4}{\pi}\left(\sin\omega t + \frac{1}{3}\sin 3\omega t + \frac{1}{5}\sin 5\omega t + \cdots\right).$$

Thus, for example, the amplitude of the third harmonic of the plant output is

$$\frac{4}{3\pi}|G(j6)| = \frac{1}{45}\frac{4}{\pi}.$$

The higher harmonics are even more attenuated compared with the fundamental harmonic.

Notes

The system of differential equations (2.43) in Exercise 2.6 comes from Boyce and DiPrima [32, p. 523]. Al-Khafaji and Tooley [6] provide an introductory treatment of numerical solutions of differential equations. For further discussion of solving differential equations using numerical methods and analyses of their computational efficiency, the reader is referred to Salvadori and Baron [252], Conte and de Boor [53, Chapter 8], or Parker and Chua [222]. For an in-depth treatment of the describing function method, we refer to Graham and McRuer [105] and Chapters 6 and 7 of Hsu and Meyer [128]. Graham and McRuer [105] make an interesting remark about the origins of the describing function method. They note on page 92 of their 1961 book: "While the notion of representing a nonlinearity by an 'equivalent' linear element is quite old and has been used by many writers, the first instance of a major systematic exploitation of the technique was probably by N. M. Kryloff and N. N. Bogoliuboff, *Introduction to Non-linear Mechanics* (a free translation of the Russian edition of 1937, by S. Lefschetz), Princeton University Press, Princeton, N.J., 1943. See also N. Minorsky, *Introduction to Non-linear Mechanics,* J. W. Edwards, Ann Arbor, Mich., 1947." The so-called sinusoidal describing function that we analyzed in the previous section, was developed almost simultaneously in several different countries during and just after World War II. The first to introduce the method in the open literature appears to be A. Tustin of England, who published his paper in the *Journal of the IEE* in 1947.

The behavior of numerous dynamical systems is often modeled by a set of linear, first-order, differential equations. Frequently, however, a linear model is a result of linearization of a nonlinear model. Linear models seem to dominate the controls' literature. Yet nature is nonlinear. "All physical systems are nonlinear and have time-varying parameters in some degree" [105, p. 1]. How can one reconcile this paradox? Graham and McRuer [105, p. 1] have the following answer: "Where the effect of the nonlinearity is very small, or if the parameters vary only slowly with time, linear constant-parameter methods of analysis can be applied to give an approximate answer which is adequate for engineering purposes. The analysis and synthesis of physical systems, predicated on the study of linear constant-parameter mathematical models, has been, in fact, an outstandingly successful enterprise. A system represented as linear, however, is to some extent a mathematical abstraction that can never be encountered in a real world. Either by design or because of nature's ways it is often true that experimental facts do not, or would not, correspond with any prediction of linear constant-parameter theory. In this case nonlinear or time-varying-parameter theory is essential to the description and understanding of physical phenomena."

EXERCISES

2.1 Sketch a phase-plane portrait of the differential equation

$$\ddot{x} - \dot{x} + 2x + 2\text{sign}(2x + \dot{x}) = 0.$$

2.2 For the system of differential equations

$$\dot{x}_1 = -\frac{7}{4}x_1 + \frac{1}{4}x_2,$$
$$\dot{x}_2 = \frac{3}{4}x_1 - \frac{5}{4}x_2,$$

find the phase-plane asymptotes.

2.3 For a dynamical system modeled by

$$\dot{x}_1 = x_2,$$
$$\dot{x}_2 = -x_1,$$

find the time T it takes the system trajectory to move from the state $\boldsymbol{x}(0) = [0 \;\; 1]^T$ to the state $\boldsymbol{x}(T) = [1 \;\; 0]^T$.

2.4 Consider a trajectory in the state plane shown in Figure 2.25. Find the time it takes the representative point to move from A to B.

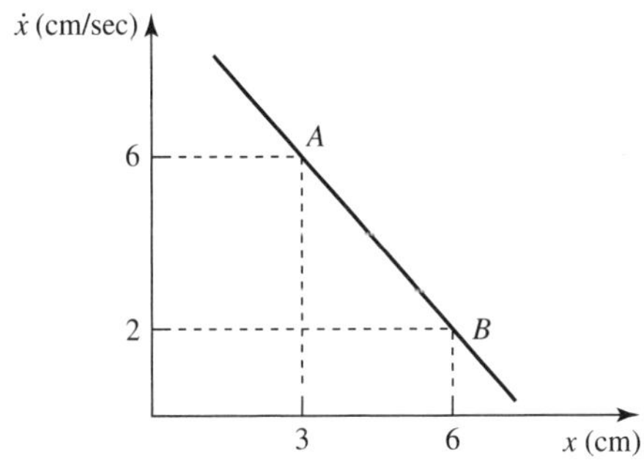

Figure 2.25 A trajectory for Exercise 2.4.

2.5 Given the following initial value problem:

$$\dot{x} = f(t, x), \qquad x(t_0) = x_0.$$

(a) Derive the Euler formula for solving the above initial value problem using a Taylor series.

(b) Make use of a Taylor series with a remainder to compute the local error formula assuming that the data at the nth step are correct.

2.6 Use Bendixson's theorem to investigate the existence of limit cycles for the system

$$\begin{aligned} \dot{x}_1 &= x_1 + x_2 - x_1\left(x_1^2 + x_2^2\right), \\ \dot{x}_2 &= -x_1 + x_2 - x_2\left(x_1^2 + x_2^2\right). \end{aligned} \tag{2.43}$$

Sketch a phase-plane portrait for this system.

2.7 Use Bendixson's theorem to investigate the existence of limit cycles for the system

$$\begin{aligned} \dot{x}_1 &= x_2 - x_1\left(x_1^2 + x_2^2 - c\right), \\ \dot{x}_2 &= -x_1 - x_2\left(x_1^2 + x_2^2 - c\right), \end{aligned} \tag{2.44}$$

where c is a constant. Sketch phase-plane portraits for this system for $c = -0.2$ and $c = 0.2$. The above nonlinear system was analyzed in reference 222, p. 204.

2.8 Verify that the Runge method agrees with the Taylor series solution through terms of degree three.

2.9 Verify that the Runge–Kutta method agrees with the Taylor series solution through terms of degree four.

2.10 Consider the nonlinear mass–spring–damper system depicted in Figure 2.26. The spring coefficient $K = 0.5x^2$ N/m, the damping coefficient $B = 0.1\dot{x}^2$ N·sec/m, and $M = 1$ kg.

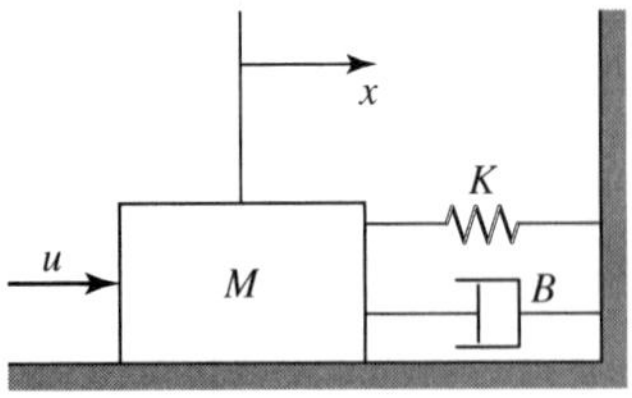

Figure 2.26 The mass–spring–damper system of Exercise 2.10.

(a) Write the second-order differential equation describing the motion of the system.

(b) Choose $x_1 = x$ and $x_2 = \dot{x}$. Represent the equation of motion derived in (a) in state-space format.

(c) For the model from (b), find the equilibrium state corresponding to $u = 0$. Then, linearize the model about the obtained equilibrium point.

2.11 (Based on Toptscheev and Tsyplyakov [282, pp. 29–31]) A cyclotron, whose operation is illustrated in Figure 2.27, accelerates charged particles into high energies so that they can be used in atom-smashing experiments. The operation of the cyclotron can be described as follows. Positive ions enter the cyclotron from the ion source, S, and are available to be accelerated. An electric oscillator establishes an accelerating potential difference across the gap of the two D-shaped objects, called dees. An emerging ion from the ion source, S, finds the dee that is negative. It will accelerate toward this dee and will enter it. Once inside the dee, the ion is screened from electric fields by the metal walls of the dees. The dees are immersed in a magnetic field so that the ion's trajectory bends in a circular path. The radius of the path depends on the ion's velocity. After some time, the ion emerges from the dee on the other side of the ion source. In

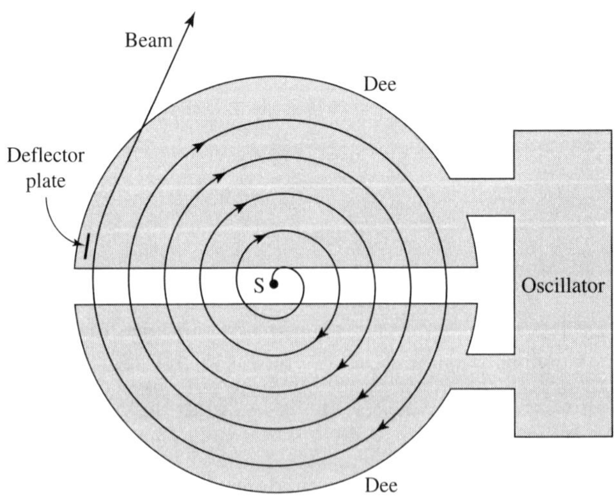

Figure 2.27 Schematic of a cyclotron of Exercise 2.11.

the meantime, the accelerating potential changes its sign. Thus the ion again faces a negative dee. The ion further accelerates and describes a semicircle of larger radius. This process goes on until the ion reaches the outer edge of one dee, where it is pulled out of the cyclotron by a negatively charged deflector plate. The moving ion is acted upon by a deflecting force of magnitude

$$q(v+\Delta v)(B+\Delta B),$$

where q is the charge of the ion, v is the ion's speed, Δv is the increase in the ion's speed, and B is the magnetic field acting on the ion. The nominal radius of the ion's orbit is r. The change in the magnetic field as the result of a change in the orbit of the ion is ΔB. As the ion speed increases, the relativistic mass m increases by Δm. The ion moving in a circular path undergoes centrifugal force of magnitude

$$\frac{(m+\Delta m)(v+\Delta v)^2}{r+\Delta r}.$$

By Newton's second law, the equation describing the motion of the ion is

$$(m+\Delta m)\frac{d^2}{dt^2}(r+\Delta r) = \frac{(m+\Delta m)(v+\Delta v)^2}{r+\Delta r} - q(v+\Delta v)(B+\Delta B).$$

Linearize the above equation. The motion of the ion along a nominal path is described by

$$m\frac{d^2 r}{dt^2} = \frac{mv^2}{r} - qvB.$$

Find the transfer function relating the change in radius of motion, Δr, of the ion and the change in the momentum of the ion:

$$\Delta p = m\Delta v + \Delta m v.$$

In other words, find the transfer function $\mathcal{L}(\Delta r)/\mathcal{L}(\Delta p)$, where $\mathcal{L}$ is the Laplace transform operator.

2.12 Linearize the permanent-magnet stepper motor model, analyzed in Subsection 1.7.3, about $x_{1eq} = \theta_{eq} = \text{constant}$.

2.13 Let $v_1, v_2, \ldots, v_n$ be nonzero, mutually orthogonal elements of the space V and let v be an arbitrary element of V.

(a) Find the component of v along v_j.

(b) Let c_j denote the component of v along v_j. Show that the element

$$v - c_1 v_1 - \cdots c_n v_n$$

is orthogonal to $v_1, v_2, \ldots, v_n$.

(c) Let c_j be the component of $v \in V$ along v_j. Show that

$$\sum_{i=1}^{n} c_i^2 \le \|v\|^2.$$

The above relation is called the Bessel inequality.

2.14 Consider characteristics of two nonlinear elements that are shown in Figure 2.28. Let

$$N_S(x) = \frac{2}{\pi}\left[\arcsin\frac{1}{x} + \frac{1}{x}\cos\left(\arcsin\frac{1}{x}\right)\right]. \tag{2.45}$$

Then, the describing function of the limiter is

$$N_l = \begin{cases} K, & M \le S \\ K N_S\left(\dfrac{M}{S}\right), & M > S. \end{cases}$$

The describing function of the dead zone is

$$N_{d-z} = \begin{cases} 0, & M \le A, \\ K\left[1 - N_S\left(\dfrac{M}{A}\right)\right], & M > A. \end{cases}$$

Derive the describing function of the nonlinear element whose characteristic is shown in Figure 2.29.

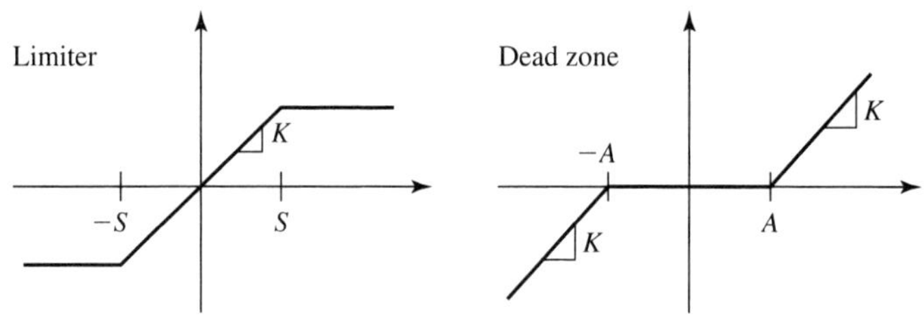

Figure 2.28 Characteristics of a limiter and dead zone nonlinear elements of Exercise 2.14.

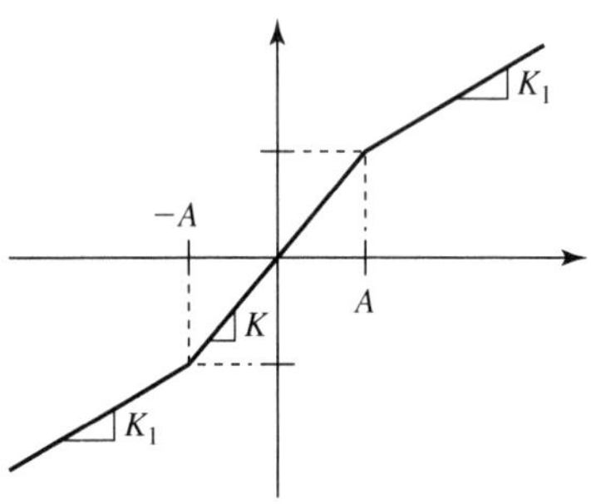

Figure 2.29 Characteristic of a nonlinear element of Exercise 2.14.

2.15 The describing function of the nonlinear element shown in Figure 2.30 is

$$N_1 = \begin{cases} 0, & M \le A, \\ K\left[1 - N_S\left(\dfrac{M}{A}\right)\right] + \dfrac{4B}{\pi M}\sqrt{1 - \left(\dfrac{A}{M}\right)^2}, & M > A, \end{cases}$$

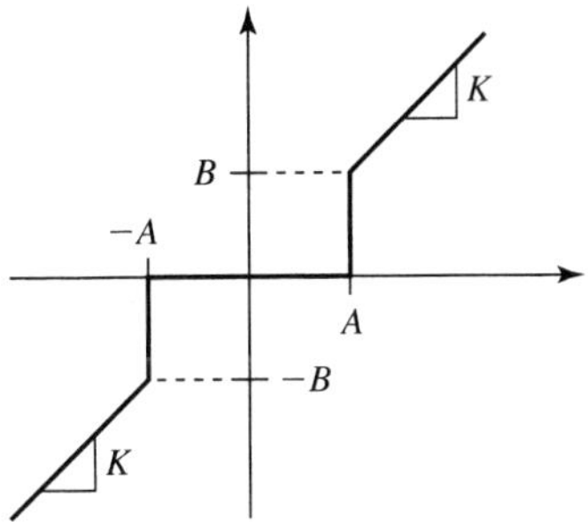

Figure 2.30 Nonlinear element with the describing function N_1 of Exercise 2.15.

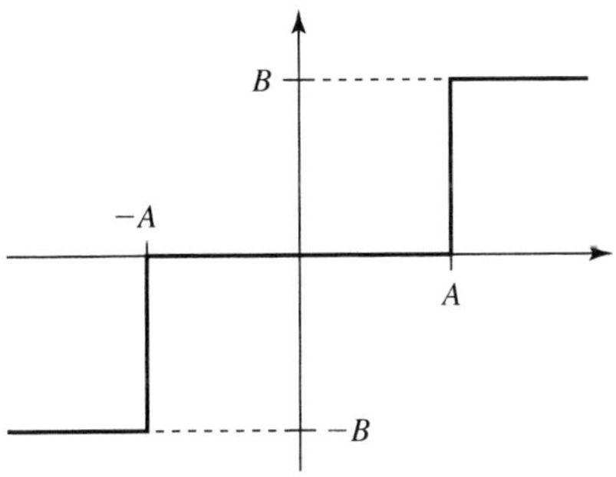

Figure 2.31 Characteristic of the ideal relay with dead zone.

where $N_S(x)$ is given by (2.45). The describing function of the dead-zone nonlinearity is given in Exercise 2.14 and its characteristic is shown in Figure 2.28. Derive the describing function of the ideal relay with dead zone whose characteristic is shown in Figure 2.31.

CHAPTER 3 Linear Systems

In bear country, wear bells to avoid surprising bruins. Don't approach a bear, but don't run away from one either. Movement may incite a charge, and a bear can outrun you.

If a bear approaches you, talk to it calmly in a low, firm voice. Slowly back away until the bear is out of sight. If you climb a tree, get at least 12 feet up.

If a bear senses that its food may be stolen or that its cub is in danger, it may charge. If this happens, drop to the ground and assume a fetal position, protecting your head with your arms, and play dead. Keep your pack on; it may serve as armor.

—*Reader's Digest Practical Problem Solver* [241]

We begin our discussion of linear system theory with a description of the concept of linearity by Mitchell J. Feigenbaum [227, p. 3], who is a researcher of nonlinear systems. "Linearity means that the rule that determines what a piece of a system is going to do next is not influenced by what it is doing now. More precisely, this is intended in a differential or incremental sense: For a linear spring, the increase of its tension is proportional to the increment whereby it is stretched, with the ratio of these increments exactly independent of how much it has already been stretched. Such a spring can be stretched arbitrarily far, and in particular will never snap or break. Accordingly, no real spring is linear." The real world is nonlinear. However, there are a number of reasons to investigate linear systems. Linear models are often used to approximate nonlinear systems. Many times, this approximation is sufficient for the controller design for the underlying nonlinear system. Knowledge of linear system theory helps to understand the intricate theory of nonlinear systems.

In this chapter we acquaint ourselves with basic concepts from linear systems that will be used in subsequent chapters.

3.1 Reachability and Controllability

We first consider a linear, time-invariant, discrete-time system modeled by

$$\boldsymbol{x}(k+1) = \boldsymbol{A}\boldsymbol{x}(k) + \boldsymbol{B}\boldsymbol{u}(k), \qquad \boldsymbol{x}(0) = \boldsymbol{x}_0, \tag{3.1}$$

where $\boldsymbol{A} \in \mathbb{R}^{n\times n}$, $\boldsymbol{B} \in \mathbb{R}^{n\times m}$. The solution to (3.1) is

$$\boldsymbol{x}(i) = \boldsymbol{A}^i \boldsymbol{x}(0) + \sum_{k=0}^{i-1} \boldsymbol{A}^{i-k-1} \boldsymbol{B}\boldsymbol{u}(k),$$

which can be represented as

$$\boldsymbol{x}(i) = \boldsymbol{A}^i \boldsymbol{x}(0) + [\boldsymbol{B} \quad \boldsymbol{A}\boldsymbol{B} \quad \cdots \quad \boldsymbol{A}^{i-1}\boldsymbol{B}] \begin{bmatrix} \boldsymbol{u}(i-1) \\ \boldsymbol{u}(i-2) \\ \vdots \\ \boldsymbol{u}(0) \end{bmatrix}. \tag{3.2}$$

Let

$$\boldsymbol{U}_i = [\boldsymbol{B} \quad \boldsymbol{A}\boldsymbol{B} \quad \cdots \quad \boldsymbol{A}^{i-1}\boldsymbol{B}], \qquad i = 1, 2, \ldots.$$

Note that $\boldsymbol{U}_i \in \mathbb{R}^{n\times mi}$. If for some positive l we have

$$\operatorname{rank} \boldsymbol{U}_l = \operatorname{rank} \boldsymbol{U}_{l+1},$$

then all the columns of $\boldsymbol{A}^l\boldsymbol{B}$ are linearly dependent on those of $\boldsymbol{U}_l$. This, in turn, implies that all the columns of $\boldsymbol{A}^{l+1}\boldsymbol{B}, \boldsymbol{A}^{l+2}\boldsymbol{B}, \ldots$ must also be linearly dependent on those of $\boldsymbol{U}_l$. Hence,

$$\operatorname{rank} \boldsymbol{U}_l = \operatorname{rank} \boldsymbol{U}_{l+1} = \operatorname{rank} \boldsymbol{U}_{l+2} = \cdots. \tag{3.3}$$

The above implies that the rank of $\boldsymbol{U}_i$ increases by at least one when i is increased by one, until the maximum value of rank $\boldsymbol{U}_i$ is attained. The maximum value of the rank $\boldsymbol{U}_i$ is guaranteed to be achieved for $i = n$, which means that we always have

$$\operatorname{rank} \boldsymbol{U}_n = \operatorname{rank} \boldsymbol{U}_{n+1} = \operatorname{rank} \boldsymbol{U}_{n+2} = \cdots. \tag{3.4}$$

It follows from (3.3) that if rank $\boldsymbol{B} = p$, then

$$\operatorname{rank}[\boldsymbol{B} \quad \boldsymbol{A}\boldsymbol{B} \quad \cdots \quad \boldsymbol{A}^{n-1}\boldsymbol{B}] = \operatorname{rank}[\boldsymbol{B} \quad \boldsymbol{A}\boldsymbol{B} \quad \cdots \quad \boldsymbol{A}^{n-p}\boldsymbol{B}]. \tag{3.5}$$

We say that the system (3.1) is *reachable* if for any $\boldsymbol{x}_f$ there exists an integer $q > 0$ and a control sequence, $\{\boldsymbol{u}(i) : i = 0, 1, \ldots, q-1\}$, that transfers $\boldsymbol{x}(0) = \boldsymbol{0}$ to $\boldsymbol{x}(q) = \boldsymbol{x}_f$.

Theorem 3.1 The system (3.1) is reachable if and only if

$$\operatorname{rank}[\boldsymbol{B} \quad \boldsymbol{A}\boldsymbol{B} \quad \cdots \quad \boldsymbol{A}^{n-1}\boldsymbol{B}] = n.$$

Proof Using (3.2) where $\boldsymbol{x}(0) = \boldsymbol{0}$ and $\boldsymbol{x}(q) = \boldsymbol{x}_f$, we obtain

$$\begin{aligned}\boldsymbol{x}_f &= \sum_{k=0}^{q-1} \boldsymbol{A}^{q-k-1}\boldsymbol{B}\boldsymbol{u}(k)\\ &= [\boldsymbol{B} \quad \boldsymbol{A}\boldsymbol{B} \quad \cdots \quad \boldsymbol{A}^{q-1}\boldsymbol{B}]\begin{bmatrix}\boldsymbol{u}(q-1)\\ \boldsymbol{u}(q-2)\\ \vdots\\ \boldsymbol{u}(0)\end{bmatrix}.\end{aligned} \tag{3.6}$$

It follows from the above that for any $\boldsymbol{x}_f$ there exists a control sequence $\{\boldsymbol{u}(i) : i = 0, 1, \ldots, q-1\}$ that transfers $\boldsymbol{x}(0) = \boldsymbol{0}$ to $\boldsymbol{x}(q) = \boldsymbol{x}_f$ if and only if

$$\operatorname{rank}[\boldsymbol{B} \quad \boldsymbol{A}\boldsymbol{B} \quad \cdots \quad \boldsymbol{A}^{q-1}\boldsymbol{B}] = \operatorname{rank} \boldsymbol{U}_q = n.$$

By (3.4) the maximum value of the rank of $\boldsymbol{U}_q$ is guaranteed to be attained for $q = n$, which proves the theorem.

A notion closely related to reachability is the notion of *controllability*. The notion of controllability is related to the property of a controlled system being transferable from any given state to the origin $\boldsymbol{0}$ of the state space by means of an appropriately chosen control law.

Theorem 3.2 The system (3.1) is controllable if and only if

$$\operatorname{range}(\boldsymbol{A}^n) \subset \operatorname{range}[\boldsymbol{B} \quad \boldsymbol{A}\boldsymbol{B} \quad \cdots \quad \boldsymbol{A}^{n-1}\boldsymbol{B}].$$

Proof Using the solution formula for the system (3.1), for the case when $\boldsymbol{x}(0) = \boldsymbol{x}_0$ and $\boldsymbol{x}_f = \boldsymbol{x}(q) = \boldsymbol{0}$, we obtain

$$\begin{aligned}\boldsymbol{A}^q\boldsymbol{x}_0 &= -\sum_{k=0}^{q-1} \boldsymbol{A}^{q-k-1}\boldsymbol{B}\boldsymbol{u}(k)\\ &= -[\boldsymbol{B} \quad \boldsymbol{A}\boldsymbol{B} \quad \cdots \quad \boldsymbol{A}^{q-1}\boldsymbol{B}]\begin{bmatrix}\boldsymbol{u}(q-1)\\ \boldsymbol{u}(q-2)\\ \vdots\\ \boldsymbol{u}(0)\end{bmatrix}.\end{aligned} \tag{3.7}$$

The system (3.1) is controllable if and only if for some $q > 0$ and arbitrary initial condition $\boldsymbol{x}_0$, the vector $\boldsymbol{A}^q\boldsymbol{x}_0$ is in the range of $[\boldsymbol{B} \quad \boldsymbol{A}\boldsymbol{B} \quad \cdots \quad \boldsymbol{A}^{q-1}\boldsymbol{B}] = \boldsymbol{U}_q$. Taking into account that the maximal range of $\boldsymbol{U}_q$ is guaranteed to be attained for $q = n$, the system (3.1) is controllable if and only if the condition of the theorem is satisfied.

Note that the condition of the above theorem can be equivalently represented as

$$\operatorname{rank}[\boldsymbol{B} \quad \boldsymbol{A}\boldsymbol{B} \quad \cdots \quad \boldsymbol{A}^{n-1}\boldsymbol{B}] = \operatorname{rank}[\boldsymbol{B} \quad \boldsymbol{A}\boldsymbol{B} \quad \cdots \quad \boldsymbol{A}^{n-1}\boldsymbol{B} \quad \boldsymbol{A}^n]. \tag{3.8}$$

We illustrate the above discussion with a simple example that we found in Kaczorek [144, p. 482].

◆ Example 3.1

Consider a discrete-time system modeled by

$$
\begin{aligned}
\boldsymbol{x}(k+1) &= \boldsymbol{A}\boldsymbol{x}(k) + \boldsymbol{b}u(k) \\
&= \begin{bmatrix} 0 & a \\ 0 & 0 \end{bmatrix} \boldsymbol{x}(k) + \begin{bmatrix} 1 \\ 0 \end{bmatrix} u(k), \qquad a \in \mathbb{R}.
\end{aligned} \tag{3.9}
$$

The above system model is not reachable because

$$
\operatorname{rank}[\boldsymbol{b} \quad \boldsymbol{A}\boldsymbol{b}] = \operatorname{rank} \begin{bmatrix} 1 & 0 \\ 0 & 0 \end{bmatrix} = 1 < 2.
$$

Using (3.6), it is easy to check that, for example, the state $\boldsymbol{x}_f = [0 \;\; 1]^T$ is not reachable from $\boldsymbol{x}(0) = \boldsymbol{0}$ because $\boldsymbol{x}_f$ does not belong to the range of $[\boldsymbol{b} \;\; \boldsymbol{A}\boldsymbol{b}]$. In fact, nonreachable states form the set

$$
\left\{ c \begin{bmatrix} 0 \\ 1 \end{bmatrix}, \qquad c \in \mathbb{R} \right\}.
$$

However, the system (3.9) is controllable because for any $a \in \mathbb{R}$,

$$
\operatorname{rank}[\boldsymbol{b} \quad \boldsymbol{A}\boldsymbol{b}] = \operatorname{rank} \begin{bmatrix} 1 & 0 \\ 0 & 0 \end{bmatrix} = \operatorname{rank}[\boldsymbol{b} \quad \boldsymbol{A}\boldsymbol{b} \quad \boldsymbol{A}^2\,] = \operatorname{rank} \begin{bmatrix} 1 & 0 & 0 & 0 \\ 0 & 0 & 0 & 0 \end{bmatrix};
$$

that is,

$$
\operatorname{range}(\boldsymbol{A}^2) = \operatorname{range}(\boldsymbol{O}) \subset \operatorname{range}[\boldsymbol{b} \quad \boldsymbol{A}\boldsymbol{b}].
$$

(See appendix, Section A.3, for an introduction to the $\boldsymbol{O}$ notation.) Using (3.7), we can easily see that arbitrary initial state $\boldsymbol{x}(0) \in \mathbb{R}^2$ of the system (3.9) can be transferred to the origin of $\mathbb{R}^2$ using, for example, the zero control sequence, $u(0) = u(1) = 0$. This is because $\boldsymbol{A}^2 = \boldsymbol{O}$ and hence $\boldsymbol{A}^2\boldsymbol{x}(0) = \boldsymbol{0}$ for any $\boldsymbol{x}(0) \in \mathbb{R}^2$.

In summary, for discrete-time linear systems, reachability implies controllability, and the two notions are equivalent if the matrix $\boldsymbol{A}$ of the given discrete system is nonsingular.

We next discuss the notions of reachability and controllability for continuous-time linear systems.

We say that the system $\dot{\boldsymbol{x}}(t) = \boldsymbol{A}\boldsymbol{x}(t) + \boldsymbol{B}\boldsymbol{u}(t)$, where $\boldsymbol{A} \in \mathbb{R}^{n\times n}$, $\boldsymbol{B} \in \mathbb{R}^{n\times m}$, or equivalently the pair $(\boldsymbol{A}, \boldsymbol{B})$, is reachable if for any $\boldsymbol{x}_f \in \mathbb{R}^n$ there is t_1 and a control law $\boldsymbol{u}(t)$ that transfers $\boldsymbol{x}(t_0) = \boldsymbol{0}$ to $\boldsymbol{x}(t_1) = \boldsymbol{x}_f$.

We say that the system $\dot{\boldsymbol{x}}(t) = \boldsymbol{A}\boldsymbol{x}(t) + \boldsymbol{B}\boldsymbol{u}(t)$, or equivalently the pair $(\boldsymbol{A}, \boldsymbol{B})$, is *controllable* if there is a control law $\boldsymbol{u}(t)$ that transfers any initial state $\boldsymbol{x}_0 = \boldsymbol{x}(t_0)$ at any time t_0 to the origin $\boldsymbol{0}$ of the state space at some time $t_1 > t_0$.

It turns out that in the case of continuous, time-invariant systems the two notions are equivalent. Indeed, from the solution formula for the controlled system (A.151), for $\boldsymbol{x}(0) = \boldsymbol{0}$ and

$x(t_1) = x_f$, we obtain

$$x_f = \int_0^{t_1} e^{A(t_1-t)} Bu(t)\, dt.$$

Premultiplying the above equation by e^{-At_1} and rearranging, we get

$$v = \int_0^{t_1} e^{-At} Bu(t)\, dt, \tag{3.10}$$

where $v = e^{-At_1} x_f$. On the other hand, from the solution formula for the controlled system (A.151) for $x(0) = x_0$ and $x(t_1) = \mathbf{0}$, we obtain

$$\mathbf{0} = e^{At_1} x_0 + \int_0^{t_1} e^{A(t_1-t)} Bu(t)\, dt.$$

Premultiplying the above equation by e^{-At_1} and rearranging yields

$$-x_0 = \int_0^{t_1} e^{-At} Bu(t)\, dt. \tag{3.11}$$

Comparing (3.10) and (3.11), we conclude that reachability of the system $\dot{x}(t) = Ax(t)+Bu(t)$ is equivalent to its controllability. This is not true in the case of discrete-time systems, where reachability implies controllability, and the two notions are equivalent if the matrix A of the given discrete system is nonsingular.

There are many criteria for checking if a given system is reachable/controllable or not. We present a few of them in this section.

Theorem 3.3 The following are equivalent:

1. The system $\dot{x}(t) = Ax(t) + Bu(t)$ is reachable.
2. $\operatorname{rank}[B \;\; AB \;\; \cdots \;\; A^{n-1}B] = n$.
3. The n rows of the matrix $e^{-At}B$ are linearly independent over the set of real numbers for all $t \in [0, \infty)$.
4. The matrix

$$W(t_0, t_1) = \int_{t_0}^{t_1} e^{-At} BB^T e^{-A^T t}\, dt$$

is nonsingular for all $t_1 > t_0$.

Proof We first prove the implication 1 ⇒ 2; that is, if the pair (A, B) is reachable, then the so-called controllability matrix

$$[B \quad AB \quad \cdots \quad A^{n-1}B] \in \mathbb{R}^{n\times mn}$$

is of full rank n. We prove this statement by contraposition. Thus, we assume that

$$\operatorname{rank}[B \quad AB \quad \cdots \quad A^{n-1}B] < n;$$

then, there is a constant n-vector $\boldsymbol{q} \neq \boldsymbol{0}$ so that

$$\boldsymbol{q}^T \boldsymbol{B} = \boldsymbol{0}^T, \qquad \boldsymbol{q}^T \boldsymbol{A}\boldsymbol{B} = \boldsymbol{0}^T, \ldots, \boldsymbol{q}^T \boldsymbol{A}^{n-1} \boldsymbol{B} = \boldsymbol{0}^T.$$

By the Cayley–Hamilton theorem the matrix $\boldsymbol{A}$ satisfies its own characteristic equation. Hence

$$\boldsymbol{A}^n = -a_{n-1} \boldsymbol{A}^{n-1} - \cdots - a_1 \boldsymbol{A} - a_0 \boldsymbol{I}_n.$$

Therefore

$$\boldsymbol{q}^T \boldsymbol{A}^n \boldsymbol{B} = \boldsymbol{q}^T(-a_{n-1} \boldsymbol{A}^{n-1} \boldsymbol{B} - \cdots - a_1 \boldsymbol{A}\boldsymbol{B} - a_0 \boldsymbol{B}) = \boldsymbol{0}^T.$$

By induction we obtain

$$\boldsymbol{q}^T \boldsymbol{A}^i \boldsymbol{B} = \boldsymbol{0}^T \qquad \text{for } i = n+1, n+2, \ldots.$$

Now let $\boldsymbol{x}(0) = \boldsymbol{0}$ and $\boldsymbol{x}(t_1) = \boldsymbol{q}$. We will now show that there is no control law that can transfer the system from $\boldsymbol{x}(0) = \boldsymbol{0}$ to $\boldsymbol{x}(t_1) = \boldsymbol{q}$. From the solution formula for the controlled system, we obtain

$$\boldsymbol{q} = \int_0^{t_1} e^{\boldsymbol{A}(t_1-t)} \boldsymbol{B}\boldsymbol{u}(t)\, dt.$$

We now premultiply the resulting equation by $\boldsymbol{q}^T$ to get

$$0 \neq \|\boldsymbol{q}\|^2 = \int_0^{t_1} \boldsymbol{q}^T e^{\boldsymbol{A}(t_1-t)} \boldsymbol{B}\boldsymbol{u}(t)\, dt = 0$$

because

$$\boldsymbol{q}^T e^{\boldsymbol{A}(t_1-t)} \boldsymbol{B} = \boldsymbol{q}^T \left(\boldsymbol{B} + (t_1 - t)\boldsymbol{A}\boldsymbol{B} + \frac{(t_1-t)^2}{2!} \boldsymbol{A}^2 \boldsymbol{B} - \cdots \right) = \boldsymbol{0}^T.$$

Thus the system cannot be transferred from $\boldsymbol{x}(0) = \boldsymbol{0}$ to $\boldsymbol{x}(t_1) = \boldsymbol{q}$, and hence the pair $(\boldsymbol{A}, \boldsymbol{B})$ is not reachable.

We will now prove the implication 2 $\Rightarrow$ 3; that is, if the controllability matrix is of full rank, then the n rows of $e^{-\boldsymbol{A}t}\boldsymbol{B}$ are linearly independent over $\mathbb{R}$ for all $t \in [0, \infty)$. We prove this statement by contraposition. So assume that for some nonzero n-vector $\boldsymbol{q}$ we have

$$\boldsymbol{q}^T e^{-\boldsymbol{A}t} \boldsymbol{B} = \boldsymbol{0}^T, \qquad \text{for all } t \in [0, \infty).$$

If we set $t = 0$, then we obtain

$$\boldsymbol{q}^T \boldsymbol{B} = \boldsymbol{0}^T.$$

Differentiating $\boldsymbol{q}^T e^{-\boldsymbol{A}t} \boldsymbol{B} = \boldsymbol{0}$ and letting $t = 0$ yields

$$\boldsymbol{q}^T \boldsymbol{A}\boldsymbol{B} = \boldsymbol{0}^T.$$

Continuing this process gives

$$\boldsymbol{q}^T [\boldsymbol{B} \quad \boldsymbol{A}\boldsymbol{B} \quad \cdots \quad \boldsymbol{A}^{n-1}\boldsymbol{B}] = \boldsymbol{0}^T,$$

which means that $\operatorname{rank}[\boldsymbol{B} \ \ \boldsymbol{A}\boldsymbol{B} \ \cdots \ \boldsymbol{A}^{n-1}\boldsymbol{B}] < n$.

We will now prove, by contraposition, the implication 3 $\Rightarrow$ 4; that is, if the n rows of $e^{-\boldsymbol{A}t}\boldsymbol{B}$ are linearly independent over $\mathbb{R}$ for all $t \in [0, \infty)$, then $\det \boldsymbol{W}(t_0, t_1) \neq 0$ for any $t_1 > t_0$. So assume that $\det \boldsymbol{W}(t_0, t_1) = 0$. Then, there is a nonzero n-vector

$\boldsymbol{q} \in \mathbb{R}^n$ such that

$$\boldsymbol{q}^T \int_{t_0}^{t_1} e^{-At} \boldsymbol{B} \boldsymbol{B}^T e^{-A^T t} \, dt = \boldsymbol{0}^T$$

for all $t \in [t_0, t_1]$. We postmultiply the above equation by $\boldsymbol{q}$ to get

$$\int_{t_0}^{t_1} (\boldsymbol{q}^T e^{-At} \boldsymbol{B})(\boldsymbol{B}^T e^{-A^T t} \boldsymbol{q}) \, dt = 0.$$

Let $\boldsymbol{v}(t) = \boldsymbol{B}^T e^{-A^T t} \boldsymbol{q}$; then the above equation can be written as

$$\int_{t_0}^{t_1} \|\boldsymbol{v}(t)\|^2 \, dt = 0.$$

However, the above is true if and only if $\boldsymbol{v}(t) = \boldsymbol{0}$—that is, if and only if the rows of $e^{-At}\boldsymbol{B}$ are linearly dependent over the set of real numbers.

We now prove, constructively, the implication $4 \Rightarrow 1$. Let $\boldsymbol{x}(t_0) = \boldsymbol{x}_0$ be an arbitrary initial state and let $\boldsymbol{x}(t_1) = \boldsymbol{x}_f$ be an arbitrary final state. Then, we claim that the control law

$$\boldsymbol{u}(t) = \boldsymbol{B}^T e^{-A^T t} \boldsymbol{W}^{-1}(t_0, t_1)(e^{-At_1} \boldsymbol{x}(t_1) - e^{-At_0} \boldsymbol{x}_0) \tag{3.12}$$

will transfer the system from $\boldsymbol{x}_0$ to $\boldsymbol{x}_f$. Note that to construct the above control law the matrix $\boldsymbol{W}(t_0, t_1)$ must be invertible. Substituting (3.12) into

$$\boldsymbol{x}(t_1) = e^{A(t_1 - t_0)} \boldsymbol{x}_0 + \int_{t_0}^{t_1} e^{A(t_1 - \tau)} \boldsymbol{B} \boldsymbol{u}(\tau) \, d\tau$$

yields

$$\begin{aligned} \boldsymbol{x}(t_1) &= e^{A(t_1 - t_0)} \boldsymbol{x}_0 \\ &\quad + e^{At_1} \int_{t_0}^{t_1} e^{-A\tau} \boldsymbol{B} \boldsymbol{B}^T e^{-A^T \tau} d\tau \, \boldsymbol{W}^{-1}(t_0, t_1)(e^{-At_1} \boldsymbol{x}(t_1) - e^{-At_0} \boldsymbol{x}_0). \end{aligned}$$

Performing easy manipulations gives $\boldsymbol{x}(t_1) = \boldsymbol{x}(t_1)$, which means that indeed the proposed control law does the required job. Thus, we have established equivalence between the four statements of the theorem.

◆ Example 3.2

Consider an armature-controlled DC motor system whose schematic is shown in Figure 3.1, where the system parameters are $R_a = 1\ \Omega$, $L_a \approx 0$ H, $K_b = 5$ V/rad/sec, $K_i = 5$ Nm/A, and the effective moment of inertia is $I_{eq} = 0.1$ kg·m^2. The gear inertia and friction are negligible.

Applying Kirchhoff's voltage law to the armature circuit gives

$$R_a i_a + K_b \dot{\theta} = e_a. \tag{3.13}$$

The armature-controlled DC motor uses a constant field current, and therefore the motor torque is

$$\tau_m = K_i i_a.$$

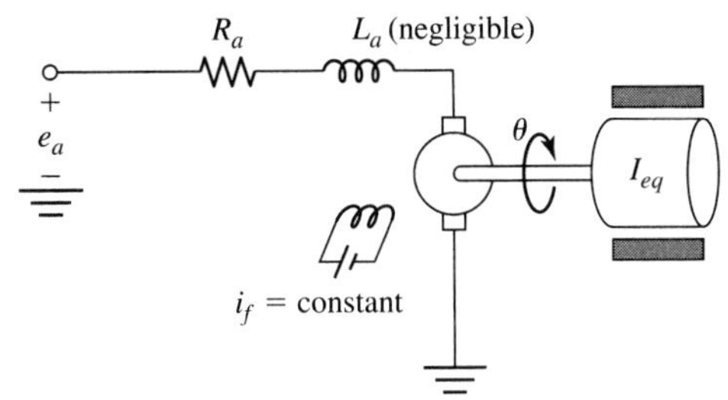

Figure 3.1 Schematic of an armature-controlled DC motor system of Example 3.2.

The motor torque is equal to the torque delivered to the load. Hence,

$$I_{eq}\ddot{\theta} = K_i i_a. \tag{3.14}$$

Substituting i_a from (3.13) into the above equation and dividing both sides by I_{eq} yields

$$\begin{aligned}\ddot{\theta} &= \frac{K_i}{I_{eq}}\left(\frac{e_a - K_b\dot{\theta}}{R_a}\right)\\ &= -\frac{K_i K_b}{I_{eq}R_a}\dot{\theta} + \frac{K_i}{I_{eq}R_a}e_a.\end{aligned}$$

Let $x_1 = \theta$, $x_2 = \dot{\theta}$, and $u = e_a$. Then taking into account the system parameters, we represent the above equation in a state-space format:

$$\begin{bmatrix}\dot{x}_1\\ \dot{x}_2\end{bmatrix} = \begin{bmatrix}0 & 1\\ 0 & -250\end{bmatrix}\begin{bmatrix}x_1\\ x_2\end{bmatrix} + \begin{bmatrix}0\\ 50\end{bmatrix}u.$$

We will now find a control u that transfers the system from the initial state $\boldsymbol{x}(t_0) = \boldsymbol{x}(0) = [10\ \ 1]^T$ to the final state $\boldsymbol{x}(2) = [0\ \ 0]^T$. Recall that the solution of a controlled system has the form

$$\boldsymbol{x}(t) = e^{\boldsymbol{A}(t-t_0)}\boldsymbol{x}(t_0) + \int_{t_0}^{t} e^{\boldsymbol{A}(t-\tau)}\boldsymbol{B}\boldsymbol{u}(\tau)\,d\tau.$$

Our goal is to find u so that

$$\boldsymbol{0} = e^{2\boldsymbol{A}}\boldsymbol{x}(0) + \int_0^2 e^{\boldsymbol{A}(2-\tau)}\boldsymbol{B}\boldsymbol{u}(\tau)\,d\tau.$$

One can verify that one such u has the form

$$u(t) = -\boldsymbol{B}^T e^{\boldsymbol{A}^T(2-t)}\left(\int_0^2 e^{\boldsymbol{A}(2-\tau)}\boldsymbol{B}\boldsymbol{B}^T e^{\boldsymbol{A}^T(2-\tau)}\,d\tau\right)^{-1} e^{2\boldsymbol{A}}\boldsymbol{x}(0),$$

where

$$e^{\boldsymbol{A}t} = \begin{bmatrix}1 & \frac{1}{250}(1 - e^{-250t})\\ 0 & e^{-250t}\end{bmatrix}.$$

Hence,

$$\int_0^2 e^{\boldsymbol{A}(2-\tau)}\boldsymbol{B}\boldsymbol{B}^T e^{\boldsymbol{A}^T(2-\tau)}\,d\tau = \int_0^2 \begin{bmatrix} 1 & \frac{1}{250}\left(1-e^{-250(2-\tau)}\right) \\ 0 & e^{-250(2-\tau)} \end{bmatrix} \begin{bmatrix} 0 \\ 50 \end{bmatrix} [0 \quad 50]$$

$$\times \begin{bmatrix} 1 & 0 \\ \frac{1}{250}\left(1-e^{-250(2-\tau)}\right) & e^{-250(2-\tau)} \end{bmatrix} d\tau$$

$$= \begin{bmatrix} 0.07976 & 0.02 \\ 0.02 & 5 \end{bmatrix}.$$

The inverse of the above matrix is

$$\begin{bmatrix} 12.5502 & -0.0502 \\ -0.0502 & 0.2002 \end{bmatrix}.$$

Hence,

$$u(t) = -[0 \quad 50] \begin{bmatrix} 1 & 0 \\ \frac{1}{250}\left(1-e^{-250(2-t)}\right) & e^{-250(2-t)} \end{bmatrix}$$

$$\times \begin{bmatrix} 12.5502 & -0.0502 \\ -0.0502 & 0.2002 \end{bmatrix} e^{2\boldsymbol{A}} \begin{bmatrix} 10 \\ 1 \end{bmatrix}$$

$$= 50.22e^{(250t-500)} - 25.11.$$

The above control strategy transfers the system from $\boldsymbol{x}(0) = [10 \ \ 1]^T$ to $\boldsymbol{x}(2) = [0 \ \ 0]^T$. Plots of x_1, x_2 versus time are shown in Figure 3.2. A plot of u versus time is given in Figure 3.3.

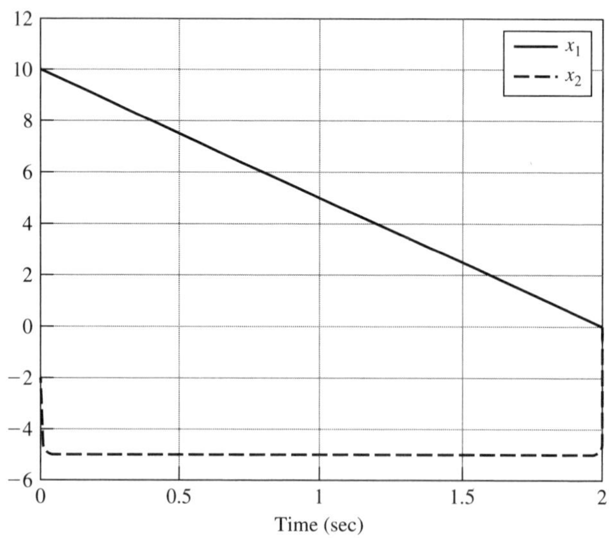

Figure 3.2 Plots of state variables versus time of Example 3.2.

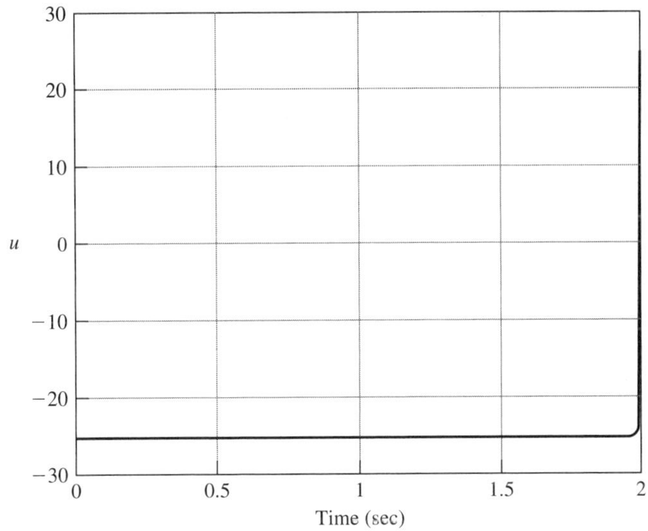

Figure 3.3 A plot of the control action versus time of Example 3.2.

We now give a necessary and sufficient condition for the pair $(\boldsymbol{A}, \boldsymbol{B})$ to be nonreachable.

Theorem 3.4 The pair $(\boldsymbol{A}, \boldsymbol{B})$ is nonreachable if and only if there is a similarity transformation $\boldsymbol{z} = \boldsymbol{T}\boldsymbol{x}$ such that

$$\tilde{\boldsymbol{A}} = \boldsymbol{T}\boldsymbol{A}\boldsymbol{T}^{-1} = \begin{bmatrix} \boldsymbol{A}_1 & \boldsymbol{A}_2 \\ \boldsymbol{O} & \boldsymbol{A}_4 \end{bmatrix}, \qquad \tilde{\boldsymbol{B}} = \boldsymbol{T}\boldsymbol{B} = \begin{bmatrix} \boldsymbol{B}_1 \\ \boldsymbol{O} \end{bmatrix},$$

where the pair $(\boldsymbol{A}_1, \boldsymbol{B}_1)$ is reachable, $\boldsymbol{A}_1 \in \mathbb{R}^{r\times r}$, $\boldsymbol{B}_1 \in \mathbb{R}^{r\times m}$, and the rank of the controllability matrix of the pair $(\boldsymbol{A}, \boldsymbol{B})$ equals r.

Proof We first prove necessity ($\Rightarrow$). After forming the controllability matrix of the pair $(\boldsymbol{A}, \boldsymbol{B})$, we use elementary row operations to get

$$\begin{aligned}
\boldsymbol{T}[\boldsymbol{B} \quad \boldsymbol{A}\boldsymbol{B} \quad \cdots \quad \boldsymbol{A}^{n-1}\boldsymbol{B}] &= [\boldsymbol{T}\boldsymbol{B} \quad (\boldsymbol{T}\boldsymbol{A}\boldsymbol{T}^{-1})(\boldsymbol{T}\boldsymbol{B}) \quad \cdots \quad (\boldsymbol{T}\boldsymbol{A}\boldsymbol{T}^{-1})^{n-1}(\boldsymbol{T}\boldsymbol{B})] \\
&= \begin{bmatrix} x & x & x & \cdots & x & x & x \\ 0 & x & x & \cdots & x & x & x \\ \vdots & & & \vdots & & & \vdots \\ 0 & 0 & 0 & \cdots & x & x & x \\ 0 & 0 & 0 & 0 & 0 & 0 & 0 \\ \vdots & & & \vdots & & & \vdots \\ 0 & 0 & 0 & 0 & 0 & 0 & 0 \end{bmatrix} \\
&= \begin{bmatrix} \boldsymbol{B}_1 & \boldsymbol{A}_1\boldsymbol{B}_1 & \boldsymbol{A}_1^2\boldsymbol{B}_1 & \cdots & \boldsymbol{A}_1^{n-1}\boldsymbol{B}_1 \\ \boldsymbol{O} & \boldsymbol{O} & \boldsymbol{O} & \cdots & \boldsymbol{O} \end{bmatrix}, \qquad (3.15)
\end{aligned}$$

where the symbol x denotes a "don't care"—that is, an unspecified scalar. We now show that indeed

$$\tilde{A} = TAT^{-1} = \begin{bmatrix} A_1 & A_2 \\ A_3 & A_4 \end{bmatrix}, \qquad \tilde{B} = TB = \begin{bmatrix} B_1 \\ O \end{bmatrix},$$

where $A_3 = O$. It is clear that TB must have the above form. If A_3 were not a zero matrix, then we would have

$$\tilde{A}\tilde{B} = \begin{bmatrix} A_1 B_1 \\ A_3 B_1 \end{bmatrix}.$$

Comparing the above with (3.15), we conclude that $A_3 B_1 = O$. Thus,

$$\tilde{A}\tilde{B} = \begin{bmatrix} A_1 B_1 \\ O \end{bmatrix}.$$

Next we consider

$$\tilde{A}^2 \tilde{B} = \begin{bmatrix} A_1^2 B_1 \\ A_3 A_1 B_1 \end{bmatrix}.$$

Again, comparing the above with (3.15), we conclude that

$$A_3 A_1 B_1 = O,$$

and thus

$$\tilde{A}^2 \tilde{B} = \begin{bmatrix} A_1^2 B_1 \\ O \end{bmatrix}.$$

Continuing in this manner, we conclude that $A_3 A_1^{n-1} B_1 = O$. Thus it follows that

$$A_3 \begin{bmatrix} B_1 & A_1 B_1 & \cdots & A_1^{n-1} B_1 \end{bmatrix} = O.$$

Because $\text{rank}[B_1 \;\; A_1 B_1 \;\; \cdots \;\; A_1^{n-1} B_1] = r$—that is, the controllability matrix of the pair (A_1, B_1) is of full rank—we have to have $A_3 = O$, and the proof of necessity is complete.

To prove sufficiency ($\Leftarrow$), note that if there is a similarity transformation such that

$$\tilde{A} = TAT^{-1} = \begin{bmatrix} A_1 & A_2 \\ O & A_4 \end{bmatrix}, \qquad \tilde{B} = TB = \begin{bmatrix} B_1 \\ O \end{bmatrix},$$

then the controllability matrix of the pair $(\tilde{A}, \tilde{B})$ has the form

$$\begin{bmatrix} B_1 & A_1 B_1 & A_1^2 B_1 & \cdots & A_1^{n-1} B_1 \\ O & O & O & \cdots & O \end{bmatrix}.$$

Hence, the pair $(\tilde{A}, \tilde{B})$ is nonreachable. Because the similarity transformation preserves the reachability property, the pair (A, B) is nonreachable as well.

◆ **Example 3.3**

For the system model,

$$\dot{x} = \begin{bmatrix} -1 & 1 & 0 & 1 \\ 1 & 1 & 0 & 0 \\ 0 & 0 & 4 & 0 \\ 1 & 1 & -1 & 2 \end{bmatrix} x + \begin{bmatrix} 1 & 1 \\ 0 & 0 \\ 0 & 0 \\ 0 & 1 \end{bmatrix} u,$$

construct a state variable transformation so that in the new coordinates the reachable part is separated and distinct from the nonreachable part.

We first form the controllability matrix of the pair (A, B) to get

$$\mathcal{C} = [\, B \quad AB \quad A^2B \quad A^3B \,] = \begin{bmatrix} 1 & 1 & -1 & 0 & 3 & 4 & -1 & 4 \\ 0 & 0 & 1 & 1 & 0 & 1 & 3 & 5 \\ 0 & 0 & 0 & 0 & 0 & 0 & 0 & 0 \\ 0 & 1 & 1 & 3 & 2 & 7 & 7 & 19 \end{bmatrix}.$$

We then use the MATLAB command `[Q,R]=qr(C)` to obtain

$$Q = \begin{bmatrix} -1 & 0 & 0 & 0 \\ 0 & 0 & -1 & 0 \\ 0 & 0 & 0 & 1 \\ 0 & -1 & 0 & 0 \end{bmatrix}, \qquad R = \begin{bmatrix} -1 & -1 & 1 & 0 & -3 & -4 & 1 & -4 \\ 0 & -1 & -1 & -3 & -2 & -7 & -7 & -19 \\ 0 & 0 & -1 & -1 & 0 & -1 & -3 & -5 \\ 0 & 0 & 0 & 0 & 0 & 0 & 0 & 0 \end{bmatrix},$$

where the matrices Q and R satisfy $\mathcal{C} = QR$, and R is upper row triangular such that rank R = rank $\mathcal{C}$. Thus, premultiplying the controllability matrix, $\mathcal{C}$, by Q^{-1} reduces this matrix to an upper row triangular matrix R because $Q^{-1}\mathcal{C} = R$. Let $z = Q^{-1}x$ be the state variable transformation. Then in the new coordinates the matrices A and B take the form

$$Q^{-1}AQ = \left[\begin{array}{ccc|c} -1 & 1 & 1 & 0 \\ 1 & 2 & 1 & 1 \\ 1 & 0 & 1 & 0 \\ \hline 0 & 0 & 0 & 4 \end{array}\right] \quad \text{and} \quad Q^{-1}B = \left[\begin{array}{cc} -1 & -1 \\ 0 & -1 \\ 0 & 0 \\ \hline 0 & 0 \end{array}\right].$$

The dimension of the nonreachable part is 1.

If the eigenvalues of the noncontrollable part of a given pair (A, B) are all in the open left-hand complex plane—that is, the nonreachable part is asymptotically stable—then the pair (A, B) is called *stabilizable*.

We now give yet another test for reachability. This test is also known as the Popov–Belevitch–Hautus (PBH) eigenvector test. We denote the set of eigenvalues of a matrix A as eig(A).

Theorem 3.5 The pair (A, B) is reachable if and only if

$$\text{rank}[\, s I_n - A \quad B\,] = n \qquad \text{for all } s \in \text{eig}(A).$$

Proof We first prove the necessity part ($\Rightarrow$) by contraposition. If there is a vector $q \neq 0$ such that for some $s \in \text{eig}(A)$,

$$q^T[\, s I_n - A \quad B\,] = 0^T,$$

that is,

$$q^T A = s q^T, \qquad q^T B = 0^T,$$

then

$$q^T AB = s q^T B = 0^T,$$
$$q^T A^2 B = s q^T AB = 0^T,$$

and so on, until

$$q^T A^{n-1} B = s q^T A^{n-2} B = 0^T.$$

We write the above equations in the form

$$q^T[\, B \quad AB \quad \cdots \quad A^{n-1} B\,] = 0^T,$$

which means that the pair (A, B) is nonreachable.

We now prove the sufficiency part ($\Leftarrow$) also by contraposition. So we assume that $\text{rank}[B \;\; AB \;\cdots\; A^{n-1} B] = r < n$. By Theorem 3.4, there exists a state variable transformation such that in the new basis the pair (A, B) takes the form

$$\tilde{A} = T A T^{-1} = \begin{bmatrix} A_1 & A_2 \\ O & A_4 \end{bmatrix}, \qquad \tilde{B} = T B = \begin{bmatrix} B_1 \\ O \end{bmatrix}.$$

We choose q of the form

$$q^T = [\, 0 \quad z\,],$$

where $z \neq 0$ is a left eigenvector of A_4; that is,

$$z A_4 = s z,$$

where $s \in \text{eig}(A_4)$. Thus

$$\begin{aligned} q^T \tilde{A} &= [\, 0 \quad z\,] \tilde{A} \\ &= [\, 0 \quad s z\,] \\ &= s q^T. \end{aligned}$$

Clearly,

$$q^T[\, s I_n - \tilde{A} \quad \tilde{B}\,] = 0^T,$$

which means that the rank of the matrix $[\, s I_n - \tilde{A} \;\; \tilde{B}\,]$ is less than n, and therefore the proof of the sufficiency part is complete.

3.2 Observability and Constructability

As we have seen in the previous section, system reachability implies the ability to completely control the entire state of the system. Suppose now that the system state is not directly accessible. Instead, we have the output of the system,

$$y = Cx + Du,$$

where $C \in \mathbb{R}^{p\times n}$, $p \le n$, and $D \in \mathbb{R}^{p\times m}$. However, we still want to know the dynamical behavior of the entire state.

We say that the system

$$\dot{x} = Ax + Bu,$$
$$y = Cx + Du,$$

or equivalently the pair (A, C), is *observable* if there is a finite $t_1 > t_0$ such that for arbitrary $u(t)$ and resulting $y(t)$ over $[t_0, t_1]$, we can determine $x(t_0)$ from complete knowledge of the system input u and output y.

Note that once $x(t_0)$ is known, we can determine $x(t)$ from knowledge of $u(t)$ and $y(t)$ over any finite time interval $[t_0, t_1]$. We now show how to constructively determine $x(t_0)$, given $u(t)$ and $y(t)$. While doing so we will establish a sufficiency condition for the system to be observable. The condition is also necessary for the system observability, as we show next. To proceed with constructive determination of the state vector, we note that the solution $y(t)$ is given by

$$y(t) = Ce^{A(t-t_0)}x_0 + \int_{t_0}^{t} Ce^{A(t-\tau)}Bu(\tau)\,d\tau + Du(t).$$

We subtract $\int_{t_0}^{t} Ce^{A(t-\tau)}Bu(\tau)\,d\tau + Du(t)$ from both sides of the above equation. Let

$$g(t) = y(t) - \int_{t_0}^{t} Ce^{A(t-\tau)}Bu(\tau)\,d\tau - Du(t),$$

then we have

$$g(t) = Ce^{A(t-t_0)}x(t_0),$$

where g is known to us. Our goal now is to determine $x(t_0)$. Having $x(t_0)$, we can determine the entire $x(t)$ for all $t \in [t_0, t_1]$ from the formula

$$x(t) = e^{A(t-t_0)}x_0 + \int_{t_0}^{t} e^{A(t-\tau)}Bu(\tau)\,d\tau.$$

Premultiplying both side of $g(t)$ by $e^{A^T(t-t_0)}C^T$ and integrating the resulting expression between the limits t_0 and t_1, we get

$$\int_{t_0}^{t_1} e^{A^T(t-t_0)}C^T Ce^{A(t-t_0)}\,dt\, x_0 = \int_{t_0}^{t_1} e^{A^T(t-t_0)}C^T g(t)\,dt.$$

Let

$$V(t_0, t_1) = \int_{t_0}^{t_1} e^{A^T t} C^T C e^{At}\, dt.$$

Then, after performing some manipulations and assuming that the matrix $V(t_0, t_1)$ is invertible, we obtain

$$x_0 = e^{At_0} V^{-1}(t_0, t_1) e^{A^T t_0} \int_{t_0}^{t_1} e^{A^T(t-t_0)} C^T g(t)\, dt.$$

Knowledge of x_0 allows us to reconstruct the entire state $x(t)$ over the interval $[t_0, t_1]$. We conclude that if the matrix $V(t_0, t_1)$ is invertible, then the system is observable.

◆ Example 3.4

For a dynamical system modeled by

$$\dot{x} = \begin{bmatrix} -1/2 & 1/2 \\ -1/2 & 1/2 \end{bmatrix} x + \begin{bmatrix} -1 \\ 1 \end{bmatrix} u = Ax + bu,$$
$$y = [1 \quad 1]x = cx,$$

we shall determine $x(t)$ over the time interval $[0, 10]$ knowing that $u(t) = 1$ for $t \geq 0$ and $y(t) = t^2 + 2t$ for $t \geq 0$.

We first find $x(0)$. Knowledge of $x(0)$ will allow us to find the entire state vector for all $t \in [0, 10]$. We have $t_0 = 0$ and $t_1 = 10$. Hence,

$$\begin{aligned} x(t_0) = x(0) &= e^{At_0} V^{-1}(t_0, t_1) e^{A^T t_0} \int_{t_0}^{t_1} e^{A^T(t-t_0)} c^T g(t)\, dt \\ &= V^{-1}(0, 10) \int_0^{10} e^{A^T t} c^T g(t)\, dt, \end{aligned}$$

where

$$g(t) = y(t) - c\int_0^t e^{A(t-\tau)} bu(\tau)\, d\tau.$$

To evaluate the above expression, we first use formula (A.148) to compute

$$\begin{aligned} e^{At} = \mathcal{L}^{-1}([sI_2 - A]^{-1}) &= \mathcal{L}^{-1}\left(\begin{bmatrix} \frac{1}{s} - \frac{1}{2}\frac{1}{s^2} & \frac{1}{2}\frac{1}{s^2} \\ -\frac{1}{2}\frac{1}{s^2} & \frac{1}{s} + \frac{1}{2}\frac{1}{s^2} \end{bmatrix}\right) \\ &= \begin{bmatrix} 1 - \frac{1}{2}t & \frac{1}{2}t \\ -\frac{1}{2}t & 1 + \frac{1}{2}t \end{bmatrix}. \end{aligned}$$

Next, we compute

$$
\begin{aligned}
\boldsymbol{V}(0,10) &= \int_0^{10} e^{\boldsymbol{A}^T t}\boldsymbol{c}^T \boldsymbol{c} e^{\boldsymbol{A}t}\,dt \\
&= \int_0^{10} \left(\begin{bmatrix} 1-t \\ 1+t \end{bmatrix}\begin{bmatrix} 1-t & 1+t \end{bmatrix}\right) dt = \int_0^{10} \begin{bmatrix} 1-2t+t^2 & 1-t^2 \\ 1-t^2 & 1+2t+t^2 \end{bmatrix} dt \\
&= \left.\begin{bmatrix} t-t^2+\dfrac{t^3}{3} & t-\dfrac{t^3}{3} \\ t-\dfrac{t^3}{3} & t+t^2+\dfrac{t^3}{3} \end{bmatrix}\right|_0^{10} = \begin{bmatrix} 243.333 & -323.333 \\ -323.333 & 443.333 \end{bmatrix}.
\end{aligned}
$$

The inverse of $\boldsymbol{V}(0,10)$ is

$$
\boldsymbol{V}^{-1}(0,10) = \begin{bmatrix} 0.133 & 0.097 \\ 0.097 & 0.073 \end{bmatrix}.
$$

We can now compute $g(t)$ to get

$$
\begin{aligned}
g(t) &= y(t) - \boldsymbol{c}\int_0^t e^{\boldsymbol{A}(t-\tau)}\boldsymbol{b}u(\tau)\,d\tau \\
&= t^2+2t-\begin{bmatrix} 1 & 1 \end{bmatrix}\int_0^t \begin{bmatrix} -1+t-\tau \\ 1+t+\tau \end{bmatrix} d\tau = t^2+2t-\int_0^t (2t-2\tau)\,d\tau \\
&= t^2+2t-2t(\tau)\Big|_0^t + \tau^2\Big|_0^t = t^2+2t-t^2 \\
&= 2t.
\end{aligned}
$$

Hence,

$$
\int_0^{10} e^{\boldsymbol{A}^T t}\boldsymbol{c}^T g(t)\,dt = \int_0^{10} \begin{bmatrix} (1-t)2t \\ (1+t)2t \end{bmatrix} dt = \left.\begin{bmatrix} t^2-\frac{2}{3}t^3 \\ t^2+\frac{2}{3}t^3 \end{bmatrix}\right|_0^{10} = \begin{bmatrix} -566.6667 \\ 766.6667 \end{bmatrix}.
$$

We can now compute

$$
\boldsymbol{x}(0) = \boldsymbol{V}^{-1}(0,10)\int_0^{10} e^{\boldsymbol{A}^T t}\boldsymbol{c}^T g(t)\,dt = \begin{bmatrix} -1 \\ 1 \end{bmatrix}.
$$

Alternatively, we could use the Laplace transform method to find $\boldsymbol{x}(0)$. Knowing $\boldsymbol{x}(0)$, we can find $\boldsymbol{x}(t)$ for $t \geq 0$,

$$
\begin{aligned}
\boldsymbol{x}(t) &= e^{\boldsymbol{A}t}\boldsymbol{x}(0) + \int_0^t e^{\boldsymbol{A}(t-\tau)}\boldsymbol{b}u(\tau)\,d\tau \\
&= \begin{bmatrix} -1+t \\ 1+t \end{bmatrix} + \begin{bmatrix} (-1+t)t-\dfrac{t^2}{2} \\ (1+t)t-\dfrac{t^2}{2} \end{bmatrix} = \begin{bmatrix} \dfrac{t^2}{2}-1 \\ \dfrac{t^2}{2}+2t+1 \end{bmatrix}.
\end{aligned}
$$

We will now show that the invertibility of $V(t_0, t_1)$ is also necessary for observability. We first note that one can use similar arguments to those in the previous section to show that $V(t_0, t_1)$ is nonsingular if and only if the n columns of the matrix

$$Ce^{At}$$

are linearly independent for all $t \in [0, \infty)$ over the set of real numbers. Hence, if $V(t_0, t_1)$ is singular, there exists a nonzero constant vector, say x_a, such that $V(t_0, t_1)x_a = \mathbf{0}$. This implies that

$$g(t) = Ce^{At}x_0 = Ce^{At}(x_0 + x_a).$$

Thus, $x(0) = x_0 + x_a$ yields the same response as $x(0) = x_0$, which means that we cannot determine the system state, or in other words, the state fails to be observable. In view of the above result, it follows that by repeating essentially the same arguments used in the previous section we can prove the following theorem:

Theorem 3.6 The following statements are equivalent:

1. The pair (A, C) is observable.
2. The matrix $V(t_0, t_1)$ is nonsingular for all $t_1 > t_0$.
3. The n columns of Ce^{At} are linearly independent for all $t \in [0, \infty)$ over the real numbers.
4. The observability matrix

$$\begin{bmatrix} C \\ CA \\ \vdots \\ CA^{n-1} \end{bmatrix} \in \mathbb{R}^{pn \times n}$$

is of full rank n.

5.

$$\operatorname{rank} \begin{bmatrix} sI_n - A \\ C \end{bmatrix} = n \qquad \text{for all } s \in \operatorname{eig}(A).$$

◆ Example 3.5

For the dynamical system model,

$$\dot{x} = Ax = \begin{bmatrix} 2 & 0 \\ 3 & -1 \end{bmatrix} x,$$

$$y = cx = [-1 \quad 1]x,$$

we will find a nonzero initial vector $x(0) = x_0$ such that $y(t) = 0$ for all $t \geq 0$.

We have $\boldsymbol{x}(t) = e^{\boldsymbol{A}t}\boldsymbol{x}_0$, where

$$e^{\boldsymbol{A}t} = \mathcal{L}^{-1}((s\boldsymbol{I} - \boldsymbol{A})^{-1}) = \begin{bmatrix} e^{2t} & 0 \\ e^{2t} - e^{-t} & e^{-t} \end{bmatrix}.$$

We want

$$0 = y(t) = [-1 \quad 1]e^{\boldsymbol{A}t}\boldsymbol{x}_0 \qquad \text{for all } t \geq 0.$$

For example, if

$$\boldsymbol{x}_0 = [1 \quad 1]^T,$$

then $y(t) = 0$ for all $t \geq 0$. Note that we were able to find $\boldsymbol{x}_0 \neq \boldsymbol{0}$ such that $y(t) = 0$ for all $t \geq 0$, because for all $t \geq 0$ the columns of $\boldsymbol{c}e^{\boldsymbol{A}t}$ are linearly dependent over the real numbers because the pair $(\boldsymbol{A}, \boldsymbol{c})$ is nonobservable. The vector $\boldsymbol{x}_0$ is in the null space of the observability matrix

$$\begin{bmatrix} \boldsymbol{c} \\ \boldsymbol{c}\boldsymbol{A} \end{bmatrix}.$$

We can also prove a result analogous to that of Theorem 3.4 for the given pair $(\boldsymbol{A}, \boldsymbol{C})$; that is, using a similarity transformation we can separate the observable part from the nonobservable. If the eigenvalues of the nonobservable part of a given pair $(\boldsymbol{A}, \boldsymbol{C})$ are all in the open left-hand complex plane—that is, the nonobservable part is asymptotically stable—then we say that the pair $(\boldsymbol{A}, \boldsymbol{C})$ is *detectable.*

A notion related to observability is the notion of constructability. We say that the system

$$\dot{\boldsymbol{x}} = \boldsymbol{A}\boldsymbol{x} + \boldsymbol{B}\boldsymbol{u},$$
$$\boldsymbol{y} = \boldsymbol{C}\boldsymbol{x} + \boldsymbol{D}\boldsymbol{u},$$

or equivalently the pair $(\boldsymbol{A}, \boldsymbol{C})$, is *constructable* if there is a finite $t_1 > t_0$ such that for arbitrary $\boldsymbol{u}(t)$ and resulting $\boldsymbol{y}(t)$ over $[t_0, t_1]$ we can determine $\boldsymbol{x}(t_1)$ from complete knowledge of the system input $\boldsymbol{u}$ and output $\boldsymbol{y}$.

One can verify that in the case of continuous, time-invariant systems, observability is equivalent to constructability. In the case of discrete-time systems, observability implies constructability, and the two notions are equivalent if the matrix $\boldsymbol{A}$ of the given discrete system is nonsingular.

3.3 Companion Forms

In this section we discuss two types of special forms of a given linear dynamical system model represented by the triple $(\boldsymbol{A}, \boldsymbol{B}, \boldsymbol{C})$. These forms are obtained by a change of state variables and reveal structural properties of the model. We will use the above-mentioned forms to construct algorithms for pole placement. The pole placement algorithms will then be used to design state-feedback controllers, as well as state estimators. We begin by introducing the controller companion form, or controller form for short.

3.3.1 Controller Form

We begin by considering a single-input, continuous-time model represented by

$$\dot{\boldsymbol{x}}(t) = \boldsymbol{A}\boldsymbol{x}(t) + \boldsymbol{b}u(t),$$

where the pair $(\boldsymbol{A}, \boldsymbol{b})$ is assumed to be reachable, or, equivalently, controllable. This means that

$$\operatorname{rank}[\boldsymbol{b} \quad \boldsymbol{A}\boldsymbol{b} \quad \cdots \quad \boldsymbol{A}^{n-1}\boldsymbol{b}] = n.$$

We select the last row of the inverse of the controllability matrix. Let $\boldsymbol{q}_1$ be that row. Next, we form the matrix

$$\boldsymbol{T} = \begin{bmatrix} \boldsymbol{q}_1 \\ \boldsymbol{q}_1\boldsymbol{A} \\ \vdots \\ \boldsymbol{q}_1\boldsymbol{A}^{n-1} \end{bmatrix}.$$

Note that $\boldsymbol{T}$ is invertible. Indeed, because

$$\boldsymbol{T}[\boldsymbol{b} \quad \boldsymbol{A}\boldsymbol{b} \quad \cdots \quad \boldsymbol{A}^{n-1}\boldsymbol{b}] = \begin{bmatrix} 0 & 0 & \cdots & 0 & 1 \\ 0 & 0 & \cdots & 1 & x \\ \vdots & \vdots & & \vdots & \vdots \\ 1 & x & \cdots & x & x \end{bmatrix},$$

this implies that $\boldsymbol{T}$ must be nonsingular. In the above, we used the symbol x to denote "don't care," unspecified scalars in our present discussion. Consider now the state variable transformation, $\tilde{\boldsymbol{x}} = \boldsymbol{T}\boldsymbol{x}$. Then, in the new coordinates, the system model is

$$\begin{aligned} \dot{\tilde{\boldsymbol{x}}}(t) &= \boldsymbol{T}\boldsymbol{A}\boldsymbol{T}^{-1}\tilde{\boldsymbol{x}}(t) + \boldsymbol{T}\boldsymbol{b}u(t) \\ &= \tilde{\boldsymbol{A}}\tilde{\boldsymbol{x}}(t) + \tilde{\boldsymbol{b}}u(t). \end{aligned}$$

The matrices $\tilde{\boldsymbol{A}}$ and $\tilde{\boldsymbol{b}}$ have particular structures that we now discuss. We first analyze the structure of $\tilde{\boldsymbol{b}}$. Because $\boldsymbol{q}_1$ is the last row of the inverse of the controllability matrix, we have

$$\boldsymbol{q}_1\boldsymbol{b} = \boldsymbol{q}_1\boldsymbol{A}\boldsymbol{b} = \cdots = \boldsymbol{q}_1\boldsymbol{A}^{n-2}\boldsymbol{b} = 0$$

and

$$\boldsymbol{q}_1\boldsymbol{A}^{n-1}\boldsymbol{b} = 1.$$

Hence,

$$\tilde{\boldsymbol{b}} = \boldsymbol{T}\boldsymbol{b} = \begin{bmatrix} \boldsymbol{q}_1\boldsymbol{b} \\ \boldsymbol{q}_1\boldsymbol{A}\boldsymbol{b} \\ \vdots \\ \boldsymbol{q}_1\boldsymbol{A}^{n-2}\boldsymbol{b} \\ \boldsymbol{q}_1\boldsymbol{A}^{n-1}\boldsymbol{b} \end{bmatrix} = \begin{bmatrix} 0 \\ 0 \\ \vdots \\ 0 \\ 1 \end{bmatrix}.$$

The structure of $\tilde{\boldsymbol{A}}$ is revealed by considering the relation $\boldsymbol{T}\boldsymbol{A}\boldsymbol{T}^{-1} = \tilde{\boldsymbol{A}}$ represented as

$$\boldsymbol{T}\boldsymbol{A} = \tilde{\boldsymbol{A}}\boldsymbol{T}.$$

The left-hand side of the above matrix equation is

$$TA = \begin{bmatrix} q_1 A \\ q_1 A^2 \\ \vdots \\ q_1 A^{n-1} \\ q_1 A^n \end{bmatrix}.$$

By the Cayley–Hamilton theorem, we have

$$A^n = -a_0 I_n - a_1 A - \cdots - a_{n-1} A^{n-1},$$

and hence

$$q_1 A^n = -a_0 q_1 - a_1 q_1 A - \cdots - a_{n-1} q_1 A^{n-1}.$$

Comparing both sides of the equation $TA = \tilde{A}T$ and taking into account the Cayley–Hamilton theorem, we get

$$\tilde{A} = \begin{bmatrix} 0 & 1 & 0 & \cdots & 0 & 0 \\ 0 & 0 & 1 & \cdots & 0 & 0 \\ \vdots & \vdots & & & & \vdots \\ 0 & 0 & 0 & \cdots & 0 & 1 \\ -a_0 & -a_1 & -a_2 & \cdots & -a_{n-2} & -a_{n-1} \end{bmatrix}.$$

Note that the coefficients of the characteristic polynomial of A are immediately apparent by inspecting the last row of $\tilde{A}$. There are other benefits of the pair (A, b) being given in its controller form that we discuss in Section 3.4. We now illustrate the above algorithm for reducing a single-input system into the controller form with a numerical example.

◆ Example 3.6

Consider the following dynamical system model:

$$\dot{x} = \begin{bmatrix} 0 & 0 & 1 & 0 \\ 0 & 0 & 0 & 1 \\ -2 & 1 & 0 & 0 \\ 1 & -1 & 0 & 0 \end{bmatrix} x + \begin{bmatrix} 0 \\ 0 \\ 1 \\ 0 \end{bmatrix} u.$$

The controllability matrix of the pair (A, b) is

$$[b \quad Ab \quad A^2 b \quad A^3 b] = \begin{bmatrix} 0 & 1 & 0 & -2 \\ 0 & 0 & 0 & 1 \\ 1 & 0 & -2 & 0 \\ 0 & 0 & 1 & 0 \end{bmatrix}.$$

The pair (A, b) is controllable because the controllability matrix is nonsingular. We find the last row of the inverse of the controllability matrix to be

$q_1 = [0\ \ 1\ \ 0\ \ 0]$. Hence the transformation matrix is

$$T = \begin{bmatrix} q_1 \\ q_1 A \\ q_1 A^2 \\ q_1 A^3 \end{bmatrix} = \begin{bmatrix} 0 & 1 & 0 & 0 \\ 0 & 0 & 0 & 1 \\ 1 & -1 & 0 & 0 \\ 0 & 0 & 1 & -1 \end{bmatrix}.$$

The matrices A and b in the new coordinates have the form

$$TAT^{-1} = \begin{bmatrix} 0 & 1 & 0 & 0 \\ 0 & 0 & 0 & 1 \\ 1 & -1 & 0 & 0 \\ 0 & 0 & 1 & -1 \end{bmatrix} \begin{bmatrix} 0 & 0 & 1 & 0 \\ 0 & 0 & 0 & 1 \\ -2 & 1 & 0 & 0 \\ 1 & -1 & 0 & 0 \end{bmatrix} \begin{bmatrix} 1 & 0 & 1 & 0 \\ 1 & 0 & 0 & 0 \\ 0 & 1 & 0 & 1 \\ 0 & 1 & 1 & 0 \end{bmatrix} = \begin{bmatrix} 0 & 1 & 0 & 0 \\ 0 & 0 & 1 & 0 \\ 0 & 0 & 0 & 1 \\ -1 & 0 & -3 & 0 \end{bmatrix}$$

and

$$Tb = \begin{bmatrix} 0 \\ 0 \\ 0 \\ 1 \end{bmatrix}.$$

We now generalize the above method of transforming the given pair (A, B) into the controller form for multi-input systems. We consider a controllable pair (A, B), where $A \in \mathbb{R}^{n\times n}$, $B \in \mathbb{R}^{n\times m}$, $m \le n$, and rank $B = m$. The last assumption we made is just for convenience to simplify further manipulations. This assumption implies that all inputs are mutually independent, which is usually the case in practical applications. We first form the controllability matrix of the pair (A, B) and represent it as

$$[\,b_1 \quad b_2 \quad \cdots \quad b_m \quad Ab_1 \quad Ab_2 \quad \cdots \quad Ab_m \quad \cdots \quad A^{n-1}b_1 \quad \cdots \quad A^{n-1}b_m\,], \tag{3.16}$$

where b_i, $i = 1, 2, \ldots, m$, is the ith column of B. The above controllability matrix is of full rank because we assumed that the pair (A, B) is controllable. Thus, we can select n linearly independent columns from (3.16). We select these columns proceeding from left to right. Then, we rearrange the selected columns to form a $n \times n$ nonsingular matrix L of the form

$$L = [\,b_1 \quad Ab_1 \quad \cdots \quad A^{d_1-1}b_1 \quad b_2 \quad \cdots \quad A^{d_2-1}b_2 \quad \cdots \quad b_m \quad \cdots \quad A^{d_m-1}b_m\,].$$

We call the m integers d_i the *controllability indices* of the pair (A, B). Note that

$$\sum_{i=1}^{m} d_i = n.$$

We call the integer

$$d = \max\{d_i : i = 1, 2, \ldots, m\}$$

the *controllability index* of the pair (A, B). Observe that all m columns of B are present in L because we assumed that B is of full rank, and observe the way we selected mutually independent columns from the controllability matrix (3.16) to construct the matrix L. Furthermore, if the column $\boldsymbol{A}^k\boldsymbol{b}_j$ is present in $\boldsymbol{L}$, then so is $\boldsymbol{A}^{k-1}\boldsymbol{b}_j$. To proceed further, we let

$$\sigma_k = \sum_{i=1}^{k} d_i \qquad \text{for } k = 1, 2, \ldots, m.$$

Note that $\sigma_1 = d_1$, $\sigma_2 = d_1 + d_2, \ldots, \sigma_m = \sum_{i=1}^{m} d_i = n$. We now select m rows from $\boldsymbol{L}^{-1}$. We denote the selected rows by $\boldsymbol{q}_k$, where $\boldsymbol{q}_k$ is the σ_kth row of $\boldsymbol{L}^{-1}$. Next, we form the $n \times n$ matrix

$$\boldsymbol{T} = \begin{bmatrix} \boldsymbol{q}_1 \\ \boldsymbol{q}_1\boldsymbol{A} \\ \vdots \\ \boldsymbol{q}_1\boldsymbol{A}^{d_1-1} \\ \boldsymbol{q}_2 \\ \vdots \\ \boldsymbol{q}_2\boldsymbol{A}^{d_2-1} \\ \vdots \\ \boldsymbol{q}_m \\ \vdots \\ \boldsymbol{q}_m\boldsymbol{A}^{d_m-1} \end{bmatrix}.$$

We can show that $\boldsymbol{T}$ is nonsingular by computing the product $\boldsymbol{TL}$ and noticing that $\det(\boldsymbol{TL}) = \pm 1$. We define a state-variable transformation $\tilde{\boldsymbol{x}} = \boldsymbol{Tx}$; and then by using similar arguments as in the single-input case, we arrive at the special structure of the pair

$$(\tilde{\boldsymbol{A}}, \tilde{\boldsymbol{B}}) = (\boldsymbol{TAT}^{-1}, \boldsymbol{TB}),$$

where

$$\tilde{\boldsymbol{A}} = \left[\begin{array}{ccccc|ccccc|c|ccccc}
0 & 1 & 0 & \cdots & 0 & 0 & 0 & & \cdots & 0 & \cdots & 0 & 0 & & \cdots & 0 \\
0 & 0 & 1 & \cdots & 0 & 0 & 0 & & \cdots & 0 & & 0 & 0 & & \cdots & 0 \\
 & & & \ddots & & \vdots & & & & \vdots & & \vdots & & & & \vdots \\
0 & 0 & 0 & \cdots & 1 & 0 & 0 & & \cdots & 0 & & 0 & 0 & & \cdots & 0 \\
x & x & x & \cdots & x & x & x & & \cdots & x & \cdots & x & x & & \cdots & x \\
\hline
0 & 0 & 0 & \cdots & 0 & 0 & 1 & & \cdots & 0 & \cdots & 0 & 0 & & \cdots & 0 \\
\vdots & & & & \vdots & \vdots & & & \ddots & \vdots & & \vdots & & & & \vdots \\
0 & 0 & 0 & \cdots & 0 & 0 & 0 & & \cdots & 1 & & 0 & 0 & & \cdots & 0 \\
x & x & x & \cdots & x & x & x & & \cdots & x & \cdots & x & x & & \cdots & x \\
\hline
\vdots & & & & & & & \vdots & & & \ddots & & & & \vdots & \\
\hline
0 & 0 & 0 & \cdots & 0 & 0 & 0 & & \cdots & 0 & \cdots & 0 & 1 & & \cdots & 0 \\
\vdots & & & & \vdots & \vdots & & & & \vdots & & \vdots & & & \ddots & \vdots \\
0 & 0 & 0 & \cdots & 0 & 0 & 0 & & \cdots & 0 & & 0 & 0 & & \cdots & 1 \\
x & x & x & \cdots & x & x & x & & \cdots & x & \cdots & x & x & & \cdots & x
\end{array}\right].$$

Note that the m diagonal blocks of $\tilde{A}$ have the structure of the controller form as in the single-input case. The dimensions of the diagonal blocks are $d_i \times d_i$, $i = 1, 2, \ldots, m$. The off-diagonal blocks consist of zero elements except for their respective last rows. Thus, all information about $\tilde{A}$, as well as A, can be derived from knowledge of the controllability indices, as well as the m nonzero rows containing the elements marked using x, of the matrix $\tilde{A}$. The matrix $\tilde{B}$ has the form

$$\tilde{\boldsymbol{B}} = \left[\begin{array}{ccccc} 0 & 0 & \cdots & 0 & 0 \\ 0 & 0 & \cdots & 0 & 0 \\ \vdots & & & & \vdots \\ 1 & x & x & x & x \\ \hline 0 & 0 & \cdots & 0 & 0 \\ 0 & 0 & \cdots & 0 & 0 \\ \vdots & & & & \vdots \\ 0 & 1 & x & x & x \\ \hline \vdots & & & & \vdots \\ \hline 0 & 0 & \cdots & 0 & 0 \\ 0 & 0 & \cdots & 0 & 0 \\ \vdots & & & & \vdots \\ 0 & 0 & 0 & 0 & 1 \end{array}\right].$$

The nonzero rows of $\tilde{B}$ correspond to the rows of $\tilde{A}$ containing the elements marked using x. We now illustrate the above algorithm for reducing a multi-input system model into the controller form with a numerical example.

◆ Example 3.7

Consider the pair (A, B), where

$$\boldsymbol{A} = \begin{bmatrix} -1 & 1 & 0 & 1 \\ 1 & 1 & 0 & 0 \\ 0 & 0 & 4 & 0 \\ 1 & 1 & -1 & 2 \end{bmatrix}, \qquad \boldsymbol{B} = \begin{bmatrix} 1 & 1 \\ 2 & 0 \\ 1 & 0 \\ 0 & 1 \end{bmatrix}.$$

We transform the above pair into its controller companion form. Following the algorithm, we find

$$\operatorname{rank}[\,\boldsymbol{b}_1 \quad \boldsymbol{b}_2 \quad \boldsymbol{A}\boldsymbol{b}_1 \quad \boldsymbol{A}\boldsymbol{b}_2\,] = 4.$$

Hence $d_1 = d_2 = 2$. We next form the matrix $\boldsymbol{L}$;

$$\boldsymbol{L} = [\,\boldsymbol{b}_1 \quad \boldsymbol{A}\boldsymbol{b}_1 \quad \boldsymbol{b}_2 \quad \boldsymbol{A}\boldsymbol{b}_2\,] = \begin{bmatrix} 1 & 1 & 1 & 0 \\ 2 & 3 & 0 & 1 \\ 1 & 4 & 0 & 0 \\ 0 & 2 & 1 & 3 \end{bmatrix}$$

and

$$\boldsymbol{L}^{-1} = \begin{bmatrix} 0.20 & 0.60 & 0.40 & -0.20 \\ -0.05 & -0.15 & 0.35 & 0.05 \\ 0.85 & -0.45 & 0.05 & 0.15 \\ -0.25 & 0.25 & -0.25 & 0.25 \end{bmatrix}.$$

Hence

$$\boldsymbol{q}_1 = [-0.05 \quad -0.15 \quad 0.35 \quad 0.05]$$

and

$$\boldsymbol{q}_2 = [-0.25 \quad 0.25 \quad -0.25 \quad 0.25].$$

The transformation matrix is

$$\boldsymbol{T} = \begin{bmatrix} \boldsymbol{q}_1 \\ \boldsymbol{q}_1\boldsymbol{A} \\ \boldsymbol{q}_2 \\ \boldsymbol{q}_2\boldsymbol{A} \end{bmatrix} = \begin{bmatrix} -0.05 & -0.15 & 0.35 & 0.05 \\ 0.05 & -0.15 & 1.35 & 0.05 \\ -0.25 & 0.25 & -0.25 & 0.25 \\ 0.75 & 0.25 & -1.25 & 0.25 \end{bmatrix}.$$

The matrices $\boldsymbol{A}$ and $\boldsymbol{B}$ in the new coordinates have the form

$$\boldsymbol{T}\boldsymbol{A}\boldsymbol{T}^{-1} = \left[\begin{array}{cc|cc} 0 & 1 & 0 & 0 \\ -4 & 5 & 0 & 0 \\ \hline 0 & 0 & 0 & 1 \\ 3 & -3 & 4 & 1 \end{array}\right] \quad \text{and} \quad \boldsymbol{T}\boldsymbol{B} = \left[\begin{array}{cc} 0 & 0 \\ 1 & 0 \\ \hline 0 & 0 \\ 0 & 1 \end{array}\right].$$

3.3.2 Observer Form

In this subsection, we use duality and the algorithms for reducing reachable systems into controller forms to derive analogous results for observable systems. We consider an observable dynamical system model

$$\begin{aligned} \dot{\boldsymbol{x}}(t) &= \boldsymbol{A}\boldsymbol{x}(t), \\ \boldsymbol{y}(t) &= \boldsymbol{C}\boldsymbol{x}(t), \end{aligned} \tag{3.17}$$

or, equivalently, the observable pair $(\boldsymbol{A}, \boldsymbol{C})$, where $\boldsymbol{A} \in \mathbb{R}^{n\times n}$, $\boldsymbol{C} \in \mathbb{R}^{p\times n}$, $p \le n$, and rank $\boldsymbol{C} = p$. We omitted the input variable $\boldsymbol{u}$ for convenience because it plays no role in the subsequent discussion. We next consider the *dual system* model associated with the system (3.17),

$$\dot{\boldsymbol{z}}(t) = \boldsymbol{A}^T\boldsymbol{z}(t) + \boldsymbol{C}^T\boldsymbol{v}(t).$$

Because the pair $(\boldsymbol{A}, \boldsymbol{C})$ is observable, the dual pair $(\boldsymbol{A}^T, \boldsymbol{C}^T)$ is reachable. Therefore, the pair $(\boldsymbol{A}^T, \boldsymbol{C}^T)$ can be transformed into the controller form using the algorithm from the previous subsection. We then take the dual of the result to get

$$\begin{aligned} \dot{\hat{\boldsymbol{x}}}(t) &= \hat{\boldsymbol{A}}\hat{\boldsymbol{x}}, \\ \boldsymbol{y}(t) &= \hat{\boldsymbol{C}}\hat{\boldsymbol{x}}, \end{aligned}$$

where

$$\hat{A} = \begin{bmatrix} \hat{A}_{11} & \hat{A}_{12} & \cdots & \hat{A}_{1p} \\ \hat{A}_{21} & \hat{A}_{22} & \cdots & \hat{A}_{2p} \\ \vdots & & & \vdots \\ \hat{A}_{p1} & \hat{A}_{p2} & \cdots & \hat{A}_{pp} \end{bmatrix}.$$

Each diagonal submatrix $\hat{A}_{ii} \in \mathbb{R}^{\hat{d}_i \times \hat{d}_i}$ of $\hat{A}$ has the form

$$\hat{A}_{ii} = \begin{bmatrix} 0 & 0 & \cdots & 0 & x \\ 1 & 0 & \cdots & 0 & x \\ \vdots & & \ddots & & \vdots \\ 0 & 0 & \cdots & 1 & x \end{bmatrix},$$

which is just the transpose of the controller form for the single-input case. The integers $\hat{d}_i$, $i = 1, 2, \ldots, p$ are the controllability indices of the controllable pair (A^T, C^T). The off-diagonal blocks, $j \neq i$, have the form

$$\hat{A}_{ji} = \begin{bmatrix} 0 & 0 & \cdots & 0 & x \\ 0 & 0 & \cdots & 0 & x \\ \vdots & & & & \vdots \\ 0 & 0 & \cdots & 0 & x \end{bmatrix};$$

that is, they consist of zeros except for their last columns. The matrix $\hat{C}$ consists of p blocks and has the form

$$\hat{C} = \left[\begin{array}{cccc|cccc|c|cccc} 0 & 0 & \cdots & 1 & 0 & 0 & \cdots & 0 & \cdots & 0 & 0 & \cdots & 0 \\ 0 & 0 & \cdots & x & 0 & 0 & \cdots & 1 & \cdots & 0 & 0 & \cdots & 0 \\ \vdots & & & \vdots & \vdots & & & \vdots & & \vdots & & & \vdots \\ 0 & 0 & \cdots & x & 0 & 0 & \cdots & x & \cdots & 0 & 0 & \cdots & 1 \end{array}\right].$$

The p controllability indices of the pair (A^T, C^T) are called the *observability indices* of the pair (A, C). Note that

$$\sum_{i=1}^{p} \hat{d}_i = n.$$

Similarly as in the controller form, we define the *observability index* of the pair (A, C) to be the integer

$$\hat{d} = \max\{\hat{d}_i : i = 1, 2, \ldots, p\}.$$

We now illustrate the above algorithm for transforming an observable pair into the observer form with a numerical example.

◆ **Example 3.8**

Given the observable pair $(\boldsymbol{A}, \boldsymbol{C})$,

$$\boldsymbol{A} = \begin{bmatrix} 0 & 0 & 1 & 0 \\ 1 & 0 & 2 & 0 \\ 0 & 1 & 3 & 0 \\ 0 & 0 & -21 & 5 \end{bmatrix}, \qquad \boldsymbol{C} = \begin{bmatrix} 1 & 0 & 0 & 0 \\ 0 & 0 & 0 & 1 \end{bmatrix},$$

we transform the pair $(\boldsymbol{A}^T, \boldsymbol{C}^T)$ into the controller form. Thus, we first form the controllability matrix for $(\boldsymbol{A}^T, \boldsymbol{C}^T)$; then proceeding from left to right, we select the first $n = 4$ linearly independent columns. They are

$$\boldsymbol{c}_1^T, \boldsymbol{c}_2^T, \boldsymbol{A}^T \boldsymbol{c}_1^T, (\boldsymbol{A}^T)^2 \boldsymbol{c}_1^T.$$

Therefore, the observability indices are: $\hat{d}_1 = 3$ and $\hat{d}_2 = 1$. Hence,

$$\boldsymbol{L} = \begin{bmatrix} \boldsymbol{c}_1^T & \boldsymbol{A}^T \boldsymbol{c}_1^T & (\boldsymbol{A}^T)^2 \boldsymbol{c}_1^T & \boldsymbol{c}_2^T \end{bmatrix} = \begin{bmatrix} 1 & 0 & 0 & 0 \\ 0 & 0 & 1 & 0 \\ 0 & 1 & 3 & 0 \\ 0 & 0 & 0 & 1 \end{bmatrix}$$

and

$$\boldsymbol{L}^{-1} = \begin{bmatrix} 1 & 0 & 0 & 0 \\ 0 & -3 & 1 & 0 \\ 0 & 1 & 0 & 0 \\ 0 & 0 & 0 & 1 \end{bmatrix}.$$

The two vectors that we need to form the desired transformation are

$$\boldsymbol{q}_1 = [0 \quad 1 \quad 0 \quad 0] \quad \text{and} \quad \boldsymbol{q}_2 = [0 \quad 0 \quad 0 \quad 1].$$

The transformation matrix is

$$\hat{\boldsymbol{T}} = \begin{bmatrix} 0 & 1 & 0 & 0 \\ 0 & -0 & 1 & 0 \\ 1 & 2 & 3 & -21 \\ 0 & 0 & 0 & 1 \end{bmatrix}.$$

The observer form of the pair $(\boldsymbol{A}, \boldsymbol{C})$ is given by the matrices

$$\hat{\boldsymbol{A}} = \hat{\boldsymbol{T}}^{-T} \boldsymbol{A} \hat{\boldsymbol{T}}^T = \left[\begin{array}{ccc|c} 0 & 0 & 1 & 0 \\ 1 & 0 & 2 & 0 \\ 0 & 1 & 3 & 0 \\ \hline 0 & 0 & -105 & 5 \end{array}\right] \quad \text{and} \quad \hat{\boldsymbol{C}} = \boldsymbol{C} \hat{\boldsymbol{T}}^T = \left[\begin{array}{ccc|c} 0 & 0 & 1 & 0 \\ 0 & 0 & -21 & 1 \end{array}\right].$$

3.4 Linear State-Feedback Control

In the following sections we discuss methods for a feedback controller design for linear systems. The design process involves three steps. The first step is the design of a state-feedback control law. In this step we assume that we have all states available to us for feedback purposes. In general, this may not be the case. The assumption that all the states are available allows us to proceed with the state-feedback control law design. We then proceed with the second step; this involves the design of an estimator, also referred to as an observer of the state vector. The estimator should use only the input and output of the plant to be controlled to generate an estimate of the plant state. The last step is to combine the first two steps in that the control law, designed in the first step, uses the state estimate instead of the true state vector. The result of this step is a combined controller–estimator compensator. We first discuss the problem of designing a linear state-feedback control law. In the following sections we present methods for constructing estimators and combined controller–estimator compensators.

The linear state-feedback control law, for a system modeled by $\dot{\boldsymbol{x}} = \boldsymbol{A}\boldsymbol{x} + \boldsymbol{B}\boldsymbol{u}$, is the feedback of a linear combination of all the state variables and has the form

$$\boldsymbol{u} = -\boldsymbol{K}\boldsymbol{x},$$

where $\boldsymbol{K} \in \mathbb{R}^{m\times n}$ is a constant matrix. The closed-loop system then is

$$\dot{\boldsymbol{x}} = (\boldsymbol{A} - \boldsymbol{B}\boldsymbol{K})\boldsymbol{x}.$$

The poles of the closed-loop system are the roots of the characteristic equation

$$\det(s\boldsymbol{I}_n - \boldsymbol{A} + \boldsymbol{B}\boldsymbol{K}) = 0.$$

The linear state-feedback control law design consists of selecting the gains

$$k_{ij}, \qquad i = 1, 2, \ldots, m, \quad j = 1, 2, \ldots, n,$$

so that the roots of the closed-loop characteristic equation

$$\det(s\boldsymbol{I}_n - \boldsymbol{A} + \boldsymbol{B}\boldsymbol{K}) = 0$$

are in desirable locations in the complex plane. We assume that a designer has made the selection of the desired poles of the closed-loop system, and they are

$$s_1, s_2, \ldots, s_n.$$

The desired closed-loop poles can be real or complex. If they are complex, then they must come in complex conjugate pairs. This is because we use only real gains k_{ij}. Having selected the desired closed-loop poles, we can form the desired closed-loop characteristic polynomial,

$$\begin{aligned}\alpha_c(s) &= (s - s_1)(s - s_2)\cdots(s - s_n)\\ &= s^n + \alpha_{n-1}s^{n-1} + \cdots + \alpha_1 s + \alpha_0.\end{aligned}$$

Our goal is to select a feedback matrix $\boldsymbol{K}$ such that

$$\det(s\boldsymbol{I}_n - \boldsymbol{A} + \boldsymbol{B}\boldsymbol{K}) = s^n + \alpha_{n-1}s^{n-1} + \cdots + \alpha_1 s + \alpha_0.$$

The above problem is also referred to as the *pole placement* problem. We first discuss the pole placement problem for the single-input plants. In this case, $\boldsymbol{K} = \boldsymbol{k} \in \mathbb{R}^{1\times n}$. The solution to the problem is easily obtained if the pair $(\boldsymbol{A}, \boldsymbol{b})$ is already in the controller companion form. In such a case we have

$$\boldsymbol{A} - \boldsymbol{b}\boldsymbol{k} = \begin{bmatrix} 0 & 1 & 0 & \cdots & 0 & 0 \\ 0 & 0 & 1 & \cdots & 0 & 0 \\ \vdots & & & & & \vdots \\ 0 & 0 & 0 & \cdots & 0 & 1 \\ -a_0 - k_1 & -a_1 - k_2 & -a_2 - k_3 & \cdots & -a_{n-2} - k_{n-1} & -a_{n-1} - k_n \end{bmatrix}.$$

Hence, the desired gains are

$$\begin{aligned} k_1 &= \alpha_0 - a_0, \\ k_2 &= \alpha_1 - a_1, \\ &\vdots \\ k_n &= \alpha_{n-1} - a_{n-1}. \end{aligned}$$

If the pair $(\boldsymbol{A}, \boldsymbol{b})$ is not in the controller companion form, we first transform it into the companion form and then compute the gain vector $\tilde{\boldsymbol{k}}$ such that

$$\det(s\boldsymbol{I}_n - \tilde{\boldsymbol{A}} + \tilde{\boldsymbol{b}}\tilde{\boldsymbol{k}}) = s^n + \alpha_{n-1}s^{n-1} + \cdots + \alpha_1 s + \alpha_0.$$

Thus,

$$\tilde{\boldsymbol{k}} = [\alpha_0 - a_0 \quad \alpha_1 - a_1 \quad \cdots \quad \alpha_{n-1} - a_{n-1}].$$

Then,

$$\boldsymbol{k} = \tilde{\boldsymbol{k}}\boldsymbol{T},$$

where $\boldsymbol{T}$ is the transformation that brings the pair $(\boldsymbol{A}, \boldsymbol{b})$ into the controller companion form.

We can represent the above formula for the gain matrix in an alternative way. For this, note that

$$\begin{aligned} \tilde{\boldsymbol{k}}\boldsymbol{T} &= [\alpha_0 - a_0 \quad \alpha_1 - a_1 \quad \cdots \quad \alpha_{n-1} - a_{n-1}] \begin{bmatrix} \boldsymbol{q}_1 \\ \boldsymbol{q}_1\boldsymbol{A} \\ \vdots \\ \boldsymbol{q}_1\boldsymbol{A}^{n-1} \end{bmatrix} \\ &= \boldsymbol{q}_1(\alpha_0\boldsymbol{I}_n + \alpha_1\boldsymbol{A} + \cdots + \alpha_{n-1}\boldsymbol{A}^{n-1}) - \boldsymbol{q}_1(a_0\boldsymbol{I}_n + a_1\boldsymbol{A} + \cdots + a_{n-1}\boldsymbol{A}^{n-1}). \end{aligned}$$

By the Cayley–Hamilton theorem, we have

$$\boldsymbol{A}^n = -(a_0\boldsymbol{I}_n + a_1\boldsymbol{A} + \cdots + a_{n-1}\boldsymbol{A}^{n-1}).$$

Hence,

$$\boldsymbol{k} = \boldsymbol{q}_1\alpha_c(\boldsymbol{A}).$$

The above expression for the gain row vector was proposed by Ackermann in 1972 and is now referred to as *Ackermann's formula* for pole placement.

◆ Example 3.9

For the linear dynamical system

$$\dot{\boldsymbol{x}} = \begin{bmatrix} 1 & -1 \\ 1 & -2 \end{bmatrix} \boldsymbol{x} + \begin{bmatrix} 2 \\ 1 \end{bmatrix} u,$$

we will use Ackermann's formula to design a state-feedback controller, $u = -\boldsymbol{kx}$, such that the closed-loop poles are located at $\{-1, -2\}$.

To use Ackermann's formula, we first form the controllability matrix of the pair $(\boldsymbol{A}, \boldsymbol{b})$ and then find the last row of its inverse, denoted $\boldsymbol{q}_1$. The controllability matrix is

$$[\boldsymbol{b} \quad \boldsymbol{Ab}] = \begin{bmatrix} 2 & 1 \\ 1 & 0 \end{bmatrix}.$$

The inverse of the above matrix is

$$\begin{bmatrix} 0 & 1 \\ 1 & -2 \end{bmatrix}.$$

Hence, $\boldsymbol{q}_1 = [1 \quad -2]$. The desired closed-loop characteristic polynomial is

$$\alpha_c(s) = (s+1)(s+2) = s^2 + 3s + 2.$$

Therefore,

$$\begin{aligned}
\boldsymbol{k} &= \boldsymbol{q}_1 \alpha_c(\boldsymbol{A}) \\
&= \boldsymbol{q}_1(\boldsymbol{A}^2 + 3\boldsymbol{A} + 2\boldsymbol{I}_2) \\
&= \boldsymbol{q}_1 \left(\begin{bmatrix} 1 & -1 \\ 1 & -2 \end{bmatrix} \begin{bmatrix} 1 & -1 \\ 1 & -2 \end{bmatrix} + 3 \begin{bmatrix} 1 & -1 \\ 1 & -2 \end{bmatrix} + 2 \begin{bmatrix} 1 & 0 \\ 0 & 1 \end{bmatrix} \right) \\
&= \boldsymbol{q}_1 \left(\begin{bmatrix} 0 & 1 \\ -1 & 3 \end{bmatrix} + \begin{bmatrix} 3 & -3 \\ 3 & -6 \end{bmatrix} + \begin{bmatrix} 2 & 0 \\ 0 & 2 \end{bmatrix} \right) \\
&= [1 \quad -2] \begin{bmatrix} 5 & -2 \\ 2 & -1 \end{bmatrix} \\
&= [1 \quad 0]
\end{aligned}$$

We now present a pole placement algorithm for multi-input systems. If the pair $(\boldsymbol{A}, \boldsymbol{B})$ is already in the controller companion form, then we may proceed as follows. First, represent the matrix $\boldsymbol{B}$ as

$$
\boldsymbol{B}=\left[\begin{array}{ccccc}
0 & 0 & \cdots & 0 & 0\\
0 & 0 & \cdots & 0 & 0\\
\vdots & & & & \vdots\\
1 & x & x & x & x\\
\hline
0 & 0 & \cdots & 0 & 0\\
0 & 0 & \cdots & 0 & 0\\
\vdots & & & & \vdots\\
0 & 1 & x & x & x\\
\hline
\vdots & & & & \vdots\\
\hline
0 & 0 & \cdots & 0 & 0\\
0 & 0 & \cdots & 0 & 0\\
\vdots & & & & \vdots\\
0 & 0 & 0 & 0 & 1
\end{array}\right]
=\left[\begin{array}{ccccc}
0 & 0 & \cdots & 0 & 0\\
0 & 0 & \cdots & 0 & 0\\
\vdots & & & & \vdots\\
1 & 0 & 0 & 0 & 0\\
\hline
0 & 0 & \cdots & 0 & 0\\
0 & 0 & \cdots & 0 & 0\\
\vdots & & & & \vdots\\
0 & 1 & 0 & 0 & 0\\
\hline
\vdots & & & & \vdots\\
\hline
0 & 0 & \cdots & 0 & 0\\
0 & 0 & \cdots & 0 & 0\\
\vdots & & & & \vdots\\
0 & 0 & 0 & 0 & 1
\end{array}\right]
\left[\begin{array}{ccccc}
1 & x & \cdots & x & x\\
0 & 1 & \cdots & x & x\\
\vdots & & \ddots & & \vdots\\
0 & 0 & \cdots & 1 & x\\
0 & 0 & \cdots & 0 & 1
\end{array}\right]
=\hat{\boldsymbol{B}}\boldsymbol{\Gamma},
$$

where the square and nonsingular matrix $\boldsymbol{\Gamma}$ consists of the nonzero rows of $\boldsymbol{B}$. Next, let

$$
\hat{\boldsymbol{K}}=\boldsymbol{\Gamma}\boldsymbol{K}.
$$

Notice that

$$
\hat{\boldsymbol{B}}\hat{\boldsymbol{K}}=\left[\begin{array}{ccccc}
0 & 0 & \cdots & 0 & 0\\
0 & 0 & \cdots & 0 & 0\\
\vdots & & & & \vdots\\
\hat{k}_{11} & \hat{k}_{12} & \cdots & \hat{k}_{1\,n-1} & \hat{k}_{1n}\\
\hline
0 & 0 & \cdots & 0 & 0\\
0 & 0 & \cdots & 0 & 0\\
\vdots & & & & \vdots\\
\hat{k}_{21} & \hat{k}_{22} & \cdots & \hat{k}_{2\,n-1} & \hat{k}_{2n}\\
\hline
\vdots & & & & \vdots\\
\hline
0 & 0 & \cdots & 0 & 0\\
0 & 0 & \cdots & 0 & 0\\
\vdots & & & & \vdots\\
\hat{k}_{m1} & \hat{k}_{m2} & \cdots & \hat{k}_{m\,n-1} & \hat{k}_{mn}
\end{array}\right];
$$

that is, the nonzero rows in the product $\hat{\boldsymbol{B}}\hat{\boldsymbol{K}}$ match the nonzero rows of the matrix $\boldsymbol{A}$ in the controller companion form. We can then, for example, select the gains k_{ij}, $i=1,2,\ldots,m$,

$j = 1, 2, \ldots, n$, so that

$$A - BK = A - \hat{B}\hat{K} = \begin{bmatrix} 0 & 1 & 0 & \cdots & 0 & 0 \\ 0 & 0 & 1 & \cdots & 0 & 0 \\ \vdots & & & & & \vdots \\ 0 & 0 & 0 & \cdots & 0 & 1 \\ -\alpha_0 & -\alpha_1 & -\alpha_2 & \cdots & -\alpha_{n-2} & -\alpha_{n-1} \end{bmatrix},$$

where

$$K = \Gamma^{-1}\hat{K}.$$

If the pair (A, B) is not in the controller companion form, we first transform it into this form and then compute the gain matrix that allocates the closed-loop poles into the desired locations for the pair (A, B) in the controller form. The gain matrix K that allocates the closed-loop poles into the prespecified locations for the pair (A, B) in the original coordinates is then given by

$$K = \Gamma^{-1}\hat{K}T,$$

where T is the transformation matrix that brings the pair (A, B) into the controller companion form.

◆ Example 3.10

For the pair

$$A = \begin{bmatrix} 0 & 0 & 1 & 0 \\ 1 & 0 & 2 & 0 \\ 0 & 1 & 3 & 0 \\ 0 & 0 & -21 & 5 \end{bmatrix}, \qquad B = \begin{bmatrix} 1 & 0 \\ 0 & 0 \\ 0 & 0 \\ 0 & 1 \end{bmatrix},$$

we will use its controller companion form to find the matrix $K \in \mathbb{R}^{2\times 4}$ so that closed-loop poles are located at

$$-2, -3 + j, -3 - j, -4.$$

We first transform the pair (A, B) into its controller companion form. We begin by forming the controllability matrix

$$[b_1 \quad b_2 \quad Ab_1 \quad Ab_2 \quad A^2b_1 \quad \cdots] = \begin{bmatrix} 1 & 0 & 0 & 0 & 0 & \cdots \\ 0 & 0 & 1 & 0 & 0 & \cdots \\ 0 & 0 & 0 & 0 & 1 & \cdots \\ 0 & 1 & 0 & 5 & 0 & \cdots \end{bmatrix}.$$

We then select, proceeding from left to right, the first four linearly independent columns from the above controllability matrix. We obtain

$$[b_1 \quad b_2 \quad Ab_1 \quad A^2b_1].$$

Thus, the controllability indices are $d_1 = 3$ and $d_2 = 1$. We rearrange the above columns and form the matrix $\boldsymbol{L}$ of the form

$$\boldsymbol{L} = [\boldsymbol{b}_1 \quad \boldsymbol{A}\boldsymbol{b}_1 \quad \boldsymbol{A}^2\boldsymbol{b}_1 \quad \boldsymbol{b}_2] = \boldsymbol{I}_4 = \boldsymbol{L}^{-1}.$$

The two row vectors that we need in constructing the transformation are

$$\boldsymbol{q}_1 = [0 \quad 0 \quad 1 \quad 0] \quad \text{and} \quad \boldsymbol{q}_2 = [0 \quad 0 \quad 0 \quad 1].$$

The transformation matrix is

$$\boldsymbol{T} = \begin{bmatrix} \boldsymbol{q}_1 \\ \boldsymbol{q}_1\boldsymbol{A} \\ \boldsymbol{q}_1\boldsymbol{A}^2 \\ \boldsymbol{q}_2 \end{bmatrix} = \begin{bmatrix} 0 & 0 & 1 & 0 \\ 0 & 1 & 3 & 0 \\ 1 & 3 & 11 & 0 \\ 0 & 0 & 0 & 1 \end{bmatrix}.$$

The pair $(\boldsymbol{A}, \boldsymbol{B})$ in the new coordinates has the form

$$\boldsymbol{A} = \boldsymbol{T}\boldsymbol{A}\boldsymbol{T}^{-1} = \left[\begin{array}{ccc|c} 0 & 1 & 0 & 0 \\ 0 & 0 & 1 & 0 \\ 1 & 2 & 3 & 0 \\ \hline -21 & 0 & 0 & 5 \end{array}\right] \quad \text{and} \quad \tilde{\boldsymbol{B}} = \boldsymbol{T}\boldsymbol{B} = \left[\begin{array}{cc} 0 & 0 \\ 0 & 0 \\ 1 & 0 \\ \hline 0 & 1 \end{array}\right].$$

The desired closed-loop characteristic polynomial is

$$\alpha_c(s) = (s+2)(s+3-j)(s+3+j)(s+4) = s^4 + 12s^3 + 54s^2 + 108s + 80.$$

A possible choice of the gain matrix $\tilde{\boldsymbol{K}}$, from among many, that does the job is $\tilde{\boldsymbol{K}}$ such that

$$\tilde{\boldsymbol{A}} - \tilde{\boldsymbol{B}}\tilde{\boldsymbol{K}} = \begin{bmatrix} 0 & 1 & 0 & 0 \\ 0 & 0 & 1 & 0 \\ 0 & 0 & 0 & 1 \\ -80 & -108 & -54 & -12 \end{bmatrix}.$$

Note that $\boldsymbol{\Gamma} = \boldsymbol{I}_2$. Therefore,

$$\tilde{\boldsymbol{K}} = \begin{bmatrix} 1 & 2 & 3 & -1 \\ 59 & 108 & 54 & 17 \end{bmatrix},$$

and thus

$$\boldsymbol{K} = \tilde{\boldsymbol{K}}\boldsymbol{T} = \begin{bmatrix} 3 & 11 & 40 & -1 \\ 54 & 270 & 977 & 17 \end{bmatrix}.$$

The above algorithm for pole placement for multi-input systems is more of theoretical value rather than practical. This algorithm is not well-suited for numerical implementations because the transformation of a given pair $(\boldsymbol{A}, \boldsymbol{B})$ into the controller companion form suffers from poor numerical properties. For a discussion of pole placement numerically stable algorithms for multi-input systems, we refer the reader to Jamshidi, Tarokh, and Shafai [141].

Unlike in the single-input case, a solution to the pole placement problem for a multi-input system is not unique. Therefore, one can use the remaining degrees of freedom to achieve secondary goals. In Subsection 5.3.3 we discuss a method for constructing a linear state-feedback control law that allocates the closed-loop poles into prespecified locations and at the same time minimizes a quadratic performance index.

Using the results preceding the above discussion, we can state and prove a fundamental theorem of linear systems.

Theorem 3.7 The pole placement problem is solvable for all choices of n desired closed-loop poles, symmetric with respect to the real axis, if and only if the given pair $(\boldsymbol{A}, \boldsymbol{B})$ is reachable.

Proof The proof of necessity ($\Leftarrow$) follows immediately from the discussion preceding Example 3.10. Indeed, the pair $(\boldsymbol{A}, \boldsymbol{B})$ is reachable, if and only if it can be transformed into the controller canonical form. Once the transformation is performed, we can solve the pole placement problem for the given set of desired closed-loop poles, symmetric with respect to the real axis. We then transform the pair $(\boldsymbol{A}, \boldsymbol{B})$ and the gain matrix back into the original coordinates. Because the similarity transformation affects neither controllability nor the characteristic polynomial, the eigenvalues of $\boldsymbol{A} - \boldsymbol{B}\boldsymbol{K}$ are precisely the desired prespecified closed-loop poles. This completes the proof of the necessity part.

We use a proof by contraposition to prove the sufficiency part ($\Rightarrow$). Assume that the pair $(\boldsymbol{A}, \boldsymbol{B})$ is nonreachable. By Theorem 3.4, there is a similarity transformation $\boldsymbol{z} = \boldsymbol{T}\boldsymbol{x}$ such that the pair $(\boldsymbol{A}, \boldsymbol{B})$ in the new coordinates has the form

$$\tilde{\boldsymbol{A}} = \boldsymbol{T}\boldsymbol{A}\boldsymbol{T}^{-1} = \begin{bmatrix} \boldsymbol{A}_1 & \boldsymbol{A}_2 \\ \boldsymbol{O} & \boldsymbol{A}_4 \end{bmatrix}, \qquad \tilde{\boldsymbol{B}} = \boldsymbol{T}\boldsymbol{B} = \begin{bmatrix} \boldsymbol{B}_1 \\ \boldsymbol{O} \end{bmatrix},$$

where the pair $(\boldsymbol{A}_1, \boldsymbol{B}_1)$ is reachable, $\boldsymbol{A}_1 \in \mathbb{R}^{r\times r}$, and $\boldsymbol{B}_1 \in \mathbb{R}^{r\times m}$. Let

$$\boldsymbol{u} = -\tilde{\boldsymbol{K}}\boldsymbol{z} = -[\,\boldsymbol{K}_1 \quad \boldsymbol{K}_2\,]\boldsymbol{z},$$

where $\boldsymbol{K}_1 \in \mathbb{R}^{m\times r}$ and $\boldsymbol{K}_2 \in \mathbb{R}^{m\times(n-r)}$. Then,

$$\tilde{\boldsymbol{A}} - \tilde{\boldsymbol{B}}\tilde{\boldsymbol{K}} = \begin{bmatrix} \boldsymbol{A}_1 - \boldsymbol{B}_1\boldsymbol{K}_1 & \boldsymbol{A}_2 - \boldsymbol{B}_1\boldsymbol{K}_2 \\ \boldsymbol{O} & \boldsymbol{A}_4 \end{bmatrix}. \tag{3.18}$$

It follows from (3.18) that the nonreachable portion of the system is not affected by the state feedback; that is, the eigenvalues of $\boldsymbol{A}_4$ cannot be allocated. Hence, the pole placement problem cannot be solved if the pair $(\tilde{\boldsymbol{A}}, \tilde{\boldsymbol{B}})$, or equivalently the pair $(\boldsymbol{A}, \boldsymbol{B})$, is nonreachable, which completes the proof of the sufficiency part.

For an alternative proof of this important result, the reader may consult Wonham [304].

3.5 State Estimators

In the previous section, we presented algorithms for solving the pole placement problem using the linear state-feedback control law. To implement such a control law, one needs availability of all the state variables. Often, this requirement is not met, either because measuring all the state variables would require excessive number of sensors or because the state variables are not accessible for direct measurement. Instead, only a subset of state variables or their combination may be available. To implement the control law constructed in the previous section, we will use estimates of state variables rather than the true states. We now discuss the problem of constructing state estimators. Much of the literature refers to state estimators as "observers." However, we agree with Franklin and Powell [89, p. 139] that "estimator" is much more descriptive in its function because "observer" implies a direct measurement. We present two types of state estimators. First, full-order state estimators will be discussed and then a reduced-order state estimator will be introduced.

3.5.1 Full-Order Estimator

We now discuss the problem of constructing a full-order state estimator for dynamical systems modeled by

$$\dot{\boldsymbol{x}}(t) = \boldsymbol{A}\boldsymbol{x}(t) + \boldsymbol{B}\boldsymbol{u}(t),$$
$$\boldsymbol{y}(t) = \boldsymbol{C}\boldsymbol{x}(t),$$

where $\boldsymbol{C} \in \mathbb{R}^{p\times n}$, $p \leq n$. We assume that the pair $(\boldsymbol{A}, \boldsymbol{C})$ is observable. Our goal is to construct a dynamical system that will estimate the state vector $\boldsymbol{x}$ based on the plant input $\boldsymbol{u}$ and the plant output $\boldsymbol{y}$. One could consider constructing a model of the plant dynamics and connecting the resulting dynamical system, referred to as an *open-loop estimator,* as in Figure 3.4. This open-loop estimator is described by

$$\dot{\tilde{\boldsymbol{x}}}(t) = \boldsymbol{A}\tilde{\boldsymbol{x}}(t) + \boldsymbol{B}\boldsymbol{u}(t),$$

where $\tilde{\boldsymbol{x}}(t)$ is the estimate of $\boldsymbol{x}(t)$. Let

$$\boldsymbol{e}(t) = \tilde{\boldsymbol{x}}(t) - \boldsymbol{x}(t)$$

be the estimation error. Then, the dynamics of the estimation error are described by

$$\dot{\boldsymbol{e}}(t) = \dot{\tilde{\boldsymbol{x}}}(t) - \dot{\boldsymbol{x}}(t) = \boldsymbol{A}\boldsymbol{e}(t),$$

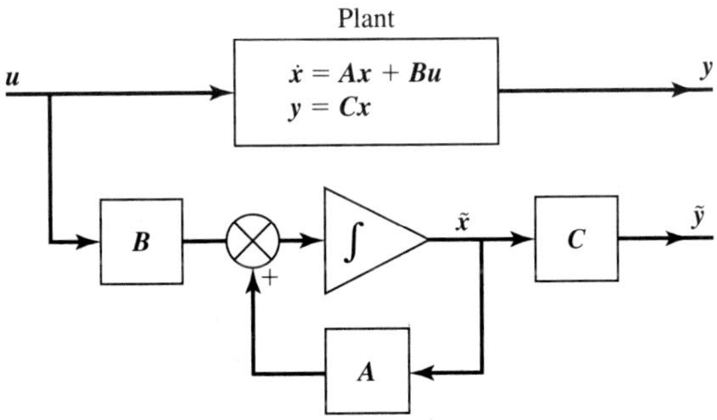

Figure 3.4 Open-loop estimator.

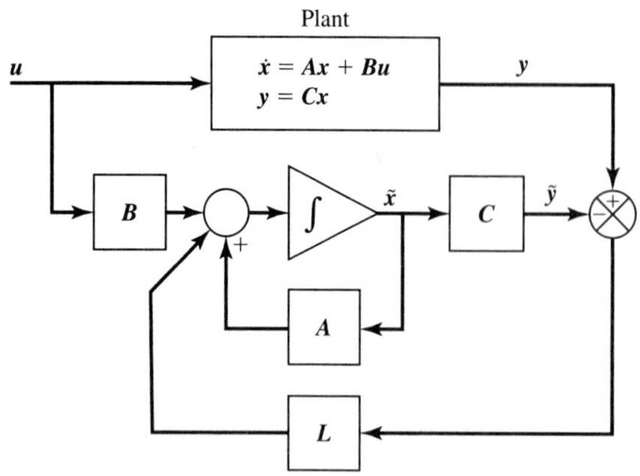

Figure 3.5 Closed-loop estimator.

with the initial estimation error

$$e(0) = \tilde{x}(0) - x(0).$$

If the eigenvalues of the matrix A are in the open left-hand plane, then the error converges to zero. However, we have no control over the convergence rate. Furthermore, the matrix A does not have to have all its eigenvalues in the open left-hand plane. Thus, the open-loop estimator is impractical. We modify this estimator by adding a feedback term to it. The resulting structure, depicted in Figure 3.5, is called the *closed-loop estimator,* the *Luenberger observer,* or the asymptotic full-order estimator. The dynamics of the closed-loop estimator are described by

$$\dot{\tilde{x}}(t) = A\tilde{x}(t) + Bu(t) + L(y(t) - C\tilde{x}(t)),$$

and the dynamics of the estimation error are governed by

$$\begin{aligned}\dot{e}(t) &= \dot{\tilde{x}}(t) - \dot{x}(t)\\ &= (A - LC)\,e(t), \qquad e(0) = \tilde{x}(0) - x(0).\end{aligned}$$

The pair (A, C) is observable if and only if the dual pair (A^T, C^T) is reachable. By assumption, the pair (A, C) is observable, and therefore the pair (A^T, C^T) is reachable. By Theorem 3.7, we can solve the pole placement problem for the dual pair (A^T, C^T); that is, for any set of prespecified n complex numbers, symmetric with respect to the real axis, there is a matrix, call it L^T, such that the eigenvalues of $A^T - C^T L^T$ and hence of $A - LC$ are in the prespecified locations. It follows from the above that if the pair (A, C) is observable, then in addition to forcing the estimation error to converge to zero, we can also control its rate of convergence by appropriately selecting the eigenvalues of the matrix $A - LC$. We also note that the selection of the closed-loop estimator gain matrix L can be approached in exactly the same fashion as the construction of the gain matrix K in the linear state-feedback control law design. We illustrate the above point with a numerical example.

◆ Example 3.11

For the given observable pair

$$A = \begin{bmatrix} 0 & 0 & 1 & 0 \\ 1 & 0 & 2 & 0 \\ 0 & 1 & 3 & 0 \\ 0 & 0 & -21 & 5 \end{bmatrix}, \qquad C = \begin{bmatrix} 1 & 0 & 0 & 0 \\ 0 & 0 & 0 & 1 \end{bmatrix},$$

construct the matrix $L \in \mathbb{R}^{4\times 2}$ so that the eigenvalues of $A - LC$ are located at

$$\{-2, -3 + j, -3 - j, -4\}.$$

In our construction of the gain matrix L we use the observer companion form of the pair (A, C) that we computed in Example 3.8. Our goal then is to construct a matrix L so that the characteristic polynomial of the matrix $A - LC$ is

$$\det(sI_4 - A + LC) = s^4 + 12s^3 + 54s^2 + 108s + 80.$$

Using the observer form of the pair (A, C), we select $\hat{L}$ so that

$$\hat{A} - \hat{L}\hat{C} = \begin{bmatrix} 0 & 0 & 0 & -80 \\ 1 & 0 & 0 & -108 \\ 0 & 1 & 0 & -54 \\ 0 & 0 & 1 & -12 \end{bmatrix}.$$

We have

$$\hat{L} = \begin{bmatrix} 1681 & 80 \\ 2270 & 108 \\ 1137 & 54 \\ 251 & 17 \end{bmatrix},$$

and hence

$$L = \hat{T}^T \hat{L} = \begin{bmatrix} 1137 & 54 \\ 3955 & 188 \\ 5681 & 270 \\ -23{,}626 & -1117 \end{bmatrix}.$$

We note that there are other possible gain matrices L that could be used to allocate the eigenvalues of $A - LC$ into desired locations.

We now present an alternative approach to the closed-loop estimator design. This approach will then be used to construct lower-order estimators that we discuss in the subsequent section. We assume that the dynamics of an estimator are given by

$$\dot{z}(t) = Dz(t) + Eu(t) + Gy(t), \tag{3.19}$$

where $z(t) \in \mathbb{R}^n$. To begin with the development, suppose that we would like to obtain

$$z = \boldsymbol{T}\boldsymbol{x},$$

where $\boldsymbol{T} \in \mathbb{R}^{n\times n}$ is invertible. Premultiplying $\dot{\boldsymbol{x}}(t) = \boldsymbol{A}\boldsymbol{x}(t) + \boldsymbol{B}\boldsymbol{u}(t)$ by $\boldsymbol{T}$ yields

$$\boldsymbol{T}\dot{\boldsymbol{x}}(t) = \boldsymbol{T}\boldsymbol{A}\boldsymbol{x}(t) + \boldsymbol{T}\boldsymbol{B}\boldsymbol{u}(t). \tag{3.20}$$

Substituting into (3.19) the relation $\boldsymbol{y}(t) = \boldsymbol{C}\boldsymbol{x}(t)$ gives

$$\dot{z}(t) = \boldsymbol{D}z(t) + \boldsymbol{E}\boldsymbol{u}(t) + \boldsymbol{G}\boldsymbol{C}\boldsymbol{x}(t). \tag{3.21}$$

Subtracting (3.21) from (3.20) and taking into account that $z = \boldsymbol{T}\boldsymbol{x}$ and $\dot{z} = \boldsymbol{T}\dot{\boldsymbol{x}}$ yields

$$\boldsymbol{0} = (\boldsymbol{T}\boldsymbol{A} - \boldsymbol{D}\boldsymbol{T} - \boldsymbol{G}\boldsymbol{C})\,\boldsymbol{x}(t) + (\boldsymbol{T}\boldsymbol{B} - \boldsymbol{E})\,\boldsymbol{u}(t)$$

for all t and for arbitrary input $\boldsymbol{u}$. Hence, we must have

$$\begin{aligned} \boldsymbol{T}\boldsymbol{A} - \boldsymbol{D}\boldsymbol{T} &= \boldsymbol{G}\boldsymbol{C}, \\ \boldsymbol{E} &= \boldsymbol{T}\boldsymbol{B}. \end{aligned}$$

If $z \neq \boldsymbol{T}\boldsymbol{x}$ but the two above equations hold, then subtracting (3.20) from (3.21) and taking into account the above relations yields

$$\begin{aligned} \frac{d}{dt}(z - \boldsymbol{T}\boldsymbol{x}) &= \boldsymbol{D}z - \boldsymbol{T}\boldsymbol{A}\boldsymbol{x} + \boldsymbol{G}\boldsymbol{C}\boldsymbol{x} \\ &= \boldsymbol{D}z - \boldsymbol{D}\boldsymbol{T}\boldsymbol{x} \\ &= \boldsymbol{D}(z - \boldsymbol{T}\boldsymbol{x}). \end{aligned}$$

If the eigenvalues of $\boldsymbol{D}$ all have negative real parts, then

$$z - \boldsymbol{T}\boldsymbol{x} \to \boldsymbol{0} \qquad \text{as } t \to \infty.$$

Note that if we try $\boldsymbol{T} = \boldsymbol{I}_n$, then we have the previously analyzed case, where $\boldsymbol{G} = \boldsymbol{L}$, and

$$\begin{aligned} \boldsymbol{E} &= \boldsymbol{B}, \\ \boldsymbol{A} - \boldsymbol{L}\boldsymbol{C} &= \boldsymbol{D}. \end{aligned}$$

The resulting estimator, when $\boldsymbol{T} = \boldsymbol{I}_n$, was called the identity observer by its developer, Luenberger [188, 189]. Given the desired eigenvalues, symmetric with respect to the real axis, the equation $\boldsymbol{A} - \boldsymbol{L}\boldsymbol{C} = \boldsymbol{D}$ can be solved for $\boldsymbol{L}$ so that the resulting $\boldsymbol{D}$ has its eigenvalues in the desired locations, if and only if the pair $(\boldsymbol{A}, \boldsymbol{C})$ is observable. However, it is not true that given an arbitrary $\boldsymbol{D}$ we can solve for $\boldsymbol{L}$ so that $\boldsymbol{A} - \boldsymbol{L}\boldsymbol{C} = \boldsymbol{D}$ even if the pair $(\boldsymbol{A}, \boldsymbol{C})$ is observable. We will use the results of the above discussion to construct estimators of reduced order in the following section.

3.5.2 Reduced-Order Estimator

If the output $\boldsymbol{y}$ is a p vector with $p < n$, we would like to use the p outputs to determine p of the x_i's and design an estimator of order $n - p$ to estimate the rest of the state variables. Specifically, if rank $\boldsymbol{C} = p$, then we can use the equation $\boldsymbol{y} = \boldsymbol{C}\boldsymbol{x}$ to solve for p of the x_i's in terms of the y_i's and the remaining $n - p$ state variables x_k's. Then, we construct an estimator

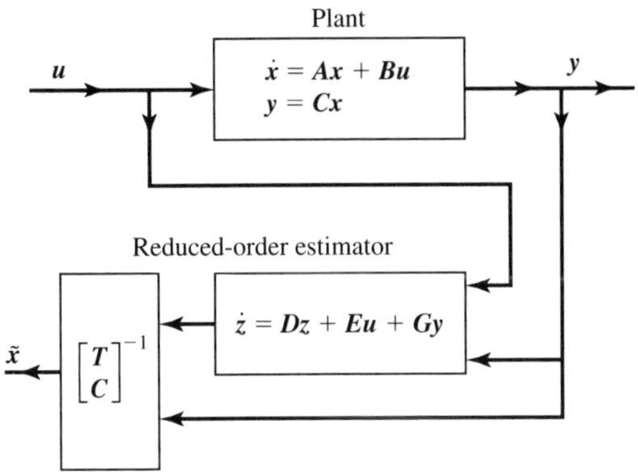

Figure 3.6 Reduced-order estimator.

of order $n - p$. To recover all states, we should like to solve for $\boldsymbol{x}$ the equations

$$z = \boldsymbol{T}\boldsymbol{x},$$
$$y = \boldsymbol{C}\boldsymbol{x},$$

where now $\boldsymbol{T} \in \mathbb{R}^{(n-p)\times n}$. This can be done if and only if

$$\operatorname{rank}\begin{bmatrix} \boldsymbol{T} \\ \boldsymbol{C} \end{bmatrix} = n. \tag{3.22}$$

Given $\boldsymbol{D} \in \mathbb{R}^{(n-p)\times(n-p)}$, $\boldsymbol{G} \in \mathbb{R}^{(n-p)\times p}$, and the triple $(\boldsymbol{A}, \boldsymbol{B}, \boldsymbol{C})$, we can use the Kronecker product to show that the equation

$$\boldsymbol{T}\boldsymbol{A} - \boldsymbol{D}\boldsymbol{T} = \boldsymbol{G}\boldsymbol{C}$$

can be uniquely solved for $\boldsymbol{T}$ if and only if no eigenvalue of $\boldsymbol{D}$ is also an eigenvalue of $\boldsymbol{A}$. Indeed, the above equation is a special case of the matrix equation (A.54) that we analyze in Section A.5. Note that when choosing the matrices $\boldsymbol{D}$ and $\boldsymbol{G}$, we must make sure that the condition (3.22) is satisfied. The structure of the resulting reduced-order estimator is shown in Figure 3.6. The last step in the control design process is to combine the control law that we constructed in the previous section with an estimator.

3.6 Combined Controller–Estimator Compensator

We start by combining a control law with a full-order estimator. We write the equations that govern the behavior of the dynamical system consisting of the plant model and the full-order estimator. The resulting system's order is $2n$, and the equations that describe the system are

$$\left.\begin{aligned} \begin{bmatrix} \dot{\boldsymbol{x}}(t) \\ \dot{\tilde{\boldsymbol{x}}}(t) \end{bmatrix} &= \begin{bmatrix} \boldsymbol{A} & \boldsymbol{O} \\ \boldsymbol{L}\boldsymbol{C} & \boldsymbol{A} - \boldsymbol{L}\boldsymbol{C} \end{bmatrix} \begin{bmatrix} \boldsymbol{x}(t) \\ \tilde{\boldsymbol{x}}(t) \end{bmatrix} + \begin{bmatrix} \boldsymbol{B} \\ \boldsymbol{B} \end{bmatrix} \boldsymbol{u}(t) \\ \boldsymbol{y}(t) &= [\,\boldsymbol{C} \quad \boldsymbol{O}\,] \begin{bmatrix} \boldsymbol{x}(t) \\ \tilde{\boldsymbol{x}}(t) \end{bmatrix} \end{aligned}\right\} \tag{3.23}$$

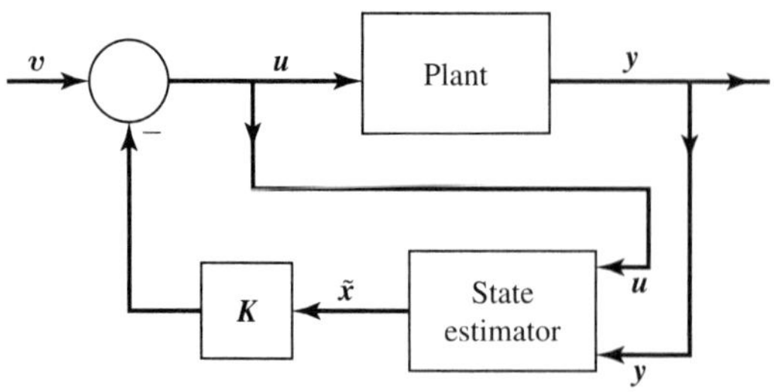

Figure 3.7 Closed-loop system driven by the combined controller–estimator compensator.

Suppose that we now apply the control law

$$\boldsymbol{u}(t) = -\boldsymbol{K}\tilde{\boldsymbol{x}}(t) + \boldsymbol{v}(t) \tag{3.24}$$

instead of the actual state-feedback control law. In the above, the vector $\boldsymbol{v}$ denotes an external input signal. A schematic of a closed-loop system driven by the combined controller-estimator compensator is shown in Figure 3.7. A model of the closed-loop system is obtained by combining equations (3.23) and (3.24):

$$\begin{bmatrix} \dot{\boldsymbol{x}}(t) \\ \dot{\tilde{\boldsymbol{x}}}(t) \end{bmatrix} = \begin{bmatrix} \boldsymbol{A} & -\boldsymbol{B}\boldsymbol{K} \\ \boldsymbol{L}\boldsymbol{C} & \boldsymbol{A} - \boldsymbol{L}\boldsymbol{C} - \boldsymbol{B}\boldsymbol{K} \end{bmatrix} \begin{bmatrix} \boldsymbol{x}(t) \\ \tilde{\boldsymbol{x}}(t) \end{bmatrix} + \begin{bmatrix} \boldsymbol{B} \\ \boldsymbol{B} \end{bmatrix} \boldsymbol{v}(t),$$

$$\boldsymbol{y}(t) = [\,\boldsymbol{C} \quad \boldsymbol{O}\,] \begin{bmatrix} \boldsymbol{x}(t) \\ \tilde{\boldsymbol{x}}(t) \end{bmatrix}.$$

To analyze the above closed-loop system, it is convenient to perform a change of state variables using the following transformation:

$$\begin{bmatrix} \boldsymbol{x} \\ \tilde{\boldsymbol{x}} - \boldsymbol{x} \end{bmatrix} = \begin{bmatrix} \boldsymbol{I}_n & \boldsymbol{O} \\ -\boldsymbol{I}_n & \boldsymbol{I}_n \end{bmatrix} \begin{bmatrix} \boldsymbol{x} \\ \tilde{\boldsymbol{x}} \end{bmatrix}.$$

Note that

$$\begin{bmatrix} \boldsymbol{I}_n & \boldsymbol{O} \\ -\boldsymbol{I}_n & \boldsymbol{I}_n \end{bmatrix}^{-1} = \begin{bmatrix} \boldsymbol{I}_n & \boldsymbol{O} \\ \boldsymbol{I}_n & \boldsymbol{I}_n \end{bmatrix}.$$

In the new coordinates the equations describing the closed-loop system are

$$\begin{bmatrix} \dot{\boldsymbol{x}}(t) \\ \dot{\tilde{\boldsymbol{x}}}(t) - \dot{\boldsymbol{x}}(t) \end{bmatrix} = \begin{bmatrix} \boldsymbol{A} - \boldsymbol{B}\boldsymbol{K} & -\boldsymbol{B}\boldsymbol{K} \\ \boldsymbol{O} & \boldsymbol{A} - \boldsymbol{L}\boldsymbol{C} \end{bmatrix} \begin{bmatrix} \boldsymbol{x}(t) \\ \tilde{\boldsymbol{x}}(t) - \boldsymbol{x}(t) \end{bmatrix} + \begin{bmatrix} \boldsymbol{B} \\ \boldsymbol{O} \end{bmatrix} \boldsymbol{v}(t),$$

$$\boldsymbol{y}(t) = [\,\boldsymbol{C} \quad \boldsymbol{O}\,] \begin{bmatrix} \boldsymbol{x}(t) \\ \tilde{\boldsymbol{x}}(t) - \boldsymbol{x}(t) \end{bmatrix}.$$

Note that in the above system the subsystem corresponding to the error component $\boldsymbol{e}(t) = \tilde{\boldsymbol{x}}(t) - \boldsymbol{x}(t)$ is uncontrollable. Furthermore, the $2n$ poles of the closed-loop system are equal to the individual eigenvalues of both $\boldsymbol{A} - \boldsymbol{B}\boldsymbol{K}$ and $\boldsymbol{A} - \boldsymbol{L}\boldsymbol{C}$, which means that the design of the control law is separated from the construction of the estimator. This is what is known as the *separation principle*. The closed-loop transfer function relating $\boldsymbol{Y}(s)$ and $\boldsymbol{V}(s)$ is

$$\begin{aligned}\boldsymbol{Y}(s) &= [\boldsymbol{C} \quad \boldsymbol{O}]\begin{bmatrix} s\boldsymbol{I}_n - \boldsymbol{A} + \boldsymbol{B}\boldsymbol{K} & \boldsymbol{B}\boldsymbol{K} \\ \boldsymbol{O} & s\boldsymbol{I}_n - \boldsymbol{A} + \boldsymbol{L}\boldsymbol{C}\end{bmatrix}^{-1}\begin{bmatrix}\boldsymbol{B} \\ \boldsymbol{O}\end{bmatrix}\boldsymbol{V}(s) \\ &= \boldsymbol{C}(s\boldsymbol{I}_n - \boldsymbol{A} + \boldsymbol{B}\boldsymbol{K})^{-1}\boldsymbol{B}\boldsymbol{V}(s).\end{aligned}$$

The above expression is identical to that for the closed-loop system if we applied the state-feedback control law $\boldsymbol{u}(t) = -\boldsymbol{K}\boldsymbol{x}(t) + \boldsymbol{v}(t)$. Thus, the combined estimator-controller compensator yields the same closed-loop transfer function as the actual state-feedback control law.

We now briefly discuss the issue of the estimator pole selection. Franklin, Powell, and Emami-Naeini [90, p. 552] recommend that the real parts of the estimator poles—that is, the real parts of the eigenvalues of the matrix $\boldsymbol{A} - \boldsymbol{L}\boldsymbol{C}$—be a factor of 2 to 6 times deeper in the open left-half plane than the real parts of the controller poles, which are the eigenvalues of the matrix $\boldsymbol{A} - \boldsymbol{B}\boldsymbol{K}$. Such a choice ensures a faster decay of the estimator error $\boldsymbol{e}(t) = \tilde{\boldsymbol{x}}(t) - \boldsymbol{x}(t)$ compared with the desired controller dynamics. This, in turn, causes the controller poles to dominate the closed-loop system response. Because the estimator poles represent a measure of the speed with which the error $\boldsymbol{e}(t) = \tilde{\boldsymbol{x}}(t) - \boldsymbol{x}(t)$ decays to zero, one would tend to assign estimator poles deep in the left-hand plane. However, fast decay requires large gains that may lead to saturation of some signals and unpredictable nonlinear effects. If the estimator poles were slower than the controller poles, the closed-loop system response would be dominated by the estimator, which is undesirable. As is usual in engineering practice, the term *compromise* can be used to describe the process of constructing the final compensator structure.

We now analyze the combined estimator–controller compensator that uses the reduced-order estimator. The closed-loop system driven by this type of compensator also enjoys the separation property as in the case of the full-order estimator. To prove the above statement, we begin with writing the equations describing the closed-loop system:

$$\begin{aligned}\dot{\boldsymbol{x}}(t) &= \boldsymbol{A}\boldsymbol{x}(t) + \boldsymbol{B}\boldsymbol{u}(t), \\ \dot{\boldsymbol{z}}(t) &= \boldsymbol{D}\boldsymbol{z}(t) + \boldsymbol{T}\boldsymbol{B}\boldsymbol{u}(t) + \boldsymbol{G}\boldsymbol{C}\boldsymbol{x}(t), \\ \boldsymbol{y}(t) &= \boldsymbol{C}\boldsymbol{x}(t), \\ \boldsymbol{u}(t) &= -\boldsymbol{K}\tilde{\boldsymbol{x}}(t) + \boldsymbol{v}(t),\end{aligned}$$

where

$$\tilde{\boldsymbol{x}} = \begin{bmatrix}\boldsymbol{T} \\ \boldsymbol{C}\end{bmatrix}^{-1}\begin{bmatrix}\boldsymbol{z} \\ \boldsymbol{y}\end{bmatrix}.$$

Let

$$\begin{bmatrix}\boldsymbol{T} \\ \boldsymbol{C}\end{bmatrix}^{-1} = [\boldsymbol{M} \quad \boldsymbol{N}],$$

where $\boldsymbol{M} \in \mathbb{R}^{n\times(n-p)}$ and $\boldsymbol{N} \in \mathbb{R}^{n\times p}$. Note that

$$\boldsymbol{M}\boldsymbol{T} + \boldsymbol{N}\boldsymbol{C} = \boldsymbol{I}_n.$$

Hence, the control input can be represented as

$$\begin{aligned} \boldsymbol{u}(t) &= -\boldsymbol{K}[\boldsymbol{M} \quad \boldsymbol{N}] \begin{bmatrix} z(t) \\ y(t) \end{bmatrix} + \boldsymbol{v}(t) \\ &= -\boldsymbol{K}\boldsymbol{M}z(t) - \boldsymbol{K}\boldsymbol{N}\boldsymbol{y}(t) + \boldsymbol{v}(t). \end{aligned}$$

Using the above equations and noting that $\boldsymbol{y}(t) = \boldsymbol{C}\boldsymbol{x}(t)$, we represent equations that describe the closed-loop system as

$$\begin{aligned} \begin{bmatrix} \dot{\boldsymbol{x}}(t) \\ \dot{z}(t) \end{bmatrix} &= \begin{bmatrix} \boldsymbol{A} & \boldsymbol{O} \\ \boldsymbol{G}\boldsymbol{C} & \boldsymbol{D} \end{bmatrix} \begin{bmatrix} \boldsymbol{x}(t) \\ z(t) \end{bmatrix} + \begin{bmatrix} \boldsymbol{B} \\ \boldsymbol{T}\boldsymbol{B} \end{bmatrix} (-\boldsymbol{K}\boldsymbol{M}z(t) - \boldsymbol{K}\boldsymbol{N}\boldsymbol{y}(t) + \boldsymbol{v}(t)) \\ &= \begin{bmatrix} \boldsymbol{A} - \boldsymbol{B}\boldsymbol{K}\boldsymbol{N}\boldsymbol{C} & -\boldsymbol{B}\boldsymbol{K}\boldsymbol{M} \\ \boldsymbol{G}\boldsymbol{C} - \boldsymbol{T}\boldsymbol{B}\boldsymbol{K}\boldsymbol{N}\boldsymbol{C} & \boldsymbol{D} - \boldsymbol{T}\boldsymbol{B}\boldsymbol{K}\boldsymbol{M} \end{bmatrix} \begin{bmatrix} \boldsymbol{x}(t) \\ z(t) \end{bmatrix} + \begin{bmatrix} \boldsymbol{B} \\ \boldsymbol{T}\boldsymbol{B} \end{bmatrix} \boldsymbol{v}(t), \\ y(t) &= [\boldsymbol{C} \quad \boldsymbol{O}] \begin{bmatrix} \boldsymbol{x}(t) \\ z(t) \end{bmatrix}. \end{aligned}$$

To proceed further, we perform a transformation of the state variable of the above system:

$$\begin{bmatrix} \boldsymbol{x} \\ z - \boldsymbol{T}\boldsymbol{x} \end{bmatrix} = \begin{bmatrix} \boldsymbol{I}_n & \boldsymbol{O} \\ -\boldsymbol{T} & \boldsymbol{I}_{n-p} \end{bmatrix} \begin{bmatrix} \boldsymbol{x} \\ z \end{bmatrix}.$$

Note that

$$\begin{bmatrix} \boldsymbol{I}_n & \boldsymbol{O} \\ -\boldsymbol{T} & \boldsymbol{I}_{n-p} \end{bmatrix}^{-1} = \begin{bmatrix} \boldsymbol{I}_n & \boldsymbol{O} \\ \boldsymbol{T} & \boldsymbol{I}_{n-p} \end{bmatrix}.$$

In the new coordinates the equations describing the closed-loop system become

$$\begin{aligned} \begin{bmatrix} \dot{\boldsymbol{x}}(t) \\ \dot{z}(t) - \boldsymbol{T}\dot{\boldsymbol{x}}(t) \end{bmatrix} &= \begin{bmatrix} \boldsymbol{A} - \boldsymbol{B}\boldsymbol{K}\boldsymbol{N}\boldsymbol{C} - \boldsymbol{B}\boldsymbol{K}\boldsymbol{M}\boldsymbol{T} & -\boldsymbol{B}\boldsymbol{K}\boldsymbol{M} \\ -\boldsymbol{T}\boldsymbol{A} + \boldsymbol{G}\boldsymbol{C} + \boldsymbol{D}\boldsymbol{T} & \boldsymbol{D} \end{bmatrix} \begin{bmatrix} \boldsymbol{x}(t) \\ z(t) - \boldsymbol{T}\boldsymbol{x}(t) \end{bmatrix} + \begin{bmatrix} \boldsymbol{B} \\ \boldsymbol{O} \end{bmatrix} \boldsymbol{v}(t) \\ y(t) &= [\boldsymbol{C} \quad \boldsymbol{O}] \begin{bmatrix} \boldsymbol{x}(t) \\ z(t) - \boldsymbol{T}\boldsymbol{x}(t) \end{bmatrix}. \end{aligned}$$

Taking into account the equations

$$\boldsymbol{T}\boldsymbol{A} - \boldsymbol{D}\boldsymbol{T} - \boldsymbol{G}\boldsymbol{C} = \boldsymbol{0} \quad \text{and} \quad \boldsymbol{M}\boldsymbol{T} = \boldsymbol{I}_n - \boldsymbol{N}\boldsymbol{C},$$

we obtain

$$\begin{aligned} \begin{bmatrix} \dot{\boldsymbol{x}}(t) \\ \dot{z}(t) - \boldsymbol{T}\dot{\boldsymbol{x}}(t) \end{bmatrix} &= \begin{bmatrix} \boldsymbol{A} - \boldsymbol{B}\boldsymbol{K} & -\boldsymbol{B}\boldsymbol{K}\boldsymbol{M} \\ \boldsymbol{O} & \boldsymbol{D} \end{bmatrix} \begin{bmatrix} \boldsymbol{x}(t) \\ z(t) - \boldsymbol{T}\boldsymbol{x}(t) \end{bmatrix} + \begin{bmatrix} \boldsymbol{B} \\ \boldsymbol{O} \end{bmatrix} \boldsymbol{v}(t) \\ y(t) &= [\boldsymbol{C} \quad \boldsymbol{O}] \begin{bmatrix} \boldsymbol{x}(t) \\ z(t) - \boldsymbol{T}\boldsymbol{x}(t) \end{bmatrix}. \end{aligned}$$

It is now apparent that the separation principle also holds if we use the reduced-order estimator in constructing the compensator; that is, the selection of the gain matrix $\boldsymbol{K}$ that assigns closed-loop poles is independent, separate from the selection of the estimator matrix $\boldsymbol{D}$ that assigns the estimator poles. Furthermore, the transfer function matrix relating $\boldsymbol{Y}(s)$ and $\boldsymbol{V}(s)$ in the closed-loop system with the reduced-order estimator is exactly the same as the one that would be obtained using the actual state-feedback control law $\boldsymbol{u}(t) = -\boldsymbol{K}\boldsymbol{x}(t) + \boldsymbol{v}(t)$.

◆ Example 3.12

Consider a schematic of an armature-controlled DC motor system, shown in Figure 3.8, where the system parameters are: $R_a = 5\,\Omega$, $L_a = 200\,\text{mH}$, $K_b =$ 0.1 V/rad/sec, $K_i = 0.1$ Nm/A, the gear ratio $N_1/N_2 = 1/50$. The armature inertia is $I_{armature} = 2 \times 10^{-3}\ \text{kg}\cdot\text{m}^2$. The load of 10 kg is located at an effective radius of 0.2 m. The gear inertia and friction are negligible.

We first construct the state-space model of the DC motor system. We start with writing the equation relating torque to angular acceleration:

$$T_m = I_{eq}\ddot{\theta}, \tag{3.25}$$

where

$$\begin{aligned} I_{eq} &= I_{armature} + (N_1/N_2)^2 I_l \\ &= 2 \times 10^{-3}\ \text{kg}\cdot\text{m}^2 + (1/50)^2 \times 10 \times (0.2)^2\ \text{kg}\cdot\text{m}^2 \\ &= 2.16 \times 10^{-3}\ \text{kg}\cdot\text{m}^2. \end{aligned}$$

Kirchhoff's voltage law applied to the armature circuit gives

$$R_a i_a + L_a \frac{di_a}{dt} + e_b = e_a,$$

where $e_b = K_b \frac{d\theta}{dt}$. The equation for the developed torque is

$$T_m = K_i i_a. \tag{3.26}$$

Combining (3.25) and (3.26) gives

$$\ddot{\theta} = \frac{K_i}{I_{eq}} i_a.$$

Let $x_1 = i_a$, $x_2 = \theta$, $x_3 = \dot{\theta} = \omega$, $u = e_a$, and $y = \theta_l$. Taking into account the definition of the state variables, we represent the above modeling equations in

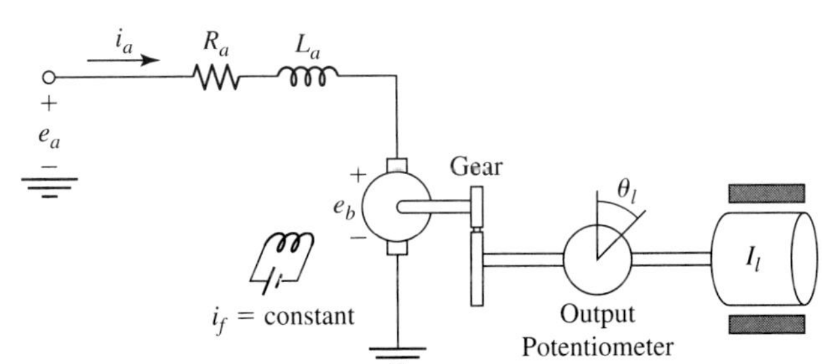

Figure 3.8 Schematic of an armature-controlled DC motor system of Example 3.12.

state-space format:

$$\begin{bmatrix} \dot{x}_1 \\ \dot{x}_2 \\ \dot{x}_3 \end{bmatrix} = \begin{bmatrix} -\frac{R_a}{L_a} & 0 & -\frac{K_b}{L_a} \\ 0 & 0 & 1 \\ \frac{K_i}{I_{eq}} & 0 & 0 \end{bmatrix} \begin{bmatrix} x_1 \\ x_2 \\ x_3 \end{bmatrix} + \begin{bmatrix} \frac{1}{L_a} \\ 0 \\ 0 \end{bmatrix} u$$

$$y = \begin{bmatrix} 0 & \frac{N_1}{N_2} & 0 \end{bmatrix} \begin{bmatrix} x_1 \\ x_2 \\ x_3 \end{bmatrix}.$$

Substituting the given parameter values, we get

$$\begin{bmatrix} \dot{x}_1 \\ \dot{x}_2 \\ \dot{x}_3 \end{bmatrix} = \begin{bmatrix} -25 & 0 & -0.5 \\ 0 & 0 & 1 \\ 46.296 & 0 & 0 \end{bmatrix} \begin{bmatrix} x_1 \\ x_2 \\ x_3 \end{bmatrix} + \begin{bmatrix} 5 \\ 0 \\ 0 \end{bmatrix} u = \boldsymbol{A}\boldsymbol{x} + \boldsymbol{b}u,$$

$$y = [0 \quad 0.02 \quad 0] \begin{bmatrix} x_1 \\ x_2 \\ x_3 \end{bmatrix} = \boldsymbol{c}\boldsymbol{x}.$$

Our next step is to design a state-feedback controller $u = -\boldsymbol{k}\boldsymbol{x} + v$, such that the eigenvalues of $\boldsymbol{A} - \boldsymbol{b}\boldsymbol{k}$ are

$$-1 + j, \quad -1 - j, \quad -10.$$

To allocate the poles into the desired locations, we first transform the system model into the controller companion form. We first form the controllability matrix:

$$[\boldsymbol{b} \quad \boldsymbol{A}\boldsymbol{b} \quad \boldsymbol{A}^2\boldsymbol{b}] = \begin{bmatrix} 5 & -125 & 3009.3 \\ 0 & 0 & 231.5 \\ 0 & 231.5 & -5787 \end{bmatrix}.$$

The system is controllable. The last row of the inverse of the controllability matrix is

$$\boldsymbol{q}_1 = [0 \quad 0.0043 \quad 0],$$

and hence the transformation matrix we are seeking has the form

$$\boldsymbol{T} = \begin{bmatrix} \boldsymbol{q}_1 \\ \boldsymbol{q}_1\boldsymbol{A} \\ \boldsymbol{q}_1\boldsymbol{A}^2 \end{bmatrix} = \begin{bmatrix} 0 & 0.0043 & 0 \\ 0 & 0 & 0.0043 \\ 0.2 & 0 & 0 \end{bmatrix}.$$

The system's matrices in the new coordinates are

$$\tilde{\boldsymbol{A}} = \boldsymbol{T}\boldsymbol{A}\boldsymbol{T}^{-1} = \begin{bmatrix} 0 & 1 & 0 \\ 0 & 0 & 1 \\ 0 & -23.148 & -25 \end{bmatrix},$$

$$\tilde{\boldsymbol{b}} = \boldsymbol{T}\boldsymbol{b} = \begin{bmatrix} 0 \\ 0 \\ 1 \end{bmatrix},$$

and

$$\tilde{\boldsymbol{c}} = \boldsymbol{c}\boldsymbol{T}^{-1} = [4.63 \quad 0 \quad 0].$$

The desired closed-loop characteristic polynomial $\alpha_c(s)$ is

$$\alpha_s(s) = (s+1-j)(s+1+j)(s+10) = s^3 + 12s^2 + 22s + 20.$$

We now find $\tilde{\boldsymbol{k}}$ so that

$$\tilde{\boldsymbol{A}} - \tilde{\boldsymbol{b}}\tilde{\boldsymbol{k}} = \begin{bmatrix} 0 & 1 & 0 \\ 0 & 0 & 1 \\ -20 & -22 & -12 \end{bmatrix}.$$

We have

$$\tilde{\boldsymbol{k}} = [\,20 \quad -1.148 \quad -13\,]$$

and

$$\boldsymbol{k} = \tilde{\boldsymbol{k}}\boldsymbol{T} = [\,-2.6 \quad 0.0864 \quad -0.005\,].$$

Hence,

$$\boldsymbol{A} - \boldsymbol{b}\boldsymbol{k} = \begin{bmatrix} -12 & -0.432 & -0.4752 \\ 0 & 0 & 1 \\ 46.294 & 0 & 0 \end{bmatrix}.$$

The transfer function of the closed-loop system is

$$\frac{\mathcal{L}(y(t))}{\mathcal{L}(v(t))} = \frac{Y(s)}{V(s)} = \boldsymbol{c}(s\boldsymbol{I} - \boldsymbol{A} + \boldsymbol{b}\boldsymbol{k})^{-1}\boldsymbol{b} = \frac{4.6296}{s^3 + 12s^2 + 22s + 20}.$$

The unit-step response of the closed-loop system is shown in Figure 3.9.

Next, we design a full-order state estimator placing the estimator poles at $\{-4, -5+2j, -5-2j\}$ and then synthesize the combined controller–estimator

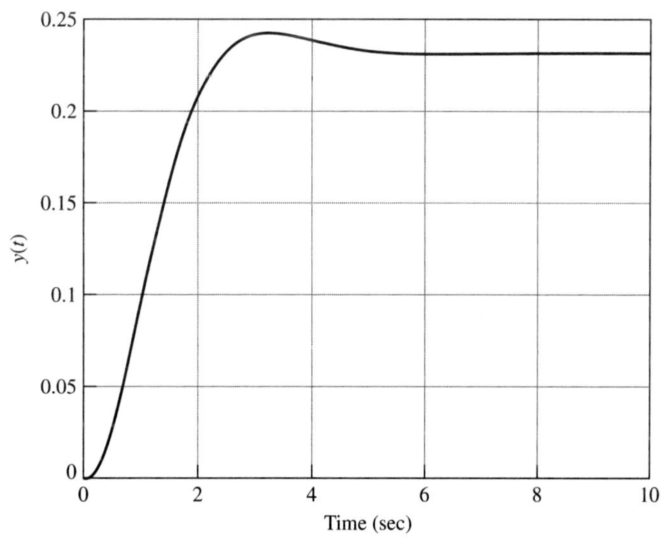

Figure 3.9 Step response of the closed-loop DC motor system with state feedback.

compensator. To compute the estimator gain vector $\boldsymbol{l}$, we first transform the pair $(\boldsymbol{A}, \boldsymbol{c})$ into the observer companion form. This is equivalent to transforming the pair $(\boldsymbol{A}^T, \boldsymbol{c}^T)$ into the controller companion form. The transformation matrix that brings the pair $(\boldsymbol{A}, \boldsymbol{c})$ into its observer companion form is

$$\hat{\boldsymbol{Q}} = \hat{\boldsymbol{T}}^T = \begin{bmatrix} 1.1 & -27 & 650 \\ 0 & 0 & 50 \\ 0 & 50 & -1250 \end{bmatrix}.$$

The matrices $\boldsymbol{A}$ and $\boldsymbol{c}$ in the new coordinates have the form

$$\hat{\boldsymbol{A}} = \hat{\boldsymbol{Q}}^{-1}\boldsymbol{A}\hat{\boldsymbol{Q}} = \begin{bmatrix} 0 & 0 & 0 \\ 1 & 0 & -23.1481 \\ 0 & 1 & -25 \end{bmatrix} \quad \text{and} \quad \hat{\boldsymbol{c}} = \boldsymbol{c}\hat{\boldsymbol{Q}} = [0 \quad 0 \quad 1].$$

The desired characteristic polynomial of the estimator is

$$\det(\boldsymbol{I}_3 - \boldsymbol{A} + \boldsymbol{l}\boldsymbol{c}) = (s+4)(s+5-2j)(s+5+2j) = s^3 + 14s^2 + 69s + 116.$$

We now find $\hat{\boldsymbol{l}}$ so that

$$\hat{\boldsymbol{A}} - \hat{\boldsymbol{l}}\hat{\boldsymbol{c}} = \begin{bmatrix} 0 & 0 & -116 \\ 1 & 0 & -69 \\ 0 & 1 & -14 \end{bmatrix}.$$

We have

$$\hat{\boldsymbol{l}} = \begin{bmatrix} 116 \\ 45.8519 \\ -11 \end{bmatrix} \quad \text{and therefore} \quad \boldsymbol{l} = \hat{\boldsymbol{Q}}\hat{\boldsymbol{l}} = \begin{bmatrix} -8263 \\ -550 \\ 16043 \end{bmatrix}.$$

The dynamics of the full-order state estimator are described by

$$\dot{\tilde{\boldsymbol{x}}} = (\boldsymbol{A} - \boldsymbol{l}\boldsymbol{c})\tilde{\boldsymbol{x}} + \boldsymbol{b}u + \boldsymbol{l}y;$$

that is,

$$\dot{\tilde{\boldsymbol{x}}} = \begin{bmatrix} -25 & 165.2544 & -0.5 \\ 0 & 11 & 1 \\ 46.2963 & -320.8519 & 0 \end{bmatrix}\tilde{\boldsymbol{x}} + \begin{bmatrix} 5 \\ 0 \\ 0 \end{bmatrix}u + \begin{bmatrix} -8263 \\ -550 \\ 16043 \end{bmatrix}y.$$

We connect the full-order estimator to the DC motor system, thus obtaining a closed-loop system with the combined controller–full-order-estimator compensator in the loop as in Figure 3.7. In Figure 3.10 we show a plot of x_1 and its estimate for the closed-loop system with the estimator in the loop, where $v = 0$, $\boldsymbol{x}(0) = [1 \;\; 0.2 \;\; -0.1]^T$, and $\tilde{\boldsymbol{x}}(0) = \boldsymbol{0}$.

Our last part of this example is to design the reduced-order state estimator. We choose the reduced-order estimator poles at $\{-5+2j, -5-2j\}$. The reduced-order estimator dynamics are given by

$$\dot{\boldsymbol{z}} = \boldsymbol{D}\boldsymbol{z} + \boldsymbol{E}u + \boldsymbol{G}y.$$

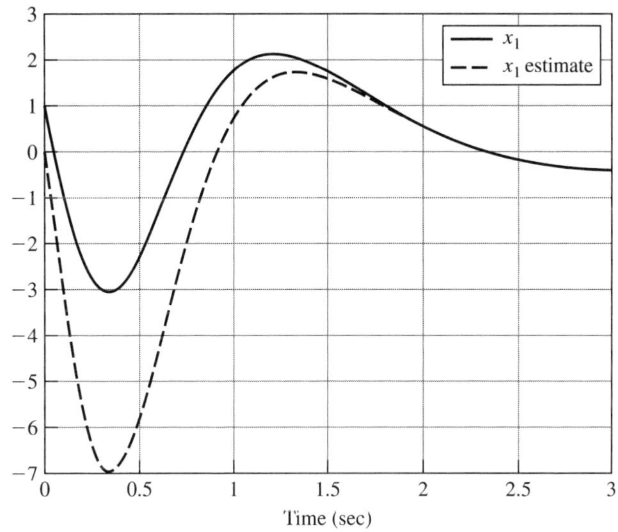

Figure 3.10 A plot of x_1 and its estimate versus time for the closed-loop system with the full-order estimator in the loop.

Let

$$\boldsymbol{G} = \begin{bmatrix} 100 \\ 100 \end{bmatrix}.$$

The estimator's poles are supposed to be $-5 + 2j$ and $-5 - 2j$. Thus, the desired characteristic polynomial of $\boldsymbol{D}$ should be

$$\det(s\boldsymbol{I}_2 - \boldsymbol{D}) = s^2 + 10s + 29.$$

Let

$$\boldsymbol{D} = \begin{bmatrix} 0 & 1 \\ -29 & -10 \end{bmatrix}.$$

We now solve the matrix equation,

$$-\boldsymbol{DT} + \boldsymbol{TA} = \boldsymbol{G}c,$$

for $\boldsymbol{T}$. We first convert the above equation in the format similar to that given by (A.55) in Section A.5 and then solve the resulting equation. Employing the following MATLAB's commands,

```
Gc=G*c;
t=(-kron(eye(3),D)+kron(A',eye(2)))\Gc(:);
T=[t(1:2) t(3:4) t(5:6)]
```

we obtain

$$\boldsymbol{T} = \begin{bmatrix} -0.5493 & 0.7586 & -0.2507 \\ 2.1271 & -2.0000 & 1.0332 \end{bmatrix}.$$

Note that

$$\det \begin{bmatrix} \boldsymbol{T} \\ \boldsymbol{c} \end{bmatrix} = 6.8691 \times 10^{-4} \neq 0$$

and

$$\begin{bmatrix} \boldsymbol{T} \\ \boldsymbol{c} \end{bmatrix}^{-1} = \begin{bmatrix} -30.0840 & -7.2980 & 411.3200 \\ 0 & 0 & 50 \\ 61.9338 & 15.9921 & -750 \end{bmatrix}.$$

We also have

$$\boldsymbol{E} = \boldsymbol{T}\boldsymbol{b} = \begin{bmatrix} -2.7463 \\ 10.6357 \end{bmatrix}.$$

Hence, the dynamics of the reduced-order estimator are described by

$$\dot{\boldsymbol{z}} = \begin{bmatrix} 0 & 1 \\ -29 & -10 \end{bmatrix} \boldsymbol{z} + \begin{bmatrix} -2.7463 \\ 10.6357 \end{bmatrix} u + \begin{bmatrix} 100 \\ 100 \end{bmatrix} y.$$

An estimate of the entire state can be obtained from

$$\hat{\boldsymbol{x}} = \begin{bmatrix} \boldsymbol{T} \\ \boldsymbol{c} \end{bmatrix}^{-1} \begin{bmatrix} \boldsymbol{z} \\ y \end{bmatrix} = \begin{bmatrix} -30.0840 & -7.2980 & 411.3200 \\ 0 & 0 & 50 \\ 61.9338 & 15.9921 & -750 \end{bmatrix} \begin{bmatrix} \boldsymbol{z} \\ y \end{bmatrix}.$$

A plot of x_1 and its estimate versus time for the closed-loop system with the combined controller–reduced-order-estimator compensator is depicted in Figure 3.11, where $v = 0$, $\boldsymbol{x}(0) = [1 \;\; 0.2 \;\; -0.1]^T$, and $\tilde{\boldsymbol{z}}(0) = \boldsymbol{0}$. Of course, for zero initial

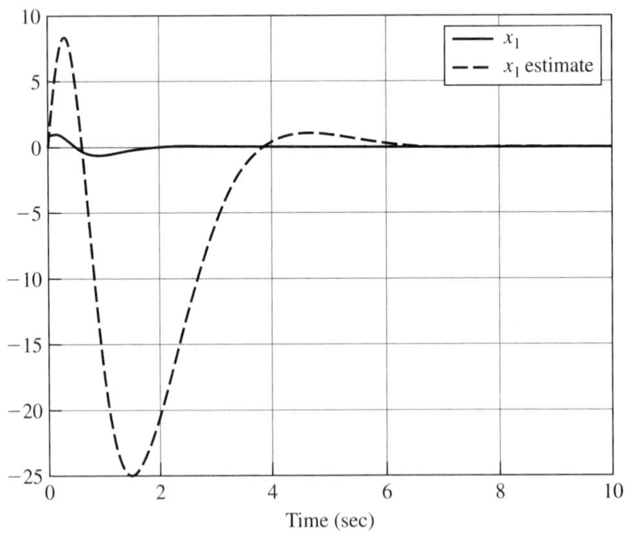

Figure 3.11 A plot of x_1 and its estimate versus time for the closed-loop system with the reduced-order estimator in the loop.

conditions, the step responses of the closed-loop systems with the full-order as well as with the reduced-order estimators in the loop are the same as the step response of the closed-loop system with the state-feedback controller $u = -\boldsymbol{k}\boldsymbol{x} + v$. This follows from our analyses of the combined controller–estimator compensators.

Notes

For further reading on the subject of linear system theory, see Brockett [34], Wolovich [301], Barnett [20], Kailath [146], Skelton [261], DeCarlo [59], Rugh [245], Antsaklis and Michel [12], or Chen [48]. For more information on canonical forms, see the papers by Popov [236] and Denham [62].

One may argue that Maxwell's [197] analysis of Watt's governor constitutes the first work in linear system theory: the physical object, its linear model, and a method of analysis. For example, MacFarlane [192, p. 251] states that "Maxwell's fundamentally important contribution lay in recognizing that the behavior of an automatic feedback control system in the vicinity of an equilibrium condition could be approximated by a linear differential equation, and hence the stability of the control system could be discussed in terms of the location of the roots of an associated algebraic equation." On the other hand, L. Markus in Elworthy et al. [76, p. xxvii] makes the following statement: "But certainly Watt's application of the centrifugal governor for regulating steam engines (1775) can be taken as the start of modern control engineering."

In Kalman's [147] well-known 1963 paper, there is a theorem which states that the state space, $\mathcal{X}$, of a linear system model may be decomposed as a direct sum:

$$\mathcal{X} = \mathcal{X}_A \oplus \mathcal{X}_B \oplus \mathcal{X}_C \oplus \mathcal{X}_D,$$

where

$$\begin{aligned}
\mathcal{X}_A &= \text{states that are reachable but not observable,}\\
\mathcal{X}_B &= \text{states that are reachable and observable,}\\
\mathcal{X}_C &= \text{states that are neither reachable nor observable,}\\
\mathcal{X}_D &= \text{states that are not reachable but observable.}
\end{aligned}$$

This decomposition became, to a certain extent, an icon of linear systems. For example, the logo on the covers of the old issues of the *International Journal of Systems Science* was the above decomposition. Many authors included Kalman's original decomposition algorithm in their textbooks on linear systems. Yet almost twenty years after Kalman's publication of the decomposition theorem, Boley [29] generated an example that revealed some shortcomings of Kalman's original algorithm. In his reply, Kalman [148] revised his decomposition algorithm and fixed his original proof of the decomposition theorem.

EXERCISES

3.1 (Adapted from Barnett [20, p. 94]) Consider the system, shown in Figure 3.12, which is composed of two masses moving on a smooth—that is, frictionless—horizontal plane.

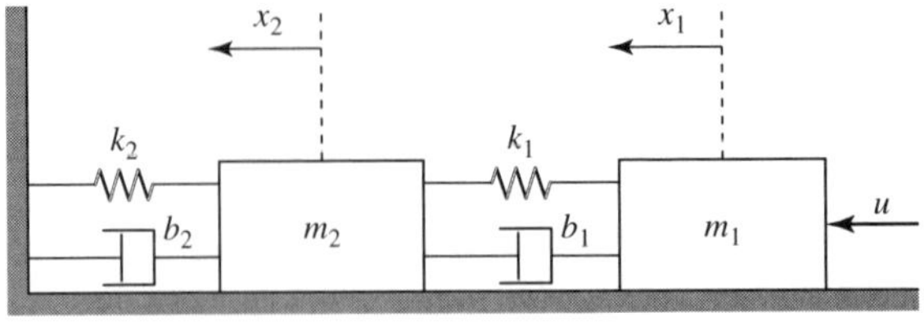

Figure 3.12 A mechanical system for Exercise 3.1.

The springs are linear; that is, they obey Hooke's law:

$$\text{force} = k \times \text{extension},$$

where k is a constant called the force constant of the spring. In addition to the springs, the masses are joined by linear dampers. In a linear damper, the force is proportional to the velocity of one end of the damper relative to the other end. The proportionality constant is called the damping coefficient. In this exercise, damping coefficients are b_1 and b_2, respectively. A force u is applied to the right end mass. Obtain the system's state-space model in the form $\dot{\boldsymbol{x}} = \boldsymbol{A}\boldsymbol{x} + \boldsymbol{b}u$. Determine under what condition, in terms of k_1, the system model is controllable for $m_1 = m_2 = 1$ kg, $b_1 = b_2 = 1$ Nsec/m, and $k_2 = 1/4$ N/m.

3.2 In the circuit shown in Figure 3.13, take the state variables to be the voltage across the capacitor and the current through the inductor. The source voltage v_s is the control variable, and the current i is the system output. Obtain a state-space model of this circuit. Obtain the condition, in terms of R_1, R_2, C, and L, under which the model is uncontrollable and unobservable. Find the transfer function of the circuit model.

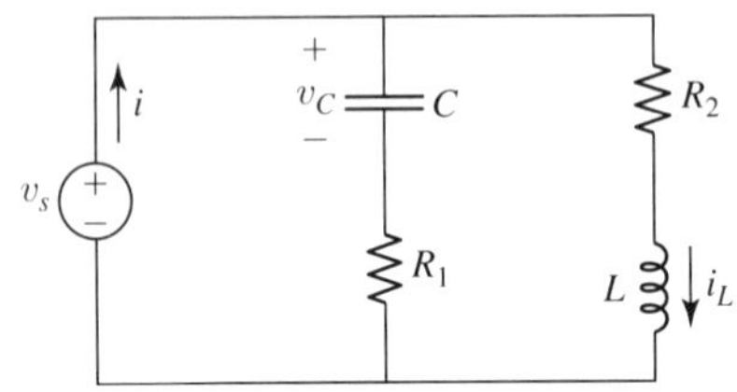

Figure 3.13 Electric circuit for Exercise 3.2.

3.3 In the circuit shown in Figure 3.14, take the state variables to be the voltage across the capacitor C_1, the current through the inductor L_1, and the current through the inductor L_2. The source current i_s is the control variable, and the current through the inductor

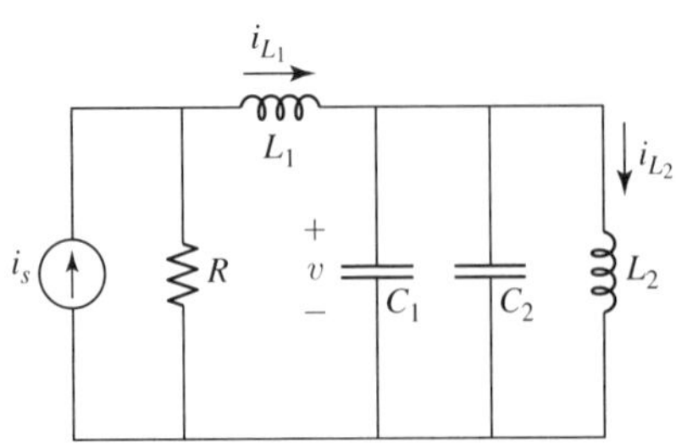

Figure 3.14 Electric circuit for Exercise 3.3.

L_2 is the system output. Obtain a state-space model of this circuit. Analyze the circuit model controllability and observability properties. Find the transfer function of the circuit model.

3.4 For the system

$$\dot{x} = x + u,$$

find the set of points that can be steered to the origin $x = 0$ by an admissible control $|u| \leq 1$.

3.5 (a) Find e^{At} for

$$A = \begin{bmatrix} 0 & 1 \\ -ab & -a-b \end{bmatrix}.$$

(b) Given a dynamical system modeled by the differential equation

$$\ddot{x} + 2\dot{x} + x = \dot{r} + r,$$

where $r = r(t) = \sin t$, and the initial conditions are $x(0) = 1$ and $\dot{x}(0) = 0$, choose the state variables $x_1 = x$ and $x_2 = \dot{x}_1$. Represent the differential equation in state-space format. Solve for $x(t)$ for $t \geq 0$.

In both cases, verify your calculations using MATLAB's Symbolic Math Toolbox.

3.6 Find sufficient conditions, in terms of a, b, c, and d, for reachability of the following pairs:

(a)

$$A = \begin{bmatrix} -1 & 1 & 0 & 0 & 0 \\ 0 & -1 & 1 & 0 & 0 \\ 0 & 0 & -1 & 0 & 0 \\ 0 & 0 & 0 & -2 & 1 \\ 0 & 0 & 0 & 0 & -2 \end{bmatrix}, \quad B = \begin{bmatrix} 0 & 0 \\ 0 & 0 \\ a & b \\ 0 & 0 \\ c & d \end{bmatrix};$$

(b)

$$A = \begin{bmatrix} -1 & 1 & 0 & 0 & 0 \\ 0 & -1 & 1 & 0 & 0 \\ 0 & 0 & -1 & 0 & 0 \\ 0 & 0 & 0 & -1 & 1 \\ 0 & 0 & 0 & 0 & -1 \end{bmatrix}, \quad B = \begin{bmatrix} 0 & 0 \\ 0 & 0 \\ a & b \\ 0 & 0 \\ c & d \end{bmatrix}.$$

3.7 A system composed of two subsystems connected in cascade is depicted in the block diagram shown in Figure 3.15. Take the state variables of the system to be $x_1 = y_1 - u_1$ and $x_2 = y_2$. Assume that the system input is $u = u_1$ and the system output is $y = y_2$. Obtain the state-space model of the interconnected system. When is the system uncontrollable? Find the transfer function of the uncontrollable system.

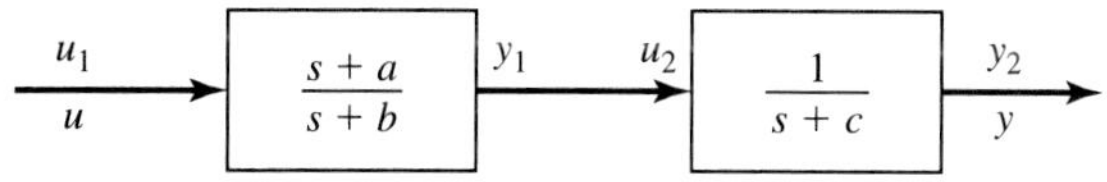

Figure 3.15 A block diagram for Exercise 3.7.

3.8 Derive a state-space description of a system represented by the block diagram shown in Figure 3.16. Construct a state-variable transformation $z = Tx$ such that the matrix TAT^{-1} has a diagonal form. Find the eigenvalue of TAT^{-1} corresponding to the noncontrollable state, and find the eigenvalue corresponding to the nonobservable state. Find the transfer function of the system.

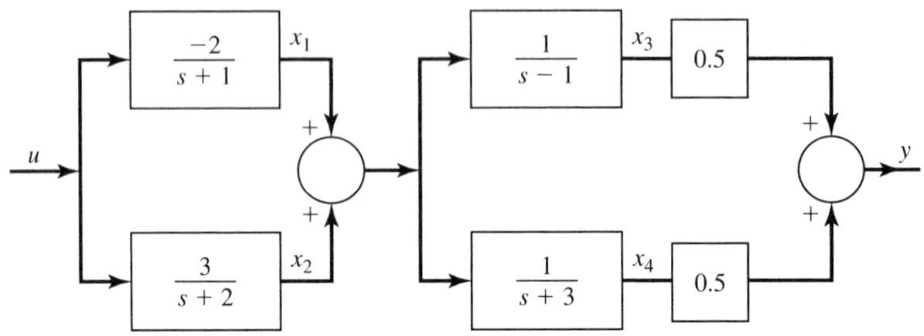

Figure 3.16 A block diagram for Exercise 3.8.

3.9 (Adapted from Barnett [20, p. 95]) A simple representation of a vehicle suspension system is shown in Figure 3.17. Assume that the forces produced by the springs and dampers act at the end points P and Q and that x_1 and x_2 are the displacements of these points from equilibrium. The force exerted by a damper is proportional to the velocity of one end of the damper relative to the other end, and the springs obey Hooke's law. Assume that the mass of the platform can be neglected so that the spring motions can be regarded as independent. The control force $4u$ is applied one-quarter of the way from one end of the platform. Derive the two-dimensional state-space model of this simple representation of a vehicle suspension system. Next, if the ends of the platform are each given an initial displacement of 10 units, find a control law that returns the platform to equilibrium at $t = 1$ sec.

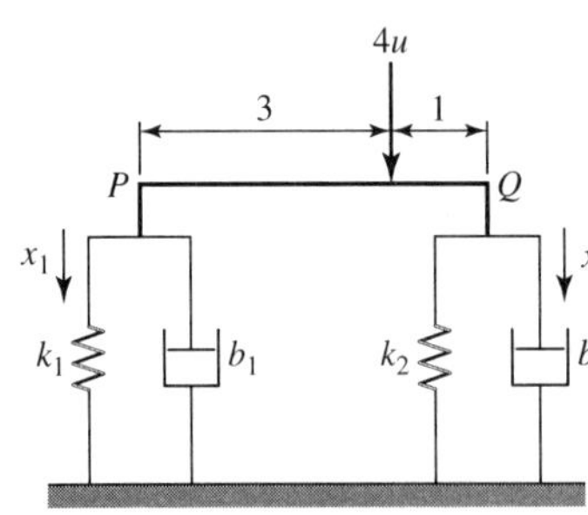

Figure 3.17 A simple representation of a vehicle suspension system of Exercise 3.9.

3.10 For the dynamical system model, $\dot{x} = Ax + bu$, where

$$A = \begin{bmatrix} 0 & 1 \\ 1 & 0 \end{bmatrix} \quad \text{and} \quad b = \begin{bmatrix} 0 \\ 1 \end{bmatrix},$$

design a state-feedback controller that stabilizes the system about the operating state corresponding to the constant input $u = 2$. The desired closed-loop poles are $-2, -3$.

3.11 Consider the vehicle model given in Subsection 1.7.2. Let the input signal be $u = \delta_f$ (the front-wheel steer angle) and $\delta_r = 0$. Compute the transfer function $C(s)/U(s)$. Plot the vehicle response Y versus X for X changing from 0 m to 100 m, when the input δ_f, in the vehicle coordinates, has the form $\delta_f(t) = 0.0872\, h(t)$ rad, where $h(t)$ is the unit step function.

Suppose now that the plot of δ_f versus time, in the car coordinates, has the form shown in Figure 3.18. Note that the values of δ_f are given in deg. The value of δ_f changes form -2.5 deg to 2.5 deg. In your simulations you will need to convert deg to rad. Plot the vehicle response Y versus X for X changing from 0 m to 100 m.

Construct the state-feedback controller $u = -\boldsymbol{kx} + r$ such that $\text{eig}(\boldsymbol{A} - \boldsymbol{bk}) = \{-2, -3 + j, -3 - j, -4\}$. Plot u and r versus time, and Y versus X, when $r = 0.0872\, h(t)$ rad.

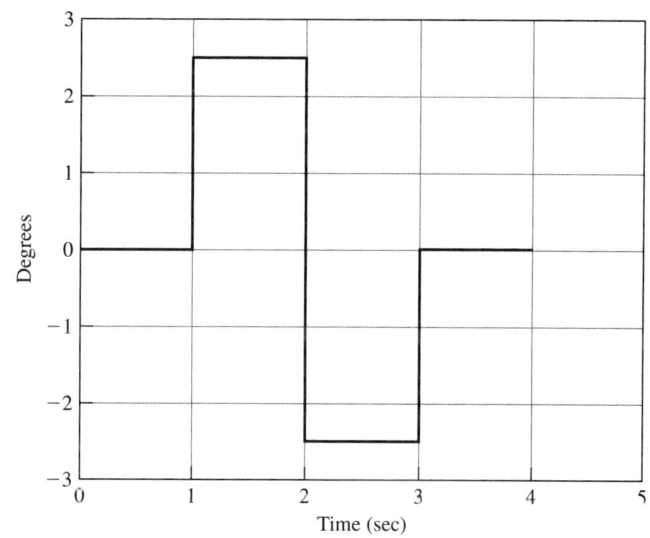

Figure 3.18 Time history of the front steer angle in the car coordinates of Exercise 3.11.

3.12 Show that the controllability property is invariant under the following set of transformations:

(a) Nonsingular coordinate transformation in the state space

(b) Nonsingular transformation of the inputs

(c) State feedback

(d) Output feedback

3.13 Given the following model of a dynamical system,

$$\dot{\boldsymbol{x}}(t) = \begin{bmatrix} 0 & 2 \\ 1 & 0 \end{bmatrix} \boldsymbol{x}(t) + \begin{bmatrix} 1 \\ 0 \end{bmatrix} u(t),$$

$$y(t) = [\,0 \quad 1\,]\boldsymbol{x}(t),$$

do the following:

(a) Design a full-order asymptotic state estimator for the above system with the estimator poles located at -2 and -3. Write the equations of the estimator dynamics.

Note that the pair $(\boldsymbol{A}, \boldsymbol{c})$ is already in the observer form, which should facilitate your estimator design.

(b) Denote the estimator state vector by z. Let $u = -[2 \;\; 3]z + v$. Find the transfer function of the closed-loop system $Y(s)/V(s)$.

3.14 Let

$$\dot{\boldsymbol{x}} = \boldsymbol{A}\boldsymbol{x} + \boldsymbol{b}u = \begin{bmatrix} -1 & 1 \\ 0 & 2 \end{bmatrix} \boldsymbol{x} + \begin{bmatrix} 1 \\ 1 \end{bmatrix} u,$$

$$y = \boldsymbol{c}\boldsymbol{x} = [1 \quad 0]\boldsymbol{x}.$$

(a) Find the state transformation that transforms the pair $(\boldsymbol{A}, \boldsymbol{c})$ into the observer companion form.

(b) Find the representation of the dynamical system in the new coordinates.

3.15 For the following model of a dynamical system,

$$\dot{\boldsymbol{x}} = \begin{bmatrix} 0 & 1 \\ 2 & 0 \end{bmatrix} \boldsymbol{x} + \begin{bmatrix} 1 \\ 0 \end{bmatrix} u,$$

$$y = [1 \quad 0]\boldsymbol{x} + 2u,$$

do the following:

(a) Design a full-order asymptotic state estimator with the estimator poles located at -3 and -4. Write the equations of the estimator dynamics.

(b) Denote the estimator state by z. Let $u = -[3 \;\; 2]\, z + r$. Find the transfer function, $Y(s)/R(s)$, of the closed-loop system.

3.16 Given a dynamical system modeled by

$$\dot{\boldsymbol{x}} = \boldsymbol{A}\boldsymbol{x} + \boldsymbol{b}u,$$

$$y = \boldsymbol{c}\boldsymbol{x},$$

where

$$\boldsymbol{A} = \begin{bmatrix} 1 & -3 \\ 4 & 2 \end{bmatrix}, \qquad \boldsymbol{b} = \begin{bmatrix} 1 \\ 1 \end{bmatrix}, \qquad \boldsymbol{c} = [2 \quad 3],$$

do the following:

(a) Design a reduced-order state estimator

$$\dot{z} = dz + eu + gy,$$

where $d = -5$ and $g = 1$.

(b) Write an expression for the estimated states.

3.17 For the plant

$$\begin{bmatrix} \dot{x}_1 \\ \dot{x}_2 \end{bmatrix} = \begin{bmatrix} 1 & -1 \\ 1 & -2 \end{bmatrix} \begin{bmatrix} x_1 \\ x_2 \end{bmatrix} + \begin{bmatrix} 2 \\ 1 \end{bmatrix} u$$

$$y = [0 \quad 1] \begin{bmatrix} x_1 \\ x_2 \end{bmatrix},$$

do the following:

(a) Design the reduced-order state estimator for $\boldsymbol{G} = [1]$ and $\boldsymbol{D} = [-2]$.

(b) Write the equations of the estimated state in terms of the plant's output and the estimator state.

3.18 Consider a dynamical system that consists of a cart with an inverted pendulum (point mass on a massless shaft) attached to it as depicted in Figure 3.19.

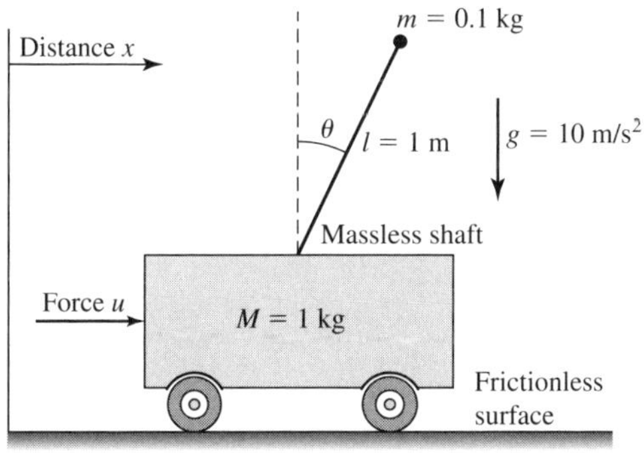

Figure 3.19 An inverted pendulum moving on a cart of Exercise 3.18.

(a) Obtain a nonlinear state-space model of the system using the state variables

$$x_1 = x, \qquad x_2 = \dot{x}, \qquad x_3 = \theta, \qquad x_4 = \dot{\theta}.$$

Designate x_1 as the system output.

(b) Linearize the model about the operating point $\boldsymbol{x} = \boldsymbol{0}$, $u = 0$.

(c) Design a state feedback control law $u = -\boldsymbol{k}\boldsymbol{x}$ such that the closed-loop poles are located at

$$-1, \quad -2, \quad -1 \pm j$$

(d) Write a simple MATLAB script that animates the motion of the linearized closed-loop model of the inverted pendulum on the cart along with animation of the displacement x and the angle θ data. The basic handle graphics of MATLAB can be used to write the script. A script contained in Example 4 in Pratap [238, p. 194] can be helpful here. Alternatively, you can employ SIMULINK.

3.19 Consider the nonlinear model of the one-link robot manipulator of Example 5.13. In the derivation, assume that $L_a = 100$ mH and $g = 9.8$ m/s^2. Represent the obtained third-order model in the form $\dot{\boldsymbol{x}} = \boldsymbol{f}(\boldsymbol{x}, u)$, where $x_1 = \theta_p$, $x_2 = \dot{\theta}_p$, and $x_3 = i_a$.

(a) Find the constant value of $u = u_e$ and x_{3e} so that the state $\boldsymbol{x}_e = [30\pi/180 \;\; 0 \;\; x_{3e}]^T$ becomes the equilibrium state; that is, the obtained $\boldsymbol{x}_e$ and u_e satisfy the equation

$$\boldsymbol{0} = \boldsymbol{f}(\boldsymbol{x}_e, u_e).$$

(b) Linearize the nonlinear model, $\dot{\boldsymbol{x}} = \boldsymbol{f}(\boldsymbol{x}, u)$, about the operating point $[\boldsymbol{x}_e^T \;\; u_e]^T$ to obtain

$$\frac{d}{dt}\Delta\boldsymbol{x} = \boldsymbol{A}\Delta\boldsymbol{x} + \boldsymbol{b}\Delta u,$$

where

$$\Delta \boldsymbol{x} = \boldsymbol{x} - \boldsymbol{x}_e, \qquad \Delta u = u - u_e,$$

and

$$\boldsymbol{A} = \frac{\partial \boldsymbol{f}}{\partial \boldsymbol{x}}(\boldsymbol{x}_e, u_e) \quad \text{and} \quad \boldsymbol{b} = \frac{\partial \boldsymbol{f}}{\partial u}(\boldsymbol{x}_e, u_e).$$

(c) Design a state feedback control law $\Delta u = -\boldsymbol{k}\Delta\boldsymbol{x}$ such that the closed-loop poles are located at

$$-2, \quad -2 \pm j$$

(d) Implement the obtained control law on the nonlinear model $\dot{\boldsymbol{x}} = \boldsymbol{f}(\boldsymbol{x}, u)$. Write a MATLAB script that animates the dynamical behavior of the nonlinear closed-loop system along with animation of the angle θ_p and the input u data.

(e) Design a full-order state estimator for the linearized model. The estimator's poles should be located at

$$-5, \quad -5 \pm j$$

(f) Implement the combined controller–estimator compensator on the nonlinear model and write a MATLAB script that animates the dynamical behavior of the nonlinear closed-loop system along with animation of the angle θ_p and the input u data.

Stability

> Examine each question in terms of what is ethically and aesthetically right, as well as what is economically expedient. A thing is right when it tends to preserve the integrity, stability, and beauty of the biotic community. It is wrong when it tends otherwise.
>
> —Aldo Leopold

4.1 Informal Introduction to Stability

In general, it can be difficult to obtain an explicit solution of nonlinear differential equations modeling a dynamical system. However, often we do not need to know the explicit solution of the modeling equations. Instead, we are interested in the behavior of the solution as time tends to infinity. Furthermore, in many models of physical systems a "small" change in the initial conditions results in a "small" change in the solution. Specifically, if a system is perturbed from its rest, or equilibrium position, then it starts to move. We can roughly say that the equilibrium position is stable if the system does not go far from this position for small initial perturbations. This concept may be illustrated, as in Willems [300], Brogan [35], or Barnett [20], by means of a ball resting in equilibrium on a sheet of metal bent into various shapes with cross sections depicted in Figure 4.1. This figure illustrates four possible scenarios. If we neglect frictional forces, a small perturbation from the equilibrium leads to the following:

1. Oscillatory motion about equilibrium irrespective of the magnitude of the initial perturbation. In this case the equilibrium position is globally stable.
2. The new position is also an equilibrium, and the ball will remain in it. Here the equilibrium position is stable.
3. The ball moving away without returning to the equilibrium. This equilibrium position is unstable.
4. Oscillatory motion about the equilibrium, unless the initial perturbation is so large that the ball is forced to oscillate about the new equilibrium position. This equilibrium position is locally stable.

If friction is taken into account, then the oscillatory motion steadily decreases until the ball returns to the equilibrium positions as in items 1 and 4 above. We refer to such equilibrium positions as being asymptotically stable. In our discussion of stability concepts, we implicitly

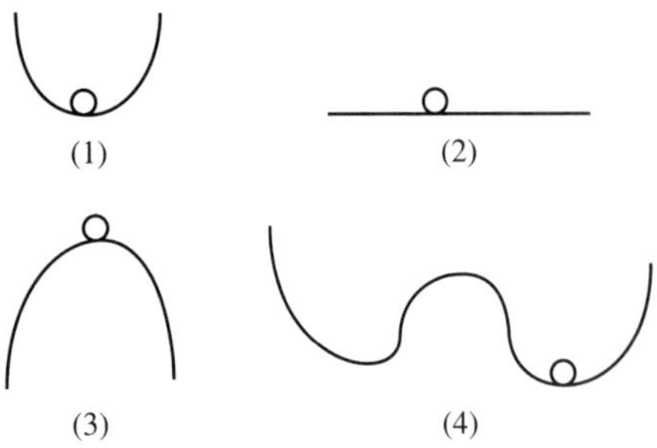

Figure 4.1 Illustration of stable and unstable equilibrium positions.

assume, as in item 4, that there is some "small" region for allowed initial perturbations for which subsequent motions converge to the equilibrium position. If however, as in item 1, any initial perturbation yields a motion that converges to the equilibrium position, then we have global asymptotic stability, which is also referred to as asymptotic stability in the large.

The above informal discussion should serve as a preparation for an in-depth analysis of stability of dynamical systems. We begin our analysis by formally defining the equilibrium state. We then define stability, asymptotic stability, and other relevant notions.

4.2 Basic Definitions of Stability

We consider a class of dynamical systems modeled by the equation:

$$\dot{\boldsymbol{x}}(t) = \boldsymbol{f}(t, \boldsymbol{x}(t)), \qquad \boldsymbol{x}(t_0) = \boldsymbol{x}_0, \tag{4.1}$$

where $\boldsymbol{x}(t) \in \mathbb{R}^n$ is the state vector and $\boldsymbol{f} : \mathbb{R} \times \mathbb{R}^n \to \mathbb{R}$ is a vector-valued function with the components

$$f_i(t, x_1, x_2, \ldots, x_n) : \mathbb{R} \times \mathbb{R}^n \to \mathbb{R}, \qquad i = 1, 2, \ldots, n.$$

We assume that f_i's are continuous and have continuous first partial derivatives. Under these assumptions, the solution to $\dot{\boldsymbol{x}}(t) = \boldsymbol{f}(t, \boldsymbol{x}(t))$, $\boldsymbol{x}(t_0) = \boldsymbol{x}_0$, exists and is unique. We denote this solution as $\boldsymbol{x}(t) = \boldsymbol{x}(t; t_0, \boldsymbol{x}_0)$. If the f_i's do not depend explicitly on t, then the system represented by $\dot{\boldsymbol{x}}(t) = \boldsymbol{f}(t, \boldsymbol{x}(t)) = \boldsymbol{f}(\boldsymbol{x}(t))$ is called *autonomous;* otherwise it is called *nonautonomous*. An *equilibrium point* or state is a constant vector, say $\boldsymbol{x}_e$, such that

$$\boldsymbol{f}(t, \boldsymbol{x}_e) = \boldsymbol{0} \qquad \text{for all } t.$$

Observe that an equilibrium state $\boldsymbol{x}_e$ is a constant solution to (4.1).

If $\boldsymbol{x}_e$ is an equilibrium point, we can arrange for this equilibrium point to be transferred to the origin of $\mathbb{R}^n$ by introducing the new variable:

$$\tilde{\boldsymbol{x}} = \boldsymbol{x} - \boldsymbol{x}_e.$$

We assume that this has been done for an equilibrium point under consideration. Therefore, we have

$$\boldsymbol{f}(t, \boldsymbol{0}) = \boldsymbol{0} \qquad \text{for all } t.$$

We now introduce basic definitions of stability in the sense of A. M. Lyapunov* (1857–1918). In the following considerations we use the standard Euclidean norm of a vector. Thus, if $\boldsymbol{x} \in \mathbb{R}^n$, then

$$\|\boldsymbol{x}\| = \|\boldsymbol{x}\|_2 = (\boldsymbol{x}^T\boldsymbol{x})^{1/2}.$$

Definition 4.1 An equilibrium state $\boldsymbol{x}_e$ is said to be stable if for any given t_0 and any positive scalar ε, there exists a positive scalar $\delta = \delta(t_0, \varepsilon)$ such that if $\|\boldsymbol{x}(t_0) - \boldsymbol{x}_e\| < \delta$, then $\|\boldsymbol{x}(t; t_0, \boldsymbol{x}_0) - \boldsymbol{x}_e\| < \varepsilon$ for all $t \geq t_0$.

Friedland [91, p. 43] makes the following comment regarding the above definition of stability: "If this 'epsilonics' definition leaves you cold, think of a contest between you, the system designer, and an adversary (nature?). The adversary picks a region in state space of radius ε and challenges you to find another region, of radius δ, such that if the initial state starts out inside your region, it remains inside his region. If you can do this, your system design is stable." We add that you have to find an appropriate δ for any choice of ε for the equilibrium state to be stable.

Definition 4.2 An equilibrium state $\boldsymbol{x}_e$ is said to be *convergent*, or *attractive*, if for any given t_0 there is a positive scalar $\delta_1 = \delta_1(t_0)$ such that if $\|\boldsymbol{x}(t_0) - \boldsymbol{x}_e\| < \delta_1$, then

$$\lim_{t\to\infty} \boldsymbol{x}(t; t_0, \boldsymbol{x}_0) = \boldsymbol{x}_e.$$

We can restate Definition 4.2 as follows. An equilibrium state $\boldsymbol{x}_e$ is convergent if for any t_0 there is a positive scalar $\delta_1 = \delta_1(t_0)$ such that if $\|\boldsymbol{x}(t_0) - \boldsymbol{x}_e\| < \delta_1$, then for any given $\varepsilon_1 > 0$ there exists a $T = T(\varepsilon_1, t_0, \boldsymbol{x}_0)$ such that $\|\boldsymbol{x}(t; t_0, \boldsymbol{x}_0) - \boldsymbol{x}_e\| < \varepsilon_1$ for all $t > t_0 + T$.

Definition 4.3 An equilibrium state $\boldsymbol{x}_e$ is asymptotically stable if it is stable and convergent.

Definition 4.4 An equilibrium state $\boldsymbol{x}_e$ is said to be uniformly stable if the positive scalar $\delta = \delta(t_0, \varepsilon)$ introduced in Definition 4.1 can be taken independent of the initial time t_0.

◆ Example 4.1

The following example that comes from Rugh [245, p. 100] nicely illustrates the notion of uniform stability. The solution to the differential equation,

$$\dot{x}(t) = (4t\sin t - 2t)x(t), \qquad x(t_0) = x_0, \tag{4.2}$$

is

$$x(t) = x_0 \exp\left(4\sin t - 4t\cos t - t^2 - 4\sin t_0 + 4t_0\cos t_0 + t_0^2\right). \tag{4.3}$$

One can verify this using MATLAB's Symbolic Toolbox by typing the command

```
x=dsolve('Dx=(4*t*sin(t)-2*t)*x, x(t_0)=x_0','t')
```

It is easy to show that for fixed t_0 there is a γ such that (4.3) is bounded by $\gamma|x_0|$ for all $t \geq t_0$. This is because the term $(-t^2)$ dominates the exponent as t increases. Therefore, for a given ε, it is enough to choose x_0 so that $\gamma|x_0| < \varepsilon$, or equivalently $\delta < \varepsilon/\gamma$, to show that the equilibrium solution $x_e = 0$ is stable. However, the

*Sometimes transliterated in the literature as Liapunov or Liapunoff.

equilibrium solution $x_e = 0$ is not uniformly stable. To show this, fix initial state x_0 and consider a sequence of initial times $t_0 = 2k\pi$, where $k = 0, 1, \ldots$, and the values of the respective solutions at times π units later:

$$x(2k\pi + \pi) = x_0 \exp((4k + 1)\pi(4 - \pi)).$$

One can see that there is no bound on the exponential factor and hence a candidate bound γ increases as k increases, where increasing k corresponds to the increasing initial time.

Definition 4.5 An equilibrium state $\boldsymbol{x}_e$ is said to be uniformly convergent, or uniformly attractive, if the positive scalars $\delta_1 = \delta_1(t_0)$ and $T = T(\varepsilon_1, t_0, \boldsymbol{x}_0)$ introduced in Definition 4.2 are independent of t_0.

The last two definitions allow us to introduce the notion of uniform asymptotic stability.

Definition 4.6 An equilibrium state $\boldsymbol{x}_e$ is said to be uniformly asymptotically stable if it is uniformly stable and uniformly convergent.

Notice that the solution $\boldsymbol{x}(t) = \boldsymbol{x}_e$ must first be stable in order to qualify to be asymptotically stable. This is to prevent a system trajectory from straying arbitrarily far from the equilibrium state before converging toward it. Note also that the concepts of Lyapunov stability and asymptotic stability are local concepts. They apply to the system's behavior in a small neighborhood about the equilibrium point (Hsu and Meyer [128]).

A special case of uniform asymptotic stability is *uniform exponential stability* that implies uniform stability, and it imposes an additional requirement that all solutions originating "close" to the given equilibrium, $\boldsymbol{x}_e = \boldsymbol{0}$, approach this equilibrium exponentially as $t \to \infty$. More rigorously, we say that $\boldsymbol{x}_e = \boldsymbol{0}$ is uniformly exponentially stable if it is stable and there exist finite positive constants γ and λ such that for any t_0 and $\boldsymbol{x}_0$, any solution starting in a neighborhood of $\boldsymbol{0}$ satisfies

$$\|\boldsymbol{x}(t)\| \leq \gamma e^{-\lambda(t-t_0)} \|\boldsymbol{x}_0\|.$$

Note that γ is no less than unity, and uniform implies that γ and λ are independent of t_0 (see Figure 4.2).

An equilibrium state is unstable if it is not stable. An ε–δ definition of instability is given next.

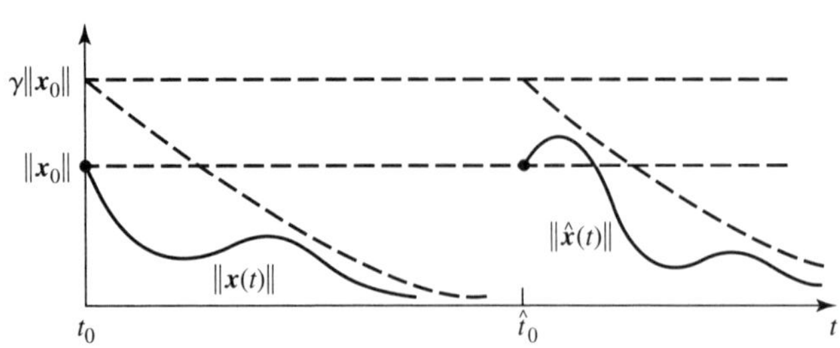

Figure 4.2 Illustration of uniform exponential stability: A decaying-exponential bound is independent of the initial time t_0.

Definition 4.7 An equilibrium state $\boldsymbol{x}_e$ is said to be unstable if there is an $\varepsilon > 0$ so that for any $\delta > 0$ there exists $\boldsymbol{x}(t_0)$ such that if $\|\boldsymbol{x}(t_0) - \boldsymbol{x}_e\| < \delta$, then $\|\boldsymbol{x}(t_1) - \boldsymbol{x}_e\| \geq \varepsilon$ for some $t_1 > t_0$.

A two-dimensional geometric interpretation of the above definitions is shown in Figure 4.3. We assume that $\boldsymbol{x}_e = 0$ is the equilibrium state. If the equilibrium state is stable, as in Figure 4.3(a), then given an outer circle of radius ε, there exists an inner circle of radius δ such that trajectories starting within the δ-circle never leave the ε-circle. If the equilibrium state is asymptotically stable, then trajectories tend to the equilibrium state as t tends to ∞. This situation is depicted in Figure 4.3(b). Finally, if the equilibrium state is unstable, as in Figure 4.3(c), then there exists an ε-circle such that for every δ-circle there exists a trajectory starting within it that leaves the ε-circle at some later time. A geometric interpretation of the above described scenarios in one dimension is given in Figure 4.4.

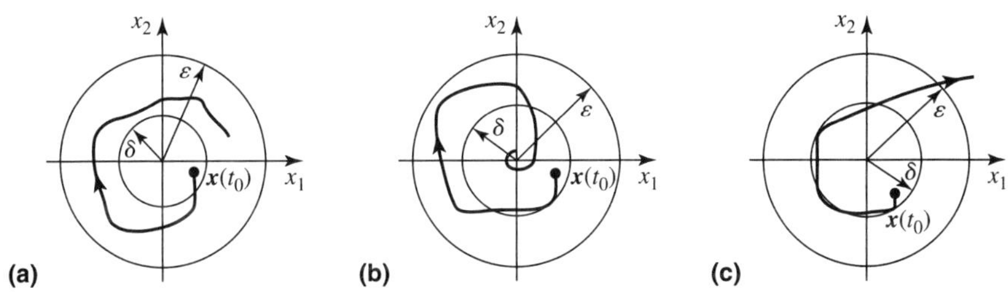

Figure 4.3 Illustration of stability concepts in two dimensions. (a) Stable equilibrium state in the sense of Lyapunov. (b) Asymptotically stable equilibrium state. (c) Unstable equilibrium state.

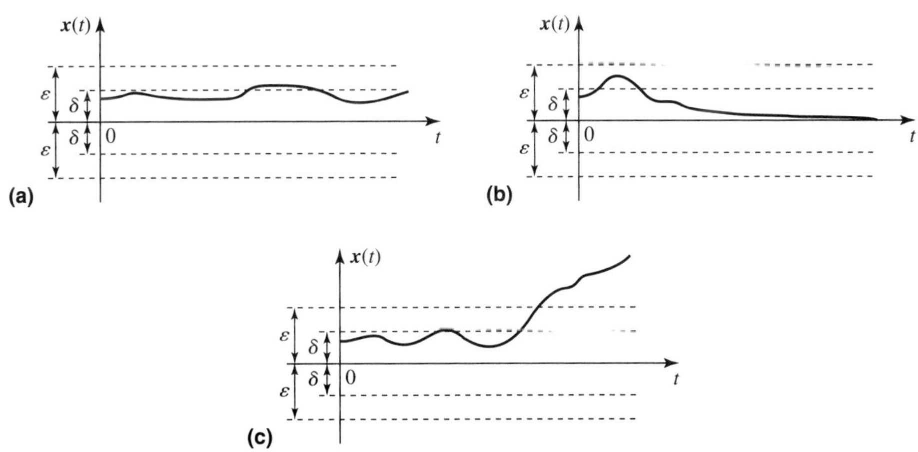

Figure 4.4 Illustration of stability concepts in one dimension. (a) Stable equilibrium state in the sense of Lyapunov. (b) Asymptotically stable equilibrium state. (c) Unstable equilibrium state.

4.3 Stability of Linear Systems

It follows from the discussion in the previous sections that a linear time-invariant dynamical system modeled by the equations

$$\dot{\boldsymbol{x}} = \boldsymbol{A}\boldsymbol{x}, \qquad \boldsymbol{x}(t_0) = \boldsymbol{x}_0 \tag{4.4}$$

is asymptotically stable if and only if the solution to equation (4.4) decays to zero as $t \to \infty$ for any initial state $\boldsymbol{x}_0$. We can view the vector $\boldsymbol{x}(t) \in \mathbb{R}^n$ as defining the coordinates of a point moving in an n-dimensional state space $\mathbb{R}^n$. In an asymptotically stable system modeled by equation (4.4), this point converges to the origin of $\mathbb{R}^n$. We can argue that if a trajectory is converging to the origin of the state space, then it should be possible to find a family of nested surfaces described by $V(x_1, x_2, \ldots, x_n) = c$, $c \geq 0$, such that monotonically decreasing values of c give surfaces progressively shrinking in on the origin with the limiting surface $V(x_1, x_2, \ldots, x_n) = 0$ being the origin $\boldsymbol{x} = \boldsymbol{0}$. Furthermore, along every trajectory of (4.4), the parameter c should steadily decrease. We now introduce some notation and definitions that we use in the ensuing discussion.

Let $V(\boldsymbol{x}) = V(x_1, x_2, \ldots, x_n) : \mathbb{R}^n \to \mathbb{R}$ be a real-valued function and let $\mathcal{S} \subseteq \mathbb{R}^n$ be a compact region containing the origin $\boldsymbol{x} = \boldsymbol{0}$ in its interior.

Definition 4.8 We say that the function $V = V(\boldsymbol{x})$ is *positive semidefinite* (p.s.d.) in $\mathcal{S}$, with respect to $\boldsymbol{x} = \boldsymbol{0}$, if:

1. V is continuously differentiable, that is, $V \in \mathcal{C}^1$.
2. $V(\boldsymbol{0}) = 0$.
3. $V(\boldsymbol{x}) \geq 0$ for all $\boldsymbol{x} \in \mathcal{S}$.

Definition 4.9 We say that the function $V = V(\boldsymbol{x})$ is *positive definite* (p.d.) in $\mathcal{S}$, with respect to $\boldsymbol{x} = \boldsymbol{0}$, if:

1. $V \in \mathcal{C}^1$.
2. $V(\boldsymbol{0}) = 0$.
3. $V(\boldsymbol{x}) > 0$ for all $\boldsymbol{x} \in \mathcal{S} \setminus \{\boldsymbol{0}\}$.

The symbol $\mathcal{S} \setminus \{\boldsymbol{0}\}$ in item 3 above denotes the set $\{\boldsymbol{x} : \boldsymbol{x} \in \mathcal{S}, \boldsymbol{x} \neq \boldsymbol{0}\}$.

Negative semidefinite (n.s.d.) and *negative definite* (n.d.) functions are defined in a similar fashion by reversing inequality signs in item 3 in both definitions above.

There are many functions that satisfy the above definitions. In the stability analysis of dynamical systems, one class is particularly useful. This is the positive definite quadratic form, a label attached to a nested family of ellipses and their extensions to higher-dimensional spaces—ellipsoids for $n = 3$ and hyperellipsoids for $n \geq 4$. Quadratic forms are reviewed in Section A.4. We consider the quadratic form

$$V(\boldsymbol{x}) = \boldsymbol{x}^T \boldsymbol{P} \boldsymbol{x},$$

where $\boldsymbol{P}$ is a real symmetric $n \times n$ matrix. The time derivative of $V(\boldsymbol{x}(t))$ evaluated on a solution of $\dot{\boldsymbol{x}} = \boldsymbol{A}\boldsymbol{x}$ is

$$\begin{aligned}\frac{dV(\boldsymbol{x}(t))}{dt} &= \dot{V}(t)\\ &= \dot{\boldsymbol{x}}^T(t)\boldsymbol{P}\boldsymbol{x}(t) + \boldsymbol{x}^T(t)\boldsymbol{P}\dot{\boldsymbol{x}}(t).\end{aligned}$$

Substituting $\dot{x} = Ax$ and $\dot{x}^T = x^T A^T$ into the above equation yields

$$\begin{aligned} \dot{V} &= x^T A^T P x + x^T P A x \\ &= x^T (A^T P + PA) x. \end{aligned}$$

Let

$$Q = -(A^T P + PA).$$

Then,

$$\dot{V} = -x^T Q x. \tag{4.5}$$

We are now ready to state and prove the following theorem concerning the stability of the system $\dot{x} = Ax$, or, equivalently, the matrix $A \in \mathbb{R}^n$.

Theorem 4.1 (A. M. Lyapunov, 1892) The system $\dot{x} = Ax$, $x(t_0) = x_0$, is asymptotically stable if and only if for any given real symmetric positive definite (r.s.p.d.) matrix Q the solution P to the continuous Lyapunov matrix equation,

$$A^T P + PA = -Q \tag{4.6}$$

is also (real symmetric) positive definite.

Proof ($\Rightarrow$) We prove necessity, for asymptotic stability, by contradiction. We have $-x^T Qx < 0$ for all $x \neq 0$. We assume that A is an asymptotically stable matrix and that for some nonzero x_0 we have $x_0^T P x_0 \le 0$. From the discussion before the theorem, and in particular from (4.5), it follows that

$$\frac{d}{dt}(x^T(t) P x(t)) = -x^T(t) Q x(t) < 0.$$

The above means that $V(x(t)) = x^T(t) P x(t)$ decreases on the trajectories of the system $\dot{x}(t) = Ax(t)$ as $t \to \infty$. On the other hand, because A is asymptotically stable, we have

$$\lim_{t\to\infty} x(t) = 0.$$

Hence,

$$\lim_{t\to\infty} x^T(t) P x(t) = \lim_{x(t)\to 0} x^T(t) P x(t) = 0.$$

Thus, we arrived at a contradiction. This completes the proof of the first part of the theorem that states that asymptotic stability of A implies that for any r.s.p.d matrix Q the solution P to the equation $A^T P + PA = -Q$ is r.s.p.d..

($\Leftarrow$) We also prove sufficiency by contradiction. We assume that

$$x^T P x > 0 \quad \text{and} \quad -x^T Q x < 0 \qquad \text{for all } x \neq 0,$$

and that the matrix A is not asymptotically stable. This means that the solution to $\dot{x}(t) = Ax(t)$, $x_0 \neq 0$, which has the form $x(t) = e^{A(t-t_0)} x_0$, is bounded away from the origin $x = 0$. Thus, there is a constant $b > 0$ such that

$$x^T(t) Q x(t) > b > 0 \qquad \text{for all } t > t_0.$$

Integrating $\frac{d}{dt}(\boldsymbol{x}^T(t)\,\boldsymbol{P}\boldsymbol{x}(t)) = -\boldsymbol{x}^T(t)\,\boldsymbol{Q}\boldsymbol{x}(t)$ yields

$$\boldsymbol{x}^T(t)\,\boldsymbol{P}\boldsymbol{x}(t) = \boldsymbol{x}^T(t_0)\,\boldsymbol{P}\boldsymbol{x}(t_0) + \int_{t_0}^{t}(-\boldsymbol{x}^T(s)\,\boldsymbol{Q}\boldsymbol{x}(s))\,ds.$$

Taking into account that $-\boldsymbol{x}^T(t)\,\boldsymbol{Q}\boldsymbol{x}(t) < -b < 0$ for all $t > t_0$, we obtain for $t > t_0$,

$$\begin{aligned}\boldsymbol{x}^T(t)\,\boldsymbol{P}\boldsymbol{x}(t) &= \boldsymbol{x}^T(t_0)\,\boldsymbol{P}\boldsymbol{x}(t_0) + \int_{t_0}^{t}(-\boldsymbol{x}^T(s)\,\boldsymbol{Q}\boldsymbol{x}(s))\,ds \\ &< \boldsymbol{x}^T(t_0)\,\boldsymbol{P}\boldsymbol{x}(t_0) - b(t - t_0).\end{aligned}$$

The above implies that for a sufficiently large t we would have $\boldsymbol{x}^T(t)\,\boldsymbol{P}\boldsymbol{x}(t) < 0$, which contradicts the assumption that $\boldsymbol{x}^T(t)\,\boldsymbol{P}\boldsymbol{x}(t) > 0$. This means that $\boldsymbol{A}$ must be asymptotically stable, and the proof is complete.

◆ Example 4.2

Let

$$\boldsymbol{A} = \begin{bmatrix} 0 & 1 \\ -1 & -2 \end{bmatrix},$$

and $\boldsymbol{Q} = \boldsymbol{I}_2$. We solve the Lyapunov equation $\boldsymbol{A}^T\boldsymbol{P} + \boldsymbol{P}\boldsymbol{A} = -\boldsymbol{Q}$ for $\boldsymbol{P}$. Assume $\boldsymbol{P}$ to have the form

$$\boldsymbol{P} = \begin{bmatrix} p_1 & p_2 \\ p_2 & p_3 \end{bmatrix} = \boldsymbol{P}^T.$$

Then, we have

$$\begin{aligned}\boldsymbol{A}^T\boldsymbol{P} + \boldsymbol{P}\boldsymbol{A} &= \begin{bmatrix} 0 & -1 \\ 1 & -2 \end{bmatrix}\begin{bmatrix} p_1 & p_2 \\ p_2 & p_3 \end{bmatrix} + \begin{bmatrix} p_1 & p_2 \\ p_2 & p_3 \end{bmatrix}\begin{bmatrix} 0 & 1 \\ -1 & -2 \end{bmatrix} \\ &= \begin{bmatrix} -2p_2 & -p_3 + p_1 - 2p_2 \\ -p_3 + p_1 - 2p_2 & 2p_2 - 4p_3 \end{bmatrix} \\ &= -\boldsymbol{Q} = \begin{bmatrix} -1 & 0 \\ 0 & -1 \end{bmatrix}.\end{aligned}$$

Solving the corresponding linear equations yields

$$\boldsymbol{P} = \begin{bmatrix} 3/2 & 1/2 \\ 1/2 & 1/2 \end{bmatrix}.$$

The solution matrix $\boldsymbol{P}$ is positive definite. Hence, the matrix $\boldsymbol{A}$ must have its eigenvalues in the open left-half plane. Indeed, the eigenvalues of $\boldsymbol{A}$ are both located at -1.

Definition 4.10 A positive definite function $V(\boldsymbol{x})$ whose time derivative evaluated on the solutions of the system $\dot{\boldsymbol{x}} = \boldsymbol{A}\boldsymbol{x}$ is negative definite (or negative semidefinite) is called a Lyapunov function for this system.

Lyapunov's theorem gives us a necessary and sufficient condition for asymptotic stability of linear time-invariant dynamical systems. To check if a given matrix $\boldsymbol{A}$ is asymptotically stable or not, it is enough to take an arbitrary r.s.p.d. matrix $\boldsymbol{Q}$, solve the equation $\boldsymbol{A}^T\boldsymbol{P} + \boldsymbol{P}\boldsymbol{A} = -\boldsymbol{Q}$ for real symmetric $\boldsymbol{P}$, and check if the obtained $\boldsymbol{P}$ is positive definite, or not. If $\boldsymbol{P} \succ 0$, then $\boldsymbol{A}$ is asymptotically stable. If $\boldsymbol{P}$ is not positive definite, then $\boldsymbol{A}$ cannot be asymptotically stable. Because $\boldsymbol{Q}$ can be an arbitrary r.s.p.d. matrix, $\boldsymbol{Q} = \boldsymbol{I}_n$ (the $n \times n$ identity matrix) is a good choice. It is important to note that there would be no use, as the following example shows, in first choosing a matrix $\boldsymbol{P}$ and then calculating $\boldsymbol{Q}$ from the equation $\boldsymbol{A}^T\boldsymbol{P} + \boldsymbol{P}\boldsymbol{A} = -\boldsymbol{Q}$.

◆ Example 4.3

Let

$$\boldsymbol{A} = \begin{bmatrix} -1 & 3 \\ 0 & -1 \end{bmatrix},$$

which is asymptotically stable. Let

$$\boldsymbol{P} = \begin{bmatrix} 1 & 0 \\ 0 & 1 \end{bmatrix} = \boldsymbol{I}_2 > 0.$$

Then,

$$\boldsymbol{Q} = -(\boldsymbol{A}^T\boldsymbol{P} + \boldsymbol{P}\boldsymbol{A}) = \begin{bmatrix} 2 & -3 \\ -3 & 2 \end{bmatrix},$$

which is indefinite. Thus, unless $\boldsymbol{Q}$ turns out to be definite, nothing can be inferred about asymptotic stability from the Lyapunov theorem if we start with $\boldsymbol{P}$ and then calculate $\boldsymbol{Q}$.

We now give a geometric interpretation of the Lyapunov theorem. First, note that the function

$$\dot{V} = \frac{dV(\boldsymbol{x}(t))}{dt}$$

can be expressed, using the chain rule, as

$$\dot{V} = \nabla V^T \dot{\boldsymbol{x}} = \|\nabla V\| \, \|\dot{\boldsymbol{x}}\| \cos\theta,$$

where θ is the angle between the vectors ∇V and $\dot{\boldsymbol{x}}$. If $\|\dot{\boldsymbol{x}}\| = 1$, then $\nabla V^T \dot{\boldsymbol{x}}$ can be viewed as the rate of increase of V at $\boldsymbol{x}$ in the direction $\dot{\boldsymbol{x}}$. Thus, $\dot{V} < 0$ if and only if $90^\circ < \theta < 270^\circ$. In such a case the velocity vector $\dot{\boldsymbol{x}}(t)$ points in the direction of decreasing V. We illustrate the above discussion in Figure 4.5.

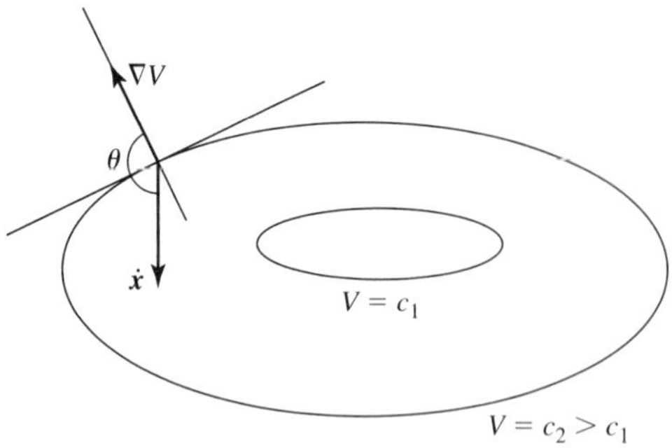

Figure 4.5 Geometric interpretation of the Lyapunov theorem.

We now prove a variant of Theorem 4.1, which is very useful in numerical verification of stability.

Theorem 4.2 A matrix $\boldsymbol{A}$ is asymptotically stable if and only if for any $\boldsymbol{Q}=\boldsymbol{C}^T\boldsymbol{C}$, $\boldsymbol{C} \in \mathbb{R}^{p\times n}$, such that the pair $(\boldsymbol{A}, \boldsymbol{C})$ is observable, the solution $\boldsymbol{P}$ to the Lyapunov matrix equation $\boldsymbol{A}^T\boldsymbol{P} + \boldsymbol{P}\boldsymbol{A} = -\boldsymbol{C}^T\boldsymbol{C}$ is positive definite.

Proof We first prove the necessity condition ($\Rightarrow$) for asymptotic stability. If $\boldsymbol{A}$ is asymptotically stable—that is, $\Re\lambda_i(\boldsymbol{A}) < 0$—then we will show that

$$\boldsymbol{P} = \int_0^\infty e^{\boldsymbol{A}^T t}\boldsymbol{C}^T\boldsymbol{C}e^{\boldsymbol{A}t}\,dt \tag{4.7}$$

is the symmetric positive definite solution to the Lyapunov equation. Indeed, the integrand of (4.7) is a sum of terms of the form

$$t^k e^{\lambda_i t},$$

where λ_i is the ith eigenvalue of $\boldsymbol{A}$. Since $\Re\lambda_i(\boldsymbol{A}) < 0$, the integral in (4.7) exists because the terms in $\boldsymbol{P}$ do not diverge when $t \to \infty$. Indeed, if we let $-\alpha_i = \Re\lambda_i$, where $\alpha_i > 0$, then

$$\lim_{t\to\infty} t^k e^{-\alpha_i t} = \lim_{t\to\infty} \frac{t^k}{e^{\alpha_i t}},$$

We differentiate the numerator and denominator of the above k times and use L'Hôpital's rule to get

$$\begin{aligned}
\lim_{t\to\infty} \frac{t^k}{e^{\alpha_i t}} &= \lim_{t\to\infty} \frac{kt^{k-1}}{\alpha_i e^{\alpha_i t}} \\
&= \lim_{t\to\infty} \frac{k(k-1)t^{k-2}}{\alpha_i^2 e^{\alpha_i t}} \\
&\;\;\vdots \\
&= \lim_{t\to\infty} \frac{k!}{\alpha_i^k e^{\alpha_i t}} \\
&= 0.
\end{aligned} \tag{4.8}$$

Thus, by (4.8), the formula for P given by (4.7) is well-defined. Next, note that $P = P^T$. The matrix P is also positive definite because the pair (A, C) is observable. Substituting the expression for P into the Lyapunov equation and performing manipulations yields

$$\begin{aligned} A^T P + PA &= \int_0^\infty A^T e^{A^T t} C^T C e^{At}\, dt + \int_0^\infty e^{A^T t} C^T C e^{At} A\, dt \\ &= \int_0^\infty \frac{d}{dt}\left(e^{A^T t} C^T C e^{At}\right) dt \\ &= e^{A^T t} C^T C e^{At} \Big|_0^\infty \\ &= -C^T C. \end{aligned}$$

The sufficiency part ($\Leftarrow$) can be proven using the arguments of the proof of the sufficiency part of the Lyapunov theorem and noting that the observability of the pair (A, C) implies that $x^T(t) C^T C x(t)$ is not identically zero along any nonzero solution of $\dot{x}(t) = Ax(t)$.

We now show that if A is asymptotically stable, then the Lyapunov equation (4.6) has a unique solution for every Q. Uniqueness can be proven in many ways. We use the Kronecker product in our proof. Using the results of Section A.5, we represent the Lyapunov equation (4.6) as

$$(I_n \otimes A^T + A^T \otimes I_n)s(P) = -s(Q). \tag{4.9}$$

The condition for a solution $s(P)$ of the above equation to be unique is

$$\lambda_i(A^T) + \lambda_j(A) \neq 0, \qquad i, j = 1, 2, \ldots, n.$$

Because $\lambda_i(A^T) = \lambda_i(A)$, and A is asymptotically stable by assumption, the condition $\lambda_i(A^T) + \lambda_j(A) \neq 0$ is met, and hence the Lyapunov equation has a unique solution.

4.4 Evaluating Quadratic Indices

In this section we apply the Lyapunov theory to evaluate the quadratic indices

$$J_r = \int_0^\infty t^r x^T(t) Q x(t)\, dt, \qquad r = 0, 1, \ldots$$

subject to

$$\dot{x}(t) = Ax(t), \qquad x(0) = x_0,$$

where $Q = C^T C$ such that the pair (A, C) is observable, and the matrix A is asymptotically stable. We can reformulate the above problem as follows. Evaluate the quadratic performance indices

$$J_r = \int_0^\infty t^r y^T(t) y(t)\, dt, \qquad r = 0, 1, \ldots$$

subject to

$$\dot{\boldsymbol{x}}(t) = \boldsymbol{A}\boldsymbol{x}(t), \qquad \boldsymbol{x}(0) = \boldsymbol{x}_0,$$
$$\boldsymbol{y}(t) = \boldsymbol{C}\boldsymbol{x}(t),$$

where the pair $(\boldsymbol{A}, \boldsymbol{C})$ is observable, and the matrix $\boldsymbol{A}$ is asymptotically stable.

To evaluate

$$J_0 = \int_0^\infty \boldsymbol{y}^T(t)\boldsymbol{y}(t)\,dt = \int_0^\infty \boldsymbol{x}^T(t)\boldsymbol{C}^T\boldsymbol{C}\boldsymbol{x}(t)\,dt = \int_0^\infty \boldsymbol{x}^T(t)\boldsymbol{Q}\boldsymbol{x}(t)\,dt,$$

we note that

$$\frac{d}{dt}(\boldsymbol{x}^T(t)\boldsymbol{P}\boldsymbol{x}(t)) = -\boldsymbol{x}^T(t)\boldsymbol{Q}\boldsymbol{x}(t), \tag{4.10}$$

where $\boldsymbol{P}$ and $\boldsymbol{Q} = \boldsymbol{C}^T\boldsymbol{C}$ satisfy the Lyapunov matrix equation, $\boldsymbol{A}^T\boldsymbol{P} + \boldsymbol{P}\boldsymbol{A} = -\boldsymbol{Q}$. Hence,

$$J_0 = -\int_0^\infty \frac{d}{dt}(\boldsymbol{x}^T(t)\boldsymbol{P}\boldsymbol{x}(t))\,dt. \tag{4.11}$$

Integrating both sides of (4.10) with respect to t yields

$$\begin{aligned} J_0 &= -\int_0^\infty \frac{d}{dt}(\boldsymbol{x}^T(t)\boldsymbol{P}\boldsymbol{x}(t))\,dt \\ &= -\left(\boldsymbol{x}^T(t)\boldsymbol{P}\boldsymbol{x}(t)\right)\Big|_0^\infty \\ &= \lim_{t\to\infty}(-\boldsymbol{x}^T(t)\boldsymbol{P}\boldsymbol{x}(t)) + \boldsymbol{x}^T(0)\boldsymbol{P}\boldsymbol{x}(0) \\ &= \boldsymbol{x}^T(0)\boldsymbol{P}\boldsymbol{x}(0) \end{aligned}$$

because, by assumption, $\boldsymbol{A}$ is asymptotically stable and therefore $\lim_{t\to\infty}\boldsymbol{x}(t) = \boldsymbol{0}$ for all $\boldsymbol{x}(0)$. Furthermore, $\boldsymbol{P} = \boldsymbol{P}^T > 0$, and thus $J_0 > 0$ for all $\boldsymbol{x}(0) \neq \boldsymbol{0}$.

Repeating the above arguments leads to a similar expression for J_1. First observe that

$$\begin{aligned} \frac{d}{dt}(t\boldsymbol{x}^T(t)\boldsymbol{P}\boldsymbol{x}(t)) &= \boldsymbol{x}^T(t)\boldsymbol{P}\boldsymbol{x}(t) + t\frac{d}{dt}(\boldsymbol{x}^T(t)\boldsymbol{P}\boldsymbol{x}(t)) \\ &= \boldsymbol{x}^T(t)\boldsymbol{P}\boldsymbol{x}(t) - t\boldsymbol{x}^T(t)\boldsymbol{Q}\boldsymbol{x}(t), \end{aligned}$$

where $\boldsymbol{P}$ satisfies the Lyapunov matrix equation, $\boldsymbol{A}^T\boldsymbol{P} + \boldsymbol{P}\boldsymbol{A} = -\boldsymbol{Q}$. Hence,

$$\begin{aligned} J_1 &= \int_0^\infty t\boldsymbol{x}^T(t)\boldsymbol{Q}\boldsymbol{x}(t)\,dt \\ &= \int_0^\infty \boldsymbol{x}^T(t)\boldsymbol{P}\boldsymbol{x}(t)\,dt - \int_0^\infty \frac{d}{dt}(t\boldsymbol{x}^T(t)\boldsymbol{P}\boldsymbol{x}(t))\,dt. \end{aligned} \tag{4.12}$$

Because $\boldsymbol{P} = \boldsymbol{P}^T > 0$, we can solve

$$\boldsymbol{A}^T\boldsymbol{P}_1 + \boldsymbol{P}_1\boldsymbol{A} = -\boldsymbol{P}$$

for $\boldsymbol{P}_1 = \boldsymbol{P}_1^T > 0$. The above equation can also be represented as

$$\frac{d}{dt}(\boldsymbol{x}^T(t)\boldsymbol{P}_1\boldsymbol{x}(t)) = -\boldsymbol{x}^T(t)\boldsymbol{P}\boldsymbol{x}(t). \tag{4.13}$$

Substituting (4.13) into (4.12) yields

$$\begin{aligned} J_1 &= -\int_0^\infty \frac{d}{dt}(\boldsymbol{x}^T(t)\boldsymbol{P}_1\boldsymbol{x}(t))dt - (t\boldsymbol{x}^T(t)\boldsymbol{P}\boldsymbol{x}(t))\Big|_0^\infty \\ &= \lim_{t\to\infty}(-\boldsymbol{x}^T(t)\boldsymbol{P}_1\boldsymbol{x}(t)) + \boldsymbol{x}^T(0)\boldsymbol{P}_1\boldsymbol{x}(0) - 0 + 0 \\ &= \boldsymbol{x}^T(0)\boldsymbol{P}_1\boldsymbol{x}(0). \end{aligned}$$

Using similar arguments, we can evaluate J_r, $r > 1$, which involves successive solutions to the Lyapunov matrix equations.

In many control problems a dynamical system model is given in terms of a transfer function rather than in terms of a system of differential equations. We now discuss a method of converting a given rational function into a system of first-order differential equations. After this process is completed, we use the Lyapunov theory to evaluate a performance index of interest. Suppose we are given a rational function of the form

$$E(s) = \frac{b_{n-1}s^{n-1} + \cdots + b_1 s + b_0}{s^n + a_{n-1}s^{n-1} + \cdots + a_1 s + a_0},$$

where $E(s) = \mathcal{L}(e(t))$ is the Laplace transform of $e(t)$. We now show that $e(t)$ can be generated as the solution of the differential equation

$$\frac{d^n e(t)}{dt^n} + a_{n-1}\frac{d^{n-1}e(t)}{dt^{n-1}} + \cdots + a_1\frac{de(t)}{dt} + a_0 e(t) = 0, \tag{4.14}$$

subject to a set of initial conditions that we shall derive. We follow the method of Power and Simpson [237, pp. 220–221]. We first take the Laplace transform of the differential equation and then rearrange the resulting terms in an appropriate matrix equation. Recall that

$$\mathcal{L}\left(\frac{d^i e(t)}{dt^i}\right) = s^i E(s) - s^{i-1}e(0) - s^{i-2}\frac{de(0)}{dt} - \cdots - \frac{d^{i-1}e(0)}{dt^{i-1}}. \tag{4.15}$$

Applying the above to (4.14) and comparing corresponding terms with $E(s)$ leads to the equation for the initial conditions:

$$\begin{bmatrix} a_1 & a_2 & \cdots & a_{n-1} & 1 \\ a_2 & a_3 & \cdots & 1 & 0 \\ \vdots & \vdots & & \vdots & \vdots \\ a_{n-1} & 1 & \cdots & 0 & 0 \\ 1 & 0 & \cdots & 0 & 0 \end{bmatrix} \begin{bmatrix} e(0) \\ \dfrac{de(0)}{dt} \\ \vdots \\ \dfrac{d^{n-2}e(0)}{dt^{n-2}} \\ \dfrac{d^{n-1}e(0)}{dt^{n-1}} \end{bmatrix} = \begin{bmatrix} b_0 \\ b_1 \\ \vdots \\ b_{n-2} \\ b_{n-1} \end{bmatrix}.$$

The coefficient matrix in the above equation is nonsingular. Thus, there is a unique solution to this equation, that is, we can uniquely determine the initial conditions. We next define the state vector

$$\boldsymbol{x} = \left[e \quad \frac{de}{dt} \quad \cdots \quad \frac{d^{n-1}e}{dt^{n-1}} \right]^T,$$

and represent the differential equation (4.14) as

$$\dot{\boldsymbol{x}} = \boldsymbol{A}\boldsymbol{x},$$

where

$$\boldsymbol{A} = \begin{bmatrix} 0 & 1 & 0 & \cdots & 0 & 0 \\ 0 & 0 & 1 & \cdots & 0 & 0 \\ \vdots & \vdots & \vdots & & \vdots & \vdots \\ 0 & 0 & 0 & \cdots & 0 & 1 \\ -a_0 & -a_1 & -a_2 & \cdots & -a_{n-2} & -a_{n-1} \end{bmatrix}.$$

The vector of initial conditions is

$$\boldsymbol{x}(0) = \begin{bmatrix} e(0) \\ \frac{de(0)}{dt} \\ \vdots \\ \frac{d^{n-2}e(0)}{dt^{n-2}} \\ \frac{d^{n-1}e(0)}{dt^{n-1}} \end{bmatrix} = \begin{bmatrix} a_1 & a_2 & \cdots & a_{n-1} & 1 \\ a_2 & a_3 & \cdots & 1 & 0 \\ \vdots & \vdots & & \vdots & \vdots \\ a_{n-1} & 1 & \cdots & 0 & 0 \\ 1 & 0 & \cdots & 0 & 0 \end{bmatrix}^{-1} \begin{bmatrix} b_0 \\ b_1 \\ \vdots \\ b_{n-2} \\ b_{n-1} \end{bmatrix}.$$

Note that $e = [1 \;\; 0 \;\; \cdots \;\; 0 \;\; 0]\boldsymbol{x}$.

The above procedure for converting a rational function description of a process into its equivalent state-space representation shows a relation between the zeros of the given rational function and the initial conditions of the associated state-space realization.

We will now use the above method of constructing a state-space realization of a given rational function in the following example that involves evaluation of a quadratic index.

◆ Example 4.4

Consider a block diagram of a dynamical control system model shown in Figure 4.6. This system comes from Dorf and Bishop [66, p. 257]. The values of the parameters are $K_1 = 0.5$ and $K_1 K_2 K_p = 2.5$. We shall use the Lyapunov theory to select the gain K_3 that minimizes the effect of the disturbance $u(t) = \mathcal{L}^{-1}(U(s))$ in the sense of the integral of the square error (ISE) criterion. In other words, we wish to find K_3 that leads to minimization of the performance index

$$J_0 = \int_0^\infty c^2(t)\, dt,$$

where $c(t) = \mathcal{L}^{-1}(C(s))$. We assume a unit step disturbance and $R(s) = 0$.

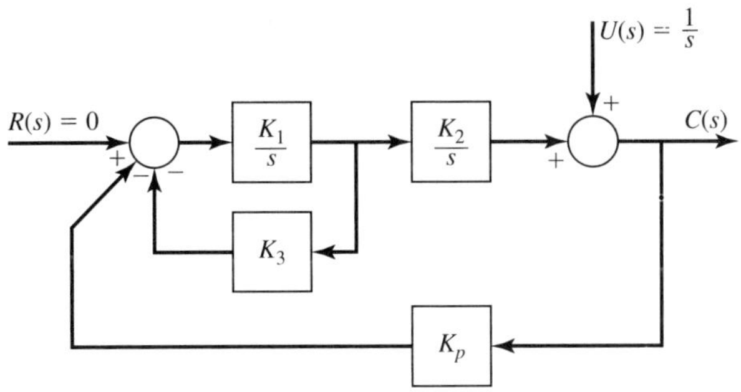

Figure 4.6 Block diagram of the optimized control system model of Example 4.4.

We first find a rational function that models $C(s)$. From the block diagram shown in Figure 4.6, we obtain

$$C(s) = U(s) - \frac{K_2}{s}\frac{K_1}{s + K_1 K_3} K_p C(s).$$

Hence

$$C(s) = \frac{s + 0.5K_3}{s^2 + 0.5K_3 s + 2.5}.$$

The next step is to determine the state-space representation corresponding to $C(s)$. We use the method given before this example. We first form the differential equation for $c(t)$:

$$\frac{d^2c(t)}{dt^2} + 0.5K_3\frac{dc(t)}{dt} + 2.5c(t) = 0$$

with initial conditions

$$\begin{bmatrix} 0.5K_3 & 1 \\ 1 & 0 \end{bmatrix} \begin{bmatrix} c(0) \\ \dfrac{dc(0)}{dt} \end{bmatrix} = \begin{bmatrix} 0.5K_3 \\ 1 \end{bmatrix}.$$

Solving the above equation yields

$$\begin{bmatrix} c(0) \\ \dfrac{dc(0)}{dt} \end{bmatrix} = \begin{bmatrix} 1 \\ 0 \end{bmatrix}.$$

Let $x_1 = c$, $x_2 = \frac{dc}{dt}$. The differential equation in vector–matrix form corresponding to $C(s)$ is

$$\dot{\mathbf{x}}(t) = \begin{bmatrix} 0 & 1 \\ -2.5 & -0.5K_3 \end{bmatrix} \mathbf{x}(t), \qquad \mathbf{x}(0) = \begin{bmatrix} 1 \\ 0 \end{bmatrix}$$

$$c(t) = \mathbf{c}\mathbf{x}(t),$$

where $\mathbf{c} = [1 \quad 0]$. We are now ready to evaluate the performance index J_0:

$$J_0 = \int_0^\infty c^2(t)\,dt = \int_0^\infty \mathbf{x}^T(t) \begin{bmatrix} 1 & 0 \\ 0 & 0 \end{bmatrix} \mathbf{x}(t)\,dt = \int_0^\infty \mathbf{x}^T(t)\,\mathbf{c}^T\,\mathbf{c}\,\mathbf{x}(t)\,dt.$$

The pair $(\boldsymbol{A}, \mathbf{c})$ is observable. Therefore, we can solve the Lyapunov equation

$$\boldsymbol{A}^T\boldsymbol{P} + \boldsymbol{P}\boldsymbol{A} = -\mathbf{c}^T\mathbf{c}$$

for $\boldsymbol{P}$ to obtain

$$\boldsymbol{P} = \begin{bmatrix} \dfrac{K_3}{10} + \dfrac{1}{K_3} & \dfrac{1}{5} \\ \dfrac{1}{5} & \dfrac{2}{5K_3} \end{bmatrix}.$$

Hence

$$J_0 = \mathbf{x}^T(0)\,\boldsymbol{P}\,\mathbf{x}(0) = \frac{K_3}{10} + \frac{1}{K_3}.$$

Differentiating J_0 and equating the result to zero gives

$$\frac{dJ_0}{dK_3} = \frac{1}{10} - \frac{1}{K_3^2} = 0.$$

Selecting

$$K_3 = \sqrt{10}$$

yields a positive definite $\boldsymbol{P}$. Note that because

$$\left.\frac{d^2J_0}{dK_3^2}\right|_{K_3=\sqrt{10}} = \left.\frac{2}{K_3^3}\right|_{K_3=\sqrt{10}} > 0,$$

that is, the second-order sufficiency condition for the minimum is satisfied, we conclude that $K_3 = \sqrt{10} \approx 3.16$ minimizes J_0.

We now wish to determine the gain K_3 that leads to minimization of the effect of the disturbance $u(t)$ in the sense of the integral of time multiplied by the squared error (ITSE) criterion. That is, we wish to select K_3 to minimize the performance index:

$$J_1 = \int_0^\infty t c^2(t)\,dt.$$

We first solve the Lyapunov equation,

$$\boldsymbol{A}^T\boldsymbol{P}_1 + \boldsymbol{P}_1\boldsymbol{A} = -\boldsymbol{P},$$

to get

$$P_1 = \begin{bmatrix} \frac{2}{K_3^2} + \frac{K_3^2}{100} & \frac{1}{5}\left(\frac{1}{K_3} + \frac{K_3}{10}\right) \\ \frac{1}{5}\left(\frac{1}{K_3} + \frac{K_3}{10}\right) & \frac{4}{5K_3^2} + \frac{1}{25} \end{bmatrix}.$$

Hence

$$J_1 = x^T(0)\, P_1\, x(0) = \frac{2}{K_3^2} + \frac{K_3^2}{100}.$$

Differentiating J_1 and equating the result to zero gives

$$\frac{dJ_1}{dK_3} = -\frac{4}{K_3^3} - \frac{K_3}{50} = 0.$$

We select $K_3 = \sqrt[4]{200}$ to keep P_1 positive definite. For this choice of K_3 the second-order sufficiency condition for the minimum of J_1 is satisfied. Thus, $K_3 \approx 3.76$ minimizes J_1.

4.5 Discrete-Time Lyapunov Equation

We will now use the Lyapunov approach to the stability analysis of the discrete-time linear time-invariant systems modeled by

$$x(k+1) = Ax(k), \qquad x(k_0) = x_0. \tag{4.16}$$

By analogy with the continuous-time case, consider a quadratic positive definite form,

$$V(x(k)) = x^T(k)Px(k).$$

We evaluate the first difference of V, defined as $\Delta V = V(x(k+1)) - V(x(k))$, on the trajectories of (4.16) to obtain

$$\begin{aligned} \Delta V(x(k)) &= V(x(k+1)) - V(x(k)) \\ &= x^T(k+1)Px(k+1) - x^T(k)Px(k) \\ &= x^T(k)A^T PAx(k) - x^T(k)Px(k) \\ &= x^T(k)(A^T PA - P)x(k), \end{aligned} \tag{4.17}$$

where we replaced $x(k+1)$ in the above with $Ax(k)$. For the solutions of (4.16) to decrease the "energy function" V, at each $k > k_0$, it is necessary and sufficient that $\Delta V(x(k)) < 0$, or, equivalently, that for some positive definite Q,

$$\Delta V(x(k)) = -x^T(k)Qx(k). \tag{4.18}$$

Combining (4.16) and (4.18), we obtain the so-called *discrete-time Lyapunov equation,*

$$\boxed{A^T PA - P = -Q.} \tag{4.19}$$

Theorem 4.3 The matrix $\boldsymbol{A}$ has its eigenvalues in the open unit disk if and only if for any positive definite $\boldsymbol{Q}$ the solution $\boldsymbol{P}$ to (4.19) is positive definite.

Proof ($\Rightarrow$) We can use the Kronecker product, and in particular relation (A.62), to represent the discrete-time Lyapunov equation as

$$(\boldsymbol{I}_{n^2} - \boldsymbol{A}^T \otimes \boldsymbol{A}^T)\boldsymbol{s}(\boldsymbol{P}) = s(\boldsymbol{Q}), \tag{4.20}$$

where $\otimes$ denotes the Kronecker product and $\boldsymbol{s}(\cdot)$ is the stacking operator. It is easy to see that the above equation has a unique solution if and only if

$$1 - \lambda_i \lambda_j \neq 0 \qquad \text{for all } i, j = 1, 2, \ldots, n,$$

where $\{\lambda_i\}$ is the set of eigenvalues of $\boldsymbol{A}$. By assumption, $|\lambda_i| < 1$, which is a sufficient condition for (4.20) and hence (4.19) to have a unique solution. The solution $\boldsymbol{P}$ can be expressed as

$$\boldsymbol{P} = \sum_{k=0}^{\infty} (\boldsymbol{A}^T)^k \boldsymbol{Q}\boldsymbol{A}^k. \tag{4.21}$$

The above expression is well-defined because all eigenvalues of $\boldsymbol{A}$ have magnitude (strictly) less than unity. Substituting (4.21) into (4.19) gives

$$\begin{aligned}
\boldsymbol{A}^T \boldsymbol{P} \boldsymbol{A} - \boldsymbol{P} &= \boldsymbol{A}^T \left(\sum_{k=0}^{\infty} (\boldsymbol{A}^T)^k \boldsymbol{Q}\boldsymbol{A}^k \right) \boldsymbol{A} - \sum_{k=0}^{\infty} (\boldsymbol{A}^T)^k \boldsymbol{Q}\boldsymbol{A}^k \\
&= \sum_{k=1}^{\infty} (\boldsymbol{A}^T)^k \boldsymbol{Q}\boldsymbol{A}^k - \sum_{k=1}^{\infty} (\boldsymbol{A}^T)^k \boldsymbol{Q}\boldsymbol{A}^k - \boldsymbol{Q} \\
&= -\boldsymbol{Q}.
\end{aligned}$$

Thus, indeed (4.21) satisfies the discrete-time Lyapunov equation. Note that if $\boldsymbol{Q}$ is symmetric, so is $\boldsymbol{P}$. If $\boldsymbol{Q}$ is positive definite, so is $\boldsymbol{P}$.

($\Leftarrow$) Let λ_i be an eigenvalue of $\boldsymbol{A}$ and let $\boldsymbol{v}_i$ be a corresponding eigenvector; that is, $\boldsymbol{A}\boldsymbol{v}_i = \lambda_i \boldsymbol{v}_i$. Observe that λ_i and $\boldsymbol{v}_i$ may be complex. We denote by $\boldsymbol{v}_i^*$ the complex conjugate transpose of $\boldsymbol{v}_i$. Note that $\boldsymbol{v}_i^* \boldsymbol{P} \boldsymbol{v}_i$ is a scalar. We will show that its complex conjugate equals itself, which means that $\boldsymbol{v}_i^* \boldsymbol{P} \boldsymbol{v}_i$ is real. Indeed,

$$\overline{\boldsymbol{v}_i^* \boldsymbol{P} \boldsymbol{v}_i} = \boldsymbol{v}_i^* \boldsymbol{P}^* \boldsymbol{v}_i = \boldsymbol{v}_i^* \boldsymbol{P}^T \boldsymbol{v}_i = \boldsymbol{v}_i^* \boldsymbol{P} \boldsymbol{v}_i,$$

where we used the fact that the complex conjugate transpose of a real matrix reduces to its transpose. Thus, $\boldsymbol{v}_i^* \boldsymbol{P} \boldsymbol{v}_i$ is real for any complex $\boldsymbol{v}_i$.

Premultiplying now (4.19) by $\boldsymbol{v}_i^*$ and postmultiplying by $\boldsymbol{v}_i$ gives

$$\boldsymbol{v}_i^* (\boldsymbol{A}^T \boldsymbol{P} \boldsymbol{A} - \boldsymbol{P}) \boldsymbol{v}_i = -\boldsymbol{v}_i^* \boldsymbol{Q} \boldsymbol{v}_i. \tag{4.22}$$

The left-hand side of the above equality evaluates to

$$\begin{aligned}
\boldsymbol{v}_i^* \boldsymbol{A}^T \boldsymbol{P} \boldsymbol{A} \boldsymbol{v}_i - \boldsymbol{v}_i^* \boldsymbol{P} \boldsymbol{v}_i &= \bar{\lambda}_i \boldsymbol{v}_i^* \boldsymbol{P} \boldsymbol{v}_i \lambda_i - \boldsymbol{v}_i^* \boldsymbol{P} \boldsymbol{v}_i \\
&= (|\lambda_i|^2 - 1) \boldsymbol{v}_i^* \boldsymbol{P} \boldsymbol{v}_i.
\end{aligned}$$

Comparing the above with the right-hand side of (4.22) yields

$$(1 - |\lambda_i|^2) \boldsymbol{v}_i^* \boldsymbol{P} \boldsymbol{v}_i = \boldsymbol{v}_i^* \boldsymbol{Q} \boldsymbol{v}_i > 0$$

because $\boldsymbol{Q}$ is positive definite. Because $\boldsymbol{P}$ is positive definite, we have to have $|\lambda_i| < 1$. The proof is complete.

We will now concern ourselves with evaluating performance indices on the trajectories of (4.16). Specifically, we wish to evaluate

$$K_r = \sum_{k=0}^{\infty} k^r \boldsymbol{x}^T(k)\boldsymbol{Q}\boldsymbol{x}(k), \qquad r = 0, 1, \ldots \tag{4.23}$$

where $\boldsymbol{Q} = \boldsymbol{Q}^T > 0$ subject to (4.16). We assume that the matrix $\boldsymbol{A}$ has its eigenvalues in the open unit disk; that is, the matrix $\boldsymbol{A}$ is *convergent*. We first consider the case when $k = 0$. For this case, K_r becomes

$$K_0 = \sum_{k=0}^{\infty} \boldsymbol{x}^T(k)\boldsymbol{Q}\boldsymbol{x}(k). \tag{4.24}$$

Because the matrix $\boldsymbol{A}$ is convergent, the solution $\boldsymbol{P}$ to (4.19) is positive definite. Using (4.17), (4.18), and (4.16), we write

$$\begin{aligned} \boldsymbol{x}^T(k)\boldsymbol{Q}\boldsymbol{x}(k) &= -\boldsymbol{x}^T(k)(\boldsymbol{A}^T\boldsymbol{P}\boldsymbol{A} - \boldsymbol{P})\boldsymbol{x}(k) \\ &= \boldsymbol{x}^T(k)\boldsymbol{P}\boldsymbol{x}(k) - \boldsymbol{x}^T(k+1)\boldsymbol{P}\boldsymbol{x}(k+1). \end{aligned} \tag{4.25}$$

Summing both sides of the above form $k = 0$ to $k = \infty$ and using the fact that $\boldsymbol{x}(\infty) = \boldsymbol{0}$ because the matrix $\boldsymbol{A}$ is convergent, we obtain

$$K_0 = \boldsymbol{x}^T(0)\boldsymbol{P}\boldsymbol{x}(0). \tag{4.26}$$

We will now evaluate

$$K_1 = \sum_{k=0}^{\infty} k\boldsymbol{x}^T(k)\boldsymbol{Q}\boldsymbol{x}(k),$$

subject to (4.16). To proceed, we multiply both sides of (4.25) by k to get

$$k\boldsymbol{x}^T(k)\boldsymbol{Q}\boldsymbol{x}(k) = k\boldsymbol{x}^T(k)\boldsymbol{P}\boldsymbol{x}(k) - k\boldsymbol{x}^T(k+1)\boldsymbol{P}\boldsymbol{x}(k+1).$$

Summing the above from $k = 0$ to $k = \infty$ gives

$$\begin{aligned} K_1 &= \boldsymbol{x}^T(1)\boldsymbol{P}\boldsymbol{x}(1) - \boldsymbol{x}^T(2)\boldsymbol{P}\boldsymbol{x}(2) + 2\boldsymbol{x}^T(2)\boldsymbol{P}\boldsymbol{x}(2) - 2\boldsymbol{x}^T(3)\boldsymbol{P}\boldsymbol{x}(3) + \cdots \\ &= \sum_{k=0}^{\infty} \boldsymbol{x}^T(k)\boldsymbol{P}\boldsymbol{x}(k) - \boldsymbol{x}^T(0)\boldsymbol{P}\boldsymbol{x}(0). \end{aligned} \tag{4.27}$$

Using the method we employed to find the value of K_0, we now evaluate the first term on the right-hand side of (4.27) to get

$$\sum_{k=0}^{\infty} \boldsymbol{x}^T(k)\boldsymbol{P}\boldsymbol{x}(k) = \boldsymbol{x}^T(0)\boldsymbol{P}_1\boldsymbol{x}(0),$$

where $\boldsymbol{P}_1$ satisfies

$$\boldsymbol{A}^T\boldsymbol{P}_1\boldsymbol{A} - \boldsymbol{P}_1 = -\boldsymbol{P}.$$

Taking the above into account, we represent (4.27) as

$$K_1 = \boldsymbol{x}^T(0)(\boldsymbol{P}_1 - \boldsymbol{P})\boldsymbol{x}(0).$$

Proceeding in a similar manner, we can find expressions for K_r, where $r > 1$.

4.6 Constructing Robust Linear Controllers

In this section we use the Lyapunov theorem to construct robust state-feedback controllers. One of the basic issues in the control of dynamical systems is the effect of uncertainties, or neglected nonlinearities, on the stability of the closed-loop system. A controller is viewed as robust if it maintains stability for all uncertainties in an expected range. We consider a class of dynamical systems modeled by

$$\dot{\boldsymbol{x}}(t) = \boldsymbol{A}\boldsymbol{x}(t) + \boldsymbol{B}\boldsymbol{u}(t) + \boldsymbol{f}(t, \boldsymbol{x}(t)) + \boldsymbol{B}\boldsymbol{h}(t, \boldsymbol{x}(t), \boldsymbol{u}(t)), \tag{4.28}$$

where $\boldsymbol{x}(t) \in \mathbb{R}^n$, $\boldsymbol{u}(t) \in \mathbb{R}^m$, and the functions $\boldsymbol{h}$ and $\boldsymbol{f}$ model uncertainties, or nonlinearities, in the system. We refer to $\boldsymbol{h}$ as the *matched uncertainty* because it affects the system dynamics via the input matrix $\boldsymbol{B}$ in the same fashion as the input $\boldsymbol{u}$ does. In other words, the uncertainty $\boldsymbol{h}$ matches the system input $\boldsymbol{u}$. The vector $\boldsymbol{f}$ models the *unmatched uncertainty*. We assume that the uncertain elements $\boldsymbol{h}$ and $\boldsymbol{f}$ satisfy the following norm bounds:

1. $\|\boldsymbol{h}(t, \boldsymbol{x}, \boldsymbol{u})\| \le \gamma_h \|\boldsymbol{u}\| + \alpha_h \|\boldsymbol{x}\|$,
2. $\|\boldsymbol{f}(t, \boldsymbol{x})\| \le \alpha_f \|\boldsymbol{x}\|$,

where γ_h, α_h, and α_f are known nonnegative constants. We further assume that the matrix $\boldsymbol{A}$ is asymptotically stable. If this is not the case, we apply a preliminary state-feedback controller $\boldsymbol{u} = -\boldsymbol{K}\boldsymbol{x}$ such that $\boldsymbol{A} - \boldsymbol{B}\boldsymbol{K}$ is asymptotically stable. Such a feedback exists provided that the pair $(\boldsymbol{A}, \boldsymbol{B})$ is stabilizable. Our goal is to construct a linear state-feedback controller that would make the closed-loop system asymptotically stable for arbitrary $\boldsymbol{f}$ and $\boldsymbol{h}$ that satisfy the above norm bounds.

Theorem 4.4 Suppose that $\boldsymbol{A}$ is asymptotically stable and that $\boldsymbol{P} = \boldsymbol{P}^T > 0$ is the solution to the Lyapunov matrix equation $\boldsymbol{A}^T\boldsymbol{P} + \boldsymbol{P}\boldsymbol{A} = -2\boldsymbol{Q}$ for some $\boldsymbol{Q} = \boldsymbol{Q}^T > 0$. Suppose also that

$$\alpha_f < \frac{\lambda_{\min}(\boldsymbol{Q})}{\lambda_{\max}(\boldsymbol{P})} \quad \text{and} \quad \gamma_h < 1.$$

Then, the state-feedback controller $\boldsymbol{u} = -\gamma\boldsymbol{B}^T\boldsymbol{P}\boldsymbol{x}$, where

$$\gamma > \frac{\alpha_h^2}{4(\lambda_{\min}(\boldsymbol{Q}) - \alpha_f\lambda_{\max}(\boldsymbol{P}))(1 - \gamma_h)}$$

stabilizes the uncertain system model (4.28) for arbitrary $\boldsymbol{f}$ and $\boldsymbol{h}$ that satisfy the norm bounds.

Proof The time derivative of the positive definite function $V = \boldsymbol{x}^T\boldsymbol{P}\boldsymbol{x}$ evaluated on any trajectory of the closed-loop system is

$$\begin{aligned}\dot{V} &= 2\boldsymbol{x}^T\boldsymbol{P}\dot{\boldsymbol{x}}\\ &= -2\boldsymbol{x}^T\boldsymbol{Q}\boldsymbol{x} - 2\gamma\boldsymbol{x}^T\boldsymbol{P}\boldsymbol{B}\boldsymbol{B}^T\boldsymbol{P}\boldsymbol{x} + 2\boldsymbol{x}^T\boldsymbol{P}\boldsymbol{f} + 2\boldsymbol{x}^T\boldsymbol{P}\boldsymbol{B}\boldsymbol{h}.\end{aligned}$$

Our goal is to determine $\tilde{\gamma} > 0$ such that if $\gamma > \tilde{\gamma}$, then $\dot{V} < 0$, which in turn implies the asymptotic stability of the closed-loop system. To proceed further, recall that if $\boldsymbol{Q}$ is a symmetric matrix, then

$$\lambda_{\min}(\boldsymbol{Q})\|\boldsymbol{x}\|^2 \le \boldsymbol{x}^T\boldsymbol{Q}\boldsymbol{x} \le \lambda_{\max}(\boldsymbol{Q})\|\boldsymbol{x}\|^2,$$

and therefore

$$-\boldsymbol{x}^T\boldsymbol{Q}\boldsymbol{x} \le -\lambda_{\min}(\boldsymbol{Q})\|\boldsymbol{x}\|^2.$$

Furthermore,

$$\boldsymbol{x}^T\boldsymbol{P}\boldsymbol{B}\boldsymbol{B}^T\boldsymbol{P}\boldsymbol{x} = \boldsymbol{x}^T\boldsymbol{P}\boldsymbol{B}\left(\boldsymbol{x}^T\boldsymbol{P}\boldsymbol{B}\right)^T = \|\boldsymbol{x}^T\boldsymbol{P}\boldsymbol{B}\|^2.$$

For a symmetric positive definite matrix $\boldsymbol{P}$, its induced 2-norm is

$$\|\boldsymbol{P}\| = \lambda_{\max}(\boldsymbol{P}).$$

Taking the above relations into account, using the norm bounds on the uncertain elements, and performing some manipulations yields

$$\begin{aligned}\dot{V} \le &-2\lambda_{\min}(\boldsymbol{Q})\|\boldsymbol{x}\|^2 - 2\gamma\|\boldsymbol{x}^T\boldsymbol{P}\boldsymbol{B}\|^2 + 2\alpha_f\lambda_{\max}(\boldsymbol{P})\|\boldsymbol{x}\|^2\\ &+2\|\boldsymbol{x}^T\boldsymbol{P}\boldsymbol{B}\|(\alpha_h\|\boldsymbol{x}\| + \gamma_h\gamma\|\boldsymbol{x}^T\boldsymbol{P}\boldsymbol{B}\|).\end{aligned} \tag{4.29}$$

Let

$$\tilde{\boldsymbol{Q}} = \begin{bmatrix} \lambda_{\min}(\boldsymbol{Q}) - \alpha_f\lambda_{\max}(\boldsymbol{P}) & -\dfrac{\alpha_h}{2} \\ -\dfrac{\alpha_h}{2} & \gamma(1-\gamma_h)\end{bmatrix} \in \mathbb{R}^{2\times 2}. \tag{4.30}$$

Then, we can represent (4.29) as

$$\dot{V} \le -2[\,\|\boldsymbol{x}\| \quad \|\boldsymbol{x}^T\boldsymbol{P}\boldsymbol{B}\|\,]\,\tilde{\boldsymbol{Q}}\begin{bmatrix}\|\boldsymbol{x}\| \\ \|\boldsymbol{x}^T\boldsymbol{P}\boldsymbol{B}\|\end{bmatrix}. \tag{4.31}$$

By Theorem A.6 on page 634, for $\dot{V}$ to be negative it is enough that the leading principal minors of the matrix $\tilde{\boldsymbol{Q}}$ given by (4.30) be positive. The first-order leading principal minor is positive by assumption. For the second-order leading principal minor to be positive it is sufficient that $\gamma_h < 1$ and

$$\gamma > \frac{\alpha_h^2}{4(\lambda_{\min}(\boldsymbol{Q}) - \alpha_f\lambda_{\max}(\boldsymbol{P}))(1-\gamma_h)}.$$

The proof is now complete.

Note that if $\gamma_h = 0$, then the closed-loop system is asymptotically stable if

$$\alpha_f < \frac{\lambda_{\min}(\boldsymbol{Q})}{\lambda_{\max}(\boldsymbol{P})} \quad \text{and} \quad \gamma > \frac{\alpha_h^2}{4(\lambda_{\min}(\boldsymbol{Q}) - \alpha_f\lambda_{\max}(\boldsymbol{P}))}.$$

◆ Example 4.5

Consider a dynamical system model

$$\dot{\boldsymbol{x}} = \boldsymbol{A}\boldsymbol{x} + \boldsymbol{b}(u + h(\boldsymbol{x}, u)) = \begin{bmatrix} -2 & 0 \\ 0 & -1 \end{bmatrix} \boldsymbol{x} + \begin{bmatrix} 1 \\ 1 \end{bmatrix} (u + h(\boldsymbol{x}, u)).$$

Let $\boldsymbol{P}$ be the solution of the matrix Lyapunov equation $\boldsymbol{A}^T\boldsymbol{P} + \boldsymbol{P}\boldsymbol{A} = -2\boldsymbol{I}_2$. We will design a linear state-feedback controller $u = -\gamma \boldsymbol{b}^T\boldsymbol{P}\boldsymbol{x}$ so that the origin of the closed-loop system is uniformly asymptotically stable for any $h(\boldsymbol{x}, u)$ such that

$$|h(\boldsymbol{x}, u)| \le 4\|\boldsymbol{x}\| + \tfrac{1}{2}|u|. \tag{4.32}$$

In particular, we will find a bound $\tilde{\gamma} > 0$ so that for any $\gamma > \tilde{\gamma}$ the closed-loop system is uniformly asymptotically stable in the face of uncertainties satisfying the norm bound (4.32).

Solving the Lyapunov equation $\boldsymbol{A}^T\boldsymbol{P} + \boldsymbol{P}\boldsymbol{A} = -2\boldsymbol{I}_2$ yields

$$\boldsymbol{P} = \begin{bmatrix} \frac{1}{2} & 0 \\ 0 & 1 \end{bmatrix}.$$

Hence,

$$u = -\gamma \boldsymbol{b}^T\boldsymbol{P}\boldsymbol{x} = -\gamma\left(\tfrac{1}{2}x_1 + x_2\right).$$

We obtain a bound on γ by evaluating $\frac{d}{dt}(\boldsymbol{x}^T\boldsymbol{P}\boldsymbol{x})$ on the trajectories of the closed-loop system. First, we evaluate

$$\begin{aligned} \frac{d}{dt}\left(\boldsymbol{x}^T\boldsymbol{P}\boldsymbol{x}\right) &= 2\boldsymbol{x}^T\boldsymbol{P}\dot{\boldsymbol{x}} \\ &= 2\boldsymbol{x}^T\boldsymbol{P}(\boldsymbol{A}\boldsymbol{x} + \boldsymbol{b}(u + h(\boldsymbol{x}, u))) \\ &= -2\boldsymbol{x}^T\boldsymbol{x} - 2\gamma\boldsymbol{x}^T\boldsymbol{P}\boldsymbol{b}\boldsymbol{b}^T\boldsymbol{P}\boldsymbol{x} + 2\boldsymbol{x}^T\boldsymbol{P}\boldsymbol{b}h. \end{aligned}$$

Taking the norms and performing manipulations yields

$$\begin{aligned} \frac{d}{dt}\left(\boldsymbol{x}^T\boldsymbol{P}\boldsymbol{x}\right) &\le -2\boldsymbol{x}^T\boldsymbol{x} - 2\gamma|\boldsymbol{x}^T\boldsymbol{P}\boldsymbol{b}|^2 + 2|\boldsymbol{x}^T\boldsymbol{P}\boldsymbol{b}|\left(4\|\boldsymbol{x}\| + \tfrac{1}{2}\gamma|\boldsymbol{x}^T\boldsymbol{P}\boldsymbol{b}|\right) \\ &= -2\|\boldsymbol{x}\|^2 - \gamma|\boldsymbol{x}^T\boldsymbol{P}\boldsymbol{b}|^2 + 8|\boldsymbol{x}^T\boldsymbol{P}\boldsymbol{b}|\|\boldsymbol{x}\| \\ &= -2[\,\|\boldsymbol{x}\| \quad |\boldsymbol{x}^T\boldsymbol{P}\boldsymbol{b}|\,]\begin{bmatrix} 1 & -2 \\ -2 & \gamma/2 \end{bmatrix}\begin{bmatrix} \|\boldsymbol{x}\| \\ |\boldsymbol{x}^T\boldsymbol{P}\boldsymbol{b}| \end{bmatrix}. \end{aligned}$$

For $\frac{d}{dt}(\boldsymbol{x}^T\boldsymbol{P}\boldsymbol{x})$ to be negative definite, it is enough that the matrix

$$\begin{bmatrix} 1 & -2 \\ -2 & \gamma/2 \end{bmatrix}$$

be positive definite. This is the case whenever

$$\gamma > \tilde{\gamma} = 8.$$

4.7 Hurwitz and Routh Stability Criteria

In this section we present two tests for determining if the zeros of an nth-degree polynomial are located in the open left-half complex plane. In deriving these tests, we will use the Lyapunov matrix equation given by (4.6). We consider the nth-degree polynomial

$$p(s) = a_n s^n + a_{n-1} s^{n-1} + \cdots + a_1 s + a_0 \tag{4.33}$$

with real coefficients. Associated with the polynomial $p(s)$ is the $n \times n$ real Hurwitz matrix,

$$\boldsymbol{H} = \begin{bmatrix} a_{n-1} & a_{n-3} & a_{n-5} & a_{n-7} & \cdots \\ a_n & a_{n-2} & a_{n-4} & a_{n-6} & \cdots \\ 0 & a_{n-1} & a_{n-3} & a_{n-5} & \cdots \\ 0 & a_n & a_{n-2} & a_{n-4} & \cdots \\ \vdots & & & \vdots & \ddots \end{bmatrix}. \tag{4.34}$$

In our further discussion we assume, without loss of generality, that $a_n = 1$. The leading principal minors of the Hurwitz matrix are

$$H_1 = a_{n-1}, \quad H_2 = \det \begin{bmatrix} a_{n-1} & a_{n-3} \\ 1 & a_{n-2} \end{bmatrix},$$

$$H_3 = \det \begin{bmatrix} a_{n-1} & a_{n-3} & a_{n-5} \\ 1 & a_{n-2} & a_{n-4} \\ 0 & a_{n-1} & a_{n-3} \end{bmatrix}, \ldots, H_n = \det \boldsymbol{H}. \tag{4.35}$$

We next define

$$b_1 = H_1, \qquad b_2 = \frac{H_2}{H_1}, \qquad b_3 = \frac{H_3}{H_2 H_1},$$

and

$$b_k = \frac{H_{k-3} H_k}{H_{k-2} H_{k-1}} \qquad \text{for } k = 4, 5, \ldots, n.$$

Thus, for example, $b_4 = \frac{H_1 H_4}{H_2 H_3}$. Let

$$\boldsymbol{B} = \begin{bmatrix} 0 & 1 & 0 & 0 & \cdots & 0 & 0 & 0 \\ -b_n & 0 & 1 & 0 & \cdots & 0 & 0 & 0 \\ 0 & -b_{n-1} & 0 & 1 & \cdots & 0 & 0 & 0 \\ \vdots & & & & & & & \vdots \\ 0 & 0 & 0 & 0 & \cdots & -b_3 & 0 & 1 \\ 0 & 0 & 0 & 0 & \cdots & 0 & -b_2 & -b_1 \end{bmatrix} \in \mathbb{R}^{n \times n}.$$

For example, for $n = 4$, we have

$$B = \begin{bmatrix} 0 & 1 & 0 & 0 \\ -b_4 & 0 & 1 & 0 \\ 0 & -b_3 & 0 & 1 \\ 0 & 0 & -b_2 & -b_1 \end{bmatrix}.$$

Let

$$A = \begin{bmatrix} 0 & 1 & 0 & \cdots & 0 & 0 \\ 0 & 0 & 1 & \cdots & 0 & 0 \\ \vdots & & & & & \vdots \\ 0 & 0 & 0 & \cdots & 0 & 1 \\ -a_0 & -a_1 & -a_2 & \cdots & -a_{n-2} & -a_{n-1} \end{bmatrix} \in \mathbb{R}^{n\times n}.$$

Lemma 4.1 The matrices $\boldsymbol{B}$ and $\boldsymbol{A}$ are similar; that is, there is a nonsingular matrix $\boldsymbol{T} \in \mathbb{R}^{n\times n}$ such that

$$\boldsymbol{B} = \boldsymbol{T}^{-1}\boldsymbol{A}\boldsymbol{T}. \tag{4.36}$$

Proof The similarity matrix is lower triangular and has the form

$$\boldsymbol{T} = \begin{bmatrix} 1 & 0 & 0 & 0 & 0 & 0 & \cdots & 0 \\ 0 & 1 & 0 & 0 & 0 & 0 & \cdots & 0 \\ t_{31} & 0 & 1 & 0 & 0 & 0 & \cdots & 0 \\ 0 & t_{42} & 0 & 1 & 0 & 0 & \cdots & 0 \\ t_{51} & 0 & t_{53} & 0 & 1 & 0 & \cdots & 0 \\ 0 & t_{62} & 0 & t_{64} & 0 & 1 & \cdots & 0 \\ \vdots & & & & & & & \vdots \end{bmatrix}. \tag{4.37}$$

To determine the nonfixed coefficients t_{ij} of the matrix $\boldsymbol{T}$, we represent (4.36) as

$$\boldsymbol{T}\boldsymbol{B} = \boldsymbol{A}\boldsymbol{T}, \tag{4.38}$$

and compare both sides of (4.38) to determine the coefficients t_{ij}. We obtain, for example, $t_{31} = -b_n$, $t_{42} = -b_n - b_{n-1}$, and so on.

Let $\boldsymbol{P}$ be an $n \times n$ diagonal matrix defined as

$$\boldsymbol{P} = \operatorname{diag}[\, b_1 b_2 \times \cdots \times b_n \quad b_1 b_2 \times \cdots \times b_{n-1} \quad \cdots \quad b_1 b_2 \quad b_1 \,], \tag{4.39}$$

and $\boldsymbol{Q}$ be an $n \times n$ diagonal matrix defined as

$$\boldsymbol{Q} = \operatorname{diag}\begin{bmatrix} 0 & 0 & \cdots & 0 & 2b_1^2 \end{bmatrix}.$$

We represent the matrix $\boldsymbol{Q}$ as $\boldsymbol{Q} = \boldsymbol{c}^T\boldsymbol{c}$, where

$$\boldsymbol{c} = [\, 0 \quad 0 \quad \cdots \quad 0 \quad \sqrt{2}b_1 \,].$$

Note that the pair $(\boldsymbol{B}, \boldsymbol{c})$ is observable. The matrices $\boldsymbol{B}$, $\boldsymbol{P}$, and $\boldsymbol{Q}$ satisfy the Lyapunov matrix equation,

$$\boldsymbol{B}^T\boldsymbol{P} + \boldsymbol{P}\boldsymbol{B} = -\boldsymbol{Q}. \tag{4.40}$$

By Theorem 4.2, on page 158, applied to (4.40), the matrix $\boldsymbol{B}$ is asymptotically stable if and only if the symmetric matrix $\boldsymbol{P}$ is positive definite. By Theorem A.6, the matrix $\boldsymbol{P}$ is positive definite if and only if its leading principal minors are positive. Using the above results, we can prove the following theorem:

Theorem 4.5 (Hurwitz) The real polynomial $p(s)$ of degree n with positive coefficients a_i has its zeros in the open left-half plane if and only if the leading principal minors of the Hurwitz matrix are positive.

Proof The polynomial $p(s)$ has its zeros in the open left plane if and only if the matrix $\boldsymbol{B}$ is asymptotically stable. This in turn is true if and only if the diagonal elements of $\boldsymbol{P}$, given by (4.39), are positive. We have

$$b_1 = H_1, \qquad b_1 b_2 = H_2,$$

and for $k = 3, \ldots, n$ we obtain

$$b_1 b_2 \times \cdots \times b_k = \frac{H_k}{H_{k-2}}.$$

Thus, the diagonal elements of $\boldsymbol{P}$ are all positive, if and only if the leading principal minors of the Hurwitz matrix are positive, which completes the proof.

We will now prove the Routh necessary and sufficient condition for the real polynomial (4.33) to have its zeros in the open left-half plane. We first use the polynomial coefficients to form the following $(n+1)$-row Routh array:

$$\begin{array}{ll|lllll}
s^n & \text{row} & a_n & a_{n-2} & a_{n-4} & a_{n-6} & \cdots \\
s^{n-1} & \text{row} & a_{n-1} & a_{n-3} & a_{n-5} & a_{n-7} & \cdots \\
s^{n-2} & \text{row} & r_{21} & r_{22} & r_{23} & r_{24} & \cdots \\
s^{n-3} & \text{row} & r_{31} & r_{32} & r_{33} & r_{34} & \cdots \\
\vdots & & \vdots & & & \vdots & \\
s^0 & \text{row} & r_{n1} & & & &
\end{array} \tag{4.41}$$

The elements that comprise the s^n and s^{n-1} rows are appropriately ordered coefficients of $p(s)$. The elements of the s^{n-2} row are formed from its preceding two rows as follows:

$$r_{21} = -\frac{1}{a_{n-1}} \det \begin{bmatrix} a_n & a_{n-2} \\ a_{n-1} & a_{n-3} \end{bmatrix} = \frac{a_{n-1}a_{n-2} - a_n a_{n-3}}{a_{n-1}},$$

$$r_{22} = -\frac{1}{a_{n-1}} \det \begin{bmatrix} a_n & a_{n-4} \\ a_{n-1} & a_{n-5} \end{bmatrix} = \frac{a_{n-1}a_{n-4} - a_n a_{n-5}}{a_{n-1}},$$

$$r_{23} = -\frac{1}{a_{n-1}} \det \begin{bmatrix} a_n & a_{n-6} \\ a_{n-1} & a_{n-7} \end{bmatrix} = \frac{a_{n-1}a_{n-6} - a_n a_{n-7}}{a_{n-1}},$$

$$\vdots \qquad \vdots$$

The elements of the s^{n-3} row of the Routh array are generated using its preceding two rows:

$$
\begin{aligned}
r_{31} &= -\frac{1}{r_{21}} \det \begin{bmatrix} a_{n-1} & a_{n-3} \\ r_{21} & r_{22} \end{bmatrix} = \frac{r_{21}a_{n-3} - r_{22}a_{n-1}}{r_{21}}, \\
r_{32} &= -\frac{1}{r_{21}} \det \begin{bmatrix} a_{n-1} & a_{n-5} \\ r_{21} & r_{23} \end{bmatrix} = \frac{r_{21}a_{n-5} - r_{23}a_{n-1}}{r_{21}}, \\
r_{33} &= -\frac{1}{r_{21}} \det \begin{bmatrix} a_{n-1} & a_{n-7} \\ r_{21} & r_{24} \end{bmatrix} = \frac{r_{21}a_{n-7} - r_{24}a_{n-1}}{r_{21}}, \\
&\vdots \qquad \vdots
\end{aligned}
$$

Subsequent rows of the array are formed from their preceding rows in an analogous fashion. Undefined positions can be filled with zeros. The last two rows of the array will have one term each. We will now formulate the stability test that in some texts (for example, in Rohrs et al. [243] or Wolovich [302]) is also called the Routh–Hurwitz criterion.

Theorem 4.6 (Routh) The zeros of the polynomial $p(s) = a_n s^n + a_{n-1}s^{n-1} + \cdots + a_1 s + a_0$, whose coefficients are all nonzero and have the same sign, are all in the open left-half plane if and only if the elements of the first column of the associated Routh array all have the same sign and are nonzero.

Proof Without loss of generality, we assume that the polynomial coefficients are all positive. If they are negative, we can multiply both sides of the equation $p(s) = 0$ by -1, thus making the coefficients positive. We then perform elementary row operations on the associated Hurwitz matrix, $\boldsymbol{H}$. These operations are equivalent to premultiplying the Hurwitz matrix by lower triangular matrices with ones on the diagonal. For example,

$$
\begin{bmatrix} 1 & 0 & 0 & 0 & \cdots \\ -a_n/a_{n-1} & 1 & 0 & 0 & \cdots \\ 0 & 0 & 1 & 0 & \cdots \\ \vdots & & & \ddots & \vdots \\ 0 & 0 & 0 & 0 & 1 \end{bmatrix}
\begin{bmatrix} a_{n-1} & a_{n-3} & a_{n-5} & a_{n-7} \\ a_n & a_{n-2} & a_{n-4} & a_{n-6} \\ 0 & a_{n-1} & a_{n-3} & a_{n-5} \\ 0 & a_n & a_{n-2} & a_{n-4} \\ \vdots & & \vdots & \ddots \end{bmatrix}
=
\begin{bmatrix} a_{n-1} & a_{n-3} & a_{n-5} & a_{n-7} \\ 0 & r_{21} & r_{22} & r_{23} \\ 0 & a_{n-1} & a_{n-3} & a_{n-5} \\ 0 & a_n & a_{n-2} & a_{n-4} \\ \vdots & & \vdots & \ddots \end{bmatrix}.
$$

Let $\boldsymbol{U}$ be the product of the matrices corresponding to the above described elementary operations on $\boldsymbol{H}$. Then,

$$
\boldsymbol{U} = \begin{bmatrix} 1 & 0 & \cdots & 0 \\ u_{12} & 1 & \cdots & 0 \\ \vdots & & \ddots & \vdots \\ u_{n1} & u_{n2} & \cdots & 1 \end{bmatrix},
$$

where, for example, $u_{12} = -a_n/a_{n-1}$. Premultiplying the Hurwitz matrix (4.34) by the matrix $\boldsymbol{U}$ given above yields

$$\boldsymbol{UH} = \begin{bmatrix} a_{n-1} & a_{n-3} & a_{n-5} & \cdots & \cdots \\ 0 & r_{21} & r_{22} & \cdots & r_{2,n-1} \\ 0 & 0 & r_{31} & \cdots & r_{3,n-2} \\ \vdots & & \vdots & \ddots & \vdots \\ 0 & 0 & 0 & \cdots & r_{n1} \end{bmatrix}. \tag{4.42}$$

Because the matrix $\boldsymbol{U}$ is lower triangular with the ones on the diagonal, the leading principal minors of $\boldsymbol{H}$ and $\boldsymbol{UH}$ are the same, where

$$H_1 = a_{n-1}, \quad H_2 = a_{n-1}r_{21}, \ldots, \quad H_i = a_{n-1}r_{21} \times \cdots \times r_{i1}, \qquad i = 1, 2, \ldots, n.$$

Hence,

$$a_{n-1} = H_1 \quad \text{and} \quad r_{i1} = \frac{H_i}{H_{i-1}} \qquad \text{for } i = 2, \ldots, n.$$

By the Hurwitz theorem, the zeros of the polynomial $p(s)$ are in the open left-half plane if and only if the coefficients r_{i1}, $i = 1, 2 \ldots, n$, which are the elements of the first column of the Routh array, are all positive. The proof is complete.

It is easy to verify that a necessary and sufficient condition for the first- and second-order polynomials to have their zeros in the open left-half plane is that all their coefficients be nonzero and have the same sign. However, this condition is only necessary and not sufficient for third- and higher-order polynomials to have all their zeros in the open left-half plane.

It is possible to extend the above Routh stability criterion so that it can be used to find the number of polynomial zeros with positive real parts. The number of zeros of $p(s)$ with positive real parts is equal to the number of sign changes in the first column of the Routh array. For a proof of this statement using the law of inertia of quadratic forms, see, for example, Kaczorek [143, Section 8.4.3].

◆ Example 4.6

We will use the Routh stability criterion to analyze stability of the linearized model of the flyball governor. The governor's linearized model is given by (2.26) on page 77. The model's characteristic equation is

$$\det \begin{bmatrix} s & -1 & 0 \\ \dfrac{g\sin^2 x_{1e}}{\cos x_{1e}} & s + \dfrac{b}{m} & -\dfrac{2g\sin x_{1e}}{x_{3e}} \\ \dfrac{\kappa}{I}\sin x_{1e} & 0 & s \end{bmatrix} = s^3 + \frac{b}{m}s^2 + \frac{g\sin^2 x_{1e}}{\cos x_{1e}}s + \frac{2g\kappa\sin^2 x_{1e}}{I x_{3e}}$$

$$= s^3 + a_2 s^2 + a_1 s + a_0. \tag{4.43}$$

The coefficients of the above third-order polynomial are all positive. Therefore, the necessary condition for stability is satisfied. We next construct the Routh array for (4.43) to get

$$\begin{array}{c|cc} s^3 & 1 & a_1 \\ s^2 & a_2 & a_0 \\ s^1 & \frac{1}{a_2}(a_2 a_1 - a_0) & \\ s^0 & a_0 & \end{array} \tag{4.44}$$

Because the characteristic polynomial coefficients are all positive, the linearized model is asymptotically stable if and only if

$$a_2 a_1 - a_0 = \frac{b}{m}\frac{g \sin^2 x_{1e}}{\cos x_{1e}} - \frac{2g\kappa \sin^2 x_{1e}}{I x_{3e}} > 0. \tag{4.45}$$

Let

$$\nu = \frac{x_{3e}}{2\kappa \cos x_{1e}}.$$

Using the above notation, we express (4.45) as

$$\frac{bI}{m}\nu > 1. \tag{4.46}$$

It follows from condition (4.46) that:

1. An increase of the mass m of the weights of the governor has a deteriorating effect on the governor's stability.
2. A decrease of the friction coefficient b has also a harmful effect on stability.
3. A decrease of the moment of inertia I of the flywheel harms stability.
4. A decrease of the coefficient ν harms stability.

The above deductions are attributed by Pontryagin [235, p. 219] to the Russian engineer Vyshnegradskiy. In 1876, Vyshnegradskiy used these observations to propose design rules for governors to regulate engines without engine speeds fluctuating.

4.8 Stability of Nonlinear Systems

In this section we will be concerned with stability analysis of nonlinear time-varying dynamical system models described by (4.1)—that is, models of the form

$$\dot{\boldsymbol{x}}(t) = \boldsymbol{f}(t, \boldsymbol{x}(t)), \qquad \boldsymbol{x}(t_0) = \boldsymbol{x}_0,$$

where $\boldsymbol{x}(t) \in \mathbb{R}^n$ is the state vector and $\boldsymbol{f}$ is a vector-valued function.

To proceed, we need the following lemma. Let $B_q(\mathbf{0})$ denote an open ball with the center at $\mathbf{0}$ and radius q, and let $\mathcal{S} = \bar{B}_R(\mathbf{0}) \subset \mathbb{R}^n$ be a closed ball, a compact set, that contains the origin $\boldsymbol{x} = \mathbf{0}$ in its interior.

Lemma 4.2 For a continuously differentiable function $V : \mathbb{R}^n \to \mathbb{R}$, where $V(\mathbf{0}) = 0$, we have $V(\boldsymbol{x}) > 0$ for all $\boldsymbol{x} \in \mathcal{S} \setminus \{\mathbf{0}\}$, if and only if there is a continuous, strictly increasing function $\psi : \mathbb{R} \to \mathbb{R}$ such that $\psi(0) = 0$, $\psi(r) > 0$ for $r > 0$, and $V(\boldsymbol{x}) \geq \psi(r)$, where $r = \|\boldsymbol{x}\|$ and $r \leq R$.

Proof Sufficiency ($\Leftarrow$) is clear.

To prove necessity ($\Rightarrow$), we define, following Willems [300, p. 21], the function $\kappa : \mathbb{R} \to \mathbb{R}$,

$$\kappa(y) = \min_{\substack{\boldsymbol{x} \in \mathcal{S} \\ \|\boldsymbol{x}\| = y}} V(\boldsymbol{x}). \tag{4.47}$$

Note that the function κ is well-defined because the set

$$\{\boldsymbol{x} : \|\boldsymbol{x}\| = y\} \cap \mathcal{S} \tag{4.48}$$

is compact and V is continuous. Therefore, by the Weierstrass theorem on page 650, the function V achieves its infimum over the set (4.48).

We will now show that κ is continuous on $\mathcal{S} = B_R(\mathbf{0})$. Then, we will construct an appropriate function ψ satisfying the hypotheses of the lemma. To show that κ is continuous on $\mathcal{S}$, we will use the fact that V is uniformly continuous in $\mathcal{S}$. To this end let $\boldsymbol{x}^{(1)} \in \mathcal{S}$ be such that $\|\boldsymbol{x}^{(1)}\| = y$ and $\kappa(y) = V(\boldsymbol{x}^{(1)})$. Next, let Δy be an increment of y and let $\boldsymbol{x}^{(2)} \in \mathcal{S}$ be a vector such that $y + \Delta y = \|\boldsymbol{x}^{(2)}\|$. For example, we can take

$$\boldsymbol{x}^{(2)} = \boldsymbol{x}^{(1)} + \frac{\boldsymbol{x}^{(1)}}{\|\boldsymbol{x}^{(1)}\|} \Delta y.$$

Observe that

$$V(\boldsymbol{x}^{(2)}) \geq \kappa(y + \Delta y).$$

Choose now a vector $\boldsymbol{x}^{(3)}$ such that

$$\|\boldsymbol{x}^{(3)}\| = y + \Delta y \quad \text{and} \quad V(\boldsymbol{x}^{(3)}) = \kappa(y + \Delta y).$$

Note that for any vector, say $\boldsymbol{x}^{(4)}$, such that $\|\boldsymbol{x}^{(4)}\| = y$ we have $V(\boldsymbol{x}^{(4)}) \geq \kappa(y)$ (see Figure 4.7). We can take

$$\boldsymbol{x}^{(4)} = \boldsymbol{x}^{(3)} - \frac{\boldsymbol{x}^{(3)}}{\|\boldsymbol{x}^{(3)}\|} \Delta y.$$

By the uniform continuity of V, for any $\epsilon > 0$ there exists a $\delta > 0$ such that if

$$\|\boldsymbol{x}^{(1)} - \boldsymbol{x}^{(2)}\| < \delta \quad \text{and} \quad \|\boldsymbol{x}^{(3)} - \boldsymbol{x}^{(4)}\| < \delta,$$

then

$$\left| V(\boldsymbol{x}^{(1)}) - V(\boldsymbol{x}^{(2)}) \right| < \epsilon \quad \text{and} \quad \left| V(\boldsymbol{x}^{(3)}) - V(\boldsymbol{x}^{(4)}) \right| < \epsilon.$$

From $\left| V(\boldsymbol{x}^{(1)}) - V(\boldsymbol{x}^{(2)}) \right| < \epsilon$, we obtain

$$-\epsilon < V(\boldsymbol{x}^{(1)}) - V(\boldsymbol{x}^{(2)}) < \epsilon.$$

Thus

$$V(\boldsymbol{x}^{(1)}) > V(\boldsymbol{x}^{(2)}) - \epsilon.$$

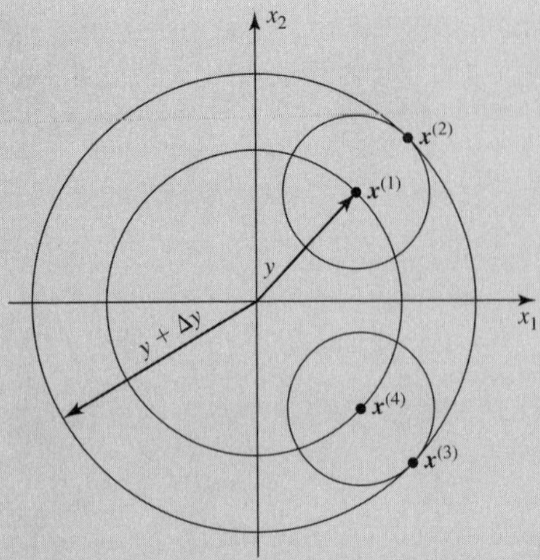

Figure 4.7 A two-dimensional illustration of the proof of the continuity of the function κ given by (4.47).

From $|V(\boldsymbol{x}^{(3)}) - V(\boldsymbol{x}^{(4)})| < \epsilon$, we get

$$V(\boldsymbol{x}^{(3)}) > V(\boldsymbol{x}^{(4)}) - \epsilon.$$

From the definition of the function κ it follows that

$$V(\boldsymbol{x}^{(2)}) \geq V(\boldsymbol{x}^{(3)}) \quad \text{and} \quad V(\boldsymbol{x}^{(4)}) \geq V(\boldsymbol{x}^{(1)}).$$

Because $V(\boldsymbol{x}^{(1)}) > V(\boldsymbol{x}^{(2)}) - \epsilon$, we obtain

$$V(\boldsymbol{x}^{(1)}) > V(\boldsymbol{x}^{(3)}) - \epsilon$$

and therefore

$$-\epsilon < V(\boldsymbol{x}^{(1)}) - V(\boldsymbol{x}^{(3)}). \tag{4.49}$$

Because $V(\boldsymbol{x}^{(3)}) > V(\boldsymbol{x}^{(4)}) - \epsilon$, we have

$$V(\boldsymbol{x}^{(3)}) > V(\boldsymbol{x}^{(1)}) - \epsilon,$$

and thus

$$V(\boldsymbol{x}^{(1)}) - V(\boldsymbol{x}^{(3)}) < \epsilon. \tag{4.50}$$

Combining (4.49) and (4.50), we get

$$\left| V(\boldsymbol{x}^{(1)}) - V(\boldsymbol{x}^{(3)}) \right| < \epsilon,$$

or, equivalently,

$$|\kappa(y) - \kappa(y + \Delta y)| < \epsilon$$

if

$$\Delta y < \delta,$$

which proves the continuity of κ.

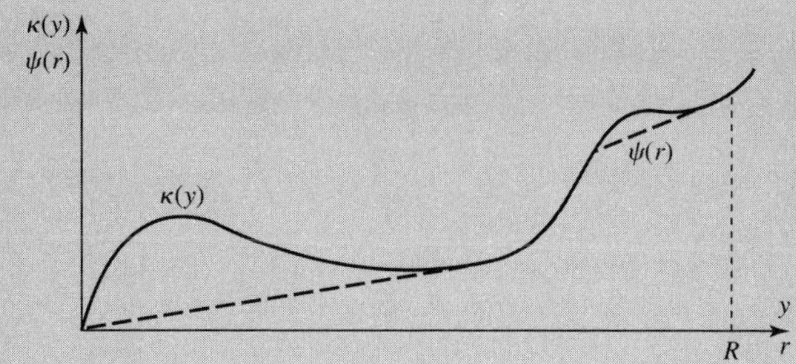

Figure 4.8 Constructing the function ψ given by (4.51).

We are now ready to construct a function ψ that satisfies the hypotheses of the lemma. This function comes from Willems [300, p. 22] and has the form

$$\boxed{\psi(r) = r \min_{r \le y \le R} \frac{\kappa(y)}{y}.} \tag{4.51}$$

Note that ψ is continuous, $\psi(0) = 0$, and it is positive and strictly increasing; see Figure 4.8 for a graphical illustration of the way ψ is constructed from κ. The proof of the lemma is now complete.

We can characterize a positive definite function V in another way as in the following lemma.

Lemma 4.3 A continuously differentiable function $V: \mathbb{R}^n \to \mathbb{R}$ is positive definite in $\mathcal{S} = \bar{B}_R(\mathbf{0}) \subset \mathbb{R}^n$, with respect to $\boldsymbol{x} = \mathbf{0}$, if and only if there exists a continuous nondecreasing function $\alpha: \mathbb{R} \to \mathbb{R}$, such that $\alpha(0) = 0$, $\alpha(r) > 0$ for $r > 0$, and $V(\boldsymbol{x}) \geq \alpha(r)$, where $r = \|\boldsymbol{x}\|$ and $r \leq R$.

A possible choice for α is

$$\alpha(r) = \min_{r \le \|\boldsymbol{x}\| \le R} V(\boldsymbol{x}).$$

Note that the function α is nondecreasing because as r increases, the minimum is taken over a smaller region. Furthermore, $\alpha(r) > 0$ whenever $r > 0$ because $V(\boldsymbol{x}) > 0$ for $\boldsymbol{x} \in \mathcal{S} \setminus \{\mathbf{0}\}$. The function α is continuous because V is continuous.

The main difference between the above two characterizations of positive definite functions is that in Lemma 4.2 the function ψ is strictly increasing rather than just nondecreasing. In our further discussion, we will exploit extensively the existence of such a function ψ associated with a positive definite function V.

So far we only considered positive definite, or negative definite functions, that are time-invariant. We now extend our definition of positive definiteness to time-varying functions of the form $W(t, \boldsymbol{x})$. We first define a time-varying positive semidefinite function. As before, we let $\mathcal{S}$ to denote a compact subset of $\mathbb{R}^n$ containing the origin $\boldsymbol{x} = \mathbf{0}$ in its interior.

Definition 4.11 A real-valued function $W : \mathbb{R} \times \mathbb{R}^n \to \mathbb{R}$ is positive semidefinite, with respect to $\boldsymbol{x} = \mathbf{0}$, in a region $\mathcal{S} \subseteq \mathbb{R}^n$ if for all t and all $\boldsymbol{x}$ in $\mathcal{S}$:

1. W is continuous and has continuous partial derivatives with respect to all of its arguments; that is, $W \in \mathcal{C}^1$.
2. $W(t, \mathbf{0}) = 0$.
3. $W(t, \boldsymbol{x}) \geq 0$.

Definition 4.12 A real-valued function $W : \mathbb{R} \times \mathbb{R}^n \to \mathbb{R}$ is positive definite, with respect to $\boldsymbol{x} = \mathbf{0}$, in a region $\mathcal{S} \subseteq \mathbb{R}^n$ if for all t and all $\boldsymbol{x}$ in $\mathcal{S}$:

1. $W \in \mathcal{C}^1$.
2. $W(t, \mathbf{0}) = 0$.
3. There exists a positive definite time-invariant function V such that $W(t, \boldsymbol{x}) \geq V(\boldsymbol{x})$.

We next introduce a notion of a *decrescent* function.

Definition 4.13 A time-varying function $W : \mathbb{R} \times \mathbb{R}^n \to \mathbb{R}$ is decrescent in a region $\mathcal{S}$ if a positive definite function $\tilde{V}$ exists such that for all $\boldsymbol{x}$ in $\mathcal{S}$ and all t:

$$|W(t, \boldsymbol{x})| \leq \tilde{V}(\boldsymbol{x});$$

that is, W tends to zero uniformly with respect to t as $\|\boldsymbol{x}\| \to 0$.

Now consider an equilibrium state of the dynamical system modeled by (4.1)—that is, a constant state $\boldsymbol{x}_e$ such that $\boldsymbol{f}(t, \boldsymbol{x}_e) = \mathbf{0}$ for all t. Let $W(t, \boldsymbol{x})$ be a time-dependent real-valued function whose time derivative along the solution $\boldsymbol{x}(t) = \boldsymbol{x}(t; t_0, \boldsymbol{x}_0)$ of the given differential equation is

$$\frac{dW(t, \boldsymbol{x})}{dt} = \dot{W}(t, \boldsymbol{x}) = \frac{\partial W}{\partial t} + \nabla W^T \dot{\boldsymbol{x}}.$$

Then, we have the following theorem concerning the stability of the equilibrium point in the sense of Lyapunov.

Theorem 4.7 If in some neighborhood $\mathcal{S}$ of the equilibrium state $\boldsymbol{x}_e$ there is a positive definite function W, with respect to $\boldsymbol{x}_e$, such that its time derivative along any solution of $\dot{\boldsymbol{x}}(t) = \boldsymbol{f}(t, \boldsymbol{x}(t))$ is negative semidefinite in $\mathcal{S}$, then the equilibrium state $\boldsymbol{x}_e$ is stable. If W is also decrescent, then the equilibrium state is uniformly stable.

Proof Without loss of generality, we assume that $\boldsymbol{x}_e = \mathbf{0}$ and follow the arguments of Willems [300, pp. 29, 30]. Because W is positive definite, by Definition 4.12 and Lemma 4.2 there is a strictly increasing function $\psi_1(r)$, $\psi(0) = 0$, such that

$$W(t, \boldsymbol{x}) \geq \psi_1(\|\boldsymbol{x}\|),$$

where $\|\boldsymbol{x}\| \leq R$, and R is the radius of the largest ball contained in $\mathcal{S}$. Let

$$\psi_2(t_0, r) = \max_{\|\boldsymbol{x}\| \leq r} W(t_0, \boldsymbol{x}).$$

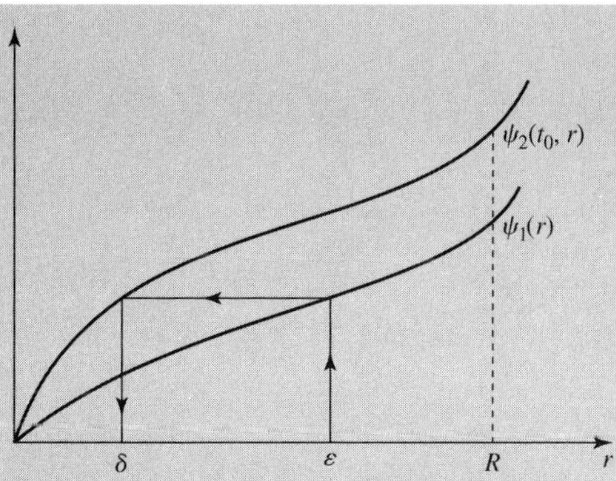

Figure 4.9 Determining a δ given an ϵ.

Then

$$\psi_1(r) \leq \psi_2(t_0, r).$$

Thus, for any $\epsilon > 0$ there is a positive δ that depends on ϵ and t_0, such that

$$\psi_2(t_0, \delta) < \min\{\psi_1(\epsilon), \psi_1(R)\}.$$

Figure 4.9 illustrates how to determine a δ given an ϵ. By assumption, $dW(t, \boldsymbol{x})/dt \leq 0$ along any solution of the system. Hence $W(t, \boldsymbol{x})$ is a nonincreasing function and therefore

$$W(t, \boldsymbol{x}(t; t_0, \boldsymbol{x}_0)) \leq W(t_0, \boldsymbol{x}_0) \qquad \text{for all } t \geq t_0.$$

If $\|\boldsymbol{x}_0\| < \delta$, then for $t \geq t_0$ we have

$$\psi_1(\epsilon) > \psi_2(t_0, \delta) \geq W(t_0, \boldsymbol{x}_0) \geq W(t, \boldsymbol{x}(t; t_0, \boldsymbol{x}_0)) \geq \psi_1(\|\boldsymbol{x}(t; t_0, \boldsymbol{x}_0)\|).$$

Therefore,

$$\psi_1(\epsilon) > \psi_1(\|\boldsymbol{x}(t; t_0, \boldsymbol{x}_0)\|).$$

Because ψ_1 is strictly increasing, we have to have

$$\|\boldsymbol{x}(t; t_0, \boldsymbol{x}_0)\| < \epsilon$$

for all $t \geq t_0$. This means that the null solution is (locally) stable, which completes the proof of the first part of the theorem.

If W is also decrescent, then there is a nondecreasing function $\tilde{\psi}_2 : \mathbb{R} \rightarrow \mathbb{R}$, $\tilde{\psi}_2(0) = 0$, and

$$W(t, \boldsymbol{x}) \leq \tilde{\psi}_2(\|\boldsymbol{x}\|)$$

for all $\boldsymbol{x}$ such that $\|\boldsymbol{x}\| \leq R$ and all t. Indeed, because $W(t, \boldsymbol{x})$ is positive definite and decrescent, there exists a positive definite $V(\boldsymbol{x})$ such that $W(t, \boldsymbol{x}) \leq V(\boldsymbol{x})$. Then, we can take $\tilde{\psi}_2$ defined as

$$\tilde{\psi}_2(r) = \max_{\|\boldsymbol{x}\| \leq r} V(\boldsymbol{x}).$$

Let

$$\tilde{\psi}_2(\delta) < \min\{\psi_1(\epsilon), \psi_1(R)\}.$$

Note that δ depends now only on ϵ, and not on t_0 as before. To show the uniform stability of the null solution, we repeat the argument above with $\tilde{\psi}_2$ replacing ψ_2. This completes the proof of the theorem.

The next theorem gives us a sufficiency condition for an equilibrium point to be uniformly asymptotically stable.

Theorem 4.8 If in some neighborhood $\mathcal{S}$ of the given equilibrium point $\boldsymbol{x}_e$ there is a positive definite, with respect to $\boldsymbol{x}_e$, and decrescent function W such that its time derivative along any solution of the system model (4.1) is negative definite in $\mathcal{S}$, then the equilibrium state is uniformly asymptotically stable.

Proof As before, without loss of generality, we assume that $\boldsymbol{x}_e = \boldsymbol{0}$ and then follow the arguments of Willems [300, pp. 30, 31]. The previous theorem ensures uniform stability of the equilibrium state $\boldsymbol{x}_e$. Thus, only the uniform convergence remains to be shown. From the assumptions of the theorem it follows that there exist strictly increasing functions ψ_1 and ψ_3, along with a nondecreasing function ψ_2 such that

$$\begin{aligned}\psi_1(0) &= \psi_2(0) = \psi_3(0),\\ \psi_1(\|\boldsymbol{x}\|) &\le W(t, \boldsymbol{x}) \le \psi_2(\|\boldsymbol{x}\|),\\ \psi_3(\|\boldsymbol{x}\|) &\le -\dot{W}(t, \boldsymbol{x}),\end{aligned}$$

for all t and all $\boldsymbol{x}$ such that $\|\boldsymbol{x}\| \le R$, where R is the radius of the largest ball contained in $\mathcal{S}$. First select δ_1 so that

$$\psi_2(\delta_1) < \psi_1(R), \tag{4.52}$$

Next, given an $\epsilon_1 > 0$, we define positive constants δ_2 and T as follows:

$$\psi_2(\delta_2) < \min\{\psi_1(\epsilon_1), \psi_2(\delta_1)\}, \tag{4.53}$$

$$T = \psi_1(R)/\psi_3(\delta_2). \tag{4.54}$$

Note that δ_1 and T are independent of t_0.

We shall prove the theorem in two stages. First, we shall show that for some $t_2 \in [t_0, t_0 + T]$,

$$\|\boldsymbol{x}(t_2; t_0, \boldsymbol{x}_0)\| < \delta_2$$

whenever $\|\boldsymbol{x}_0\| < \delta_1$. Then, we shall show that once $\|\boldsymbol{x}(t_2; t_0, \boldsymbol{x}_0)\| < \delta_2$, then

$$\|\boldsymbol{x}(t; t_0, \boldsymbol{x}_0)\| < \epsilon_1 \qquad \text{for all } t \ge t_2.$$

We begin with proving the claim that for some $t_2 \in [t_0, t_0 + T]$,

$$\|\boldsymbol{x}(t_2; t_0, \boldsymbol{x}_0)\| < \delta_2$$

whenever $\|\boldsymbol{x}_0\| < \delta_1$. We prove this claim by contradiction. We assume that for all $t \in [t_0, t_0 + T]$,

$$\|\boldsymbol{x}(t; t_0, \boldsymbol{x}_0)\| \ge \delta_2 \qquad \text{if } \|\boldsymbol{x}_0\| < \delta_1. \tag{4.55}$$

Using (4.55), (4.54), and (4.52), we obtain

$$\begin{aligned}
0 < \psi_1(\delta_2) &\le \psi_1(\|\boldsymbol{x}(t; t_0, \boldsymbol{x}_0)\|) \\
&\le W[t_0 + T, \boldsymbol{x}(t_0 + T; t_0, \boldsymbol{x}_0)] \\
&= W(t_0, \boldsymbol{x}_0) + \int_{t_0}^{t_0+T} \dot{W}[t, \boldsymbol{x}(t; t_0, \boldsymbol{x}_0)]\, dt \\
&\le W(t_0, \boldsymbol{x}_0) - \int_{t_0}^{t_0+T} \psi_3(\|\boldsymbol{x}(t; t_0, \boldsymbol{x}_0)\|)\, dt \\
&\le W(t_0, \boldsymbol{x}_0) - T\psi_3(\delta_2) \\
&\le \psi_2(\|\boldsymbol{x}_0\|) - T\psi_3(\delta_2) \\
&< \psi_2(\delta_1) - \psi_1(R) \\
&< 0,
\end{aligned}$$

which is a contradiction, and thus we proved the first part of the theorem. We now proceed with the second part. It follows from the first part that if $\|\boldsymbol{x}_0\| < \delta_1$, then for some $t_2 \in [t_0, t_0 + T]$ we have $\|\boldsymbol{x}(t_2; t_0, \boldsymbol{x}_0)\| < \delta_2$. Hence, for $t \ge t_2$,

$$\psi_1(\|\boldsymbol{x}(t; t_0, \boldsymbol{x}_0)\|) \le W(t, \boldsymbol{x}(t; t_0, \boldsymbol{x}_0)) \le W(t_2, \boldsymbol{x}(t_2; t_0, \boldsymbol{x}_0)) < \psi_2(\delta_2) < \psi_1(\epsilon_1).$$

Therefore,

$$\psi_1(\|\boldsymbol{x}(t; t_0, \boldsymbol{x}_0)\|) < \psi_1(\epsilon_1),$$

and because ψ_1 is strictly increasing we obtain

$$\|\boldsymbol{x}(t; t_0, \boldsymbol{x}_0)\| < \epsilon_1$$

for all $t \ge t_2$, which proves the uniform convergence. The proof of the theorem is thus complete.

◆ Example 4.7

Given a dynamical system model

$$\dot{\boldsymbol{x}}(t) = \begin{bmatrix} -0.75 & 0.3536 \\ 0.3536 & -0.5 \end{bmatrix} \boldsymbol{x}(t), \qquad \boldsymbol{x}(0) = \boldsymbol{x}_0.$$

The solution $\boldsymbol{P} = \boldsymbol{P}^T$ to the Lyapunov matrix equation $\boldsymbol{A}^T\boldsymbol{P} + \boldsymbol{P}\boldsymbol{A} = -\boldsymbol{Q}$, where $\boldsymbol{Q} = 2\boldsymbol{I}_2$, is

$$\boldsymbol{P} = \begin{bmatrix} 2 & \sqrt{2} \\ \sqrt{2} & 3 \end{bmatrix}.$$

The eigenvalues of $\boldsymbol{P}$ are

$$\lambda_1 = \lambda_{\min}(\boldsymbol{P}) = 1, \qquad \lambda_2 = \lambda_{\max}(\boldsymbol{P}) = 4.$$

In general, for a symmetric matrix $\boldsymbol{P}$,

$$\lambda_{\min}(\boldsymbol{P})\|\boldsymbol{x}\|^2 \le \boldsymbol{x}^T\boldsymbol{P}\boldsymbol{x} \le \lambda_{\max}(\boldsymbol{P})\|\boldsymbol{x}\|^2.$$

Therefore, we can take

$$\psi_1(\|\boldsymbol{x}\|) = \lambda_{\min}(\boldsymbol{P})\|\boldsymbol{x}\|^2, \quad \psi_2(\|\boldsymbol{x}\|) = \lambda_{\max}(\boldsymbol{P})\|\boldsymbol{x}\|^2, \quad \text{and} \quad \psi_3(\|\boldsymbol{x}\|) = \lambda_{\min}(\boldsymbol{Q})\|\boldsymbol{x}\|^2.$$

In our example, $\lambda_{\min}(\boldsymbol{Q}) = 2$.

- Given $\varepsilon = 4$. We will find $\delta > 0$ such that if $\|\boldsymbol{x}_0\| \le \delta$, then $\|\boldsymbol{x}(t)\| \le \varepsilon$ for $t \ge 0$. We obtain $\delta = \delta(\varepsilon)$ by solving the equation,

$$\psi_1(\varepsilon) = \psi_2(\delta),$$

  to get

$$\delta = \sqrt{\frac{\lambda_{\min}(\boldsymbol{P})}{\lambda_{\max}(\boldsymbol{P})}}\varepsilon = 2.$$

- Given $\boldsymbol{x}_0 = [8 \;\; 6]^T$ and $\varepsilon_1 = 2$. We will find T such that $\|\boldsymbol{x}(t)\| \le \varepsilon_1$ for all $t \ge T$. We have $\|\boldsymbol{x}_0\| = 10$. Let $\delta_1 = 10$. We then compute R, δ_2, and T from the equations

$$\begin{aligned}\psi_2(\delta_1) &= \psi_1(R),\\ \psi_2(\delta_2) &= \min(\psi_1(\varepsilon_1), \psi_2(\delta_1)),\\ T &= \frac{\psi_1(R)}{\psi_3(\delta_2)}.\end{aligned}$$

  Using the first equation, we find

$$R = \sqrt{\frac{\lambda_{\max}(\boldsymbol{P})}{\lambda_{\min}(\boldsymbol{P})}}\delta_1 = 20.$$

  We then solve for $\delta_2 \le \delta_1$ to get

$$\delta_2 = \sqrt{\frac{\lambda_{\min}(\boldsymbol{P})}{\lambda_{\max}(\boldsymbol{P})}}\varepsilon_1 = 1.$$

  Hence,

$$T = \frac{\lambda_{\min}(\boldsymbol{P})R^2}{\lambda_{\min}(\boldsymbol{Q})\delta_2^2} = \frac{\lambda_{\min}(\boldsymbol{P})\frac{\lambda_{\max}(\boldsymbol{P})}{\lambda_{\min}(\boldsymbol{P})}\delta_1^2}{\lambda_{\min}(\boldsymbol{Q})\frac{\lambda_{\min}(\boldsymbol{P})}{\lambda_{\max}(\boldsymbol{P})}\varepsilon_1^2} = \frac{\lambda_{\max}^2(\boldsymbol{P})\delta_1^2}{\lambda_{\min}(\boldsymbol{Q})\lambda_{\min}(\boldsymbol{P})\varepsilon_1^2} = 200.$$

Proposition 4.1 If in some neighborhood $\mathcal{S}$ of $\boldsymbol{x}_e = \mathbf{0}$ the positive definite and decrescent function W satisfies

$$c_1\|\boldsymbol{x}\|^2 \le W(t, \boldsymbol{x}) \le c_2\|\boldsymbol{x}\|^2$$

for some positive constants c_1 and c_2 and its time derivative along any solution of the system (4.1) is negative definite in $\mathcal{S}$ and such that for some positive constant c_3,

$$\dot{W}(t, \boldsymbol{x}) \le -c_3\|\boldsymbol{x}\|^2,$$

then the equilibrium state $\boldsymbol{x}_e = \mathbf{0}$ is uniformly exponentially stable.

Proof We have

$$\begin{aligned}\dot{W}(t, \boldsymbol{x}) &\le -c_3\|\boldsymbol{x}\|^2 \\ &= -\frac{c_3}{c_2}(c_2\|\boldsymbol{x}\|^2) \\ &\le -\frac{c_3}{c_2}W(t, \boldsymbol{x}).\end{aligned}$$

Applying to the above Theorem A.20, which is called the comparison theorem, we obtain

$$W(t, \boldsymbol{x}) \le W(t_0, \boldsymbol{x}_0)\exp\left(-\frac{c_3}{c_2}(t - t_0)\right).$$

Hence,

$$\begin{aligned}\|\boldsymbol{x}\| &\le \sqrt{\frac{W(t, \boldsymbol{x})}{c_1}} \\ &\le \sqrt{\frac{W(t_0, \boldsymbol{x}_0)\exp\left(-\frac{c_3}{c_2}(t - t_0)\right)}{c_1}} \\ &\le \left(\frac{c_2}{c_1}\right)^{\frac{1}{2}}\|\boldsymbol{x}(t_0)\|\exp\left(-\frac{c_3}{2c_2}(t - t_0)\right),\end{aligned}$$

which means that the equilibrium state $\boldsymbol{x}_e = \mathbf{0}$ is uniformly exponentially stable.

The above theorems were concerned with local stability. We now proceed to prove sufficiency conditions for global stability. For this we need the following definitions. We first define a *radially unbounded* or *coercive* function.

Definition 4.14 A time-varying function $W : \mathbb{R} \times \mathbb{R}^n \to \mathbb{R}$ is radially unbounded, or coercive, if as $\|\boldsymbol{x}\| \to \infty$ the function W tends to infinity uniformly with respect to t; that is, for any given positive real number M, there is a positive real number N so that $W(t, \boldsymbol{x}) > M$ for all t and all $\boldsymbol{x}$ whenever $\|\boldsymbol{x}\| > N$.

The next two definitions are concerned with *Lagrange stable equilibrium states*.

Definition 4.15 The equilibrium state $\boldsymbol{x}_e$ is called Lagrange stable, or bounded, if for every real number $d > 0$ there exists a positive real number $b = b(t_0, d)$ such that if

$$\|\boldsymbol{x}_0 - \boldsymbol{x}_e\| < d,$$

then

$$\|\boldsymbol{x}(t; t_0, \boldsymbol{x}_0) - \boldsymbol{x}_e\| < b$$

for all $t \geq t_0$.

Definition 4.16 The equilibrium state $\boldsymbol{x}_e$ is uniformly bounded if the number b in Definition 4.15 can be taken independent of the initial time t_0.

We now define *uniformly asymptotically stable in the large* equilibrium state.

Definition 4.17 If the equilibrium state $\boldsymbol{x}_e$ is uniformly stable and uniformly bounded and all trajectories are uniformly convergent to $\boldsymbol{x}_e$, then this equilibrium state is called uniformly asymptotically stable in the large, or globally uniformly asymptotically stable.

The following theorem gives a sufficiency condition for an equilibrium state to be globally uniformly asymptotically stable.

Theorem 4.9 If there is a function W that is positive definite for all $\boldsymbol{x} \neq \boldsymbol{x}_e$, $W(t, \boldsymbol{x}_e) = 0$, as well as decrescent and radially unbounded, and if its time derivative along any solution of the system is negative definite for all $\boldsymbol{x}$, then the equilibrium state $\boldsymbol{x}_e$ is globally uniformly asymptotically stable.

Proof We assume, for simplicity, that $\boldsymbol{x}_e = \boldsymbol{0}$ and follow the arguments of Willems [300, pp. 31, 32]. By Theorem 4.7, the equilibrium state $\boldsymbol{x}_e$ is uniformly stable. Similarly as in Theorem 4.8, we construct the functions ψ_1, ψ_2, and ψ_3. Since, by assumption, W is radially unbounded, $\psi_1(\|\boldsymbol{x}\|) \to \infty$ as $\|\boldsymbol{x}\| \to \infty$. Therefore, for any constant $d > 0$, there is a constant $b > 0$ such that $\psi_1(b) > \psi_2(d)$. See Figure 4.10 for an illustration of the procedure that can be used to find the constant b. If $\|\boldsymbol{x}_0\| < d$, then $\|\boldsymbol{x}(t; t_0, \boldsymbol{x}_0)\| < b$ for all $t \geq t_0$. Indeed, because

$$\psi_1(b) > \psi_2(d) \geq W(t_0, \boldsymbol{x}_0) \geq W(t, \boldsymbol{x}(t; t_0, \boldsymbol{x}_0)) \geq \psi_1(\|\boldsymbol{x}(t; t_0, \boldsymbol{x}_0)\|),$$

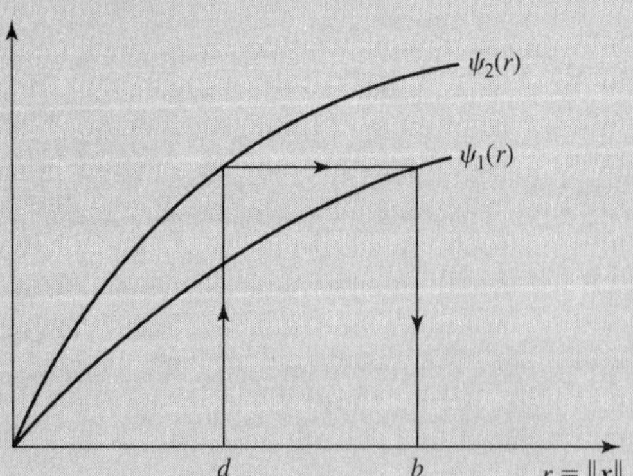

Figure 4.10 Determining a constant $b > 0$ given $d > 0$ in the proof of Theorem 4.9.

then

$$\psi_1(b) > \psi_1(\|\boldsymbol{x}(t; t_0, \boldsymbol{x}_0)\|).$$

The above implies that

$$\|\boldsymbol{x}(t; t_0, \boldsymbol{x}_0)\| < b$$

because ψ_1 is strictly increasing. Hence all trajectories of the system are uniformly bounded. To prove their convergence to the equilibrium state, we define δ_2, T, and b, given d and ϵ_1, as

$$\psi_2(d) < \psi_1(b),$$
$$\psi_2(\delta_2) < \psi_1(\epsilon_1),$$
$$T = \psi_1(b)/\psi_3(\delta_2).$$

If $\|\boldsymbol{x}_0\| < d$, then for all $t > t_0 + T$, we have

$$\|\boldsymbol{x}(t; t_0, \boldsymbol{x}_0)\| < \epsilon_1,$$

which can be shown using the same arguments as in the proof of Theorem 4.8 with b playing the role of R, and d playing the role of δ_1. This proves that all trajectories of the system converge uniformly to the equilibrium state. Hence, the equilibrium state is globally uniformly asymptotically stable.

We now give a theorem that can be used as a test for checking the instability of nonlinear dynamical systems. This instability test is illustrated in Figure 4.11.

Theorem 4.10 Let $\boldsymbol{x}_e$ be an equilibrium state of the system $\dot{\boldsymbol{x}} = \boldsymbol{f}(t, \boldsymbol{x})$, that is, $\boldsymbol{f}(t, \boldsymbol{x}_e) = \boldsymbol{0}$. If there exists a function $W: \mathbb{R} \times \mathbb{R}^n \to \mathbb{R}$ such that in the region $S = \{\boldsymbol{x} : \|\boldsymbol{x} - \boldsymbol{x}_e\| \le R\}$

1. $W \in C^1$,
2. $W(t, \boldsymbol{x}_e) = 0$,

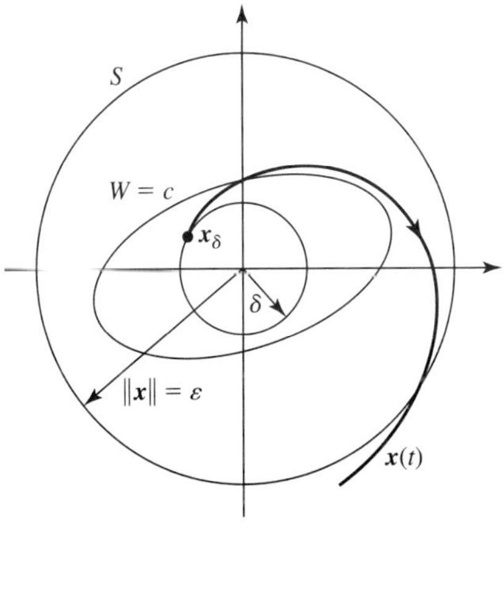

Figure 4.11 An illustration of Theorem 4.10.

3. $\dot{W}(t, \boldsymbol{x}) > 0$ for all $\boldsymbol{x} \in \mathcal{S} \setminus \{\boldsymbol{x}_e\}$, and
4. in every neighborhood of $\boldsymbol{x}_e$ in $\mathcal{S}$, for any t_0 there is at least one point $\boldsymbol{x}$ at which $W(t_0, \boldsymbol{x}) > 0$,

then the equilibrium state $\boldsymbol{x}_e$ is unstable.

Proof Without loss of generality, assume that $\boldsymbol{x}_e = \boldsymbol{0}$. By assumption 4, for any t_0 and any δ, $0 < \delta < R$, there exists $\boldsymbol{x}_\delta$, $0 < \|\boldsymbol{x}_\delta\| < \delta$, such that $W(t_0, \boldsymbol{x}_\delta) > 0$. The system trajectory $\boldsymbol{x}(t) = \boldsymbol{x}(t; t_0, \boldsymbol{x}_\delta)$ cannot approach the equilibrium state $\boldsymbol{x}_e = \boldsymbol{0}$ as $t \to \infty$ because $W(t, \boldsymbol{0}) = 0$ but $W(t_0, \boldsymbol{x}_\delta) > 0$ and $\dot{W}(t, \boldsymbol{x}) > 0$ for $\boldsymbol{x} \neq \boldsymbol{x}_e$. Since $\dot{W}$ is continuous, because $W \in \mathcal{C}^1$, and the trajectory is bounded away from the equilibrium state, there is a constant $m > 0$ such that for all $t \geq t_0$,

$$\dot{W}(t, \boldsymbol{x}(t)) \geq m > 0,$$

as long as

$$\|\boldsymbol{x}(t)\| \leq R, \qquad \boldsymbol{x} \neq \boldsymbol{0}.$$

Hence,

$$W(t, \boldsymbol{x}) = W(t_0, \boldsymbol{x}_\delta) + \int_{t_0}^{t} \dot{W}(s, \boldsymbol{x}(s))\, ds \geq W(t_0, \boldsymbol{x}_\delta) + m(t - t_0). \tag{4.56}$$

The function W is continuous and therefore bounded on the compact set $\mathcal{S}$. However, the right-hand side of (4.56) is unbounded. Therefore the trajectory $\boldsymbol{x}(t; t_0, \boldsymbol{x}_\delta)$ cannot remain infinitely long in $\mathcal{S}$. Thus, for some $t_1 > t_0$ the trajectory reaches the boundary of the region $\|\boldsymbol{x}\| \leq R$. Hence, given any ϵ, $0 < \epsilon < R$, then for any $\delta > 0$, no matter how small, there is at least one trajectory such that $\|\boldsymbol{x}_\delta\| < \delta$ but $\|\boldsymbol{x}(t; t_0, \boldsymbol{x}_\delta)\| \geq \epsilon$ for some $t_1 > t_0$. Therefore, the equilibrium state is unstable.

Corollary 4.1 If there exists a function $W : \mathbb{R} \times \mathbb{R}^n \to \mathbb{R}$ such that in the region $\mathcal{S} = \{\boldsymbol{x} : \|\boldsymbol{x} - \boldsymbol{x}_e\| \leq R\}$

1. $W \in \mathcal{C}^1$,
2. $W(t, \boldsymbol{x}_e) = 0$,
3. $\dot{W}(t, \boldsymbol{x}) < 0$ for all $\boldsymbol{x} \in \mathcal{S} \setminus \{\boldsymbol{x}_e\}$, and
4. in every neighborhood of $\boldsymbol{x}_e$ in $\mathcal{S}$, for any t_0 there is at least one point $\boldsymbol{x}$ at which $W(t_0, \boldsymbol{x}) < 0$,

then the equilibrium state $\boldsymbol{x}_e$ is unstable.

4.9 Lyapunov's Indirect Method

In this section we are concerned with the problem of investigating stability properties of an equilibrium state of a nonlinear system based on its linearization about the given equilibrium. We devise a method that allows one to determine whether the equilibrium of the nonlinear system is asymptotically stable or unstable based on the dynamics of the linearized system about this equilibrium. The method is sometimes referred to as the *Lyapunov's first method* or *Lyapunov's indirect method*. For simplicity of notation, we assume that the equilibrium point to be tested for stability is the origin. We consider a nonlinear system model,

$$\dot{\boldsymbol{x}} = \boldsymbol{f}(\boldsymbol{x}), \tag{4.57}$$

where $\boldsymbol{f} : \mathcal{D} \rightarrow \mathbb{R}^n$ is a continuously differentiable function from a domain $\mathcal{D} \subseteq \mathbb{R}^n$ into $\mathbb{R}^n$. We assume that the origin $\boldsymbol{x} = \boldsymbol{0}$ is in the interior of $\mathcal{D}$ and is an equilibrium state of (4.57), that is, $\boldsymbol{f}(\boldsymbol{0}) = \boldsymbol{0}$. Let

$$\boldsymbol{A} = \left.\frac{\partial \boldsymbol{f}}{\partial \boldsymbol{x}}\right|_{\boldsymbol{x}=\boldsymbol{0}} = \left.\begin{bmatrix} \frac{\partial f_1}{\partial x_1} & \frac{\partial f_1}{\partial x_2} & \cdots & \frac{\partial f_1}{\partial x_n} \\ \vdots & & \ddots & \vdots \\ \frac{\partial f_n}{\partial x_1} & \frac{\partial f_n}{\partial x_2} & \cdots & \frac{\partial f_n}{\partial x_n} \end{bmatrix}\right|_{\boldsymbol{x}=\boldsymbol{0}} \tag{4.58}$$

be the Jacobian matrix of $\boldsymbol{f}$ evaluated at $\boldsymbol{x} = \boldsymbol{0}$. Then, applying Taylor's theorem to the right-hand side of (4.57) yields

$$\dot{\boldsymbol{x}} = \boldsymbol{A}\boldsymbol{x} + \boldsymbol{g}(\boldsymbol{x}), \tag{4.59}$$

where the function $\boldsymbol{g}$ represents higher-order terms and has the property

$$\lim_{\|\boldsymbol{x}\| \to 0} \frac{\|\boldsymbol{g}(\boldsymbol{x})\|}{\|\boldsymbol{x}\|} = 0. \tag{4.60}$$

The above suggests that in a "small" neighborhood of the origin, we can approximate the nonlinear system (4.57) with its linearized system about the origin, $\dot{\boldsymbol{x}} = \boldsymbol{A}\boldsymbol{x}$. The following theorems give conditions under which we can infer about stability properties of the origin of the nonlinear system based on stability properties of the linearized system.

Theorem 4.11 Let the origin, $\boldsymbol{x} = 0$, be an equilibrium state of the nonlinear system, $\dot{\boldsymbol{x}} = \boldsymbol{f}(\boldsymbol{x})$. Then, the origin is an asymptotically stable equilibrium of the nonlinear system if $\boldsymbol{A}$, the Jacobian matrix of $\boldsymbol{f}$ evaluated at the origin, has all its eigenvalues in the open left-half plane.

Proof Our proof is based on that of Khalil [157, pp. 127–128]. To begin with, we suppose that all the eigenvalues of $\boldsymbol{A}$ are in the open left-half plane. Then, by Theorem 4.1, for any positive definite symmetric matrix $\boldsymbol{Q}$, the solution $\boldsymbol{P}$ to the Lyapunov equation (4.6) is also positive definite. We employ the positive definite quadratic form $V(\boldsymbol{x}) = \boldsymbol{x}^T\boldsymbol{P}\boldsymbol{x}$ as a Lyapunov function candidate for the nonlinear system $\dot{\boldsymbol{x}} = \boldsymbol{f}(\boldsymbol{x})$. We use (4.59) and the Lyapunov matrix equation (4.6) when evaluating the time derivative of $V(\boldsymbol{x})$ on the trajectories of the nonlinear system. We obtain

$$\begin{aligned} \dot{V}(\boldsymbol{x}) &= 2\boldsymbol{x}^T\boldsymbol{P}\,\boldsymbol{f}(\boldsymbol{x}) \\ &= 2\boldsymbol{x}^T\boldsymbol{P}(\boldsymbol{A}\boldsymbol{x} + \boldsymbol{g}(\boldsymbol{x})) \\ &= \boldsymbol{x}^T(\boldsymbol{A}^T\boldsymbol{P} + \boldsymbol{P}\boldsymbol{A})\boldsymbol{x} + 2\boldsymbol{x}^T\boldsymbol{P}\boldsymbol{g}(\boldsymbol{x}) \\ &= -\boldsymbol{x}^T\boldsymbol{Q}\boldsymbol{x} + 2\boldsymbol{x}^T\boldsymbol{P}\boldsymbol{g}(\boldsymbol{x}). \end{aligned} \tag{4.61}$$

The first term on the right-hand side of the above equation is negative definite, while the second term is, in general, indefinite. The function $\boldsymbol{g}$ satisfies

$$\frac{\|\boldsymbol{g}(\boldsymbol{x})\|_2}{\|\boldsymbol{x}\|_2} \to 0 \qquad \text{as } \|\boldsymbol{x}\|_2 \to 0.$$

Therefore, for any $\gamma > 0$ there exists $r > 0$ such that

$$\|\boldsymbol{g}(\boldsymbol{x})\|_2 < \gamma \|\boldsymbol{x}\|_2 \qquad \text{for all } \|\boldsymbol{x}\|_2 < r. \tag{4.62}$$

Substituting the above into (4.61) gives

$$\begin{aligned}\dot{V}(\boldsymbol{x}) &< -\boldsymbol{x}^T \boldsymbol{Q}\boldsymbol{x} + 2\gamma \|\boldsymbol{P}\|_2 \|\boldsymbol{x}\|_2^2 \\ &\le -(\lambda_{\min}(\boldsymbol{Q}) - 2\gamma\lambda_{\max}(\boldsymbol{P})) \|\boldsymbol{x}\|_2^2 \qquad \text{for all } \|\boldsymbol{x}\|_2 < r.\end{aligned} \tag{4.63}$$

Thus, selecting

$$\gamma < \frac{\lambda_{\min}(\boldsymbol{Q})}{2\lambda_{\max}(\boldsymbol{P})}$$

ensures that $\dot{V}(\boldsymbol{x})$ is negative definite. By Theorem 4.8, we conclude that the origin, $\boldsymbol{x} = \boldsymbol{0}$, is an asymptotically stable equilibrium state of the nonlinear system, which completes the proof.

Theorem 4.12 Let the origin, $\boldsymbol{x} = \boldsymbol{0}$, be an equilibrium state of the nonlinear system, $\dot{\boldsymbol{x}} = \boldsymbol{f}(\boldsymbol{x})$. Then, the origin is unstable if $\boldsymbol{A}$ has one or more eigenvalues in the right-half plane.

Proof Our proof follows that of Khalil [157, pp. 128–130]. We first consider the special case when $\boldsymbol{A}$ has no eigenvalues on the imaginary axis. Suppose that the eigenvalues cluster into a group in the open right-half plane and a group in the open left-half plane. Then, we can find a nonsingular matrix $\boldsymbol{T}$ such that

$$\boldsymbol{T}\boldsymbol{A}\boldsymbol{T}^{-1} = \begin{bmatrix} -\boldsymbol{A}_1 & \boldsymbol{O} \\ \boldsymbol{O} & \boldsymbol{A}_2 \end{bmatrix},$$

where both $\boldsymbol{A}_1$ and $\boldsymbol{A}_2$ have their eigenvalues in the open left-half plane. A possible candidate for $\boldsymbol{T}$ would be the matrix that transforms $\boldsymbol{A}$ into its real Jordan form. Now, let

$$\boldsymbol{z} = \boldsymbol{T}\boldsymbol{x} = \begin{bmatrix} \boldsymbol{z}_1 \\ \boldsymbol{z}_2 \end{bmatrix},$$

where the partition of $\boldsymbol{z}$ is compatible with the dimensions of $\boldsymbol{A}_1$ and $\boldsymbol{A}_2$. The change of variables, $\boldsymbol{z} = \boldsymbol{T}\boldsymbol{x}$, transforms the nonlinear system (4.59) into the form

$$\begin{aligned}\dot{\boldsymbol{z}}_1 &= -\boldsymbol{A}_1\boldsymbol{z}_1 + \boldsymbol{g}_1(\boldsymbol{z}), \\ \dot{\boldsymbol{z}}_2 &= \boldsymbol{A}_2\boldsymbol{z}_2 + \boldsymbol{g}_2(\boldsymbol{z}),\end{aligned} \tag{4.64}$$

where the functions g_i, $i = 1, 2$, have the property that for any $\gamma > 0$, there exists $r > 0$ such that for $i = 1, 2$, we have

$$\|g_i(z)\|_2 \le \gamma \|z\|_2 \qquad \text{for all } \|z\|_2 \le r. \tag{4.65}$$

The origin $z = 0$ is an equilibrium state for the nonlinear system (4.64). Because the matrix T is nonsingular, stability properties of $z = 0$ in the z coordinates are the same as the stability properties of $x = 0$ in the x coordinates. We will now show that $z = 0$ is an unstable equilibrium of the system (4.64), which implies that $x = 0$ is an unstable equilibrium of the system (4.57). We use Theorem 4.10, which entails constructing a function V satisfying the hypotheses of that theorem. Let Q_1 and Q_2 be positive definite symmetric matrices of the dimensions of A_1 and A_2, respectively. Because A_1 and A_2 have their eigenvalues in the open left-half plane, the Lyapunov matrix equations,

$$A_i^T P_i + P_i A_i = -Q_i, \qquad i = 1, 2,$$

have unique positive definite solutions P_1 and P_2. Let

$$V(z) = z_1^T P_1 z_1 - z_2^T P_2 z_2 = z^T \begin{bmatrix} P_1 & O \\ O & -P_2 \end{bmatrix} z. \tag{4.66}$$

In the subspace, $\{z = [\, z_1^T \;\; z_2^T \,]^T : z_2 = 0\}$, $V(z) > 0$ in every neighborhood of the origin. Let

$$\mathcal{U} = \{z \in \mathbb{R}^n : \|z\|_2 \le r\}.$$

The time derivative of V evaluated on the trajectories of the system (4.64) is

$$\begin{aligned}
\dot V(z) &= 2z_1^T P_1 \dot z_1 - 2z_2^T P_2 \dot z_2 \\
&= 2z_1^T P_1(-A_1 z_1 + g_1(z)) - 2z_2^T P_2(A_2 z_2 + g_2(z)) \\
&= z_1^T Q_1 z_1 + 2z_1^T P_1 g_1(z) + z_2^T Q_2 z_2 - 2z_2^T P_2 g_2(z) \\
&\ge \lambda_{\min}(Q_1)\|z_1\|_2^2 + \lambda_{\min}(Q_2)\|z_2\|_2^2 + 2z^T \begin{bmatrix} P_1 g_1(z) \\ -P_2 g_2(z) \end{bmatrix} \\
&\ge \lambda_{\min}(Q_1)\|z_1\|_2^2 + \lambda_{\min}(Q_2)\|z_2\|_2^2 \\
&\quad - 2\|z\|_2 \sqrt{\|P_1\|_2^2 \|g_1(z)\|_2^2 + \|P_2\|_2^2 \|g_2(z)\|_2^2}.
\end{aligned} \tag{4.67}$$

Let

$$\alpha = \min(\lambda_{\min}(Q_1), \lambda_{\min}(Q_2)) \quad \text{and} \quad \beta = \max(\|P_1\|_2, \|P_2\|_2).$$

With (4.65) and the above taken into account, (4.67) evaluates to

$$\dot V(z) \ge (\alpha - 2\sqrt{2}\beta\gamma)\|z\|_2^2 \qquad \text{for all } z \in \mathcal{U}. \tag{4.68}$$

If we select

$$\gamma < \frac{\alpha}{2\sqrt{2}\beta},$$

then $\dot{V}(z) > 0$ in $\mathcal{U}\backslash\{0\}$. Thus, all hypotheses of Theorem 4.10 are satisfied and hence the origin is an unstable equilibrium of the system (4.57).

In the general case when A has eigenvalues on the imaginary axis, in addition to eigenvalues in the open right-half plane, we shift the imaginary axis in order to reduce this case to the previous one. Specifically, suppose that A has m eigenvalues in the open right-half plane with $\Re\lambda_i > \delta > 0$. Then, the matrix

$$A - \frac{\delta}{2} I$$

still has m eigenvalues in open right-half plane, but no eigenvalues on the imaginary axis. We can proceed similarly as in the previous case to conclude that the origin is an unstable equilibrium of the corresponding nonlinear system (4.57). For details of this part of the proof, we refer to Khalil [157, p. 130].

In summary, the previous two theorems provide us with a simple method for determining asymptotic stability or instability of an equilibrium state of the nonlinear system (4.57). We compute the Jacobian matrix (4.58) of the nonlinear system, evaluate it at the equilibrium state of interest, and test its eigenvalues. If $\Re\lambda_i < 0$ for all i, then the equilibrium state is asymptotically stable. If for some i, $\Re\lambda_i > 0$, then the equilibrium is unstable. Finally, if $\Re\lambda_i \le 0$ for all i with $\Re\lambda_i = 0$ for some i, linearization fails to determine asymptotic stability of the equilibrium of the nonlinear system. In this case, the stability of the nonlinear system depends on the nature of the higher-order terms.

◆ Example 4.8

For the nonlinear system model

$$\begin{bmatrix} \dot{x}_1 \\ \dot{x}_2 \end{bmatrix} = \begin{bmatrix} -1.1x_1 + g_2(x_2) \\ g_1(x_1) - 1.1x_2 \end{bmatrix}, \tag{4.69}$$

where

$$g_i(x_i) = \frac{2}{\pi}\arctan(0.7\pi x_i), \qquad i = 1, 2,$$

we will find its equilibrium states and determine their stability properties using the linearization method. We mention here that the above system was analyzed in references 126 and 203.

To find the system equilibrium states, we solve the algebraic equations $\dot{x}_1 = 0$ and $\dot{x}_2 = 0$, that is,

$$\begin{aligned} -1.1x_1 + \frac{2}{\pi}\arctan(0.7\pi x_2) &= 0, \\ \frac{2}{\pi}\arctan(0.7\pi x_1) - 1.1x_2 &= 0. \end{aligned}$$

We obtain three equilibrium states:

$$\boldsymbol{x}^{(1)} = \begin{bmatrix} 0.454 \\ 0.454 \end{bmatrix}, \quad \boldsymbol{x}^{(2)} = \begin{bmatrix} -0.454 \\ -0.454 \end{bmatrix}, \quad \text{and} \quad \boldsymbol{x}^{(3)} = \begin{bmatrix} 0 \\ 0 \end{bmatrix}.$$

Let

$$\Delta \boldsymbol{x} = \boldsymbol{x} - \boldsymbol{x}^{(i)}, \qquad i = 1, 2, 3.$$

Then, a linearized model about an equilibrium state has the form

$$\frac{d}{dt}\Delta \boldsymbol{x} = \boldsymbol{A}\Delta \boldsymbol{x},$$

where

$$\boldsymbol{A} = \left.\begin{bmatrix} \dfrac{\partial f_1(\boldsymbol{x})}{\partial x_1} & \dfrac{\partial f_1(\boldsymbol{x})}{\partial x_2} \\ \dfrac{\partial f_2(\boldsymbol{x})}{\partial x_1} & \dfrac{\partial f_2(\boldsymbol{x})}{\partial x_2} \end{bmatrix}\right|_{\boldsymbol{x}=\boldsymbol{x}^{(i)}} = \left.\begin{bmatrix} -1.1 & \dfrac{1.4}{1 + (0.7\pi x_2)^2} \\ \dfrac{1.4}{1 + (0.7\pi x_2)^2} & -1.1 \end{bmatrix}\right|_{\boldsymbol{x}=\boldsymbol{x}^{(i)}}. \tag{4.70}$$

We have three linearized models:

1. About $\boldsymbol{x}^{(1)}$,

$$\frac{d}{dt}\Delta \boldsymbol{x} = \begin{bmatrix} -1.1 & 0.7073 \\ 0.7073 & -1.1 \end{bmatrix} \Delta \boldsymbol{x},$$

2. About $\boldsymbol{x}^{(2)}$,

$$\frac{d}{dt}\Delta \boldsymbol{x} = \begin{bmatrix} -1.1 & 0.7073 \\ 0.7073 & -1.1 \end{bmatrix} \Delta \boldsymbol{x},$$

3. About $\boldsymbol{x}^{(3)}$,

$$\frac{d}{dt}\Delta \boldsymbol{x} = \begin{bmatrix} -1.1 & 1.4 \\ 1.4 & -1.1 \end{bmatrix} \Delta \boldsymbol{x}.$$

By Theorem 4.11, the states $\boldsymbol{x}^{(1)}$ and $\boldsymbol{x}^{(2)}$ are asymptotically stable because the corresponding linearized models have their eigenvalues in the open left-half plane: -1.8073 and -0.3927. By Theorem 4.12, the equilibrium state $\boldsymbol{x}^{(3)}$ is unstable because the linearized model has an eigenvalue in the open right-half plane, at 0.3 (the other eigenvalue is at -2.5). In Figure 4.12, we show a phase portrait of the above nonlinear system, where one can see clearly the stability properties of the equilibrium states.

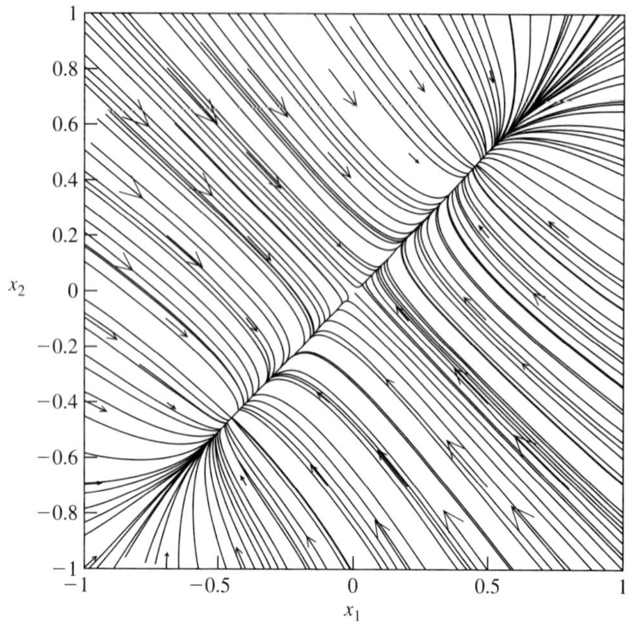

Figure 4.12 Phase portrait of the nonlinear system of Example 4.8.

4.10 Discontinuous Robust Controllers

In this section we consider the problem of constructing stabilizing state-feedback controllers for nonlinear dynamical systems whose mathematical models contain uncertainties. The design of the controllers is based on the Lyapunov theory presented in the previous sections. The class of dynamical systems considered here is modeled by

$$\dot{\boldsymbol{x}}(t) = \boldsymbol{f}(t, \boldsymbol{x}(t)) + \boldsymbol{G}(t, \boldsymbol{x}(t))\left(\boldsymbol{u}(t) + \boldsymbol{\xi}(t, \boldsymbol{x}(t))\right), \qquad \boldsymbol{x}(t_0) = \boldsymbol{x}_0, \tag{4.71}$$

where $\boldsymbol{x}(t) \in \mathbb{R}^n$, $\boldsymbol{u}(t) \in \mathbb{R}^m$, and $\boldsymbol{\xi}(t, \boldsymbol{x}(t)) \in \mathbb{R}^m$ represent the lumped uncertainties. We assume that the uncertain element $\boldsymbol{\xi}$ is bounded by a known continuous nonnegative function ρ, that is,

$$\|\boldsymbol{\xi}(t, \boldsymbol{x}(t))\| \leq \rho(t, \boldsymbol{x}(t)).$$

We also assume that the equilibrium state $\boldsymbol{x} = \boldsymbol{0}$ of the uncontrolled nominal system

$$\dot{\boldsymbol{x}}(t) = \boldsymbol{f}(t, \boldsymbol{x}(t)) \tag{4.72}$$

is globally uniformly asymptotically stable and that we were able to construct a Lyapunov function $W : \mathbb{R} \times \mathbb{R}^n \to \mathbb{R}$ corresponding to this equilibrium state. The Lyapunov function W is assumed to satisfy the following conditions: There exist three continuous and strictly increasing functions ψ_i, $i = 1, 2, 3$, such that

$$\begin{aligned} \psi_i(0) &= 0, \qquad i = 1, 2, 3, \\ \lim_{r\to\infty} \psi_1(r) &= \infty, \\ \psi_1(\|\boldsymbol{x}\|) &\le W(t, \boldsymbol{x}) \le \psi_2(\|\boldsymbol{x}\|), \\ \psi_3(\|\boldsymbol{x}\|) &\le -\dot{W}(t, \boldsymbol{x}) \end{aligned} \tag{4.73}$$

for all t and all $\boldsymbol{x}$. We will show that if we apply the control strategy

$$\boldsymbol{u}(t, \boldsymbol{x}) = \begin{cases} -\dfrac{\boldsymbol{\mu}(t, \boldsymbol{x})}{\|\boldsymbol{\mu}(t, \boldsymbol{x})\|}\rho(t, \boldsymbol{x}) & \text{if } (t, \boldsymbol{x}) \notin \Omega \\ \boldsymbol{p} \in \mathbb{R}^m,\ \|\boldsymbol{p}\| \le \rho(t, \boldsymbol{x}) & \text{if } (t, \boldsymbol{x}) \in \Omega, \end{cases} \tag{4.74}$$

where $\boldsymbol{\mu}(t, \boldsymbol{x}) = \boldsymbol{G}^T(t, \boldsymbol{x})\nabla_x W(t, \boldsymbol{x})$ and $\Omega = \{(t, \boldsymbol{x}) : \boldsymbol{\mu} = \boldsymbol{0}\}$, then the equilibrium state $\boldsymbol{x} = \boldsymbol{0}$ is a globally uniformly asymptotically stable solution of the closed-loop system. For this, consider the Lyapunov function W for the uncontrolled nominal system. The time derivative of W evaluated on the solutions of the closed-loop system is

$$\begin{aligned} \dot{W}(t, \boldsymbol{x}) &= \frac{\partial W}{\partial t} + \nabla_x W^T \left(\boldsymbol{f}(t, \boldsymbol{x}) + \boldsymbol{G}(t, \boldsymbol{x})\left(\boldsymbol{u} + \boldsymbol{\xi}(t, \boldsymbol{x})\right)\right) \\ &\le -\psi_3(\|\boldsymbol{x}\|) + \nabla_x W^T \boldsymbol{G}(t, \boldsymbol{x})\boldsymbol{u} + \nabla_x W^T \boldsymbol{G}(t, \boldsymbol{x})\boldsymbol{\xi}(t, \boldsymbol{x}) \\ &\le -\psi_3(\|\boldsymbol{x}\|) + \boldsymbol{\mu}^T(t, \boldsymbol{x})\boldsymbol{u} + \|\boldsymbol{\mu}(t, \boldsymbol{x})\|\rho(t, \boldsymbol{x}). \end{aligned} \tag{4.75}$$

For $(t, \boldsymbol{x}) \notin \Omega$,

$$\begin{aligned} \dot{W}(t, \boldsymbol{x}) &\le -\psi_3(\|\boldsymbol{x}\|) - \boldsymbol{\mu}^T(t, \boldsymbol{x})\frac{\boldsymbol{\mu}(t, \boldsymbol{x})}{\|\boldsymbol{\mu}(t, \boldsymbol{x})\|}\rho(t, \boldsymbol{x}) + \|\boldsymbol{\mu}(t, \boldsymbol{x})\|\rho(t, \boldsymbol{x}) \\ &= -\psi_3(\|\boldsymbol{x}\|) - \frac{\|\boldsymbol{\mu}(t, \boldsymbol{x})\|^2}{\|\boldsymbol{\mu}(t, \boldsymbol{x})\|}\rho(t, \boldsymbol{x}) + \|\boldsymbol{\mu}(t, \boldsymbol{x})\|\rho(t, \boldsymbol{x}) \\ &= -\psi_3(\|\boldsymbol{x}\|) - \|\boldsymbol{\mu}(t, \boldsymbol{x})\|\rho(t, \boldsymbol{x}) + \|\boldsymbol{\mu}(t, \boldsymbol{x})\|\rho(t, \boldsymbol{x}) \\ &= -\psi_3(\|\boldsymbol{x}\|) \\ &< 0. \end{aligned} \tag{4.76}$$

For $(t, \boldsymbol{x}) \in \Omega$, we have $\boldsymbol{\mu}(t, \boldsymbol{x}) = \boldsymbol{0}$; and using (4.75); we obtain

$$\dot{W}(t, \boldsymbol{x}) \le -\psi_3(\|\boldsymbol{x}\|) < 0. \tag{4.77}$$

Thus, W is also a Lyapunov function for the closed-loop system, which means that the uncertain system (4.71) driven by the control law (4.74) is indeed globally uniformly asymptotically stable.

◆ Example 4.9

For the uncertain system model

$$\dot{\boldsymbol{x}} = \boldsymbol{A}\boldsymbol{x} + \boldsymbol{\xi}_1(\boldsymbol{x}) + \boldsymbol{b}(u + \xi_2(\boldsymbol{x})),$$

where

$$A = \begin{bmatrix} -2 & 1 \\ 0 & -3 \end{bmatrix}, \quad b = \begin{bmatrix} 0 \\ 1 \end{bmatrix}, \quad \|\xi_1(x)\| \leq 1.75\|x\|, \quad \text{and} \quad |\xi_2| \leq 10\|x\| + 0.5|u|,$$

design stabilizing control strategy of the form (4.74).

The uncontrolled nominal system $\dot{x} = Ax$ is asymptotically stable. Let

$$Q = \begin{bmatrix} 4 & -1 \\ -1 & 6 \end{bmatrix}.$$

The eigenvalues of Q are $\lambda_{min}(Q) = 3.5858$ and $\lambda_{max}(Q) = 6.4142$. Solving the Lyapunov matrix equation $A^T P + PA = -Q$ yields

$$P = I_2.$$

The proposed controller has the form

$$u = -\rho(x)\frac{\mu(x)}{|\mu(x)|} = -\rho(x)\frac{b^T P x}{|b^T P x|} = -\rho(x)\frac{b^T x}{|b^T x|} = -\rho(x)\operatorname{sign}(b^T x). \tag{4.78}$$

Our goal is to determine a bound $\rho^*(x) > 0$ such that for any $\rho(x) \geq \rho^*(x)$ the uncertain closed-loop system is globally uniformly asymptotically stable. To begin, consider

$$V(x) = x^T P x = x^T x$$

as the Lyapunov function candidate for the closed-loop system. Evaluating the time derivative of V on the trajectories of the closed-loop system yields

$$\begin{aligned} \frac{d}{dt}V(x) &= 2x^T \dot{x} \\ &= 2x^T(Ax + \xi_1(x) + b(u + \xi_2(x))) \\ &= -x^T Qx + 2x^T \xi_1(x) + 2x^T bu(x) + 2x^T b\xi_2(x). \end{aligned} \tag{4.79}$$

Substituting into the above the expression for the controller given by (4.78), taking the norms, and using bounds on the uncertain elements, we obtain

$$\begin{aligned} \frac{d}{dt}V(x) &\leq -\lambda_{min}(Q)\|x\|^2 + 2\|x\|\|\xi_1(x)\| - 2\rho(x)\frac{|x^T b|^2}{|x^T b|} + 2|x^T b||\xi_2(x)| \\ &\leq (-3.5858 + 3.5)\|x\|^2 - 2\rho(x)|x^T b| + 2(10\|x\| + 0.5\rho(x))|x^T b|. \end{aligned} \tag{4.80}$$

In the above, we used the fact that

$$|u| = \rho(x).$$

Collecting the terms in (4.80) yields

$$\frac{d}{dt}V(x) \leq (-3.5858 + 3.5)\|x\|^2 + (20\|x\| - \rho(x))\,|x^T b|. \tag{4.81}$$

Therefore, $\frac{dV}{dt} < 0$ for any

$$\rho(\boldsymbol{x}) \geq \rho^*(\boldsymbol{x}) = 20\|\boldsymbol{x}\|. \tag{4.82}$$

In summary, the uncertain closed-loop system will be globally uniformly asymptotically stable if the controller gain $\rho(\boldsymbol{x})$ is chosen to satisfy (4.82).

We now use the Lyapunov approach to design a model-reference tracking controller. A block diagram of the model-reference control system is shown in Figure 4.13. Model-reference control can be used for plants with nonlinearities and time-varying parameters. One of the assumptions that is made while designing a model-reference controller is that a plant model is in a controller companion form. Here, we propose a method for designing model-reference controllers for a class of multi-input systems that does not require a plant model to be in a specific form. The crux of our method is in an appropriate selection of the model-reference system. Consider a plant model of the form

$$\dot{\boldsymbol{x}} = \boldsymbol{A}\boldsymbol{x} + \boldsymbol{B}\boldsymbol{u}, \tag{4.83}$$

where $\boldsymbol{A} \in \mathbb{R}^{n\times n}$ and $\boldsymbol{B} \in \mathbb{R}^{n\times m}$. The model-reference signal generator is described by

$$\dot{\boldsymbol{x}}_d = \boldsymbol{A}_d\boldsymbol{x}_d + \boldsymbol{B}_d\boldsymbol{r}, \tag{4.84}$$

where $\boldsymbol{A}_d \in \mathbb{R}^{n\times n}$, $\boldsymbol{B}_d \in \mathbb{R}^{n\times m}$, and $\boldsymbol{r} \in \mathbb{R}^m$ is a command signal that is chosen so that $\boldsymbol{x}_d$ represents a desired state trajectory of the plant. The matrix $\boldsymbol{A}_d$ of the reference system is chosen from the set

$$\{\boldsymbol{A}_d : \boldsymbol{A}_d = \boldsymbol{A} - \boldsymbol{B}\boldsymbol{K}_d,\ \boldsymbol{K}_d \in \mathbb{R}^{m\times n}\}, \tag{4.85}$$

and $\boldsymbol{B}_d = \boldsymbol{B}$. Thus, the reference system is state-feedback equivalent with the plant model (4.83). We assume that the pair $(\boldsymbol{A}, \boldsymbol{B})$ is controllable. Therefore, we can select $\boldsymbol{A}_d$ to have its eigenvalues in prescribed locations in the open left-hand plane. To proceed further, we define the tracking error $\boldsymbol{e}$ as

$$\boldsymbol{e}(t) = \boldsymbol{x}(t) - \boldsymbol{x}_d(t). \tag{4.86}$$

Our goal is to construct a state-feedback control law so that

$$\lim_{t\to\infty} \|\boldsymbol{e}(t)\| = 0. \tag{4.87}$$

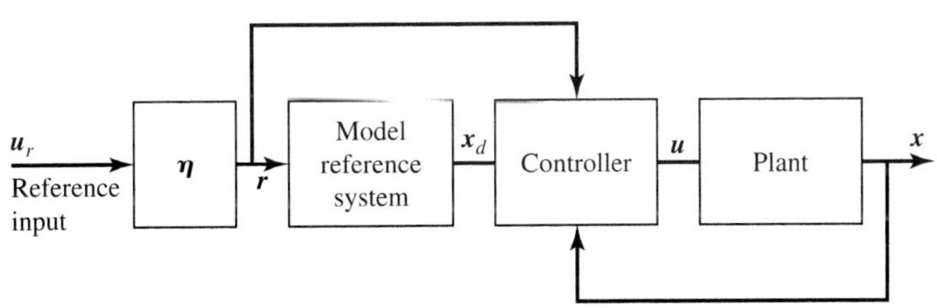

Figure 4.13 Model-reference control system block diagram with input scaling.

We now form the differential equation that describes the error dynamics. For this we evaluate the time derivative of the error vector to get

$$\begin{aligned}\dot{\boldsymbol{e}}(t) &= \dot{\boldsymbol{x}}(t) - \dot{\boldsymbol{x}}_d(t)\\ &= \boldsymbol{A}\boldsymbol{x}(t) + \boldsymbol{B}\boldsymbol{u} - \boldsymbol{A}_d\boldsymbol{x}_d(t) - \boldsymbol{B}\boldsymbol{r}(t).\end{aligned} \tag{4.88}$$

Recall that $\boldsymbol{A}_d = \boldsymbol{A} - \boldsymbol{B}\boldsymbol{K}_d$, hence $\boldsymbol{A} = \boldsymbol{A}_d + \boldsymbol{B}\boldsymbol{K}_d$. Substituting this expression for $\boldsymbol{A}$ into the error equation (4.88), we obtain

$$\begin{aligned}\dot{\boldsymbol{e}}(t) &= \boldsymbol{A}_d(\boldsymbol{x}(t) - \boldsymbol{x}_d(t)) + \boldsymbol{B}(\boldsymbol{u} + \boldsymbol{K}_d\boldsymbol{x}(t) - \boldsymbol{r}(t))\\ &= \boldsymbol{A}_d\boldsymbol{e}(t) + \boldsymbol{B}(\boldsymbol{u} + \boldsymbol{K}_d\boldsymbol{x}(t) - \boldsymbol{r}(t)).\end{aligned} \tag{4.89}$$

We will use a Lyapunov function approach in our further analysis that will result in a control law forcing the error $\boldsymbol{e}(t)$ to go asymptotically to zero. As a Lyapunov function candidate for (4.89), we take

$$V(\boldsymbol{e}) = \boldsymbol{e}^T\boldsymbol{P}\boldsymbol{e}, \tag{4.90}$$

where $\boldsymbol{P}$ solves the Lyapunov matrix equation, $\boldsymbol{A}_d^T\boldsymbol{P} + \boldsymbol{P}\boldsymbol{A}_d = -\boldsymbol{Q}$, for some $\boldsymbol{Q} = \boldsymbol{Q}^T > 0$. Because by construction $\boldsymbol{A}_d$ is asymptotically stable, the solution to the Lyapunov matrix equation is positive definite, that is, $\boldsymbol{P} = \boldsymbol{P}^T > 0$. We next evaluate the time derivative of $V(\boldsymbol{e}(t))$ on the trajectories of (4.89),

$$\begin{aligned}\dot{V}(\boldsymbol{e}) &= 2\boldsymbol{e}^T\boldsymbol{P}\dot{\boldsymbol{e}}\\ &= 2\boldsymbol{e}^T\boldsymbol{P}(\boldsymbol{A}_d\boldsymbol{e} + \boldsymbol{B}(\boldsymbol{u} + \boldsymbol{K}_d\boldsymbol{x} - \boldsymbol{r}))\\ &= \boldsymbol{e}^T\left(\boldsymbol{A}_d^T\boldsymbol{P} + \boldsymbol{P}\boldsymbol{A}_d\right)\boldsymbol{e} + 2\boldsymbol{e}^T\boldsymbol{P}\boldsymbol{B}(\boldsymbol{u} + \boldsymbol{K}_d\boldsymbol{x} - \boldsymbol{r}).\end{aligned} \tag{4.91}$$

Using the Lyapunov matrix equation and taking the norms of the above yields

$$\dot{V}(\boldsymbol{e}) \le -\boldsymbol{e}^T\boldsymbol{Q}\boldsymbol{e} + 2\boldsymbol{e}^T\boldsymbol{P}\boldsymbol{B}\boldsymbol{u} + 2\|\boldsymbol{e}^T\boldsymbol{P}\boldsymbol{B}\|\|\boldsymbol{K}_d\boldsymbol{x}\| + 2\|\boldsymbol{e}^T\boldsymbol{P}\boldsymbol{B}\|\|\boldsymbol{r}\|. \tag{4.92}$$

If $\boldsymbol{e}$ is such that $\boldsymbol{B}^T\boldsymbol{P}\boldsymbol{e} = \boldsymbol{0}$, then

$$\dot{V}(\boldsymbol{e}) \le -\boldsymbol{e}^T\boldsymbol{Q}\boldsymbol{e} < 0. \tag{4.93}$$

We now analyze the case when $\boldsymbol{B}^T\boldsymbol{P}\boldsymbol{e} \ne \boldsymbol{0}$. Let the proposed control law be

$$\boldsymbol{u} = -\rho(\boldsymbol{x}, \boldsymbol{r})\frac{\boldsymbol{B}^T\boldsymbol{P}\boldsymbol{e}}{\|\boldsymbol{B}^T\boldsymbol{P}\boldsymbol{e}\|}, \tag{4.94}$$

where $\rho(\boldsymbol{x}, \boldsymbol{r}) = \|\boldsymbol{K}_d\boldsymbol{x}\| + \|\boldsymbol{r}\|$. Substituting (4.94) into (4.92), we obtain

$$\begin{aligned}\dot{V}(\boldsymbol{e}) &\le -\boldsymbol{e}^T\boldsymbol{Q}\boldsymbol{e} - 2\|\boldsymbol{e}^T\boldsymbol{P}\boldsymbol{B}\|\left(\rho(\boldsymbol{x}, \boldsymbol{r}) - \|\boldsymbol{K}_d\boldsymbol{x}\| - \|\boldsymbol{r}\|\right)\\ &= -\boldsymbol{e}^T\boldsymbol{Q}\boldsymbol{e}\\ &< 0.\end{aligned}$$

Since for all $\boldsymbol{e} \in \mathbb{R}^n$ we have $V(\boldsymbol{e}) > 0$ and $\dot{V}(\boldsymbol{e}) < 0$, the equilibrium point $\boldsymbol{e} = \boldsymbol{0}$ of the error equation (4.89) is globally uniformly asymptotically stable, which means that $\boldsymbol{x}(t) \to \boldsymbol{x}_d(t)$.

◆ Example 4.10

We now apply the above tracking controller to the vehicle model derived in Subsection 1.7.2, where $\delta_r = 0$. That is, we wish to construct a tracking controller for a vehicle with front-wheel steering. The first step is to select the poles of the model reference system. We choose the model reference system to have the following set of asymptotically stable poles:

$$\{-4 - 2j, -4 + 2j, -4.5, -5.0\}. \tag{4.95}$$

The pair $(\boldsymbol{A}, \boldsymbol{b})$ is controllable; hence we can find $\boldsymbol{k}_d$ such that the eigenvalues of the matrix $\boldsymbol{A}_d = \boldsymbol{A} - \boldsymbol{b}\boldsymbol{k}_d$ are at the locations in (4.95). The feedback gain is

$$\boldsymbol{k}_d = [0.0329 \quad 2.2854 \quad 0.2429 \quad -0.1653], \tag{4.96}$$

and the resulting model reference plant matrix is

$$\boldsymbol{A}_d = \begin{bmatrix} -6.6638 & -107.1281 & -29.6350 & 7.7479 \\ 0 & 0 & 1 & 0 \\ -0.9205 & -65.8195 & -10.8362 & 4.7603 \\ -1 & -18.3 & 0 & 0 \end{bmatrix}. \tag{4.97}$$

The next step is to find $\boldsymbol{P}$ that solves the Lyapunov equation, $\boldsymbol{A}_d^T \boldsymbol{P} + \boldsymbol{P}\boldsymbol{A}_d = -\boldsymbol{Q}$. We take $\boldsymbol{Q} = \boldsymbol{I}_4$. Solving the Lyapunov equation, we get

$$\boldsymbol{P} = \begin{bmatrix} 0.1145 & 0.7530 & -0.1809 & -0.0966 \\ 0.7530 & 34.6174 & -0.2883 & -3.3437 \\ -0.1809 & -0.2883 & 0.5144 & 0.0522 \\ -0.0966 & -3.3437 & 0.0522 & 0.6216 \end{bmatrix}. \tag{4.98}$$

Now that we have $\boldsymbol{P}$, we are ready to construct the controller. We have

$$\begin{aligned} u &= -\rho(\boldsymbol{x}, r)\frac{\boldsymbol{b}^T \boldsymbol{P}\boldsymbol{e}}{|\boldsymbol{b}^T \boldsymbol{P}\boldsymbol{e}|} \\ &= -(|\boldsymbol{k}_d \boldsymbol{x}| + |r|)\operatorname{sign}(\boldsymbol{b}^T \boldsymbol{P}(\boldsymbol{x} - \boldsymbol{x}_d)) \\ &= -\left(\left|\begin{bmatrix} 0.0329 \\ 2.2854 \\ 0.2429 \\ -0.1653 \end{bmatrix}^T \boldsymbol{x}\right| + |r|\right)\operatorname{sign}\left(\begin{bmatrix} 0.1571 \\ 26.9931 \\ 6.3324 \\ -3.0250 \end{bmatrix}^T (\boldsymbol{x} - \boldsymbol{x}_d)\right). \end{aligned} \tag{4.99}$$

Our goal is to have the reference input signal u_r to correspond to a desired vehicle path. This requires scaling the reference input. The scaling factor is

$$\eta = \left(\lim_{s \to 0} \boldsymbol{c}(s\boldsymbol{I}_4 - \boldsymbol{A}_d)^{-1}\boldsymbol{b}\right)^{-1}, \tag{4.100}$$

where $\boldsymbol{c} = [0 \ 0 \ 0 \ 1]$. Note that η is well-defined because the matrix $\boldsymbol{A}_d$ is asymptotically stable. In our example, $\eta = -0.1653$. Thus, we have $r = \eta u_r$. The reference input is $u_r = 4h(t - 1 \text{ sec})$, where $h(t)$ is the unit step function.

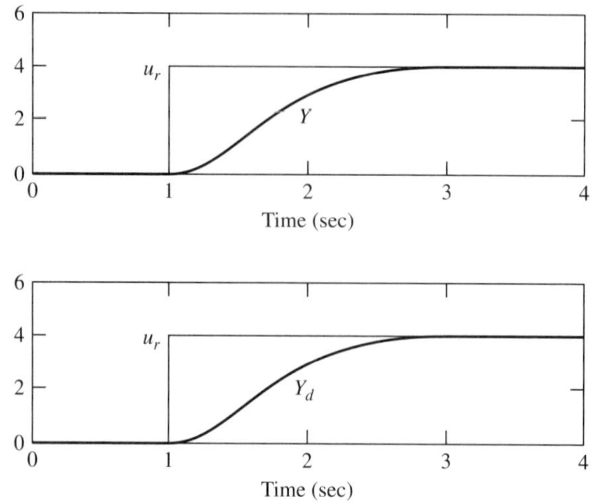

Figure 4.14 Plots of the actual output of the closed-loop system and the desired output in Example 4.10.

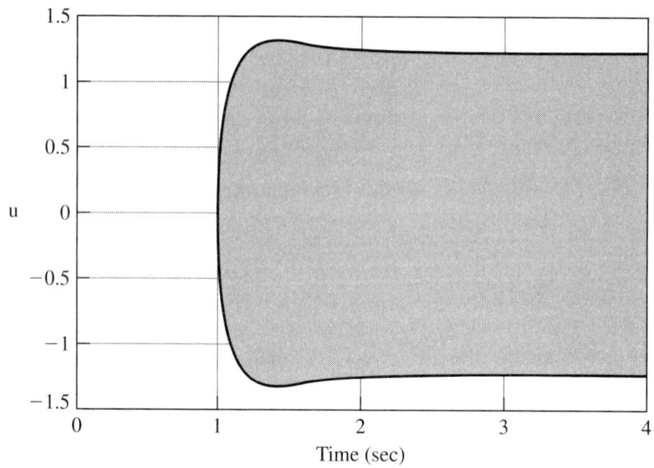

Figure 4.15 Plot of the controller action versus time in Example 4.10.

The performance of the controller, for zero initial conditions, is shown in Figures 4.14 and 4.15. In Figure 4.14, the closed-loop system output is shown in the upper subplot, while the desired output is shown in the bottom subplot. The vehicle tracks the desired path with no noticeable steady-state error. The control action is shown in Figure 4.15.

The proposed tracking controller is discontinuous in the state, which results in high activity referred to as chattering. Chattering can be reduced by using a continuous version of the above controller. The continuous controller is obtained by introducing a so-called boundary layer that smooths out the discontinuous controller. We discuss the problem of constructing a boundary layer controller in the following section.

4.11 Uniform Ultimate Boundedness

A drawback to the discontinuous controllers, discussed in the previous section, is that rapid switching tends to excite high-frequency modes of the plant, and this may degrade the performance of the controller. These problems can be alleviated by adding a so-called *boundary layer* to the discontinuous controller resulting in a continuous control strategy. However, there is a price for making a discontinuous controller a continuous one. The closed-loop system driven by a boundary layer controller may only be *uniformly ultimately bounded* rather than asymptotically stable. The notions we are concerned with in this section are described in the following three definitions.

Definition 4.18 The closed ball $\bar{B}_q(\mathbf{0}) = \{\boldsymbol{x} : \|\boldsymbol{x}\| \le q\}$ is uniformly stable if for any $\varepsilon > q$, there is a $\delta = \delta(\varepsilon) > 0$ so that if $\|\boldsymbol{x}(t_0)\| \le \delta$, then $\|\boldsymbol{x}(t)\| \le \varepsilon$ for $t \ge t_0$.

For the benefit of the reader, we now rephrase the definition of uniform boundedness from Section 4.8.

Definition 4.19 The solutions of $\dot{\boldsymbol{x}} = \boldsymbol{f}(t, \boldsymbol{x})$ are uniformly bounded if for any given $d > 0$, there exists $b = b(d) < \infty$ so that for each solution $\boldsymbol{x}(t) = \boldsymbol{x}(t; t_0, \boldsymbol{x}_0)$ with $\|\boldsymbol{x}_0\| \le d$, we have $\|\boldsymbol{x}(t)\| \le b$ for all $t \ge t_0$.

We next define the notion of uniform ultimate boundedness (UUB) of a system.

Definition 4.20 The solutions of $\dot{\boldsymbol{x}} = \boldsymbol{f}(t, \boldsymbol{x})$ are uniformly ultimately bounded (UUB) with respect to the ball $\bar{B}_q(\mathbf{0})$ if for every $d > 0$, there exists $T(q, d)$ such that for each solution $\boldsymbol{x}(t) = \boldsymbol{x}(t; t_0, \boldsymbol{x}_0)$ with $\|\boldsymbol{x}_0\| \le d$, we have $\boldsymbol{x}(t) \in \bar{B}_q(\mathbf{0})$ for $t \ge t_0 + T(q, d)$.

We now consider a dynamical system model (4.71):

$$\dot{\boldsymbol{x}}(t) = \boldsymbol{f}(t, \boldsymbol{x}(t)) + \boldsymbol{G}(t, \boldsymbol{x}(t))\left(\boldsymbol{u}(t) + \boldsymbol{\xi}(t, \boldsymbol{x}(t))\right), \qquad \boldsymbol{x}(t_0) = \boldsymbol{x}_0,$$

where $\boldsymbol{x}(t) \in \mathbb{R}^n$, $\boldsymbol{u}(t) \in \mathbb{R}^m$, and $\boldsymbol{\xi}(t, \boldsymbol{x}(t)) \in \mathbb{R}^m$ represents the lumped uncertainties. We assume that the uncertain element $\boldsymbol{\xi}$ is bounded by a known continuous nonnegative function ρ, that is, $\|\boldsymbol{\xi}(t, \boldsymbol{x}(t))\| \le \rho(t, \boldsymbol{x}(t))$. As in Section 4.10, we assume that the equilibrium state $\boldsymbol{x} = \mathbf{0}$ of the uncontrolled nominal system, $\dot{\boldsymbol{x}}(t) = \boldsymbol{f}(t, \boldsymbol{x}(t))$, is globally uniformly asymptotically stable and that we were able to construct a Lyapunov function $W : \mathbb{R} \times \mathbb{R}^n \to \mathbb{R}$ satisfying the conditions (4.73) for all t and all $\boldsymbol{x}$.

We showed in the previous section that if we apply the discontinuous control strategy

$$\boldsymbol{u}(t, \boldsymbol{x}) = \begin{cases} -\dfrac{\boldsymbol{\mu}(t, \boldsymbol{x})}{\|\boldsymbol{\mu}(t, \boldsymbol{x})\|}\rho(t, \boldsymbol{x}) & \text{if } \boldsymbol{\mu}(t, \boldsymbol{x}) \ne \mathbf{0} \\ \boldsymbol{p} \in \mathbb{R}^m, \quad \|\boldsymbol{p}\| \le \rho(t, \boldsymbol{x}) & \text{if } \boldsymbol{\mu}(t, \boldsymbol{x}) = \mathbf{0}, \end{cases}$$

where $\boldsymbol{\mu}(t, \boldsymbol{x}) = \boldsymbol{G}^T(t, \boldsymbol{x})\nabla_{\boldsymbol{x}} W(t, \boldsymbol{x})$, then the equilibrium state $\boldsymbol{x} = \mathbf{0}$ is a globally uniformly asymptotically stable solution of the closed-loop system. However, the above controller is discontinuous in $\boldsymbol{x}$. Following Corless and Leitmann [56], we modify it to make it continuous by introducing a boundary layer. The proposed boundary layer controller is

$$\boldsymbol{u}(t,\boldsymbol{x}) = \begin{cases} -\dfrac{\boldsymbol{\mu}(t,\boldsymbol{x})\rho(t,\boldsymbol{x})}{\|\boldsymbol{\mu}(t,\boldsymbol{x})\rho(t,\boldsymbol{x})\|}\rho(t,\boldsymbol{x}) & \text{if } \|\boldsymbol{\mu}(t,\boldsymbol{x})\rho(t,\boldsymbol{x})\| > \nu \\ \boldsymbol{p}(t,\boldsymbol{x}), \quad \|\boldsymbol{p}(t,\boldsymbol{x})\| \le \rho(t,\boldsymbol{x}) & \text{if } \|\boldsymbol{\mu}(t,\boldsymbol{x})\rho(t,\boldsymbol{x})\| \le \nu, \end{cases} \tag{4.101}$$

where ν is the thickness of the boundary layer about the naturally defined switching surface $\{(t,\boldsymbol{x}) : \|\boldsymbol{\mu}(t,\boldsymbol{x})\rho(t,\boldsymbol{x})\| = 0\}$, and $\boldsymbol{p}$ is a continuous function such that

$$\boldsymbol{p} = -\frac{\boldsymbol{\mu}(t,\boldsymbol{x})\rho(t,\boldsymbol{x})}{\|\boldsymbol{\mu}(t,\boldsymbol{x})\rho(t,\boldsymbol{x})\|}\rho(t,\boldsymbol{x}) \qquad \text{if } \|\boldsymbol{\mu}(t,\boldsymbol{x})\rho(t,\boldsymbol{x})\| = \nu.$$

A particular example of the above type of a controller is

$$\boldsymbol{u}(t,\boldsymbol{x}) = \begin{cases} -\dfrac{\boldsymbol{\mu}(t,\boldsymbol{x})\rho(t,\boldsymbol{x})}{\|\boldsymbol{\mu}(t,\boldsymbol{x})\rho(t,\boldsymbol{x})\|}\rho(t,\boldsymbol{x}) & \text{if } \|\boldsymbol{\mu}(t,\boldsymbol{x})\rho(t,\boldsymbol{x})\| > \nu \\ -\dfrac{\boldsymbol{\mu}(t,\boldsymbol{x})}{\nu}\rho^2(t,\boldsymbol{x}) & \text{if } \|\boldsymbol{\mu}(t,\boldsymbol{x})\rho(t,\boldsymbol{x})\| \le \nu. \end{cases} \tag{4.102}$$

We will now show that the closed-loop system driven by the boundary layer controller (4.101) is UUB with respect to a neighborhood of the zero state. This neighborhood can be made arbitrarily small by letting $\nu \to 0$. Before we state the results, we first compute the time derivative of a Lyapunov function candidate on the trajectories of the closed-loop system. Consider the Lyapunov function W for the uncontrolled nominal system. The time derivative of W evaluated on the solutions of the closed-loop system is

$$\begin{aligned} \dot{W}(t,\boldsymbol{x}) &= \frac{\partial W}{\partial t} + \nabla_{\boldsymbol{x}} W^T \left(\boldsymbol{f}(t,\boldsymbol{x}) + \boldsymbol{G}(t,\boldsymbol{x})\left(\boldsymbol{u} + \boldsymbol{\xi}(t,\boldsymbol{x})\right)\right) \\ &\le -\psi_3(\|\boldsymbol{x}\|) + \nabla_{\boldsymbol{x}} W^T \boldsymbol{G}(t,\boldsymbol{x})\boldsymbol{u} + \|\boldsymbol{\mu}(t,\boldsymbol{x})\rho(t,\boldsymbol{x})\|. \end{aligned}$$

We now consider two cases. The first case is when $\|\boldsymbol{\mu}(t,\boldsymbol{x})\rho(t,\boldsymbol{x})\| > \nu$. Taking into account the controller structure for this case yields

$$\begin{aligned} \dot{W}(t,\boldsymbol{x}) &\le -\psi_3(\|\boldsymbol{x}\|) - \nabla_{\boldsymbol{x}} W^T \boldsymbol{G}(t,\boldsymbol{x}) \frac{\boldsymbol{\mu}(t,\boldsymbol{x})\rho(t,\boldsymbol{x})}{\|\boldsymbol{\mu}(t,\boldsymbol{x})\rho(t,\boldsymbol{x})\|}\rho(t,\boldsymbol{x}) + \|\boldsymbol{\mu}(t,\boldsymbol{x})\rho(t,\boldsymbol{x})\| \\ &= -\psi_3(\|\boldsymbol{x}\|) - \frac{\|\boldsymbol{\mu}(t,\boldsymbol{x})\rho(t,\boldsymbol{x})\|^2}{\|\boldsymbol{\mu}(t,\boldsymbol{x})\rho(t,\boldsymbol{x})\|} + \|\boldsymbol{\mu}(t,\boldsymbol{x})\rho(t,\boldsymbol{x})\| \\ &= -\psi_3(\|\boldsymbol{x}\|). \end{aligned}$$

If, on the other hand, $\|\boldsymbol{\mu}(t,\boldsymbol{x})\rho(t,\boldsymbol{x})\| \le \nu$, then by using (4.101) we obtain

$$\begin{aligned} \dot{W}(t,\boldsymbol{x}) &\le -\psi_3(\|\boldsymbol{x}\|) + \nabla_{\boldsymbol{x}} W^T \boldsymbol{G}(t,\boldsymbol{x})\boldsymbol{u} + \|\boldsymbol{\mu}(t,\boldsymbol{x})\rho(t,\boldsymbol{x})\| \\ &\le -\psi_3(\|\boldsymbol{x}\|) + \|\nabla_{\boldsymbol{x}} W^T \boldsymbol{G}(t,\boldsymbol{x})\| \, \|\boldsymbol{u}\| + \|\boldsymbol{\mu}(t,\boldsymbol{x})\rho(t,\boldsymbol{x})\| \\ &\le -\psi_3(\|\boldsymbol{x}\|) + 2\|\boldsymbol{\mu}(t,\boldsymbol{x})\rho(t,\boldsymbol{x})\| \\ &\le -\psi_3(\|\boldsymbol{x}\|) + 2\nu. \end{aligned}$$

We conclude that for all t and all $\boldsymbol{x}$, we have

$$\boxed{\dot{W}(t, \boldsymbol{x}) \leq -\psi_3(\|\boldsymbol{x}\|) + 2\nu.} \tag{4.103}$$

To proceed further, we need an assumption on the function ψ_3; if ψ_3 is bounded and

$$\lim_{r \to \infty} \psi_3(r) = l < \infty, \tag{4.104}$$

then we assume

$$2\nu < l. \tag{4.105}$$

Proposition 4.2 The solutions of the system (4.71) driven by the boundary layer controller (4.101) are uniformly bounded.

Proof Let R be such that

$$\psi_3(R) = 2\nu.$$

Note that R is well-defined by virtue of (4.105) and the fact that ψ_3 is strictly increasing. Because ψ_3 is strictly increasing, we can write

$$R = \psi_3^{-1}(2\nu).$$

Figure 4.16 illustrates the above construction.

For a given $d > 0$, where $\|\boldsymbol{x}_0\| \leq d$, let

$$\tilde{d} = \max\{d, R\}.$$

We claim that

$$b = b(d) = \psi_1^{-1}(\psi_2(\tilde{d}));$$

that is, the solutions of (4.71) driven by the boundary layer controller (4.101) are uniformly bounded. In other words, if $\|\boldsymbol{x}_0\| \leq d$, then $\|\boldsymbol{x}(t; t_0, \boldsymbol{x}_0)\| \leq b$ for $t \geq t_0$. First note that $\|\boldsymbol{x}_0\| \leq \tilde{d}$ and that $\psi_1(b) = \psi_2(\tilde{d})$. Hence, $\tilde{d} \leq b$ (see Figure 4.17 for an

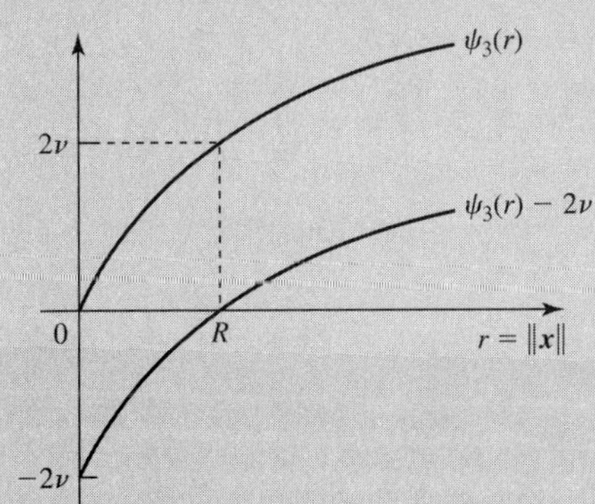

Figure 4.16 Finding R for a given ν in the proof of Proposition 4.2.

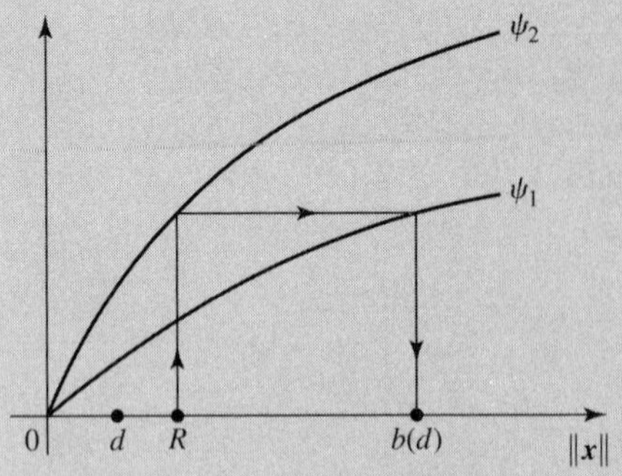

Figure 4.17 Finding b for a given d in the proof of Proposition 4.2.

illustration of this construction), and thus

$$\begin{aligned}\dot{W}(t, \mathbf{x}(t)) &\le -\psi_3(\|\mathbf{x}(t)\|) + 2\nu \\ &= -\psi_3(\|\mathbf{x}(t)\|) + \psi_3(R). \qquad (4.106)\end{aligned}$$

Suppose first that $d > R$. Then for $\|\mathbf{x}_0\| > R$ and $\|\mathbf{x}(t; t_0, \mathbf{x}_0)\| > R$, $\dot{W} < 0$ and therefore

$$\begin{aligned}\psi_1(b) &= \psi_2(\tilde{d}) \\ &\ge W(t_0, \mathbf{x}(t_0; t_0, \mathbf{x}_0)) \\ &\ge W(t, \mathbf{x}(t; t_0, \mathbf{x}_0)) \\ &\ge \psi_1(\|\mathbf{x}(t; t_0, \mathbf{x}_0)\|),\end{aligned}$$

which implies that $\|\mathbf{x}(t; t_0, \mathbf{x}_0)\| \le b$ for $t \ge t_0$.

If, on the other hand, $\|\mathbf{x}_0\| \le R$, then because $\dot{W}(t, \mathbf{x}(t)) \le -\psi_3(\|\mathbf{x}(t)\|) + 2\nu$, the solution may reach the sphere $\|\mathbf{x}\| = R$ at some time $t_2 \ge t_0$. However, by (4.106), $\dot{W} < 0$ outside $\|\mathbf{x}\| = R$. Therefore, $\|\mathbf{x}(t; t_0, \mathbf{x}_0)\| \le b$. Thus the solution is uniformly bounded.

We illustrate Proposition 4.2 with Figure 4.18. The initial condition in Figure 4.18 is such that $\|\boldsymbol{x}_0\| < d$ and $d < R$. The trajectory is confined to stay within the ball of radius $b = \psi_1^{-1}(\psi_2(R))$.

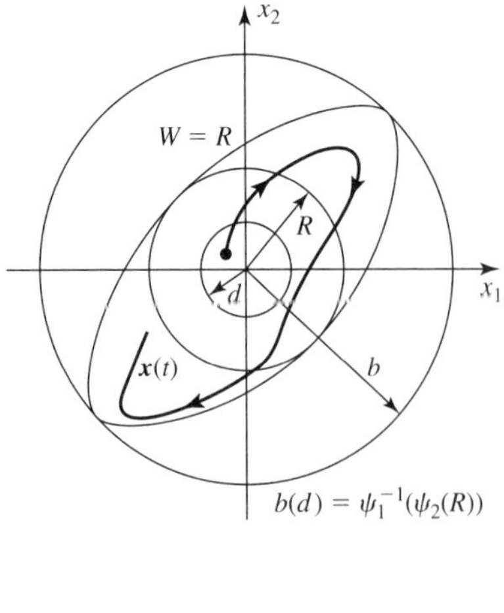

Figure 4.18 A two-dimensional illustration of an uniformly ultimately bounded (UUB) solution.

We now show, using arguments similar to those of Corless and Leitmann [56], that the solutions of the closed-loop system are UUB with respect to any ball of radius $\tilde{b} > \psi_1^{-1}(\psi_2(R))$.

Theorem 4.13 Let $\boldsymbol{x}(t) = \boldsymbol{x}(t; t_0, \boldsymbol{x}_0)$ be a solution of the closed-loop system for $[t_0, \infty)$ such that (4.103) is satisfied. If $\|\boldsymbol{x}_0\| \leq d$, then for a given

$$\tilde{b} > \psi_1^{-1}(\psi_2(R)), \tag{4.107}$$

$\|\boldsymbol{x}(t; t_0, \boldsymbol{x}_0)\| \leq \tilde{b}$ for all $t \geq t_0 + T(\tilde{b}, d)$, where

$$T(\tilde{b}, d) = \begin{cases} 0 & \text{if } d \leq \tilde{R} \\ \dfrac{\psi_2(d) - \psi_1(\tilde{R})}{\psi_3(\tilde{R}) - 2\nu} & \text{if } d > \tilde{R}, \end{cases}$$

and

$$\psi_2(\tilde{R}) = \psi_1(\tilde{b}). \tag{4.108}$$

Proof From assumption (4.107),

$$\psi_1(\tilde{b}) > \psi_2(R),$$

and by (4.108),

$$\psi_1(\tilde{b}) = \psi_2(\tilde{R}).$$

Hence, by (4.73), where we assumed that ψ_2 is strictly increasing,

$$\tilde{R} > R. \tag{4.109}$$

We consider two cases. The first case is when $d \leq \tilde{R}$. By uniform boundedness, for all $t \geq t_0$,

$$\|\boldsymbol{x}(t)\| \leq \tilde{b},$$

for any $\tilde{b}$ such that $\psi_1(\tilde{b}) > \psi_2(\tilde{R})$. Hence in this case $T(\tilde{b}, d) = 0$.

We now consider the case when $d > \tilde{R}$. We show, by contradiction, that for some $t_1 \in [t_0, t_0 + T(\tilde{b}, d)]$,

$$\|\boldsymbol{x}(t_1)\| \leq \tilde{R}.$$

From this we shall conclude the result by referring to the previous case. So suppose that $\|\boldsymbol{x}(t)\| > \tilde{R}$ for all $t \in [t_0, t_0 + T(\tilde{b}, d)]$. Then,

$$\begin{aligned} \psi_1(\|\boldsymbol{x}(t_0 + T(\tilde{b}, d); t_0, \boldsymbol{x}_0)\|) &\leq W(t_0 + T(\tilde{b}, d), \boldsymbol{x}(t_0 + T(\tilde{b}, d); t_0, \boldsymbol{x}_0)) \\ &= W(t_0, \boldsymbol{x}_0) + \int_{t_0}^{t_0+T(\tilde{b},d)} \dot{W} dt \\ &\leq W(t_0, \boldsymbol{x}_0) + \int_{t_0}^{t_0+T(\tilde{b},d)} (-\psi_3(\tilde{R}) + 2\nu) dt \\ &\leq \psi_2(d) - \int_{t_0}^{t_0+T(\tilde{b},d)} (\psi_3(\tilde{R}) - 2\nu) dt. \end{aligned}$$

Using the fact that $\psi_3(\tilde{R}) - 2\nu > 0$ because by (4.109), $\tilde{R} > R$, we obtain

$$\begin{aligned}\psi_1(\|\boldsymbol{x}(t_0 + T(\tilde{b}, d); t_0, \boldsymbol{x}_0)\|) &\le \psi_2(d) - (\psi_3(\tilde{R}) - 2\nu)\, T(\tilde{b}, d) \\ &= \psi_2(d) - (\psi_3(\tilde{R}) - 2\nu)\frac{\psi_2(d) - \psi_1(\tilde{R})}{\psi_3(\tilde{R}) - 2\nu} \\ &= \psi_1(\tilde{R}).\end{aligned}$$

Thus, because ψ_1 is strictly increasing, we obtain

$$\|\boldsymbol{x}(t_0 + T(\tilde{b}, d); t_0, \boldsymbol{x}_0)\| \le \tilde{R},$$

which contradicts our assumption that for all $t \in [t_0, t_0 + T(\tilde{b}, d)]$ it should have been $\|\boldsymbol{x}(t)\| > \tilde{R}$. Hence, there exists $t_1 \in [t_0, t_0 + T(\tilde{b}, d)]$ such that $\|\boldsymbol{x}(t_1)\| \le \tilde{R}$. Then using the same argument as in the previous case we can show that

$$\|\boldsymbol{x}(t)\| \le \tilde{b}$$

for all $t \ge t_1$ and the proof is complete.

We have the following consequence of the above theorem.

Corollary 4.2 The ball

$$\{\boldsymbol{x} : \|\boldsymbol{x}\| \le \psi_1^{-1}(\psi_2(R))\}$$

is uniformly stable.

Proof Let

$$\varepsilon > \psi_1^{-1}(\psi_2(R)),$$

and consider $\delta = \delta(\varepsilon)$ obtained from

$$\psi_1(\varepsilon) = \psi_2(\delta).$$

Obviously $\delta \ge R$. Figure 4.19 illustrates the above relations. Hence, by uniform boundedness if $\|\boldsymbol{x}_0\| \le \delta$, then $\|\boldsymbol{x}(t)\| \le \varepsilon$ for all $t \ge t_0$, which means that the ball is uniformly stable.

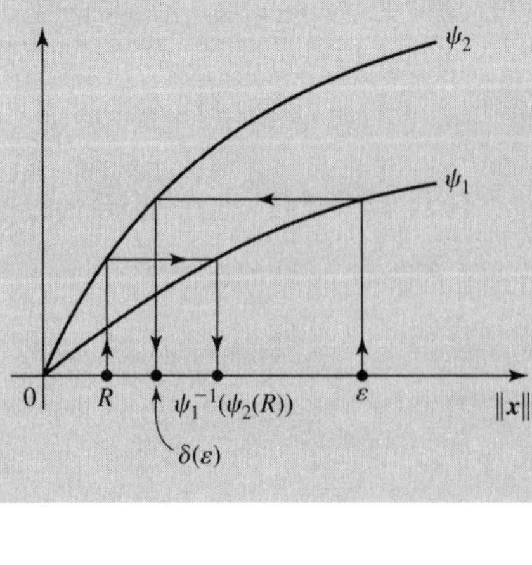

Figure 4.19 Finding δ given ε for an uniformly stable ball in the proof of Corollary 4.2.

◆ Example 4.11

Consider a dynamical system model

$$\dot{\boldsymbol{x}} = \boldsymbol{A}\boldsymbol{x} + \boldsymbol{b}(u + h(t, \boldsymbol{x})),$$

where

$$\boldsymbol{A} = \begin{bmatrix} 0 & 1 \\ -1 & -1 \end{bmatrix}, \quad \boldsymbol{b} = \begin{bmatrix} 0 \\ 1 \end{bmatrix}, \quad \text{and} \quad |h(t, \boldsymbol{x})| \le 2.$$

Let $\boldsymbol{P}$ be the solution of the matrix Lyapunov equation $\boldsymbol{A}^T\boldsymbol{P} + \boldsymbol{P}\boldsymbol{A} = -2\boldsymbol{I}_2$, and let

$$u = \begin{cases} -2\,\text{sign}(\boldsymbol{b}^T\boldsymbol{P}\boldsymbol{x}) & \text{if } |\boldsymbol{b}^T\boldsymbol{P}\boldsymbol{x}| \ge \nu \\ -2\dfrac{\boldsymbol{b}^T\boldsymbol{P}\boldsymbol{x}}{\nu} & \text{if } |\boldsymbol{b}^T\boldsymbol{P}\boldsymbol{x}| < \nu, \end{cases}$$

where $\nu = 1$. Suppose that the initial condition, $\boldsymbol{x}(0)$, is such that $\|\boldsymbol{x}(0)\|_2 \le 10$. Find the radius, b, of a ball such that $\|\boldsymbol{x}(t)\|_2 \le b$ for all $t \ge 0$.

Solving the Lyapunov equation $\boldsymbol{A}^T\boldsymbol{P} + \boldsymbol{P}\boldsymbol{A} = -2\boldsymbol{I}_2$ yields

$$\boldsymbol{P} = \begin{bmatrix} 3 & 1 \\ 1 & 2 \end{bmatrix}.$$

We next evaluate $d/dt(\boldsymbol{x}^T\boldsymbol{P}\boldsymbol{x})$ on the trajectories of the closed-loop system. For the case when $|\boldsymbol{b}^T\boldsymbol{P}\boldsymbol{x}| \ge \nu$, we get

$$\frac{d}{dt}(\boldsymbol{x}^T\boldsymbol{P}\boldsymbol{x}) \le -2\|\boldsymbol{x}\|^2.$$

For the case when $|\boldsymbol{b}^T\boldsymbol{P}\boldsymbol{x}| < \nu$, we obtain

$$\begin{aligned} \frac{d}{dt}(\boldsymbol{x}^T\boldsymbol{P}\boldsymbol{x}) &= -2\|\boldsymbol{x}\|^2 + 2\boldsymbol{x}^T\boldsymbol{P}\boldsymbol{b}(u+h) \\ &\le -2\|\boldsymbol{x}\|^2 - 4\frac{|\boldsymbol{x}^T\boldsymbol{P}\boldsymbol{b}|^2}{\nu} + 4|\boldsymbol{x}^T\boldsymbol{P}\boldsymbol{b}| \\ &\le -2\|\boldsymbol{x}\|^2 + 4\nu \\ &= -2\|\boldsymbol{x}\|^2 + 4. \end{aligned}$$

Hence, $d/dt(\boldsymbol{x}^T\boldsymbol{P}\boldsymbol{x}) < 0$ outside of the ball

$$\bar{B}_R(\boldsymbol{0}) = \{\boldsymbol{x} : \|\boldsymbol{x}\| \le R = \sqrt{2}\}.$$

To find the radius, b, of a ball, such that $\|\boldsymbol{x}(t)\|_2 \le b$ for all $t \ge 0$ and $\|\boldsymbol{x}(0)\|_2 \le 10 = d$, we first observe that

$$\lambda_{\min}(\boldsymbol{P})\|\boldsymbol{x}\|^2 = \psi_1(\|\boldsymbol{x}\|) \le \boldsymbol{x}^T\boldsymbol{P}\boldsymbol{x} \le \psi_2(\|\boldsymbol{x}\|) = \lambda_{\max}(\boldsymbol{P})\|\boldsymbol{x}\|^2.$$

Let $\tilde{d} = \max\{d, R\} = 10$. Then, the radius b is computed by solving the equation

$$\psi_1(b) = \psi_2(\tilde{d}),$$

that is,

$$\lambda_{\min}(\boldsymbol{P})b^2 = \lambda_{\max}(\boldsymbol{P})\tilde{d}^2.$$

We obtain

$$\begin{aligned} b &= \tilde{d}\sqrt{\frac{\lambda_{\max}(\boldsymbol{P})}{\lambda_{\min}(\boldsymbol{P})}} \\ &= 10\sqrt{\frac{(5+\sqrt{5})/2}{(5-\sqrt{5})/2}} = 16.18. \end{aligned}$$

◆ Example 4.12

In this example, we investigate the performance of a boundary layer tracking controller on the tracking problem considered in Example 4.10. Specifically, we replace the discontinuous controller given by (4.99) with a boundary layer continuous controller obtained by replacing the sign nonlinearity with a linear saturation nonlinearity. A linear saturation nonlinearity, denoted sat_ν, can be described by

$$\text{sat}_\nu(x) = \frac{1}{2\nu}\left(|x+\nu| - |x-\nu|\right), \tag{4.110}$$

where the parameter ν controls the width of the boundary layer. In other words,

$$\text{sat}_\nu(x) = \begin{cases} 1 & \text{if } x \geq \nu, \\ \frac{1}{\nu}x & \text{if } -\nu < x < \nu, \\ -1 & \text{if } x \leq \nu. \end{cases} \tag{4.111}$$

A plot of (4.110) for $\nu = 0.5$ is shown in Figure 4.20. A boundary layer version of the discontinuous tracking controller (4.99) is

$$\begin{aligned} u &= -\rho(\boldsymbol{x}, r)\text{sat}_\nu(\boldsymbol{b}^T\boldsymbol{P}\boldsymbol{e}) \\ &= -(|\boldsymbol{k}_d\boldsymbol{x}| + |r|)\text{sat}_\nu(\boldsymbol{b}^T\boldsymbol{P}(\boldsymbol{x}-\boldsymbol{x}_d)) \\ &= -\left(\left|\begin{bmatrix} 0.0329 \\ 2.2854 \\ 0.2429 \\ -0.1653 \end{bmatrix}^T \boldsymbol{x}\right| + |r|\right)\text{sat}_\nu\left(\begin{bmatrix} 0.1571 \\ 26.9931 \\ 6.3324 \\ -3.0250 \end{bmatrix}^T (\boldsymbol{x}-\boldsymbol{x}_d)\right). \end{aligned} \tag{4.112}$$

The plot of the actual output, Y, of the closed-loop system driven by (4.112) for $\nu = 0.5$ and zero initial conditions was indistinguishable from that shown in Figure 4.14 when the discontinuous controller was used. The boundary layer controller action versus time is shown in Figure 4.21. Chattering is basically eliminated in this case.

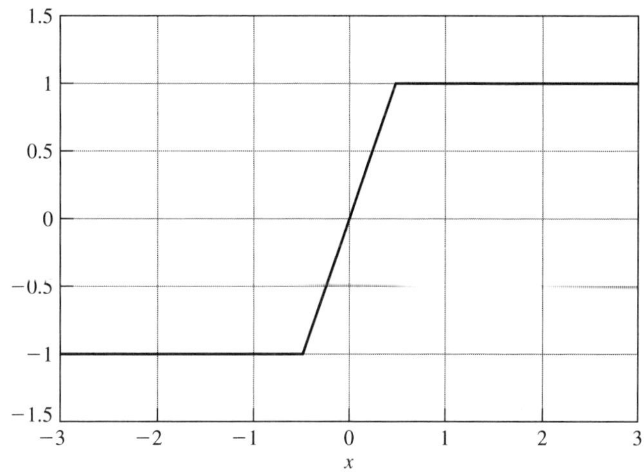

Figure 4.20 Plot of the linear saturation function described by (4.111).

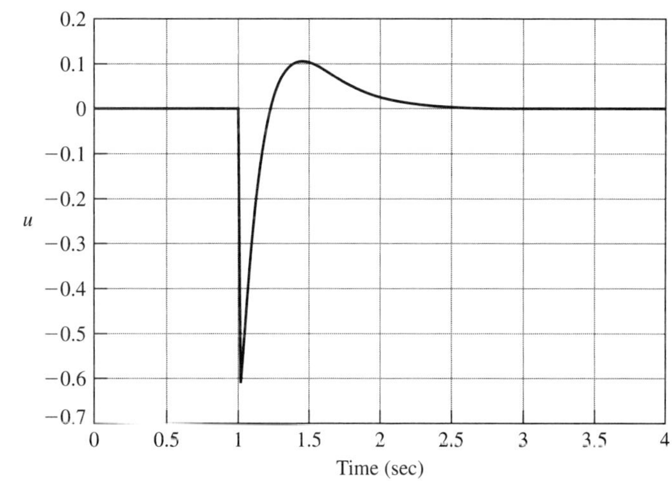

Figure 4.21 Plot of the boundary layer controller action versus time in Example 4.12.

4.12 Lyapunov-Like Analysis

In this section we present a number of results that are very useful in the analysis and design of adaptive controllers.

Lemma 4.4 (Barbalat) Given a differentiable function $f = f(t) : \mathbb{R} \to \mathbb{R}$ such that $\dot{f}$ is uniformly continuous. If $\lim_{t\to\infty} f(t)$ exists, then

$$\lim_{t\to\infty} \dot{f}(t) = 0.$$

Proof We prove the lemma by contraposition. We assume that $\dot{f}(t)$ does not converge to 0 as $t \to \infty$. Therefore, there exists an $\varepsilon > 0$ such that for every $T > 0$ we can find $t_i \geq T$ so that $|\dot{f}(t_i)| \geq \varepsilon$. By assumption $\dot{f}$ is uniformly continuous. Hence, given $\varepsilon/2 > 0$ there is $\delta_{\varepsilon/2} > 0$ so that for every t_i and every $\tau \in [0, \delta_{\varepsilon/2}]$, we have

$$|\dot{f}(t_i + \tau) - \dot{f}(t_i)| < \frac{\varepsilon}{2}. \tag{4.113}$$

Hence, for all $t \in [t_i, t_i + \delta_{\varepsilon/2}]$, we have

$$\begin{aligned} |\dot{f}(t)| &= |\dot{f}(t) - \dot{f}(t_i) + \dot{f}(t_i)| \\ &\geq |\dot{f}(t_i)| - |\dot{f}(t) - \dot{f}(t_i)| \\ &> \varepsilon - \frac{\varepsilon}{2} \\ &= \frac{\varepsilon}{2}. \end{aligned}$$

Suppose that $\dot{f}(t_i) > 0$. (The other case is when $\dot{f}(t_i) < 0$ and its analysis is similar to the case we analyze now.) It follows from the above that for any $t \in [t_i, t_i + \delta_{\varepsilon/2}]$,

$$\dot{f}(t) > \frac{\varepsilon}{2}.$$

Using the above, we obtain

$$\int_{t_i}^{t_i+\delta_{\varepsilon/2}} \dot{f}(t)\, dt > \int_{t_i}^{t_i+\delta_{\varepsilon/2}} \frac{\varepsilon}{2}\, dt = \frac{\varepsilon}{2}\delta_{\varepsilon/2}. \tag{4.114}$$

However,

$$\int_{t_i}^{t_i+\delta_{\varepsilon/2}} \dot{f}(t)\, dt = f(t_i + \delta_{\varepsilon/2}) - f(t_i). \tag{4.115}$$

Combining (4.114) and (4.115) yields

$$f(t_i + \delta_{\varepsilon/2}) - f(t_i) > \frac{\varepsilon}{2}\delta_{\varepsilon/2} \qquad \text{for every } i,$$

which contradicts the assumption that $f(t)$ has a finite limit as $t \to \infty$. The proof of the lemma is complete.

A simple sufficient condition for a differentiable function to be uniformly continuous is boundedness of its derivative. Thus if a function f is twice differentiable, then its derivative $\dot{f}$ is uniformly continuous if its second derivative $\ddot{f}$ is bounded.

The following lemma, called by Slotine and Li [262, p. 125] the "Lyapunov-like lemma," follows immediately from Barbalat's lemma.

Lemma 4.5 ("Lyapunov-Like Lemma") Given a real-valued function $W(t, \boldsymbol{x})$ such that

1. $W(t, \boldsymbol{x})$ is bounded below,
2. $\dot{W}(t, \boldsymbol{x})$ is negative semidefinite, and
3. $\dot{W}(t, \boldsymbol{x})$ is uniformly continuous in time,

then

$$\dot{W}(t, \boldsymbol{x}) \to 0 \qquad \text{as } t \to \infty.$$

◆ Example 4.13

Investigate the stability of the following dynamical system model:

$$\begin{bmatrix} \dot{x}_1 \\ \dot{x}_2 \end{bmatrix} = \begin{bmatrix} -2x_1 + x_2 u \\ -x_1 u \end{bmatrix}, \tag{4.116}$$

where the input u is bounded.

Consider the Lyapunov function candidate

$$V = \|\boldsymbol{x}\|^2 = x_1^2 + x_2^2.$$

Note that V is bounded below. The time derivative of V evaluated on the trajectories of the system model (4.116) is

$$\dot{V} = 2x_1\dot{x}_1 + 2x_2\dot{x}_2 = 2x_1(-2x_1 + x_2 u) - 2x_2 x_1 u = -4x_1^2 \le 0.$$

Thus, the system modeled by (4.116) is uniformly stable, and hence x_1 and x_2 are bounded. We now use Lemma 4.5 to show that in addition, $x_1(t) \to 0$ as $t \to \infty$. To use this lemma, we have to show that $\dot{V}$ is uniformly continuous in t. For this it is enough to show that the derivative of $\dot{V}$ is bounded. The derivative of $\dot{V}$ is

$$\ddot{V} = -8x_1\dot{x}_1 = -8x_1(-2x_1 + x_2 u),$$

which is bounded because u is bounded by assumption and we showed that x_1 and x_2 are also bounded. Hence, $\dot{V}$ is uniformly continuous. The function V satisfies all the conditions of Lemma 4.5, and therefore $x_1(t) \to 0$ as $t \to \infty$.

◆ Example 4.14

Consider a dynamical system model,

$$\dot{\boldsymbol{x}} = \boldsymbol{A}\boldsymbol{x} + \boldsymbol{b}u, \qquad e = \boldsymbol{c}\boldsymbol{x}, \tag{4.117}$$

where

$$u = \boldsymbol{\phi}^T \boldsymbol{\xi}. \tag{4.118}$$

The components $\phi_i = \phi_i(t)$ of the vector-valued function $\boldsymbol{\phi} = \boldsymbol{\phi}(t) \in \mathbb{R}^K$ are adjustable parameters according to the adaptation law

$$\dot{\boldsymbol{\phi}} = -\boldsymbol{\Gamma} e \boldsymbol{\xi}, \tag{4.119}$$

where $\boldsymbol{\Gamma} = \boldsymbol{\Gamma}^T > 0$ is a design parameter matrix. The vector-valued function $\boldsymbol{\xi} = \boldsymbol{\xi}(t) \in \mathbb{R}^K$ is bounded for all t. The matrix $\boldsymbol{A} \in \mathbb{R}^{n\times n}$ is asymptotically stable, and $\boldsymbol{b} \in \mathbb{R}^n$. The row vector $\boldsymbol{c} \in \mathbb{R}^{1\times n}$ satisfies the equation

$$\boldsymbol{c} = \boldsymbol{b}^T \boldsymbol{P}, \tag{4.120}$$

where $\boldsymbol{P} = \boldsymbol{P}^T > 0$ is the solution of the matrix Lyapunov equation

$$\boldsymbol{A}^T \boldsymbol{P} + \boldsymbol{P}\boldsymbol{A} = -2\boldsymbol{Q}$$

for some $\boldsymbol{Q} = \boldsymbol{Q}^T > 0$. We will use the Lyapunov function candidate,

$$V(\boldsymbol{x}, \boldsymbol{\phi}) = \tfrac{1}{2}\boldsymbol{x}^T \boldsymbol{P}\boldsymbol{x} + \tfrac{1}{2}\boldsymbol{\phi}^T \boldsymbol{\Gamma}^{-1} \boldsymbol{\phi}, \tag{4.121}$$

and the Lyapunov-like lemma to show that $\boldsymbol{x}$ and $\boldsymbol{\phi}$ are both bounded and that

$$\lim_{t\to\infty} \boldsymbol{x}(t) = \boldsymbol{0}.$$

We first evaluate the time derivative of the Lyapunov function candidate $V = V(\boldsymbol{x}, \boldsymbol{\phi})$ on the trajectories of the closed-loop system,

$$\dot{\boldsymbol{x}} = \boldsymbol{A}\boldsymbol{x} + \boldsymbol{b}(\boldsymbol{\phi}^T\boldsymbol{\xi}), \qquad \dot{\boldsymbol{\phi}} = -\boldsymbol{\Gamma} e \boldsymbol{\xi}. \tag{4.122}$$

We have

$$\begin{aligned}
\dot{V} &= \boldsymbol{x}^T \boldsymbol{P}\dot{\boldsymbol{x}} + \boldsymbol{\phi}^T \boldsymbol{\Gamma}^{-1} \dot{\boldsymbol{\phi}} \\
&= \boldsymbol{x}^T \boldsymbol{P}(\boldsymbol{A}\boldsymbol{x} + \boldsymbol{b}(\boldsymbol{\phi}^T\boldsymbol{\xi})) - \boldsymbol{\phi}^T e \boldsymbol{\Gamma}^{-1}\boldsymbol{\Gamma}\boldsymbol{\xi} \\
&= -\boldsymbol{x}^T \boldsymbol{Q}\boldsymbol{x} + (e\boldsymbol{\phi}^T - e\boldsymbol{\phi}^T)\boldsymbol{\xi} \\
&= -\boldsymbol{x}^T \boldsymbol{Q}\boldsymbol{x},
\end{aligned}$$

where we used the fact that $\boldsymbol{c} = \boldsymbol{b}^T \boldsymbol{P}$ and hence $e = \boldsymbol{c}\boldsymbol{x} = \boldsymbol{x}^T \boldsymbol{c}^T = \boldsymbol{x}^T \boldsymbol{P}\boldsymbol{b}$. Thus, $\dot{V}$ is negative semidefinite in the $\begin{bmatrix}\boldsymbol{x}\\ \boldsymbol{\phi}\end{bmatrix}$-space, which implies that $\boldsymbol{x}$ and $\boldsymbol{\phi}$ are bounded.

To show the convergence of $\boldsymbol{x}$, we use the Lyapunov-like lemma. For this, we first show that $\dot{V}$ is uniformly continuous. Indeed, because

$$\ddot{V} = -2\boldsymbol{x}^T \boldsymbol{Q}\dot{\boldsymbol{x}}$$

is bounded, since $\boldsymbol{x}$ and $\dot{\boldsymbol{x}}$ are bounded, $\dot{V}$ is uniformly continuous. Hence, $\dot{V} \to 0$ as $t \to \infty$. Since $\dot{V} = -\boldsymbol{x}^T \boldsymbol{Q}\boldsymbol{x}$ and $\boldsymbol{Q} = \boldsymbol{Q}^T > 0$, we have to have $\boldsymbol{x}(t) \to \boldsymbol{0}$ as $t \to 0$.

We can re-state Barbalat's lemma in alternative form as follows (see, for example, Khalil [157, p. 192]).

Lemma 4.6 Let $\phi : \mathbb{R} \to \mathbb{R}$ be a uniformly continuous function on $[0, \infty)$. Suppose that $\lim_{t\to\infty} \int_0^t \phi(\tau)\, d\tau$ exists and is finite. Then,

$$\phi(t) \to 0 \qquad \text{as } t \to \infty.$$

4.13 LaSalle's Invariance Principle

In this section we generalize the Lyapunov function concept using LaSalle's invariant set. We first present this generalization, following the exposition in LaSalle [176], for nonlinear dynamical systems modeled by the difference equation

$$\boldsymbol{x}(k+1) = \boldsymbol{f}(\boldsymbol{x}(k)), \tag{4.123}$$

where $k \in \mathcal{Z}_+$, the set of nonnegative integers, and $\boldsymbol{x} : \mathcal{Z}_+ \to \mathbb{R}^n$; that is, $\boldsymbol{x}(k)$ is the state of the system model (4.123) at time k, and $\boldsymbol{f} : \mathbb{R}^n \to \mathbb{R}^n$. We assume that the vector-valued function $\boldsymbol{f}$ is continuous on $\mathbb{R}^n$; that is, if $\boldsymbol{x}^j \to \boldsymbol{y}$, then $\boldsymbol{f}(\boldsymbol{x}^j) \to \boldsymbol{f}(\boldsymbol{y})$. If $\boldsymbol{x}$ is a vector, then $\boldsymbol{f}(\boldsymbol{x})$ is a vector and is the value of the function $\boldsymbol{f}$ at $\boldsymbol{x}$.

The solution to the initial value problem

$$\boldsymbol{x}(k+1) = \boldsymbol{f}(\boldsymbol{x}(k)), \qquad \boldsymbol{x}(0) = \boldsymbol{x}^0$$

is

$$\boldsymbol{x}(k) = \boldsymbol{x}^k = \boldsymbol{f}^k(\boldsymbol{x}^0),$$

where $\boldsymbol{f}^k$ is the kth iterate of $\boldsymbol{f}$; that is, $\boldsymbol{f}^0 = \boldsymbol{I}_n$, and $\boldsymbol{f}^k = \boldsymbol{f}\boldsymbol{f}^{k-1}$.

A state $\boldsymbol{x}_e \in \mathbb{R}^n$ is a *fixed point* or an *equilibrium state* of the system $\boldsymbol{x}(k+1) = \boldsymbol{f}(\boldsymbol{x}(k))$ if $\boldsymbol{x}_e = \boldsymbol{f}(\boldsymbol{x}_e)$. Thus the solution of the system that starts at $\boldsymbol{x}_e$ stays at $\boldsymbol{x}_e$ for all $k \geq 0$.

A subset $\mathcal{A} \subset \mathbb{R}^n$ is called *positively invariant* if $\boldsymbol{f}(\mathcal{A}) \subset \mathcal{A}$; that is, if $\boldsymbol{x} \in \mathcal{A}$, then $\boldsymbol{f}(\boldsymbol{x}) \in \mathcal{A}$. It is clear that an equilibrium solution is invariant. The adjective "positive" is added to emphasize the fact that we will be interested in the behavior of the system for $k \geq 0$.

The solution through $\boldsymbol{x}^0$ is bounded if there is a constant M such that $\|\boldsymbol{f}^k(\boldsymbol{x}^0)\| \leq M$ for all $k \geq 0$.

Consider again a particular solution $\boldsymbol{f}^k(\boldsymbol{x}^0)$. A point $\boldsymbol{y} \in \mathbb{R}^n$ is a *positive limit point* or a *positive point of accumulation* of $\boldsymbol{f}^k(\boldsymbol{x}^0)$ if there is a subsequence k_i such that $\boldsymbol{f}^{k_i}(\boldsymbol{x}^0) \to \boldsymbol{y}$. The set of all positive limit points of $\boldsymbol{f}^k(\boldsymbol{x}^0)$ is called the *positive limit set* and denoted $\Omega(\boldsymbol{x}^0)$.

A sequence $\{\boldsymbol{x}^k\}$ is said to approach a set $\mathcal{A}$ if

$$\text{dist}(\boldsymbol{x}^k, \mathcal{A}) = \inf\{\|\boldsymbol{x}^k - \boldsymbol{y}\| : \boldsymbol{y} \in \mathcal{A}\} \to 0,$$

where $\text{dist}(\boldsymbol{p}, \mathcal{S})$ denotes the distance from a point $\boldsymbol{p}$ to a set $\mathcal{S}$.

Lemma 4.7 If $\boldsymbol{f}$ is continuous and $\boldsymbol{f}^k(\boldsymbol{x}^0)$ is bounded for all $k \geq 0$, then $\Omega(\boldsymbol{x}^0)$ is non-empty, compact, positively invariant set. Furthermore, $\boldsymbol{f}^k(\boldsymbol{x}^0) \to \Omega(\boldsymbol{x}^0)$ as $k \to \infty$.

Proof Because the sequence $\{\boldsymbol{x}^k\}$ is bounded by the Bolzano–Weierstrass theorem, stated on page 645, there is at least one positive limit point of $\{\boldsymbol{x}^k\}$ and so $\Omega(\boldsymbol{x}^0)$ is nonempty and bounded.

We now show that the set $\Omega(\boldsymbol{x}^0)$ is closed; that is, if $\boldsymbol{y}^k \to \boldsymbol{y}$ with $\boldsymbol{y}^k \in \Omega(\boldsymbol{x}^0)$, then $\boldsymbol{y} \in \Omega(\boldsymbol{x}^0)$. Because $\Omega(\boldsymbol{x}^0)$ is nonempty and bounded, there is a sequence, say $\{\boldsymbol{y}^l\}$, $\boldsymbol{y}^l \in \Omega(\boldsymbol{x}^0)$, such that $\boldsymbol{y}^l \to \boldsymbol{y}$. Because each $\boldsymbol{y}^l$ is in $\Omega(\boldsymbol{x}^0)$, for each $\boldsymbol{y}^l$ there is a subsequence, $\{\boldsymbol{x}_l^{k_i}\}$, of $\{\boldsymbol{x}^k\}$ such that $\boldsymbol{x}_l^{k_i} \to \boldsymbol{y}^l$. The sequences $\{\boldsymbol{x}_l^{k_i}\}$ are convergent. Therefore, we can construct a sequence $\{\boldsymbol{x}_l\}$, a subsequence of $\{\boldsymbol{x}^k\}$, such that for any $\varepsilon > 0$ there exist N_1 so that

$$\|\boldsymbol{x}_l - \boldsymbol{y}^l\| < \varepsilon/2 \qquad \text{for } l \geq N_1.$$

The construction of such a sequence can be accomplished as follows. For $l = 1$ select a term $\boldsymbol{x}_1$ from $\{\boldsymbol{x}_1^{k_i}\}$ such that

$$\|\boldsymbol{x}_1 - \boldsymbol{y}^1\| < 1.$$

Next, for $l = 2$ select a term $\boldsymbol{x}_2$ form $\{\boldsymbol{x}_2^{k_i}\}$ such that

$$\|\boldsymbol{x}_2 - \boldsymbol{y}^2\| < 1/2,$$

and so on. The sequence $\{\boldsymbol{y}^l\}$ is also convergent. Hence, for any $\varepsilon > 0$ there is N_2 such that

$$\|\boldsymbol{y}^l - \boldsymbol{y}\| < \varepsilon/2 \qquad \text{for } l \geq N_2.$$

Combining the above gives

$$\|\boldsymbol{x}_l - \boldsymbol{y}\| \leq \|\boldsymbol{x}_l - \boldsymbol{y}^l\| + \|\boldsymbol{y}^l - \boldsymbol{y}\| < \varepsilon \qquad \text{for } l \geq \max\{N_1, N_2\},$$

which implies that $\boldsymbol{y} \in \Omega(\boldsymbol{x}^0)$. That is, the limit of any convergent sequence with elements in $\Omega(\boldsymbol{x}^0)$ is also in this set, which means that the set $\Omega(\boldsymbol{x}^0)$ is closed. Because $\Omega(\boldsymbol{x}^0)$ is also bounded, it is compact.

We now show that $\Omega(\boldsymbol{x}^0)$ is positively invariant. For this let $\boldsymbol{y} \in \Omega(\boldsymbol{x}^0)$. Hence, there is a subsequence k_i such that $\boldsymbol{x}^{k_i} \to \boldsymbol{y}$. Because $\boldsymbol{f}$ is continuous, $\boldsymbol{x}^{k_i+1} = \boldsymbol{f}(\boldsymbol{x}^{k_i}) \to \boldsymbol{f}(\boldsymbol{y})$ as $i \to \infty$. Therefore $\boldsymbol{f}(\boldsymbol{y}) \in \Omega(\boldsymbol{x}^0)$, which implies that $\Omega(\boldsymbol{x}^0)$ is invariant. Note that we do not have to have $\boldsymbol{y} = \boldsymbol{f}(\boldsymbol{y})$.

Finally, we will show that $\boldsymbol{f}^k(\boldsymbol{x}^0) \to \Omega(\boldsymbol{x}^0)$ as $k \to \infty$. We shall prove the statement by contradiction. First observe that $\text{dist}(\boldsymbol{x}^k, \Omega(\boldsymbol{x}^0))$ is bounded because both $\{\boldsymbol{x}^k\}$ and $\Omega(\boldsymbol{x}^0)$ are. Assume that $\text{dist}(\boldsymbol{x}^k, \Omega(\boldsymbol{x}^0))$ does not converge to zero. Then, there is a subsequence k_i such that $\boldsymbol{x}^{k_i} \to \boldsymbol{y}$ and $\rho(\boldsymbol{x}^{k_i}, \Omega(\boldsymbol{x}^0)) \to a > 0$. But because $\boldsymbol{y} \in \Omega(\boldsymbol{x}^0)$,

$$\text{dist}(\boldsymbol{x}^{k_i}, \Omega(\boldsymbol{x}^0)) \leq \|\boldsymbol{x}^{k_i} - \boldsymbol{y}\| \to 0,$$

and thus $\text{dist}(\boldsymbol{x}^{k_i}, \Omega(\boldsymbol{x}^0)) \to 0$, which is a contradiction. The proof of the lemma is now complete.

Let $V : \mathbb{R}^n \to \mathbb{R}$. Define the *Lyapunov difference* of V relative to $\boldsymbol{x}(k+1) = \boldsymbol{f}(\boldsymbol{x}(k))$, where $\boldsymbol{x}(k) \in \mathcal{G} \subseteq \mathbb{R}^n$, as

$$\Delta V(\boldsymbol{x}) = V(\boldsymbol{f}(\boldsymbol{x})) - V(\boldsymbol{x});$$

that is,

$$\Delta V(\boldsymbol{x}(k)) = V(\boldsymbol{x}(k+1)) - V(\boldsymbol{x}(k)).$$

If $\Delta V(\boldsymbol{x}(k)) \leq 0$, then V is nonincreasing along solutions of the system.

We say that V is a *Lyapunov* function for $\boldsymbol{x}(k+1) = \boldsymbol{f}(\boldsymbol{x}(k))$ on a set $\mathcal{G}$ if

1. V is continuous on $\mathbb{R}^n$ and
2. $\Delta V(\boldsymbol{x}(k)) \leq 0$ for all $\boldsymbol{x}(k) \in \mathcal{G}$.

A c-level surface of V is the set

$$V^{-1}(c) = \{\boldsymbol{x} : V(\boldsymbol{x}) = c, \ \boldsymbol{x} \in \mathbb{R}^n\}.$$

Theorem 4.14 (LaSalle's Invariance Principle) If

1. V is a Lyapunov function for $\boldsymbol{x}(k+1) = \boldsymbol{f}(\boldsymbol{x}(k))$ in a compact set $\mathcal{G}$ and
2. the solution $\boldsymbol{f}^k(\boldsymbol{x}^0)$ is bounded and in $\mathcal{G}$,

then there is a number c such that $\boldsymbol{f}^k(\boldsymbol{x}^0) \to \mathcal{M} \cap V^{-1}(c)$, where $\mathcal{M}$ is the largest positively invariant set contained in the set $\mathcal{E} = \{\boldsymbol{x} \in \mathbb{R}^n : \Delta V = 0\}$.

Proof Because $\boldsymbol{x}^k = \boldsymbol{f}^k(\boldsymbol{x}^0)$ is bounded and in $\mathcal{G}$, by Lemma 4.7 the positive limit set $\Omega(\boldsymbol{x}^0)$ is nonempty and in $\mathcal{G}$. Furthermore, $\boldsymbol{x}^k$ tends to $\Omega(\boldsymbol{x}^0)$. Now $\{V(\boldsymbol{x}^k)\}$ is nonincreasing and bounded below, so $V(\boldsymbol{x}^k) \to c$ for some c. If $\boldsymbol{y} \in \Omega(\boldsymbol{x}^0)$, then there is a subsequence k_i such that $\boldsymbol{x}^{k_i} \to \boldsymbol{y}$, and hence $V(\boldsymbol{x}^{k_i}) \to V(\boldsymbol{y})$; that is, $V(\boldsymbol{y}) = c$. Thus $V(\Omega(\boldsymbol{x}^0)) = c$, and hence $\Omega(\boldsymbol{x}^0) \subset V^{-1}(c)$. Because $V(\Omega(\boldsymbol{x}^0)) = c$ and $\Omega(\boldsymbol{x}^0)$ is positively invariant, for arbitrary $\boldsymbol{y} \in \Omega(\boldsymbol{x}^0)$ we have

$$\begin{aligned} \Delta V(\boldsymbol{y}) &= V(F(\boldsymbol{y})) - V(\boldsymbol{y}) \\ &= c - c = 0; \end{aligned}$$

that is, $\Delta V(\Omega(\boldsymbol{x}^0)) = 0$. In summary,

$$\boldsymbol{x}^k \to \Omega(\boldsymbol{x}^0) \subset \{\boldsymbol{x} \in \mathbb{R}^n : \Delta V = 0\} \cap V^{-1}(c).$$

Because $\Omega(\boldsymbol{x}^0)$ is positively invariant, we have $\Omega(\boldsymbol{x}^0) \subset \mathcal{M}$, and the proof of the theorem is complete.

Using the preceding theorem we can study stability properties of difference equations.

LaSalle's invariance principle can be used to prove a result concerning the instability of equilibrium states.

Theorem 4.15 Let $\boldsymbol{x}_e$ be an equilibrium state of $\boldsymbol{x}(k+1) = \boldsymbol{f}(\boldsymbol{x}(k))$, where $\{\boldsymbol{x}(k)\}$ is bounded. Let ΔV be positive definite with respect to $\boldsymbol{x}_e$, and let V take positive values arbitrarily close to $\boldsymbol{x}_e$. If $V(\boldsymbol{x}_e) = 0$, then $\boldsymbol{x}_e$ is unstable.

Proof We prove the theorem by contradiction. First observe that $-V$ is a Lyapunov function because $\Delta(-V) < 0$. Because $\boldsymbol{x}^k = \boldsymbol{f}^k(\boldsymbol{x}^0)$ is bounded, we can apply the invariance principle to conclude that $\boldsymbol{x}^k$ tends to $\{\boldsymbol{x} : \Delta(-V(\boldsymbol{x})) = 0\} = \{\boldsymbol{x}_e\}$, that is, $\boldsymbol{x}^k \to \boldsymbol{x}_e$. Using the above, we conclude that $V(\boldsymbol{x}^k) \to V(\boldsymbol{x}_e) = 0$. However, $\Delta V(\boldsymbol{x}^k) > 0$ and so

$$V(\boldsymbol{x}^k) > V(\boldsymbol{x}^{k-1}) > \cdots > V(\boldsymbol{x}^0) > 0.$$

Hence, for some positive ε, we have $V(\boldsymbol{x}^k) > \varepsilon > 0$, and thus we would have to have $V(\boldsymbol{x}_e) > 0$. The above contradiction completes the proof.

We now present LaSalle's invariance principle for continuous-time dynamical systems models of the form

$$\dot{\boldsymbol{x}} = \boldsymbol{f}(\boldsymbol{x}), \tag{4.124}$$

where $\boldsymbol{f} : \mathcal{D} \to \mathbb{R}^n$ with $\mathcal{D} \subseteq \mathbb{R}^n$ being open and connected. The function $\boldsymbol{f}$ is assumed to be locally Lipschitz on $\mathcal{D}$; that is, each point of $\mathcal{D}$ has a neighborhood such that $\boldsymbol{f}$ satisfies the Lipschitz condition in this neighborhood with some Lipschitz constant. Each neighborhood can have a different Lipschitz constant. The above assumptions are made in order to ensure the existence and uniqueness of a solution to (4.124) for a given initial condition $\boldsymbol{x}(0)$.

To proceed, we need a few definitions. Suppose that $\boldsymbol{x}(t)$ is a solution of (4.124). Then, a point $\boldsymbol{p}$ is a *positive limit point* of $\boldsymbol{x}(t)$ if there is a sequence $\{t_n\}$ with $t_n \to \infty$ as $n \to \infty$ such that $\boldsymbol{x}(t_n) \to \boldsymbol{p}$ as $n \to \infty$. The set of all positive limit points of $\boldsymbol{x}(t)$ is called the positive limit set of $\boldsymbol{x}(t)$. A set $\mathcal{A}$ is a positively invariant set with respect to (4.124) if

$$\boldsymbol{x}(0) \in \mathcal{A} \quad \text{implies} \quad \boldsymbol{x}(t) \in \mathcal{A} \qquad \text{for all } t \geq 0.$$

We say that $\boldsymbol{x}(t)$ approaches a set $\mathcal{A}$, as $t \to \infty$, if for each $\varepsilon > 0$ there is $T > 0$ such that

$$\operatorname{dist}(\boldsymbol{x}(t), \mathcal{A}) < \varepsilon \qquad \text{for all } t > T.$$

Employing arguments similar to those used in the proof of Lemma 4.7, we can prove the following lemma.

Lemma 4.8 If a solution $\boldsymbol{x}(t)$ to (4.124) is bounded and belongs to $\mathcal{D}$ for $t \geq 0$, then its positive limit set Ω is non-empty, compact, and invariant. Furthermore,

$$\boldsymbol{x}(t) \to \Omega \qquad \text{as } t \to \infty.$$

Using the above lemma, we can prove the following version of the LaSalle's invariance principle for the continuous-time case:

Theorem 4.16 If $V : \mathbb{R}^n \to \mathbb{R}$ is a continuously differentiable function such that $\dot{V}(x) \leq 0$ on the solutions of (4.124) that reside in a compact set $\mathcal{G} \subset \mathcal{D}$, then there is a number c such that $\boldsymbol{x}(t) \to \mathcal{M} \cap V^{-1}(c)$, where $\mathcal{M}$ is the largest positively invariant set contained in the set $\mathcal{E} = \{\boldsymbol{x} \in \mathbb{R}^n : \dot{V} = 0\}$.

For further discussion of the continuous-time LaSalle's invariance principle, we refer to Khalil [157, pp. 114–115]. Applications of LaSalle's invariance principle to the stability analysis of various neural networks can be found in references 43 and 287, while in references 156, and 256 positively invariant sets were used to analyze stability of a class of nonlinear systems driven by adaptive feedback controllers.

Notes

Hahn [112] is a classical text on stability of dynamical systems. Stability of discrete dynamical systems is presented by Hahn [111] and LaSalle [176]. Extensive coverage of stability issues can also be found in Willems [300], Slotine and Li [262], Vidyasagar [289], Krstić, Kanellakopoulos, and Kokotović [171], and Khalil [157].

There is an interesting story behind the origins of the Routh test. The problem of investigating the stability of a liner dynamical system in terms of the location of the roots of its characteristic equation was posed by Maxwell [197]. On page 271 of his 1868 paper, Maxwell writes, "This condition is mathematically equivalent to the condition that all the possible roots, and all the possible parts of the impossible roots, of a certain equation shall be negative." By impossible roots, Maxwell means complex roots, and by possible parts he means real parts in today's terminology. He continues, "I have not been able completely to determine these conditions for equations of a higher degree than the third; but I hope that the subject will obtain the attention of mathematicians." A complete general solution to the problem was published by Routh in 1877. When Maxwell and Routh were classmates at Cambridge University in 1854, Routh was placed first in their final math examination ahead of Maxwell. A story says (Smith [265, p. 55]) that apparently Maxwell was so sure of coming in first on this exam that he did not even bother to rise early to hear the exam results read out in the Senate House. Maxwell sent his servant instead. When the servant came back, Maxwell asked, "Well, tell me who's second!" Obviously, he did not expect to receive the reply, "You are, sir." In 1877, Routh received the prestigious Adams Prize for solving the problem posed by Maxwell in 1868. According to an account by Truxal [283, p. 527], Routh presented his solution with the opening remark, "It has recently come to my attention that my good friend James Clerk Maxwell has had difficulty with a rather trivial problem . . ."

EXERCISES

4.1 (a) Consider a negative unity feedback closed-loop system whose open-loop transfer function is

$$G(s) = \frac{6s^3 + 11s^2}{s^4 + 6s + K}.$$

For what values of the parameter K the closed-loop system is asymptotically stable?

(b) Figure 4.22 shows a closed-loop system in which the error, measured by a transducer that has a time constant of 1 sec, is applied to a proportional-plus-derivative (PD) controller. The output of the controller drives a third-order plant. Use the

Routh–Hurwitz criterion to find the region of asymptotic stability, of the closed-loop feedback system, in the (a, K)-plane.

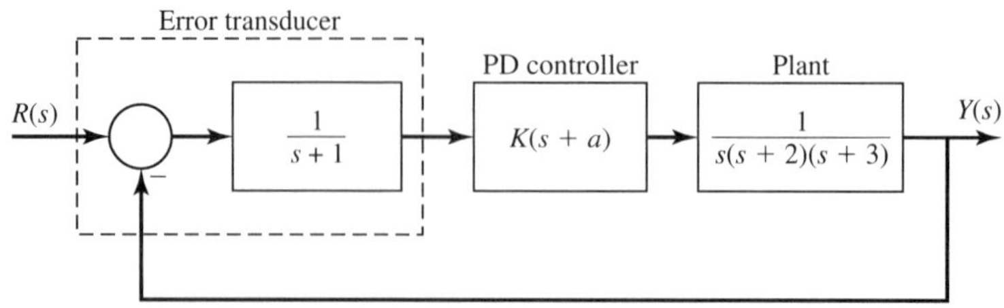

Figure 4.22 A closed-loop feedback system of Exercise 4.1.

4.2 (a) Determine if the polynomial

$$P(s) = s^7 + 2s^6 + 5s^5 + 7s^4 + 2s^3 - 5s^2 - 8s - 4$$

is Hurwitz. Use the Routh–Hurwitz criterion to find the number of the polynomial RHP zeros and determine the $j\omega$-axis zeros.

(b) Consider a negative unity feedback system with the forward loop transfer function

$$G(s) = \frac{K}{s(s+3)(s+10)}.$$

Use the Routh–Hurwitz criterion to find the interval of K for which the closed-loop system is asymptotically stable.

(c) Consider a negative unity feedback system with the forward loop transfer function

$$G(s) = \frac{K}{s(Ts+1)(0.5\,s+1)}.$$

Use the Routh–Hurwitz criterion to derive the domain of asymptotic stability of the closed-loop system in the (T, K)-plane, and plot it using MATLAB.

4.3 For a dynamical system model

$$\begin{bmatrix} \dot{x}_1 \\ \dot{x}_2 \end{bmatrix} = \begin{bmatrix} x_2 \\ x_1 - x_2 \sin x_1 \end{bmatrix},$$

(a) find an equilibrium point for this system;

(b) linearize the system about the found equilibrium point;

(c) determine if the linearized system is stable, asymptotically stable, or unstable.

4.4 Consider the following model of a dynamical system,

$$\begin{aligned} \dot{x}_1 &= x_2, \\ \dot{x}_2 &= -x_1^5 - x_2^3. \end{aligned} \tag{4.125}$$

(a) Find the equilibrium state of the system. Then, consider the following Lyapunov function candidate,

$$V(x_1, x_2) = \frac{1}{2}x_2^2 + \int_0^{x_1} s^5\, ds.$$

Evaluate $d/dt V(x_1, x_2)$ on the trajectories of the system (4.125).

(b) Use the results obtained in (a) to decide whether the equilibrium state is stable, asymptotically stable, or unstable.

4.5 For a dynamical system modeled by $\dot{\boldsymbol{x}} = \boldsymbol{A}\boldsymbol{x} + \boldsymbol{\zeta}(t, \boldsymbol{x})$, where

$$\boldsymbol{A} = \begin{bmatrix} 0 & 1 \\ -1/2 & -3/2 \end{bmatrix} \quad \text{and} \quad \|\boldsymbol{\zeta}(t, \boldsymbol{x})\| \le k\|\boldsymbol{x}\|,$$

find $k^* > 0$ such that for any $k < k^*$ the system is globally uniformly asymptotically stable.

4.6 (a) Find the value of

$$J_0 = 2\int_0^{\infty} \|\boldsymbol{x}(t)\|^2\, dt$$

subject to

$$\dot{\boldsymbol{x}} = \begin{bmatrix} 0 & 1 \\ -1 & -1 \end{bmatrix}\boldsymbol{x}, \qquad \boldsymbol{x}(0) = \begin{bmatrix} 0 \\ -1 \end{bmatrix}.$$

(b) Find the value of

$$J_0 = \sum_0^{\infty} \|\boldsymbol{x}(k)\|^2$$

subject to

$$\boldsymbol{x}(k+1) = \begin{bmatrix} 1/2 & 0 \\ 0 & 1/4 \end{bmatrix}\boldsymbol{x}(k), \qquad \boldsymbol{x}(0) = \begin{bmatrix} 3 \\ 0 \end{bmatrix}.$$

4.7 Determine the stability or instability of the zero solution of the system

$$\dot{x}_1 = x_2,$$
$$\dot{x}_2 = x_1 - x_2 \sin x_1.$$

using the second method of Lyapunov and compare the result of your analysis with the solution to Exercise 4.3.

4.8 (Based on Toptscheev and Tsyplyakov [282, p. 329]) Construct a Lyapunov function for the system modeled by

$$\dot{x}_1 = x_2,$$
$$\dot{x}_2 = -x_2 - x_1^3.$$

Show that the origin is a globally stable equilibrium state of the system.

4.9 Investigate the stability of the following dynamical system model:

$$\dot{x}_1 = x_2 - 2x_1x_2^2,$$
$$\dot{x}_2 = -x_1 - x_2 - 2x_2^3.$$

Is the null solution, $x_1(t) = x_2(t) = 0$, $t \geq 0$, locally stable, globally stable, locally asymptotically stable, globally asymptotically stable, or unstable? Justify your answer.

4.10 Use the second method of Lyapunov to estimate the region of asymptotic stability of the equilibrium state $[2/3 \;\; 2/3]^T$ of the dynamical system model

$$\dot{x}_1 = (1 - x_1 - 0.5x_2)\,x_1,$$
$$\dot{x}_2 = (1 - 0.5x_1 - x_2)\,x_2.$$

4.11 For the nonlinear system model $\dot{\boldsymbol{x}} = \boldsymbol{f}(\boldsymbol{x})$, where

$$\boldsymbol{f}(\boldsymbol{x}) = \begin{bmatrix} f_1(\boldsymbol{x}) \\ f_2(\boldsymbol{x}) \end{bmatrix} = \begin{bmatrix} 4x_1 - 2x_1^2x_2 \\ -16x_2 + 4x_1^2x_2 \end{bmatrix}, \tag{4.126}$$

find its equilibrium points and determine their stability properties using the linearization method.

4.12 Consider a dynamical system modeled by

$$\dot{\boldsymbol{x}} = \boldsymbol{A}\boldsymbol{x},$$

where $\boldsymbol{A} \in \mathbb{R}^{n\times n}$. The eigenvalues λ_i, $i = 1, 2, \ldots, n$, of the matrix $\boldsymbol{A}$ are all negative, and the matrix $\boldsymbol{A}$ is diagonalizable; that is, there exists a nonsingular matrix $\boldsymbol{T} \in \mathbb{R}^{n\times n}$ such that

$$\boldsymbol{T}\boldsymbol{A}\boldsymbol{T}^{-1} = \boldsymbol{D} = \operatorname{diag}\{\lambda_1, \lambda_2, \ldots, \lambda_n\}.$$

Let

$$V(\boldsymbol{x}) = \tfrac{1}{2}\boldsymbol{x}^T\boldsymbol{T}^T\boldsymbol{T}\boldsymbol{x} = \boldsymbol{x}^T\boldsymbol{P}\boldsymbol{x}.$$

(a) Find $\dot{V}$; that is, find the derivative of $V(\boldsymbol{x}(t))$ with respect to time along the solutions of the system $\dot{\boldsymbol{x}} = \boldsymbol{A}\boldsymbol{x}$.

(b) Find the upper bound on $\dot{V}$ such that

$$\dot{V}(\boldsymbol{x}) \leq c\|\boldsymbol{x}\|,$$

where the constant $c < 0$ depends on an eigenvalue of $\boldsymbol{A}$ and an eigenvalue of $\boldsymbol{P}$.

4.13 Consider a system of linear equations of the form

$$\boldsymbol{A}\boldsymbol{x} = \boldsymbol{b},$$

where $\boldsymbol{A} \in \mathbb{R}^{m\times n}$, $m \geq n$, $\operatorname{rank}(\boldsymbol{A}) = n$, and $\boldsymbol{x}$ and $\boldsymbol{b}$ are of appropriate dimensions. Consider also an associated iterative algorithm,

$$\boldsymbol{x}(k+1) = \boldsymbol{x}(k) + \alpha\boldsymbol{A}^T(\boldsymbol{b} - \boldsymbol{A}\boldsymbol{x}(k)), \qquad \boldsymbol{x}(0) = \boldsymbol{x}_0,$$

where $\alpha > 0$. Treat the above algorithm as a discrete-time dynamical system.

(a) Find the equilibrium state of the algorithm.

(b) Use the conditions for asymptotic stability of discrete-time systems to find the largest range of values of α for which the algorithm converges to its equilibrium state.

(c) Show that the algorithm, for the range of α found in (b), minimizes the objective function

$$J(\boldsymbol{x}) = (\boldsymbol{b} - \boldsymbol{A}\boldsymbol{x})^T(\boldsymbol{b} - \boldsymbol{A}\boldsymbol{x}).$$

4.14 Consider a system of linear equations of the form

$$\boldsymbol{A}\boldsymbol{x} = \boldsymbol{b},$$

where $\boldsymbol{A} \in \mathbb{R}^{m \times n}$, $m \geq n$, rank$(\boldsymbol{A}) = n$, and $\boldsymbol{x}$ and $\boldsymbol{b}$ are of appropriate dimensions. Consider also associated objective function

$$J(\boldsymbol{x}) = (\boldsymbol{b} - \boldsymbol{A}\boldsymbol{x})^T(\boldsymbol{b} - \boldsymbol{A}\boldsymbol{x}).$$

Write down an iterative steepest descent algorithm, in terms of $\boldsymbol{A}$ and $\boldsymbol{b}$, for minimizing the above objective function. Carefully derive an expression for the step size α_k in terms of $\nabla J(\boldsymbol{x}^{(k)})$ and $\boldsymbol{A}$ only.

4.15 (a) Given two symmetric positive definite matrices $\boldsymbol{P}$ and $\tilde{\boldsymbol{P}}$. Show, by contradiction, that if for all $\boldsymbol{x}$,

$$\boldsymbol{x}^T\boldsymbol{P}\boldsymbol{x} \geq \boldsymbol{x}^T\tilde{\boldsymbol{P}}\boldsymbol{x},$$

then

$$\lambda_{\max}(\boldsymbol{P}) \geq \lambda_{\max}(\tilde{\boldsymbol{P}}).$$

(b) Let $\boldsymbol{P} = \boldsymbol{P}^T > 0$ be the unique solution to

$$\boldsymbol{A}^T\boldsymbol{P} + \boldsymbol{P}\boldsymbol{A} = -2\boldsymbol{Q}, \tag{4.127}$$

where $\boldsymbol{Q} = \boldsymbol{Q}^T > 0$, and let $\hat{\boldsymbol{P}} = \hat{\boldsymbol{P}}^T > 0$ be the unique solution to

$$\boldsymbol{A}^T\hat{\boldsymbol{P}} + \hat{\boldsymbol{P}}\boldsymbol{A} = -2\hat{\boldsymbol{Q}}, \tag{4.128}$$

where $\hat{\boldsymbol{Q}} = q\boldsymbol{Q}$, $q > 0$. Show that

$$\frac{\lambda_{\min}(\hat{\boldsymbol{Q}})}{\lambda_{\max}(\hat{\boldsymbol{P}})} = \frac{\lambda_{\min}(\boldsymbol{Q})}{\lambda_{\max}(\boldsymbol{P})}.$$

(c) Prove the following lemma of Patel and Toda [223].

Lemma 4.9 Let $\boldsymbol{P} = \boldsymbol{P}^T > 0$ be the unique solution to

$$\boldsymbol{A}^T\boldsymbol{P} + \boldsymbol{P}\boldsymbol{A} = -2\boldsymbol{Q},$$

where $\boldsymbol{Q} = \boldsymbol{Q}^T > 0$. Then, the ratio

$$\frac{\lambda_{\min}(\boldsymbol{Q})}{\lambda_{\max}(\boldsymbol{P})}$$

is maximized if $\boldsymbol{Q} = \boldsymbol{I}$.

4.16 Consider the Lyapunov matrix equation,

$$A^T P + PA = -2I_n,$$

where $A \in \mathbb{R}^{n\times n}$ is asymptotically stable.

(a) Let $\sigma_{\min}$ be the smallest singular value of A; that is,

$$\sigma_{\min} = \sqrt{\lambda_{\min}(AA^T)},$$

where $\lambda_{\min}\left(AA^T\right)$ denotes the smallest eigenvalue of AA^T. Show that $\sigma_{\min} > 0$.

(b) Let $\lambda_{\max}(P)$ be the largest eigenvalue of P, the solution of the above Lyapunov equation. Show that

$$\lambda_{\max}(P) \geq \frac{1}{\sigma_{\min}}.$$

4.17 (Based on Theorem 1 in Brockett [34]) Consider an asymptotically stable strictly proper transfer function

$$T(s) = \frac{n(s)}{d(s)}.$$

Suppose that the triple (c, A, b) is a controllable and observable realization of the transfer function $T(s)$. Thus A must be asymptotically stable. This and the observability assumption guarantees that we can solve the Lyapunov equation

$$A^T P + PA = -c^T c$$

for $P = P^T > 0$. Let

$$q(s) = d(s)b^T P(sI - A)^{-1}b.$$

Show that

$$d(s)q(-s) + q(s)d(-s) = n(s)n(-s).$$

4.18 Consider a dynamical system model

$$\dot{x} = Ax, \qquad x(0) = x_0,$$

where $A = \begin{bmatrix} 0 & 1 \\ -1 & -1 \end{bmatrix}$ and $\|x_0\| \leq 10$. Use the Lyapunov approach to find $T \geq 0$, given $\varepsilon_1 = 20$, such that

$$\|x(t; 0, x_0)\| \leq \varepsilon_1 \qquad \text{for all } t \geq T.$$

You can use, for example,

$$Q = \begin{bmatrix} 2 & 1 \\ 1 & 2 \end{bmatrix}$$

when solving the matrix Lyapunov equation $A^T P + PA = -Q$.

4.19 The role of an automotive suspension is to support the vehicle body on the axles. A simple linear model of the suspension assembly associated with a single wheel, called the quarter-car model, is shown in Figure 1.27. We represent the vehicle body by the "sprung" mass, while the tire and axle are represented by the "unsprung" mass. We model the tire as a spring with spring coefficient k_t, and we assume its damping to be negligible. The suspension consists of the spring, the shock absorber, and the variable force element. The role of the variable force element is to generate a force that compensates for the uneven road, which is the source of the system disturbances. We assume that the variable force element F_a can instantaneously provide any desired force; thus, we have a linear active suspension. The equations modeling the dynamics of the two-degree-of-freedom system depicted in Figure 4.23 are derived by applying Newton's law to each mass of the system and by taking into account the positive direction of the z coordinates. We obtain

$$
\begin{aligned}
m_s \ddot{z}_s &= -k_s(z_s - z_u) - b_s(\dot{z}_s - \dot{z}_u) + F_a,\\
m_u \ddot{z}_u &= k_s(z_s - z_u) + b_s(\dot{z}_s - \dot{z}_u) - F_a - k_t(z_u - z_r).
\end{aligned}
$$

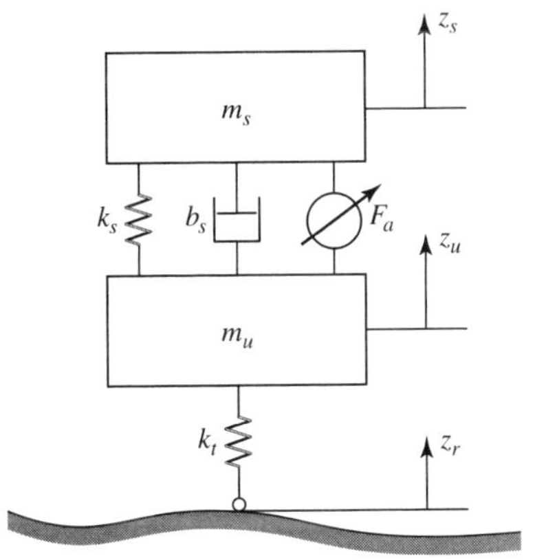

Figure 4.23 Linear quarter-car model.

Define the following state variables:

$$
\begin{aligned}
x_1 &= z_s - z_u = \text{suspension stroke},\\
x_2 &= \dot{z}_s = \text{sprung mass velocity},\\
x_3 &= z_u - z_r = \text{tire deflection},\\
x_4 &= \dot{z}_u = \text{unsprung mass velocity}.
\end{aligned}
$$

Let $u = F_a$ be the system input, and let $\delta(t) = \dot{z}_r(t)$ be the disturbance input due to the road roughness.

(a) Represent the active suspension model in state-space format.

(b) Substitute the vehicle data, given in Table 4.1, into the modeling equations.

(c) Our goal is to isolate the sprung mass from road disturbances. Specifically, we wish to design a control law such that the sprung mass velocity $\dot{z}_s(t) = x_2$ is as close to 0 as possible in spite of road disturbances. Thus, we assume that we can measure the sprung mass velocity, which means that the output is

$$
y(t) = \boldsymbol{c}\boldsymbol{x}(t) = [0 \quad 1 \quad 0 \quad 0]\boldsymbol{x}(t).
$$

Table 4.1 Data for a Vehicle Suspension Taken from [272]

m_s	355 kg
m_u	46 kg
k_s	14384 N/m
b_s	1860 N/m/sec
k_t	300000 N/m

To investigate the effect of road disturbances on the vertical velocity of the sprung mass, set $u = 0$ and $\delta(t) = d\sin(6t)$. Generate a plot of y versus t for $d = 0.5$.

(d) Construct a state-feedback Lyapunov-based controller of Theorem 4.4. Generate a plot of y of the closed-loop system versus t for $d = 0.5$. Discuss your strategy of the above state-feedback controller design that would minimize the influence of the disturbance $\delta(t)$ on the sprung mass velocity y.

4.20 Consider a dynamical system model

$$\dot{\boldsymbol{x}} = \begin{bmatrix} -2 & 0 \\ 0 & -1 \end{bmatrix} \boldsymbol{x} + \begin{bmatrix} 1 \\ 1 \end{bmatrix} (u + h(t, \boldsymbol{x})),$$

where

$$|h(t, \boldsymbol{x})| \le 3.$$

Let $\boldsymbol{P}$ be the solution of the matrix Lyapunov equation $\boldsymbol{A}^T\boldsymbol{P} + \boldsymbol{P}\boldsymbol{A} = -2\boldsymbol{I}_2$, and let

$$u = \begin{cases} -3\,\text{sign}(\boldsymbol{b}^T\boldsymbol{P}\boldsymbol{x}) & \text{if } |\boldsymbol{b}^T\boldsymbol{P}\boldsymbol{x}| \ge \nu \\ -3\dfrac{\boldsymbol{b}^T\boldsymbol{P}\boldsymbol{x}}{\nu} & \text{if } |\boldsymbol{b}^T\boldsymbol{P}\boldsymbol{x}| < \nu, \end{cases}$$

where $\nu = 0.5$. Find the radius of the ball, in $\mathbb{R}^2$, that is uniformly stable for the closed-loop system.

Optimal Control

BEFORE me grew the human soul,
And after I am dead and gone,
Through grades of effort and control
The marvellous work shall still go on

—Archibald Lampman

If everything seems under control, you're just not going fast enough.

—Mario Andretti

One of the essential elements of the control problem, discussed in Section 1.1, is a means of testing the performance of any proposed control law. Whenever we use the term "best" or "optimal" to describe the effectiveness of a given control strategy, we do so with respect to some numerical index of performance called the performance index. We assume that the value of the performance index decreases as the quality of the given admissible control law increases. The admissible controller that ensures the completion of the system objective and at the same time minimizes the performance index is called an *optimal controller* for the system. In this chapter, we discuss various methods of constructing optimal controllers.

5.1 Performance Indices

Constructing a performance index—that is, choosing a means to measure the system performance—can be considered a part of the system modeling. Here, we discuss some typical choices of performance indices.

First, suppose that the objective is to control a dynamical system modeled by the equations

$$\dot{\boldsymbol{x}}(t) = \boldsymbol{A}\boldsymbol{x}(t) + \boldsymbol{B}\boldsymbol{u}(t), \qquad \boldsymbol{x}(t_0) = \boldsymbol{x}_0, \tag{5.1}$$

$$\boldsymbol{y}(t) = \boldsymbol{C}\boldsymbol{x}(t) \tag{5.2}$$

on a fixed interval $[t_0, t_f]$ so that the components of the state vector are "small." A suitable performance index to be minimized would be

$$J_1 = \int_{t_0}^{t_f} \boldsymbol{x}^T(t)\boldsymbol{x}(t)\,dt. \tag{5.3}$$

Obviously, if J_1 is small, then the state vector norm, $\|\boldsymbol{x}(t)\|$, is small in the sense of the above performance index.

If the objective is to control the system so that the components of the output, $\boldsymbol{y}(t)$, are to be small, then we could use the performance index

$$\begin{aligned} J_2 &= \int_{t_0}^{t_f} \boldsymbol{y}^T(t)\boldsymbol{y}(t)\,dt \\ &= \int_{t_0}^{t_f} \boldsymbol{x}^T(t)\boldsymbol{C}^T\boldsymbol{C}\boldsymbol{x}(t)\,dt \\ &= \int_{t_0}^{t_f} \boldsymbol{x}^T(t)\boldsymbol{Q}\boldsymbol{x}(t)\,dt, \end{aligned} \tag{5.4}$$

where the weight matrix $\boldsymbol{Q} = \boldsymbol{C}^T\boldsymbol{C}$ is symmetric positive semidefinite.

In the case when we wish to control the system in such a manner that the components of the input, $\boldsymbol{u}(t)$, are "not too large," a suitable performance index to be minimized is

$$J_3 = \int_{t_0}^{t_f} \boldsymbol{u}^T(t)\boldsymbol{u}(t)\,dt, \tag{5.5}$$

or

$$J_4 = \int_{t_0}^{t_f} \boldsymbol{u}^T(t)\boldsymbol{R}\boldsymbol{u}(t)\,dt, \tag{5.6}$$

where the weight matrix $\boldsymbol{R}$ is symmetric positive definite. There is no loss of generality in assuming the weight matrix $\boldsymbol{R}$ to be symmetric. If $\boldsymbol{R}$ was not symmetric, we could represent the quadratic term $\boldsymbol{u}^T\boldsymbol{R}\boldsymbol{u}$ equivalently as

$$\boldsymbol{u}^T\boldsymbol{R}\boldsymbol{u} = \boldsymbol{u}^T\left(\frac{\boldsymbol{R}+\boldsymbol{R}^T}{2}\right)\boldsymbol{u},$$

where the matrix $\frac{1}{2}(\boldsymbol{R}+\boldsymbol{R}^T)$ is symmetric.

As pointed out by Owens [218, p. 186], we cannot simultaneously minimize the performance indices (5.3) and (5.5) because minimization of (5.3) requires large control signals, while minimization of (5.5) requires small control signals. To solve the dilemma, we could compromise between the two conflicting objectives by minimizing the performance index that is a convex combination of J_1 and J_3,

$$\begin{aligned} J &= \lambda J_1 + (1-\lambda)J_3 \\ &= \int_{t_0}^{t_f} (\lambda\boldsymbol{x}^T(t)\boldsymbol{x}(t) + (1-\lambda)\boldsymbol{u}^T(t)\boldsymbol{u}(t))\,dt, \end{aligned} \tag{5.7}$$

where λ is a parameter in the range $[0, 1]$. If $\lambda = 1$, then $J = J_1$; and if $\lambda = 0$, then $J = J_3$. By trial and error, we could select λ from the interval $[0, 1]$ to compromise between the two extremes. A generalization of the performance index (5.7) is

$$J = \int_{t_0}^{t_f} (\boldsymbol{x}^T(t)\boldsymbol{Q}\boldsymbol{x}(t) + \boldsymbol{u}^T(t)\boldsymbol{R}\boldsymbol{u}(t))\,dt.$$

In certain applications, we may wish the final state $\boldsymbol{x}(t_f)$ to be as close as possible to $\mathbf{0}$. Then, a possible performance measure to be minimized is

$$\boldsymbol{x}^T(t_f)\boldsymbol{F}\boldsymbol{x}(t_f), \tag{5.8}$$

where $\boldsymbol{F}$ is a symmetric positive definite matrix.

We can combine the performance measures (5.4), (5.6), and (5.8) when our control aim is to keep the state "small," the control "not too large," and the final state as near to $\mathbf{0}$ as possible. The resulting performance index is

$$J = \frac{1}{2}\boldsymbol{x}^T(t_f)\boldsymbol{F}\boldsymbol{x}(t_f) + \frac{1}{2}\int_{t_0}^{t_f}(\boldsymbol{x}^T(t)\boldsymbol{Q}\boldsymbol{x}(t) + \boldsymbol{u}^T(t)\boldsymbol{R}\boldsymbol{u}(t))\,dt, \tag{5.9}$$

where the factor $\frac{1}{2}$ is placed to simplify subsequent algebraic manipulations. Minimizing (5.9) subject to (5.1) is called the *linear quadratic regulator* problem, or the LQR problem for short.

In some cases, the controller's goal is to force the system state to track a desired state trajectory, $\boldsymbol{x}_d(t)$, throughout the interval $[t_0, t_f]$ while maintaining the deviations of the actual state $\boldsymbol{x}(t)$ "small" from the desired trajectory with the control effort $\boldsymbol{u}(t)$ "not too large" and the final state $\boldsymbol{x}(t_f)$ being as near as possible to some desired state $\boldsymbol{x}_d(t_f)$. A suitable performance index to be minimized in this case might be

$$\begin{aligned} J = {} & \frac{1}{2}(\boldsymbol{x}(t_f) - \boldsymbol{x}_d(t_f))^T\boldsymbol{F}(\boldsymbol{x}(t_f) - \boldsymbol{x}_d(t_f)) \\ & + \frac{1}{2}\int_{t_0}^{t_f}((\boldsymbol{x}(t) - \boldsymbol{x}_d(t))^T\boldsymbol{Q}(\boldsymbol{x}(t) - \boldsymbol{x}_d(t)) + \boldsymbol{u}^T(t)\boldsymbol{R}\boldsymbol{u}(t))\,dt. \end{aligned} \tag{5.10}$$

5.2 A Glimpse at the Calculus of Variations

The goal of this section is to provide a brief introduction to the subject of the *calculus of variations*. The calculus of variations deals with minimization or maximization of *functionals*. A functional is a mapping or a transformation that depends on one or more functions and the values of the functional are numbers. Examples of functionals are performance indices that we introduced in the previous section. We now give two more examples of functionals.

◆ Example 5.1

The length l of the arc connecting two given points $A(x_0, y_0)$ and $B(x_1, y_1)$ in the x–y plane, as depicted in Figure 5.1, is an example of a functional. We know that

$$l = l(y(x)) = \int_{x_0}^{x_1}\sqrt{1 + (dy(x)/dx)^2}\,dx,$$

subject to the constraints $y(x_0) = y_0$ and $y(x_1) = y_1$.

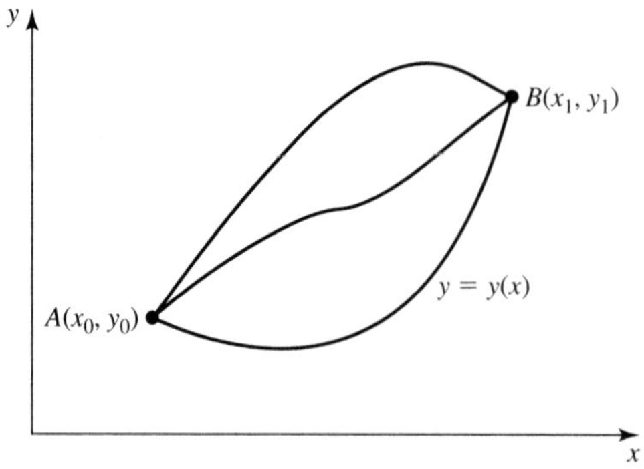

Figure 5.1 The length of arcs joining two given points in the plane is an example of a functional.

◆ Example 5.2

Another example of a functional is the area of the surface of revolution formed by a surface connecting two circular disks as illustrated in Figure 5.2. In particular, the surface of revolution can be formed by a soap film clinging to two circular disks. When a wire circle is dipped into a soap solution and withdrawn, a circular disk of soap film bounded by the circle is formed. If a second smaller circle is made to touch the disk and then moved away, the two circles will be joined by a surface of film that is a surface of revolution in the particular case when the circles are parallel and have their centers on the same axis perpendicular to their planes. If we denote the arc of intersection with the x–y plane by $y = y(x)$, then the area of the surface of revolution is

$$v = v(y(x)) = 2\pi \int_{x_0}^{x_1} y(x)\sqrt{1 + (dy(x)/dx)^2}\, dx,$$

subject to the constraints $y(x_0) = y_0$, $y(x_1) = y_1$. One may then want to minimize $v = v(y(x))$ to find $y = y(x)$ resulting in the surface of revolution of minimum area.

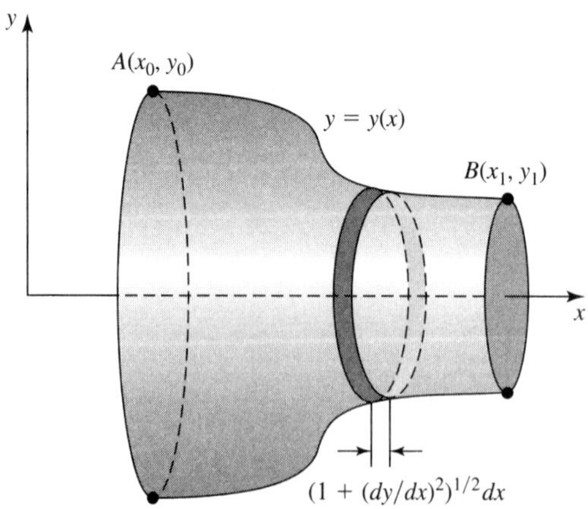

Figure 5.2 The area of surfaces of revolution connecting two circular disks is an example of a functional.

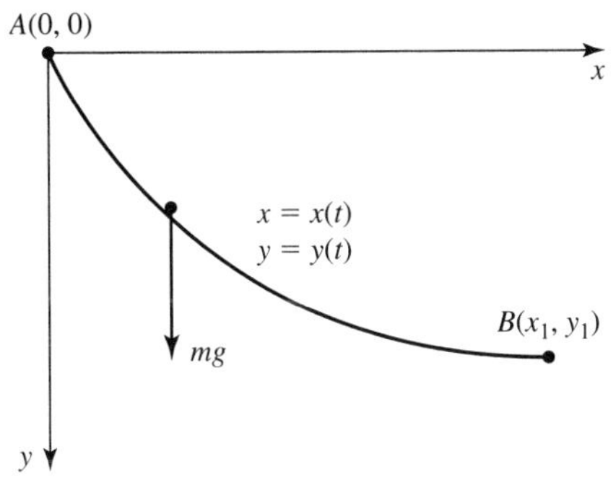

Figure 5.3 Illustration of the brachistochrone problem.

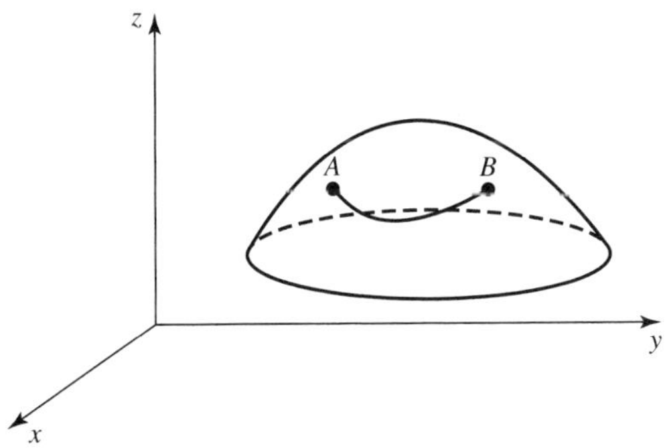

Figure 5.4 The problem of finding geodesics.

The problem of optimization of functionals dates back to Johann Bernoulli. In 1696 he published a letter in which he posed the problem, which is now known as the *brachistochrone problem*. Brachistochrone translates from Greek as the shortest time. The brachistochrone problem can be described as follows. A particle of mass m is to slide down along a frictionless track connecting two given points. The motion of the particle is to be constrained only by gravity and the shape of the curve of the track. Find the curve of the track that minimizes the time travel between the two given points. We illustrate the problem in Figure 5.3. The brachistochrone problem was solved by Johann Bernoulli himself, his brother Jacob, Isaac Newton, the Marquis de L'Hôpital, Gottfried Leibniz, and Tschirnhaus. A solution to the brachistochrone problem is presented in Example 5.23. In 1697 Johann Bernoulli posed and solved the *geodesics* problem, which can be formulated as follows: Find a line of shortest length that connects two points on a given surface; see Figure 5.4 for an illustration of the problem. The curves of the least length connecting pairs of points on a surface are called *geodesics*. A general procedure for solving the above type of problems, using the calculus of variations, was given by Leonhard Euler (who flourished between 1707 and 1783).

To acquaint ourselves with the methods of the calculus of variations, we first need to introduce concepts concerning functionals.

5.2.1 Variation and Its Properties

A *function* can be viewed as a rule, or a mapping, that assigns to each element of some set a unique element of a possibly different set. In particular, a function $x : \mathbb{R} \to \mathbb{R}$ of a real variable t

assigns to each real number a uniqe real number. An increment of the argument of a function of one variable t is $\Delta t = t - t_1$.

Similarly, a *functional* is a mapping that assigns to each function, from some class of functions, a unique number. Thus, we can say that a functional is a "function of a function." Let $x : \mathbb{R} \to \mathbb{R}$ be an argument of a functional. By a *variation* δx of an argument x of a functional v we mean the difference of two functions

$$\delta x = x - x_1.$$

We assume that x can change in an arbitrary way in some class of functions.

Recall that a function $x : \mathbb{R} \to \mathbb{R}$ is continuous if a small change of its argument t corresponds to a small change of the value $x(t)$ of the function. Similarly, a functional v is said to be continuous if a "small" change of its argument x corresponds to a small change of the value of the functional.

A functional v is called linear if

$$v(ax_1 + x_2) = av(x_1) + v(x_2),$$

where a is a constant.

We will now introduce the notion of the *variation of a functional*. This notion is analogous to the notion of a function differential. To connect the two, we first consider the case of a function of one variable. Let x be a differentiable function defined on an open interval U, and let $t \in U$. The derivative of x at t is defined as

$$x'(t) = \lim_{\Delta t \to 0} \frac{x(t + \Delta t) - x(t)}{\Delta t}. \tag{5.11}$$

Let

$$\varphi(\Delta t) = \frac{x(t + \Delta t) - x(t)}{\Delta t} - x'(t). \tag{5.12}$$

The function φ is not defined at $\Delta t = 0$; however,

$$\lim_{\Delta t \to 0} \varphi(\Delta t) = 0.$$

We rewrite (5.12) as

$$x(t + \Delta t) - x(t) = x'(t)\Delta t + \varphi(\Delta t)\,\Delta t. \tag{5.13}$$

The above relation has meaning only when $\Delta t \neq 0$. To make it hold at $\Delta t = 0$, we define $\varphi(\Delta t)|_{\Delta t = 0} = 0$. To proceed, let

$$\begin{aligned} \beta(\Delta t) &= \varphi(\Delta t) \qquad &\text{if } \Delta t > 0 \\ \beta(\Delta t) &= -\varphi(\Delta t) \qquad &\text{if } \Delta t < 0. \end{aligned} \tag{5.14}$$

Then, if x is differentiable, there exists a function β such that

$$\begin{aligned} x(t + \Delta t) - x(t) = \Delta x &= x'(t)\Delta t + \beta(\Delta t)|\Delta t| \\ &= L(t, \Delta t) + \beta(\Delta t)|\Delta t|, \end{aligned} \tag{5.15}$$

where $\lim_{\Delta t \to 0} \beta(\Delta t) = 0$, and $L(t, \Delta t) = x'(t)\Delta t$ is a linear function in Δt.

For a real-valued function $f = f(\boldsymbol{x}) : \mathbb{R}^n \to \mathbb{R}$, relation (5.15) takes the form

$$\begin{aligned} f(\boldsymbol{x} + \Delta \boldsymbol{x}) - f(\boldsymbol{x}) = \Delta f &= Df(\boldsymbol{x})\Delta \boldsymbol{x} + \beta(\Delta \boldsymbol{x})\|\Delta \boldsymbol{x}\| \\ &= L(\boldsymbol{x}, \Delta \boldsymbol{x}) + \beta(\Delta \boldsymbol{x})\|\Delta \boldsymbol{x}\|, \end{aligned} \tag{5.16}$$

where $\lim_{\Delta \boldsymbol{x}\to \boldsymbol{0}} \beta(\Delta \boldsymbol{x}) = 0$, and

$$L(\boldsymbol{x}, \Delta \boldsymbol{x}) = Df(\boldsymbol{x})\, \Delta \boldsymbol{x} = \nabla f(\boldsymbol{x})^T \Delta \boldsymbol{x}$$

is a linear function in $\Delta \boldsymbol{x}$. Relation (5.16) holds for operators defined over Banach spaces with scalars being either real or complex numbers. The functionals we consider are operators over a Banach space of continuous functions, $\mathcal{C}([t_0, t_1])$, with the norm given by (A.99) on page 656.

If an increment $\Delta v = v(x + \delta x) - v(x)$ of a functional v can be represented as

$$\Delta v = L(x, \delta x) + \beta(\delta x)\|\delta x\|,$$

where $L(x, \delta x)$ is a linear functional with respect to δx, the term $\|\delta x\| = \max_{t\in[t_0,t_1]} |\delta x|$ denotes the maximal value of $|\delta x|$, and $\beta(\delta x) \to 0$ if $\|\delta x\| \to 0$, then the linear part L of Δv is called the *variation of the functional* and is denoted δv, that is,

$$\delta v = L(x, \delta x).$$

◆ Example 5.3

We find the variation of the functional

$$v = \int_0^1 (2x^2(t) + x(t))\, dt.$$

For this we first calculate its increment to get

$$\begin{aligned}
\Delta v &= v(x + \delta x) - v(x)\\
&= \int_0^1 (2(x + \delta x)^2 + (x + \delta x))\, dt - \int_0^1 (2x^2 + x)\, dt\\
&= \int_0^1 (2x^2 + 4x\delta x + 2(\delta x)^2 + x + \delta x - 2x^2 - x)\, dt\\
&= \int_0^1 (4x + 1)\delta x\, dt + 2\int_0^1 (\delta x)^2\, dt.
\end{aligned}$$

The linear part of Δv is

$$\delta v = \int_0^1 (4x + 1)\delta x\, dt,$$

which is the variation of the given functional.

To proceed, note that the linear part on the right-hand side of (5.16) can be computed as

$$Df(\boldsymbol{x})\Delta \boldsymbol{x} = L(\boldsymbol{x}, \Delta \boldsymbol{x}) = \left. \frac{d}{d\alpha} f(\boldsymbol{x} + \alpha \Delta \boldsymbol{x}) \right|_{\alpha=0} \tag{5.17}$$

As mentioned before, the above also holds for operators over Banach spaces other than Euclidean spaces. Using this observation, we can state and prove a lemma that will allow us to calculate the variation of a given functional more efficiently than the method we used in Example 5.3.

Lemma 5.1

$$\delta v = \frac{d}{d\alpha} v(x + \alpha\delta x)\bigg|_{\alpha=0}.$$

Proof Suppose that for a given functional v there exists its variation. This means that we can represent Δv as

$$\Delta v = v(x + \alpha\delta x) - v(x) = L(x, \alpha\delta x) + \beta(\alpha\delta x)|\alpha|\|\delta x\|.$$

Then, the derivative of $v(x + \alpha\delta x)$ with respect to α evaluated at $\alpha = 0$ is equal to

$$\begin{aligned}
\lim_{\alpha\to 0} \frac{\Delta v}{\alpha} &= \lim_{\alpha\to 0} \frac{L(x, \alpha\delta x) + \beta(\alpha\delta x)|\alpha|\|\delta x\|}{\alpha} \\
&= \lim_{\alpha\to 0} \frac{L(x, \alpha\delta x)}{\alpha} + \lim_{\alpha\to 0} \frac{\beta(\alpha\delta x)|\alpha|\|\delta x\|}{\alpha} \\
&= L(x, \delta x),
\end{aligned}$$

because $L(\cdot, \cdot)$ is linear with respect to the second argument, and hence

$$L(x, \alpha\delta x) = \alpha L(x, \delta x).$$

Furthermore,

$$\lim_{\alpha\to 0} \frac{\beta(\alpha\delta x)|\alpha|\|\delta x\|}{\alpha} = \lim_{\alpha\to 0} \beta(\alpha\delta x)\|\delta x\| = 0.$$

This completes the proof.

◆ Example 5.4

We will now use Lemma 5.1 to find the variation of the functional considered in Example 5.3. We have

$$\begin{aligned}
\delta v &= \frac{d}{d\alpha} v(x + \alpha\delta x)|_{\alpha=0} \\
&= \frac{d}{d\alpha}\left(\int_0^1 (2(x + \alpha\delta x)^2 + (x + \alpha\delta x))\, dt\right)\bigg|_{\alpha=0} \\
&= \int_0^1 (4(x + \alpha\delta x)\delta x + \delta x)\, dt\bigg|_{\alpha=0} \\
&= \int_0^1 (4x + 1)\delta x\, dt,
\end{aligned}$$

which is the variation of the given functional in Example 5.3.

◆ Example 5.5

We will find the variation of the functional

$$v(x_1, x_2) = v(x_1(t), x_2(t)) = \int_0^1 x_1 e^{-x_2}\, dt,$$

using the formula

$$\delta v = \frac{d}{d\alpha} v(\mathbf{x} + \alpha \delta \mathbf{x}) \bigg|_{\alpha=0}.$$

We have

$$\begin{aligned}
\delta v &= \frac{d}{d\alpha}\left(\int_0^1 (x_1 + \alpha\delta x_1) e^{-(x_2 + \alpha \delta x_2)}\, dt\right)\Bigg|_{\alpha=0} \\
&= \int_0^1 \frac{d}{d\alpha}\left((x_1 + \alpha\delta x_1) e^{-(x_2+\alpha\delta x_2)}\, dt\right)\Bigg|_{\alpha=0} \\
&= \int_0^1 \left(\delta x_1 e^{-(x_2+\alpha\delta x_2)} - \delta x_2 e^{-(x_2+\alpha\delta x_2)}(x_1 + \alpha\delta x_1)\right) dt \Bigg|_{\alpha=0} \\
&= \int_0^1 (e^{-x_2}\delta x_1 - x_1 e^{-x_2}\delta x_2)\, dt.
\end{aligned}$$

Thus,

$$\delta v = \int_0^1 \left(e^{-x_2}\delta x_1 - x_1 e^{\ x_2}\delta x_2\right) dt.$$

Definition 5.1 A functional v is maximal on some function x_0 if the value of the functional on any function x close to x_0 is less than or equal to that on x_0, that is,

$$\Delta v = v(x) - v(x_0) \leq 0.$$

We will now give a first order necessary condition for a given functional to achieve a maximum or minimum on a curve.

Theorem 5.1 If for a functional v there exists a variation and the functional reaches a maximum or a minimum on a curve x_0, then

$$\delta v(x_0) = 0.$$

Proof Suppose that x_0 and δx are fixed. Then

$$v(x_0 + \alpha\delta x) = \Phi(\alpha);$$

that is, $v(x_0+\alpha\delta x)$ is a function of α that for $\alpha = 0$ achieves its maximum or minimum. This implies that $\frac{d}{d\alpha}\Phi = 0$, or, equivalently,

$$\left.\frac{d}{d\alpha}v(x_0 + \alpha\delta x)\right|_{\alpha=0} = 0.$$

Hence, by Lemma 5.1, we have $\delta v = 0$. Therefore, a functional that achieves its maximum or a minimum on a function x_0 has its variation evaluated at x_0 equal to zero.

The above results will be used in the next section to determine conditions that a curve has to satisfy to be a maximizer or a minimizer, that is, to be an extremal for a given functional.

5.2.2 Euler–Lagrange Equation

In this section, we consider the following optimization problem.

Problem 1 Suppose that we are given two points (t_0, x_0) and (t_1, x_1) in the (t, x)-plane. We wish to find a curve, or trajectory, joining the given points such that the functional

$$v(x) = \int_{t_0}^{t_1} F(t, x, \dot{x})\, dt \tag{5.18}$$

along this trajectory can achieve its extremal—that is, maximal or minimal—value. The problem is illustrated in Figure 5.5. In solving Problem 1, we use the following lemma.

Lemma 5.2 Suppose that a function $\phi : \mathbb{R} \to \mathbb{R}$ is continuous on the interval $[t_0, t_1]$. Then,

$$\int_{t_0}^{t_1} \phi(t)\delta x(t)\, dt = 0$$

if and only if $\phi(t) = 0$ at every point of the interval $[t_0, t_1]$.

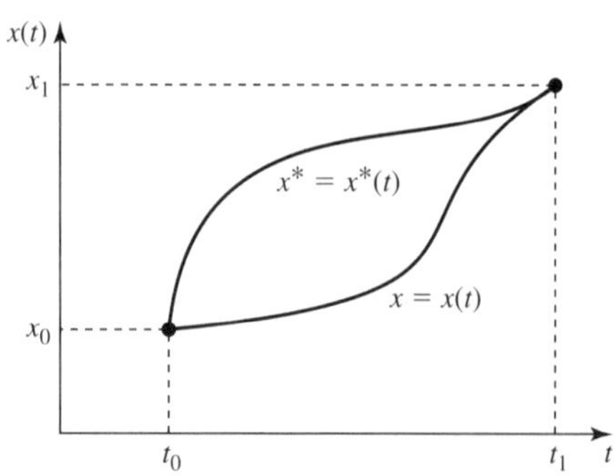

Figure 5.5 Illustration of Problem 1.

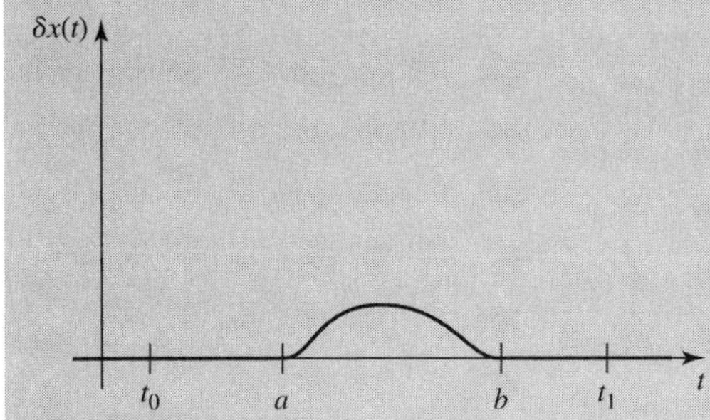

Figure 5.6 Illustration of the proof of Lemma 5.2.

Proof Sufficiency is obvious because we can satisfy $\int_{t_0}^{t_1} \phi(t)\delta x(t)\,dt = 0$ by choosing ϕ that vanishes at every point of the interval $[t_0, t_1]$.

We prove necessity by contraposition. We thus prove that if ϕ is nonzero on some subinterval of $[t_0, t_1]$, then $\int_{t_0}^{t_1} \phi(t)\delta x(t)\,dt \neq 0$. Specifically, assume, without loss of generality, that ϕ is positive on the interval $[a, b]$, where $t_0 \le a \le b \le t_1$. Note that if ϕ were nonzero at only one point, then because it is continuous, it would have to be nonzero in some neighborhood of that point. Hence, we can assume that $a < b$. Consider now a variation given by

$$\delta x = \delta x(t) = \begin{cases} k^2(t-a)^2(t-b)^2 & \text{for } a \le t \le b, \\ 0 & \text{elsewhere.} \end{cases} \tag{5.19}$$

If we assumed ϕ to be negative on the interval $[a, b]$, then we would use δx to be negative of (5.19). We illustrate δx in Figure 5.6. Because δx is positive in $[a, b]$ and zero elsewhere in the interval $[t_0, t_1]$, the integrand of $\int_{t_0}^{t_1} \phi(t)\delta x(t)\,dt$ will have the same sign as ϕ in $[a, b]$ and will be zero outside this subinterval. Thus

$$\int_{t_0}^{t_1} \phi(t)\delta x(t)\,dt = \int_a^b \phi(t)\delta x(t)\,dt \neq 0,$$

and the proof is complete.

Solution to Problem 1 We assume that there is an optimal trajectory $x = x(t)$ joining the points (t_0, x_0) and (t_1, x_1) such that $\delta v(x(t)) = 0$. Let us then consider an arbitrary acceptable curve $x^* = x^*(t)$ that is close to x and from a family of curves

$$x(t, \alpha) = x(t) + \alpha(x^*(t) - x(t)) = x(t) + \alpha\delta x(t).$$

Note that for $\alpha = 0$ we get $x(t)$, and for $\alpha = 1$ we obtain $x^*(t)$. Furthermore,

$$\frac{d}{dt}(\delta x(t)) = \frac{d}{dt}(x^*(t) - x(t)) = \delta x',$$

and analogously

$$\frac{d^2}{dt^2}(\delta x(t)) = \frac{d^2}{dt^2}(x^*(t) - x(t)) = \delta x''.$$

We are now ready to investigate the behavior of the functional v given by (5.18) on the curves from the family $x(t, \alpha)$. Note that the functional v can be considered as a function of α, that is, $v(x(t, \alpha)) = \Phi(\alpha)$. It follows from the first-order necessary condition for a point to be an extremizer of a function of one variable that $\alpha = 0$ is a candidate to be an extremizer of Φ if

$$\left.\frac{d\Phi(\alpha)}{d\alpha}\right|_{\alpha=0} = 0,$$

where

$$\Phi(\alpha) = \int_{t_0}^{t_1} F(t, x(t, \alpha), x'(t, \alpha))\, dt,$$

and $x'(t, \alpha) = \frac{d}{dt}x(t, \alpha)$. We now evaluate $\frac{d\Phi(\alpha)}{d\alpha}$:

$$\frac{d\Phi(\alpha)}{d\alpha} = \int_{t_0}^{t_1} \left(F_x \frac{d}{d\alpha}x(t, \alpha) + F_{x'}\frac{d}{d\alpha}x'(t, \alpha)\right) dt,$$

where $F_x = \frac{\partial}{\partial x}F(t, x(t, \alpha), x'(t, \alpha))$, and $F_{x'} = \frac{\partial}{\partial x'}F(t, x(t, \alpha), x'(t, \alpha))$. Because

$$\frac{d}{d\alpha}x(t, \alpha) = \frac{d}{d\alpha}(x(t) + \alpha\delta x(t)) = \delta x(t)$$

and

$$\frac{d}{d\alpha}x'(t, \alpha) = \frac{d}{d\alpha}(x'(t) + \alpha\delta x'(t)) = \delta x'(t),$$

we can write

$$\frac{d}{d\alpha}\Phi(\alpha) = \int_{t_0}^{t_1} (F_x(t, x(t, \alpha), x'(t, \alpha))\delta x(t) + F_{x'}(t, x(t, \alpha), x'(t, \alpha))\delta x'(t))\, dt.$$

By Lemma 5.1,

$$\frac{d}{d\alpha}\Phi(0) = \delta v.$$

Therefore,

$$\delta v = \int_{t_0}^{t_1} (F_x\delta x + F_{x'}\delta x')\, dt,$$

where for convenience we dropped the argument t. Integrating by parts the last term, along with taking into account that $\delta x' = (\delta x)'$, yields

$$\delta v = (F_{x'}\delta x)|_{t_0}^{t_1} + \int_{t_0}^{t_1} \left(F_x - \frac{d}{dt}F_{x'}\right) \delta x\, dt.$$

The end points (t_0, x_0) and (t_1, x_1) are fixed. This implies that

$$\delta x|_{t=t_0} = x^*(t_0) - x(t_0) = 0,$$

and

$$\delta x|_{t=t_1} = x^*(t_1) - x(t_1) = 0.$$

Hence

$$\delta v = \int_{t_0}^{t_1} \left(F_x - \frac{d}{dt} F_{x'} \right) \delta x \, dt.$$

By Lemma 5.2,

$$\delta v = 0 \quad \text{if and only if} \quad F_x - \frac{d}{dt} F_{x'} = 0.$$

The equation $F_x - \frac{d}{dt} F_{x'} = 0$ can also be represented as

$$F_x - F_{x't} - F_{x'x} x' - F_{x'x'} x'' = 0.$$

The equation

$$\boxed{F_x - \frac{d}{dt} F_{x'} = 0} \tag{5.20}$$

is known as the Euler–Lagrange equation. The curves $x = x(t, C_1, C_2)$ that are solutions to the Euler–Lagrange equation are called *extremals*. The constants C_1 and C_2 are the integration constants. Thus, a functional can only achieve its extremum on the extremals.

◆ Example 5.6

We find the extremal for the functional

$$v = \int_0^{\pi/2} ((x')^2 - x^2)\, dt, \tag{5.21}$$

where $x(0) = 0$, and $x(\pi/2) = 1$. The Euler–Lagrange equation for the problem takes the form

$$\begin{aligned} F_x - \frac{d}{dt} F_{x'} &= \frac{\partial}{\partial x}((x')^2 - x^2) - \frac{d}{dt}\left(\frac{\partial}{\partial x'}((x')^2 - x^2) \right) \\ &= -2x - \frac{d}{dt}(2x') \\ &= -2x - 2x'' \\ &= 0. \end{aligned}$$

Equivalently, the Euler–Lagrange equation for the functional (5.21) can be represented as

$$x'' + x = 0.$$

The characteristic equation of the above differential equation is $s^2 + 1 = 0$.

Therefore, the general solution of the Euler–Lagrange equation has the form

$$x(t) = C_1 \cos t + C_2 \sin t.$$

Taking into account the end-point conditions gives

$$x(0) = C_1 = 0,$$
$$x\left(\frac{\pi}{2}\right) = C_2 = 1.$$

Thus, we have one extremal

$$x(t) = \sin t.$$

The functional (5.21) is minimized or maximized only on the extremal $x(t) = \sin t$.

◆ Example 5.7

We now find the extremal for the functional

$$v = \int_0^1 ((x')^2 + 12xt)\,dt, \tag{5.22}$$

where $x(0) = 0$, and $x(1) = 1$. The Euler–Lagrange equation for the this functional is

$$\begin{aligned} F_x - \frac{d}{dt}F_{x'} &= \frac{\partial}{\partial x}((x')^2 + 12xt) - \frac{d}{dt}\left(\frac{\partial}{\partial x'}((x')^2 + 12xt)\right) \\ &= 12t - \frac{d}{dt}(2x') \\ &= 12t - 2x'' \\ &= 0. \end{aligned}$$

Equivalently, the Euler–Lagrange equation for the functional (5.22) can be represented as

$$x'' = 6t.$$

The general solution to the above differential equation is

$$x(t) = C_1 t + C_2 + t^3.$$

From the end-point conditions, we get

$$x(0) = C_2 = 0,$$
$$x(1) = C_1 + 1 = 1.$$

Hence, $C_1 = 0$, and the extremal has the form

$$x(t) = t^3.$$

Suppose now that $A = A(t_0, x_{10}, \ldots, x_{n0})$ and $B = B(t_1, x_{11}, \ldots, x_{n1})$ are two given points in the $(n+1)$-dimensional space $(t, x_1, \ldots, x_n)$, along with the functional

$$v = v(x_1, \ldots, x_n) = \int_{t_0}^{t_1} F(t, x_1, \ldots, x_n, \dot{x}_1, \ldots, \dot{x}_n)\, dt.$$

To find a trajectory along which $\delta v = 0$, we consider only the variation in one coordinate, say x_i, while other coordinates are fixed. We perform this process for each coordinate. The functional v will depend only upon x_i, and we write

$$v(x_1, \ldots, x_i, \ldots, x_n) = \tilde{v}(x_i).$$

Thus, we reduced the given n-dimensional problem to n one-dimensional problems. Proceeding as before, we conclude that the functional $v = v(x_1, \ldots, x_n)$ can only be extremized on the trajectories that satisfy n Euler–Lagrange equations,

$$F_{x_i} - \frac{d}{dt} F_{\dot{x}_i} = 0, \qquad i = 1, \ldots, n. \tag{5.23}$$

Observe that if in (5.23) we set $F = L = K - U$, where L is the Lagrangian, $K = K(\dot{q}_1, \ldots, \dot{q}_n)$ is the kinetic energy of the particle, $U = U(q_1, \ldots, q_n)$ is its potential energy, and $x_i = q_i$, then (5.23) becomes

$$\frac{\partial L}{\partial q_i} - \frac{d}{dt}\left(\frac{\partial L}{\partial \dot{q}_i}\right) = 0, \qquad i = 1, 2, \ldots, n,$$

which are the Lagrange equations of motion for a particle. Thus a particle moves along a path on which the functional

$$J = \int_{t_1}^{t_2} L\, dt$$

is extremized, which agrees with the Hamilton principle saying that a particle moves along the path that minimizes the integral J.

In our discussion up to now, both end points were fixed. We now would like to solve a more general optimization problem, where the end points need not be fixed. For simplicity, we assume first that one of the end points is fixed, while the other is free. After we solve this problem, we can easily generalize to the case when both end points are free.

Problem 2 In the (t, x) plane the left end point A is fixed, while the right end point B is free (see Figure 5.7). It is required to find a trajectory $x = x(t)$ starting at A on which the functional $v = \int_{t_0}^{t_1} F(t, x, x')\, dt$ can achieve its extremal value.

Solution to Problem 2 Following Elsgolc [75, pp. 64–72], we begin by assuming that a solution to Problem 2 exists; that is, there is a trajectory $x = x(t)$ joining (t_0, x_0) and (t_1, x_1) such that $\delta v = 0$. This trajectory must satisfy the Euler–Lagrange equation. Consider an adjoining

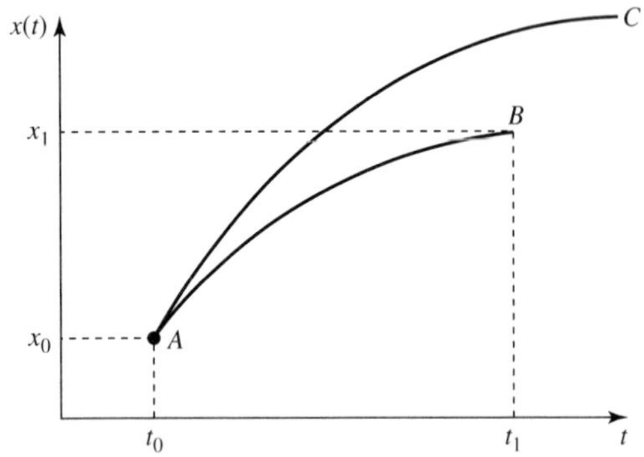

Figure 5.7 Illustration of Problem 2.

trajectory specified by $x + \delta x$ joining (t_0, x_0) and $(t_1 + \delta t_1, x_1 + \delta x_1)$. Then,

$$\begin{aligned}\Delta v &= v(x + \delta x) - v(x) \\ &= \int_{t_0}^{t_1+\delta t_1} F(t, x + \delta x, x' + \delta x')\, dt - \int_{t_0}^{t_1} F(t, x, x')\, dt \\ &= \int_{t_1}^{t_1+\delta t_1} F(t, x + \delta x, x' + \delta x')\, dt + \int_{t_0}^{t_1} (F(t, x + \delta x, x' + \delta x') - F(t, x, x'))\, dt.\end{aligned}$$

Our goal now is to extract from Δv its linear portion in δx, which we know is the variation of the given functional. We begin by approximating the first integral as

$$\int_{t_1}^{t_1+\delta t_1} F(t, x + \delta x, x' + \delta x')\, dt \approx F(t, x, x')|_{t=t_1} \delta t_1.$$

Expanding the integrand of the second integral into a Taylor series and retaining only first-order terms yields

$$\int_{t_0}^{t_1} (F(t, x + \delta x, x' + \delta x') - F(t, x, x'))\, dt = \int_{t_0}^{t_1} (F_x(t, x, x')\delta x + F_{x'}(t, x, x')\delta x')\, dt.$$

Integrating the second term by parts and combining with the first term gives

$$\int_{t_0}^{t_1} (F_x \delta x + F_{x'} \delta x')\, dt = (F_{x'} \delta x)\Big|_{t_0}^{t_1} + \int_{t_0}^{t_1} \left(F_x - \frac{d}{dt} F_{x'} \right) \delta x\, dt,$$

where for simplicity of notation we suppressed the arguments of F_x and $F_{x'}$. We assumed that a solution to the problem exits. Hence, an extremal must satisfy the Euler–Lagrange equation $F_x - \frac{d}{dt} F_{x'} = 0$. Because the left end point is fixed, we obtain

$$\delta x|_{t=t_0} = 0.$$

Taking into account the results of the above manipulations yields

$$\Delta v \approx F(t, x, x')|_{t=t_1} \delta t_1 + (F_{x'} \delta x)|_{t=t_1}. \tag{5.24}$$

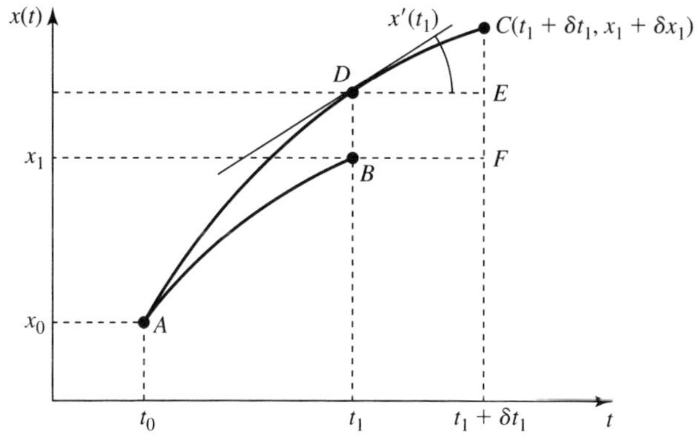

Figure 5.8 A geometric interpretation of calculating the variation of the functional for Problem 2.

The final step in obtaining δv is to evaluate $(F_{x'}\delta x)|_{t_1}$. The problem is two-dimensional, so we can illustrate it graphically as shown in Figure 5.8. Referring to Figure 5.8, we obtain

$$\begin{aligned} DB &= \delta x|_{t=t_1}, \\ FC &= \delta x_1, \\ EC &\approx x'(t_1)\delta t_1, \\ DB &= FC - EC. \end{aligned}$$

Hence,

$$\delta x|_{t=t_1} \approx \delta x_1 - x'(t_1)\delta t_1.$$

Substituting the above into (5.24) gives

$$\begin{aligned} \delta v &= F(t, x, x')|_{t=t_1}\delta t_1 + F_{x'}|_{t=t_1}(\delta x_1 - x'(t_1)\delta t_1) \\ &= (F - x'F_{x'})|_{t=t_1}\delta t_1 + F_{x'}|_{t=t_1}\delta x_1. \end{aligned}$$

If the variations δt_1 and δx_1 are arbitrary, we obtain $\delta v = 0$ if in addition to the Euler–Lagrange equation the following conditions are also satisfied:

$$\boxed{(F - x'F_{x'})|_{t=t_1} = 0 \quad \text{and} \quad F_{x'}|_{t=t_1} = 0.}$$

The above end-point relations are called the *transversality conditions*. In conclusion, the optimal trajectory must satisfy both the Euler–Lagrange equation and the transversality conditions.

◆ Example 5.8

Consider the functional of Example 5.7, where now the left end point is fixed as before—that is, $x(0) = 0$—while the right end point $x(1)$ is free. Here, we have $t_1 = 1$. Because $\delta t_1 = 0$ and δx_1 is arbitrary, the transversality conditions for the right end point are reduced to one condition,

$$F_{x'}|_{t=t_1} = 0.$$

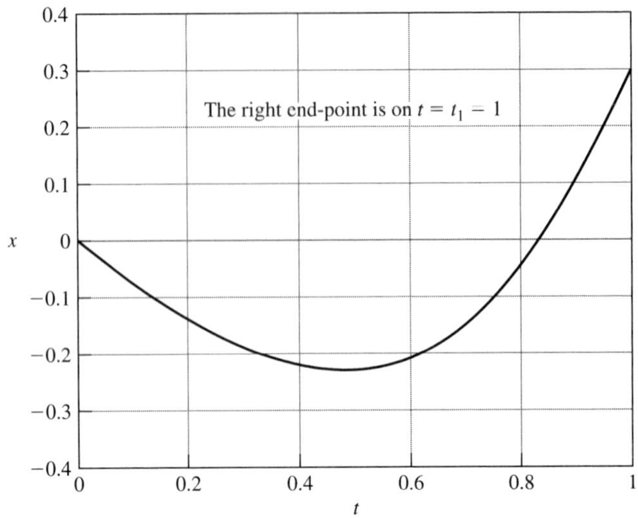

Figure 5.9 A plot of the extremal, $x(t) = -3t + t^3$, of Example 5.8.

Taking into account the solution to the Euler–Lagrange equation obtained in Example 5.7 and then using the fact that the left end point is fixed, we obtain

$$\begin{aligned} 0 &= F_{x'}|_{t=t_1} \\ &= 2x'(t_1) \\ &= 2(C_1 + 3t^2)|_{t=1} \\ &= 2(C_1 + 3). \end{aligned}$$

Hence, $C_1 = -3$ and the extremal for this case is

$$x(t) = -3t + t^3.$$

In Figure 5.9, we show a plot of the above extremal.

◆ Example 5.9

We consider again the same functional as in Example 5.7, where, this time, the left end point is fixed, as before, while at the right end point t_1 is free and $x(t_1) = 1$. Hence, in this case, $\delta x_1 = 0$ while δt_1 is arbitrary. The transversality conditions reduce to one condition,

$$(F - x' F_{x'})|_{t=t_1} = 0.$$

We perform simple calculations to obtain

$$F - x' F_{x'} = 12xt - (x')^2,$$

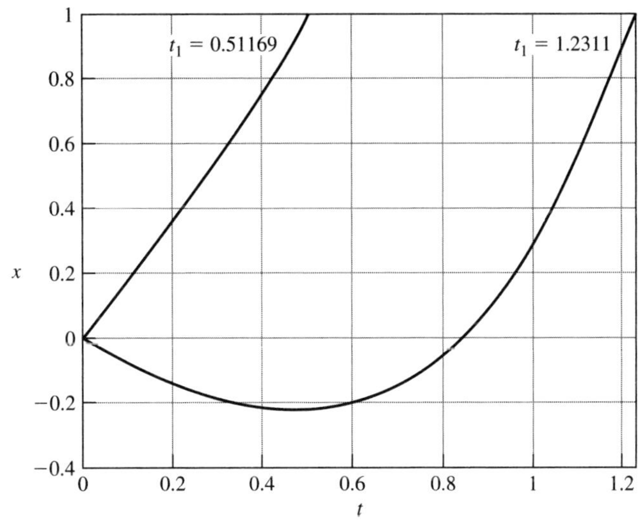

Figure 5.10 Plots of the extremals of Example 5.9.

where

$$x' = (C_1 t + t^3)' = C_1 + 3t^2.$$

Note that because $x(t_1) = C_1 t_1 + t_1^3 = 1$, we have

$$C_1 = \frac{1 - t_1^3}{t_1}.$$

Taking into account the above and the fact that $x(t_1) = 1$, we perform simple algebraic manipulations to arrive at the following equation for t_1:

$$(F - x' F_{x'})|_{t=t_1} = (12x(t_1)t_1 - (x'(t_1))^2) = -4t_1^6 + 8t_1^3 - 1 = 0.$$

Solving the above equation, we obtain the following two solutions:

$$t_1 = \left(1 \pm \sqrt{\frac{3}{4}}\right)^{1/3}.$$

Plots of the two extremals for this example are shown in Figure 5.10.

If the position of the end point B is restricted to a curve $x = g_1(t)$, then

$$\delta x_1 \approx g_1'(t_1)\delta t_1,$$

and hence

$$\delta v = (F + (g_1' - x')F_{x'})\delta t_1 = 0$$

if

$$(F + (g_1' - x')F_{x'})|_{t=t_1} = 0. \tag{5.25}$$

The above is the transversality condition for the case when the end point B is restricted to the curve $x = g_1(t)$.

In the next section we discuss a specific problem of minimizing a functional subject to constraints in the form of differential equations.

5.3 Linear Quadratic Regulator

5.3.1 Algebraic Riccati Equation (ARE)

Consider a linear system modeled by

$$\dot{\boldsymbol{x}} = \boldsymbol{A}\boldsymbol{x} + \boldsymbol{B}\boldsymbol{u}, \qquad \boldsymbol{x}(0) = \boldsymbol{x}_0$$

and the associated performance index

$$J = \int_0^\infty (\boldsymbol{x}^T \boldsymbol{Q}\boldsymbol{x} + \boldsymbol{u}^T \boldsymbol{R}\boldsymbol{u})\, dt,$$

where $\boldsymbol{Q} = \boldsymbol{Q}^T \geq 0$ and $\boldsymbol{R} = \boldsymbol{R}^T > 0$. Our goal is to construct a stabilizing linear state-feedback controller of the form $\boldsymbol{u} = -\boldsymbol{K}\boldsymbol{x}$ that minimizes the performance index J. We denote such a linear control law by $\boldsymbol{u}^*$.

First, we assume that a linear state-feedback optimal controller exists such that the optimal closed-loop system

$$\dot{\boldsymbol{x}} = (\boldsymbol{A} - \boldsymbol{B}\boldsymbol{K})\boldsymbol{x}$$

is asymptotically stable. The above assumption implies that there is a Lyapunov function $V = \boldsymbol{x}^T \boldsymbol{P}\boldsymbol{x}$ for the closed-loop system; that is, for some positive definite matrix $\boldsymbol{P}$ the time derivative dV/dt evaluated on the trajectories of the closed-loop system is negative definite.

Theorem 5.2 If the state-feedback controller $\boldsymbol{u}^* = -\boldsymbol{K}\boldsymbol{x}$ is such that

$$\min_{\boldsymbol{u}} \left(\frac{dV}{dt} + \boldsymbol{x}^T \boldsymbol{Q}\boldsymbol{x} + \boldsymbol{u}^T \boldsymbol{R}\boldsymbol{u} \right) = 0, \tag{5.26}$$

for some $V = \boldsymbol{x}^T \boldsymbol{P}\boldsymbol{x}$, then the controller is optimal.

Proof We can rewrite the condition of the theorem as

$$\left.\frac{dV}{dt}\right|_{\boldsymbol{u}=\boldsymbol{u}^*} + \boldsymbol{x}^T \boldsymbol{Q}\boldsymbol{x} + \boldsymbol{u}^{*T} \boldsymbol{R}\boldsymbol{u}^* = 0.$$

Hence,

$$\left.\frac{dV}{dt}\right|_{\boldsymbol{u}=\boldsymbol{u}^*} = -\boldsymbol{x}^T \boldsymbol{Q}\boldsymbol{x} - \boldsymbol{u}^{*T} \boldsymbol{R}\boldsymbol{u}^*.$$

Integrating both sides of the resulting equation with respect to time from 0 to ∞, we obtain

$$V(\boldsymbol{x}(\infty)) - V(\boldsymbol{x}(0)) = -\int_0^\infty (\boldsymbol{x}^T \boldsymbol{Q}\boldsymbol{x} + \boldsymbol{u}^{*T} \boldsymbol{R}\boldsymbol{u}^*)\, dt.$$

Because by assumption the closed-loop system is asymptotically stable, $\boldsymbol{x}(\infty) = \boldsymbol{0}$, and therefore

$$V(\boldsymbol{x}(0)) = \boldsymbol{x}_0^T \boldsymbol{P}\boldsymbol{x}_0 = \int_0^\infty (\boldsymbol{x}^T \boldsymbol{Q}\boldsymbol{x} + \boldsymbol{u}^{*T} \boldsymbol{R}\boldsymbol{u}^*)\, dt.$$

Thus, we showed that if a linear state-feedback controller satisfies the assumption of the theorem, then the value of the performance index for such a controller is

$$J(\boldsymbol{u}^*) = \boldsymbol{x}_0^T \boldsymbol{P}\boldsymbol{x}_0.$$

To show that such a controller is indeed optimal, we use a proof by contradiction. We assume that (5.26) holds and that $\boldsymbol{u}^*$ is not optimal. Suppose that $\tilde{\boldsymbol{u}}$ yields a smaller value of J, that is,

$$J(\tilde{\boldsymbol{u}}) < J(\boldsymbol{u}^*).$$

It follows from (5.26) that

$$\left.\frac{dV}{dt}\right|_{u=\tilde{u}} + \boldsymbol{x}^T \boldsymbol{Q}\boldsymbol{x} + \tilde{\boldsymbol{u}}^T \boldsymbol{R}\tilde{\boldsymbol{u}} \geq 0,$$

that is,

$$\left.\frac{dV}{dt}\right|_{u=\tilde{u}} \geq -\boldsymbol{x}^T \boldsymbol{Q}\boldsymbol{x} - \tilde{\boldsymbol{u}}^T \boldsymbol{R}\tilde{\boldsymbol{u}}.$$

Integrating the above with respect to time from 0 to ∞ yields

$$V(\boldsymbol{x}(0)) \leq \int_0^\infty (\boldsymbol{x}^T \boldsymbol{Q}\boldsymbol{x} + \tilde{\boldsymbol{u}}^T \boldsymbol{R}\tilde{\boldsymbol{u}})\, dt$$

implying that

$$J(\boldsymbol{u}^*) \leq J(\tilde{\boldsymbol{u}}),$$

which is a contradiction, and the proof is complete.

It follows from the above theorem that the synthesis of the optimal control law involves finding an appropriate Lyapunov function, or equivalently, the matrix $\boldsymbol{P}$. The appropriate $\boldsymbol{P}$ is found by minimizing

$$f(\boldsymbol{u}) = \frac{dV}{dt} + \boldsymbol{x}^T \boldsymbol{Q}\boldsymbol{x} + \boldsymbol{u}^T \boldsymbol{R}\boldsymbol{u}, \tag{5.27}$$

We first apply to (5.27) the necessary condition for unconstrained minimization,

$$\left.\frac{\partial}{\partial \boldsymbol{u}}\left(\frac{dV}{dt} + \boldsymbol{x}^T \boldsymbol{Q}\boldsymbol{x} + \boldsymbol{u}^T \boldsymbol{R}\boldsymbol{u}\right)\right|_{u=u^*} = \boldsymbol{0}^T.$$

Differentiating the above yields

$$\begin{aligned}\frac{\partial}{\partial \boldsymbol{u}}\left(\frac{dV}{dt} + \boldsymbol{x}^T\boldsymbol{Q}\boldsymbol{x} + \boldsymbol{u}^T\boldsymbol{R}\boldsymbol{u}\right) &= \frac{\partial}{\partial \boldsymbol{u}}(2\boldsymbol{x}^T\boldsymbol{P}\dot{\boldsymbol{x}} + \boldsymbol{x}^T\boldsymbol{Q}\boldsymbol{x} + \boldsymbol{u}^T\boldsymbol{R}\boldsymbol{u})\\ &= \frac{\partial}{\partial \boldsymbol{u}}(2\boldsymbol{x}^T\boldsymbol{P}\boldsymbol{A}\boldsymbol{x} + 2\boldsymbol{x}^T\boldsymbol{P}\boldsymbol{B}\boldsymbol{u} + \boldsymbol{x}^T\boldsymbol{Q}\boldsymbol{x} + \boldsymbol{u}^T\boldsymbol{R}\boldsymbol{u})\\ &= 2\boldsymbol{x}^T\boldsymbol{P}\boldsymbol{B} + 2\boldsymbol{u}^T\boldsymbol{R}\\ &= \boldsymbol{0}^T.\end{aligned}$$

Hence, the candidate for an optimal control law has the form

$$\boldsymbol{u}^* = -\boldsymbol{R}^{-1}\boldsymbol{B}^T\boldsymbol{P}\boldsymbol{x} = -\boldsymbol{K}\boldsymbol{x}, \tag{5.28}$$

where $\boldsymbol{K} = \boldsymbol{R}^{-1}\boldsymbol{B}^T\boldsymbol{P}$. Note that

$$\begin{aligned}\frac{\partial^2}{\partial \boldsymbol{u}^2}\left(\frac{dV}{dt} + \boldsymbol{x}^T\boldsymbol{Q}\boldsymbol{x} + \boldsymbol{u}^T\boldsymbol{R}\boldsymbol{u}\right) &= \frac{\partial^2}{\partial \boldsymbol{u}^2}(2\boldsymbol{x}^T\boldsymbol{P}\boldsymbol{A}\boldsymbol{x} + 2\boldsymbol{x}^T\boldsymbol{P}\boldsymbol{B}\boldsymbol{u} + \boldsymbol{x}^T\boldsymbol{Q}\boldsymbol{x} + \boldsymbol{u}^T\boldsymbol{R}\boldsymbol{u})\\ &= \frac{\partial}{\partial \boldsymbol{u}}(2\boldsymbol{x}^T\boldsymbol{P}\boldsymbol{B} + 2\boldsymbol{u}^T\boldsymbol{R})\\ &= 2\boldsymbol{R}\\ &> 0.\end{aligned}$$

Thus the second-order sufficiency condition for $\boldsymbol{u}^*$ to minimize (5.27) is satisfied.

We now turn our attention to finding an appropriate matrix $\boldsymbol{P}$. The optimal closed-loop system has the form

$$\dot{\boldsymbol{x}} = (\boldsymbol{A} - \boldsymbol{B}\boldsymbol{R}^{-1}\boldsymbol{B}^T\boldsymbol{P})\boldsymbol{x}, \qquad \boldsymbol{x}(0) = \boldsymbol{x}_0.$$

Our optimal controller satisfies the equation

$$\left.\frac{dV}{dt}\right|_{u=u^*} + \boldsymbol{x}^T\boldsymbol{Q}\boldsymbol{x} + \boldsymbol{u}^{*T}\boldsymbol{R}\boldsymbol{u}^* = 0,$$

that is,

$$2\boldsymbol{x}^T\boldsymbol{P}\boldsymbol{A}\boldsymbol{x} + 2\boldsymbol{x}^T\boldsymbol{P}\boldsymbol{B}\boldsymbol{u}^* + \boldsymbol{x}^T\boldsymbol{Q}\boldsymbol{x} + \boldsymbol{u}^{*T}\boldsymbol{R}\boldsymbol{u}^* = 0.$$

We substitute the expression for $\boldsymbol{u}^*$, given by (5.28), into the above equation and represent it as

$$\boldsymbol{x}^T(\boldsymbol{A}^T\boldsymbol{P} + \boldsymbol{P}\boldsymbol{A})\boldsymbol{x} - 2\boldsymbol{x}^T\boldsymbol{P}\boldsymbol{B}\boldsymbol{R}^{-1}\boldsymbol{B}^T\boldsymbol{P}\boldsymbol{x} + \boldsymbol{x}^T\boldsymbol{Q}\boldsymbol{x} + \boldsymbol{x}^T\boldsymbol{P}\boldsymbol{B}\boldsymbol{R}^{-1}\boldsymbol{B}^T\boldsymbol{P}\boldsymbol{x} = 0.$$

Factoring out $\boldsymbol{x}$ yields

$$\boldsymbol{x}^T(\boldsymbol{A}^T\boldsymbol{P} + \boldsymbol{P}\boldsymbol{A} + \boldsymbol{Q} - \boldsymbol{P}\boldsymbol{B}\boldsymbol{R}^{-1}\boldsymbol{B}^T\boldsymbol{P})\boldsymbol{x} = 0.$$

The above equation should hold for any $\boldsymbol{x}$. For this to be true we have to have

$$\boxed{\boldsymbol{A}^T\boldsymbol{P} + \boldsymbol{P}\boldsymbol{A} + \boldsymbol{Q} - \boldsymbol{P}\boldsymbol{B}\boldsymbol{R}^{-1}\boldsymbol{B}^T\boldsymbol{P} = \boldsymbol{O}.} \tag{5.29}$$

The above equation is referred to as the *algebraic Riccati equation* (ARE). In conclusion, the synthesis of the optimal linear state feedback controller minimizing the performance index

$$J = \int_0^\infty (\boldsymbol{x}^T\boldsymbol{Q}\boldsymbol{x} + \boldsymbol{u}^T\boldsymbol{R}\boldsymbol{u})\,dt$$

subject to

$$\dot{\boldsymbol{x}} = \boldsymbol{A}\boldsymbol{x} + \boldsymbol{B}\boldsymbol{u}, \quad \boldsymbol{x}(0) = \boldsymbol{x}_0$$

requires solving the ARE given by (5.29).

◆ Example 5.10

Consider the following model of a dynamical system:

$$\dot{x} = 2u_1 + 2u_2, \qquad x(0) = 3,$$

along with the associated performance index

$$J = \int_0^\infty (x^2 + ru_1^2 + ru_2^2)\, dt,$$

where $r > 0$ is a parameter.

1. We first find the solution to the ARE corresponding to the linear state-feedback optimal controller. We have

$$\boldsymbol{A} = 0, \qquad \boldsymbol{B} = [2 \quad 2], \qquad \boldsymbol{Q} = 1, \qquad \boldsymbol{R} = r\boldsymbol{I}_2.$$

   The ARE for this problem is

$$\boldsymbol{O} = \boldsymbol{A}^T\boldsymbol{P} + \boldsymbol{P}\boldsymbol{A} + \boldsymbol{Q} - \boldsymbol{P}\boldsymbol{B}\boldsymbol{R}^{-1}\boldsymbol{B}^T\boldsymbol{P} = 1 - \frac{8}{r}p^2,$$

   whose solution is

$$p = \sqrt{\frac{r}{8}}.$$

2. We now write the equation of the closed-loop system driven by the optimal controller. The optimal controller has the form

$$\boldsymbol{u} = -\boldsymbol{R}^{-1}\boldsymbol{B}^T\boldsymbol{P}\boldsymbol{x} = -\frac{1}{\sqrt{2r}}\begin{bmatrix}1\\1\end{bmatrix}x.$$

   Hence, the closed-loop optimal system is described by

$$\dot{x} = [2 \quad 2]\boldsymbol{u} = -\frac{4}{\sqrt{2r}}x.$$

3. Finally, we find the value of J for the optimal closed-loop system. We have

$$J = \boldsymbol{x}^T(0)\boldsymbol{P}\boldsymbol{x}(0) = \frac{9}{2}\sqrt{\frac{r}{2}}.$$

We can directly verify that $J = \frac{9}{2}\sqrt{\frac{r}{2}}$ as follows:

$$\begin{aligned}\min J &= \min \int_0^\infty (x^2 + ru_1^2 + ru_2^2)\,dt \\ &= \int_0^\infty 9\exp\left(-\frac{8t}{\sqrt{2r}}\right)(1+1)\,dt = 9\frac{\sqrt{2r}}{4} = \frac{9}{2}\sqrt{\frac{r}{2}}.\end{aligned}$$

A major difficulty when solving the ARE is that it is a nonlinear equation. In the next section we present an efficient method for solving the ARE.

5.3.2 Solving the ARE Using the Eigenvector Method

We now present a method for solving the ARE referred to as the MacFarlane and Potter method or the eigenvector method. In our discussion of the method, we follow the development of Power and Simpson [237]. We begin by representing the ARE in the form

$$[\boldsymbol{P} \quad -\boldsymbol{I}_n]\begin{bmatrix} \boldsymbol{A} & -\boldsymbol{B}\boldsymbol{R}^{-1}\boldsymbol{B}^T \\ -\boldsymbol{Q} & -\boldsymbol{A}^T \end{bmatrix}\begin{bmatrix} \boldsymbol{I}_n \\ \boldsymbol{P} \end{bmatrix} = \boldsymbol{O}. \tag{5.30}$$

The $2n \times 2n$ matrix in the middle is called the *Hamiltonian matrix*. We use the symbol $\boldsymbol{H}$ to denote the Hamiltonian matrix, that is,

$$\boldsymbol{H} = \begin{bmatrix} \boldsymbol{A} & -\boldsymbol{B}\boldsymbol{R}^{-1}\boldsymbol{B}^T \\ -\boldsymbol{Q} & -\boldsymbol{A}^T \end{bmatrix}.$$

Thus, the ARE can be represented as

$$[\boldsymbol{P} \quad -\boldsymbol{I}_n]\boldsymbol{H}\begin{bmatrix} \boldsymbol{I}_n \\ \boldsymbol{P} \end{bmatrix} = \boldsymbol{O}.$$

We premultiply the above equation by $\boldsymbol{X}^{-1}$ and then postmultiply it by $\boldsymbol{X}$, where $\boldsymbol{X}$ is a nonsingular $n \times n$ matrix,

$$[\boldsymbol{X}^{-1}\boldsymbol{P} \quad -\boldsymbol{X}^{-1}]\boldsymbol{H}\begin{bmatrix} \boldsymbol{X} \\ \boldsymbol{P}\boldsymbol{X} \end{bmatrix} = \boldsymbol{O}. \tag{5.31}$$

Observe that if we can find matrices $\boldsymbol{X}$ and $\boldsymbol{P}\boldsymbol{X}$ such that

$$\boldsymbol{H}\begin{bmatrix} \boldsymbol{X} \\ \boldsymbol{P}\boldsymbol{X} \end{bmatrix} = \begin{bmatrix} \boldsymbol{X} \\ \boldsymbol{P}\boldsymbol{X} \end{bmatrix}\boldsymbol{\Lambda},$$

then equation (5.31) becomes

$$[\boldsymbol{X}^{-1}\boldsymbol{P} \quad -\boldsymbol{X}^{-1}]\begin{bmatrix} \boldsymbol{X} \\ \boldsymbol{P}\boldsymbol{X} \end{bmatrix}\boldsymbol{\Lambda} = \boldsymbol{O}.$$

We have thus reduced the problem of solving the ARE to that of constructing appropriate matrices $\boldsymbol{X}$ and $\boldsymbol{PX}$. To proceed further, let $\boldsymbol{v}_i$ be an eigenvector of $\boldsymbol{H}$ and let s_i be the corresponding eigenvalue; then

$$\boldsymbol{H}\boldsymbol{v}_i = s_i \boldsymbol{v}_i.$$

In our further discussion, we assume that $\boldsymbol{H}$ has at least n distinct real eigenvalues among its $2n$ eigenvalues. (The results obtained can be generalized for the case when the eigenvalues of $\boldsymbol{H}$ are complex or nondistinct.) Thus, we can write

$$\boldsymbol{H}[\boldsymbol{v}_1 \quad \boldsymbol{v}_2 \quad \cdots \quad \boldsymbol{v}_n] = [\boldsymbol{v}_1 \quad \boldsymbol{v}_2 \quad \cdots \quad \boldsymbol{v}_n] \begin{bmatrix} s_1 & 0 & \cdots & 0 \\ 0 & s_2 & \cdots & 0 \\ \vdots & \vdots & & \vdots \\ 0 & 0 & \cdots & s_n \end{bmatrix}.$$

Let

$$\begin{bmatrix} \boldsymbol{X} \\ \boldsymbol{PX} \end{bmatrix} = [\boldsymbol{v}_1 \quad \boldsymbol{v}_2 \quad \cdots \quad \boldsymbol{v}_n]$$

and

$$\boldsymbol{\Lambda} = \begin{bmatrix} s_1 & 0 & \cdots & 0 \\ 0 & s_2 & \cdots & 0 \\ \vdots & \vdots & & \vdots \\ 0 & 0 & \cdots & s_n \end{bmatrix}.$$

The above choice of $\boldsymbol{X}$ and $\boldsymbol{PX}$ constitutes a possible solution to the equation

$$[\boldsymbol{X}^{-1}\boldsymbol{P} \quad -\boldsymbol{X}^{-1}] \begin{bmatrix} \boldsymbol{X} \\ \boldsymbol{PX} \end{bmatrix} \boldsymbol{\Lambda} = \boldsymbol{O}.$$

To construct $\boldsymbol{P}$, we partition the $2n \times n$ eigenvector matrix $[\boldsymbol{v}_1 \; \boldsymbol{v}_2 \; \cdots \; \boldsymbol{v}_n]$ into two $n \times n$ submatrices as follows:

$$[\boldsymbol{v}_1 \quad \boldsymbol{v}_2 \quad \cdots \quad \boldsymbol{v}_n] = \begin{bmatrix} \boldsymbol{W} \\ \boldsymbol{Z} \end{bmatrix}.$$

Thus,

$$\begin{bmatrix} \boldsymbol{X} \\ \boldsymbol{PX} \end{bmatrix} = \begin{bmatrix} \boldsymbol{W} \\ \boldsymbol{Z} \end{bmatrix}.$$

Taking $\boldsymbol{X} = \boldsymbol{W}$ and $\boldsymbol{PX} = \boldsymbol{Z}$ and assuming that $\boldsymbol{W}$ is invertible, we obtain

$$\boxed{\boldsymbol{P} = \boldsymbol{Z}\boldsymbol{W}^{-1}.} \tag{5.32}$$

We now have to decide what set of n eigenvalues should be chosen from those of $\boldsymbol{H}$ to construct the particular $\boldsymbol{P}$. In the case when all $2n$ eigenvalues of $\boldsymbol{H}$ are distinct, the number of different matrices P generated by the above-described method is given by

$$\frac{(2n)!}{(n!)^2}.$$

Let $Q = C^T C$ be a full rank factorization of Q, as defined on page 636. It follows from Theorem 3 of Kučera [172, p. 346] that the Hamiltonian matrix H has n eigenvalues in the open left-half complex plane and n in the open right-half plane if and only if the pair (A, B) is stabilizable and the pair (A, C) is detectable. The matrix P that we seek corresponds to the asymptotically stable eigenvalues of H. The eigenvalues of the Hamiltonian matrix come in pairs $\pm s_i$ (see Exercise 5.15). The characteristic polynomial of H contains only even powers of s. With such constructed matrix P, we have the following result:

Theorem 5.3 The poles of the closed-loop system

$$\dot{x}(t) = (A - BR^{-1}B^T P)x(t)$$

are those eigenvalues of H having negative real parts.

Proof Because $[v_1 \; v_2 \; \cdots \; v_n] = \begin{bmatrix} W \\ Z \end{bmatrix}$, we can write

$$\begin{bmatrix} A & -BR^{-1}B^T \\ -Q & -A^T \end{bmatrix} \begin{bmatrix} W \\ Z \end{bmatrix} = \begin{bmatrix} W \\ Z \end{bmatrix} \Lambda.$$

Performing appropriate multiplication of $n \times n$ block matrices yields

$$AW - BR^{-1}B^T Z = W\Lambda,$$

or

$$A - BR^{-1}B^T ZW^{-1} = A - BR^{-1}B^T P = W\Lambda W^{-1},$$

because $P = ZW^{-1}$. Thus, the matrix $A - BR^{-1}B^T P$ is similar to the matrix Λ whose eigenvalues are the asymptotically stable eigenvalues of H. The proof is complete.

◆ Example 5.11

Consider the following model of a dynamical system:

$$\dot{x} = 2x + u,$$

along with the associated performance index

$$J = \int_0^\infty (x^2 + ru^2)\,dt.$$

Find the value of r such that the optimal closed-loop system has its pole at -3.

We form the associated Hamiltonian matrix

$$H = \begin{bmatrix} A & -BR^{-1}B^T \\ -Q & -A^T \end{bmatrix} = \begin{bmatrix} 2 & -\frac{1}{r} \\ -1 & -2 \end{bmatrix}.$$

The characteristic equation of $\boldsymbol{H}$ is

$$\det(s\boldsymbol{I}_2 - \boldsymbol{H}) = s^2 - 4 - \frac{1}{r} = 0.$$

Hence,

$$r = \frac{1}{5}$$

results in the optimal closed-loop system having its pole located at -3.

The eigenvector method for solving the ARE can be generalized to the case when the eigenvalues of the Hamiltonian matrix $\boldsymbol{H}$ are not necessarily distinct or real.

5.3.3 Optimal Systems with Prescribed Poles

We can always find a state feedback $u = -\boldsymbol{K}\boldsymbol{x}$ such that the closed-loop system $\dot{\boldsymbol{x}} = (\boldsymbol{A}-\boldsymbol{B}\boldsymbol{K})\boldsymbol{x}$ will have its poles in prescribed locations (symmetric with respect to the real axis) if and only if the pair $(\boldsymbol{A}, \boldsymbol{B})$ is reachable. In the multi-input systems the gain matrix $\boldsymbol{K}$ that shifts the poles to the prescribed locations is not unique. We would like to use the remaining degrees of freedom to construct a gain matrix $\boldsymbol{K}$ that shifts the system poles to the desired locations and at the same time minimizes a performance index. Specifically, we are given a system model,

$$\dot{\boldsymbol{x}}(t) = \boldsymbol{A}\boldsymbol{x}(t) + \boldsymbol{B}\boldsymbol{u}(t),$$

the desired set of closed-loop poles,

$$\{s_1, s_2, \ldots, s_n\},$$

and an $m \times m$ positive definite matrix $\boldsymbol{R} = \boldsymbol{R}^T$. We wish to construct a state-feedback control law $\boldsymbol{u}(t) = -\boldsymbol{K}\boldsymbol{x}(t)$ and an $n \times n$ positive semidefinite matrix $\boldsymbol{Q} = \boldsymbol{Q}^T$ such that the closed-loop system has its poles in the desired locations, and at the same time the performance index

$$J = \int_0^\infty (\boldsymbol{x}^T(t)\boldsymbol{Q}\boldsymbol{x}(t) + \boldsymbol{u}^T(t)\boldsymbol{R}\boldsymbol{u}(t))\,dt$$

is minimized. In other words, we are interested in designing an optimal control system with prescribed poles. We illustrate the problem with a very simple example.

◆ **Example 5.12**

For a dynamical system modeled by

$$\begin{aligned}\dot{x} &= 2x + u\\ &= \boldsymbol{A}x + \boldsymbol{B}u,\end{aligned}$$

and the performance index

$$J = \int_0^\infty (qx^2 + u^2)\,dt$$
$$= \int_0^\infty (\boldsymbol{x}^T \boldsymbol{Q}\boldsymbol{x} + \boldsymbol{u}^T \boldsymbol{R}\boldsymbol{u})\,dt,$$

find q so that the eigenvalue of

$$\boldsymbol{A} - \boldsymbol{B}\boldsymbol{R}^{-1}\boldsymbol{B}^T\boldsymbol{P}$$

is equal to -3.

The corresponding Hamiltonian matrix is

$$\boldsymbol{H} = \begin{bmatrix} 2 & -1 \\ -q & -2 \end{bmatrix}.$$

The characteristic polynomial of $\boldsymbol{H}$ is

$$\det[s\boldsymbol{I}_2 - \boldsymbol{H}] = \det\begin{bmatrix} s-2 & 1 \\ q & s+2 \end{bmatrix} = s^2 - 4 - q.$$

We want

$$\det[s\boldsymbol{I}_2 - \boldsymbol{H}] = (s+3)(s-3) = s^2 - 9.$$

Hence, the desired value of q is $q = 5$.

We now discuss a method for designing optimal control systems with prescribed poles. In constructing the gain matrix $\boldsymbol{K}$ and the weight matrix $\boldsymbol{Q}$, we repeatedly use a state variable transformation. To proceed, we analyze the system model, the performance index, and the Hamiltonian matrix in a new basis. Let

$$\boldsymbol{x} = \boldsymbol{T}\tilde{\boldsymbol{x}},$$

where $\det \boldsymbol{T} \neq 0$. Then, the system model in the new coordinates has the form

$$\dot{\tilde{\boldsymbol{x}}}(t) = \boldsymbol{T}^{-1}\boldsymbol{A}\boldsymbol{T}\tilde{\boldsymbol{x}}(t) + \boldsymbol{T}^{-1}\boldsymbol{B}\boldsymbol{u}(t)$$
$$= \tilde{\boldsymbol{A}}\tilde{\boldsymbol{x}}(t) + \tilde{\boldsymbol{B}}\boldsymbol{u}(t).$$

The performance index J, after state variable transformation, takes the form

$$\tilde{J} = \int_0^\infty (\tilde{\boldsymbol{x}}^T(t)\boldsymbol{T}^T\boldsymbol{Q}\boldsymbol{T}\tilde{\boldsymbol{x}}(t) + \boldsymbol{u}^T(t)\boldsymbol{R}\boldsymbol{u}(t))\,dt.$$

The Hamiltonian matrix is transformed to

$$\tilde{\boldsymbol{H}} = \begin{bmatrix} \boldsymbol{T}^{-1}\boldsymbol{A}\boldsymbol{T} & -\boldsymbol{T}^{-1}\boldsymbol{B}\boldsymbol{R}^{-1}\boldsymbol{B}^T\boldsymbol{T}^{-T} \\ -\boldsymbol{T}^T\boldsymbol{Q}\boldsymbol{T} & -\boldsymbol{T}^T\boldsymbol{A}^T\boldsymbol{T}^{-T} \end{bmatrix}$$
$$= \begin{bmatrix} \tilde{\boldsymbol{A}} & -\tilde{\boldsymbol{B}}\boldsymbol{R}^{-1}\tilde{\boldsymbol{B}}^T \\ -\tilde{\boldsymbol{Q}} & -\tilde{\boldsymbol{A}}^T \end{bmatrix},$$

where $\boldsymbol{T}^{-T} = (\boldsymbol{T}^{-1})^T = (\boldsymbol{T}^T)^{-1}$. Observe that

$$\tilde{\boldsymbol{H}} = \begin{bmatrix} \boldsymbol{T}^{-1} & \boldsymbol{0} \\ \boldsymbol{0} & \boldsymbol{T}^T \end{bmatrix} \boldsymbol{H} \begin{bmatrix} \boldsymbol{T} & \boldsymbol{0} \\ \boldsymbol{0} & \boldsymbol{T}^{-T} \end{bmatrix};$$

that is, $\tilde{\boldsymbol{H}}$ and $\boldsymbol{H}$ are similarity matrices because

$$\begin{bmatrix} \boldsymbol{T}^{-1} & \boldsymbol{0} \\ \boldsymbol{0} & \boldsymbol{T}^T \end{bmatrix} \begin{bmatrix} \boldsymbol{T} & \boldsymbol{0} \\ \boldsymbol{0} & \boldsymbol{T}^{-T} \end{bmatrix} = \boldsymbol{I}_{2n}.$$

Hence, the eigenvalues of $\boldsymbol{H}$ and $\tilde{\boldsymbol{H}}$ are identical. Our next step is to compute the characteristic equation of $\tilde{\boldsymbol{H}}$ as a function of $\tilde{\boldsymbol{Q}}$. Computing $\det(s\boldsymbol{I}_{2n} - \tilde{\boldsymbol{H}})$ is facilitated after we perform some manipulations on the characteristic matrix $s\boldsymbol{I}_{2n} - \tilde{\boldsymbol{H}}$ that do not change its determinant. Specifically, we premultiply the characteristic matrix $s\boldsymbol{I}_{2n} - \tilde{\boldsymbol{H}}$ by a $2n \times 2n$ matrix whose determinant equals 1. Then, the determinant of the product is the same as $\det(s\boldsymbol{I}_{2n} - \tilde{\boldsymbol{H}})$, but it is much easier to evaluate. We have

$$\begin{aligned} &\begin{bmatrix} \boldsymbol{I}_n & \boldsymbol{0} \\ -\tilde{\boldsymbol{Q}}(s\boldsymbol{I}_n - \tilde{\boldsymbol{A}})^{-1} & \boldsymbol{I}_n \end{bmatrix} \begin{bmatrix} s\boldsymbol{I}_n - \tilde{\boldsymbol{A}} & \tilde{\boldsymbol{B}}\boldsymbol{R}^{-1}\tilde{\boldsymbol{B}}^T \\ \tilde{\boldsymbol{Q}} & s\boldsymbol{I}_n + \tilde{\boldsymbol{A}}^T \end{bmatrix} \\ &\quad = \begin{bmatrix} s\boldsymbol{I}_n - \tilde{\boldsymbol{A}} & \tilde{\boldsymbol{B}}\boldsymbol{R}^{-1}\tilde{\boldsymbol{B}}^T \\ \boldsymbol{0} & s\boldsymbol{I}_n + \tilde{\boldsymbol{A}}^T - \tilde{\boldsymbol{Q}}(s\boldsymbol{I}_n - \tilde{\boldsymbol{A}})^{-1}\tilde{\boldsymbol{B}}\boldsymbol{R}^{-1}\tilde{\boldsymbol{B}}^T \end{bmatrix}. \end{aligned}$$

Let

$$\boldsymbol{F} = \tilde{\boldsymbol{B}}\boldsymbol{R}^{-1}\tilde{\boldsymbol{B}}^T,$$

then

$$\begin{aligned} \det(s\boldsymbol{I}_{2n} - \tilde{\boldsymbol{H}}) &= \det \begin{bmatrix} s\boldsymbol{I}_n - \tilde{\boldsymbol{A}} & \boldsymbol{F} \\ \boldsymbol{0} & s\boldsymbol{I}_n + \tilde{\boldsymbol{A}}^T - \boldsymbol{Q}(s\boldsymbol{I}_n - \tilde{\boldsymbol{A}})^{-1}\boldsymbol{F} \end{bmatrix} \\ &= \det(s\boldsymbol{I}_n - \tilde{\boldsymbol{A}}) \det(s\boldsymbol{I}_n + \tilde{\boldsymbol{A}}^T - \tilde{\boldsymbol{Q}}(s\boldsymbol{I}_n - \tilde{\boldsymbol{A}})^{-1}\boldsymbol{F}). \end{aligned}$$

Suppose now that the matrix $\boldsymbol{A}$ has n linearly independent eigenvectors; hence it is diagonalizable. Let

$$\boldsymbol{T} = [\boldsymbol{t}_1 \quad \boldsymbol{t}_2 \quad \cdots \quad \boldsymbol{t}_n]$$

be a matrix composed of the eigenvectors of $\boldsymbol{A}$. Therefore,

$$\tilde{\boldsymbol{A}} = \boldsymbol{T}^{-1}\boldsymbol{A}\boldsymbol{T} = \begin{bmatrix} \lambda_1 & 0 & \cdots & 0 \\ 0 & \lambda_2 & \cdots & 0 \\ \vdots & \vdots & & \vdots \\ 0 & 0 & \cdots & \lambda_n \end{bmatrix}.$$

Note that $\tilde{\boldsymbol{A}}^T = \tilde{\boldsymbol{A}}$. Assume that

$$\tilde{\boldsymbol{Q}} = \tilde{\boldsymbol{Q}}_j = \begin{bmatrix} 0 & 0 & 0 & \cdots & 0 \\ \vdots & \vdots & \vdots & & \vdots \\ 0 & 0 & 0 & \cdots & 0 \\ 0 & 0 & \tilde{q}_{jj} & \cdots & 0 \\ \vdots & \vdots & \vdots & & \vdots \\ 0 & 0 & 0 & \cdots & 0 \end{bmatrix};$$

that is, $\tilde{\boldsymbol{Q}}_j$ has only one nonzero element $\tilde{q}_{jj}$. Under the above assumptions the matrix $(s\boldsymbol{I}_n + \tilde{\boldsymbol{A}}^T - \tilde{\boldsymbol{Q}}_j(s\boldsymbol{I}_n - \tilde{\boldsymbol{A}})^{-1}\boldsymbol{F})$ takes the form

$$s\boldsymbol{I}_n + \tilde{\boldsymbol{A}}^T - \tilde{\boldsymbol{Q}}_j(s\boldsymbol{I}_n - \tilde{\boldsymbol{A}})^{-1}\boldsymbol{F} = \begin{bmatrix} s+\lambda_1 & 0 & \cdots & 0 \\ 0 & s+\lambda_2 & \cdots & 0 \\ \vdots & \vdots & \ddots & \vdots \\ 0 & 0 & \cdots & s+\lambda_n \end{bmatrix}$$

$$- \begin{bmatrix} 0 & 0 & 0 & \cdots & 0 \\ \vdots & \vdots & \vdots & & \vdots \\ 0 & 0 & \tilde{q}_{jj} & & 0 \\ \vdots & \vdots & \vdots & & \vdots \\ 0 & 0 & 0 & \cdots & 0 \end{bmatrix} \begin{bmatrix} \frac{1}{s-\lambda_1} & 0 & \cdots & 0 \\ 0 & \frac{1}{s-\lambda_2} & \cdots & 0 \\ \vdots & \vdots & \ddots & \vdots \\ 0 & 0 & \cdots & \frac{1}{s-\lambda_n} \end{bmatrix} \begin{bmatrix} f_{11} & f_{12} & \cdots & f_{1n} \\ f_{21} & f_{22} & \cdots & f_{2n} \\ \vdots & \vdots & \ddots & \vdots \\ f_{n1} & f_{n2} & \cdots & f_{nn} \end{bmatrix}$$

$$= \begin{bmatrix} s+\lambda_1 & 0 & 0 & \cdots & 0 \\ 0 & s+\lambda_2 & 0 & \cdots & 0 \\ \vdots & \vdots & \vdots & & \vdots \\ 0 & 0 & s+\lambda_j & & 0 \\ \vdots & \vdots & \vdots & \ddots & \vdots \\ 0 & 0 & 0 & \cdots & s+\lambda_n \end{bmatrix} - \begin{bmatrix} 0 & 0 & 0 & \cdots & 0 \\ 0 & 0 & 0 & \cdots & 0 \\ \vdots & \vdots & \vdots & & \vdots \\ \frac{\tilde{q}_{jj}f_{j1}}{s-\lambda_j} & \frac{\tilde{q}_{jj}f_{j2}}{s-\lambda_j} & \frac{\tilde{q}_{jj}f_{jj}}{s-\lambda_j} & \cdots & \frac{\tilde{q}_{jj}f_{jn}}{s-\lambda_j} \\ \vdots & \vdots & \vdots & & \vdots \\ 0 & 0 & 0 & \cdots & 0 \end{bmatrix}$$

$$= \begin{bmatrix} s+\lambda_1 & 0 & 0 & \cdots & 0 & 0 \\ 0 & s+\lambda_2 & 0 & \cdots & 0 & 0 \\ \vdots & \vdots & \vdots & & \vdots & \vdots \\ -\frac{\tilde{q}_{jj}f_{j1}}{s-\lambda_j} & -\frac{\tilde{q}_{jj}f_{j2}}{s-\lambda_j} & \cdots & \cdots & s+\lambda_j-\frac{\tilde{q}_{jj}f_{jj}}{s-\lambda_j} & -\frac{\tilde{q}_{jj}f_{jn}}{s-\lambda_j} \\ \vdots & \vdots & \vdots & \ddots & \vdots & \vdots \\ 0 & 0 & 0 & \cdots & 0 & s+\lambda_n \end{bmatrix}.$$

Hence,

$$\det(s\boldsymbol{I}_n + \tilde{\boldsymbol{A}}^T - \tilde{\boldsymbol{Q}}_j(s\boldsymbol{I}_n - \tilde{\boldsymbol{A}})^{-1}\boldsymbol{F}) = \left(\prod_{\substack{i=1 \\ i\neq j}}^{n}(s+\lambda_i)\right)\left(s+\lambda_j-\frac{\tilde{q}_{jj}f_{jj}}{s-\lambda_j}\right).$$

We can now evaluate the characteristic polynomial of the Hamiltonian matrix,

$$\begin{aligned}\det(s\boldsymbol{I}_{2n} - \boldsymbol{H}) &= \det(s\boldsymbol{I}_{2n} - \tilde{\boldsymbol{H}}) \\ &= \left(\prod_{i=1}^{n}(s-\lambda_i)\right)\left(\prod_{\substack{i=1\\ i\neq j}}^{n}(s+\lambda_i)\right)\left(s+\lambda_j - \frac{\tilde{q}_{jj}f_{jj}}{s-\lambda_j}\right) \\ &= \prod_{\substack{i=1\\ i\neq j}}^{n}(s+\lambda_i)(s-\lambda_i)((s+\lambda_j)(s-\lambda_j) - \tilde{q}_{jj}f_{jj}).\end{aligned}$$

From the form of the characteristic polynomial of the Hamiltonian matrix, we conclude that selecting an appropriate value of $\tilde{q}_{jj}$ allows us to shift the jth pole of the closed-loop system into a prespecified location while leaving the other poles in their original positions. To see this we represent the term that involves $\tilde{q}_{jj}$ as

$$\begin{aligned}(s+\lambda_j)(s-\lambda_j) - \tilde{q}_{jj}f_{jj} &= s^2 - \lambda_j^2 - \tilde{q}_{jj}f_{jj} \\ &= s^2 - \left(\lambda_j^2 + \tilde{q}_{jj}f_{jj}\right) \\ &= (s - s_j)(s + s_j),\end{aligned}$$

where

$$s_j = \pm\sqrt{\lambda_j^2 + \tilde{q}_{jj}f_{jj}}.$$

From the above expression for s_j, we can calculate $\tilde{q}_{jj}$ that results in the jth pole being shifted to $-\sqrt{\lambda_j^2 + \tilde{q}_{jj}f_{jj}}$. We obtain

$$\boxed{\tilde{q}_{jj} = \frac{s_j^2 - \lambda_j^2}{f_{jj}}.} \tag{5.33}$$

Note that while selecting the desired location for s_j, we have to make sure that $\tilde{q}_{jj} > 0$ to ensure positive semidefiniteness of the weight matrix $\tilde{\boldsymbol{Q}}_j$. Having $\tilde{\boldsymbol{Q}}_j$, we compute $\boldsymbol{Q}_j = \boldsymbol{T}^{-T}\tilde{\boldsymbol{Q}}_j\boldsymbol{T}^{-1}$ and solve the Riccati equation

$$\boldsymbol{A}^T\boldsymbol{P}_j + \boldsymbol{P}_j\boldsymbol{A} + \boldsymbol{Q}_j - \boldsymbol{P}_j\boldsymbol{B}\boldsymbol{R}^{-1}\boldsymbol{B}^T\boldsymbol{P}_j = \boldsymbol{O} \tag{5.34}$$

for $\boldsymbol{P}_j$. We compute $\boldsymbol{u} = -\boldsymbol{R}^{-1}\boldsymbol{B}^T\boldsymbol{P}_j\boldsymbol{x} + \boldsymbol{u}_j = -\boldsymbol{K}_j\boldsymbol{x} + \boldsymbol{u}_j$ that shifts only one pole of the closed-loop system

$$\begin{aligned}\dot{\boldsymbol{x}} &= (\boldsymbol{A} - \boldsymbol{B}\boldsymbol{K}_j)\boldsymbol{x} + \boldsymbol{B}\boldsymbol{u}_j \\ &= \boldsymbol{A}_j\boldsymbol{x} + \boldsymbol{B}\boldsymbol{u}_j\end{aligned}$$

to its specified position. We then start with the new system

$$\dot{\boldsymbol{x}} = \boldsymbol{A}_j\boldsymbol{x} + \boldsymbol{B}\boldsymbol{u}_j$$

and shift the next pole to a specified location, and then we combine the obtained feedback gains. The above method is based on the result, established by Solheim [267], that different feedback matrices $\boldsymbol{K}_j$ and the corresponding weight matrices $\boldsymbol{Q}_j$ can be combined to give the final optimal gain $\boldsymbol{K}$ and the corresponding weight matrix $\boldsymbol{Q}$. Indeed, suppose we apply the above method to the system $\dot{\boldsymbol{x}} = (\boldsymbol{A} - \boldsymbol{B}\boldsymbol{K}_j)\boldsymbol{x} + \boldsymbol{B}\boldsymbol{u}_j$. After finding an appropriate weight matrix $\boldsymbol{Q}_{j+1}$, we solve the Riccati equation

$$(\boldsymbol{A} - \boldsymbol{B}\boldsymbol{K}_j)^T\boldsymbol{P}_{j+1} + \boldsymbol{P}_{j+1}(\boldsymbol{A} - \boldsymbol{B}\boldsymbol{K}_j) + \boldsymbol{Q}_{j+1} - \boldsymbol{P}_{j+1}\boldsymbol{B}\boldsymbol{R}^{-1}\boldsymbol{B}^T\boldsymbol{P}_{j+1} = \boldsymbol{O} \tag{5.35}$$

for $\boldsymbol{P}_{j+1}$ to compute the gain $\boldsymbol{K}_{j+1}$. Adding the two Riccati equations (5.34) and (5.35), and performing simple manipulations yields

$$\boldsymbol{A}^T(\boldsymbol{P}_j + \boldsymbol{P}_{j+1}) + (\boldsymbol{P}_j + \boldsymbol{P}_{j+1})\boldsymbol{A} + \boldsymbol{Q}_j + \boldsymbol{Q}_{j+1} - (\boldsymbol{P}_j + \boldsymbol{P}_{j+1})\boldsymbol{B}\boldsymbol{R}^{-1}\boldsymbol{B}^T(\boldsymbol{P}_j + \boldsymbol{P}_{j+1}) = \boldsymbol{O}.$$

Let $\boldsymbol{P} = \boldsymbol{P}_j + \boldsymbol{P}_{j+1}$ and $\boldsymbol{Q} = \boldsymbol{Q}_j + \boldsymbol{Q}_{j+1}$, then

$$\begin{aligned}\boldsymbol{K} &= \boldsymbol{K}_j + \boldsymbol{K}_{j+1}\\ &= \boldsymbol{R}^{-1}\boldsymbol{B}^T\boldsymbol{P}_j + \boldsymbol{R}^{-1}\boldsymbol{B}^T\boldsymbol{P}_{j+1}.\end{aligned}$$

In a similar manner we can show that the method works for more than just two iterations.

◆ Example 5.13

Consider a simple model of one-link robot manipulator shown in Figure 5.11. The motion of the robot arm is controlled by a DC motor via a gear. The DC motor is armature-controlled and its schematic is shown in Figure 5.12. We assume that the motor moment of inertia is negligible compared with that of the robot arm. We model the arm as a point mass m attached to the end of a massless rod of length l. Hence the arm inertia is $I_a = ml^2$. We assume that the gear train has no backlash, and all connecting shafts are rigid. As can be seen from Figure 5.11, counterclockwise rotation of the arm is defined as positive, and clockwise rotation

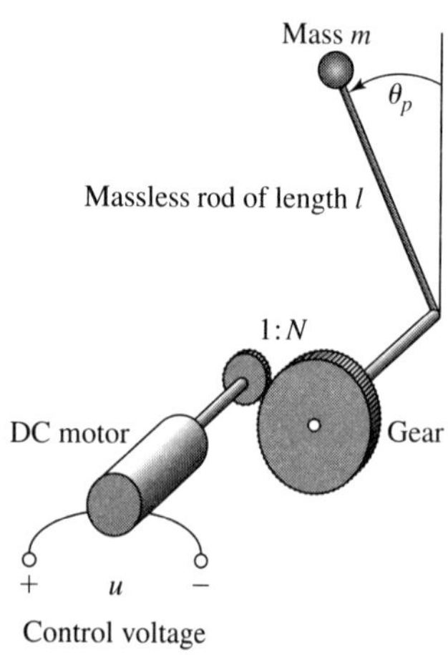

Figure 5.11 One-link manipulator controlled by a DC motor via a gear.

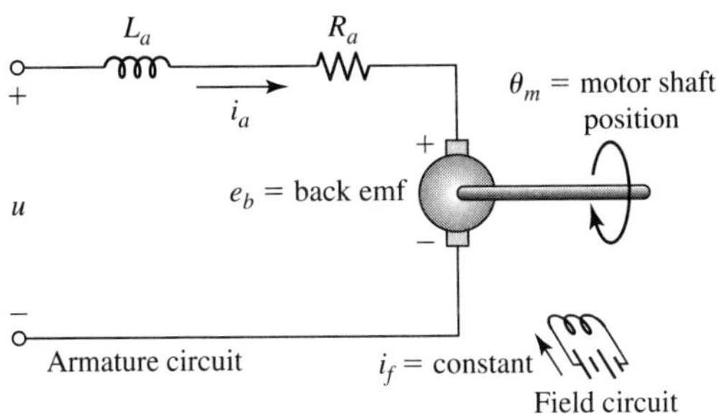

Figure 5.12 Schematic of an armature-controlled DC motor of Example 5.13.

of the arm is defined as negative, whereas counterclockwise rotation of the motor shaft is defined as negative, and clockwise rotation of the shaft is defined as positive. The torque delivered by the motor is

$$T_m = K_m i_a,$$

where K_m is the motor-torque constant, and i_a is the armature current. Let N denote the gear ratio. Then we have

$$\frac{\theta_p}{\theta_m} = \frac{\text{radius of motor gear}}{\text{radius of arm gear}} = \frac{\text{number of teeth of motor gear}}{\text{number of teeth of arm gear}} = \frac{1}{N}.$$

This is because the gears are in contact and thus

$$\theta_p \times \text{radius of arm gear} = \theta_m \times \text{radius of motor gear},$$

and the radii of the gears are proportional to their number of teeth. The work done by the gears must be equal. Let T_p denote the torque applied to the robot arm. Then,

$$T_p\theta_p = T_m\theta_m.$$

Thus, the torque applied to the pendulum is

$$T_p = NT_m = NK_m i_a.$$

We use Newton's second law to write the equation modeling the arm dynamics,

$$I_a\frac{d^2\theta_p}{dt^2} = mgl\sin\theta_p + T_p. \tag{5.36}$$

Substituting into (5.36) the expressions for I_a and T_p and then rearranging yields

$$ml^2\frac{d^2\theta_p}{dt^2} = mgl\sin\theta_p + NK_m i_a, \tag{5.37}$$

where $g = 9.8$ m/sec^2 is the gravitational constant. Applying Kirchhoff's voltage law to the armature circuit yields

$$L_a \frac{di_a}{dt} + R_a i_a + K_b N \frac{d\theta_p}{dt} = u,$$

where K_b is the back emf constant. We assume that $L_a \approx 0$. Hence,

$$u = R_a i_a + K_b N \frac{d\theta_p}{dt}, \tag{5.38}$$

We next compute i_a from (5.38) and substitute the result into (5.37) to get

$$ml^2 \frac{d^2\theta_p}{dt^2} = mgl \sin\theta_p + NK_m \left(\frac{u}{R_a} - \frac{K_b N \frac{d\theta_p}{dt}}{R_a} \right). \tag{5.39}$$

We can now construct a state-space model of the one-link robot. For this we choose the following state and output variables:

$$x_1 = \theta_p, \quad x_2 = \frac{d\theta_p}{dt} = \omega_p, \quad \text{and} \quad y = x_1.$$

Then, using (5.39), we obtain the following simple state-space model of the robot manipulator:

$$\begin{bmatrix} \dot{x}_1 \\ \dot{x}_2 \end{bmatrix} = \begin{bmatrix} x_2 \\ \frac{g}{l} \sin x_1 - \frac{K_b K_m N^2}{ml^2 R_a} x_2 + \frac{NK_m}{ml^2 R_a} u \end{bmatrix},$$
$$y = x_1.$$

Reasonable parameters for the robot are: $l = 1$ m, $m = 1$ kg, $N = 10$, $K_m = 0.1$ Nm/A, $K_b = 0.1$ Vsec/rad, $R_a = 1\ \Omega$. With the above parameters the robot model takes the form:

$$\begin{bmatrix} \dot{x}_1 \\ \dot{x}_2 \end{bmatrix} = \begin{bmatrix} x_2 \\ 9.8 \sin x_1 - x_2 + u \end{bmatrix},$$
$$y = x_1.$$

Time histories of state trajectories of the uncontrolled nonlinear system, $u = 0$, are shown in Figure 5.13 for the initial conditions $x_1(0) = 1$ and $x_2(0) = 0$. A phase portrait of the uncontrolled nonlinear system is shown in Figure 5.14. The linearized model about $\boldsymbol{x} = \boldsymbol{0}$, $u = 0$ has the form

$$\frac{d}{dt}\Delta\boldsymbol{x} = \begin{bmatrix} 0 & 1 \\ 9.8 & -1 \end{bmatrix} \Delta\boldsymbol{x} + \begin{bmatrix} 0 \\ 1 \end{bmatrix} \Delta u,$$
$$\Delta y = [1 \quad 0] \Delta\boldsymbol{x}. \tag{5.40}$$

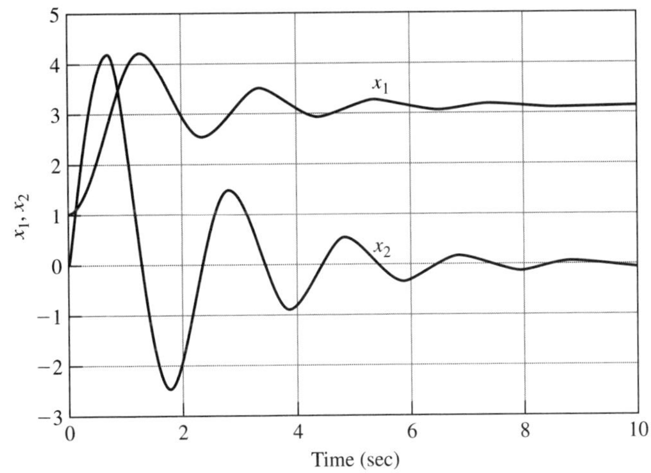

Figure 5.13 Plots of $y = x_1$ and x_2 versus time of the uncontrolled nonlinear system of Example 5.13.

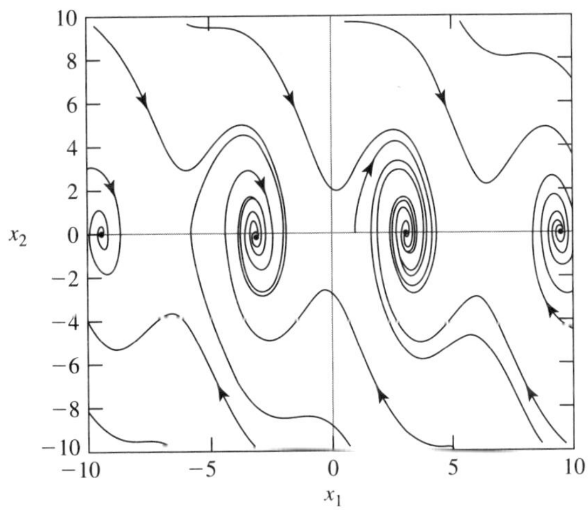

Figure 5.14 A phase portrait of the uncontrolled nonlinear system of Example 5.13.

Plots of $y = x_1$ and x_2 versus time are shown in Figure 5.15. A phase portrait of the linearized system is shown in Figure 5.16. Let

$$J = \int_0^\infty (y^2 + u^2)\, dt.$$

We shall find the linear state-feedback control law $u = -\boldsymbol{kx}$ that minimizes J subject to the equations given by (5.40). We have

$$\boldsymbol{Q} = \boldsymbol{c}^T \boldsymbol{c} = \begin{bmatrix} 1 & 0 \\ 0 & 0 \end{bmatrix} \quad \text{and} \quad \boldsymbol{R} = [1].$$

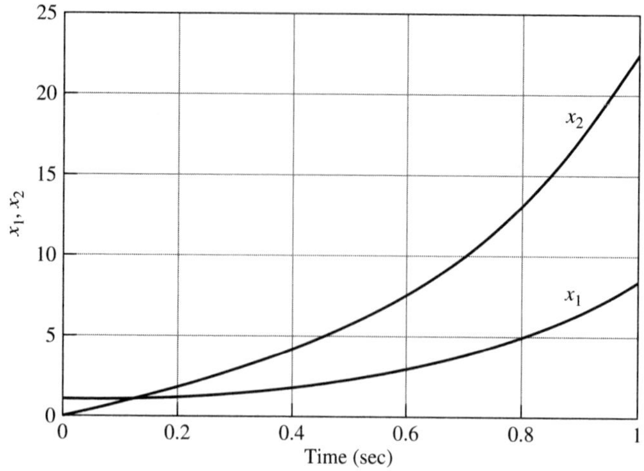

Figure 5.15 Plots of x_1 and x_2 versus time of the linearized system of Example 5.13.

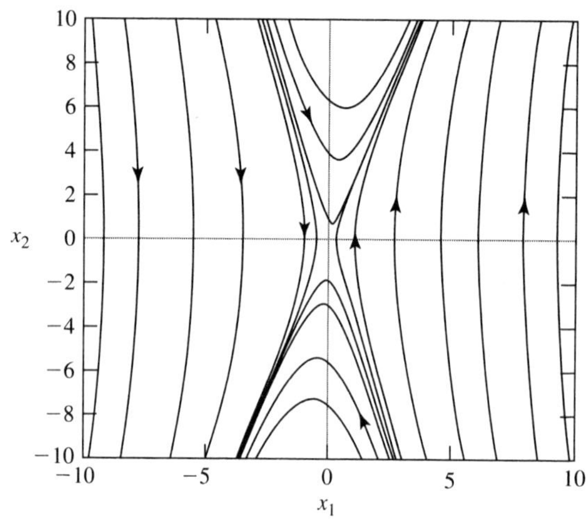

Figure 5.16 A phase portrait of the linearized system of Example 5.13.

Solving the associated Riccati equation yields

$$\boldsymbol{P} = \begin{bmatrix} 72.3371 & 19.6509 \\ 19.6509 & 5.3484 \end{bmatrix}. \tag{5.41}$$

Hence,

$$\boldsymbol{k} = [19.6509 \quad 5.3484] \tag{5.42}$$

and

$$\boldsymbol{A}_c = \begin{bmatrix} 0 & 1 \\ -9.8509 & -6.3484 \end{bmatrix}.$$

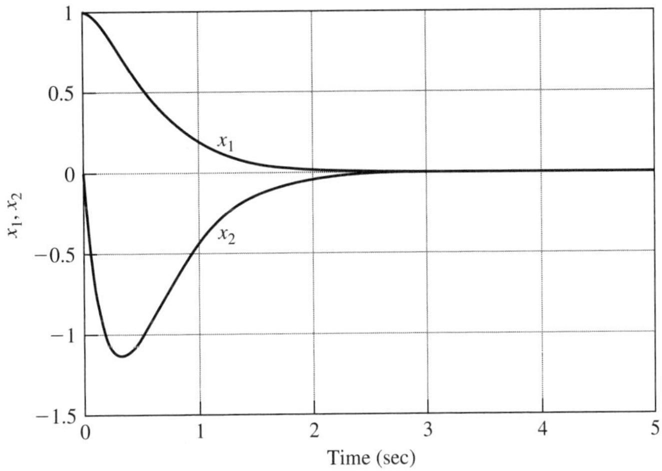

Figure 5.17 Plots of x_1 and x_2 versus time of the linear closed-loop system of Example 5.13.

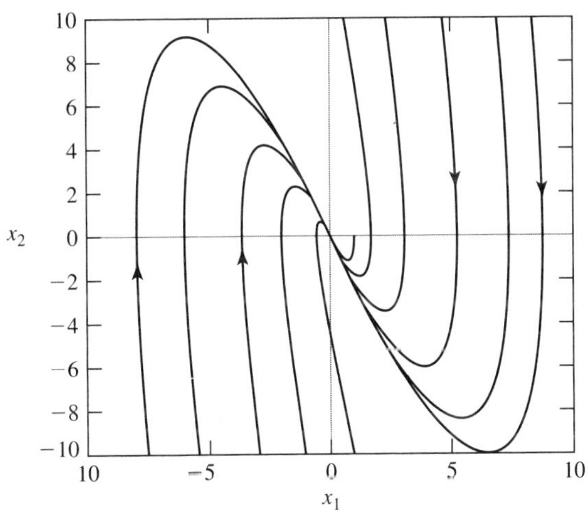

Figure 5.18 A phase portrait of the linear closed-loop system of Example 5.13.

Plots of x_1 and x_2 versus time of the closed-loop linear system

$$\dot{\boldsymbol{x}} = (\boldsymbol{A} - \boldsymbol{bk})\,\boldsymbol{x} = \boldsymbol{A}_c\boldsymbol{x}$$

$$y = x_1,$$

when the initial conditions are $x_1(0) = 1$ and $x_2(0) = 0$, are shown in Figure 5.17. A phase portrait of the closed-loop linear system is shown in Figure 5.18. We now apply the above optimal controller to the nonlinear model. Plots of x_1 and x_2 versus time of the closed-loop nonlinear systems are shown in Figure 5.19. A phase portrait of the nonlinear closed-loop system is shown in Figure 5.20. The closed-loop poles of the linearized system—that is, the eigenvalues of $\boldsymbol{A}_c$—are $\lambda_1 = -2.7003$ and $\lambda_2 = -3.6481$.

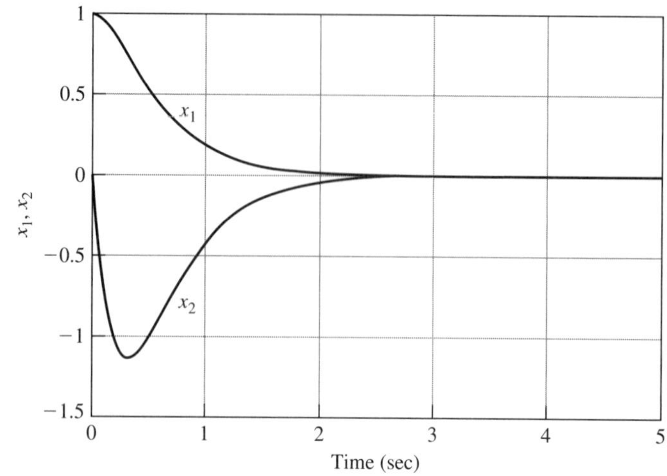

Figure 5.19 Plots of x_1 and x_2 of the nonlinear closed-loop system of Example 5.13.

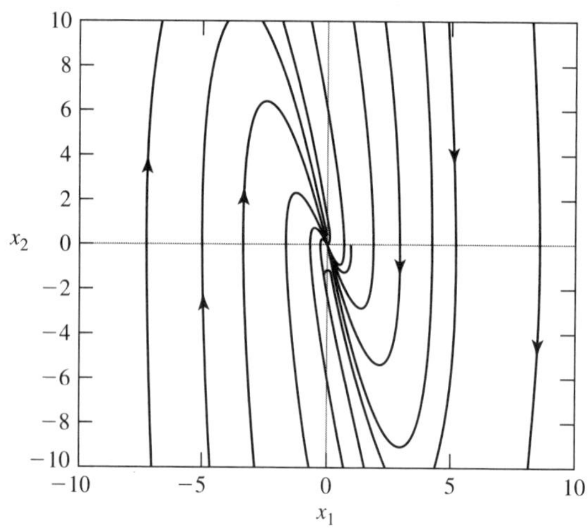

Figure 5.20 A phase portrait of the nonlinear closed-loop system of Example 5.13.

◆ Example 5.14

We use the data from the previous example to find the weight matrix $\boldsymbol{Q}_2$ and the corresponding $\boldsymbol{k}_2$ so that λ_2 is shifted to $s_2 = -5$; that is, the new poles of the closed-loop system

$$\dot{\boldsymbol{x}} = (\boldsymbol{A}_C - \boldsymbol{b}\boldsymbol{k}_2)\boldsymbol{x}$$

are $s_1 = \lambda_1$ and $s_2 = -5$. For the control law $u = -\boldsymbol{k}\boldsymbol{x} + r$, the linear closed-loop system is

$$\begin{aligned}\dot{\boldsymbol{x}} &= \begin{bmatrix} 0 & 1 \\ -9.8509 & -6.3484 \end{bmatrix} \boldsymbol{x} + \begin{bmatrix} 0 \\ 1 \end{bmatrix} u \\ &= \boldsymbol{A}_c \boldsymbol{x} + \boldsymbol{b} r,\end{aligned}$$

where for simplicity we used $\boldsymbol{x}$ rather than $\Delta \boldsymbol{x}$, and u instead of Δu. The eigenvalues of $\boldsymbol{A}_c$ are $\lambda_1 = -2.7003$ and $\lambda_2 = -3.6481$. We have $\lambda_1 \neq \lambda_2$ and thus the matrix $\boldsymbol{A}_c$ is diagonalizable. Note that different eigenvalues are only a sufficient condition for diagonalizability. We construct the similarity transformation using the eigenvectors of $\boldsymbol{A}_c$. The eigenvectors of $\boldsymbol{A}_c$ are

$$\boldsymbol{v}_1 = \begin{bmatrix} 0.3473 \\ -0.9378 \end{bmatrix} \quad \text{and} \quad \boldsymbol{v}_2 = \begin{bmatrix} -0.2644 \\ -0.9644 \end{bmatrix}.$$

Let $\boldsymbol{x} = \boldsymbol{T}\boldsymbol{z}$, where $\boldsymbol{T} = [\boldsymbol{v}_1 \ \ \boldsymbol{v}_2]$. Then, $\boldsymbol{A}_c$ in the new coordinates has the form

$$\tilde{\boldsymbol{A}}_c = \boldsymbol{T}^{-1} \boldsymbol{A}_c \boldsymbol{T} = \begin{bmatrix} -2.7003 & 0 \\ 0 & -3.6481 \end{bmatrix}$$

and

$$\tilde{\boldsymbol{b}} = \boldsymbol{T}^{-1} \boldsymbol{b} = \begin{bmatrix} 3.0383 \\ 3.9912 \end{bmatrix}.$$

Next, we compute

$$\boldsymbol{T}^{-1} \boldsymbol{b} \boldsymbol{R}^{-1} \boldsymbol{b}^T \boldsymbol{T}^{-T} = \begin{bmatrix} 9.2312 & 12.1264 \\ 12.1264 & 15.9296 \end{bmatrix}.$$

Hence,

$$\tilde{q}_{22} = \frac{s_2^2 - \lambda_2^2}{15.9296} = 0.7339.$$

Therefore,

$$\tilde{\boldsymbol{Q}}_2 = \begin{bmatrix} 0 & 0 \\ 0 & 0.7339 \end{bmatrix} = \boldsymbol{T}^T \boldsymbol{Q}_2 \boldsymbol{T}$$

and

$$\boldsymbol{Q}_2 = \boldsymbol{T}^{-T} \tilde{\boldsymbol{Q}}_2 \boldsymbol{T}^{-1} = \begin{bmatrix} 85.2450 & 31.5686 \\ 31.5686 & 11.6907 \end{bmatrix}.$$

Solving the ARE with $\boldsymbol{A}_c$, $\boldsymbol{b}$, $\boldsymbol{Q}_2$, and $\boldsymbol{R}$ yields

$$\boldsymbol{P}_2 = \begin{bmatrix} 9.8572 & 3.6504 \\ 3.6504 & 1.3518 \end{bmatrix}$$

and

$$\boldsymbol{k}_2 = [3.6504 \quad 1.3518].$$

As a check, if we take

$$Q_{new} = Q + Q_2 = \begin{bmatrix} 86.2450 & 31.5686 \\ 31.5686 & 11.6907 \end{bmatrix},$$

where $Q = c^T c$ as in Example 5.13, and $R_{new} = R$, then solving the associated ARE using A, rather then A_c, we obtain

$$P_{new} = \begin{bmatrix} 82.1943 & 23.3013 \\ 23.3013 & 6.7002 \end{bmatrix} = P + P_2,$$

where P is given by (5.41), and

$$k_{new} = [23.3013 \quad 6.7002] = k + k_2,$$

where k is the optimal gain given by (5.42).

◆ Example 5.15

For the linearized model given by (5.40) and $R = [2]$, we will now determine Q and the corresponding k that result in the closed-loop system $\dot{x} = (A - bk)x$ with poles located at $s_1 = -4$ and $s_2 = -5$, where

$$A = \begin{bmatrix} 0 & 1 \\ 9.8 & -1 \end{bmatrix}, \qquad b = \begin{bmatrix} 0 \\ 1 \end{bmatrix}.$$

We shift poles into the desired locations one by one. We first diagonalize A using similarity transformation constructed from eigenvectors of A. The eigenvalues of A are $\lambda_1 = 2.6702$ and $\lambda_2 = -3.6702$. We denote the corresponding eigenvectors by v_1 and v_2, respectively. We obtain

$$T^{-1}AT = \begin{bmatrix} 2.6702 & 0 \\ 0 & -3.6702 \end{bmatrix},$$

where

$$T = [v_1 \quad v_2] = \begin{bmatrix} 0.3507 & -0.2629 \\ 0.9365 & 0.9648 \end{bmatrix}.$$

We will now construct k_1 so that λ_1 is shifted to $s_1 = -4$, while λ_2 remains unchanged; that is, the eigenvalues of $A - bk_1$ become s_1 and λ_2. For this, we first compute

$$F = T^{-1}bR^{-1}T^T b^T = \begin{bmatrix} 0.1011 & 0.1349 \\ 0.1349 & 0.1800 \end{bmatrix}.$$

Hence, $f_{11} = 0.1011$,

$$\tilde{q}_{11} = \frac{4^2 - (2.6702)^2}{0.1011} = 87.7352,$$

and

$$\tilde{\boldsymbol{Q}}_1 = \begin{bmatrix} 87.7352 & 0 \\ 0 & 0 \end{bmatrix} = \boldsymbol{T}^T \boldsymbol{Q}_1 \boldsymbol{T}.$$

Hence,

$$\boldsymbol{Q}_1 = \boldsymbol{T}^{-T} \tilde{\boldsymbol{Q}}_1 \boldsymbol{T}^{-1} = \begin{bmatrix} 239.0024 & 65.1202 \\ 65.1202 & 17.7431 \end{bmatrix}.$$

Solving the associated ARE, using $\boldsymbol{R} = [2]$, yields

$$\boldsymbol{P}_1 = \begin{bmatrix} 179.7014 & 48.9626 \\ 48.9626 & 13.3407 \end{bmatrix}.$$

Therefore,

$$\boldsymbol{k}_1 = [24.4813 \quad 6.6703]$$

and

$$\boldsymbol{A}_1 = \boldsymbol{A} - \boldsymbol{b}\boldsymbol{k}_1 = \begin{bmatrix} 0 & 1 \\ -14.6813 & -7.6703 \end{bmatrix}.$$

We will now construct $\boldsymbol{k}_2$ that shifts λ_2 to s_2 leaving s_1 unchanged. We first diagonalize $\boldsymbol{A}_1$ using its eigenvectors. Note that $\boldsymbol{A}_1$ and $\boldsymbol{A}$ share the common eigenvector $\boldsymbol{v}_2$. The other eigenvector of $\boldsymbol{A}_1$ corresponding to its eigenvalue s_1 is $\boldsymbol{u}_1 = [-0.2425 \ \ 0.9701]^T$. We have

$$\boldsymbol{T}^{-1} \boldsymbol{A}_1 \boldsymbol{T} = \begin{bmatrix} s_1 & 0 \\ 0 & \lambda_2 \end{bmatrix},$$

where

$$\boldsymbol{T} = [\boldsymbol{u}_1 \quad \boldsymbol{v}_2] = \begin{bmatrix} -0.2425 & 0.2629 \\ 0.9701 & -0.9648 \end{bmatrix}.$$

We next compute

$$\boldsymbol{T}^{-1} \boldsymbol{b} \boldsymbol{R}^{-1} \boldsymbol{b}^T \boldsymbol{T}^{-T} = \begin{bmatrix} 78.0607 & 72.0157 \\ 72.0157 & 66.4389 \end{bmatrix}.$$

Thus, $\tilde{r}_{22} = 66.4389$,

$$\tilde{q}_{22} = \frac{5^2 - (-3.6702)^2}{66.4389} = 0.1735,$$

and

$$\tilde{\boldsymbol{Q}}_2 = \begin{bmatrix} 0 & 0 \\ 0 & 0.1735 \end{bmatrix} = \boldsymbol{T}^T \boldsymbol{Q}_2 \boldsymbol{T}.$$

From the above, we obtain

$$\boldsymbol{Q}_2 = \boldsymbol{T}^{-T} \tilde{\boldsymbol{Q}}_2 \boldsymbol{T}^{-1} = \begin{bmatrix} 368.9003 & 92.2211 \\ 92.2211 & 23.0543 \end{bmatrix}.$$

Solving the associated ARE, we get

$$P_2 = \begin{bmatrix} 42.5495 & 10.6369 \\ 10.6369 & 2.6591 \end{bmatrix}$$

and

$$k_2 = [5.3185 \quad 1.3296].$$

The final weight matrix Q that results in shifting poles into the desired locations is

$$Q = Q_1 + Q_2 = \begin{bmatrix} 607.9027 & 157.3413 \\ 157.3413 & 40.7974 \end{bmatrix}.$$

The corresponding P is

$$P = P_1 + P_2 = \begin{bmatrix} 222.2509 & 59.5996 \\ 59.5996 & 15.9998 \end{bmatrix}.$$

The corresponding optimal gain vector k is

$$k = k_1 + k_2 = [29.7998 \quad 7.9999].$$

5.3.4 Optimal Saturating Controllers

Asymptotic stability of an equilibrium state of a nonlinear system is a local property. This means that the system's trajectories starting in a neighborhood of the equilibrium state will approach this equilibrium asymptotically. Often, we are interested in determining how far from the equilibrium state the system's trajectories can start and still converge to the equilibrium asymptotically. A region around an equilibrium state within which all trajectories approach the equilibrium is called a *region of attraction,* or *region of asymptotic stability* (RAS). Finding the exact RAS analytically can be a challenging task. A tool often used to estimate the exact RAS is a Lyapunov function. Specifically, if $V = V(x)$ is a Lyapunov function of the equilibrium state of interest—that is, V is positive definite with respect to the equilibrium state in a domain $\mathcal{S}$ and the time derivative of V evaluated on the trajectories of the system is negative definite in $\mathcal{S}$—then the set

$$\Omega_c = \{x : x \in \mathcal{S},\, V(x) \le c\}$$

is an estimate of the exact RAS. In other words, a region of attraction is a set of states x for which $\dot{V}(x) < 0$ and $V(x) \le c$ for some positive constant c. The set Ω_c may be a conservative estimate, often a crude underestimate, of the true RAS. In order to obtain the least conservative estimate of the domain of attraction, corresponding to the given Lyapunov function, we could try to maximize the size of Ω_c by finding the largest V contour that can just fit on the equilibrium side of the surface described by $\dot{V}(x) = 0$. This can be accomplished by solving the optimization problem for c such that

$$c = \min_{\dot{V}(x)=0} V(x). \tag{5.43}$$

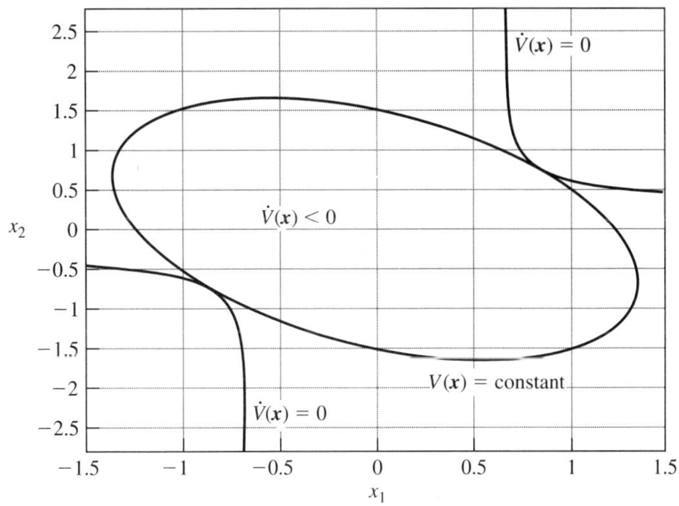

Figure 5.21 Estimating the region of asymptotic stability.

Solving the above optimization problem corresponds to maximizing the size of the region $\{\boldsymbol{x} : V(\boldsymbol{x}) \leq c\}$ in which $\dot{V}(\boldsymbol{x}) < 0$ by making the boundary of $\{\boldsymbol{x} : V(\boldsymbol{x}) \leq c\}$ to be tangent to the boundary of $\{\boldsymbol{x} : \dot{V}(\boldsymbol{x}) < 0\}$. This is illustrated in Figure 5.21.

We next estimate the RAS for single-input linear dynamical systems with saturating linear optimal control. Our development follows that of Friedland [91, Section 5.1]. We will use such controllers in Chapter 11 to stabilize chaotic systems. We consider a single-input linear system model,

$$\dot{\boldsymbol{x}} = \boldsymbol{A}\boldsymbol{x} + \boldsymbol{b}U_{max}\,\text{sat}(u/U_{max}), \tag{5.44}$$

where $\boldsymbol{x} \in \mathbb{R}^n$ and

$$\text{sat}(z) = \begin{cases} 1 & \text{if } \ z \geq 1 \\ z & \text{if } \ |z| < 1 \\ -1 & \text{if } \ z \leq -1. \end{cases}$$

In the linear region, $|u| \leq U_{max}$, (5.44) becomes $\dot{\boldsymbol{x}} = \boldsymbol{A}\boldsymbol{x} + \boldsymbol{b}u$. We will consider using the control law of the form

$$u = U_{max}\,\text{sat}(-\boldsymbol{k}\boldsymbol{x}/U_{max}). \tag{5.45}$$

When $|\boldsymbol{k}\boldsymbol{x}| \leq U_{max}$, the control law given by (5.45) becomes a linear state-feedback controller, $u = -\boldsymbol{k}\boldsymbol{x}$. For simplicity of notation, we can assume that $U_{max} = 1$. Indeed, we can always scale the input vector $\boldsymbol{b}$ so that $U_{max} = 1$. The gain vector $\boldsymbol{k}$ is obtained by solving the associated LQR problem with the performance index

$$J = \int_0^\infty (\boldsymbol{x}^T \boldsymbol{Q}\boldsymbol{x} + ru^2)\, dt,$$

where we assume that the pair $(\boldsymbol{A}, \boldsymbol{b})$ is reachable, $\boldsymbol{Q} = \boldsymbol{Q}^T > 0$, and $r > 0$. Solving the algebraic

Riccati equation,

$$A^T P + P A + Q - P b b^T P / r = O, \tag{5.46}$$

we use its solution, $P = P^T > 0$, to construct the gain vector,

$$k = b^T P / r. \tag{5.47}$$

We then proceed with evaluating of $\frac{d}{dt} V = \frac{d}{dt} x^T P x$ on the trajectories of the closed-loop system,

$$\begin{aligned} \dot{V} &= 2x^T P \dot{x} \\ &= x^T (A^T P + P A) x - 2x^T P b \operatorname{sat}(k x). \end{aligned} \tag{5.48}$$

By (5.47),

$$b^T P = r k. \tag{5.49}$$

Using the above and (5.46) gives

$$A^T P + P A = r k^T k - Q. \tag{5.50}$$

Substituting (5.50) and (5.49) into (5.48) yields

$$\boxed{\dot{V} = -x^T Q x - 2 r k x \operatorname{sat}(k x) + r (k x)^2} \tag{5.51}$$

For x such that $|kx| \leq 1$, the above becomes

$$\dot{V} = -x^T Q x - r (k x)^2,$$

which is negative definite. Thus, the origin is (locally) asymptotically stable for the closed-loop system.

An estimate of the RAS is a set of states bounded by the surface V = constant for which $\dot{V} < 0$. To find as large an estimate as possible, we will attempt to find the surface $\dot{V} = 0$ and find $c = \min V(x)$ on that surface. Then, the region bounded by $V(x) = c$ is an estimate of the RAS. Solving the above problem amounts to solving the optimization problem,

$$\min_x x^T P x$$

$$\text{subject to} \quad -x^T Q x - 2 r k x \operatorname{sat}(k x) + r (k x)^2 = 0. \tag{5.52}$$

Solving the above problem is not a simple task. Instead, we will solve an easier problem resulting in a more conservative estimate of the RAS. We first rewrite (5.52) as

$$y^2 - 2y \operatorname{sat}(y) - x^T Q x / r = 0, \tag{5.53}$$

where $y = kx$. We will investigate the case when $|y| \geq 1$ because for $|y| < 1$, we have $\dot{V} < 0$ and we are interested in the solution to the above when $\dot{V} = 0$. When $y \geq 1$, (5.53) becomes

$$y^2 - 2y - x^T Q x / r = 0. \tag{5.54}$$

because for $y \geq 1$, $\operatorname{sat}(y) = 1$. The solution to (5.54) for $y \geq 1$ is

$$y = 1 + \sqrt{1 + x^T Q x / r} > 2, \tag{5.55}$$

while for $y \leq -1$ the solution to (5.53) has the form

$$y = -1 - \sqrt{1 + \boldsymbol{x}^T \boldsymbol{Q}\boldsymbol{x}/r} < -2. \tag{5.56}$$

The above solutions satisfy the inequality

$$|y| \geq 2.$$

Thus, we have $\dot{V} < 0$ for $|y| = |\boldsymbol{k}\boldsymbol{x}| \leq 2$ and hence the RAS includes the region

$$\{\boldsymbol{x} : V(\boldsymbol{x}) \leq \tilde{c}\},$$

where $\tilde{c}$ is obtained by solving the optimization problem

$$\boxed{\tilde{c} = \max_{|\boldsymbol{k}\boldsymbol{x}| \leq 2} \boldsymbol{x}^T \boldsymbol{P}\boldsymbol{x}}$$

We will use the Lagrange multipliers method to solve the above constrained optimization problem. We will solve the equivalent problem,

$$\max \boldsymbol{x}^T \boldsymbol{P}\boldsymbol{x}$$
$$\text{subject to} \quad (\boldsymbol{k}\boldsymbol{x})^2 = 4.$$

The Lagrangian for the above problem is

$$l(\boldsymbol{x}, \lambda) = \boldsymbol{x}^T \boldsymbol{P}\boldsymbol{x} + \lambda(4 - (\boldsymbol{k}\boldsymbol{x})^2).$$

Applying to the Lagrangian l the first-order necessary conditions for unconstrained minimization gives

$$D_{\boldsymbol{x}} l(\boldsymbol{x}, \lambda) = 2\boldsymbol{x}^T \boldsymbol{P} - 2\lambda(\boldsymbol{k}\boldsymbol{x})\boldsymbol{k} = \boldsymbol{0}^T, \tag{5.57}$$

$$D_{\lambda} l(\boldsymbol{x}, \lambda) = 4 - (\boldsymbol{k}\boldsymbol{x})^2 = 0. \tag{5.58}$$

From (5.57), we obtain

$$\boldsymbol{x}^T \boldsymbol{P}\boldsymbol{x} = \lambda(\boldsymbol{k}\boldsymbol{x})^2.$$

Because $(\boldsymbol{k}\boldsymbol{x})^2 = 4$,

$$\lambda = \tfrac{1}{4}\boldsymbol{x}^T \boldsymbol{P}\boldsymbol{x},$$

and therefore

$$\tilde{c} = \boldsymbol{x}^T \boldsymbol{P}\boldsymbol{x} = 4\lambda.$$

Again, using (5.57), we obtain

$$\boldsymbol{x}^T \boldsymbol{P} = \lambda(\boldsymbol{k}\boldsymbol{x})\boldsymbol{k}.$$

Postmultiplying both sides of the above by $\boldsymbol{P}^{-1}\boldsymbol{k}^T$ gives

$$\boldsymbol{x}^T \boldsymbol{k}^T = \boldsymbol{k}\boldsymbol{x} = \lambda(\boldsymbol{k}\boldsymbol{x})\boldsymbol{k}\boldsymbol{P}^{-1}\boldsymbol{k}^T. \tag{5.59}$$

Note that $\boldsymbol{k}\boldsymbol{x} \neq 0$ because $(\boldsymbol{k}\boldsymbol{x})^2 = 4$. Hence, from (5.59),

$$\lambda = \frac{1}{\boldsymbol{k}\boldsymbol{P}^{-1}\boldsymbol{k}^T}.$$

Substituting (5.47) into the above gives

$$\lambda = \frac{1}{\boldsymbol{k}\boldsymbol{P}^{-1}\boldsymbol{k}^T} = \frac{r^2}{\boldsymbol{b}^T\boldsymbol{P}\boldsymbol{b}}.$$

Hence,

$$\boxed{\tilde{c} = 4\lambda = \frac{4r^2}{\boldsymbol{b}^T\boldsymbol{P}\boldsymbol{b}}}$$

5.3.5 Linear Quadratic Regulator for Discrete Systems on an Infinite Time Interval

Consider a discrete linear plant model,

$$\boldsymbol{x}(k+1) = \boldsymbol{A}\boldsymbol{x}(k) + \boldsymbol{B}\boldsymbol{u}(k), \qquad k = 0, 1, 2, \ldots, \tag{5.60}$$

with a specified initial condition $\boldsymbol{x}(0) = \boldsymbol{x}_0$, where $\boldsymbol{x}(k) \in \mathbb{R}^n$ and $\boldsymbol{u}(k) \in \mathbb{R}^m$. We assume that the pair $(\boldsymbol{A}, \boldsymbol{B})$ is reachable. Our objective is to construct a stabilizing linear state-feedback controller,

$$\boldsymbol{u}(k) = -\boldsymbol{K}\boldsymbol{x}(k),$$

that minimizes the quadratic performance index

$$J(\boldsymbol{u}) = \sum_{k=0}^{\infty} \{\boldsymbol{x}^T(k)\boldsymbol{Q}\boldsymbol{x}(k) + \boldsymbol{u}^T(k)\boldsymbol{R}\boldsymbol{u}(k)\}, \tag{5.61}$$

where $\boldsymbol{Q} = \boldsymbol{Q}^T \geq 0$, and $\boldsymbol{R} = \boldsymbol{R}^T > 0$. We assume that the components of the control vector are unconstrained. An optimal control law will be denoted $\boldsymbol{u}^*$.

Because we seek the controller among the ones that stabilize the system (5.60), the closed-loop system,

$$\boldsymbol{x}(k+1) = (\boldsymbol{A} - \boldsymbol{B}\boldsymbol{K})\boldsymbol{x}(k), \tag{5.62}$$

is asymptotically stable; that is, the eigenvalues of the matrix $\boldsymbol{A} - \boldsymbol{B}\boldsymbol{K}$ are in the open unit disk. This implies that there exists a Lyapunov function, $V(\boldsymbol{x}(k)) = \boldsymbol{x}^T(k)\boldsymbol{P}\boldsymbol{x}(k)$, for the system (5.62). Therefore the first forward difference, $\Delta V(\boldsymbol{x}(k)) = V(\boldsymbol{x}(k+1)) - V(\boldsymbol{x}(k))$, evaluated on the trajectories of (5.62) is negative definite. We now state and prove a sufficient condition for $\boldsymbol{u}(k) = -\boldsymbol{K}\boldsymbol{x}(k)$ to be optimal.

Theorem 5.4 If the state feedback controller, $\boldsymbol{u}(k) = -\boldsymbol{K}\boldsymbol{x}(k)$, is such that

$$\min_{\boldsymbol{u}}(\Delta V(\boldsymbol{x}(k)) + \boldsymbol{x}^T(k)\boldsymbol{Q}\boldsymbol{x}(k) + \boldsymbol{u}^T(k)\boldsymbol{R}\boldsymbol{u}(k)) = 0, \tag{5.63}$$

then it is optimal.

Proof We can represent (5.63) as

$$\Delta V(\boldsymbol{x}(k))|_{u=u^*} + \boldsymbol{x}^T(k)\,\boldsymbol{Q}\boldsymbol{x}(k) + \boldsymbol{u}^{*T}(k)\,\boldsymbol{R}\boldsymbol{u}^*(k) = 0.$$

Hence,

$$\Delta V(\boldsymbol{x}(k))|_{u=u^*} = (V(\boldsymbol{x}(k+1)) - V(\boldsymbol{x}(k))|_{u=u^*} = -\boldsymbol{x}^T(k)\,\boldsymbol{Q}\boldsymbol{x}(k) - \boldsymbol{u}^{*T}(k)\,\boldsymbol{R}\boldsymbol{u}^*(k).$$

We sum both sides from $k = 0$ to $k = \infty$ to get

$$V(\boldsymbol{x}(\infty)) - V(\boldsymbol{x}(0)) = -\sum_{k=0}^{\infty}(\boldsymbol{x}^T(k)\,\boldsymbol{Q}\boldsymbol{x}(k) + \boldsymbol{u}^{*T}(k)\,\boldsymbol{R}\boldsymbol{u}^*(k)).$$

Because by assumption the closed-loop system is asymptotically stable, we have $\boldsymbol{x}(\infty) = \boldsymbol{0}$. Hence,

$$V(\boldsymbol{x}(0)) = \boldsymbol{x}_0^T\boldsymbol{P}\boldsymbol{x}_0 = \sum_{k=0}^{\infty}(\boldsymbol{x}^T(k)\,\boldsymbol{Q}\boldsymbol{x}(k) + \boldsymbol{u}^{*T}(k)\,\boldsymbol{R}\boldsymbol{u}^*(k)).$$

Thus, if a linear stabilizing controller satisfies the hypothesis of the theorem, then the value of the performance index (5.61) resulting from applying this controller is

$$J = \boldsymbol{x}_0^T\boldsymbol{P}\boldsymbol{x}_0.$$

To show that such a controller is an optimal one, we will use a proof by contradiction. We assume that the hypothesis of the theorem holds but the controller $\boldsymbol{u}^*$ that satisfies this hypothesis is not optimal; that is, there is a controller $\tilde{\boldsymbol{u}}$ such that

$$J(\tilde{\boldsymbol{u}}) < J(\boldsymbol{u}^*). \tag{5.64}$$

The hypothesis of the theorem implies that

$$\Delta V(\boldsymbol{x}(k))|_{u=\tilde{u}} + \boldsymbol{x}^T(k)\,\boldsymbol{Q}\boldsymbol{x}(k) + \tilde{\boldsymbol{u}}^T(k)\,\boldsymbol{R}\tilde{\boldsymbol{u}}(k) \geq 0. \tag{5.65}$$

We represent (5.65) as

$$\Delta V(\boldsymbol{x}(k))|_{u=\tilde{u}} \geq -\boldsymbol{x}^T(k)\,\boldsymbol{Q}\boldsymbol{x}(k) - \tilde{\boldsymbol{u}}^T(k)\,\boldsymbol{R}\tilde{\boldsymbol{u}}(k).$$

Summing the above from $k = 0$ to $k = \infty$ yields

$$V(\boldsymbol{x}(0)) \leq \sum_{k=0}^{\infty}(\boldsymbol{x}^T(k)\,\boldsymbol{Q}\boldsymbol{x}(k) + \tilde{\boldsymbol{u}}^T(k)\,\boldsymbol{R}\tilde{\boldsymbol{u}}(k));$$

that is,

$$J(\tilde{\boldsymbol{u}}) \geq J(\boldsymbol{u}^*),$$

which contradicts (5.64). The proof is complete.

Finding an optimal controller involves finding an appropriate quadratic Lyapunov function $V(\boldsymbol{x}) = \boldsymbol{x}^T\boldsymbol{P}\boldsymbol{x}$, which then is used to construct the optimal controller. We first find $\boldsymbol{u}^*$ that minimizes the function

$$f = f(\boldsymbol{u}(k)) = \Delta V(\boldsymbol{x}(k)) + \boldsymbol{x}^T(k)\boldsymbol{Q}\boldsymbol{x}(k) + \boldsymbol{u}^T(k)\boldsymbol{R}\boldsymbol{u}(k). \tag{5.66}$$

To proceed, we perform preliminary manipulations on the function f. In particular, we use the definition of ΔV and substitute into (5.66) the system equation (5.60) to obtain

$$\begin{aligned} f(\boldsymbol{u}(k)) &= \boldsymbol{x}^T(k+1)\boldsymbol{P}\boldsymbol{x}(k+1) - \boldsymbol{x}^T(k)\boldsymbol{P}\boldsymbol{x}(k) + \boldsymbol{x}^T(k)\boldsymbol{Q}\boldsymbol{x}(k) + \boldsymbol{u}^T(k)\boldsymbol{R}\boldsymbol{u}(k) \\ &= (\boldsymbol{A}\boldsymbol{x}(k) + \boldsymbol{B}\boldsymbol{u}(k))^T\boldsymbol{P}(\boldsymbol{A}\boldsymbol{x}(k) + \boldsymbol{B}\boldsymbol{u}(k)) - \boldsymbol{x}^T(k)\boldsymbol{P}\boldsymbol{x}(k) + \boldsymbol{x}^T(k)\boldsymbol{Q}\boldsymbol{x}(k) \\ &\quad + \boldsymbol{u}^T(k)\boldsymbol{R}\boldsymbol{u}(k). \end{aligned} \tag{5.67}$$

Applying to the above function the first-order necessary condition for a relative minimum gives

$$\begin{aligned} \frac{\partial f(\boldsymbol{u}(k))}{\partial \boldsymbol{u}(k)} &= 2(\boldsymbol{A}\boldsymbol{x}(k) + \boldsymbol{B}\boldsymbol{u}(k))^T\boldsymbol{P}\boldsymbol{B} + 2\boldsymbol{u}^T(k)\boldsymbol{R} \\ &= 2\boldsymbol{x}^T(k)\boldsymbol{A}^T\boldsymbol{P}\boldsymbol{B} + 2\boldsymbol{u}^T(k)(\boldsymbol{R} + \boldsymbol{B}^T\boldsymbol{P}\boldsymbol{B}) \\ &= \boldsymbol{0}^T. \end{aligned}$$

The matrix $\boldsymbol{R} + \boldsymbol{B}^T\boldsymbol{P}\boldsymbol{B}$ is positive definite because $\boldsymbol{R}$ is positive definite, and therefore the matrix $\boldsymbol{R} + \boldsymbol{B}^T\boldsymbol{P}\boldsymbol{B}$ is invertible. Hence,

$$\boldsymbol{u}^*(k) = -(\boldsymbol{R} + \boldsymbol{B}^T\boldsymbol{P}\boldsymbol{B})^{-1}\boldsymbol{B}^T\boldsymbol{P}\boldsymbol{A}\boldsymbol{x}(k) = -\boldsymbol{K}\boldsymbol{x}(k), \tag{5.68}$$

where

$$\boldsymbol{K} = (\boldsymbol{R} + \boldsymbol{B}^T\boldsymbol{P}\boldsymbol{B})^{-1}\boldsymbol{B}^T\boldsymbol{P}\boldsymbol{A}.$$

Let

$$\boldsymbol{S} = \boldsymbol{R} + \boldsymbol{B}^T\boldsymbol{P}\boldsymbol{B}.$$

Then, we represent (5.68) as

$$\boldsymbol{u}^*(k) = -\boldsymbol{S}^{-1}\boldsymbol{B}^T\boldsymbol{P}\boldsymbol{A}\boldsymbol{x}(k). \tag{5.69}$$

We next check if $\boldsymbol{u}^*$ satisfies the second-order sufficient condition for a relative minimizer of the function (5.66). We have

$$\begin{aligned} &\frac{\partial^2}{\partial \boldsymbol{u}^2(k)}(\boldsymbol{x}^T(k+1)\boldsymbol{P}\boldsymbol{x}(k+1) - \boldsymbol{x}^T(k)\boldsymbol{P}\boldsymbol{x}(k) + \boldsymbol{x}^T(k)\boldsymbol{Q}\boldsymbol{x}(k) + \boldsymbol{u}^T(k)\boldsymbol{R}\boldsymbol{u}(k)) \\ &= \frac{\partial}{\partial \boldsymbol{u}(k)}(2\boldsymbol{x}^T(k)\boldsymbol{A}^T\boldsymbol{P}\boldsymbol{B} + 2\boldsymbol{u}^T(k)(\boldsymbol{R} + \boldsymbol{B}^T\boldsymbol{P}\boldsymbol{B})) \\ &= 2(\boldsymbol{R} + \boldsymbol{B}^T\boldsymbol{P}\boldsymbol{B}) \\ &> 0, \end{aligned} \tag{5.70}$$

that is, $\boldsymbol{u}^*$ satisfies the second-order sufficient condition for a relative minimizer of the function (5.66).

The optimal controller (5.69) can be constructed if we have found an appropriate positive definite matrix $\boldsymbol{P}$. Our next task is to devise a method that would allow us to compute the desired matrix $\boldsymbol{P}$. For this, we first find the equation describing the closed-loop system driven by the optimal controller given by (5.69):

$$\boldsymbol{x}(k+1) = (\boldsymbol{A} - \boldsymbol{B}\boldsymbol{S}^{-1}\boldsymbol{B}^T\boldsymbol{P}\boldsymbol{A})\boldsymbol{x}(k). \tag{5.71}$$

Our controller satisfies the hypothesis of Theorem 5.4, that is,

$$\boldsymbol{x}^T(k+1)\boldsymbol{P}\boldsymbol{x}(k+1) - \boldsymbol{x}^T(k)\boldsymbol{P}\boldsymbol{x}(k) + \boldsymbol{x}^T(k)\boldsymbol{Q}\boldsymbol{x}(k) + \boldsymbol{u}^{*T}(k)\boldsymbol{R}\boldsymbol{u}^*(k) = 0.$$

Substituting into (5.71) the expression for $\boldsymbol{u}^*$ given by (5.69), along with performing simple manipulations, gives

$$\begin{aligned}
&\boldsymbol{x}^T(k)(\boldsymbol{A} - \boldsymbol{B}\boldsymbol{S}^{-1}\boldsymbol{B}^T\boldsymbol{P}\boldsymbol{A})^T\boldsymbol{P}(\boldsymbol{A} - \boldsymbol{B}\boldsymbol{S}^{-1}\boldsymbol{B}^T\boldsymbol{P}\boldsymbol{A})\boldsymbol{x}(k) - \boldsymbol{x}^T(k)\boldsymbol{P}\boldsymbol{x}(k)\\
&\quad + \boldsymbol{x}^T(k)\boldsymbol{Q}\boldsymbol{x}(k) + \boldsymbol{x}^T(k)\boldsymbol{A}^T\boldsymbol{P}\boldsymbol{B}\boldsymbol{S}^{-1}\boldsymbol{R}\boldsymbol{S}^{-1}\boldsymbol{B}^T\boldsymbol{P}\boldsymbol{A}\boldsymbol{x}(k)\\
&= \boldsymbol{x}^T(k)\boldsymbol{A}^T\boldsymbol{P}\boldsymbol{A}\boldsymbol{x}(k) - \boldsymbol{x}^T(k)\boldsymbol{A}^T\boldsymbol{P}\boldsymbol{B}\boldsymbol{S}^{-1}\boldsymbol{B}^T\boldsymbol{P}\boldsymbol{A}\boldsymbol{x}(k)\\
&\quad - \boldsymbol{x}^T(k)\boldsymbol{A}^T\boldsymbol{P}\boldsymbol{B}\boldsymbol{S}^{-1}\boldsymbol{B}^T\boldsymbol{P}\boldsymbol{A}\boldsymbol{x}(k) + \boldsymbol{x}^T(k)\boldsymbol{A}^T\boldsymbol{P}\boldsymbol{B}\boldsymbol{S}^{-1}\boldsymbol{B}^T\boldsymbol{P}\boldsymbol{B}\boldsymbol{S}^{-1}\boldsymbol{B}^T\boldsymbol{P}\boldsymbol{A}\boldsymbol{x}(k)\\
&\quad - \boldsymbol{x}^T(k)\boldsymbol{P}\boldsymbol{x}(k) + \boldsymbol{x}^T(k)\boldsymbol{Q}\boldsymbol{x}(k) + \boldsymbol{x}^T(k)\boldsymbol{A}^T\boldsymbol{P}\boldsymbol{B}\boldsymbol{S}^{-1}\boldsymbol{R}\boldsymbol{S}^{-1}\boldsymbol{B}^T\boldsymbol{P}\boldsymbol{A}\boldsymbol{x}(k)\\
&= \boldsymbol{x}^T(k)\boldsymbol{A}^T\boldsymbol{P}\boldsymbol{A}\boldsymbol{x}(k) - \boldsymbol{x}^T(k)\boldsymbol{P}\boldsymbol{x}(k) + \boldsymbol{x}^T(k)\boldsymbol{Q}\boldsymbol{x}(k)\\
&\quad - 2\boldsymbol{x}^T(k)\boldsymbol{A}^T\boldsymbol{P}\boldsymbol{B}\boldsymbol{S}^{-1}\boldsymbol{B}^T\boldsymbol{P}\boldsymbol{A}\boldsymbol{x}(k)\\
&\quad + \boldsymbol{x}^T(k)\boldsymbol{A}^T\boldsymbol{P}\boldsymbol{B}\boldsymbol{S}^{-1}(\boldsymbol{R} + \boldsymbol{B}^T\boldsymbol{P}\boldsymbol{B})\boldsymbol{S}^{-1}\boldsymbol{B}^T\boldsymbol{P}\boldsymbol{A}\boldsymbol{x}(k)\\
&= \boldsymbol{x}^T(k)\boldsymbol{A}^T\boldsymbol{P}\boldsymbol{A}\boldsymbol{x}(k) - \boldsymbol{x}^T(k)\boldsymbol{P}\boldsymbol{x}(k) + \boldsymbol{x}^T(k)\boldsymbol{Q}\boldsymbol{x}(k)\\
&\quad - 2\boldsymbol{x}^T(k)\boldsymbol{A}^T\boldsymbol{P}\boldsymbol{B}\boldsymbol{S}^{-1}\boldsymbol{B}^T\boldsymbol{P}\boldsymbol{A}\boldsymbol{x}(k)\\
&\quad + \boldsymbol{x}^T(k)\boldsymbol{A}^T\boldsymbol{P}\boldsymbol{B}\boldsymbol{S}^{-1}\boldsymbol{B}^T\boldsymbol{P}\boldsymbol{A}\boldsymbol{x}(k)\\
&= \boldsymbol{x}^T(k)(\boldsymbol{A}^T\boldsymbol{P}\boldsymbol{A} - \boldsymbol{P} + \boldsymbol{Q} - \boldsymbol{A}^T\boldsymbol{P}\boldsymbol{B}\boldsymbol{S}^{-1}\boldsymbol{B}^T\boldsymbol{P}\boldsymbol{A})\boldsymbol{x}(k)\\
&= 0.
\end{aligned} \tag{5.72}$$

The above equation should hold for any $\boldsymbol{x}$. For this to be true we have to have

$$\boxed{\boldsymbol{A}^T\boldsymbol{P}\boldsymbol{A} - \boldsymbol{P} + \boldsymbol{Q} - \boldsymbol{A}^T\boldsymbol{P}\boldsymbol{B}\boldsymbol{S}^{-1}\boldsymbol{B}^T\boldsymbol{P}\boldsymbol{A} = \boldsymbol{O}} \tag{5.73}$$

The above equation is called the *discrete algebraic Riccati equation.*

5.4 Dynamic Programming

The objective of this section is to introduce the reader to the method of dynamic programming invented by Richard E. Bellman (1920–1984) in 1957. Dynamic programming can be used to synthesize optimal controllers for nonlinear, time-varying dynamical systems. We first present dynamic programming for discrete-time systems and then discuss the continuous version of dynamic programming.

5.4.1 Discrete-Time Systems

Dynamic programming is based on the principle of optimality (PO), which was formulated by Bellman as follows:

> An optimal policy has the property that whatever the initial state and initial decision are, the remaining decisions must constitute an optimal policy with regard to the state resulting from the first decision.

In other words, an optimal trajectory has the property that at an intermediate point, no matter how it was reached, the rest of the trajectory must coincide with an optimal trajectory as computed from this intermediate point as the initial point. We will justify the PO by contradiction. Suppose that we are given an optimal trajectory starting from initial state $\boldsymbol{x}_0 = \boldsymbol{x}(t_0)$ at time t_0 and

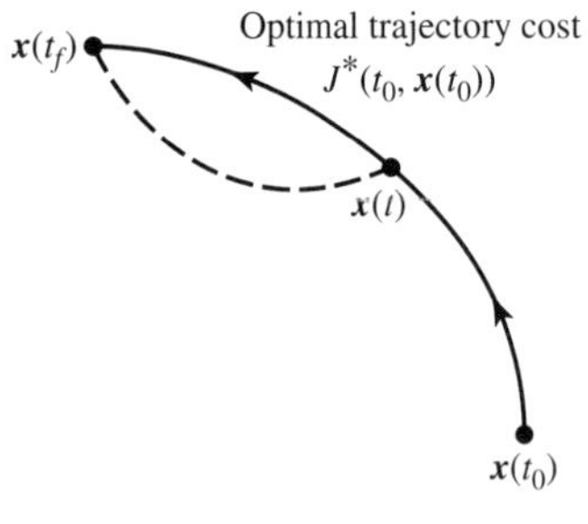

Figure 5.22 Illustration of the Principle of Optimality.

terminating at the final state $\boldsymbol{x}(t_f)$ at time t_f. Let $\boldsymbol{x}(t)$ be an intermediate state on the optimal trajectory and let $J^*(t, \boldsymbol{x}(t))$ be the optimal cost of traveling from $\boldsymbol{x}(t)$ to the final state $\boldsymbol{x}(t_f)$. Thus, $J^*(t_0, \boldsymbol{x}(t_0))$ is the optimal cost of traveling from $\boldsymbol{x}_0 = \boldsymbol{x}(t_0)$ to $\boldsymbol{x}(t_f)$. It then follows that the portion of the optimal trajectory from $\boldsymbol{x}(t)$ to $\boldsymbol{x}(t_f)$ is also optimal. For if some other path, indicated by the dashed curve in Figure 5.22, connecting $\boldsymbol{x}(t)$ and $\boldsymbol{x}(t_f)$ resulted in a smaller cost than the solid path, then this would provide a less expensive route from $\boldsymbol{x}_0$ to $\boldsymbol{x}(t_f)$, thus contradicting the optimality of the original trajectory.

We note that the PO seems to be already known to Johann Bernoulli, who stated in 1706 that "... all curves, which should give a maximum, preserve also in all their sections the laws of the same maximum." Here, the reader should interpret maximum as optimum.

Bellman's principle of optimality serves to "limit the number of potentially optimal control strategies that must be investigated" [182, p. 315]. The optimal control strategy is determined, using the PO, by working backward from the final stage.

We illustrate this "backwardness" with an example of a sequential decision-making process. We define a decision as the choice of alternative paths leaving a given node.

◆ Example 5.16

We use the PO to find the minimum cost path that starts at the node **a** and ends at any one of the nodes **k**, **l**, **m** in the routing network depicted in Figure 5.23. The travel costs are shown beside each path segment. The line segments can only be traversed from left to right. Applying the PO to this routing problem, we find the optimal path by performing a backward pass through the network. Let J_q^* be the minimum cost from a generic node **q** to any one of the three possible final nodes. Then, we have

$$J_g^* = 16, \qquad J_h^* = 8, \qquad J_i^* = 7, \qquad J_j^* = 6.$$

The next stage yields

$$J_d^* = \min\{3 + J_g^*, 11 + J_h^*\} = \min\{19, 19\} = 19.$$

Thus, in this case, we have two optimal paths from node **d**. We can take either **dgk** or **dhk** paths. We then compute optimal paths from nodes **e** and **f**. We have

$$J_e^* = \min\{6 + J_h^*, 10 + J_i^*\} = \min\{6 + 8, 10 + 7\} = 14$$

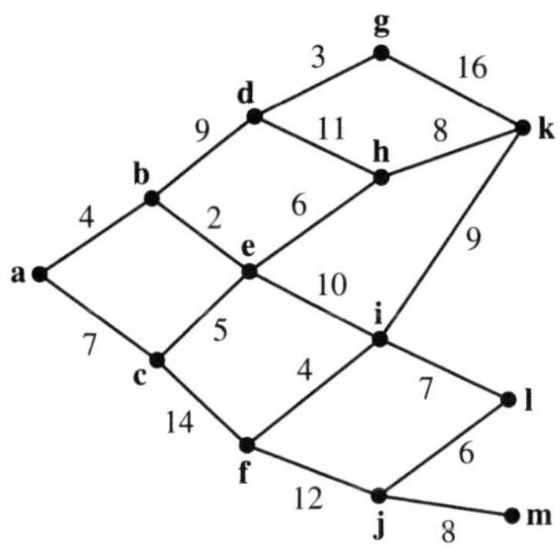

Figure 5.23 A routing network.

and

$$J_f^* = \min\{4 + J_i^*, 12 + J_j^*\} = \min\{4 + 7, 12 + 6\} = 11.$$

Thus, the optimal path emanating from node **e** is **ehk**, and the path from node **f** is **fil**. The next stage yields

$$J_b^* = \min\{9 + J_d^*, 2 + J_e^*\} = \min\{9 + 19, 2 + 14\} = 16,$$

and

$$J_c^* = \min\{5 + J_e^*, 14 + J_f^*\} = \min\{5 + 14, 14 + 11\} = 19.$$

Therefore, the optimal path starting from **b** is **behk**, and starting from **c** the optimal path is **cehk**. Finally, the initial, and hence final, stage yields

$$J_a^* = \min\{4 + J_b^*, 7 + J_c^*\} = \min\{4 + 16, 7 + 19\} = 20.$$

Hence, the optimal, minimum total cost route from **a** to any one of the three possible termination nodes **k**, **l**, or **m** is **abehk**. The total cost of this path is 20.

We now consider a class of discrete-time dynamical systems modeled by the difference equation

$$\boldsymbol{x}(k+1) = \boldsymbol{f}(k, \boldsymbol{x}(k), \boldsymbol{u}(k)), \tag{5.74}$$

with a given initial condition $\boldsymbol{x}(k_0) = \boldsymbol{x}_0$, and the associated performance index

$$J = J_0 = \Phi(N, \boldsymbol{x}(N)) + \sum_{k=0}^{N-1} F(k, \boldsymbol{x}(k), \boldsymbol{u}(k)).$$

Let $J_k^*(\boldsymbol{x}(k))$ denote the minimum cost of transferring the system from $\boldsymbol{x}(k)$ to the terminal point $\boldsymbol{x}(N)$. Because $\boldsymbol{x}(N)$ is the terminal point, we have $J_N^*(\boldsymbol{x}(N)) = \Phi(N, \boldsymbol{x}(N))$. Using the PO, we obtain the cost of transfer from $\boldsymbol{x}(N-1)$ to the terminal point:

$$\begin{aligned} J_{N-1}^*(\boldsymbol{x}(N-1)) &= \min_{\boldsymbol{u}(N-1)} \{F(N-1, \boldsymbol{x}(N-1), \boldsymbol{u}(N-1)) + J_N^*(\boldsymbol{x}(N))\} \\ &= \min_{\boldsymbol{u}(N-1)} \{F(N-1, \boldsymbol{x}(N-1), \boldsymbol{u}(N-1)) + \Phi(N, \boldsymbol{x}(N))\}. \end{aligned}$$

When applying the PO to solving optimal control problems for discrete-time dynamical systems, we usually perform two stage-by-stage passes through the time stages. We begin with a backward pass. For this, we use (5.74) to eliminate $\boldsymbol{x}(N)$. Next, we carry out the minimization to find $\boldsymbol{u}^*(N-1)$. We then repeat the process. A typical step of the backward pass is described as

$$\begin{aligned} J_k^*(\boldsymbol{x}(k)) &= \min_{\boldsymbol{u}(k)}\{F(k,\boldsymbol{x}(k),\boldsymbol{u}(k)) + J_{k+1}^*(\boldsymbol{x}(k+1))\} \\ &= \min_{\boldsymbol{u}(k)}\{F(k,\boldsymbol{x}(k),\boldsymbol{u}(k)) + J_{k+1}^*(\boldsymbol{f}(k,\boldsymbol{x}(k),\boldsymbol{u}(k)))\}. \end{aligned}$$

The backward pass is completed when the initial time k_0 is reached. Now, because $\boldsymbol{x}_0 = \boldsymbol{x}(k_0)$ is known, we find $\boldsymbol{u}_0 = \boldsymbol{u}(k_0)$ in terms of this state. We can then proceed with the *forward pass*. We use $\boldsymbol{x}_0$, $\boldsymbol{u}_0$, and (5.74) to compute $\boldsymbol{x}(k_0+1)$. The state $\boldsymbol{x}(k_0+1)$ is then used to compute $\boldsymbol{u}(k_0+1)$ from $\boldsymbol{x}(k_0+2) = \boldsymbol{f}(k_0+1, \boldsymbol{x}(k_0+1), \boldsymbol{u}(k_0+1))$, and so on.

◆ Example 5.17

Consider a scalar discrete-time dynamical system modeled by

$$x(k+1) = 2x(k) - 3u(k), \qquad x(0) = 4,$$

and the performance index

$$J = J_0 = (x(2) - 10)^2 + \frac{1}{2}\sum_{k=0}^{1}(x^2(k) + u^2(k)).$$

Note that the final state, in this example, is free. We use the PO to find optimal $u(0)$ and $u(1)$. There are no constraints on $u(k)$. We begin with the backward pass. We have

$$J^*(x(2)) = (x(2) - 10)^2.$$

Then,

$$\begin{aligned} J^*(x(1)) &= \min_{u(1)}\left\{\tfrac{1}{2}(x^2(1) + u^2(1)) + J^*(x(2))\right\} \\ &= \min_{u(1)}\left\{\tfrac{1}{2}(x^2(1) + u^2(1)) + (2x(1) - 3u(1) - 10)^2\right\}. \end{aligned}$$

Because there are no constraints on $u(1)$, we find optimal $u(1)$ as a function of $x(1)$ by applying the first-order necessary condition for unconstrained optimization to get

$$\frac{\partial}{\partial u(1)}\left\{\tfrac{1}{2}(x^2(1) + u^2(1)) + (2x(1) - 3u(1) - 10)^2\right\} = 0.$$

Hence,

$$u(1) = \tfrac{1}{19}(12x(1) - 60),$$

and therefore

$$J^*(x(1)) = \frac{1}{2}\left(x^2(1) + \left(\frac{12x(1)-60}{19}\right)^2\right) + \left(2x(1) - 3\frac{12x(1)-60}{19} - 10\right)^2.$$

Next, we compute $u(0)$. For this observe that

$$J^*(x(0)) = \min_{u(0)} \left\{\tfrac{1}{2}(x^2(0) + u^2(0)) + J^*(x(1))\right\}.$$

Taking into account that

$$x(1) = 2x(0) - 3u(0), \qquad x(0) = 4,$$

and the fact that there are no constraints on $u(0)$, we again use the first-order necessary condition for unconstrained optimization to find optimal $u(0)$. We compute

$$\frac{\partial}{\partial u(0)} \left\{\tfrac{1}{2}(x^2(0) + u^2(0)) + J^*(x(1))\right\} = 0$$

to get

$$\frac{\partial}{\partial u(0)} \left\{8 + \frac{1}{2}u^2(0) + \frac{1}{2}\left(\frac{12(8-3u(0))-60}{19}\right)^2\right\}$$
$$+ \frac{\partial}{\partial u(0)} \left\{2(8-3u(0)) - 3\frac{12(8-3u(0))-60}{19} - 10\right\}^2 = 0.$$

Performing some manipulations yields $13.7895u(0) = 27.7895$. Hence, $u(0) = 2.0153$. This completes the backward pass. We are now ready for the forward pass. We use the system difference equation, $x(0)$, and $u(0)$ to compute $x(1) = 1.9541$. Using the above, we find

$$u(1) = \frac{12x(1)-60}{19} = -1.9237.$$

Therefore,

$$x(2) = 2x(1) - 3u(1) = 9.6794,$$

which completes the forward pass.

5.4.2 Discrete Linear Quadratic Regulator Problem

In Example 5.17 above, we solved a simple discrete linear quadratic regulator problem where we found the open-loop optimal control strategy. We will now apply the PO to find the optimal control of a linear state-feedback form for the discrete linear quadratic regulator problem. This problem is stated as follows. Given a discrete linear plant model,

$$\boldsymbol{x}(k+1) = \boldsymbol{A}\boldsymbol{x}(k) + \boldsymbol{B}\boldsymbol{u}(k), \qquad 0 \leq k \leq N-1,$$

with a specified initial condition $\boldsymbol{x}(0) = \boldsymbol{x}_0$. We wish to calculate the optimal control sequence

$$\{\boldsymbol{u}^*(0), \boldsymbol{u}^*(1), \ldots, \boldsymbol{u}^*(N-1)\},$$

that minimizes the quadratic performance index

$$J = J_0 = \frac{1}{2}\boldsymbol{x}^T(N)\boldsymbol{H}_N\boldsymbol{x}(N) + \frac{1}{2}\sum_{k=0}^{N-1}\{\boldsymbol{x}^T(k)\boldsymbol{Q}\boldsymbol{x}(k) + \boldsymbol{u}^T(k)\boldsymbol{R}\boldsymbol{u}(k)\},$$

where $\boldsymbol{H}_N = \boldsymbol{H}_N^T \geq 0$, $\boldsymbol{Q} = \boldsymbol{Q}^T \geq 0$, and $\boldsymbol{R} = \boldsymbol{R}^T > 0$. We assume that the components of the control vector are unconstrained. To solve this problem we use the PO. To begin, we note that

$$J^*(\boldsymbol{x}(N)) = \tfrac{1}{2}\boldsymbol{x}^T(N)\boldsymbol{H}_N\boldsymbol{x}(N).$$

The above is the penalty for being in a state $\boldsymbol{x}(N)$ at a time N.

We now decrement k to $N-1$ to get

$$J_{N-1}(\boldsymbol{x}(N-1)) = \tfrac{1}{2}\boldsymbol{x}^T(N)\boldsymbol{H}_N\boldsymbol{x}(N) + \tfrac{1}{2}\boldsymbol{x}^T(N-1)\boldsymbol{Q}\boldsymbol{x}(N-1) + \tfrac{1}{2}\boldsymbol{u}^T(N-1)\boldsymbol{R}\boldsymbol{u}(N-1).$$

We use the state equation to eliminate $\boldsymbol{x}(N)$ from J_{N-1} to obtain

$$\begin{aligned} J_{N-1} = {} & \tfrac{1}{2}(\boldsymbol{A}\boldsymbol{x}(N-1) + \boldsymbol{B}\boldsymbol{u}(N-1))^T\boldsymbol{H}_N(\boldsymbol{A}\boldsymbol{x}(N-1) + \boldsymbol{B}\boldsymbol{u}(N-1)) \\ & + \tfrac{1}{2}\boldsymbol{x}^T(N-1)\boldsymbol{Q}\boldsymbol{x}(N-1) + \tfrac{1}{2}\boldsymbol{u}^T(N-1)\boldsymbol{R}\boldsymbol{u}(N-1). \end{aligned}$$

Because there are no constraints on control, we can apply the first-order necessary condition from static optimization to find $\boldsymbol{u}^*(N-1)$ as a function of $\boldsymbol{x}(N-1)$,

$$\begin{aligned} \frac{\partial J_{N-1}}{\partial \boldsymbol{u}(N-1)} &= (\boldsymbol{A}\boldsymbol{x}(N-1) + \boldsymbol{B}\boldsymbol{u}(N-1))^T\boldsymbol{H}_N\boldsymbol{B} + \boldsymbol{u}^T(N-1)\boldsymbol{R} \\ &= \boldsymbol{0}^T. \end{aligned}$$

Solving for the control satisfying the first-order necessary condition, we obtain

$$\boldsymbol{u}^*(N-1) = -(\boldsymbol{R} + \boldsymbol{B}^T\boldsymbol{H}_N\boldsymbol{B})^{-1}\boldsymbol{B}^T\boldsymbol{H}_N\boldsymbol{A}\boldsymbol{x}_{N-1}.$$

Let

$$\boldsymbol{K}_{N-1} = (\boldsymbol{R} + \boldsymbol{B}^T\boldsymbol{H}_N\boldsymbol{B})^{-1}\boldsymbol{B}^T\boldsymbol{H}_N\boldsymbol{A},$$

then

$$\boldsymbol{u}^*(N-1) = -\boldsymbol{K}_{N-1}\boldsymbol{x}(N-1).$$

Computing the optimal cost of transferring the system from $N-1$ to N yields

$$\begin{aligned} J_{N-1}^* = {} & \tfrac{1}{2}\boldsymbol{x}^T(N-1)(\boldsymbol{A} - \boldsymbol{B}\boldsymbol{K}_{N-1})^T\boldsymbol{H}_N(\boldsymbol{A} - \boldsymbol{B}\boldsymbol{K}_{N-1})\boldsymbol{x}(N-1) \\ & + \tfrac{1}{2}\boldsymbol{x}^T(N-1)\big(\boldsymbol{K}_{N-1}^T\boldsymbol{R}\boldsymbol{K}_{N-1} + \boldsymbol{Q}\big)\boldsymbol{x}(N-1). \end{aligned}$$

Let

$$H_{N-1} = (A - BK_{N-1})^T H_N (A - BK_{N-1}) + K_{N-1}^T R K_{N-1} + Q,$$

then

$$J_{N-1}^* = \tfrac{1}{2} x^T(N-1) H_{N-1} x(N-1).$$

Decrementing k to $N-2$ yields

$$J_{N-2} = \tfrac{1}{2} x^T(N-1) H_{N-1} x(N-1) + \tfrac{1}{2} x^T(N-2) Q x(N-2) + \tfrac{1}{2} u^T(N-2) R u(N-2).$$

Note that J_{N-2} has the same form as J_{N-1}. Thus, we obtain analogous optimal feedback gain where N is being replaced with $N-1$. Continuing in this fashion, we get the following results for each $k = N-1, N-2, \ldots, 0$:

$$\begin{aligned} K_k &= (R + B^T H_{k+1} B)^{-1} B^T H_{k+1} A, \\ u^*(k) &= -K_k x(k), \\ H_k &= (A - BK_k)^T H_{k+1} (A - BK_k) + K_k^T R K_k + Q, \end{aligned}$$

and

$$J_k^* = \tfrac{1}{2} x^T(k) H_k x(k).$$

The above control scheme can be implemented by computing the sequence of gain matrices $\{K_k\}$ off-line and stored. Then, we can implement the controller $u^*(k) = -K_k x(k)$. Note that the time-varying nature of the feedback controller may make an implementation of the control strategy quite complicated, especially for large n and N. However, as we go to the limit as $N \to \infty$, we can simplify the solution to the problem. Specifically, let us consider a (reachable) system model

$$x(k+1) = Ax(k) + Bu(k), \qquad k \geq 0, \quad x(0) = x_0$$

and the performance index

$$J = \frac{1}{2} \sum_{k=0}^{\infty} \{x^T(k) Q x(k) + u^T(k) R u(k)\}$$

obtained from the previously considered performance index by letting $N \to \infty$ and setting $H_N = 0$. Then, the optimal controller takes the form

$$u^*(k) = -K_\infty x(k), \qquad k \geq 0,$$

where

$$K_\infty = (R + B^T H_\infty B)^{-1} B^T H_\infty A \tag{5.75}$$

and H_∞ is defined as the limit obtained by continuing the recursion

$$H_k = (A - BK_k)^T H_{k+1} (A - BK_k) + K_k^T R K_k + Q \tag{5.76}$$

back to $k = -\infty$, that is,

$$\lim_{k \to -\infty} \boldsymbol{H}_k = \boldsymbol{H}_\infty.$$

Alternatively, letting $k \to -\infty$ in (5.76), we obtain

$$\boldsymbol{H}_\infty = (\boldsymbol{A} - \boldsymbol{B}\boldsymbol{K}_\infty)^T \boldsymbol{H}_\infty (\boldsymbol{A} - \boldsymbol{B}\boldsymbol{K}_\infty) + \boldsymbol{K}_\infty^T \boldsymbol{R}\boldsymbol{K}_\infty + \boldsymbol{Q}, \tag{5.77}$$

where $\boldsymbol{K}_\infty$ is given by (5.75). Substituting (5.75) into (5.77) yields

$$\begin{aligned}\boldsymbol{H}_\infty = {} & (\boldsymbol{A} - \boldsymbol{B}(\boldsymbol{R} + \boldsymbol{B}^T\boldsymbol{H}_\infty\boldsymbol{B})^{-1}\boldsymbol{B}^T\boldsymbol{H}_\infty\boldsymbol{A})^T\boldsymbol{H}_\infty(\boldsymbol{A} - \boldsymbol{B}(\boldsymbol{R} + \boldsymbol{B}^T\boldsymbol{H}_\infty\boldsymbol{B})^{-1}\boldsymbol{B}^T\boldsymbol{H}_\infty\boldsymbol{A}) \\ & + ((\boldsymbol{R} + \boldsymbol{B}^T\boldsymbol{H}_\infty\boldsymbol{B})^{-1}\boldsymbol{B}^T\boldsymbol{H}_\infty\boldsymbol{A})^T\boldsymbol{R}((\boldsymbol{R} + \boldsymbol{B}^T\boldsymbol{H}_\infty\boldsymbol{B})^{-1}\boldsymbol{B}^T\boldsymbol{H}_\infty\boldsymbol{A}) + \boldsymbol{Q}.\end{aligned} \tag{5.78}$$

Thus, we can compute $\boldsymbol{H}_\infty$ by solving the above algebraic equation. We note that the above formidable looking equation is just the discrete algebraic Riccati equation. Indeed, setting $\boldsymbol{P} = \boldsymbol{H}_\infty$, using the notation

$$\boldsymbol{S} = \boldsymbol{R} + \boldsymbol{B}^T\boldsymbol{H}_\infty\boldsymbol{B} = \boldsymbol{R} + \boldsymbol{B}^T\boldsymbol{P}\boldsymbol{B},$$

and performing simple manipulations, we obtain (5.73).

We next discuss a continuous form of dynamic programming. We begin our discussion by analyzing the minimum time regulator problem. Then, we focus our attention on a more general optimal control problem.

5.4.3 Continuous Minimum Time Regulator Problem

In this subsection we consider a control system modeled by

$$\dot{\boldsymbol{x}} = \boldsymbol{f}(t, \boldsymbol{x}, \boldsymbol{u})$$

with a given initial state $\boldsymbol{x}(t_0) = \boldsymbol{x}_0$. We restrict the controller $\boldsymbol{u}(t)$ to be restrained by the compact set $U \subset \mathbb{R}^m$. For example, we may restrict the components u_i, $i = 1, 2, \ldots, m$, of the control vector $\boldsymbol{u}$ to satisfy the amplitude constraints

$$|u_i| \leq k_i,$$

where k_i are fixed real numbers. A piecewise continuous control vector $\boldsymbol{u}$ satisfying the above constraints is called admissible. The set of admissible controllers is denoted U. Our goal is to determine an admissible control law $\boldsymbol{u}^*$ that steers the system's trajectory from a given initial state $\boldsymbol{x}(t_0) = \boldsymbol{x}_0$ to a specified final state $\boldsymbol{x}_f$ in the shortest possible time. This is equivalent to minimizing the performance index

$$J = \int_{t_0}^{t_f} dt = t_f - t_0.$$

We call such a controller *time optimal*. We assume that for any $\boldsymbol{x} \neq \boldsymbol{x}_f$ there exists admissible, time optimal control transferring the system from $\boldsymbol{x}$ to $\boldsymbol{x}_f$ in minimum time. We denote by $J^*(t, \boldsymbol{x}) = J^*(t, \boldsymbol{x}(t))$ the time of the optimal transfer from $\boldsymbol{x}(t)$ to the specified $\boldsymbol{x}_f$. Observe that $J^*(t, \boldsymbol{x}) = J^*(t, x_1, x_2 \ldots, x_n)$, that is,

$$J^* : \mathbb{R}^{n+1} \to \mathbb{R},$$

which means that J^* is a real-valued function of $n + 1$ variables. We assume that the function J^* is continuous and has continuous partial derivatives everywhere except at the point $\boldsymbol{x}_f$.

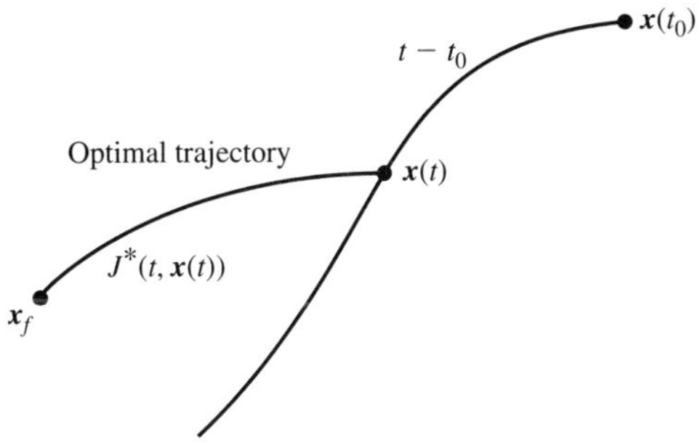

Figure 5.24 Illustration of the derivation of the condition that a controller must satisfy when transferring a dynamical system from an initial state to the final state in an optimal manner.

Now, let $\boldsymbol{x}_0 = \boldsymbol{x}(t_0) \neq \boldsymbol{x}_f$ be an arbitrary initial state in $\mathbb{R}^n$. Assume that we applied some admissible $\boldsymbol{u}_0 \in U$ on the time interval $[t_0, t]$. As a result of this control action, the system is transferred from $\boldsymbol{x}(t_0)$ to $\boldsymbol{x}(t)$. The time spent on the motion from $\boldsymbol{x}(t_0)$ to $\boldsymbol{x}(t)$ is $t - t_0$. We then apply an admissible optimal control that transfers the system in minimum time $J^*(t, \boldsymbol{x}(t))$ from $\boldsymbol{x}(t)$ to $\boldsymbol{x}_f$. The above considerations are illustrated in Figure 5.24. Denoting the optimal time of transferring the system from $\boldsymbol{x}(t_0)$ to $\boldsymbol{x}_f$ by $J^*(t_0, \boldsymbol{x}(t_0))$, we obtain

$$(t - t_0) + J^*(t, \boldsymbol{x}(t)) \geq J^*(t_0, \boldsymbol{x}(t_0)).$$

Rearranging the terms of the above equation and dividing by $(t - t_0)$ yields

$$1 + \frac{J^*(t, \boldsymbol{x}(t)) - J^*(t_0, \boldsymbol{x}(t_0))}{t - t_0} \geq 0.$$

Letting $t \to t_0$ and taking into account that $\dot{x}_i = f_i(t, \boldsymbol{x}, \boldsymbol{u})$, $i = 1, 2, \ldots, n$, we obtain

$$\begin{aligned} 1 + \frac{\partial J^*(t, \boldsymbol{x})}{\partial t} + \sum_{i=1}^{n} \frac{\partial J^*(t, \boldsymbol{x})}{\partial x_i} \dot{x}_i &= 1 + \frac{\partial J^*(t, \boldsymbol{x})}{\partial t} + \sum_{i=1}^{n} \frac{\partial J^*(t, \boldsymbol{x})}{\partial x_i} f_i \\ &= 1 + \frac{\partial J^*}{\partial t} + \frac{\partial J^*}{\partial \boldsymbol{x}} \boldsymbol{f} \\ &\geq 0. \end{aligned}$$

We conclude that for any $\boldsymbol{x}_0 \neq \boldsymbol{x}_f$ and for any $\boldsymbol{u} \in U$,

$$1 + \frac{\partial J^*}{\partial t} + \frac{\partial J^*}{\partial \boldsymbol{x}} \boldsymbol{f} \geq 0.$$

We next define the Hamiltonian function for the time optimal control problem to be

$$H(t, \boldsymbol{x}, \boldsymbol{u}, \boldsymbol{p}) = 1 + \boldsymbol{p}^T \boldsymbol{f}(t, \boldsymbol{x}, \boldsymbol{u}),$$

where

$$\boldsymbol{p} = \begin{bmatrix} \frac{\partial J^*}{\partial x_1} & \frac{\partial J^*}{\partial x_2} & \cdots & \frac{\partial J^*}{\partial x_n} \end{bmatrix}^T.$$

Thus, we can say that for any $\boldsymbol{x}_0 \neq \boldsymbol{x}_f$ and for any $\boldsymbol{u} \in U$,

$$\frac{\partial J^*(t, \boldsymbol{x})}{\partial t} + H(t, \boldsymbol{x}, \boldsymbol{u}) \geq 0.$$

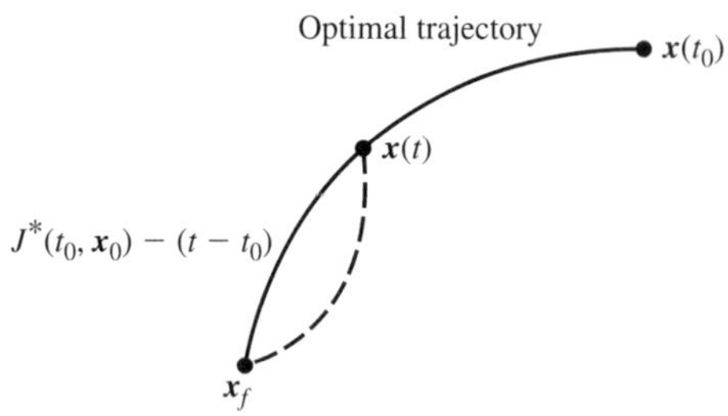

Figure 5.25 Illustration of the derivation of the optimality condition of Theorem 5.5.

We will now show that

$$\frac{\partial J^*(t, \boldsymbol{x})}{\partial t} + H(t, \boldsymbol{x}, \boldsymbol{u}) = 0$$

for any optimal transfer, that is,

$$\frac{\partial J^*(t, \boldsymbol{x})}{\partial t} + \min_{\boldsymbol{u} \in U} H(t, \boldsymbol{x}, \boldsymbol{u}) = 0 \qquad \text{for any } \boldsymbol{x} \neq \boldsymbol{x}_f.$$

In the following, we use the PO. Suppose that $\boldsymbol{u}(t) \in U$ is an optimal controller that transfers the system from the initial state $\boldsymbol{x}(t_0) \neq \boldsymbol{x}_f$ to the given final state $\boldsymbol{x}_f$ in a time t_f. Thus, we have $J^*(t_0, \boldsymbol{x}(t_0)) = t_f - t_0$. The transfer from $\boldsymbol{x}(t)$ to $\boldsymbol{x}_f$ requires $(t_f - t)$ time units. See Figure 5.25 for an illustration of the arguments. Hence,

$$t_f - t = J^*(t_0, \boldsymbol{x}(t_0)) - (t - t_0),$$

and it is not possible to go from $\boldsymbol{x}(t)$ to $\boldsymbol{x}_f$ in a time less than $J^*(t_0, \boldsymbol{x}(t_0)) - (t - t_0)$. For if such a more rapid motion were to exist, indicated by the dashed curve in Figure 5.25, then by moving from $\boldsymbol{x}_0$ to $\boldsymbol{x}(t)$ during the time $t - t_0$ and then moving from $\boldsymbol{x}(t)$ to $\boldsymbol{x}_f$ during a time less than $J^*(t_0, \boldsymbol{x}(t_0)) - (t - t_0)$, we would accomplish the transfer from $\boldsymbol{x}_0$ to $\boldsymbol{x}_f$ during a time less than $J^*(t_0, \boldsymbol{x}(t_0))$, which is impossible. Therefore, we can write

$$\begin{aligned} t_f - t &= J^*(t, \boldsymbol{x}(t)) \\ &= J^*(t_0, \boldsymbol{x}(t_0)) - (t - t_0). \end{aligned}$$

Rearranging terms, dividing by $t - t_0$, and letting $t \to t_0$ yields

$$\frac{\partial J^*(t, \boldsymbol{x})}{\partial t} + \min_{\boldsymbol{u} \in U} H(t, \boldsymbol{x}, \boldsymbol{u}) = 0 \qquad \text{for any } \boldsymbol{x} \neq \boldsymbol{x}_f.$$

We summarize our considerations in the following theorem.

Theorem 5.5 Let $\dot{\boldsymbol{x}} = \boldsymbol{f}(t, \boldsymbol{x}, \boldsymbol{u})$ be a control system model, let U be a set of admissible controllers, and let $\boldsymbol{x}_f$ be a prescribed final state. If we suppose that

1. for any $\boldsymbol{x} \neq \boldsymbol{x}_f$ there exists an admissible optimal control transferring $\boldsymbol{x}$ to $\boldsymbol{x}_f$, and
2. the function $J^*(t, \boldsymbol{x}(t))$, which is the time of optimal transfer of the system from $\boldsymbol{x}$, at time t, to $\boldsymbol{x}_f$, is continuously differentiable except at the point $\boldsymbol{x}_f$,

then

$$\frac{\partial J^*(t, \boldsymbol{x})}{\partial t} + H(t, \boldsymbol{x}, \boldsymbol{u}) \geq 0 \qquad \text{for any } \boldsymbol{u} \in U \text{ and } \boldsymbol{x} \neq \boldsymbol{x}_f,$$

$$\frac{\partial J^*(t, \boldsymbol{x})}{\partial t} + \min_{\boldsymbol{u} \in U} H(t, \boldsymbol{x}, \boldsymbol{u}) = 0 \qquad \text{for any optimal transfer where } \boldsymbol{x} \neq \boldsymbol{x}_f.$$

The above theorem is a special case of a more general result known as the continuous form of the dynamic programming. We now use the PO to derive a continuous form of dynamic programming for a class of problems with more general performance indices.

5.4.4 The Hamilton–Jacobi–Bellman Equation

We consider a control process model of the form

$$\dot{\boldsymbol{x}} = \boldsymbol{f}(t, \boldsymbol{x}, \boldsymbol{u}), \qquad \boldsymbol{x}(t_0) = \boldsymbol{x}_0, \tag{5.79}$$

and we consider the performance index to minimize

$$J(t_0, \boldsymbol{x}(t_0), \boldsymbol{u}) = \Phi(t_f, \boldsymbol{x}(t_f)) + \int_{t_0}^{t_f} F(\tau, \boldsymbol{x}(\tau), \boldsymbol{u}(\tau))\, d\tau. \tag{5.80}$$

Define

$$J(t, \boldsymbol{x}(t), \boldsymbol{u}(\tau)) = \Phi(t_f, \boldsymbol{x}(t_f)) + \int_{t}^{t_f} F(\tau, \boldsymbol{x}(\tau), \boldsymbol{u}(\tau))\, d\tau$$

with $t \leq \tau \leq t_f$, and let

$$J^*(t, \boldsymbol{x}(t)) = \min_{\boldsymbol{u}} J(t, \boldsymbol{x}(t), \boldsymbol{u}(\tau)).$$

We subdivide the interval $[t, t_f]$ as

$$[t, t_f] = [t, t + \Delta t] \cup [t + \Delta t, t_f].$$

Then, we can write

$$J^*(t, \boldsymbol{x}(t), \boldsymbol{u}(\tau)) = \min_{\boldsymbol{u}} \left\{ \int_{t}^{t+\Delta t} F\, d\tau + \int_{t+\Delta t}^{t_f} F\, d\tau + \Phi(t_f, \boldsymbol{x}(t_f)) \right\}.$$

However, by the PO the trajectory on the interval $[t + \Delta t, t_f]$ must be optimal. Hence,

$$J^*(t, \boldsymbol{x}(t)) = \min_{\boldsymbol{u}} \left\{ \int_{t}^{t+\Delta t} F\, d\tau + J^*(t + \Delta t, \boldsymbol{x}(t + \Delta t)) \right\}.$$

We next expand $J^*(t + \Delta t, \boldsymbol{x}(t + \Delta t))$ in a Taylor series about the point $(t, \boldsymbol{x}(t))$ to get

$$J^*(t, \boldsymbol{x}(t)) = \min_{\boldsymbol{u}} \left\{ \int_{t}^{t+\Delta t} F\, d\tau + J^* + \frac{\partial J^*}{\partial t}\Delta t + \frac{\partial J^*}{\partial \boldsymbol{x}}(\boldsymbol{x}(t + \Delta t) - \boldsymbol{x}(t)) + \text{H.O.T.} \right\},$$

where H.O.T. stands for higher-order terms. Because $J^*(t, \boldsymbol{x}(t))$ is independent of $\boldsymbol{u}$, we cancel the terms $J^*(t, \boldsymbol{x}(t))$ out and use that fact that $\boldsymbol{x}(t + \Delta t) - \boldsymbol{x}(t) \approx \dot{\boldsymbol{x}}\Delta t$ to obtain

$$0 = \min_{\boldsymbol{u}} \left\{ \int_t^{t+\Delta t} F\, d\tau + \frac{\partial J^*}{\partial t}\Delta t + \frac{\partial J^*}{\partial \boldsymbol{x}}\dot{\boldsymbol{x}}\,\Delta t + \text{H.O.T.} \right\}.$$

By assumption, Δt is small; hence we can write

$$0 = \min_{\boldsymbol{u}} \left\{ F\,\Delta t + \frac{\partial J^*}{\partial t}\Delta t + \frac{\partial J^*}{\partial \boldsymbol{x}}\boldsymbol{f}\,\Delta t + \text{H.O.T.} \right\}.$$

Dividing the above equation by Δt and letting $\Delta t \to 0$ yields

$$0 = \frac{\partial J^*}{\partial t} + \min_{\boldsymbol{u}} \left\{ F + \frac{\partial J^*}{\partial \boldsymbol{x}}\boldsymbol{f} \right\}. \tag{5.81}$$

Let $H = F + \frac{\partial J^*}{\partial \boldsymbol{x}}\boldsymbol{f}$ denote the Hamiltonian function. Then, we write the partial differential equation (5.81) as

$$\boxed{0 = \frac{\partial J^*}{\partial t} + \min_{\boldsymbol{u}} H} \tag{5.82}$$

subject to the boundary condition

$$J^*(t_f, \boldsymbol{x}(t_f)) = \Phi(t_f, \boldsymbol{x}(t_f)).$$

The partial differential equation (5.82) for the optimal cost $J^*(t, \boldsymbol{x}(t))$ is called the *Hamilton–Jacobi–Bellman* (HJB) equation. It provides the solution to the optimal control problem for general nonlinear dynamical systems. However, analytical solution to the HJB equation is difficult to obtain in most cases. We now illustrate, with three examples, how the HJB equation can be used to construct optimal controllers.

◆ Example 5.18

We consider a general class of dynamical systems modeled by the scalar differential equation

$$\dot{x} = ax + bu, \qquad x(t_0) = x_0,$$

with the associated cost index

$$J(t_0) = \frac{1}{2}fx^2(t_f) + \frac{1}{2}\int_{t_0}^{t_f} (qx^2(t) + ru^2(t))\,dt,$$

where the initial time t_0 equals zero, the final time $t_f < \infty$ is fixed, and the final state $x(t_f)$ is free. Our goal is construct the control u^* that minimizes $J(0)$.

We first form the Hamiltonian function for the problem:

$$H\left(t, x, u, \frac{\partial J^*}{\partial x}\right) = \frac{1}{2}qx^2 + \frac{1}{2}ru^2 + \frac{\partial J^*}{\partial x}(ax + bu).$$

Because there are no constraints on the control u, we can determine the form of the optimal control by applying the first-order necessary condition for static optimization. We compute

$$\frac{\partial H}{\partial u} = ru + b\frac{\partial J^*}{\partial x} = 0.$$

The optimal control, therefore, has the form

$$u^* = -\frac{1}{r}b\frac{\partial J^*}{\partial x}.$$

This control indeed minimizes the Hamiltonian function H because

$$\frac{\partial^2 H}{\partial u^2} = r > 0;$$

that is, the second-order sufficient condition for static minimization is satisfied. We thus reduced the problem of constructing the optimal control to that of finding $\partial J^*/\partial x$. Substituting u^* into the HJB equation yields

$$\begin{aligned}
&\frac{\partial J^*}{\partial t} + \frac{1}{2}qx^2 + \frac{b^2}{2r}\left(\frac{\partial J^*}{\partial x}\right)^2 + ax\frac{\partial J^*}{\partial x} - \frac{b^2}{r}\left(\frac{\partial J^*}{\partial x}\right)^2 \\
&= \frac{\partial J^*}{\partial t} + \frac{1}{2}qx^2 - \frac{b^2}{2r}\left(\frac{\partial J^*}{\partial x}\right)^2 + ax\frac{\partial J^*}{\partial x} \\
&= 0.
\end{aligned}$$

Let us now assume that J^* is a quadratic function of the state for all $t \le t_f$, that is,

$$J^* = \tfrac{1}{2}p(t)x^2,$$

where $p(t)$ is a scalar function of time to be determined. Partial derivatives of J^* are

$$\frac{\partial J^*}{\partial t} = \frac{1}{2}\dot{p}(t)x^2 \quad \text{and} \quad \frac{\partial J^*}{\partial x} = p(t)x.$$

Substituting the above into the HJB equation and performing some manipulations gives

$$\left(\frac{1}{2}\dot{p}(t) + \frac{1}{2}q + ap(t) - \frac{b^2p^2(t)}{2r}\right)x^2 = 0.$$

The above holds for all $x = x(t)$ if and only if

$$\frac{1}{2}\dot{p}(t) + \frac{1}{2}q + ap(t) - \frac{b^2p^2(t)}{2r} = 0,$$

or, equivalently,

$$\dot{p}(t) + q + 2ap(t) - \frac{b^2 p^2(t)}{r} = 0 \tag{5.83}$$

subject to the boundary condition $p(t_f) = f$ that we obtain by comparing $J^* = \frac{1}{2} p(t) x^2$ and $J(t_f) = \frac{1}{2} f x^2(t_f)$. To solve the above equation we assume that

$$p(t) = \rho \frac{\dot{w}(t)}{w(t)},$$

where the parameter ρ is to be determined. Note that

$$\dot{p}(t) = \rho \frac{\ddot{w}(t) w(t) - \dot{w}^2(t)}{w^2(t)}.$$

Substituting the above into (5.83), we obtain

$$\begin{aligned}
\dot{p} + q + 2ap - \frac{b^2 p^2}{r} \\
&= \rho \frac{\ddot{w}w - \dot{w}^2}{w^2} + q\frac{w^2}{w^2} + 2a\rho\frac{\dot{w}}{w} - \frac{b^2\rho^2\dot{w}^2/r}{w^2} \\
&= \frac{\rho\ddot{w}w - \rho\dot{w}^2 + qw^2 + 2a\rho\dot{w}w - b^2\rho^2\dot{w}^2/r}{w^2} \\
&= 0.
\end{aligned}$$

We now choose ρ so that the terms nonlinear in $\dot{w}$ cancel out. For this it is enough that

$$-\rho\dot{w}^2 - b^2\rho^2\dot{w}^2/r = 0.$$

Hence, if we set

$$\rho = -\frac{r}{b^2},$$

then the equation to be solved takes the form

$$\frac{\left(-\frac{r}{b^2}\ddot{w} - \frac{2ar}{b^2}\dot{w} + qw\right) w}{w^2} = 0.$$

The above will be satisfied if

$$-\frac{r}{b^2}\ddot{w} - \frac{2ar}{b^2}\dot{w} + qw = 0,$$

or, equivalently,

$$\ddot{w} + 2a\dot{w} - \frac{qb^2}{r}w = 0,$$

which is just a second-order linear differential equation. Its characteristic equation has two real roots:

$$s_{1,2} = -a \pm \sqrt{a^2 + \frac{qb^2}{r}}.$$

Hence

$$w(t) = C_1 e^{s_1 t} + C_2 e^{s_2 t},$$

and

$$p(t) = \rho \frac{\dot{w}(t)}{w(t)} = \rho \frac{s_1 C_1 e^{s_1 t} + s_2 C_2 e^{s_2 t}}{C_1 e^{s_1 t} + C_2 e^{s_2 t}}.$$

Using the boundary value $p(t_f) = f$, we express C_1 in terms of C_2 to obtain

$$C_1 = \frac{\rho s_2 C_2 e^{s_1 t_f} - f C_2 e^{s_2 t_f}}{f e^{s_1 t_f} - \rho s_1 e^{s_1 t_f}}.$$

We substitute the above expression for C_1 into $p(t)$ and cancel out C_2. Then, we use the obtained result to construct the optimal state-feedback controller:

$$\boxed{u^* = -\frac{1}{r} b \frac{\partial J^*}{\partial x} = -\frac{1}{r} b p(t) x}$$

◆ Example 5.19

Consider the schematic of an armature controlled DC motor, given in Figure 5.26, where the system parameters are $R_a = 2\,\Omega$, $L_a \approx 0$ H, $K_b = 2$ V/rad/sec, and $K_i = 2$ Nm/A, and the equivalent moment of inertia referred to the motor shaft is $I_{eq} = 1\ \text{kg}\cdot\text{m}^2$. The friction is assumed to be negligible.

1. Construct the first-order differential equation modeling the DC motor.
2. Use the Hamilton–Jacobi–Bellman equation to find the optimal state-feedback

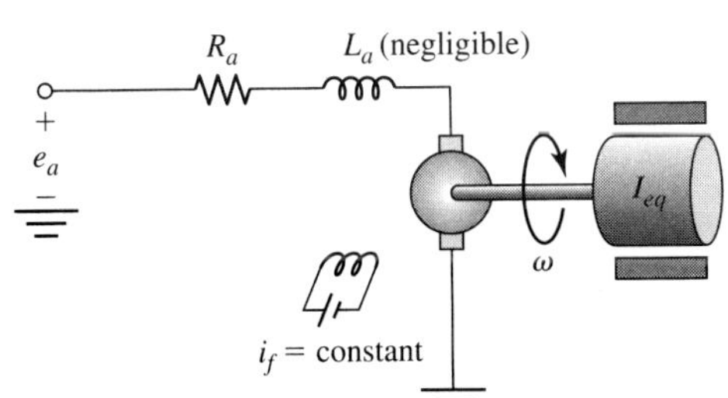

Figure 5.26 Schematic of an armature-controlled DC motor system of Example 5.19.

controller, $e_a = k(t)\omega$, which minimizes the performance index

$$J = \frac{1}{2}\omega^2(10) + \int_0^{10} R_a i_a^2(t)\,dt,$$

where i_a is the armature current. The final state is free. There are no constraints on e_a. Assume $J^* = \frac{1}{2}p(t)\omega^2$.

Applying Kirchhoff's voltage law to the armature circuit yields

$$R_a i_a + K_b \omega = e_a.$$

The torque developed by the motor, T_m, is

$$I_{eq}\dot{\omega} = T_m = K_i i_a.$$

Substituting i_a from the first equation into the second and dividing both sides by I_{eq} yields

$$\dot{\omega} = -\frac{K_i K_b}{R_a I_{eq}}\omega + \frac{K_i}{I_{eq} R_a} e_a.$$

Substituting into the above the parameter values, we obtain

$$\dot{\omega} = -2\omega + e_a. \tag{5.84}$$

We now represent the performance index, J, in terms of ω and e_a. Applying Ohm's law to the armature circuit, we get

$$e_a - K_b\omega = R_a i_a.$$

Hence,

$$J = \frac{1}{2}\omega^2(10) + \int_0^{10} \frac{(e_a - K_b\omega)^2}{R_a}\,dt.$$

The Hamiltonian function is

$$\begin{aligned} H &= \frac{(e_a - K_b\omega)^2}{R_a} + \frac{\partial J^*}{\partial \omega}\left(-\frac{K_i K_b}{R_a I_{eq}}\omega + \frac{K_i}{I_{eq} R_a} e_a\right) \\ &= \frac{(e_a - 2\omega)^2}{2} + \frac{\partial J^*}{\partial \omega}(-2\omega + e_a). \end{aligned}$$

Because there are no constraints on e_a, we can find the optimal control by solving the equation

$$0 - \frac{\partial H}{\partial e_a} = \frac{2(e_a - K_b\omega)}{R_a} + \frac{K_i}{I_{eq} R_a}\frac{\partial J^*}{\partial \omega}.$$

Hence,

$$e_a = 2\omega - \frac{\partial J^*}{\partial \omega}. \tag{5.85}$$

Substituting (5.85) into the HJB equation yields

$$0 = \frac{\partial J^*}{\partial t} + \frac{1}{2}\left(-\frac{\partial J^*}{\partial \omega}\right)^2 + \frac{\partial J^*}{\partial \omega}\left(-\frac{\partial J^*}{\partial \omega}\right) = \frac{\partial J^*}{\partial t} - \frac{1}{2}\left(\frac{\partial J^*}{\partial \omega}\right)^2.$$

By assumption $J^* = \frac{1}{2}p(t)\omega^2$. Hence,

$$\frac{\partial J^*}{\partial t} = \frac{1}{2}\dot{p}\omega^2 \quad \text{and} \quad \frac{\partial J^*}{\partial \omega} = p\omega.$$

Substituting the above into the HJB equation, we get

$$\tfrac{1}{2}(\dot{p} - p^2)\omega^2 = 0.$$

We next solve the nonlinear differential equation

$$\dot{p} - p^2 = 0. \tag{5.86}$$

We assume that $p = \rho \frac{\dot{w}}{w}$. Therefore,

$$\dot{p} = \rho \frac{\ddot{w}w - \dot{w}^2}{w^2}.$$

We substitute the expressions for p and $\dot{p}$ into (5.86) to get

$$\frac{\rho\ddot{w}w - \rho\dot{w}^2 - \rho^2\dot{w}^2}{w^2} = 0.$$

Selecting $\rho = -1$ eliminates the nonlinear terms in the numerator, and we are left with the equation

$$\ddot{w} = 0,$$

whose solution is

$$w = w(t) = C_1 t + C_2,$$

where C_1 and C_2 are integration constants. Hence,

$$p = -\frac{\dot{w}}{w} = -\frac{C_1}{C_1 t + C_2}.$$

Because

$$J^*(10) = \tfrac{1}{2}\omega^2(10) = \tfrac{1}{2}p(10)\omega^2(10),$$

we conclude that $p(10) = 1$. We use this information to eliminate one of the integration constants. We obtain

$$p(10) = 1 = -\frac{C_1}{10C_1 + C_2}.$$

Hence,

$$C_2 = -11C_1,$$

and

$$p = p(t) = -\frac{C_1}{C_1 t + C_2} = -\frac{C_1}{C_1 t - 11C_1} = -\frac{1}{t-11}.$$

Therefore,

$$\boxed{e_a^* = 2\omega - \frac{\partial J^*}{\partial \omega} = 2\omega - p\omega = \left(2 + \frac{1}{t-11}\right)\omega}$$

In the above examples, the final time t_f was fixed and $t_f < \infty$. In the following example, we consider the linear quadratic regulator design for $t_f = \infty$ using the HJB equation.

◆ Example 5.20

For the dynamical system model

$$\begin{bmatrix} \dot{x}_1 \\ \dot{x}_2 \end{bmatrix} = \begin{bmatrix} 0 & 1 \\ 0 & -1 \end{bmatrix} \begin{bmatrix} x_1 \\ x_2 \end{bmatrix} + \begin{bmatrix} 0 \\ 1 \end{bmatrix} u$$

and the performance index

$$J = \frac{1}{2} \int_0^\infty \left(x_1^2 + x_2^2 + u^2\right) dt,$$

we will use the HJB equation to find the optimal state-feedback controller. We first form the Hamiltonian function

$$H = \frac{1}{2}\left(x_1^2 + x_2^2 + u^2\right) + \frac{\partial J^*}{\partial x_1} x_2 + \frac{\partial J^*}{\partial x_2}(-x_2 + u).$$

We assume that J^* has the form

$$J^* = \tfrac{1}{2}\left(\alpha x_1^2 + 2\beta x_1 x_2 + \gamma x_2^2\right).$$

Hence,

$$\frac{\partial J^*}{\partial t} = 0,$$

and the HJB equation takes the form

$$\min_u H = 0,$$

with the boundary condition

$$J^*(\boldsymbol{x}(t_f)) = J^*(\boldsymbol{x}(\infty)) = 0.$$

Because there are no constraints on u, we apply the first-order necessary condition for unconstrained optimization to find u minimizing H:

$$\frac{\partial H}{\partial u} = 0.$$

We denote by u^* the solution to the above equation. Differentiating, we obtain

$$u^* + \frac{\partial J^*}{\partial x_2} = 0.$$

Hence,

$$u^* = -\frac{\partial J^*}{\partial x_2}.$$

We substitute u^* into the HJB equation to get

$$\frac{1}{2}\left(x_1^2 + x_2^2 + \left(\frac{\partial J^*}{\partial x_2}\right)^2\right) + \frac{\partial J^*}{\partial x_1}x_2 - \frac{\partial J^*}{\partial x_2}x_2 - \left(\frac{\partial J^*}{\partial x_2}\right)^2 = 0.$$

Note that

$$\frac{\partial J^*}{\partial x_1} = \alpha x_1 + \beta x_2 \quad \text{and} \quad \frac{\partial J^*}{\partial x_2} = \beta x_1 + \gamma x_2.$$

Substituting the above into the HJB equation yields

$$\tfrac{1}{2}\left(x_1^2 + x_2^2 + (\beta x_1 + \gamma x_2)^2\right) + (\alpha x_1 + \beta x_2)x_2 - (\beta x_1 + \gamma x_2)x_2 - (\beta x_1 + \gamma x_2)^2$$
$$= 0.$$

We rearrange the above equation to obtain

$$\tfrac{1}{2}x_1^2(1-\beta^2) + x_1 x_2(\alpha - \beta - \beta\gamma) + \tfrac{1}{2}x_2^2(1 - 2\gamma - \gamma^2 + 2\beta) = 0.$$

For this equation to hold, we have to have

$$\begin{aligned} 1 - \beta^2 &= 0,\\ \alpha - \beta - \beta\gamma &= 0,\\ 1 - 2\gamma - \gamma^2 + 2\beta &= 0. \end{aligned}$$

There are three solutions to the above system of equations:

1. $\alpha = 2$, $\beta = \gamma = 1$.
2. $\alpha = -2$, $\beta = 1$, $\gamma = -3$.
3. $\alpha = 0$, $\beta = \gamma = -1$.

Thus, we have three candidate solutions to the problem:

$$u^1 = -x_1 - x_2,$$
$$u^2 = -x_1 + 3x_2,$$
$$u^3 = x_1 + x_2.$$

However, only u^1 stabilizes the plant. Hence, the optimal control law for the problem is

$$u^* = -x_1 - x_2.$$

Note that only for this controller the boundary condition, $J^*(\boldsymbol{x}(t_f)) = J^*(\boldsymbol{x}(\infty)) = 0$, is satisfied.

5.5 Pontryagin's Minimum Principle

5.5.1 Optimal Control with Constraints on Inputs

In this section we provide an informal derivation of the minimum principle of Pontryagin, which is a generalization of the Euler–Lagrange equations that also includes problems with constraints on the control inputs. Only a special case of the minimum principle will be stated, but this special case covers a large class of control problems. We consider the problem of minimizing the performance index given by

$$J = \Phi(t_f, \boldsymbol{x}(t_f)) + \int_{t_0}^{t_f} F(\boldsymbol{x}(t), \boldsymbol{u}(t))\, dt \tag{5.87}$$

subject to

$$\dot{\boldsymbol{x}} = \boldsymbol{f}(\boldsymbol{x}, \boldsymbol{u}), \quad \boldsymbol{x}(t_0) = \boldsymbol{x}_0, \quad \text{and} \quad \boldsymbol{x}(t_f) = \boldsymbol{x}_f. \tag{5.88}$$

We consider two cases: fixed final state and free final state.

It follows from our discussion of the HJB equation that the optimal controller, $\boldsymbol{u}^*$, minimizes the Hamiltonian function, that is,

$$\boldsymbol{u}^* = \arg\min_{\boldsymbol{u} \in U} H(\boldsymbol{x}, \boldsymbol{u}, \boldsymbol{p}),$$

where $H(\boldsymbol{x}, \boldsymbol{u}, \boldsymbol{p}) = H = F + \boldsymbol{p}^T \boldsymbol{f}$, and

$$\boldsymbol{p} = [p_1 \quad p_2 \quad \cdots \quad p_n]^T = \begin{bmatrix} \frac{\partial J^*}{\partial x_1} & \frac{\partial J^*}{\partial x_2} & \cdots & \frac{\partial J^*}{\partial x_n} \end{bmatrix}^T.$$

To proceed, we need one more assumption in addition to the assumptions we made while discussing the HJB equation. We assume that $J^*(\boldsymbol{x})$ has continuous second partial derivatives $\partial^2 J^*/\partial x_i \partial x_j$; that is, the functions p_i have continuous first derivatives $\partial p_i/\partial x_j$. This implies that

$$\frac{\partial \boldsymbol{p}}{\partial \boldsymbol{x}} = \left(\frac{\partial \boldsymbol{p}}{\partial \boldsymbol{x}}\right)^T. \tag{5.89}$$

We also assume that the components of $\boldsymbol{f}$ have continuous first derivatives $\partial f_i/\partial x_j$. We then compute

$$\frac{\partial H}{\partial \boldsymbol{x}} = \frac{\partial}{\partial \boldsymbol{x}}(F + \boldsymbol{p}^T \boldsymbol{f})$$
$$= \frac{\partial F}{\partial \boldsymbol{x}} + \boldsymbol{f}^T \frac{\partial \boldsymbol{p}}{\partial \boldsymbol{x}} + \boldsymbol{p}^T \frac{\partial \boldsymbol{f}}{\partial \boldsymbol{x}}, \tag{5.90}$$

where $\partial F/\partial \boldsymbol{x} = \nabla_x^T F$. From our previous discussion it follows that the Hamiltonian function attains its minimum on the optimal trajectory. Taking this into account, we conclude that for the optimal process we obtain

$$\frac{\partial H}{\partial \boldsymbol{x}} = \boldsymbol{0}^T \quad \text{for } (\boldsymbol{x}, \boldsymbol{u}) = (\boldsymbol{x}^*, \boldsymbol{u}^*).$$

Hence, for the optimal process $(\boldsymbol{x}^*, \boldsymbol{u}^*)$ we have

$$\frac{\partial F}{\partial \boldsymbol{x}} + \boldsymbol{f}^T \frac{\partial \boldsymbol{p}}{\partial \boldsymbol{x}} + \boldsymbol{p}^T \frac{\partial \boldsymbol{f}}{\partial \boldsymbol{x}} = \boldsymbol{0}^T. \tag{5.91}$$

We now differentiate $\boldsymbol{p}$ with respect to time to get

$$\dot{\boldsymbol{p}} = \frac{\partial \boldsymbol{p}}{\partial \boldsymbol{x}}\dot{\boldsymbol{x}} = \frac{\partial \boldsymbol{p}}{\partial \boldsymbol{x}}\boldsymbol{f}. \tag{5.92}$$

Combining (5.91) and (5.92) and taking into account (5.89), we obtain

$$\dot{\boldsymbol{p}} = -\left(\frac{\partial F}{\partial \boldsymbol{x}}\right)^T - \left(\frac{\partial \boldsymbol{f}}{\partial \boldsymbol{x}}\right)^T \boldsymbol{p}. \tag{5.93}$$

Thus, $\boldsymbol{p}$ is the solution to the differential equation (5.93), and hence $\boldsymbol{p} = \boldsymbol{p}(t)$. Therefore $\partial \boldsymbol{p}/\partial \boldsymbol{x} = \boldsymbol{0}$, and by (5.90) we have

$$\frac{\partial H}{\partial \boldsymbol{x}} = \frac{\partial F}{\partial \boldsymbol{x}} + \boldsymbol{p}^T \frac{\partial \boldsymbol{f}}{\partial \boldsymbol{x}}. \tag{5.94}$$

Comparing (5.93) and (5.94) gives

$$\dot{\boldsymbol{p}} = -\left(\frac{\partial H}{\partial \boldsymbol{x}}\right)^T \tag{5.95}$$

The above equation is called in the literature the *adjoint* or *costate* equation.

We summarize our development in the following theorem.

Theorem 5.6 Necessary conditions for $\boldsymbol{u} \in U$ to minimize (5.87) subject to (5.88) are

$$\dot{\boldsymbol{p}} = -\left(\frac{\partial H}{\partial \boldsymbol{x}}\right)^T,$$

where $H = H(\boldsymbol{x}, \boldsymbol{u}, \boldsymbol{p}) = F(\boldsymbol{x}, \boldsymbol{u}) + \boldsymbol{p}^T \boldsymbol{f}(\boldsymbol{x}, \boldsymbol{u})$, and

$$H(\boldsymbol{x}^*, \boldsymbol{u}^*, \boldsymbol{p}^*) = \min_{\boldsymbol{u} \in U} H(\boldsymbol{x}, \boldsymbol{u}, \boldsymbol{p}).$$

If the final state, $\boldsymbol{x}(t_f)$, is free, then in addition to the above conditions it is required that the following end-point condition is satisfied:

$$\boldsymbol{p}(t_f) = \nabla_{\boldsymbol{x}} \Phi|_{t=t_f}.$$

We now illustrate the above theorem with the following well-known example.

◆ Example 5.21

We consider a controlled object modeled by

$$\begin{aligned}\begin{bmatrix}\dot{x}_1\\ \dot{x}_2\end{bmatrix} &= \begin{bmatrix}0 & 1\\ 0 & 0\end{bmatrix}\begin{bmatrix}x_1\\ x_2\end{bmatrix} + \begin{bmatrix}0\\ 1\end{bmatrix}u\\ &= \boldsymbol{A}\boldsymbol{x} + \boldsymbol{b}u\end{aligned}$$

with the performance index

$$J = \int_0^{t_f} dt.$$

The control is required to satisfy

$$|u(t)| \le 1$$

for all $t \in [0, t_f]$. This constraint means that the control must have magnitude no greater than 1. Our objective is to find admissible control that minimizes J and transfers the system from a given initial state $\boldsymbol{x}_0$ to the origin.

We begin by finding the Hamiltonian function for the problem:

$$\begin{aligned}H &= 1 + \boldsymbol{p}^T(\boldsymbol{A}\boldsymbol{x} + \boldsymbol{b}u)\\ &= 1 + [p_1 \quad p_2]\left(\begin{bmatrix}0 & 1\\ 0 & 0\end{bmatrix}\begin{bmatrix}x_1\\ x_2\end{bmatrix} + \begin{bmatrix}0\\ 1\end{bmatrix}u\right)\\ &= 1 + p_1 x_2 + p_2 u.\end{aligned}$$

The costate equations are

$$\begin{bmatrix}\dot{p}_1\\ \dot{p}_2\end{bmatrix} = -\begin{bmatrix}\dfrac{\partial H}{\partial x_1}\\ \dfrac{\partial H}{\partial x_2}\end{bmatrix} = \begin{bmatrix}0\\ -p_1\end{bmatrix}.$$

Solving the costate equations yields

$$p_1 = d_1 \quad \text{and} \quad p_2 = -d_1 t + d_2,$$

where d_1 and d_2 are integration constants. We now find an admissible control minimizing the Hamiltonian:

$$\begin{aligned}\arg_u \min H &= \arg_u \min(1 + p_1 x_2 + p_2 u)\\ &= \arg_u \min(p_2 u)\\ &= \begin{cases}u(t) = 1 & \text{if } p_2 < 0,\\ u(t) = -1 & \text{if } p_2 > 0.\end{cases}\end{aligned}$$

Hence,

$$u^*(t) = -\text{sign}(p_2^*) = -\text{sign}(-d_1 t + d_2),$$

where

$$\text{sign}(z) = \begin{cases}1 & \text{if } z > 0,\\ -1 & \text{if } z < 0.\end{cases}$$

Thus, the optimal control law is piecewise constant taking the values 1 or -1. This control law has at most two intervals of constancy because the argument of the sign function is a linear function, $-d_1 t + d_2$, that changes its sign at most once. This type of control is called a *bang-bang control* because it switches back and forth between its extreme values. System trajectories for $u = 1$ and $u = -1$ are families of parabolas, and their equations were derived in Example 2.2. A phase portrait for $u = 1$ is shown in Figure 2.4, while that for $u = -1$ is shown in Figure 2.5. Note that only one parabola from each family passes through the specified terminal point $\boldsymbol{x} = [0 \;\; 0]^T$ in the state plane. Segments of the two parabolas through the origin form the switching curve,

$$x_1 = -\tfrac{1}{2} x_2^2 \,\mathrm{sign}(x_2).$$

This means that if an initial state is above the switching curve, then $u = -1$ is used until the switching curve is reached. Then, $u = 1$ is used to reach the origin. For an initial state below the switching curve, the control $u = 1$ is used first to reach the switching curve, and then the control is switched to $u = -1$. The above control action can be described as

$$u = \begin{cases} 1 & \text{if } v < 0, \\ -\mathrm{sign}(x_2) & \text{if } v = 0, \\ -1 & \text{if } v > 0, \end{cases}$$

where $v = v(x_1, x_2) = x_1 + \frac{1}{2} x_2^2 \mathrm{sign}(x_2) = 0$ is the equation describing the switching curve. We can use the above equation to synthesize a closed-loop system such that starting at an arbitrary initial state in the state plane, the trajectory will always be moving in an optimal fashion toward the origin. Once the origin is reached, the trajectory will stay there. A phase portrait of the closed-loop time optimal system is given in Figure 5.27.

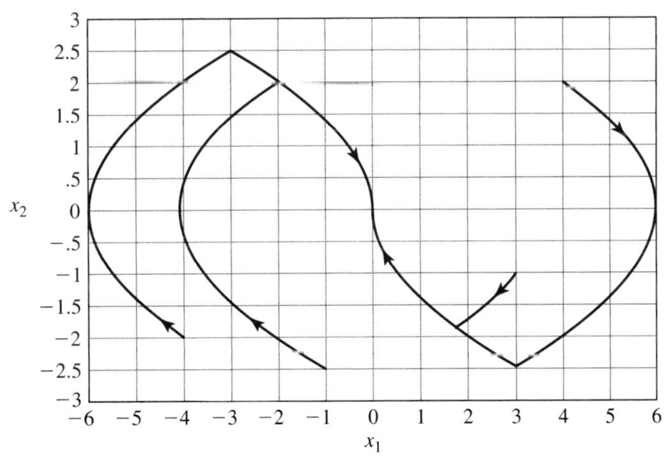

Figure 5.27 A phase-plane portrait of the time optimal closed-loop system of Example 5.21.

◆ Example 5.22

Consider the schematic of an armature-controlled DC motor, given in Figure 5.26. Use the minimum principle to find the armature voltage $e_a(t)$ such that the motor angular velocity ω changes from $\omega(0)=0$ rad/sec at $t=0$ to $\omega(1)=10$ rad/sec while minimizing the energy dissipated in the armature resistor, R_a. There are no constraints on e_a. Note that the energy dissipated in the armature resistor is

$$J = \int_0^1 R_a i_a^2(t)\,dt,$$

where i_a is the armature current.

The modeling equation of the DC motor was derived in Example 5.19. It is given by (5.84) on page 288. We now represent the expression for the energy dissipated in the armature resistor in terms of ω and e_a. Applying Ohm's law to the armature circuit, we get

$$e_a - K_b\omega = R_a i_a.$$

Hence,

$$J = \int_0^1 R_a i_a^2(t)\,dt = \int_0^1 \frac{(e_a - K_b\omega)^2}{R_a}\,dt.$$

The Hamiltonian function is

$$H = \frac{(e_a - K_b\omega)^2}{R_a} + p\left(-\frac{K_i K_b}{R_a I_{eq}}\omega + \frac{K_i}{I_{eq} R_a} e_a\right).$$

Because there are no constraints on e_a, we can find the optimal control by solving the equation

$$0 = \frac{\partial H}{\partial e_a} = \frac{2(e_a - K_b\omega)}{R_a} + \frac{K_i}{I_{eq} R_a} p.$$

Hence,

$$e_a = 2\omega - p. \tag{5.96}$$

The costate equation is

$$\dot{p} = -\frac{\partial H}{\partial \omega} = \frac{2K_b(e_a - K_b\omega)}{R_a} + \frac{K_i K_b}{R_a I_{eq}} p = 2(e_a - 2\omega) + 2p. \tag{5.97}$$

We substitute (5.96) into (5.97) to obtain

$$\dot{p} = 0.$$

Hence,

$$p(t) = \text{constant} = c.$$

Therefore,

$$e_a = 2\omega - c. \tag{5.98}$$

Substituting (5.98) into (5.84), we get

$$\dot{\omega} = -c.$$

Solving the above yields

$$\omega(t) = -ct + \omega(0) = -ct,$$

because $\omega(0) = 0$. We next use the boundary condition $\omega(1) = 10$ to find $c = -10$. Hence, the optimal armature voltage is

$$\boxed{e_a^*(t) = 20t + 10}$$

◆ Example 5.23

In this example, we will solve Johann Bernoulli's brachistochrone problem, introduced in Section 5.2, using the minimum principle. We first recall the problem as restated by Johann Bernoulli in 1697 (quoted from Dunham [70, p. 200]):

"Among the infinitely many curves which join the two given points ... choose the one such that, if the curve is replaced by a thin tube or groove, and a small sphere placed in it and released, then this will pass from one point to the other in the shortest time."

It is difficult not to mention the reward that one obtains for solving the problem (quoted from Dunham [70, p. 200]):

"Let who can seize quickly the prize which we have promised to the solver. Admittedly this prize is neither of gold nor silver, for these appeal only to base and venal souls. ... Rather, since virtue itself is its own most desirable reward and a fame is a powerful incentive, we offer the prize fitting for the man of noble blood, compounded of honor, praise, and approbation ..." We add that Johann Bernoulli wrote the above after he already knew the solution to the problem!

Model and the Performance Index

The field is conservative, hence the sums of the kinetic and potential energy is constant,

$$\tfrac{1}{2}mv^2(t) - mgy(t) = \tfrac{1}{2}mv^2(t_0) - mgy(t_0) = 0,$$

where we assumed that the particle is at rest at t_0. Using the above, we can find the particle's speed at any time $t \geq t_0$:

$$v(t) = \sqrt{2gy(t)}.$$

Then, the brachistochrone problem can be formulated as an optimal control problem of the form

$$\text{minimize} \quad J(t_0) = \int_{t_0}^{T} dt$$

$$\text{subject to} \quad \begin{cases} \dot{x} = v\cos\theta, \\ \dot{y} = v\sin\theta, \end{cases}$$

where θ is the angle that the particle velocity vector forms with the positive x axis, while $\dot{x}$ and $\dot{y}$ are the horizontal and vertical components of the particle velocity vector. The path angle θ is considered as the control input to be determined.

Particle Trajectory

To proceed, we form the Hamiltonian function

$$H = 1 + p_1 v \cos\theta + p_2 v \sin\theta, \tag{5.99}$$

where p_1 and p_2 are the costate variables. The costate equations are

$$\begin{bmatrix} \dot{p}_1 \\ \dot{p}_2 \end{bmatrix} = - \begin{bmatrix} \dfrac{\partial H}{\partial x} \\ \dfrac{\partial H}{\partial y} \end{bmatrix} = \begin{bmatrix} 0 \\ -\dfrac{g}{v}(p_1 \cos\theta + p_2 \sin\theta) \end{bmatrix}.$$

It follows from the above that $p_1 = K = \text{constant}$. Because there are no constraints on the control input θ, the necessary condition for the control optimality is

$$\frac{\partial H}{\partial \theta} = 0,$$

where

$$\frac{\partial H}{\partial \theta} = -p_1 v \sin\theta + p_2 v \cos\theta.$$

Hence,

$$p_2 = p_1 \tan\theta = K \tan\theta.$$

We next find an expression for $\dot{\theta}$. First note that

$$\dot{p}_2 = \frac{\partial p_2}{\partial \theta}\dot{\theta} = \frac{\partial (K \tan\theta)}{\partial \theta}\dot{\theta} = \frac{K}{\cos^2\theta}\dot{\theta}. \tag{5.100}$$

On the other hand,

$$\begin{aligned} \dot{p}_2 &= -\frac{g}{v}(p_1 \cos\theta + p_2 \sin\theta) \\ &= -\frac{g}{v}(K \cos\theta + K \tan\theta \sin\theta) \\ &= -\frac{Kg}{v}\left(\cos\theta + \frac{\sin^2\theta}{\cos\theta}\right) \\ &= -\frac{Kg}{v}\left(\frac{\cos^2\theta + \sin^2\theta}{\cos\theta}\right) \\ &= -\frac{Kg}{v\cos\theta}. \end{aligned} \tag{5.101}$$

Combining (5.100) and (5.101) yields

$$\dot{\theta} = -\frac{g}{v}\cos\theta. \tag{5.102}$$

We now use the boundary conditions to express x and y as functions of time and θ. First, using the above we can write

$$\dot{y} = \frac{dy}{d\theta}\dot{\theta} = \frac{dy}{d\theta}\left(-\frac{g}{v}\cos\theta\right).$$

On the other hand,

$$\dot{y} = v\sin\theta.$$

Equating the right-hand sides of the above two equations yields

$$\frac{dy}{v^2} = \frac{dy}{2gy} = -\frac{1}{g}\tan\theta\, d\theta.$$

Integrating both sides gives

$$\frac{1}{2g}\int\frac{dy}{y} = -\frac{1}{g}\int\tan\theta\, d\theta,$$

that is,

$$\frac{1}{2g}\ln|y| = \frac{1}{g}\ln|\cos\theta|.$$

Performing manipulations, we obtain

$$y = C\cos^2\theta,$$

where C is an integration constant. Evaluating the above at $t = T$ gives

$$y(T) = y_1 = C\cos^2\theta(T).$$

Hence,

$$C = \frac{y_1}{\cos^2\theta(T)}$$

and

$$\boxed{y(t) = \frac{y_1}{\cos^2\theta(T)}\cos^2\theta(t)} \tag{5.103}$$

In a similar way, we can obtain an expression for x as a function of time and θ. Indeed, on the one hand we have

$$\dot{x} = \frac{dx}{d\theta}\dot{\theta} = -\frac{dx}{d\theta}\frac{g}{v}\cos\theta,$$

while on the other hand we obtain

$$\dot{x} = v\cos\theta.$$

Comparing the right-hand sides of the above two equations gives

$$\frac{dx}{d\theta} = -\frac{v^2}{g} = -2y = -2\frac{y_1}{\cos^2\theta(T)}\cos^2\theta.$$

We represent the above as

$$dx = -2\frac{y_1}{\cos^2\theta(T)}\cos^2\theta\, d\theta.$$

Integrating both sides, using the identity, $\cos^2\theta = \frac{1}{2}(1+\cos 2\theta)$, yields

$$x = -2\frac{y_1}{\cos^2\theta(T)}\left(\frac{\theta}{2} + \frac{\sin 2\theta}{4}\right) + \tilde{C},$$

where $\tilde{C}$ is the integration constant. Evaluating the above at $t = T$ gives

$$x_1 = -2\frac{y_1}{\cos^2\theta(T)}\left(\frac{\theta(T)}{2} + \frac{\sin 2\theta(T)}{4}\right) + \tilde{C}.$$

Hence,

$$\tilde{C} = x_1 + \frac{y_1}{\cos^2\theta(T)}\left(\theta(T) + \frac{\sin 2\theta(T)}{2}\right).$$

Using the above, we obtain

$$\boxed{x(t) = x_1 + \frac{y_1}{2\cos^2\theta(T)}\left(2(\theta(T) - \theta(t)) + \sin 2\theta(T) - \sin 2\theta(t)\right)} \tag{5.104}$$

We point out that the equation for $x(t)$, $y(t)$ and $\dot{\theta}$ specify an implicit feedback control law in the sense that the control θ can be calculated given the state $(x(t), y(t))$.

We now use the boundary conditions to find the description of the time-optimal closed loop system. We can assume, without loss of generality, that $x(t_0) = y(t_0) = x(0) = y(0) = 0$. Then, using the expression for $y(t)$ we note that $y(t_0) = 0 = \frac{y_1}{\cos^2\theta(T)}\cos^2\theta(t_0)$. Thus, the angle of departure is $\theta(t_0) = 90°$. With the above initial path angle, y, and hence v, is increased at the outset as quickly as possible. Using these data, we next evaluate x at $t = t_0$ to form an equation that can be used to find the angle of arrival $\theta(T)$. The equation is

$$2x_1\cos^2\theta(T) + y_1(2\theta(T) - \pi + \sin 2\theta(T)) = 0.$$

The above equation can be solved easily using MATLAB's Symbolic Toolbox or the MATLAB's `fsolve` function. Once $\theta(T)$ is found, we use the expression for y to find $\cos\theta$, where

$$\cos\theta(t) = \sqrt{\frac{y(t)}{y_1}}\cos\theta(T).$$

Hence,

$$\sin\theta(t) = \sqrt{1 - \cos^2\theta(t)}.$$

We substitute the above expression into $\dot{x} = v\cos\theta$ and $\dot{y} = v\sin\theta$ to obtain the state-space model of the time optimal motion of the particle moving under the force of gravity from a given point to another specified point. The equations are

$$\dot{x}(t) = \sqrt{2gy(t)}\sqrt{\frac{y(t)}{y_1}}\cos\theta(T),$$
$$\dot{y}(t) = \sqrt{2gy(t)}\sqrt{1 - \frac{y(t)}{y_1}\cos^2\theta(T)}$$

subject to the boundary conditions $[x(t_0) \quad y(t_0)] = [x_0 \quad y_0]$ and $[x(T) \quad y(T)] = [x_1 \quad y_1]$.

The Time of Quickest Descent

We now derive an expression that can be used to compute the time of the particle travel. We first establish an auxiliary result that we will use in our derivation. Specifically, we will show that $\dot{\theta}(t)$ is constant on the interval $[t_0, T]$. Differentiating both sides of equation (5.102) gives

$$\begin{aligned}\frac{d\dot{\theta}}{dt} &= \frac{d}{dt}\left(-\frac{g}{\sqrt{2gy}}\cos\theta\right)\\ &= -\frac{g}{\sqrt{2g}}\left(\frac{d}{dt}\left(\frac{1}{\sqrt{y}}\cos\theta\right)\right)\\ &= -\frac{g}{\sqrt{2g}}\left(-\frac{1}{2}\frac{\dot{y}\cos\theta}{y^{3/2}} - \frac{\sin\theta}{\sqrt{y}}\dot{\theta}\right)\\ &= \frac{g}{\sqrt{2g}}\left(\frac{1}{2}\frac{\dot{y}\cos\theta}{y^{3/2}} + \frac{\sin\theta}{\sqrt{y}}\dot{\theta}\right).\end{aligned}$$

Substituting into the above the expressions for $\dot{\theta}$ and $\dot{y}$ gives

$$\begin{aligned}\frac{d\dot{\theta}}{dt} &= \frac{g}{\sqrt{2g}}\left(\frac{1}{2}\frac{\sqrt{2gy}\sin\theta\cos\theta}{y^{3/2}} - g\frac{\sin\theta\cos\theta}{\sqrt{y}\sqrt{2gy}}\right)\\ &= \frac{g}{\sqrt{2g}}\left(\frac{\sqrt{g}}{y\sqrt{2}}\sin\theta\cos\theta - \frac{\sqrt{g}}{y\sqrt{2}}\sin\theta\cos\theta\right)\\ &= 0.\end{aligned}$$

Because $\dot{\theta}$ is constant on the interval $[t_0, T]$, we use (5.102) to write

$$\frac{d\theta}{dt} = -\frac{g}{\sqrt{2gy(T)}}\cos\theta(T).$$

Separating the variables yields

$$d\theta = -\frac{g}{\sqrt{2gy(T)}}\cos\theta(T)\,dt.$$

Integrating both sides of the above, we obtain

$$\int_{\theta(t)}^{\theta(T)} d\theta = -\frac{g}{\sqrt{2gy(T)}} \cos\theta(T) \int_t^T dt.$$

Hence,

$$\theta(T) - \theta(t) = -\frac{g}{\sqrt{2gy(T)}} \cos\theta(T)(T - t).$$

To obtain an expression for the particle travel time, we set in the above equation $t = t_0 = 0$ and perform simple manipulations to obtain

$$\boxed{T = \frac{\pi/2 - \theta(T)}{\cos\theta(T)} \sqrt{2y(T)/g}}$$

It's an Upside-Down Cycloid!

The path of quickest descent is an upside-down cycloid. A cycloid is a curve described by a point on the circumference of a circle that rolls without slipping (see Figure 5.28). The equations describing a cycloid are

$$\begin{aligned} x(\phi) &= R(\phi - \sin\phi), \\ y(\phi) &= R(1 - \cos\phi). \end{aligned}$$

We will now show that equations (5.104) and (5.103) describe a cycloid. To proceed, we define

$$R = \frac{y_1}{2\cos^2\theta(T)} \quad \text{and} \quad \phi(t) = \pi - 2\theta(t). \tag{5.105}$$

Using the above notation and taking into account that $\sin(\pi - \phi) = \sin\phi$, we rewrite (5.104) as

$$x(t) - x_1 + R(\phi(T) - \sin\phi(T)) = R(\phi(t) - \sin\phi(t)).$$

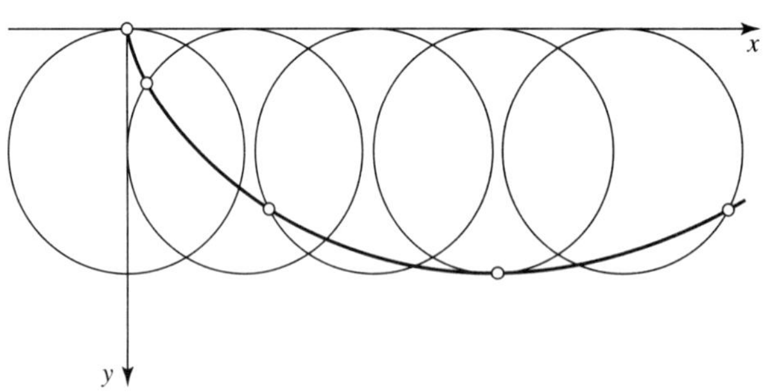

Figure 5.28 Constructing an upside-down cycloid—the solution to the brachistochrone problem.

Using the notation given in (5.105) and the fact that $\cos 2\phi = 2\cos^2\phi - 1$, we represent (5.103) as

$$\begin{aligned} y(t) &= \frac{y_1}{\cos^2\theta(T)}\cos^2\theta(t) \\ &= \frac{y_1}{\cos^2\theta(T)}\frac{\cos 2\theta(t)+1}{2} \\ &= R(1+\cos(\pi-\phi(t))) \\ &= R(1-\cos\phi(t)), \end{aligned}$$

because $\cos(\pi - \phi(t)) = -\cos\phi$.

In the following, we discuss a general minimum time regulator problem for a class of multi-input dynamical systems modeled by

$$\dot{\boldsymbol{x}} = \boldsymbol{A}\boldsymbol{x} + \boldsymbol{B}\boldsymbol{u}.$$

The goal is to transfer the system from some initial state to the origin in minimum time subject to

$$|u_i| \le k_i, \quad k_i > 0, \qquad i = 1, 2, \ldots, m.$$

Thus, the performance index for the problem is $J = \int_{t_0}^{t_f} dt$. We begin our analysis by forming the associated Hamiltonian function

$$\begin{aligned} H &= 1 + \boldsymbol{p}^T(\boldsymbol{A}\boldsymbol{x} + \boldsymbol{B}\boldsymbol{u}) \\ &= 1 + \boldsymbol{p}^T\boldsymbol{A}\boldsymbol{x} + \sum_{i=1}^{m}(\boldsymbol{p}^T\boldsymbol{b}_i)u_i, \end{aligned}$$

where $\boldsymbol{b}_i$ is the ith column of $\boldsymbol{B}$. Let

$$s_i(t) = \boldsymbol{p}^{*T}(t)\boldsymbol{b}_i.$$

Then, applying Pontryagin's minimum principle, we obtain the following form of the optimal control:

$$u_i^*(t) = -k_i \operatorname{sign}(s_i(t)), \qquad i = 1, 2, \ldots, m.$$

Thus, the optimal control for the ith control channel has the bang-bang form, $u_i^* = \pm k_i$. Note that $s_i(t) = \boldsymbol{p}^{*T}(t)\boldsymbol{b}_i = 0$ is the switching curve for the ith control channel. We compute the costate vector from the costate equation,

$$\begin{aligned} \dot{\boldsymbol{p}}^{*T} &= -\frac{\partial H}{\partial \boldsymbol{x}} \\ &= -\frac{\partial}{\partial \boldsymbol{x}}(\boldsymbol{p}^{*T}\boldsymbol{A}\boldsymbol{x}) \\ &= -\boldsymbol{p}^{*T}\boldsymbol{A}, \end{aligned}$$

or

$$\dot{\boldsymbol{p}}^* = -\boldsymbol{A}^T\boldsymbol{p}^*.$$

The solution to the costate equation is

$$p^*(t) = e^{-A^T t} p^*(0).$$

Thus, the ith switching curve can be represented as

$$s_i(t) = p^{*T}(0)e^{-At} b_i = 0.$$

Observe that if $s_i(t) = 0$ on some time interval, then $u_i^*(t)$ is indefinite on this interval. We now show that under some mild assumptions there is no time interval on which $s_i(t) = 0$; that is, there is no time interval on which $u_i^*(t)$ is indefinite. First, we assume that $b_i \neq 0$. Note that in our problem here, $\min_{u \in U} H(x, u) = 0$ because $J^*(t, x) = J^*(x)$ and therefore $\partial J^*/\partial t = 0$. Hence, $1 + p^{*T}(t)(Ax^* + Bu^*) = 0$ for all t for the optimal process. Thus, the optimal costate vector $p^*(t)$ cannot be zero for any value of t. Finally, if $s_i = p^{*T} b_i$ is zero on some time interval—that is, s_i is constant on a time interval—then this implies that on the above time interval, we obtain

$$\dot{s}_i = \dot{p}^{*T} b_i = -p^{*T} A b_i = 0.$$

Similarly, higher derivatives of $s_i(t) = p^{*T} b_i$ will also be zero on the time interval on which $s_i(t) = p^{*T} b_i$ is zero, that is, $s_i^{(j)}(t) = (-1)^j p^{*T} A^j b_i = 0$, $j = 1, 2, \ldots$. We represent the above relations as

$$p^{*T} [b_i \quad Ab_i \quad \cdots \quad A^{n-1} b_i \quad \cdots] = 0^T.$$

If the system is controllable by the ith input acting alone, then

$$\operatorname{rank}[b_i \quad Ab_i \quad \cdots \quad A^{n-1} b_i \quad \cdots] = n,$$

and we would have to have $p^* = p^*(t) = 0$. However, we already ruled out this case and thus $s_i(t) = p^{*T} b_i$ cannot be zero on any time interval. In conclusion, if the system is controllable by each input acting alone, then there is no time interval on which optimal control vector is indefinite.

5.5.2 A Two-Point Boundary-Value Problem

We consider the problem of minimizing the quadratic performance index

$$J = \frac{1}{2} x^T(t_f) F x(t_f) + \frac{1}{2} \int_{t_0}^{t_f} (x^T(t) Q x(t) + u^T(t) R u(t))\, dt \tag{5.106}$$

subject to

$$\dot{x}(t) = Ax(t) + Bu(t), \qquad x(t_0) = x_0, \qquad x(t_f) \text{ is free.} \tag{5.107}$$

We assume that there are no constraints on the control input u. The weight matrices F and Q are real symmetric positive semidefinite, and the matrix R is real symmetric positive definite. The Hamiltonian function for the above linear quadratic control problem is

$$H = \tfrac{1}{2} x^T Q x + \tfrac{1}{2} u^T R u + p^T (Ax + Bu), \tag{5.108}$$

where $p \in \mathbb{R}^n$ is the costate vector. It follows from Theorem 5.6 that the optimal controller must minimize the Hamiltonian function. Because the control vector is unconstrained, the necessary condition for optimality of the control $\boldsymbol{u}$ is

$$\frac{\partial H}{\partial \boldsymbol{u}} = \mathbf{0}^T. \tag{5.109}$$

Evaluating (5.109) yields

$$\frac{\partial H}{\partial \boldsymbol{u}} = \boldsymbol{u}^T \boldsymbol{R} + \boldsymbol{p}^T \boldsymbol{B} = \mathbf{0}^T. \tag{5.110}$$

Hence, the optimal controller has the form

$$\boldsymbol{u} = -\boldsymbol{R}^{-1} \boldsymbol{B}^T \boldsymbol{p}. \tag{5.111}$$

Substituting the above into (5.107), we obtain

$$\dot{\boldsymbol{x}} = \boldsymbol{A}\boldsymbol{x} - \boldsymbol{B}\boldsymbol{R}^{-1}\boldsymbol{B}^T \boldsymbol{p}, \qquad \boldsymbol{x}(t_0) = \boldsymbol{x}_0. \tag{5.112}$$

We next find the costate equation,

$$\begin{aligned} \dot{\boldsymbol{p}} &= -\left(\frac{\partial H}{\partial \boldsymbol{x}}\right)^T \\ &= -\boldsymbol{Q}\boldsymbol{x} - \boldsymbol{A}^T \boldsymbol{p}. \end{aligned} \tag{5.113}$$

Combining (5.112) and the above yields $2n$ linear differential equations:

$$\begin{bmatrix} \dot{\boldsymbol{x}} \\ \dot{\boldsymbol{p}} \end{bmatrix} = \begin{bmatrix} \boldsymbol{A} & -\boldsymbol{B}\boldsymbol{R}^{-1}\boldsymbol{B}^T \\ -\boldsymbol{Q} & -\boldsymbol{A}^T \end{bmatrix} \begin{bmatrix} \boldsymbol{x} \\ \boldsymbol{p} \end{bmatrix}. \tag{5.114}$$

By Theorem 5.6, the state vector at time t_f must satisfy

$$\boldsymbol{p}(t_f) = \tfrac{1}{2}(\nabla \boldsymbol{x}^T \boldsymbol{F} \boldsymbol{x})|_{t=t_f} = \boldsymbol{F}\boldsymbol{x}(t_f). \tag{5.115}$$

In summary, we reduced the linear quadratic control problem to solving $2n$ linear differential equations with mixed boundary conditions, which is an example of a *two-point boundary value problem,* or TPBVP for short. We shall now solve the above TPBVP. If we had the initial conditions $\boldsymbol{x}(t_0)$ and $\boldsymbol{p}(t_0)$, then the solution to (5.114) would have the form

$$\begin{bmatrix} \boldsymbol{x}(t_f) \\ \boldsymbol{p}(t_f) \end{bmatrix} = e^{\boldsymbol{H}(t_f - t_0)} \begin{bmatrix} \boldsymbol{x}(t_0) \\ \boldsymbol{p}(t_0) \end{bmatrix}, \tag{5.116}$$

where

$$\boldsymbol{H} = \begin{bmatrix} \boldsymbol{A} & -\boldsymbol{B}\boldsymbol{R}^{-1}\boldsymbol{B}^T \\ -\boldsymbol{Q} & -\boldsymbol{A}^T \end{bmatrix}.$$

However, in this particular TPBVP, $\boldsymbol{p}(t_0)$ is unknown and instead we are given $\boldsymbol{p}(t_f)$. To proceed, let

$$\begin{bmatrix} \boldsymbol{x}(t_f) \\ \boldsymbol{p}(t_f) \end{bmatrix} = e^{\boldsymbol{H}(t_f - t)} \begin{bmatrix} \boldsymbol{x}(t) \\ \boldsymbol{p}(t) \end{bmatrix}, \tag{5.117}$$

where

$$e^{H(t_f-t)} = \begin{bmatrix} \Phi_{11}(t_f,t) & \Phi_{12}(t_f,t) \\ \Phi_{21}(t_f,t) & \Phi_{22}(t_f,t) \end{bmatrix}.$$

Each of the blocks $\Phi_{ij}(t_f, t)$, $i, j = 1, 2$, is $n \times n$. Thus we have

$$\begin{bmatrix} x(t_f) \\ p(t_f) \end{bmatrix} = \begin{bmatrix} \Phi_{11}(t_f,t) & \Phi_{12}(t_f,t) \\ \Phi_{21}(t_f,t) & \Phi_{22}(t_f,t) \end{bmatrix} \begin{bmatrix} x(t) \\ p(t) \end{bmatrix}. \tag{5.118}$$

The above matrix equation can be represented as

$$x(t_f) = \Phi_{11}(t_f,t)x(t) + \Phi_{12}(t_f,t)p(t), \tag{5.119}$$

$$p(t_f) = \Phi_{21}(t_f,t)x(t) + \Phi_{22}(t_f,t)p(t). \tag{5.120}$$

Using the boundary condition (5.115) and (5.119) gives

$$p(t_f) = Fx(t_f) = F\Phi_{11}(t_f,t)x(t) + F\Phi_{12}(t_f,t)p(t). \tag{5.121}$$

Subtracting (5.120) from (5.121), we obtain

$$0 = (F\Phi_{11}(t_f,t) - \Phi_{21}(t_f,t))x(t) + (F\Phi_{12}(t_f,t) - \Phi_{22}(t_f,t))p(t). \tag{5.122}$$

Using the above, we compute

$$\begin{aligned} p(t) &= (\Phi_{22}(t_f,t) - F\Phi_{12}(t_f,t))^{-1}(F\Phi_{11}(t_f,t) - \Phi_{21}(t_f,t))x(t) \\ &= P(t)x(t), \end{aligned} \tag{5.123}$$

where

$$P(t) = (\Phi_{22}(t_f,t) - F\Phi_{12}(t_f,t))^{-1}(F\Phi_{11}(t_f,t) - \Phi_{21}(t_f,t)). \tag{5.124}$$

Substituting (5.123) into (5.111) yields the optimal state-feedback controller

$$u(t) = -R^{-1}B^T P(t)x(t). \tag{5.125}$$

We will now show that $P(t)$ satisfies a matrix differential equation. Indeed, differentiating (5.123) gives

$$\dot{P}x + P\dot{x} - \dot{p} = 0. \tag{5.126}$$

Substituting into (5.112), (5.113), and (5.123) yields

$$(\dot{P} + PA - PBR^{-1}B^T P + Q + A^T P)x = 0. \tag{5.127}$$

The preceding must hold throughout $t_0 \le t \le t_f$. Hence, P must satisfy

$$\boxed{\dot{P} = PBR^{-1}B^T P - A^T P - PA - Q} \tag{5.128}$$

subject to the boundary condition

$$P(t_f) = F. \tag{5.129}$$

Equation (5.128) is called the matrix *Riccati differential equation*. Note that because F is symmetric, so is $\boldsymbol{P} = \boldsymbol{P}(t)$. Observe that $\boldsymbol{P} = \boldsymbol{P}(t)$ given by (5.124) is a solution to (5.128) subject to the boundary condition (5.129).

The Riccati differential equation (5.128) associated with a class of one-dimensional plants was solved in Example 5.18.

Notes

Sussmann and Willems [273] made the following statement, in 1997, regarding the origins of optimal control: "Optimal control was born in 1697—300 years ago—in Groningen, a university town in the north of The Netherlands, when Johann Bernoulli, professor of mathematics at the local university from 1695 to 1705, published his solution of the *brachystochrone problem*."

For a serious student of optimal control, the seminal text by Lee and Markus [177] is recommended. Less advanced than Lee and Markus are Kirk [160], Bryson and Ho [36], Owens [218], and Pierre [234]. The classical text by Kwakernaak and Sivan [173] is also worth perusing because of its clarity and its treatment of the stochastic aspects of control problems. Informative, and at the same time entertaining, accounts of the origins of the calculus of variations are given by Dunham [70, Chapter 8] and by Sussmann and Willems [273]. Elsgolc [75] provides a lucid introduction, illustrated with many examples, to the calculus of variations. Applications of the calculus of variations in physics and engineering are provided by Weinstock [297]. Interrelations between the calculus of variations and optimal control are discussed by Hestenes [121]. Frequency-domain aspects of optimal control and robust optimal control design are discussed by Burl [39]. Informal discussion of the Pontryagin minimum principle in Section 5.5 was adapted from Boltyanskii [30]. For more numerical examples related to optimal control, see Lewis and Syrmos [182], and Bryson [37]. Exciting applications of optimal control to a number of control problems of technology and science can be found in Lee and Markus [177] .

EXERCISES

5.1 Find the variations of the functionals:

(a) $v(x_1, x_2) = \int_0^{\pi/2}(x_1^4 + 2x_1^2x_2^2 + x_2^2)\,dt$,

(b) $v(x_1, x_2) = \int_0^1(e^{x_1-x_2} - e^{x_1+x_2})\,dt$.

5.2 Find the equations of the curves that are extremals for the functional

$$v(x) = \int_0^{t_f}\left(\tfrac{1}{4}\dot{x}^2 - t\dot{x} - x + \tfrac{1}{2}x\dot{x}\right)dt$$

for the boundary conditions

(a) $x(0) = 1$, $t_f = 2$, and $x(2) = 10$;

(b) $x(0) = 1$, $t_f = 10$, and $x(10)$ is free.

5.3 Find the equation of the curve that is an extremal for the functional

$$v(x) = \int_0^{t_f}\left(\tfrac{1}{2}\dot{x}^2 - t\dot{x} + 2x^2\right)dt$$

subject to the boundary conditions

$$x(0) = 3/4, \quad t_f = 1, \quad \text{and} \quad x(1) \text{ is free.}$$

5.4 Given a dynamical system model

$$\dot{\boldsymbol{x}} = \begin{bmatrix} 0 & 1 \\ 0 & 0 \end{bmatrix} \boldsymbol{x} + \begin{bmatrix} 0 \\ 1 \end{bmatrix} u,$$
$$y = [2 \quad 0]\boldsymbol{x},$$

and the associated performance index

$$J = \int_0^\infty (y^2 + u^2)\, dt.$$

Assuming that $u = -\boldsymbol{k}\boldsymbol{x}$ is such that J is minimized, find the poles of the closed-loop system.

5.5 Given a dynamical system model

$$\dot{\boldsymbol{x}} = \begin{bmatrix} 0 & 1 \\ 0 & 0 \end{bmatrix} \boldsymbol{x} + \begin{bmatrix} 0 \\ 1 \end{bmatrix} u,$$
$$y = [2 \quad 0]\boldsymbol{x}$$

and the performance index

$$J = \int_0^\infty (y^2 + u^2)\, dt.$$

Use the Kronecker product to represent the associated algebraic Riccati equation as the nonlinear vector algebraic equation. Solve it, and construct the optimal state-feedback control law $u = -\boldsymbol{k}\boldsymbol{x}$. Assume that

$$\boldsymbol{x}(0) = \begin{bmatrix} 1 \\ 2 \end{bmatrix}.$$

Find the optimal value of J.

5.6 Find a state-feedback control law u so as to minimize

$$J = \int_0^1 (x^2 + u^2)\, dt$$

subject to

$$\dot{x} = -x + u, \qquad x(0) = 2.$$

Assume $J^* = p(t)x^2$.

5.7 Use the principle of optimality to find the set of numbers $x_1, x_2, \ldots, x_n$ satisfying $x_1 + x_2 + \cdots + x_n = c$, where c is a positive constant such that the product $x_1 x_2 \cdots x_n$ is maximized (see [20]).

5.8 Use dynamic programming to find the maximum cost path between node **a** and node **t** in the routing network depicted in Figure 5.29, where the travel costs are shown

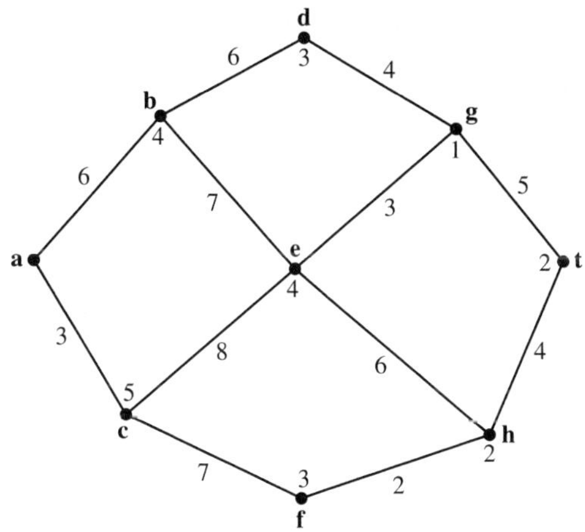

Figure 5.29 A routing network of Exercise 5.8.

beside each segment and toll charges are shown by each node. The network can only be traversed from left to right.

5.9 Use dynamic programming to find $u(0)$ and $u(1)$ that minimize

$$J = (x(2) - 2)^2 + \sum_{k=0}^{1} u^2(k)$$

subject to

$$x(k+1) = x(k) + u(k), \qquad x(0) = 1.$$

5.10 Consider the following optimal control problem:

$$\text{minimize} \quad J_2(u) = \frac{1}{2} \sum_{k=0}^{1} u(k)^2$$
$$\text{subject to} \quad x(k+1) = x(k) + 2u(k), \qquad x(0) = 3, \qquad x(2) = 0.$$

Convert the above optimal control problem into a quadratic optimization problem with an equality constraint using the composite input vector

$$z = \begin{bmatrix} u(0) \\ u(1) \end{bmatrix}.$$

Then, employ Lagrange's multiplier theorem to find the optimal sequence $\{u^*(0), u^*(1)\}$.

5.11 Consider the following optimal control problem:

$$\text{minimize} \quad J_2(u) = \frac{1}{2} x(2)^2 + \frac{1}{2} \sum_{k=0}^{1} (x(k)^2 + u(k)^2)$$
$$\text{subject to} \quad x(k+1) = x(k) + 2u(k), \qquad x(0) = 3.$$

Convert the above optimal control problem into a quadratic optimization problem with an equality constraint using the composite input-state vector

$$z = \begin{bmatrix} u(0) \\ x(1) \\ u(1) \\ x(2) \end{bmatrix}.$$

Then, employ Lagrange's multiplier theorem to find the optimal sequence $\{u^*(0), u^*(1)\}$.

5.12 Find $u = u(t)$ so as to minimize

$$J = \int_0^1 u^2\, dt$$

subject to

$$\dot{x} = -2x + u, \quad x(0) = 10, \quad x(1) = 0.$$

5.13 For the circuit shown in Figure 5.30, find the input voltage, $u(t)$, that changes the current in the inductor from $i = 10$ A at $t = 0$ to $i = 0$ at $t = 1$ sec while minimizing the performance index,

$$J = \int_0^1 u^2(t)\, dt.$$

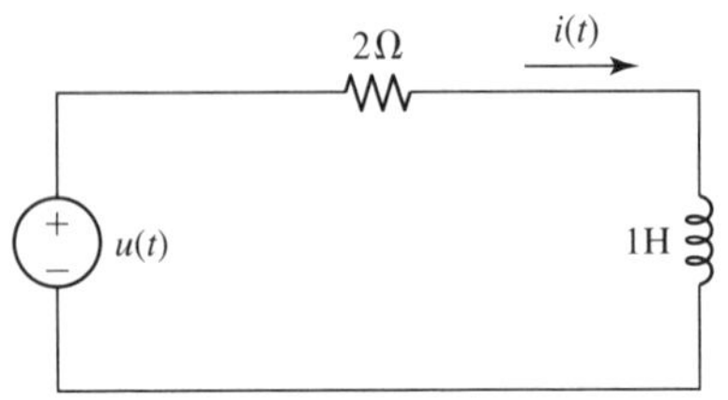

Figure 5.30 A circuit for Exercise 5.13.

5.14 Use the minimum principle to find the input voltage $u(t)$ that charges the capacitor, in Figure 5.31, from $x(0) = 2$ V at $t = 0$ to $x(10) = 12$ V while minimizing the energy dissipated in the resistor. There are no constraints on $u(t)$. Note that, by Kirchhoff's voltage law,

$$Ri(t) + x(t) = u(t),$$

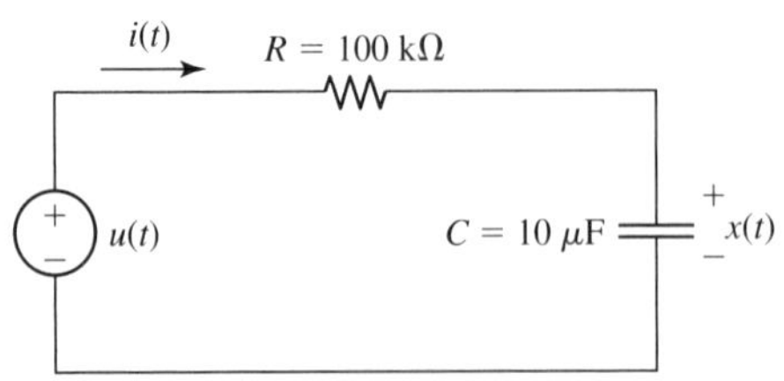

Figure 5.31 A capacitor charging circuit of Exercise 5.14.

where $i(t) = C\frac{dx(t)}{dt}$, and the energy dissipated is

$$J = \int_0^{10} Ri^2(t)\, dt.$$

5.15 Show that if μ is an eigenvalue of the Hamiltonian matrix $\boldsymbol{H}$, then so is $-\mu$.

5.16 Consider the Hamiltonian function defined by

$$H = F + \boldsymbol{p}^T \boldsymbol{f}.$$

Show that if the functions $\boldsymbol{f}$ and F do not explicitly depend upon t, then

$$H|_{\boldsymbol{u}=\boldsymbol{u}^*} = \text{constant}, \qquad t_0 \le t \le t_f.$$

5.17 Given the Hamiltonian matrix corresponding to a dynamical system and its associated performance index. The Hamiltonian matrix is

$$\boldsymbol{H} = \begin{bmatrix} 2 & -5 \\ -1 & -2 \end{bmatrix}.$$

The MATLAB command `[V,D] = eig(H)` resulted in the following data:

$$\boldsymbol{V} = \begin{bmatrix} 0.9806 & 0.7071 \\ -0.1961 & 0.7071 \end{bmatrix}, \qquad \boldsymbol{D} = \begin{bmatrix} 3 & 0 \\ 0 & -3 \end{bmatrix}.$$

(a) Write the equation of the closed loop system driven by the optimal controller $u = -kx$.

(b) Find the solution to the algebraic Riccati equation corresponding to the optimal controller.

5.18 (a) Suppose that s_i is an eigenvalue of the matrix

$$\boldsymbol{A} - \boldsymbol{B}\boldsymbol{R}^{-1}\boldsymbol{B}^T\boldsymbol{P},$$

and $\boldsymbol{v}_i \ne \boldsymbol{0}$ is the corresponding eigenvector, where $\boldsymbol{P}$ is the symmetric positive definite solution of the algebraic Riccati equation. Show that

$$\begin{bmatrix} \boldsymbol{v}_i \\ \boldsymbol{P}\boldsymbol{v}_i \end{bmatrix}$$

is an eigenvector of the associated Hamiltonian matrix corresponding to the eigenvalue s_i.

(b) Show that if $\boldsymbol{Q} = \boldsymbol{Q}^T > 0$, then the real parts of the eigenvalues of the matrix

$$\boldsymbol{A} - \boldsymbol{B}\boldsymbol{R}^{-1}\boldsymbol{B}^T\boldsymbol{P}$$

are all negative, where $\boldsymbol{P} = \boldsymbol{P}^T > 0$ is the solution of the associated algebraic Riccati equation.

5.19 Given the following model of a dynamical system:

$$\dot{x}_1 = x_2 + 5,$$
$$\dot{x}_2 = u,$$

where

$$|u| \le 1.$$

The performance index to be minimized is

$$J = \int_0^{t_f} dt.$$

Find the state-feedback control law $u = u(x_1, x_2)$ that minimizes J and drives the system from a given initial condition $\boldsymbol{x}(0) = [x_1(0), x_2(0)]^T$ to the final state $\boldsymbol{x}(t_f) = \boldsymbol{0}$. Proceed as indicated below.

(a) Derive the equations of the optimal trajectories.

(b) Derive the equation of the switching curve.

(c) Write the expression for the optimal state-feedback controller.

5.20 (Based on Rugh and Murphy [246]) Given a dynamical system model

$$\dot{\boldsymbol{x}} = \boldsymbol{A}\boldsymbol{x} + \boldsymbol{b}u,$$

where

$$\boldsymbol{A} = \begin{bmatrix} 0 & 1 & 0 & \cdots & 0 & 0 \\ 0 & 0 & 1 & \cdots & 0 & 0 \\ \vdots & & & \cdots & & \vdots \\ 0 & 0 & 0 & \cdots & 0 & 1 \\ -a_0 & -a_1 & -a_2 & \cdots & a_{n-2} & a_{n-1} \end{bmatrix}, \qquad \boldsymbol{b} = \begin{bmatrix} 0 \\ 0 \\ \vdots \\ 0 \\ 1 \end{bmatrix},$$

and the associated performance index

$$J = \frac{1}{2}\int_0^{\infty} (\boldsymbol{x}^T\boldsymbol{Q}\boldsymbol{x} + u^2)\,dt, \qquad \boldsymbol{Q} = \boldsymbol{Q}^T \geq 0.$$

We label the above problem as $(\boldsymbol{A}, \boldsymbol{b})$. Consider now the dynamical system model

$$\dot{\tilde{\boldsymbol{x}}} = \tilde{\boldsymbol{A}}\tilde{\boldsymbol{x}} + \tilde{\boldsymbol{b}}\tilde{u},$$

where $\boldsymbol{b} = \tilde{\boldsymbol{b}}$,

$$\tilde{\boldsymbol{A}} = \begin{bmatrix} 0 & 1 & 0 & \cdots & 0 & 0 \\ 0 & 0 & 1 & \cdots & 0 & 0 \\ \vdots & & & \cdots & & \vdots \\ 0 & 0 & 0 & \cdots & 0 & 1 \\ 0 & 0 & 0 & \cdots & 0 & 0 \end{bmatrix},$$

and the associated performance index has the form

$$\tilde{J} = \frac{1}{2}\int_0^{\infty} (\tilde{\boldsymbol{x}}^T\tilde{\boldsymbol{Q}}\tilde{\boldsymbol{x}} + \tilde{u}^2)\,dt,$$

where $\tilde{\boldsymbol{Q}} = \boldsymbol{Q} + \boldsymbol{D}\tilde{\boldsymbol{A}} + \tilde{\boldsymbol{A}}^T\boldsymbol{D} + \boldsymbol{D}\boldsymbol{b}\boldsymbol{b}^T\boldsymbol{D}$, and

$$\boldsymbol{D} = \begin{bmatrix} 0 & 1 & 0 & \cdots & 0 & -a_0 \\ 0 & 0 & 1 & \cdots & 0 & -a_1 \\ \vdots & & & \cdots & & \vdots \\ 0 & 0 & 0 & \cdots & 0 & -a_{n-2} \\ -a_0 & -a_1 & -a_2 & \cdots & -a_{n-2} & -a_{n-1} \end{bmatrix}.$$

We refer to the second problem as $(\tilde{A}, \tilde{b})$.

(a) Show that the symmetric matrix P is a solution to the (A, b) problem if, and only if, the symmetric matrix $\tilde{P} = P - D$ is a solution to the $(\tilde{A}, \tilde{b})$ problem.

(b) Show that $A - bb^T P = \tilde{A} - bb^T \tilde{P}$.

Hint Note that $\tilde{A} = A - bb^T D$ and $\tilde{b} = b$. Write down the algebraic Riccati equation for the $(\tilde{A}, \tilde{b})$ problem and perform appropriate substitutions.

5.21 Consider the following variation of the brachistochrone problem: A particle of mass m is to slide down a frictionless track connecting two given points that are not located on the vertical line. The particle's initial speed is v_0. The motion of the particle is constrained only by gravity and the shape of the track. Find the curve of the track that minimizes the time travel between the two given points. Solve this brachistochrone problem using the minimum principle of Pontryagin. Illustrate your derivations with a numerical example and simulations using MATLAB.

5.22 (a) Consider the ground vehicle model described in Subsection 1.7.2. Let the input signal be $u = \delta_f$ (the front wheel steer angle) and $\delta_r = 0$. Write a MATLAB script that animates the lateral motion Y versus the longitudinal motion X of the vehicle, the steer angle δ_f, and the heading angle ψ. Use the steer angle shown in Figure 5.32.

(b) Use the linear model and the data from Table 1.1 to design a linear quadratic regulator (LQR). In this part, the control has the form

$$\delta_f = -kx + Y_{ref},$$

where Y_{ref} is the command input shown in Figure 5.33. That is, the closed-loop

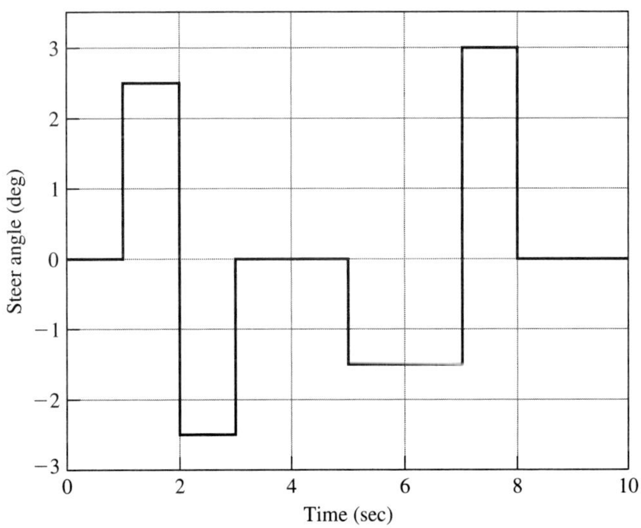

Figure 5.32 Plot of the steer angle δ_f versus time for Exercise 5.22.

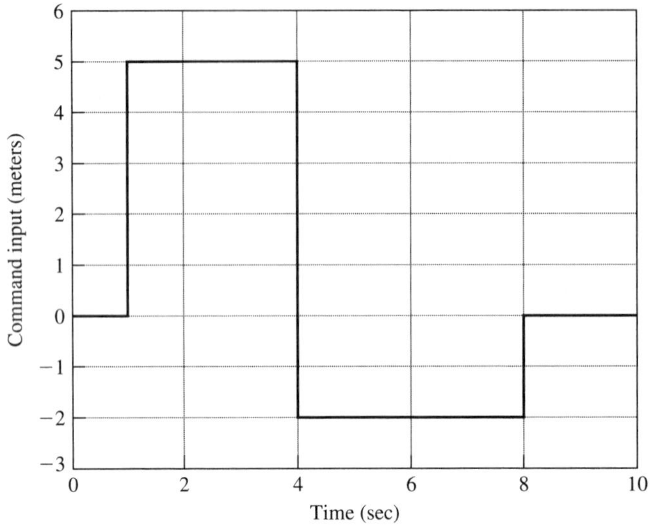

Figure 5.33 Plot of the command input Y_{ref} versus time for Exercise 5.22.

system here will have the form

$$\dot{\boldsymbol{x}} = (\boldsymbol{A} - \boldsymbol{bk})\boldsymbol{x} + \boldsymbol{b}Y_{ref}.$$

You as a designer will select the weights $\boldsymbol{Q}$ and r. See how the linear vehicle tracks the command input. Observe the sign difference in the command input and the heading angle. Your LQR gain should be calculated for each chosen v_x. In other words, if in your interactive simulation a specific value of v_x is chosen, then your program should calculate the corresponding gain for this particular value of v_x. Implement the LQR state feedback employing the vehicle model that uses the nonlinear cornering characteristic shown in Figure 1.15 and test the performance of the closed-loop system utilizing the command input shown in Figure 5.33. You may need to impose constraints on δ_f—for example, $|\delta_f| \leq 20°$. (Do not forget about converting degrees into radians.)

CHAPTER 6 Sliding Modes

As philosopher of science Karl Popper has emphasized, a good theory is characterized by the fact that it makes a number of predictions that could in principle be disproved or falsified by observation. Each time new experiments are observed to agree with the predictions the theory survives, and our confidence in it is increased; but if ever a new observation is found to disagree, we have to abandon or modify the theory. At least that is what is supposed to happen, but you can always question the competence of the person who carried out the observation.

—Stephen Hawking, *A Brief History of Time* [117]

6.1 Simple Variable Structure Systems

The purpose of this section is to informally introduce the reader to variable structure sliding mode control systems. A formal analysis is presented in the following sections.

A *variable structure system* is a dynamical system whose structure changes in accordance with the current value of its state. A variable structure system can be viewed as a system composed of independent structures together with a switching logic between each of the structures. With appropriate switching logic, a variable structure system can exploit the desirable properties of each of the structures the system is composed of. Even more, a variable structure system may have a property that is not a property of any of its structures. We illustrate the above ideas with two numerical examples.

◆ Example 6.1

This example was adapted from Utkin [284]. We consider a dynamical system model

$$\begin{aligned} \dot{x}_1 &= x_2, \\ \dot{x}_2 &= -ux_1, \end{aligned} \tag{6.1}$$

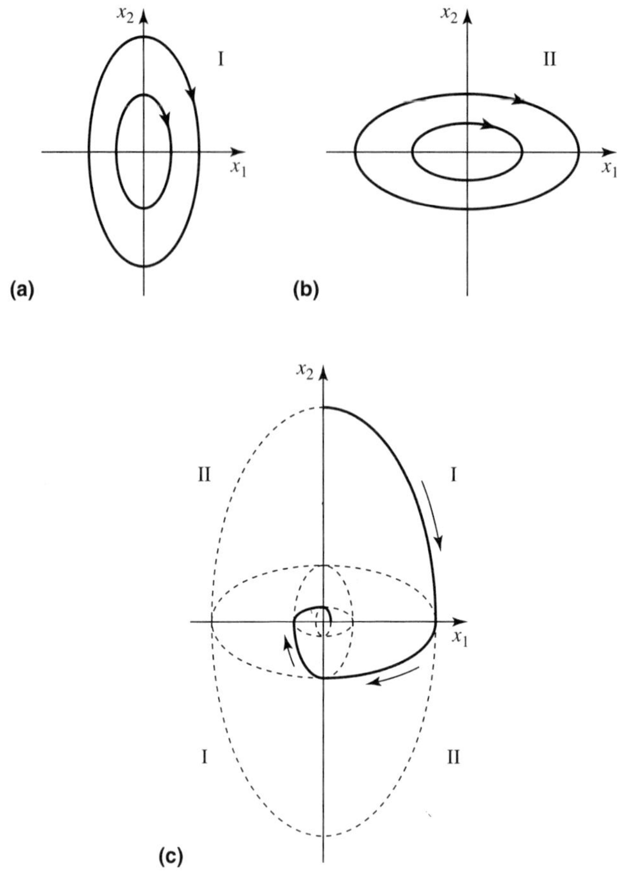

Figure 6.1 (a) and (b) Phase-plane portraits of structures making up a variable structure system. (c) Phase-plane portrait of the variable structure system itself, where the structure of the system is changed any time the system's trajectory crosses either of the coordinate axis. (Adapted from the survey paper by Utkin [284].)

having two structures corresponding to $u = 1/a$ and $u = a$, where a is a positive constant greater than 1; in our simulations we take $a = 5$. The phase-plane portraits of the structures are shown in Figures 6.1(a) and 6.1(b). The phase-plane portraits of the individual structures are families of ellipses. Neither of the structures is asymptotically stable—they are only stable. However, by choosing a suitable switching logic between the structures, we can make the resulting variable structure system asymptotically stable. Indeed, suppose the structure of the system is changed any time the system's trajectory crosses either of the coordinate axes of the state plane, that is,

$$a = \begin{cases} 1/5 & \text{if } x_1 x_2 < 0, \\ 5 & \text{if } x_1 x_2 > 0. \end{cases} \tag{6.2}$$

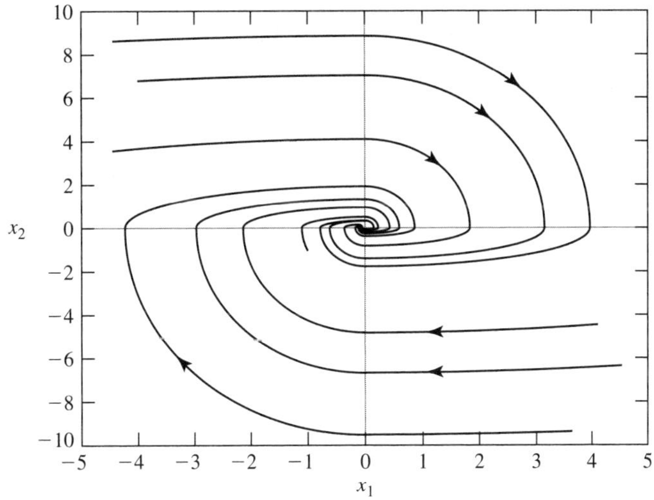

Figure 6.2 Phase-plane portrait of the variable structure system described by (6.1) and (6.2).

A phase plane portrait of the system (6.1) with the switching logic specified by (6.2) is shown in Figure 6.2. We add that the trajectories spiral clockwise towards the origin.

◆ Example 6.2

Consider the double integrator

$$\ddot{x} = u,$$

whose state space model is

$$\begin{aligned}\dot{x}_1 &= x_2,\\ \dot{x}_2 &= u.\end{aligned} \tag{6.3}$$

Let $u = \pm 2$. Typical trajectories of the above system are shown in Figure 6.3. We thus have two independent structures, neither of them being stable. Let us now use the following switching logic between each of the structures:

$$u = -2\,\text{sign}(x_1 + x_2). \tag{6.4}$$

With the switching logic in (6.4), the closed-loop system changes its structure anytime the state trajectory crosses the line

$$x_1 + x_2 = 0.$$

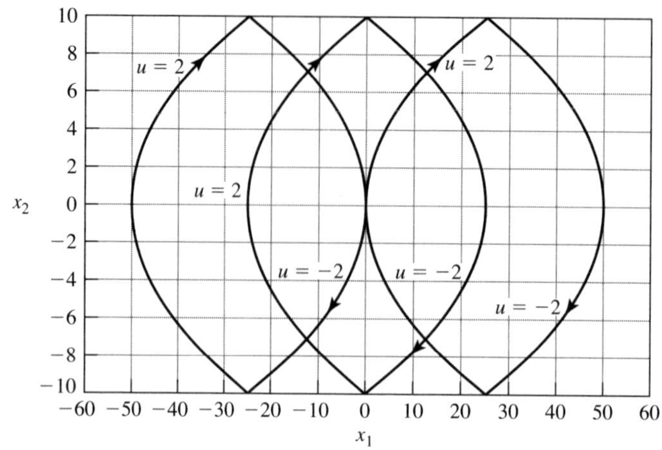

Figure 6.3 Typical trajectories of the double integrator system.

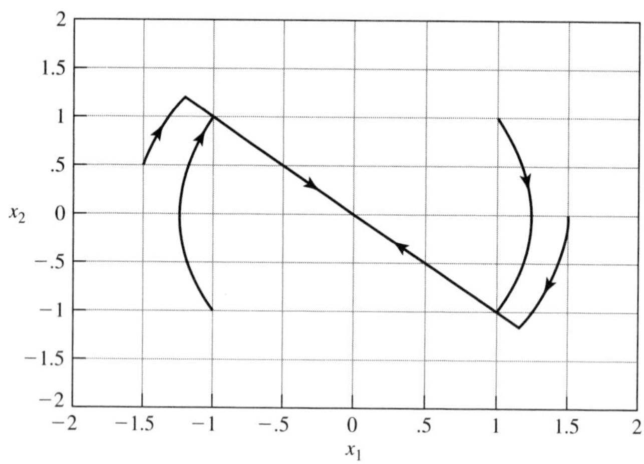

Figure 6.4 Typical trajectories of the closed-system (6.3), (6.4) for initial conditions "close" to the origin of the state-plane. The applied control law is given by (6.4).

We introduce the notation

$$\sigma(\boldsymbol{x}) = \sigma(x_1, x_2) = x_1 + x_2.$$

Then, we can describe the switching logic action as

$$u = \begin{cases} -2 & \text{if } \sigma(\boldsymbol{x}) > 0, \\ 2 & \text{if } \sigma(\boldsymbol{x}) < 0. \end{cases} \tag{6.5}$$

Typical trajectories of the closed-loop system described by (6.3) and (6.4) for initial conditions located "close" to the origin of the state-plane are shown in Figure 6.4. Close to the origin, once the trajectories intercept the switching line, they are confined to it thereafter. If we increase the integration step size, then we can observe oscillations, or *chattering*, of the trajectory about the switching surface as it is explained in Figure 6.5. A plot of two trajectories with an increased integration time is shown in Figure 6.6.

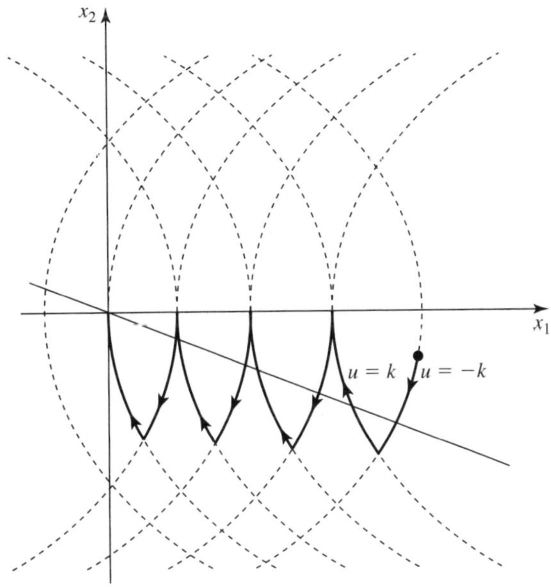

Figure 6.5 Chattering in a variable structure sliding mode system—a magnified view.

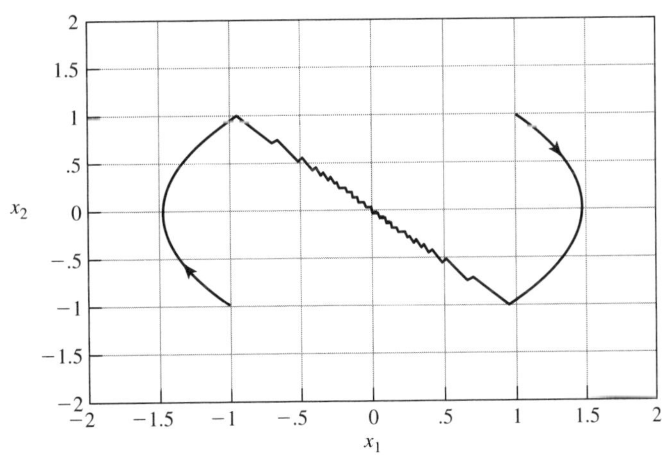

Figure 6.6 An illustration of chattering in a variable structure sliding mode system consisting of the double integrator and the control law, $u = -\text{sign}(x_1 + x_2)$. Two trajectories starting from two different initial conditions are plotted.

6.2 Sliding Mode Definition

The motion of the system while confined to the switching line or a surface is referred to as *sliding*. A sliding mode will exist if in the vicinity of the switching surface the state velocity vectors are directed toward the surface. In such a case, the switching surface attracts trajectories when they are in its vicinity; and once a trajectory intersects the switching surface, it will stay on it thereafter. A surface $\sigma(\boldsymbol{x}) = 0$ is attractive if

1. any trajectory starting on the surface remains there, and
2. any trajectory starting outside the surface tends to it at least asymptotically.

Thus, for a sliding motion to occur we need

$$\lim_{\sigma\to 0^+} \dot{\sigma} < 0 \quad \text{and} \quad \lim_{\sigma\to 0^-} \dot{\sigma} > 0.$$

The above conditions are illustrated in Figure 6.7. They ensure that the motion of the state trajectory $\boldsymbol{x}$ on either side of the switching surface $\sigma(\boldsymbol{x}) = 0$ is toward the switching surface. The two conditions may be combined to give

$$\sigma\dot{\sigma} < 0 \tag{6.6}$$

in the neighborhood of the switching surface. Note, however, that the closed-loop system is now modeled by differential equations with discontinuous right-hand sides. The modeling differential equations are discontinuous on the switching surface $\sigma(\boldsymbol{x}) = 0$. This is the reason we often refer to the switching surface as the *discontinuity surface*. A rigorous analysis of such systems cannot be carried out using methods of classical theory of differential equations. A solution to the differential equation $\dot{\boldsymbol{x}} = \boldsymbol{f}(t, \boldsymbol{x})$ with a continuous right-hand side is a function $\boldsymbol{x}(t)$ that has a derivative and satisfies the differential equation everywhere on a given interval. In the case when the right-hand side of $\dot{\boldsymbol{x}} = \boldsymbol{f}(t, \boldsymbol{x})$ is discontinuous, we have to use more general methods such as one, for example, developed by Filippov [84]. We illustrate these points with a numerical example taken from Filippov [84, p. 1].

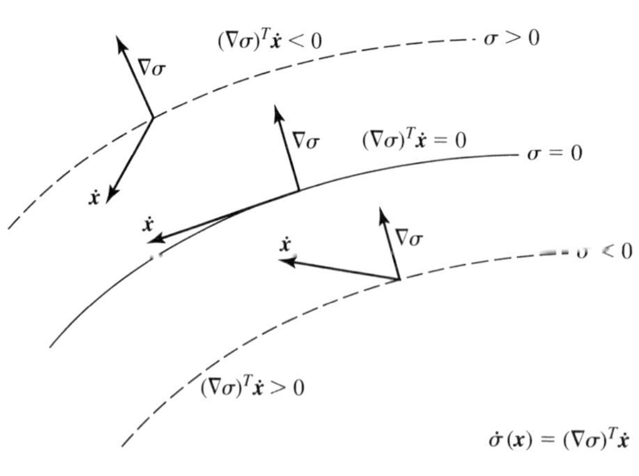

Figure 6.7 An illustration of the sufficient condition of the attractivity to the switching surface.

◆ Example 6.3

Consider the following differential equation with the discontinuous right-hand side:

$$\dot{x} = 1 - 2\,\mathrm{sign}(x). \tag{6.7}$$

For $x < 0$, we have $\dot{x} = 3$, while for $x > 0$, $\dot{x} = -1$. In Figure 6.8, plots of x versus time for different initial conditions are given. As t increases, the solution x of (6.7) reaches the line $x = 0$ and stays there. The function $x(t)$, however, does not satisfy (6.7) in the usual sense because if $\dot{x}(t) = 0$ and $x(t) = 0$, the right-hand side of (6.7) becomes

$$1 - 2\,\mathrm{sign}(0) = 1 \neq 0,$$

where we defined $\mathrm{sign}(0) = 0$.

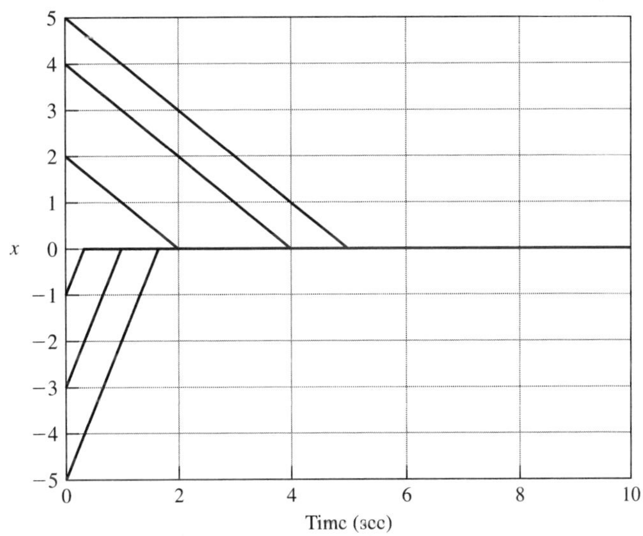

Figure 6.8 Plots of x versus time for different initial conditions for the differential equation (6.7).

We now briefly describe Filippov's approach. Filippov [84, pp. 2–3] recommends that the generalized solution to an ordinary differential equation with a discontinuous right-hand side should meet the following criteria:

1. For a differential equation with a continuous right-hand side, the generalized solution should be equivalent to the usual one.
2. For the equation $\dot{x} = f(t)$, the generalized solution should be of the form

$$x(t) = \int f(t)\,dt + \text{constant}.$$

3. For any initial condition $x(t_0) = x_0$, the solution must exist at least for $t > t_0$ and should be continuable.
4. The generalized solution should be applicable to a broad class of equations that model dynamical systems.
5. The limit of an uniformly convergent sequence of solutions should be a solution.
6. After applying to a solution a common change of variables, the result of the transformation should be a solution.

We next describe the proposed generalized solution in the sense of Filippov. We begin with an example. Suppose that we are given a single-input dynamical system model, $\dot{\boldsymbol{x}} = \boldsymbol{f}(\boldsymbol{x}, u)$ along with a switching surface $\sigma(\boldsymbol{x}) = 0$. Let $\boldsymbol{f}^- = \boldsymbol{f}(\boldsymbol{x}, u^-)$ and $\boldsymbol{f}^+ = \boldsymbol{f}(\boldsymbol{x}, u^+)$, where $u = u^-$ if $\sigma < 0$ and $u = u^+$ if $\sigma > 0$. We next join the velocity vectors $\boldsymbol{f}^-$ and $\boldsymbol{f}^+$. The resulting velocity vector $\boldsymbol{f}^0$ is obtained by drawing the line tangent to the switching surface that intersects the line segment joining the velocity vectors $\boldsymbol{f}^-$ and $\boldsymbol{f}^+$. Thus,

$$\boldsymbol{f}^0 = \alpha \boldsymbol{f}^- + (1 - \alpha)\boldsymbol{f}^+,$$

where $\alpha \in [0, 1]$. The above construction is illustrated in Figure 6.9. Observe that $\boldsymbol{f}^0$ belongs to the smallest convex set containing $\boldsymbol{f}^-$ and $\boldsymbol{f}^+$. Recall that a set $\mathcal{S}$ is said to be convex if for every $\boldsymbol{x}, \boldsymbol{y} \in \mathcal{S}$ and every real number α, $0 < \alpha < 1$, we have $\alpha\boldsymbol{x} + (1 - \alpha)\boldsymbol{y} \in \mathcal{S}$.

Following Itkis [140], we give a simple interpretation of the above construction. The number α may be interpreted as the time the system's trajectory remains below the switching surface, while $(1 - \alpha)$ may be interpreted as the time the trajectory remains above the switching surface during the trajectory oscillation period about the switching surface. Specifically, if the trajectory performs one oscillation about $\sigma = 0$ in a "small" time interval $\Delta\tau$, then it will traverse a distance $\|\boldsymbol{f}^-\|\alpha\Delta\tau$ in time $\alpha\Delta\tau$ with velocity $\boldsymbol{f}^-$. During the remaining time of oscillation the trajectory will traverse a distance $\|\boldsymbol{f}^+\|(1 - \alpha)\Delta\tau$ in time $(1 - \alpha)\Delta\tau$ with velocity $\boldsymbol{f}^+$. The average displacement of the trajectory during the time $\Delta\tau$ will then be $\|\alpha\boldsymbol{f}^- + (1 - \alpha)\boldsymbol{f}^+\|\Delta\tau$ with the average velocity $\boldsymbol{f}^0$.

Suppose now that there are m discontinuity surfaces of the form

$$\sigma_i = \{\boldsymbol{x} : \sigma_i(\boldsymbol{x}) = 0\}, \qquad i = 1, 2, \ldots, m.$$

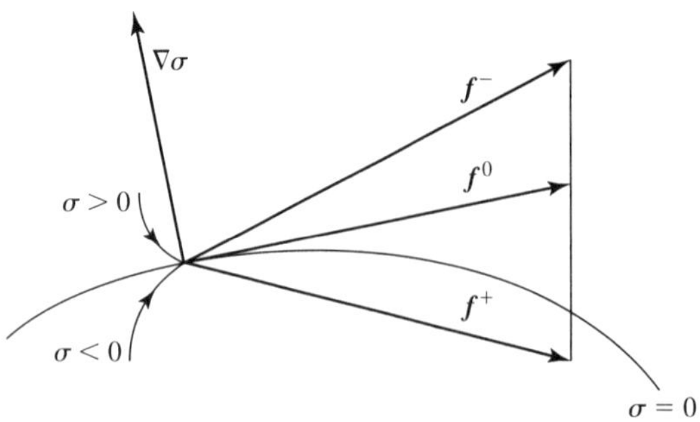

Figure 6.9 Constructing a solution in the sense of Filippov in $\mathbb{R}^2$.

Let

$$H = \bigcup_{i=1}^{m} \sigma_i .$$

Then, we take a solution of the differential equation $\dot{\boldsymbol{x}} = \boldsymbol{f}(\boldsymbol{x})$ with the discontinuous right-hand side to be a solution of the differential inclusion

$$\dot{\boldsymbol{x}}(t) \in G(\boldsymbol{x}(t)),$$

where for each $\boldsymbol{x}$ the set $G(\boldsymbol{x})$ is the smallest closed convex set containing the cluster values, limits, of the function $\boldsymbol{f}(\boldsymbol{y})$ as $\boldsymbol{y} \to \boldsymbol{x}$, $\boldsymbol{y} \notin H$; that is, a solution of the dynamical system $\dot{\boldsymbol{x}} = \boldsymbol{f}(\boldsymbol{x})$ is an absolutely continuous function $\boldsymbol{x}(t)$, in the sense of Definition A.2, defined on an interval, or segment, L for which $\dot{\boldsymbol{x}}(t) \in G(\boldsymbol{x}(t))$ almost everywhere on L (Filippov [84]). Observe that if the function $\boldsymbol{f}$ is continuous at a point $\boldsymbol{x}$—that is, $\boldsymbol{x} \notin H$—then $G(\boldsymbol{x})$ consists of a single point, namely $\boldsymbol{f}(\boldsymbol{x})$, and the solution satisfies $\dot{\boldsymbol{x}} = \boldsymbol{f}(\boldsymbol{x})$ in the usual sense. However, if $\boldsymbol{x} \in H$, then $\boldsymbol{x}$ lies on one or more of the surfaces $\sigma_1, \sigma_2, \ldots, \sigma_m$. Suppose that $\boldsymbol{x}$ lies on the surfaces $\sigma_{i_1}, \sigma_{i_2}, \ldots, \sigma_{i_k}$. It then follows that for a sufficiently small $\delta > 0$ the surfaces $\sigma_{i_1}, \sigma_{i_2}, \ldots, \sigma_{i_k}$ partition the δ-neighborhood of $\boldsymbol{x}$ into $\ell \leq 2^k$ regions $\mathcal{R}_1, \mathcal{R}_2, \ldots, \mathcal{R}_\ell$, in each of which the function $\boldsymbol{f}$ is continuous. Let $\boldsymbol{f}_j$ denote the function $\boldsymbol{f}$ restricted to the region $\mathcal{R}_j$. Then, $G(\boldsymbol{x})$ is the smallest closed convex set containing the set of points

$$\left\{ \lim_{\substack{\tilde{\boldsymbol{x}} \to \boldsymbol{x} \\ \tilde{\boldsymbol{x}} \in \mathcal{R}_j}} \boldsymbol{f}_j(\tilde{\boldsymbol{x}}) : j = 1, 2, \ldots, \ell \right\}. \tag{6.8}$$

The velocity vector, $\dot{\boldsymbol{x}} = \boldsymbol{f}^0$, lies then on the intersection of the tangent plane to the surfaces of discontinuity and the convex set described by (6.8). This is illustrated in Figures 6.9 and 6.10.

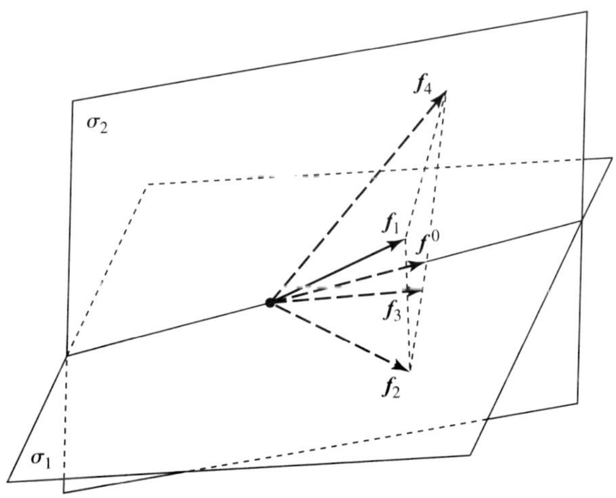

Figure 6.10 Constructing a solution in the sense of Filippov in $\mathbb{R}^3$.

6.3 A Simple Sliding Mode Controller

In this section we discuss the problem of designing sliding mode controllers for single-input plants. The design of a variable structure sliding mode controller consists of two phases:

- Sliding (switching) surface design so as to achieve the desired system behavior, like the stability to the origin, when restricted to the surface.
- Selecting feedback gains of the controller so that the closed-system is stable to the sliding surface.

There are a number of benefits when the system is in sliding. In particular:

- While in sliding, the system is not affected by matched uncertainties.
- While in sliding, the system behavior is governed by a reduced set of differential equations. This is what we refer to as the *order reduction*. It is very useful in designing variable structure sliding mode controllers.

We now discuss a sliding surface design phase for linear single-input plants. Without loss of generality, we assume that the plant model is already in its controller companion form,

$$\dot{\boldsymbol{x}} = \begin{bmatrix} 0 & 1 & 0 & \cdots & 0 \\ 0 & 0 & 1 & \cdots & 0 \\ \vdots & & & & \vdots \\ 0 & 0 & 0 & \cdots & 1 \\ -a_0 & -a_1 & -a_2 & \cdots & -a_{n-1} \end{bmatrix} \boldsymbol{x} + \begin{bmatrix} 0 \\ 0 \\ \vdots \\ 0 \\ 1 \end{bmatrix} u.$$

Let the sliding surface be

$$\begin{aligned} \sigma(\boldsymbol{x}) &= \boldsymbol{s}\boldsymbol{x} \\ &= [s_1 \quad s_2 \quad \cdots \quad s_{n-1} \quad 1] \begin{bmatrix} x_1 \\ \vdots \\ x_n \end{bmatrix} \\ &= s_1 x_1 + s_2 x_2 + \cdots + s_{n-1} x_{n-1} + x_n \\ &= 0. \end{aligned}$$

Note that, without loss of generality, we assumed that $s_n = 1$. If this was not the case, we could divide both sides of $\boldsymbol{s}\boldsymbol{x} = 0$ by $s_n \neq 0$ to ensure that $s_n = 1$. Using the above equation of sliding surface, we can write

$$x_n = -s_1 x_1 - s_2 x_2 - \cdots - s_{n-1} x_{n-1}.$$

We now combine the equations of the plant model and that of the sliding surface. First note that

$$\dot{x}_{n-1} = x_n.$$

Hence, we can write

$$\begin{aligned} \dot{x}_1 &= x_2, \\ \dot{x}_2 &= x_3, \\ &\vdots \\ \dot{x}_{n-1} &= -s_1 x_1 - s_2 x_2 - \cdots - s_{n-1} x_{n-1}, \end{aligned} \tag{6.9}$$

or, equivalently in matrix form,

$$\dot{z} = \begin{bmatrix} 0 & 1 & 0 & \cdots & 0 \\ 0 & 0 & 1 & \cdots & 0 \\ \vdots & & & & \vdots \\ 0 & 0 & 0 & \cdots & 1 \\ -s_1 & -s_2 & -s_3 & \cdots & -s_{n-1} \end{bmatrix} z, \tag{6.10}$$

where $z = [x_1 \ x_2 \ \cdots \ x_{n-1}]^T$. The above equation describes the system dynamics in sliding. Observe the order reduction of the system dynamics in sliding. We now have $n - m = n - 1$ first-order differential equations. The poles of the reduced-order system are the roots of the equation

$$\lambda^{n-1} + s_{n-1}\lambda^{n-2} + \cdots + s_2\lambda + s_1 = 0.$$

The response of the system in sliding is completely specified by an appropriate choice of the components $s_1, s_2, \ldots, s_{n-1}$ of the switching surface.

Consider again Example 6.2. The sliding surface for this example can be represented as

$$\sigma(\boldsymbol{x}) = [s_1 \quad 1]\boldsymbol{x} = 0,$$

or, equivalently,

$$s_1 x_1 + x_2 = 0.$$

The dynamics of the system in sliding are described by

$$\dot{z} = -s_1 z,$$

which is just a first-order system. Note that to ensure the reduced-system stability, we select $s_1 > 0$.

After designing a sliding surface, we construct a feedback controller. In this section we discuss a controller design for linear single-input plants. The controller objective is to drive the plant state to the sliding surface and maintain it on the surface for all subsequent time. We use a generalized Lyapunov approach in constructing the controller. Specifically, we use a distance, $V = 0.5\sigma^2$, from the sliding surface σ as a Lyapunov function candidate. Then, we select the controllers' gains so that the time derivative of the chosen Lyapunov function candidate evaluated on the solution of the controlled system is negative definite with respect to the switching surface, thus ensuring the motion of the state trajectory to the surface as it is illustrated in Figure 6.7. Our goal is to find $u = u(\boldsymbol{x})$ so that

$$\frac{d}{dt}\left(\frac{1}{2}\sigma^2\right) = \sigma\dot{\sigma} < 0.$$

We consider the controller structure of the form

$$u = k_1^{\pm} x_1 + k_2^{\pm} x_2 + \cdots + k_n^{\pm} x_n, \tag{6.11}$$

where $k_i^{\pm}$ are the feedback gains to be determined so that the condition $\sigma\dot{\sigma} < 0$ is satisfied. To determine the feedback gains, we substitute u, given by (6.11), into the expression $\sigma\dot{\sigma}$ and utilize (6.9) to get

$$\begin{aligned}
\sigma\dot{\sigma} &= \sigma[s_1 \quad s_2 \quad \cdots \quad 1]\dot{\boldsymbol{x}} \\
&= \sigma(s_1\dot{x}_1 + \cdots + s_{n-1}\dot{x}_{n-1} + \dot{x}_n) \\
&= \sigma(s_1 x_2 + \cdots + s_{n-1}x_n + \dot{x}_n) \\
&= \sigma(s_1 x_2 + \cdots + s_{n-1}x_n - a_0 x_1 - \cdots - a_{n-1}x_n + u) \\
&= \sigma\left(s_1 x_2 + \cdots + s_{n-1}x_n - a_0 x_1 - \cdots - a_{n-1}x_n + k_1^{\pm}x_1 + \cdots + k_n^{\pm}x_n\right).
\end{aligned}$$

Rearranging the terms yields

$$\sigma\dot{\sigma} = \sigma x_1\left(-a_0 + k_1^{\pm}\right) + \sigma x_2\left(-a_1 + s_1 + k_2^{\pm}\right) + \cdots + \sigma x_n\left(-a_{n-1} + s_{n-1} + k_n^{\pm}\right).$$

A possible way to ensure that $\sigma\dot{\sigma} < 0$ is to select the gains to force each term in $\sigma\dot{\sigma}$ to be negative, that is,

$$\begin{aligned}
\sigma x_1\left(-a_0 + k_1^{\pm}\right) &< 0, \\
\sigma x_2\left(-a_1 + s_1 + k_2^{\pm}\right) &< 0, \\
&\vdots \\
\sigma x_n\left(-a_{n-1} + s_{n-1} + k_n^{\pm}\right) &< 0.
\end{aligned}$$

A possible choice of gains that leads to satisfaction of the attractivity to the surface condition, $\sigma\dot{\sigma} < 0$, is

$$\begin{aligned}
k_1^{+} &< -|a_0| && \text{if } \sigma x_1 > 0, \\
k_1^{-} &> |a_0| && \text{if } \sigma x_1 < 0, \\
&\vdots \\
k_n^{+} &< -(|a_{n-1}| + |s_{n-1}|) && \text{if } \sigma x_n > 0, \\
k_n^{-} &> |a_{n-1}| + |s_{n-1}| && \text{if } \sigma x_n < 0.
\end{aligned} \tag{6.12}$$

We can also use the following controller gains:

$$\begin{aligned}
k_1^{\pm} &= -(|a_0| + \varepsilon)\operatorname{sign}(x_1\sigma), \\
&\vdots \\
k_n^{\pm} &= -(|a_{n-1}| + |s_{n-1}| + \varepsilon)\operatorname{sign}(x_n\sigma),
\end{aligned} \tag{6.13}$$

where $\varepsilon > 0$ is a design parameter. Note that the controllers' gains given by (6.13) satisfy conditions (6.12). The larger the value of ε the faster the trajectory converges to the sliding surface.

6.4 Sliding in Multi-Input Systems

In this section we describe a sliding mode in linear multi-input plants. When the number of the plant inputs is m, we can use sliding surface of the form

$$\{\boldsymbol{x} : \boldsymbol{S}\boldsymbol{x} = \boldsymbol{0},\ \boldsymbol{S} \in \mathbb{R}^{m\times n}\}.$$

We also use $\boldsymbol{S}\boldsymbol{x} = \boldsymbol{0}$ to denote the above sliding surface. In many applications, nonlinear sliding surfaces arise. Here, we use linear sliding surfaces only, although one can use nonlinear sliding surfaces as well. Let $\boldsymbol{\sigma}(\boldsymbol{x}) = \boldsymbol{S}\boldsymbol{x}$. Then,

$$\boldsymbol{\sigma}(\boldsymbol{x}) = \begin{bmatrix} \sigma_1(\boldsymbol{x}) \\ \sigma_2(\boldsymbol{x}) \\ \vdots \\ \sigma_m(\boldsymbol{x}) \end{bmatrix} = \begin{bmatrix} \boldsymbol{s}_1\boldsymbol{x} \\ \boldsymbol{s}_2\boldsymbol{x} \\ \vdots \\ \boldsymbol{s}_m\boldsymbol{x} \end{bmatrix},$$

where $\boldsymbol{s}_i \in \mathbb{R}^{1\times n}$. Recall, that the system is in sliding mode if $\boldsymbol{\sigma}(\boldsymbol{x}(t)) = \boldsymbol{0}$ for $t \geq t_0$, where $\boldsymbol{x}$ is the state trajectory and t_0 is a specified time. It follows that in a sliding mode the velocity vector $\dot{\boldsymbol{x}}(t)$ is tangent to the switching surface. Equivalently,

$$(\nabla\sigma_i)^T\dot{\boldsymbol{x}}(t) = \boldsymbol{s}_i\dot{\boldsymbol{x}}(t) = 0 \qquad \text{for } i = 1, 2, \ldots, m,$$

that is,

$$\dot{\boldsymbol{\sigma}}(\boldsymbol{x}(t)) = \boldsymbol{S}\dot{\boldsymbol{x}}(t) = \boldsymbol{0}.$$

We can thus characterize the system in sliding mode as

$$\begin{aligned} \boldsymbol{\sigma}(\boldsymbol{x}(t)) &= \boldsymbol{0}, \\ \dot{\boldsymbol{\sigma}}(\boldsymbol{x}(t)) &= \boldsymbol{0}. \end{aligned} \tag{6.14}$$

Consider now the plant model $\dot{\boldsymbol{x}}(t) = \boldsymbol{A}\boldsymbol{x}(t) + \boldsymbol{B}\boldsymbol{u}(t)$. Combining $\boldsymbol{S}\dot{\boldsymbol{x}}(t) = \boldsymbol{0}$ and the modeling equation of the plant, we get

$$\boldsymbol{S}\boldsymbol{A}\boldsymbol{x} + \boldsymbol{S}\boldsymbol{B}\boldsymbol{u} = \boldsymbol{0}.$$

If the matrix $\boldsymbol{S}\boldsymbol{B}$ is nonsingular, then we can solve the above algebraic equation for $\boldsymbol{u}$. The result is called the *equivalent control,* and it is denoted using the symbol $\boldsymbol{u}_{eq}$. We have

$$\boldsymbol{u}_{eq} = -(\boldsymbol{S}\boldsymbol{B})^{-1}\boldsymbol{S}\boldsymbol{A}\boldsymbol{x}.$$

Substituting $\boldsymbol{u}_{eq}$ into the plant model equation yields

$$\dot{\boldsymbol{x}} = (\boldsymbol{I}_n - \boldsymbol{B}(\boldsymbol{S}\boldsymbol{B})^{-1}\boldsymbol{S})\boldsymbol{A}\boldsymbol{x}. \tag{6.15}$$

The above system is sometimes referred to as the *equivalent system*. Thus, we can describe the behavior of the system in sliding alternatively as

$$\begin{aligned} \dot{\boldsymbol{x}} &= (\boldsymbol{I}_n - \boldsymbol{B}(\boldsymbol{S}\boldsymbol{B})^{-1}\boldsymbol{S})\boldsymbol{A}\boldsymbol{x}, \\ \boldsymbol{S}\boldsymbol{x} &= \boldsymbol{0}. \end{aligned} \tag{6.16}$$

Note that in sliding, the system is governed by a reduced set of differential equations that are obtained by combining the above two sets of equations.

6.5 Sliding Mode and System Zeros

In this section we show that the poles of the reduced-order system in sliding are the so-called *system zeros* of the dynamical system described by

$$\begin{aligned} \dot{x} &= Ax + Bu, \\ \sigma &= Sx, \end{aligned} \tag{6.17}$$

where $S \in \mathbb{R}^{m \times n}$ and we assume that $\det SB \neq 0$. The system zeros are defined using the *system matrix*. The system matrix, $P(s)$, corresponding to the plant (6.17) has the form

$$P(s) = \begin{bmatrix} sI_n - A & -B \\ S & O \end{bmatrix}. \tag{6.18}$$

The system matrix $P(s)$ corresponding to the plant (6.17) is a square matrix of order $n+m$. The zeros of the determinant of the system matrix (6.18) are the system zeros of (6.17). The system zeros are invariant under the following set of transformations:

1. Nonsingular state variable transformation
2. Nonsingular transformation of the inputs
3. Nonsingular transformation of the outputs
4. State feedback to the inputs
5. Output feedback to the rates of change of the states

We now evaluate the determinant of the system matrix (6.18). To facilitate this task, we premultiply the system matrix by the matrix

$$\begin{bmatrix} I_n & O \\ -S(sI_n - A)^{-1} & I_m \end{bmatrix}.$$

The determinant of the above matrix is unity and so the determinant of the product is the same as the determinant of the system matrix,

$$\begin{aligned} \det P(s) &= \det\left(\begin{bmatrix} I_n & O \\ -S(sI_n - A)^{-1} & I_m \end{bmatrix} \begin{bmatrix} sI_n - A & -B \\ S & O \end{bmatrix} \right) \\ &= \det \begin{bmatrix} sI_n - A & -B \\ O & S(sI_n - A)^{-1}B \end{bmatrix} \\ &= \det(sI_n - A)\det(S(sI_n - A)^{-1}B). \end{aligned} \tag{6.19}$$

We next evaluate the characteristic polynomial of the equivalent system (6.15). While performing manipulations, we use the result of Exercise A.3 to obtain

$$\begin{aligned} &\det(sI_n - A + B(SB)^{-1}SA) \\ &\quad = \det(sI_n - A)\det(I_n + (sI_n - A)^{-1}B(SB)^{-1}SA) \\ &\quad = \det(sI_n - A)\det(I_m + SA(sI_n - A)^{-1}B(SB)^{-1}). \end{aligned} \tag{6.20}$$

In our further manipulations, we use the identity

$$A(\operatorname{Adj}(sI_n - A)) = s\operatorname{Adj}(sI_n - A) - \det(sI_n - A)I_n. \tag{6.21}$$

The above follows from the fact that

$$(s\boldsymbol{I}_n - \boldsymbol{A})(s\boldsymbol{I}_n - \boldsymbol{A})^{-1} = \boldsymbol{I}_n,$$

which is equivalent to

$$(s\boldsymbol{I}_n - \boldsymbol{A})\operatorname{Adj}(s\boldsymbol{I}_n - \boldsymbol{A}) = \det(s\boldsymbol{I}_n - \boldsymbol{A})\boldsymbol{I}_n. \tag{6.22}$$

We use (6.21) to get

$$\begin{aligned}
&\det(\boldsymbol{I}_m + \boldsymbol{S}\boldsymbol{A}(s\boldsymbol{I}_n - \boldsymbol{A})^{-1}\boldsymbol{B}(\boldsymbol{S}\boldsymbol{B})^{-1}) \\
&\quad = \det\left(\frac{\det(s\boldsymbol{I}_n - \boldsymbol{A})\boldsymbol{I}_m + \boldsymbol{S}(s\operatorname{Adj}(s\boldsymbol{I}_n - \boldsymbol{A}) - \det(s\boldsymbol{I}_n - \boldsymbol{A})\boldsymbol{I}_n)\boldsymbol{B}(\boldsymbol{S}\boldsymbol{B})^{-1}}{\det(s\boldsymbol{I}_n - \boldsymbol{A})}\right) \\
&\quad = \det\left(\frac{s\boldsymbol{S}\operatorname{Adj}(s\boldsymbol{I}_n - \boldsymbol{A})\boldsymbol{B}(\boldsymbol{S}\boldsymbol{B})^{-1}}{\det(s\boldsymbol{I}_n - \boldsymbol{A})}\right) \\
&\quad = s^m \det(\boldsymbol{S}(s\boldsymbol{I}_n - \boldsymbol{A})^{-1}\boldsymbol{B})\det(\boldsymbol{S}\boldsymbol{B})^{-1}.
\end{aligned}$$

Using (6.19) and the above expression, we represent (6.20) as

$$\begin{aligned}
&\det(s\boldsymbol{I}_n - \boldsymbol{A} + \boldsymbol{B}(\boldsymbol{S}\boldsymbol{B})^{-1}\boldsymbol{S}\boldsymbol{A}) \\
&\quad = s^m \det(\boldsymbol{S}\boldsymbol{B})^{-1}\det(s\boldsymbol{I}_n - \boldsymbol{A})\det(\boldsymbol{S}(s\boldsymbol{I}_n - \boldsymbol{A})^{-1}\boldsymbol{B}),
\end{aligned}$$

and hence

$$\boxed{\det(s\boldsymbol{I}_n - \boldsymbol{A} + \boldsymbol{B}(\boldsymbol{S}\boldsymbol{B})^{-1}\boldsymbol{S}\boldsymbol{A}) = \det(\boldsymbol{S}\boldsymbol{B})^{-1}s^m \det \boldsymbol{P}(s)} \tag{6.23}$$

We conclude that the dynamics—that is, the poles—of the system modeled by (6.17) in sliding along $\boldsymbol{\sigma} = \boldsymbol{S}\boldsymbol{x}$ are determined by its system zeros. This fact leads to the observation that the switching surface design can be viewed as choosing an output matrix $\boldsymbol{S}$ so that the system (6.17) has a desired set of system zero locations, which in turn govern the dynamics of the system in sliding along $\boldsymbol{S}\boldsymbol{x} = \boldsymbol{0}$.

6.6 Nonideal Sliding Mode

We consider a nonlinear dynamical system modeled by

$$\dot{\boldsymbol{x}} = \boldsymbol{f}(t, \boldsymbol{x}) + \boldsymbol{B}(t, \boldsymbol{x})\boldsymbol{u}, \tag{6.24}$$

where $\boldsymbol{x}(t) \in \mathbb{R}^n$, $\boldsymbol{f}(t, \boldsymbol{x}(t)) \in \mathbb{R}^n$, $\boldsymbol{B}(t, \boldsymbol{x}(t)) \in \mathbb{R}^{n\times m}$, and $\boldsymbol{u} \in \mathbb{R}^m$. The functions $\boldsymbol{f}$ and $\boldsymbol{B}$ are continuous functions of all their arguments. The switching surface is $\{\boldsymbol{x} : \boldsymbol{\sigma}(\boldsymbol{x}) = \boldsymbol{0}\}$, where $\boldsymbol{\sigma}(\boldsymbol{x}(t)) \in \mathbb{R}^m$. Let

$$\boldsymbol{S} = \frac{\partial \boldsymbol{\sigma}}{\partial \boldsymbol{x}},$$

where we assume that $\boldsymbol{S}\boldsymbol{B}$ is invertible. Ideally, once the system enters sliding mode, an appropriately constructed controller is switched with an infinite frequency to keep the system on the surface $\{\boldsymbol{\sigma} = \boldsymbol{0}\}$. Thus, in the ideal case, once the state trajectory hits the switching surface, say at t_0, then for $t \geq t_0$ we have $\boldsymbol{\sigma} = \boldsymbol{0}$ and $\dot{\boldsymbol{\sigma}}(\boldsymbol{x}) = \boldsymbol{0}$. Solving the algebraic equation

$$\dot{\boldsymbol{\sigma}}(\boldsymbol{x}) = \boldsymbol{S}\dot{\boldsymbol{x}} = \boldsymbol{S}(\boldsymbol{f}(t, \boldsymbol{x}) + \boldsymbol{B}(t, \boldsymbol{x})\boldsymbol{u}) = \boldsymbol{0}$$

for $\boldsymbol{u}$ yields

$$\boldsymbol{u} = -(\boldsymbol{S}\boldsymbol{B})^{-1}\boldsymbol{S}\boldsymbol{f}.$$

Recall that the above expression for $\boldsymbol{u}$ is called the equivalent control and is denoted $\boldsymbol{u}_{eq}$. Substituting $\boldsymbol{u}_{eq}$ into to the modeling equation (6.24), we obtain

$$\dot{\boldsymbol{x}} = \boldsymbol{f} - \boldsymbol{B}(\boldsymbol{S}\boldsymbol{B})^{-1}\boldsymbol{S}\boldsymbol{f} = (\boldsymbol{I}_n - \boldsymbol{B}(\boldsymbol{S}\boldsymbol{B})^{-1}\boldsymbol{S})\boldsymbol{f}.$$

The above equation together with the condition $\boldsymbol{\sigma} = \mathbf{0}$ describes the system in ideal sliding.

We now study a nonideal sliding mode. In nonideal sliding mode we no longer have $\dot{\boldsymbol{\sigma}}(\boldsymbol{x}) = \mathbf{0}$. Hence, solving the equation

$$\dot{\boldsymbol{\sigma}}(\boldsymbol{x}) = \boldsymbol{S}(\boldsymbol{f}(t, \boldsymbol{x}) + \boldsymbol{B}(t, \boldsymbol{x})\boldsymbol{u})$$

for $\boldsymbol{u}$, which we now refer to as $\tilde{\boldsymbol{u}}$, yields

$$\tilde{\boldsymbol{u}} = -(\boldsymbol{S}\boldsymbol{B})^{-1}\boldsymbol{S}\boldsymbol{f} + (\boldsymbol{S}\boldsymbol{B})^{-1}\dot{\boldsymbol{\sigma}}.$$

Substituting the above into the plant dynamics gives

$$\dot{\tilde{\boldsymbol{x}}} = \boldsymbol{f} - \boldsymbol{B}(\boldsymbol{S}\boldsymbol{B})^{-1}\boldsymbol{S}\boldsymbol{f} + \boldsymbol{B}(\boldsymbol{S}\boldsymbol{B})^{-1}\dot{\boldsymbol{\sigma}}, \tag{6.25}$$

where we used the notation $\tilde{\boldsymbol{x}}$ to indicate the state of the system in nonideal sliding mode. The above describes the plant modeled by (6.24) in nonideal sliding mode. The term $\boldsymbol{B}(\boldsymbol{S}\boldsymbol{B})^{-1}\dot{\boldsymbol{\sigma}}$ on the right-hand side of (6.25) can be viewed as representing nonidealities. In real sliding the state trajectory $\tilde{\boldsymbol{x}}$ will reside in some vicinity of the manifold $\{\tilde{\boldsymbol{x}} : \boldsymbol{\sigma}(\tilde{\boldsymbol{x}}) = \mathbf{0}\}$ so that

$$\|\boldsymbol{\sigma}(\tilde{\boldsymbol{x}})\| \le \Delta,$$

where $\Delta > 0$. We now show, following Utkin [285, pp. 45–48], that if $\Delta \to 0$, then the real sliding mode behavior approaches that of the ideal sliding mode behavior.

Theorem 6.1 (Utkin, 1974) Suppose the following:

1. A solution to $\dot{\tilde{\boldsymbol{x}}} = \boldsymbol{f}(t, \tilde{\boldsymbol{x}}) + \boldsymbol{B}(t, \tilde{\boldsymbol{x}})\tilde{\boldsymbol{u}}$, on the interval $[0, T]$, satisfies the condition

$$\|\boldsymbol{\sigma}(\tilde{\boldsymbol{x}})\| \le \Delta,$$

   where $\tilde{\boldsymbol{u}}$ may incorporate nonidealities corresponding to $\dot{\boldsymbol{\sigma}} \ne \mathbf{0}$.
2. There is a Lipschitz constant L for the right-hand side of

$$\dot{\boldsymbol{x}} = \boldsymbol{f} - \boldsymbol{B}(\boldsymbol{S}\boldsymbol{B})^{-1}\boldsymbol{S}\boldsymbol{f}.$$

3. Partial derivatives of the function

$$\boldsymbol{B}(\boldsymbol{S}\boldsymbol{B})^{-1}$$

   exist and are bounded in any bounded domain of $\mathbb{R}^n$.

4. For the function

$$\boldsymbol{f}(t, \tilde{\boldsymbol{x}}) + \boldsymbol{B}(t, \tilde{\boldsymbol{x}})\tilde{\boldsymbol{u}},$$

denoted $\boldsymbol{f} + \boldsymbol{B}\tilde{\boldsymbol{u}}$, there exist nonnegative constants M_1 and M_2 such that

$$\|\boldsymbol{f} + \boldsymbol{B}\tilde{\boldsymbol{u}}\| \le M_1 + M_2\|\tilde{\boldsymbol{x}}\|.$$

Then, for any pair of state trajectories that are solutions to the ideal and nonideal sliding models for which the initial conditions satisfy

$$\|\boldsymbol{x}(0) - \tilde{\boldsymbol{x}}(0)\| \le P\Delta, \qquad P \ge 0,$$

there exists a nonnegative H such that

$$\|\boldsymbol{x}(t) - \tilde{\boldsymbol{x}}(t)\| \le H\Delta$$

for all $t \in [0, T]$.

Proof The trajectory of the system in an ideal sliding mode satisfies

$$\boldsymbol{x}(t) = \boldsymbol{x}(0) + \int_0^t (\boldsymbol{f} - \boldsymbol{B}(\boldsymbol{S}\boldsymbol{B})^{-1}\boldsymbol{S}\boldsymbol{f})(\boldsymbol{x}(\tau))\,d\tau, \tag{6.26}$$

while the trajectory of the system in nonideal sliding satisfies

$$\tilde{\boldsymbol{x}}(t) = \tilde{\boldsymbol{x}}(0) + \int_0^t (\boldsymbol{f} - \boldsymbol{B}(\boldsymbol{S}\boldsymbol{B})^{-1}\boldsymbol{S}\boldsymbol{f})(\tilde{\boldsymbol{x}}(\tau))\,d\tau + \int_0^t \boldsymbol{B}(\boldsymbol{S}\boldsymbol{B})^{-1}\dot{\boldsymbol{\sigma}}(\tilde{\boldsymbol{x}}(\tau))\,d\tau.$$

Integrating the last term of the above by parts gives

$$\begin{aligned}\tilde{\boldsymbol{x}}(t) = \tilde{\boldsymbol{x}}(0) &+ \int_0^t (\boldsymbol{f} - \boldsymbol{B}(\boldsymbol{S}\boldsymbol{B})^{-1}\boldsymbol{S}\boldsymbol{f})(\tilde{\boldsymbol{x}}(\tau))\,d\tau \\ &+ \boldsymbol{B}(\boldsymbol{S}\boldsymbol{B})^{-1}\boldsymbol{\sigma}(\tilde{\boldsymbol{x}}(\tau))\big|_0^t - \int_0^t \left(\frac{d}{d\tau}\boldsymbol{B}(\boldsymbol{S}\boldsymbol{B})^{-1}\right)\boldsymbol{\sigma}(\tilde{\boldsymbol{x}}(\tau))\,d\tau.\end{aligned}$$

Subtracting the above from (6.26), taking the norms of both sides, and applying the triangle inequality yields

$$\begin{aligned}\|\boldsymbol{x}(t) - \tilde{\boldsymbol{x}}(t)\| \le\ & \|\boldsymbol{x}(0) - \tilde{\boldsymbol{x}}(0)\| + \|\boldsymbol{B}(\boldsymbol{S}\boldsymbol{B})^{-1}\boldsymbol{\sigma}(\tilde{\boldsymbol{x}}(\tau))|_0^t\| \\ &+ \int_0^t \left\|\left(\frac{d}{d\tau}\boldsymbol{B}(\boldsymbol{S}\boldsymbol{B})^{-1}\right)\boldsymbol{\sigma}(\tilde{\boldsymbol{x}}(\tau))\right\|\,d\tau \\ &+ \int_0^t \|(\boldsymbol{f} - \boldsymbol{B}(\boldsymbol{S}\boldsymbol{B})^{-1}\boldsymbol{S}\boldsymbol{f})(\boldsymbol{x}(\tau)) - (\boldsymbol{f} - \boldsymbol{B}(\boldsymbol{S}\boldsymbol{B})^{-1}\boldsymbol{S}\boldsymbol{f})(\tilde{\boldsymbol{x}}(\tau))\|\,d\tau.\end{aligned}$$

Taking into account the hypothesis $\|\boldsymbol{x}(0) - \tilde{\boldsymbol{x}}(0)\| \le P\Delta$ and assumption 2, we obtain

$$\begin{aligned}\|\boldsymbol{x}(t) - \tilde{\boldsymbol{x}}(t)\| \le\ & P\Delta + \|\boldsymbol{B}(\boldsymbol{S}\boldsymbol{B})^{-1}\boldsymbol{\sigma}(\tilde{\boldsymbol{x}}(\tau))\,|_0^t\| \\ &+ \int_0^t \left\|\left(\frac{d}{d\tau}\boldsymbol{B}(\boldsymbol{S}\boldsymbol{B})^{-1}\right)\boldsymbol{\sigma}(\tilde{\boldsymbol{x}}(\tau))\right\|\,d\tau + \int_0^t L\,\|\boldsymbol{x}(\tau) - \tilde{\boldsymbol{x}}(\tau)\|\,d\tau.\end{aligned}$$

We next use assumption 1 to get

$$\|\boldsymbol{x}(t)-\tilde{\boldsymbol{x}}(t)\| \leq \Bigg(P+\|\boldsymbol{B}(\boldsymbol{S}\boldsymbol{B})^{-1}(\tilde{\boldsymbol{x}}(t))\|+\|\boldsymbol{B}(\boldsymbol{S}\boldsymbol{B})^{-1}(\tilde{\boldsymbol{x}}(0))\|$$
$$+\int_0^t \left\|\left(\frac{d}{d\tau}\boldsymbol{B}(\boldsymbol{S}\boldsymbol{B})^{-1}\right)(\tilde{\boldsymbol{x}}(\tau))\right\| d\tau\Bigg)\Delta$$
$$+\int_0^t L\|\boldsymbol{x}(\tau)-\tilde{\boldsymbol{x}}(\tau)\|d\tau. \tag{6.27}$$

The trajectory of the system in nonideal sliding is bounded on the interval $[0, T]$. Indeed, using assumption 4 to the system $\dot{\tilde{\boldsymbol{x}}} = \boldsymbol{f}(t, \tilde{\boldsymbol{x}}) + \boldsymbol{B}(t, \tilde{\boldsymbol{x}})\tilde{\boldsymbol{u}}$, we get

$$\begin{aligned}\|\tilde{\boldsymbol{x}}(t)\| &\leq \|\tilde{\boldsymbol{x}}(0)\|+\int_0^t \|\boldsymbol{f}(\tau,\tilde{\boldsymbol{x}})+\boldsymbol{B}(\tau,\tilde{\boldsymbol{x}})\tilde{\boldsymbol{u}}\|d\tau\\ &\leq \|\tilde{\boldsymbol{x}}(0)\|+\int_0^t (M_1+M_2\|\tilde{\boldsymbol{x}}\|)d\tau\\ &\leq \|\tilde{\boldsymbol{x}}(0)\|+M_1T+\int_0^t M_2\|\tilde{\boldsymbol{x}}\|d\tau.\end{aligned}$$

Applying now the Bellman–Gronwall lemma (from 663), to the above integral inequality, we obtain

$$\begin{aligned}\|\tilde{\boldsymbol{x}}(t)\| &\leq (\|\tilde{\boldsymbol{x}}(0)\|+M_1T)e^{M_2t}\\ &\leq (\|\tilde{\boldsymbol{x}}(0)\|+M_1T)e^{M_2T}, \qquad t\in[0,T].\end{aligned}$$

The above means that $\tilde{\boldsymbol{x}}$ is bounded on the interval $[0, T]$ and hence the function $\boldsymbol{B}(\boldsymbol{S}\boldsymbol{B})^{-1}$ is bounded on the same interval. By assumption 3, partial derivatives of the function $\boldsymbol{B}(\boldsymbol{S}\boldsymbol{B})^{-1}$ exist and are bounded. Applying the above to (6.27), we conclude that for some $S \geq 0$ that depends on the initial conditions, time T, and constant P, we have

$$\|\boldsymbol{x}(t)-\tilde{\boldsymbol{x}}(t)\| \leq S\Delta + L\int_0^t \|\boldsymbol{x}(\tau)-\tilde{\boldsymbol{x}}(\tau)\|d\tau.$$

Application of the Bellman–Gronwall lemma to the above inequality yields

$$\|\boldsymbol{x}(t)-\tilde{\boldsymbol{x}}(t)\| \leq S\Delta e^{LT}, \qquad t\in[0,T].$$

Let $H = Se^{LT}$. Then, we can write

$$\|\boldsymbol{x}(t)-\tilde{\boldsymbol{x}}(t)\| \leq H\Delta,$$

which completes the proof of the theorem.

6.7 Sliding Surface Design

We now prove a sufficient condition for the existence of a sliding surface $\sigma(\boldsymbol{x}) = \boldsymbol{S}\boldsymbol{x} = \boldsymbol{0}$ such that the plant model $\dot{\boldsymbol{x}} = \boldsymbol{A}\boldsymbol{x} + \boldsymbol{B}\boldsymbol{u}$ restricted to it has arbitrarily prescribed poles. A general method for designing a sliding surface is contained implicitly in the proof of the following theorem that can be found in Hui and Żak [132]. In our discussion, we use the following notation. For a full-column-rank matrix $\boldsymbol{X}$, its left inverse will be denoted $\boldsymbol{X}^g$. Thus, $\boldsymbol{X}^g\boldsymbol{X} = \boldsymbol{I}$, where $\boldsymbol{I}$ is the identity matrix whose size is equal to the number of columns of $\boldsymbol{X}$. The imaginary part of a complex number z is denoted $\Im(z)$. The set of eigenvalues of a square matrix $\boldsymbol{A}$, also called its spectrum, is denoted $\operatorname{eig}(\boldsymbol{A})$.

Theorem 6.2 Consider the plant model

$$\dot{\boldsymbol{x}} = \boldsymbol{A}\boldsymbol{x} + \boldsymbol{B}\boldsymbol{u}, \tag{6.28}$$

where $\boldsymbol{x}(t) \in \mathbb{R}^n$, $\boldsymbol{u}(t) \in \mathbb{R}^m$, $\operatorname{rank}(\boldsymbol{B}) = m$, and the pair $(\boldsymbol{A}, \boldsymbol{B})$ is controllable. Let

$$\{\lambda_1, \lambda_2, \ldots, \lambda_{n-m}\} \in \mathbb{C},$$

where $\mathbb{C}$ denotes the set of complex numbers, be an arbitrary set of $n-m$ complex numbers such that any λ_i with $\Im(\lambda_i) \neq 0$ appears in this set in conjugate pair. Then, there exist full-rank matrices $\boldsymbol{W} \in \mathbb{R}^{n\times(n-m)}$ and $\boldsymbol{W}^g \in \mathbb{R}^{(n-m)\times n}$ so that $\boldsymbol{W}^g\boldsymbol{W} = \boldsymbol{I}_{n-m}$, $\boldsymbol{W}^g\boldsymbol{B} = \boldsymbol{O}$, and the spectrum of the matrix $\boldsymbol{W}^g\boldsymbol{A}\boldsymbol{W}$ is

$$\operatorname{eig}(\boldsymbol{W}^g\boldsymbol{A}\boldsymbol{W}) = \{\lambda_1, \lambda_2, \ldots, \lambda_{n-m}\}.$$

Furthermore, there exists a matrix $\boldsymbol{S} \in \mathbb{R}^{m\times n}$ such that $\boldsymbol{S}\boldsymbol{B} = \boldsymbol{I}_m$ and the system (6.28) restricted to the surface $\boldsymbol{S}\boldsymbol{x} = \boldsymbol{0}$ has the poles $\lambda_1, \lambda_2, \ldots, \lambda_{n-m}$.

Proof We first prove the existence of the desired matrices $\boldsymbol{W}$ and $\boldsymbol{W}^g$. We can find $\boldsymbol{T} \in \mathbb{R}^{n\times(n-m)}$ so that $[\boldsymbol{T} \;\; \boldsymbol{B}]$ is invertible because $\boldsymbol{B}$ has full rank. Let

$$[\boldsymbol{T} \;\; \boldsymbol{B}]^{-1} = \begin{bmatrix} \boldsymbol{T}^g \\ \boldsymbol{B}^g \end{bmatrix}.$$

We next introduce a new coordinate $\boldsymbol{z}$ such that $\boldsymbol{x} = [\boldsymbol{T} \;\; \boldsymbol{B}]\boldsymbol{z}$. Then, in the new coordinates the system $\dot{\boldsymbol{x}} = \boldsymbol{A}\boldsymbol{x} + \boldsymbol{B}\boldsymbol{u}$ has the form

$$\dot{\boldsymbol{z}} = \begin{bmatrix} \boldsymbol{T}^g\boldsymbol{A}\boldsymbol{T} & \boldsymbol{T}^g\boldsymbol{A}\boldsymbol{B} \\ \boldsymbol{B}^g\boldsymbol{A}\boldsymbol{T} & \boldsymbol{B}^g\boldsymbol{A}\boldsymbol{B} \end{bmatrix}\boldsymbol{z} + \begin{bmatrix} \boldsymbol{O} \\ \boldsymbol{I}_m \end{bmatrix}\boldsymbol{u}.$$

Because the pair $(\boldsymbol{A}, \boldsymbol{B})$ is controllable, we can use the PBH rank test, given on page 106, to conclude that

$$\operatorname{rank}[\lambda\boldsymbol{I}_{n-m} - \boldsymbol{T}^g\boldsymbol{A}\boldsymbol{T} \quad \boldsymbol{T}^g\boldsymbol{A}\boldsymbol{B}] = n - m \qquad \text{for all } \lambda \in \mathbb{C},$$

which means that the pair $(T^g AT, T^g AB)$ is controllable. The set $\{\lambda_1, \lambda_2, \ldots, \lambda_{n-m}\}$ is symmetric with respect to the real axis; therefore we can find a matrix $F \in \mathbb{R}^{m\times(n-m)}$ such that

$$\operatorname{eig}(T^g AT - T^g ABF) = \{\lambda_1, \lambda_2, \ldots, \lambda_{n-m}\}.$$

Let $W^g = T^g$ and let $W = T - BF$. Then, because by construction $T^g T = I_{n-m}$ and $T^g B = O$, we have $W^g W = I_{n-m}$ and $W^g B = O$. Note that

$$W^g AW = T^g AT - T^g ABF.$$

Hence,

$$\operatorname{eig}(W^g AW) = \operatorname{eig}(T^g AT - T^g ABF) = \{\lambda_1, \lambda_2, \ldots, \lambda_{n-m}\}.$$

We next prove the existence of the sliding surface with the desired properties. We first note that the matrix $[W \;\; B]$ is invertible because $W^g W = I_{n-m}$, $W^g B = O$, and W, B are full rank. Hence, Range(W) $\cap$ Range(B) $= \{\mathbf{0}\}$ and we write

$$[W \;\; B]^{-1} = \begin{bmatrix} W^g \\ S \end{bmatrix}$$

for some $S \in \mathbb{R}^{m\times n}$. Note that $SB = I_m$ and $SW = O$. We now introduce new coordinates

$$z = \begin{bmatrix} z_1 \\ z_2 \end{bmatrix} = \begin{bmatrix} W^g \\ S \end{bmatrix} x,$$

where $z_2 = Sx$. In the new coordinates, the equivalent system $\dot{x} = (I_n - B(SB)^{-1}S)Ax$ takes the form

$$\dot{z} = \begin{bmatrix} W^g \\ S \end{bmatrix} (I_n - B(SB)^{-1}S)A[W \;\; B]z$$

$$= \begin{bmatrix} W^g AW & W^g AB \\ O & O \end{bmatrix} z.$$

The restriction to $z_2 = Sx = \mathbf{0}$ is

$$\dot{z}_1 = W^g AWz_1,$$

and hence the system restricted to the sliding surface $Sx = \mathbf{0}$ has the same eigenvalues as the matrix $W^g AW$. The proof of the theorem is now complete.

We use the results of the proof of the above theorem to propose the following sliding surface design algorithm.

SWITCHING SURFACE DESIGN ALGORITHM

Given a controllable pair $(\boldsymbol{A}, \boldsymbol{B})$ with rank $\boldsymbol{B} = m$, along with the set of desired poles $\{\lambda_i, \ i = 1, 2, \ldots, n-m\}$ of the system restricted to a sliding surface, do the following:

STEP 1 Find a matrix $\boldsymbol{T} \in \mathbb{R}^{(n-m)\times n}$ such that $[\boldsymbol{T} \quad \boldsymbol{B}]$ is invertible.

STEP 2 Compute

$$[\boldsymbol{T} \quad \boldsymbol{B}]^{-1} = \begin{bmatrix} \boldsymbol{T}^g \\ \boldsymbol{B}^g \end{bmatrix}.$$

STEP 3 Find a matrix $\boldsymbol{F} \in \mathbb{R}^{m\times(n-m)}$ such that

$$\operatorname{eig}(\boldsymbol{T}^g\boldsymbol{A}\boldsymbol{T} - \boldsymbol{T}^g\boldsymbol{A}\boldsymbol{B}\boldsymbol{F}) = \{\lambda_1, \lambda_2, \ldots, \lambda_{n-m}\}.$$

STEP 4 Take $\boldsymbol{W} = \boldsymbol{T} - \boldsymbol{B}\boldsymbol{F}$.

STEP 5 Compute

$$[\boldsymbol{W} \quad \boldsymbol{B}]^{-1} - \begin{bmatrix} \boldsymbol{W}^g \\ \boldsymbol{S} \end{bmatrix}.$$

Form sliding surface $\boldsymbol{S}\boldsymbol{x} = \boldsymbol{0}$ using the last m rows of $[\boldsymbol{W} \quad \boldsymbol{B}]^{-1}$.

We illustrate the above algorithm with a simple numerical example.

◆ Example 6.4

Given a dynamical system model of the form (6.28), where

$$\boldsymbol{A} = \begin{bmatrix} 0 & 1 & 0 \\ 0 & 0 & 0 \\ 0 & 0 & 0 \end{bmatrix}, \qquad \boldsymbol{B} = \begin{bmatrix} 0 & 0 \\ 1 & 0 \\ 0 & 1 \end{bmatrix}.$$

We design a sliding surface

$$\sigma = \boldsymbol{S}\boldsymbol{x} = \boldsymbol{0}$$

so that the system restricted to it will behave like a linear system described by

$$\dot{z} = -z.$$

Following the steps of the of the design algorithm, we first find a matrix $\boldsymbol{T}$ such that the augmented matrix $[\boldsymbol{T} \quad \boldsymbol{B}]$ is invertible. For example, we can select

$$\boldsymbol{T} = [1 \quad 0 \quad 0]^T.$$

Note that

$$[\boldsymbol{T} \quad \boldsymbol{B}] = \boldsymbol{I}_3.$$

Hence, the inverse is

$$[T \quad B]^{-1} = \begin{bmatrix} T^g \\ B^g \end{bmatrix} = I_3.$$

Next, we find $F \in \mathbb{R}^{m\times(n-m)}$ such that

$$T^g A T - T^g A B F = -1.$$

We have

$$F = \begin{bmatrix} 1 \\ 0 \end{bmatrix}.$$

We then compute $W = T - BF = [1 \quad -1 \quad 0]^T$ and find

$$[W \quad B]^{-1} = \begin{bmatrix} 1 & 0 & 0 \\ -1 & 1 & 0 \\ 0 & 0 & 1 \end{bmatrix}^{-1} = \begin{bmatrix} 1 & 0 & 0 \\ 1 & 1 & 0 \\ 0 & 0 & 1 \end{bmatrix} = \begin{bmatrix} W^g \\ S \end{bmatrix}.$$

Hence,

$$S = \begin{bmatrix} 1 & 1 & 0 \\ 0 & 0 & 1 \end{bmatrix}.$$

6.8 State Estimation of Uncertain Systems

6.8.1 Discontinuous Estimators

So far in our discussion of variable structure sliding mode control system design we assumed that the entire state vector of the plant to be controlled is available to us through measurement. If the state vector cannot be measured, then we can try to determine its suitable estimate that can be used by the controller instead of the true state vector. If we use a state vector estimate for the inaccessible state, then we split a control design problem into three phases. The first phase is the design of a controller assuming the availability of the entire state vector of the plant. In the second phase, we construct a state estimator based on the plant input and output. Finally, in the third phase, we combine the controller and the estimator into the combined controller–estimator compensator, where the controller uses the plant state estimate instead of the actual state.

We discuss the problem of constructing a state estimator for a class of dynamical systems with matched uncertainties modeled by

$$\dot{x} = Ax + B(u + \xi),$$
$$y = Cx,$$

where $C \in \mathbb{R}^{p\times n}$, and $p \geq m$, which means that the number of input channels does not exceed the number of output channels. The vector function ξ models the lumped uncertainties or nonlinearities in the plant. We assume that there is a nonnegative real-valued function $\rho = \rho(\boldsymbol{y})$ such that

$$\|\boldsymbol{\xi}\| \leq \rho.$$

We assume that the pair $(\boldsymbol{A}, \boldsymbol{C})$ is observable. Our goal is to construct the plant state estimator. Let $\hat{\boldsymbol{x}}$ be an estimate of $\boldsymbol{x}$. Let $\boldsymbol{e}$ denote the estimation error, that is,

$$\boldsymbol{e}(t) = \hat{\boldsymbol{x}}(t) - \boldsymbol{x}(t).$$

The observability of $(\boldsymbol{A}, \boldsymbol{C})$ implies the existence of a matrix $\boldsymbol{L} \in \mathbb{R}^{n\times p}$ such that $\boldsymbol{A} - \boldsymbol{LC}$ has prescribed (symmetric with respect to the real axis) eigenvalues in the open left-half plane. Because $\boldsymbol{A} - \boldsymbol{LC}$ is asymptotically stable, by the Lyapunov theorem, for any $\boldsymbol{Q} = \boldsymbol{Q}^T > 0$, there is a unique $\boldsymbol{P} = \boldsymbol{P}^T > 0$ such that

$$(\boldsymbol{A} - \boldsymbol{LC})^T\boldsymbol{P} + \boldsymbol{P}(\boldsymbol{A} - \boldsymbol{LC}) = -\boldsymbol{Q}.$$

We choose $\boldsymbol{Q}$, if possible, so that for some $\boldsymbol{F} \in \mathbb{R}^{m\times p}$ we have

$$\boldsymbol{FC} = \boldsymbol{B}^T\boldsymbol{P}.$$

We need this technical condition for the construction of the state estimator. This condition has an interesting system theoretic interpretation. Its satisfaction means that the transfer function matrix

$$\boldsymbol{G}(s) = \boldsymbol{FC}(s\boldsymbol{I} - \boldsymbol{A} + \boldsymbol{LC})^{-1}\boldsymbol{B}$$

is positive real; that is, the elements of $\boldsymbol{G}(s)$ are analytic in the open right-half plane and $\boldsymbol{G}(s) + \boldsymbol{G}^*(s) \geq 0$ in the open right-half plane. Here the superscript asterisk denotes the conjugate transpose.

To proceed, define the vector function

$$\boldsymbol{E}(\boldsymbol{e}, \eta) = \begin{cases} \dfrac{\boldsymbol{FCe}}{\|\boldsymbol{FCe}\|}\eta & \text{for } \boldsymbol{FCe} \neq \boldsymbol{0}, \\ \boldsymbol{v} \in \mathbb{R}^m,\ \|\boldsymbol{v}\| \leq \eta & \text{for } \boldsymbol{FCe} = \boldsymbol{0}, \end{cases}$$

where $\eta \geq 0$ is a design parameter. Note that if we have a single input single-output plant, then we can write

$$\boldsymbol{E}(\boldsymbol{e}, \eta) = \eta\,\mathrm{sign}(\boldsymbol{FCe}).$$

We now introduce a state estimator described in Walcott and Żak [292]. It has the form

$$\boxed{\dot{\hat{\boldsymbol{x}}} = (\boldsymbol{A} - \boldsymbol{LC})\hat{\boldsymbol{x}} - \boldsymbol{BE}(\boldsymbol{e}, \eta) + \boldsymbol{Ly} + \boldsymbol{Bu}.}$$

If $\eta \geq \rho$, the function $\hat{\boldsymbol{x}}$ is an asymptotic estimate of the state $\boldsymbol{x}$, that is,

$$\lim_{t\to\infty} \boldsymbol{e}(t) = \lim_{t\to\infty} (\hat{\boldsymbol{x}}(t) - \boldsymbol{x}(t)) = \boldsymbol{0}.$$

To prove the above statement using Lyapunov's type of arguments, construct the differential equation satisfied by the estimation error $\boldsymbol{e}$:

$$\dot{\boldsymbol{e}} = \dot{\hat{\boldsymbol{x}}} - \dot{\boldsymbol{x}} = (\boldsymbol{A} - \boldsymbol{LC})\boldsymbol{e} - \boldsymbol{BE}(\boldsymbol{e}, \eta) - \boldsymbol{B\xi}. \tag{6.29}$$

Then, show that

$$\frac{d}{dt}(\boldsymbol{e}^T \boldsymbol{P}\boldsymbol{e}) = -\boldsymbol{e}^T \boldsymbol{Q}\boldsymbol{e} < 0,$$

which implies

$$\lim_{t\to\infty} \boldsymbol{e}(t) = \boldsymbol{0}.$$

It follows from the above that the estimation error is insensitive to matched uncertainty.

6.8.2 Boundary Layer Estimators

We can view the error differential equation (6.29) as a model of a closed-loop system, with matched uncertainties, controlled by $-\boldsymbol{E}(\boldsymbol{e}, \eta)$. This "controller" is discontinuous and varies its structure with respect to the switching surface $\{\boldsymbol{e} : \boldsymbol{F}\boldsymbol{C}\boldsymbol{e} = \boldsymbol{0}\}$. In our analysis of (6.29), we assumed that $\boldsymbol{E}(\boldsymbol{e}, \eta)$ can switch with infinitely high frequency. In practice, the switching frequency is finite, albeit very high. High-frequency switching is not always desirable because it may degrade the system performance. In mechanical systems, it may lead to excessive wear and tear of the system's components. To alleviate the problem, we smooth out the discontinuous part of the estimator by introducing a boundary layer around the surface $\{\boldsymbol{e} : \boldsymbol{F}\boldsymbol{C}\boldsymbol{e} = \boldsymbol{0}\}$. This results in replacing $\boldsymbol{E}(\boldsymbol{e}, \eta)$ with its continuous approximation $\tilde{\boldsymbol{E}}(\boldsymbol{e}, \eta)$. An example of such a continuous approximation is

$$\tilde{\boldsymbol{E}}(\boldsymbol{e}, \eta) = \begin{cases} \dfrac{\boldsymbol{F}\boldsymbol{C}\boldsymbol{e}}{\|\boldsymbol{F}\boldsymbol{C}\boldsymbol{e}\|}\eta & \text{for } \|\boldsymbol{F}\boldsymbol{C}\boldsymbol{e}\| \geq \nu, \\ \dfrac{\boldsymbol{F}\boldsymbol{C}\boldsymbol{e}}{\nu}\eta & \text{for } \|\boldsymbol{F}\boldsymbol{C}\boldsymbol{e}\| < \nu, \end{cases}$$

where ν is the width of the boundary layer. However, the error equation, or more precisely $\boldsymbol{e} = \boldsymbol{0}$, is not anymore asymptotically stable when a boundary layer estimator is used. To describe the behavior of the error differential equation resulting from the application of a boundary layer estimator, we recall the following definitions.

Definition 6.1 The closed ball $\bar{B}_q(\boldsymbol{0}) = \{\boldsymbol{e} : \|\boldsymbol{e}\| < q\}$ is uniformly stable if for any $\varepsilon > q$, there is $\delta(\varepsilon) > 0$ so that if $\|\boldsymbol{e}(t_0)\| \leq \delta(\varepsilon)$, then $\|\boldsymbol{e}(t)\| \leq \varepsilon$ for $t \geq t_0$.

Definition 6.2 The equilibrium $\boldsymbol{e} = \boldsymbol{0}$ is uniformly bounded if for any given $r > 0$, there exists $d(r) < \infty$ so that for each solution $\boldsymbol{e}(t) = \boldsymbol{e}(t; t_0, \boldsymbol{e}_0)$ with $\|\boldsymbol{e}_0\| \leq r$, we have $\|\boldsymbol{e}(t)\| \leq d(r)$ for all $t \geq t_0$.

We next adapt the definition of uniform ultimate boundedness, abbreviated UUB, to the estimation error, $\boldsymbol{e}$.

Definition 6.3 The error $\boldsymbol{e}$ is uniformly ultimately bounded (UUB) with respect to a closed ball $\bar{B}$ if for every $r > 0$, there is $T(r) \geq 0$ such that for each solution $\boldsymbol{e}(t)$ with $\|\boldsymbol{e}(t_0)\| \leq r$, we have $\boldsymbol{e}(t) \in \bar{B}$ for $t \geq t_0 + T(r)$.

We can show that the error of the boundary layer estimator is UUB. Specifically, the estimation error of the boundary layer estimator ends up in a ball whose radius depends on ν. The ball radius tends to zero with ν tending to zero. To see this consider, as before, the Lyapunov function candidate $V(\boldsymbol{e}) = \boldsymbol{e}^T \boldsymbol{P}\boldsymbol{e}$. For $\|\boldsymbol{F}\boldsymbol{C}\boldsymbol{e}\| \geq \nu$, we have

$$\begin{aligned}
\frac{d}{dt}V(\boldsymbol{e}) &= 2\boldsymbol{e}^T \boldsymbol{P}\dot{\boldsymbol{e}} \\
&= 2\boldsymbol{e}^T \boldsymbol{P}((\boldsymbol{A} - \boldsymbol{L}\boldsymbol{C})\boldsymbol{e} - \boldsymbol{B}\tilde{\boldsymbol{E}}(\boldsymbol{e}, \eta) - \boldsymbol{B}\boldsymbol{\xi}) \\
&= -\boldsymbol{e}^T \boldsymbol{Q}\boldsymbol{e} - 2\boldsymbol{e}^T \boldsymbol{P}\boldsymbol{B}\frac{\boldsymbol{F}\boldsymbol{C}\boldsymbol{e}}{\|\boldsymbol{F}\boldsymbol{C}\boldsymbol{e}\|}\eta - 2\boldsymbol{e}^T \boldsymbol{P}\boldsymbol{B}\boldsymbol{\xi}.
\end{aligned} \tag{6.30}$$

Because by assumption, $\boldsymbol{F}\boldsymbol{C} = \boldsymbol{B}^T \boldsymbol{P}$, therefore

$$\boldsymbol{e}^T \boldsymbol{P}\boldsymbol{B}\boldsymbol{F}\boldsymbol{C}\boldsymbol{e} = \|\boldsymbol{F}\boldsymbol{C}\boldsymbol{e}\|^2 = \|\boldsymbol{e}^T \boldsymbol{P}\boldsymbol{B}\|^2.$$

Substituting the above into (6.30) and taking into account that $\|\boldsymbol{e}^T \boldsymbol{P}\boldsymbol{B}\boldsymbol{\xi}\| \leq \|\boldsymbol{e}^T \boldsymbol{P}\boldsymbol{B}\|\rho$, we obtain

$$\frac{d}{dt}V(\boldsymbol{e}) \leq -\boldsymbol{e}^T \boldsymbol{Q}\boldsymbol{e} - 2\|\boldsymbol{F}\boldsymbol{C}\boldsymbol{e}\|\eta + 2\|\boldsymbol{e}^T \boldsymbol{P}\boldsymbol{B}\|\eta = -\boldsymbol{e}^T \boldsymbol{Q}\boldsymbol{e}.$$

For $\|\boldsymbol{F}\boldsymbol{C}\boldsymbol{e}\| < \nu$, we have

$$\begin{aligned}
\frac{d}{dt}V(\boldsymbol{e}) &= 2\boldsymbol{e}^T \boldsymbol{P}\dot{\boldsymbol{e}} \\
&= -\boldsymbol{e}^T \boldsymbol{Q}\boldsymbol{e} - 2\boldsymbol{e}^T \boldsymbol{P}\boldsymbol{B}\frac{\boldsymbol{F}\boldsymbol{C}\boldsymbol{e}}{\nu}\eta - 2\boldsymbol{e}^T \boldsymbol{P}\boldsymbol{B}\boldsymbol{\xi} \\
&\leq -\boldsymbol{e}^T \boldsymbol{Q}\boldsymbol{e} - 2\frac{\|\boldsymbol{F}\boldsymbol{C}\boldsymbol{e}\|^2}{\nu}\eta + 2\|\boldsymbol{e}^T \boldsymbol{P}\boldsymbol{B}\|\rho \\
&\leq -\boldsymbol{e}^T \boldsymbol{Q}\boldsymbol{e} + 2\nu\rho \\
&\leq -\lambda_{\min}(\boldsymbol{Q})\|\boldsymbol{e}\|^2 + 2\nu\rho.
\end{aligned} \tag{6.31}$$

In summary,

$$\begin{aligned}
\frac{d}{dt}V(\boldsymbol{e}) &= -\boldsymbol{e}^T \boldsymbol{Q}\boldsymbol{e} - 2\boldsymbol{e}^T \boldsymbol{P}\boldsymbol{B}\tilde{\boldsymbol{E}}(\boldsymbol{e}, \eta) - 2\boldsymbol{e}^T \boldsymbol{P}\boldsymbol{B}\boldsymbol{\xi} \\
&\leq \begin{cases} -\boldsymbol{e}^T \boldsymbol{Q}\boldsymbol{e} & \text{for } \|\boldsymbol{F}\boldsymbol{C}\boldsymbol{e}\| \geq \nu \\ -\lambda_{\min}(\boldsymbol{Q})\|\boldsymbol{e}\|^2 + 2\nu\rho & \text{for } \|\boldsymbol{F}\boldsymbol{C}\boldsymbol{e}\| < \nu. \end{cases}
\end{aligned}$$

Hence, we can conclude that $\frac{d}{dt}V(\boldsymbol{e}) < 0$ for all $\boldsymbol{e} \in \mathbb{R}^n$ such that

$$\|\boldsymbol{e}\| > \sqrt{\frac{2\nu\rho}{\lambda_{\min}(\boldsymbol{Q})}} \stackrel{\text{def}}{=} R.$$

We illustrate the regions used in the discussion of the boundary layer state estimator in Figure 6.11. Referring to Figure 6.11 and Definition 6.2, we have for $r \leq R$,

$$d(r) = \sqrt{\frac{\lambda_{\max}(\boldsymbol{P})}{\lambda_{\min}(\boldsymbol{P})}}\,R.$$

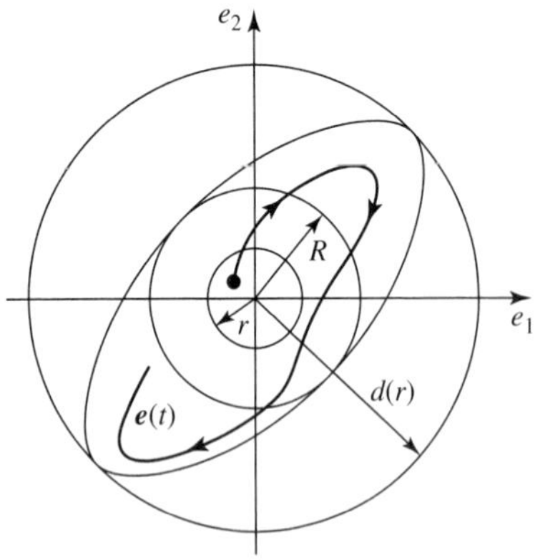

Figure 6.11 A two-dimensional illustration of an UUB error.

If, on the other hand, $r > R$, then

$$d(r) = \sqrt{\frac{\lambda_{\max}(\boldsymbol{P})}{\lambda_{\min}(\boldsymbol{P})}}\, r.$$

We conclude that the estimation error is UUB with respect to any ball with a radius greater than

$$\sqrt{\frac{\lambda_{\max}(\boldsymbol{P})}{\lambda_{\min}(\boldsymbol{P})}}\, R.$$

6.9 Sliding Modes in Solving Optimization Problems

In this section we use sliding modes to analyze dynamics of an analog circuit, proposed by Rodríguez-Vázquez et al. [242], that can solve convex optimization problems with linear constraints. This class of networks for solving optimization problems is particularly attractive because it can be can be realized using switched-capacitor technology. The network we analyze is modeled by a set of nonlinear ordinary differential equations with discontinuous right-hand sides. In carrying out the analysis, we use concepts from the theory of differential equations with discontinuous right-hand sides and the Lyapunov stability theory. We show that the equilibrium points of the differential equations modeling the network coincide with the optimizers of the underlying optimization problem. We then prove that irrespective of the initial state of the network, its state converges to the solution set of the optimization problem. Our analysis is based on the results of the paper by Glazos, Hui, and Żak [98].

6.9.1 Optimization Problem Statement

We consider the constrained optimization problem

$$\begin{array}{ll} \text{minimize} & f(\boldsymbol{x}) \\ \text{subject to} & \boldsymbol{A}\boldsymbol{x} \leq \boldsymbol{b}, \end{array} \tag{6.32}$$

where $\boldsymbol{x} \in \mathbb{R}^n$, $f : \mathbb{R}^n \to \mathbb{R}$, $\boldsymbol{A} \in \mathbb{R}^{m\times n}$, and $\boldsymbol{b} \in \mathbb{R}^m$. Let

$$\Omega = \{\boldsymbol{x} : \boldsymbol{A}\boldsymbol{x} \leq \boldsymbol{b}\}. \tag{6.33}$$

denote the feasible region for the above optimization problem. The feasible set Ω is the intersection of m closed half-spaces and is therefore a convex polytope. Let

$$h_i = \{\boldsymbol{x} : \boldsymbol{a}_i \boldsymbol{x} = b_i\}, \qquad i = 1, 2, \ldots, m,$$

denote the hyperplanes associated with the above m closed half-spaces, where $\boldsymbol{a}_i$ is the ith row of $\boldsymbol{A}$ and b_i is the ith component of the vector $\boldsymbol{b}$. Also, we define

$$H = \bigcup_{i=1}^{m} h_i.$$

We let Γ denote the collection of all local minimizers of the optimization problem (6.32). Lastly, we introduce the index sets

$$I(\boldsymbol{x}) = \{i : \boldsymbol{a}_i \boldsymbol{x} = b_i\}$$

and

$$J(\boldsymbol{x}) = \{i : \boldsymbol{a}_i \boldsymbol{x} > b_i\}.$$

Recall that a function $f : \mathbb{R}^n \to \mathbb{R}$ defined on a convex set Ω is said to be convex if, for every $\boldsymbol{x}, \boldsymbol{y} \in \Omega$ and every $0 \le \alpha \le 1$, there holds

$$f(\alpha \boldsymbol{x} + (1-\alpha)\boldsymbol{y}) \le \alpha f(\boldsymbol{x}) + (1-\alpha) f(\boldsymbol{y}).$$

Let $f \in \mathcal{C}^1$. Then, f is convex over a convex set Ω if and only if

$$f(\boldsymbol{y}) \ge f(\boldsymbol{x}) + \nabla f^T(\boldsymbol{x})(\boldsymbol{y} - \boldsymbol{x})$$

for all $\boldsymbol{x}, \boldsymbol{y} \in \Omega$.

We make the following assumptions:

A1. The objective function f is convex over $\mathbb{R}^n$ and has continuous first partial derivatives on $\mathbb{R}^n$; that is, $f \in \mathcal{C}^1$.

A2. The feasible set Ω is a nonempty, bounded, and closed polyhedron, which means that Ω is a compact set.

A3. Each point in Ω is a regular point of the constraints; that is, for any $\boldsymbol{x} \in \Omega$ the vectors $\boldsymbol{a}_i$, $i \in I(\boldsymbol{x})$, are linearly independent.

It follows from Assumption A1 that the optimization problem (6.32) is a convex programming problem. Indeed, we are asked to minimize a convex objective function f over a polyhedron Ω.

Problem (6.32) is well-posed in the sense that the feasible set Ω is nonempty and the optimal value of f is not $-\infty$ because by assumptions A1 and A2 the set $f(\Omega)$ is compact.

An elementary overview of results concerning convex optimization problems can be found, for example, in Chong and Żak [50, Chapter 21]. We state two results related to problem (6.32) that are relevant in our analysis.

Lemma 6.1 Any local minimizer of the optimization problem (6.32) is a global minimizer.

Furthermore, the Karush–Kuhn–Tucker condition for optimality is both necessary and sufficient for problem (6.32). Specifically, we have the following lemma:

Lemma 6.2 Suppose that $x^* \in \Omega$. Then, $x^* \in \Gamma$ if and only if there exist nonnegative constants $\mu_1, \mu_2, \ldots, \mu_m$ such that

$$\nabla f(x^*) + \sum_{i \in I(x^*)} \mu_i a_i^T = 0.$$

The network applied to the optimization problem (6.32) has the form

$$\begin{aligned} \tau \dot{x}(t) &= g(x(t)) \\ &= \begin{cases} -\mu \sum_{i \in J(x(t))} a_i^T & \text{if } x(t) \notin \Omega \\ -\nabla f(x(t)) & \text{if } x(t) \in \Omega, \end{cases} \\ x(0) &= x_0, \end{aligned} \tag{6.34}$$

where τ and μ are positive design constants. The function $g(x)$ is discontinuous on the surfaces $h_1, h_2, \ldots h_m$ and is continuous elsewhere. A phenomenon commonly occurring in systems such as (6.34) is the so-called sliding mode that we discussed in Section 6.2. An analysis of the behavior of the above dynamical system using a sliding mode approach is presented in the following sections.

We remark that we can view (6.34) as a dynamical gradient system for the unconstrained problem

$$\text{minimize} \quad \kappa_0 f(x) + \mu \sum_{i=1}^{m} \kappa_i (a_i x - b_i),$$

where

$$\kappa_i = \begin{cases} 0 & \text{if } a_i x \le b_i, \\ 1 & \text{if } a_i x > b_i \end{cases}$$

and

$$\kappa_0 = \begin{cases} 1 & \text{if } \kappa_i = 0 \text{ for all } i = 1, 2, \ldots, m, \\ 0 & \text{otherwise;} \end{cases}$$

that is, $\kappa_0 = \prod_{i=1}^{m}(1 - \kappa_i)$. Using the preceding notation, we can represent system (6.34) as

$$\tau \dot{x}(t) = g(x(t)) = -\kappa_0 \nabla f(x(t)) - \mu \sum_{i=1}^{m} \kappa_i a_i^T, \qquad x(0) = x_0. \tag{6.35}$$

6.9.2 Penalty Function Method

An approach that can be used to construct an analog optimizer to solve a constrained optimization problem is to first convert, or approximate, the constrained optimization problem into an associated unconstrained optimization problem and then design an analog network that solves the unconstrained problem. Such a network is typically an implementation of the dynamical gradient system for minimizing the objective function of the associated unconstrained problem.

One method of approximating a constrained optimization problem into an associated unconstrained optimization problem is the penalty function method. For more details on the subject of the penalty method see, for example, Luenberger [190]. The idea behind the penalty function method is to approximate the constrained optimization problem

$$\begin{array}{ll} \text{minimize} & f(\boldsymbol{x}) \\ \text{subject to} & \boldsymbol{x} \in \Omega, \end{array} \tag{6.36}$$

where $\Omega \subset \mathbb{R}^n$ is the feasible set, with an associated unconstrained problem of the form

$$\text{minimize} \quad f(\boldsymbol{x}) + \zeta p(\boldsymbol{x}), \tag{6.37}$$

where the objective function $f : \mathbb{R}^n \to \mathbb{R}$ is continuous on $\mathbb{R}^n$ and ζ is a positive constant. The function $p : \mathbb{R}^n \to \mathbb{R}$ is continuous on $\mathbb{R}^n$ and satisfies the following:

1. $p(\boldsymbol{x}) \geq 0$ for all $\boldsymbol{x} \in \mathbb{R}^n$.
2. $p(\boldsymbol{x}) = 0$ if and only if $\boldsymbol{x} \in \Omega$.

If for a finite value of the penalty parameter ζ the solution of the unconstrained problem (6.37) coincides with the solution of the constrained problem (6.36), then the penalty function p is said to be exact. Bertsekas [26] has shown that except for trivial cases, an exact penalty function cannot be everywhere differentiable—see Exercise 6.9 for an illustration of this statement. Quite often we can find an exact penalty function that results in the function $f(\boldsymbol{x}) + \zeta p(\boldsymbol{x})$ being piecewise smooth. However, the dynamical gradient system for such a function will have a discontinuous right-hand side. As such, a rigorous analysis of its behavior cannot be carried out using only methods derived from the classical theory of differential equations. Rather, the analysis of the system must be carried out using other methods—for example, those reported by Filippov [84] and by Aubin and Cellina [14]. Such an approach was undertaken in 1967 by Karpinskaya [150] and was later undertaken by Korovin and Utkin [164], Utkin [286], Flåm [87], Flåm and Zowe [86], Żak et al. [314], and Chong, Hui, and Żak [49], among others. A differential geometric approach to the analysis of dynamical systems that solve optimization problems is reported by Faybusovich [80] and by Helmke and Moore [119].

6.9.3 Dynamical Gradient Circuit Analysis

We will now show that every trajectory of system (6.34), irrespective of the initial state, converges to the solution set of the optimization problem (6.32). We do this by showing that every trajectory of system (6.34) has the following properties:

1. Irrespective of the initial state $\boldsymbol{x}_0$, the resulting trajectory reaches the feasible set Ω in finite time and is thereafter confined to Ω. This portion of the trajectory is hereafter referred to as the *reaching phase*.
2. While confined to the feasible set Ω, the trajectory converges asymptotically to the solution set Γ. This portion of the trajectory is hereafter referred to as the *convergence phase*.

Reaching Phase We begin by introducing the function $V : \mathbb{R}^n \to \mathbb{R}$ of the form

$$V(\boldsymbol{x}) = \sum_{i=1}^{m} \max\{\boldsymbol{a}_i \boldsymbol{x} - b_i, 0\},$$

or, equivalently,

$$V(\boldsymbol{x}) = \begin{cases} \sum_{i\in J(\boldsymbol{x})} (\boldsymbol{a}_i\boldsymbol{x} - b_i) & \text{if } \boldsymbol{x} \notin \Omega, \\ 0 & \text{if } \boldsymbol{x} \in \Omega. \end{cases} \tag{6.38}$$

Observe that $V(\boldsymbol{x}) > 0$ for all $\boldsymbol{x} \notin \Omega$, and $V(\boldsymbol{x}) = 0$ for all $\boldsymbol{x} \in \Omega$. We can view $V(\boldsymbol{x})$ as a distance measure from the point $\boldsymbol{x}$ to the feasible set Ω. We shall show that V is strictly decreasing when evaluated along any trajectory of system (6.34) outside Ω. Specifically, we will prove the existence of a positive constant, η, with the property that

$$\frac{d}{dt}V(\boldsymbol{x}(t)) \le -\eta$$

when evaluated along any trajectory of system (6.34) outside Ω. In doing so, we use the following result:

Lemma 6.3 Suppose $\hat{\boldsymbol{x}} \notin \Omega$ and $\beta_1, \beta_2, \ldots, \beta_m$ are nonnegative constants. If $\beta_j > 0$ for $j \in J(\hat{\boldsymbol{x}})$, then $\sum_{i\in I(\hat{\boldsymbol{x}})\cup J(\hat{\boldsymbol{x}})} \beta_i \boldsymbol{a}_i \neq \boldsymbol{0}$.

Proof Let the function $q: \mathbb{R}^n \to \mathbb{R}$ be defined as

$$q(\boldsymbol{x}) = \sum_{i\in I(\hat{\boldsymbol{x}})\cup J(\hat{\boldsymbol{x}})} \beta_i(\boldsymbol{a}_i\boldsymbol{x} - b_i).$$

Clearly, $q \in C^1$ and

$$\nabla q^T(\boldsymbol{x}) = \sum_{i\in I(\hat{\boldsymbol{x}})\cup J(\hat{\boldsymbol{x}})} \beta_i \boldsymbol{a}_i.$$

Because $\beta_1, \beta_2, \ldots, \beta_m$ are all nonnegative, it follows that

1. the function q is convex over $\mathbb{R}^n$, and
2. $q(\boldsymbol{x}) \le 0$ for all $\boldsymbol{x} \in \Omega$.

Furthermore, because $\beta_j > 0$ for $j \in J(\hat{\boldsymbol{x}})$, we also have $q(\hat{\boldsymbol{x}}) > 0$. Hence, we must have

$$\nabla q^T(\hat{\boldsymbol{x}}) = \sum_{i\in I(\hat{\boldsymbol{x}})\cup J(\hat{\boldsymbol{x}})} \beta_i \boldsymbol{a}_i \neq \boldsymbol{0}.$$

We must consider two cases when analyzing the behavior of system (6.34). The first case is when the motion of the system is not confined to any of the surfaces $h_1, h_2, \ldots, h_m$. The second case is when the system is in sliding mode on one or more of the surfaces $h_1, h_2, \ldots, h_m$. We only need to consider these two cases because every trajectory of system (6.34) is composed of these two types of motion. That is, every trajectory $\boldsymbol{x}(t)$ of system (6.34) can be broken into a countable number of intervals, $(t_0, t_1), \ldots, (t_{l-1}, t_l), \ldots$, on each of which both index sets $I(\boldsymbol{x}(t))$ and $J(\boldsymbol{x}(t))$ are constant. If $I(\boldsymbol{x}(t))$ is not empty, then the system is in a sliding mode on that interval, and if $I(\boldsymbol{x}(t))$ is empty, then the system is not in a sliding mode.

Case 1 Let $\boldsymbol{x}(t)$ denote any particular trajectory of system (6.34), and let $t' \in (t_{l-1}, t_l)$. Suppose that on the interval (t_{l-1}, t_l) the trajectory $\boldsymbol{x}(t)$ does not intersect Ω or any of the surfaces $h_1, h_2, \ldots, h_m$, that is,

$$\boldsymbol{x}(t) \notin \Omega \cup H \qquad \text{for all } t \in (t_{l-1}, t_l). \tag{6.39}$$

Denoting the index set $J(\boldsymbol{x}(t'))$ by J', it follows from (6.39) and the fact that $\boldsymbol{x}(t)$ is absolutely continuous that $I(\boldsymbol{x}(t)) = \emptyset$ and $J' \neq \emptyset$ for all $t \in (t_{l-1}, t_l)$. This in turn implies that on the interval (t_{l-1}, t_l) the trajectory $\boldsymbol{x}(t)$ must satisfy (6.34) in the usual sense. Hence, we have

$$\tau \dot{\boldsymbol{x}}(t) = -\mu \sum_{i \in J'} \boldsymbol{a}_i^T$$

almost everywhere on the interval (t_{l-1}, t_l). Also, it follows from (6.38) that

$$V(\boldsymbol{x}(t)) = \sum_{i \in J'} (\boldsymbol{a}_i \boldsymbol{x}(t) - b_i) \tag{6.40}$$

for all $t \subset (t_{l-1}, t_l)$. Applying the chain rule to (6.40), we obtain

$$\frac{d}{dt} V(\boldsymbol{x}(t)) = \nabla V^T(\boldsymbol{x}(t)) \dot{\boldsymbol{x}}(t) = -\frac{\mu}{\tau} \left\| \sum_{i \in J'} \boldsymbol{a}_i \right\|_2^2$$

almost everywhere on the interval (t_{l-1}, t_l). However, Lemma 6.3 along with the fact that there is a finite number of constraints—that is, m is finite—implies the existence of a positive constant, η_1, with the property that

$$\eta_1 \leq \left\| \sum_{i \in J(\boldsymbol{x})} \boldsymbol{a}_i \right\|_2^2$$

for all $\boldsymbol{x} \notin \Omega$. Hence,

$$\frac{d}{dt} V(\boldsymbol{x}(t)) \leq -\frac{\mu \eta_1}{\tau} < 0 \tag{6.41}$$

almost everywhere on the interval (t_{l-1}, t_l), which concludes our analysis of the first case.

Case 2 Suppose now that on an interval (t_{l-1}, t_l) a particular trajectory $\boldsymbol{x}(t)$ is confined to one or more of the surfaces $h_1, h_2, \ldots, h_m$ and does not intersect Ω. Specifically, suppose that there exist nonempty index sets I' and J' with the property that

$$I(\boldsymbol{x}(t)) = I' \qquad \text{for all } t \in (t_{l-1}, t_l), \tag{6.42}$$

and

$$J(\boldsymbol{x}(t)) = J' \qquad \text{for all } t \in (t_{l-1}, t_l). \tag{6.43}$$

Let $i_1, i_2, \ldots, i_k$ denote the elements of I', and let $S = \bigcap_{i \in I'} h_i$ denote the surface to which the trajectory $\boldsymbol{x}(t)$ is confined on the interval (t_{l-1}, t_l), and let

$$\tilde{\boldsymbol{a}} = \sum_{i \in J'} \boldsymbol{a}_i^T.$$

It follows from (6.42) and (6.43) that for a sufficiently small $\delta > 0$ the surfaces $h_{i_1}, \ldots, h_{i_k}$ partition the δ-neighborhood of $\boldsymbol{x}(t')$ into $\ell \leq 2^k$ regions, $\mathcal{R}_1, \mathcal{R}_2, \ldots, \mathcal{R}_\ell$, in each of which the function $\boldsymbol{g}$ is continuous. Then, $G(\boldsymbol{x}(t'))$ is the set of all vectors $\boldsymbol{v}$ of the form

$$\boldsymbol{v} = -\mu \sum_{j=1}^{\ell} \alpha_j \left(\tilde{\boldsymbol{a}} + \sum_{i \in I'} U(i, j) \boldsymbol{a}_i^T \right), \tag{6.44}$$

where

$$\alpha_j \geq 0, \qquad j = 1, 2, \ldots, \ell, \tag{6.45}$$

$$\sum_{j=1}^{\ell} \alpha_j = 1, \tag{6.46}$$

and

$$U(i, j) = \begin{cases} 1 & \text{if } \boldsymbol{a}_i \boldsymbol{x} > b_i \quad \text{for all } \boldsymbol{x} \in \mathcal{R}_j \\ 0 & \text{if } \boldsymbol{a}_i \boldsymbol{x} \leq b_i \quad \text{for all } \boldsymbol{x} \in \mathcal{R}_j. \end{cases}$$

Using (6.46), we rewrite (6.44) as

$$\boldsymbol{v} = -\mu \tilde{\boldsymbol{a}} - \mu \sum_{j=1}^{\ell} \alpha_j \left(\sum_{i \in I'} U(i, j) \boldsymbol{a}_i^T \right). \tag{6.47}$$

For notational simplicity, we let

$$\boldsymbol{r} = \sum_{j=1}^{\ell} \alpha_j \left(\sum_{i \in I'} U(i, j) \boldsymbol{a}_i^T \right).$$

Then, (6.47) becomes

$$\boldsymbol{v} = -\mu \tilde{\boldsymbol{a}} - \mu \boldsymbol{r}. \tag{6.48}$$

Let $T(\boldsymbol{x})$ denote the tangent plane to the surface S at the point $\boldsymbol{x}$. Observe that because the trajectory $\boldsymbol{x}(t)$ is confined to the surface S on the interval (t_{l-1}, t_l), then $\tau \dot{\boldsymbol{x}}(t) \in T(\boldsymbol{x}(t))$ almost everywhere on the interval (t_{l-1}, t_l). Thus, $\boldsymbol{x}(t)$ is an absolutely continuous function that satisfies

$$\tau \dot{\boldsymbol{x}}(t) \in G(\boldsymbol{x}(t)) \cap T(\boldsymbol{x}(t))$$

almost everywhere on the interval (t_{l-1}, t_l). We note that because the trajectory $\boldsymbol{x}(t)$ is confined to the surface S on the interval (t_{l-1}, t_l), the set $G(\boldsymbol{x}(t)) \cap T(\boldsymbol{x}(t))$ is nonempty for all (t_{l-1}, t_l). In particular, $G(\boldsymbol{x}(t')) \cap T(\boldsymbol{x}(t'))$ is the set of all $\boldsymbol{v}$ of the form (6.48) that also lie on the tangent plane $T(\boldsymbol{x}(t'))$. However,

$$\boldsymbol{r} \in \operatorname{Span}\{\boldsymbol{a}_{i_1}^T, \boldsymbol{a}_{i_2}^T, \ldots, \boldsymbol{a}_{i_k}^T\},$$

and thus $\boldsymbol{r}$ is orthogonal to $T(\boldsymbol{x}(t'))$. Therefore, the set $G(\boldsymbol{x}(t')) \cap T(\boldsymbol{x}(t'))$ consists of exactly one element, the orthogonal projection of the vector $-\mu \tilde{\boldsymbol{a}}$ onto the tangent plane $T(\boldsymbol{x}(t'))$. Hence, for (t_{l-1}, t_l) the set $G(\boldsymbol{x}(t)) \cap T(\boldsymbol{x}(t))$ consists of exactly one element, and thus $\dot{\boldsymbol{x}}(t) = -(\mu/\tau) \boldsymbol{P} \tilde{\boldsymbol{a}}$ is uniquely determined almost everywhere on the interval (t_{l-1}, t_l), where $\boldsymbol{P}$ is the orthogonal projector onto the tangent plane $T(\boldsymbol{x}(t))$. Note that

$$\boldsymbol{P} = \boldsymbol{I}_n - \boldsymbol{B}(\boldsymbol{B}^T\boldsymbol{B})^{-1}\boldsymbol{B}^T, \tag{6.49}$$

where $\boldsymbol{I}_n$ denotes the $n \times n$ identity matrix and

$$\boldsymbol{B} = \begin{bmatrix} \boldsymbol{a}_{i_1}^T & \boldsymbol{a}_{i_2}^T & \cdots & \boldsymbol{a}_{i_k}^T \end{bmatrix}.$$

By (6.38) and (6.43),

$$V(\boldsymbol{x}(t)) = \sum_{i \in J'} (\boldsymbol{a}_i \boldsymbol{x}(t) - b_i) \tag{6.50}$$

for all $t \in (t_{l-1}, t_l)$. Applying the chain rule to (6.50) yields

$$\frac{d}{dt}V(\boldsymbol{x}(t)) = \nabla V^T(\boldsymbol{x}(t))\dot{\boldsymbol{x}}(t) = -\frac{\mu}{\tau}\tilde{\boldsymbol{a}}^T\boldsymbol{P}\tilde{\boldsymbol{a}} \tag{6.51}$$

almost everywhere on the interval (t_{l-1}, t_l). Observing that $\boldsymbol{P} = \boldsymbol{P}^T = \boldsymbol{P}^2$, we rewrite (6.51) as

$$\frac{d}{dt}V(\boldsymbol{x}(t)) = -\frac{\mu}{\tau}\|\boldsymbol{P}\tilde{\boldsymbol{a}}\|_2^2. \tag{6.52}$$

We can show, using Lemma 6.3, that $\boldsymbol{P}\tilde{\boldsymbol{a}} \neq \boldsymbol{0}$ (see Exercise 6.17). Combining this with the fact that the number of constraints is finite, we conclude that there exists a positive constant, η_2, such that

$$\eta_2 \leq \|\boldsymbol{P}\tilde{\boldsymbol{a}}\|_2^2 \qquad \text{for all } \boldsymbol{x} \in H \setminus \Omega.$$

Hence,

$$\frac{d}{dt}V(\boldsymbol{x}(t)) \leq -\frac{\mu\eta_2}{\tau} < 0 \tag{6.53}$$

almost everywhere on the interval (t_{l-1}, t_l), which concludes our analysis of the second case.

Using the above results, we can now state and prove the following theorem.

Theorem 6.3 Every trajectory of system (6.34) reaches Ω in finite time and is thereafter confined to Ω.

Proof Let $\boldsymbol{x}(t)$ be any particular trajectory of system (6.34) and let

$$\eta = \min\left\{\frac{\mu\eta_1}{\tau}, \frac{\mu\eta_2}{\tau}\right\}.$$

It follows directly from (6.41) and (6.53) that

$$\frac{d}{dt}V(\boldsymbol{x}(t)) \leq -\eta$$

for almost all t when $\boldsymbol{x}(t) \notin \Omega$. Hence, $\boldsymbol{x}(t) \in \Omega$ for all $t \geq V(\boldsymbol{x}(0))/\eta$.

Convergence Phase We now analyze dynamical behavior of system (6.34) when its motion is confined to the feasible set Ω. We begin by introducing the function $F:\mathbb{R}^n \to \mathbb{R}$ such that

$$F(\boldsymbol{x}) = f(\boldsymbol{x}) - f^*,$$

where f^* is the optimal value of f for the optimization problem (6.32). Clearly, $F \in \mathcal{C}^1$. Furthermore, we see from Lemma 6.1 that $F(\boldsymbol{x}) = 0$ for all $\boldsymbol{x} \in \Gamma$ and $F(\boldsymbol{x}) > 0$ for all $\boldsymbol{x} \in \Omega \setminus \Gamma$.

Similar to the previous subsection, we consider two cases when analyzing the behavior of system (6.34) in the region Ω. We first consider the case when the motion of system (6.34) is confined to the interior of Ω. Then, we consider the case when the motion of system (6.34) is confined to the boundary of Ω—that is, when system (6.34) is in a sliding mode on one or more of the surfaces $h_1, \ldots, h_m$.

Case 1 Let $\boldsymbol{x}(t)$ be any particular trajectory of system (6.34). Suppose that on an interval (t_{l-1}, t_l) the trajectory $\boldsymbol{x}(t)$ is confined to the interior of Ω. Then, $I(\boldsymbol{x}(t)) = \emptyset$ and $J(\boldsymbol{x}(t)) = \emptyset$ for all $t \in (t_{l-1}, t_l)$. This implies that on the interval (t_{l-1}, t_l), the trajectory $\boldsymbol{x}(t)$ satisfies $\tau\dot{\boldsymbol{x}}(t) = -\nabla f(\boldsymbol{x}(t))$ almost everywhere on the interval (t_{l-1}, t_l) in the usual sense. Applying the chain rule to $F(\boldsymbol{x}(t))$, we obtain

$$\frac{d}{dt}F(\boldsymbol{x}(t)) = \nabla F^T(\boldsymbol{x}(t))\dot{\boldsymbol{x}}(t) = -\frac{1}{\tau}\|\nabla f(\boldsymbol{x}(t))\|_2^2 \tag{6.54}$$

almost everywhere on the interval (t_{l-1}, t_l). Using the fact that $I(\boldsymbol{x}(t)) = \emptyset$ for all $t \in (t_{l-1}, t_l)$, we see from Lemma 6.2 that on the interval (t_{l-1}, t_l), we have $\nabla f(\boldsymbol{x}(t)) = \boldsymbol{0}$ if and only if $\boldsymbol{x}(t) \in \Gamma$. Hence, we have

$$\frac{d}{dt}F(\boldsymbol{x}(t)) = 0 \qquad \text{if and only if } \boldsymbol{x}(t) \in \Gamma, \tag{6.55}$$

and

$$\frac{d}{dt}F(\boldsymbol{x}(t)) < 0 \qquad \text{if and only if } \boldsymbol{x}(t) \notin \Gamma, \tag{6.56}$$

almost everywhere on the interval (t_{l-1}, t_l). This concludes our analysis of the first case.

Case 2 Let $\boldsymbol{x}(t)$ denote any particular trajectory of system (6.34) and let $t \in (t_{l-1}, t_l)$. Suppose that on the interval (t_{l-1}, t_l) the trajectory $\boldsymbol{x}(t)$ is confined to the boundary of Ω. Thus:

1. There exists a nonempty index set I' with the property that $I(\boldsymbol{x}(t)) = I'$ for all $t \in (t_{l-1}, t_l)$.
2. $J(\boldsymbol{x}(t)) = \emptyset$ for all $t \in (t_{l-1}, t_l)$.

Let $i_1, \ldots, i_k$ denote the elements of I', and let $S = \bigcap_{i \in I'} h_i$ denote the surface to which the trajectory $\boldsymbol{x}(t)$ is confined on the interval (t_{l-1}, t_l). Let $\boldsymbol{B} \in \mathbb{R}^{n\times k}$ be the matrix with columns $\boldsymbol{a}_{i_1}^T, \ldots, \boldsymbol{a}_{i_k}^T$. By assumption A3 the vectors $\boldsymbol{a}_{i_1}^T, \ldots, \boldsymbol{a}_{i_k}^T$ are linearly independent, and hence the matrix $\boldsymbol{B}$ has full rank.

It follows from the above that for a sufficiently small $\delta > 0$ the surfaces $h_{i_1}, \ldots, h_{i_k}$ partition the δ-neighborhood of $\boldsymbol{x}(t')$ into 2^k regions, in each of which the function $\boldsymbol{g}$ is continuous. Using a similar argument to the one in the previous subsection, we can show that the set $G(\boldsymbol{x}(t'))$ is the set of all $\boldsymbol{v}$ of the form

$$\boldsymbol{v} = -\alpha_0 \nabla f(\boldsymbol{x}(t')) - \mu \sum_{j=1}^{2^k-1} \alpha_j \boldsymbol{B}\boldsymbol{u}_j, \tag{6.57}$$

where

$$\alpha_j \geq 0, \qquad j = 0, 1, \ldots, 2^k - 1, \tag{6.58}$$

$$\sum_{j=0}^{2^k-1} \alpha_j = 1, \tag{6.59}$$

and $\boldsymbol{u}_j \in \mathbb{R}^k$, $j = 1, \ldots, 2^k - 1$, are defined as

$$\begin{aligned}
\boldsymbol{u}_1 &= [1 \quad 0 \quad 0 \quad \cdots \quad 0 \quad 0 \quad 0]^T, \\
\boldsymbol{u}_2 &= [0 \quad 1 \quad 0 \quad \cdots \quad 0 \quad 0 \quad 0]^T, \\
\boldsymbol{u}_3 &= [1 \quad 1 \quad 0 \quad \cdots \quad 0 \quad 0 \quad 0]^T, \\
&\vdots \\
\boldsymbol{u}_{2^k-3} &= [1 \quad 1 \quad 1 \quad \cdots \quad 1 \quad 0 \quad 1]^T, \\
\boldsymbol{u}_{2^k-2} &= [1 \quad 1 \quad 1 \quad \cdots \quad 1 \quad 1 \quad 0]^T, \\
\boldsymbol{u}_{2^k-1} &= [1 \quad 1 \quad 1 \quad \cdots \quad 1 \quad 1 \quad 1]^T.
\end{aligned}$$

We illustrate this case in Figure 6.12 for a hypothetical situation of sliding occurring on the intersection of two hyperplanes. In the figure, the components of the vector $\boldsymbol{v}$ as given by (6.57) are shown.

Let $T(\boldsymbol{x})$ denote the plane tangent to the surface S at the point $\boldsymbol{x}$. Observe that $\dot{\boldsymbol{x}}(t) \in T(\boldsymbol{x}(t))$ almost everywhere on the interval (t_{l-1}, t_l) because the trajectory $\boldsymbol{x}(t)$ is confined to the surface S on the interval (t_{l-1}, t_l). Thus, $\boldsymbol{x}(t)$ is an absolutely continuous function that satisfies

$$\tau \dot{\boldsymbol{x}}(t) \in G(\boldsymbol{x}(t)) \cap T(\boldsymbol{x}(t))$$

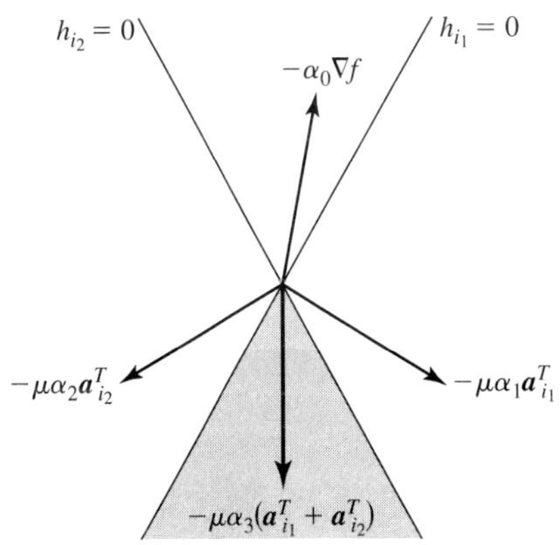

Figure 6.12 Components of the vector $\boldsymbol{v}$ as given by (6.57) for sliding occurring on the intersection of two hyperplanes h_{i_1} and h_{i_2}.

almost everywhere on the interval (t_{l-1}, t_l). We note that because the trajectory $\boldsymbol{x}(t)$ is confined to the surface S on the interval (t_{l-1}, t_l), the set $G(\boldsymbol{x}(t)) \cap T(\boldsymbol{x}(t))$ is by assumption nonempty for all $t \in (t_{l-1}, t_l)$. In particular, we see that $G(\boldsymbol{x}(t')) \cap T(\boldsymbol{x}(t'))$ is the set of all $\boldsymbol{v}$ that have the form (6.57) and also lie on the tangent plane $T(\boldsymbol{x}(t'))$. Let

$$\boldsymbol{r} = \sum_{j=1}^{2^k-1} \alpha_j \boldsymbol{B}\boldsymbol{u}_j.$$

Then, we represent (6.57) as

$$\boldsymbol{v} = -\alpha_0 \nabla f(\boldsymbol{x}(t')) - \mu \boldsymbol{r}. \tag{6.60}$$

Obviously, $\boldsymbol{r} \in \text{Span}\{\boldsymbol{a}_{i_1}^T, \ldots, \boldsymbol{a}_{i_k}^T\}$, and thus $\boldsymbol{r}$ is orthogonal to $T(\boldsymbol{x}(t'))$. However, $\boldsymbol{v} \in T(\boldsymbol{x}(t'))$ by assumption. Therefore, $\boldsymbol{v}$ is the orthogonal projection of the vector $-\alpha_0 \nabla f(\boldsymbol{x}(t'))$ onto the tangent plane $T(\boldsymbol{x}(t'))$; that is,

$$\boldsymbol{v} = -\alpha_0 \boldsymbol{P} \nabla f(\boldsymbol{x}(t')), \tag{6.61}$$

where $\boldsymbol{P} = \boldsymbol{I}_n - \boldsymbol{B}(\boldsymbol{B}^T\boldsymbol{B})^{-1}\boldsymbol{B}^T$.

Note that by virtue of assumptions A1 and A2, the set $\nabla f(\Omega)$ is compact. This implies the existence of a nonnegative constant, M, with the property that

$$M \geq \|\nabla f(\boldsymbol{x})\|_2 \qquad \text{for all } \boldsymbol{x} \in \Omega.$$

Let

$$\gamma = \frac{\mu}{\mu + M\sqrt{k}\|(\boldsymbol{B}^T\boldsymbol{B})^{-1}\boldsymbol{B}^T\|_2}.$$

Proposition 6.1

$$\gamma \leq \alpha_0 \leq 1.$$

Proof It follows from (6.58) and (6.59) that $\alpha_0 \leq 1$. Thus, it remains to show that $\alpha_0 \geq \gamma$. It follows from (6.60) and (6.61) that $\mu \boldsymbol{r}$ is the orthogonal projection of the vector $-\alpha_0 \nabla f(\boldsymbol{x}(t'))$ onto the subspace $T(\boldsymbol{x}(t'))^\perp$, where $T(\boldsymbol{x}(t'))^\perp$ is the orthogonal complement of $T(\boldsymbol{x}(t'))$ in $\mathbb{R}^n$. Therefore,

$$-\alpha_0(\boldsymbol{I}_n - \boldsymbol{P})\nabla f(\boldsymbol{x}(t')) = \mu \sum_{j=1}^{2^k-1} \alpha_j \boldsymbol{B}\boldsymbol{u}_j. \tag{6.62}$$

Let $\boldsymbol{U} \in \mathbb{R}^{k \times (2^k-1)}$ be the matrix with columns $\boldsymbol{u}_1, \ldots, \boldsymbol{u}_{2^k-1}$, and let $\boldsymbol{\alpha} \in \mathbb{R}^{2^k-1}$ be the column vector with the components $\alpha_1, \ldots, \alpha_{2^k-1}$. Then, we can represent (6.62) as

$$-\frac{\alpha_0}{\mu}(\boldsymbol{I}_n - \boldsymbol{P})\nabla f(\boldsymbol{x}(t')) = \boldsymbol{B}\boldsymbol{U}\boldsymbol{\alpha}. \tag{6.63}$$

Premultiplying both sides of (6.63) by $(\boldsymbol{B}^T\boldsymbol{B})^{-1}\boldsymbol{B}^T$, we obtain

$$-\frac{\alpha_0}{\mu}(\boldsymbol{B}^T\boldsymbol{B})^{-1}\boldsymbol{B}^T(\boldsymbol{I}_n - \boldsymbol{P})\nabla f(\boldsymbol{x}(t')) = \boldsymbol{U}\boldsymbol{\alpha},$$

and hence

$$\frac{\alpha_0}{\mu}\|(\boldsymbol{B}^T\boldsymbol{B})^{-1}\boldsymbol{B}^T(\boldsymbol{I}_n-\boldsymbol{P})\nabla f(\boldsymbol{x}(t'))\|_2 = \|\boldsymbol{U}\boldsymbol{\alpha}\|_2. \tag{6.64}$$

Applying the Cauchy–Schwarz inequality to (6.64) and using the fact that $\|\boldsymbol{I}_n-\boldsymbol{P}\| = 1$, we have

$$\frac{M\alpha_0}{\mu}\|(\boldsymbol{B}^T\boldsymbol{B})^{-1}\boldsymbol{B}^T\|_2 \geq \|\boldsymbol{U}\boldsymbol{\alpha}\|_2 \geq \frac{1}{\sqrt{k}}\|\boldsymbol{U}\boldsymbol{\alpha}\|_1$$

since for $\boldsymbol{z}\in\mathbb{R}^k$, we have $\|\boldsymbol{z}\|_2 \geq \frac{1}{\sqrt{k}}\|\boldsymbol{z}\|_1$. However, using (6.58) and (6.59), we see that

$$\|\boldsymbol{U}\boldsymbol{\alpha}\|_1 \geq \sum_{j=1}^{2^k-1}\alpha_j = 1-\alpha_0.$$

Thus,

$$\frac{M\alpha_0\sqrt{k}}{\mu}\|(\boldsymbol{B}^T\boldsymbol{B})^{-1}\boldsymbol{B}^T\|_2 \geq 1-\alpha_0. \tag{6.65}$$

It follows from (6.65) that $\alpha_0 > 0$. Dividing both sides of (6.65) by α_0 yields

$$\frac{M\sqrt{k}}{\mu}\|(\boldsymbol{B}^T\boldsymbol{B})^{-1}\boldsymbol{B}^T\|_2 \geq \frac{1-\alpha_0}{\alpha_0} = \frac{1}{\alpha_0}-1.$$

Hence,

$$\alpha_0 \geq \frac{1}{1+\dfrac{M\sqrt{k}}{\mu}\|(\boldsymbol{B}^T\boldsymbol{B})^{-1}\boldsymbol{B}^T\|_2} = \gamma.$$

The proof of the proposition is complete.

We have the following two corollaries.

Corollary 6.1

$$G(\boldsymbol{x}(t'))\cap T(\boldsymbol{x}(t'))\subset\{-\alpha\boldsymbol{P}\nabla f(\boldsymbol{x}(t')):\gamma\leq\alpha\leq 1\}.$$

Proof This is directly from (6.61) and Proposition 6.1.

Corollary 6.2

$$\boldsymbol{P}\nabla f(\boldsymbol{x}(t')) = \boldsymbol{0} \qquad \text{if and only if } \boldsymbol{0}\in G(\boldsymbol{x}(t'))\cap T(\boldsymbol{x}(t')).$$

Proof If $\boldsymbol{P}\nabla f(\boldsymbol{x}(t')) = \boldsymbol{0}$, then $\boldsymbol{0}\in G(\boldsymbol{x}(t'))\cap T(\boldsymbol{x}(t'))$ because, by Corollary 6.1,

$$\boldsymbol{P}\nabla f(\boldsymbol{x}(t'))\in G(\boldsymbol{x}(t'))\cap T(\boldsymbol{x}(t')).$$

Conversely, suppose that $\mathbf{0} \in G(\mathbf{x}(t')) \cap T(\mathbf{x}(t'))$. By Corollary 6.1, $-\alpha \mathbf{P}\nabla f(\mathbf{x}(t')) = \mathbf{0}$ for some $\alpha \in [\gamma, 1]$. Since $\gamma > 0$ for all $\mu > 0$, $-\alpha \mathbf{P}\nabla f(\mathbf{x}(t')) = \mathbf{0}$ is possible only if $\mathbf{P}\nabla f(\mathbf{x}(t')) = \mathbf{0}$, which completes the proof.

Proposition 6.2

$$\mathbf{P}\nabla f(\mathbf{x}(t')) = \mathbf{0} \qquad \text{if and only if } \mathbf{x}(t') \in \Gamma.$$

Proof We first prove the sufficiency part of the proposition. Suppose $\mathbf{P}\nabla f(\mathbf{x}(t')) = \mathbf{0}$. It follows from (6.61) and (6.57) that

$$\alpha_0 \nabla f(\mathbf{x}(t')) + \mu \sum_{j=1}^{2^k-1} \alpha_j \mathbf{B}\mathbf{u}_j = \mathbf{0}. \tag{6.66}$$

Let

$$\mathbf{d} = \frac{\mu}{\alpha_0} \sum_{j=1}^{2^k-1} \alpha_j \mathbf{u}_j.$$

Note that because $\alpha_0 \neq 0$, $\mathbf{d}$ is well-defined. It then follows from (6.66) that

$$\nabla f(\mathbf{x}(t')) + \mathbf{B}\mathbf{d} = \mathbf{0}.$$

We see from (6.58) and the definition of $\mathbf{u}_1, \ldots, \mathbf{u}_{2^k-1}$ that each element of $\mathbf{d}$ is nonnegative. By Lemma 6.2, the point $\mathbf{x}(t')$ satisfies the Karush–Kuhn–Tucker conditions for problem (6.32) and hence $\mathbf{x}(t') \in \Gamma$. The sufficiency condition is thus proven.

We now prove the necessity condition. Suppose that $\mathbf{x}(t') \in \Gamma$. By Lemma 6.2, the point $\mathbf{x}(t')$ satisfies the Karush–Kuhn–Tucker optimality conditions for problem (6.32). This implies the existence of nonnegative constants $\mu_1, \ldots, \mu_k$ with the property that

$$\nabla f(\mathbf{x}(t')) + \mathbf{B}\boldsymbol{\mu} = \mathbf{0}, \tag{6.67}$$

where $\boldsymbol{\mu} = [\mu_1, \ldots, \mu_k]^T$. Let

$$\nu = 1 + \frac{1}{\mu} \sum_{i=1}^{k} \mu_i, \tag{6.68}$$

$$\tilde{\alpha}_0 = \frac{1}{\nu}, \tag{6.69}$$

$$\tilde{\alpha}_{2^{i-1}} = \frac{\mu_i}{\mu\nu}, \qquad i = 1, \ldots, k, \tag{6.70}$$

$$\tilde{\alpha}_j = 0, \qquad j = 1, \ldots, 2^k - 1, \quad j \neq 2^{i-1} \quad \text{for } i = 1, \ldots, k, \tag{6.71}$$

$$\tilde{\mathbf{v}} = -\tilde{\alpha}_0 \nabla f(\mathbf{x}(t')) - \mu \sum_{j=1}^{2^k-1} \tilde{\alpha}_j \mathbf{B}\mathbf{u}_j. \tag{6.72}$$

It follows from (6.68)–(6.71) that

$$\tilde{\alpha}_j \geq 0, \qquad j = 0, 1, \ldots, 2^k - 1 \quad \text{and} \quad \sum_{j=0}^{2^k-1} \tilde{\alpha}_j = 1.$$

Therefore, $\tilde{\boldsymbol{v}} \in G(\boldsymbol{x}(t'))$. Note that

$$\tilde{\boldsymbol{v}} = -\frac{\mu}{\mu + \sum_{i=1}^{k} \mu_i}(\nabla f(\boldsymbol{x}(t')) + \boldsymbol{B}\boldsymbol{\mu}).$$

Using (6.67), we conclude that $\boldsymbol{v} = \boldsymbol{0}$, and thus $\tilde{\boldsymbol{v}} \in T(\boldsymbol{x}(t'))$. Hence,

$$\boldsymbol{0} \in G(\boldsymbol{x}(t')) \cap T(\boldsymbol{x}(t')).$$

Using Corollary 6.2, we get $\boldsymbol{P}\nabla f(\boldsymbol{x}(t')) = \boldsymbol{0}$, which completes the proof of the necessary condition and the proof of the proposition.

Applying now the chain rule to F, taking into account (6.61), Proposition 6.1, and the fact that $\boldsymbol{P} = \boldsymbol{P}^T = \boldsymbol{P}^2$, we obtain

$$\frac{d}{dt}F(\boldsymbol{x}(t)) \leq -\frac{\gamma}{\tau}\|\boldsymbol{P}\nabla f(\boldsymbol{x}(t))\|_2^2 \tag{6.73}$$

almost everywhere on the interval (t_{l-1}, t_l). By Proposition 6.2,

$$\frac{d}{dt}F(\boldsymbol{x}(t)) = 0 \qquad \text{if and only if } \boldsymbol{x}(t) \in \Gamma, \tag{6.74}$$

and

$$\frac{d}{dt}F(\boldsymbol{x}(t)) < 0 \qquad \text{if and only if } \boldsymbol{x}(t) \notin \Gamma, \tag{6.75}$$

almost everywhere on the interval (t_{l-1}, t_l), which concludes our analysis for the second case.

We now show that any trajectory of system (6.34) that is confined to Ω converges asymptotically to the region Γ.

Theorem 6.4 Let $\boldsymbol{x}(t)$ be any trajectory of system (6.34). If $\boldsymbol{x}(t) \in \Omega$ for all t, then the trajectory $\boldsymbol{x}(t)$ converges asymptotically to Γ.

Proof Recall that $F(\boldsymbol{x}) = 0$ if $\boldsymbol{x} \in \Gamma$ and $F(\boldsymbol{x}) > 0$ if $\boldsymbol{x} \in \Omega \setminus \Gamma$. Therefore, to prove the theorem it is enough to show that

$$\lim_{t\to\infty} F(\boldsymbol{x}(t)) = 0,$$

or equivalently, that for any $\varepsilon > 0$ there exists a finite time, $T(\boldsymbol{x}(0), \varepsilon)$, such that $F(\boldsymbol{x}(t)) < \varepsilon$ for all $t \geq T(\boldsymbol{x}(0), \varepsilon)$. We see from (6.56) and (6.75) that

$$\frac{d}{dt} F(\boldsymbol{x}(t)) < 0$$

for almost all t when $\boldsymbol{x}(t) \notin \Gamma$. Consequently $F(\boldsymbol{x}(t)) \leq F(\boldsymbol{x}(0))$ for all $t \geq 0$. Therefore, we only need to consider the case when $F(\boldsymbol{x}(0)) \geq \varepsilon$. Let

$$\theta(\varepsilon) = \left\{ \boldsymbol{x} : \boldsymbol{x} \in \Omega,\ F(\boldsymbol{x}) \geq \frac{\varepsilon}{2} \right\}.$$

Because $F(\boldsymbol{x}(0)) \geq \varepsilon$, we have $\boldsymbol{x}(0) \in \theta(\varepsilon)$ and $\theta(\varepsilon) \subset \Omega \setminus \Gamma$. Furthermore, because the function F is continuous, the set $\theta(\varepsilon)$ is compact. We will now show that there exists a positive constant, $\rho(\varepsilon)$, such that

$$\frac{d}{dt} F(\boldsymbol{x}(t)) \leq -\rho(\varepsilon)$$

for almost all t when $\boldsymbol{x}(t) \in \theta(\varepsilon)$, which guarantees that $F(\boldsymbol{x}(t)) \leq \varepsilon/2$ for all

$$t \geq \frac{F(\boldsymbol{x}(0)) - \varepsilon/2}{\rho(\varepsilon)}.$$

We analyze two cases. The first case is when the trajectory $\boldsymbol{x}(t)$ is in $\theta(\varepsilon)$ and is not in a sliding mode on any of the surfaces h_i. Then, we analyze the case when the trajectory is in $\theta(\varepsilon)$ in a sliding mode on a surface that is an intersection of some of the hyperplanes h_i.

Case 1 Because the function f is convex and $f \in C^1$, we have $\nabla f(\boldsymbol{x}) \neq \boldsymbol{0}$ for all $\boldsymbol{x} \in \Omega \setminus \Gamma$. Hence,

$$-\frac{1}{\tau} \|\nabla f(\boldsymbol{x})\|_2^2 < 0 \qquad \text{for all } \boldsymbol{x} \in \theta(\varepsilon) \tag{6.76}$$

because $\theta \subset \Omega \setminus \Gamma$ and $\nabla f(\boldsymbol{x}) \neq \boldsymbol{0}$ in $\theta(\varepsilon)$. The fact that the function $-1/\tau \|\nabla f(\boldsymbol{x})\|_2^2$ is continuous on the compact set $\theta(\varepsilon)$ implies the existence of a positive constant, $\rho_1(\varepsilon)$, such that

$$-\frac{1}{\tau} \|\nabla f(\boldsymbol{x})\|_2^2 \leq -\rho_1(\varepsilon) \qquad \text{for all } \boldsymbol{x} \in \theta(\varepsilon).$$

Combining the above with (6.54) yields

$$\frac{d}{dt} F(\boldsymbol{x}(t)) \leq -\rho_1(\varepsilon)$$

for almost all t when $\boldsymbol{x}(t) \in \theta(\varepsilon)$, which concludes our analysis of the first case.

Case 2 Let $i_1, \ldots, i_k$ denote the elements of I', and let $S = \bigcap_{i \in I'} h_i$ denote the surface to which the trajectory $\boldsymbol{x}(t)$ is confined on the interval (t_{l-1}, t_l). Suppose that $S \cap \theta(\varepsilon) \neq \emptyset$. Clearly,

$$S \cap \theta(\varepsilon) \subset \Omega \setminus \Gamma.$$

Using arguments similar to those of Glazos et al. [98, p. 692], we can show that

$$-\frac{1}{\tau}\| \boldsymbol{P}\nabla f(\boldsymbol{x})\|_2^2 < 0 \qquad \text{for all } \boldsymbol{x} \in S \cap \theta(\varepsilon), \tag{6.77}$$

where $\boldsymbol{P}$ is the orthogonal projector onto the tangent plane to the surface S. By the above and the fact that the function $-\frac{1}{\tau}\| \boldsymbol{P}\nabla f(\boldsymbol{x})\|_2^2$ is continuous on the compact set $S \cap \theta(\varepsilon)$, there exists a positive constant, $\hat{\rho}(\varepsilon)$, such that

$$-\frac{1}{\tau}\| \boldsymbol{P}\nabla f(\boldsymbol{x})\|_2^2 < -\hat{\rho}(\varepsilon) \qquad \text{for all } \boldsymbol{x} \in S \cap \theta(\varepsilon).$$

It then follows directly from (6.73) that there exists a positive constant $\hat{\gamma}$ such that

$$\frac{d}{dt}F(\boldsymbol{x}(t)) \le -\hat{\gamma}\hat{\rho}(\varepsilon)$$

for almost all t when $\boldsymbol{x}(t) \in S \cap \theta(\varepsilon)$. The above along with the fact that the number of constraints is finite implies the existence of a positive constant, $\rho_2(\varepsilon)$, such that

$$\frac{d}{dt}F(\boldsymbol{x}(t)) \le -\rho_2(\varepsilon) \tag{6.78}$$

for almost all t when $\boldsymbol{x}(t) \in S \cap \theta(\varepsilon)$. This concludes our analysis for the second case.

Let

$$\rho(\varepsilon) = \min\{\rho_1(\varepsilon), \rho_2(\varepsilon)\}.$$

Then, using (6.77) and (6.78) yields

$$\frac{d}{dt}F(\boldsymbol{x}(t)) \le -\rho(\varepsilon)$$

for almost all t when $\boldsymbol{x}(t) \in \theta(\varepsilon)$. Hence, $F(\boldsymbol{x}(t)) \le \varepsilon/2 < \varepsilon$ for all

$$t \ge \frac{F(\boldsymbol{x}(0)) - \varepsilon/2}{\rho(\varepsilon)}.$$

Therefore,

$$T(\boldsymbol{x}(0), \varepsilon) = \begin{cases} 0 & \text{if } F(\boldsymbol{x}(0)) < \varepsilon, \\ \dfrac{F(\boldsymbol{x}(0)) - \varepsilon/2}{\rho(\varepsilon)} & \text{if } F(\boldsymbol{x}(0)) \ge \varepsilon. \end{cases}$$

The proof of the theorem is complete.

Theorem 6.5 Every trajectory of system (6.34), irrespective of the initial condition, $\boldsymbol{x}(0)$, converges to a solution of problem (6.32).

Proof This statement follows directly from Theorems 6.3 and 6.4.

◆ **Example 6.5** (Based on Glazos et al. [98])

Consider the quadratic programming problem

$$\begin{array}{ll} \text{minimize} & \frac{1}{2}\boldsymbol{x}^T \boldsymbol{Q}\boldsymbol{x} + \boldsymbol{c}^T \boldsymbol{x}, \qquad \boldsymbol{x} \in \mathbb{R}^2 \\ \text{subject to} & \boldsymbol{A}\boldsymbol{x} \le \boldsymbol{b}, \end{array}$$

where

$$\boldsymbol{Q} = \begin{bmatrix} 4 & 1 \\ 1 & 2 \end{bmatrix}, \quad \boldsymbol{c} = \begin{bmatrix} -12 \\ -10 \end{bmatrix}, \quad \boldsymbol{A} = \begin{bmatrix} 5 & 1 \\ -5 & 1 \\ -1 & -2 \end{bmatrix}, \quad \text{and} \quad \boldsymbol{b} = \begin{bmatrix} 0 \\ 10 \\ 10 \end{bmatrix}.$$

The above optimization problem clearly satisfies assumptions A1–A4, and one can easily verify that the point $\boldsymbol{x}^* = [-19/22 \;\; 95/22]^T$ is a unique minimizer for the problem, that is, $\Gamma = \{\boldsymbol{x}^*\}$. Note that $\boldsymbol{x}^* \in h_1$. A phase-plane portrait for system (6.34) that solves the above optimization problem is shown in Figure 6.13. The feasible region is depicted by the shaded area. Also shown are level sets of the quadratic objective function.

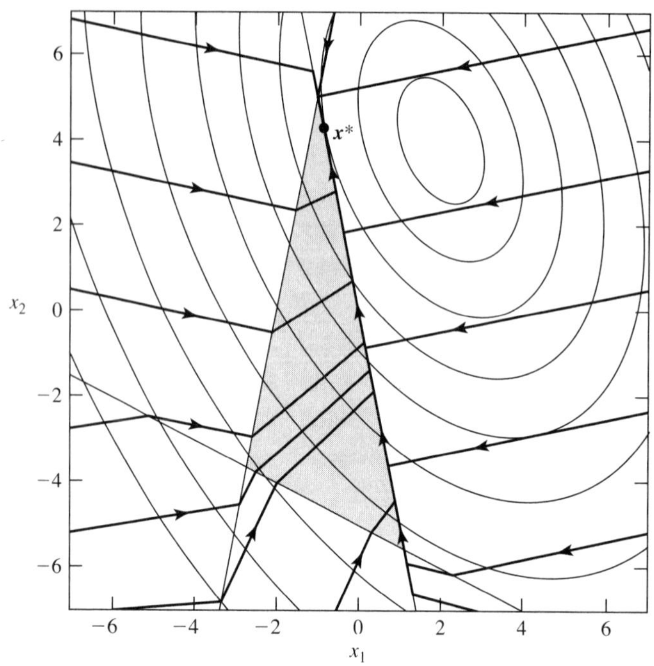

Figure 6.13 Phase-plane portrait for the system in Example 6.5.

Notes

The idea of controlling a dynamical system by varying its structure was originated in the 1950s. Early pioneers in the field of variable structure systems were S. V. Emelyanov and his co-workers at the Institute of Control Sciences in Moscow in the former Soviet Union. A large body of literature was produced by these studies; unfortunately, most of the work was published in Russian only. The titles of books and papers on the subject of variable structure control systems published in the years 1957–1989 were collected by Emelyanov [77]. The first widely accessible books in English on variable structure sliding mode systems were by Itkis [140] and Utkin [285]. A survey paper by Utkin [284] refers to many early contributions to the field; see also a tutorial by DeCarlo, Żak, and Matthews [60] and a survey by Hung, Gao, and Hung [135]. At about the same time, A. F. Filippov was developing his theory of differential equations with discontinuous right-hand sides [83]. Filippov's approach is now accepted as the tool for the analysis of sliding mode systems. Sliding modes also found applications in the state estimation of dynamical uncertain systems as we show in Section 6.8. For further analysis of discontinuous state estimators, we refer the reader to Edwards and Spurgeon [71, 73]. For a stability analysis of uncertain closed-loop systems with the combined variable structure estimator–controller compensators we refer to Hui and Żak [131]. Output feedback sliding mode controllers are presented in references 64, 72, 133, 174, 310, and 311.

In 1967, Karpinskaya [150] identified the occurrence of sliding modes in an analog circuit for solving optimization problems. The idea of using analog circuits to solve constrained optimization problems first appeared in the literature in 1956 when Pyne [239] proposed a method for solving linear programming problems using an electronic analog computer. Soon after, other methods were proposed for solving various mathematical programming problems on an analog computer—see, for example, Rybashov [248–251] and Karpinskaya [150], among others. More recently, many new analog architectures have been proposed for solving constrained optimization problems. In particular, Chua and Lin [51] developed a canonical nonlinear programming circuit for solving general nonlinear programming problems. Tank and Hopfield [278] later proposed an analog circuit for solving linear programming problems. The results of Tank and Hopfield were then extended to nonlinear programming problems by Kennedy and Chua [155]. Rodríguez-Vázquez et al. [242] and Cichocki and Unbehauen [52] proposed analog networks that use switched-capacitor integrated circuit technology for solving certain constrained optimization problems. We analyzed dynamical behavior of a neural network of Rodríguez-Vázquez et al. [242] for solving a class of convex programming problems. In carrying out the analysis, we used concepts from the theory of differential equations with discontinuous right-hand sides and Lyapunov stability theory. Our method can also be used to analyze other classes of analog dynamical optimizers whose designs are based on exact penalty functions. An open problem is to extend the results obtained to a more general class of mathematical programming problems. Preliminary results in this direction can be found in Glazos et al. [98] and Chong et al. [49].

EXERCISES

6.1 Consider the following variable structure system that was analyzed in Hung et al. [135] (see also Utkin [284] for a similar system):

$$\begin{aligned}\dot{x}_1 &= x_2,\\ \dot{x}_2 &= -x_1 + 2x_2 + u,\end{aligned} \tag{6.79}$$

where

$$u = -4x_1 \operatorname{sign} \sigma(x_1, x_2), \qquad \sigma(x_1, x_2) = x_1(0.5x_1 + x_2).$$

(a) Sketch phase plane portraits of two linear subsystems that constitute the given variable structure system (6.79).

(b) Sketch a phase-plane portrait of the variable structure system (6.79).

6.2 (a) For the dynamical system model,

$$\dot{x} = \begin{bmatrix} 1 & 1 \\ 2 & 1 \end{bmatrix} x + \begin{bmatrix} 1 \\ -1 \end{bmatrix} u,$$

construct the switching surface

$$\sigma(x) = sx = 0$$

such that the system in sliding along the surface has its pole at -2.

(b) Consider the following model of a dynamical system:

$$\dot{x} = \begin{bmatrix} 2 & 0 & -1 \\ 0 & -2 & 2 \\ -1 & -2 & 3 \end{bmatrix} x + \begin{bmatrix} 1 \\ 1 \\ 1 \end{bmatrix} u.$$

Design a switching surface $\sigma(x) = sx = 0$ such that the system restricted to it has poles at -4, -5.

(c) Given the following dynamical system model,

$$\dot{x} = \begin{bmatrix} 2 & 0 & -1 \\ 0 & -2 & 2 \\ -1 & -2 & 3 \end{bmatrix} x + \begin{bmatrix} 1 \\ 1 \\ 1 \end{bmatrix} u,$$

along with the switching surface

$$\sigma(x) = sx = [8 \quad 8 \quad -15]x = 0.$$

Determine the reduced-order equations that describe the system behavior in a sliding mode.

6.3 Consider a dynamical system modeled by the equations

$$\begin{bmatrix} \dot{x}_1 \\ \dot{x}_2 \end{bmatrix} = \begin{bmatrix} 0 & 1 \\ a_1(t) & a_2(t) \end{bmatrix} \begin{bmatrix} x_1 \\ x_2 \end{bmatrix} + \begin{bmatrix} 0 \\ 1 \end{bmatrix} u,$$

where

$$|a_1(t)| \le 3, \quad |a_2(t)| \le 4 \qquad \text{for all } t.$$

(a) Design a switching surface

$$\sigma(x_1, x_2) = s_1 x_1 + x_2 = 0$$

such that the system restricted to it has its pole at -2.

(b) Design a sliding mode controller

$$u = k_1^{\pm} x_1 + k_2^{\pm} x_2,$$

so that the switching surface $\sigma = 0$ is globally attractive. The gains k_1 and k_2 are functions of $x_1\sigma$ and $x_2\sigma$, respectively.

6.4 Consider the following model of an armature controlled DC motor controlling an inverted pendulum:

$$\begin{bmatrix} \dot{x}_1 \\ \dot{x}_2 \\ \dot{x}_3 \end{bmatrix} = \begin{bmatrix} x_2 \\ 9.8 \sin x_1 + x_3 \\ -\dfrac{1}{L_a} x_2 - \dfrac{1}{L_a} x_3 \end{bmatrix} + \begin{bmatrix} 0 \\ 0 \\ \dfrac{1}{L_a} \end{bmatrix} u$$

$$y = x_1,$$

where L_a is the armature inductance of the motor.

(a) Derive the reduced-order model assuming L_a is negligible.

(b) For the obtained reduced-order model from the first part, design the switching surface $\sigma = 0$ of the form

$$\sigma = \sigma(x_1, x_2) = [s_1 \quad 1] \begin{bmatrix} x_1 \\ x_2 \end{bmatrix} = 0$$

so that the system in sliding will have its pole at -3.

(c) Use the above sliding surface to design a variable structure controller of the form

$$u = -k_1 x_1 \operatorname{sign}(x_1\sigma) - k_2 x_2 \operatorname{sign}(x_2\sigma).$$

The smaller the gains k_1, k_2, the better.

6.5 Consider a dynamical system model,

$$\begin{bmatrix} \dot{\boldsymbol{x}}_1 \\ \dot{x}_2 \end{bmatrix} = \begin{bmatrix} \boldsymbol{A}_1 & \boldsymbol{A}_2 \\ \boldsymbol{A}_3 & A_4 \end{bmatrix} \begin{bmatrix} \boldsymbol{x}_1 \\ x_2 \end{bmatrix} + \begin{bmatrix} \boldsymbol{0} \\ b \end{bmatrix} u, \tag{6.80}$$

$$y = [\boldsymbol{C}_1 \quad C_2] \begin{bmatrix} \boldsymbol{x}_1 \\ x_2 \end{bmatrix}, \tag{6.81}$$

where b and C_2 are nonzero scalars. Let

$$\sigma = [\boldsymbol{S}_1 \quad S_2] \begin{bmatrix} \boldsymbol{x}_1 \\ x_2 \end{bmatrix} = F \left[\boldsymbol{C}_1 \quad C_2\right] \begin{bmatrix} \boldsymbol{x}_1 \\ x_2 \end{bmatrix} = Fy, \tag{6.82}$$

where S_2 and F are nonzero scalars.

(a) Find the reduced-order system that models (6.80) when restricted to the surface $\sigma = 0$.

(b) Find the transfer function $H(s) = \mathcal{L}(y(t))/\mathcal{L}(u(t)) = Y(s)/U(s)$. What is the relation between the zeros of $H(s)$ and the poles of the reduced-order system found in part (a)?

6.6 (Based on Kwan [174]) Prove the following lemma.

Lemma 6.4 Let

$$\dot{\boldsymbol{x}} = \boldsymbol{A}\boldsymbol{x} + \boldsymbol{B}\boldsymbol{u},$$

where the matrix $\boldsymbol{A}$ has negative real distinct eigenvalues, that is,

$$\text{eig}(\boldsymbol{A}) = \{-\lambda_1, \quad -\lambda_2, \quad \ldots, \quad -\lambda_n\}, \qquad \lambda_i > 0, \lambda_i \neq \lambda_j \text{ for } i \neq j.$$

Let $\lambda_{\min} = \min\{\lambda_1, \quad \lambda_2, \quad \ldots, \quad \lambda_n\}$ and K be a positive constant such that

$$\|e^{\boldsymbol{A}t}\| \leq Ke^{-\lambda_{\min} t}$$

Then,

$$\|\boldsymbol{x}(t)\| \leq w(t),$$

where $w(t)$ is a solution of

$$\dot{w} = -\lambda_{\min} w + K\|\boldsymbol{B}\|\|\boldsymbol{u}\|, \qquad \text{subject to} \quad w(0) \geq K\|\boldsymbol{x}(0)\|.$$

6.7 (Based on Kwan [174]) Consider an nth-order model of an uncertain dynamical system,

$$\begin{bmatrix} \dot{\boldsymbol{x}}_1 \\ \dot{x}_2 \end{bmatrix} = \begin{bmatrix} \boldsymbol{A}_1 & \boldsymbol{A}_2 \\ \boldsymbol{A}_3 & A_4 \end{bmatrix} \begin{bmatrix} \boldsymbol{x}_1 \\ x_2 \end{bmatrix} + \begin{bmatrix} \boldsymbol{0} \\ b \end{bmatrix} (u + \xi), \tag{6.83}$$

$$y = [\,\boldsymbol{C}_1 \quad C_2\,] \begin{bmatrix} \boldsymbol{x}_1 \\ x_2 \end{bmatrix},$$

where b and C_2 are nonzero scalars and ξ models system's uncertainty, where

$$\xi = \boldsymbol{D}_1\boldsymbol{x}_1 + D_2x_2 + D_3u + D_4(t).$$

In the above, $\boldsymbol{D}_1$ is a row vector whose size is equal to the dimension of $\boldsymbol{x}_1 \in \mathbb{R}^{n-1}$, and D_2 and D_3 are scalars, where $|D_3| < 1$ and D_4 is an unknown bounded function of time whose bound is known to us; that is, $|D_4(t)| \leq d_4$, where d_4 is known to us. Let $\sigma = Fy$ be as defined in Exercise 6.5. Assume that the switching surface,

$$\sigma = Fy = F[\boldsymbol{C}_1 \quad C_2] \begin{bmatrix} \boldsymbol{x}_1 \\ x_2 \end{bmatrix} = 0,$$

was chosen in such a way that the nominal system (6.83) restricted to it has the eigenvalues

$$\{-\lambda_1, \quad -\lambda_2, \quad \ldots, \quad -\lambda_{n-1}\}, \qquad \lambda_i > 0, \lambda_i \neq \lambda_j \text{ for } i \neq j.$$

(a) Evaluate $\dot{\sigma}$ on the trajectories of the uncertain system model (6.83).

(b) For the controller

$$u = (S_2b)^{-1}(1 - |D_3|)^{-1}(-Hw(t) - G|\sigma| - \delta)\text{sign}(\sigma),$$

where $w(t)$ is a solution of a first-order differential equation such that $\|\boldsymbol{x}_1(t)\| \leq w(t)$, find lower bounds for H, G, and δ such that

$$\sigma\dot{\sigma} \leq -\eta|\sigma|$$

for some $\eta > 0$. Use the lemma of Exercise 6.6 to find the form of the differential equation whose solution satisfies the condition $\|\boldsymbol{x}_1(t)\| \leq w(t)$.

6.8 For the dynamical system model

$$\dot{\boldsymbol{x}} = \boldsymbol{A}\boldsymbol{x} + \boldsymbol{b}(u + f(y)),$$
$$y = \boldsymbol{c}\boldsymbol{x},$$

where

$$\boldsymbol{A} = \begin{bmatrix} -2 & 2 \\ 0 & 1 \end{bmatrix}, \quad \boldsymbol{b} = \begin{bmatrix} 0 \\ 1 \end{bmatrix}, \quad \boldsymbol{c} = [0 \quad 1], \quad \text{and} \quad |f(y)| \leq 10y,$$

design a discontinuous state-estimator. The estimator poles should be placed at -2 and 3.

Hint After constructing the required estimator gain $\boldsymbol{l}$, find a symmetric positive definite matrix $\boldsymbol{Q}$ such that the solution, $\boldsymbol{P}$, to the Lyapunov matrix equation, $(\boldsymbol{A} - \boldsymbol{l}\boldsymbol{c})^T\boldsymbol{P} + \boldsymbol{P}(\boldsymbol{A} - \boldsymbol{l}\boldsymbol{c}) = -\boldsymbol{Q}$, satisfies the condition, $\boldsymbol{b}^T\boldsymbol{P} = \boldsymbol{c}$.

6.9 The notion of *feasible direction,* which we present next, plays an important role in the analysis of constrained optimization problems such as that given by (6.36). We say that a vector $\boldsymbol{d}$ is a feasible direction at $\boldsymbol{x} \in \Omega$ if there is an $\bar{\alpha} > 0$ such that $\boldsymbol{x} + \alpha\boldsymbol{d} \in \Omega$ for all $\alpha \in [0, \bar{\alpha}]$. Suppose now that the minimizer $\boldsymbol{x}^*$ of the optimization problem (6.36) lies on the boundary of Ω and there is a feasible direction $\boldsymbol{d}$ at $\boldsymbol{x}^*$ such that $\boldsymbol{d}^T\nabla f(\boldsymbol{x}^*) > 0$. Show, using contraposition, that if p is an exact penalty function for the problem, then p is not differentiable at $\boldsymbol{x}^*$ (see [50]).

6.10 Consider an uncertain, or nonlinear, dynamical system model,

$$\dot{\boldsymbol{x}}(t) = \boldsymbol{A}\boldsymbol{x}(t) + \boldsymbol{f}(t, \boldsymbol{x}(t)) + \boldsymbol{B}(\boldsymbol{u}(t) + \boldsymbol{h}(t, \boldsymbol{x}(t), \boldsymbol{u}(t))), \tag{6.84}$$

where $\boldsymbol{x}(t) \in \mathbb{R}^n$, $\boldsymbol{u}(t) \in \mathbb{R}^m$, the constant matrices $\boldsymbol{A} \in \mathbb{R}^{n\times n}$ and $\boldsymbol{B} \in \mathbb{R}^{n\times m}$ are known, and the function $\boldsymbol{f}$ and $\boldsymbol{h}$ model uncertainties, or nonlinearities, in the system. We refer to $\boldsymbol{h}$ as the matched uncertainty, while we refer to $\boldsymbol{f}$ as the unmatched uncertainty. The reason for calling $\boldsymbol{h}$ the matched uncertainty is that it affects the system dynamics through the column space of the matrix $\boldsymbol{B}$ in a similar fashion as the control $\boldsymbol{u}$ does. In other words, $\boldsymbol{h}$ matches the control action $\boldsymbol{u}$. We assume that

1. $\|\boldsymbol{f}(t, \boldsymbol{x})\| \leq \alpha_f\|\boldsymbol{x}\| + \beta_f$ and
2. $\|\boldsymbol{h}(t, \boldsymbol{x}, \boldsymbol{u})\| \leq \gamma_h\|\boldsymbol{u}\| + \alpha_h\|\boldsymbol{x}\| + \beta_h$,

where α_f, β_f, γ_h, α_h, and β_h are known nonnegative constants, and $\|\cdot\| = \|\cdot\|_2$ denotes the Euclidean norm for vectors and the induced spectral norm for matrices. We also assume that

3. the pair $(\boldsymbol{A}, \boldsymbol{B})$ is controllable and
4. rank $\boldsymbol{B} = m$.

Let

$$\tilde{x} = Mx = \begin{bmatrix} W^g \\ S \end{bmatrix} x,$$

where W^g and S are the matrices obtained from the Switching Surface Design Algorithm. Transform the uncertain system model (6.84) into the $\tilde{x}$ coordinates.

6.11 The presence of unmatched uncertainties has a deteriorating effect on the stability properties of a closed-loop system. The above combined with the imposed hard bounds on the control action may make it impossible to globally asymptotically stabilize the closed-loop system. The best it can be achieved in such a situation may be UUB with some region of attraction $\Gamma \subset \mathbb{R}^n$ and uniformly stable set Σ containing the origin of $\mathbb{R}^n$. The goal of this exercise is to estimate the "size" of the set Γ and that of Σ.

It follows from Exercise 6.10 that the uncertain system (6.84) confined to the switching surface $\sigma = Sx = \mathbf{0}$ is described, in the new coordinates, by

$$\dot{z} = A_{11}z + W^g \tilde{f}(t, z, \mathbf{0}), \tag{6.85}$$

where $\|\tilde{f}(t, z, \mathbf{0})\| \leq \alpha_f \|W\| \|z\| + \beta_f$. We refer to the above system as the *reduced-order system*. The matrix A_{11} is asymptotically stable by construction. Therefore, for any symmetric and positive definite matrix $Q \in \mathbb{R}^{(n-m)\times(n-m)}$ the solution P to the Lyapunov equation

$$A_{11}^T P + P A_{11} = -2Q$$

is (symmetric) positive definite. Consider now the positive definite function:

$$V_1 = (z^T P z)^{1/2}.$$

The matrix P is symmetric positive definite, hence it can be represented as

$$P = P^{1/2} P^{1/2},$$

where $P^{1/2} \in \mathbb{R}^{(n-m)\times(n-m)}$ is symmetric and nonsingular. We refer to $P^{1/2}$ as a square root of P—see page 635 for a discussion of a matrix square root. Let

$$\zeta = P^{1/2} z.$$

Then,

$$V_1 = (z^T P z)^{1/2} = (\zeta^T \zeta)^{1/2} = \|\zeta\|.$$

Evaluate the time derivative of V_1 on the trajectories of the reduced-order system (6.85). Specifically, let

$$\rho = \frac{\lambda_{\min}(Q)}{\lambda_{\max}(P)} - \alpha_f \frac{\lambda_{\max}(P)}{\lambda_{\min}(P)} \|W^g\| \|W\|,$$

and

$$r = \frac{\beta_f}{\rho} \frac{\lambda_{\max}(P)}{\lambda_{\min}^{\frac{1}{2}}(P)} \|W^g\|.$$

Assume that $\rho > 0$. Using the above notation, show that

$$\frac{d}{dt}\|\zeta\| \leq -\rho(\|\zeta\| - r).$$

6.12 Prove the following theorem concerning stability of the motion on the switching surface.

Theorem 6.6 The reduced-order system, analyzed in Exercise 6.11,

$$\dot{z} = A_{11} z + W^g \tilde{f}(t, z, 0)$$

is globally UUB, with the uniformly stable set

$$\Sigma = \left\{ z : \|\zeta\| = \|P^{\frac{1}{2}} z\| \leq r \right\} \subset \mathbb{R}^{n-m}.$$

In the proof of the above theorem, use the following lemma, which we recommend that you prove before using it.

Lemma 6.5 Suppose $\phi(t)$, a, $b \in \mathbb{R}$, and $a \neq 0$. If

$$\dot{\phi}(t) - a\phi(t) \leq -b,$$

then, for $t \geq t_0$,

$$\phi(t) \leq \frac{b}{a} + \left(\phi(t_0) - \frac{b}{a}\right) e^{(t-t_0)}.$$

6.13 Let $\sigma = Sx$, $\|\cdot\|_p$ be the p norm of a vector, and let ∇_σ denote the gradient operation with respect to σ. Consider the following class of controllers:

$$\boxed{u^{(p)} = -\eta \nabla_\sigma \|\sigma\|_p, \qquad 1 \leq p \leq \infty} \tag{6.86}$$

where

$$\nabla_\sigma \|\sigma\|_p = \begin{bmatrix} \frac{\partial}{\partial \sigma_1}\|\sigma\|_p & \frac{\partial}{\partial \sigma_2}\|\sigma\|_p & \cdots & \frac{\partial}{\partial \sigma_m}\|\sigma\|_p \end{bmatrix}^T.$$

(a) Write down the expressions for the controllers when $p = 1, 2, \infty$.

(b) Find the bounds of $u^{(p)}$ for $p = 1, 2, \infty$.

6.14 The goal of this exercise is to obtain certain estimates that can be used to investigate stability of the closed-loop system

$$\begin{aligned} \dot{x} &= Ax + f + B(u + h), \\ u &= u^{(p)} = -\eta \nabla_\sigma \|\sigma(x)\|_p \qquad \text{for } p = 1, 2, \text{ or } \infty. \end{aligned} \tag{6.87}$$

Let

$$\begin{bmatrix} \zeta \\ \sigma \end{bmatrix} = \begin{bmatrix} P^{1/2} W^g \\ S \end{bmatrix} x.$$

Note that

$$x = [\boldsymbol{W}\boldsymbol{P}^{-1/2} \quad \boldsymbol{B}] \begin{bmatrix} \boldsymbol{\zeta} \\ \boldsymbol{\sigma} \end{bmatrix}.$$

In the new coordinates the system takes the form

$$\begin{bmatrix} \dot{\boldsymbol{\zeta}} \\ \dot{\boldsymbol{\sigma}} \end{bmatrix} = \begin{bmatrix} \boldsymbol{P}^{1/2}\boldsymbol{A}_{11}\boldsymbol{P}^{-1/2} & \boldsymbol{P}^{1/2}\boldsymbol{A}_{12} \\ \boldsymbol{A}_{21}\boldsymbol{P}^{-1/2} & \boldsymbol{A}_{22} \end{bmatrix} \begin{bmatrix} \boldsymbol{\zeta} \\ \boldsymbol{\sigma} \end{bmatrix} + \begin{bmatrix} \boldsymbol{O} \\ \boldsymbol{I}_m \end{bmatrix} (\boldsymbol{u} + \tilde{\boldsymbol{h}}) + \begin{bmatrix} \boldsymbol{P}^{1/2}\boldsymbol{W}^g \tilde{\boldsymbol{f}} \\ \boldsymbol{S}\tilde{\boldsymbol{f}} \end{bmatrix}. \tag{6.88}$$

Obtain estimates for $\frac{d}{dt}\|\boldsymbol{\sigma}\|$ ($\boldsymbol{\sigma} \neq \boldsymbol{0}$) and $\frac{d}{dt}\|\boldsymbol{\zeta}\|$ ($\boldsymbol{\zeta} \neq \boldsymbol{0}$) evaluated on any solution of the closed-loop system (6.88). Specifically, show that under some assumptions we obtain

$$\boxed{\frac{d}{dt}\|\boldsymbol{\sigma}\| \le a_{21}\|\boldsymbol{\zeta}\| + a_{22}\|\boldsymbol{\sigma}\| - \mu^{(p)}}$$

and

$$\boxed{\frac{d}{dt}\|\boldsymbol{\zeta}\| \le -\rho\|\boldsymbol{\zeta}\| + a_{12}\|\boldsymbol{\sigma}\| + \tilde{\beta}_f}$$

where

$$\begin{aligned}
a_{21} &= \|\boldsymbol{A}_{21}\boldsymbol{P}^{-1/2}\| + \alpha_h\|\boldsymbol{W}\boldsymbol{P}^{-1/2}\| + \alpha_f\|\boldsymbol{S}\|\|\boldsymbol{W}\boldsymbol{P}^{-1/2}\|, \\
a_{22} &= \|\boldsymbol{A}_{22}\| + \alpha_h\|\boldsymbol{B}\| + \alpha_f\|\boldsymbol{S}\|\|\boldsymbol{B}\|, \\
a_{12} &= \|\boldsymbol{P}^{1/2}\boldsymbol{A}_{12}\| + \alpha_f\|\boldsymbol{P}^{1/2}\boldsymbol{W}^g\|\|\boldsymbol{B}\|, \\
\tilde{\beta}_f &= \beta_f\|\boldsymbol{P}^{1/2}\boldsymbol{W}^g\|, \\
\mu^{(1)} &= \eta(1 - \gamma_h\sqrt{m}) - \beta_h - \beta_f\|\boldsymbol{S}\|, \\
\mu^{(2)} &= \eta(1 - \gamma_h) - \beta_h - \beta_f\|\boldsymbol{S}\|, \\
\mu^{(\infty)} &= \frac{\eta}{\sqrt{m}}(1 - \gamma_h\sqrt{m}) - \beta_h - \beta_f\|\boldsymbol{S}\|.
\end{aligned}$$

6.15 In the following lemma, a definition of sliding mode domain is used, which we provide below. This definition can be found, for example, in Utkin [284, p. 218].

Definition 6.4 A domain Δ in the manifold $\{\boldsymbol{x} : \boldsymbol{\sigma}(\boldsymbol{x}) = \boldsymbol{0}\}$ is a sliding mode domain if for each $\varepsilon > 0$ there exists a $\delta > 0$ such that any trajectory starting in the n-dimensional δ-neighborhood of Δ may leave the n-dimensional ε neighborhood of Δ only through the n-dimensional ε neighborhood of the boundaries of Δ.

The previous definition is illustrated in Figure 6.14. Prove the following lemma:

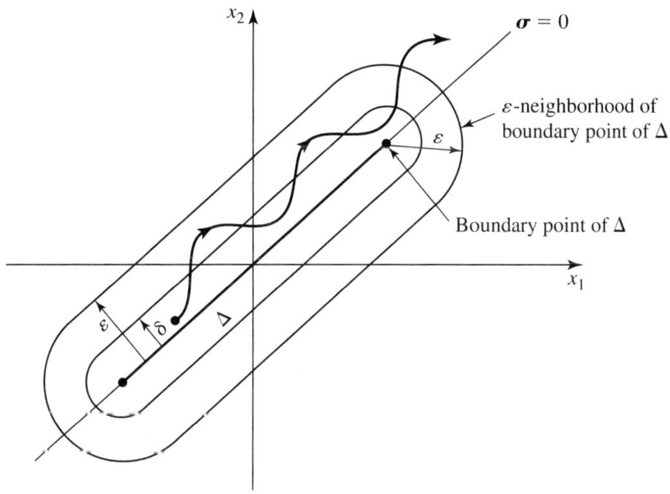

Figure 6.14 Two-dimensional illustration of a sliding mode domain.

Lemma 6.6 If $a_{21} \neq 0$, then the region

$$\Delta = \left\{ (\boldsymbol{\zeta}, \boldsymbol{\sigma}) : \|\boldsymbol{\zeta}\| < \frac{\mu^{(p)}}{a_{21}}, \boldsymbol{\sigma} = \mathbf{0} \right\}$$

is a sliding mode domain for the closed-loop system (6.87) of Exercise 6.14.

The notation of the lemma is explained in Exercise 6.14.

6.16 Prove the following result concerning UUB of the closed-loop system (6.87) of Exercise 6.14. Consult Glazos and Żak [99] for a method of proving this result.

Theorem 6.7 Let $a_{21} \neq 0$ and

$$\Gamma = \{(\boldsymbol{\zeta}, \boldsymbol{\sigma}) : \alpha\|\boldsymbol{\zeta}\| + \beta\|\boldsymbol{\sigma}\| < \gamma\},$$

where

$$\alpha = \frac{\beta a_{21}}{\beta + \rho},$$

$$\beta = \frac{1}{2}\left(a_{22} - \rho + \sqrt{(a_{22} + \rho)^2 + 4a_{12}a_{21}}\right),$$

$$\gamma = \mu^{(p)} - \frac{\alpha \tilde{\beta}_f}{\beta}.$$

Then, the closed-loop system is UUB with respect to the uncertainty region

$$\Sigma \times \{\mathbf{0}\} = \{(\boldsymbol{\zeta}, \mathbf{0}) : \|\boldsymbol{\zeta}\| \leq r\}$$

with region of attraction Γ, where

$$r = \frac{\beta_f \lambda_{\max}(\boldsymbol{P}) \|\boldsymbol{W}^g\|}{\lambda_{\min}^{1/2}(\boldsymbol{P})\rho}.$$

The set Σ is defined in Exercise 6.12.

For an illustration of the above mentioned region see Figure 6.15.

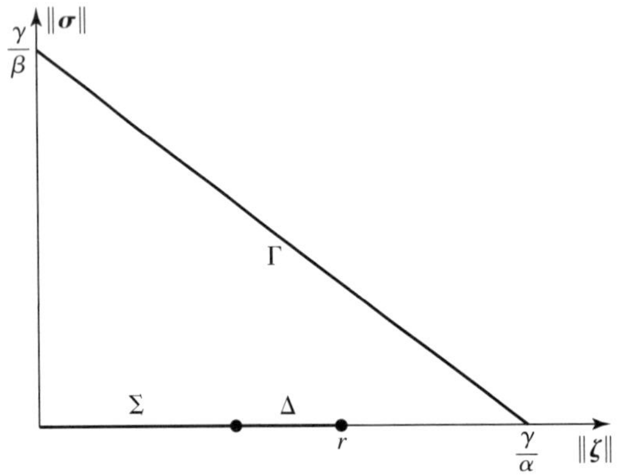

Figure 6.15 Illustration of the region of attraction and the uncertainty region of Theorem 6.7.

6.17 Show, using Lemma 6.3, that $\boldsymbol{P}\tilde{\boldsymbol{a}} \neq \mathbf{0}$, where $\boldsymbol{P}$ and $\tilde{\boldsymbol{a}}$ are defined on page 347.

CHAPTER 7 Vector Field Methods

1. If your postage stamps get stuck together, place them in the freezer for an hour or two. They'll usually come apart and be usable.
2. To remove a stamp from an envelope for collecting purposes, put it in the freezer overnight; then take it out and slide a knife between it and the envelope.
3. Or cut off the corner of the envelope with the stamp and float—don't immerse—it in lukewarm water until the paper comes off. Dry the stamp on blotting paper, then flatten it between the pages of a book. Or lay a thin paper over the stamp and run moderately hot iron over it.

Caution: Never put an antique stamp in water because the ink might run. And don't use steam to loosen such a stamp: it might be ruined as a collector's item.

—*Reader's Digest Practical Problem Solver* [241]

In this chapter we discuss methods that use vector fields in the controller construction for a class of nonlinear dynamical systems. The control design process, as in the case of linear systems, involves three steps. The first step is the design of a state-feedback control law, the second step involves the design of a state estimator, and the third step combines the first two steps to obtain a combined controller–estimator compensator. We use the so-called *feedback linearization* method to construct a state-feedback control strategy in the first stage of the above design algorithm. The basic idea of the feedback linearization approach is to use a control consisting of two components: one component cancels out the plant's nonlinearities, and the other controls the resulting linear system. The method is limited to a class of dynamical systems, plants, whose models are sufficiently smooth; that is, the plants whose right-hand sides of the modeling differential equations are sufficiently many times differentiable. In the next section we describe the class of plants for which the feedback linearization method is applicable.

7.1 A Nonlinear Plant Model

Dynamical systems that we consider in this chapter are modeled by the equations

$$\dot{\boldsymbol{x}} = \boldsymbol{f}(\boldsymbol{x}) + \boldsymbol{G}(\boldsymbol{x})\boldsymbol{u}, \tag{7.1}$$

$$\boldsymbol{y} = \boldsymbol{h}(\boldsymbol{x}), \tag{7.2}$$

where $\boldsymbol{f}:\mathbb{R}^n \rightarrow \mathbb{R}^n$, $\boldsymbol{G}:\mathbb{R}^n \rightarrow \mathbb{R}^{n\times m}$, and the output map $\boldsymbol{h}:\mathbb{R}^n \rightarrow \mathbb{R}^p$. Note that $\boldsymbol{G}(\boldsymbol{x})$ may be written as $\boldsymbol{G}(\boldsymbol{x}) = [\boldsymbol{g}_1(\boldsymbol{x})\ \boldsymbol{g}_2(\boldsymbol{x})\ \cdots\ \boldsymbol{g}_m(\boldsymbol{x})]$, where $\boldsymbol{g}_i:\mathbb{R}^n \rightarrow \mathbb{R}^n$, $i = 1, 2, \ldots, m$. We write $\boldsymbol{f} \in \mathcal{C}^\infty$ if $\boldsymbol{f}$ is in all $\mathcal{C}^k$ for $k = 1, 2, \ldots$. The functions $\boldsymbol{f}, \boldsymbol{g}_i$, and $\boldsymbol{h}$ are assumed to be $\mathcal{C}^\infty$ vector fields on $\mathbb{R}^n$. We also assume that $\boldsymbol{f}(\boldsymbol{0}) = \boldsymbol{0}$ and $\boldsymbol{h}(\boldsymbol{0}) = \boldsymbol{0}$. We note that the linear system model $\dot{\boldsymbol{x}} = \boldsymbol{A}\boldsymbol{x} + \boldsymbol{B}\boldsymbol{u}$, $\boldsymbol{y} = \boldsymbol{C}\boldsymbol{x}$ is a special case of the above more general model. Indeed, we have $\boldsymbol{f}(\boldsymbol{x}) = \boldsymbol{A}\boldsymbol{x}$, $\boldsymbol{G}(\boldsymbol{x}) = \boldsymbol{B}$, and $\boldsymbol{h}(\boldsymbol{x}) = \boldsymbol{C}\boldsymbol{x}$. We now give a simple example of a nonlinear dynamical system whose model fits into the above class of systems.

◆ Example 7.1

The example comes from Żak and MacCarley [313]. It is a simple model of a one-link robot manipulator whose reduced-order model we analyzed in Example 5.13. The motion of the robot arm is controlled by a DC motor via a gear as shown in Figure 5.11. The DC motor is armature-controlled, and its schematic is shown in Figure 5.12. We used Newton's second law to write the equation modeling the arm dynamics,

$$ml^2\frac{d^2\theta_p}{dt^2} = mgl\sin\theta_p + NK_m i_a,$$

where $g = 9.8\ \text{m/sec}^2$ is the gravitational constant. We next applied Kirchhoff's voltage law to the armature circuit to obtain

$$L_a\frac{di_a}{dt} + Ri_a + e_b = L_a\frac{di_a}{dt} + Ri_a + K_b N\frac{d\theta_p}{dt} = u,$$

where K_b is the back emf constant. We can now construct a third-order state-space model of the one-link robot. For this we choose the following state variables:

$$x_1 = \theta_p, \qquad x_2 = \frac{d\theta_p}{dt} = \omega_p, \qquad x_3 = i_a.$$

Then, the model in state-space format is

$$\begin{bmatrix}\dot{x}_1\\ \dot{x}_2\\ \dot{x}_3\end{bmatrix} = \begin{bmatrix} x_2 \\ \dfrac{g}{l}\sin x_1 + \dfrac{NK_m}{ml^2}x_3 \\ -\dfrac{K_b N}{L_a}x_2 - \dfrac{R_a}{L_a}x_3\end{bmatrix} + \begin{bmatrix}0\\0\\ \dfrac{1}{L_a}\end{bmatrix}u, \tag{7.3}$$

Reasonable parameters of the robot are: $l = 1$ m, $m = 1$ kg, $N = 10$, $K_m = 0.1$ Nm/A, $K_b = 0.1$ V sec/rad, $R_a = 1\,\Omega$, $L_a = 100$ mH. With the above parameters the robot model takes the form

$$\begin{bmatrix} \dot{x}_1 \\ \dot{x}_2 \\ \dot{x}_3 \end{bmatrix} = \begin{bmatrix} x_2 \\ 9.8 \sin x_1 + x_3 \\ -10\, x_2 - 10\, x_3 \end{bmatrix} + \begin{bmatrix} 0 \\ 0 \\ 10 \end{bmatrix} u. \tag{7.4}$$

We assume that the output, y, is

$$y = x_1. \tag{7.5}$$

7.2 Controller Form

One of the objectives of this chapter is to devise a method for constructing state-feedback stabilizing control law for a class of dynamical systems modeled by (7.1). To construct such a law it is advantageous to work with an equivalent system model rather than with the original one. We thus first discuss a method for reducing a nonlinear system model into an equivalent form that is a generalization of the controller form known from linear system theory. In our subsequent discussion, we will be using three types of *Lie derivatives*. They are as follows:

1. *Derivative of a vector field with respect to a vector field, also known as the* ***Lie bracket***. Given the vector-valued functions $\boldsymbol{f}: \mathbb{R}^n \to \mathbb{R}^n$ and $\boldsymbol{g}: \mathbb{R}^n \to \mathbb{R}^n$, where $\boldsymbol{f}$ and $\boldsymbol{g}$ are C^∞ vector fields, their Lie bracket is defined as

$$[\boldsymbol{f}, \boldsymbol{g}] = \frac{\partial \boldsymbol{f}}{\partial \boldsymbol{x}} \boldsymbol{g} - \frac{\partial \boldsymbol{g}}{\partial \boldsymbol{x}} \boldsymbol{f}, \tag{7.6}$$

where $\frac{\partial \boldsymbol{f}}{\partial \boldsymbol{x}}$ and $\frac{\partial \boldsymbol{g}}{\partial \boldsymbol{x}}$ are the Jacobian matrices of $\boldsymbol{f}$ and $\boldsymbol{g}$, respectively. Sometimes in the literature the negative of the above definition is used. For convenience we will also be using an alternative notation of the form

$$[\boldsymbol{f}, \boldsymbol{g}] = (\mathrm{ad}^1 \boldsymbol{f}, \boldsymbol{g}). \tag{7.7}$$

We define

$$(\mathrm{ad}^k \boldsymbol{f}, \boldsymbol{g}) = [\boldsymbol{f}, (\mathrm{ad}^{k-1} \boldsymbol{f}, \boldsymbol{g})],$$

where

$$(\mathrm{ad}^0 \boldsymbol{f}, \boldsymbol{g}) = \boldsymbol{g}.$$

For a single-input linear system model, $\boldsymbol{f}(\boldsymbol{x}) = \boldsymbol{A}\boldsymbol{x}$ and $\boldsymbol{g}(\boldsymbol{x}) = \boldsymbol{b}$. Hence,

$$\begin{aligned} (\mathrm{ad}^0 \boldsymbol{f}, \boldsymbol{g}) &= \boldsymbol{b}, \\ (\mathrm{ad}^1 \boldsymbol{f}, \boldsymbol{g}) &= \boldsymbol{A}\boldsymbol{b}, \\ &\vdots \\ (\mathrm{ad}^k \boldsymbol{f}, \boldsymbol{g}) &= \boldsymbol{A}^k \boldsymbol{b}. \end{aligned}$$

Thus, the controllability matrix of the linear system can be expressed in terms of Lie brackets as

$$[\boldsymbol{b} \quad A\boldsymbol{b} \quad \cdots \quad A^{n-1}\boldsymbol{b}] = [(\mathrm{ad}^0 \boldsymbol{f}, \boldsymbol{g}) \quad (\mathrm{ad}^1 \boldsymbol{f}, \boldsymbol{g}) \quad \cdots \quad (\mathrm{ad}^{n-1} \boldsymbol{f}, \boldsymbol{g})].$$

We will call the $n \times n$ matrix

$$[(\mathrm{ad}^0 \boldsymbol{f}, \boldsymbol{g}) \quad (\mathrm{ad}^1 \boldsymbol{f}, \boldsymbol{g}) \quad \cdots \quad (\mathrm{ad}^{n-1} \boldsymbol{f}, \boldsymbol{g})], \tag{7.8}$$

the controllability matrix of the single input nonlinear system modeled by

$$\dot{\boldsymbol{x}} = \boldsymbol{f}(\boldsymbol{x}) + \boldsymbol{g}(\boldsymbol{x})u. \tag{7.9}$$

2. *Derivative of a function with respect to a vector field.* Let $h : \mathbb{R}^n \to \mathbb{R}$ be a $\mathcal{C}^\infty$ function on $\mathbb{R}^n$. Let $Dh = \nabla h^T$, where ∇h is the gradient (a column vector) of h with respect to $\boldsymbol{x}$. Then, the Lie derivative of the function h with respect to the vector field $\boldsymbol{f}$, denoted $L_{\boldsymbol{f}} h$ or $L_{\boldsymbol{f}}(h)$, is defined as

$$L_{\boldsymbol{f}} h = L_{\boldsymbol{f}}(h) = \langle \nabla h, \boldsymbol{f} \rangle = Dh\, \boldsymbol{f} = \frac{\partial h}{\partial x_1} f_1 + \frac{\partial h}{\partial x_2} f_2 + \cdots + \frac{\partial h}{\partial x_n} f_n.$$

We use the following notation:

$$\begin{aligned} L_{\boldsymbol{f}}^0 h &= h, \\ L_{\boldsymbol{f}}^1 h &= L_{\boldsymbol{f}} h, \\ L_{\boldsymbol{f}}^2 h &= L_{\boldsymbol{f}} L_{\boldsymbol{f}} h, \\ &\vdots \\ L_{\boldsymbol{f}}^k h &= L_{\boldsymbol{f}} L_{\boldsymbol{f}}^{k-1} h. \end{aligned}$$

3. *Derivative of Dh with respect to the vector field.* The Lie derivative of Dh with respect to the vector field $\boldsymbol{f}$, denoted $L_{\boldsymbol{f}}(Dh)$, is defined as

$$L_{\boldsymbol{f}}(Dh) = \left(\frac{\partial \nabla h}{\partial x} \boldsymbol{f} \right)^T + Dh \frac{\partial \boldsymbol{f}}{\partial x}. \tag{7.10}$$

Note that

$$L_{\boldsymbol{f}}(Dh) = DL_{\boldsymbol{f}} h = \nabla L_{\boldsymbol{f}}^T h. \tag{7.11}$$

The above three Lie derivatives satisfy the *Leibniz formula,*

$$\boxed{L_{[\boldsymbol{f},\boldsymbol{g}]} h = \langle \nabla h, [\boldsymbol{f}, \boldsymbol{g}] \rangle = L_{\boldsymbol{g}} L_{\boldsymbol{f}} h - L_{\boldsymbol{f}} L_{\boldsymbol{g}} h} \tag{7.12}$$

Our goal now is to construct a $\mathcal{C}^\infty$ state variable transformation

$$z = \boldsymbol{T}(\boldsymbol{x}), \qquad \boldsymbol{T}(\boldsymbol{0}) = \boldsymbol{0},$$

for which there is a $\mathcal{C}^\infty$ inverse $\boldsymbol{x} = \boldsymbol{T}^{-1}(z)$, such that system model $\dot{\boldsymbol{x}} = \boldsymbol{f}(\boldsymbol{x}) + \boldsymbol{g}(\boldsymbol{x})u$ in the new coordinates has the form

$$\begin{bmatrix} \dot{z}_1 \\ \dot{z}_2 \\ \vdots \\ \dot{z}_{n-1} \\ \dot{z}_n \end{bmatrix} = \begin{bmatrix} z_2 \\ z_3 \\ \vdots \\ z_n \\ \tilde{f}_n(z_1, z_2, \ldots, z_n) \end{bmatrix} + \begin{bmatrix} 0 \\ 0 \\ \vdots \\ 0 \\ 1 \end{bmatrix} u. \tag{7.13}$$

A $\mathcal{C}^\infty$ transformation $z = \boldsymbol{T}(\boldsymbol{x})$ such that $\boldsymbol{T}(\boldsymbol{0}) = \boldsymbol{0}$ and for which there is a $\mathcal{C}^\infty$ inverse $\boldsymbol{x} = \boldsymbol{T}^{-1}(z)$ is called a *diffeomorphism*. We devise a method for constructing the desired transformation. Suppose that such a transformation exists. Differentiating $z(t)$ with respect to time t gives

$$\dot{z}(t) = \frac{\partial \boldsymbol{T}}{\partial \boldsymbol{x}}\dot{\boldsymbol{x}} = \frac{\partial \boldsymbol{T}}{\partial \boldsymbol{x}}(\boldsymbol{f}(\boldsymbol{x}) + \boldsymbol{g}(\boldsymbol{x})u). \tag{7.14}$$

Comparing the right-hand sides of (7.13) and (7.14), we obtain

$$\left.\frac{\partial \boldsymbol{T}}{\partial \boldsymbol{x}}\boldsymbol{f}(\boldsymbol{x})\right|_{\boldsymbol{x}=\boldsymbol{T}^{-1}(z)} = \begin{bmatrix} z_2 \\ z_3 \\ \vdots \\ z_n \\ \tilde{f}_n(z_1, z_2, \ldots, z_n) \end{bmatrix}, \tag{7.15}$$

and

$$\left.\frac{\partial \boldsymbol{T}}{\partial \boldsymbol{x}}\boldsymbol{g}(\boldsymbol{x})\right|_{\boldsymbol{x}=\boldsymbol{T}^{-1}(z)} = \begin{bmatrix} 0 \\ 0 \\ \vdots \\ 0 \\ 1 \end{bmatrix}. \tag{7.16}$$

We first consider (7.15). We have

$$\begin{bmatrix} z_2 \\ z_3 \\ \vdots \\ z_n \\ \tilde{f}_n(z_1, z_2, \ldots, z_n) \end{bmatrix} = \begin{bmatrix} T_2 \\ T_3 \\ \vdots \\ T_n \\ \tilde{f}_n(z_1, z_2, \ldots, z_n) \end{bmatrix}$$

because $z = \boldsymbol{T}(\boldsymbol{x})$. We conclude that

$$\frac{\partial T_i}{\partial \boldsymbol{x}}\boldsymbol{f}(\boldsymbol{x}) = T_{i+1}, \qquad i = 1, 2, \ldots, n-1,$$

or equivalently

$$T_{i+1} = L_{\boldsymbol{f}} T_i, \qquad i = 1, 2, \ldots, n-1.$$

Therefore, the transformation $z = \boldsymbol{T}(\boldsymbol{x})$ has the form

$$\boldsymbol{T} = \begin{bmatrix} T_1 \\ L_f T_1 \\ \vdots \\ L_f^{n-2} T_1 \\ L_f^{n-1} T_1 \end{bmatrix}. \tag{7.17}$$

The above means that the problem of constructing the desired transformation $z = \boldsymbol{T}(\boldsymbol{x})$ is reduced to finding its first component T_1. The remaining components of $\boldsymbol{T}$ can be successively computed using T_1. To specify T_1, we use (7.16), which we represent componentwise as

$$L_g T_i = 0, \qquad i = 1, 2, \ldots, n-1,$$

and

$$L_g T_n = 1.$$

Note that $L_g T_1 = 0$ can be represented as

$$\frac{\partial T_1}{\partial \boldsymbol{x}}(\mathrm{ad}^0 \boldsymbol{f}, \boldsymbol{g}) = 0.$$

Next, consider the relation $L_g T_2 = 0$. We will show that

$$L_g T_2 = \frac{\partial T_1}{\partial \boldsymbol{x}}(\mathrm{ad}^1 \boldsymbol{f}, \boldsymbol{g}) = 0. \tag{7.18}$$

By the Leibniz formula, we obtain

$$\begin{aligned} L_g T_2 &= L_g L_f T_1 \\ &= L_f L_g T_1 + L_{[f,g]} T_1 \\ &= L_{[f,g]} T_1, \end{aligned} \tag{7.19}$$

because $L_g T_1 = 0$, and hence $L_f L_g T_1 = 0$. Using (7.19), we obtain (7.18). Continuing in this manner, we get

$$\frac{\partial T_1}{\partial \boldsymbol{x}}(\mathrm{ad}^i \boldsymbol{f}, \boldsymbol{g}) = 0, \qquad i = 0, 1, \ldots, n-2,$$

and

$$\frac{\partial T_1}{\partial \boldsymbol{x}}(\mathrm{ad}^{n-1} \boldsymbol{f}, \boldsymbol{g}) = 1.$$

We represent the above n equations in matrix form as

$$\begin{aligned} &\frac{\partial T_1}{\partial \boldsymbol{x}}[(\mathrm{ad}^0 \boldsymbol{f}, \boldsymbol{g}) \quad (\mathrm{ad}^1 \boldsymbol{f}, \boldsymbol{g}) \quad \cdots \quad (\mathrm{ad}^{n-1} \boldsymbol{f}, \boldsymbol{g})] \\ &= \left[L_g T_1 \quad L_{(\mathrm{ad}^1 f, g)} T_1 \quad \cdots \quad L_{(\mathrm{ad}^{n-1} f, g)} T_1\right] \\ &= [0 \quad 0 \quad \cdots \quad 1]. \end{aligned}$$

Thus the derivative of T_1—that is, the row vector $\frac{\partial T_1}{\partial \boldsymbol{x}}$—is the last row of the inverse of the controllability matrix, provided the controllability matrix is invertible. We denote the last row of the inverse of the controllability matrix by $\boldsymbol{q}(\boldsymbol{x})$. Then, the problem we have to solve is to find $T_1 : \mathbb{R}^n \to \mathbb{R}$ such that

$$\frac{\partial T_1(\boldsymbol{x})}{\partial \boldsymbol{x}} = \boldsymbol{q}(\boldsymbol{x}), \qquad T_1(\boldsymbol{0}) = 0, \tag{7.20}$$

or, equivalently,

$$\nabla T_1 = \boldsymbol{q}^T(\boldsymbol{x}), \qquad T_1(\boldsymbol{0}) = 0. \tag{7.21}$$

The above problem can be viewed as finding a potential function T_1 for a given vector field $\boldsymbol{q}^T$. A necessary and sufficient condition for a given vector field to have a potential function is given by Theorem A.21 on page 672. We will show that if $\boldsymbol{q}^T$ has a potential function, then the resulting state variable transformation is a diffeomorphism. Before we proceed with the proof, we first illustrate the results obtained so far with a numerical example.

◆ Example 7.2

Consider the nonlinear system model from Example 7.1. The controllability matrix of this system model is

$$[(\text{ad}^0\, \boldsymbol{f}, \boldsymbol{g}) \quad (\text{ad}^1\, \boldsymbol{f}, \boldsymbol{g}) \quad (\text{ad}^2\, \boldsymbol{f}, \boldsymbol{g})] = \begin{bmatrix} 0 & 0 & 10 \\ 0 & 10 & -100 \\ 10 & -100 & 900 \end{bmatrix}.$$

The above controllability matrix is of full rank on $\mathbb{R}^3$. The last row of its inverse is

$$\boldsymbol{q} = [0.1 \quad 0 \quad 0].$$

Hence,

$$T_1 = 0.1 x_1.$$

Having obtained T_1, we construct the desired transformation $\boldsymbol{z} = \boldsymbol{T}(\boldsymbol{x})$, where

$$\boldsymbol{T}(\boldsymbol{x}) = \begin{bmatrix} T_1 \\ L_{\boldsymbol{f}} T_1 \\ L_{\boldsymbol{f}}(L_{\boldsymbol{f}} T_1) \end{bmatrix} = \begin{bmatrix} 0.1 x_1 \\ 0.1 x_2 \\ 0.98 \sin x_1 + 0.1 x_3 \end{bmatrix}.$$

Note that the inverse transformation $\boldsymbol{x} = \boldsymbol{T}^{-1}(\boldsymbol{z})$ exists and has the form

$$\boldsymbol{T}^{-1}(\boldsymbol{z}) = \begin{bmatrix} 10 z_1 \\ 10 z_2 \\ 10 z_3 - 9.8 \sin(10 z_1) \end{bmatrix}.$$

Furthermore,

$$\frac{\partial \boldsymbol{T}}{\partial \boldsymbol{x}} = \begin{bmatrix} 0.1 & 0 & 0 \\ 0 & 0.1 & 0 \\ 0.98\cos x_1 & 0 & 0.1 \end{bmatrix}.$$

Applying the transformation $\boldsymbol{z} = \boldsymbol{T}(\boldsymbol{x})$ to the model of the one-link robot manipulator yields

$$\begin{aligned} \dot{\boldsymbol{z}} &= \frac{\partial \boldsymbol{T}}{\partial \boldsymbol{x}} \boldsymbol{f}(\boldsymbol{x})\Bigg|_{\boldsymbol{x}=\boldsymbol{T}^{-1}(\boldsymbol{z})} + \frac{\partial \boldsymbol{T}}{\partial \boldsymbol{x}} \boldsymbol{g}(\boldsymbol{x})\Bigg|_{\boldsymbol{x}=\boldsymbol{T}^{-1}(\boldsymbol{z})} u \\ &= \begin{bmatrix} z_2 \\ z_3 \\ 9.8\sin(10z_1) + (9.8\cos(10z_1) - 10)z_2 - 10z_3 \end{bmatrix} + \begin{bmatrix} 0 \\ 0 \\ 1 \end{bmatrix} u. \end{aligned} \tag{7.22}$$

We will now show that if there is a potential function for $\boldsymbol{q}^T(\boldsymbol{x})$, then the resulting transformation is indeed a diffeomorphism. By construction, $\boldsymbol{T}$ is of $\mathcal{C}^\infty$ class, and $\boldsymbol{T}(\boldsymbol{0}) = \boldsymbol{0}$. We now show that $\boldsymbol{T}$ is locally $\mathcal{C}^\infty$-invertible. By the inverse function theorem, stated in Exercise 7.3, if the Jacobian matrix of $\boldsymbol{T}$ evaluated at $\boldsymbol{0}$ is invertible, then $\boldsymbol{T}$ is locally $\mathcal{C}^\infty$-invertible. The Jacobian matrix of $\boldsymbol{T}$ is

$$D\boldsymbol{T} = \begin{bmatrix} DT_1 \\ DL_f T_1 \\ \vdots \\ DL_f^{n-2} T_1 \\ DL_f^{n-1} T_1 \end{bmatrix}.$$

We compute the product of the above Jacobian matrix and the controllability matrix

$$[(\mathrm{ad}^0 \boldsymbol{f}, \boldsymbol{g}) \quad (\mathrm{ad}^1 \boldsymbol{f}, \boldsymbol{g}) \quad \cdots \quad (\mathrm{ad}^{n-1} \boldsymbol{f}, \boldsymbol{g})]$$

to get

$$\begin{bmatrix} DT_1 \\ DL_f T_1 \\ \vdots \\ DL_f^{n-2} T_1 \\ DL_f^{n-1} T_1 \end{bmatrix} [(\mathrm{ad}^0 \boldsymbol{f}, \boldsymbol{g}) \quad (\mathrm{ad}^1 \boldsymbol{f}, \boldsymbol{g}) \quad \cdots \quad (\mathrm{ad}^{n-1} \boldsymbol{f}, \boldsymbol{g})].$$

Performing multiplications yields

$$
\begin{aligned}
&\boldsymbol{DT}[(\mathrm{ad}^0 \boldsymbol{f}, \boldsymbol{g}) \quad (\mathrm{ad}^1 \boldsymbol{f}, \boldsymbol{g}) \quad \cdots \quad (\mathrm{ad}^{n-1} \boldsymbol{f}, \boldsymbol{g})] \\
&= \begin{bmatrix} L_g T_1 & L_{(\mathrm{ad}^1 f, g)} T_1 & \cdots & L_{(\mathrm{ad}^{n-1} f, g)} T_1 \\ L_g L_f T_1 & L_{(\mathrm{ad}^1 f, g)} L_f T_1 & \cdots & L_{(\mathrm{ad}^{n-1} f, g)} L_f T_1 \\ \vdots & \vdots & & \vdots \\ L_g L_f^{n-1} T_1 & L_{(\mathrm{ad}^1 f, g)} L_f^{n-1} T_1 & \cdots & L_{(\mathrm{ad}^{n-1} f, g)} L_f^{n-1} T_1 \end{bmatrix} \\
&= \begin{bmatrix} L_g T_1 & L_{(\mathrm{ad}^1 f, g)} T_1 & \cdots & L_{(\mathrm{ad}^{n-1} f, g)} T_1 \\ L_g T_2 & L_{(\mathrm{ad}^1 f, g)} T_2 & \cdots & L_{(\mathrm{ad}^{n-1} f, g)} T_2 \\ \vdots & \vdots & & \vdots \\ L_g T_n & L_{(\mathrm{ad}^1 f, g)} T_n & \cdots & L_{(\mathrm{ad}^{n-1} f, g)} T_n \end{bmatrix} \\
&= \begin{bmatrix} L_g T_1 & L_{(\mathrm{ad}^1 f, g)} T_1 & \cdots & L_{(\mathrm{ad}^{n-1} f, g)} T_1 \\ L_{(\mathrm{ad}^1 f, g)} T_1 & L_{(\mathrm{ad}^2 f, g)} T_1 & \cdots & L_{(\mathrm{ad}^{n} f, g)} T_1 \\ \vdots & \vdots & & \vdots \\ L_{(\mathrm{ad}^{n-1} f, g)} T_1 & L_{(\mathrm{ad}^n f, g)} T_1 & \cdots & L_{(\mathrm{ad}^{2n-2} f, g)} T_1 \end{bmatrix} \\
&= \begin{bmatrix} 0 & 0 & \cdots & 0 & 1 \\ 0 & 0 & \cdots & 1 & x \\ \vdots & \vdots & & \vdots & \vdots \\ 1 & x & \cdots & x & x \end{bmatrix},
\end{aligned}
$$

where we use x to denote "don't care" elements. It follows from the above that at $\boldsymbol{x} = \boldsymbol{0}$ the Jacobian matrix $\boldsymbol{DT}$ is invertible, that is,

$$\operatorname{rank} \boldsymbol{DT}(\boldsymbol{0}) = n.$$

In conclusion, a diffeomorphic state variable transformation exists, in a neighborhood of $\boldsymbol{x} = \boldsymbol{0}$, if

1. in a neighborhood of $\boldsymbol{x} = \boldsymbol{0}$,

$$\operatorname{rank}[(\mathrm{ad}^0 \boldsymbol{f}, \boldsymbol{g}) \quad (\mathrm{ad}^1 \boldsymbol{f}, \boldsymbol{g}) \quad \cdots \quad (\mathrm{ad}^{n-1} \boldsymbol{f}, \boldsymbol{g})] = n,$$

2. the equation

$$\boldsymbol{q}(\boldsymbol{x}) = [0 \quad 0 \quad \cdots \quad 0 \quad 1][(\mathrm{ad}^0 \boldsymbol{f}, \boldsymbol{g}) \quad (\mathrm{ad}^1 \boldsymbol{f}, \boldsymbol{g}) \quad \cdots \quad (\mathrm{ad}^{n-1} \boldsymbol{f}, \boldsymbol{g})]^{-1}$$

has a potential function in a neighborhood of $\boldsymbol{x} = \boldsymbol{0}$; that is, $\boldsymbol{q}^T(\boldsymbol{x})$ is a conservative vector field in a neighborhood of $\boldsymbol{x} = \boldsymbol{0}$.

The above method of transforming a class of single-input nonlinear systems into the controller form can be generalized to multi-input systems in a way similar to that in the linear case discussed in Subsection 3.3.1. We will illustrate the method with a numerical example.

◆ Example 7.3

Consider the model of the permanent stepper motor given by (1.61). We first represent the model as

$$\dot{\boldsymbol{x}} = \boldsymbol{f}(\boldsymbol{x}) + \boldsymbol{g}_1 u_1 + \boldsymbol{g}_2 u_2, \tag{7.23}$$

where

$$\boldsymbol{f}(\boldsymbol{x}) = \begin{bmatrix} x_2 \\ -K_1 x_2 - K_2 x_3 \sin(K_3 x_1) + K_2 x_4 \cos(K_3 x_1) \\ K_4 x_2 \sin(K_3 x_1) - K_5 x_3 \\ -K_4 x_2 \cos(K_3 x_1) - K_5 x_4 \end{bmatrix},$$

$$\boldsymbol{g}_1 = \begin{bmatrix} 0 \\ 0 \\ 1 \\ 0 \end{bmatrix}, \quad \text{and} \quad \boldsymbol{g}_2 = \begin{bmatrix} 0 \\ 0 \\ 0 \\ 1 \end{bmatrix}.$$

We next form the controllability matrix for the above model to get

$$[\boldsymbol{g}_1 \ \boldsymbol{g}_2 \ (\mathrm{ad}^1 \boldsymbol{f}, \boldsymbol{g}_1) \ (\mathrm{ad}^1 \boldsymbol{f}, \boldsymbol{g}_2) \ (\mathrm{ad}^2 \boldsymbol{f}, \boldsymbol{g}_1) \ (\mathrm{ad}^2 \boldsymbol{f}, \boldsymbol{g}_2) \ (\mathrm{ad}^3 \boldsymbol{f}, \boldsymbol{g}_1) \ (\mathrm{ad}^3 \boldsymbol{f}, \boldsymbol{g}_2)], \tag{7.24}$$

where

$$(\mathrm{ad}^1 \boldsymbol{f}, \boldsymbol{g}_1) = \begin{bmatrix} 0 \\ -K_2 \sin(K_3 x_1) \\ -K_5 \\ 0 \end{bmatrix}, \qquad (\mathrm{ad}^1 \boldsymbol{f}, \boldsymbol{g}_2) = \begin{bmatrix} 0 \\ K_2 \cos(K_3 x_1) \\ 0 \\ -K_5 \end{bmatrix},$$

$$\begin{aligned}
&(\mathrm{ad}^2 \boldsymbol{f}, \boldsymbol{g}_1) \\
&\quad = [\boldsymbol{f}, (\mathrm{ad}^1 \boldsymbol{f}, \boldsymbol{g}_1)] = \frac{\partial \boldsymbol{f}}{\partial \boldsymbol{x}}(\mathrm{ad}^1 \boldsymbol{f}, \boldsymbol{g}_1) - \frac{\partial(\mathrm{ad}^1 \boldsymbol{f}, \boldsymbol{g}_1)}{\partial \boldsymbol{x}} \boldsymbol{f} \\
&\quad = \begin{bmatrix} -K_2 \sin(K_3 x_1) \\ K_1 K_2 \sin(K_3 x_1) + K_2 K_5 \sin(K_3 x_1) + K_2 K_3 x_2 \cos(K_3 x_1) \\ -K_4 K_2 \sin^2(K_3 x_1) + K_5^2 \\ K_4 K_2 \cos(K_3 x_1) \sin(K_3 x_1) \end{bmatrix},
\end{aligned} \tag{7.25}$$

and

$$(\text{ad}^2\, \boldsymbol{f}, \boldsymbol{g}_2) = [\, \boldsymbol{f}, (\text{ad}^1\, \boldsymbol{f}, \boldsymbol{g}_2)] = \frac{\partial \boldsymbol{f}}{\partial \boldsymbol{x}}(\text{ad}^1\, \boldsymbol{f}, \boldsymbol{g}_2) - \frac{\partial(\text{ad}^1\, \boldsymbol{f}, \boldsymbol{g}_2)}{\partial \boldsymbol{x}}\, \boldsymbol{f}$$

$$= \begin{bmatrix} K_2 \cos(K_3 x_1) \\ -K_1 K_2 \cos(K_3 x_1) - K_2 K_5 \cos(K_3 x_1) + K_2 K_3 x_2 \sin(K_3 x_1) \\ K_4 K_2 \cos(K_3 x_1) \sin(K_3 x_1) \\ -K_4 K_2 \cos^2(K_3 x_1) + K_5^2 \end{bmatrix}. \tag{7.26}$$

Observe that

$$(\text{ad}^1\, \boldsymbol{f}, \boldsymbol{g}_2) = -K_5 \cot(K_3 x_1)\, \boldsymbol{g}_1 - K_5\, \boldsymbol{g}_2 - \cot(K_3 x_1)(\text{ad}^1\, \boldsymbol{f}, \boldsymbol{g}_1); \tag{7.27}$$

that is, $(\text{ad}^1\, \boldsymbol{f}, \boldsymbol{g}_2)$ is a linear combination of $\boldsymbol{g}_1$, $\boldsymbol{g}_2$, and $(\text{ad}^1\, \boldsymbol{f}, \boldsymbol{g}_1)$. We will use this fact when selecting columns to construct a transformation into a controller form. Proceeding from left to right and using (7.27), we select linearly independent columns from the controllability matrix (7.24). The PM stepper motor system is controllable because we have found four linearly independent columns—$\boldsymbol{g}_1$, $(\text{ad}^1\, \boldsymbol{f}, \boldsymbol{g}_1)$, $(\text{ad}^2\, \boldsymbol{f}, \boldsymbol{g}_1)$, and $\boldsymbol{g}_2$—in its controllability matrix. Thus, the controllability indices of the PM stepper motor system are $d_1 = 3$ and $d_2 = 1$. Therefore, the controller form of the PM stepper motor model is

$$\begin{bmatrix} \dot{z}_1 \\ \dot{z}_2 \\ \dot{z}_3 \\ \dot{z}_4 \end{bmatrix} = \begin{bmatrix} z_2 \\ z_3 \\ \tilde{f}_3(z_1, z_2, z_3, z_4) \\ \tilde{f}_4(z_1, z_2, z_3, z_4) \end{bmatrix} + \begin{bmatrix} 0 & 0 \\ 0 & 0 \\ \tilde{g}_{31} & \tilde{g}_{32} \\ \tilde{g}_{41} & \tilde{g}_{42} \end{bmatrix} \begin{bmatrix} u_1 \\ u_2 \end{bmatrix}, \tag{7.28}$$

where the elements $\tilde{g}_{ij}$ may depend on z_k, $k = 1, 2, 3, 4$, and

$$\det \begin{bmatrix} \tilde{g}_{31} & \tilde{g}_{32} \\ \tilde{g}_{41} & \tilde{g}_{42} \end{bmatrix} \neq 0. \tag{7.29}$$

The transformation we are looking for has the form

$$\boldsymbol{z} = \boldsymbol{T}(\boldsymbol{x}) = \begin{bmatrix} T_1 \\ T_2 \\ T_3 \\ T_4 \end{bmatrix} = \begin{bmatrix} T_1 \\ L_{\boldsymbol{f}} T_1 \\ L_{\boldsymbol{f}}^2 T_1 \\ T_4 \end{bmatrix}, \tag{7.30}$$

where T_1 and T_4 are functions that we have to find. There is no unique solution for T_1 and T_4. Differentiating $\boldsymbol{z} = \boldsymbol{T}(\boldsymbol{x})$ and using (7.23) yields

$$\begin{aligned} \dot{\boldsymbol{z}} &= \frac{\partial \boldsymbol{T}}{\partial \boldsymbol{x}} \dot{\boldsymbol{x}} \\ &= \left. \frac{\partial \boldsymbol{T}}{\partial \boldsymbol{x}} \boldsymbol{f}(\boldsymbol{x}) \right|_{\boldsymbol{x}=\boldsymbol{T}^{-1}(\boldsymbol{z})} + u_1 \left. \frac{\partial \boldsymbol{T}}{\partial \boldsymbol{x}} \boldsymbol{g}_1(\boldsymbol{x}) \right|_{\boldsymbol{x}=\boldsymbol{T}^{-1}(\boldsymbol{z})} + u_2 \left. \frac{\partial \boldsymbol{T}}{\partial \boldsymbol{x}} \boldsymbol{g}_2(\boldsymbol{x}) \right|_{\boldsymbol{x}=\boldsymbol{T}^{-1}(\boldsymbol{z})}, \end{aligned} \tag{7.31}$$

where

$$\frac{\partial \boldsymbol{T}}{\partial \boldsymbol{x}} = D\boldsymbol{T} = \begin{bmatrix} DT_1 \\ DT_2 \\ DT_3 \\ DT_4 \end{bmatrix}.$$

Comparing (7.28) and (7.31), we obtain

$$z_2 = T_2 = DT_1\, \boldsymbol{f} = L_{\boldsymbol{f}} T_1, \qquad z_3 = T_3 = DT_2\, \boldsymbol{f} = L_{\boldsymbol{f}} T_2, \tag{7.32}$$

and

$$L_{\boldsymbol{g}_1} T_1 = 0, \qquad L_{\boldsymbol{g}_1} T_2 = L_{\boldsymbol{g}_1} L_{\boldsymbol{f}} T_1 = 0, \tag{7.33}$$

$$L_{\boldsymbol{g}_2} T_1 = 0, \qquad L_{\boldsymbol{g}_2} T_2 = L_{\boldsymbol{g}_2} L_{\boldsymbol{f}} T_1 = 0. \tag{7.34}$$

Applying the Leibniz formula (7.12) to the right-hand sides of (7.33) and (7.34) gives

$$L_{\boldsymbol{f}} L_{\boldsymbol{g}_1} T_1 = L_{[\boldsymbol{f},\boldsymbol{g}_1]} T_1 \quad \text{and} \quad L_{\boldsymbol{f}} L_{\boldsymbol{g}_2} T_1 = L_{[\boldsymbol{f},\boldsymbol{g}_2]} T_1.$$

Because by (7.33) and (7.34), $L_{\boldsymbol{g}_1} T_1 = 0$ and $L_{\boldsymbol{g}_2} T_1 = 0$, we can write

$$DT_1 [\boldsymbol{g}_1 \quad (\text{ad}^1\, \boldsymbol{f}, \boldsymbol{g}_1) \quad \boldsymbol{g}_2 \quad (\text{ad}^1\, \boldsymbol{f}, \boldsymbol{g}_2)] = \boldsymbol{0}^T, \tag{7.35}$$

that is,

$$\begin{bmatrix} \frac{\partial T_1}{\partial x_1} & \frac{\partial T_1}{\partial x_2} & \frac{\partial T_1}{\partial x_3} & \frac{\partial T_1}{\partial x_4} \end{bmatrix} \begin{bmatrix} 0 & 0 & 0 & 0 \\ 0 & -K_2 \sin(K_3 x_1) & 0 & K_2 \cos(K_3 x_1) \\ 1 & -K_5 & 0 & 0 \\ 0 & 0 & 1 & -K_5 \end{bmatrix} = [0 \;\; 0 \;\; 0 \;\; 0].$$

By inspection, a nontrivial solution to the above is

$$DT_1 = \begin{bmatrix} \frac{\partial T_1}{\partial x_1} & 0 & 0 & 0 \end{bmatrix}, \tag{7.36}$$

where the function T_1 may only be a function of x_1 for DT_1 to be integrable. We will specify T_1 later. To proceed, we note that by (7.30) and (7.31), we obtain

$$\begin{bmatrix} \tilde{g}_{31} & \tilde{g}_{32} \\ \tilde{g}_{41} & \tilde{g}_{42} \end{bmatrix} = \begin{bmatrix} (DT_3)\boldsymbol{g}_1 & (DT_3)\boldsymbol{g}_2 \\ (DT_4)\boldsymbol{g}_1 & (DT_4)\boldsymbol{g}_2 \end{bmatrix}, \tag{7.37}$$

Using the structure of the state transformation (7.30) and the Leibniz formula (7.12), we express the element $\tilde{g}_{31} = (DT_3)\boldsymbol{g}_1$ as

$$\begin{aligned} (DT_3)\boldsymbol{g}_1 &= L_{\boldsymbol{g}_1} L_{\boldsymbol{f}} T_2 \\ &= L_{[\boldsymbol{f},\boldsymbol{g}_1]} T_2 + L_{\boldsymbol{f}} L_{\boldsymbol{g}_1} T_2 \\ &= L_{[\boldsymbol{f},\boldsymbol{g}_1]} T_2, \end{aligned} \tag{7.38}$$

because $L_{\boldsymbol{g}_1} T_2 = 0$ and therefore $L_{\boldsymbol{f}} L_{\boldsymbol{g}_1} T_2 = 0$. Observing that $T_2 = L_{\boldsymbol{f}} T_1$, employing the Leibniz formula, and taking into account the fact that $L_{\boldsymbol{g}_1} T_1 = 0$, we get

$$\begin{aligned} L_{[\boldsymbol{f},\boldsymbol{g}_1]} T_2 &= L_{[\boldsymbol{f},\boldsymbol{g}_1]} L_{\boldsymbol{f}} T_1 \\ &= L_{[\boldsymbol{f},[\boldsymbol{f},\boldsymbol{g}_1]]} T_1 + L_{\boldsymbol{f}} L_{[\boldsymbol{f},\boldsymbol{g}_1]} T_1 \\ &= L_{[\boldsymbol{f},[\boldsymbol{f},\boldsymbol{g}_1]]} T_1 + L_{\boldsymbol{f}}(L_{\boldsymbol{g}_1} L_{\boldsymbol{f}} T_1 - L_{\boldsymbol{f}} L_{\boldsymbol{g}_1} T_1) \\ &= L_{[\boldsymbol{f},[\boldsymbol{f},\boldsymbol{g}_1]]} T_1 \\ &= DT_1(\mathrm{ad}^2\, \boldsymbol{f}, \boldsymbol{g}_1). \end{aligned} \tag{7.39}$$

Hence,

$$\tilde{g}_{31} = DT_1(\mathrm{ad}^2\, \boldsymbol{f}, \boldsymbol{g}_1). \tag{7.40}$$

Similarly, we find

$$\tilde{g}_{32} = DT_1(\mathrm{ad}^2\, \boldsymbol{f}, \boldsymbol{g}_2). \tag{7.41}$$

Substituting (7.25), (7.26), and (7.36) into (7.40) and (7.41), we find

$$\tilde{g}_{31} = -\frac{\partial T_1}{\partial x_1} K_2 \sin(K_3 x_1), \tag{7.42}$$

and

$$\tilde{g}_{32} = \frac{\partial T_1}{\partial x_1} K_2 \cos(K_3 x_1). \tag{7.43}$$

Substituting the above expressions into (7.29) yields

$$\begin{aligned} \tilde{g}_{31}\tilde{g}_{42} - \tilde{g}_{32}\tilde{g}_{41} &= (DT_1(\mathrm{ad}^2\, \boldsymbol{f}, \boldsymbol{g}_1))((DT_4)\boldsymbol{g}_2) - (DT_1(\mathrm{ad}^2\, \boldsymbol{f}, \boldsymbol{g}_2))((DT_4)\boldsymbol{g}_1) \\ &= -\frac{\partial T_1}{\partial x_1} K_2 \sin(K_3 x_1) \frac{\partial T_4}{\partial x_4} - \frac{\partial T_1}{\partial x_1} K_2 \cos(K_3 x_1) \frac{\partial T_4}{\partial x_3} \\ &\neq 0. \end{aligned} \tag{7.44}$$

Suppose that we choose

$$\frac{\partial T_1}{\partial x_1} = \frac{1}{K_2}, \quad \frac{\partial T_4}{\partial x_3} = \cos(K_3 x_1), \quad \text{and} \quad \frac{\partial T_4}{\partial x_4} = \sin(K_3 x_1). \tag{7.45}$$

Substituting (7.45) into (7.44) gives

$$\det \begin{bmatrix} \tilde{g}_{31} & \tilde{g}_{32} \\ \tilde{g}_{41} & \tilde{g}_{42} \end{bmatrix} = -1.$$

Using (7.36), (7.45), and (7.30), we obtain

$$\begin{aligned} T_1 &= x_1/K_2, \\ T_2 &= L_{\boldsymbol{f}} T_1 = x_2/K_2, \\ T_3 &= L_{\boldsymbol{f}} T_2 = -\tfrac{K_1}{K_2} x_2 - x_3 \sin(K_3 x_1) + x_4 \cos(K_3 x_1), \\ T_4 &= x_3 \cos(K_3 x_1) + x_4 \sin(K_3 x_1). \end{aligned} \tag{7.46}$$

The Jacobian matrix of $\boldsymbol{T}$ is

$$D\boldsymbol{T} = \begin{bmatrix} 1/K_2 & 0 & 0 & 0 \\ 0 & 1/K_2 & 0 & 0 \\ -K_3x_3\cos(K_3x_1) - K_3x_4\sin(K_3x_1) & -K_1/K_2 & -\sin(K_3x_1) & \cos(K_3x_1) \\ -K_3x_3\sin(K_3x_1) + K_3x_4\cos(K_3x_1) & 0 & \cos(K_3x_1) & \sin(K_3x_1) \end{bmatrix}. \tag{7.47}$$

We have

$$\boldsymbol{T}(\boldsymbol{0}) = \boldsymbol{0} \quad \text{and} \quad \det D\boldsymbol{T} = -1/K_2^2 \neq 0 \qquad \text{for all } \boldsymbol{x} \in \mathbb{R}^4, \tag{7.48}$$

which means that we were able to find a global transformation into the controller form. The inverse, $\boldsymbol{x} = \boldsymbol{T}^{-1}(\boldsymbol{z})$, is

$$\begin{aligned} x_1 &= K_2 z_1, \\ x_2 &= K_2 z_2, \\ x_3 &= -K_1 z_2 \sin(K_3K_2z_1) - z_3\sin(K_3K_2z_1) + z_4\cos(K_3K_2z_1), \\ x_4 &= K_1 z_2 \cos(K_3K_2z_1) + z_3\cos(K_3K_2z_1) + z_4\sin(K_3K_2z_1). \end{aligned} \tag{7.49}$$

We transform the PM stepper motor model into the new coordinates using (7.30), (7.46), and (7.46) to obtain

$$\begin{bmatrix} \dot z_1 \\ \dot z_2 \\ \dot z_3 \\ \dot z_4 \end{bmatrix} = \begin{bmatrix} z_2 \\ z_3 \\ -K_2K_3z_2z_4 - K_1z_3 - K_2K_4z_2 - K_5z_3 - K_5K_1z_2 \\ K_2K_3z_2z_3 + K_1K_2K_3z_2^2 - K_5z_4 \end{bmatrix} + \begin{bmatrix} 0 & 0 \\ 0 & 0 \\ -\sin(K_3K_2z_1) & \cos(K_3K_2z_1) \\ \cos(K_3K_2z_1) & \sin(K_3K_2z_1) \end{bmatrix} \begin{bmatrix} u_1 \\ u_2 \end{bmatrix}. \tag{7.50}$$

7.3 Linearizing State-Feedback Control

Once the plant model is transformed into the controller form, we can construct a state-feedback controller in the new coordinates and then, using the inverse transformation, represent the controller in the original coordinates. While constructing the controller in the new coordinates, a part of the controller is used to cancel nonlinearities, thus resulting in a linear system in the new coordinates. This procedure, as we mentioned at the beginning of this chapter, is called feedback linearization. Then, we proceed to construct the other part of the controller. This part can be designed using linear control methods because the feedback linearized system is linear.

We illustrate the above method of state feedback design on a numerical example involving the one-link manipulator whose model was derived in Example 7.1. The state-space model of the manipulator, given by (7.4), was transformed into the controller form in Example 7.2. The controller form of the one-link manipulator model is given by (7.22). It is easy to design a stabilizing state-feedback controller in the new coordinates. It takes the form

$$u = -\tilde{f}_3(z_1, z_2, z_3) - (k_1 z_1 + k_2 z_2 + k_3 z_3), \tag{7.51}$$

where $\tilde{f}_3(z_1, z_2, z_3) = 9.8\sin(10z_1) + (9.8\cos(10z_1) - 10)\,z_2 - 10z_3$. Suppose that the desired closed-loop poles of the feedback linearized one-link manipulator are to be located at $\{-3.5, -4, -5\}$. Then, the linear feedback gains k_i, $i = 1, 2, 3$, that shift the poles to these desired locations are

$$k_1 = 70, \qquad k_2 = 51.5, \qquad k_3 = 12.5.$$

Applying (7.51) with the above values of the linear feedback gains to the model (7.22) gives

$$\dot{z} = \begin{bmatrix} 0 & 1 & 0 \\ 0 & 0 & 1 \\ -70 & -51.5 & -12.5 \end{bmatrix} z.$$

Next, applying the inverse transformation yields the controller in the original coordinates,

$$u = -7x_1 - 4.15x_2 - 0.25x_3 - 0.98x_2\cos(x_1) - 12.25\sin(x_1). \tag{7.52}$$

In Figure 7.1, three plots of the manipulator's link angle, $x_1 = \theta$, versus time are presented. The initial conditions have the form $\boldsymbol{x}(0) = [\theta(0) \;\; 0 \;\; 0]^T$. The control law applied to the manipulator is given by (7.52).

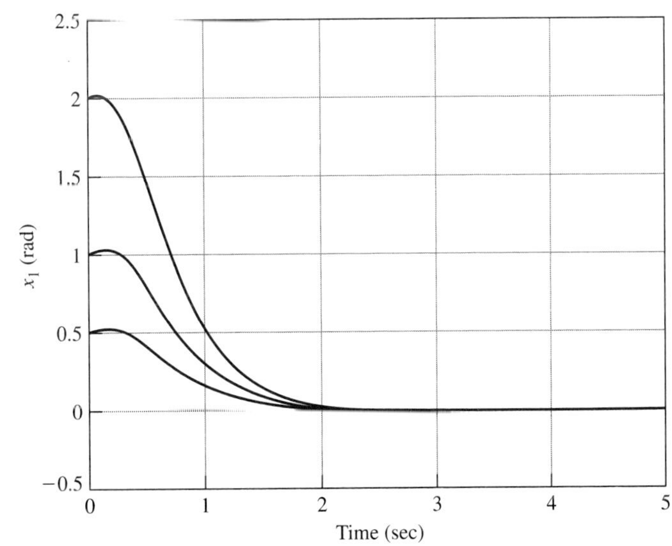

Figure 7.1 Plots of the one-link manipulator's link angle, $x_1 = \theta$, versus time for three different initial angles $\theta(0)$. The manipulator is driven by the state-feedback control law (7.52).

7.4 Observer Form

In this section we discuss the problem of transforming a class of nonlinear system models into an observer form, which is then used to construct state estimators for these systems. We consider a single output nonlinear system model of the form

$$\begin{aligned}\dot{\boldsymbol{x}} &= \boldsymbol{f}(\boldsymbol{x}) + \boldsymbol{G}(\boldsymbol{x})\boldsymbol{u},\\ y &= h(\boldsymbol{x}).\end{aligned} \tag{7.53}$$

First, we analyze the case of unforced system when $\boldsymbol{u} = \boldsymbol{0}$. In such a case, (7.53) reduces to

$$\begin{aligned}\dot{\boldsymbol{x}} &= \boldsymbol{f}(\boldsymbol{x}),\\ y &= h(\boldsymbol{x}).\end{aligned} \tag{7.54}$$

We desire to find a state variable transformation, written as $\boldsymbol{x} = \boldsymbol{T}(\boldsymbol{z})$, such that the model (7.54) in the $\boldsymbol{z}$ coordinates is

$$\begin{aligned}\dot{\boldsymbol{z}} &= \begin{bmatrix} 0 & 0 & \cdots & 0 & 0\\ 1 & 0 & \cdots & 0 & 0\\ \vdots & & \ddots & & \vdots\\ 0 & 0 & \cdots & 1 & 0\end{bmatrix}\boldsymbol{z} + \begin{bmatrix}\alpha_0(y)\\ \alpha_1(y)\\ \vdots\\ \alpha_{n-1}(y)\end{bmatrix} = \boldsymbol{A}\boldsymbol{z} + \boldsymbol{\alpha}(y) = \tilde{\boldsymbol{f}}(\boldsymbol{z}),\\ y &= [0\ \ 0\ \ \cdots\ \ 0\ \ 1]\boldsymbol{z} = \boldsymbol{c}\boldsymbol{z}.\end{aligned} \tag{7.55}$$

It is easy to verify that the state $\tilde{\boldsymbol{z}}$ of the dynamical system

$$\dot{\tilde{\boldsymbol{z}}} = (\boldsymbol{A} - \boldsymbol{l}\boldsymbol{c})\tilde{\boldsymbol{z}} + \boldsymbol{\alpha}(y) + \boldsymbol{l}y, \tag{7.56}$$

will asymptotically converge to the state $\boldsymbol{z}$ of (7.55) if the matrix $\boldsymbol{A} - \boldsymbol{l}\boldsymbol{c}$ is asymptotically stable. This is because the dynamics of the error, $\boldsymbol{e} = \tilde{\boldsymbol{z}} - \boldsymbol{z}$, are described by

$$\dot{\boldsymbol{e}} = (\boldsymbol{A} - \boldsymbol{l}\boldsymbol{c})\boldsymbol{e}, \qquad \boldsymbol{e}(t_0) = \tilde{\boldsymbol{z}}(t_0) - \boldsymbol{z}(t_0). \tag{7.57}$$

We will now derive a sufficiency condition for the existence of a state variable transformation that brings (7.54) into the observer form (7.55).

Taking the derivative of $\boldsymbol{x} = \boldsymbol{T}(\boldsymbol{z})$ with respect to time yields

$$\dot{\boldsymbol{x}} = \frac{\partial \boldsymbol{T}}{\partial \boldsymbol{z}}\dot{\boldsymbol{z}} = \frac{\partial \boldsymbol{T}}{\partial \boldsymbol{z}}\tilde{\boldsymbol{f}}(\boldsymbol{z}), \tag{7.58}$$

where we represent the Jacobian matrix of $\boldsymbol{T}(\boldsymbol{z})$, denoted $\dfrac{\partial \boldsymbol{T}}{\partial \boldsymbol{z}}$, as

$$\frac{\partial \boldsymbol{T}}{\partial \boldsymbol{z}} = D\boldsymbol{T} = \begin{bmatrix}\dfrac{\partial \boldsymbol{T}}{\partial z_1} & \cdots & \dfrac{\partial \boldsymbol{T}}{\partial z_n}\end{bmatrix}. \tag{7.59}$$

Equating the right-hand sides of $\dot{\boldsymbol{x}} = \boldsymbol{f}(\boldsymbol{x})$ and (7.58) yields

$$\boldsymbol{f}(\boldsymbol{x}) = \boldsymbol{f}(\boldsymbol{T}(\boldsymbol{z})) = \frac{\partial \boldsymbol{T}}{\partial \boldsymbol{z}}\tilde{\boldsymbol{f}}(\boldsymbol{z}). \tag{7.60}$$

Taking the partial derivative of (7.60) with respect to z_k yields

$$\boxed{\frac{\partial \boldsymbol{f}(\boldsymbol{T}(\boldsymbol{z}))}{\partial z_k} = \frac{\partial\left(\frac{\partial \boldsymbol{T}}{\partial \boldsymbol{z}}\right)}{\partial z_k}\tilde{\boldsymbol{f}}(\boldsymbol{z}) + \frac{\partial \boldsymbol{T}}{\partial \boldsymbol{z}}\frac{\partial \tilde{\boldsymbol{f}}(\boldsymbol{z})}{\partial z_k}} \tag{7.61}$$

On the other hand, using the chain rule gives

$$\frac{\partial \boldsymbol{f}(\boldsymbol{T}(\boldsymbol{z}))}{\partial z_k} = \frac{\partial \boldsymbol{f}}{\partial \boldsymbol{x}}\frac{\partial \boldsymbol{T}(\boldsymbol{z})}{\partial z_k}. \tag{7.62}$$

Using (7.60) and the fact that $\boldsymbol{x} = \boldsymbol{T}(\boldsymbol{z})$, the first term on the right-hand side of (7.61) evaluates to

$$\begin{aligned}
\frac{\partial\left(\frac{\partial \boldsymbol{T}}{\partial \boldsymbol{z}}\right)}{\partial z_k}\tilde{\boldsymbol{f}}(\boldsymbol{z}) &= \frac{\partial\left(\frac{\partial \boldsymbol{T}}{\partial z_k}\right)}{\partial \boldsymbol{z}}\tilde{\boldsymbol{f}}(\boldsymbol{z}) = \frac{\partial\left(\frac{\partial \boldsymbol{T}}{\partial z_k}\right)}{\partial \boldsymbol{z}}\frac{\partial \boldsymbol{x}}{\partial \boldsymbol{x}}\tilde{\boldsymbol{f}}(\boldsymbol{z}) = \frac{\partial\left(\frac{\partial \boldsymbol{T}}{\partial z_k}\right)}{\partial \boldsymbol{x}}\frac{\partial \boldsymbol{T}(\boldsymbol{z})}{\partial \boldsymbol{z}}\tilde{\boldsymbol{f}}(\boldsymbol{z}) \\
&= \frac{\partial\left(\frac{\partial \boldsymbol{T}}{\partial z_k}\right)}{\partial \boldsymbol{x}}\boldsymbol{f}(\boldsymbol{x}).
\end{aligned} \tag{7.63}$$

Employing (7.55) and (7.59), we evaluate the second term on the right-hand side of (7.61) to get

$$\frac{\partial \boldsymbol{T}}{\partial \boldsymbol{z}}\frac{\partial \tilde{\boldsymbol{f}}(\boldsymbol{z})}{\partial z_k} = \frac{\partial \boldsymbol{T}}{\partial z_{k+1}} \qquad \text{for } k = 1, 2, \ldots, n-1, \tag{7.64}$$

Using the above and substituting (7.62) and (7.63) into (7.61) gives

$$\boxed{\frac{\partial \boldsymbol{T}}{\partial z_{k+1}} = \frac{\partial \boldsymbol{f}}{\partial \boldsymbol{x}}\frac{\partial \boldsymbol{T}(\boldsymbol{z})}{\partial z_k} - \frac{\partial\left(\frac{\partial \boldsymbol{T}}{\partial z_k}\right)}{\partial \boldsymbol{x}}\boldsymbol{f}(\boldsymbol{x}), \qquad k = 1, 2, \ldots, n-1} \tag{7.65}$$

With the help of the Lie-derivative notation introduced in Section 7.2, we express (7.65) in an alternative way. Specifically, using (7.6) and (7.7), we represent (7.65) as

$$\frac{\partial \boldsymbol{T}}{\partial z_{k+1}} = \left[\boldsymbol{f}, \frac{\partial \boldsymbol{T}}{\partial z_k}\right] = \left(\operatorname{ad}^1 \boldsymbol{f}, \frac{\partial \boldsymbol{T}}{\partial z_k}\right), \qquad k = 1, 2, \ldots, n-1. \tag{7.66}$$

Observe that

$$\begin{aligned}
\frac{\partial \boldsymbol{T}}{\partial z_2} &= \left(\operatorname{ad}^1 \boldsymbol{f}, \frac{\partial \boldsymbol{T}}{\partial z_1}\right), \\
\frac{\partial \boldsymbol{T}}{\partial z_3} &= \left(\operatorname{ad}^1 \boldsymbol{f}, \frac{\partial \boldsymbol{T}}{\partial z_2}\right) = \left(\operatorname{ad}^2 \boldsymbol{f}, \frac{\partial \boldsymbol{T}}{\partial z_1}\right), \\
&\ \ \vdots \qquad \vdots
\end{aligned}$$

and in general

$$\boxed{\frac{\partial \boldsymbol{T}}{\partial z_k} = \left(\operatorname{ad}^{k-1} \boldsymbol{f}, \frac{\partial \boldsymbol{T}}{\partial z_1}\right), \qquad k = 1, 2, \ldots, n.} \tag{7.67}$$

With the aid of the above, we express all the columns of the Jacobian matrix (7.59) of $\boldsymbol{T}(z)$ in terms of the starting vector $\partial \boldsymbol{T}/\partial z_1$, that is,

$$\frac{\partial \boldsymbol{T}}{\partial z} = \left[\left(\operatorname{ad}^0 \boldsymbol{f}, \frac{\partial \boldsymbol{T}}{\partial z_1}\right) \quad \left(\operatorname{ad}^1 \boldsymbol{f}, \frac{\partial \boldsymbol{T}}{\partial z_1}\right) \quad \cdots \quad \left(\operatorname{ad}^{n-1} \boldsymbol{f}, \frac{\partial \boldsymbol{T}}{\partial z_1}\right)\right] \tag{7.68}$$

To obtain an expression for the starting vector $\partial \boldsymbol{T}/\partial z_1$, we use the output equation

$$y = h(\boldsymbol{x}) = z_n. \tag{7.69}$$

We use the chain rule when taking the partial derivative of (7.69) with respect to z to get

$$\frac{\partial h(\boldsymbol{x})}{\partial \boldsymbol{x}} \frac{\partial \boldsymbol{T}}{\partial z} = [0 \quad 0 \quad \cdots \quad 0 \quad 1]. \tag{7.70}$$

We utilize (7.10) to express the first component on the left-hand side of the above as

$$\frac{\partial h(\boldsymbol{x})}{\partial \boldsymbol{x}} \frac{\partial \boldsymbol{T}}{\partial z_1} = L_f^0(Dh)\frac{\partial \boldsymbol{T}}{\partial z_1} = 0. \tag{7.71}$$

Using the Leibniz formula, we represent the second component of the left-hand side of (7.70) as

$$\begin{aligned} 0 &= \frac{\partial h(\boldsymbol{x})}{\partial \boldsymbol{x}} \frac{\partial \boldsymbol{T}}{\partial z_2} \\ &= \left\langle \nabla h, \left[\boldsymbol{f}, \frac{\partial \boldsymbol{T}}{\partial z_1}\right]\right\rangle \\ &= L_{\frac{\partial \boldsymbol{T}}{\partial z_1}} L_{\boldsymbol{f}} h - L_{\boldsymbol{f}} L_{\frac{\partial \boldsymbol{T}}{\partial z_1}} h \\ &= L_{\frac{\partial \boldsymbol{T}}{\partial z_1}} L_{\boldsymbol{f}} h, \end{aligned} \tag{7.72}$$

because $\dfrac{\partial h(\boldsymbol{x})}{\partial \boldsymbol{x}} \dfrac{\partial \boldsymbol{T}}{\partial z_1} = 0$ by (7.71). We use (7.11) to represent (7.72) as

$$\begin{aligned} 0 &= L_{\frac{\partial \boldsymbol{T}}{\partial z_1}} L_{\boldsymbol{f}} h \\ &= DL_{\boldsymbol{f}} h \frac{\partial \boldsymbol{T}}{\partial z_1} \\ &= L_{\boldsymbol{f}}^1(Dh)\frac{\partial \boldsymbol{T}}{\partial z_1}. \end{aligned} \tag{7.73}$$

By repeated application of the Leibniz formula and (7.68), we express (7.70) equivalently as

$$\begin{bmatrix} L_f^0(Dh) \\ L_f^1(Dh) \\ \vdots \\ L_f^{n-2}(Dh) \\ L_f^{n-1}(Dh) \end{bmatrix} \frac{\partial \boldsymbol{T}}{\partial z_1} = \begin{bmatrix} 0 \\ 0 \\ \vdots \\ 0 \\ 1 \end{bmatrix}. \tag{7.74}$$

The matrix

$$\begin{bmatrix} L_f^0(Dh) \\ L_f^1(Dh) \\ \vdots \\ L_f^{n-2}(Dh) \\ L_f^{n-1}(Dh) \end{bmatrix} \tag{7.75}$$

is called the observability matrix of the nonlinear system model (7.54). It follows from (7.74) that the starting vector $\partial \boldsymbol{T}/\partial z_1$ is equal to the last column of the inverse of the observability matrix given by (7.75).

We now discuss the case of transforming a nonlinear system model with inputs, described by (7.53), into its observer form. We proceed as recommended by Krener and Respondek [170, Section 6]. First transform the unforced system into the observer form (7.55). Then, check if the vector field $\boldsymbol{G}(\boldsymbol{x})\boldsymbol{u}$ is transformed into the form

$$\left(\frac{\partial \boldsymbol{T}}{\partial \boldsymbol{z}}\right)^{-1} \boldsymbol{G}(\boldsymbol{x})|_{\boldsymbol{x}=\boldsymbol{T}(\boldsymbol{z})}\, \boldsymbol{u} = \boldsymbol{\beta}(y, \boldsymbol{u}). \tag{7.76}$$

If this is the case, then set

$$\boldsymbol{\gamma}(y, \boldsymbol{u}) = \boldsymbol{\alpha}(y) + \boldsymbol{\beta}(y, \boldsymbol{u}).$$

The above means that the system with inputs is transformed into the observer form

$$\begin{aligned} \dot{\boldsymbol{z}} &= \boldsymbol{A}\boldsymbol{z} + \boldsymbol{\gamma}(y, \boldsymbol{u}), \\ y &= \boldsymbol{c}\boldsymbol{z}, \end{aligned} \tag{7.77}$$

where the matrices $\boldsymbol{A}$ and $\boldsymbol{c}$ are as in (7.55). If the unforced system cannot be transformed into the observer form or $\boldsymbol{G}(\boldsymbol{x})\boldsymbol{u}$ does not transform into (7.76), then the original system (7.53) cannot be transformed into the observer form (7.77).

◆ Example 7.4

Consider the system consisting of an inverted pendulum controlled by an armature-controlled DC motor that we analyzed in Example 7.1. The observability matrix for the system is

$$\begin{bmatrix} L_f^0(Dh) \\ L_f^1(Dh) \\ L_f^2(Dh) \end{bmatrix} = \begin{bmatrix} 1 & 0 & 0 \\ 0 & 1 & 0 \\ 9.8\cos(x_1) & 0 & 1 \end{bmatrix}. \tag{7.78}$$

The starting vector $\partial \boldsymbol{T}/\partial z_1$ is the last column of the inverse of the above observability matrix,

$$\frac{\partial \boldsymbol{T}}{\partial z_1} = \begin{bmatrix} 0 \\ 0 \\ 1 \end{bmatrix}. \tag{7.79}$$

We will now find the Jacobian matrix of the desired transformation:

$$\frac{\partial \boldsymbol{T}}{\partial \boldsymbol{z}} = \left[\left(\mathrm{ad}^0 \boldsymbol{f}, \frac{\partial \boldsymbol{T}}{\partial z_1} \right) \quad \left(\mathrm{ad}^1 \boldsymbol{f}, \frac{\partial \boldsymbol{T}}{\partial z_1} \right) \quad \left(\mathrm{ad}^2 \boldsymbol{f}, \frac{\partial \boldsymbol{T}}{\partial z_1} \right) \right] = \begin{bmatrix} 0 & 0 & 1 \\ 0 & 1 & -10 \\ 1 & -10 & 90 \end{bmatrix}. \tag{7.80}$$

Therefore, we can take

$$\boldsymbol{x} = \boldsymbol{T}(\boldsymbol{z}) = \begin{bmatrix} 0 & 0 & 1 \\ 0 & 1 & -10 \\ 1 & -10 & 90 \end{bmatrix} \boldsymbol{z}. \tag{7.81}$$

Note that $\boldsymbol{T}(\boldsymbol{0}) = \boldsymbol{0}$. The inverse transformation is

$$\boldsymbol{z} = \boldsymbol{T}^{-1}\boldsymbol{x} = \begin{bmatrix} 10 & 10 & 1 \\ 10 & 1 & 0 \\ 1 & 0 & 0 \end{bmatrix} \boldsymbol{x}. \tag{7.82}$$

The system in the new coordinates has the form

$$\dot{\boldsymbol{z}} = \left(\frac{\partial \boldsymbol{T}}{\partial \boldsymbol{z}} \right)^{-1} \boldsymbol{f}(\boldsymbol{x}) \Bigg|_{\boldsymbol{x}=\boldsymbol{T}(\boldsymbol{z})} + \left(\frac{\partial \boldsymbol{T}}{\partial \boldsymbol{z}} \right)^{-1} \boldsymbol{g}(\boldsymbol{x}) \Bigg|_{\boldsymbol{x}=\boldsymbol{T}(\boldsymbol{z})} u = \boldsymbol{A}\boldsymbol{z} + \boldsymbol{\gamma}(y, u).$$

$$y = \boldsymbol{c}\boldsymbol{z}.$$

Performing simple manipulations yields

$$\dot{\boldsymbol{z}} = \begin{bmatrix} 0 & 0 & 0 \\ 1 & 0 & 0 \\ 0 & 1 & 0 \end{bmatrix} \boldsymbol{z} + \begin{bmatrix} 98\sin(y) + 10u \\ 9.8\sin(y) - 10y \\ -10y \end{bmatrix}, \tag{7.83}$$

$$y = [0 \quad 0 \quad 1]\boldsymbol{z} = z_3. \tag{7.84}$$

7.5 Asymptotic State Estimator

Once the nonlinear system model (7.53) is in the observer form, we can proceed with the construction of an asymptotic state estimator using the technique from the linear systems. The state estimator dynamics in the new coordinates are

$$\dot{\tilde{z}} = (\boldsymbol{A} - \boldsymbol{l}\boldsymbol{c})\tilde{z} + \boldsymbol{l}y + \boldsymbol{\gamma}(y, \boldsymbol{u}), \tag{7.85}$$

where the estimator gain vector $\boldsymbol{l}$ is chosen so that the matrix $(\boldsymbol{A} - \boldsymbol{l}\boldsymbol{c})$ has its eigenvalues in the desired locations in the open left-hand complex plane. It is easy to verify that the state estimation error, $z - \tilde{z}$, satisfies the linear differential equation

$$\frac{d}{dt}(z - \tilde{z}) = (\boldsymbol{A} - \boldsymbol{l}\boldsymbol{c})\,(z - \tilde{z})\,.$$

Applying the inverse transformation to (7.85), we obtain the state estimator dynamics in the original coordinates.

We illustrate the above method of constructing a state estimator using the model of the one-link manipulator from Example 7.1. We first construct a state estimator for the one-link manipulator model in the observer form given by (7.83) and (7.84). Suppose that we wish the eigenvalues of the state estimator matrix, $\boldsymbol{A} - \boldsymbol{l}\boldsymbol{c}$, to be located at $\{-9, -10, -11\}$. Then, the estimator gain vector is $\boldsymbol{l} = [\,990 \;\; 299 \;\; 30\,]^T$. With the above choice of the state estimator design parameters, its dynamics in the new coordinates become

$$\dot{\tilde{z}} = \begin{bmatrix} 0 & 0 & -990 \\ 1 & 0 & -299 \\ 0 & 1 & -30 \end{bmatrix} \tilde{z} + \begin{bmatrix} 990 \\ 299 \\ 30 \end{bmatrix} y + \begin{bmatrix} 98\sin(y) + 10u \\ 9.8\sin(y) - 10y \\ -10y \end{bmatrix}, \tag{7.86}$$

$$y = [0 \quad 0 \quad 1]\, z = z_3. \tag{7.87}$$

Using transformations (7.81) and (7.82), we represent the state estimator dynamics in the original coordinates,

$$\dot{\tilde{x}} = \begin{bmatrix} -20 & 1 & 0 \\ -89 & 0 & 1 \\ 100 & -10 & -10 \end{bmatrix} \tilde{x} + \begin{bmatrix} 30 \\ -1 \\ 700 \end{bmatrix} y + \begin{bmatrix} -10y \\ 9.8\sin(y) + 90y \\ -800y + 10u \end{bmatrix}, \tag{7.88}$$

$$y = [1 \quad 0 \quad 0]\, \boldsymbol{x} = x_1. \tag{7.89}$$

Performing simple manipulations, we represent the above state estimator as

$$\dot{\tilde{x}} = \begin{bmatrix} 0 & 1 & 0 \\ 0 & 0 & 1 \\ 0 & -10 & -10 \end{bmatrix} \tilde{x} + \begin{bmatrix} 0 \\ 9.8\sin(y) \\ 0 \end{bmatrix} - \begin{bmatrix} 20 \\ 89 \\ -100 \end{bmatrix} (\tilde{y} - y) + \begin{bmatrix} 0 \\ 0 \\ 10 \end{bmatrix} u, \tag{7.90}$$

$$y = [1 \quad 0 \quad 0]\, \boldsymbol{x} = x_1, \tag{7.91}$$

$$\tilde{y} = [1 \quad 0 \quad 0]\, \tilde{\boldsymbol{x}} = \tilde{x}_1. \tag{7.92}$$

7.6 Combined Controller–Estimator Compensator

Having constructed a state estimator, we use the state estimates, rather than the states themselves, in the state-feedback control law implementation. The question now is how the incorporation of the state estimator affects the stability of the closed-loop system with the combined estimator–controller compensator in the loop. Recall that in the case of linear systems, we were able to show that the closed-loop system was asymptotically stable. In the present case, one has to be careful when analyzing the stability of the closed-loop system with the combined estimator–controller compensator in the loop. The plant's nonlinearities may cause difficulties in stabilizing the closed-loop system using the combined controller–estimator compensator. However, for certain "nice" nonlinearities—for example, the ones for which the Lipschitz condition is satisfied—one should be able to find stability conditions in terms of the nonlinearity's Lipschitz constant and the location of the estimator's poles rather easily.

We now illustrate the implementation of the combined estimator–controller compensator on the one-link manipulator model. In this example, the controller is

$$u = -7\tilde{x}_1 - 4.15\tilde{x}_2 - 0.25\tilde{x}_3 - 0.98\tilde{x}_2\cos(\tilde{x}_1) - 12.25\sin(\tilde{x}_1). \tag{7.93}$$

In Figure 7.2, plots of the one-link manipulator output for three different initial conditions are shown. The manipulator is driven by the combined controller-estimator compensator consisting of the state estimator (7.90), (7.91), and (7.92) and the control law (7.93). The initial conditions of the estimator were set to zero while the initial conditions of the plant were the same as when we implemented the state-feedback control alone. Comparing plots of Figures 7.1 and 7.2, we can see that the presence of the estimator alters the system output in a noticeable way.

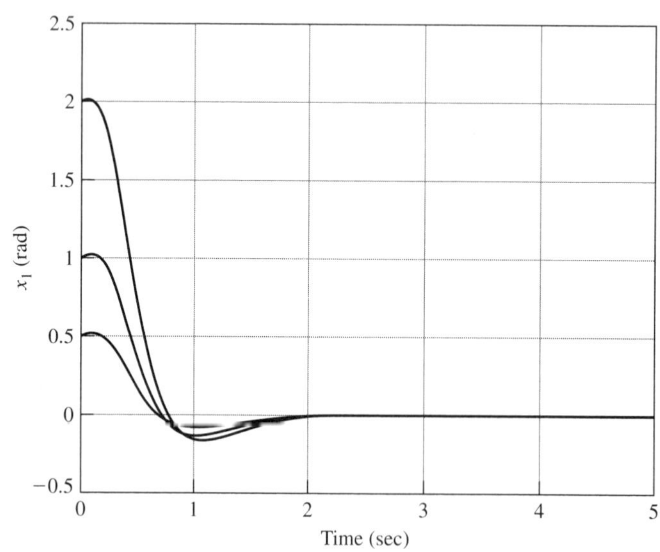

Figure 7.2 Plots of the one-link manipulator's link angle, $x_1 = \theta$, versus time for three different initial angles $\theta(0)$. The manipulator is driven by the combined controller–estimator compensator.

Notes

Friedland [91, p. 123] has the following remarks to offer on the method of feedback linearization: "The idea underlying feedback linearization is fairly obvious. The method probably was used for many years in practical control system design without anyone having reported it. Concurrently, a number of mathematical investigations on the use of differential geometry in dynamical systems were conducted without much notice by control system designers."

Our exposition of feedback linearization is based on the papers of Sommer [268] and Żak and MacCarley [313]. The research in applying differential geometric methods to the control design intensified after the papers by Su [270] and by Hunt, Su, and Meyer [136] appeared. Bakker and Nijmeijer [17] showed the equivalence of the Sommer's approach and that of Hunt, Su, and Meyer. Our treatment of the subject of state estimation follows the approach of Bestle and Zeitz [27], which in turn was presented, using the differential geometry notation, in the paper by Walcott, Corless, and Żak [291]. Zeitz [316] offers a generalization of the method proposed by Walcott et al. [291]. For further study of differential geometric techniques in nonlinear feedback design, we recommend books by Isidori [139] and by Marino and Tomei [196].

EXERCISES

7.1 Verify the Leibniz formula (7.12).

7.2 Consider the one-link robot manipulator of Example 7.1 whose model is given by (7.4) and (7.5). Treat this model as a simulation model. In order to obtain a design model, eliminate the armature inductance from the design model. In other words, neglect high-frequency parasitics represented by L_a in the simulation model. For this set $L_a = 0$, compute x_3 from the equation obtained by setting $L_a = 0$ and substitute the result into the second of the state equations. The result will be a second-order design model. Represent the design model in the form

$$\dot{\boldsymbol{x}} = \boldsymbol{A}\boldsymbol{x} + \boldsymbol{b}(u + \xi),$$

where $\xi = 9.8\sin(x_1)$ represents the model nonlinearity. Use the design model to do the following:

(a) Construct the sliding surface, $\sigma = \boldsymbol{s}\boldsymbol{x} = 0$, so that the design model restricted to this surface will have a pole at -3.

(b) Construct a control law of the form

$$u = -k_1 x_1 \operatorname{sign}(x_1\sigma) - k_2 x_2 \operatorname{sign}(x_2\sigma)$$

so that $\sigma\dot{\sigma} < 0$ for all $\boldsymbol{x} \in \mathbb{R}^2$ for which $\sigma(\boldsymbol{x}) \neq 0$. While constructing the controller u use the fact that

$$-1 \le \frac{\sin(x_1)}{x_1} \le 1.$$

(c) Apply the above control law to the simulation model in which $L_a = 10\,\text{mH}$. Plot x_1, x_2, and x_3 versus time and compare these plots with the plots of x_1 and x_2 versus time when the above controller was applied to the design model. Investigate a destabilizing effect of increasing L_a on the behavior of the closed-loop system composed of the simulation model and the control law from part (b).

7.3 Let $\mathcal{S}$ be an open set in $\mathbb{R}^n$ and let $\boldsymbol{f} : \mathcal{S} \to \mathbb{R}^n$, where

$$\boldsymbol{f}(\boldsymbol{x}) = [\, f_1(\boldsymbol{x}) \quad f_2(\boldsymbol{x}) \quad \cdots \quad f_n(\boldsymbol{x})\,]^T .$$

Recall that if all partial derivatives of all functions f_i exist and are continuous in $\mathcal{S}$, then we say that $\boldsymbol{f}$ is a $\mathcal{C}^1$ function, map, in $\mathcal{S}$. We say that $\boldsymbol{f}$ is a $\mathcal{C}^p$ function, map, in $\mathcal{S}$ if each component f_i is continuous and has continuous partial derivatives in $\mathcal{S}$ for all $k = 1, 2, \ldots, p$.

Let $\boldsymbol{f} : \mathbb{R}^n \to \mathbb{R}^n$ be a $\mathcal{C}^p$ map, $p > 0$, on an open subset $\mathcal{S}$ of $\mathbb{R}^n$, and let $\boldsymbol{a}$ be a point of $\mathcal{S}$. We say that $\boldsymbol{f}$ is locally $\mathcal{C}^p$-invertible at $\boldsymbol{a}$ if there is an open set $\mathcal{U} \subset \mathcal{S}$ containing $\boldsymbol{a}$ and an open set $\mathcal{V}$ containing $\boldsymbol{f}(\boldsymbol{a})$ such that $\boldsymbol{f} : \mathcal{U} \to \mathcal{V}$ has a $\mathcal{C}^p$ inverse $\boldsymbol{h} : \mathcal{V} \to \mathcal{U}$ so that for all $\boldsymbol{x} \in \mathcal{U}$,

$$\boldsymbol{h}(\boldsymbol{f}(\boldsymbol{x})) = \boldsymbol{x}.$$

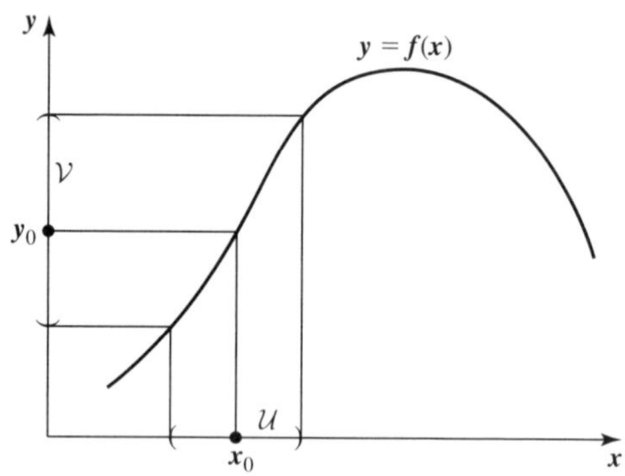

Figure 7.3 An illustration of the sets $\mathcal{U}$ and $\mathcal{V}$ that appear in the inverse function theorem.

Figure 7.3 illustrates the sets $\mathcal{U}$ and $\mathcal{V}$ for one variable case. The Jacobian matrix of $\boldsymbol{f} : \mathbb{R}^n \to \mathbb{R}^n$, denoted $\boldsymbol{J}(\boldsymbol{x})$, is defined as

$$\boldsymbol{J}(\boldsymbol{x}) = D\boldsymbol{f}(\boldsymbol{x}) = \begin{bmatrix} \nabla f_1^T(\boldsymbol{x}) \\ \nabla f_2^T(\boldsymbol{x}) \\ \vdots \\ \nabla f_n^T(\boldsymbol{x}) \end{bmatrix} = \begin{bmatrix} Df_1(\boldsymbol{x}) \\ Df_2(\boldsymbol{x}) \\ \vdots \\ Df_n(\boldsymbol{x}) \end{bmatrix}.$$

Prove the inverse function theorem for maps:

Theorem 7.1 Let $\boldsymbol{f} : \mathbb{R}^n \to \mathbb{R}^n$ be a $\mathcal{C}^p$ map, $p > 0$, on an open subset $\mathcal{S}$ of $\mathbb{R}^n$, and let $\boldsymbol{a}$ be a point of $\mathcal{S}$ such that the Jacobian matrix $\boldsymbol{J}(\boldsymbol{a})$ is invertible, and $\boldsymbol{b} = \boldsymbol{f}(\boldsymbol{a})$. Then, there exist open sets $\mathcal{U}$ and $\mathcal{V}$ in $\mathbb{R}^n$ such that $\boldsymbol{a} \in \mathcal{U}$, $\boldsymbol{b} \in \mathcal{V}$, $\boldsymbol{f}$ is one-to-one on $\mathcal{U}$, $\boldsymbol{f}(\mathcal{U}) = \mathcal{V}$, and $\boldsymbol{f}$ is locally $\mathcal{C}^p$-invertible at $\boldsymbol{a}$.

7.4 For the dynamical system model

$$\begin{bmatrix}\dot{x}_1\\ \dot{x}_2\end{bmatrix} = \begin{bmatrix}x_1x_2 + x_2\\ x_1 + x_2\end{bmatrix} + \begin{bmatrix}1\\0\end{bmatrix}u,$$

do the following:

(a) Construct a smooth state-space transformation, $z = \boldsymbol{T}(\boldsymbol{x})$, that transforms the system into the controller form.

(b) Transform the system into the controller form.

(c) In the z coordinates, design a linearizing state-feedback controller so that the linear closed-loop system has poles at $-2 \pm 2j$.

(d) Represent the control law in the $\boldsymbol{x}$ coordinates.

7.5 Consider the following model of a dynamical system:

$$\begin{bmatrix}\dot{x}_1\\ \dot{x}_2\end{bmatrix} = \begin{bmatrix}x_1^2 + x_2\\ x_1^3 + x_2\end{bmatrix} + \begin{bmatrix}0\\1\end{bmatrix}u.$$

(a) Transform the system into the controller form.

(b) Design the linearizing state-feedback controller such that the closed-loop system in the new coordinates has its poles at -1 and -2.

(c) Linearize the closed-loop system about $\boldsymbol{x} = \boldsymbol{0}$ and find the poles of the linearized model.

(d) Perform numerical simulations of the dynamical behavior of the closed-loop system for a few initial conditions. Plot x_1 versus time as well as x_2 versus time.

7.6 Consider the following model of a dynamical system:

$$\dot{\boldsymbol{x}} = \begin{bmatrix}x_2 + 2x_2^2\\ 2x_2\end{bmatrix} + \begin{bmatrix}x_2\\1\end{bmatrix}u.$$

(a) Construct a smooth state-space transformation $z = \boldsymbol{T}(\boldsymbol{x})$ that transforms the system into the controller form, and then transform the system into the controller form.

(b) In the z coordinates, design a linearizing state-feedback controller so that the linear closed-loop system has poles at $-2 \pm 2j$, and then represent the control law in the $\boldsymbol{x}$ coordinates.

(c) Perform numerical simulations of the dynamical behavior of the closed-loop system for a few initial conditions. Plot x_1 versus time as well as x_2 versus time.

7.7 Consider a dynamical system modeled by the equations

$$\begin{bmatrix}\dot{x}_1\\ \dot{x}_2\end{bmatrix} = \begin{bmatrix}-x_1\\ x_1 - x_2 - x_2^3\end{bmatrix} + \begin{bmatrix}1\\0\end{bmatrix}u,$$

$$y = x_2.$$

(a) Find a smooth state-space transformation $z = \boldsymbol{T}(\boldsymbol{x})$ that brings the above system model into the controller form.

(b) Find the system's equations in the new coordinates.

(c) Find a smooth state-space transformation $\boldsymbol{x} = \boldsymbol{T}(\boldsymbol{z})$ that brings the above system model into the observer form.

(d) Find the system's observer form.

7.8 Consider a dynamical system modeled by the following equations:

$$\begin{bmatrix} \dot{x}_1 \\ \dot{x}_2 \end{bmatrix} = \begin{bmatrix} x_1^2 + x_2 \\ x_1^3 + x_2 \end{bmatrix} + \begin{bmatrix} 0 \\ 1 \end{bmatrix} u,$$

$$y = \begin{bmatrix} 1 & 0 \end{bmatrix} \begin{bmatrix} x_1 \\ x_2 \end{bmatrix}.$$

Note that this is the same plant as in Exercise 7.5 plus the output map.

(a) Transform the above model into the observer form.

(b) Design a state estimator so that the poles of the linearized error dynamics are located at -3 and -4.

(c) Use the control law developed in Exercise 7.5 and the state estimator from the previous part of this exercise to construct a combined controller-estimator compensator. Simulate the behavior of the closed-loop system for a few different initial conditions. Note that because the state variables of the state estimator are available to you, you can always set zero initial conditions on the estimator if you wish to do so. Plot the output of the plant versus time.

CHAPTER 8 Fuzzy Systems

> It is in the response to mistakes, of technique or of judgment, that one sees most clearly the human side of science's practitioners: a mix of pride, stubbornness, humility, honesty, dismay, and regret that one could expect to encounter in any section of the population. There are the inevitable diehards, whose pride forces them to concoct ever-more-elaborate explanations for the facts in order to avoid confession of error. There are the crusaders, determined at all costs to expose the foolishness. And there are the "told-you-so's," generally members of the scientific elite, who make little contribution while the debate is raging but are ready to heap scorn on the perpetrators afterward for their "obvious" mistakes. All this, of course, is the antithesis of objective science, but follows as inevitably as night follows day from the fact that science is practiced by people and not by automatons.
>
> —*A Biography of Water: Life's Matrix* [18, p. 291]

8.1 Motivation and Basic Definitions

Conventional control algorithms require a mathematical model of the dynamical system to be controlled. The mathematical model is then used to construct a controller. In many practical situations, however, it is not always feasible to obtain an accurate mathematical model of the controlled system. *Fuzzy logic control* offers a way of dealing with modeling problems by implementing linguistic, nonformally expressed control laws derived from expert knowledge. This approach is especially successful in the case of processes such as washing machines, camera focusing, or drying processes that are being controlled well by human operators. The control of such processes can be captured in the form of rules

IF (process state) THEN (control action).

Thus, by a rule we will understand an IF-THEN statement that defines the set of facts that must be true (the IF part) before a set of actions (the THEN part) can be executed. The basic notion of fuzzy logic is a *fuzzy set*. An ordinary set, or simply a set, is a collection of objects. Ordinary sets have two membership states: inclusion or exclusion. An object is either in the set or it is

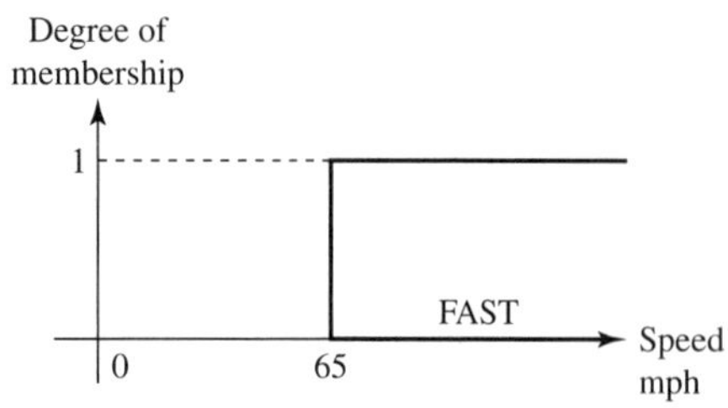

Figure 8.1 An example of a crisp set—fast cars.

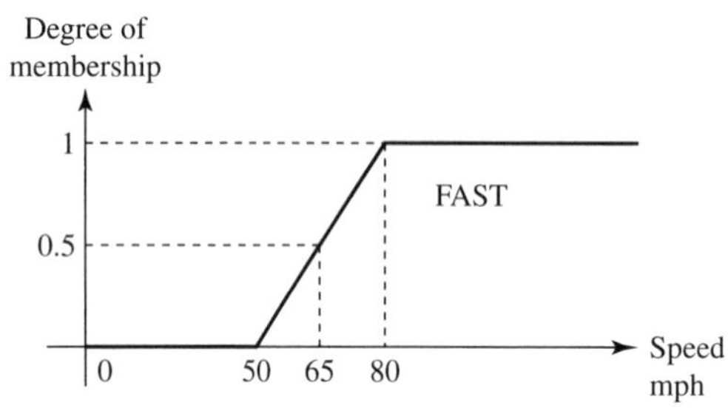

Figure 8.2 A fuzzy set FAST.

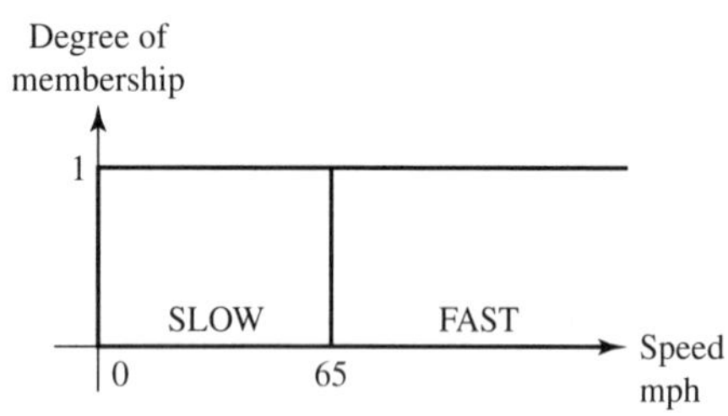

Figure 8.3 Crisp sets for SLOW and FAST.

not. Thus the degree to which an object is a member of a set is either 0 or 1. For example, a car is fast if it is driven faster than 65 mph, and it is slow if it is driven slower than 65 mph (see Figure 8.1). This is the reason we sometimes refer to ordinary sets as *crisp* sets. Fuzzy sets, on the other hand, allow for partial membership in a set. The degree to which an object is a member of a fuzzy set can vary between 0 and 1. With a fuzzy set, we have a gradual transition from membership to nonmembership. For example, a car moving with velocity of 65 mph is a member of the fuzzy set FAST to degree 0.5. A car being driven 57.5 mph is fast to degree 0.25 (see Figure 8.2). We next illustrate the crisp and fuzzy versions of the sets SLOW and FAST. Observe the sudden change of membership in the crisp SLOW and FAST sets at a speed of 65 mph in Figure 8.3. The fuzzy SLOW and FAST sets allow for a gradual change from SLOW to FAST for a car moving between 50 and 80 mph, as illustrated in Figure 8.4. Notice that a car moving at a speed between 50 and 80 mph has partial membership in SLOW and partial membership in FAST at the same time. In the above example, the variable SPEED can have two possible linguistic values: SLOW and FAST. For this reason, we refer to the variable SPEED as the *linguistic variable* because its possible values are the linguistic values: in this example they are SLOW and FAST.

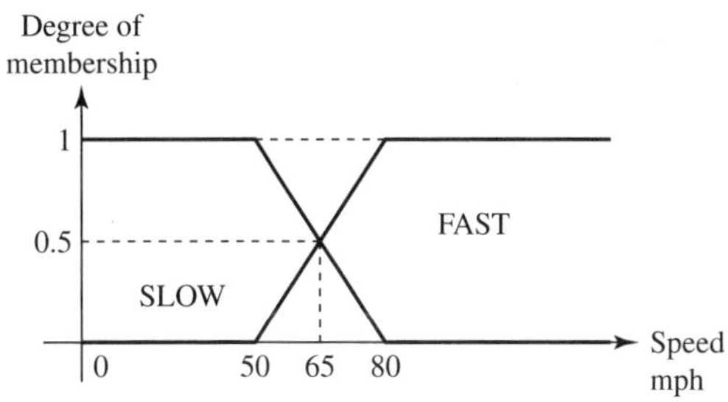

Figure 8.4 Fuzzy sets for SLOW and FAST.

With the above examples in mind, we now formally define a fuzzy set. For this, we need to introduce some notation. Let U be a set. We use $u \in U$ to denote that u is an element of U. We call U the *universe of discourse,* or just the *universe*. We first recall the definition of a function. A function A from a set X to a set Y is a set of ordered pairs in $X \times Y$ with the property that if (x, y) and (x, z) are elements of A, then $y = z$.

Definition 8.1 (Fuzzy Set) A fuzzy set A on U is a function that maps U into the interval $[0, 1]$. Equivalently,

$$A = \{(u, \mu_A(u)) : u \in U\},$$

where μ_A is commonly called the membership function of the fuzzy set A.

Note that the concept of a fuzzy set generalizes that of a crisp set in the following way. Let $V \subset U$. Define $\mu_V : U \to [0, 1]$ by $\mu_V(u) = 1$ if $u \in V$ and $\mu_V(u) = 0$ otherwise. Then V can be identified with the fuzzy set associated with μ_V. In other words, the subsets of U are identified with the functions from U to the two point set $\{0, 1\}$.

◆ Example 8.1

Consider the following universe $U = \{4, 5, 6, 7, 8\}$. Let

$$A = \{(4, 0.0), (5, 0.5), (6, 1.0), (7, 0.5), (8, 0.0)\}.$$

Then A is a fuzzy set. One interpretation of A is closeness to 6.

Let A and B be fuzzy sets on U. We define the *union, intersection,* and *complement* using the following membership functions:

Union: $\mu_{A \cup B}(u) = \max\{\mu_A(u), \mu_B(u)\}$ for all $u \in U$;

Intersection: $\mu_{A \cap B}(u) = \min\{\mu_A(u), \mu_B(u)\}$ for all $u \in U$;

Complement: $\mu_{\bar{A}}(u) = 1 - \mu_A(u)$ for all $u \in U$.

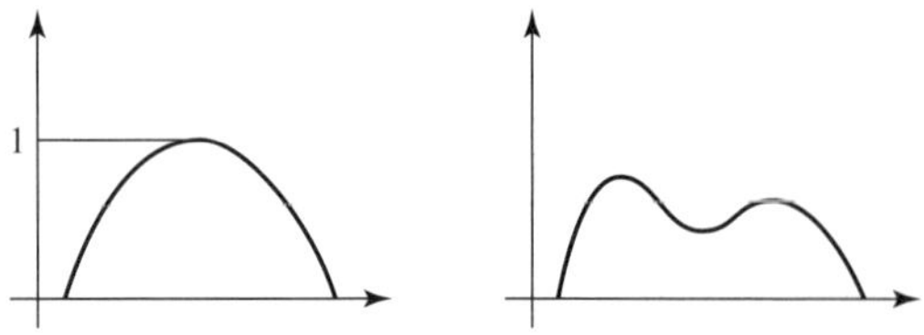

Figure 8.5 The first fuzzy set is a fuzzy number, while the second one is not.

◆ Example 8.2

Consider another fuzzy set B, "number close to 4," given by

$$B = \{(4, 1.0), (5, 0.8), (6, 0.5), (7, 0.2), (8, 0.0)\}.$$

Then,

$$A \cup B = \{(4, 1.0), (5, 0.8), (6, 1.0), (7, 0.5), (8, 0.0)\}.$$

The intersection of the two fuzzy sets A and B is

$$A \cap B = \{(4, 0.0), (5, 0.5), (6, 0.5), (7, 0.2), (8, 0.0)\}.$$

The complement of the fuzzy set A is

$$\bar{A} = \{(4, 1.0), (5, 0.5), (6, 0.0), (7, 0.5), (8, 1.0)\}.$$

Definition 8.2 (Fuzzy Number) A fuzzy number is a fuzzy set F on a universe $U \subset \mathbb{R}$ such that

$$\max_{u \in U} \mu_F(u) = 1$$

and

$$\mu_F(\lambda u_1 + (1 - \lambda)u_2) \geq \min(\mu_F(u_1), \mu_F(u_2))$$

for all $u_1, u_2 \in U$ and all $\lambda \in [0, 1]$.

A fuzzy set that satisfies the conditions $\max_{u \in U} \mu_F(u) = 1$ and $\mu_F(\lambda u_1 + (1 - \lambda)u_2) \geq \min(\mu_F(u_1), \mu_F(u_2))$ is called *normal* and *convex*, respectively. Thus, a fuzzy number is a fuzzy set that is both normal and convex. Examples of convex and nonconvex fuzzy sets are given in Figure 8.5.

8.2 Fuzzy Arithmetic and Fuzzy Relations

8.2.1 Interval Arithmetic

We begin our discussion by introducing the concept of the *interval of confidence*. For this, imagine an experiment where a measurement was performed, resulting in a data point a. The measurement was performed in nonideal conditions. As a result of the presence of the noise measurement in the experiment, we can say that the data point a lies in the interval $[a_1, \ a_2]$, that is,

$$a \in [a_1, \quad a_2].$$

We refer to the closed interval $A = [a_1, \ a_2]$ as the interval of confidence. A fuzzy number can be viewed as an extension of the concept of the interval of confidence. This extension can be used to devise useful interpretations to various operations on fuzzy numbers. We now discuss operations that we can perform on the intervals of confidence. We start with the operation of addition.

Let $A = [a_1, \ a_2]$ and $B = [b_1, \ b_2]$ be two intervals of confidence. If $x \in [a_1, \ a_2]$ and $y \in [b_1, \ b_2]$, then $x + y \in [a_1 + b_1, \ a_2 + b_2]$. We write this symbolically as

$$\begin{aligned} A\,(+)\,B &= [a_1, \quad a_2]\,(+)\,[b_1, \quad b_2] \\ &= [a_1 + b_1, \quad a_2 + b_2]. \end{aligned}$$

To prove the above, note that if $a_1 \leq x$ and $b_1 \leq y$, then $a_1 + b_1 \leq x + y$. Similarly, if $x \leq a_2$ and $y \leq b_2$, then $x + y \leq a_2 + b_2$.

We now define the operation of subtraction on the intervals of confidence. Let $A = [a_1, \ a_2]$ and $B = [b_1, \ b_2]$ be two intervals of confidence. If $x \in [a_1, \ a_2]$ and $y \in [b_1, \ b_2]$, then $x - y \in [a_1 - b_2, \ a_2 - b_1]$. We write this symbolically as

$$\begin{aligned} A\,(-)\,B &= [a_1, \quad a_2]\,(-)\,[b_1, \quad b_2] \\ &= [a_1 - b_2, \quad a_2 - b_1]. \end{aligned}$$

We can also multiply the given intervals of confidence. We first discuss the simple case when the intervals of confidence belong to the set of nonnegative real numbers denoted $\mathbb{R}^+$. Let $A = [a_1, \ a_2]$ and $B = [b_1, \ b_2]$ be two intervals of confidence such that $A, B \subset \mathbb{R}^+$. If $x \in [a_1, \ a_2]$, and $y \in [b_1, \ b_2]$, then $xy \in [a_1 b_1, \ a_2 b_2]$, written symbolically as

$$\begin{aligned} A\,(\cdot)\,B &= [a_1, \quad a_2]\,(\cdot)\,[b_1, \quad b_2] \\ &= [a_1 b_1, \quad a_2 b_2]. \end{aligned}$$

When the intervals of confidence are arbitrary intervals of the set of real numbers $\mathbb{R}$, we have nine possible combinations that are discussed in Kaufmann and Gupta [151]. We list the above-mentioned nine cases in Table 8.1.

Table 8.1 Multiplication of the Intervals of Confidence in $\mathbb{R}$

A	B	$A\,(\cdot)\,B$
$0 \leq a_1 \leq a_2$	$0 \leq b_1 \leq b_2$	$[a_1 b_1, a_2 b_2]$
$0 \leq a_1 \leq a_2$	$b_1 \leq 0 \leq b_2$	$[a_2 b_1, a_2 b_2]$
$0 \leq a_1 \leq a_2$	$b_1 \leq b_2 \leq 0$	$[a_2 b_1, a_1 b_2]$
$a_1 \leq 0 \leq a_2$	$0 \leq b_1 \leq b_2$	$[a_1 b_1, a_2 b_2]$
$a_1 \leq 0 \leq a_2$	$b_1 \leq 0 \leq b_2$	$[\min(a_1 b_2, a_2 b_1), \max(a_1 b_1, a_2 b_2)]$
$a_1 \leq 0 \leq a_2$	$b_1 \leq b_2 \leq 0$	$[a_2 b_1, a_1 b_1]$
$a_1 \leq a_2 \leq 0$	$0 \leq b_1 \leq b_2$	$[a_1 b_2, a_2 b_1]$
$a_1 \leq a_2 \leq 0$	$b_1 \leq 0 \leq b_2$	$[a_1 b_2, a_1 b_1]$
$a_1 \leq a_2 \leq 0$	$b_1 \leq b_2 \leq 0$	$[a_2 b_2, a_1 b_1]$

Observe that the multiplication of an interval of confidence $A \subset \mathbb{R}^+$ by a real number $k \in \mathbb{R}$ can be easily deduced from the operation of multiplication of two intervals of confidence. First, assume that $k \in \mathbb{R}^+$ and represent it as

$$k = [k, \quad k].$$

Then,

$$\begin{aligned} k \cdot A &= k \cdot [a_1, \quad a_2] \\ &= [k, \quad k] \, (\cdot) \, [a_1, \quad a_2] \\ &= [ka_1, \, ka_2]. \end{aligned}$$

If $k < 0$, then

$$k \cdot [a_1, \quad a_2] = [ka_2, \quad ka_1].$$

We now discuss the operations of maximum and minimum of two intervals of confidence. First, we explain the meaning of the symbols we are going to use. Consider two real numbers a and b. By definition, we have

$$\begin{aligned} a \wedge b &= \min(a, b) \\ &= \begin{cases} a & \text{if } a \leq b, \\ b & \text{if } b < a, \end{cases} \end{aligned}$$

and

$$\begin{aligned} a \vee b &= \max(a, b) \\ &= \begin{cases} b & \text{if } a \leq b, \\ a & \text{if } b < a. \end{cases} \end{aligned}$$

Consider now two intervals of confidence $A, B \subset \mathbb{R}$, where $A = [a_1, \; a_2]$ and $B = [b_1, \; b_2]$. We introduce two operations $(\wedge)$ and $(\vee)$ on the intervals of confidence. We define

$$\begin{aligned} A \, (\wedge) \, B &= [a_1, \quad a_2] \, (\wedge) \, [b_1, \quad b_2] \\ &= [a_1 \wedge b_1, \quad a_2 \wedge b_2] \end{aligned}$$

and

$$\begin{aligned} A \, (\vee) \, B &= [a_1, \quad a_2] \, (\vee) \, [b_1, \quad b_2] \\ &= [a_1 \vee b_1, \quad a_2 \vee b_2]. \end{aligned}$$

The above operations on the intervals will be used in the next section to define the corresponding operations on fuzzy numbers.

8.2.2 Manipulating Fuzzy Numbers

By relating the concept of the interval of confidence with the new concept of the *level of presumption,* we arrive at the concept of fuzzy number. Recall that a fuzzy number is a fuzzy set that is convex and normal. The fuzzy set $A \subset \mathbb{R} \times \mathbb{R}$ is convex if for any $x, y \in \mathbb{R}$ and $\lambda \in [0, \; 1]$

$$\mu_A(\lambda x + (1 - \lambda)y) \geq \mu_A(x) \wedge \mu_A(y).$$

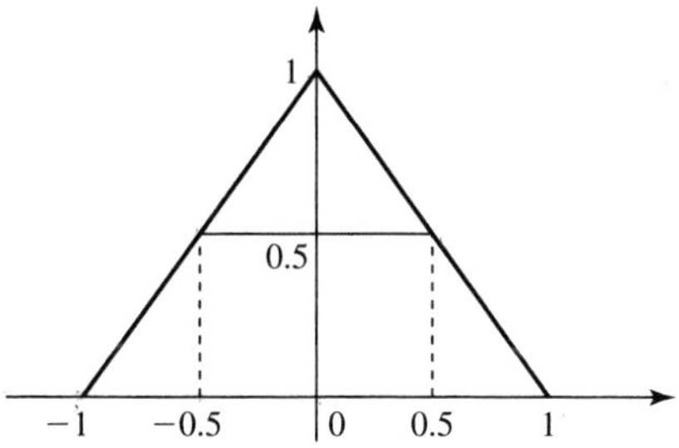

Figure 8.6 A fuzzy zero.

The fuzzy set $A \subset \mathbb{R} \times \mathbb{R}$ is normal if for all $x \in \mathbb{R}$

$$\bigvee_x \mu_A(x) = 1,$$

where $\bigvee$ denotes the operation of taking a maximum. Thus, the fuzzy set $A \subset \mathbb{R} \times \mathbb{R}$ is normal if the largest value of $\mu_A(x)$ is equal to 1.

We consider now a fuzzy zero as depicted in Figure 8.6. The interval of confidence corresponding to the level of presumption of 0 is $[-1,\ 1]$. The interval of confidence corresponding to the level of presumption of 0.5 is $[-0.5,\ 0.5]$, and so on. Thus the level of presumption $\alpha \in [0,\ 1]$ gives an interval of confidence

$$A_\alpha = \left[a_1^{(\alpha)},\quad a_2^{(\alpha)}\right].$$

Note that if $\alpha_1 < \alpha_2$, then

$$\left[a_1^{(\alpha_2)},\quad a_2^{(\alpha_2)}\right] \subseteq \left[a_1^{(\alpha_1)},\quad a_2^{(\alpha_1)}\right],$$

or, equivalently, $A_{\alpha_2} \subseteq A_{\alpha_1}$. Thus, if the presumption level α is increasing, then the corresponding interval of confidence is nonincreasing; that is, A_α is a nonincreasing function of α. In the previous section we discussed how to add intervals of confidence. The addition of fuzzy numbers can be performed following the same process, level by level. For example, let A and B be two fuzzy numbers, and let A_α and B_α be their intervals of confidence for the level of presumption $\alpha \in [0,\ 1]$; that is, $A_\alpha = \{x : \mu_A(x) \geq \alpha\}$ and $B_\alpha = \{x : \mu_B(x) \geq \alpha\}$. Then,

$$\begin{aligned} A_\alpha\,(+)\,B_\alpha &= \left[a_1^{(\alpha)},\quad a_2^{(\alpha)}\right](+)\left[b_1^{(\alpha)},\quad b_2^{(\alpha)}\right] \\ &= \left[a_1^{(\alpha)} + b_1^{(\alpha)},\quad a_2^{(\alpha)} + b_2^{(\alpha)}\right]. \end{aligned}$$

Similarly, we can subtract two given fuzzy numbers,

$$\begin{aligned} A_\alpha\,(-)\,B_\alpha &= \left[a_1^{(\alpha)},\quad a_2^{(\alpha)}\right](-)\left[b_1^{(\alpha)},\quad b_2^{(\alpha)}\right] \\ &= \left[a_1^{(\alpha)} - b_2^{(\alpha)},\quad a_2^{(\alpha)} - b_1^{(\alpha)}\right]. \end{aligned}$$

◆ Example 8.3

Consider two triangular fuzzy numbers whose membership functions are

$$\mu_A(x) = \begin{cases} 0 & \text{if } x \le 0.5, \\ 2/5x - 1/5 & \text{if } 0.5 < x < 3, \\ -1/2x + 5/2 & \text{if } 3 \le x < 5, \\ 0 & \text{if } x \ge 5 \end{cases}$$

and

$$\mu_B(x) = \begin{cases} 0 & \text{if } x \le 1, \\ x - 1 & \text{if } 1 < x < 2, \\ -1/5x + 7/5 & \text{if } 2 \le x < 7, \\ 0 & \text{if } x \ge 7. \end{cases}$$

We now compute the intervals of confidence for the above fuzzy numbers for each level of presumption $\alpha \in [0, 1]$. We begin with the fuzzy number A. In a sense, our goal is to express the triangular shape of the fuzzy number A in the form $A_\alpha = [a_1^{(\alpha)}, a_2^{(\alpha)}]$ in terms of α. From the description of A above, we obtain

$$\alpha = 2/5a_1^{(\alpha)} - 1/5$$

and

$$\alpha = -1/2a_2^{(\alpha)} + 5/2.$$

Hence, the interval of confidence at the level α for the fuzzy number A is

$$\begin{aligned} A_\alpha &= \left[a_1^{(\alpha)}, \quad a_2^{(\alpha)}\right] \\ &= [5/2\alpha + 1/2, \quad -2\alpha + 5]. \end{aligned}$$

Proceeding in a similar manner, we express the triangular shape of the fuzzy number B in the form $B_\alpha = [b_1^{(\alpha)}, b_2^{(\alpha)}]$ in terms of α. From the description of B above, we obtain

$$\alpha = b_1^{(\alpha)} - 1,$$

and

$$\alpha = -1/5b_2^{(\alpha)} + 7/5.$$

To find the interval of confidence at the level α for the fuzzy number B, we perform simple manipulations to get

$$\begin{aligned} B_\alpha &= \left[b_1^{(\alpha)}, \quad b_2^{(\alpha)}\right] \\ &= [\alpha + 1, \quad -5\alpha + 7]. \end{aligned}$$

We next compute the sum of the above intervals of confidence,

$$\begin{aligned} A_\alpha\,(+)\,B_\alpha &= \left[a_1^{(\alpha)} + b_1^{(\alpha)}, \quad a_2^{(\alpha)} + b_2^{(\alpha)}\right] \\ &= [7/2\alpha + 3/2, \quad -7\alpha + 12]. \end{aligned}$$

From the above, we can find analytic description of the fuzzy number $A(+)B$ in terms of its membership function $\mu_{A(+)B}$. Because $a_1^{(\alpha)} + b_1^{(\alpha)} = 7/2\alpha + 3/2$, and $a_2^{(\alpha)} + b_2^{(\alpha)} = -7\alpha + 12$, solving for α yields

$$\mu_{A(+)B}(x) = \begin{cases} 0 & \text{if } x \le 3/2, \\ 2/7x - 3/7 & \text{if } 3/2 < x \le 5, \\ -1/7x + 12/7 & \text{if } 5 < x \le 12, \\ 0 & \text{if } x > 12. \end{cases}$$

We now compute the fuzzy number $A(-)B$. First, we find the expression for the interval of confidence $A_\alpha\,(-)\,B_\alpha$ corresponding to the level α:

$$\begin{aligned} A_\alpha\,(-)\,B_\alpha &= \left[a_1^{(\alpha)} - b_2^{(\alpha)},\quad a_2^{(\alpha)} - b_1^{(\alpha)}\right] \\ &= [15/2\alpha - 13/2, \quad -3\alpha + 4]. \end{aligned}$$

Thus,

$$a_1^{(\alpha)} - b_2^{(\alpha)} = 15/2\alpha - 13/2$$

and

$$a_2^{(\alpha)} - b_1^{(\alpha)} = -3\alpha + 4.$$

We use the above two equations to solve for α, which then allows us to find analytic expression for the membership function $\mu_{A(-)B}$. Performing simple manipulations yields

$$\mu_{A(-)B}(x) = \begin{cases} 0 & \text{if } x \le -13/2, \\ 2/15x + 13/15 & \text{if } -13/2 < x \le 1, \\ -1/3x + 4/3 & \text{if } 1 < x \le 4, \\ 0 & \text{if } x > 4. \end{cases}$$

As far as the multiplication is concerned, we consider only the case when for $\alpha \in [0, 1]$ the intervals of confidence $A_\alpha, B_\alpha \subset \mathbb{R}^+$. In such a case the multiplication of the two fuzzy numbers A and B can be performed level by level on the intervals of confidence following the process of multiplication of the intervals of confidence in $\mathbb{R}^+$, where

$$\begin{aligned} A_\alpha\,(\cdot)\,B_\alpha &= \left[a_1^{(\alpha)},\quad a_2^{(\alpha)}\right](\cdot)\left[b_1^{(\alpha)},\quad b_2^{(\alpha)}\right] \\ &= \left[a_1^{(\alpha)} b_1^{(\alpha)},\quad a_2^{(\alpha)} b_2^{(\alpha)}\right]. \end{aligned}$$

From the above we can deduce the way we can multiply a fuzzy number by a nonnegative real number k. We have

$$\begin{aligned} k \cdot A_\alpha &= [k,\quad k]\,(\cdot)\left[a_1^{(\alpha)},\quad a_2^{(\alpha)}\right] \\ &= \left[k a_1^{(\alpha)},\quad k a_2^{(\alpha)}\right]. \end{aligned}$$

The multiplication of fuzzy numbers in $\mathbb{R}$ is discussed in detail in Kaufmann and Gupta [151, pp. 325–334].

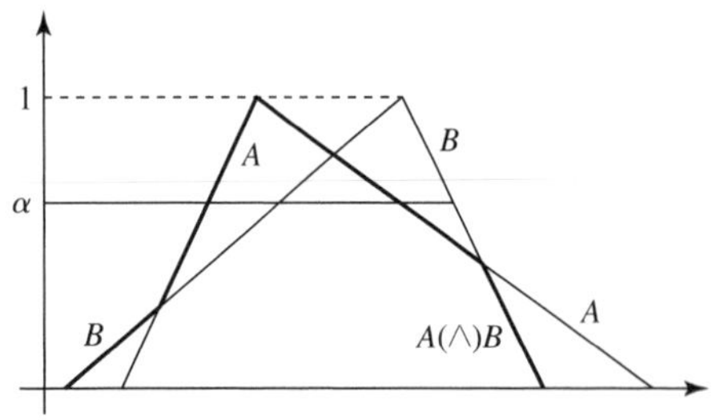

Figure 8.7 Calculating the fuzzy minimum.

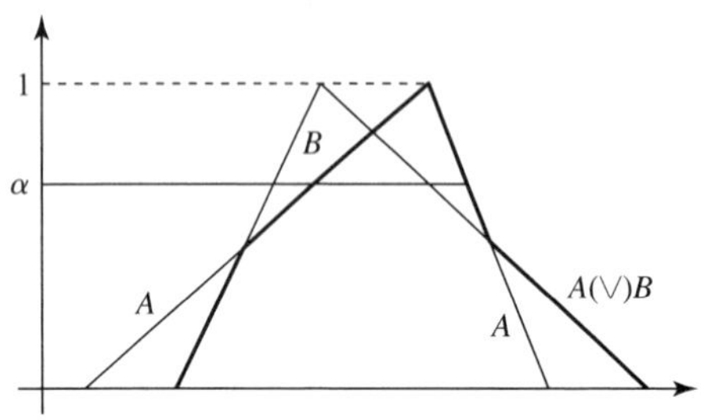

Figure 8.8 Calculating the fuzzy maximum of two fuzzy numbers.

The fuzzy minimum of two fuzzy numbers A and B is the fuzzy number that can be obtained by performing the following level by level operation:

$$\begin{aligned} A_\alpha\,(\wedge)\,B_\alpha &= \left[a_1^{(\alpha)},\quad a_2^{(\alpha)}\right](\wedge)\left[b_1^{(\alpha)},\quad b_2^{(\alpha)}\right] \\ &= \left[a_1^{(\alpha)} \wedge b_1^{(\alpha)},\quad a_2^{(\alpha)} \wedge b_2^{(\alpha)}\right]. \end{aligned}$$

We illustrate a graphical procedure for calculating the fuzzy minimum in Figure 8.7.

The fuzzy maximum of the two fuzzy numbers is a fuzzy number that can be obtained by performing the following level-by-level operation:

$$\begin{aligned} A_\alpha\,(\vee)\,B_\alpha &= \left[a_1^{(\alpha)},\quad a_2^{(\alpha)}\right](\vee)\left[b_1^{(\alpha)},\quad b_2^{(\alpha)}\right] \\ &= \left[a_1^{(\alpha)} \vee b_1^{(\alpha)},\quad a_2^{(\alpha)} \vee b_2^{(\alpha)}\right]. \end{aligned}$$

The process of calculating the fuzzy maximum is illustrated in Figure 8.8. We now present a numerical example to illustrate analytical method for calculating the fuzzy minimum and maximum of two fuzzy numbers.

◆ Example 8.4

Consider the two triangular fuzzy numbers A and B from Example 8.3. Recall that

$$A_\alpha = \left[a_1^{(\alpha)},\quad a_2^{(\alpha)}\right] = [5/2\alpha + 1/2,\quad -2\alpha + 5]$$

and

$$B_\alpha = \left[b_1^{(\alpha)},\quad b_2^{(\alpha)}\right] = [\alpha + 1,\quad -5\alpha + 7].$$

Therefore, for the fuzzy minimum

$$A_\alpha\,(\wedge)\,B_\alpha = [(5/2\alpha + 1/2) \wedge (\alpha + 1), \quad (-2\alpha + 5) \wedge (-5\alpha + 7)]$$

we find that

$$A_\alpha\,(\wedge)\,B_\alpha = \begin{cases} [5/2\alpha + 1/2, \ -2\alpha + 5] & \text{if } 0 \le \alpha \le 1/3, \\ [\alpha + 1, \ -2\alpha + 5] & \text{if } 1/3 < \alpha \le 2/3, \\ [\alpha + 1, \ -5\alpha + 7] & \text{if } 2/3 < \alpha \le 1. \end{cases}$$

From the above we can find an analytic expression for the membership function of the fuzzy number $A(\wedge)\,B$:

$$\mu_{A(\wedge)B}(x) = \begin{cases} 0 & \text{if } x \le 1/2, \\ 2/5x - 1/5 & \text{if } 1/2 < x \le 4/3, \\ x - 1 & \text{if } 4/3 < x \le 2, \\ -1/5x + 7/5 & \text{if } 2 < x \le 11/3, \\ -1/2x + 5/2 & \text{if } 11/3 < x \le 5, \\ 0 & \text{if } 5 < x. \end{cases}$$

We next find the fuzzy maximum of the fuzzy numbers A and B:

$$\begin{aligned} A_\alpha\,(\vee)\,B_\alpha &= \left[a_1^{(\alpha)} \vee b_1^{(\alpha)}, \quad a_2^{(\alpha)} \vee b_2^{(\alpha)}\right] \\ &= [(5/2\alpha + 1/2) \vee (\alpha + 1), \quad (-2\alpha + 5) \vee (-5\alpha + 7)]. \end{aligned}$$

Performing simple manipulations yields

$$A_\alpha\,(\vee)\,B_\alpha = \begin{cases} [\alpha + 1, \ -5\alpha + 7] & \text{if } 0 \le \alpha \le 1/3, \\ [5/2\alpha + 1/2, \ -5\alpha + 7] & \text{if } 1/3 < \alpha \le 2/3, \\ [5/2\alpha + 1/2, \ -2\alpha + 5] & \text{if } 2/3 < \alpha \le 1. \end{cases}$$

Using the above, we compute

$$\mu_{A(\vee)B}(x) = \begin{cases} 0 & \text{if } x \le 1, \\ x - 1 & \text{if } 1 < x \le 4/3, \\ 2/5x - 1/5 & \text{if } 4/3 < x \le 3, \\ -1/2x + 5/2 & \text{if } 3 < x \le 11/3, \\ -1/5x + 7/5 & \text{if } 11/3 < x \le 7, \\ 0 & \text{if } 7 < x. \end{cases}$$

8.2.3 Fuzzy Relations

In this subsection we discuss fuzzy relations. We begin with the definition of a fuzzy relation.

Definition 8.3 Let U and V be two universes of discourse. A fuzzy relation R between U and V is a fuzzy set in the product space

$$U \times V = \{(u, v) : u \in U, \ v \in V\}$$

characterized by a membership function

$$\mu_R : U \times V \to [0,\ 1].$$

Suppose we are given two universes of discourse:

$$U = \{u_1, u_2, \ldots, u_m\} \quad \text{and} \quad V = \{v_1, v_2, \ldots, v_n\}.$$

A fuzzy relation in $U \times V$ can be represented by an $m \times n$ matrix:

$$\boldsymbol{R} = \begin{bmatrix} \mu_R(u_1, v_1) & \mu_R(u_1, v_2) & \cdots & \mu_R(u_1, v_n) \\ \mu_R(u_2, v_1) & \mu_R(u_2, v_2) & \cdots & \mu_R(u_2, v_n) \\ \vdots & \vdots & \ddots & \vdots \\ \mu_R(u_m, v_1) & \mu_R(u_m, v_2) & \cdots & \mu_R(u_m, v_n) \end{bmatrix}.$$

Terano, Asai, and Sugeno [280] refer to the matrix $\boldsymbol{R}$ as a fuzzy matrix. Because $\mu_R : U \times V \to [0,\ 1]$, the values of the elements of the fuzzy matrix $\boldsymbol{R}$ are in the interval $[0,\ 1]$.

◆ Example 8.5

Let $U = \{-1, 0, 1\}$ and $V = \{1, 2, 3, 4\}$. An example of a fuzzy matrix corresponding to a fuzzy relation in $U \times V$ is

$$\boldsymbol{R} = \begin{bmatrix} 0.2 & 0.6 & 0.7 & 0.9 \\ 0.0 & 0.3 & 0.6 & 1 \\ 0.1 & 0.4 & 0.5 & 0.8 \end{bmatrix}.$$

Definition 8.3 is concerned with a binary fuzzy relation. We can generalize this definition to the n-ary fuzzy relation R in $X_1 \times X_2 \times \cdots \times X_n$ as a fuzzy set in the product space $X_1 \times X_2 \times \cdots \times X_n$ with a membership function:

$$\mu_R : X_1 \times X_2 \times \cdots \times X_n \to [0,\ 1].$$

A special kind of a fuzzy relation is a fuzzy implication that we discuss next. Let A and B be fuzzy sets in U and V, respectively. A fuzzy implication from A to B, denoted $A \to B$, is a fuzzy relation in $U \times V$ that is characterized by a specific membership function. Among different types of membership functions that can characterize a fuzzy implication, we list two classes:

- A fuzzy conjunction denoted $\mu_A(u) \star \mu_B(v)$
- A fuzzy disjunction denoted $\mu_A(u) \dot{+} \mu_B(v)$

A fuzzy conjunction, also called a t-norm, represents a generalized intersection operator, while a fuzzy disjunction, also called a t-conorm, represents a generalized union operator. Examples of a t-norm are as follows:

- Fuzzy intersection $= \min\{\mu_A(u), \mu_B(v)\}$
- Algebraic product $= \mu_A(u)\mu_B(v)$

- Bounded product $= \max\{0, \mu_A + \mu_B - 1\}$
- Drastic product $= \begin{cases} \mu_A & \text{if } \mu_B = 1, \\ \mu_B & \text{if } \mu_A = 1, \\ 0 & \text{if } \mu_A, \mu_B < 1 \end{cases}$

Examples of a t-conorm are as follows:

- Fuzzy union $= \max\{\mu_A(u), \mu_B(v)\}$
- Algebraic sum $= \mu_A(u) + \mu_B(v) - \mu_A(u)\mu_B(v)$
- Bounded sum $= \min\{1, \mu_A + \mu_B\}$
- Drastic sum $= \begin{cases} \mu_A & \text{if } \mu_B = 0 \\ \mu_B & \text{if } \mu_A = 0 \\ 1 & \text{if } \mu_A, \mu_B > 0 \end{cases}$

A fuzzy implication $A \to B$ can be viewed as a fuzzy IF–THEN rule of the form

$$\text{IF } x \text{ is } A, \quad \text{THEN } y \text{ is } B,$$

where x and y are linguistic variables and A and B are the corresponding values of the linguistic variables.

8.2.4 Composition of Fuzzy Relations

Suppose we are given a fuzzy relation in $U \times V$ labeled R, along with a fuzzy relation in $V \times W$ labeled S. The max-star composition of R and S is a fuzzy relation, denoted $R \circ S$, that is characterized by the membership function

$$\mu_{R \circ S}(u, w) = \bigvee_{v \in V} \{\mu_R(u, v) \star \mu_S(v, w)\},$$

where $\bigvee$ denotes the maximum operator, and $\star$ refers to any operator in the class of t norms. Note that $R \circ S$ is a fuzzy set in $U \times W$.

◆ Example 8.6

Let $U = \{u_1, u_2\}$, $V = \{v_1, v_2, v_3\}$, and $W = \{w_1, w_2\}$ be the given three universes of discourse. Let

$$\boldsymbol{R} = \begin{bmatrix} 0.5 & 0.7 & 0.0 \\ 0.8 & 1.0 & 0.2 \end{bmatrix}$$

be the fuzzy matrix that represents the relation R in $U \times V$, and let

$$\boldsymbol{S} = \begin{bmatrix} 0.4 & 0.6 \\ 0.2 & 0.9 \\ 0.0 & 0.5 \end{bmatrix}$$

be the given fuzzy matrix representing the relation S in $V \times W$. Let the $\star$ operator be the minimum operator $\wedge$. Then the fuzzy matrix representing the max–min fuzzy relation $R \circ S$ has the form

$$\boldsymbol{R} \circ \boldsymbol{S}$$

$$= \begin{bmatrix} 0.5 & 0.7 & 0.0 \\ 0.8 & 1.0 & 0.2 \end{bmatrix} \circ \begin{bmatrix} 0.4 & 0.6 \\ 0.2 & 0.9 \\ 0.1 & 0.5 \end{bmatrix}$$

$$= \begin{bmatrix} (0.5 \wedge 0.4) \vee (0.7 \wedge 0.2) \vee (0.0 \wedge 0.1) & (0.5 \wedge 0.6) \vee (0.7 \wedge 0.9) \vee (0.0 \wedge 0.5) \\ (0.8 \wedge 0.4) \vee (1.0 \wedge 0.2) \vee (0.2 \wedge 0.1) & (0.8 \wedge 0.6) \vee (1.0 \wedge 0.9) \vee (0.2 \wedge 0.5) \end{bmatrix}$$

$$= \begin{bmatrix} 0.4 & 0.7 \\ 0.4 & 0.9 \end{bmatrix}.$$

Other kinds of composition of fuzzy relations are possible. In particular, we can define the dual composition to the max $-\star$ as the min $-\dot{+}$. The min $-\dot{+}$ composition of the fuzzy relations R and S is characterized by the membership function

$$\mu_{R \circ S}(u, w) = \bigwedge_{v \in V} \{\mu_R(u, v) \dot{+} \mu_S(v, w)\},$$

where $\bigwedge$ refers to the minimum operator, and $\dot{+}$ can be any operator from the class of t conorm that we discussed in the previous section.

A particular type of the composition of fuzzy relations is the max $-\star$ composition when S is just a fuzzy set A in V, rather than a fuzzy relation in the product space $V \times W$. In this case the fuzzy matrix representing S is a column vector. This means that the fuzzy matrix representing the composition $R \circ A$ is also a column vector. A general form of the membership function that characterizes such a composition is

$$\mu_{R \circ A}(w) = \bigvee_{v \in V} \{\mu_R(u, v) \star \mu_A(v)\}.$$

Note that the composition of the fuzzy relation R and the fuzzy set A is a fuzzy set in U.

◆ Example 8.7

Let $U = \{u_1, u_2, u_3\}$ and $V = \{v_1, v_2, v_3\}$ be given two universes of discourse. Let A be a fuzzy set in V described by the fuzzy column vector $\boldsymbol{A} = [0.2\ \ 0.8\ \ 1.0]^T$. The fuzzy matrix $\boldsymbol{R}$ that describes the fuzzy relation R in $U \times V$ is

$$\boldsymbol{R} = \begin{bmatrix} 0.5 & 0.4 & 0.8 \\ 0.8 & 0.6 & 0.4 \\ 0.9 & 0.7 & 0.1 \end{bmatrix}.$$

Assume the $\wedge$ operator for the $\star$ operator—that is, the membership function of the composition of R and A—has the form

$$\mu_{R \circ A}(w) = \bigvee_{v \in V} \{\mu_R(u, v) \wedge \mu_A(v)\}.$$

The fuzzy column vector corresponding to the composition $R \circ A$ is

$$\boldsymbol{R} \circ \boldsymbol{A} = \begin{bmatrix} 0.5 & 0.4 & 0.8 \\ 0.8 & 0.6 & 0.4 \\ 0.9 & 0.7 & 0.1 \end{bmatrix} \circ \begin{bmatrix} 0.2 \\ 0.8 \\ 1.0 \end{bmatrix}$$

$$= \begin{bmatrix} (0.5 \wedge 0.2) \vee (0.4 \wedge 0.8) \vee (0.8 \wedge 1.0) \\ (0.8 \wedge 0.2) \vee (0.6 \wedge 0.8) \vee (0.4 \wedge 1.0) \\ (0.9 \wedge 0.2) \vee (0.7 \wedge 0.8) \vee (0.1 \wedge 1.0) \end{bmatrix}$$

$$= \begin{bmatrix} 0.8 \\ 0.6 \\ 0.7 \end{bmatrix}.$$

We can think of A as the fuzzy input to the fuzzy system modeled by the fuzzy relation R. Let $B = R \circ A$. Then the output of the fuzzy system is B, which is a fuzzy set. In other words, the fuzzy system modeled by a fuzzy relation R yields a fuzzy output in response to a fuzzy input. We illustrate the above in Figure 8.9.

8.3 Standard Additive Model

Let X and Y be nonempty sets and let I and J be nonempty index sets. Suppose we are given collections of fuzzy sets $\{A_\alpha : \alpha \in I\}$ and $\{B_\beta : \beta \in J\}$ on X and Y, respectively, and rules of the form

$$\text{IF } A_\alpha, \quad \text{THEN } B_\beta.$$

The IF–THEN rules along with the collections of fuzzy sets form a function R from $\{A_\alpha : \alpha \in I\}$ to $\{B_\beta : \beta \in J\}$, which we call the *fuzzy system* associated with the rules.

The *support* of a function into $\mathbb{R}$ is the subset of the domain where the function is nonzero. The support of the fuzzy sets A_α and B_β in the IF–THEN rules describe subsets, or patches, in the Cartesian product space $X \times Y$, as illustrated in Figure 8.10. Suppose the support of each A_α and each B_β is a one-point set, say $x_\alpha \in X$ and $y_\beta \in Y$, respectively. If $X = \bigcup\{x_\alpha : \alpha \in I\}$, then R can be identified with a function $\tilde{R}$ from X to Y by

$$\tilde{R}(x_\alpha) = y_\alpha$$

if

$$R(A_\alpha) = B_\alpha.$$

One can think of this case as having complete knowledge and the patches are just points.

A fuzzy system can be thought of as a "fuzzified" function from X to Y with fuzzy sets replacing points in the domain. The size of a patch is usually inversely proportional to the

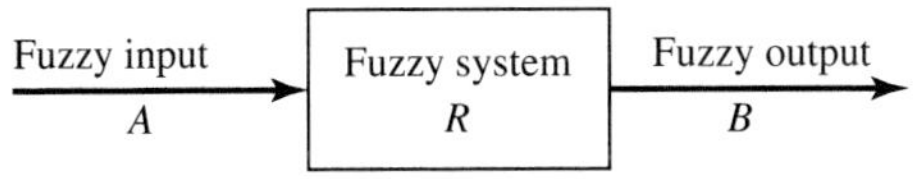

Figure 8.9 A fuzzy system produces a fuzzy output in response to a fuzzy input.

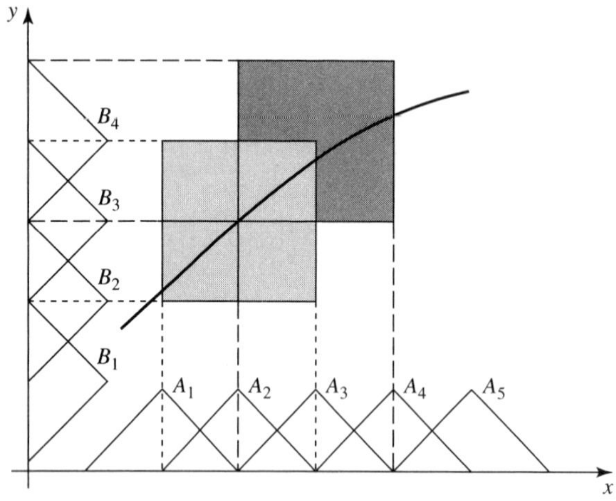

Figure 8.10 Function approximation using IF–THEN rules.

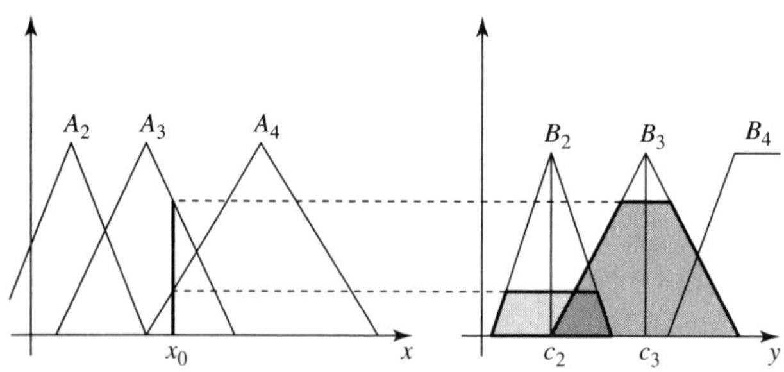

Figure 8.11 Generating output fuzzy sets from THEN-part sets of IF–THEN rules.

knowledge or confidence of the situation. A point $x \in X$ is usually in the support of more than one A_α with confidence represented by the values of the membership functions μ_α. The fuzzy set $R(A_\alpha)$ represents the desired action corresponding to A_α. This is illustrated in Figure 8.11. Suppose we are given the following two rules:

$$\text{IF } A_3, \quad \text{THEN } B_3,$$
$$\text{IF } A_4, \quad \text{THEN } B_2.$$

The point x_0 is contained in the support of A_3 and A_4. Thus, according to the given rules, both B_3 and B_2 need to be taken into account. The process by which we combine these actions is called *defuzzification*. The process of defuzzification is also known as fuzzy reasoning in the literature. The outcome of defuzzification is a numerical value $y \in Y$. An example of a defuzzification algorithm is a centroidal defuzzifier of the form

$$y = \frac{\sum_\alpha \mu_{A_\alpha}(x)\xi(x, A_\alpha, R(A_\alpha))}{\sum_\alpha \mu_{A_\alpha}(x)}, \tag{8.1}$$

where $\xi(x, A_\alpha, R(A_\alpha))$ is a function into the real numbers. We now give an example of ξ.

Let $\xi(x, A_\alpha, R(A_\alpha))$ be the centroid of the support of $R(A_\alpha)$. For example, the centroids of the support of $R(A_3) = B_3$ is c_3, and that of $R(A_4) = B_2$ is c_2. A fuzzy system with the centroidal

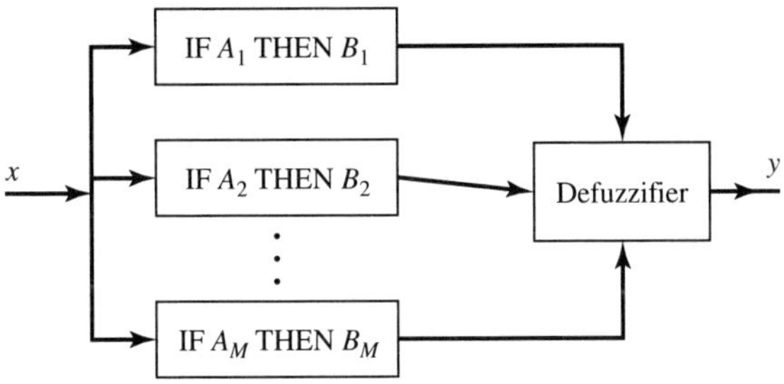

Figure 8.12 Single-input, single-output standard additive model schematic representation.

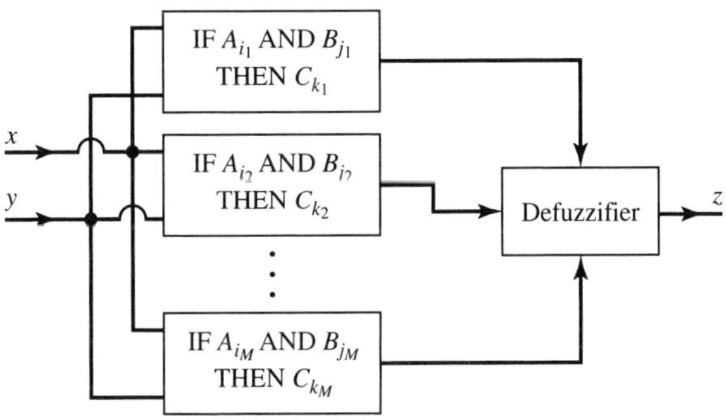

Figure 8.13 Two-input, single-output standard additive model representation.

defuzzifier is called the *standard additive model* (SAM). The term standard additive model was coined by Kosko [165]. A schematic representation of a SAM is shown in Figure 8.12. In the above description of the SAM, we were concerned with a single-input, single-output case. The above construction can be extended to a many-inputs case. Suppose we are given IF–THEN rules of the form

$$\text{IF } A_i \text{ AND } B_i, \quad \text{THEN } C_i.$$

The SAM corresponding to this case is illustrated in Figure 8.13. We could use a defuzzification algorithm of the form

$$z = \frac{\sum_{l=1}^{M} \mu_{A_l}(x)\mu_{B_l}(y)c_l}{\sum_{l=1}^{M} \mu_{A_l}(x)\mu_{B_l}(y)}, \tag{8.2}$$

where M is the number of IF–THEN rules, and c_l is the centroid of the fuzzy set C_l, $l = 1, 2, \ldots, M$. This type of defuzzifier is called the *center average defuzzifier*. Another type of defuzzifier that we could use is

$$z = \frac{\sum_{l=1}^{M} \min\{\mu_{A_l}(x), \mu_{B_l}(y)\}c_l}{\sum_{l=1}^{M} \min\{\mu_{A_l}(x)\mu_{B_l}(y)\}}. \tag{8.3}$$

An illustration of fuzzy reasoning using defuzzification scheme (8.3) is given in Figure 8.14. In general, for q inputs the center average defuzzifier takes the form

$$z = \frac{\sum_{l=1}^{M} \prod_{j=1}^{q} \mu_{A_j^l}(x_j)c_l}{\sum_{l=1}^{M} \prod_{j=1}^{q} \mu_{A_j^l}(x_j)}. \tag{8.4}$$

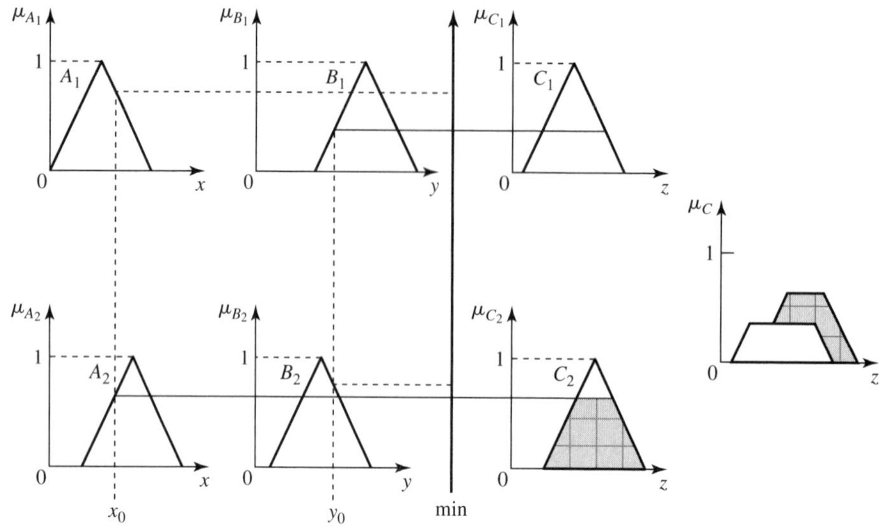

Figure 8.14 An illustration of fuzzy reasoning in a two-input fuzzy system.

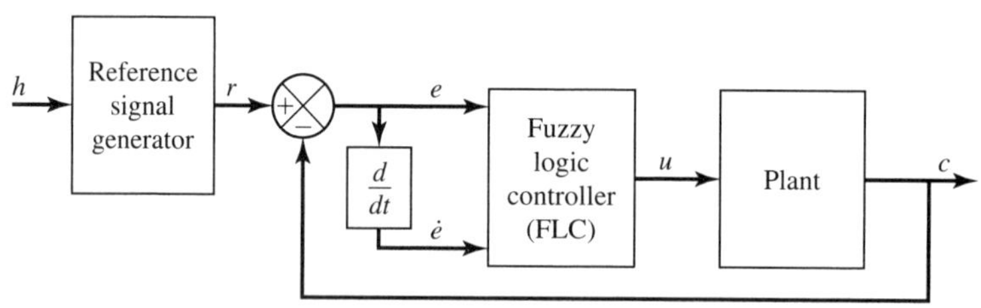

Figure 8.15 A diagram of a closed-loop system with a fuzzy logic controller.

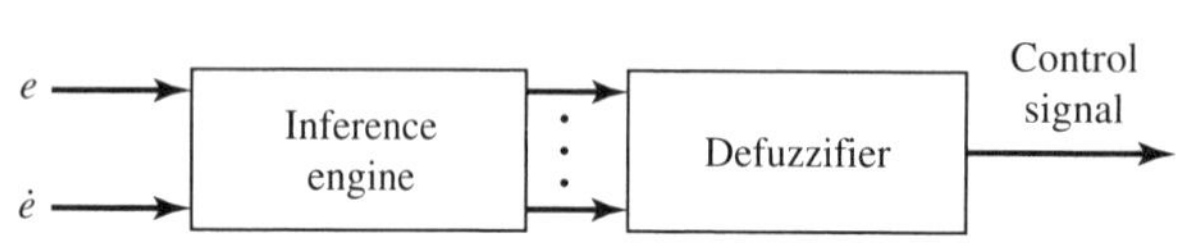

Figure 8.16 Standard additive model as a fuzzy logic controller.

8.4 Fuzzy Logic Control

A block diagram of a closed-loop system with a fuzzy logic controller (FLC) is shown in Figure 8.15. A typical fuzzy logic control architecture is depicted in Figure 8.16. The controller uses the control error e and the rate of change in the control error, $\dot{e}$ or Δe, as its inputs. A typical fuzzy logic control architecture has the structure of a SAM; hence, it is composed of two basic components: a decision-making logic module, also referred to as an inference engine, and a defuzzifier. The inference engine processes the inputs using a knowledge base. The outputs of the inference engine are converted into crisp values, defuzzified, by a defuzzifier. Designing a fuzzy logic controller can be summarized in the following six-step algorithm:

- Identify ranges of the controller inputs.
- Identify ranges of the controller outputs.
- Create fuzzy sets for each input and output.

Table 8.2 The Rules Matrix

Error \ Change-in-error	LN	N	ZE	P	LP
LN	LN	LN	LN	N	ZE
N	LN	LN	N	ZE	P
ZE	N	N	ZE	P	P
P	N	ZE	P	LP	LP
LP	ZE	P	LP	LP	LP

- Translate the interaction of the inputs and outputs into IF–THEN rules, and then form the rules matrix.
- Decide on the inference engine and the defuzzifier, and then apply them to crisp inputs to produce crisp outputs.
- Implement the controller, test it, and modify it if necessary.

As an example, suppose we use five membership functions for the controller's inputs and its output. We then generate the rules. A possible set of rules is shown in Table 8.2, where LN denotes large negative, N means negative, ZE is zero, P is positive, and LP refers to large positive. The next step involves choosing an inference engine and a defuzzifier. We can use the product inference rule and the modified center average defuzzifier to get

$$u = \frac{\sum_{k=1}^{25} m_k^o p_k / s_k^o}{\sum_{k=1}^{25} p_k / s_k^o},$$

where for $k = 1, 2, \ldots, 25$, following the rules shown in Table 8.2, we compute

$$p_k = \mu_{i_k}^e \mu_{j_k}^{ce}, \qquad i, j = 1, 2, \ldots, 5.$$

The functions μ_i^e, μ_j^{ec} are the membership functions corresponding to the linguistic variables ERROR and CHANGE-IN-ERROR, respectively. The parameters m_k^o and s_k^o are the m and s parameters of the controller output membership functions—see figures below for their definition. The modified center average defuzzifier takes into account the shape of the membership functions, and in particular their "spread." As pointed out by Wang [295], the sharper the shape of μ, the stronger is our belief that the value should be closer to its m. This observation is reflected in the formula for the defuzzifier we use, where we divide the weights p_k by s_k^o—the spread of the corresponding controller output membership function. We can use triangular membership functions whose equations are given next. The expression for the function shown in Figure 8.17 is

$$\mu(x, m, s) = \max\left(0, 1 - \frac{|x - m|}{s}\right).$$

The expression for the function shown in Figure 8.18 is

$$\mu(x, m, s) = \min\left(1, \max\left(0, 1 + \frac{x - m}{s}\right)\right).$$

Finally, the expression for the function shown in Figure 8.19 is

$$\mu(x, m, s) = \min\left(1, \max\left(0, 1 - \frac{x - m}{s}\right)\right).$$

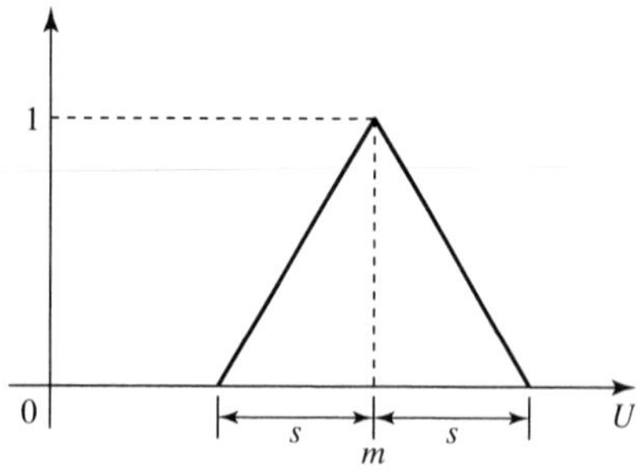

Figure 8.17 A triangular shaped fuzzy set.

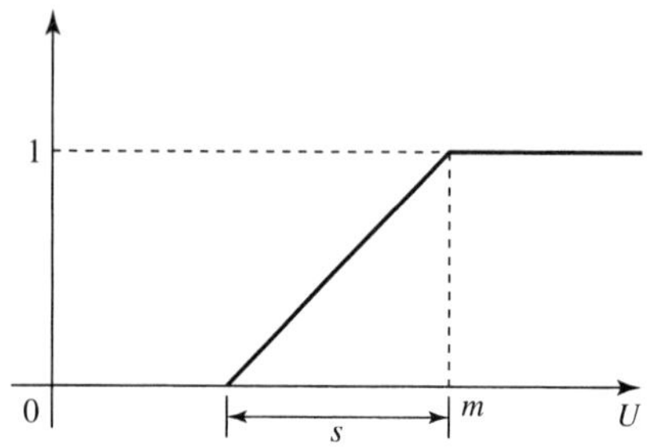

Figure 8.18 A fuzzy set at the right end of a universe of discourse.

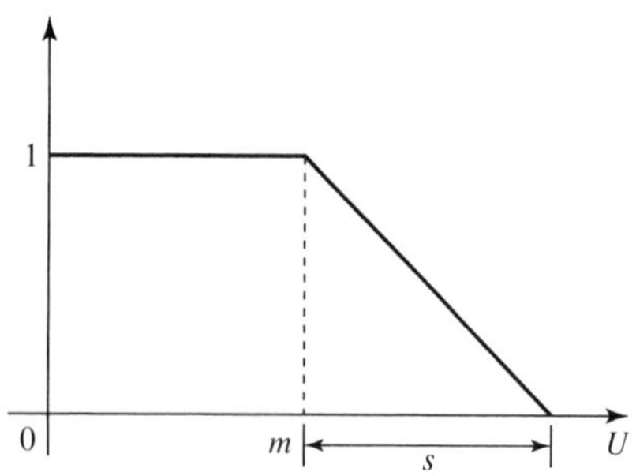

Figure 8.19 A fuzzy set at the left end of a universe of discourse.

◆ Example 8.8

Consider the vehicle model given in Subsection 2.4.2. Let the input signal be $u = \delta_f$ (the front-wheel steer angle) and $\delta_r = 0$. We will design a simple fuzzy logic controller for the above vehicle model. The range for the error is $[-0.5, 0.5]$. The range for the change-in-error is $[-0.06, 0.06]$. The range for the controller output is $[-3^\circ, 3^\circ]$. The rules matrix is as in Table 8.2. The parameters of the membership functions are given in Table 8.3.

Plots of the error, the change-in-error, and controller's output membership functions are given in Figures 8.20, 8.21, and 8.22, respectively. We created a nonlinear mapping from the controller's input space into its output space and formed a look-up table that implements the obtained nonlinear mapping representing fuzzy logic controller. We used SIMULINK's 2-D Look-Up Table block to implement the

Table 8.3 The *m* and *s* Parameters of the Membership Functions

ERROR		CHANGE-IN-ERROR		OUTPUT	
m	*s*	*m*	*s*	*m*	*s*
$m_1 = -0.4$	$s_1 = 0.1$	$m_1 = -0.05$	$s_1 = 0.025$	$m_1 = -2.5$	$s_1 = 1$
$m_2 = -0.2$	$s_2 = 0.2$	$m_2 = -0.025$	$s_2 = 0.025$	$m_2 = -1$	$s_2 = 1$
$m_3 = 0$	$s_3 = 0.2$	$m_3 = 0$	$s_3 = 0.025$	$m_3 = 0$	$s_3 = 1$
$m_4 = 0.2$	$s_4 = 0.2$	$m_4 = 0.025$	$s_4 = 0.025$	$m_4 = 1$	$s_4 = 1$
$m_5 = 0.4$	$s_5 = 0.1$	$m_5 = 0.05$	$s_5 = 0.025$	$m_5 = 2.5$	$s_5 = 1$

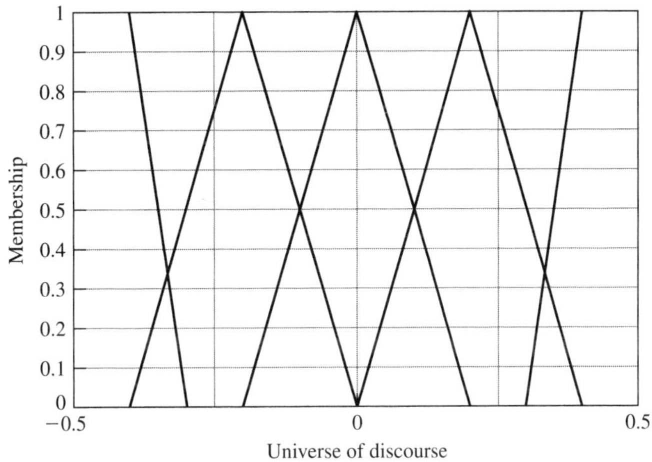

Figure 8.20 Fuzzy numbers of the linguistic variable ERROR in Example 8.8.

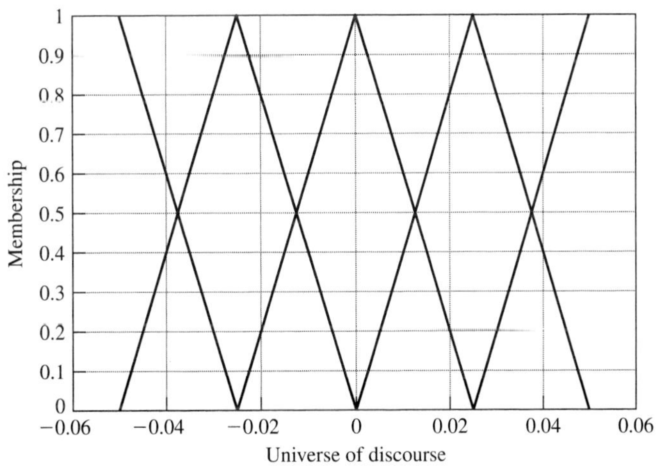

Figure 8.21 Fuzzy numbers of the linguistic variable CHANGE-IN-ERROR in Example 8.8.

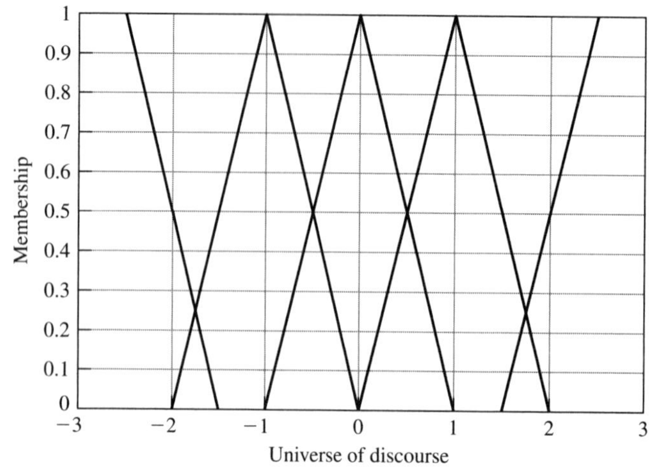

Figure 8.22 Fuzzy numbers of the linguistic variable CONTROL in Example 8.8.

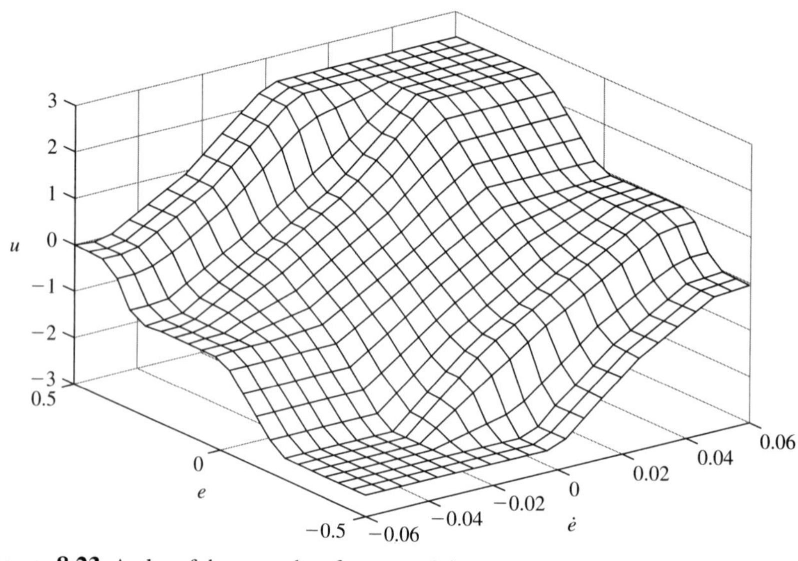

Figure 8.23 A plot of the control surface $u = f(\dot{e}, e)$ in Example 8.8.

controller. The look-up table size was 22×22. The output of the look-up table was scaled by the factor $-2\pi/180$ before using it as the vehicle's steering input. The negative sign is to take into account the opposite directions of the lateral axes of the vehicle coordinate system and the fixed system. The factor $\pi/180$ converts degrees into radians. We multiplied the output of the look-up table by 2 to increase the control authority. This allows us to use the same controller to track step inputs with large amplitudes. A plot of the control surface is given in Figure 8.23. Next, a reference signal generator was constructed. We selected the following transfer function of the signal generator:

$$\frac{R(s)}{H(s)} = \frac{1}{s^3 + 1.75s^2 + 2.15s + 1},$$

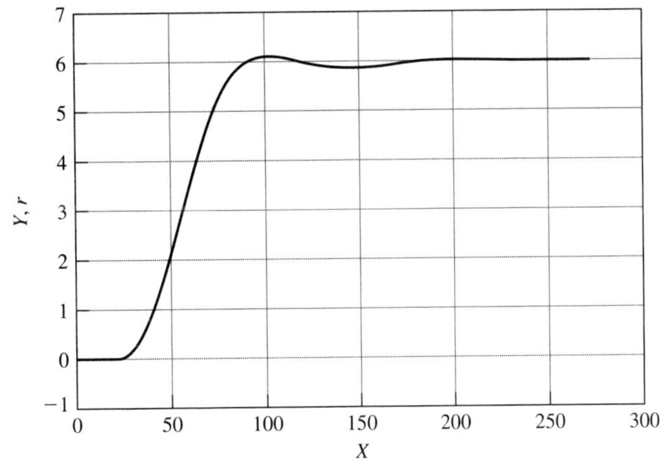

Figure 8.24 Plots of Y and r versus X in Example 8.8.

where $R(s) = \mathcal{L}(r(t))$, $H(s) = \mathcal{L}(6h(t))$, and $h(t)$ is the unit step function. The closed-loop system was simulated using SIMULINK. Plots of Y and r versus X are shown in Figure 8.24.

◆ Example 8.9

This example is based on Example 6.1 found in Driankov, Hellendoorn, and Reinfrank [67, pp. 266–272]. We consider a nonlinear dynamical system modeled by the differential equation

$$\ddot{x} + a_1(1 - x^2)\dot{x} + a_2 x = bu, \tag{8.5}$$

where $|u| \le 15$, $a_1 = 2.2165$, and $a_2 = b = 12.7388$. Let $x_1 = x$ and $x_2 = \dot{x}_1$. Then, the above system model in state-space format has the form

$$\begin{bmatrix} \dot{x}_1 \\ \dot{x}_2 \end{bmatrix} = \begin{bmatrix} x_1 \\ -a_2 x_1 - a_1\left(1 - x_1^2\right)x_2 \end{bmatrix} + \begin{bmatrix} 0 \\ b \end{bmatrix} u.$$

A phase-plane portrait of the uncontrolled system is shown in Figure 8.25. Plots of x_1 and x_2 versus time for the initial conditions $x_1(0) = -1$ and $x_2(0) = 0$ are shown in Figure 8.26. The relevant membership functions are shown in Figure 8.27, 8.28, 8.29, and 8.30. The following MATLAB function can be used to generate all the relevant membership functions.

```
function y = trap_a(x,a1,a2,c,s)
%TRAP_A is an asymmetric trapezoid type membership
%  function,
% also called a ''pi'' function,
% where x is the function domain, a1 is the length of the
```

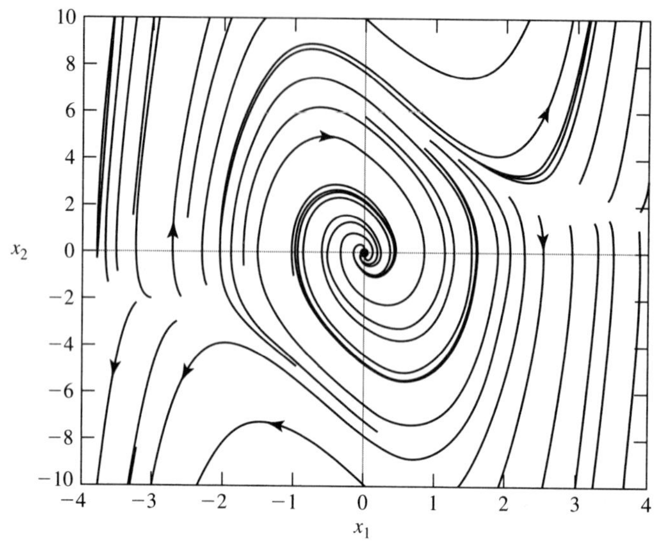

Figure 8.25 A phase-plane portrait of the uncontrolled system in Example 8.9.

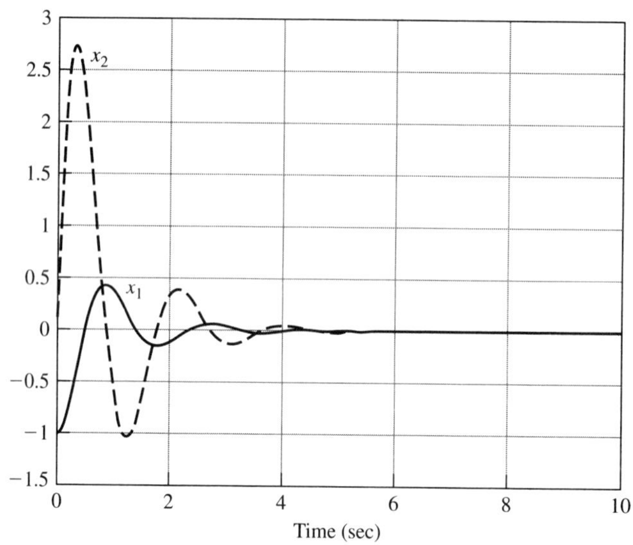

Figure 8.26 Plots of x_1 and x_2 versus time of the uncontrolled system of Example 8.9.

```
% left trapezoid's side
% at the base, a2 is the length of the right side at the
%  base,
% c is the trapezoid's center, and s is a half of the
% trapezoid upper side. Thus the length of the trapezoid
%  upper
```

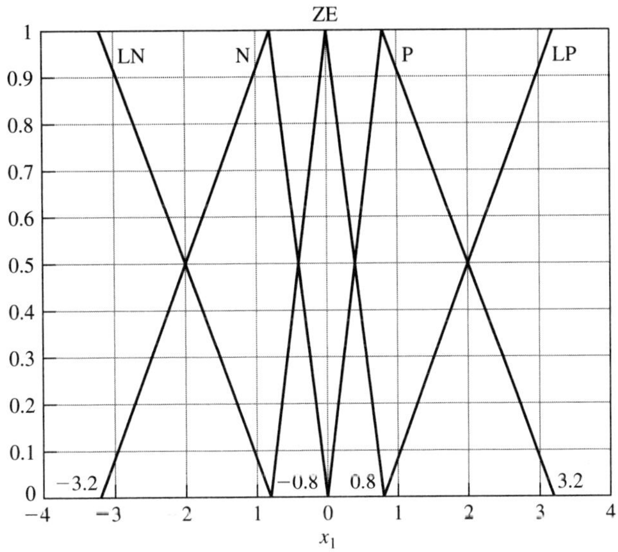

Figure 8.27 Membership functions of x_1 of Example 8.9.

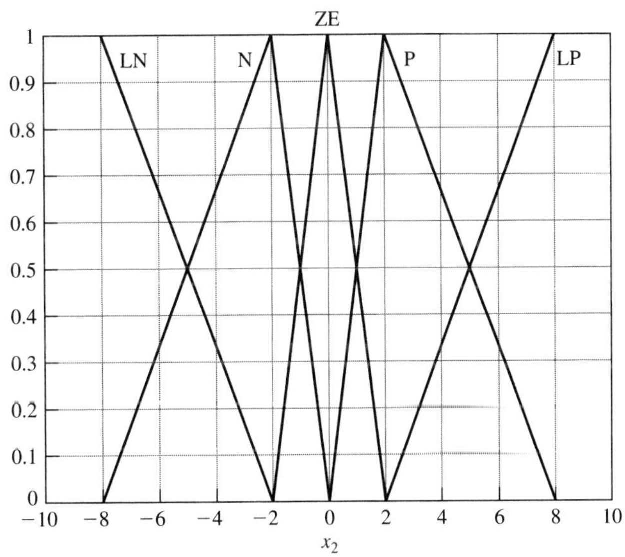

Figure 8.28 Membership functions of x_2 of Example 8.9.

```
% side is 2*s. The length of the trapezoid base is then
%  2*s+a1+a2.
% Set s=0 to get a triangular membership function.

y1=min(0.5,(max((-0.5),(0.5 +((x-c+s)./a1)))));
y2=min(0.5,(max((-0.5),(0.5 -((x-c-s)./a2)))));
y=y1+y2;
```

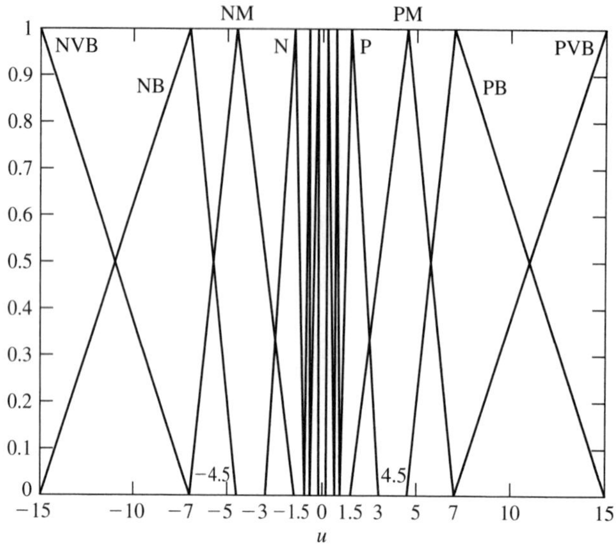

Figure 8.29 Membership functions of u of Example 8.9.

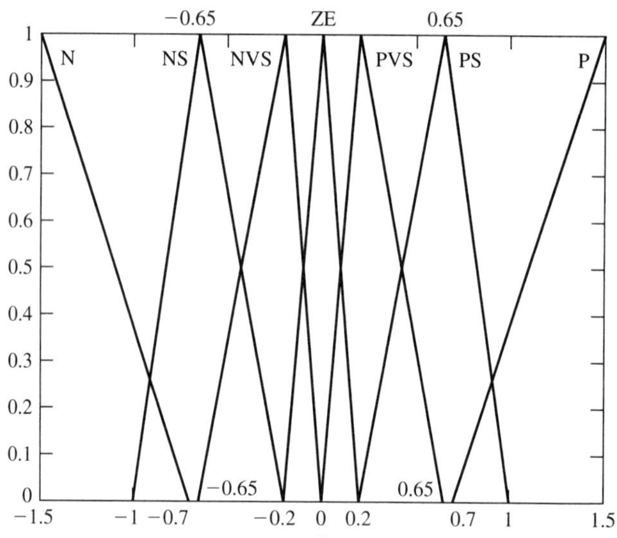

Figure 8.30 Magnification of the membership functions of u on the interval $[-1.5, 1.5]$ of Example 8.9.

We used the rules matrix given in Table 8.4 to produce a nonlinear mapping $u = \Phi(x_1, x_2)$ and to construct a look-up table that implements the obtained mapping representing a fuzzy logic controller. The 2-D Look-Up Table block to implement the controller was used. The look-up table size was 22×22. A plot

Table 8.4 The Rules Matrix of Example 8.9

x_2 \ x_1	LN	N	ZE	P	LP
LN	PVB	PM	PM	PM	PVB
N	PB	P	PS	PS	PM
ZE	P	PVS	ZE	NVS	N
P	NM	NS	NS	N	NB
LP	NVB	NM	NM	NM	NVB

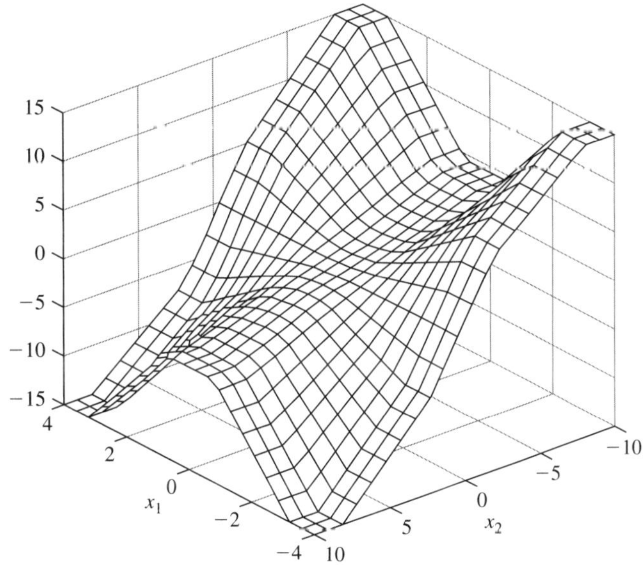

Figure 8.31 A plot of $u = \Phi(x_1, x_2)$ of Example 8.9.

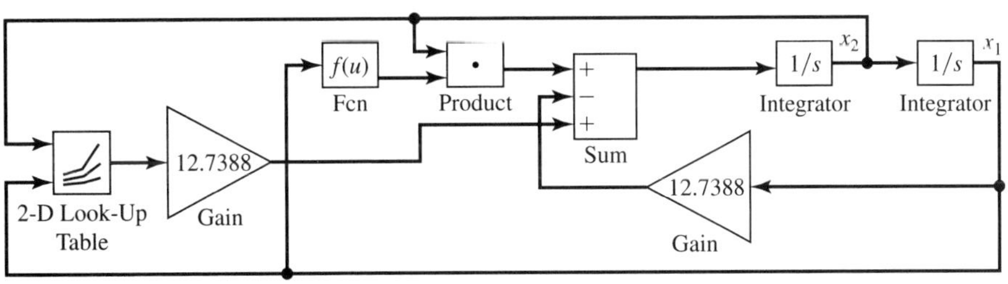

Figure 8.32 A SIMULINK diagram of the closed-loop system of Example 8.9.

of $u = \Phi(x_1, x_2)$ is shown in Figure 8.31. A SIMULINK diagram of the closed-loop system is given in Figure 8.32. A phase portrait of the closed-loop system is depicted in Figure 8.33. Plots of x_1 and x_2 versus time for $x_1(0) = -1$ and $x_2(0) = 0$ for the closed-loop system are shown in Figure 8.34.

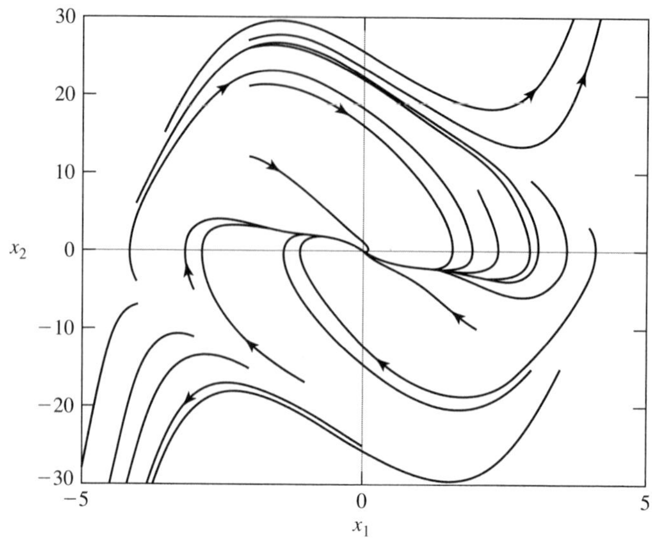

Figure 8.33 A phase-plane portrait of the closed-loop system of Example 8.9.

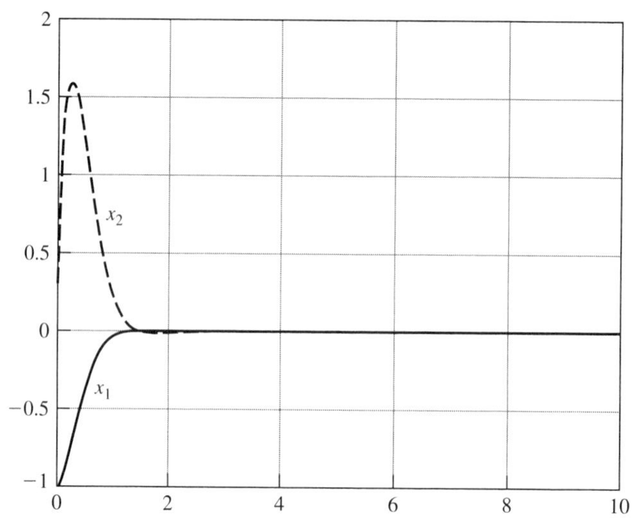

Figure 8.34 Plots of x_1 and x_2 versus time for $x_1(0) = -1$ and $x_2(0) = 0$ for the closed-loop system in Example 8.9.

8.5 Stabilization Using Fuzzy Models

In this section a systematic Lyapunov-based stabilizing control design method for uncertain nonlinear dynamical systems using fuzzy models is presented. The controller is constructed using a design model of the dynamical process to be controlled. Recall that the dynamical process to be controlled is also called the plant. We obtain the design model from the truth model using a fuzzy modeling approach.

8.5.1 Fuzzy Modeling

In many situations there may be human experts that can provide a linguistic description, in terms of IF–THEN rules, of the plant dynamical behavior. Combining available mathematical description of the plant with its linguistic description results in a fuzzy system model. Such an approach to modeling of continuous-time dynamical processes was proposed by Takagi and Sugeno [275] and further developed by Sugeno and Kang [271]. This type of model is called the Takagi–Sugeno (T–S) or Takagi–Sugeno–Kang (TSK) fuzzy model. We mention here that the discrete-time version of the TSK model was analyzed by Tanaka and Sugeno [277], and Tanaka [276], while Wang, Tanaka, and Griffin [293], Cao, Rees, and Feng [42], and Feng et al. [82] used TSK fuzzy models to design stabilizing controllers for nonlinear systems. One can construct a TSK model if local description of the plant is available in terms of local linear models,

$$\dot{\boldsymbol{x}}(t) = \boldsymbol{A}_i\boldsymbol{x}(t) + \boldsymbol{B}_i\boldsymbol{u}(t), \qquad i = 1, 2, \ldots, r,$$

where the state vector $\boldsymbol{x}(t) \in \mathbb{R}^n$, the control input $\boldsymbol{u}(t) \in \mathbb{R}^m$, and the matrices $\boldsymbol{A}_i$ and $\boldsymbol{B}_i$ are of appropriate dimensions. The above information is then fused with the available IF–THEN rules, where the ith rule can have the following form:

Rule i: IF $x_1(t)$ is F_1^i AND $\cdots$ AND $x_n(t)$ is F_n^i,
THEN $\dot{\boldsymbol{x}}(t) = \boldsymbol{A}_i\boldsymbol{x}(t) + \boldsymbol{B}_i\boldsymbol{u}(t)$,

where F_j^i, $j = 1, 2, \ldots, n$, is the jth fuzzy set of the ith rule. Let $\mu_j^i(x_j)$ be the membership function of the fuzzy set F_j^i, and let

$$w^i = w^i(\boldsymbol{x}) = \prod_{j=1}^{n} \mu_j^i(x_j).$$

Then, the resulting fuzzy system model is constructed as the weighted average of the local models and has the form

$$\begin{aligned}
\dot{\boldsymbol{x}} &= \frac{\sum_{i=1}^{r} w^i(\boldsymbol{A}_i\boldsymbol{x} + \boldsymbol{B}_i\boldsymbol{u})}{\sum_{i=1}^{r} w^i} \\
&= \sum_{i=1}^{r} \alpha_i(\boldsymbol{A}_i\boldsymbol{x} + \boldsymbol{B}_i\boldsymbol{u}) \\
&= \left(\sum_{i=1}^{r} \alpha_i\boldsymbol{A}_i\right)\boldsymbol{x} + \left(\sum_{i=1}^{r} \alpha_i\boldsymbol{B}_i\right)\boldsymbol{u} \\
&= \boldsymbol{A}(\boldsymbol{\alpha})\boldsymbol{x} + \boldsymbol{B}(\boldsymbol{\alpha})\boldsymbol{u},
\end{aligned} \tag{8.6}$$

where for $i = 1, 2, \ldots, r$,

$$\alpha_i = \frac{w^i}{\sum_{i=1}^{r} w^i}. \tag{8.7}$$

Note that for $i = 1, 2, \ldots, r$, we have

$$\alpha_i \geq 0, \quad \sum_{i=1}^{r} \alpha_i = 1, \quad \text{and} \quad \boldsymbol{\alpha} = [\alpha_1 \quad \alpha_2 \quad \cdots \quad \alpha_r]^T \in [0, 1]^r. \tag{8.8}$$

A system model given by (8.6) and (8.8) is also referred to as in the literature as the *polytopic system*—see, for example, Takaba [274].

8.5.2 Constructing a Fuzzy Design Model Using a Nonlinear Model

In the stabilization problem of a nonlinear plant, we are concerned with constructing a controller so that starting from an arbitrary point, in some neighborhood of the operating point, the controller forces the closed-loop system trajectory to converge to this operating point. On the other hand, if the starting point coincides with the operating point, the closed-loop system trajectory is expected to stay at this point for all subsequent time. When designing a controller, we use a TSK fuzzy design model obtained by combining a mathematical description of the plant in terms of its truth model and a linguistic description of the plant. The essential components of the TSK model are linear local models, which describe the plant dynamical behavior at its different operating points. We now show that using a common linearization approach to construct local models may result in a fuzzy design model different from what the designer had in his or her mind. For example, suppose that the truth model of a plant has the form

$$\dot{\boldsymbol{x}} = \boldsymbol{f}(\boldsymbol{x}) + \boldsymbol{G}(\boldsymbol{x})\boldsymbol{u}. \tag{8.9}$$

For simplicity of notation, let

$$\boldsymbol{F}(\boldsymbol{x}, \boldsymbol{u}) = \boldsymbol{f}(\boldsymbol{x}) + \boldsymbol{G}(\boldsymbol{x})\boldsymbol{u}.$$

Then, we can represent (8.9) as

$$\dot{\boldsymbol{x}} = \boldsymbol{F}(\boldsymbol{x}, \boldsymbol{u}). \tag{8.10}$$

Expanding $\boldsymbol{F}$ into a Taylor series around $(\boldsymbol{x}_0, \boldsymbol{u}_0)$ yields

$$\dot{\boldsymbol{x}} = \boldsymbol{F}(\boldsymbol{x}_0, \boldsymbol{u}_0) + \left.\frac{\partial \boldsymbol{F}}{\partial \boldsymbol{x}}\right|_{\substack{\boldsymbol{x}=\boldsymbol{x}_0\\ \boldsymbol{u}=\boldsymbol{u}_0}} (\boldsymbol{x} - \boldsymbol{x}_0) + \left.\frac{\partial \boldsymbol{F}}{\partial \boldsymbol{u}}\right|_{\substack{\boldsymbol{x}=\boldsymbol{x}_0\\ \boldsymbol{u}=\boldsymbol{u}_0}} (\boldsymbol{u} - \boldsymbol{u}_0) + \text{higher-order terms}, \tag{8.11}$$

where

$$\boldsymbol{F}(\boldsymbol{x}, \boldsymbol{u}) = \boldsymbol{f}(\boldsymbol{x}) + \boldsymbol{G}(\boldsymbol{x})\boldsymbol{u} = \boldsymbol{f}(\boldsymbol{x}) + \sum_{k=1}^{m} u_k \boldsymbol{g}_k(\boldsymbol{x}), \tag{8.12}$$

where $\boldsymbol{g}_k$ is the kth column of $\boldsymbol{G}$. To write expressions for the second and the third terms of (8.11) as functions of $\boldsymbol{f}$ and $\boldsymbol{G}$, let g_{ij} be the (i, j)th element of the matrix $\boldsymbol{G}$. Then,

$$\left.\frac{\partial \boldsymbol{F}}{\partial \boldsymbol{x}}\right|_{\substack{\boldsymbol{x}=\boldsymbol{x}_0\\ \boldsymbol{u}=\boldsymbol{u}_0}} = \left.\frac{\partial \boldsymbol{f}}{\partial \boldsymbol{x}}\right|_{\boldsymbol{x}=\boldsymbol{x}_0} + \sum_{k=1}^{m} u_k \left.\frac{\partial \boldsymbol{g}_k}{\partial \boldsymbol{x}}\right|_{\substack{\boldsymbol{x}=\boldsymbol{x}_0\\ \boldsymbol{u}=\boldsymbol{u}_0}} = \left.\frac{\partial \boldsymbol{f}}{\partial \boldsymbol{x}}\right|_{\boldsymbol{x}=\boldsymbol{x}_0} + \boldsymbol{H}(\boldsymbol{x}_0, \boldsymbol{u}_0), \tag{8.13}$$

where the (i, j)th element of the $n \times n$ matrix $\boldsymbol{H}$ has the form

$$\sum_{k=1}^{m} u_k \left.\frac{\partial g_{ik}(\boldsymbol{x})}{\partial x_j}\right|_{\substack{\boldsymbol{x}=\boldsymbol{x}_0\\ \boldsymbol{u}=\boldsymbol{u}_0}}.$$

Finally,

$$\left.\frac{\partial \boldsymbol{F}}{\partial \boldsymbol{u}}\right|_{\substack{\boldsymbol{x}=\boldsymbol{x}_0\\ \boldsymbol{u}=\boldsymbol{u}_0}} = \boldsymbol{G}(\boldsymbol{x}_0). \tag{8.14}$$

A point $(\boldsymbol{x}_0^T, \boldsymbol{u}_0^T)^T \in \mathbb{R}^{n+m}$ is an equilibrium point of (8.10) if $\boldsymbol{F}(\boldsymbol{x}_0, \boldsymbol{u}_0) = \mathbf{0}$—that is, if at $(\boldsymbol{x}_0, \boldsymbol{u}_0)$ we have $\dot{\boldsymbol{x}} = \mathbf{0}$. Let $\Delta\boldsymbol{x} = \boldsymbol{x} - \boldsymbol{x}_0$ and $\Delta\boldsymbol{u} = \boldsymbol{u} - \boldsymbol{u}_0$. Note that

$$\frac{d\boldsymbol{x}_0}{dt} = \mathbf{0}.$$

Then, the linearized model about the equilibrium $(\boldsymbol{x}_0, \boldsymbol{u}_0)$ is obtained by neglecting higher-order terms and observing that at the equilibrium point $\boldsymbol{F}(\boldsymbol{x}_0, \boldsymbol{u}_0) = \mathbf{0}$. The linearized model has the form

$$\frac{d}{dt}\Delta\boldsymbol{x} = \boldsymbol{A}\Delta\boldsymbol{x} + \boldsymbol{B}\Delta\boldsymbol{u}, \tag{8.15}$$

where

$$\boldsymbol{A} = \left.\frac{\partial \boldsymbol{F}}{\partial \boldsymbol{x}}\right|_{\substack{x=x_0\\u=u_0}} \quad \text{and} \quad \boldsymbol{B} = \left.\frac{\partial \boldsymbol{F}}{\partial \boldsymbol{u}}\right|_{\substack{x=x_0\\u=u_0}}.$$

Suppose now that we wish to construct a stabilizing controller for the truth model $\dot{\boldsymbol{x}} = \boldsymbol{f}(\boldsymbol{x}) + \boldsymbol{G}(\boldsymbol{x})\boldsymbol{u}$, where $\boldsymbol{f}(\mathbf{0}) = \mathbf{0}$, so that the pair $(\boldsymbol{x} = \mathbf{0}, \boldsymbol{u} = \mathbf{0})$ constitutes an asymptotically stable equilibrium point of the closed-loop system. Our first step is to construct a fuzzy design model. We start with generating linear local models describing plant's behavior at selected operating points. It is natural to first construct a local linear model that describes plant's behavior about the equilibrium state $\boldsymbol{x} = \mathbf{0}$. We can obtain this first local linear model using the above-described linearization technique. The resulting model is given by

$$\dot{\boldsymbol{x}} = \boldsymbol{A}_1\boldsymbol{x} + \boldsymbol{B}_1\boldsymbol{u}.$$

Next, we construct local linear models describing plant's behavior at the remaining operating points. Suppose that $\boldsymbol{x} = \boldsymbol{x}_j$ is the next operating state of interest. The result of Taylor's linearization of a nonlinear model about an operating nonequilibrium point is an affine, rather than linear, model. Even if the operating point is an equilibrium point, Taylor series linearization, in general, will not yield a local model that is linear in $\boldsymbol{x}$ and $\boldsymbol{u}$. Indeed, suppose that the operating point $(\boldsymbol{x}_j, \boldsymbol{u}_j)$ is an equilibrium point, that is,

$$\boldsymbol{f}(\boldsymbol{x}_j) + \boldsymbol{G}(\boldsymbol{x}_j)\boldsymbol{u}_j = \mathbf{0}. \tag{8.16}$$

By (8.11), the resulting linearized model is

$$\frac{d}{dt}(\boldsymbol{x} - \boldsymbol{x}_j) = \boldsymbol{f}(\boldsymbol{x}_j) + \boldsymbol{G}(\boldsymbol{x}_j)\boldsymbol{u}_j + \boldsymbol{A}_j(\boldsymbol{x} - \boldsymbol{x}_j) + \boldsymbol{B}_j(\boldsymbol{u} - \boldsymbol{u}_j) \tag{8.17}$$

$$= \boldsymbol{A}_j(\boldsymbol{x} - \boldsymbol{x}_j) + \boldsymbol{B}_j(\boldsymbol{u} - \boldsymbol{u}_j). \tag{8.18}$$

We can represent model (8.18) in the form

$$\dot{\boldsymbol{x}} = \boldsymbol{A}_j\boldsymbol{x} + \boldsymbol{B}_j\boldsymbol{u} - (\boldsymbol{A}_j\boldsymbol{x}_j + \boldsymbol{B}_j\boldsymbol{u}_j). \tag{8.19}$$

The term $(\boldsymbol{A}_j\boldsymbol{x}_j + \boldsymbol{B}_j\boldsymbol{u}_j)$ does not have to be equal to zero, and hence (8.19) is not a linear model in $\boldsymbol{x}$ and $\boldsymbol{u}$, but rather an affine model. We illustrate, on a numerical example, that the term $(\boldsymbol{A}_j\boldsymbol{x}_j + \boldsymbol{B}_j\boldsymbol{u}_j)$ is indeed, in general, not equal to zero.

◆ Example 8.10

Consider a subsystem of model (1.78) of the form

$$\begin{bmatrix}\dot{x}_1\\ \dot{x}_2\end{bmatrix}=\begin{bmatrix}x_2\\ \dfrac{g\sin(x_1)-mlax_2^2\sin(2x_1)/2}{4l/3-mla\cos^2(x_1)}\end{bmatrix}+\begin{bmatrix}0\\ -\dfrac{a\cos(x_1)}{4l/3-mla\cos^2(x_1)}\end{bmatrix}u, \tag{8.20}$$

where $g = 9.8$ m/sec^2, $m = 2$ kg, $M = 8$ kg, $a = 1/(m + M)$, and $l = 0.5$ m. We linearize (8.20) about $x_1 = \pi/4$. For $x_1 = \pi/4$ to be a component of an equilibrium state, we have to have $x_2 = 0$. Hence, the equilibrium state is

$$\boldsymbol{x}_e = [\pi/4 \quad 0]^T.$$

We next compute $u = u_e$ so that equation (8.16) holds. Performing simple calculations, we get $u_e = 98$. Then, using (8.13) we obtain

$$\boldsymbol{A} = \begin{bmatrix}0 & 1\\ 22.4745 & 0\end{bmatrix}. \tag{8.21}$$

Next, using (8.14) gives

$$\boldsymbol{B} = \begin{bmatrix}0\\ -0.1147\end{bmatrix}. \tag{8.22}$$

Note that

$$\boldsymbol{A}\boldsymbol{x}_e + \boldsymbol{B}u_e = \begin{bmatrix}0\\ 6.4108\end{bmatrix} \neq \boldsymbol{0}.$$

The above example illustrates the fact that the system being linear in $\Delta\boldsymbol{x} = \boldsymbol{x} - \boldsymbol{x}_0$ and $\Delta\boldsymbol{u} = \boldsymbol{u} - \boldsymbol{u}_0$ is, in general, only affine and not linear in $\boldsymbol{x}$ and $\boldsymbol{u}$ because of the presence of the offset term $-(\boldsymbol{A}\boldsymbol{x}_0 + \boldsymbol{B}\boldsymbol{u}_0)$. The designer must then be careful when constructing local models using Taylor series linearization because this approach, in general, will not yield a local model that is linear in $\boldsymbol{x}$ and $\boldsymbol{u}$. We note that Taylor's linearization will always yield a linear system in $\boldsymbol{x}$ and $\boldsymbol{u}$ if the equilibrium point is $(\boldsymbol{x}_0^T, \boldsymbol{u}_0^T) = (\boldsymbol{0}^T, \boldsymbol{0}^T)$.

We now describe our method of constructing local models using the truth model of the given plant. We assume that the truth model has the form of (8.9). Suppose that we are given an operating state $\boldsymbol{x}_0$ that does not have to be an equilibrium state of (8.9). Our goal is to construct a linear model in $\boldsymbol{x}$ and $\boldsymbol{u}$ that approximates the behavior of (8.9) in the vicinity of the operating state $\boldsymbol{x}_0$; that is, we wish to find constant matrices $\boldsymbol{A}$ and $\boldsymbol{B}$ such that in a neighborhood of $\boldsymbol{x}_0$,

$$\boldsymbol{f}(\boldsymbol{x}) + \boldsymbol{G}(\boldsymbol{x})\boldsymbol{u} \approx \boldsymbol{A}\boldsymbol{x} + \boldsymbol{B}\boldsymbol{u} \tag{8.23}$$

and

$$\boldsymbol{f}(\boldsymbol{x}_0) + \boldsymbol{G}(\boldsymbol{x}_0)\boldsymbol{u} = \boldsymbol{A}\boldsymbol{x}_0 + \boldsymbol{B}\boldsymbol{u} \qquad \text{for all } \boldsymbol{u}. \tag{8.24}$$

Because $\boldsymbol{u}$ is arbitrary, we have to have

$$\boldsymbol{G}(\boldsymbol{x}_0) = \boldsymbol{B}. \tag{8.25}$$

Thus, we are left with finding a constant matrix $\boldsymbol{A}$ such that in a neighborhood of $\boldsymbol{x}_0$,

$$\boldsymbol{f}(\boldsymbol{x}) \approx \boldsymbol{A}\boldsymbol{x} \tag{8.26}$$

and

$$\boldsymbol{f}(\boldsymbol{x}_0) = \boldsymbol{A}\boldsymbol{x}_0. \tag{8.27}$$

Let $\boldsymbol{a}_i^T$ denote the ith row of the matrix $\boldsymbol{A}$. Then, for the purpose of further analysis, we represent condition (8.26) as

$$f_i(\boldsymbol{x}) \approx \boldsymbol{a}_i^T \boldsymbol{x}, \qquad i = 1, 2, \ldots, n, \tag{8.28}$$

and we represent (8.27) as

$$f_i(\boldsymbol{x}_0) = \boldsymbol{a}_i^T \boldsymbol{x}_0, \qquad i = 1, 2, \ldots, n, \tag{8.29}$$

where the ith component of $\boldsymbol{f}$ is $f_i : \mathbb{R}^n \to \mathbb{R}$. Expanding the left-hand side of (8.28) about $\boldsymbol{x}_0$ and neglecting second- and higher-order terms yields

$$f_i(\boldsymbol{x}_0) + \nabla^T f_i(\boldsymbol{x}_0)(\boldsymbol{x} - \boldsymbol{x}_0) \approx \boldsymbol{a}_i^T \boldsymbol{x}, \tag{8.30}$$

where $\nabla f_i(\boldsymbol{x}) : \mathbb{R}^n \to \mathbb{R}^n$ is the gradient (a column vector) of f_i at $\boldsymbol{x}$. Now, using (8.29), we represent (8.30) as

$$\nabla^T f_i(\boldsymbol{x}_0)(\boldsymbol{x} - \boldsymbol{x}_0) \approx \boldsymbol{a}_i^T(\boldsymbol{x} - \boldsymbol{x}_0), \tag{8.31}$$

where $\boldsymbol{x}$ is arbitrary but "close" to $\boldsymbol{x}_0$. Our task now is to determine a constant vector $\boldsymbol{a}_i$ that is as "close as possible" to $\nabla f_i(\boldsymbol{x}_0)$ and satisfies the constraint $\boldsymbol{a}_i^T \boldsymbol{x}_0 = f_i(\boldsymbol{x}_0)$. Let

$$E = \tfrac{1}{2} \|\nabla f_i(\boldsymbol{x}_0) - \boldsymbol{a}_i\|_2^2 .$$

Then, we can formulate our task as a constrained optimization problem of the form

$$\begin{aligned} &\underset{\boldsymbol{a}_i}{\text{minimize}}\ E \\ &\text{subject to } \boldsymbol{a}_i^T \boldsymbol{x}_0 = f_i(\boldsymbol{x}_0). \end{aligned} \tag{8.32}$$

We note that (8.32) is a convex constrained optimization problem. This means that the first-order necessary condition for a minimum of E is also sufficient; see, for example, Luenberger [190, Section 6.5] or Chong and Żak [50, Chapter 21]. The first-order conditions for the optimization problem given by (8.32) are

$$\nabla_{\boldsymbol{a}_i} E + \lambda \nabla_{\boldsymbol{a}_i} \left(\boldsymbol{a}_i^T \boldsymbol{x}_0 - f_i(\boldsymbol{x}_0) \right) = \boldsymbol{0}, \tag{8.33}$$

$$\boldsymbol{a}_i^T \boldsymbol{x}_0 = f_i(\boldsymbol{x}_0), \tag{8.34}$$

where λ is the Lagrange multiplier and the subscript $\boldsymbol{a}_i$ in $\nabla_{\boldsymbol{a}_i}$ indicates that the gradient, ∇ is computed with respect to $\boldsymbol{a}_i$. Performing the required differentiation in (8.33) yields

$$\boldsymbol{a}_i - \nabla f_i(\boldsymbol{x}_0) + \lambda \boldsymbol{x}_0 = \boldsymbol{0}, \tag{8.35}$$

$$\boldsymbol{a}_i^T \boldsymbol{x}_0 = f_i(\boldsymbol{x}_0). \tag{8.36}$$

Recall that we consider now the case when $\boldsymbol{x}_0 \neq \mathbf{0}$. Premultiplying (8.35) by $\boldsymbol{x}_0^T$ and substituting (8.36) into the resulting equation yields

$$\lambda = \frac{\boldsymbol{x}_0^T \nabla f_i(\boldsymbol{x}_0) - f_i(\boldsymbol{x}_0)}{\|\boldsymbol{x}_0\|^2}. \tag{8.37}$$

Substituting λ given by (8.37) into (8.35), we obtain

$$\boxed{\boldsymbol{a}_i = \nabla f_i(\boldsymbol{x}_0) + \frac{f_i(\boldsymbol{x}_0) - \boldsymbol{x}_0^T \nabla f_i(\boldsymbol{x}_0)}{\|\boldsymbol{x}_0\|^2}\boldsymbol{x}_0, \qquad \boldsymbol{x}_0 \neq \mathbf{0}} \tag{8.38}$$

◆ Example 8.11

We illustrate the above procedure for constructing local linear models on the nonlinear model (8.20) that we analyzed in Example 8.10. This model was also considered by Wang et al. [293, p. 20]. Suppose that, as in Wang et al. [293], we wish to construct two local models; one describing the plant operation about $x_1 = 0$ and the other local model when x_1 is about $\pm\pi/2$. In both models, we assume that $x_2 = 0$ and u is arbitrary. Note that the nonlinear model (8.20) is in the format given by (8.9). We first construct a local model about the operating state $x_1 = x_2 = 0$. The local model for this case can be obtained using Taylor's linearization about the equilibrium point $(\boldsymbol{x}_0, \boldsymbol{u}_0) = (\mathbf{0}, \mathbf{0})$.

We next construct the second local model that corresponds to $x_1 = 88°\pi/180°$ and $x_2 = 0$. Let $\beta = \cos(88°)$. Then, the input matrix $\boldsymbol{B}$ can be obtained from (8.20) using (8.25):

$$\boldsymbol{B} = \begin{bmatrix} 0 \\ -\dfrac{a\beta}{4l/3 - mla\beta^2} \end{bmatrix}.$$

We then compute the matrix $\boldsymbol{A}$ of the second local model using (8.38). We first compute the first row $\boldsymbol{A}$. We note that $f_1(\boldsymbol{x}) = x_2$, and hence

$$\nabla f_1 = [0 \quad 1]^T.$$

The operating state is $\boldsymbol{x}_0 = [88°\pi/180°\ 0]^T$. We compute $\boldsymbol{a}_1^T$, the first row of $\boldsymbol{A}$, using (8.38) to get

$$\boldsymbol{a}_1^T = \nabla^T f_1(\boldsymbol{x}_0) + [0 \quad 0] = [0 \quad 1].$$

We again use (8.38) to compute $\boldsymbol{a}_2^T$, the second row of the matrix $\boldsymbol{A}$, of the second local model:

$$\boldsymbol{a}_2^T = [\, f_2(\boldsymbol{x}_0)/(88°\pi/180°) \quad 0] \approx \begin{bmatrix} \dfrac{g}{4l/3 - mla\beta^2}\left(\dfrac{2}{\pi}\right) & 0 \end{bmatrix}.$$

The above-obtained local models are the same as the ones that were proposed by Wang et al. [293] using a heuristic approach.

8.5.3 Stabilizability of Fuzzy Models

In this subsection we first present sufficient conditions for asymptotic stability in the large of the unforced continuous-time fuzzy model

$$\dot{\boldsymbol{x}} = \sum_{i=1}^{r} \alpha_i \boldsymbol{A}_i \boldsymbol{x}, \tag{8.39}$$

where $\alpha_i = \alpha_i(\boldsymbol{x}(t)) \geq 0$ for $i = 1, 2, \ldots, r$, and $\sum_{i=1}^{r} \alpha_i = 1$. The above model can be viewed as an example of unforced continuous-time polytopic model. We next discuss the stabilizability problem of forced fuzzy models using fuzzy state-feedback control strategies.

It is straightforward to show that a sufficient condition for asymptotic stability in the large of the equilibrium state $\boldsymbol{x} = \boldsymbol{0}$ of system (8.39) is that there exists a symmetric positive definite matrix $\boldsymbol{P}$ such that for $i = 1, 2, \ldots, r$,

$$\boldsymbol{A}_i^T \boldsymbol{P} + \boldsymbol{P}\boldsymbol{A}_i < 0. \tag{8.40}$$

It is obvious that a necessary condition for the existence of a common symmetric positive definite $\boldsymbol{P}$ satisfying (8.40) is that each $\boldsymbol{A}_i$ be asymptotically stable—that is, the eigenvalues of each $\boldsymbol{A}_i$ be in the open left-hand complex plane. We now give another necessary condition for the existence of a common $\boldsymbol{P}$ that satisfies (8.40).

Theorem 8.1 Suppose that each $\boldsymbol{A}_i$, $i = 1, 2, \ldots, r$, in the model (8.39) is asymptotically stable and there exists a symmetric positive definite $\boldsymbol{P}$ such that conditions (8.40) hold. Then, the matrices

$$\sum_{k=1}^{s} \boldsymbol{A}_{i_k},$$

where $i_k \in \{1, 2, \ldots, r\}$ and $s = 1, 2, 3, \ldots, r$ are asymptotically stable.

Proof Suppose that conditions (8.40) hold. Let

$$\boldsymbol{A}_{i_k}^T \boldsymbol{P} + \boldsymbol{P}\boldsymbol{A}_{i_k} = -\boldsymbol{Q}_{i_k}, \tag{8.41}$$

where $\boldsymbol{Q}_{i_k} = \boldsymbol{Q}_{i_k}^T > 0$ for $i_k \in \{1, 2, \ldots, r\}$. Summing both sides of (8.41) yields

$$\left(\sum_{k=1}^{s} \boldsymbol{A}_{i_k}^T\right) \boldsymbol{P} + \boldsymbol{P}\left(\sum_{k=1}^{s} \boldsymbol{A}_{i_k}\right) = -\sum_{k=1}^{s} \boldsymbol{Q}_{i_k}. \tag{8.42}$$

By assumption $\boldsymbol{P} = \boldsymbol{P}^T > 0$ and because $\boldsymbol{Q}_{i_k} = \boldsymbol{Q}_{i_k}^T > 0$ for $i_k \in \{1, 2, \ldots, r\}$, the symmetric matrices

$$\sum_{k=1}^{s} \boldsymbol{Q}_{i_k}, \qquad s = 2, 3, \ldots, r,$$

are also positive definite. Hence, by the Lyapunov theorem, the matrices $\sum_{k=1}^{s} \boldsymbol{A}_{i_k}$, $s = 2, 3, \ldots, r$, are asymptotically stable.

The following corollary can be viewed as a continuous-time counterpart of Theorem 4.3 of Tanaka and Sugeno [277, pp. 146–147].

Corollary 8.1 Suppose that each A_i, $i = 1, 2, \ldots, r$ in fuzzy model (8.39) is asymptotically stable and there exists a symmetric positive definite P such that conditions (8.40) hold. Then, the matrices

$$A_i + A_j, \qquad i, j = 1, 2, \ldots, r,$$

are asymptotically stable.

We mention that Theorem 4.3 in Tanaka and Sugeno [277] can be generalized to correspond to Theorem 8.1 rather than to the above corollary.

The above corollary can be restated in an equivalent way as follows. If there exist i and j such that a matrix $A_i + A_j$ is not asymptotically stable, then there is no positive definite matrix P such that $A_i^T P + P A_i < 0$ for $i = 1, 2, \ldots, r$. The following example illustrates this fact.

◆ Example 8.12

Given the unforced fuzzy system model described by (8.39), where $r = 2$ and

$$A_1 = \begin{bmatrix} -1 & 4 \\ 0 & -2 \end{bmatrix}, \qquad A_2 = \begin{bmatrix} -1 & 0 \\ 4 & -2 \end{bmatrix}.$$

Obviously each A_i, $i = 1, 2$, is asymptotically stable. However, there is no common positive definite P such that $A_i^T P + P A_i < 0$ for $i = 1, 2$ because the matrix

$$A_1 + A_2 = \begin{bmatrix} -1 & 4 \\ 0 & -2 \end{bmatrix} + \begin{bmatrix} -1 & 0 \\ 4 & -2 \end{bmatrix} = \begin{bmatrix} -2 & 4 \\ 4 & -4 \end{bmatrix}$$

is unstable since its eigenvalues are

$$\{1.1231, -7.1231\}.$$

Theorem 8.1 can be used to check if there does not exist a positive definite matrix P satisfying conditions (8.40) because this theorem gives us a sufficient condition for nonexistence of a common P that satisfies (8.40).

We now present a sufficiency condition for asymptotic stabilizability in the large of the model given by (8.6) and (8.8) using fuzzy state feedback control law:

$$u = -\sum_{j=1}^{r} \alpha_j K_j x. \tag{8.43}$$

The closed-loop system has the form

$$\boxed{\dot{x} = \sum_{i=1}^{r} \sum_{j=1}^{r} \alpha_i \alpha_j (A_i - B_i K_j) x.} \tag{8.44}$$

To obtain the above model, we used the fact that $\sum_{j=1}^{r} \alpha_j = 1$. Suppose that the gain matrices, K_i, were chosen so that the matrices

$$A_i - B_i K_i, \qquad i = 1, 2, \ldots, r,$$

have their eigenvalues in the open left-hand complex plane; that is, they are asymptotically stable. Let

$$G_{ij} = (A_i - B_i K_j) + (A_j - B_j K_i) \qquad \text{for } i < j \le r.$$

We now present sufficient conditions of Wang et al. [293] for the closed-loop system modeled by (8.44) to be asymptotically stable in the large.

Theorem 8.2 If for some symmetric positive definite P the conditions

$$(A_i - B_i K_i)^T P + P(A_i - B_i K_i) < 0, \qquad i = 1, 2, \ldots, r, \tag{8.45}$$

and

$$G_{ij}^T P + P G_{ij} < 0, \qquad i < j \le r, \tag{8.46}$$

are satisfied, then the closed-loop system modeled by (8.44) is asymptotically stable in the large.

Proof Let P be such that (8.45) and (8.46) are satisfied. We evaluate the time derivative of $x^T(t) P x(t)$ on the trajectories of (8.44) to get

$$\begin{aligned}
\frac{d}{dt}(x^T(t) P x(t)) &= 2x^T P \dot{x} \\
&= 2x^T P \left(\sum_{i=1}^{r} \sum_{j=1}^{r} \alpha_i \alpha_j (A_i - B_i K_j) x \right) \\
&= \sum_{i=1}^{r} \sum_{j=1}^{r} \alpha_i \alpha_j x^T ((A_i - B_i K_j)^T P + P(A_i - B_i K_j)) x \\
&= \sum_{r=1}^{r} \alpha_i^2 x^T ((A_i - B_i K_i)^T P + P(A_i - B_i K_i)) x \\
&\quad + \sum_{i=1}^{r} \sum_{j>i}^{r} \alpha_i \alpha_j x^T \left(G_{ij}^T P + P G_{ij} \right) x.
\end{aligned} \tag{8.47}$$

Hence, when (8.45) and (8.46) are satisfied, the positive definite quadratic function $x^T P x$ becomes a Lyapunov function for the system (8.44), which implies asymptotic stability in the large of (8.44). The proof of the theorem is complete.

We now give a sufficient condition for the stability of system (8.44) for the case when conditions (8.46) are not satisfied. To proceed, we need the following notation. Suppose that there is a positive definite P such that conditions (8.45) are satisfied. Represent conditions (8.45) as

$$(A_i - B_i K_i)^T P + P(A_i - B_i K_i) = -Q_i, \qquad i = 1, 2, \ldots, r, \tag{8.48}$$

where each Q_i is symmetric positive definite. Let λ_i denote the minimum eigenvalue of Q_i. We have

$$\lambda_i > 0, \qquad i = 1, 2, \ldots, r, \tag{8.49}$$

because $Q_i = Q_i^T > 0$. Let

$$G_{ij}^T P + P G_{ij} = -Q_{ij}, \qquad i < j \le r, \tag{8.50}$$

where now we do not require Q_{ij} to be positive definite. Note that $Q_{ij} = Q_{ij}^T$, and hence the eigenvalues of Q_{ij} are real. Let λ_{ij} denote the minimum eigenvalue of Q_{ij}. We are ready to state and prove the following theorem.

Theorem 8.3 Suppose that each $(A_i - B_i K_i)$, $i = 1, 2, \ldots, r$, in (8.44) is asymptotically stable and there exists a symmetric positive definite P such that conditions (8.45) hold. Then, the closed-loop fuzzy model (8.44) is asymptotically stable in the large if the matrix

$$\Lambda = \begin{bmatrix} \lambda_1 & \lambda_{12}/2 & \cdots & \lambda_{1r}/2 \\ \lambda_{12}/2 & \lambda_2 & & \lambda_{2r}/2 \\ \vdots & & \ddots & \vdots \\ \lambda_{1r}/2 & \lambda_{2r}/2 & \cdots & \lambda_r \end{bmatrix} \tag{8.51}$$

is positive definite.

Proof Let

$$W = W(x(t)) = x^T(t) P x(t)$$

be the Lyapunov function candidate for the closed-loop system modeled by (8.44). We then evaluate the time derivative of W on the trajectories of (8.44) to get

$$\begin{aligned} \dot{W} &= 2x^T P \dot{x} \\ &= 2x^T P \left(\sum_{i=1}^{r} \sum_{j=1}^{r} \alpha_i \alpha_j (A_i - B_i K_j) x \right) \\ &= \sum_{i=1}^{r} \sum_{j=1}^{r} \alpha_i \alpha_j x^T ((A_i - B_i K_j)^T P + P(A_i - B_i K_j)) x \\ &= -\sum_{i=1}^{r} \alpha_i^2 x^T Q_i x - \sum_{i=1}^{r} \sum_{j>i}^{r} \alpha_i \alpha_j x^T Q_{ij} x. \end{aligned} \tag{8.52}$$

Recall that for any symmetric matrix $M = M^T$, we have

$$\lambda_{\min}(M) \|x\|^2 \le x^T M x,$$

where $\lambda_{\min}(M)$ is the smallest eigenvalue of M. Using this fact in (8.52) yields

$$\begin{aligned} \dot{W} &\le -\left(\sum_{i=1}^{r} \alpha_i^2 \lambda_i + \sum_{i=1}^{r} \sum_{j>i}^{r} \alpha_i \alpha_j \lambda_{ij} \right) \|x\|^2 \\ &= -\left([\alpha_1 \ \ \alpha_2 \ \ \cdots \ \ \alpha_r] \begin{bmatrix} \lambda_1 & \lambda_{12}/2 & \cdots & \lambda_{1r}/2 \\ \lambda_{12}/2 & \lambda_2 & & \lambda_{2r}/2 \\ \vdots & & \ddots & \vdots \\ \lambda_{1r}/2 & \lambda_{2r}/2 & \cdots & \lambda_r \end{bmatrix} \begin{bmatrix} \alpha_1 \\ \alpha_2 \\ \vdots \\ \alpha_r \end{bmatrix} \right) \|x\|^2 \\ &= -(\alpha^T \Lambda \alpha) \|x\|^2. \end{aligned} \tag{8.53}$$

The Lyapunov derivative is negative definite if the matrix Λ is positive definite, which completes the proof.

8.5.4 A Lyapunov-Based Stabilizer

In this subsection we apply a Lyapunov approach to the fuzzy control design for uncertain dynamical systems of the form

$$\dot{\boldsymbol{x}} = \boldsymbol{A}(\boldsymbol{\alpha})\boldsymbol{x} + \boldsymbol{B}(\boldsymbol{\alpha})(\boldsymbol{u} + \boldsymbol{h}(t, \boldsymbol{x})), \tag{8.54}$$

where $\boldsymbol{\alpha}$ satisfies (8.8) and the vector-valued function $\boldsymbol{h}$ represents modeling uncertainties or disturbances. The only information about uncertainties available to the designer are their bounds. Specifically, we assume that the uncertainty modeling function is bounded by a nonnegative function $\eta = \eta(t, \boldsymbol{x})$; that is,

$$\|\boldsymbol{h}(t, \boldsymbol{x})\| \leq \eta(t, \boldsymbol{x}),$$

where $\|\cdot\|$ denotes the vector p-norm. Observe that $\boldsymbol{h}$ affects the system dynamics via the input matrix $\boldsymbol{B}(\boldsymbol{\alpha})$ in a similar fashion as the control input $\boldsymbol{u}$ does. For this reason, we say that the uncertainty $\boldsymbol{h}$ satisfies the matching condition.

Some of the Lyapunov-based control strategies, using nonfuzzy controllers, for uncertain systems are documented in Zinober [318]. The method of constructing stabilizing controllers, presented in this section, requires that each of the matrices $\boldsymbol{A}_i$, $i = 1, 2, \ldots, r$, is already asymptotically stable and that there exists a common symmetric positive definite $\boldsymbol{P} = \boldsymbol{P}^T > 0$ such that, for $i = 1, 2, \ldots, r$, we have

$$\boldsymbol{A}_i^T \boldsymbol{P} + \boldsymbol{P}\boldsymbol{A}_i < 0. \tag{8.55}$$

If this is not the case, we break the design of stabilizing control law, $\boldsymbol{u}$, into two parts $\boldsymbol{u} = \boldsymbol{u}_1 + \boldsymbol{u}_2$. We first design $\boldsymbol{u}_1$ of the form

$$\boldsymbol{u}_1 = -\sum_{j=1}^{r} \alpha_j \boldsymbol{K}_j \boldsymbol{x} \tag{8.56}$$

so that the system

$$\dot{\boldsymbol{x}} = \sum_{i=1}^{r}\sum_{j=1}^{r} \alpha_i \alpha_j (\boldsymbol{A}_i - \boldsymbol{B}_i \boldsymbol{K}_j)\boldsymbol{x}$$

is asymptotically stable in the large and there exists a common $\boldsymbol{P}$ such that conditions (8.45) and (8.46) or conditions of Theorem 8.3 are satisfied. Note that the system modeled by (8.54) and driven by (8.56) has the form

$$\begin{aligned}\dot{\boldsymbol{x}} &= \sum_{i=1}^{r} \alpha_i \boldsymbol{A}_i \boldsymbol{x} - \sum_{i=1}^{r} \alpha_i \boldsymbol{B}_i \left(\sum_{j=1}^{r} \alpha_j \boldsymbol{K}_j \boldsymbol{x} \right) + \sum_{i=1}^{r} \alpha_i \boldsymbol{B}_i (\boldsymbol{u}_2 + \boldsymbol{h}) \\ &= \sum_{i=1}^{r}\sum_{j=1}^{r} \alpha_i \alpha_j (\boldsymbol{A}_i - \boldsymbol{B}_i \boldsymbol{K}_j)\boldsymbol{x} + \sum_{i=1}^{r} \alpha_i \boldsymbol{B}_i (\boldsymbol{u}_2 + \boldsymbol{h}).\end{aligned} \tag{8.57}$$

We next proceed with the design of $\boldsymbol{u}_2$. The role of $\boldsymbol{u}_2$ is to reject the uncertainty $\boldsymbol{h}$. Then, $\boldsymbol{u} = \boldsymbol{u}_1 + \boldsymbol{u}_2$ will asymptotically stabilize in the large the uncertain system (8.54). We propose a family of fuzzy controllers $\boldsymbol{u}_2$ that reject the uncertainty $\boldsymbol{h}$. These controllers are obtained by "fuzzifying" the controllers proposed in Hui and Żak [133]. To proceed, we need the following notation. Let

$$\boldsymbol{\mu} = [\mu_1 \quad \mu_2 \quad \dots \quad \mu_m]^T = \sum_{i=1}^{r} \alpha_i \boldsymbol{B}_i^T \boldsymbol{P}\boldsymbol{x}. \tag{8.58}$$

Let $\|\cdot\|_p$ denote the p-norm of a vector; that is, if a vector $\boldsymbol{v} \in \mathbb{R}^m$, then its p-norm is defined as

$$\|\boldsymbol{v}\|_p = \left(|v_1|^p + |v_2|^p + \cdots + |v_m|^p\right)^{1/p}, \qquad p \ge 1.$$

A well-known result concerning p-norms is the Hölder inequality,

$$\left|\boldsymbol{v}^T\boldsymbol{w}\right| \le \|\boldsymbol{v}\|_p \|\boldsymbol{w}\|_q, \qquad \frac{1}{p} + \frac{1}{q} = 1.$$

All norms on $\mathbb{R}^m$ are equivalent in the sense that if $\|\cdot\|_\alpha$ and $\|\cdot\|_\beta$ are norms on $\mathbb{R}^m$, then there exist positive constants C_1 and C_2 such that

$$C_1\|\boldsymbol{v}\|_\alpha \le \|\boldsymbol{v}\|_\beta \le C_2\|\boldsymbol{v}\|_\alpha \qquad \text{for all } \boldsymbol{v} \in \mathbb{R}^m. \tag{8.59}$$

Finally, the gradient of $\|\boldsymbol{\mu}\|_p$, where it is defined, has the form

$$\nabla\|\boldsymbol{\mu}\|_p = \left[\frac{\partial\|\boldsymbol{\mu}\|_p}{\partial\mu_1} \quad \frac{\partial\|\boldsymbol{\mu}\|_p}{\partial\mu_2} \quad \cdots \quad \frac{\partial\|\boldsymbol{\mu}\|_p}{\partial\mu_m}\right]^T, \qquad p \ge 1. \tag{8.60}$$

Applying the chain rule, we compute

$$\nabla\|\boldsymbol{\mu}\|_p = \frac{1}{\|\boldsymbol{\mu}\|_p^{p-1}} \begin{bmatrix} |\mu_1|^{p-1}\operatorname{sign}(\mu_1) \\ |\mu_2|^{p-1}\operatorname{sign}(\mu_2) \\ \vdots \\ |\mu_m|^{p-1}\operatorname{sign}(\mu_m) \end{bmatrix}. \tag{8.61}$$

Let $\|\boldsymbol{h}(t, \boldsymbol{x})\|_q \le \eta_q$, where $\eta_q \ge 0$ is a constant. We then show that when

$$\boldsymbol{u}_2 = -\eta_q \nabla\|\boldsymbol{\mu}\|_p, \quad p \ge 1, \quad \text{and} \quad \frac{1}{p} + \frac{1}{q} = 1, \tag{8.62}$$

then the closed-loop system driven by $\boldsymbol{u} = \boldsymbol{u}_1 + \boldsymbol{u}_2$ is asymptotically stable in the large for any uncertainty $\boldsymbol{h}$ such that $\|\boldsymbol{h}\|_q \le \eta_q$.

Theorem 8.4 Suppose that $\boldsymbol{u}_1$, given by (8.56), is such that there exists a common $\boldsymbol{P}$ satisfying conditions (8.45) and (8.46) or conditions of Theorem 8.3. Then, the system (8.54) driven by $\boldsymbol{u} = \boldsymbol{u}_1 + \boldsymbol{u}_2$, where $\boldsymbol{u}_2$ is given by (8.62), is asymptotically stable in the large for any uncertainty $\boldsymbol{h}$ such that $\|\boldsymbol{h}\|_q \leq \eta_q$.

Proof Suppose that $\boldsymbol{u}_1$, given by (8.56), was constructed such that (8.45) and (8.46) or the conditions of Theorem 8.3 hold. Then, (8.54) driven by $\boldsymbol{u}_1$ takes the form (8.57); that is,

$$\dot{\boldsymbol{x}} = \sum_{i=1}^{r}\sum_{j=1}^{r} \alpha_i\alpha_j(\boldsymbol{A}_i - \boldsymbol{B}_i\boldsymbol{K}_j)\boldsymbol{x} + \sum_{i=1}^{r}\alpha_i\boldsymbol{B}_i(\boldsymbol{u}_2 + \boldsymbol{h}).$$

We now show that

$$W = \boldsymbol{x}^T\boldsymbol{P}\boldsymbol{x}, \tag{8.63}$$

is the Lyapunov function for the closed-loop system, where $\boldsymbol{P} = \boldsymbol{P}^T > 0$ satisfies (8.45) and (8.46) or the conditions of Theorem 8.3. For this, we evaluate the Lyapunov derivative of W—that is, we evaluate $\frac{d}{dt}W$ on the trajectories of (8.57)—and show that the Lyapunov derivative of W is negative definite. Before we proceed with evaluating $\frac{d}{dt}W$, observe that using the fact that

$$\operatorname{sign}(\mu_j) = \frac{|\mu_j|}{\mu_j}$$

and using (8.61), we obtain

$$\boldsymbol{\mu}^T\nabla\|\boldsymbol{\mu}\|_p = \|\boldsymbol{\mu}\|_p.$$

Hence,

$$\begin{aligned}\frac{d}{dt}W &= 2\boldsymbol{x}^T\boldsymbol{P}\left(\sum_{i=1}^{r}\sum_{j=1}^{r}\alpha_i\alpha_j(\boldsymbol{A}_i - \boldsymbol{B}_i\boldsymbol{K}_j)\boldsymbol{x} + \sum_{i=1}^{r}\alpha_i\boldsymbol{B}_i(\boldsymbol{u}_2 + \boldsymbol{h})\right)\\ &< -2\eta_q\boldsymbol{\mu}^T\nabla\|\boldsymbol{\mu}\|_p + 2\boldsymbol{\mu}^T\boldsymbol{h}\\ &\leq -2\eta_q\|\boldsymbol{\mu}\|_p + 2\left|\boldsymbol{\mu}^T\boldsymbol{h}\right|.\end{aligned} \tag{8.64}$$

Applying Hölder's inequality to the above yields

$$\begin{aligned}\frac{d}{dt}W &< -2\eta_q\|\boldsymbol{\mu}\|_p + 2\|\boldsymbol{\mu}\|_p\|\boldsymbol{h}\|_q\\ &\leq -2\eta_q\|\boldsymbol{\mu}\|_p + 2\|\boldsymbol{\mu}\|_p\,\eta_q.\end{aligned} \tag{8.65}$$

Therefore $\frac{d}{dt}W < 0$, and by Lyapunov's theorem the closed loop-system is asymptotically stable in the large.

◆ Example 8.13

Using system model (1.79), we propose the following two rules describing the plant dynamics:

Rule 1: IF $x_1(t)$ is about 0,

THEN $\dot{\boldsymbol{x}}(t) = \boldsymbol{A}_1\boldsymbol{x}(t) + \boldsymbol{B}_1(u(t) + h)$

and

Rule 2: IF $x_1(t)$ is about $\pm\pi/4$,
THEN $\dot{\boldsymbol{x}}(t) = \boldsymbol{A}_2\boldsymbol{x}(t) + \boldsymbol{B}_2(u(t) + h)$,

where $h = -f_c + mlx_2^2\sin(x_1)$. Local linear models are constructed using the procedure described in the previous section, and in particular relations (8.25) and (8.38). We have

$$\boldsymbol{A}_1 = \begin{bmatrix} 0 & 1 & 0 & 0 \\ \dfrac{g}{4l/3 - aml} & 0 & 0 & 0 \\ 0 & 0 & 0 & 1 \\ -\dfrac{amg}{4/3 - am} & 0 & 0 & 0 \end{bmatrix}, \qquad \boldsymbol{B}_1 = \begin{bmatrix} 0 \\ -\dfrac{a}{4l/3 - aml} \\ 0 \\ \dfrac{4a/3}{4/3 - ma} \end{bmatrix}.$$

The matrices $\boldsymbol{A}_2$ and $\boldsymbol{B}_2$ of the local model corresponding to the second rule are:

$$\boldsymbol{A}_2 = \begin{bmatrix} 0 & 1 & 0 & 0 \\ \dfrac{g\sin(x_1)}{4l/3 - mla\cos^2(x_1)}\left(\dfrac{1}{x_1}\right) & 0 & 0 & 0 \\ 0 & 0 & 0 & 1 \\ \dfrac{-mag\sin(2x_1)/2}{4/3 - ma\cos^2(x_1)}\left(\dfrac{1}{x_1}\right) & 0 & 0 & 0 \end{bmatrix}_{|x_1=\pm\pi/4}$$

and

$$\boldsymbol{B}_2 = \begin{bmatrix} 0 \\ \dfrac{-a\cos(x_1)}{4l/3 - mla\cos^2(x_1)} \\ 0 \\ \dfrac{4a/3}{4/3 - ma\cos^2(x_1)} \end{bmatrix}_{|x_1=\pm\pi/4}.$$

The parameter values are the same as in Example 8.10. We use weighting functions of the form:

$$w_1(x_1) = \frac{1 - 1/(1+\exp(-14(x_1 - \pi/8)))}{1+\exp(-14(x_1+\pi/8))} \quad \text{and} \quad w_2(x_1) = 1 - w_1(x_1).$$

The corresponding fuzzy system model is

$$\dot{\boldsymbol{x}} = (\alpha_1\boldsymbol{A}_1 + \alpha_2\boldsymbol{A}_2)\boldsymbol{x} + (\alpha_1\boldsymbol{B}_1 + \alpha_2\boldsymbol{B}_2)\left(u - \mu_c\operatorname{sign}(x_4) + mlx_2^2\sin(x_1)\right), \tag{8.66}$$

where $\alpha_i = w_i$, $i = 1, 2$. Note that $w_1 + w_2 = 1$ for all t. The friction coefficient $\mu_c = 0.005$. We then proceed to construct a stabilizing controller. We first construct

$$u_1 = -(\alpha_1\boldsymbol{K}_1 + \alpha_2\boldsymbol{K}_2)\boldsymbol{x}$$

such that the system

$$\dot{\boldsymbol{x}} = \sum_{i=1}^{2}\sum_{j=1}^{2}\alpha_i\alpha_j(\boldsymbol{A}_i - \boldsymbol{B}_i\boldsymbol{K}_j)\boldsymbol{x} \tag{8.67}$$

is asymptotically stable in the large. To test our stabilization algorithm, we randomly chose the set of desired poles for $(\boldsymbol{A}_1 - \boldsymbol{B}_1\boldsymbol{K}_1)$ to be in the interval $[-4, -1]$. The set of desired poles is

$$\{-1.0970, -2.1263, -2.5553, -3.9090\}.$$

The desired poles for $(\boldsymbol{A}_2 - \boldsymbol{B}_2\boldsymbol{K}_2)$ were also chosen in random fashion to be in the interval $[-4, -1]$. In this case, we generated the set of poles of the form

$$\{-1.5794, -1.6857, -2.7908, -3.6895\}.$$

For this choice of the closed-loop poles of the local models, we obtain the following feedback gain matrices:

$$\boldsymbol{K}_1 = [-294.8766 \quad -73.1208 \quad -13.4726 \quad -27.3362]$$

and

$$\boldsymbol{K}_2 = [-440.3915 \quad -118.5144 \quad -19.1611 \quad -35.5575].$$

Let

$$\boldsymbol{G}_{12} = \boldsymbol{A}_1 - \boldsymbol{B}_1\boldsymbol{K}_2 + \boldsymbol{A}_2 - \boldsymbol{B}_2\boldsymbol{K}_1$$

and

$$\boldsymbol{P} = \begin{bmatrix} 54.9580 & 15.6219 & 6.9389 & 12.1165 \\ 15.6219 & 4.5429 & 2.1011 & 3.5488 \\ 6.9389 & 2.1011 & 1.3972 & 1.7978 \\ 12.1165 & 3.5488 & 1.7978 & 2.9375 \end{bmatrix}.$$

Then, it is easy to check that sufficient conditions, given by (8.45) and (8.46), for asymptotic stability in the large of system (8.67) are satisfied. This completes the design of u_1.

We now construct u_2. We first use (8.58) to construct

$$\mu = \left(\alpha_1 \boldsymbol{B}_1^T \boldsymbol{P} + \alpha_2 \boldsymbol{B}_2^T \boldsymbol{P}\right)\boldsymbol{x}.$$

In this example, the number of inputs is $m = 1$, and hence (8.62) reduces to

$$u_2 = -\eta_q \text{sign}(\mu),$$

where we can take $\eta_q = |h|$. We combine u_1 and u_2 to obtain the following stabilizing controller:

$$u = u_1 + u_2 = -(\alpha_1\boldsymbol{K}_1 + \alpha_2\boldsymbol{K}_2)\boldsymbol{x} - \eta_q \text{sign}(\mu). \tag{8.68}$$

The above controller globally stabilizes the design model. In our simulations we connected the controller to the truth model given by (1.79). Typical responses are shown in Figures 8.35 and 8.36. The controller locally stabilizes the truth model. We observed that when $|x_1(0)| > 1.3$ rads and $x_2(0) = x_3(0) = x_4(0) = 0$, then the equilibrium state $\boldsymbol{x} = \boldsymbol{0}$ of the closed-loop system consisting of the truth model driven by (8.68) becomes unstable.

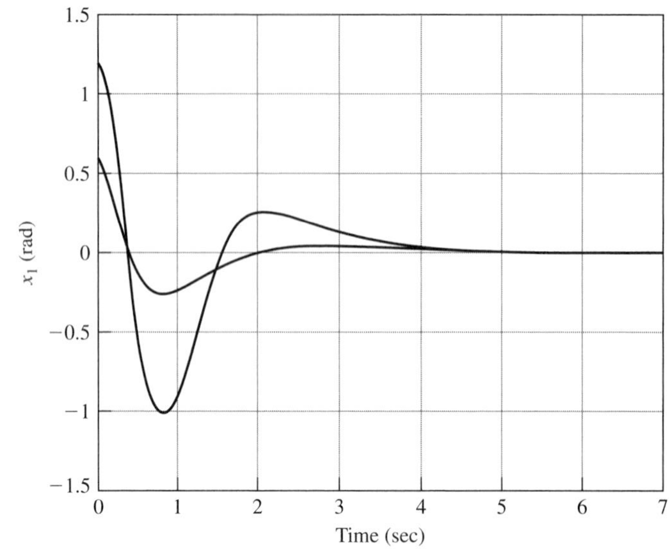

Figure 8.35 Plots of x_1 versus time for two different initial values of x_1: $x_1(0) = 1.2$ and $x_1(0) = 0.6$. In both cases, $x_2(0) = x_3(0) = x_4(0) = 0$.

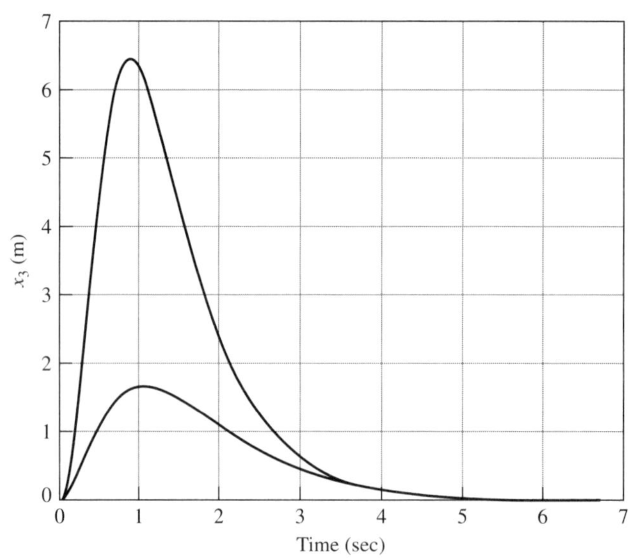

Figure 8.36 Plots of x_3 versus time for two different initial values of x_1: $x_1(0) = 1.2$ and $x_1(0) = 0.6$. In both cases, $x_2(0) = x_3(0) = x_4(0) = 0$.

We summarize this section as follows. A common approach to the control system design is to use a design model to construct a controller. The design model is a simplified version of the truth model and usually neglects high-frequency dynamics of the plant. Thus, the designer has to cope with an incomplete mathematical model of the process to be controlled. To account for modeling uncertainties, we used a two-part controller. The first part of the controller stabilized the nominal plant—that is, the plant model that does not include modeling uncertainties. The role of the second part of the controller was to reject modeling uncertainties. In both cases, fuzzy state-feedback control strategies were used. In order to be able to apply the proposed controllers in real-life systems, we have to account for limited authority of actuators, unmatched uncertainties, and inaccessibility of state variables.

8.6 Stability of Discrete Fuzzy Models

We analyze the model of the form

$$\boldsymbol{x}(k+1) = \sum_{i=1}^{r} w_i(\boldsymbol{x}(k))\boldsymbol{A}_i\boldsymbol{x}(k), \tag{8.69}$$

where

$$w_i(\boldsymbol{x}(k)) \geq 0 \quad \text{and} \quad \sum_{i=1}^{r} w_i(\boldsymbol{x}(k)) = 1.$$

To proceed, we need the following notation. Given two $n \times n$ matrices $\boldsymbol{A}$ and $\boldsymbol{B}$. Then, $\boldsymbol{A} < \boldsymbol{B}$ means that

$$\boldsymbol{x}^T\boldsymbol{A}\boldsymbol{x} < \boldsymbol{x}^T\boldsymbol{B}\boldsymbol{x} \qquad \text{for all } \boldsymbol{x} \in \mathbb{R}^n, \boldsymbol{x} \neq \boldsymbol{0};$$

that is, the quadratic form $\boldsymbol{x}^T(\boldsymbol{A} - \boldsymbol{B})\boldsymbol{x}$ is negative definite.

We now prove the following matrix inequality lemma that is useful in the stability analysis of discrete fuzzy models.

Lemma 8.1 Let $\boldsymbol{A}, \boldsymbol{B}, \boldsymbol{P} \in \mathbb{R}^{n\times n}$. If $\boldsymbol{P} = \boldsymbol{P}^T > 0$ such that

$$\boldsymbol{A}^T\boldsymbol{P}\boldsymbol{A} - \boldsymbol{P} < 0 \quad \text{and} \quad \boldsymbol{B}^T\boldsymbol{P}\boldsymbol{B} - \boldsymbol{P} < 0,$$

then

$$\boldsymbol{A}^T\boldsymbol{P}\boldsymbol{B} + \boldsymbol{B}^T\boldsymbol{P}\boldsymbol{A} - 2\boldsymbol{P} < 0.$$

Proof We first represent $\boldsymbol{A}^T\boldsymbol{P}\boldsymbol{B} + \boldsymbol{B}^T\boldsymbol{P}\boldsymbol{A} - 2\boldsymbol{P}$ as

$$\begin{aligned}\boldsymbol{A}^T\boldsymbol{P}\boldsymbol{B} + \boldsymbol{B}^T\boldsymbol{P}\boldsymbol{A} - 2\boldsymbol{P} &= -(\boldsymbol{A} - \boldsymbol{B})^T\boldsymbol{P}(\boldsymbol{A} - \boldsymbol{B}) + \boldsymbol{A}^T\boldsymbol{P}\boldsymbol{A} + \boldsymbol{B}^T\boldsymbol{P}\boldsymbol{B} - 2\boldsymbol{P}\\ &= -(\boldsymbol{A} - \boldsymbol{B})^T\boldsymbol{P}(\boldsymbol{A} - \boldsymbol{B}) + (\boldsymbol{A}^T\boldsymbol{P}\boldsymbol{A} - \boldsymbol{P})\\ &\quad +(\boldsymbol{B}^T\boldsymbol{P}\boldsymbol{B} - \boldsymbol{P}).\end{aligned}$$

The first term, $-(\boldsymbol{A} - \boldsymbol{B})^T\boldsymbol{P}(\boldsymbol{A} - \boldsymbol{B})$, is nonpositive definite because $\boldsymbol{P} = \boldsymbol{P}^T > 0$ by assumption, while the second and the third terms are negative definite by assumption. Hence,

$$\boldsymbol{A}^T\boldsymbol{P}\boldsymbol{B} + \boldsymbol{B}^T\boldsymbol{P}\boldsymbol{A} - 2\boldsymbol{P} < 0,$$

and the lemma is proven.

Using Lemma 8.1, we now prove a sufficient condition for global asymptotic stability of model (8.69).

Theorem 8.5 The equilibrium state, $\boldsymbol{x}_e = \boldsymbol{0}$, of model (8.69) is globally asymptotically stable if there exists a positive definite matrix $\boldsymbol{P} = \boldsymbol{P}^T > 0$ such that for $i = 1, 2, \ldots, r$,

$$\boldsymbol{A}_i^T \boldsymbol{P} \boldsymbol{A}_i - \boldsymbol{P} < 0. \tag{8.70}$$

Proof Suppose that there exists a symmetric positive definite matrix $\boldsymbol{P} = \boldsymbol{P}^T > 0$ such that for $i = 1, 2, \ldots, r$, inequalities given by (8.70) are satisfied. Consider then the Lyapunov function candidate

$$V(\boldsymbol{x}(k)) = \boldsymbol{x}^T(k)\, \boldsymbol{P} \boldsymbol{x}(k). \tag{8.71}$$

To prove the theorem, it is enough to show that

$$\Delta V(\boldsymbol{x}(k)) < 0.$$

We have

$$\begin{aligned}\Delta V(\boldsymbol{x}(k)) &= V(\boldsymbol{x}(k+1)) - V(\boldsymbol{x}(k)) \\ &= \boldsymbol{x}^T(k+1)\, \boldsymbol{P} \boldsymbol{x}(k+1) - \boldsymbol{x}^T(k)\, \boldsymbol{P} \boldsymbol{x}(k) \\ &= \left(\sum_{i=1}^{r} w_i \boldsymbol{x}^T \boldsymbol{A}_i^T\right) \boldsymbol{P} \left(\sum_{i=1}^{r} w_i \boldsymbol{A}_i \boldsymbol{x}\right) - \boldsymbol{x}^T \boldsymbol{P} \boldsymbol{x} \\ &= \boldsymbol{x}^T \left(\left(\sum_{i=1}^{r} w_i \boldsymbol{A}_i^T\right) \boldsymbol{P} \left(\sum_{i=1}^{r} w_i \boldsymbol{A}_i\right) - \boldsymbol{P}\right) \boldsymbol{x}.\end{aligned} \tag{8.72}$$

To proceed, we represent (8.72) in an equivalent form as

$$\Delta V(\boldsymbol{x}) = \boldsymbol{x}^T \left(\left(\sum_{i=1}^{r} w_i \boldsymbol{A}_i^T\right) \boldsymbol{P} \left(\sum_{j=1}^{r} w_j \boldsymbol{A}_j\right) - \boldsymbol{P}\right) \boldsymbol{x}. \tag{8.73}$$

By assumption we have

$$\sum_{i=1}^{r} w_i = 1,$$

and hence

$$\sum_{i=1}^{r} w_i \sum_{j=1}^{r} w_j = \sum_{i=1}^{r} \sum_{j=1}^{r} w_i w_j = 1. \tag{8.74}$$

Substituting (8.74) into (8.73) yields

$$\begin{aligned}\Delta V(\boldsymbol{x}) &= \boldsymbol{x}^T \left(\left(\sum_{i=1}^{r} w_i \boldsymbol{A}_i^T\right) \boldsymbol{P} \left(\sum_{j=1}^{r} w_j \boldsymbol{A}_j\right) - \sum_{i=1}^{r} \sum_{j=1}^{r} w_i w_j \boldsymbol{P}\right) \boldsymbol{x} \\ &= \boldsymbol{x}^T \left(\sum_{i=1}^{r} \sum_{j=1}^{r} w_i w_j \left(\boldsymbol{A}_i^T \boldsymbol{P} \boldsymbol{A}_j - \boldsymbol{P}\right)\right) \boldsymbol{x}.\end{aligned} \tag{8.75}$$

We rearrange (8.75) to get

$$\Delta V(\boldsymbol{x}(k)) = \sum_{i=1}^{r} w_i^2 \boldsymbol{x}^T(\boldsymbol{A}_i^T \boldsymbol{P} \boldsymbol{A}_i - \boldsymbol{P})\boldsymbol{x} + \sum_{i=1}^{r} \sum_{\substack{j=1 \\ j \neq i}}^{r} w_i w_j \boldsymbol{x}^T(\boldsymbol{A}_i^T \boldsymbol{P} \boldsymbol{A}_j - \boldsymbol{P})\boldsymbol{x}$$

$$= \sum_{i=1}^{r} w_i^2 \boldsymbol{x}^T(\boldsymbol{A}_i^T \boldsymbol{P} \boldsymbol{A}_i - \boldsymbol{P})\boldsymbol{x}$$

$$+ \sum_{i=1}^{r} \sum_{j>i}^{r} w_i w_j \boldsymbol{x}^T(\boldsymbol{A}_i^T \boldsymbol{P} \boldsymbol{A}_j + \boldsymbol{A}_j^T \boldsymbol{P} \boldsymbol{A}_i - 2\boldsymbol{P})\boldsymbol{x}. \tag{8.76}$$

The first term in (8.76) is negative definite because of the assumption and the fact that at least one of the weights, w_i, is positive because $w_i \geq 0$ and $\sum_{i=1}^{r} w_i = 1$. The second term in (8.76) is nonpositive definite by virtue of Lemma 8.1. Hence, $V = \boldsymbol{x}^T \boldsymbol{P} \boldsymbol{x}$ is a Lyapunov function for the fuzzy model (8.69), and by the Lyapunov theorem the equilibrium state $\boldsymbol{x}_e = \boldsymbol{0}$ of the model is globally asymptotically stable.

The above theorem gives no method for finding a common positive definite matrix $\boldsymbol{P}$ that satisfies r inequalities (8.70). All of the matrices $\boldsymbol{A}_i$, $i = 1, 2, \ldots, r$, are asymptotically stable if there exists a common positive definite $\boldsymbol{P}$ that satisfies (8.70). However, the asymptotic stability of all the matrices $\boldsymbol{A}_i$ is only necessary but not sufficient for the existence of a common symmetric positive definite $\boldsymbol{P}$ that satisfies (8.70). The following theorem provides a necessary, but not sufficient, condition for the existence of positive definite $\boldsymbol{P}$ satisfying (8.70).

Theorem 8.6 Given r square $n \times n$ matrices $\boldsymbol{A}_i$ that are asymptotically stable and nonsingular. If there exists a symmetric positive definite matrix $\boldsymbol{P} = \boldsymbol{P}^T > 0$ such that for $i = 1, 2, \ldots, r$,

$$\boldsymbol{A}_i^T \boldsymbol{P} \boldsymbol{A}_i - \boldsymbol{P} < 0, \tag{8.77}$$

then for $i, j = 1, 2, \ldots, r$, the matrices $\boldsymbol{A}_i \boldsymbol{A}_j$ are asymptotically stable.

Proof By assumption, there exists a symmetric positive definite matrix $\boldsymbol{P}$ such that for $j = 1, 2, \ldots, r$, we have

$$\boldsymbol{A}_j^T \boldsymbol{P} \boldsymbol{A}_j - \boldsymbol{P} < 0. \tag{8.78}$$

Premultiplying the above by $(\boldsymbol{A}_j^T)^{-1}$ and postmultiplying by $\boldsymbol{A}_j^{-1}$ yields

$$\boldsymbol{P} - (\boldsymbol{A}_j^T)^{-1} \boldsymbol{P} \boldsymbol{A}_j^{-1} < 0. \tag{8.79}$$

Hence,

$$\boldsymbol{P} < (\boldsymbol{A}_j^T)^{-1} \boldsymbol{P} \boldsymbol{A}_j^{-1}. \tag{8.80}$$

On the other hand, using (8.77) we get

$$\boldsymbol{A}_i^T \boldsymbol{P} \boldsymbol{A}_i < \boldsymbol{P}. \tag{8.81}$$

Combining (8.80) and (8.81), we obtain

$$\boldsymbol{A}_i^T \boldsymbol{P} \boldsymbol{A}_i < (\boldsymbol{A}_j^T)^{-1} \boldsymbol{P} \boldsymbol{A}_j^{-1}. \tag{8.82}$$

Premultiplying both sides of (8.82) by A_j^T and postmultiplying by A_j, we get, for $i, j = 1, 2, \ldots, r$,

$$(A_i A_j)^T P(A_i A_j) - P < 0. \tag{8.83}$$

Therefore, for $i, j = 1, 2, \ldots, r$, the matrices $A_i A_j$ are asymptotically stable, which completes the proof.

The above theorem can be stated in an equivalent way as follows. Assume that for $i = 1, 2, \ldots, r$, the matrices A_i are asymptotically stable and nonsingular. If for some $i, j = 1, 2, \ldots, r$, the matrix $A_i A_j$ is unstable, then there does not exist a common positive definite P such that

$$A_i^T P A_i - P < 0 \qquad \text{for } i = 1, 2, \ldots, r.$$

◆ Example 8.14

Let

$$A_1 = \begin{bmatrix} 1 & -0.5 \\ 1 & 0 \end{bmatrix} \quad \text{and} \quad A_2 = \begin{bmatrix} -1 & -0.5 \\ 1 & 0 \end{bmatrix}.$$

Both matrices, A_1 and A_2, are asymptotically stable. Indeed, the eigenvalues of A_1 are located at $\{0.5 + j0.5, 0.5 - j0.5\}$, while the eigenvalues of A_2 are located at $\{-0.5 + j0.5, -0.5 - j0.5\}$. However, the matrix $A_1 A_2$ has its eigenvalues located at $\{-1.866, -0.134\}$, and hence is unstable. Thus, it follows from Theorem 8.6, that there is no positive definite P such that

$$A_i^T P A_i - P < 0 \qquad \text{for } i = 1, 2. \tag{8.84}$$

8.7 Fuzzy Estimator

In this section we are concerned with the problem of designing a state estimator using a TSK model of the plant. We first discuss the issue of constructing a TSK model using input–output data. To construct a TSK model, we need local description of the plant in terms of linear models,

$$\dot{x}(t) = A_i x(t) + B_i u(t),$$
$$y(t) = C_i x(t),$$

where $i = 1, 2, \ldots, r$, the state vector $x(t) \in \mathbb{R}^n$, the control input $u(t) \in \mathbb{R}^m$, the output $y(t) \in \mathbb{R}^p$, and the matrices A_i, B_i, and C_i are of appropriate dimensions. The above information is then fused with the available IF–THEN rules, where the ith rule can have the following form:

$$\textbf{Rule i:} \quad \text{IF } z_1(t) \text{ is } F_1^i \text{ AND } \cdots \text{ AND } z_q(t) \text{ is } F_q^i,$$

$$\text{THEN } \begin{cases} \dot{x}(t) = A_i x(t) + B_i u(t), \\ y = C_i x(t), \end{cases}$$

where F_j^i, $j = 1, 2, \ldots, q$, is the jth fuzzy set of the ith rule, and $z_1(t), \ldots, z_q(t)$ are measurable system variables. Let $\mu_j^i(z_j)$ be the membership function of the fuzzy set F_j^i, and let

$$w^i = w^i(z) = \prod_{j=1}^{q} \mu_j^i(z_j), \qquad i = 1, 2, \ldots, r,$$

where r is the number of fuzzy rules. The resulting fuzzy system model is the weighted average of the local models and has the form of (8.6), that is,

$$\begin{aligned}\dot{\boldsymbol{x}} &= \frac{\sum_{i=1}^{r} w^i(\boldsymbol{A}_i\boldsymbol{x} + \boldsymbol{B}_i\boldsymbol{u})}{\sum_{i=1}^{r} w^i} \\ &= \sum_{i=1}^{r} \alpha_i(\boldsymbol{A}_i\boldsymbol{x} + \boldsymbol{B}_i\boldsymbol{u}) \\ &= \left(\sum_{i=1}^{r} \alpha_i \boldsymbol{A}_i\right)\boldsymbol{x} + \left(\sum_{i=1}^{r} \alpha_i \boldsymbol{B}_i\right)\boldsymbol{u} \\ &= \boldsymbol{A}(\boldsymbol{\alpha})\boldsymbol{x} + \boldsymbol{B}(\boldsymbol{\alpha})\boldsymbol{u},\end{aligned}$$

where for $i = 1, 2, \ldots, r$ we have

$$\alpha_i = \frac{w^i}{\sum_{i=1}^{r} w^i}.$$

Similarly,

$$\begin{aligned}\boldsymbol{y} &= \frac{\sum_{i=1}^{r} w^i \boldsymbol{C}_i\boldsymbol{x}}{\sum_{i=1}^{r} w^i} \\ &= \left(\sum_{i=1}^{r} \alpha_i \boldsymbol{C}_i\right)\boldsymbol{x} \\ &= \boldsymbol{C}(\boldsymbol{\alpha})\boldsymbol{x}.\end{aligned} \tag{8.85}$$

The coefficients α_i satisfy (8.8); that is,

$$\alpha_i \geq 0, \quad \sum_{i=1}^{r} \alpha_i = 1 \quad \text{and} \quad \boldsymbol{\alpha} = [\alpha_1 \quad \alpha_2 \quad \cdots \quad \alpha_r]^T \in [0, 1]^r.$$

A fuzzy state estimator for the system modeled by (8.6) and (8.85) can have the form

$$\dot{\tilde{\boldsymbol{x}}} = \boldsymbol{A}(\boldsymbol{\alpha})\tilde{\boldsymbol{x}} + \boldsymbol{B}(\boldsymbol{\alpha})\boldsymbol{u} + \boldsymbol{L}(\boldsymbol{\alpha})(\boldsymbol{y} - \tilde{\boldsymbol{y}}), \tag{8.86}$$

where $\tilde{\boldsymbol{x}}$ is the estimate of the plant state $\boldsymbol{x}$ and

$$\boldsymbol{L}(\boldsymbol{\alpha}) = \sum_{i=1}^{r} \alpha_i \boldsymbol{L}_i.$$

The state estimator (8.86) can be represented equivalently as

$$\boxed{\dot{\tilde{\boldsymbol{x}}} = \sum_{i=1}^{r} \alpha_i \boldsymbol{A}_i\tilde{\boldsymbol{x}} + \sum_{i=1}^{r} \alpha_i \boldsymbol{B}_i\boldsymbol{u} + \sum_{i=1}^{r} \alpha_i \boldsymbol{L}_i(\boldsymbol{y} - \tilde{\boldsymbol{y}})} \tag{8.87}$$

We note that the state estimator (8.87) can be viewed as a result of implementing the following r fuzzy rules:

$$\textbf{Rule i:}\quad \text{IF } z_1(t) \text{ is } F_1^i \text{ AND } \cdots \text{ AND } z_q(t) \text{ is } F_q^i,$$
$$\text{THEN } \begin{cases} \dot{\tilde{\boldsymbol{x}}} = \boldsymbol{A}_i\tilde{\boldsymbol{x}} + \boldsymbol{B}_i\boldsymbol{u} + \boldsymbol{L}_i(\boldsymbol{y} - \tilde{\boldsymbol{y}}), \\ \tilde{\boldsymbol{y}} = \boldsymbol{C}_i\tilde{\boldsymbol{x}}. \end{cases}$$

We use the state estimate in the controller implementation to get

$$\boldsymbol{u} = -\sum_{j=1}^{r} \alpha_j \boldsymbol{K}_j \tilde{\boldsymbol{x}} + \boldsymbol{v}. \tag{8.88}$$

Substituting the above into the plant model yields

$$\dot{\boldsymbol{x}} = \sum_{i=1}^{r} \alpha_i \boldsymbol{A}_i \boldsymbol{x} - \sum_{i=1}^{r}\sum_{j=1}^{r} \alpha_i\alpha_j \boldsymbol{B}_i\boldsymbol{K}_j\tilde{\boldsymbol{x}} + \sum_{i=1}^{r}\alpha_i\boldsymbol{B}_i\boldsymbol{v}. \tag{8.89}$$

Substituting (8.85) and (8.88) into the estimator equation (8.87), we obtain

$$\begin{aligned} \dot{\tilde{\boldsymbol{x}}} &= \sum_{i=1}^{r} \alpha_i \boldsymbol{A}_i\tilde{\boldsymbol{x}} - \sum_{i=1}^{r}\sum_{j=1}^{r} \alpha_i\alpha_j\boldsymbol{B}_i\boldsymbol{K}_j\tilde{\boldsymbol{x}} + \sum_{i=1}^{r}\sum_{j=1}^{r}\alpha_i\alpha_j\boldsymbol{L}_i\boldsymbol{C}_j(\boldsymbol{x} - \tilde{\boldsymbol{x}}) + \sum_{i=1}^{r}\alpha_i\boldsymbol{B}_i\boldsymbol{v} \\ &= \sum_{i=1}^{r}\sum_{j=1}^{r}\alpha_i\alpha_j(\boldsymbol{A}_i - \boldsymbol{B}_i\boldsymbol{K}_j)\tilde{\boldsymbol{x}} + \sum_{i=1}^{r}\sum_{j=1}^{r}\alpha_i\alpha_j\boldsymbol{L}_i\boldsymbol{C}_j(\boldsymbol{x} - \tilde{\boldsymbol{x}}) + \sum_{i=1}^{r}\alpha_i\boldsymbol{B}_i\boldsymbol{v}. \end{aligned} \tag{8.90}$$

Equations (8.89) and (8.90) constitute the closed-loop system driven by the combined fuzzy controller–estimator compensator. We represent the above two equations modeling the closed-loop system in matrix form as

$$\begin{aligned} \begin{bmatrix} \dot{\boldsymbol{x}} \\ \dot{\tilde{\boldsymbol{x}}} \end{bmatrix} &= \begin{bmatrix} \sum_{i=1}^{r}\alpha_i\boldsymbol{A}_i & -\sum_{i=1}^{r}\sum_{j=1}^{r}\alpha_i\alpha_j\boldsymbol{B}_i\boldsymbol{K}_j \\ \sum_{i=1}^{r}\sum_{j=1}^{r}\alpha_i\alpha_j\boldsymbol{L}_i\boldsymbol{C}_j & \sum_{i=1}^{r}\sum_{j=1}^{r}\alpha_i\alpha_j(\boldsymbol{A}_i - \boldsymbol{B}_i\boldsymbol{K}_j - \boldsymbol{L}_i\boldsymbol{C}_j) \end{bmatrix} \begin{bmatrix} \boldsymbol{x} \\ \tilde{\boldsymbol{x}} \end{bmatrix} \\ &\quad + \begin{bmatrix} \sum_{i=1}^{r}\alpha_i\boldsymbol{B}_i \\ \sum_{i=1}^{r}\alpha_i\boldsymbol{B}_i \end{bmatrix}\boldsymbol{v} \end{aligned} \tag{8.91}$$

$$\boldsymbol{y} = \begin{bmatrix} \sum_{i=1}^{r}\alpha_i\boldsymbol{C}_i & \boldsymbol{O} \end{bmatrix} \begin{bmatrix} \boldsymbol{x} \\ \tilde{\boldsymbol{x}} \end{bmatrix}. \tag{8.92}$$

8.7.1 The Comparison Method for Linear Systems

We will now present the comparison method for linear systems. This method will then be used to investigate stability properties of the closed-loop system, modeled by (8.91) and (8.92), consisting of a fuzzy model of a given plant and a fuzzy controller–estimator compensator.

Let $\boldsymbol{A}$ be a constant $n \times n$ matrix and let $e^{\boldsymbol{A}t}$ denote the matrix exponential. We will now give a necessary and sufficient condition for the matrix exponential to have all its elements nonnegative for $t \geq 0$. This result appears in Bellman [24, p. 171].

Lemma 8.2 A necessary and sufficient condition that all elements of e^{At} be nonnegative for $t \geq 0$ is that

$$a_{ij} \geq 0, \qquad i \neq j. \tag{8.93}$$

Proof We follow the arguments of Bellman [24, p. 172]; see also Beckenbach and Bellman [23]. We first prove the necessary condition. Because

$$e^{At} = \boldsymbol{I}_n + \boldsymbol{A}t + \tfrac{1}{2}\boldsymbol{A}^2t^2 + \cdots$$

it is clear that (8.93) is necessary for the result to hold for small t, which establishes the necessity condition.

To prove the sufficiency part, let c_1 be a scalar so that all the elements of $\boldsymbol{A} + c_1\boldsymbol{I}_n$ are nonnegative. Then, obviously, all the elements of $e^{(\boldsymbol{A}+c_1\boldsymbol{I}_n)t}$ are nonnegative. Also, the elements of $e^{-c_1\boldsymbol{I}_nt}$ are nonnegative because exponentials are always nonnegative. The matrices $\boldsymbol{A} + c_1\boldsymbol{I}_n$ and $-c_1\boldsymbol{I}_n$ commute. Hence,

$$\begin{aligned} e^{At} &= e^{(\boldsymbol{A}+c_1\boldsymbol{I}_n)t - c_1\boldsymbol{I}_nt} \\ &= e^{(\boldsymbol{A}+c_1\boldsymbol{I}_n)t}e^{-c_1\boldsymbol{I}_nt}, \end{aligned}$$

which establishes the desired nonnegativity condition and the sufficiency condition.

Using the above lemma, we can prove the following lemma due to Bellman [25]. We use the following notation: $\boldsymbol{x} \leq \boldsymbol{y}$ implies component-by-component majorization; that is, $x_i \leq y_i$ for $i = 1, 2, \ldots, n$, where x_i and y_i are the ith components of $\boldsymbol{x}$ and $\boldsymbol{y}$, respectively.

Lemma 8.3 If $a_{ij} \geq 0$, $i \neq j$, then for $t \geq 0$,

$$\frac{d\boldsymbol{x}}{dt} \leq \boldsymbol{A}\boldsymbol{x}, \qquad \boldsymbol{x}(0) = \boldsymbol{c}, \tag{8.94}$$

implies $\boldsymbol{x} \leq \boldsymbol{y}$, where

$$\frac{d\boldsymbol{y}}{dt} = \boldsymbol{A}\boldsymbol{y}, \qquad \boldsymbol{y}(0) = \boldsymbol{c}. \tag{8.95}$$

Proof We can represent (8.94) as

$$\frac{d\boldsymbol{x}(t)}{dt} = \boldsymbol{A}\boldsymbol{x}(t) - \boldsymbol{f}(t), \qquad \boldsymbol{x}(0) = \boldsymbol{c}, \tag{8.96}$$

where $\boldsymbol{f}(t)$ is a vector-valued function whose components are nonnegative for all $t \geq 0$. The solution of (8.96) can be represented as

$$\begin{aligned} \boldsymbol{x}(t) &= e^{At}\boldsymbol{c} - \int_0^t e^{A(t-s)}\boldsymbol{f}(s)\,ds \\ &= \boldsymbol{y}(t) - \int_0^t e^{A(t-s)}\boldsymbol{f}(s)\,ds. \end{aligned} \tag{8.97}$$

In the above, s is the integration variable and $t \geq 0$. Hence, $t \geq s \geq 0$. Using Lemma 8.2, we conclude that the components of the integrand of the second term in (8.97) are all nonnegative for $t \geq s \geq 0$. Therefore,

$$\int_0^t e^{\boldsymbol{A}(t-s)} \boldsymbol{f}(s)\, ds \geq \boldsymbol{0} \qquad \text{for } t \geq 0,$$

which implies

$$\boldsymbol{x} \leq \boldsymbol{y}.$$

The proof is complete.

Equation (8.95) is referred to as the *equation of comparison* for (8.94).

8.7.2 Stability Analysis of the Closed-Loop System

We now perform a stability analysis of the closed-loop system given by (8.91) and (8.92). We first apply a state variable transformation to bring the system into a form convenient for further analysis. The transformation has the form

$$\begin{bmatrix} \boldsymbol{x} \\ \tilde{\boldsymbol{x}} - \boldsymbol{x} \end{bmatrix} = \begin{bmatrix} \boldsymbol{I}_n & \boldsymbol{O} \\ -\boldsymbol{I}_n & \boldsymbol{I}_n \end{bmatrix} \begin{bmatrix} \boldsymbol{x} \\ \tilde{\boldsymbol{x}} \end{bmatrix}. \tag{8.98}$$

The closed-loop system model in the new coordinates is

$$\begin{bmatrix} \dot{\boldsymbol{x}} \\ \dot{\tilde{\boldsymbol{x}}} - \dot{\boldsymbol{x}} \end{bmatrix} = \begin{bmatrix} \sum_{i=1}^r \sum_{j=1}^r \alpha_i \alpha_j (\boldsymbol{A}_i - \boldsymbol{B}_i \boldsymbol{K}_j) & -\sum_{i=1}^r \sum_{j=1}^r \alpha_i \alpha_j \boldsymbol{B}_i \boldsymbol{K}_j \\ \boldsymbol{O} & \sum_{i=1}^r \sum_{j=1}^r \alpha_i \alpha_j (\boldsymbol{A}_i - \boldsymbol{L}_i \boldsymbol{C}_j) \end{bmatrix}$$

$$\times \begin{bmatrix} \boldsymbol{x} \\ \tilde{\boldsymbol{x}} - \boldsymbol{x} \end{bmatrix} + \begin{bmatrix} \sum_{i=1}^r \alpha_i \boldsymbol{B}_i \\ \boldsymbol{O} \end{bmatrix} \boldsymbol{v}, \tag{8.99}$$

$$\boldsymbol{y} = \begin{bmatrix} \sum_{i=1}^r \alpha_i \boldsymbol{C}_i & \boldsymbol{O} \end{bmatrix} \begin{bmatrix} \boldsymbol{x} \\ \tilde{\boldsymbol{x}} - \boldsymbol{x} \end{bmatrix}. \tag{8.100}$$

Let $\boldsymbol{e} = \tilde{\boldsymbol{x}} - \boldsymbol{x}$. We now present a sufficient condition for the unforced closed-loop system to be globally uniformly asymptotically stable; that is, we will give a condition for uniform asymptotic stability in the large of the system

$$\begin{bmatrix} \dot{\boldsymbol{x}} \\ \dot{\boldsymbol{e}} \end{bmatrix} = \begin{bmatrix} \sum_{i=1}^r \sum_{j=1}^r \alpha_i \alpha_j (\boldsymbol{A}_i - \boldsymbol{B}_i \boldsymbol{K}_j) & -\sum_{i=1}^r \sum_{j=1}^r \alpha_i \alpha_j \boldsymbol{B}_i \boldsymbol{K}_j \\ \boldsymbol{O} & \sum_{i=1}^r \sum_{j=1}^r \alpha_i \alpha_j (\boldsymbol{A}_i - \boldsymbol{L}_i \boldsymbol{C}_j) \end{bmatrix} \begin{bmatrix} \boldsymbol{x} \\ \boldsymbol{e} \end{bmatrix}. \tag{8.101}$$

This condition was proposed by Ma, Sun, and He [191].

Theorem 8.7 If there exist two scalar functions $V(\boldsymbol{x}) : \mathbb{R}^n \to \mathbb{R}$ and $\tilde{V}(\boldsymbol{e}) : \mathbb{R}^n \to \mathbb{R}$ and positive real numbers γ_i, $\tilde{\gamma}_i$, $i = 1, \ldots, 4$, such that

$$\gamma_1 \|\boldsymbol{x}\|^2 \le V(\boldsymbol{x}) \le \gamma_2 \|\boldsymbol{x}\|^2, \qquad \tilde{\gamma}_1 \|\boldsymbol{e}\|^2 \le \tilde{V}(\boldsymbol{e}) \le \tilde{\gamma}_2 \|\boldsymbol{e}\|^2, \tag{8.102}$$

$$\begin{aligned} &\frac{\partial V(\boldsymbol{x})}{\partial \boldsymbol{x}} \sum_{i=1}^{r} \sum_{j=1}^{r} \alpha_i \alpha_j (\boldsymbol{A}_i - \boldsymbol{B}_i \boldsymbol{K}_j) \boldsymbol{x} \le -\gamma_3 \|\boldsymbol{x}\|^2, \\ &\frac{\partial \tilde{V}(\boldsymbol{e})}{\partial \boldsymbol{e}} \sum_{i=1}^{r} \sum_{j=1}^{r} \alpha_i \alpha_j (\boldsymbol{A}_i - \boldsymbol{L}_i \boldsymbol{C}_j) \boldsymbol{e} \le -\tilde{\gamma}_3 \|\boldsymbol{e}\|^2, \end{aligned} \tag{8.103}$$

and

$$\left\| \frac{\partial V(\boldsymbol{x})}{\partial \boldsymbol{x}} \right\| \le \gamma_4 \|\boldsymbol{x}\|, \qquad \left\| \frac{\partial \tilde{V}(\boldsymbol{e})}{\partial \boldsymbol{c}} \right\| \le \tilde{\gamma}_4 \|\boldsymbol{e}\|, \tag{8.104}$$

then the system (8.101) is globally uniformly asymptotically stable.

Proof We evaluate the time derivative of V on the trajectories of the system (8.101),

$$\begin{aligned} \dot{V}(\boldsymbol{x}) &= \frac{\partial V(\boldsymbol{x})}{\partial \boldsymbol{x}} \dot{\boldsymbol{x}} \\ &= \frac{\partial V(\boldsymbol{x})}{\partial \boldsymbol{x}} \sum_{i=1}^{r} \sum_{j=1}^{r} \alpha_i \alpha_j (\boldsymbol{A}_i - \boldsymbol{B}_i \boldsymbol{K}_j) \boldsymbol{x} - \frac{\partial V(\boldsymbol{x})}{\partial \boldsymbol{x}} \sum_{i=1}^{r} \sum_{j=1}^{r} \alpha_i \alpha_j \boldsymbol{B}_i \boldsymbol{K}_j \boldsymbol{e} \\ &\le -\gamma_3 \|\boldsymbol{x}\|^2 + \left\| \frac{\partial V(\boldsymbol{x})}{\partial \boldsymbol{x}} \right\| \sum_{i=1}^{r} \sum_{j=1}^{r} \|\boldsymbol{B}_i \boldsymbol{K}_j\| \|\boldsymbol{e}\| \\ &\le -\frac{\gamma_3}{2} \|\boldsymbol{x}\|^2 - \frac{\gamma_3}{2} \|\boldsymbol{x}\|^2 + \gamma_4 \sum_{i=1}^{r} \sum_{j=1}^{r} \|\boldsymbol{B}_i \boldsymbol{K}_j\| \|\boldsymbol{e}\| \|\boldsymbol{x}\|. \end{aligned} \tag{8.105}$$

To proceed, we observe that

$$\frac{\gamma_3}{2} \|\boldsymbol{x}\|^2 - \gamma_4 \sum_{i=1}^{r} \sum_{j=1}^{r} \|\boldsymbol{B}_i \boldsymbol{K}_j\| \|\boldsymbol{e}\| \|\boldsymbol{x}\| + \frac{\gamma_4^2}{2\gamma_3} \left(\sum_{i=1}^{r} \sum_{j=1}^{r} \|\boldsymbol{B}_i \boldsymbol{K}_j\| \right)^2 \|\boldsymbol{e}\|^2 \ge 0. \tag{8.106}$$

Indeed, we can represent the left-hand side of the above as

$$[\|\boldsymbol{x}\| \quad \|\boldsymbol{e}\|] \begin{bmatrix} \dfrac{\gamma_3}{2} & -\dfrac{\gamma_4}{2} \displaystyle\sum_{i=1}^{r} \sum_{j=1}^{r} \|\boldsymbol{B}_i \boldsymbol{K}_j\| \\ -\dfrac{\gamma_4}{2} \displaystyle\sum_{i=1}^{r} \sum_{j=1}^{r} \|\boldsymbol{B}_i \boldsymbol{K}_j\| & \dfrac{\gamma_4^2}{2\gamma_3} \left(\displaystyle\sum_{i=1}^{r} \sum_{j=1}^{r} \|\boldsymbol{B}_i \boldsymbol{K}_j\| \right)^2 \end{bmatrix} \begin{bmatrix} \|\boldsymbol{x}\| \\ \|\boldsymbol{e}\| \end{bmatrix}. \tag{8.107}$$

It is easy to check that the above quadratic form is positive semidefinite—all three principle minors are nonnegative. We use this fact and (8.102) to further evaluate (8.105) as

$$\begin{aligned}\dot{V}(\boldsymbol{x}) &\le -\frac{\gamma_3}{2}\|\boldsymbol{x}\|^2 + \frac{\gamma_4^2}{2\gamma_3}\left(\sum_{i=1}^{r}\sum_{j=1}^{r}\|\boldsymbol{B}_i\boldsymbol{K}_j\|\right)^2\|\boldsymbol{e}\|^2 \\ &\le -\frac{\gamma_3}{2\gamma_2}V(\boldsymbol{x}) + \frac{\gamma_4^2}{2\gamma_3\tilde{\gamma}_1}\left(\sum_{i=1}^{r}\sum_{j=1}^{r}\|\boldsymbol{B}_i\boldsymbol{K}_j\|\right)^2\tilde{V}(\boldsymbol{e}). \end{aligned} \tag{8.108}$$

Let $a = \left(\sum_{i=1}^{r}\sum_{j=1}^{r}\|\boldsymbol{B}_i\boldsymbol{K}_j\|\right)^2$. Then, (8.108) can be written as

$$\dot{V}(\boldsymbol{x}) \le -\frac{\gamma_3}{2\gamma_2}V(\boldsymbol{x}) + \frac{a\gamma_4^2}{2\gamma_3\tilde{\gamma}_1}\tilde{V}(\boldsymbol{e}). \tag{8.109}$$

We now evaluate the time derivative of $\tilde{V}(\boldsymbol{e})$ on the trajectories of (8.101),

$$\begin{aligned}\dot{\tilde{V}}(\boldsymbol{e}) &= \frac{\partial\tilde{V}(\boldsymbol{e})}{\partial\boldsymbol{e}}\dot{\boldsymbol{e}} \\ &\le \frac{\partial\tilde{V}(\boldsymbol{e})}{\partial\boldsymbol{e}}\sum_{i=1}^{r}\sum_{j=1}^{r}\alpha_i\alpha_j(\boldsymbol{A}_i - \boldsymbol{L}_i\boldsymbol{C}_j)\boldsymbol{e} \\ &\le -\tilde{\gamma}_3\|\boldsymbol{e}\|^2 \\ &\le -\frac{\tilde{\gamma}_3}{\tilde{\gamma}_2}\tilde{V}(\boldsymbol{e}). \end{aligned} \tag{8.110}$$

We combine (8.109) and (8.110) into a single matrix differential inequality of the form

$$\begin{aligned}\begin{bmatrix}\dot{V}(\boldsymbol{x}) \\ \dot{\tilde{V}}(\boldsymbol{e})\end{bmatrix} &\le \begin{bmatrix}-\dfrac{\gamma_3}{2\gamma_2} & \dfrac{a\gamma_4^2}{2\gamma_3\tilde{\gamma}_1} \\ 0 & -\dfrac{\tilde{\gamma}_3}{\tilde{\gamma}_2}\end{bmatrix}\begin{bmatrix}V(\boldsymbol{x}) \\ \tilde{V}(\boldsymbol{e})\end{bmatrix} \\ &= \boldsymbol{A}\begin{bmatrix}V(\boldsymbol{x}) \\ \tilde{V}(\boldsymbol{e})\end{bmatrix}. \end{aligned} \tag{8.111}$$

We then refer to Lemma 8.3 to conclude that the system (8.101) is globally uniformly asymptotically stable. Indeed, because the equation of comparison for (8.111) is globally uniformly asymptotically stable, we have to have

$$V(\boldsymbol{x}) \to 0 \quad \text{and} \quad \tilde{V}(\boldsymbol{e}) \to 0 \qquad \text{as } t \to \infty.$$

Because V and $\tilde{V}$ are both positive definite decrescent and radially unbounded functions, the above can hold provided that $\boldsymbol{x} \to \boldsymbol{0}$ and $\boldsymbol{e} \to \boldsymbol{0}$ uniformly with respect to t and irrespective of the initial conditions $\boldsymbol{x}(0)$ and $\boldsymbol{e}(0)$. Therefore, the closed-loop system is globally uniformly asymptotically stable.

8.8 Adaptive Fuzzy Control

In this section we analyze adaptive fuzzy robust tracking controllers for a class of uncertain dynamical systems. The controllers' construction and their analysis involves sliding modes.

8.8.1 Plant Model and Control Objective

We consider a class of plants modeled by the differential equation

$$x^{(n)} = f(\boldsymbol{x}) + g(\boldsymbol{x})u + \eta(t, \boldsymbol{x}), \tag{8.112}$$

where $\boldsymbol{x} = [x \ \dot{x} \ \cdots \ x^{(n-1)}]^T$, and the functions $f = f(\boldsymbol{x})$ and $g = g(\boldsymbol{x})$ are unknown to us. The function $\eta = \eta(t, \boldsymbol{x})$ models the plant's disturbances. We assume that

$$|\eta(t, \boldsymbol{x})| \leq d, \tag{8.113}$$

where $d > 0$ is known to us. We further assume that the function $g(\boldsymbol{x})$ has a positive lower bound; that is, there exists a positive constant $\underline{g}$ such that $g(\boldsymbol{x}) \geq \underline{g} > 0$. Let

$$\boldsymbol{x}_d = \boldsymbol{x}_d(t) = \begin{bmatrix} x_d(t) & \dot{x}_d(t) & \cdots & x_d^{(n-1)}(t) \end{bmatrix}^T$$

denote the desired state trajectory and let

$$\begin{aligned} \boldsymbol{e} &= \boldsymbol{x} - \boldsymbol{x}_d \\ &= \begin{bmatrix} x - x_d & \dot{x} - \dot{x}_d & \cdots & x^{(n-1)} - x_d^{(n-1)} \end{bmatrix}^T \\ &= \begin{bmatrix} e & \dot{e} & \cdots & e^{(n-1)} \end{bmatrix}^T. \end{aligned}$$

We wish to construct a controller u such that

$$\lim_{t\to\infty} \boldsymbol{e}(t) = \boldsymbol{0}.$$

We analyze two different fuzzy adaptive tracking controllers and show that the plant (8.112) driven by these controllers is stable and the adaptation parameters are bounded.

In the next subsection we present background results used in the adaptive fuzzy controllers' construction.

8.8.2 Background Results

Fuzzy Logic System Recall, from Section 8.3, that a fuzzy logic controller with q inputs, v_1, $v_2, \ldots, v_q$, and the center average defuzzifier can be represented as

$$u = \frac{\sum_{l=1}^{m} \prod_{j=1}^{q} \mu_{F_j^l}(v_j)\theta_l}{\sum_{l=1}^{m} \prod_{j=1}^{q} \mu_{F_j^l}(v_j)}, \tag{8.114}$$

where m is the number of fuzzy rules used to construct the controller, F_j^l is the jth fuzzy set corresponding to the lth fuzzy rule, and θ_l can be taken as a centroid of the lth fuzzy set corresponding to the controller's output, u. To proceed, we represent (8.114) as

$$u = \sum_{l=1}^{m} \left(\frac{\prod_{j=1}^{q} \mu_{F_j^l}(v_j)}{\sum_{l=1}^{m} \prod_{j=1}^{q} \mu_{F_j^l}(v_j)} \right) \theta_l. \tag{8.115}$$

Let

$$\boldsymbol{\theta} = [\theta_1 \quad \theta_2 \quad \cdots \quad \theta_m]^T$$

and

$$\boldsymbol{\xi} = [\xi_1 \quad \xi_2 \quad \cdots \quad \xi_m]^T = \left[\frac{\prod_{j=1}^q \mu_{F_j^1}(v_j)}{\sum_{l=1}^m \prod_{j=1}^q \mu_{F_j^l}(v_j)} \quad \cdots \quad \frac{\prod_{j=1}^q \mu_{F_j^m}(v_j)}{\sum_{l=1}^m \prod_{j=1}^q \mu_{F_j^l}(v_j)} \right]^T.$$

Then, we can represent (8.115) as

$$u = \boldsymbol{\theta}^T \boldsymbol{\xi}, \tag{8.116}$$

where the vector $\boldsymbol{\xi}$ is referred to as the *regressor* vector in the control literature.

Projection Operator Consider a vector-valued function of time,

$$\boldsymbol{\theta}(t) = [\theta_1(t) \quad \cdots \quad \theta_m(t)]^T \in \mathbb{R}^m,$$

where $\dot{\theta}_i(t) = \alpha_i(t)$, $i = 1, \ldots, m$. We wish the components $\theta_i(t)$ to lie between $\underline{\theta}_i$ and $\bar{\theta}_i$, that is, $\underline{\theta}_i \le \theta_i(t) \le \bar{\theta}_i$. One way to achieve this goal is by setting $\dot{\theta}_i(t) = 0$ when $\theta_i(t)$ reaches either of the bounds and tends to go beyond the bound, that is,

$$\dot{\theta}_i(t) = \begin{cases} 0 & \text{if } \theta_i = \underline{\theta}_i \text{ and } \alpha_i(t) < 0, \\ 0 & \text{if } \theta_i = \bar{\theta}_i \text{ and } \alpha_i(t) > 0, \\ \alpha_i(t) & \text{otherwise.} \end{cases} \tag{8.117}$$

For compactness, we use the notation *Proj* for the right-hand side of (8.117). Then, we write

$$\dot{\theta}_i(t) = Proj_{\theta_i}(\alpha_i), \qquad i = 1, \ldots, m, \tag{8.118}$$

or in vector notation,

$$\dot{\boldsymbol{\theta}}(t) = Proj_{\boldsymbol{\theta}}(\boldsymbol{\alpha}). \tag{8.119}$$

Sliding Modes Let

$$\boldsymbol{s} = [s_1 \quad s_2 \quad \cdots \quad s_{n-1} \quad 1] \in \mathbb{R}^{1\times n}$$

and $\sigma(\boldsymbol{e}) = \boldsymbol{se}$. Then, $\{\boldsymbol{e} : \sigma(\boldsymbol{e}) = 0\}$ represents a sliding surface in the tracking error, $\boldsymbol{e}$, space. The coefficients s_i of the sliding surface are chosen so that the system (8.112) restricted to the sliding surface is asymptotically stable—see Section 6.7 for a sliding surface design algorithm. Our objective is to construct fuzzy tracking controllers that force the tracking error to approach the sliding surface, $\{\sigma(\boldsymbol{e}) = 0\}$, and then stay in some neighborhood of the sliding surface for all subsequent time. If f and g were known exactly to us and $\eta = 0$, then the controller

$$u^* = \frac{1}{g}\left(-f - \mu\sigma + x_d^{(n)} - \boldsymbol{k}^T\boldsymbol{e}\right), \tag{8.120}$$

where $\boldsymbol{k} = [0 \;\; s_1 \;\; \cdots \;\; s_{n-1}]^T$ and $\mu > 0$ is a design parameter, would force the tracking error, $\boldsymbol{e}$, to behave as desired. We can verify this using standard arguments from sliding mode control practice—see Section 6.2. We consider the function $V = 0.5\sigma^2$. Note that V is positive definite with respect to the sliding surface $\{\sigma = 0\}$. The function V can be viewed as a distance measure to the sliding surface $\{\sigma = 0\}$. The time derivative of V evaluated on the trajectories of the closed-loop system (8.112), (8.120) is

$$\begin{aligned}\dot{V} &= \sigma\dot{\sigma} \\ &= \sigma \boldsymbol{s}\dot{\boldsymbol{e}} \\ &= \sigma\left(e^{(n)} + s_{n-1}e^{(n-1)} + \cdots + s_1\dot{e}\right) \\ &= \sigma\left(x^{(n)} - x_d^{(n)} + \boldsymbol{k}^T\boldsymbol{e}\right) \\ &= \sigma\left(f + gu - x_d^{(n)} + \boldsymbol{k}^T\boldsymbol{e}\right).\end{aligned} \tag{8.121}$$

Substituting into the equation above the control law given by (8.120) yields

$$\dot{V} = -\mu\sigma^2. \tag{8.122}$$

Thus, the sliding surface, $\{\sigma = 0\}$, is asymptotically attractive and the system restricted to the sliding surface can be made asymptotically stable with respect to the origin by an appropriate choice of the parameters, s_i, of the sliding surface.

The control law (8.120) requires perfect knowledge of the plant's model. However, the plant model components f and g are unknown to us. In the following subsection, we describe how to use adaptive fuzzy logic controllers to approximate, as closely as possible, the performance of (8.120). In the following, we consider two cases. In the first case, the function $f = f(\boldsymbol{x})$ is unknown to us while $g = g(\boldsymbol{x}) > 0$ is known to us. In the second case, both f and g are unknown to us. In both cases, we consider the presence of a bounded disturbance modeled by η.

8.8.3 Controllers

Unknown $f(\boldsymbol{x})$ When in our plant model (8.112) only $g(\boldsymbol{x})$ is known, $\eta \neq 0$, and $f(\boldsymbol{x})$ is unknown, we propose to approximate $f(\boldsymbol{x})$ with fuzzy logic system of the form $\boldsymbol{\theta}_f^T\boldsymbol{\xi}_f(\boldsymbol{x})$, where the functions $\boldsymbol{\theta}_f$ and $\boldsymbol{\xi}_f$ will be described after defining relevant symbols. We use $\boldsymbol{\theta}_f^*$ to denote an "optimal" vector such that

$$\boldsymbol{\theta}_f^* = \operatorname{argmin}_{\boldsymbol{\theta}_f} \sup_{\boldsymbol{x}\in\Omega}\left|f(\boldsymbol{x}) - \boldsymbol{\theta}_f^T\boldsymbol{\xi}_f(\boldsymbol{x})\right|, \tag{8.123}$$

where $\Omega \subseteq \mathbb{R}^n$ is a region to which the state $\boldsymbol{x}$ is constrained to reside. We assume that

$$\left|f(\boldsymbol{x}) - \boldsymbol{\theta}_f^{*T}\boldsymbol{\xi}_f(\boldsymbol{x})\right| \le d_f \qquad \text{for all } \boldsymbol{x} \in \Omega, \tag{8.124}$$

where $d_f > 0$ and each element of $\boldsymbol{\theta}_f^*$ is constant and bounded below and above as follows:

$$\underline{\theta}_{fi} \le \theta_{fi}^* \le \bar{\theta}_{fi} \qquad \text{for all } i = 1, \ldots, m, \tag{8.125}$$

or in vector notation,

$$\underline{\boldsymbol{\theta}}_f \le \boldsymbol{\theta}_f^* \le \bar{\boldsymbol{\theta}}_f. \tag{8.126}$$

We define the adaptation parameter error as

$$\boldsymbol{\phi}_f = \boldsymbol{\theta}_f - \boldsymbol{\theta}_f^*. \tag{8.127}$$

We will use the adaptation law

$$\dot{\boldsymbol{\phi}}_f(t) = \dot{\boldsymbol{\theta}}_f(t) = Proj_{\boldsymbol{\theta}_f}(\gamma_f\sigma\boldsymbol{\xi}_f(t)), \tag{8.128}$$

where $\gamma_f > 0$ is a design parameter. It is easy to verify, using the definition of *Proj* and the fact that $\underline{\theta}_{fi} \le \theta^*_{fi} \le \bar{\theta}_{fi}$ for each i, that

$$\boldsymbol{\phi}_f^T \left(\frac{1}{\gamma_f} Proj_{\boldsymbol{\theta}_f}(\gamma_f \sigma \boldsymbol{\xi}_f) - \sigma \boldsymbol{\xi}_f \right) \le 0. \tag{8.129}$$

We now analyze the following fuzzy adaptive control law,

$$\boxed{u = \frac{1}{g}\left(-\boldsymbol{\theta}_f^T \boldsymbol{\xi}_f - \mu\sigma + x_d^{(n)} - \boldsymbol{k}^T \boldsymbol{e} + u_s\right)} \tag{8.130}$$

with u_s satisfying

$$\sigma\left(f - \boldsymbol{\theta}_f^T \boldsymbol{\xi}_f + u_s + \eta\right) \le \epsilon, \tag{8.131}$$

$$\sigma u_s \le 0, \tag{8.132}$$

where $\epsilon > 0$ is a design parameter. We state and prove a theorem concerning the dynamical behavior of the closed-loop system driven by the control law (8.130).

Theorem 8.8 For the closed-loop system,

$$\begin{aligned} x^{(n)} &= f + gu + \eta, \\ u &= \frac{1}{g}\left(-\boldsymbol{\theta}_f^T \boldsymbol{\xi}_f - \mu\sigma + x_d^{(n)} - \boldsymbol{k}^T \boldsymbol{e} + u_s\right), \\ \dot{\boldsymbol{\theta}}_f &= Proj_{\boldsymbol{\theta}_f}(\gamma_f \sigma \boldsymbol{\xi}_f), \end{aligned} \tag{8.133}$$

where u_s satisfies (8.131) and (8.132), we have the following:

1.
$$\sigma^2(t) \le e^{-2\mu t}\sigma^2(0) + \frac{\epsilon}{\mu}. \tag{8.134}$$

2. If $\eta = 0$ and there exists $\boldsymbol{\theta}_f^*$ such that $f(\boldsymbol{x}) = \boldsymbol{\theta}_f^{*T} \boldsymbol{\xi}_f(\boldsymbol{x})$, then the origin of the $(\sigma, \boldsymbol{\phi}_f)$ space is stable, and hence $\sigma(t)$ and $\boldsymbol{\phi}_f(t)$ are bounded for all $t \ge 0$ and $e(t) \to 0$ as $t \to \infty$.

Proof To prove 1, consider $V = 0.5\sigma^2$. The time derivative of V evaluated on the solutions of the closed-loop system (8.112), (8.130) is

$$\begin{aligned} \dot{V} &= \sigma\dot{\sigma} \\ &= -\mu\sigma^2 + \sigma\left(f - \boldsymbol{\theta}_f^T \boldsymbol{\xi}_f + u_s + \eta\right). \end{aligned}$$

Taking into account (8.131) yields

$$\dot{V} \le -\mu\sigma^2 + \epsilon = -2\mu V + \epsilon.$$

Applying to the above the comparison theorem, presented in Subsection A.14.2, we obtain

$$\begin{aligned} V(t) &\le e^{-2\mu t} V(0) + \frac{\epsilon}{2\mu}(1 - e^{-2\mu t}) \\ &\le e^{-2\mu t} V(0) + \frac{\epsilon}{2\mu}. \end{aligned}$$

Noting in the above that $V = 0.5\sigma^2$, we obtain (8.134), which proves the first part of the theorem.

To prove 2, consider the following Lyapunov function candidate:

$$V = \frac{1}{2}\left(\sigma^2 + \frac{1}{\gamma_f}\boldsymbol{\phi}_f^T\boldsymbol{\phi}_f\right). \tag{8.135}$$

We evaluate the time derivative of the above Lyapunov function candidate on the solutions of the closed-loop system (8.133), where $\eta = 0$. We obtain

$$\dot{V} = \sigma\left(-\mu\sigma + f - \boldsymbol{\theta}_f^T\boldsymbol{\xi}_f + u_s\right) + \frac{1}{\gamma_f}\boldsymbol{\phi}_f^T\dot{\boldsymbol{\phi}}_f.$$

Taking into account the assumption that $f = \boldsymbol{\theta}_f^{*T}\boldsymbol{\xi}_f$ and using (8.127), (8.128), (8.129), and (8.132) gives

$$\begin{aligned}
\dot{V} &= -\mu\sigma^2 + \sigma u_s - (\boldsymbol{\theta}_f^T\boldsymbol{\xi}_f - \boldsymbol{\theta}_f^{*T}\boldsymbol{\xi}_f)\sigma + \frac{1}{\gamma_f}\boldsymbol{\phi}_f^T\dot{\boldsymbol{\phi}}_f \\
&= -\mu\sigma^2 + \sigma u_s - \boldsymbol{\phi}_f^T\boldsymbol{\xi}_f\sigma + \boldsymbol{\phi}_f^T\frac{1}{\gamma_f}Proj_{\boldsymbol{\theta}_f}(\gamma_f\sigma\boldsymbol{\xi}_f) \\
&\le -\mu\sigma^2 + \sigma u_s \\
&\le -\mu\sigma^2.
\end{aligned} \tag{8.136}$$

Thus, $\dot{V} \le 0$ in the $[\sigma \;\; \boldsymbol{\phi}_f]^T$ space, which implies that the zero solution of the closed-loop system (8.133) is stable; therefore, $\sigma(t)$ and $\boldsymbol{\phi}_f(t)$ are bounded for $t \ge 0$.

To show that $\sigma(t) \to 0$ as $t \to \infty$, we integrate (8.136) from 0 to t to obtain

$$\int_0^t \dot{V}(\tau)\, d\tau = V(t) - V(0) \le \int_0^t (-\mu\sigma^2)\, d\tau, \tag{8.137}$$

or, equivalently,

$$\int_0^t \mu\sigma^2 d\tau \le V(0) - V(t) \le V(0). \tag{8.138}$$

Thus, $\lim_{t\to\infty}\int_0^t \sigma^2(\tau)\,d\tau$ exists. It follows from the first part of the theorem that σ^2 is bounded and because

$$\frac{d}{dt}\sigma^2 = 2\sigma\dot{\sigma} \le -\mu\sigma^2 + \epsilon, \tag{8.139}$$

we conclude that $\frac{d}{dt}\sigma^2$ is also bounded. Therefore, σ^2 is uniformly continuous. By Barbalat's lemma (see Section 4.12), $\sigma^2 \to 0$ as $t \to \infty$. This implies the following for tracking error: $e(t) \to 0$ as $t \to \infty$, which concludes the proof of the theorem.

Selecting u_s in Control Law (8.130) We now discuss a method that can be used to select the component u_s while implementing the control law (8.130). Let h be a constant such that

$$h \ge \|\boldsymbol{\xi}_f\|\|\bar{\boldsymbol{\theta}}_f - \underline{\boldsymbol{\theta}}_f\|.$$

Given the design parameter ϵ, choose positive numbers ϵ_1, ϵ_2, and ϵ_3 such that $\epsilon = \epsilon_1 + \epsilon_2 + \epsilon_3$. Then, u_s can be chosen as

$$u_s = -k_s\sigma, \tag{8.140}$$

where

$$k_s \geq \frac{d_f^2}{4\epsilon_1} + \frac{h^2}{4\epsilon_2} + \frac{d^2}{4\epsilon_3}. \tag{8.141}$$

We will now verify that the above choice of u_s guarantees that (8.140) satisfies (8.131) and (8.132). We first consider the left-hand side of (8.131). Adding and subtracting $\boldsymbol{\theta}_f^{*T}\boldsymbol{\xi}_f$ and using (8.127) yields

$$\begin{aligned}\sigma\left(f - \boldsymbol{\theta}_f^T\boldsymbol{\xi}_f + u_s + \eta\right) &= \sigma\left(f - \boldsymbol{\theta}_f^{*T}\boldsymbol{\xi}_f + \boldsymbol{\theta}_f^{*T}\boldsymbol{\xi}_f - \boldsymbol{\theta}_f^T\boldsymbol{\xi}_f + u_s + \eta\right)\\ &= \sigma\left(f - \boldsymbol{\theta}_f^{*T}\boldsymbol{\xi}_f - \boldsymbol{\phi}_f^T\boldsymbol{\xi}_f - k_s\sigma + \eta\right).\end{aligned} \tag{8.142}$$

We now use (8.141) and group the terms to obtain

$$\begin{aligned}\sigma\left(f - \boldsymbol{\theta}_f^T\boldsymbol{\xi}_f + u_s + \eta\right) \leq{}& \sigma\left(f - \boldsymbol{\theta}_f^{*T}\boldsymbol{\xi}_f - \frac{d_f^2}{4\epsilon_1}\sigma\right)\\ &-\sigma\left(\boldsymbol{\phi}_f^T\boldsymbol{\xi}_f + \frac{h^2}{4\epsilon_2}\sigma\right) + \sigma\left(\eta - \frac{d^2}{4\epsilon_3}\sigma\right).\end{aligned} \tag{8.143}$$

Next, we consider each term on the right-hand side of the above inequality separately. We start with the first term. Completing the squares, we obtain

$$\sigma\left(f - \boldsymbol{\theta}_f^{*T}\boldsymbol{\xi}_f - \frac{d_f^2}{4\epsilon_1}\sigma\right) = -\left(\frac{d_f}{2\sqrt{\epsilon_1}}\sigma - \frac{f - \boldsymbol{\theta}_f^{*T}\boldsymbol{\xi}_f}{d_f/\sqrt{\epsilon_1}}\right)^2 + \left(\frac{f - \boldsymbol{\theta}_f^{*T}\boldsymbol{\xi}_f}{d_f/\sqrt{\epsilon_1}}\right)^2.$$

Neglecting the first term in the above and using the assumption (8.124), we get

$$\sigma\left(f - \boldsymbol{\theta}_f^{*T}\boldsymbol{\xi}_f - \frac{d_f^2}{4\epsilon_1}\sigma\right) \leq \frac{d_f^2}{d_f^2/\epsilon_1} \leq \epsilon_1. \tag{8.144}$$

Completing the squares in the second term on the right-hand side of (8.143) gives

$$\begin{aligned}-\sigma\left(\boldsymbol{\phi}_f^T\boldsymbol{\xi}_f + \frac{h^2}{4\epsilon_2}\sigma\right) &= -\left(\frac{h}{2\sqrt{\epsilon_2}}\sigma + \frac{\boldsymbol{\phi}_f^T\boldsymbol{\xi}_f}{h/\sqrt{\epsilon_2}}\right)^2 + \left(\frac{\boldsymbol{\phi}_f^T\boldsymbol{\xi}_f}{h/\sqrt{\epsilon_2}}\right)^2\\ &\leq \left(\frac{\boldsymbol{\phi}_f^T\boldsymbol{\xi}_f}{h/\sqrt{\epsilon_2}}\right)^2\\ &= \left(\frac{(\boldsymbol{\theta}_f - \boldsymbol{\theta}_f^*)^T\boldsymbol{\xi}_f}{h}\right)^2\epsilon_2\\ &\leq \frac{\|\boldsymbol{\theta}_f - \boldsymbol{\theta}_f^*\|^2\|\boldsymbol{\xi}_f\|^2}{h^2}\epsilon_2\\ &\leq \frac{\|\boldsymbol{\theta}_f - \boldsymbol{\theta}_f^*\|^2\|\boldsymbol{\xi}_f\|^2}{\|\boldsymbol{\xi}_f\|^2\|\bar{\boldsymbol{\theta}}_f - \underline{\boldsymbol{\theta}}_f\|^2}\epsilon_2 && (8.145)\\ &\leq \epsilon_2. && (8.146)\end{aligned}$$

Completing the squares in the third term on the right-hand side of (8.143) yields

$$\sigma\left(\eta - \frac{d^2}{4\epsilon_3}\sigma\right) = -\left(\frac{d}{2\sqrt{\epsilon_3}}\sigma - \frac{\eta}{d/\sqrt{\epsilon_3}}\right)^2 + \frac{\eta^2}{d^2}\epsilon_3$$
$$\leq \epsilon_3.$$

Thus,

$$\sigma\left(f - \boldsymbol{\theta}_f^T\boldsymbol{\xi}_f + u_s + \eta\right) \leq \epsilon_1 + \epsilon_2 + \epsilon_3 = \epsilon.$$

Hence (8.131) is satisfied.

Also, $u_s = -k_s\sigma$, $k_s > 0$, satisfies (8.132).

Unknown $f(x)$ and $g(x)$ We now assume that $g(\boldsymbol{x})$ and $f(\boldsymbol{x})$ are unknown to us and $\eta \neq 0$. We approximate $f(\boldsymbol{x})$ and $g(\boldsymbol{x})$ with fuzzy logic systems, $\boldsymbol{\theta}_f^T\boldsymbol{\xi}_f(\boldsymbol{x})$ and $\boldsymbol{\theta}_g^T\boldsymbol{\xi}_g(\boldsymbol{x})$, respectively. The conditions on $\boldsymbol{\theta}_f^T\boldsymbol{\xi}_f(\boldsymbol{x})$ are the same as in the previous case, and the conditions on $\boldsymbol{\theta}_g^T\boldsymbol{\xi}_g(\boldsymbol{x})$ are similar; that is, the optimal parameter vector, $\boldsymbol{\theta}_g^*$, satisfies

$$\boldsymbol{\theta}_g^* = \text{argmin}_{\boldsymbol{\theta}_g} \sup_{\boldsymbol{x}\in\Omega} \left|g(\boldsymbol{x}) - \boldsymbol{\theta}_g^T\boldsymbol{\xi}_g(\boldsymbol{x})\right|. \tag{8.147}$$

We assume that

$$\left|g(\boldsymbol{x}) - \boldsymbol{\theta}_g^{*T}\boldsymbol{\xi}_g(\boldsymbol{x})\right| \leq d_g \qquad \text{for all } \boldsymbol{x} \in \Omega, \tag{8.148}$$

where

$$\underline{\theta}_{gi} \leq \theta_{gi}^* \leq \bar{\theta}_{gi} \qquad \text{for all } i = 1, \ldots, m. \tag{8.149}$$

The parameter error, $\boldsymbol{\phi}_g = \boldsymbol{\theta}_g - \boldsymbol{\theta}_g^*$, is adapted according to the law

$$\dot{\boldsymbol{\phi}}_g = \dot{\boldsymbol{\theta}}_g = Proj_{\boldsymbol{\theta}_g}(\gamma_g\sigma\boldsymbol{\xi}_g u_a),$$

where

$$u_a = \frac{1}{\boldsymbol{\theta}_g^T\boldsymbol{\xi}_g}\left(-\boldsymbol{\theta}_f^T\boldsymbol{\xi}_f + x_d^{(n)} - \boldsymbol{k}^T\boldsymbol{e}\right). \tag{8.150}$$

The initial values of the components of $\boldsymbol{\theta}_g$ are chosen to satisfy (8.149), where for each $i = 1, \ldots, m$ we have $\theta_{gi} > 0$. This ensures that $\boldsymbol{\theta}_g^T\boldsymbol{\xi}_g > 0$ because we choose fuzzy sets so that at any t at least one rule fires and so at least one component of $\boldsymbol{\xi}$ is nonzero and in fact positive. Using the definition of *Proj*, one can verify that

$$\boldsymbol{\phi}_g^T\left(\frac{1}{\gamma_g}Proj_{\boldsymbol{\theta}_g}(\gamma_g\sigma\boldsymbol{\xi}_g u_a) - \sigma\boldsymbol{\xi}_g u_a\right) \leq 0. \tag{8.151}$$

We now analyze the following fuzzy adaptive control law,

$$\boxed{u = \frac{1}{\boldsymbol{\theta}_g^T\boldsymbol{\xi}_g}\left(-\boldsymbol{\theta}_f^T\boldsymbol{\xi}_f + x_d^{(n)} - \boldsymbol{k}^T\boldsymbol{e}\right) - \frac{1}{\underline{g}}\mu\sigma + u_s} \tag{8.152}$$

with u_s satisfying the following conditions:

$$\sigma\left(f - x_d^{(n)} + \boldsymbol{k}^T\boldsymbol{e} + gu_a + gu_s + \eta\right) \leq \epsilon, \tag{8.153}$$

$$\sigma u_s \leq 0, \tag{8.154}$$

where $\epsilon > 0$ is a design parameter. Using the definition of u_a, given by (8.150), we represent (8.152) as

$$u = u_a - \frac{1}{\underline{g}}\mu\sigma + u_s. \tag{8.155}$$

Theorem 8.9 For the closed-loop system,

$$\begin{aligned} x^{(n)} &= f + gu + \eta, \\ u &= \frac{1}{\boldsymbol{\theta}_g^T \boldsymbol{\xi}_g}\left(-\boldsymbol{\theta}_f^T \boldsymbol{\xi}_f + x_d^{(n)} - \boldsymbol{k}^T \mathbf{e}\right) - \frac{1}{\underline{g}}\mu\sigma + u_s, \\ \dot{\boldsymbol{\theta}}_f &= Proj_{\boldsymbol{\theta}_f}(\gamma_f \sigma \boldsymbol{\xi}_f), \\ \dot{\boldsymbol{\theta}}_g &= Proj_{\boldsymbol{\theta}_g}(\gamma_g \sigma \boldsymbol{\xi}_g u_a), \end{aligned} \tag{8.156}$$

where u_s satisfies (8.153) and (8.154), we have the following:

1.

$$\sigma^2(t) \le e^{-2\mu t}\sigma^2(0) + \frac{\epsilon}{\mu}. \tag{8.157}$$

2. If $\eta = 0$ and there exist $\boldsymbol{\theta}_f^*$ and $\boldsymbol{\theta}_g^*$ such that $f(\boldsymbol{x}) = \boldsymbol{\theta}_f^{*T} \boldsymbol{\xi}_f(\boldsymbol{x})$ and $g(\boldsymbol{x}) = \boldsymbol{\theta}_g^{*T} \boldsymbol{\xi}_g(\boldsymbol{x})$, then the origin of the $[\sigma \ \ \boldsymbol{\phi}_f \ \ \boldsymbol{\phi}_g]^T$-space is stable and hence $\sigma(t)$, $\boldsymbol{\phi}_f(t)$, and $\boldsymbol{\phi}_g(t)$ are bounded and $e(t) \to 0$ as $t \to \infty$.

Proof To prove 1, consider $V = 0.5\sigma^2$. The time derivative of V evaluated on the solutions of the closed-loop system (8.112), (8.152) is

$$\begin{aligned} \dot{V} &= \sigma\dot{\sigma} \\ &= \sigma\left(f + gu + \eta - x_d^{(n)} + \boldsymbol{k}^T \mathbf{e}\right). \end{aligned}$$

Substituting into the above the alternative expression for the control law given by (8.155) and applying (8.153) gives

$$\begin{aligned} \dot{V} &= \sigma\left(f + gu_a - \frac{g}{\underline{g}}\mu\sigma + gu_s + \eta - x_d^{(n)} + \boldsymbol{k}^T \mathbf{e}\right) \\ &\le -\frac{g}{\underline{g}}\mu\sigma^2 + \epsilon \\ &= -2\mu V + \epsilon. \end{aligned}$$

Invoking now the comparison theorem, we obtain (8.157).

To prove 2, consider the following Lyapunov function candidate:

$$V = \frac{1}{2}\left(\sigma^2 + \frac{1}{\gamma_f}\boldsymbol{\phi}_f^T \boldsymbol{\phi}_f + \frac{1}{\gamma_g}\boldsymbol{\phi}_g^T \boldsymbol{\phi}_g\right). \tag{8.158}$$

By assumption, $f = \boldsymbol{\theta}_f^{*T} \boldsymbol{\xi}_f$, $g = \boldsymbol{\theta}_g^{*T} \boldsymbol{\xi}_g$, $\eta = 0$ as well as conditions (8.129), (8.151), and (8.154) hold. In addition, it follows from (8.150) that

$$x_d^{(n)} - \boldsymbol{k}^T \mathbf{e} = \boldsymbol{\theta}_f^T \boldsymbol{\xi}_f + \boldsymbol{\theta}_g^T \boldsymbol{\xi}_g u_a.$$

Taking into account all of the above in evaluating the time derivative of V on the trajectories of the closed-loop system (8.156), we obtain

$$\begin{aligned}\dot{V} &= \sigma\left(-\frac{g}{\underline{g}}\mu\sigma + f - \boldsymbol{\theta}_f^T\boldsymbol{\xi}_f + (g - \boldsymbol{\theta}_g^T\boldsymbol{\xi}_g)u_a + gu_s\right) + \frac{1}{\gamma_f}\boldsymbol{\phi}_f^T\dot{\boldsymbol{\phi}}_f + \frac{1}{\gamma_g}\boldsymbol{\phi}_g^T\dot{\boldsymbol{\phi}}_g \\ &= \sigma(f - \boldsymbol{\theta}_f^T\boldsymbol{\xi}_f + (g - \boldsymbol{\theta}_g^T\boldsymbol{\xi}_g)u_a + gu_s) - \frac{g}{\underline{g}}\mu\sigma^2 \\ &\quad + \boldsymbol{\phi}_f^T\frac{1}{\gamma_f}Proj_{\boldsymbol{\theta}_f}(\gamma_f\sigma\boldsymbol{\xi}_f) + \boldsymbol{\phi}_g^T\frac{1}{\gamma_g}Proj_{\boldsymbol{\theta}_g}(\gamma_g\sigma\boldsymbol{\xi}_g u_a) \\ &\le -\frac{g}{\underline{g}}\mu\sigma^2 + \sigma g u_s - \boldsymbol{\phi}_f^T\boldsymbol{\xi}_f\sigma + \boldsymbol{\phi}_f^T\frac{1}{\gamma_f}Proj_{\boldsymbol{\theta}_f}(\gamma_f\sigma\boldsymbol{\xi}_f) \\ &\quad - \sigma\boldsymbol{\phi}_g^T\boldsymbol{\xi}_g u_a + \boldsymbol{\phi}_g^T\frac{1}{\gamma_g}Proj_{\boldsymbol{\theta}_g}(\gamma_g\sigma\boldsymbol{\xi}_g u_a) \\ &\le -\mu\sigma^2 + g\sigma u_s \\ &\le -\mu\sigma^2.\end{aligned}$$

It follows from the above that the closed-loop system (8.156) is stable; therefore, $\sigma(t)$, $\boldsymbol{\phi}_f(t)$, and $\boldsymbol{\phi}_g(t)$ are bounded for $t \ge 0$.

Applying now Barbalat's lemma, similar to the way we did in the proof of Theorem 8.8, we conclude that $\sigma(t) \to 0$ as $t \to \infty$, which completes the proof of the theorem.

Selecting u_s in Control Law (8.152) The component u_s appearing in the control law (8.152) can be chosen similarly as in the previous case. We now give a lower bound on the gain k_s. Let

$$h_f \ge \|\boldsymbol{\xi}_f\|\,\|\bar{\boldsymbol{\theta}}_f - \underline{\boldsymbol{\theta}}_f\| \quad \text{and} \quad h_g \ge \|\boldsymbol{\xi}_g\|\,\|\bar{\boldsymbol{\theta}}_g - \underline{\boldsymbol{\theta}}_g\|\,|u_a|.$$

Given the design parameter ϵ, select positive numbers ϵ_1, ϵ_2, ϵ_3, and ϵ_4 such that $\epsilon = \epsilon_1 + \epsilon_2 + \epsilon_3 + \epsilon_4$. Then, k_s should satisfy the following condition:

$$k_s \ge \frac{1}{\underline{g}}\left(\frac{d_f^2 + |u_a|^2 d_g^2}{2\epsilon_1} + \frac{h_f^2}{4\epsilon_2} + \frac{h_g^2}{4\epsilon_3} + \frac{d^2}{4\epsilon_4}\right). \tag{8.159}$$

When the gain k_s satisfies the above condition, $u_s = -k_s\sigma$ guarantees that (8.153) and (8.154) are satisfied. To show that this is indeed the case, consider the left-hand side of (8.153). Use $\boldsymbol{\theta}_g^T\boldsymbol{\xi}_g u_a + \boldsymbol{\theta}_f^T\boldsymbol{\xi}_f = x_d^{(n)} - \boldsymbol{k}^T\boldsymbol{e}$ from (8.150), and then add and subtract $\boldsymbol{\theta}_f^{*T}\boldsymbol{\xi}_f$ and $\boldsymbol{\theta}_g^{*T}\boldsymbol{\xi}_g u_a$ to obtain

$$\begin{aligned}\sigma\left(f - x_d^{(n)} + \boldsymbol{k}^T\boldsymbol{e} + gu_a + gu_s + \eta\right) &= \sigma\left(f - \boldsymbol{\theta}_f^T\boldsymbol{\xi}_f - \boldsymbol{\theta}_g^T\boldsymbol{\xi}_g u_a + gu_a + gu_s + \eta\right) \\ &= \sigma\big(f - \boldsymbol{\theta}_f^{*T}\boldsymbol{\xi}_f - \boldsymbol{\phi}_f^T\boldsymbol{\xi}_f + (g - \boldsymbol{\theta}_g^{*T}\boldsymbol{\xi}_g)u_a \\ &\quad - \boldsymbol{\phi}_g^T\boldsymbol{\xi}_g u_a + gu_s + \eta\big).\end{aligned}$$

Taking into account (8.159) and rearranging terms gives

$$\begin{aligned}
&\sigma\left(f - x_d^{(n)} + \boldsymbol{k}^T\boldsymbol{e} + gu_a + gu_s + \eta\right)\\
&\le \sigma\left(f - \boldsymbol{\theta}_f^T\boldsymbol{\xi}_f - \frac{d_f^2}{2\epsilon_1}\sigma\right) + \sigma\left(\left(g - \boldsymbol{\theta}_g^{*T}\boldsymbol{\xi}_g\right)u_a - \frac{|u_a|^2 d_g^2}{2\epsilon_1}\sigma\right)\\
&\quad - \sigma\left(\boldsymbol{\phi}_f^T\boldsymbol{\xi}_f + \frac{h_f^2}{4\epsilon_2}\sigma\right) - \sigma\left(\boldsymbol{\phi}_g^T\boldsymbol{\xi}_g u_a + \frac{h_g^2}{4\epsilon_3}\sigma\right) + \sigma\left(\eta - \frac{d^2}{4\epsilon_4}\sigma\right).
\end{aligned}$$

Completing the squares, similarly as in the previous case, yields

$$\begin{aligned}
&\sigma\left(f - x_d^{(n)} + \boldsymbol{k}^T\boldsymbol{e} + gu_a + gu_s + \eta\right)\\
&\le \left(\frac{f - \boldsymbol{\theta}_f^T\boldsymbol{\xi}_f}{\sqrt{2}d_f/\sqrt{\epsilon_1}}\right)^2 + \left(\frac{\left(g - \boldsymbol{\theta}_g^{*T}\boldsymbol{\xi}_g\right)u_a}{\sqrt{2}|u_a|d_g/\sqrt{\epsilon_1}}\right)^2 + \left(\frac{\boldsymbol{\phi}_f^T\boldsymbol{\xi}_f}{h_f^2/\sqrt{\epsilon_2}}\right)^2 + \left(\frac{\boldsymbol{\phi}_g^T\boldsymbol{\xi}_g u_a}{h_g^2/\sqrt{\epsilon_3}}\right)^2 + \left(\frac{\eta}{d/\sqrt{\epsilon_4}}\right)^2\\
&\le \frac{1}{2}\epsilon_1 + \frac{1}{2}\epsilon_1 + \epsilon_2 + \epsilon_3 + \epsilon_4\\
&= \epsilon.
\end{aligned}$$

Hence, (8.153) is satisfied, and so is (8.154) by the fact that $u_s = -k_s\sigma$, where $k_s > 0$.

8.8.4 Examples

In this subsection we present simulation results illustrating the performance of the control strategies discussed above. The controllers are tested on a simplified model of an electric drive system used by Fischle and Schröder [85] to test their fuzzy adaptive controllers. The plant is modeled by the differential equation

$$\ddot{x} = -\frac{5}{J}\arctan(5\dot{x}) + \frac{c_T}{J}u = f(x, \dot{x}) + gu, \tag{8.160}$$

where x is the angular position, $c_T = 10$ Nm/A, and $J = 0.1$ kg·m^2. We generate the reference signal x_d using the reference system whose transfer function is given by $\frac{400}{s^2+40s+400}$. The input signal, $w(t)$, to the reference system is changing its value randomly between -1.5 and 1.5 every 0.5 seconds.

◆ Example 8.15

In this example, we assume that we know g, where $g = 100$, and that there are no disturbances affecting the system; that is, $\eta = 0$. We use the control law given by (8.130), where $f(\boldsymbol{x})$ is approximated by a fuzzy logic system. We use fuzzy sets for x and $\dot{x}$ as in Fischle and Schröder [85]. They are shown in Figure 8.37. There are two fuzzy sets for x and six fuzzy sets for $\dot{x}$. Thus, we can have twelve fuzzy rules possible. They have the following form:

Rule 1: IF x is N AND $\dot{x}$ is LN, THEN $y = \theta_1$

⋮

Rule 12: IF x is P AND $\dot{x}$ is LP, THEN $y = \theta_{12}$, where θ_{fi}, $i = 1, \ldots, 12$ are the adaptation parameters. The bounds on the components of the adaptation parameter vector, $\boldsymbol{\theta}_f$, are chosen to be $\bar{\theta}_{fi} = 200$ and $\underline{\theta}_{fi} = -200$. All the initial values of the components of $\boldsymbol{\theta}_f$ are set to zero. We chose $d_f = 50$, $d_g = 10$, and $\gamma_f = 5000$. The values of the remaining design parameters are: $\mu = 250$, $\epsilon_1 = 50$, $\epsilon_2 = 50$. As $\eta(t) = 0$, we do not need to worry about ϵ_3.

SIMULINK is used to simulate the closed-loop system. Its block diagram is depicted in Figure 8.38. The regressor generator block produces $\boldsymbol{\xi}$ using the input x and the parameter adaptation block updates $\boldsymbol{\theta}_f$. It is made up of sets of integrators as shown in Figure 8.39. Adaptation rates of the parameters θ_{fi} are computed with the `MATLAB function` block, depicted in Figure 8.39, using the m-file shown in Figure 8.40. The vector $\boldsymbol{s}$ that defines the sliding surface, $\{\boldsymbol{e} : \boldsymbol{se} = 0\}$, was chosen as $\boldsymbol{s} = [40 \;\; 1]$. The simulation results are shown in Figures 8.41 and 8.42.

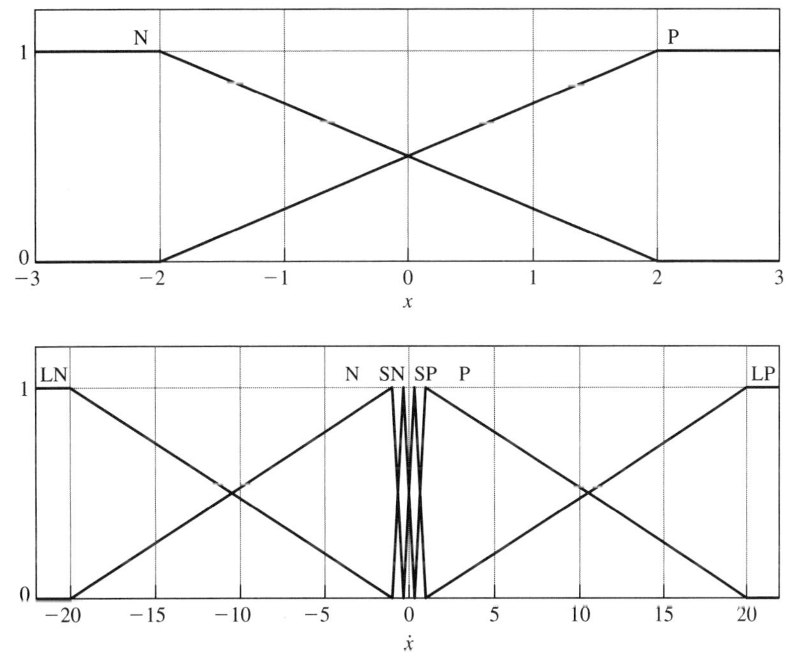

Figure 8.37 Fuzzy sets for x and $\dot{x}$.

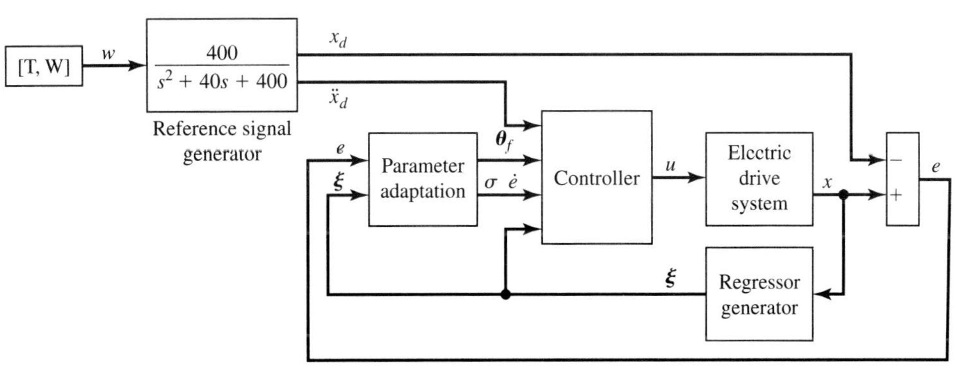

Figure 8.38 SIMULINK block diagram of fuzzy adaptive robust control system of Example 8.15.

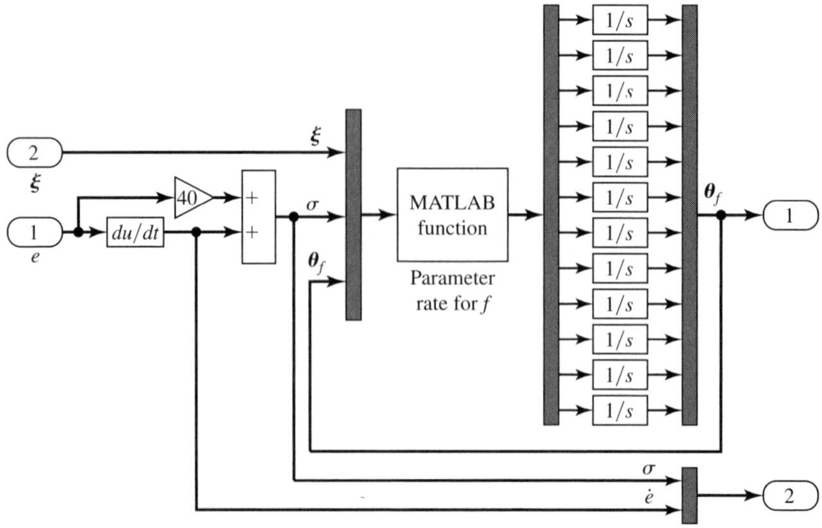

Figure 8.39 Parameter adaptation block of Example 8.15.

```
%FARC1f_UPDATE.M calculates adapation rates
% (needs 'farc1.mdl' 'farc1_fuzzifier.m', 'af_input_gen.m')

function rate = farc1f_update(x)
% x(1)--x(12) are components of xi,
% x(13) is sigma,  x(14)--x(25) are components of theta
rate = zeros(12,1);
r = 5000; %adaptation gain(learning rate)
theta_lb = -200;
theta_ub =  200;

for i = 1:12;
   alpha = r*x(13)*x(i);
   rate(i) = alpha*(((x(i+13)>theta_lb)|(alpha>0))&((x(i+13)<...
             theta_ub)|(alpha<0)));
end;
```

Figure 8.40 An m-file used in the block diagram of Figure 8.39 to compute adaptation rates.

The output $x(t)$, the reference signal $x_d(t)$, and the tracking error $e(t)$ are shown in Figure 8.41. The tracking error, $e(t)$, is so small that one cannot distinguish between the actual state $x(t)$ and the desired state $x_d(t)$. The plots of the control signal u and the components of $\boldsymbol{\theta}_f$ versus time are depicted in Figure 8.42. As the fuzzy logic system adapts the parameter $\boldsymbol{\theta}_f$, the error, $e(t)$, gets smaller and smaller. If we used smaller ϵ, we would achieve even smaller tracking error, at the expense of higher control effort.

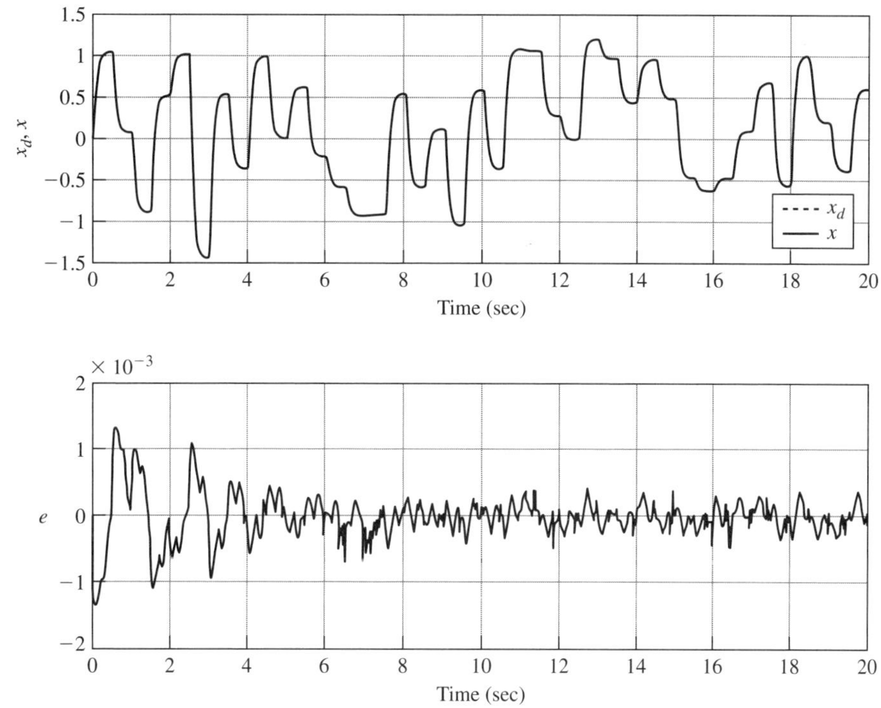

Figure 8.41 Plots of $x_d(t)$ and error $e(t)$ in Example 8.15.

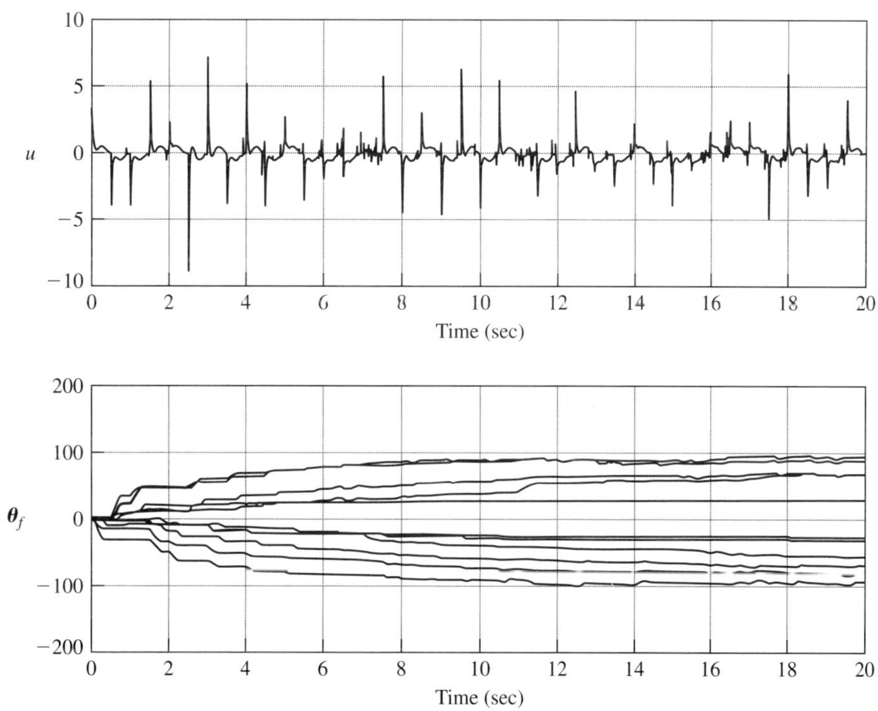

Figure 8.42 Plots of control effort $u(t)$ and the components of the parameter vector $\boldsymbol{\theta}_f$ versus time in Example 8.15.

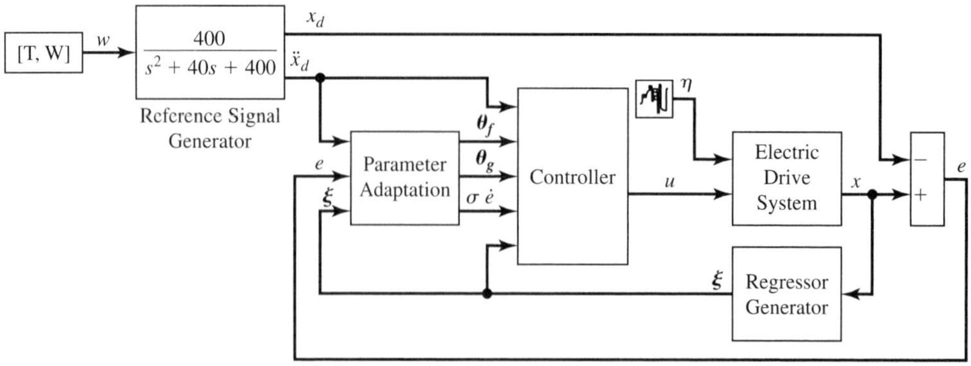

Figure 8.43 SIMULINK block diagram of the fuzzy adaptive robust control system of Example 8.16.

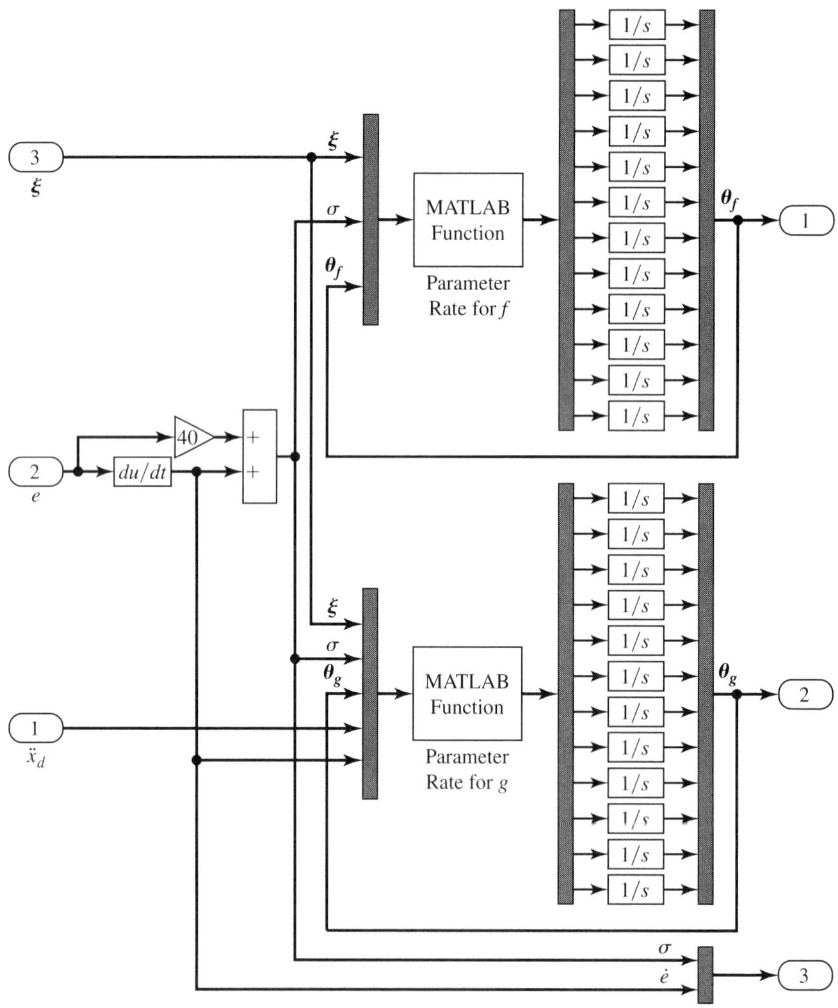

Figure 8.44 Parameter adaptation block of Example 8.16.

◆ Example 8.16

In this example, we assume that $f(\boldsymbol{x})$ and g are unknown to us and we employ the control law (8.152). Unknown $f(\boldsymbol{x})$ and g are approximated by two separate fuzzy logic systems using fuzzy sets shown in Figure 8.37 for both fuzzy logic systems.

The bounds on the adaptation parameters are chosen as follows: $\bar{\theta}_{fi} = 200$, $\underline{\theta}_{fi} = -200$, $\bar{\theta}_{gi} = 150$, and $\underline{\theta}_{gi} = 50$. All the initial values of the components of the adaptation parameter vector $\boldsymbol{\theta}_f$ are set to zero. The initial values of $\boldsymbol{\theta}_g$ are set to 150 to avoid excessively large control signal in the early stages of the simulation run. The selected gains are $\gamma_f = 5000$ and $\gamma_g = 1000$. The other design parameters are $\epsilon_1 = 50$, $\epsilon_2 = 100$, $\epsilon_3 = 50$, and $\epsilon_4 = 50$. The remaining parameters are the same as in the previous example.

SIMULINK block diagrams of the closed-loop system and the parameter adaptation law are depicted in Figure 8.43 and Figure 8.44, respectively. The disturbance, η, is a random signal whose plot versus time is shown in Figure 8.45. We used $d = 100$ so that the condition $|\eta(t)| \le d$ is satisfied. We chose, as before, $\boldsymbol{s} = [40 \;\; 1]$. The simulation results are shown in Figures 8.45, 8.46, and 8.47. As can be seen in Figure 8.46, the tracking error, $e(t)$, remains very small even in the presence of the disturbance, η. Plots of the time history of the adaptation parameters are shown in Figure 8.47.

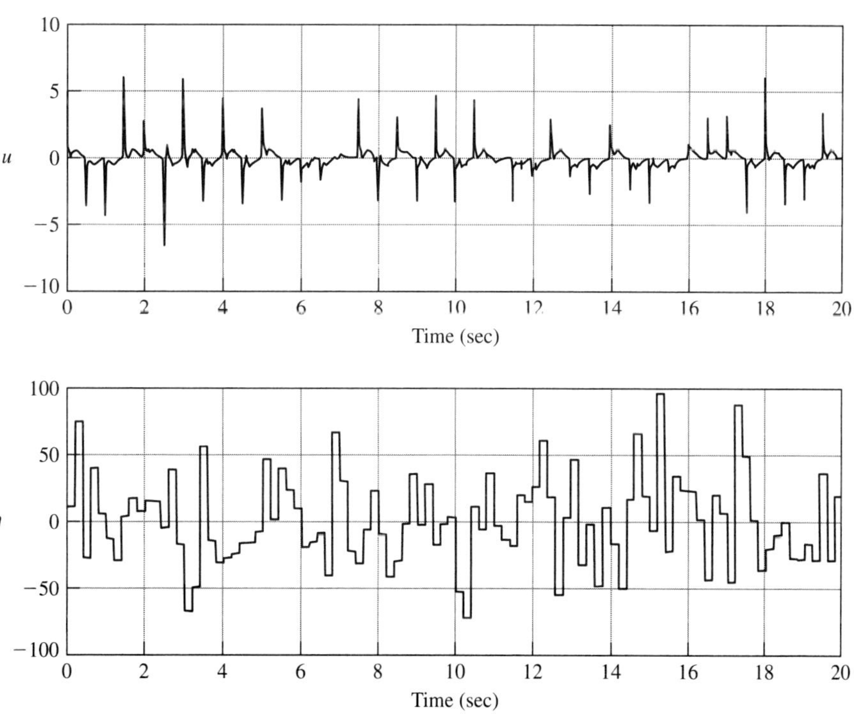

Figure 8.45 Plots of the control effort $u(t)$ and disturbance $\eta(t)$ versus time in Example 8.16.

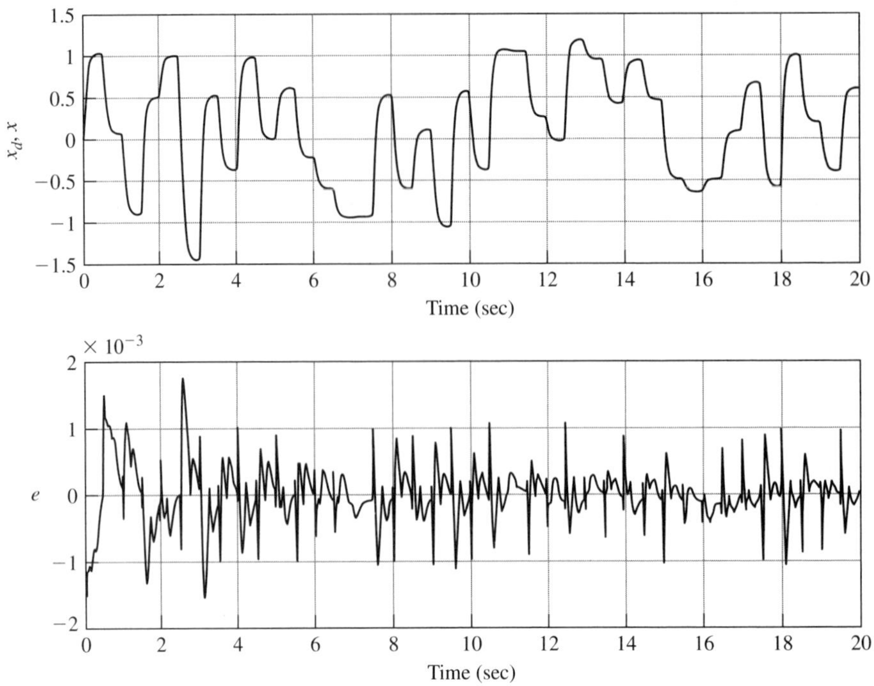

Figure 8.46 Plots of $x_d(t)$ and error $e(t)$ versus time in Example 8.16.

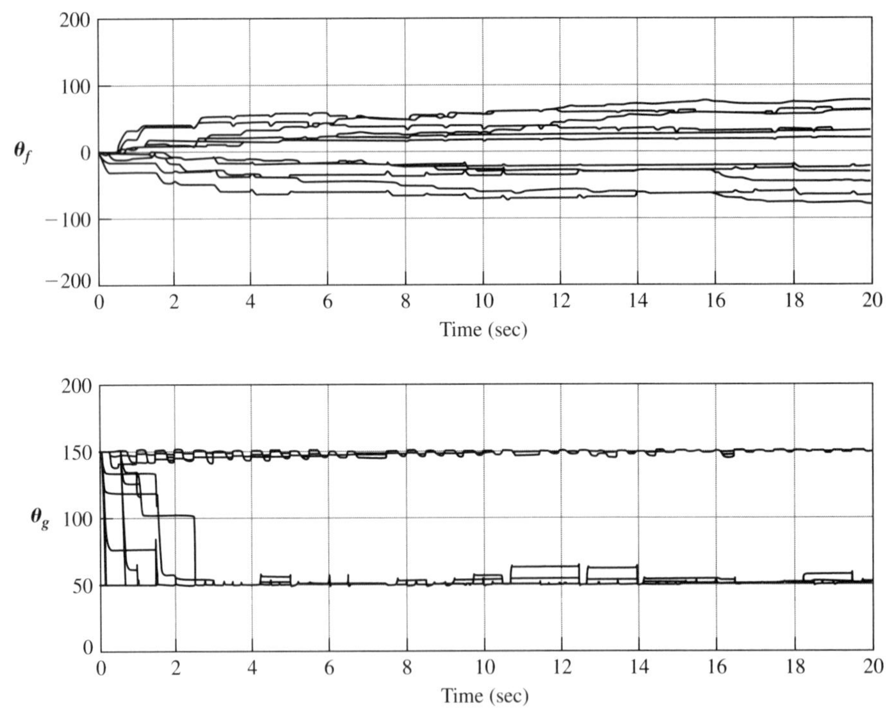

Figure 8.47 Plots of the components of the parameter vectors $\boldsymbol{\theta}_f$ and $\boldsymbol{\theta}_g$ versus time in Example 8.16.

Notes

The reader is encouraged to peruse a text on classical logic to become familiar with basic concepts of logic before undertaking his or her further study of fuzzy logic. An excellent introduction to logic is Copi [54]. For an introduction to fuzzy set theory the reader is referred to Klir, St.Clair, and Yuan [161], while for an introduction to fuzzy control, see Driankov, Hellendoorn, and Reinfrank [67]. Kosko [165] is the first book on fuzzy engineering. Graduate texts on fuzzy systems and control are Pedrycz [225] and Wang [296]. Paradigms and practice of fuzzy modeling are the topics of the edited book by Pedrycz [226]. Lin and Lee [185], and Jang, Sun, and Mizutani [142] cover applications of fuzzy systems as well as issues of merging fuzzy logic and neural networks. Mendel [201], on the other hand, presents extensions of the original fuzzy logic to deal with uncertainties.

The first chapter of Kosko's book [165] contains a thrilling account of the beginnings of fuzzy logic and fuzzy engineering. Kosko [165, p. 5] observes that "Fuzziness began as vagueness in the late nineteenth century." A concept is vague when it has blurred boundaries. For example, the concept *mountain* is vague because we do not know where a mountain ends and a hill begins. It was Bertrand Russell who in 1923 published a paper in which he described the concept of vagueness in terms of symbolic logic. In the 1920s, Polish mathematician Jan Lukasiewicz invented the multivalued logic. Max Black applied multivalued logic to sets of objects in the 1930s. In the 1950s, a number of logicians worked out a vague set algebra. In his 1965 paper, Zadeh [309] used the term *fuzzy* to mean *vague*. According to Kosko [165, p. 8], "Zadeh's 1965 paper applied Lukasiewicz's logic to each object in a set to work out a complete fuzzy set algebra and to extend the convex separation theorem to pattern recognition." The beginning of fuzzy engineering, on the other hand, can be marked by the work of Mamdani and Assilian [194], who first used fuzzy calculus to construct a fuzzy logic controller.

The method of constructing fuzzy design models described in Section 8.5 as well as the stabilizability conditions of fuzzy models along with a Lyapunov-based stabilizer were first reported by Teixeira and Żak in [279]. The controllers analyzed in Section 8.8 come from Lee et al. [178]. Other fuzzy adaptive control strategies were developed by Wang [294] and by Fischle and Schröder [85]. Fischle and Schröder [85, p. 38] observed that a major limitation of all existing stable adaptive fuzzy controllers, including the ones analyzed by us, is that they were developed for systems with unlimited actuators' authority—that is, for systems with no constraints on the effort level of the actuators. If we are serious about real-life applications of fuzzy adaptive controllers, then the case of the actuators with limited authority has to be rigorously analyzed. Another issue of practical importance is the convergence of the adaptation parameters to their "correct" values. Friedland [91, p. 345] states: "One may argue that the failure of a parameter estimate to converge to its correct value is a subsidiary issue as long as the adaptive control works (even with the incorrect parameter estimate). This reasoning might be valid if you can ascertain that the algorithm works for every input to which the system might be subjected. But in most applications the performance of the algorithm is tested only with a limited set of inputs. Satisfactory performance for this set of inputs is not a reliable indicator that its performance will be satisfactory for inputs outside the test set." In the current state-of-the-art adaptive control algorithms, the convergence of the adaptation parameters is usually ensured through the so-called *persistently exciting* inputs to the adaptation algorithms. However, the injection of the persistently exciting signals into the adaptation algorithms may not be feasible in real life applications. Thus, an alternative approach to the issues of the convergence of the adaptation parameters is needed.

EXERCISES

8.1 Consider a process whose input u is a function of two variables e and $\dot{e}$. The membership functions for e, $\dot{e}$, and u are given in Table 8.5. The control rules are:

Table 8.5 Membership Matrix Table for e, $\dot{e}$, and u for Exercise 8.1

Fuzzy Numbers	UNIVERSE OF DISCOURSE						
	−3	−2	−1	0	1	2	3
LN	1	0.5	0.25	0	0	0	0
SN	0.5	1	0.5	0.25	0	0	0
ZE	0	0.25	0.5	1	0.5	0.25	0
SP	0	0	0	0.25	0.5	1	0.5
LP	0	0	0	0	0.25	0.5	1

1. If e is ZE and $\dot{e}$ is SP, then u is SN.
2. If e is ZE and $\dot{e}$ is ZE, then u is ZE.
3. If e is SN and $\dot{e}$ is SN, then u is LP.
4. If e is SP and $\dot{e}$ is ZE, then u is SN.

Determine the value of u, for $e = 1$ and $\dot{e} = -2$, using

(a) the algebraic product defuzzifier;
(b) the center-of-gravity defuzzifier.

8.2 Use Lemma 8.3 to prove the following result due to Bailey [16]:
Let $\boldsymbol{B}$ be a matrix with negative diagonal elements and nonnegative off-diagonal elements. If $\boldsymbol{x}(t; \boldsymbol{x}_0, t_0)$ and $\boldsymbol{y}(t; \boldsymbol{y}_0, t_0)$ are solutions of $\dot{\boldsymbol{x}} \leq \boldsymbol{B}\boldsymbol{x}$ and $\dot{\boldsymbol{y}} = \boldsymbol{B}\boldsymbol{y}$, respectively, and $\boldsymbol{x}_0 = \boldsymbol{y}_0$, then $\boldsymbol{x}(t; \boldsymbol{x}_0, t_0) \leq \boldsymbol{y}(t; \boldsymbol{y}_0, t_0)$ for all $t \geq t_0$.

8.3 Consider the one-link robot manipulator of Exercise 7.2, whose model is given by (7.4) and (7.5). Let

$$\boldsymbol{P} = \boldsymbol{s}^T(\boldsymbol{s}\boldsymbol{s}^T)^{-1}\boldsymbol{s},$$

where $\boldsymbol{s} \in \mathbb{R}^{1\times 2}$ is the row vector that was found in Exercise 7.2. Then, $(\boldsymbol{I}_n - \boldsymbol{P})$ is the orthogonal projector of $\boldsymbol{x}$ onto the sliding surface $\{\boldsymbol{x} : \boldsymbol{s}\boldsymbol{x} = 0\}$. The Euclidean distance of the point $\boldsymbol{x}$ to the surface $\{\boldsymbol{x} : \boldsymbol{s}\boldsymbol{x} = 0\}$ is

$$\|\boldsymbol{P}\boldsymbol{x}\|_2.$$

Design a fuzzy sliding mode controller whose inputs are

$$\|\boldsymbol{P}\boldsymbol{x}(t)\|_2 \text{sign}(\sigma(\boldsymbol{x}(t))) \quad \text{and} \quad \Delta\|\boldsymbol{P}\boldsymbol{x}(t)\|_2 = \|\boldsymbol{P}\boldsymbol{x}(t)\|_2 - \|\boldsymbol{P}\boldsymbol{x}(t - T_d)\|_2,$$

where T_d is the time delay that is a design parameter. Use the rules matrix given in Table 8.6.

Table 8.6 The Rules Matrix for Exercise 8.3

Error \ Change-in-error	LN	N	ZE	P	LP
LN	LP	LP	LP	LP	LP
N	P	P	P	LP	LP
ZE	ZE	ZE	ZE	ZE	ZE
P	LN	LN	N	N	N
LP	LN	LN	LN	LN	LN

8.4 Consider a dynamical system modeled by (8.6)–(8.8). Let

$$\alpha_j(t) = 1 - \sum_{\substack{i=1\\ i\neq j}}^{r} \alpha_i(t). \tag{8.161}$$

Represent (8.6) as

$$\begin{aligned}\dot{\boldsymbol{x}} &= \left(\alpha_j \boldsymbol{A}_j + \sum_{\substack{i=1\\ i\neq j}}^{r} \alpha_i \boldsymbol{A}_i\right)\boldsymbol{x} + \left(\alpha_j \boldsymbol{B}_j + \sum_{\substack{i=1\\ i\neq j}}^{r} \alpha_i \boldsymbol{B}_i\right)\boldsymbol{u}\\ &= \left(\boldsymbol{A}_j + \sum_{\substack{i=1\\ i\neq j}}^{r} \alpha_i (\boldsymbol{A}_i - \boldsymbol{A}_j)\right)\boldsymbol{x} + \left(\boldsymbol{B}_j + \sum_{\substack{i=1\\ i\neq j}}^{r} \alpha_i (\boldsymbol{B}_i - \boldsymbol{B}_j)\right)\boldsymbol{u}. \end{aligned} \tag{8.162}$$

Let

$$\boldsymbol{\alpha}^j(t) = \boldsymbol{\alpha}^j = [\alpha_1 \quad \cdots \quad \alpha_{j-1} \quad \alpha_{j+1} \quad \cdots \quad \alpha_r]^T.$$

Observe that $\boldsymbol{\alpha}^j \in [0, 1]^{r-1}$; that is, the values of the vector-valued function $\boldsymbol{\alpha}^j(\cdot)$ are in the $(r-1)$-dimensional unit hypercube. Next, let

$$\Delta \boldsymbol{A}(\boldsymbol{\alpha}^j) = \sum_{\substack{i=1\\ i\neq j}}^{r} \alpha_i (\boldsymbol{A}_i - \boldsymbol{A}_j) \quad \text{and} \quad \Delta \boldsymbol{B}(\boldsymbol{\alpha}^j) = \sum_{\substack{i=1\\ i\neq j}}^{r} \alpha_i (\boldsymbol{B}_i - \boldsymbol{B}_j).$$

Using the above notation, represent model (8.162), in an equivalent way, as

$$\begin{aligned}\dot{\boldsymbol{x}} &= (\boldsymbol{A}_j + \Delta \boldsymbol{A}(\boldsymbol{\alpha}^j))\boldsymbol{x} + (\boldsymbol{B}_j + \Delta \boldsymbol{B}(\boldsymbol{\alpha}^j))\boldsymbol{u}\\ &= \boldsymbol{A}_j \boldsymbol{x} + \boldsymbol{B}_j \boldsymbol{u} + \boldsymbol{F}(\boldsymbol{\alpha}^j, \boldsymbol{x}, \boldsymbol{u}), \end{aligned} \tag{8.163}$$

where $\boldsymbol{F}(\boldsymbol{\alpha}^j, \boldsymbol{x}, \boldsymbol{u}) = \Delta \boldsymbol{A}(\boldsymbol{\alpha}^j)\boldsymbol{x} + \Delta \boldsymbol{B}(\boldsymbol{\alpha}^j)\boldsymbol{u}$. Let

$$\boldsymbol{F}(\cdot) = \boldsymbol{f}(\cdot) + \boldsymbol{B}_j \boldsymbol{h}(\cdot). \tag{8.164}$$

We refer to $\boldsymbol{f}$ as the *unmatched uncertainty,* while $\boldsymbol{h}$ is referred to as the *matched uncertainty*. Hence, (8.163) takes the form

$$\dot{\boldsymbol{x}} = \boldsymbol{A}_j \boldsymbol{x} + \boldsymbol{B}_j \boldsymbol{u} + \boldsymbol{f} + \boldsymbol{B}_j \boldsymbol{h}. \tag{8.165}$$

Assume that

$$\begin{aligned}\|\boldsymbol{f}\| &= \|\boldsymbol{f}(\boldsymbol{\alpha}^j, \boldsymbol{x})\| \leq \alpha_f \|\boldsymbol{x}\|,\\ \|\boldsymbol{h}\| &= \|\boldsymbol{h}(\boldsymbol{\alpha}^j, \boldsymbol{x}, \boldsymbol{u})\| \leq \beta_x \|\boldsymbol{x}\| + \beta_u \|\boldsymbol{u}\|. \end{aligned} \tag{8.166}$$

Prove the following theorem:

Theorem 8.10 Suppose that $\boldsymbol{A}_j$ is asymptotically stable, and $\boldsymbol{P} = \boldsymbol{P}^T > 0$ is the solution to the Lyapunov matrix equation $\boldsymbol{A}_j^T \boldsymbol{P} + \boldsymbol{P}\boldsymbol{A}_j = -2\boldsymbol{Q}$ for some $\boldsymbol{Q} = \boldsymbol{Q}^T > 0$. Suppose also that

$$\alpha_f < \frac{\lambda_{\min}(\boldsymbol{Q})}{\lambda_{\max}(\boldsymbol{P})} \quad \text{and} \quad \beta_u < 1.$$

Then, the state feedback controller $\boldsymbol{u} = -\gamma \boldsymbol{B}_j^T \boldsymbol{P}\boldsymbol{x}$, where

$$\gamma > \frac{\beta_x^2}{4(\lambda_{\min}(\boldsymbol{Q}) - \alpha_f \lambda_{\max}(\boldsymbol{P}))(1 - \beta_u)}$$

globally asymptotically stabilizes the uncertain system for arbitrary $\boldsymbol{f}$ and $\boldsymbol{h}$ that satisfy the norm bounds (8.166).

8.5 In this exercise, the following definition is used:

Definition 8.4 We say that system (8.163) satisfies the matching conditions if there exist continuous matrix functions $\boldsymbol{D}(\cdot) : \mathbb{R}^{r-1} \to \mathbb{R}^{m\times n}$ and $\boldsymbol{E}(\cdot) : \mathbb{R}^{r-1} \to \mathbb{R}^{m\times m}$ such that

$$\begin{aligned} \Delta \boldsymbol{A}(\boldsymbol{\alpha}^j) &= \boldsymbol{B}_j \boldsymbol{D}(\boldsymbol{\alpha}^j), \\ \Delta \boldsymbol{B}(\boldsymbol{\alpha}^j) &= \boldsymbol{B}_j \boldsymbol{E}(\boldsymbol{\alpha}^j) \end{aligned} \tag{8.167}$$

for all $\boldsymbol{\alpha}^j \in [0, 1]^{r-1}$.

Consider the following algorithm for linear controller design for fuzzy system models of the form (8.162) that is the subject of Exercise 8.4. The algorithm is a consequence of Theorem 8.10 of Exercise 8.4.

LINEAR CONTROLLER DESIGN ALGORITHM FOR FUZZY SYSTEMS

Given fuzzy model (8.162):

STEP 1. Set $j = 1$

STEP 2. Check matching conditions (8.167). If the matching conditions are satisfied, calculate matrices $\boldsymbol{D}$ and $\boldsymbol{E}$, and check if $\alpha_u < 1$. If the matching conditions are not satisfied, express the uncertain element as $\boldsymbol{F} = \boldsymbol{f} + \boldsymbol{B}_j \boldsymbol{h}$, calculate β_u, and check if $\beta_u < 1$. If $\beta_u > 1$, set $j = j + 1$ and repeat the above computations.

STEP 3. Stabilize $\boldsymbol{A}_j$, if necessary. Calculate $\boldsymbol{u}_1 = -\boldsymbol{K}_1 \boldsymbol{x}$ so that the matrix $\boldsymbol{A}_j - \boldsymbol{B}_j \boldsymbol{K}_1$ has its eigenvalues in the desired locations. Update the matched uncertainty involving the state vector; that is, replace $\boldsymbol{D}$ with $\boldsymbol{D} - \boldsymbol{E}\boldsymbol{K}_1$.

STEP 4. Solve the Lyapunov equation $\boldsymbol{A}_j^T \boldsymbol{P} + \boldsymbol{P}\boldsymbol{A}_j = -2\boldsymbol{Q}$ for some $\boldsymbol{Q} = \boldsymbol{Q}^T > 0$. Choosing $\boldsymbol{Q} = \boldsymbol{I}_n$ maximizes the ratio $\lambda_{\min}(\boldsymbol{Q})/\lambda_{\max}(\boldsymbol{P})$. Check if $\alpha_f < \lambda_{\min}(\boldsymbol{Q})/\lambda_{\max}(\boldsymbol{P})$. If α_f does not satisfy the condition, set $j = j + 1$ and go to step 2.

STEP 5. Construct $\boldsymbol{u}_2 = -\gamma \boldsymbol{B}_j^T \boldsymbol{P}\boldsymbol{x}$.

STEP 6. Construct the resulting state feedback linear control law $\boldsymbol{u} = \boldsymbol{u}_1 + \boldsymbol{u}_2$.

Table 8.7 Parameter Numerical Values of the Model of Exercise 8.5

m	= mass of pendulum	2.0 kg
M	= mass of cart	8.0 kg
$2l$	= length of pendulum	1.0 m
a	= $1/(m+M)$	

Illustrate the above algorithm on a reduced-order model of an inverted pendulum mounted on a cart. The equations of motion for the pendulum are

$$\begin{cases} \dot{x}_1 = x_2, \\ \dot{x}_2 = \dfrac{g\sin(x_1) - amlx_2^2\sin(2x_1)/2 - a\cos(x_1)u}{4l/3 - aml\cos^2(x_1)}, \end{cases} \tag{8.168}$$

where x_1 is the angle of the pendulum from the vertical line, x_2 is the angular velocity of the pendulum, u is the control force applied to the cart, and $g = 9.8$ m/sec^2 is the magnitude of the acceleration due to gravity. Numerical values of the parameters are given in Table 8.7. Consider modeling equations (8.168) as a truth model. Use the truth model in your simulations to evaluate the performance of the design. Design a linear stabilizing controller using the above algorithm utilizing a design model. Assume that, by using insight and experience, you obtained the following two rules describing the plant dynamics:

Rule 1: IF $x_1(t)$ is about 0,
THEN $\dot{\boldsymbol{x}}(t) = \boldsymbol{A}_1\boldsymbol{x}(t) + \boldsymbol{b}_1u(t)$,

and

Rule 2: IF $x_1(t)$ is about $\pm\pi/2$,
THEN $\dot{\boldsymbol{x}}(t) = \boldsymbol{A}_2\boldsymbol{x}(t) + \boldsymbol{b}_2u(t)$,

where

$$\boldsymbol{A}_1 = \begin{bmatrix} 0 & 1 \\ \dfrac{g}{4l/3 - aml} & 0 \end{bmatrix}, \qquad \boldsymbol{b}_1 = \begin{bmatrix} 0 \\ -\dfrac{a}{4l/3 - aml} \end{bmatrix},$$

and

$$\boldsymbol{A}_2 = \begin{bmatrix} 0 & 1 \\ \dfrac{2g}{\pi(4l/3 - amlb^2)} & 0 \end{bmatrix}, \qquad \boldsymbol{b}_2 = \begin{bmatrix} 0 \\ -\dfrac{ab}{4l/3 - amlb^2} \end{bmatrix},$$

where $b = \cos(88°)$. Use the following membership functions:

$$\mu_1(x_1) = \frac{1 - 1/(1+\exp(-7(x_1 - \pi/4)))}{1+\exp(-7(x_1+\pi/4))} \quad \text{and} \quad \mu_2(x_1) = 1 - \mu_1(x_1).$$

The above membership functions are the same as the ones employed by Cao et al. [42]. The corresponding fuzzy system model is

$$\dot{\boldsymbol{x}} = (\alpha_1 \boldsymbol{A}_1 + \alpha_2 \boldsymbol{A}_2)\boldsymbol{x} + (\alpha_1 \boldsymbol{b}_1 + \alpha_2 \boldsymbol{b}_2)u, \tag{8.169}$$

where $\alpha_i = \mu_i, i = 1, 2$, because $\mu_1 + \mu_2 = 1$ for all t. Proceed to construct a stabilizing linear controller according to the above algorithm. First, represent model (8.169) as

$$\dot{\boldsymbol{x}} = (\boldsymbol{A}_1 + \alpha_2(\boldsymbol{A}_2 - \boldsymbol{A}_1))\boldsymbol{x} + (\boldsymbol{b}_1 + \alpha_2(\boldsymbol{b}_2 - \boldsymbol{b}_1))u. \tag{8.170}$$

Check that the uncertain elements of system model (8.170) satisfy the matching conditions. Suppose that the desired eigenvalues of $\boldsymbol{A}_1 - \boldsymbol{b}_1\boldsymbol{k}_1$ are $\{-2, -2\}$. Construct the controller, and test its performance on the truth model. Generate plots of x_1 versus time for different initial angular displacements $x_1(0)$. You can set $x_2(0) = 0$.

8.6 The objective of this exercise is to understand better the dynamical behavior of the nonlinear projector used by us in adaptive fuzzy control and defined by (8.117). For example, take $\alpha(t) = \sin(0.5\pi t)$, $\underline{\theta} = -0.3$, and $\bar{\theta} = 0.5$. Compare solutions of $\dot{\theta} = Proj_\theta(\alpha)$ and $\dot{\theta} = \alpha$ for a number of initial conditions. Below is an example of MATLAB's function that implements the differential equation, $\dot{\theta} = Proj_\theta(\alpha)$, where low $= \underline{\theta}$ and upper $= \bar{\theta}$.

```
function theta_dot = Proj(t,theta)
low=-0.3;
upper=0.5;
alpha=sin(0.5*pi*t);
if (theta <= low)&(alpha<0)
   theta_dot=0;
elseif (theta >= upper)&(alpha>0)
   theta_dot=0;
else
   theta_dot=alpha;
   end
```

8.7 For the fuzzy model the stick balancer given in Example 8.13 by equation (8.66), where

$$\boldsymbol{y} = \begin{bmatrix} 1 & 0 & 0 & 0 \\ 0 & 0 & 1 & 0 \end{bmatrix} \begin{bmatrix} x_1 \\ x_2 \\ x_3 \\ x_4 \end{bmatrix},$$

construct a fuzzy state estimator. (Note that we have $\boldsymbol{C}_1 = \boldsymbol{C}_2 = \boldsymbol{C}(\boldsymbol{\alpha})$.) Implement the combined controller–estimator compensator using first the fuzzy model and then the nonlinear model of the stick balancer. Write MATLAB scripts that animate the dynamical behavior of the closed-loop systems along with animation of the angle θ and the input u data.

CHAPTER 9

Neural Networks

> We tend to regard forgetting as a failure of memory, the result of an inadequacy in the machinery for remembering, but a moment's reflection shows that that cannot always be right. If we were to remember every telephone number we have ever used, every scene we have ever looked at, every conversation we have ever heard, every sentence we have ever read, we should be overwhelmed with memories. In fact we transfer to our long-term memory stores only a tiny fraction of the contents of our short-term stores. We retain only what is likely to be useful.
>
> —*An Anatomy of Thought* [101, p. 331]

This chapter is concerned with neurocomputing, which is also called brainlike computation. The motivation behind studying artificial neural networks, commonly referred to as neural networks, or just neural nets, is to build computers whose construction would mimic the organization of the brain. The brain is the most complex natural information processing system. It is capable of organizing its building blocks, called neurons, to perform computations in a very efficient way. A drawing of a typical biological neuron is given in Figure 9.1. The neuron collects signals from other neurons through its dendrites and receptor sites. The signals have the form of electrical pulses. The pulses received by the dendrites travel to the neuron's cell body where they are being processed. The surface of the cell body of the neuron is dotted with numerous receptor sites that can receive messages from other neurons in a fashion similar to the dendrites. The neuron may produce its own pulses in response to excitation from other neurons. The neuron's own pulses travel through a long, thin strand called an axon, which splits into many end branches. The neuron's message to the other neurons is delivered via the end branches. The key to the transmission of the neuron's message is the synapse, shown in Figure 9.2. The end branches of the neuron's axon terminate in synaptic knobs. The synaptic knob of the first neuron is separated from the receptor site of another neuron by a microscopic distance. The cell body of the first neuron produces chemical substances called neurotransmitters, which are delivered down to the synaptic vesicles. The neurotransmitters are stored in the synaptic vesicles until the neuron fires and a burst of neurotransmitters is released by the vesicles. The neurotransmitters then flow across the synaptic cleft and act on the second neuron. The neurotransmitters of the first neuron may stimulate or inhibit the second neuron activity. The first neuron has many other synaptic

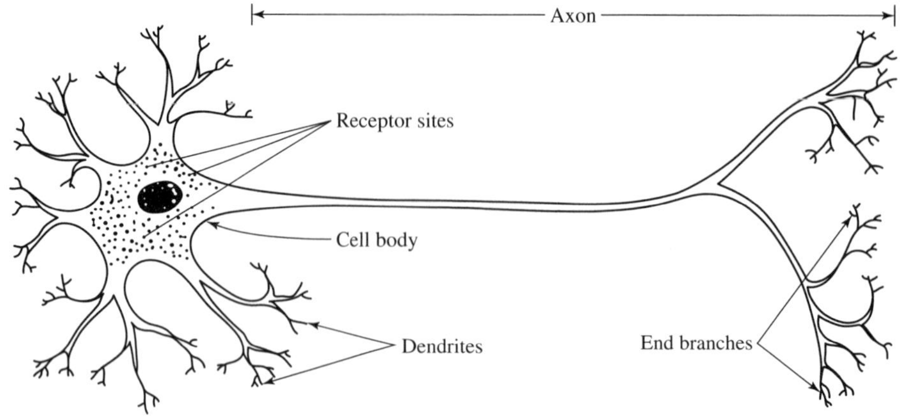

Figure 9.1 A typical biological neuron.

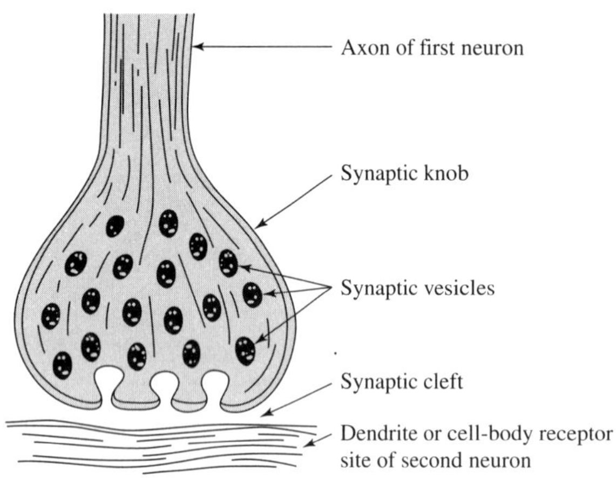

Figure 9.2 The synapse is a junction point where neuron meets neuron.

knobs elsewhere, while the second neuron has many other receptor sites that receive messages from many other neurons. Each neuron is receiving messages from hundreds or thousands of other neurons. The neuron sets off the nervous pulse; that is, it fires off its own nervous pulse only when it is properly stimulated. It was observed that, in general, the neuron will not fire in response to a single message. To fire, the neuron needs many messages arriving simultaneously or in quick succession from other neurons. In addition, the stimulating messages to fire must outweigh the messages that inhibit the neuron from firing.

In the following sections, we discuss a few mathematical models of the biological neuron. These models are referred to as the artificial neurons. Then, we analyze neural networks which are composed of the artificial neurons.

9.1 Threshold Logic Unit

A simple model of a biological neuron is the so called the *threshold logic unit* (TLU), shown in Figure 9.3. The TLU is also called the *perceptron*. A TLU is parameterized by a set of weights and a threshold (bias). The summation operation is followed by the sign activation function,

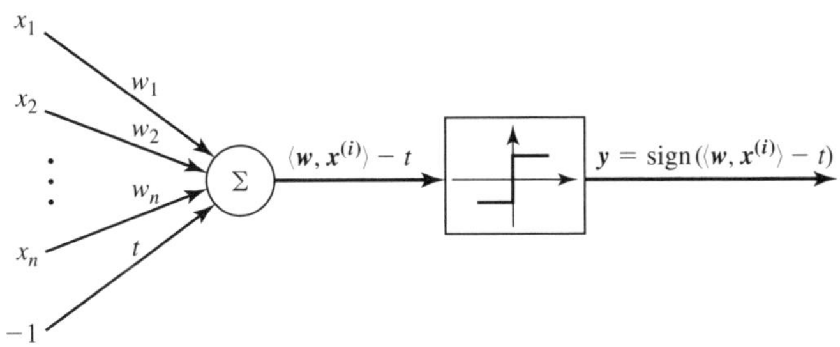

Figure 9.3 A threshold logic unit (TLU), also called the perceptron, as a model of a biological neuron.

which models the cell body of a biological neuron. The dendrites are modeled by the inputs, and their strengths are modeled by the weights. The axon corresponds to the output y. Suppose that we are given a TLU with n weights, $w_1, w_2, \ldots, w_n$, and a threshold t. Given an input vector $\boldsymbol{x}^{(i)} = [x_1^{(i)} \; x_2^{(i)} \; \cdots \; x_n^{(i)}]^T \in \mathbb{R}^n$, then the corresponding output of the TLU is

$$y_i = \text{sign}\left(\sum_{j=1}^{n} w_j x_j^{(i)} - t\right), \tag{9.1}$$

where

$$\text{sign}(z) = \begin{cases} 1 & \text{if } z \geq 0, \\ -1 & \text{if } z < 0. \end{cases}$$

We note that sometimes in the literature the sign function is undefined at 0, or it is defined as a point to a set mapping; that is, $\text{sign}(0) \in [-1, 1]$. Here, we define $\text{sign}(0) = 1$. Using the scalar product notation, we represent (9.1) as

$$y_i = \text{sign}\left(\left\langle \boldsymbol{w}, \boldsymbol{x}^{(i)}\right\rangle - t\right).$$

Note that $y_i \in \{-1, 1\}$. Specifically,

$$y_i = \begin{cases} 1 & \text{if } \left\langle \boldsymbol{w}, \boldsymbol{x}^{(i)}\right\rangle - t \geq 0, \\ -1 & \text{if } \left\langle \boldsymbol{w}, \boldsymbol{x}^{(i)}\right\rangle - t < 0. \end{cases} \tag{9.2}$$

The goal of the TLU (perceptron) is to correctly classify the set of given input vectors $\boldsymbol{x}^{(1)}, \ldots, \boldsymbol{x}^{(p)}$ into one of the two classes, C_1 or C_2. In other words, if a given vector, say $\boldsymbol{x}^{(i)}$, belongs to class C_1, then the corresponding output y_i should be $+1$. On the other hand, if $\boldsymbol{x}^{(i)}$ belongs to class C_2 then the TLU's output should be -1. Thus, we are faced with a two-class pattern classification problem. The TLU's role is to find a hyperplane that separates two decision regions. The separating hyperplane (decision boundary) has the form

$$\langle \boldsymbol{w}^*, \boldsymbol{x}\rangle - t^* = \sum_{i=1}^{n} w_i^* x_i - t^* = 0. \tag{9.3}$$

In the case of two variables x_1 and x_2 the decision boundary is a straight line. An illustration of such a case is given in Figure 9.4. In this case a point that lies above the boundary line is assigned to class C_1, while a point below the boundary line is assigned to class C_2. We will present an iterative algorithm for the decision boundary weights and threshold provided that a decision boundary exists. We assume that we are given p training pairs

$$\left(\boldsymbol{x}^{(1)}, d_1\right), \left(\boldsymbol{x}^{(2)}, d_2\right), \ldots, \left(\boldsymbol{x}^{(p)}, d_p\right),$$

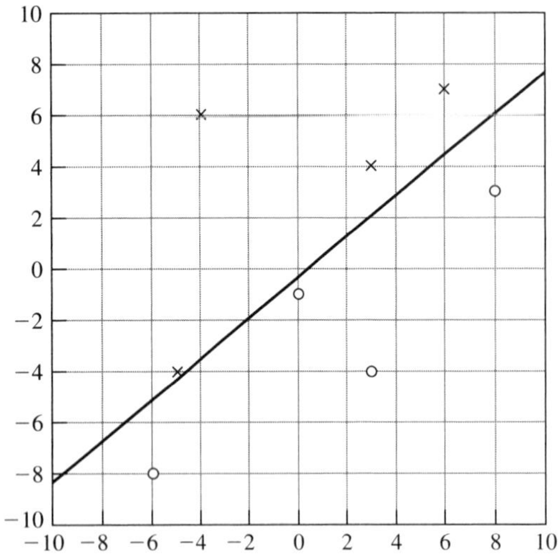

Figure 9.4 An example of separable points in two-dimensional space and a separating line found using the TLU learning algorithm.

where for $i = 1, 2, \ldots, p$, the input vectors $\boldsymbol{x}^{(i)} \in \mathbb{R}^n$ and $d_i \in \{-1, 1\}$ represent the desired responses. We would like to determine a hyperplane (9.3) such that each input vector lies on a preassigned side of the hyperplane. If such a hyperplane exists for the given training set, then the set of input vectors is referred to as *linearly separable,* and the hyperplane is referred to as a *separating hyperplane*. If no such hyperplane exists, then the training set is referred to as *linearly nonseparable*. Finding a separating hyperplane is equivalent to finding a corresponding weight vector and a threshold. If we find such a weight vector and a threshold, then we say that the TLU correctly classifies the training set. Thus, designing a TLU that correctly classifies the given training set amounts to finding a weight vector $\boldsymbol{w}^*$ and a threshold t^* such that for $i = 1, 2, \ldots, p$, we have

$$\begin{aligned} \left\langle \boldsymbol{w}^*, \boldsymbol{x}^{(i)} \right\rangle - t^* &\geq 0 \qquad \text{if } d_i = 1, \\ \left\langle \boldsymbol{w}^*, \boldsymbol{x}^{(i)} \right\rangle - t^* &< 0 \qquad \text{if } d_i = -1. \end{aligned}$$

In our discussion, we can assume the threshold value, t, to be zero. We can achieve this by increasing the dimension of every input vector, $\boldsymbol{x}^{(i)}$, by augmenting it with an entry equal to -1, and by increasing the dimension of the weight vector to $n + 1$ by augmenting it with t. Indeed,

$$y_i = \operatorname{sign}\left(\left\langle \boldsymbol{w}, \boldsymbol{x}^{(i)} \right\rangle - t\right) = \operatorname{sign}\left(\left\langle \tilde{\boldsymbol{w}}, \tilde{\boldsymbol{x}}^{(i)} \right\rangle\right),$$

where $\tilde{\boldsymbol{w}}^T = [w_1 \; w_2 \; \cdots \; w_n \; t]$ and $\tilde{\boldsymbol{x}}^{(i)T} = [x_1^{(i)} \; x_2^{(i)} \; \cdots \; x_n^{(i)} \; -1]$. In our further discussion, we assume that the given input vectors have been already augmented. In view of the above, we can formulate the training problem of a TLU as follows:

TLU Training Problem Given p training pairs $(\boldsymbol{x}^{(1)}, d_1), \ldots, (\boldsymbol{x}^{(p)}, d_p)$, where $\boldsymbol{x}^{(i)} \in \mathbb{R}^{n+1}$ and $d_i \in \{-1, 1\}$. Determine a weight vector $\boldsymbol{w} \in \mathbb{R}^{n+1}$; if there exists one, such that for $i = 1, \ldots, p$, we have

$$\operatorname{sign}\left(\left\langle \boldsymbol{w}, \boldsymbol{x}^{(i)} \right\rangle\right) = d_i.$$

Assume that the given vectors $\boldsymbol{x}^{(1)}, \ldots, \boldsymbol{x}^{(p)}$ are linearly separable—that is, there exists a weight vector $\boldsymbol{w}^*$ such that

$$\left\langle \boldsymbol{w}^*, \boldsymbol{x}^{(i)} \right\rangle \geq 0 \qquad \text{for every input vector } \boldsymbol{x}^{(i)} \text{ from class } \mathcal{C}_1$$

and

$$\left\langle \boldsymbol{w}^*, \boldsymbol{x}^{(i)} \right\rangle < 0 \qquad \text{for every input vector } \boldsymbol{x}^{(i)} \text{ from class } \mathcal{C}_2.$$

We can now formulate the algorithm for adapting the weight vector resulting in a separating hyperplane of a given set of linearly separable points. The idea is very simple and we proceed as follows. If the given vector $\boldsymbol{x}^{(i)}$ is classified correctly, do not update the weights, that is,

$$\boldsymbol{w}(k+1) = \boldsymbol{w}(k).$$

Otherwise,

$$\boldsymbol{w}(k+1) = \boldsymbol{w}(k) - \boldsymbol{x}^{(i)} \qquad \text{if} \quad \boldsymbol{w}^T(k)\boldsymbol{x}^{(i)} \geq 0 \quad \text{and} \quad \boldsymbol{x}^{(i)} \in \mathcal{C}_2 \tag{9.4}$$

and

$$\boldsymbol{w}(k+1) = \boldsymbol{w}(k) + \boldsymbol{x}^{(i)} \qquad \text{if} \quad \boldsymbol{w}^T(k)\boldsymbol{x}^{(i)} < 0 \quad \text{and} \quad \boldsymbol{x}^{(i)} \in \mathcal{C}_1. \tag{9.5}$$

Note that if $\boldsymbol{w}^T(k)\boldsymbol{x}^{(i)} \geq 0$ [that is, $y(k) = 1$] and $\boldsymbol{x}^{(i)} \in \mathcal{C}_2$ [that is, $d_i = -1$], then the error between the actual TLU output and the desired TLU output is $d_i - y(k) = -2$. Therefore, we can represent the weight update in (9.4) as

$$\boldsymbol{w}(k+1) = \boldsymbol{w}(k) + \tfrac{1}{2}(d_i - y(k))\boldsymbol{x}^{(i)}. \tag{9.6}$$

Analogously, if $\boldsymbol{w}^T(k)\boldsymbol{x}^{(i)} < 0$ [that is, $y(k) = -1$] and $\boldsymbol{x}^{(i)} \in \mathcal{C}_1$, then $d_i - y(k) = 2$. Here, as in the previous case, we can represent the weight update in (9.5) as (9.6). We now summarize the above considerations in the form of the following algorithm.

TLU (PERCEPTRON) LEARNING ALGORITHM

Given a set of training pairs $\{(\boldsymbol{x}^{(1)}, d_1), \ldots, (\boldsymbol{x}^{(i)}, d_i), \ldots, (\boldsymbol{x}^{(p)}, d_p)\}$.

STEP 1 *Initialize Weight Vector and Threshold.* Set the values of the components of the initial weight vector $\boldsymbol{w}(0)$ and threshold $t(0)$ to zero or to small random values.

STEP 2 *Present New Input Vector and Desired Output.* Present the new input vector, $\boldsymbol{x}^{(i)}$, and the corresponding desired output d_i.

STEP 3 *Calculate Actual Output:*

$$y(k) = \operatorname{sign}\left(\left\langle \boldsymbol{w}(k), \boldsymbol{x}^{(i)} \right\rangle - t(k)\right).$$

STEP 4 *Adapt Weight Vector and Threshold:*

$$\boldsymbol{w}(k+1) = \boldsymbol{w}(k) + \tfrac{1}{2}(d_i - y(k))\boldsymbol{x}^{(i)},$$
$$t(k+1) = t(k) - \tfrac{1}{2}(d_i - y(k)).$$

Note that the weight vector and threshold are unchanged if the correct decision is made by the TLU.

STEP 5 *Repeat by Going to Step 2.*

◆ Example 9.1

We used the above algorithm to generate a separating line for eight training pairs. For $i = 1, \dots, 4$ and $d_i = 1$,

$$\begin{bmatrix} x^{(1)} & x^{(2)} & x^{(3)} & x^{(4)} \end{bmatrix} = \begin{bmatrix} -5 & -4 & 3 & 6 \\ -4 & 6 & 4 & 7 \end{bmatrix}.$$

The above points are labeled "×" in Figure 9.4. The other four training pairs corresponding to $d_i = -1$, $i = 5, \dots, 8$, labeled "○" are

$$\begin{bmatrix} x^{(5)} & x^{(6)} & x^{(7)} & x^{(8)} \end{bmatrix} = \begin{bmatrix} -6 & 0 & 3 & 8 \\ -8 & -1 & -4 & 3 \end{bmatrix}.$$

The weight vector and threshold were initialized randomly, where

$$\boldsymbol{w}(0) = [-0.4860 \quad 0.8913]^T \quad \text{and} \quad t_0 = 0.7621.$$

The TLU learning algorithm generated a separating line, shown in Figure 9.4, corresponding to the weight vector $\boldsymbol{w}^* = [-5.5140 \ \ 6.8913]^T$ and threshold $t^* = -2.2379$.

Our goal now is to prove that the TLU learning algorithm can always find a separating hyperplane for linearly separable input vectors. In our proof, we use some technical results that we present next.

Lemma 9.1 Let $\boldsymbol{x}^{(1)}, \boldsymbol{x}^{(2)}, \dots, \boldsymbol{x}^{(p)} \in \mathbb{R}^n$, be a set of p linearly separable input vectors, let $\boldsymbol{w} \in \mathbb{R}^n$ be a weight vector of an n-input TLU, and let $t \in \mathbb{R}$ be its threshold. Then, there exists a threshold t' such that for $i = 1, 2, \dots, p$ we have

$$\left\langle \boldsymbol{w}, \boldsymbol{x}^{(i)} \right\rangle - t' \neq 0$$

and

$$\operatorname{sign}\left(\left\langle \boldsymbol{w}, \boldsymbol{x}^{(i)} \right\rangle - t\right) = \operatorname{sign}\left(\left\langle \boldsymbol{w}, \boldsymbol{x}^{(i)} \right\rangle - t'\right).$$

Proof Assume, without loss of generality, that the input vectors are sorted into two sets $\{\boldsymbol{x}^{(1)}, \dots, \boldsymbol{x}^{(l)}\}$ and $\{\boldsymbol{x}^{(l+1)}, \dots, \boldsymbol{x}^{(p)}\}$ such that the desired response of the vectors from the first set is $d_i = -1$, while the desired response of the vectors from the second set is $d_i = 1$. Let

$$\delta = \min\left\{\left|\langle \boldsymbol{w}, \boldsymbol{x}^{(i)} \rangle - t\right| : 1 \le i \le l\right\}.$$

It follows from (9.2) that $\delta > 0$. Let $t' = t - \delta/2$. Then,

$$\left\langle \boldsymbol{w}, \boldsymbol{x}^{(i)} \right\rangle - t' = \left\langle \boldsymbol{w}, \boldsymbol{x}^{(i)} \right\rangle - t + \delta/2 \neq 0 \qquad \text{for } i = 1, \dots, p,$$

and the output of the TLU with the modified threshold, t', is the same as the output of the TLU with the old threshold, t.

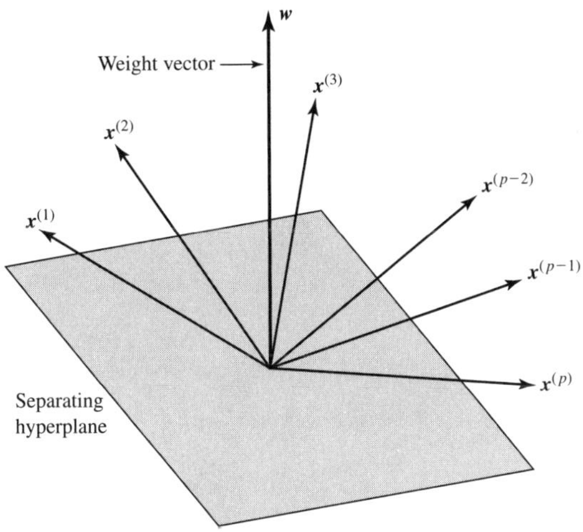

Figure 9.5 Linearly separable vectors after their preprocessing.

By the previous lemma, we can always assume that $\langle \boldsymbol{w}, \boldsymbol{x}^{(i)} \rangle - t' \neq 0$ for $i = 1, \ldots, p$ and that the output of a TLU can be defined as

$$y_i = \begin{cases} 1 & \text{if } \left\langle \boldsymbol{w}, \boldsymbol{x}^{(i)} \right\rangle - t > 0 \\ -1 & \text{if } \left\langle \boldsymbol{w}, \boldsymbol{x}^{(i)} \right\rangle - t < 0. \end{cases}$$

We further simplify the formulation of the TLU learning problem by preprocessing training pairs. Specifically, we replace every training pair, $(\boldsymbol{x}^{(i)}, d_i)$ with $d_i = -1$, with their negative, $(-\boldsymbol{x}^{(i)}, -d_i)$. After performing the above preprocessing, the TLU learning problem can be reformulated in a simple form that we use while analyzing the TLU learning algorithm.

Simplified TLU Learning Problem Given a set of input vectors $\{\boldsymbol{x}^{(1)}, \ldots, \boldsymbol{x}^{(p)}\}$. Determine a weight vector $\boldsymbol{w}$, if there exists one such that for $i = 1, \ldots, p$ we have

$$\left\langle \boldsymbol{w}, \boldsymbol{x}^{(i)} \right\rangle > 0.$$

In view of the above, for a linearly separable set of vectors, $\{\boldsymbol{x}^{(i)}\}$, after preprocessing there exists a hyperplane passing through the origin in $\mathbb{R}^{n+1}$ such that all the vectors lie on one side of the hyperplane. The normal to the hyperplane is the weight vector $\boldsymbol{w}$. This is illustrated in Figure 9.5. Taking into account the Simplified TLU learning problem, we can state the TLU learning algorithm equivalently as follows:

SIMPLIFIED TLU LEARNING ALGORITHM

STEP 1 *Initialize Weight Vector.* Set the values of the components of the initial weight vector $\boldsymbol{w}(0)$ to zero.

STEP 2 *Test.* If $\langle \boldsymbol{w}(k), \boldsymbol{x}^{(i)} \rangle \leq 0$, go to Add, else Test next input vector.

STEP 3 *Add*

$$\boldsymbol{w}(k+1) = \boldsymbol{w}(k) + \boldsymbol{x}^{(i)}, \qquad k = k + 1.$$

Go to *Test*.

The algorithm is said to *converge* if for some $k < \infty$ we have

$$\left\langle \boldsymbol{w}(k), \boldsymbol{x}^{(i)} \right\rangle > 0 \qquad \text{for } i = 1, \ldots, p.$$

The following theorem states that if the set of input vectors is linearly separable, then the TLU learning algorithm converges.

Theorem 9.1 (The TLU Convergence Theorem) If the set of vectors $\{\boldsymbol{x}^{(1)}, \ldots, \boldsymbol{x}^{(p)}\}$ is linearly separable, then the simplified TLU learning algorithm determines a weight vector $\boldsymbol{w}$ in a finite number of steps, such that $\langle \boldsymbol{w}, \boldsymbol{x}^{(i)} \rangle > 0$ for $i = 1, 2, \ldots, p$.

Proof The proof is adapted from Siu, Roychowdhury, and Kailath [260]. We begin by observing that because the set of input vectors is linearly separable, there exists a weight vector $\boldsymbol{w}^*$ and a constant $\delta > 0$ such that for $i = 1, 2, \ldots, p$, we have

$$\left\langle \boldsymbol{w}^*, \boldsymbol{x}^{(i)} \right\rangle > \delta.$$

To simplify further manipulations, we assume that $\|\boldsymbol{w}^*\| = 1$. We also assume that $\|\boldsymbol{x}^{(i)}\| \leq L$, for $i = 1, 2, \ldots, p$; that is, the lengths of all the input vectors are bounded above by L. Let k denote the number of times the weight vector $\boldsymbol{w}$ in the simplified TLU learning algorithm has been updated, and let $\boldsymbol{w}(k)$ be the value of the weight vector, $\boldsymbol{w}$, after the kth update. Our goal is to show that k is bounded above. To proceed, we note that if an input vector $\boldsymbol{x}^{(i)}$ is used for the kth update in the algorithm, then

$$\boldsymbol{w}(k) = \boldsymbol{w}(k-1) + \boldsymbol{x}^{(i)},$$

where $\langle \boldsymbol{w}(k-1), \boldsymbol{x}^{(i)} \rangle = \boldsymbol{w}^T(k-1)\boldsymbol{x}^{(i)} \leq 0$. We first derive an upper bound on $\|\boldsymbol{w}(k)\|$. We have

$$\begin{aligned}
\|\boldsymbol{w}(k)\|^2 &= \left\|\boldsymbol{w}(k-1) + \boldsymbol{x}^{(i)}\right\|^2 \\
&= \left(\boldsymbol{w}(k-1) + \boldsymbol{x}^{(i)}\right)^T \left(\boldsymbol{w}(k-1) + \boldsymbol{x}^{(i)}\right) \\
&= \boldsymbol{w}^T(k-1)\boldsymbol{w}(k-1) + \boldsymbol{x}^{(i)T}\boldsymbol{x}^{(i)} + 2\boldsymbol{w}^T(k-1)\boldsymbol{x}^{(i)} \\
&\leq \|\boldsymbol{w}(k-1)\|^2 + L^2,
\end{aligned}$$

because $\boldsymbol{w}^T(k-1)\boldsymbol{x}^{(i)} \leq 0$ and $\|\boldsymbol{x}^{(i)}\|^2 \leq L^2$. Suppose that for the $(k-1)$st update, the input vector $\boldsymbol{x}^{(j)}$ was used, that is,

$$\boldsymbol{w}(k-1) = \boldsymbol{w}(k-2) + \boldsymbol{x}^{(j)}.$$

Therefore,

$$\begin{aligned}
\|\boldsymbol{w}(k)\|^2 &\leq \|\boldsymbol{w}(k-1)\|^2 + L^2 \\
&= \left\|\boldsymbol{w}(k-2) + \boldsymbol{x}^{(j)}\right\|^2 + L^2 \\
&\leq \|\boldsymbol{w}(k-2)\|^2 + 2L^2.
\end{aligned}$$

Continuing in this manner, we obtain

$$\|\boldsymbol{w}(k)\|^2 \leq kL^2, \tag{9.7}$$

because by assumption, $\boldsymbol{w}(0) = \boldsymbol{0}$.

We next derive a lower bound on $\|\boldsymbol{w}(k)\|$. We have

$$\begin{aligned}
\langle \boldsymbol{w}^*, \boldsymbol{w}(k) \rangle &= \langle \boldsymbol{w}^*, \boldsymbol{w}(k-1) \rangle + \left\langle \boldsymbol{w}^*, \boldsymbol{x}^{(i)} \right\rangle \\
&\geq \langle \boldsymbol{w}^*, \boldsymbol{w}(k-1) \rangle + \delta,
\end{aligned}$$

because we assumed that $\langle \boldsymbol{w}^*, \boldsymbol{x}^{(i)} \rangle > \delta$. Observing that $\boldsymbol{w}(k-1) = \boldsymbol{w}(k-2) + \boldsymbol{x}^{(j)}$, we get

$$\begin{aligned}\langle \boldsymbol{w}^*, \boldsymbol{w}(k) \rangle &\geq \langle \boldsymbol{w}^*, \boldsymbol{w}(k-1) \rangle + \delta \\ &\geq \langle \boldsymbol{w}^*, \boldsymbol{w}(k-2) \rangle + 2\delta.\end{aligned}$$

Continuing in this manner, we obtain

$$\langle \boldsymbol{w}^*, \boldsymbol{w}(k) \rangle \geq k\delta. \tag{9.8}$$

Applying to the above the Cauchy–Schwarz inequality yields

$$\|\boldsymbol{w}^*\| \|\boldsymbol{w}(k)\| \geq \langle \boldsymbol{w}^*, \boldsymbol{w}(k) \rangle \geq k\delta.$$

By assumption, $\|\boldsymbol{w}^*\| = 1$. Hence,

$$\|\boldsymbol{w}(k)\| \geq k\delta.$$

Combining the obtained upper and lower bounds on $\|\boldsymbol{w}(k)\|^2$, given by (9.7) and (9.8), gives

$$k^2\delta^2 \leq \|\boldsymbol{w}(k)\|^2 \leq kL^2.$$

Therefore,

$$k \leq \frac{L^2}{\delta^2}, \tag{9.9}$$

which implies that the TLU learning algorithm converges. The proof of the TLU convergence theorem is complete.

We now present a sufficient condition for linear nonseparability.

Theorem 9.2 If there exists a positive linear combination (PLC) that equals $\boldsymbol{0}$ of the given set of input vectors, then the given set of input vectors is linearly nonseparable.

Proof We prove the theorem by contradiction. Assume that there exists a PLC of the given input vectors that is equal to $\boldsymbol{0}$, that is, there exist nonnegative constants, q_i, such that for some j, $1 \leq j \leq p$, we have $q_j > 0$, and

$$\sum_{i=1}^{p} q_i \boldsymbol{x}^{(i)} = \boldsymbol{0}. \tag{9.10}$$

Suppose now that the given input vectors are linearly separable; that is, there is a weight vector, $\boldsymbol{w}$, such that $\langle \boldsymbol{w}, \boldsymbol{x}^{(i)} \rangle > 0$ for $i = 1, \ldots, p$. We premultiply (9.10) by $\boldsymbol{w}^T$ to get

$$\boldsymbol{w}^T \sum_{i=1}^{p} q_i \boldsymbol{x}^{(i)} = \sum_{i=1}^{p} q_i \langle \boldsymbol{w}, \boldsymbol{x}^{(i)} \rangle > 0$$

because $\langle \boldsymbol{w}, \boldsymbol{x}^{(i)} \rangle > 0$ and $q_j > 0$. However, we arrived at a contradiction because by assumption we have $\sum_{i=1}^{p} q_i \boldsymbol{x}^{(i)} = \boldsymbol{0}$, which completes the proof.

◆ Example 9.2

Suppose we are given the following training set:

$$x^{(1)} = \begin{bmatrix} -1 \\ -1 \end{bmatrix}, \quad d_1 = -1; \qquad x^{(2)} = \begin{bmatrix} 1 \\ 1 \end{bmatrix}, \quad d_2 = -1;$$

$$x^{(3)} = \begin{bmatrix} -1 \\ 1 \end{bmatrix}, \quad d_3 = 1; \qquad x^{(4)} = \begin{bmatrix} 1 \\ -1 \end{bmatrix}, \quad d_4 = 1.$$

This training set represents the two-input Exclusive-OR (XOR) function. Let us now preprocess the training set so we can apply Theorem 9.2. By preprocessing we mean eliminating the threshold by augmenting the input vectors with -1 and then negating the input vectors for which the desired output is -1. The preprocessed input vectors are

$$\tilde{x}^{(1)} = \begin{bmatrix} 1 \\ 1 \\ 1 \end{bmatrix}, \quad \tilde{x}^{(2)} = \begin{bmatrix} -1 \\ -1 \\ 1 \end{bmatrix}, \quad \tilde{x}^{(3)} = \begin{bmatrix} -1 \\ 1 \\ -1 \end{bmatrix}, \quad \tilde{x}^{(4)} = \begin{bmatrix} 1 \\ -1 \\ -1 \end{bmatrix}.$$

Note that $\sum_{i=1}^{4} \tilde{x}^{(i)} = \mathbf{0}$. It follows from Theorem 9.2 that the given input vectors are not linearly separable.

9.2 Identification Using Adaptive Linear Element

The adaptive linear element, also called the adaline, is shown in Figure 9.6. We first explain the symbols used in our discussion of its operation. Let $x(t) = [x_1(t) \quad \cdots \quad x_n(t)]^T$ denote the adaline input. We assume that for all t

$$\|x(t)\| = \sqrt{x_1^2(t) + \cdots + x_n^2(t)} \le B_x,$$

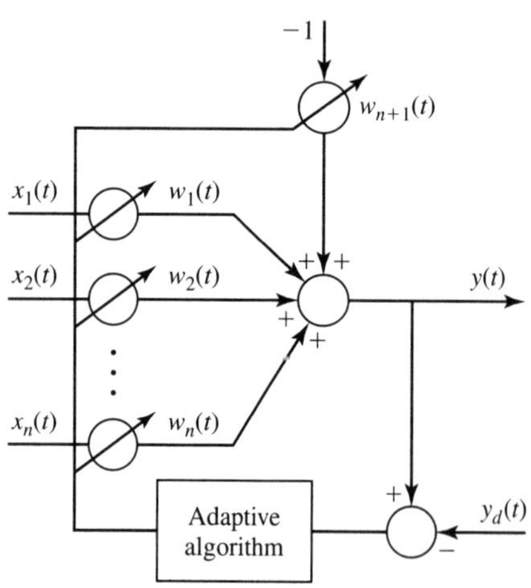

Figure 9.6 Adaline.

where $B_x > 0$ is known positive constant and $\|\cdot\| = \|\cdot\|_2$; that is, whenever we write $\|\cdot\|$ we mean the Euclidean norm $\|\cdot\|_2$. We also assume that for all t

$$\|\dot{\boldsymbol{x}}(t)\| = \sqrt{\dot{x}_1^2(t) + \cdots + \dot{x}_n^2(t)} \le B_{\dot{x}}.$$

The weight vector $\boldsymbol{w}(t) = [w_1(t) \quad \cdots \quad w_n(t)]^T$. The bias weight $w_{n+1}(t)$ is connected to a constant input equal to -1. We define $\tilde{\boldsymbol{x}}(t)$ to be an augmented input vector that includes the constant input to the bias weight, that is,

$$\tilde{\boldsymbol{x}}(t) = [x_1(t) \quad \cdots \quad x_n(t) \quad x_{n+1}(t)]^T.$$

Note that

$$\|\tilde{\boldsymbol{x}}(t)\|^2 = \tilde{\boldsymbol{x}}^T(t)\tilde{\boldsymbol{x}}(t) = \|\boldsymbol{x}(t)\|^2 + 1,$$

and hence $\tilde{\boldsymbol{x}}(t)$ is bounded away from zero at all times. The augmented vector $\tilde{\boldsymbol{x}}(t)$ is bounded, for all t, if and only if $\boldsymbol{x}(t)$ is bounded. We denote a bound of $\tilde{\boldsymbol{x}}(t)$ by $B_{\tilde{x}}$; that is, $\|\tilde{\boldsymbol{x}}(t)\| \le B_{\tilde{x}}$ for all t. Observe that $\|\dot{\tilde{\boldsymbol{x}}}(t)\| = \|\dot{\boldsymbol{x}}(t)\|$. Thus, we can take $B_{\dot{\tilde{x}}} = B_{\dot{x}}$. Similarly, we define an augmented weight vector,

$$\tilde{\boldsymbol{w}}(t) = [w_1(t) \quad \cdots \quad w_n(t) \quad w_{n+1}(t)]^T.$$

We assume that for all t

$$\|\tilde{\boldsymbol{w}}(t)\| \le B_{\tilde{w}}$$

for some constant $B_{\tilde{w}} > 0$. The output, $y(t)$, of the adaline is a linear combination of its inputs, that is,

$$y(t) = \tilde{\boldsymbol{x}}^T(t)\tilde{\boldsymbol{w}}(t).$$

Let $y_d(t)$ be the desired output of the adaline and assume that

$$|y_d(t)| \le B_y \quad \text{and} \quad |\dot{y}_d(t)| \le B_{\dot{y}}$$

for some positive constants B_y and $B_{\dot{y}}$. Let

$$e(t) = y(t) - y_d(t)$$

denote the error signal. The error dynamics are described by

$$\begin{aligned} \dot{e} &= \dot{y} - \dot{y}_d \\ &= \dot{\tilde{\boldsymbol{w}}}^T \tilde{\boldsymbol{x}} + \tilde{\boldsymbol{w}}^T \dot{\tilde{\boldsymbol{x}}} - \dot{y}_d. \end{aligned} \tag{9.11}$$

Our objective is to devise an adaptation algorithm for the weights $w_i(t)$, $i = 1, 2, \ldots, n+1$, so that the error $e(t)$ converges to zero in finite time, say t_h, and $e(t) = 0$ for $t > t_h$. Sira-Ramírez and Colina-Morles [258] showed that we can achieve our objective if we adapt the weights according to the adaptation law,

$$\boxed{\dot{\tilde{\boldsymbol{w}}} = -\eta \frac{\tilde{\boldsymbol{x}}}{\tilde{\boldsymbol{x}}^T \tilde{\boldsymbol{x}}} \operatorname{sign}(e(t))} \tag{9.12}$$

where the gain $\eta > 0$ is sufficiently large. We now state and prove the theorem of Sira-Ramírez and Colina-Morles [258].

Theorem 9.3 If the weight vector $\tilde{\boldsymbol{w}}$ is adapted according to (9.12), where

$$\eta > B_{\tilde{w}} B_{\dot{x}} + B_{\dot{y}},$$

then for arbitrary initial error $e(0)$ the error signal $e(t)$ converges to zero in finite time, say t_h, such that

$$t_h \leq \frac{|e(0)|}{\eta - B_{\tilde{w}} B_{\dot{x}} - B_{\dot{y}}}$$

and $e(t) = 0$ for $t > t_h$.

Proof Recall that the error dynamics are governed by (9.11), where the vector $\tilde{\boldsymbol{w}}(t)$ is adapted according to (9.12). Consider the following Lyapunov function candidate

$$V(e) = \tfrac{1}{2} e^2(t).$$

The time derivative of V evaluated on the solutions of (9.11) is

$$\begin{aligned} \dot{V}(e) &= e\dot{e} \\ &= e(\dot{\tilde{\boldsymbol{w}}}^T \tilde{\boldsymbol{x}} + \tilde{\boldsymbol{w}}^T \dot{\tilde{\boldsymbol{x}}} - \dot{y}_d). \end{aligned} \tag{9.13}$$

Substituting (9.12) into the above and taking into account that

$$\operatorname{sign}(e) = \frac{e}{|e|} = \frac{|e|}{e},$$

we obtain

$$\begin{aligned} \dot{V} &= -\eta|e| + e(\tilde{\boldsymbol{w}}^T \dot{\tilde{\boldsymbol{x}}} - \dot{y}_d) \\ &\leq -\eta|e| + |e|(B_{\tilde{w}} B_{\dot{\tilde{x}}} + B_{\dot{y}}) \\ &= (-\eta + B_{\tilde{w}} B_{\dot{\tilde{x}}} + B_{\dot{y}})|e| \\ &< 0 \qquad \text{for all} \quad e \neq 0 \quad \text{and} \quad \eta > B_{\tilde{w}} B_{\dot{\tilde{x}}} + B_{\dot{y}}, \end{aligned} \tag{9.14}$$

which means that $e(t) \to 0$ as t increases. It follows from our discussion above that we can take $B_{\dot{\tilde{x}}} = B_{\dot{x}}$. We now show that for some finite time t_h, we have $e(t_h) = 0$ and $e(t) = 0$ for $t > t_h$. Substitute (9.12) into (9.11) to get

$$\dot{e} = -\eta \operatorname{sign}(e) + \tilde{\boldsymbol{w}}^T \dot{\tilde{\boldsymbol{x}}} - \dot{y}_d. \tag{9.15}$$

Integrating the above from $t = 0$ to $t \leq t_h$ gives

$$e(t) - e(0) = -\eta t \operatorname{sign}(e(0)) + \int_0^t (\tilde{\boldsymbol{w}}^T(\tau)\dot{\tilde{\boldsymbol{x}}}(\tau) - \dot{y}_d(\tau))\, d\tau. \tag{9.16}$$

At $t = t_h$ we have $e(t_h) = 0$, and therefore for $t = t_h$ the above equation takes the form

$$-e(0) = -\eta t_h \operatorname{sign}(e(0)) + \int_0^{t_h} (\tilde{\boldsymbol{w}}^T(t)\dot{\tilde{\boldsymbol{x}}}(t) - \dot{y}_d(t))\, dt. \tag{9.17}$$

Multiplying both sides of the above by $-\operatorname{sign}(e(0))$ yields

$$\begin{aligned} |e(0)| &= \eta t_h - \operatorname{sign}(e(0)) \int_0^{t_h} (\tilde{\boldsymbol{w}}^T(t)\dot{\tilde{\boldsymbol{x}}}(t) - \dot{y}_d(t))\, dt \\ &\geq (\eta - (B_{\tilde{w}} B_{\dot{x}} + B_{\dot{y}}))t_h. \end{aligned} \tag{9.18}$$

Thus, the time t_h at which $e(t)$ reaches zero value for the first time satisfies the inequality

$$t_h \leq \frac{|e(0)|}{\eta - (B_{\tilde{w}} B_{\tilde{x}} + B_{\dot{y}})}.$$

Because for $e(t) \neq 0$ we have $\frac{d}{dt} e^2(t) < 0$, we are guaranteed that we also have $e(t) = 0$ for $t > t_h$. The proof of the theorem is now complete.

If the time derivative $\dot{\tilde{\boldsymbol{x}}}(t)$ is available, then we can use it in an adaptation law of $\tilde{\boldsymbol{w}}$. Using $\dot{\tilde{\boldsymbol{x}}}(t)$ in an adaptation law allows us to reduce the gain η and the bound on time t_h when the error $e(t)$ becomes zero. One such adaptation strategy that includes $\dot{\tilde{\boldsymbol{x}}}(t)$ was proposed by Sira-Ramírez and Colina-Morles [258], which we present in the following theorem.

Theorem 9.4 If the adaptation law for the augmented weight vector is chosen as

$$\dot{\tilde{\boldsymbol{w}}}(t) = -\frac{\tilde{\boldsymbol{x}}(t)\dot{\tilde{\boldsymbol{x}}}^T(t)}{\tilde{\boldsymbol{x}}^T(t)\tilde{\boldsymbol{x}}(t)}\tilde{\boldsymbol{w}}(t) - \eta \frac{\tilde{\boldsymbol{x}}(t)}{\tilde{\boldsymbol{x}}^T(t)\tilde{\boldsymbol{x}}(t)} \operatorname{sign}(e(t)) \tag{9.19}$$

where $\eta > B_{\dot{y}}$, then for any given initial error $e(0)$ the error $e(t)$ converges to zero in finite time t_h such that

$$t_h \leq \frac{|e(0)|}{\eta - B_{\dot{y}}} \tag{9.20}$$

and $e(t) = 0$ for $t > t_h$.

Proof The proof of this theorem is analogous to the proof of Theorem 9.3. The only difference is that (9.15) becomes

$$\dot{e} = -\eta \operatorname{sign}(e) - \dot{y}_d. \tag{9.21}$$

When using the adaline for control purposes, the signal to be processed by the adaline is first passed through a prefilter as shown in Figure 9.7. The prefilter is described by the equation $\dot{\boldsymbol{x}} = \boldsymbol{A}\boldsymbol{x} + \boldsymbol{b}r$, where r is the signal to be processed. The matrix $\boldsymbol{A}$ of the prefilter is chosen to be asymptotically stable. Recall that the inputs to the adaline are the components of the augmented

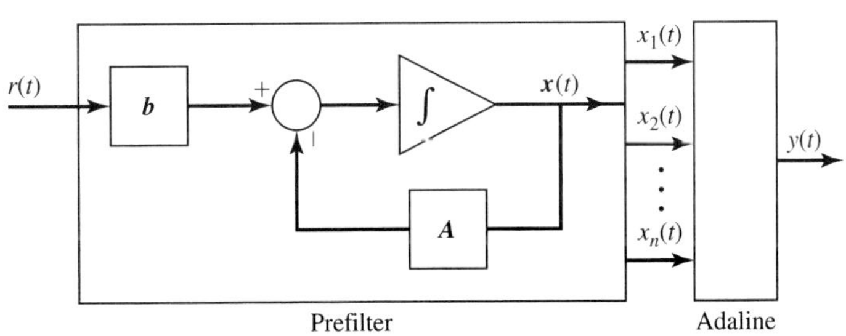

Figure 9.7 Prefilter connection to adaline.

vector $\tilde{x}$, where the $(n+1)$st component is equal to -1. Thus, $\dot{x}_{n+1} = 0$. Taking this into account, the prefiltered inputs to the adaline can be modeled as

$$\dot{\tilde{x}} = \tilde{A}\tilde{x} + \tilde{b}r, \tag{9.22}$$

where

$$\tilde{A} = \begin{bmatrix} A & 0 \\ 0^T & 0 \end{bmatrix} \quad \text{and} \quad \tilde{b} = \begin{bmatrix} b \\ 0 \end{bmatrix}.$$

Note that $\dot{\tilde{x}}$ is available to us in the architecture shown in Figure 9.7. Thus, we can employ the adaptation strategy given by (9.19), which results in the following description of the structure depicted in Figure 9.7:

$$\begin{aligned} \dot{\tilde{x}} &= \tilde{A}\tilde{x} + \tilde{b}r, \\ \dot{\tilde{w}} &= -\frac{\tilde{x}\dot{\tilde{x}}^T}{\tilde{x}^T\tilde{x}}\tilde{w} - \eta\frac{\tilde{x}}{\tilde{x}^T\tilde{x}}\,\text{sign}(e), \\ y &= \tilde{x}^T\tilde{w} \end{aligned} \tag{9.23}$$

We now demonstrate, with a numerical example, how the adaline system modeled by (9.23) can be applied to forward and inverse dynamics identification of dynamical systems. The nonlinear plant model is taken from Sira-Ramírez and Colina-Morles [258]. This plant model has the form

$$\begin{aligned} \dot{\alpha} &= p + \frac{\sin(\alpha)}{l}u, \\ \dot{p} &= \left(\frac{g}{l} - \frac{u^2}{l^2}\cos(\alpha)\right) - \frac{u}{l}p\cos(\alpha), \\ \dot{z} &= u, \\ y_p &= \alpha. \end{aligned} \tag{9.24}$$

The parameters g and l have the values: $g = 9.81$ m/sec^2 and $l = 0.7$ m. The plant input is labeled as u while the plant output is y_p. We first concern ourselves with forward dynamics identification of the above plant. A block diagram of a forward dynamics identification scheme is depicted in Figure 9.8. The input to the plant is

$$u(t) = A_1 + A_2\cos\left(\frac{t}{2\epsilon}\right) + A_3\sin\left(\frac{t}{\epsilon}\right),$$

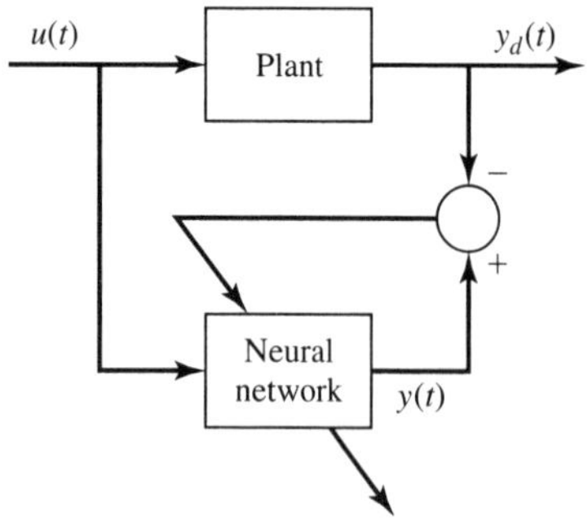

Figure 9.8 Forward dynamics identification scheme.

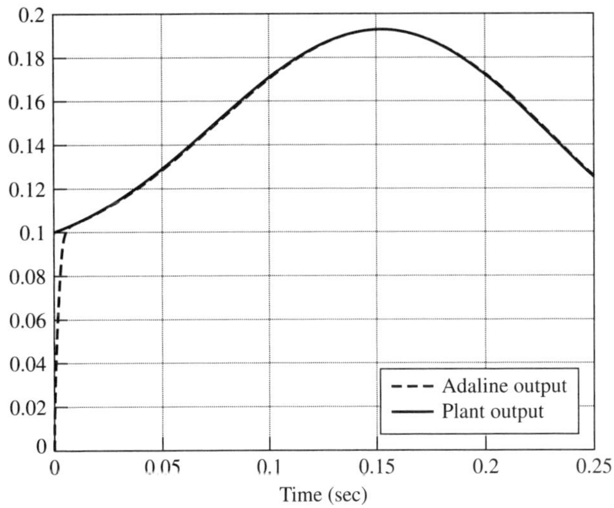

Figure 9.9 Simulation results of forward dynamics identification of (9.24).

where $A_1 = 0.4$, $A_2 = 2$, $A_3 = 3$, and $\epsilon = 0.05$. The prefilter equation (9.22) in our example has the form

$$\dot{\tilde{x}} = \begin{bmatrix} 0 & 1 & 0 & 0 \\ 0 & 0 & 1 & 0 \\ -1 & -3 & -3 & 0 \\ 0 & 0 & 0 & 0 \end{bmatrix} \tilde{x} + \begin{bmatrix} 0 \\ 0 \\ 1 \\ 0 \end{bmatrix} u.$$

Thus, the adaline has total four weights, including the bias weight. In our simulations, we use the following approximation of the sign function:

$$\text{sign}(e(t)) \approx \frac{e(t)}{|e(t)| + \delta}, \tag{9.25}$$

where $\delta = 0.05$. We used the above approximation to eliminate chattering and speed up the simulations. We used $\eta = 10$. The performance of the adaline as a forward dynamics identifier of the dynamical system modeled by (9.24) is shown in Figure 9.9. We used the following initial conditions:

$$\tilde{x}(0) = [0.1 \quad 0 \quad 0 \quad -1]^T, \qquad \tilde{w}(0) = [0.1 \quad 0 \quad 0 \quad 0]^T,$$

and $[\alpha(0) \;\; p(0) \;\; z(0)] = [0.1 \;\; 0 \;\; 0]$. Increasing the gain η results in more accurate identification; that is, the error $e(t)$ goes to zero faster. Furthermore, the effects of using the approximation (9.25) are reduced for higher gain η.

In our next experiment, we employed the adaline as an inverse dynamics identifier using the scheme shown in Figure 9.10. The results of our simulation, using the same initial conditions as before, are depicted in Figure 9.11. Here we used $\eta = 90$. Increasing the gain k decreases the identification error as well as reduces the effects of the approximation of the sign function.

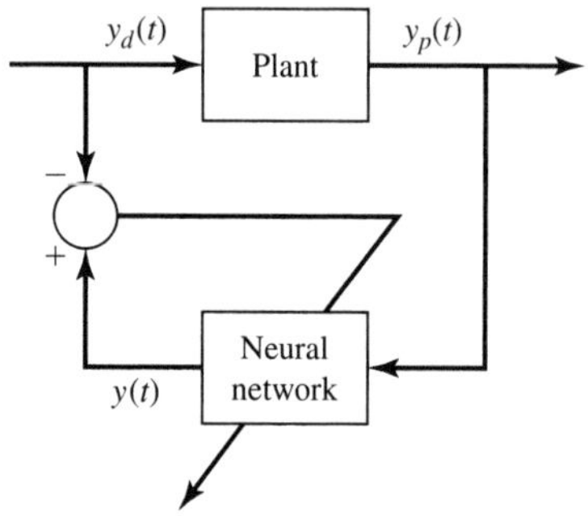

Figure 9.10 Inverse dynamics identifier scheme.

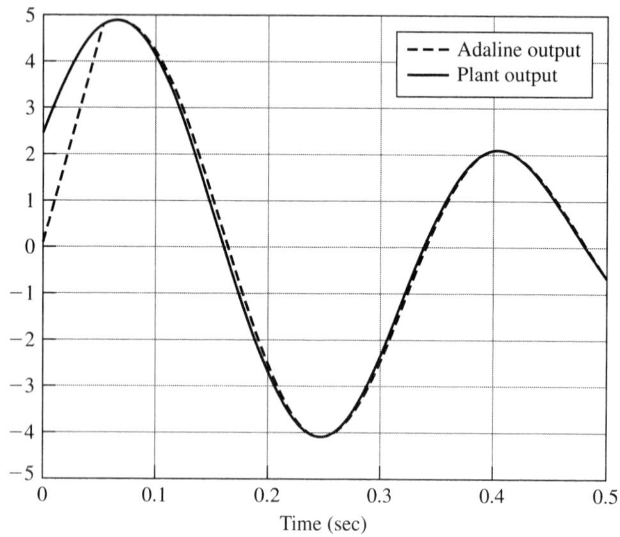

Figure 9.11 Simulation results of inverse dynamics identification of (9.24).

9.3 Backpropagation

In Section 9.1 we discussed an iterative learning algorithm for a threshold logic unit. This computational unit kept on iterating until it reached a solution, if a solution existed. In this section we present iterative learning algorithms for computational units with differentiable activation functions. This family of algorithms are called the backpropagation algorithms because they backpropagate the error between the desired and actual output of the neural network for the purpose of adjusting the network's weights in order to reduce the learning error. Specifically, a typical backpropagation algorithm consists of two passes: In the forward pass the input vectors are presented to the network, and the actual outputs of the network are calculated. Then, the error between the actual and desired outputs are calculated. In the reverse pass the error gradient with respect to the network's weights is calculated by propagating the error backwards through the network. Once the error gradient is calculated, the weights are adjusted using, for example, a descent gradient method.

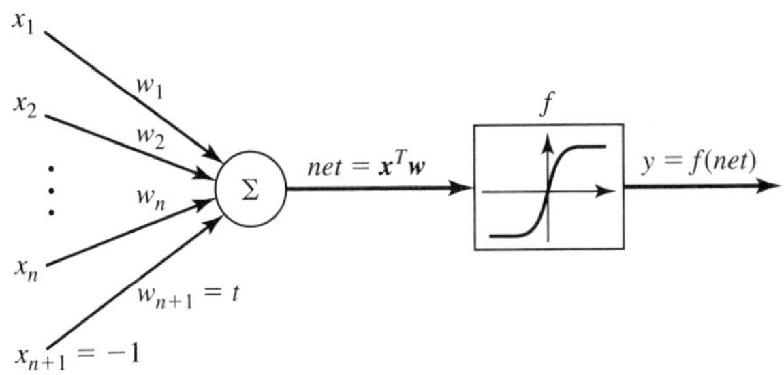

Figure 9.12 A single perceptron with a differentiable sigmoidal activation function.

We now analyze a single computational unit shown in Figure 9.12. It is also called the *perceptron*. We consider a simple case, where we train the perceptron to learn the training pair

$$(\boldsymbol{x}, d),$$

where $\boldsymbol{x} \in \mathbb{R}^{n+1}$, $x_{n+1} = (\boldsymbol{x})_{n+1} = -1$, and $d \in [-1, 1]$. Define the error function

$$\begin{aligned} E(\boldsymbol{w}) &= \tfrac{1}{2}(d - y)^2 \\ &= \tfrac{1}{2}(d - f(net))^2, \end{aligned} \tag{9.26}$$

where $net = \boldsymbol{x}^T \boldsymbol{w}$. Note that $w_{n+1} = t$, where t is a threshold, bias. The gradient of E, with respect to $\boldsymbol{w}$, is

$$\nabla E(\boldsymbol{w}) = -(d - f(net)) f'(net) \boldsymbol{x}. \tag{9.27}$$

Using the descent gradient algorithm with a fixed step size, $\alpha > 0$, to minimize E, we obtain

$$\begin{aligned} \boldsymbol{w}(k+1) &= \boldsymbol{w}(k) - \alpha \nabla E(\boldsymbol{w}(k)) \\ &= \boldsymbol{w}(k) + \alpha (d_k - f(net_k)) f'(net_k) \boldsymbol{x}^{(k)}, \end{aligned} \tag{9.28}$$

where $net_k = \boldsymbol{x}^T(k) \boldsymbol{w}(k)$. Let

$$\delta^k = (d_k - f(net_k)) f'(net_k). \tag{9.29}$$

Then, we can represent (9.28) as

$$\boxed{\boldsymbol{w}(k+1) = \boldsymbol{w}(k) + \alpha \delta^k \boldsymbol{x}(k)} \tag{9.30}$$

We now generalize the above formula for a three-layer feedforward neural network shown in Figure 9.13. We wish to train the network to learn the training pairs

$$\left(\boldsymbol{x}^{(1)}, \boldsymbol{d}^{(1)}\right), \left(\boldsymbol{x}^{(2)}, \boldsymbol{d}^{(2)}\right), \ldots, \left(\boldsymbol{x}^{(p)}, \boldsymbol{d}^{(p)}\right),$$

where $\boldsymbol{x}^{(i)} \in \mathbb{R}^n$ and $\boldsymbol{d}^{(i)} \in \mathbb{R}^m$. The network has n inputs, which is the dimension of the training vectors, and m outputs. The network's inputs are connected to a layer of hidden units, which in turn is connected to a layer of output units. To construct a network to perform some task, we must set the weights of the connections appropriately. The weights specify the strength of the influence of one unit upon another. We will derive rules for adjusting the weights w_{ji}^h and w_{kj}^o such that the error function

$$E(\boldsymbol{w}) = \frac{1}{2} \sum_{s=1}^{m} (d_s - y_s)^2$$

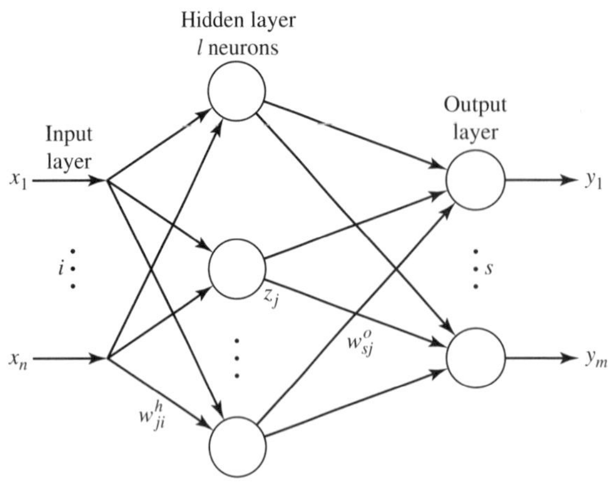

Figure 9.13 A three-layer feedforward neural network.

is minimized. The minimization is performed over the space of weights

$$\{w^h_{ji}, w^o_{sj} : i = 1, 2, \ldots, n,\ j = 1, 2, \ldots, l,\ s = 1, 2, \ldots, m\}.$$

We first give formulas for updating the weights connecting the hidden and output layers. Let

$$net_s = \sum_{j=1}^{l} w^o_{sj} z_j.$$

Using the descent gradient algorithm in a similar way as in (9.30), we obtain

$$\begin{aligned} w^o_{sj}(k+1) &= w^o_{sj}(k) - \alpha_o \frac{\partial E(\boldsymbol{w}(k))}{\partial w^o_{sj}(k)} \\ &= w^o_{sj}(k) + \alpha_o (d_s - y_s) f'_o(net_s) z_j(k), \end{aligned} \tag{9.31}$$

where f'_o is the derivative of $f_o(net_s)$ with respect to net_s, and the values of z_j are computed by feeding into the network the input vector $\boldsymbol{x}$ and then propagating it forward through the network. We have

$$z_j = f_h\left(\sum_{i=1}^{n} w^h_{ji} x_i\right) = f_h(net_j), \qquad j = 1, 2, \ldots, l.$$

To derive the update formulas for the weights in the hidden layer, we use the chain rule to compute $\partial E / \partial w^h_{ji}$, and then we apply the gradient descent algorithm:

$$\frac{\partial E}{\partial w^h_{ji}} = \frac{\partial E}{\partial z_j} \frac{\partial z_j}{\partial net_j} \frac{\partial net_j}{\partial w^h_{ji}}, \tag{9.32}$$

where

$$\frac{\partial net_j}{\partial w^h_{ji}} = x_i, \tag{9.33}$$

$$\frac{\partial z_j}{\partial net_j} = f'_h(net_j), \tag{9.34}$$

and

$$
\begin{aligned}
\frac{\partial E}{\partial z_j} &= \frac{\partial}{\partial z_j}\left(\frac{1}{2}\sum_{s=1}^{m}(d_s - f_o(net_s))^2\right) \\
&= -\sum_{s=1}^{m}(d_s - f_o(net_s))\frac{\partial f_o(net_s)}{\partial z_j} \\
&= -\sum_{s=1}^{m}(d_s - y_s) f_o'(net_s) w_{sj}^o. \qquad (9.35)
\end{aligned}
$$

Combining (9.32)–(9.35), we obtain rules for updating the hidden weights,

$$
\begin{aligned}
w_{ji}^h(k+1) &= w_{ji}^h(k) - \alpha_h \frac{\partial E(\boldsymbol{w}(k))}{\partial w_{ji}^h(k)} \\
&- w_{jl}^h(k) + \alpha_h \left(\sum_{s=1}^{m}(d_s - y_s) f_o'(net_s) w_{sj}^o(k)\right) f_h'(net_j) x_i. \qquad (9.36)
\end{aligned}
$$

A commonly used activation function is the sigmoid,

$$
f(v) = \frac{1}{1+e^{-v}}. \qquad (9.37)
$$

It is easy to check that the activation function (9.37) has the useful property,

$$
\frac{df(v)}{dv} = f(v)(1 - f(v)). \qquad (9.38)
$$

We summarize the above derivations in the following algorithm.

BACKPROPAGATION TRAINING ALGORITHM

Given a set of training pairs $\{(\boldsymbol{x}^{(1)}, \boldsymbol{d}^{(1)}), \ldots, (\boldsymbol{x}^{(i)}, \boldsymbol{d}^{(i)}), \ldots, (\boldsymbol{x}^{(p)}, \boldsymbol{d}^{(p)})\}$.

STEP 1 *Initialize Weights and Thresholds.* Initialize the weights and thresholds to small random values.

STEP 2 *Present New Input Vector and Desired Output.* Present the new input vector, $\boldsymbol{x}^{(i)}$, and the corresponding desired output, $\boldsymbol{d}^{(i)}$. Calculate actual outputs.

STEP 3 *Calculate Error Gradients.*

STEP 4 *Adapt Weight Vectors and Thresholds:*

$$
\begin{aligned}
\boldsymbol{w}(k+1) &= \boldsymbol{w}(k) - \alpha \nabla_{\boldsymbol{w}} E(\boldsymbol{w}(k), \boldsymbol{t}(k)), \\
\boldsymbol{t}(k+1) &= \boldsymbol{t}(k) - \alpha \nabla_{\boldsymbol{t}} E(\boldsymbol{w}(k), \boldsymbol{t}(k)).
\end{aligned}
$$

STEP 5 *Repeat by Returning to Step 2.*

◆ Example 9.3

We consider the two-input XOR problem. In Example 9.2 we showed that this logic function cannot be implemented using a single TLU. Here, we use a two-layer neural network with two nodes in the hidden layer and one node in the output layer. In this example, we use the unipolar representation of the two-input XOR function; that is, the elements of the training pairs are from the set {0, 1}. Therefore, the input vectors and the corresponding desired outputs have the form

$$\boldsymbol{x}^{(1)} = \begin{bmatrix} 0 \\ 0 \end{bmatrix}, \quad d_1 = 0; \qquad \boldsymbol{x}^{(2)} = \begin{bmatrix} 1 \\ 1 \end{bmatrix}, \quad d_2 = 0;$$

$$\boldsymbol{x}^{(3)} = \begin{bmatrix} 0 \\ 1 \end{bmatrix}, \quad d_3 = 1; \qquad \boldsymbol{x}^{(4)} = \begin{bmatrix} 1 \\ 0 \end{bmatrix}, \quad d_4 = 1.$$

The error function for this example takes the form

$$E = \tfrac{1}{2}(d_i - y_i)^2.$$

The activation function for all three neurons is the sigmoid given by (9.37). The outputs of the hidden neurons are

$$z_1 = f\left(w_{11}^h x_1^{(i)} + w_{12}^h x_2^{(i)} - t_1\right), \tag{9.39}$$

$$z_2 = f\left(w_{21}^h x_1^{(i)} + w_{22}^h x_2^{(i)} - t_2\right). \tag{9.40}$$

The network output is

$$y = f\left(w_{11}^o z_1 + w_{12}^o z_2 - t_o\right). \tag{9.41}$$

The threshold parameters are treated as learning weights. Thus, the above simple network requires updating of nine weights, where, in particular,

$$\frac{\partial E}{\partial t_1} = (d_i - y)\, y' w_{11} z_1' = (d_i - y)\, y(1 - y) w_{11}^o z_1(1 - z_1),$$

$$\frac{\partial E}{\partial t_2} = (d_i - y)\, y' w_{12} z_2' = (d_i - y)\, y(1 - y) w_{12}^o z_2(1 - z_2),$$

$$\frac{\partial E}{\partial t_o} = (d_i - y)\, y' = (d_i - y)\, y(1 - y).$$

In calculating the above, we used (9.38). In our simulation the weights of the network were initialized to small random values in the interval $[-0.1, 0.1]$. The four patterns were presented in a mixed order. We used a fixed learning rate of 0.75 for all nine weights. After 50,000 iterations we obtained the following weight values:

$$\boldsymbol{W}^h = \begin{bmatrix} -4.9288 & -4.9334 \\ -6.7240 & -6.7526 \end{bmatrix}, \qquad \boldsymbol{W}^o = [10.2447 \quad -10.7258],$$

and $t_1 = -7.3398$, $t_2 = -2.7853$, and $t_o = 4.8701$. In Table 9.1 we show the desired and actual responses of the trained network. In Figure 9.14 we show a plot of the function that the trained network implements.

Table 9.1 The Desired and Actual Responses of the Trained Network in Example 9.3

x_1	x_2	d_i	y_i
0	0	0	0.0087
1	1	0	0.0162
0	1	1	0.9870
1	0	1	0.9870

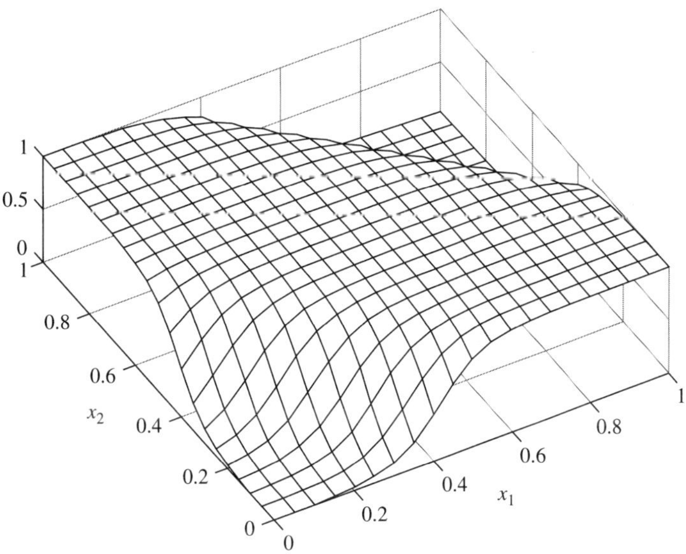

Figure 9.14 The function implemented by the trained neural network to solve the two input XOR problem in Example 9.3.

9.4 Neural Fuzzy Identifier

In this section we describe a method that can be used to construct identifiers of dynamical systems that in turn could be employed to construct neural-network-based fuzzy logic control strategies. The idea behind the method is to apply the backpropagation algorithm to a fuzzy logic system.

It follows from our discussion in Section 8.3 that a typical fuzzy logic system can be thought of as a nonlinear system described by

$$f(x_1, \ldots, x_n) = f(\boldsymbol{x}) = \frac{\sum_{l=1}^{M} v_l \left(\prod_{i=1}^{n} \mu_{il}(x_i) \right)}{\sum_{l=1}^{M} \prod_{i=1}^{n} \mu_{il}(x_i)}, \tag{9.42}$$

where, for example,

$$\mu_{il} = \exp\left\{ -\frac{(x_i - m_{il})^2}{2\sigma_{il}^2} \right\},$$

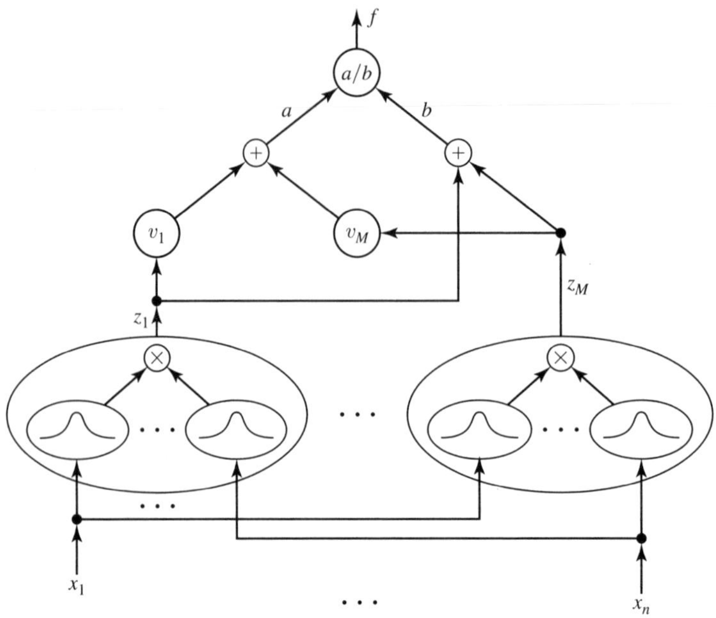

Figure 9.15 Fuzzy logic system (9.42) represented as a feedforward network.

M is the number of IF–THEN fuzzy rules, $\boldsymbol{x} = [x_1 \ x_2 \ \cdots \ x_n]^T$, and m_{il} and σ_{il} are the parameters of the Gaussian membership functions, while v_l is a weight parameter. These parameters will be determined using the backpropagation algorithm discussed in the previous section. To facilitate our discussion, we represent (9.42), following Wang [296, p. 169], as a feedforward network shown in Figure 9.15. The variables that characterize the feedforward network depicted in Figure 9.15 satisfy the following relations:

$$\begin{aligned} z_l &= \prod_{i=1}^{n} \exp\left\{-\frac{(x_i - m_{il})^2}{2\sigma_{il}^2}\right\}, \qquad l = 1, 2 \ldots, M, \\ b &= \sum_{l=1}^{M} z_l, \\ a &= \sum_{l=1}^{M} v_l z_l, \\ f &= a/b. \end{aligned} \tag{9.43}$$

We then use this feedforward network as a component of an identifier of dynamical systems. The identification scheme involving (9.42) is depicted in Figure 9.16. The adaptation algorithm for the identifier parameters will be derived using the backpropagation algorithm. The cost function to be minimized is

$$E = \tfrac{1}{2}(y - \hat{y})^2 = \tfrac{1}{2}(y - f)^2,$$

where y is the output of the identified plant and f is the output of the identifier. We use a gradient algorithm to minimize E over all

$$\boldsymbol{w} = \{v_l, \sigma_{il}, m_{il}\}.$$

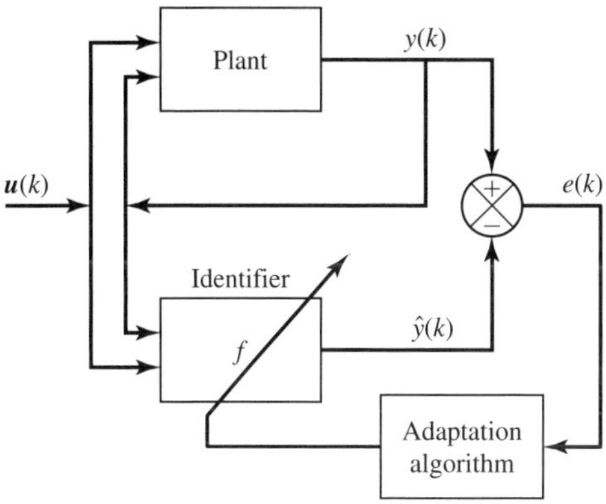

Figure 9.16 Neural fuzzy identification scheme of a nonlinear dynamical system.

The gradient algorithm that we use has the form

$$\boldsymbol{w}(k+1) = \boldsymbol{w}(k) - \alpha \nabla E(\boldsymbol{w}(k)), \tag{9.44}$$

where $\alpha > 0$ is a fixed step size. Note that (9.44) is just a simple descent gradient algorithm that we used in our discussion of the backpropagation algorithm.

In order to be able to use the above algorithm, we need to compute partial derivatives of E with respect to each component of $\boldsymbol{w}$. For this we first fix indices i, and l. We next compute the partial derivatives of E with respect to v_j. Then, the parameters v_j will be adapted accordingly as

$$v_j(k+1) = v_j(k) - \alpha \frac{\partial E(\boldsymbol{w}(k))}{\partial v_j}.$$

We note that E depends upon v_j only through a, where $f = a/b$. Hence, using the chain rule, we obtain

$$\begin{aligned}
\frac{\partial E}{\partial v_j}(\boldsymbol{w}) &= \frac{\partial}{\partial v_j}\left(\frac{1}{2}(y-f)^2\right) \\
&= \frac{\partial}{\partial v_j}\left(\frac{1}{2}(y-a/b)^2\right) \\
&= -\frac{(y-f)}{b}\frac{\partial a}{\partial v_j} \\
&= \frac{f-y}{b} z_j.
\end{aligned}$$

We next compute the partial derivative of E with respect to m_{st}. Observe that E depends upon m_{st} only through z_t. Using the chain rule again gives

$$\begin{aligned}
\frac{\partial E}{\partial m_{st}}(\boldsymbol{w}) &= -(y-f)\frac{\partial f}{\partial z_t}\frac{\partial z_t}{\partial m_{st}} \\
&= -(y-f)\left(\frac{v_t b - a}{b^2}\right) z_t \frac{(x_s - m_{st})}{\sigma_{st}^2} \\
&= (f-y)\left(\frac{v_t - f}{b}\right) z_t \frac{(x_s - m_{st})}{\sigma_{st}^2}.
\end{aligned}$$

Hence, the adaptation formula for m_{st} takes the form

$$m_{st}(k+1) = m_{st}(k) - \alpha(f - y)\left(\frac{v_t - f}{b}\right) z_t \frac{(x_s - m_{st}(k))}{\sigma_{st}^2(k)}.$$

To find the partial derivative of E with respect to σ_{st}, we note that E depends upon σ_{st} only through z_t. Applying the chain rule gives

$$\begin{aligned}\frac{\partial E}{\partial \sigma_{st}}(\boldsymbol{w}) &= -(y - f)\frac{\partial f}{\partial z_t}\frac{\partial z_t}{\partial \sigma_{st}}\\ &= (f - y)\left(\frac{v_t - f}{b}\right) z_t \frac{(x_i - m_{st})^2}{\sigma_{st}^3}.\end{aligned}$$

We are now ready to formulate an algorithm for updating on-line the weights of the neural fuzzy identifier.

ON-LINE IDENTIFICATION ALGORITHM

STEP 1 *Initialize the Identifier Parameters.* Set the values of v_l, m_{il}, and σ_{il} to small random values.

STEP 2 *Calculate the Output of the Identifier at the Time Instant k.* Compute

$$z_l = \prod_{i=1}^{n} \exp\left(-\frac{(x_i - m_{il})^2}{2\sigma_{il}^2}\right),$$

$$b = \sum_{l=1}^{M} z_l,$$

$$a = \sum_{l=1}^{M} v_l z_l,$$

$$f = a/b.$$

STEP 3 *Calculate the Identification Error:*

$$e(k) = y(k) - f(k).$$

STEP 4 *Adapt the Identifier Parameters:*

$$v_l(k+1) = v_l(k) + \alpha \frac{e(k)}{b(k)} z_l(k),$$

$$m_{il}(k+1) = m_{il}(k) + \alpha \frac{e(k)}{b(k)}(v_l(k) - f(k)) z_l(k) \frac{(x_i(k) - m_{il}(k))}{\sigma_{il}^2(k)},$$

$$\sigma_{il}(k+1) = \sigma_{il}(k) + \alpha \frac{e(k)}{b(k)}(v_l(k) - f(k)) z_l(k) \frac{(x_i(k) - m_{il}(k))^2}{\sigma_{il}^3(k)},$$

$$k = k + 1.$$

STEP 5 *Repeat by Returning to Step 2.*

◆ Example 9.4

The plant to be identified is the same as that considered by Wang [296, p. 175], and it is described by

$$y(k+1) = 0.3\,y(k) + 0.6\,y(k-1) + g(u(k)),$$

where the "unknown" function $g(u)$ has the form

$$g(u) = 0.6\sin(\pi u) + 0.3\sin(3\pi u) + 0.1\sin(5\pi u),$$

and $u(k) = \sin(2\pi k/200)$. The identifier has the structure shown in Figure 9.16 and is described by

$$\hat{y}(k+1) = 0.3\,y(k) + 0.6\,y(k-1) + f(u(k)),$$

where f has the form of (9.42) with $M=10$. We chose $\alpha = 0.5$, $y(0) = y(1) = 0$, $f(1) = f(2) = -3$, and we started with $u(1) = 0$. The identifier's parameters were initialized in a random fashion using MATLAB's command `rand`. After the identifier was trained, we tested its prediction power using the same input signal as before and the initial conditions $y(0) = y(1) = 0$, $f(1) = f(2) = -3$, and $u(1) = 0$. The results of this experiment are shown in Figures 9.17 and 9.18.

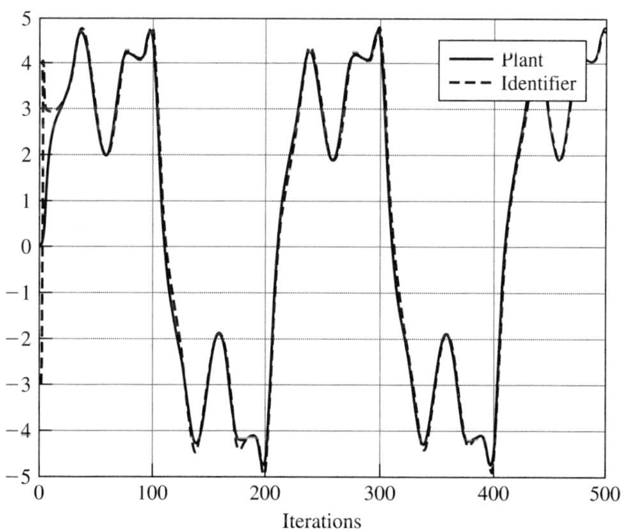

Figure 9.17 Plots of the outputs of the plant and its identifier during the learning phase in Example 9.4.

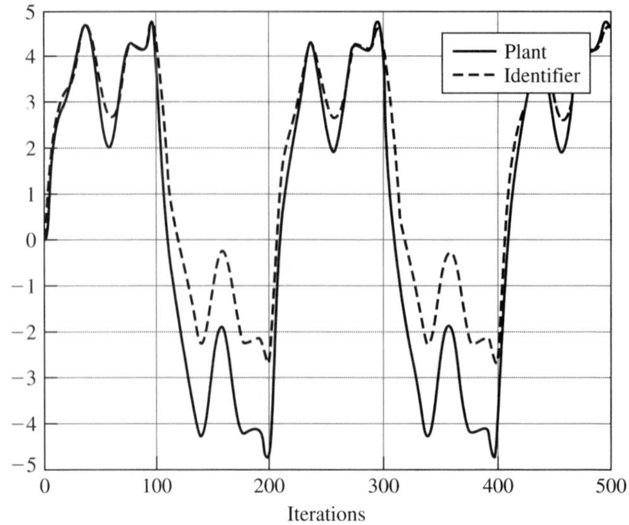

Figure 9.18 Plots of the outputs of the plant and its identifier during the testing phase in Example 9.4.

9.5 Radial-Basis Function (RBF) Networks

In this section we present radial-basis function (RBF) neural networks and their applications. We first show how an RBF network can be used for interpolation purposes. Then, we concern ourselves with the problem of identification of unknown dynamical systems using only the systems' input and output data. The RBF neural identifier is an adaptive system for which the learning algorithm is developed using Lyapunov techniques. The algorithm can add new nodes to the neural identifier to obtain the desired identification accuracy.

9.5.1 Interpolation Using RBF Networks

A diagram of a radial-basis function (RBF) network is shown in Figure 9.19. The functions $\phi(\|\boldsymbol{x} - \boldsymbol{x}^{(i)}\|)$, $i = 1, 2, \ldots, N$, are called the radial-basis functions, where $\|\cdot\|$ denotes a p-norm that is usually the Euclidean 2-norm. The real parameters w_i, $i = 1, 2, \ldots, N$, are the weights. A common choice for ϕ is

$$\phi\left(\left\|\boldsymbol{x} - \boldsymbol{x}^{(i)}\right\|\right) = \exp\left(-\frac{\left\|\boldsymbol{x} - \boldsymbol{x}^{(i)}\right\|^2}{2\sigma_i^2}\right), \tag{9.45}$$

where $\boldsymbol{x}^{(i)}$ is the center of the basis function and σ_i is sometimes called its radius. The function given by (9.45) is called the Gaussian radial-basis function.

The RBF networks are used to perform interpolation. The interpolation problem can be formulated as: For a given set of N different points

$$\left\{\boldsymbol{x}^{(i)} : \boldsymbol{x}^{(i)} \in \mathbb{R}^n, i = 1, 2, \ldots, N\right\}$$

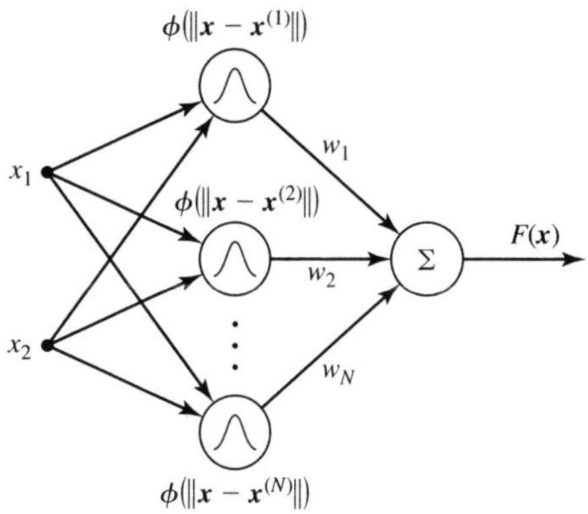

Figure 9.19 A two-input radial-basis function (RBF) network.

and a corresponding set of N real numbers

$$\{d_i : d_i \in \mathbb{R},\ i = 1, 2, \ldots, N\},$$

find a function $F : \mathbb{R}^n \to \mathbb{R}$ such that

$$F\left(\boldsymbol{x}^{(i)}\right) = d_i, \qquad i = 1, 2, \ldots, N. \tag{9.46}$$

We attempt to solve the interpolation problem using the RBF network. The function that the RBF network implements is

$$F(\boldsymbol{x}) = \sum_{j=1}^{N} w_j \phi\left(\left\|\boldsymbol{x} - \boldsymbol{x}^{(j)}\right\|\right). \tag{9.47}$$

Substituting the interpolation conditions given by (9.46) into (9.47) gives

$$\begin{bmatrix} \phi_{11} & \phi_{12} & \cdots & \phi_{1N} \\ \phi_{21} & \phi_{22} & \cdots & \phi_{2N} \\ \vdots & \vdots & \ddots & \vdots \\ \phi_{N1} & \phi_{N2} & \cdots & \phi_{NN} \end{bmatrix} \begin{bmatrix} w_1 \\ w_2 \\ \vdots \\ w_N \end{bmatrix} = \begin{bmatrix} d_1 \\ d_2 \\ \vdots \\ d_N \end{bmatrix}, \tag{9.48}$$

where $\phi_{ij} = \phi(\|\boldsymbol{x}^{(i)} - \boldsymbol{x}^{(j)}\|)$. Let $\boldsymbol{w} = [w_1\ w_2\ \cdots\ w_N]$, $\boldsymbol{d} = [d_1\ d_2\ \cdots\ d_N]$, and

$$\boldsymbol{\Phi} = \begin{bmatrix} \phi_{11} & \phi_{12} & \cdots & \phi_{1N} \\ \phi_{21} & \phi_{22} & \cdots & \phi_{2N} \\ \vdots & \vdots & \ddots & \vdots \\ \phi_{N1} & \phi_{N2} & \cdots & \phi_{NN} \end{bmatrix}.$$

Using the above notation, we represent (9.48) as

$$\boldsymbol{\Phi w} = \boldsymbol{d}. \tag{9.49}$$

The matrix $\boldsymbol{\Phi}$ is called the *interpolation matrix*. If the interpolation matrix is nonsingular, then we can solve (9.49) for $\boldsymbol{w}$ to obtain

$$\boldsymbol{w} = \boldsymbol{\Phi}^{-1}\boldsymbol{d}. \tag{9.50}$$

◆ Example 9.5

We construct an RBF network that interpolates data given in Table 9.2, where $\phi_{ij} = \exp(-\frac{(x^{(i)}-x^{(j)})^2}{2\sigma_i^2})$. We set $\sigma_i = 1$, for $i = 1, \ldots, 5$, and evaluate (9.50) to obtain

$$w = [0.8000 \quad 0.1459 \quad -0.0390 \quad 0.5737 \quad 0.2227]^T.$$

The interpolation result is shown in Figure 9.20. In Figure 9.20(a), we used a solid line to show the result of interpolating using the Gaussian radial–basis functions, while a broken line indicates the result of the straight line interpolation. The plots of the basis functions that contribute to the interpolating curve are shown in Figure 9.20(b).

Table 9.2 Interpolation Data for Example 9.5

$x^{(i)}$	d_i
0	0.8
6	0.2
7	0.4
8	0.6
10	0.3

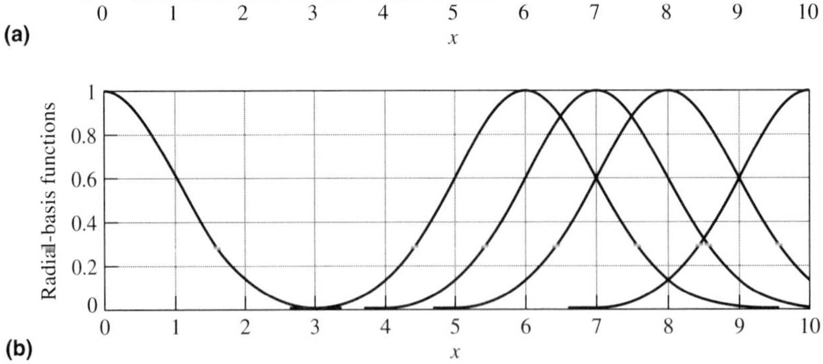

Figure 9.20 Results of interpolating data from Table 9.2 using radial-basis functions.

If the number of radial basis functions was fixed and the rows of the matrix $\boldsymbol{\Phi}$ and the elements of the vector $\boldsymbol{d}$ were obtained sequentially, we could use the recursive least squares algorithm, described in Section A.11, to compute the least squares estimate of $\boldsymbol{w}$.

9.5.2 Identification of a Single-Input, Single-State System

Here, we discuss the identification of a single-input, single-state dynamical system using neural network. We consider a dynamical system modeled by

$$\dot{x}(t) = f(x(t), u(t)), \qquad x(0) = x_0, \tag{9.51}$$

where $f : \mathbb{R} \times \mathbb{R} \to \mathbb{R}$ is an unknown, possibly nonlinear, function of x and u. We will construct a neural network to identify the function f. Adding and subtracting $ax(t)$, where $a > 0$ is a design parameter, to the right-hand side of (9.51) gives

$$\begin{aligned}\dot{x}(t) &= -ax(t) + ax(t) + f(x(t), u(t)) \\ &= -ax(t) + g(x(t), u(t)),\end{aligned} \tag{9.52}$$

where $g(x(t), u(t)) = ax(t) + f(x(t), u(t))$. The reason for the above manipulations will become apparent when we discuss learning rules of the neural network identifier of the dynamical system (9.51). This identifier is described by

$$\dot{\hat{x}}(t) = -a\hat{x}(t) + \hat{g}(x(t), u(t), \boldsymbol{m}_k, \sigma_k), \qquad \hat{x}(0) = \hat{x}_0, \tag{9.53}$$

where $\hat{x}$ is the state of the identification network, and $\boldsymbol{m}_k$ and σ_k are adjustable parameters of the identifier. The function $\hat{g}$ has the form

$$\hat{g}(x(t), u(t), \boldsymbol{m}_k, \sigma_k) = \sum_{k=1}^{K} w_k \phi(x, u, \boldsymbol{m}_k, \sigma_k), \tag{9.54}$$

where for $k = 1, 2, \ldots, K$ we have $\boldsymbol{m}_k = [m_{xk} \;\; m_{uk}]$ and

$$\phi(x(t), u(t), \boldsymbol{m}_k, \sigma_k) = \exp\left(-\frac{1}{2\sigma_k^2}\left((x(t) - m_{xk})^2 + (u(t) - m_{uk})^2\right)\right).$$

Thus, we postulate that the function $\hat{g}$ be generated using an RBF network. Suppose now that the basis functions are given to us. The problem of choosing basis functions is discussed later. Suppose also that we know the "optimal" weights w_k^*, $k = 1, 2, \ldots, K$, that result in the least identification error between g in (9.52) and $\hat{g}$ in (9.53), where the identification error is defined as

$$e(t) = g(x(t), u(t)) - \sum_{k=1}^{K} w_k^* \phi(x(t), u(t), \boldsymbol{m}_k, \sigma_k). \tag{9.55}$$

Using the above notation, we represent the dynamical system model (9.52) in the form

$$\dot{x}(t) = -ax(t) + \sum_{k=1}^{K} w_k^* \phi(x(t), u(t), \boldsymbol{m}_k, \sigma_k) + e(t). \tag{9.56}$$

The neural identifier, on the other hand, can be represented as

$$\dot{\hat{x}}(t) = -a\hat{x}(t) + \sum_{k=1}^{K} w_k \phi(x(t), u(t), \boldsymbol{m}_k, \sigma_k). \tag{9.57}$$

A schematic of the above identification architecture is depicted in Figure 9.21.

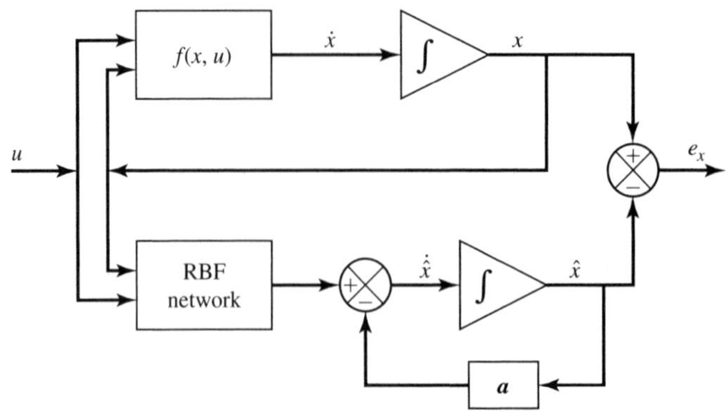

Figure 9.21 A configuration for the radial-basis function neural identifier.

9.5.3 Learning Algorithm for the RBF Identifier

We assume that both the state x and the input u of the dynamical system model (9.51), and hence that of (9.52), are bounded. We next define the state estimation error to be

$$e_x = x - \hat{x},$$

while the weights' estimation errors are

$$\xi_k = w_k^* - w_k, \qquad k = 1, \ldots, K.$$

We assume that the modeling error, e, defined in (9.55), is bounded by a finite constant e_K, that is, $|e| \leq e_K$. However, e_K is unknown to us. Next, we use (9.56) and (9.57) to form the differential equation describing the state error dynamics. For simplicity of notation, we omit the argument t of the time-varying functions. We have

$$\begin{aligned}\dot{e}_x &= \dot{x} - \dot{\hat{x}} \\ &= -ax + \sum_{k=1}^{K} w_k^* \phi(x, u, \boldsymbol{m}_k, \sigma_k) + e + a\hat{x} - \sum_{k=1}^{K} w_k \phi(x, u, \boldsymbol{m}_k, \sigma_k) \\ &= -ae_x + \sum_{k=1}^{K} \xi_k \phi(x, u, \boldsymbol{m}_k, \sigma_k) + e. \end{aligned} \tag{9.58}$$

We now use the second method of Lyapunov to devise a learning law for the RBF identifier that would yield a stable identification scheme in the sense that the state estimation error e_x decreases as the time increases and the weights do not diverge. For this purpose, we use the following Lyapunov function candidate:

$$V(e_x, \boldsymbol{\xi}) = \frac{1}{2}\left(e_x^2 + \frac{1}{\alpha}\sum_{k=1}^{K} \xi_k^2\right), \tag{9.59}$$

where $\alpha > 0$ is a design parameter and $\boldsymbol{\xi} = [\xi_1 \;\; \xi_2 \;\; \cdots \;\; \xi_K]^T$. The time derivative of V evaluated on the trajectories of (9.58) is

$$\begin{aligned}\dot{V} &= e_x\dot{e}_x + \frac{1}{\alpha}\sum_{k=1}^{K}\xi_k\dot{\xi}_k \\ &= e_x\left(-ae_x + \sum_{k=1}^{K}\xi_k\phi(x,u,\boldsymbol{m}_k,\sigma_k) + e\right) + \frac{1}{\alpha}\sum_{k=1}^{K}\xi_k\dot{\xi}_k \\ &= -ae_x^2 + \frac{1}{\alpha}\sum_{k=1}^{K}\xi_k(\alpha e_x\phi(x,u,\boldsymbol{m}_k,\sigma_k) + \dot{\xi}_k) + e_x e. \end{aligned} \tag{9.60}$$

Let

$$\dot{\xi}_k = -\alpha e_x\phi(x,u,\boldsymbol{m}_k,\sigma_k). \tag{9.61}$$

Then (9.60) simplifies to

$$\dot{V} = -ae_x^2 + e_x e. \tag{9.62}$$

Because $\dot{w}_k = -\dot{\xi}_k$, we can represent (9.61) as

$$\boxed{\dot{w}_k = \alpha e_x\phi(x,u,\boldsymbol{m}_k,\sigma_k)} \tag{9.63}$$

We further evaluate (9.62) to obtain

$$\begin{aligned}\dot{V} &\le -ae_x^2 + |e_x|\,e_K \\ &= -a|e_x|\left(|e_x| - \frac{e_K}{a}\right)\end{aligned} \tag{9.64}$$

It follows from (9.62) that if the state estimation error is zero—that is, $e_K = 0$—then

$$\dot{V} = -ae_x^2.$$

Thus in this case, $\dot{V}$ is negative semidefinite in the $[e_x\ \ \boldsymbol{\xi}^T]^T$ space, and hence the origin of the $[e_x\ \ \boldsymbol{\xi}^T]^T$ space is stable, which guarantees boundedness of e_x and the weights w_ks. We will now show, using the Lyapunov-like lemma stated on page 211, that when $e = 0$ we have $e_x(t) \to 0$ as $t \to 0$. We have to check if the assumptions of the Lyapunov-like lemma are satisfied. First, observe that V is bounded below and positive semidefinite. Next, because e_x, ξ_k as well as $\phi(x,u,\boldsymbol{m}_k,\sigma_k)$ are bounded, the derivative of the state estimation error, given by (9.58) for $e = 0$, is bounded as well, which implies that

$$\ddot{V} = -2ae_x\dot{e}_x$$

is bounded. This in turn implies that $\dot{V}$ is uniformly continuous. Hence, by the Lyapunov-like lemma, $e_x(t) \to 0$ as $t \to \infty$.

In the presence of the identification error, e, if $|e_x| < e_K/a$, then it is possible that $\dot{V} > 0$, which implies that the weights w_ks may drift to infinity with time. To avoid this problem, Liu, Kadirkamanathan, and Billings [187] proposed to set $\dot{w}_k = 0$ if $|e_x| < e_K/a$. Also, because e_K is unknown, a bound is imposed on the weights as well as on the error. The learning rule becomes

$$\dot{w}_k = \begin{cases} \alpha e_x \phi(x, u, \boldsymbol{m}_k, \sigma_k) & \text{if } |e_x| \geq e_0 \text{ and } \|\boldsymbol{w}\| \leq M \\ 0 & \text{otherwise.} \end{cases} \tag{9.65}$$

where the bounds e_0 and M are design parameters, and $\boldsymbol{w} = [w_1 \ \cdots \ w_K]^T$. With the above modified learning law, whenever $|e_x| \geq e_0 \geq e_K/a$, the Lyapunov derivative $\dot{V}$ is negative semidefinite. However, when $e_0 \leq |e_x| \leq e_K/a$, the weights may increase with time because it is possible that $\dot{V} > 0$. We impose an upper bound on the weights. We also impose an upper bound on the estimation error, e_x. When any of the two bounds is reached, more basis functions may be needed to improve the approximation of g. New basis functions are added to reduce the error until $\dot{V} < 0$. In what follows, we describe a procedure for adding radial-basis functions to the identification network.

9.5.4 Growing RBF Network

As we mentioned above, when $\dot{V} > 0$, the identification scheme is no longer stable, but the learning rule (9.65) guarantees that the adjustable parameters are bounded. Once the parameter adjustments can no longer reduce the estimation error and the bounds on the weights or on the estimation error are reached, additional basis functions in the RBF network should be added to reduce the modeling error. This continues until the approximation leads to the error reduction so that $\dot{V} < 0$.

In the proposed identification scheme by Liu, Kadirkamanathan, and Billings [187], the Gaussian radial-basis functions are used, whose areas of "influence" are determined by the functions' centers and radii (widths). As the function to be approximated moves further away from a particular basis function, the contribution of that basis function decreases. This is due to the exponential form of the Gaussian basis functions. As $|z|$ increases, $\exp(-z^2)$ decreases. The maximum contribution of a particular basis function to approximating a given function occurs when the center of the basis function and the approximated function are at the same point because in this case we have $\exp(0) = 1$. This heuristic argument leads to the following rule for determining the centers of new basis functions to be added at a time instance t:

$$\boldsymbol{m}_{K+1} = [m_{x,K+1} \quad m_{u,K+1}]^T = [\delta_x \text{round}(x(t)/\delta_x) \quad \delta_u \text{round}(u(t)/\delta_u)]^T, \tag{9.66}$$

where the round operator rounds the elements on which it operates to the nearest integer. For example, if $\boldsymbol{x} = [-1.8 \ -0.3 \ 3.7]$, then $\text{round}(\boldsymbol{x}) = [-2 \ 0 \ 4]$. The parameters δ_x and δ_u represent the spacing of the centers of the basis functions. We can think of having a grid over the input-state space where the spacing between the vertices of the grid are determined by the parameters δ_x and δ_u. Each vertex of the grid represents the potential center of a basis function. The parameter σ_k of the new basis function is determined from the relation

$$\sigma_k = \kappa \|\boldsymbol{m}_k - \boldsymbol{m}_k^*\|, \tag{9.67}$$

where $\kappa > 0$ is a design parameter and

$$\boldsymbol{m}_k^* = \arg \min_{\substack{i=1,\ldots,K+1 \\ i \neq k}} \{\|\boldsymbol{m}_k - \boldsymbol{m}_i\|\}. \tag{9.68}$$

Thus, $\boldsymbol{m}_k^*$ is the nearest center of all basis functions to the center of the kth basis function. With the above background, we can now present the rules for adding a new basis function to the

RBF network. Initially, the network contains no basis functions. Then, once the following two conditions are satisfied, a new basis function is added. The first condition is

$$\min_{k=1,\dots,K} |x(t) - m_{xk}| > \delta_x/2 \quad \text{or} \quad \min_{k=1,\dots,K} |u(t) - m_{uk}| > \delta_u/2. \tag{9.69}$$

The above will be satisfied if the state or input, at time t, is closer to an available "unoccupied" grid vertex than to any of the existing centers of the basis functions. The basis function at the new vertex would be the closest to the existing state or input and would have the greatest influence in the approximation of g. The second condition is

$$|x(t) - \hat{x}(t)| > e_{\max}, \tag{9.70}$$

where $e_{\max}$ is the maximum allowable state estimation error. If (9.70) is satisfied, a new basis function is added to reduce the state estimation error by reducing the modeling error.

◆ Example 9.6

The above technique for identification of dynamical systems using radial-basis function network is illustrated with the system

$$\dot{x} = 1.1(1 - x + 2xu - x^2)\exp(-0.5x^2 - u^2), \qquad x(0) = 0, \tag{9.71}$$

where $u = \cos(t/100)$. This model was adapted from Liu, Kadirkamanathan, and Billings [187]. There are many ways to simulate the above identification scheme. In our simulations, we discretized the above differential equation using the Euler method with the step size $dt = 0.01$ sec. The initial weights and the initial state of the identifier were set to zero. We used the following parameter values: $e_0 = 0.001$, $e_{\max} = 0.005$, $\alpha = 5$, $\kappa = 3$. In Figure 9.22, plots of the true state and its estimate

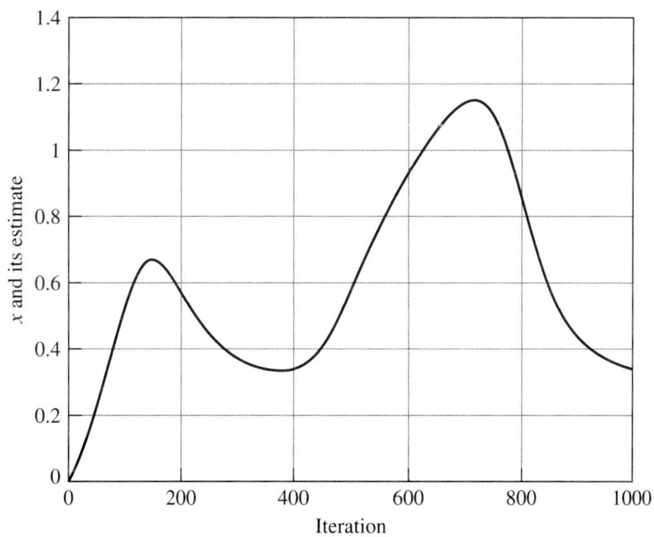

Figure 9.22 Plots of the true state and its estimate versus time. The estimated state "hugs" closely the actual state.

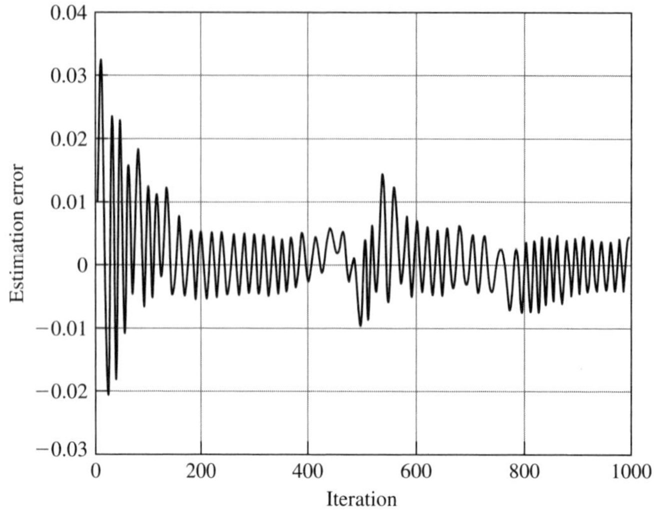

Figure 9.23 A plot of the error, e_x, between the true state and its estimate.

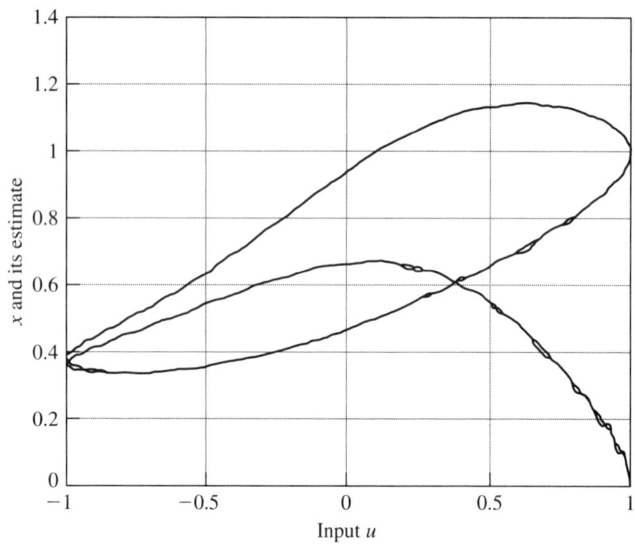

Figure 9.24 A plot of the true state and its estimate versus the input u.

versus time are shown. A plot of the state estimation error, e_x, between the true state and its estimate is shown in Figure 9.23. Initially, the error exceeds the bound e_{max}. Adding additional basis functions drives the error back down. The plots shown in Figure 9.24 depict the true state and its estimate versus the input u. Finally, in Figure 9.25, plots of the derivative of the true state versus the true state and the derivative of the estimated state versus the state estimate are shown.

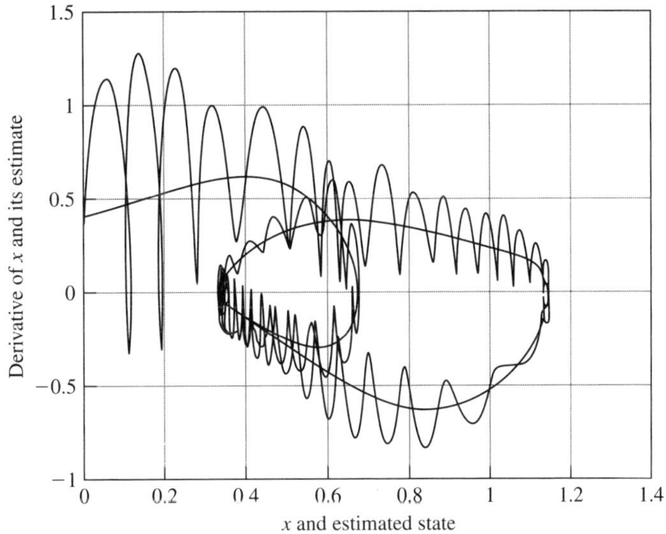

Figure 9.25 Plots of the derivative of the true state versus the true state and the derivative of the estimated state versus the state estimate of Example 9.6.

9.5.5 Identification of Multivariable Systems

We now extend the results of Subsection 9.5.2 to multi-input, multistate dynamical systems. In our derivation of the update law for the weights of the RBF network, we use a result from matrix algebra presented next. Recall the definition of the trace of a square matrix as given by (A.32). If $\boldsymbol{\Gamma} \in \mathbb{R}^{m\times n}$, then by (A.79) its Frobenius norm satisfies

$$\|\boldsymbol{\Gamma}\|_F^2 = \sum_{i=1}^{m}\sum_{j=1}^{n} \gamma_{ij}^2 = \operatorname{trace}(\boldsymbol{\Gamma}\boldsymbol{\Gamma}^T).$$

If the elements, γ_{ij}, of $\boldsymbol{\Gamma}$ are functions of time, then

$$\frac{d}{dt}(\operatorname{trace}(\boldsymbol{\Gamma}\boldsymbol{\Gamma}^T)) = 2\sum_{i=1}^{m}\sum_{j=1}^{n} \gamma_{ij}\dot{\gamma}_{ij} = 2\operatorname{trace}(\boldsymbol{\Gamma}\dot{\boldsymbol{\Gamma}}^T), \tag{9.72}$$

where $\dot{\boldsymbol{\Gamma}} = d\boldsymbol{\Gamma}/dt$. With the above background, we can proceed with the analysis of the RBF neural identifier for the multivariable systems.

The multivariable dynamical systems that we consider are modeled by

$$\dot{\boldsymbol{x}}(t) = \boldsymbol{f}(\boldsymbol{x}(t), \boldsymbol{u}(t)), \qquad \boldsymbol{x}(0) = \boldsymbol{x}_0, \tag{9.73}$$

where $\boldsymbol{u}(t) \in \mathbb{R}^m$, $\boldsymbol{x}(t) \in \mathbb{R}^n$, and $\boldsymbol{f} : \mathbb{R}^n \times \mathbb{R}^m \to \mathbb{R}^n$ is the function to be identified. Let $\boldsymbol{A} \in \mathbb{R}^{n\times n}$ be a Hurwitz matrix—that is, the matrix whose eigenvalues are all in the open left-half plane. Because $\boldsymbol{A}$ is a Hurwitz matrix, by the theorem of Lyapunov (Theorem 4.1), for any given real symmetric positive definite (r.s.p.d.) matrix $\boldsymbol{Q}$ the solution $\boldsymbol{P}$ to the continuous Lyapunov matrix equation

$$\boldsymbol{A}^T\boldsymbol{P} + \boldsymbol{P}\boldsymbol{A} = -2\boldsymbol{Q} \tag{9.74}$$

is also (real symmetric) positive definite. We use this fact later when we develop a learning law for the RBF neural identifier.

We represent the model (9.73) in an equivalent form,

$$\dot{\boldsymbol{x}}(t) = \boldsymbol{A}\boldsymbol{x}(t) + \boldsymbol{g}(\boldsymbol{x}(t), \boldsymbol{u}(t)), \qquad \boldsymbol{x}(0) = \boldsymbol{x}_0, \tag{9.75}$$

where

$$\boldsymbol{g}(\boldsymbol{x}(t), \boldsymbol{u}(t)) = -\boldsymbol{A}\boldsymbol{x}(t) + \boldsymbol{f}(\boldsymbol{x}(t), \boldsymbol{u}(t)).$$

The kth basis function can be represented as

$$\phi_k(\boldsymbol{x}, \boldsymbol{u}) = \exp\left(-\frac{1}{2\sigma_k^2}\left\|\begin{bmatrix}\boldsymbol{x}\\ \boldsymbol{u}\end{bmatrix} - \boldsymbol{m}_k\right\|^2\right).$$

Let

$$\boldsymbol{\Phi}_K(\boldsymbol{x}, \boldsymbol{u}) = [\phi_1(\boldsymbol{x}, \boldsymbol{u}) \quad \phi_2(\boldsymbol{x}, \boldsymbol{u}) \quad \cdots \quad \phi_K(\boldsymbol{x}, \boldsymbol{u})]^T$$

and

$$\hat{\boldsymbol{g}}(\boldsymbol{x}, \boldsymbol{u}, \boldsymbol{m}_k, \sigma_k) = \boldsymbol{W}_K \boldsymbol{\Phi}_K,$$

where $\boldsymbol{W}_K$ is the $n \times K$ weight matrix and $\boldsymbol{m}_k = [\boldsymbol{m}_{xk}^T \;\; \boldsymbol{m}_{uk}^T]^T$ is the center of the kth radial basis function in the $[\boldsymbol{x}^T \;\; \boldsymbol{u}^T]^T$ space. Note that $\boldsymbol{\Phi}_K$ is a column vector with K components. The RBF neural identifier has the form

$$\dot{\hat{\boldsymbol{x}}}(t) = \boldsymbol{A}\hat{\boldsymbol{x}}(t) + \hat{\boldsymbol{g}}(\boldsymbol{x}, \boldsymbol{u}, \boldsymbol{m}_k, \sigma_k), \qquad \hat{\boldsymbol{x}}(0) = \boldsymbol{x}_0. \tag{9.76}$$

Let $\boldsymbol{W}_K^*$ be the optimal weight matrix that minimizes a norm of the identification error, that is,

$$\boldsymbol{W}_K^* = \arg\min_{\boldsymbol{W}_K} \|\boldsymbol{g}(\boldsymbol{x}(t), \boldsymbol{u}(t)) - \boldsymbol{W}_K \boldsymbol{\Phi}_K(\boldsymbol{x}(t), \boldsymbol{u}(t))\|.$$

Let

$$\boldsymbol{e}(t) = \boldsymbol{g}(\boldsymbol{x}(t), \boldsymbol{u}(t)) - \boldsymbol{W}_K^* \boldsymbol{\Phi}_K(\boldsymbol{x}(t), \boldsymbol{u}(t)).$$

Using the above definition, we represent the system model (9.75) as

$$\dot{\boldsymbol{x}}(t) = \boldsymbol{A}\boldsymbol{x}(t) + \boldsymbol{W}_K^* \boldsymbol{\Phi}_K(\boldsymbol{x}(t), \boldsymbol{u}(t)) + \boldsymbol{e}(t). \tag{9.77}$$

We define the state estimation error and the weight estimation error as

$$\boldsymbol{e}_x = \boldsymbol{x} - \hat{\boldsymbol{x}} \quad \text{and} \quad \boldsymbol{\Gamma}_K = \boldsymbol{W}_K^* - \boldsymbol{W}_K.$$

Next, we form the differential equation describing the state estimation error dynamics,

$$\begin{aligned}
\dot{\boldsymbol{e}}_x &= \dot{\boldsymbol{x}} - \dot{\hat{\boldsymbol{x}}} \\
&= \boldsymbol{A}\boldsymbol{x} + \boldsymbol{W}_K^* \boldsymbol{\Phi}_K(\boldsymbol{x}, \boldsymbol{u}) + \boldsymbol{e} - \boldsymbol{A}\hat{\boldsymbol{x}} - \hat{\boldsymbol{g}}(\boldsymbol{x}, \boldsymbol{u}, \boldsymbol{m}_k, \sigma_k) \\
&= \boldsymbol{A}(\boldsymbol{x} - \hat{\boldsymbol{x}}) + \boldsymbol{W}_K^* \boldsymbol{\Phi}_K(\boldsymbol{x}, \boldsymbol{u}) - \boldsymbol{W}_K \boldsymbol{\Phi}_K(\boldsymbol{x}, \boldsymbol{u}) + \boldsymbol{e} \\
&= \boldsymbol{A}\boldsymbol{e}_x + \boldsymbol{\Gamma}_K \boldsymbol{\Phi}_K(\boldsymbol{x}, \boldsymbol{u}) + \boldsymbol{e}.
\end{aligned} \tag{9.78}$$

We employ the following Lyapunov function candidate:

$$V = V(\boldsymbol{e}_x, \boldsymbol{\Gamma}_K) = \frac{1}{2}\left(\boldsymbol{e}_x^T \boldsymbol{P} \boldsymbol{e}_x + \frac{1}{\alpha}\operatorname{trace}\left(\boldsymbol{\Gamma}_K \boldsymbol{\Gamma}_K^T\right)\right). \tag{9.79}$$

Using (9.72), we evaluate the time derivative of V on the solutions of the state estimation error equation to get

$$\begin{aligned}\dot{V} &= \boldsymbol{e}_x^T \boldsymbol{P}\dot{\boldsymbol{e}}_x + \frac{1}{\alpha}\operatorname{trace}\left(\boldsymbol{\Gamma}_K \dot{\boldsymbol{\Gamma}}_K^T\right)\\ &= \boldsymbol{e}_x^T(\boldsymbol{P}\boldsymbol{A}\boldsymbol{e}_x + \boldsymbol{P}\boldsymbol{\Gamma}_K\boldsymbol{\Phi}_K(\boldsymbol{x},\boldsymbol{u}) + \boldsymbol{P}\boldsymbol{e}) + \frac{1}{\alpha}\operatorname{trace}\left(\boldsymbol{\Gamma}_K \dot{\boldsymbol{\Gamma}}_K^T\right)\\ &= \frac{1}{2}\boldsymbol{e}_x^T(\boldsymbol{A}^T\boldsymbol{P} + \boldsymbol{P}\boldsymbol{A})\boldsymbol{e}_x + \boldsymbol{e}_x^T\boldsymbol{P}\boldsymbol{\Gamma}_K\boldsymbol{\Phi}_K(\boldsymbol{x},\boldsymbol{u}) + \frac{1}{\alpha}\operatorname{trace}\left(\boldsymbol{\Gamma}_K \dot{\boldsymbol{\Gamma}}_K^T\right) + \boldsymbol{e}_x^T\boldsymbol{P}\boldsymbol{e}.\end{aligned} \tag{9.80}$$

Using the definition of the trace operator, we obtain

$$\begin{aligned}\boldsymbol{e}_x^T\boldsymbol{P}\boldsymbol{\Gamma}_K\boldsymbol{\Phi}_K(\boldsymbol{x},\boldsymbol{u}) &= \operatorname{trace}\left(\boldsymbol{e}_x^T\boldsymbol{P}\boldsymbol{\Gamma}_K\boldsymbol{\Phi}_K(\boldsymbol{x},\boldsymbol{u})\right)\\ &= \operatorname{trace}\left(\boldsymbol{\Gamma}_K\boldsymbol{\Phi}_K(\boldsymbol{x},\boldsymbol{u})\boldsymbol{e}_x^T\boldsymbol{P}\right).\end{aligned} \tag{9.81}$$

Observe that

$$\dot{\boldsymbol{W}}_K = -\dot{\boldsymbol{\Gamma}}_K.$$

Using the above and (9.81) along with the Lyapunov equation (9.74), we write

$$\dot{V} = -\boldsymbol{e}_x^T\boldsymbol{Q}\boldsymbol{e}_x + \frac{1}{\alpha}\left(\operatorname{trace}\left(\alpha\boldsymbol{\Gamma}_K\boldsymbol{\Phi}_K(\boldsymbol{x},\boldsymbol{u})\boldsymbol{e}_x^T\boldsymbol{P}\right) - \operatorname{trace}\left(\boldsymbol{\Gamma}_K\dot{\boldsymbol{W}}_K^T\right)\right) + \boldsymbol{e}_x^T\boldsymbol{P}\boldsymbol{e}. \tag{9.82}$$

Consider the following weight matrix update law:

$$\boxed{\dot{\boldsymbol{W}}_K = \alpha\boldsymbol{P}\boldsymbol{e}_x\boldsymbol{\Phi}_K^T(\boldsymbol{x},\boldsymbol{u}).} \tag{9.83}$$

With the above update law, the Lyapunov derivative (9.82) evaluates to

$$\begin{aligned}\dot{V} &= -\boldsymbol{e}_x^T\boldsymbol{Q}\boldsymbol{e}_x + \boldsymbol{e}_x^T\boldsymbol{P}\boldsymbol{e}\\ &\le -\lambda_{\min}(\boldsymbol{Q})\|\boldsymbol{e}_x\|^2 + \lambda_{\max}(\boldsymbol{P})\|\boldsymbol{e}\|\|\boldsymbol{e}_x\|\\ &= -\lambda_{\min}(\boldsymbol{Q})\left(\|\boldsymbol{e}_x\| - \frac{\lambda_{\max}(\boldsymbol{P})}{\lambda_{\min}(\boldsymbol{Q})}\|\boldsymbol{e}\|\right)\|\boldsymbol{e}_x\|.\end{aligned} \tag{9.84}$$

Utilizing the same type of arguments as in the scalar case (analyzed in Subsection 9.5.2) and the Lyapunov-like lemma stated on page 211, we can show that when $\boldsymbol{e} = \boldsymbol{0}$, then $\boldsymbol{e}_x(t) \to \boldsymbol{0}$ as $t \to \infty$.

If the magnitude of the state estimation error is such that $\|\boldsymbol{e}_x\| < \frac{\lambda_{\max}(\boldsymbol{P})}{\lambda_{\min}(\boldsymbol{Q})}\|\boldsymbol{e}\|$, then it may happen that $\dot{V} > 0$, which in turn may lead to the weights' drift to infinity over time. To prevent this, we modify (9.83) following the analysis of the single-input single-state case. The modified weight update law is

$$\boxed{\dot{\boldsymbol{W}}_K = \begin{cases}\alpha\boldsymbol{P}\boldsymbol{e}_x\boldsymbol{\Phi}_K^T(\boldsymbol{x},\boldsymbol{u}) & \text{if } \|\boldsymbol{e}_x\| \ge e_0 \text{ and } \|\boldsymbol{W}_K\| \le M\\ \boldsymbol{O} & \text{otherwise.}\end{cases}} \tag{9.85}$$

In the above adaptation law, e_0 and M are design parameters.

9.6 A Self-Organizing Network

Up to now we discussed neural networks that must have a teacher in order to learn presented patterns. This mode of learning is referred to as the supervised mode. In this section we analyze a neural network that can learn in the unsupervised mode and the network undergoes the self-organization process. Specifically, the network is capable of clustering training patterns. It is assumed that the number of classes, p, is known and that the network's number of weight vectors is equal to the number of classes in the training set. The weights are initialized and then normalized. After initialization, the training patterns, $\boldsymbol{x}^{(1)}, \boldsymbol{x}^{(2)}, \ldots, \boldsymbol{x}^{(N)}$, are presented. In response to each presented pattern, we compute the winning neuron of the network, that is, we compute the network weight that is closest to the presented pattern. We call this computation the *similarity matching*. Let $\boldsymbol{x}$ be the presented pattern, where, for simplicity, we omitted its index. Then the similarity matching can be described as

$$\left\|\boldsymbol{x}-\boldsymbol{w}^{(i)}\right\| = \min_{1\le j\le p}\left\{\left\|\boldsymbol{x}-\boldsymbol{w}^{(j)}\right\|\right\},$$

where i is the index of the winning neuron. Thus, searching for the winning neuron corresponds to finding the maximum among the p scalar products,

$$\max_{j=1,2,\ldots,p}\left\langle \boldsymbol{w}^{(j)}, \boldsymbol{x}\right\rangle.$$

To see that this is indeed the case, note that because the weights and the patterns are normalized—that is, $\|\boldsymbol{w}^{(j)}\| = \|\boldsymbol{x}\| = 1$—we have

$$\begin{aligned}\left\|\boldsymbol{x}-\boldsymbol{w}^{(j)}\right\| &= \left(\|\boldsymbol{x}\|^2 - 2\left\langle \boldsymbol{w}^{(j)}, \boldsymbol{x}\right\rangle + \left\|\boldsymbol{w}^{(j)}\right\|^2\right)^{1/2}\\ &= \left(2-2\left\langle \boldsymbol{w}^{(j)}, \boldsymbol{x}\right\rangle\right)^{1/2}.\end{aligned}$$

Hence, minimizing $\|\boldsymbol{x}-\boldsymbol{w}^{(j)}\|$ is equivalent to maximizing the scalar product $\langle \boldsymbol{w}^{(j)}, \boldsymbol{x}\rangle$. On the other hand,

$$\left\langle \boldsymbol{w}^{(j)}, \boldsymbol{x}\right\rangle = \left\|\boldsymbol{w}^{(j)}\right\|\|\boldsymbol{x}\|\cos\left(\angle \boldsymbol{w}^{(j)}, \boldsymbol{x}\right),$$

which means that the winning neuron is "more parallel" to the input pattern $\boldsymbol{x}$ than the other neurons.

After the winning neuron is declared, it is updated, while the remaining weights are not changed. This is the reason for the name of the network—the winner-take-all (WTA) network. This update algorithm is also known as the *Kohonen learning rule* after Kohonen, who first proposed this learning rule. The role of the update is to reduce the distance between the winning neuron, $\boldsymbol{w}^{(i)}$, and the presented pattern, $\boldsymbol{x}$. One way to accomplish this task is to apply the descent gradient method to the objective function $\frac{1}{2}\|\boldsymbol{x}-\boldsymbol{w}^{(i)}\|^2$. The gradient of $\frac{1}{2}\|\boldsymbol{x}-\boldsymbol{w}^{(i)}\|^2$ with respect to $\boldsymbol{w}^{(i)}$ is

$$\tfrac{1}{2}\nabla_{\boldsymbol{w}^{(i)}}\left\|\boldsymbol{x}-\boldsymbol{w}^{(i)}\right\|^2 = -\left(\boldsymbol{x}-\boldsymbol{w}^{(i)}\right).$$

Applying the descent gradient method to minimize $\frac{1}{2}\|\boldsymbol{x}-\boldsymbol{w}^{(i)}\|^2$ with a step size α_k yields

$$\boldsymbol{w}^{(i)}(k+1) = \boldsymbol{w}^{(i)}(k) + \alpha_k\left(\boldsymbol{x}-\boldsymbol{w}^{(i)}(k)\right).$$

The weights of other neurons are unchanged, that is,

$$\boldsymbol{w}^{(j)}(k+1) = \boldsymbol{w}^{(j)}(k) \qquad \text{for } j \ne i.$$

We summarize our discussion in the form of the following algorithm.

WINNER-TAKE-ALL (WTA) LEARNING ALGORITHM

STEP 1 *Initialize Weights.* Generate the initial weights. Normalize each weight to have unit length: $\frac{\boldsymbol{w}}{\|\boldsymbol{w}\|}$.

STEP 2 *Serially Present Training Patterns.* Normalize each training pattern $\boldsymbol{x}^{(j)}$ to have unit length. The training patterns can be presented in a random order a fixed number of times.

STEP 3 *Select Winning Neuron.* For each WTA weight, compute the scalar product with the presented pattern. Select a weight that gives the largest scalar product. Call this the winning weight.

STEP 4 *Update Weight of Winning Neuron.* Let i be the index of the winning weight. Set

$$\boldsymbol{w}^{(i)}(k+1) = \boldsymbol{w}^{(i)}(k) + \alpha_k\left(\boldsymbol{x}^{(j)} - \boldsymbol{w}^{(i)}(k)\right),$$
$$\boldsymbol{w}^{(j)}(k+1) = \boldsymbol{w}^{(j)}(k) \qquad \text{for } j \neq i.$$

Normalize $\boldsymbol{w}^{(i)}(k+1)$ to have unit length.

STEP 5 *Repeat by Returning to Step 2.* Repeat until stopping criterion is met.

◆ Example 9.7

We take the same set of training patterns as in Zurada [320, p. 407]:

$$\left[\boldsymbol{x}^{(1)} \quad \boldsymbol{x}^{(2)} \quad \boldsymbol{x}^{(3)} \quad \boldsymbol{x}^{(4)} \quad \boldsymbol{x}^{(5)}\right] = \begin{bmatrix} 0.8 & 0.1736 & 0.707 & 0.342 & 0.6 \\ 0.6 & -0.9848 & 0.707 & -0.9397 & 0.8 \end{bmatrix}.$$

The above patterns are already normalized and are marked in Figure 9.26. As can be seen from Figure 9.26, we can classify the training patterns into two classes,

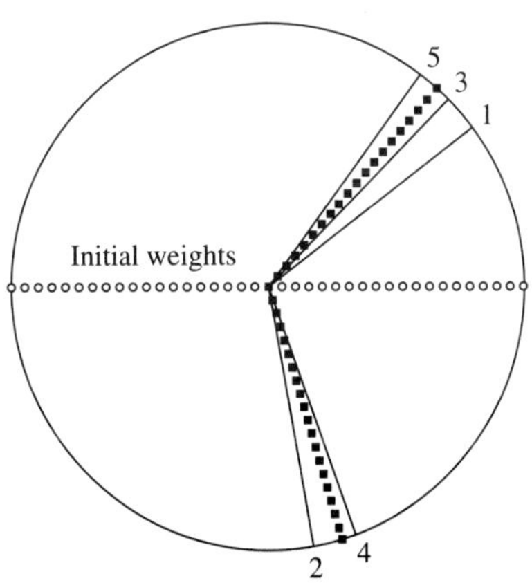

Figure 9.26 Training patterns, initial weights and the final weights of the WTA network of Example 9.7.

that is, $p = 2$. The initial normalized weights are

$$\begin{bmatrix} \boldsymbol{w}_0^{(1)} & \boldsymbol{w}_0^{(2)} \end{bmatrix} = \begin{bmatrix} 1 & -1 \\ 0 & 0 \end{bmatrix}.$$

The initial weights are also marked in Figure 9.26. The training patterns, $\boldsymbol{x}^{(1)}$, $\boldsymbol{x}^{(2)}, \ldots, \boldsymbol{x}^{(5)}$ are presented in ascending order and then recycled. After 200 presentations of the training set, we obtained the following final weights of the network:

$$\begin{bmatrix} \boldsymbol{w}^{(1)} & \boldsymbol{w}^{(2)} \end{bmatrix} = \begin{bmatrix} 0.6628 & 0.2868 \\ 0.7488 & -0.9580 \end{bmatrix}.$$

The final weights are marked in Figure 9.26 with $*$.

9.7 Hopfield Neural Network

9.7.1 Hopfield Neural Network Modeling and Analysis

The analog Hopfield neural network is a dynamical system whose schematic is shown in Figure 9.27. It is a nonlinear interconnected system proposed by Hopfield [126] as a model of biological neural networks. Each node, or neuron, in Figure 9.27 is represented by a circle. The

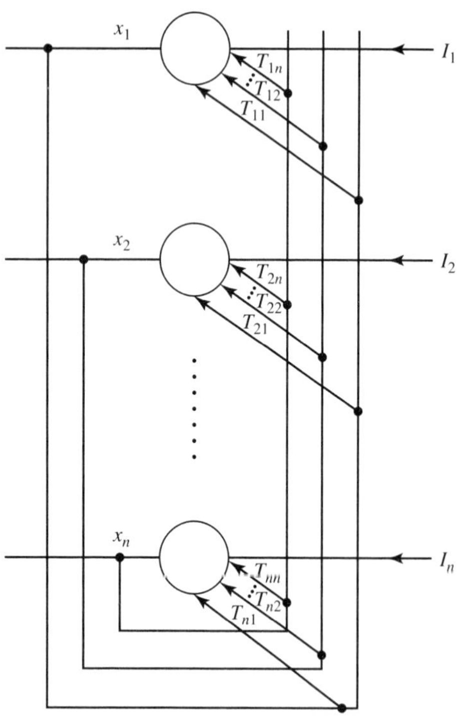

Figure 9.27 Hopfield neural network model.

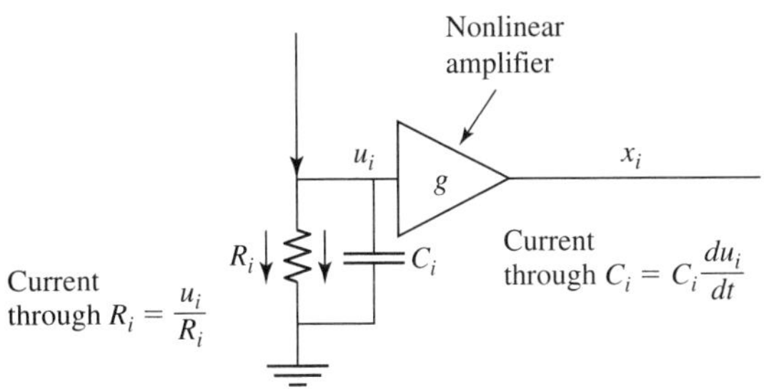

Figure 9.28 Model of a neuron in the Hopfield neural network.

structure of a neuron is depicted in Figure 9.28. The nonlinear amplifier in Figure 9.28 has an input-output characteristic described by a function $g : \mathbb{R} \to [0, 1]$. It is commonly assumed that g is continuously differentiable and strictly monotonically increasing, that is, $g(u) > g(u')$ if $u > u'$. The function g is called the *activation function*. A typical choice of g is the sigmoid function given by

$$x = g(u) = \frac{1}{1 + e^{-u/\theta}}, \tag{9.86}$$

where the positive parameter θ controls the slope of g. A circuit implementation of the Hopfield network is shown in Figure 9.29. To derive equations that model the Hopfield neural network, we use an equivalent representation of the single node of the Hopfield network shown in Figure 9.30. Applying Kirchhoff's current law at the input of each amplifier of the equivalent node representation and taking into account that the current into the amplifier is negligibly small due to the amplifier's high input resistance, we get

$$C_i \frac{du_i}{dt} + \frac{u_i}{R_i} = \frac{x_1 - u_i}{R_{i1}} + \frac{x_2 - u_i}{R_{i2}} + \cdots + \frac{x_n - u_i}{R_{in}} + I_i, \qquad i = 1, 2, \ldots, n, \tag{9.87}$$

where $x_i = g(u_i)$. Let

$$T_{ij} = \frac{1}{R_{ij}}, \qquad \frac{1}{r_i} = \sum_{j=1}^{n} T_{ij} + \frac{1}{R_i}. \tag{9.88}$$

Substituting (9.88) into (9.87) yields

$$\boxed{C_i \frac{du_i}{dt} + \frac{u_i}{r_i} = \sum_{j=1}^{n} T_{ij} x_j + I_i, \qquad i = 1, 2, \ldots, n} \tag{9.89}$$

The system of n first-order nonlinear differential equations given by (9.89) constitutes a model of the Hopfield neural network. The Hopfield neural network can be viewed as an associative memory. This is because it can have asymptotically stable states that attract the neighboring states to develop in time. This behavior can be interpreted as an evolution of an imperfect, corrupted pattern toward the correct, stored pattern—hence the term associative memory. More on neural associative memory can be found in Subsection 9.9.1.

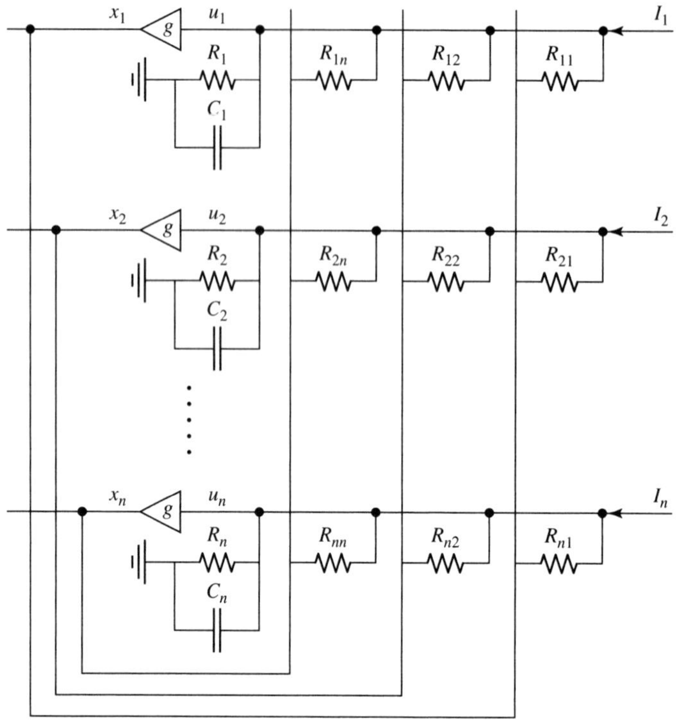

Figure 9.29 A circuit realization of the Hopfield neural network.

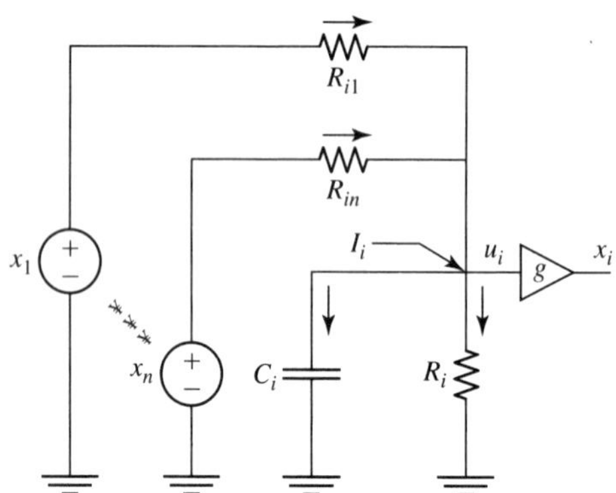

Figure 9.30 Equivalent representation of a single node in the circuit realization of the Hopfield neural network.

In further analysis, we use an equivalent representation of (9.89). To obtain this equivalent representation of the Hopfield model, we first note that the activation function g is, by assumption, strictly increasing and differentiable. Therefore g^{-1} exists and is also differentiable—see the inverse function theorem on page 390. In Figure 9.31, we show plots of $x = g(u)$ and $u = g^{-1}(x)$ for $\theta = 0.2$. We rewrite equations (9.89) as

$$C_i \frac{dg^{-1}(x_i)}{dx_i}\frac{dx_i}{dt} + \frac{g^{-1}(x_i)}{r_i} = \sum_{j=1}^{n} T_{ij}x_j + I_i, \qquad i = 1, 2, \ldots, n. \tag{9.90}$$

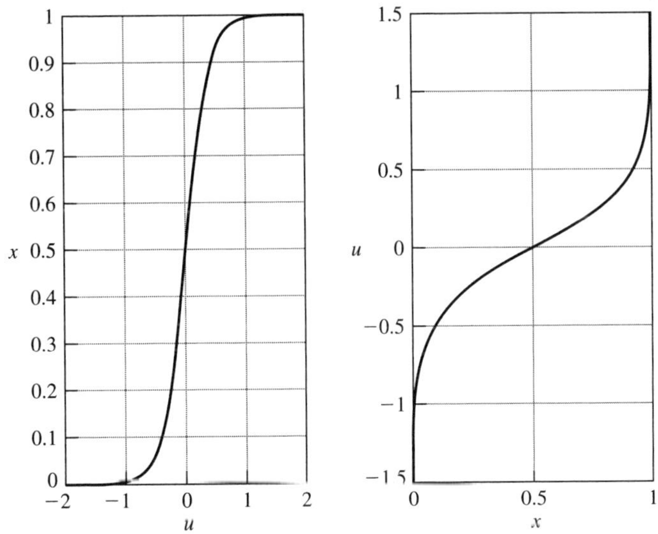

Figure 9.31 Plots of $x = g(u)$ and $u = g^{-1}(x)$.

Let $\boldsymbol{T} \in \mathbb{R}^{n\times n}$ be the matrix whose elements are T_{ij} and $\boldsymbol{v} = [I_1 \; I_2 \; \cdots \; I_n]^T \in \mathbb{R}^n$. Using this notation, we represent (9.89) in matrix form:

$$\begin{bmatrix} C_1 \dfrac{dg^{-1}(x_1)}{dx_1} & \cdots & 0 \\ \vdots & \ddots & \vdots \\ 0 & \cdots & C_n \dfrac{dg^{-1}(x_n)}{dx_n} \end{bmatrix} \dot{\boldsymbol{x}}(t) = \boldsymbol{T}\boldsymbol{x} + \boldsymbol{v} - \begin{bmatrix} \dfrac{g^{-1}(x_1)}{r_1} \\ \vdots \\ \dfrac{g^{-1}(x_n)}{r_n} \end{bmatrix}. \tag{9.91}$$

Associated with the Hopfield neural network is its computational energy function, $E : \mathbb{R}^n \to \mathbb{R}$, given by

$$\begin{aligned} E(\boldsymbol{x}) &= -\frac{1}{2}\sum_{i=1}^{n}\sum_{j=1}^{n} T_{ij}x_i x_j - \sum_{i=1}^{n} I_i x_i + \sum_{i=1}^{n}\int_0^{x_i} \frac{g^{-1}(s)}{r_i}\, ds \\ &= -\frac{1}{2}\boldsymbol{x}^T\boldsymbol{T}\boldsymbol{x} - \boldsymbol{v}^T\boldsymbol{x} + \sum_{i=1}^{n}\int_0^{x_i} \frac{g^{-1}(s)}{r_i}\, ds. \end{aligned} \tag{9.92}$$

Suppose now that $\boldsymbol{T} = \boldsymbol{T}^T$. Then, the gradient of E is

$$\nabla E(\boldsymbol{x}) = -\boldsymbol{T}\boldsymbol{x} - \boldsymbol{v} + \begin{bmatrix} \dfrac{g^{-1}(x_1)}{r_1} & \cdots & \dfrac{g^{-1}(x_n)}{r_n} \end{bmatrix}^T. \tag{9.93}$$

Let

$$S(\boldsymbol{x}) = \begin{bmatrix} C_1 \dfrac{dg^{-1}(x_1)}{dx_1} & \cdots & 0 \\ \vdots & \ddots & \vdots \\ 0 & \cdots & C_n \dfrac{dg^{-1}(x_n)}{dx_n} \end{bmatrix}. \tag{9.94}$$

We now combine (9.93) and (9.94) to obtain an equivalent form of the Hopfield network modeling equations:

$$\boxed{S(\boldsymbol{x})\dot{\boldsymbol{x}}(t) = -\nabla E(\boldsymbol{x}(t))} \tag{9.95}$$

We analyze the dynamics of the Hopfield network modeled by (9.95) using the computational energy function. We first show that E is bounded. Taking norms of (9.92) yields

$$|E(\boldsymbol{x})| \le \frac{1}{2}\|\boldsymbol{T}\|\|\boldsymbol{x}\|^2 + \|\boldsymbol{v}\|\|\boldsymbol{x}\| + \sum_{i=1}^{n} \frac{1}{r_i}\left|\int_0^{x_i} g^{-1}(s)\,ds\right|. \tag{9.96}$$

The components of the vector $\boldsymbol{x}$ are the outputs of the nonlinear amplifiers, and they take values from the interval [0, 1]. Hence, $\|\boldsymbol{x}\| = \|\boldsymbol{x}\|_2 \le \sqrt{n}$. Thus, to show that $E(\boldsymbol{x})$ is bounded, it is enough to show that the integral $\int_0^{x_i} g^{-1}(s)\,ds$ is bounded. We first find the expression for g^{-1}. Recall that $g(u)$ is given by (9.86). Therefore,

$$\frac{1}{x} = 1 + e^{-u/\theta},$$

or

$$\frac{1}{x} - 1 = e^{-u/\theta}, \tag{9.97}$$

Taking logarithms of both sides of the above yields

$$\ln\left(\frac{1-x}{x}\right) = -\frac{u}{\theta}.$$

Hence,

$$u = g^{-1}(x) = \theta(\ln(x) - \ln(1-x)). \tag{9.98}$$

Using the symmetry property of g^{-1}, we obtain

$$\begin{aligned} \left|\int_0^{x_i} g^{-1}(s)\,ds\right| &\le \int_0^{x_i} |g^{-1}(s)|ds \\ &\le -2\int_0^{1/2} g^{-1}(s)\,ds. \end{aligned} \tag{9.99}$$

Recall that

$$\int \ln(s)\,ds = s\ln(s) - s.$$

Hence,

$$\int \ln(1-s)\,ds = -(1-s)\ln(1-s) - s.$$

Using the above, we obtain

$$\begin{aligned}\int_0^{1/2} g^{-1}(s)\,ds &= \int_0^{1/2} \theta(\ln(s) - \ln(1-s))\,ds \\ &= \theta(s\ln(s) - s + (1-s)\ln(1-s) + s)|_0^{1/2} \\ &= \theta(s\ln(s) + (1-s)\ln(1-s))|_0^{1/2}. \end{aligned} \tag{9.100}$$

Note that

$$s\ln(s) = \frac{\ln(s)}{1/s}.$$

Applying L'Hôpital's rule to the above yields

$$\lim_{s\to 0}\frac{\ln(s)}{1/s} = \lim_{s\to 0}\frac{1/s}{-1/s^2} = 0.$$

Applying the above to (9.100), we get

$$\int_0^{1/2} g^{-1}(s)\,ds = \theta\ln(1/2). \tag{9.101}$$

Combining (9.99) and (9.101) yields

$$\left|\int_0^{x_i} g^{-1}(s)\,ds\right| \le -2\theta\ln(1/2), \tag{9.102}$$

which completes the proof of the statement that E is bounded.

Let $\boldsymbol{x}(t)$ denote the trajectory of the Hopfield network model (9.95). We will show that if $\boldsymbol{T} = \boldsymbol{T}^T$, then

$$\frac{d}{dt}E(\boldsymbol{x}(t)) \le 0$$

on the Hopfield network trajectories. Using the chain rule gives

$$\frac{d}{dt}E(\boldsymbol{x}(t)) = (\nabla E(\boldsymbol{x}(t)))^T\dot{\boldsymbol{x}}(t). \tag{9.103}$$

It follows from (9.95) that

$$\nabla E(\boldsymbol{x}(t)) = -\boldsymbol{S}(\boldsymbol{x}(t))\dot{\boldsymbol{x}}(t). \tag{9.104}$$

Substituting (9.104) into (9.103), we obtain

$$\begin{aligned}\frac{d}{dt}E(\boldsymbol{x}(t)) &= -\dot{\boldsymbol{x}}^T(t)\boldsymbol{S}(\boldsymbol{x}(t))\dot{\boldsymbol{x}}(t) \\ &= -\sum_{i=1}^n C_i\frac{dg^{-1}(x_i(t))}{dx_i}\dot{x}_i^2,\end{aligned} \tag{9.105}$$

where $C_i > 0$ and

$$\frac{dg^{-1}(x_i)}{dx_i} = \theta(1/x_i + 1/(1-x_i)) > 0 \tag{9.106}$$

because $x_i \in (0, 1)$. Hence,

$$\frac{d}{dt}E(\boldsymbol{x}(t)) \le 0$$

on the trajectories of the Hopfield model (9.95).

9.7.2 Analog-to-Digital Converter

We will now show how the Hopfield neural network can be used to solve some engineering problems. Specifically, we will discuss the construction of an analog-to-digital (A/D) converter using the Hopfield neural network. The idea was put forward by Tank and Hopfield [278]. In A/D conversion, an analog value is transformed into its corresponding binary representation. The conversion that we will discuss consists of two steps. The first step is to round the given analog value to the nearest integer. The second step is to convert the integer into its binary equivalent. For example, 3.4 is first rounded to 3, then 3 is converted into its binary equivalent "11" if we have a 2-bit converter, "011" if we have a 3-bit converter, or "0011" if we have a 4-bit converter. Therefore, an N-bit A/D converter can be viewed as a device that assigns to a given analog input a corresponding vertex of the hypercube $[0, 1]^N$.

The first step in A/D conversion, rounding of a given number to its nearest integer, can be organized as solving an optimization problem by a neural network. This entails formulation of a suitable cost, or penalty, function to be minimized. In fact, finding a suitable cost function will be the first step in constructing a neural A/D converter.

To better understand the idea behind solving A/D conversion problems using neural networks, we first consider constructing a 2-bit A/D converter. Let V_1 be the most significant bit (MSB), let V_0 be the least significant bit (LSB), and let x be an analog value to be converted. In the actual neural network, V_1 and V_0 are the outputs of the two neurons in the neural A/D converter. We first consider a very simple cost function,

$$E = \tfrac{1}{2}(V_1 2^1 + V_0 2^0 - x)^2. \tag{9.107}$$

A plot of the above function, for an analog input of $x = 1.5$, is shown in Figure 9.32(a). Note that E is defined over the two-dimensional hypercube $[0, 1]^2$. However, constructing a circuit minimizing the cost function (9.107) would not yield good results because V_0 and V_1 would take values from the interval $[0, 1]$ rather than the set $\{0, 1\}$. To overcome the problem, we add to (9.107) two terms whose purpose is to penalize for noninteger values of V_0 and V_1. The modified cost function is

$$E = \tfrac{1}{2}(V_1 2^1 + V_0 2^0 - x)^2 + aV_1(V_1 - 1) + bV_0(V_0 - 1), \tag{9.108}$$

where a and b are weighting coefficients. Tank and Hopfield [278] suggest selecting a and b so that the quadratic terms V_0^2 and V_1^2 are canceled out. In particular, if we set $a = -2$ and $b = -0.5$, then (9.108) becomes

$$E = \tfrac{1}{2}x^2 + 2V_0V_1 - 2V_1x - V_0x + 2V_1 + \tfrac{1}{2}V_0. \tag{9.109}$$

The term used to cancel the quadratic terms is plotted in Figure 9.32(b). The final cost function given by (9.109) is plotted in Figure 9.32(c). To convert an analog input x into its binary equivalent, we minimize the cost function (9.109). One can interpret this process as follows. A given analog input x defines a corresponding cost function E. We can view E as a landscape. Now, a ball placed somewhere on this landscape will move toward the lowest point, which should correspond to one of the vertices of the hypercube $[0, 1]^2$.

Our next task is to construct differential equations that model a neural network implementing an A/D converter. The parameters of the state equations of the network will be chosen so that the time derivative of E evaluated on the solutions of the equations describing the network is

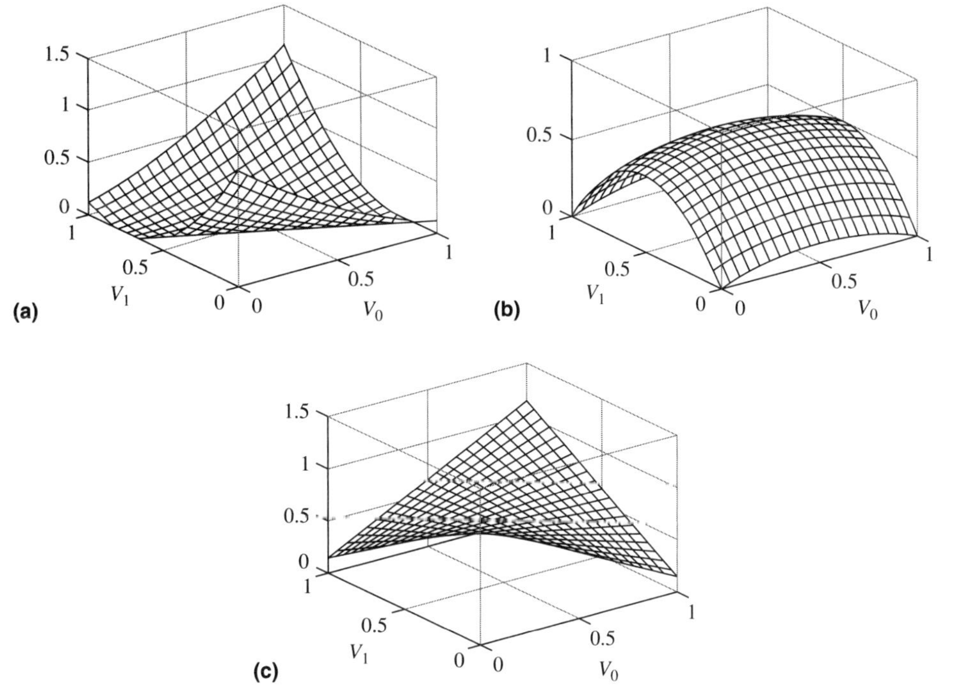

Figure 9.32 Constructing the cost function for a 2-bit A/D converter. Plots of the cost function and its components. (a) Plot of (9.107) for an analog input of $x = 1.5$. (b) Plot of the term used to cancel the quadratic terms in (9.107). (c) Plot of the cost function (9.109).

negative. We first consider a general case of an N-bit A/D converter. After this, we analyze in detail state equations of a 2-bit A/D converter. Note that

$$\frac{dE}{dt} = \frac{\partial E}{\partial V_0}\dot{V}_0 + \frac{\partial E}{\partial V_1}\dot{V}_1 + \cdots + \frac{\partial E}{\partial V_{N-1}}\dot{V}_{N-1},$$

where $V_l = g_i(u_i)$. Hence, similarly as in (9.105), it is enough to set

$$\frac{\partial E}{\partial V_i} = -C_i\dot{u}_i = -C_i\frac{dg_i^{-1}(V_i)}{dV_i}\dot{V}_i, \qquad i = 0, 1, \ldots, N-1 \tag{9.110}$$

to obtain

$$\frac{dE}{dt} = -\sum_{i=0}^{N-1} C_i\frac{dg_i^{-1}(V_i)}{dV_i}\dot{V}_i^2 \le 0, \tag{9.111}$$

whereas in (9.106) we have $\frac{dg_i^{-1}(V_i)}{dV_i} > 0$. Thus, if the state equations of an N-bit A/D converter satisfy (9.110), then $dE/dt \le 0$ on the trajectories of the converter. Using (9.110), we can write the state equations that model the 2-bit A/D converter consisting of two neurons. Let $V_0 = g_0(u_0)$ and $V_1 = g_1(u_1)$. Then, substituting (9.109) into (9.110) and performing required differentiation, we obtain

$$
\begin{aligned}
C_0\dot{u}_0 &= -\frac{\partial E}{\partial V_0} = -(2V_1 - x + 0.5) = -(2g_1(u_1) - x + 0.5),\\
C_1\dot{u}_1 &= -\frac{\partial E}{\partial V_1} = -2(V_0 - x + 1) = -2(g_0(u_0) - x + 1).
\end{aligned}
\tag{9.112}
$$

For simplicity, we assume that $C_0 = C_1 = 1$ and that $g_0 = g_1 = g$. Note that because C is present in both state equations, assigning $C_0 = C_1 = 1$ does not distort the dynamical behavior of the network; it only changes the time scale of both equations by the same factor. Observing that for $i = 1, 2$,

$$
\dot{u}_i = \frac{dg_i^{-1}(V_i)}{dV_i}\dot{V}_i,
$$

we represent (9.112) as

$$
\begin{aligned}
\dot{V}_0 &= -(2V_1 - x + 0.5)\left(\frac{dg^{-1}(V_0)}{dV_0}\right)^{-1},\\
\dot{V}_1 &= -2(V_0 - x + 1)\left(\frac{dg^{-1}(V_1)}{dV_1}\right)^{-1}.
\end{aligned}
\tag{9.113}
$$

We display a plot of the cost function E for the analog value $x = 1.5$ in Figure 9.33(a) and a phase portrait of the 2-bit A/D converter in Figure 9.33(b). We chose the activation function

$$
V_i = g(u_i) = \frac{1}{1 + e^{-4u_i}}.
$$

Using (9.98), we find

$$
u_i = g^{-1}(V_i) = \tfrac{1}{4}(\ln(V_i) - \ln(1 - V_i)).
$$

Thus,

$$
\frac{dg^{-1}(V_i)}{dV_i} = \frac{1}{4V_i(1 - V_i)}.
$$

Note that for $x = 1.5$, the cost function has two minimizers in the hypercube $[0, 1]^2$ that are separated by a "ridge." Initial conditions on each side of this ridge yield a different final value of the state vector $[V_0, V_1]$. Depending on the initial conditions of the state vector of the A/D converter, the analog value $x = 1.5$ will be converted to either the binary "01" or the binary "10." Thus, initial conditions influence the dynamical behavior of the A/D converter. Also, the activation nonlinearity influences the dynamical behavior of the A/D converter via the multiplicative term $(\frac{dg^{-1}(V_i)}{dV_i})^{-1}$ present in the state equations (9.113). One can see that the multiplicative factor influences the direction in which the converter state trajectory evolves.

Employing the above technique, we can construct higher-order A/D converters. For example, to construct a 3-bit A/D converter, we begin with the cost function

$$
E = \tfrac{1}{2}(V_2 2^2 + V_1 2^1 + V_0 2^0 - x)^2 + aV_2(V_2 - 1) + bV_1(V_1 - 1) + cV_0(V_0 - 1).
$$

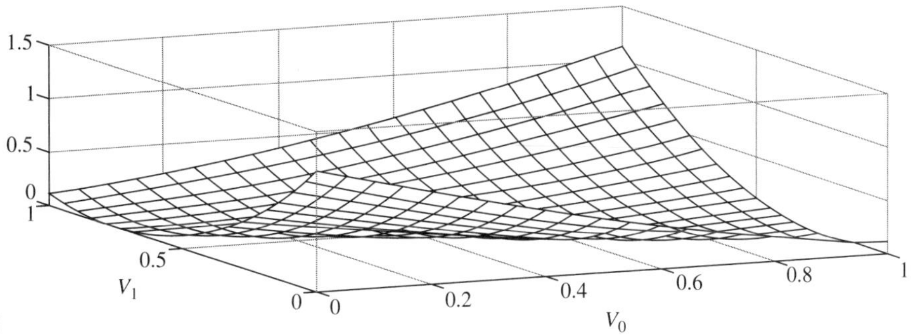

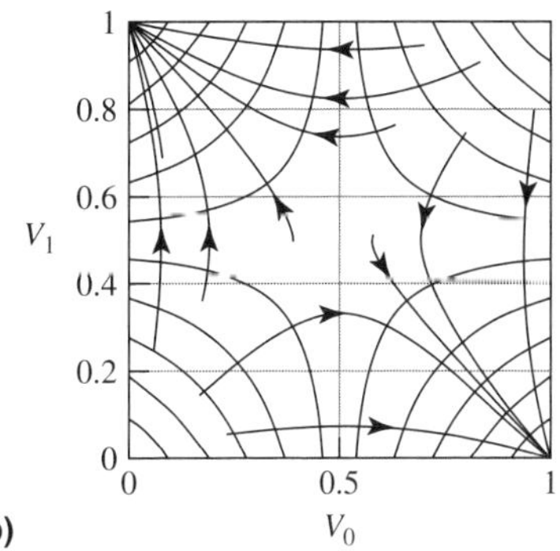

Figure 9.33 A plot of the cost function and a phase portrait of the 2-bit A/D converter for the analog value $x = 1.5$.

To cancel the quadratic terms, select $a = -8$, $b = -2$, $c = -1/2$. The state equations of the converter are obtained from the equations

$$\dot{V}_i = -\frac{1}{C_i}\frac{\partial E}{\partial V_i}\left(\frac{dg_i^{-1}(V_i)}{dV_i}\right)^{-1}, \qquad i = 0, 1, 2.$$

In our analysis of A/D neural converters, we observed that while the 2-bit converter yielded correct results, the 3-bit and 4-bit converters showed some deficiency in their performance. The errors observed for these two converters were quite significant and increased with the number of bits and consequently with network size.

9.8 Hopfield Network Stability Analysis

In this section we apply Lyapunov's second method to analyze stability of the Hopfield neural network, in the sense of Lyapunov. The neural network is viewed as a large interconnected dynamical system—that is, as an interconnection of many neurons. First we analyze stability properties of a single neuron, and then we proceed with the stability analysis of the whole network in terms of the properties of the single neurons. Interconnection information is used in conjunction with the properties of the individual neurons to obtain sufficient conditions for asymptotic stability of the whole network.

9.8.1 Hopfield Network Model Analysis

Recall the equations describing the Hopfield neural net. They are given by (9.89), that is,

$$C_i \frac{du_i}{dt} + \frac{u_i}{r_i} = \sum_{j=1}^{n} T_{ij} x_j + I_i, \qquad i = 1, 2, \ldots, n,$$

where $x_i = g_i(u_i)$. In our further analysis, we assume that the activation functions $g_i : \mathbb{R} \rightarrow (-1, 1)$, $i = 1, 2, \ldots, n$, are continuously differentiable and strictly monotonically increasing. We also assume that

$$u_i g_i(u_i) > 0 \qquad \text{for } u_i \neq 0$$

and

$$g_i(0) = 0.$$

To proceed further, we divide both sides of the above equations describing the Hopfield network by $C_i > 0$ to get

$$\dot{u}_i = -b_i u_i + \sum_{j=1}^{n} A_{ij} g_j(u_j) + U_i, \qquad i = 1, 2, \ldots, n, \tag{9.114}$$

where $b_i = 1/r_i C_i$, $A_{ij} = T_{ij}/C_i$, and $U_i = I_i/C_i$. We are interested in the behavior of solutions of the above model near its equilibrium points. For $U_i = 0$ the equilibrium points are obtained by solving the set of algebraic equations

$$-b_i u_i + \sum_{j=1}^{n} A_{ij} g_j(u_j) = 0, \qquad i = 1, 2, \ldots, n.$$

In some applications the external inputs U_i are held constant, that is, $U_i = c_i =$ constant. In such cases the equilibrium points are obtained by solving for u_i the set of the following algebraic equations:

$$-b_i u_i + \sum_{j=1}^{n} A_{ij} g_j(u_j) + c_i = 0, \qquad i = 1, 2, \ldots, n.$$

However, we can always transform the algebraic equations with nonzero constant external inputs into the algebraic equations where the external inputs are set to zero. Indeed, for $i = 1, 2, \ldots, n$ let

$$u_i = y_i + \frac{c_i}{b_i}$$

and

$$G_i(y_i) = g_i\left(y_i + \frac{c_i}{b_i}\right).$$

Substituting the above into (9.114) and assuming that $U_i = c_i$ yields

$$\dot{y}_i = -b_i y_i + \sum_{j=1}^{n} A_{ij} G_j(y_j), \qquad i = 1, 2, \ldots, n. \tag{9.115}$$

The equilibrium points of the above system of differential equations are obtained by solving the following algebraic equations:

$$-b_i y_i + \sum_{j=1}^{n} A_{ij} G_j(y_j) = -b_i y_i + \sum_{j=1}^{n} A_{ij} g_j\left(y_j + \frac{c_j}{b_j}\right)$$
$$= 0, \qquad i = 1, 2, \ldots, n.$$

Thus, after the transformation, the algebraic equations with external constant inputs used for computing the equilibrium points assume the same form as the algebraic equations corresponding to zero external inputs. Hence, without loss of generality, we can assume in our further analysis that the external inputs are zero and that the network is modeled by the equations

$$\boxed{\dot{u}_i = -b_i u_i + \sum_{j=1}^{n} A_{ij} g_j(u_j), \qquad i = 1, 2, \ldots, n} \tag{9.116}$$

Throughout, we assume that an isolated equilibrium point of interest is located at the origin of the state space $\mathbb{R}^n$. If this is not the case, then by an appropriate change of variables we can arrange for the equilibrium point of interest to be translated to the origin of $\mathbb{R}^n$. Let $\boldsymbol{u} = [u_1 \; \cdots \; u_n]^T \in \mathbb{R}^n$ and suppose $\boldsymbol{u}^* \neq \boldsymbol{0}$ is an isolated equilibrium point of interest of (9.116). For $i = 1, 2, \ldots, n$ let

$$x_i = u_i - u_i^* \quad \text{and} \quad h_i(u_i) = h_i(x_i + u_i^*) = g_i(u_i) - g_i(u_i^*).$$

Substituting the above into (9.116) yields

$$\begin{aligned}\dot{x}_i &= -b_i x_i - b_i u_i^* + \sum_{j=1}^{n} A_{ij}(h_j(x_j + u_j^*) + g_j(u_j^*)) \\ &= -b_i x_i + \sum_{j=1}^{n} A_{ij} h_j(x_j + u_j^*) + \left\{-b_i u_i^* + \sum_{j=1}^{n} A_{ij} g_j(u_j^*)\right\}.\end{aligned}$$

Note that, by the definition of the equilibrium point, the expression in the curly brackets vanishes. Thus we have

$$\dot{x}_i = -b_i x_i + \sum_{j=1}^{n} A_{ij} h_j(x_j + u_j^*), \qquad i = 1, 2, \ldots, n. \tag{9.117}$$

Let

$$G_j(x_j) = h_j(x_j + u_j^*).$$

Then, differential equations (9.117) can be written as

$$\dot{x}_i = -b_i x_i + \sum_{j=1}^{n} A_{ij} G_j(x_j), \qquad i = 1, 2, \ldots, n,$$

where $G_j : \mathbb{R} \to (c_1, c_2) \subseteq (-2, 2)$, $c_1 < 0 < c_2$, $G_j(0) = 0$, G_j is continuously differentiable and strictly monotonically increasing in x_j. Furthermore $x_j G_j(x_j) > 0$ for all $x_j \neq 0$. We now represent the above model of the Hopfield neural network in the form

$$\dot{x}_i = -b_i x_i + A_{ii} G_i(x_i) + \sum_{\substack{j=1 \\ j \neq i}}^{n} A_{ij} G_j(x_j), \qquad i = 1, 2, \ldots, n.$$

We shall view the terms

$$\sum_{\substack{j=1 \\ j \neq i}}^{n} A_{ij} G_j(x_j)$$

as the interconnections between the ith neuron and the remaining neurons in the network. We shall refer to the equations

$$\dot{x}_i = -b_i x_i + A_{ii} G_i(x_i), \qquad i = 1, 2, \ldots, n \tag{9.118}$$

as models of free isolated neurons. In what follows we analyze stability properties of an isolated single neuron.

9.8.2 Single-Neuron Stability Analysis

A single neuron in the Hopfield network, when the external input is set to zero, is modeled by a first-order differential equation:

$$\boxed{\dot{x}_i = -b_i x_i + A_{ii} G_i(x_i)} \tag{9.119}$$

Consider now, as in Michel et al. [203], the following positive definite Lyapunov function candidate

$$V_i(x_i) = \tfrac{1}{2} x_i^2.$$

Its time derivative evaluated along the trajectories of the single neuron model is

$$\begin{aligned} \dot{V}_i(x_i(t)) &= x_i \dot{x}_i \\ &= x_i(-b_i x_i + A_{ii} G_i(x_i)). \end{aligned}$$

If there exists $r_i > 0$ such that for all $x_i \in (-r_i, r_i)$

$$\begin{aligned} -b_i x_i + A_{ii} G_i(x_i) &> 0 \qquad \text{if } x_i < 0, \\ -b_i x_i + A_{ii} G_i(x_i) &= 0 \qquad \text{if } x_i = 0, \\ -b_i x_i + A_{ii} G_i(x_i) &< 0 \qquad \text{if } x_i > 0, \end{aligned}$$

then for $x_i \neq 0$ and $x_i \in (-r_i, r_i)$ we have

$$\dot{V}_i(x_i(t)) < 0$$

and $\dot{V}_i(0) = 0$. Hence the null solution $x_i = 0$ is uniformly asymptotically stable. Figure 9.34 illustrates this case. Now, if there exists $r_i > 0$ such that for $x_i \in (-r_i, r_i)$ we have

$$\begin{aligned} -b_i x_i + A_{ii} G_i(x_i) &< 0 \qquad \text{if } x_i < 0, \\ -b_i x_i + A_{ii} G_i(x_i) &= 0 \qquad \text{if } x_i = 0, \\ -b_i x_i + A_{ii} G_i(x_i) &> 0 \qquad \text{if } x_i > 0, \end{aligned}$$

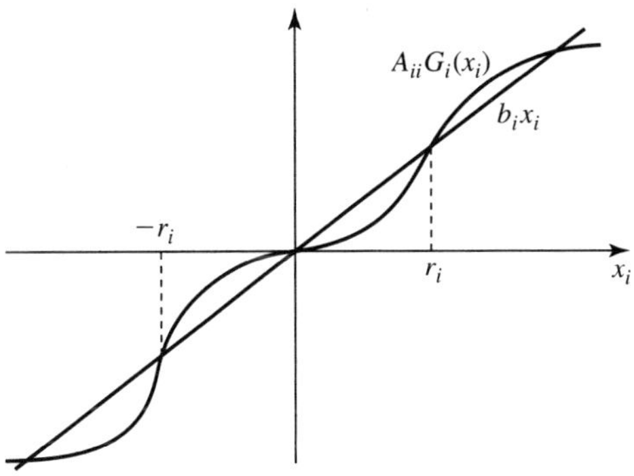

Figure 9.34 Characteristics of a single neuron model whose null solution is uniformly asymptotically stable.

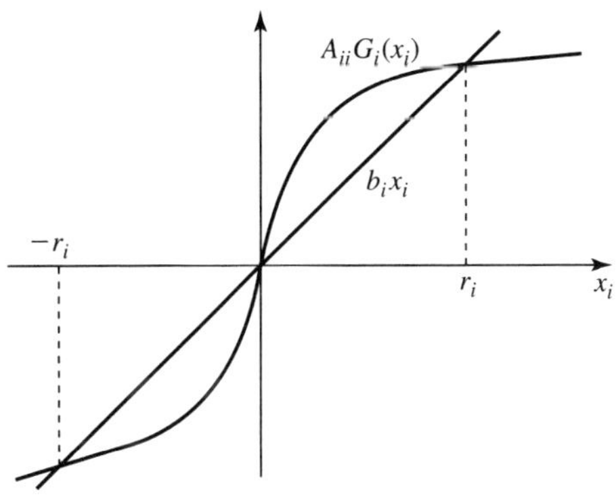

Figure 9.35 Characteristics of single-neuron model whose null solution is unstable.

then in addition to V_i being positive definite we have

$$\dot{V}_i(x_i(t)) > 0$$

for $x_i \neq 0$ and $x_i \in (-r_i, r_i)$, and $\dot{V}_i(0) = 0$. Thus the null solution $x_i = 0$ is unstable. Figure 9.35 illustrates the case when the null solution of a single neuron is unstable. Assume now that for some $r_i > 0$ there exist constants $\sigma_{i1} > 0$ and $\sigma_{i2} > 0$ such that for all $x_i \in (-r_i, r_i)$, $x_i \neq 0$ we have

$$\sigma_{i1} x_i^2 < x_i G_i(x_i) < \sigma_{i2} x_{i2}^2, \tag{9.120}$$

or, equivalently,

$$\sigma_{i1} < \frac{G_i(x_i)}{x_i} < \sigma_{i2}.$$

If in addition to the sufficient conditions for asymptotic stability we also have

$$-b_i + A_{ii}\delta_i < 0, \tag{9.121}$$

where

$$\delta_i = \begin{cases} \sigma_{i1} & \text{if } A_{ii} < 0, \\ \sigma_{i2} & \text{if } A_{ii} > 0, \end{cases} \tag{9.122}$$

then the null solution of the single-neuron model is exponentially stable. Indeed, for $x_i \neq 0$ we have

$$\begin{aligned}\dot{V}_i(x_i(t)) &= x_i\dot{x}_i \\ &= x_i(-b_i x_i + A_{ii}G_i(x_i)) \\ &= \left(-b_i + A_{ii}\frac{G_i(x_i)}{x_i}\right)x_i^2 \\ &\leq (-b_i + A_{ii}\delta_i)x_i^2 \\ &= 2(-b_i + A_{ii}\delta_i)V_i(x_i(t)).\end{aligned}$$

Applying Theorem A.20 yields

$$\begin{aligned}V_i(x_i(t)) &= \frac{x_i^2(t)}{2} \\ &\leq \exp\{2(-b_i + A_{ii}\delta_i)(t - t_0)\}V_i(x_i(t_0)) \\ &= \exp\{2(-b_i + A_{ii}\delta_i)(t - t_0)\}\frac{x_i^2(t_0)}{2}.\end{aligned}$$

Hence

$$|x_i(t)| \leq \exp\{(-b_i + A_{ii}\delta_i)(t - t_0)\}|x_i(t_0)|.$$

From the above we conclude that the null solution is not only asymptotically stable but even more, it is exponentially stable.

9.8.3 Stability Analysis of the Network

Lyapunov's second method in conjunction with interconnection information will now be used in the stability analysis of the Hopfield neural model viewed as an interconnected dynamical system. Applications of Lyapunov's second method to a general class of interconnected systems were investigated by Bailey [16]. Here, we use Lyapunov's second method to investigate stability properties of the null solution of the neural network described by

$$\boxed{\dot{x}_i = -b_i x_i + \sum_{j=1}^{n} A_{ij}G_j(x_j), \qquad i = 1, 2, \ldots, n} \tag{9.123}$$

We impose the following assumptions on the interconnections:

A1: There exist positive constants r_i, r_j, and a_{ij} such that for $|x_i| < r_i$ and $|x_j| < r_j$, $i, j = 1, 2, \ldots, n$, we have

$$x_i A_{ij} G_j(x_j) \leq |x_i| a_{ij} |x_j|.$$

A2: Let $\boldsymbol{S}$ be an $n \times n$ real matrix whose elements are

$$s_{ij} = \begin{cases} \alpha_i(-b_i + a_{ii}) & \text{if } i = j, \\ \frac{1}{2}(\alpha_i a_{ij} + \alpha_j a_{ji}) & \text{if } i \neq j. \end{cases}$$

We assume that there exists a vector $\boldsymbol{\alpha} = [\alpha_1 \cdots \alpha_n]^T$, where $\alpha_i > 0$ for $i = 1, 2, \ldots, n$, such that the matrix $\boldsymbol{S}$ is negative definite.

We now state and prove a theorem concerned with the exponential stability of the neural network model. This theorem can be found in Michel et al. [203].

Theorem 9.5 The null solution of the neural network model (9.123) is uniformly exponentially stable if assumptions A1 and A2 are satisfied.

Proof Consider the following Lyapunov function candidate:

$$V(\boldsymbol{x}) = \frac{1}{2}\sum_{i=1}^{n} \alpha_i x_i^2 = \frac{1}{2}\boldsymbol{x}^T \begin{bmatrix} \alpha_1 & & \dots & 0 \\ & \alpha_2 & & \\ \vdots & & \ddots & \vdots \\ 0 & & \dots & \alpha_n \end{bmatrix} \boldsymbol{x},$$

where $\alpha_i > 0$, $i = 1, 2, \dots, n$, are defined in assumption A2. Note that V is real symmetric positive definite. The time derivative of V evaluated along the trajectories of the neural model (9.123) is

$$\begin{aligned} \dot{V}(\boldsymbol{x}(t)) &= \sum_{i=1}^{n} \alpha_i x_i \dot{x}_i \\ &= \sum_{i=1}^{n} \alpha_i x_i \left(-b_i x_i + \sum_{j=1}^{n} A_{ij} G_j(x_j) \right) \\ &\le \sum_{i=1}^{n} \alpha_i \left(-b_i x_i^2 + |x_i| \sum_{j=1}^{n} a_{ij} |x_j| \right) \\ &= |\boldsymbol{x}|^T \boldsymbol{R} |\boldsymbol{x}|, \end{aligned}$$

where $|\boldsymbol{x}| = [|x_1| \ |x_2| \ \cdots \ |x_n|]^T$, and the matrix $\boldsymbol{R} \in \mathbb{R}^{n\times n}$ has its elements defined as

$$r_{ij} = \begin{cases} \alpha_i(-b_i + a_{ii}) & \text{if } i = j, \\ \alpha_i a_{ij} & \text{if } i \ne j. \end{cases}$$

Note that

$$\begin{aligned} |\boldsymbol{x}|^T \boldsymbol{R} |\boldsymbol{x}| &= |\boldsymbol{x}|^T \left(\frac{\boldsymbol{R} + \boldsymbol{R}^T}{2} \right) |\boldsymbol{x}| \\ &= |\boldsymbol{x}|^T \boldsymbol{S} |\boldsymbol{x}| \\ &\le \lambda_{\max}(\boldsymbol{S}) \|\boldsymbol{x}\|^2, \end{aligned}$$

where the matrix $\boldsymbol{S}$ is defined in assumption A2. By assumption A2 the matrix $\boldsymbol{S}$ is real symmetric negative definite. Hence $\lambda_{\max}(\boldsymbol{S})$ is real and negative. Thus

$$\dot{V}(\boldsymbol{x}(t)) \le \lambda_{\max}(\boldsymbol{S}) \|\boldsymbol{x}(t)\|^2 < 0$$

in a neighborhood of $\boldsymbol{x} = \boldsymbol{0}$ of radius

$$r = \min_{1\le i\le n} r_i.$$

By Proposition 4.1, the null solution is uniformly exponentially stable.

In our stability analysis of the Hopfield neural network we do not require that the parameters A_{ij}, and hence the interconnections T_{ij}, be symmetric. The symmetry requirement on the interconnections was usually imposed in the early stability analyses of this network.

Note also that a necessary condition for the matrix $\boldsymbol{S}$ to be negative definite is that its diagonal elements be negative, that is,

$$-b_i + a_{ii} < 0, \qquad i = 1, 2, \ldots, n. \tag{9.124}$$

However, this condition is not sufficient for the matrix $\boldsymbol{S}$ to be negative definite. The above necessary condition, though, can shed some light on the influence of the stability properties of the individual neurons on the stability of the whole network. To show how the stability of the individual neurons may affect the stability of the whole network, we set $i = j$ in A1 to obtain

$$x_i A_{ii} G_i(x_i) \le a_{ii}|x_i|^2. \tag{9.125}$$

Applying now the sector condition, $\sigma_{i1}x_i^2 < x_i G_i(x_i) < \sigma_{i2}x_i^2$, to $x_i A_{ii} G_i(x_i)$ gives

$$x_i A_{ii} G_i(x_i) \le A_{ii}\delta_i|x_i|^2, \tag{9.126}$$

where δ_i is defined in (9.122). Thus the sector condition (9.120) is sufficient for (9.125) to hold and we can write

$$a_{ii} = A_{ii}\delta_i.$$

Substituting the above into (9.124) yields

$$-b_i + A_{ii}\delta_i < 0, \qquad i = 1, 2, \ldots, n,$$

which are sufficient conditions for the exponential stability of the individual neurons. Thus, if we first find constants a_{ii}, $i = 1, 2, \ldots, n$, and then use them to construct the matrix $\boldsymbol{S}$, then for so-formed matrix $\boldsymbol{S}$ to be negative definite it is necessary that the above sufficient conditions for the exponential stability of the individual neurons hold.

We now present a corollary to Theorem 9.5. Before stating the corollary, we need to perform some algebraic manipulations. We first combine assumption A2 with the sector condition $\sigma_{i1}x_i^2 < x_i G_i(x_i) < \sigma_{i2}x_i^2$. It follows from the sector condition that for $x_j \ne 0$,

$$\left|\frac{G_j(x_j)}{x_j}\right| < \sigma_{j2}.$$

Applying the above when $i \ne j$ yields

$$\begin{aligned} x_i A_{ij} G_j(x_j) &\le |A_{ij}||x_i G_j(x_j)| \\ &= |A_{ij}| \left| x_i \frac{G_j(x_j)}{x_j} x_j \right| \\ &\le |A_{ij}||x_i||x_j|\sigma_{j2}. \end{aligned}$$

When $i = j$ we obtain

$$x_i A_{ii} G_i(x_i) \le \delta_i A_{ii} x_i^2,$$

where δ_i is defined as above. We use the results of the above manipulations in the following assumption:

A3: There exist positive constants r_i and r_j such that for $|x_i| < r_i$ and $|x_j| < r_j$, the interconnections of the neural network model (9.123) satisfy the conditions

$$x_i A_{ii} G_i(x_i) \le \delta_i A_{ii} x_i^2 \qquad \text{for } i = j,$$
$$x_i A_{ij} G_j(x_j) \le |x_i||A_{ij}||x_j|\sigma_{j2} \qquad \text{for } i \ne j.$$

We shall use another assumption.

A4: There exists a vector $\boldsymbol{\alpha} = [\alpha_1 \ \cdots \ \alpha_n]^T \in \mathbb{R}^n$, $\alpha_i > 0$, such that the matrix $\boldsymbol{S} \in \mathbb{R}^{n\times n}$ is negative definite, where the elements of $\boldsymbol{S}$ are defined as

$$s_{ij} = \begin{cases} \alpha_i(-b_i + \delta_i A_{ii}) & \text{if } i = j, \\ \frac{1}{2}(\alpha_i|A_{ij}|\sigma_{j2} + \alpha_j|A_{ji}|\sigma_{i2}) & \text{if } i \ne j. \end{cases}$$

We are now ready to state and prove a corollary to Theorem 9.5. This result also comes from Michel et al. [203].

Corollary 9.1 The null solution of the neural network model (9.123) is uniformly exponentially stable if the assumptions A3 and A4 are satisfied.

Proof Consider the following quadratic positive definite Lyapunov function candidate

$$V(\boldsymbol{x}) = \frac{1}{2}\sum_{i=1}^{n} \alpha_i x_i^2,$$

where $\alpha_i > 0$, $i = 1, 2, \ldots, n$, are defined in assumption A4. Taking into account the assumptions A3 and A4, we evaluate the time derivative of V along the trajectories of the neural network (9.123) to get

$$\begin{aligned}
\dot{V}(\boldsymbol{x}(t)) &= \sum_{i=1}^{n} \alpha_i x_i \left(-b_i x_i + \sum_{j=1}^{n} A_{ij} G_j(x_j) \right) \\
&= \sum_{i=1}^{n} \alpha_i x_i \left(-b_i x_i + A_{ii} G_i(x_i) + \sum_{\substack{j=1 \\ j\ne i}}^{n} A_{ij} G_j(x_j) \right) \\
&\le \sum_{i=1}^{n} \alpha_i \left((-b_i + \delta_i A_{ii})\, x_i^2 + \sum_{\substack{j=1 \\ j\ne i}}^{n} |A_{ij}||x_i||x_j|\sigma_{j2} \right) \\
&= \tfrac{1}{2}|\boldsymbol{x}|^T \boldsymbol{S} |\boldsymbol{x}| \\
&< 0 \qquad \text{for } \boldsymbol{x} \ne \boldsymbol{0}.
\end{aligned}$$

Using the above, the fact that V is a quadratic function, and Proposition 4.1, we conclude that the null solution is uniformly exponentially stable.

We illustrate the above corollary with a numerical example involving a neural network consisting of two neurons.

◆ Example 9.8

The neural net that we analyze in this example is described by the equations

$$\begin{bmatrix} \dot{u}_1 \\ \dot{u}_2 \end{bmatrix} = \begin{bmatrix} -1.4u_1 + 1.2g_2(u_2) \\ -1.1u_2 + g_1(u_1) \end{bmatrix}, \tag{9.127}$$

where

$$g_i(u_i) = \frac{2}{\pi} \arctan\left(\lambda \frac{\pi}{2} u_i\right), \qquad i = 1, 2. \tag{9.128}$$

The value of the parameter λ was set to be equal to 1.4. In Figure 9.36, we show a plot of the above activation nonlinearity. The network has three equilibrium states:

$$\boldsymbol{u}^{(1)} = \begin{bmatrix} -0.4098 \\ -0.4245 \end{bmatrix}, \quad \boldsymbol{u}^{(2)} = \begin{bmatrix} 0.4098 \\ 0.4245 \end{bmatrix}, \quad \text{and} \quad \boldsymbol{u}^{(3)} = \begin{bmatrix} 0 \\ 0 \end{bmatrix}.$$

When computing the equilibrium points, we began by finding the first coordinates of them. This can be done, for example, by combining the two algebraic equations

$$-1.4u_1 + 1.2g_2(u_2) = 0 \quad \text{and} \quad -1.1u_2 + g_1(u_1) = 0$$

and then solving them for u_1. A plot of the function obtained by substituting the second of the above equations into the first is shown in Figure 9.37. We can use Lyapunov's first method, as we did in Example 4.8, to show that the equilibrium states $\boldsymbol{u}^{(1)}$ and $\boldsymbol{u}^{(2)}$ are asymptotically stable, while the equilibrium $\boldsymbol{u}^{(3)}$ is unstable.

We now apply Corollary 9.1 to show that the equilibrium states $\boldsymbol{u}^{(1)}$ and $\boldsymbol{u}^{(2)}$ are in fact uniformly exponentially stable. We analyze the equilibrium state $\boldsymbol{u}^{(1)}$; the analysis of $\boldsymbol{u}^{(2)}$ is identical. We first translate the equilibrium state of interest

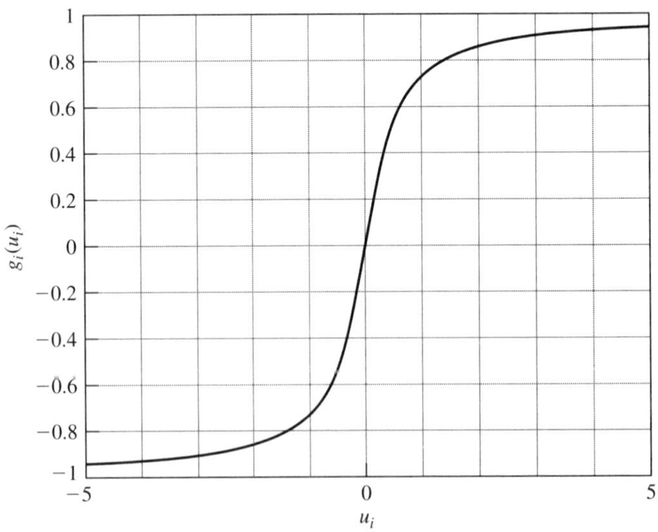

Figure 9.36 A plot of the activation nonlinearity (9.128).

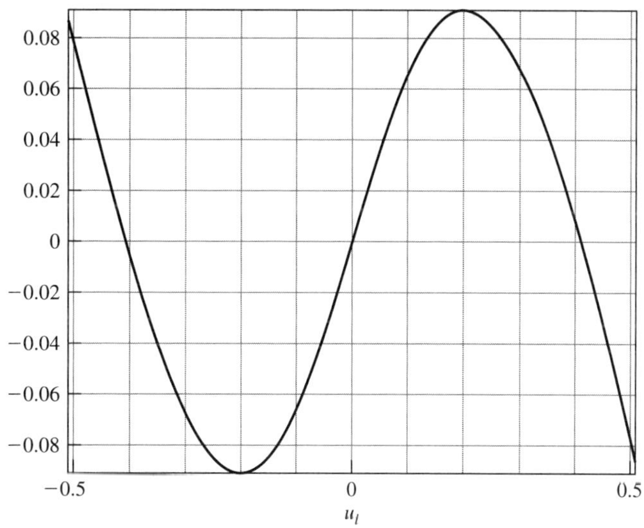

Figure 9.37 A plot of the function used to find the first coordinates of the equilibrium points of the neural net of Example 9.8.

to the origin of $\mathbb{R}^2$ using the translation

$$\boldsymbol{x} = \boldsymbol{u} - \boldsymbol{u}^{(1)}$$

Let

$$G_i(x_i) = g_i\left(x_i + u_i^{(1)}\right) - g_i\left(u_i^{(1)}\right), \qquad i = 1, 2.$$

Taking into account the above relation, the neural network model in the new coordinates takes the form

$$\begin{bmatrix} \dot{x}_1 \\ \dot{x}_2 \end{bmatrix} = \begin{bmatrix} -1.4x_1 + 1.2G_2(x_2) \\ -1.1x_2 + G_1(x_1) \end{bmatrix}. \tag{9.129}$$

The equilibrium state $\boldsymbol{x}_e = \boldsymbol{0}$ of the above net corresponds to the equilibrium state $\boldsymbol{u}_e = \boldsymbol{u}^{(1)}$ of the net in the u coordinates. To be able to apply Corollary 9.1, we need to find σ_{i2} for $i = 1, 2$ such that

$$0 < \frac{G_i(x_i)}{x_i} < \sigma_{i2} \qquad \text{for } |x_i| < r,$$

where $r > 0$. In this example, we took $r = \min\{|u_1^{(1)}|, |u_2^{(1)}|\} = 0.4098$. Graphs of $G_i(x_i)/x_i$ versus x_i, for $i = 1, 2$, are shown in Figure 9.38. We calculated $\sigma_{12} = 1.1394$ and $\sigma_{22} = 1.1165$. We next formed the matrix $\boldsymbol{S} \in \mathbb{R}^{2\times 2}$ defined in A4. Selecting $\alpha_1 = \alpha_2 = 1$ and taking into account that $A_{11} = 0$, $A_{12} = 1.4$, and $A_{21} = 1$, we obtain

$$\begin{aligned} \boldsymbol{S} &= \begin{bmatrix} -1.4\alpha_1 & (\alpha_1|A_{12}|\sigma_{22} + \alpha_2|A_{21}|\sigma_{12})/2 \\ (\alpha_2|A_{21}|\sigma_{12} + \alpha_1|A_{12}|\sigma_{22})/2 & -1.1\alpha_2 \end{bmatrix} \\ &= \begin{bmatrix} -1.4000 & 1.2396 \\ 1.2396 & -1.1000 \end{bmatrix}. \end{aligned}$$

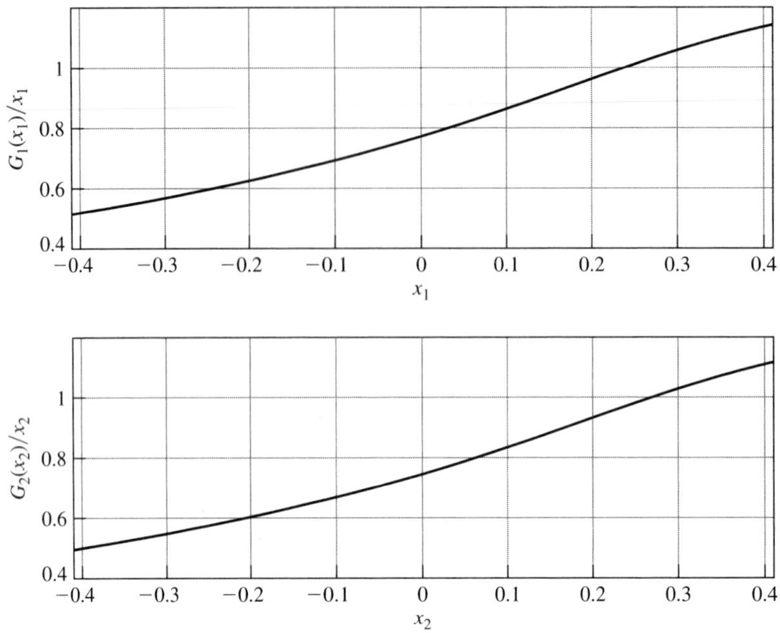

Figure 9.38 Plots of $G_i(x_i)/x_i$, $i = 1, 2$, in Example 9.8.

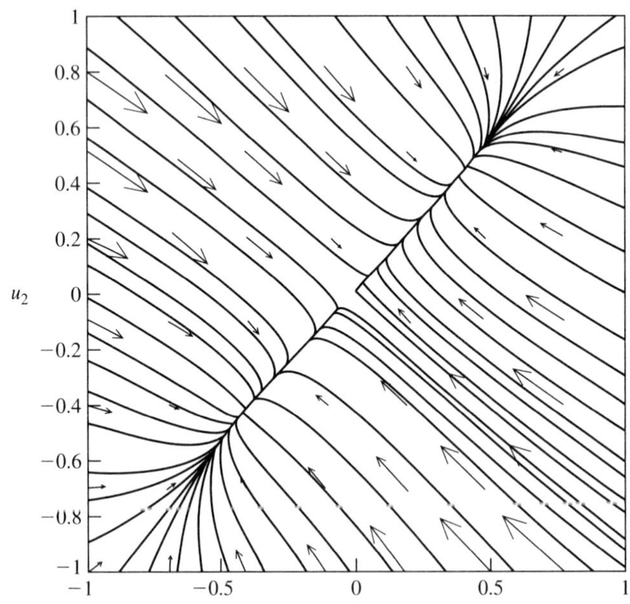

Figure 9.39 Phase plane portrait of the neural net of Example 9.8.

The eigenvalues of S are located at -2.4987 and -0.0013. Hence, the matrix $\boldsymbol{S}$ is negative definite, and therefore the equilibrium state $\boldsymbol{x}_e = \boldsymbol{0}$ of the net (9.129), and hence $\boldsymbol{u}_e = \boldsymbol{u}^{(1)}$ of the net (9.127), is uniformly exponentially stable. In Figure 9.39, we show a state plane portrait of the neural network modeled by (9.127).

9.9 Brain-State-in-a-Box (BSB) Models

9.9.1 Associative Memories

In this section we analyze a class of neural network models that can be used to construct the so-called associative memories. We distinguish between two types of association: autoassociation and heteroassociation. In autoassociation, after storing a set of patterns by a neural network, where patterns are usually represented as vectors, a distorted (noisy) pattern of a stored pattern is subsequently presented to the network. The task of the neural network is to retrieve (recall) the original stored pattern from the noisy pattern. In heteroassociation an arbitrary set of input patterns is paired with another arbitrary set of output patterns. Operation of an associative memory is characterized by two stages: storage phase, where patterns are being stored by the neural network, and recall phase, where memorized patterns are being retrieved in response to a noisy pattern presented to the network. Figure 9.40 illustrates the recall phase, where the input pattern represents a noisy version of stored pattern $\boldsymbol{x}_j$. The pattern associator produces the output pattern $\boldsymbol{y}$ in response to the input pattern $\boldsymbol{x}$. For a perfect recall in an autoassociative memory, we should have $\boldsymbol{y} = \boldsymbol{x}_j$. When $\boldsymbol{y} \neq \boldsymbol{x}_j$, the associative memory is said to have made a recall error (or error in recall). We now analyze a static linear autoassociator. For the sake of simplicity, we assume that the given patterns, $\boldsymbol{v}^{(j)}$, $j = 1, \ldots, r$, are orthogonal, that is, $\langle \boldsymbol{v}^{(j)}, \boldsymbol{v}^{(k)} \rangle = 0$ for $k \neq j$. Suppose that we form the weight matrix as the sum of the outer products of the given patterns, that is,

$$\boldsymbol{W} = \sum_{j=1}^{r} \boldsymbol{v}^{(j)} \boldsymbol{v}^{(j)T} = \boldsymbol{V}\boldsymbol{V}^T,$$

where $\boldsymbol{V} = [\boldsymbol{v}^{(1)} \ \boldsymbol{v}^{(2)} \ \cdots \ \boldsymbol{v}^{(r)}]$. Let $\boldsymbol{v}$ be one of the given patterns. We present $\boldsymbol{v}$ to the linear associator. Then, because the patterns $\boldsymbol{v}^{(j)}$, $j = 1, \ldots, r$, are mutually orthogonal, the output of the associator is

$$\boldsymbol{W}\boldsymbol{v} = \sum_{j=1}^{r} \boldsymbol{v}^{(j)} \boldsymbol{v}^{(j)T} \boldsymbol{v} = \boldsymbol{v}\boldsymbol{v}^T \boldsymbol{v} = c\boldsymbol{v},$$

where $c = \boldsymbol{v}^T \boldsymbol{v}$. Thus, we obtained back the presented pattern.

We will now show that our linear associator can reconstruct the missing part of a stored pattern. Assume that

$$\boldsymbol{v} = \boldsymbol{v}' + \boldsymbol{v}'';$$

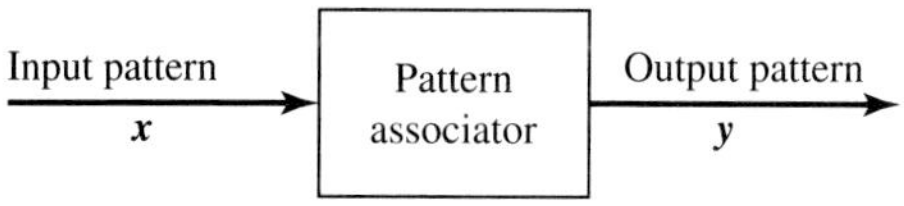

Figure 9.40 Operation of pattern associator.

that is, the given pattern $\boldsymbol{v}$ is composed of two parts. Suppose that $\boldsymbol{v}'$ and $\boldsymbol{v}''$ are orthogonal to each other and that $\boldsymbol{v}'$ is also orthogonal to the remaining patterns in the set. Taking this into account while presenting $\boldsymbol{v}'$ to our linear associator gives

$$\begin{aligned} \boldsymbol{W}\boldsymbol{v}' &= \sum_{j=1}^{r} \boldsymbol{v}^{(j)}\boldsymbol{v}^{(j)T}\boldsymbol{v}' \\ &= \boldsymbol{v}\boldsymbol{v}^T\boldsymbol{v}' \\ &= (\boldsymbol{v}' + \boldsymbol{v}'')(\boldsymbol{v}' + \boldsymbol{v}'')^T\boldsymbol{v}' \\ &= (\boldsymbol{v}' + \boldsymbol{v}'')\boldsymbol{v}'^T\boldsymbol{v}' \\ &= d\boldsymbol{v}, \end{aligned}$$

where $d = \boldsymbol{v}'^T\boldsymbol{v}'$. Thus the linear associator was able to reconstruct the missing part of the presented pattern. In the above discussion, we assumed that the patterns are mutually orthogonal. If this is not the case, the output of the associator will contain an extra component called "cross-talk." This linear static associator has very limited capability. However, we shall use the idea of the outer product to construct neural associative memories.

A neural network that can be used as an associative memory is the Brain-State-in-a-Box (BSB) neural model, which was proposed by Anderson et al. [9] in 1977. It can recognize a pattern from its given noisy version. For this reason the BSB neural model is often referred to as an associative memory. The BSB neural model is a discrete dynamical system whose dynamics are described by the difference equation

$$\boldsymbol{x}(k+1) = \boldsymbol{g}(\boldsymbol{x}(k) + \alpha\boldsymbol{W}\boldsymbol{x}(k)) \tag{9.130}$$

subject to the initial condition

$$\boldsymbol{x}(0) = \boldsymbol{x}_0,$$

where $\boldsymbol{x}(k) \in \mathbb{R}^n$ is the state of the BSB network at time k, $\alpha > 0$ is the step size, and $\boldsymbol{W} \in \mathbb{R}^{n\times n}$ is a symmetric weight matrix determined during the training procedure. The function $\boldsymbol{g} : \mathbb{R}^n \to \mathbb{R}^n$ is a vector-valued function whose ith component $(\boldsymbol{g})_i$ is defined as

$$(\boldsymbol{g}(\boldsymbol{x}))_i = \begin{cases} 1 & \text{if } x_i \geq 1 \\ x_i & \text{if } -1 < x_i < 1 \\ -1 & \text{if } x_i \leq -1. \end{cases}$$

The above nonlinear activation function is sometimes referred to as the linear saturating nonlinearity. A plot of such a function is shown in Figure 9.41. The BSB model is named from the fact that the network trajectories are constrained to be in the hypercube $H_n = [-1\ \ 1]^n$. In Figure 9.42 we show a state plane portrait of the BSB model (9.130) with the weight matrix

$$\boldsymbol{W} = \begin{bmatrix} 7 & -3 \\ -3 & 10 \end{bmatrix} \tag{9.131}$$

and the step size is $\alpha = 0.1$. The state space of this two-dimensional neural net is the square $H_2 = [-1\ \ 1]^2$. Each vertex of the square has its basin of attraction. In this example all four vertices are asymptotically stable equilibrium states of the BSB neural net. In Figure 9.43 we

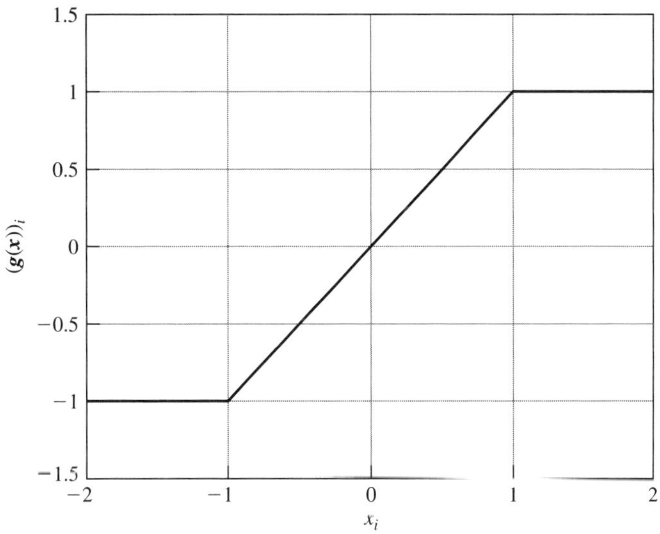

Figure 9.41 A plot of linear saturating nonlinearity used in the BSB model as an activation function.

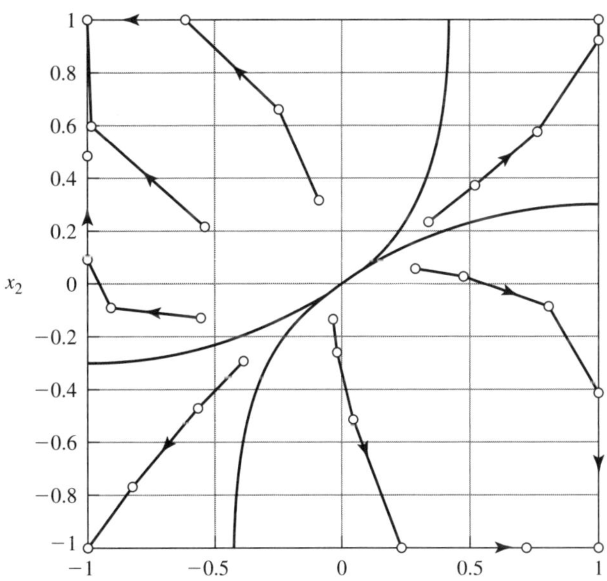

Figure 9.42 Trajectories of the BSB model with the weight matrix W given by (9.131).

show a state plane portrait of the BSB model (9.130) with the weight matrix

$$W = \begin{bmatrix} 5 & -3 \\ -3 & 2 \end{bmatrix} \tag{9.132}$$

and the step size $\alpha = 0.1$. In this example, only two vertices are asymptotically stable equilibrium states of the BSB neural net. When the BSB model is used as an associative memory,

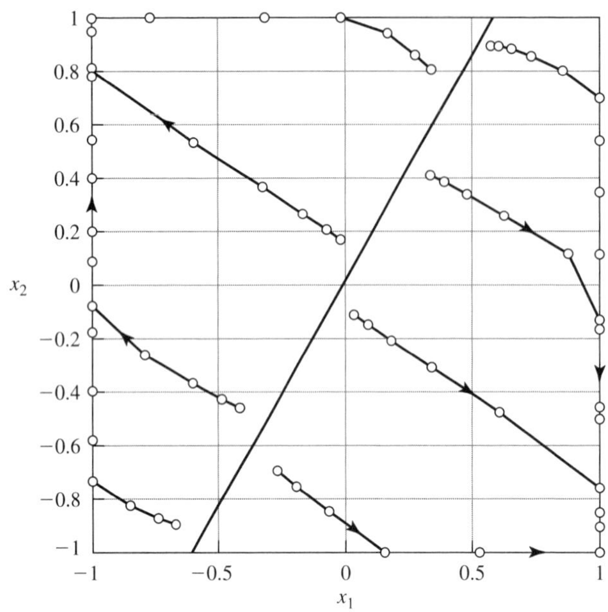

Figure 9.43 Trajectories of the BSB model with the weight matrix $\boldsymbol{W}$ given by (9.132).

asymptotically stable equilibrium states of the model represent stored patterns. Noisy patterns are represented by the states in the basins of attraction of the stable equilibrium states. Thus, suppose we are given a noisy, or imperfect, pattern represented by an initial condition $\boldsymbol{x}_0$ that is located in a basin of attraction of a stable equilibrium state corresponding to the correct pattern. Then, the corresponding trajectory of the BSB model evolves in time to this stable equilibrium state representing the correct stored pattern. An associative memory is able to recall all the information stored in the memory when it is excited with a sufficient large portion of that memory's information. This is in contrast to the address addressable memory (AAM) used in digital computers. In this type of memory, a desired set of information is recalled only when the entire correct address of the information is given. For this reason, we can view associative memories as content addressable memories (CAM).

In summary, realization of associative memories with neural networks requires storage of a set of given patterns as asymptotically stable equilibrium states of the network. The states in basins of attraction about each stable equilibrium state correspond to noisy versions of the stored patterns. When a noisy version of a stored pattern is sufficiently close to it, the network trajectory should converge to the equilibrium state corresponding to the correct pattern. Neural network associative memories should possess the following useful properties:

1. Each given pattern to be stored should be stored as an asymptotically stable equilibrium state of the network.
2. The number of asymptotically stable equilibrium states of the network that do not correspond to the given stored patterns—that is, the number of spurious states—should be minimal.
3. A nonsymmetric interconnection structure, which eases difficulties in implementing associative memory using neural networks.

4. The ability to control the extent of the basin of attraction about the equilibrium state corresponding to each stored pattern.
5. Learning capability, that is, the ability to add patterns to be stored as asymptotically stable equilibrium states to an existing set of stored patterns without affecting the existing equilibria.
6. Forgetting capability, that is, the ability to delete specified equilibria from a given set of stored equilibria without affecting the rest of the equilibria.
7. A high storage and retrieval efficiency, that is, the ability to efficiently store and retrieve a large number (compared to the order n of the network) of patterns.

9.9.2 Analysis of BSB Models

The BSB neural net has a symmetric weight matrix, which is not a desirable property of a neural associative memory. Hui and Żak [130] modified the BSB model to allow for nonsymmetric weight matrix as well as to better control the extent of basins of attraction. A modified form of the BSB neural model, which we refer to as the generalized Brain-State-in-a-Box (gBSB) neural model, is well-suited for implementing associative memories. Its dynamics are described by the equation

$$\boldsymbol{x}(k+1) = \boldsymbol{g}((\boldsymbol{I}_n + \alpha \boldsymbol{W})\boldsymbol{x}(k) + \alpha \boldsymbol{b}), \tag{9.133}$$

where $\boldsymbol{I}_n$ is the $n \times n$ identity matrix, $\boldsymbol{b} \in \mathbb{R}^n$, and the weight matrix $\boldsymbol{W} \in \mathbb{R}^{n\times n}$ need not be symmetric as in the BSB model. Thus, the original BSB model can be viewed as a special case of the gBSB model when the interconnection matrix $\boldsymbol{W}$ is symmetric and $\boldsymbol{b} = \boldsymbol{0}$. We add that the analysis of the BSB model behavior on the boundary regions of the hypercube H_n can be reduced to the study of the reduced-order gBSB-type models rather than that of the BSB type.

We now describe an algorithm for computing the weight matrix $\boldsymbol{W}$ and the vector $\boldsymbol{b}$ that result in the given patterns being stored as asymptotically stable vertices of the hypercube H_n. In further discussion, we use the following notation and definitions. Let

$$\boldsymbol{L}(\boldsymbol{x}) = (\boldsymbol{I}_n + \alpha \boldsymbol{W})\boldsymbol{x} + \alpha \boldsymbol{b}$$

and

$$\boldsymbol{T}(\boldsymbol{x}) = \boldsymbol{g}(\boldsymbol{L}(\boldsymbol{x})).$$

Definition 9.1 A point $\boldsymbol{x} \in H_n$ is an equilibrium state of $\boldsymbol{x}(k+1) = \boldsymbol{T}(\boldsymbol{x}(k))$ if $\boldsymbol{x} = \boldsymbol{T}(\boldsymbol{x})$.

Note that an equilibrium state is a constant solution to the system $\boldsymbol{x}(k+1) = \boldsymbol{T}(\boldsymbol{x}(k))$.

Let $(\boldsymbol{L}(\boldsymbol{x}))_i$ be the ith component of $\boldsymbol{L}(\boldsymbol{x})$ and let $v_i^{(j)}$ be the ith component of the vector $\boldsymbol{v}^{(j)}$. Let $\boldsymbol{v} = [\, v_1 \cdots v_n \,]^T \in \{-1, 1\}^n$; that is, $\boldsymbol{v}$ is a vertex of H_n because for $i = 1, 2, \ldots, n$ we have $v_i = \pm 1$. For a vertex $\boldsymbol{v}$ of H_n to be an equilibrium state of the gBSB net, we must have

$$v_i = (\boldsymbol{g}(\boldsymbol{L}(\boldsymbol{v})))_i, \qquad i - 1, 2, \ldots, n;$$

that is, if $v_i = 1$, then we should have

$$(\boldsymbol{L}(\boldsymbol{v}))_i \geq 1, \tag{9.134}$$

and for $v_1 = -1$ we should have

$$(\boldsymbol{L}(\boldsymbol{v}))_i \leq -1. \tag{9.135}$$

Combining (9.134) and (9.135), we obtain the following necessary and sufficient conditions for the vertex $\boldsymbol{v}$ to be an equilibrium state.

Theorem 9.6 A vertex $\boldsymbol{v}$ of the hypercube H_n is an equilibrium state of the gBSB neural net if and only if

$$(\boldsymbol{L}(\boldsymbol{v}))_i v_i \geq 1, \qquad i = 1, 2, \ldots, n.$$

Equivalently, a vertex $\boldsymbol{v}$ of the hypercube H_n is not an equilibrium state of the gBSB neural net if and only if

$$(\boldsymbol{L}(\boldsymbol{v}))_i v_i < 1 \qquad \text{for some } i = 1, 2, \ldots, n.$$

◆ Example 9.9

We verify that the vertex $\boldsymbol{v}^{(1)} = [1 \;\; 1]^T$ of H_2 is not an equilibrium state of the BSB net with the weight matrix given by (9.132) and the step size $\alpha = 0.1$. Note that for the BSB net we have

$$\boldsymbol{L}(\boldsymbol{x}) = (\boldsymbol{I}_n + \alpha \boldsymbol{W})\boldsymbol{x}.$$

We first compute $\boldsymbol{L}(\boldsymbol{v}^{(1)})$ to get

$$\boldsymbol{L}\left(\boldsymbol{v}^{(1)}\right) = (\boldsymbol{I}_2 + \alpha \boldsymbol{W})\boldsymbol{v}^{(1)} = \begin{bmatrix} 1.2 \\ 0.9 \end{bmatrix}.$$

Note that because

$$\left(\boldsymbol{L}\left(\boldsymbol{v}^{(1)}\right)\right)_2 v_2^{(1)} = 0.9 < 1,$$

the vertex $\boldsymbol{v}^{(1)}$ is not an equilibrium state of the above BSB net.

On the other hand, the vertex $\boldsymbol{v}^{(2)} = [1 \;\; -1]^T$ is an equilibrium state of this BSB model because

$$\left(\boldsymbol{L}\left(\boldsymbol{v}^{(2)}\right)\right)_1 v_1^{(2)} = 1.8 \geq 1 \quad \text{and} \quad \left(\boldsymbol{L}\left(\boldsymbol{v}^{(2)}\right)\right)_2 v_2^{(2)} = 1.5 \geq 1;$$

that is, the necessary and sufficient conditions of Theorem 9.6 are satisfied.

Using the fact that $v_i^2 = 1$ and $\alpha > 0$, we reformulate Theorem 9.6 as follows.

Theorem 9.7 A vertex $\boldsymbol{v}$ of the hypercube H_n is an equilibrium state of the gBSB neural model if and only if

$$(\boldsymbol{W}\boldsymbol{v} + \boldsymbol{b})_i v_i \geq 0, \qquad i = 1, 2, \ldots, n.$$

In our analysis we use stability definitions in the sense of Lyapunov, which we now recall.

Definition 9.2 An equilibrium state $\boldsymbol{x}^*$ of $\boldsymbol{x}(k+1) = \boldsymbol{T}(\boldsymbol{x}(k))$ is (uniformly) stable if for every $\varepsilon > 0$ there is a $\delta = \delta(\varepsilon) > 0$ such that if $\|\boldsymbol{x}(0) - \boldsymbol{x}^*\| < \delta$, then $\|\boldsymbol{x}(k) - \boldsymbol{x}^*\| < \varepsilon$ for all $k \geq 0$, where $\|\cdot\|$ may be any p-norm of a vector—for example, the Euclidean norm $\|\cdot\|_2$.

The above definition says that if an equilibrium is stable, then a trajectory starting sufficiently close to the equilibrium will stay close to the equilibrium for all future times. If the equilibrium is not stable, then it is called unstable.

Definition 9.3 The equilibrium solution $\boldsymbol{x}^*$ is asymptotically stable if it is stable and if there exists an $\eta > 0$ such that if $\|\boldsymbol{x}(0) - \boldsymbol{x}^*\| < \eta$, then $\lim_{k\to\infty} \boldsymbol{x}(k) = \boldsymbol{x}^*$.

The following theorem gives us a sufficient condition for a vertex of H_n to be an asymptotically stable equilibrium state.

Theorem 9.8 Let $\boldsymbol{v}$ be a vertex of the hypercube H_n. If

$$(\boldsymbol{L}(\boldsymbol{v}))_i v_i > 1, \qquad i = 1, 2, \ldots, n,$$

then $\boldsymbol{v}$ is an asymptotically stable equilibrium state of the gBSB neural model.

Proof Because $(\boldsymbol{L}(\boldsymbol{x}))_i x_i$ is a continuous function, there exists a neighborhood $N(\boldsymbol{v})$ about $\boldsymbol{v}$ such that

$$(\boldsymbol{I}(\boldsymbol{x}))_i x_i > 1 \qquad \text{for all} \quad i = 1, 2, \ldots, n \text{ and } \boldsymbol{x} \in N(\boldsymbol{v}).$$

Therefore, for any $\boldsymbol{x} \in N(\boldsymbol{v})$ we have

$$\boldsymbol{g}(\boldsymbol{L}(\boldsymbol{x})) = \boldsymbol{v},$$

which means that $\boldsymbol{v}$ is indeed asymptotically stable equilibrium state.

Taking into account the fact that $v_i^2 = 1$ and $\alpha > 0$, we reformulate the above theorem as follows.

Theorem 9.9 Let $\boldsymbol{v}$ be a vertex of the hypercube H_n. If

$$(\boldsymbol{W}\boldsymbol{v} + \boldsymbol{b})_i\, v_i > 0, \qquad i = 1, 2, \ldots, n,$$

then $\boldsymbol{v}$ is an asymptotically stable equilibrium state of the gBSB neural model.

9.9.3 Synthesis of Neural Associative Memory

In our further discussion, we use a special class of matrices that are characterized in the following definition.

Definition 9.4 A matrix $\boldsymbol{D} \in \mathbb{R}^{n\times n}$ is said to be row diagonal dominant if

$$d_{ii} \geq \sum_{\substack{j=1\\ j\neq i}}^{n} |d_{ij}|, \qquad i = 1, 2, \ldots, n.$$

It is strongly row diagonal dominant if

$$d_{ii} > \sum_{\substack{j=1\\ j\neq i}}^{n} |d_{ij}|, \qquad i = 1, 2, \ldots, n.$$

Now, suppose that we are given a matrix $\boldsymbol{V} \in \mathbb{R}^{n\times r}$. Its pseudo-inverse, or the Moore–Penrose inverse, is defined as follows.

Definition 9.5 Given $V \in \mathbb{R}^{n\times r}$, a matrix $V^\dagger \in \mathbb{R}^{r\times n}$ is called a pseudo-inverse of V if

1. $VV^\dagger V = V$
2. $V^\dagger V V^\dagger = V^\dagger$
3. $(VV^\dagger)^T = VV^\dagger$
4. $(V^\dagger V)^T = V^\dagger V$

The pseudo-inverse has the following useful properties:

- $(V^T)^\dagger = (V^\dagger)^T$
- $(V^\dagger)^\dagger = V$

Let $v^{(1)}, \ldots, v^{(r)}$ be given vertices of H_n that we wish to store as asymptotically stable equilibrium states of the gBSB neural net; that is, we are interested in building a neural autoassociative memory. We now present a number of technical results that will lead to formulation of a design algorithm for the weight matrix of the gBSB neural associative memory. Let $D \in \mathbb{R}^{n\times n}$ satisfy the condition

$$d_{ii} > \sum_{\substack{k=1\\k\neq i}}^{n} |d_{ik}| + |b_i|, \qquad i = 1, 2, \ldots, n. \tag{9.136}$$

We now present an expression for constructing the weight matrix W ensuring that the given patterns are stored as asymptotically stable equilibrium states of the gBSB neural net.

Theorem 9.10 Let $V = [v^{(1)} \ \cdots \ v^{(r)}] \in \{-1, 1\}^{n\times r}$. If

$$W = DVV^\dagger,$$

where the matrix D satisfies (9.136), then $v^{(1)}, \ldots, v^{(r)}$ are asymptotically stable states of the gBSB net.

Proof Using the first property of the Moore–Penrose inverse of V, we obtain

$$WV = DVV^\dagger V = DV.$$

Hence, for $j = 1, \ldots, r$ we have

$$Wv^{(j)} = Dv^{(j)}. \tag{9.137}$$

Substituting the above into the condition of Theorem 9.9 and using (9.136) gives

$$\begin{aligned}
\left(Wv^{(j)} + b\right)_i v_i^{(j)} &= \left(Dv^{(j)} + b\right)_i v_i^{(j)} \\
&= \left(d_{ii} v_i^{(j)} + \sum_{\substack{k=1\\k\neq i}}^{n} d_{ik} v_k^{(j)} + b_i \right) v_i^{(j)} \\
&\geq d_{ii} - \sum_{\substack{k=1\\k\neq i}}^{n} |d_{ik}| - |b_i| \\
&> 0.
\end{aligned}$$

By Theorem 9.9, $\boldsymbol{v}^{(j)}$ is asymptotically stable. The above argument holds for all $j = 1, \ldots, r$. The proof is complete.

Note that the weight matrix $\boldsymbol{W} = \boldsymbol{D}\boldsymbol{V}\boldsymbol{V}^{\dagger}$ does not have to be symmetric thanks to the presence of the matrix $\boldsymbol{D}$ in the formula for $\boldsymbol{W}$. The matrix $\boldsymbol{D}$ is a design parameter, and therefore it can be chosen to be nonsymmetric so that $\boldsymbol{W}$ becomes nonsymmetric.

Corollary 9.2 Every vertex that is contained in Range($\boldsymbol{V}$) is also stored as an asymptotically stable equilibrium state.

Proof Let $\boldsymbol{v} \in \{-1, 1\}^n$ be such that $\boldsymbol{v} \in$ Range($\boldsymbol{V}$). Hence, there is a vector $\boldsymbol{\gamma} \in \mathbb{R}^r$ such that $\boldsymbol{v} = \boldsymbol{V}\boldsymbol{\gamma}$. Using the above and the first property of the pseudo-inverse matrix of $\boldsymbol{V}$ yields

$$\begin{aligned}\boldsymbol{W}\boldsymbol{v} &= \boldsymbol{D}\boldsymbol{V}\boldsymbol{V}^{\dagger}\boldsymbol{v}\\ &= \boldsymbol{D}\boldsymbol{V}\boldsymbol{V}^{\dagger}\boldsymbol{V}\boldsymbol{\gamma}\\ &= \boldsymbol{D}\boldsymbol{V}\boldsymbol{\gamma}\\ &= \boldsymbol{D}\boldsymbol{v}.\end{aligned} \tag{9.138}$$

Substituting the above into the condition of Theorem 9.9 and performing manipulation similar to the ones in the proof of Theorem 9.10, we arrive at the result.

Theorem 9.11 Under the conditions of Theorem 9.10, every vertex $\boldsymbol{v}$ of H_n that is contained in Range($\boldsymbol{V}$)$^{\perp}$ such that for some $i = 1, \ldots, n$,

$$b_i v_i < 0$$

is not an equilibrium state of the gBSB net.

Proof It follows from the definition of the pseudo-inverse that if $\boldsymbol{V}^{\dagger}$ is a pseudo-inverse matrix of $\boldsymbol{V} \in \mathbb{R}^{n \times r}$, then there exists a square matrix $\boldsymbol{M}$ such that

$$\boldsymbol{V}^{\dagger} = \boldsymbol{M}\boldsymbol{V}^T. \tag{9.139}$$

Suppose that $\boldsymbol{v} \in$ Range($\boldsymbol{V}$)$^{\perp}$. Then, $\boldsymbol{V}^T\boldsymbol{v} = \boldsymbol{0}$, and therefore, by (9.139),

$$\boldsymbol{V}^{\dagger}\boldsymbol{v} = \boldsymbol{0}. \tag{9.140}$$

Hence, $\boldsymbol{W}\boldsymbol{v} = \boldsymbol{D}\boldsymbol{V}\boldsymbol{V}^{\dagger}\boldsymbol{v} = \boldsymbol{0}$, and

$$(\boldsymbol{W}\boldsymbol{v} + \boldsymbol{b})_i v_i = b_i v_i, \qquad i = 1, \ldots, n. \tag{9.141}$$

If for some i we have $b_i v_i < 0$, then

$$(\boldsymbol{W}\boldsymbol{v} + \boldsymbol{b})_i v_i < 0,$$

which implies that $\boldsymbol{v}$ is not an equilibrium point of the gBSB neural net.

The above results are clear and easy to apply to construct the weight matrix $\boldsymbol{W}$. However, if the diagonal elements of the matrix $\boldsymbol{D}$ are too large, we may end up storing all of the vertices of the hypercube H_n. This follows from the following result concerning the weight matrix.

Theorem 9.12 If the weight matrix $\boldsymbol{W}$ of the gBSB neural net is such that for all $i = 1, \ldots, n$,

$$w_{ii} > \sum_{\substack{k=1\\k\neq i}}^{n} |w_{ik}| + |b_i|,$$

then all the vertices of H_n are asymptotically stable equilibrium states of the gBSB net.

Proof Let $\boldsymbol{v}$ be an arbitrary vertex of H_n, that is, $\boldsymbol{v} \in \{-1, 1\}^n$. Then, using the assumption on the weight matrix, we obtain

$$\begin{aligned}(\boldsymbol{W}\boldsymbol{v} + \boldsymbol{b})_i v_i &= w_{ii} + \sum_{\substack{k=1\\k\neq i}}^{n} w_{ik} v_k v_i + b_i v_i \\ &\geq w_{ii} - \sum_{\substack{k=1\\k\neq i}}^{n} |w_{ik}| - |b_i| \\ &> 0.\end{aligned}$$

It follows from Theorem 9.9 that all vertices of H_n are asymptotically stable states of the gBSB net.

We now provide an "improved" method for constructing the weight matrix. The method is based on the following theorem.

Theorem 9.13 Let $\boldsymbol{V} = [\boldsymbol{v}^{(1)} \ \cdots \ \boldsymbol{v}^{(r)}] \in \{-1, 1\}^{n\times r}$. Let $\boldsymbol{B} = [\boldsymbol{b} \ \cdots \ \boldsymbol{b}] \in \mathbb{R}^{n\times r}$ be the matrix consisting of the column vector $\boldsymbol{b}$ repeated r times, $\boldsymbol{\Lambda} \in \mathbb{R}^{n\times n}$, $\beta_j = \sum_{i=1}^{r} (\boldsymbol{V}^{\dagger}\boldsymbol{v}^{(j)})_i$, and $\beta = -\max_j |\beta_j|$, where $j = 1, 2, \ldots, r$. Suppose

$$d_{ii} > \sum_{\substack{k=1\\k\neq i}}^{n} |d_{ik}| + |b_i(1-\beta)|, \qquad i = 1, 2, \ldots, n.$$

If

$$\boldsymbol{W} = (\boldsymbol{D}\boldsymbol{V} - \boldsymbol{B})\boldsymbol{V}^{\dagger} + \boldsymbol{\Lambda}(\boldsymbol{I}_n - \boldsymbol{V}\boldsymbol{V}^{\dagger}), \tag{9.142}$$

then all the patterns $\boldsymbol{v}^{(j)}$, $j = 1, 2, \ldots, r$, are stored as asymptotically stable equilibrium states of the gBSB neural net.

Proof Using the first property of the Moore–Penrose inverse of $\boldsymbol{V}$, we obtain

$$\begin{aligned}\boldsymbol{W}\boldsymbol{V} &= (\boldsymbol{D}\boldsymbol{V} - \boldsymbol{B})\boldsymbol{V}^{\dagger}\boldsymbol{V} + \boldsymbol{\Lambda}(\boldsymbol{I}_n - \boldsymbol{V}\boldsymbol{V}^{\dagger})\boldsymbol{V} \\ &= \boldsymbol{D}\boldsymbol{V} - \boldsymbol{B}\boldsymbol{V}^{\dagger}\boldsymbol{V} \\ &= \boldsymbol{D}\boldsymbol{V} - \left[\boldsymbol{B}\boldsymbol{V}^{\dagger}\boldsymbol{v}^{(1)} \quad \boldsymbol{B}\boldsymbol{V}^{\dagger}\boldsymbol{v}^{(2)} \quad \cdots \quad \boldsymbol{B}\boldsymbol{V}^{\dagger}\boldsymbol{v}^{(r)}\right] \\ &= \boldsymbol{D}\boldsymbol{V} - [\boldsymbol{b}\beta_1 \quad \boldsymbol{b}\beta_2 \quad \cdots \quad \boldsymbol{b}\beta_r].\end{aligned}$$

The jth column of the above is

$$\boldsymbol{W}\boldsymbol{v}^{(j)} = \boldsymbol{D}\boldsymbol{v}^{(j)} - \boldsymbol{b}\beta_j, \qquad j = 1, 2, \ldots, r.$$

We now show that a sufficient condition for a pattern $v^{(j)}$ to be an asymptotically stable equilibrium of $x(k+1) = T(x(k))$ is satisfied. Performing simple manipulations and taking into account the assumptions, for $j = 1, 2, \ldots, r$ we obtain

$$\begin{aligned}
(Wv^{(j)} + b)_i v_i^{(j)} &= (Dv^{(j)} - b\beta_j + b)_i v_i^{(j)} \\
&= d_{ii} + \sum_{\substack{k=1 \\ k\neq i}}^{n} d_{ik} v_k^{(j)} v_i^{(j)} + b_i(1 - \beta_j) v_i^{(j)} \\
&\geq d_{ii} - \sum_{\substack{k=1 \\ k\neq i}}^{n} |d_{ik}| - \left| b_i \left(1 - \left(-\max_j |\beta_j|\right)\right)\right| \\
&= d_{ii} - \sum_{\substack{k=1 \\ k\neq i}}^{n} |d_{ik}| - |b_i(1 - \beta)| \\
&> 0,
\end{aligned}$$

where the last inequality follows from the assumption on the matrix D. Hence, by Theorem 9.9, the given patterns are asymptotically stable equilibrium states of the gBSB model.

If the patterns are linearly independent and $r < n$, then rank $V = r$ and therefore $V^\dagger V = I_r$. In this case $\beta_j = 1$ for all j and

$$Wv^{(j)} = Dv^{(j)} - b, \qquad j = 1, 2, \ldots, r,$$

Thus in this case, a sufficient condition for the patterns $v^{(j)}$, $j = 1, 2, \ldots, r$, to be stored as asymptotically stable equilibrium states of the gBSB model is that for $i = 1, 2, \ldots, n$,

$$d_{ii} > \sum_{\substack{k=1 \\ k\neq i}}^{n} |d_{ik}|.$$

The above condition indeed guarantees that all the desired patterns are stored as asymptotically stable equilibria. As we mentioned before, if d_{ii}'s are too large, then we may end up storing many other vertices of H_n in addition to the given patterns that we want to be stored. To minimize the number of spurious patterns stored, we impose the constraints of the form

$$d_{ii} < \sum_{\substack{k=1 \\ k\neq i}}^{n} |d_{ik}| + |b_i|, \qquad i = 1, 2, \ldots, n,$$

on the matrix D while synthesizing the weight matrix W.

We now discuss the role of different terms in the expression (9.142) on the weight matrix in storing the desired patterns and minimizing the number of spurious states. First, the term $(DV - B)V^\dagger$ is responsible for the given patterns to be stored as asymptotically stable equilibrium states of the gBSB model. This is because for a given pattern, say $v^{(j)}$, we have $\Lambda(I_n - VV^\dagger)v^{(j)} = 0$, where 0 is a vector of zeros. The role of the term $\Lambda(I_n - VV^\dagger)$ is to destabilize patterns that are not in Span$[v^{(1)} \ \cdots \ v^{(r)}]$. Specifically, if a pattern y is orthogonal to Span$[\boldsymbol{v}^{(1)} \quad \cdots \quad \boldsymbol{v}^{(r)}]$, then the term $\boldsymbol{\Lambda}(\boldsymbol{I}_n - \boldsymbol{V}\boldsymbol{V}^{\dagger})$ will not only destabilize the vertex of H_n corresponding to the pattern $\boldsymbol{y}$, if it is an equilibrium state of the gBSB model, it will make this state a nonequilibrium of the gBSB model. To see this, consider the expression

$$\boldsymbol{W}\boldsymbol{y} = (\boldsymbol{D}\boldsymbol{V} - \boldsymbol{B})\boldsymbol{V}^{\dagger}\boldsymbol{y} + \boldsymbol{\Lambda}(\boldsymbol{I}_n - \boldsymbol{V}\boldsymbol{V}^{\dagger})\boldsymbol{y}.$$

Because

$$\boldsymbol{y} \perp \text{Span}\left[\boldsymbol{v}^{(1)} \quad \cdots \quad \boldsymbol{v}^{(r)}\right],$$

we have

$$\boldsymbol{W}\boldsymbol{y} = \boldsymbol{\Lambda}(\boldsymbol{I}_n - \boldsymbol{V}\boldsymbol{V}^{\dagger})\boldsymbol{y} = \boldsymbol{\Lambda}\boldsymbol{y}.$$

Therefore,

$$(\boldsymbol{W}\boldsymbol{y} + \boldsymbol{b})_i y_i = (\boldsymbol{\Lambda}\boldsymbol{y} + \boldsymbol{b})_i y_i = \lambda_{ii} + \sum_{\substack{j=1\\ j\neq i}}^{n} \lambda_{ij} y_j y_i + b_i y_i.$$

Hence, if we choose $\boldsymbol{\Lambda}$ such that for some $i = 1, 2, \ldots, n$,

$$\lambda_{ii} < -\sum_{\substack{j=1\\ j\neq i}}^{n} |\lambda_{ij}| - |b_i|,$$

then

$$(\boldsymbol{W}\boldsymbol{y} + \boldsymbol{b})_i y_i < 0 \qquad \text{for some } i = 1, 2, \ldots, n.$$

Therefore, the vertex $\boldsymbol{y}$ cannot be an equilibrium state. The inclusion of the matrix $\boldsymbol{B}$ in the synthesis formula for the weight matrix $\boldsymbol{W}$ was to prevent an automatic storage of the negatives of the given patterns as well as to somehow help to direct the gBSB model trajectories toward the given patterns. For this reason, we select $\boldsymbol{b}$ to be in Span$\{\boldsymbol{V}\}$. We recommend making $\boldsymbol{b}$ a weighted sum of $\boldsymbol{v}^{(i)}$, $i = 1, 2, \ldots, r$, with positive weight coefficients. The above results can be used as guides for selecting the components of the matrices $\boldsymbol{D}$, $\boldsymbol{\Lambda}$, the vector $\boldsymbol{b}$, and consequently the matrix $\boldsymbol{B}$. We summarize the above considerations in the form of an algorithm for synthesizing the weight matrix $\boldsymbol{W}$.

WEIGHT MATRIX CONSTRUCTION ALGORITHM

Given: $\boldsymbol{v}^{(j)}$, $j = 1, 2, \ldots, r$.

STEP 1 Form $\boldsymbol{B} = [\boldsymbol{b} \quad \boldsymbol{b} \quad \cdots \quad \boldsymbol{b}]$, where

$$\boldsymbol{b} = \sum_{j=1}^{r} \epsilon_j \boldsymbol{v}^{(j)}, \qquad \epsilon_j > 0, \quad j = 1, 2, \ldots, r.$$

STEP 2 Select $\boldsymbol{D}$ such that for $i = 1, 2, \ldots, n$ we have

$$d_{ii} > \sum_{\substack{k=1\\ k\neq i}}^{n} |d_{ik}| \quad \text{and} \quad d_{ii} < \sum_{\substack{k=1\\ k\neq i}}^{n} |d_{ik}| + |b_i|.$$

STEP 3 Choose $\boldsymbol{\Lambda}$ such that for $i = 1, 2, \ldots, n$,

$$\lambda_{ii} < -\sum_{\substack{j=1 \\ j\neq i}}^{n} |\lambda_{ij}| - |b_i|.$$

STEP 4 Compute

$$\boldsymbol{W} = (\boldsymbol{D}\boldsymbol{V} - \boldsymbol{B})\boldsymbol{V}^{\dagger} + \boldsymbol{\Lambda}(\boldsymbol{I}_n - \boldsymbol{V}\boldsymbol{V}^{\dagger}).$$

STEP 5 Identify spurious equilibrium patterns. You can use the condition of Theorem 9.7 for this purpose.

When the set of patterns to be stored is not linearly independent, we can still use the preceding algorithm to compute the weight matrix $\boldsymbol{W}$. However, the second constraint on the matrix $\boldsymbol{D}$ may lead to a violation of the condition on $\boldsymbol{D}$ in Theorem 9.13. In such a case we cannot guarantee that all the given patterns will be stored, nor we will have any control on the number of spurious states. We may then relax this condition. If attempts in synthesizing associative memory for a set of linearly dependent patterns result in unacceptable designs, we may try a different approach. We can start by picking a maximal set of patterns that are linearly independent from the given set of patterns. Let us assume that there are s patterns to be stored and that r of them form a maximal linearly independent subset. Suppose, for notational simplicity, that the patterns $\boldsymbol{v}^{(q)}$, $q = 1, \ldots, r$, form a linearly independent set. We then apply the above algorithm to generate the weight matrix $\boldsymbol{W}$ based on the linearly independent patterns $\boldsymbol{v}^{(q)}$, $q = 1, \ldots, r$. After that, we modify the obtained weight matrix and the vector $\boldsymbol{b}$ to ensure that the remaining patterns $\boldsymbol{v}^{(j)}$, $j = r+1, r+2, \ldots, s$, will also be stored. We will focus our attention on ensuring that the pattern $\boldsymbol{v}^{(r+1)}$ is stored as an asymptotically stable state. The remaining patterns $\boldsymbol{v}^{(j)}$, $j = r+2, r+3, \ldots, s$, can be stored using the same procedure. Because

$$\boldsymbol{v}^{(r+1)} \in \operatorname{Span}\{\boldsymbol{v}^{(1)}, \boldsymbol{v}^{(2)}, \ldots, \boldsymbol{v}^{(r)}\},$$

that is, $\boldsymbol{v}^{(r+1)}$ is a linear combination of $\boldsymbol{v}^{(i)}$, $i = 1, 2, \ldots, r$, it can be represented as

$$\boldsymbol{v}^{(r+1)} = \sum_{q=1}^{r} \gamma_q \boldsymbol{v}^{(q)} = \boldsymbol{V}\boldsymbol{\gamma},$$

where $\boldsymbol{\gamma} = [\gamma_1 \ \cdots \ \gamma_r]^T \in \mathbb{R}^n$. Then, using (9.142), we obtain

$$\begin{aligned}\boldsymbol{W}\boldsymbol{v}^{(r+1)} &= \boldsymbol{D}\boldsymbol{V}\boldsymbol{\gamma} - \boldsymbol{B}\boldsymbol{\gamma} \\ &= \boldsymbol{D}\boldsymbol{v}^{(r+1)} - \left(\sum_{q=1}^{r}\gamma_q\right)\boldsymbol{b}.\end{aligned}$$

Hence, if

$$\begin{aligned}(\boldsymbol{W}\boldsymbol{v}^{(r+1)} + \boldsymbol{b})v_i^{(r+1)} &= \left(\boldsymbol{D}\boldsymbol{v}^{(r+1)} - \left[\sum_{q=1}^{r}\gamma_q - 1\right]\boldsymbol{b}\right)_i v_i^{(r+1)} \\ &> 0, \qquad i = 1, 2, \ldots, n,\end{aligned}$$

then the vertex $\boldsymbol{v}^{(r+1)}$ is asymptotically stable. Observe that because of the assumptions on the matrix $\boldsymbol{D}$, we have

$$\operatorname{sign}\left(\boldsymbol{D}\boldsymbol{v}^{(r+1)}\right)_i = \operatorname{sign}\left(v_i^{(r+1)}\right).$$

With the above in mind, we obtain stronger sufficient conditions for $\boldsymbol{v}^{(r+1)}$ to be asymptotically stable. They are

$$\operatorname{sign}\left[\left(1-\sum_{q=1}^{r}\gamma_q\right)\boldsymbol{b}\right]_i = \operatorname{sign}\left(v_i^{(r+1)}\right), \qquad i = 1, 2, \ldots, n.$$

We can represent the above conditions as

$$\operatorname{sign} b_i = \operatorname{sign}\left(v_i^{(r+1)}\right)\operatorname{sign}\left(1-\sum_{q=1}^{r}\gamma_q\right), \qquad i = 1, 2, \ldots, n.$$

Thus, if we adjust the components of $\boldsymbol{b}$ so that the above conditions are satisfied, then we are guaranteed that $\boldsymbol{v}^{(r+1)}$ is stored as an asymptotically stable equilibrium state. We may try to further adjust the components of $\boldsymbol{b}$, if possible, to store the remaining patterns. Then, using the same matrices $\boldsymbol{D}$, $\boldsymbol{\Lambda}$, and $\boldsymbol{V}$ used to compute the original weight matrix $\boldsymbol{W}$, we can choose an appropriate $\boldsymbol{b}$ to generate a new weight matrix $\boldsymbol{W}$.

If $\boldsymbol{b}$ cannot be adjusted, employing the above technique, to accommodate all the patterns, we may then have to use the following alternative procedure. First, recall the stability conditions that the patterns $\boldsymbol{v}^{(r+1)}, \ldots, \boldsymbol{v}^{(s)}$ should satisfy:

$$\left(\boldsymbol{D}\boldsymbol{v}^{(k)} - \left[\sum_{q=1}^{r}\gamma_q - 1\right]\boldsymbol{b}\right)_i v_i^{(k)} > 0, \qquad i = 1, 2, \ldots, n, \quad k = r+1, r+2, \ldots, s.$$

Using the requirement for the matrix $\boldsymbol{D}$ to be diagonally dominant, we arrive at stronger conditions of the form

$$d_{ii} - \sum_{\substack{j=1\\ j\neq i}}^{n}|d_{ij}| + \left(1-\sum_{q=1}^{r}\gamma_q\right)b_i v_i^{(k)} > 0, \qquad i = 1, \ldots, n, \quad k = r+1, \ldots, s.$$

Thus, we may use the above conditions and try to decrease the magnitude of the b_i's until the conditions are met for all $k = r+1, \ldots, s$. However, we are not guaranteed to always satisfy the above conditions. If the conditions are satisfied, then employing the same matrices $\boldsymbol{D}$, $\boldsymbol{\Lambda}$, and $\boldsymbol{V}$ used to compute the original weight matrix $\boldsymbol{W}$, we choose an appropriate $\boldsymbol{b}$ to generate a new weight matrix $\boldsymbol{W}$.

◆ Example 9.10

We synthesize a gBSB-based associative memory to store the following three patterns:

$$v^{(1)} = \begin{bmatrix} -1 \\ -1 \\ 1 \\ 1 \\ 1 \end{bmatrix}, \quad v^{(2)} = \begin{bmatrix} 1 \\ 1 \\ -1 \\ -1 \\ -1 \end{bmatrix}, \quad \text{and} \quad v^{(3)} = \begin{bmatrix} 1 \\ -1 \\ 1 \\ -1 \\ 1 \end{bmatrix}.$$

We first choose matrices $\boldsymbol{D}$ and $\boldsymbol{\Lambda}$. Let

$$\boldsymbol{D} = \begin{bmatrix} 1 & 0 & 0 & 0 & -0.5 \\ 0 & 1 & 0 & 0 & 0 \\ 0 & -0.2 & 1 & 0 & 0 \\ 0 & 0 & 0 & 1 & 0 \\ 0 & 0 & 0 & 0 & 1 \end{bmatrix}, \quad \boldsymbol{\Lambda} = -2\boldsymbol{I}_5.$$

Set $\boldsymbol{V} = [v^{(1)} \ v^{(2)} \ v^{(3)}]$ and $\boldsymbol{B} = [\boldsymbol{b} \ \boldsymbol{b} \ \boldsymbol{b}]$, where

$$\boldsymbol{b} = [0.6 \quad -0.5 \quad 0.7 \quad -0.4 \quad 0.8]^T.$$

Note that rank $\boldsymbol{V} = 2$. Let the step size $\alpha = 0.9$. Using the above, we compute the weight matrix $\boldsymbol{W}$, where

$$\boldsymbol{W} = \begin{bmatrix} -0.6500 & 0.2667 & -0.2667 & -1.3500 & -0.2667 \\ 0.1250 & -1.0833 & -0.9167 & -0.1250 & -0.9167 \\ -0.1750 & -0.9500 & -1.0500 & 0.1750 & 0.9500 \\ -1.4000 & -0.0667 & 0.0667 & -0.6000 & 0.0667 \\ -0.2000 & -0.8667 & 0.8667 & 0.2000 & -1.1333 \end{bmatrix}.$$

We then check which vertices of H_5 satisfy the sufficiency condition of Theorem 9.8 for asymptotic stability. In other words, we wish to identify the spurious states. In this example, only the given patterns satisfy the condition of Theorem 9.8.

◆ Example 9.11

In this example, we use a gBSB neural net with the state in the hypercube $\{-1, 1\}^{25}$. We construct the weight matrix to store the four letters of alphabet whose bit maps are shown in Figure 9.44. The bit maps are 5×5 with elements from the set $\{-1, 1\}$, with -1 corresponding to black and 1 to white. The pattern vectors are obtained from the bit maps by stacking their columns. For example, if $A = [\boldsymbol{a}_1 \ \boldsymbol{a}_2 \ \boldsymbol{a}_3 \ \boldsymbol{a}_4 \ \boldsymbol{a}_5]$, where $\boldsymbol{a}_i$, $i = 1, 2, \ldots, 5$, is the ith column of A, then the corresponding pattern vector is obtained from A by applying the stacking

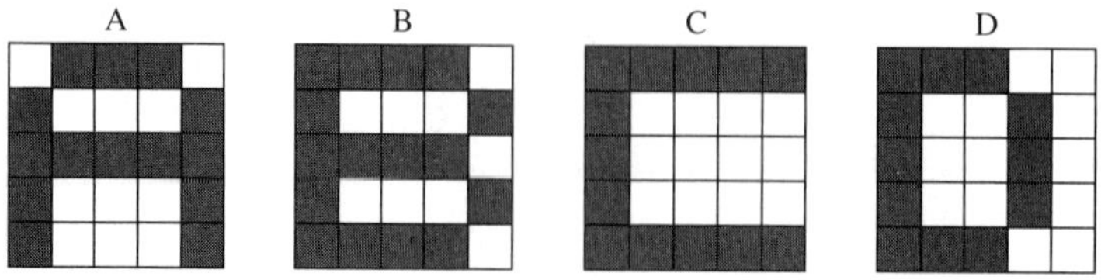

Figure 9.44 Bit maps of the characters A, B, C, and D stored in the gBSB neural net of Example 9.11.

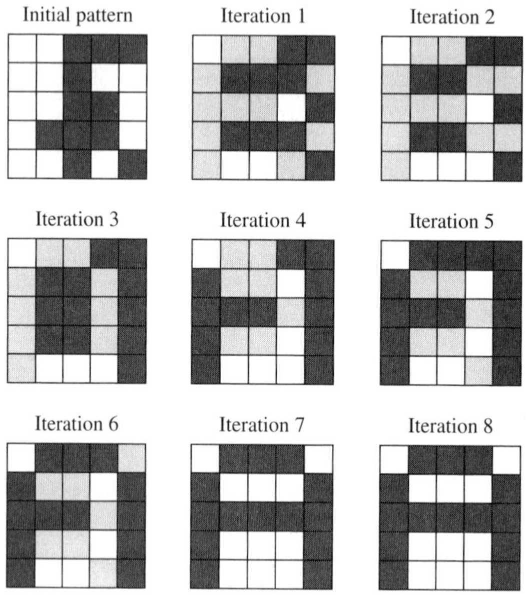

Figure 9.45 Snapshots of the iterations of the gBSB net of Example 9.11 converging toward the letter A.

operator $\boldsymbol{s} : \mathbb{R}^{5\times 5} \to \mathbb{R}^{25}$ to get

$$\boldsymbol{s}(A) = A(:) = \begin{bmatrix} \boldsymbol{a}_1 \\ \vdots \\ \boldsymbol{a}_5 \end{bmatrix},$$

where $A(:)$ is the MATLAB's command for the stacking operation on the matrix A. The weight matrix, $\boldsymbol{W}$, of the gBSB net was constructed using the weight matrix construction algorithm, where

$$\boldsymbol{V} = [A(:) \quad B(:) \quad C(:) \quad D(:)].$$

The values of other parameters chosen in this example were: $\boldsymbol{D} = 1.1\,\boldsymbol{I}_{25}$, $\boldsymbol{\Lambda} = -2\,\boldsymbol{I}_{25}$, $\boldsymbol{b} = 0.3\,A(:) + 0.1\,B(:)$, and $\alpha = 0.5$. Snapshots of the network iterates for different initial patterns are shown in Figures 9.45–9.48.

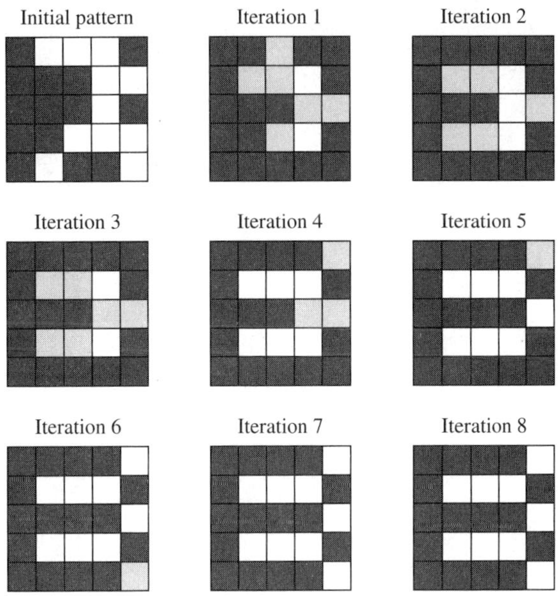

Figure 9.46 Snapshots of the iterations of the gBSB net of Example 9.11 converging toward the letter *B*.

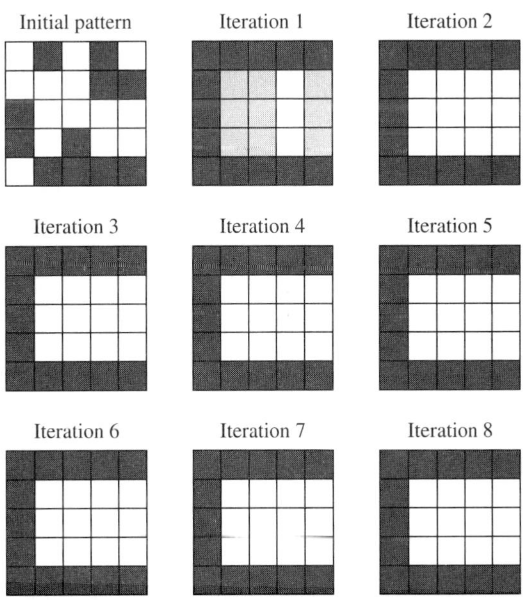

Figure 9.47 Snapshots of the iterations of the gBSB net of Example 9.11 converging toward the letter *C*.

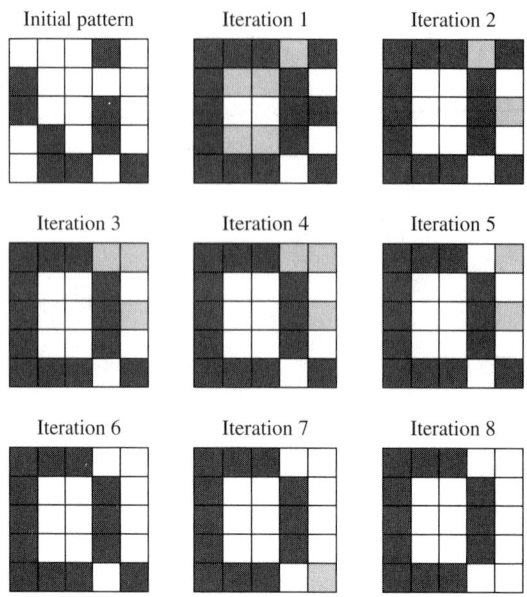

Figure 9.48 Snapshots of the iterations of the gBSB net of Example 9.11 converging toward the letter D.

9.9.4 Learning

A pattern is learned by a memory if its noisy or incomplete version presented to the memory is mapped back to this pattern. In our discussion of the proposed iterative learning, we use the following theorem, which allows one to iteratively compute the pseudo-inverse of a given matrix. The theorem can be found in Campbell and Meyer [40, p. 54].

Theorem 9.14 Let $\boldsymbol{U} \in \mathbb{R}^{n\times(r+1)}$ be partitioned by slicing off its last column, so that

$$\boldsymbol{U} = \left[\boldsymbol{V} \;\vdots\; \boldsymbol{y}\right].$$

Let

$$\begin{aligned} \boldsymbol{k} &= \boldsymbol{V}^{\dagger}\boldsymbol{y}, \\ \boldsymbol{u} &= (\boldsymbol{I}_n - \boldsymbol{V}\boldsymbol{V}^{\dagger})\,\boldsymbol{y}, \\ \boldsymbol{z} &= \begin{cases} \boldsymbol{u}^{\dagger} & \text{if } \boldsymbol{u} \neq \boldsymbol{0}, \\ \dfrac{\boldsymbol{k}^T\boldsymbol{V}^{\dagger}}{1+\boldsymbol{k}^T\boldsymbol{k}} & \text{if } \boldsymbol{u} = \boldsymbol{0}. \end{cases} \end{aligned}$$

Then,

$$\boldsymbol{U}^{\dagger} = \begin{bmatrix} \boldsymbol{V}^{\dagger} - \boldsymbol{k}\boldsymbol{z} \\ \boldsymbol{z} \end{bmatrix}.$$

Learning a new pattern requires updating the weight matrix $\boldsymbol{W}$. This, in turn, involves augmenting the matrices $\boldsymbol{V}$ and $\boldsymbol{B}$ and recomputing $\boldsymbol{V}^{\dagger}$. Augmenting $\boldsymbol{V}$ and $\boldsymbol{B}$ is straightforward, and updating $V^\dagger$ can be performed recursively using the above theorem. The above procedure guarantees that the new pattern will be stored as an asymptotically stable equilibrium state if it is linearly independent of the learned patterns. If, however, the pattern to be learned is a linear combination of the stored patterns, then we do not update the matrix V but rather modify the vector b as described at the end of the previous subsection. After that we update the weight matrix W. We now illustrate the iterative learning process on a numerical example.

◆ Example 9.12

Consider Example 9.10. Suppose that we wish to add

$$v^{(4)} = [-1 \quad -1 \quad -1 \quad -1 \quad 1]^T$$

to the previously designed memory. We augment the pattern matrix to include $v^{(4)}$ in it. Thus, we get $V = [v^{(1)} \; v^{(2)} \; v^{(3)} \; v^{(4)}]$. We also augment the matrix B by adding extra b column to it. We then update $V^\dagger$ using Theorem 9.14 and recompute the weight matrix W to obtain

$$W = \begin{bmatrix} 0.1714 & 0.8143 & 0.8286 & -0.5286 & -0.8143 \\ 0.4643 & -0.8571 & -0.4643 & 0.2143 & -1.1429 \\ 0.7786 & -0.3143 & 0.2214 & 1.1286 & 0.3143 \\ -0.8286 & 0.3143 & 0.8286 & -0.0286 & -0.3143 \\ -0.4857 & -1.0571 & 0.4857 & -0.0857 & -0.9429 \end{bmatrix}.$$

Next, we check which vertices of H_5 satisfy the condition of Theorem 9.8 for asymptotic stability. We find that only the given four patterns satisfy the condition of Theorem 9.8.

9.9.5 Forgetting

A previously stored pattern is forgotten or deleted from the memory if a stimulus that is a perturbed version of that pattern is presented to the memory and is not mapped back to the stored pattern. We now discuss the case when we wish to update the weight matrix in such a way that the specified patterns are removed from memory. We refer to the process of removing patterns from the associative memory as forgetting. Our discussion uses the following theorem, which can be considered as the inverse of Theorem 9.14. A general version of the theorem, appears in Boullion and Odell [31, p. 14].

Theorem 9.15 Let $U \in \mathbb{R}^{n\times(r+1)}$ be partitioned by slicing off its last column, so that

$$U = [V \;\vdots\; y],$$

and let its pseudo-inverse $U^\dagger \in \mathbb{R}^{(r+1)\times n}$ be partitioned by slicing off its last row, so that

$$U^\dagger = \begin{bmatrix} S \\ z \end{bmatrix}.$$

Then,

$$\boldsymbol{V}^{\dagger} = \begin{cases} \boldsymbol{S} + \boldsymbol{S}\frac{\boldsymbol{yz}}{1-\boldsymbol{zy}} & \text{if } \boldsymbol{zy} \neq 1 \\ \boldsymbol{S} - \boldsymbol{S}\boldsymbol{z}^{\dagger}\boldsymbol{z} & \text{if } \boldsymbol{zy} = 1. \end{cases}$$

Proof Referring to Theorem 9.14, we first show that $\boldsymbol{u} = \boldsymbol{0}$ if and only if $\boldsymbol{zy} \neq 1$, or, equivalently, $\boldsymbol{u} \neq \boldsymbol{0}$ if and only if $\boldsymbol{zy} = 1$. Indeed, if $\boldsymbol{u} = \boldsymbol{0}$, then

$$\boldsymbol{z} = \frac{\boldsymbol{k}^T\boldsymbol{V}^{\dagger}}{1 + \boldsymbol{k}^T\boldsymbol{k}}.$$

Hence,

$$\boldsymbol{zy} = \frac{\boldsymbol{k}^T\boldsymbol{V}^{\dagger}\boldsymbol{y}}{1 + \boldsymbol{k}^T\boldsymbol{k}} = \frac{\boldsymbol{k}^T\boldsymbol{k}}{1 + \boldsymbol{k}^T\boldsymbol{k}} < 1.$$

On the other hand, if $\boldsymbol{u} \neq \boldsymbol{0}$, then

$$\boldsymbol{z} = \boldsymbol{u}^{\dagger} = \{(\boldsymbol{I}_n - \boldsymbol{V}\boldsymbol{V}^{\dagger})\,\boldsymbol{y}\}^{\dagger}.$$

Taking into account the fact that $\boldsymbol{u} \neq \boldsymbol{0}$ and that $\boldsymbol{u} \in \mathbb{R}^n$, its pseudo-inverse is

$$\boldsymbol{u}^{\dagger} = (\boldsymbol{u}^T\boldsymbol{u})^{-1}\boldsymbol{u}^T = (\boldsymbol{y}^T(\boldsymbol{I}_n - \boldsymbol{V}\boldsymbol{V}^{\dagger})\,\boldsymbol{y})^{-1}\,\boldsymbol{y}^T(\boldsymbol{I}_n - \boldsymbol{V}\boldsymbol{V}^{\dagger}),$$

because $(\boldsymbol{I}_n - \boldsymbol{V}\boldsymbol{V}^{\dagger})$ is an orthogonal projector and hence

$$(\boldsymbol{I}_n - \boldsymbol{V}\boldsymbol{V}^{\dagger})^2 = (\boldsymbol{I}_n - \boldsymbol{V}\boldsymbol{V}^{\dagger}).$$

Therefore, $\boldsymbol{zy} = \boldsymbol{u}^{\dagger}\boldsymbol{y} = 1$.

We now prove the expressions for $\boldsymbol{V}^{\dagger}$. We start with the case when $\boldsymbol{zy} \neq 1$. It follows from Theorem 9.14 that

$$\boldsymbol{S} = \boldsymbol{V}^{\dagger} - \boldsymbol{V}^{\dagger}\boldsymbol{yz} = \boldsymbol{V}^{\dagger}(\boldsymbol{I}_n - \boldsymbol{yz}). \tag{9.143}$$

Because $\boldsymbol{zy} \neq 1$, the matrix $(\boldsymbol{I}_n - \boldsymbol{yz})$ is invertible. Using the matrix inversion formula (A.30) from Section A.3 gives

$$(\boldsymbol{I}_n - \boldsymbol{yz})^{-1} = \boldsymbol{I}_n + \frac{\boldsymbol{yz}}{1 - \boldsymbol{zy}}. \tag{9.144}$$

Applying formula (9.144) to (9.143) yields

$$\boldsymbol{V}^{\dagger} = \boldsymbol{S} + \boldsymbol{S}\frac{\boldsymbol{yz}}{1 - \boldsymbol{zy}}.$$

We now prove the expression for $\boldsymbol{V}^{\dagger}$ for the case when $\boldsymbol{zy} = 1$. It follows from Theorem 9.14 that

$$\boldsymbol{S} = \boldsymbol{V}^{\dagger} - \boldsymbol{kz}, \tag{9.145}$$

and hence, because $\boldsymbol{z} \neq \boldsymbol{0}$,

$$\boldsymbol{S}\boldsymbol{z}^{\dagger} = \boldsymbol{V}^{\dagger}\boldsymbol{z}^{\dagger} - \boldsymbol{k}. \tag{9.146}$$

But in this case, $\boldsymbol{u} \neq \boldsymbol{0}$. Therefore,

$$\boldsymbol{z}^{\dagger} = (\boldsymbol{u}^{\dagger})^{\dagger} = \boldsymbol{u} = (\boldsymbol{I}_n - \boldsymbol{V}\boldsymbol{V}^{\dagger})\,\boldsymbol{y}.$$

Substituting the above expression for $\boldsymbol{z}^{\dagger}$ into (9.146), we get

$$\begin{aligned} \boldsymbol{S}\boldsymbol{z}^{\dagger} &= \boldsymbol{V}^{\dagger}(\boldsymbol{I}_n - \boldsymbol{V}\boldsymbol{V}^{\dagger})\,\boldsymbol{y} - \boldsymbol{k} \\ &= (\boldsymbol{V}^{\dagger} - \boldsymbol{V}^{\dagger})\,\boldsymbol{y} - \boldsymbol{k} \\ &= -\boldsymbol{k}. \end{aligned}$$

We now substitute the above into (9.145) to obtain $V^{\dagger} = S - Sz^{\dagger}z$, which completes the proof of the theorem.

We can delete a pattern from the memory by first removing it from V, then updating $V^{\dagger}$ using Theorem 9.15, removing one column from B, and recomputing the weight matrix W. We illustrate the iterative forgetting process with the following numerical example.

◆ Example 9.13

We consider Example 9.12. Suppose that we wish to delete

$$v^{(3)} = [1 \quad -1 \quad 1 \quad -1 \quad 1]^T$$

from the previously designed memory. We can use Theorem 9.15 to update $V^{\dagger}$—that is, to find the pseudo-inverse of $V = [v^{(1)} \ v^{(2)} \ v^{(4)}]$. We also update the matrix B by deleting from it one column. We leave the remaining data unchanged. Recomputing the weight matrix W gives

$$W = \begin{bmatrix} -0.7333 & 1.2667 & 0.1500 & 0.1500 & -1.2667 \\ 0.9167 & -1.0833 & -0.1250 & -0.1250 & -0.9167 \\ 0.0500 & 0.0500 & -0.3250 & 1.6750 & -0.0500 \\ -0.0667 & -0.0667 & 1.4000 & -0.6000 & 0.0667 \\ -0.8667 & -0.8667 & 0.2000 & 0.2000 & -1.1333 \end{bmatrix}.$$

We then check which vertices of H_5 satisfy the condition of Theorem 9.8 for asymptotic stability. We find that only $v^{(1)}$, $v^{(2)}$, and $v^{(4)}$ satisfy the condition of Theorem 9.8, which is what we desire.

Notes

This chapter gives an introduction to computing using artificial neural networks from a dynamical system perspective. The disclaimer that we found in the book by Hertz, Krogh, and Palmer [120, p. xix] also applies to this chapter as well as to the whole field of artificial neural networks: "The adjective 'neural' is used because much of the inspiration for such networks comes from neuroscience, *not* because we are concerned with networks of real neurons. Brain modeling is a different field and, though we sometimes describe biological analogies, our prime concern is with what *artificial* networks can do, and why." The readers interested in principles underlying the organization of biological neurons into circuits and networks within different regions of the brain may consult the book edited by Shepherd [257]. For textbooks with an engineering approach to neural networks, we recommend Hassoun [116], Hagan, Demuth, and Beale [110], Zurada [320], Haykin [118], Fausett [79], Khanna [158], or Harvey [115]. Applications of artificial neural networks to signal processing and solving optimization problems are presented in the book by Cichocki and Unbehauen [52]. A psychological slant to the subject is presented in the book by Rumelhart, McClelland, and the PDP Research Group [247], and Anderson [8]. Impatient readers wishing to get a feel of artificial neural networks may want to consult a tutorial by Lippmann [186] and by Hush and Horne [137].

According to Haykin [118, p. 38], "The modern area of neural networks began with the pioneering work of McCulloch and Pitts (1943). McCulloch was a psychiatrist and neuroanatomist by training; he spent some 20 years thinking about the representation of an event in the nervous system. Pitts was a mathematical prodigy, who joined McCulloch in 1942." In their paper [200], McCulloch and Pitts, assuming that the activity of the neuron is an "all-or-none" process, developed a logical calculus of neural events. They argued that, with addition of memory, a neural network consisting of a sufficiently large number of their neurons can compute anything that can be computed using the principles of logic. McCulloch and Pitts' paper influenced John von Neumann, who used idealized switch-delay elements derived from the McCulloch–Pitts neuron model in the construction of the EDVAC (Electronic Discrete Variable Automatic Computer), which was developed out of the ENIAC (Electronic Numerical Integrator and Computer). The ENIAC was the first general-purpose computer, and it was invented by J. Presper Eckert and John Mauchly.

More recent investigators of the computational processes in the nervous system compare the brain to a hybrid, analog–digital computer [163]. According to Koch [163, p. 210], individual nerve cells communicate between each other via streams of digital pulses. Information is, on the other hand, processed in the analog domain. The resulting signal is then converted back into digital pulses and conveyed to the neurons.

For other analog learning algorithms for adaptive linear elements (adalines) in the spirit of Section 9.2, see Sira-Ramírez, Colina-Morles, and Rivas-Echeverria [259].

As we mentioned before, the Brain-State-in-a-Box (BSB) neural model was proposed by Anderson et al. in 1977 [9]. The dynamics of the BSB model and its variants were further analyzed by Greenberg [107], Golden [102, 103], Grossberg [108], Hui and Żak [130], Hakl [113], Hui, Lillo, and Żak [129], Schultz [254], Lillo et al. [184], Anderson [8], Perfetti [229], Hassoun [116], Żak, Lillo, and Hui [312], Chan and Żak [44, 45], and Park, Cho, and Park [220]. A continuous counterpart of the BSB model referred to as the linear system operating in a saturated mode was analyzed in Li, Michel, and Porod [183].

One of the first techniques used to design neural network associative memories is the *outer product method*. This method was first used by Hopfield [125, 126]. The outer product method results in a network with a symmetric interconnection structure that does not necessarily store the desired patterns as equilibrium states of the network. The network designed using this method can efficiently store up to $0.15n$ arbitrary patterns and has learning capabilities. The *projection learning rule* proposed by Personnaz, Guyon, and Dreyfus [230, 231] produces a network with a symmetric interconnection structure that always stores the desired patterns as equilibrium states, although not necessarily asymptotically stable. The network designed with this method can effectively store up to $0.5n$ desired patterns and has learning capabilities. The *eigenstructure method* used by Li, Michel, and Porod [183] yields a neural network associative memory capable of storing any given pattern as an asymptotically stable equilibrium state. The network designed with this method is capable of effectively storing a number of patterns that may exceed the order of the network. However, the resulting network will have a symmetric interconnection structure. Furthermore, there are no provisions for learning for networks designed by the eigenstructure method. In a series of papers, Michel, Farrell, and Porod [203], Michel and Farrell [202], Michel, Farrell, and Sun [204], and Yen and Michel [307] modified the eigenstructure method. The modified method can be used to design networks capable of storing on the order of $0.5n$ patterns as asymptotically stable equilibrium states. Networks designed by this method need not have symmetric interconnecting structures. Furthermore, learning is incorporated into the method. Networks designed by the previous methods use energy function techniques. A benefit of using energy functions is that the resulting networks are globally stable; that is, the networks' trajectories always tend to some equilibrium states. On the other hand, if the modified method yields a nonsymmetric interconnecting structure, the resulting network is not guaranteed to be globally stable.

EXERCISES

9.1 Write a MATLAB program that implements the TLU (Perceptron) learning algorithm for generating a separating line for points on a plane. Test your program on the following data. For $i = 1, \ldots, 4$ and $d_i = 1$,

$$\begin{bmatrix} \boldsymbol{x}^{(1)} & \boldsymbol{x}^{(2)} & \boldsymbol{x}^{(3)} & \boldsymbol{x}^{(4)} \end{bmatrix} = \begin{bmatrix} 5 & -7 & 8 & -1 \\ 5 & 6 & -3 & 2 \end{bmatrix}.$$

The above points constitute class C_1. You can label the above points using, for example, "×". Class C_2 consists of the points corresponding to $d_i = -1$. They are

$$\begin{bmatrix} \boldsymbol{x}^{(5)} & \boldsymbol{x}^{(6)} & \boldsymbol{x}^{(7)} \end{bmatrix} = \begin{bmatrix} -6 & -4 & 7 \\ 1 & -4 & -5 \end{bmatrix}$$

You can label these points using "○". The initial weight vector and threshold are

$$\boldsymbol{w}(0) = [-5 \quad -1]^T \quad \text{and} \quad t_0 = 5.$$

Plot the initial separating line as well as the final separating line.

9.2 Write a MATLAB program that implements the WTA algorithm. Test your program on the following data. The initial weights of the WTA network are

$$\begin{bmatrix} \boldsymbol{w}_0^{(1)} & \boldsymbol{w}_0^{(2)} & \boldsymbol{w}_0^{(3)} \end{bmatrix} = \begin{bmatrix} 0.4403 & 0.7411 & -0.8866 \\ 0.8979 & -0.6714 & 0.4625 \end{bmatrix}, \qquad \alpha = 0.5.$$

The training patterns are

$$X = \begin{bmatrix} -0.2 & 0.2 & -.1 & 0.98 & 0.98 & .8 & -0.6 & -0.8 & -0.7 \\ 0.98 & 0.98 & .9 & 0.2 & -0.2 & 0.1 & -0.8 & -0.6 & -0.7 \end{bmatrix}$$

Present the training set 200 times. Find the weights of the network after training.

9.3 Use the identification scheme depicted in Figure 9.16 and the neural fuzzy on-line identification algorithm from Section 9.4 to identify the plant described by

$$y(k+1) = \frac{y(k)(y(k-1)+2)(y(k)+2.5)}{8.5 + y^2(k) + y^2(k-1)} + u(k),$$

where

$$u(k) = 2\cos(2\pi k/100) \qquad \text{for } k \le 200$$

and

$$u(k) = 1.2\sin(2\pi/20) \qquad \text{for } k > 200.$$

The identifier is to be described by

$$\hat{y}(k+1) = f(k) + u(k),$$

where f has the form of (9.42) with $M = 100$. Choose $\alpha = 0.5$, $y(0) = y(1) = 0$, and $f(1) = f(2) = 0$, and start with $u(1) = 0$. (The above plant was also used by Lin and Lee [185, p. 731] to test their identification scheme.)

9.4 Write a MATLAB program that implements the learning algorithm for the RBF identifier discussed in Section 9.5.

9.5 Prove the following lemma that can be found in Park, Cho, and Park [220].

Lemma 9.2 Let $\boldsymbol{v} \in \{-1, 1\}^n$ be an asymptotically stable equilibrium state of the gBSB neural net. Let $\boldsymbol{x}$ be a vertex of H_n that differs from $\boldsymbol{v}$ in the component x_i for some $i = 1, \ldots, n$. If

$$(\boldsymbol{W}\boldsymbol{x} + \boldsymbol{b})_i v_i > 0,$$

then $\boldsymbol{x}$ is not an equilibrium state of the gBSB neural net.

9.6 Use Lemma 9.2 from Exercise 9.5 to prove the following proposition that can be found in Park, Cho, and Park [220].

Proposition 9.1 Let $\boldsymbol{v} \in \{-1, 1\}^n$ be an asymptotically stable equilibrium state of the gBSB neural net. If $w_{ii} = 0$ for $i = 1, \ldots, n$, then none of the vertices of H_n that differ from $\boldsymbol{v}$ in just one component is an equilibrium state of the gBSB neural net.

9.7 Use Lemma 9.2 from Exercise 9.5 to prove the following proposition that can be found in Park, Cho, and Park [220]. Before stating the proposition, we define the *Hamming distance* between two vectors whose components are -1 or 1 as the number of components in which the two vectors differ. Given two vectors $\boldsymbol{x}$ and $\boldsymbol{y}$, the Hamming distance between them is denoted $HD(\boldsymbol{x}, \boldsymbol{y})$.

Proposition 9.2 Let $\boldsymbol{v} \in \{-1, 1\}^n$ be an asymptotically stable equilibrium state of the gBSB neural net and let k be an integer from the set $\{1, 2, \ldots, n\}$. If

$$\left(\sum_{j=1}^{n} w_{ij} x_j + b_i \right) v_i > 2k \max_j |w_{ij}|,$$

then a vertex $\boldsymbol{x}$ of H_n, such that $0 < HD(\boldsymbol{x}, \boldsymbol{v}) \leq k$, is not an equilibrium state of the gBSB neural net.

CHAPTER 10 Genetic and Evolutionary Algorithms

> The word "algorithm" itself is quite interesting; at first glance it may look as though someone intended to write "logarithm" but jumbled up the first four letters. The word did not appear in *Webster's New World Dictionary* as late as 1957; we find only the older form "algorism" with its ancient meaning, i.e., the process of doing arithmetic using Arabic numerals." ... "Finally, historians of mathematics found the true origin of the word algorism: it comes from the name of a famous Persian textbook author, Abu Ja'far Mohammed ibn Mûsâ al-Khowârizmî (c. 825) ... Al-Khowârizmî wrote the celebrated book *Kitab al jabr w'al-muqabala* ("Rules of restoration and reduction") ... The modern meaning for algorithm is quite similar to that of *recipe, process, method, technique, procedure, routine,* except that the word "algorithm" connotes something just a little different.
>
> —*The Art of Computer Programming, Volume 1* [162, pp. 1–3]

Genetic algorithms, introduced by Holland [124] in the 1960s, have been originally proposed to model adaptive processes, both natural and man-made. In this chapter we discuss applications of genetic algorithms to solving optimization problems. Genetic algorithms were inspired by biological genetics and evolutionary processes. This is the reason for biological terms being used in the context of genetic algorithms. One of the variants of genetic algorithms, called the evolutionary algorithm, is also presented.

10.1 Genetics as an Inspiration for an Optimization Approach

The *gene* is the basic unit of inheritance. Genes are pieces of DNA, which is the material inside the cell nucleus, that carry the genetic instructions for making living organisms. Many traits are determined by genes inherited from the parents of the individual. Consider gene $\mathcal{A}$ that determines a certain trait, such as seed coat texture—round or wrinkled. Thus gene $\mathcal{A}$ has two alternative forms. These different forms of the same gene are called *alleles,* or *allelic genes*. The allelic forms of each gene are symbolized by the uppercase and lowercase of the same letter. In our example, gene $\mathcal{A}$ has two alleles: allele A and allele a. One of the alleles controls the round seed coat texture, while the other allele controls the wrinkled texture. Genes are organized into *chromosomes,* which are small packages of genes in the nucleus of a cell. Different kinds of organisms have different numbers of chromosomes. Humans have 23 pairs of chromosomes, including two sex chromosomes. In mating, each parent contributes one chromosome to each pair of the chromosomes of an offspring, so the offspring gets half of his or her chromosomes from his or her mother and half from his or her father. The assemblage of genes, encrypted as a string of DNA molecules, is called the *genome*. The human genome consists of about 35,000 genes. The particular gene constitution is called the *genotype*. The expression of the genotype is the *phenotype,* which is the way we look and behave.

Genetic algorithms are a result of an effort to model adaptation phenomena in natural and artificial systems. The term *adaptation* can be viewed as the composition of two words: *ad* + *aptare,* meaning "to fit to" [124, p. 1]. "Just what are adaptation's salient features? We can see at once that adaptation, whatever its context, involves a progressive modification of some structure or structures. These structures constitute the grist of the adaptive process, being largely determined by the field of study. Careful observation of successive structural modifications generally reveals a basic set of structural modifiers or operators; repeated action of these operators yields the observed modification sequences" [124, p. 3].

In this chapter we describe two genetic algorithms and apply them to solving optimization problems. An optimization problem of a real-valued function, $f : \Omega \to \mathbb{R}$, requires finding an element $\boldsymbol{x} \in \Omega$ such that f is maximized or minimized. The function f is called the *fitness function*. Suppose that we wish to maximize f over the parameter space Ω. The parameter space, Ω, of an underlying optimization problem is also called the *phenotype space*. A classical genetic algorithm works with binary representations of the points from the parameter space, rather than directly with the elements of Ω. We will refer to the classical genetic algorithms as the canonical algorithms. The strings of binary digits (1s and 0s) corresponding to these points are called chromosomes. The space of chromosomes is called the *genotype space*. In general, the chromosomes in the genotype space are abstract mathematical objects on which the so-called genetic operators work. After a stopping criterion for the algorithm is satisfied, a decoding function transforms a solution chromosome from the genotype space into the phenotype space. Genetic algorithms are derivative-free optimization algorithms that use only the fitness function. They do not use the fitness function derivatives. The higher the fitness of a chromosome, the better the candidate solution encoded by the chromosome. Genetic algorithms work with populations of chromosomes rather than individual chromosomes. A genetic algorithm is an iterative process that at each iteration uses the information obtained from testing a current population of chromosomes to direct its search into promising areas of the search space. Thus, generating a suitable population of chromosomes at each iteration is the key to the algorithm performance. In the simplest form, creating a new population of chromosomes to be tested involves three genetic operators: selection, crossover, and mutation. These genetic operators work on genotypes that encode phenotypes.

The role of the selection operator is to pick a population of chromosomes for further processing, called reproduction. Chromosomes are selected into a mating pool for reproduction with a probability proportional to their fitness.

The crossover operator randomly exchanges substrings between two parent chromosomes to create two offspring. The simplest crossover operator is the one-point crossover, where a crossing site is randomly picked and the corresponding substrings are exchanged, as illustrated

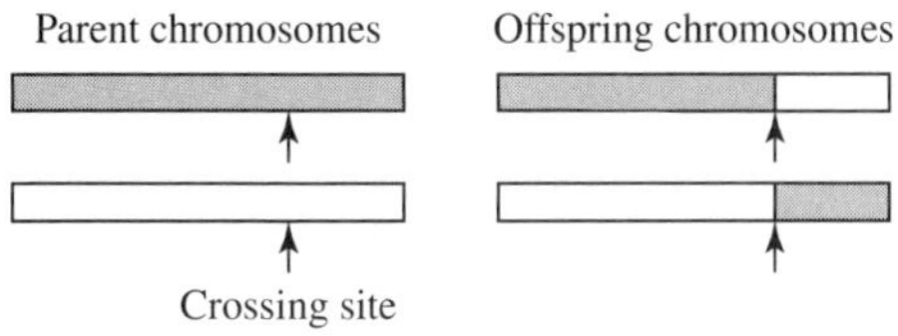

Figure 10.1 The one-point crossover.

in Figure 10.1. For example, the two parent chromosomes

$$1010101 \quad \text{and} \quad 1111010$$

could have the crossing site after the fourth bit in each to produce the two offspring

$$1010010 \quad \text{and} \quad 1111101.$$

The mutation operator randomly flips an arbitrary bit in a chromosome. For example, the chromosome 010101 might undergo a mutation in its third position to yield a new chromosome 011101.

We now provide a description of the canonical genetic algorithm using the above components.

CANONICAL GENETIC ALGORITHM (GA)

For a given optimization problem characterized by a fitness function f:

STEP 1 Choose a method of encoding candidate solutions into chromosomes; that is, decide upon a map between the phenotype and genotype spaces. (A commonly used encoding method is to concatenate the binary strings representing each real parameter value x_i.)

STEP 2 Start with a randomly generated population of chromosomes.

STEP 3 Evaluate the fitness of each chromosome in the population.

STEP 4 Replace the current population with a new population by applying selection, crossover, and mutation operations on the current population.

STEP 5 If a stopping criterion is satisfied, stop and choose the fittest chromosome from the population as a solution, else go to Step 3.

Another type of a genetic algorithm that we analyze in this chapter attempts to avoid the need for a decoding function by using the chromosome representation as close as possible to their representation in the phenotype space.

10.2 Implementing a Canonical Genetic Algorithm

We now discuss implementation issues of the canonical genetic algorithm (GA) when applied to solving optimization problems. We consider optimization problems of the form

$$\text{maximize} \quad g(\boldsymbol{x}), \qquad \boldsymbol{x} \in \Omega \subset \mathbb{R}^n. \tag{10.1}$$

A maximizing point or maximizer is denoted $\boldsymbol{x}^*$. We note that there is no loss of generality by considering only maximization problems, for if we are asked to minimize g, we can transform the minimization problem into the maximization problem by means of a simple transformation,

$$\min_{x} g(\boldsymbol{x}) = -\max_{x}(-g(\boldsymbol{x})). \tag{10.2}$$

Also note that if $\boldsymbol{x}^*$ is a minimizer of $g(\boldsymbol{x})$, then $\boldsymbol{x}^*$ is a maximizer of $-g(\boldsymbol{x})$, and vice versa. Once an objective function of the maximization problem is given, the next step is to transform the objective function into a fitness function. For example, we can use the transformation

$$f(\boldsymbol{x}) = g(\boldsymbol{x}) - g_{\min}, \tag{10.3}$$

where $f(\boldsymbol{x})$ denotes a fitness function and $g_{\min}$ is the minimum value of $g(\boldsymbol{x})$ for the current population of chromosomes. Note that the constructed fitness function is indeed nonnegative on the chromosomes of the current population.

We now discuss the implementation of Step 1 of the canonical GA presented in the previous section. We first analyze the case when the fitness function, f, is a function of only one parameter x. Suppose that

$$x \in [x_{\min}, x_{\max}] = \Omega.$$

A common encoding method, which transforms points from the phenotype space into a genotype space, is the binary encoding. In our example the phenotype space is the interval $[x_{\min}, x_{\max}]$. If we decide to use l bits to represent a point x, then this corresponds to mapping the interval $[x_{\min}, x_{\max}]$ into the discrete genotype space $[0, 2^l - 1]$ consisting of 2^l binary strings. Thus, the number of bits l determines the size of the genotype space and hence the resolution, or accuracy, of the subsequent search. We can control the resolution by appropriately choosing the parameter l. Let r denote the desired resolution. Then, the parameter l that results in the required resolution can be found by solving the equation

$$r = \frac{x_{\max} - x_{\min}}{2^l - 1}. \tag{10.4}$$

If the fitness function depends on many parameters, then we can encode the elements of the phenotype space by concatenating the binary strings representing each component of $\boldsymbol{x} = [x_1 \ x_2 \ \cdots \ x_n]^T$. For example, let $f = f(x_1, x_2)$, where $x_1 \in [x_{1,\min}, x_{1,\max}]$ and $x_2 \in [x_{2,\min}, x_{2,\max}]$. If we use 10 bits to encode each variable, then the genotype space will consists of 20-bit binary strings and the size of the genotype space is $2^{10} \times 2^{10} = 1{,}048{,}576$.

After initializing randomly a population of chromosomes, we evaluate the fitness $f(\boldsymbol{x})$ of each chromosome $\boldsymbol{x}$ in the population. This requires decoding from the genotype space into the phenotype space. Suppose, as before, that the fitness function depends on two parameters x_1 and x_2 and that we used the standard encoding method described above. We now have to convert a given binary string into two real numbers $x_1 \in [x_{1,\min}, x_{1,\max}]$ and $x_2 \in [x_{2,\min}, x_{2,\max}]$. We can perform the required decoding process in three steps. First note that the first l bits of the chromosome $\boldsymbol{x}$ correspond to x_1 while the remaining bits correspond to x_2. We first convert the strings of l bits into integers. Thus, for example, if $l = 4$ and the chromosome is 10010111, which is an 8-bit string, then the first 4 bits yield

$$x_{1,int} = 1 \times 2^3 + 0 \times 2^2 + 0 \times 2^1 + 1 \times 2^0 = 9,$$

while the remaining 4 bits yield

$$x_{2,int} = 0 \times 2^3 + 1 \times 2^2 + 1 \times 2^1 + 1 \times 2^0 = 7.$$

Then, we convert $x_{1,int}$ from the range $[0, 2^l - 1]$ into the range $[0, (x_{1,\max} - x_{1,\min})]$ by multiplying $x_{1,int}$ by

$$\frac{x_{1,\max} - x_{1,\min}}{2^l - 1}. \tag{10.5}$$

To obtain the corresponding value of x_1, we add $x_{1,\min}$ to (10.5) to get

$$x_1 = x_{1,\min} + x_{1,int}\frac{x_{1,\max} - x_{1,\min}}{2^l - 1}. \tag{10.6}$$

We obtain x_2 proceeding in an analogous manner. In general, if $\boldsymbol{x}$ has n components, the mapping from the phenotype space into the genotype space has the form

$$\boxed{x_i = x_{i,\min} + x_{i,int}\frac{x_{i,\max} - x_{i,\min}}{2^l - 1}, \qquad i = 1, 2, \ldots, n} \tag{10.7}$$

The next step of the GA algorithm is the application of the selection operator to pick chromosomes from the current population for reproduction to create the new population. In the canonical GA, a pair of chromosomes is selected from the current population with the probability of selection being proportional to the fitness values of the chromosomes. Selection is done with replacement, which means that the same chromosome can be selected more than once for reproduction. A common method of selecting parent chromosomes is the so-called *roulette-wheel method*. In this method each chromosome is given a slice of a wheel. The area of the wheel slice is proportional to the chromosome's fitness. The wheel is spun, and the chromosome is selected on whose wheel slice the ball comes to rest. The roulette-wheel method can be implemented in the following steps:

1. Calculate the fitness of each chromosome in the current population.
2. Calculate the total fitness of the current population by summing the fitnesses of each chromosome in the current population.
3. Divide the fitness of each chromosome by the total fitness.
4. Calculate the cumulative fitness of the current population using the scaled fitness of each chromosome found in the previous step. You can use MATLAB's command `cumsum` for this purpose.
5. Return the first population member whose corresponding cumulative fitness is greater than the random number from the interval (0, 1). The following MATLAB command can be used here: `find(cum_fitness-rand>0)`, where `cum_fitness` is the vector of scaled cumulative fitnesses found in step 4.
6. Repeat the above step to get the second parent chromosome.

Using the parent chromosomes picked by the selection operator, we form two offspring by crossing over, with probability p_c, at a randomly chosen crossing site. The uniform probability p_c is called the crossover probability or crossover rate. If no crossover takes place, two offspring chromosomes are formed that are exact copies of their parent chromosomes. One can also use a multipoint crossover operator. The two-point crossover operation is illustrated in Figure 10.2.

The offspring chromosomes obtained from the crossover are then acted upon by the mutation operator. Mutation is performed at each bit of the offspring chromosomes with probability p_m. The probability p_m is called the mutation probability or mutation rate. The mutation operator

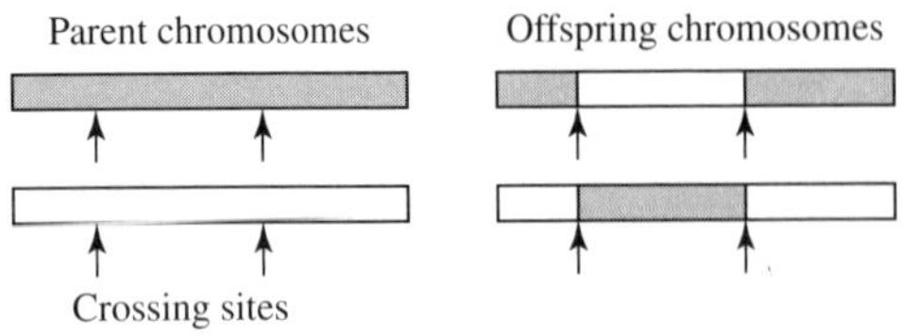

Figure 10.2 The two-point crossover.

randomly flips some of the bits in a chromosome. For example, the chromosome 101010 might be mutated in its third bit to yield 100010. The mutation probability is usually very small—for example, $p_m = 0.01$.

Each iteration, also called a *generation,* of the GA replaces the current population with a new one. A typical number of iterations of a GA algorithm is from 50 to 500 or more. The entire set of generations is called a *run*.

◆ Example 10.1

We apply the canonical GA described above to maximize the function

$$f(x) = -15\sin^2(2x) - (x-2)^2 + 160, \qquad |x| \le 10. \tag{10.8}$$

This function has many local maximizers and minimizers, as can be seen from its plot shown in Figure 10.3. The canonical GA was applied to maximize the function given by (10.8) with the chromosomes composed of 12 bits, $p_c = 0.75$, and $p_m = 0.01$. In Figure 10.4 a summary of a typical run is shown. In this run the best candidate solution was $x = 1.565324$ whose corresponding objective value is $f(x) = 159.809259$. The best solution that was observed in other runs resulted in $f(1.575092) = 159.818346$.

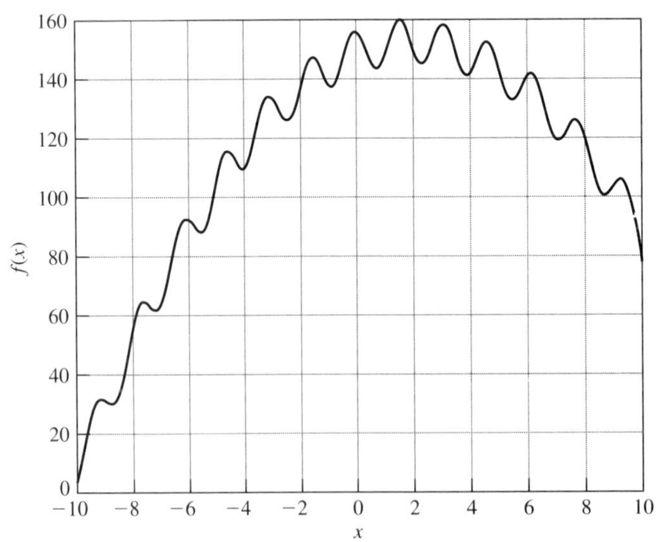

Figure 10.3 A plot of the function given by (10.8).

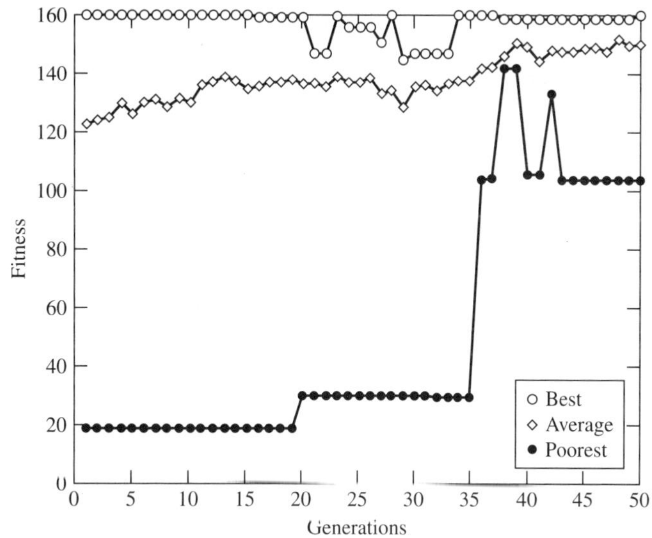

Figure 10.4 Summary of a sample run of the canonical GA when searching for a maximizer of the function given by (10.8).

In the next section we will try to get answers to some whys and hows of genetic algorithms.

10.3 Analysis of the Canonical Genetic Algorithm

In the analysis of genetic algorithms, the notion of a *schema* will be useful. A schema is a set of bit strings. It can be described using a template made up of ones, zeros, and asterisks, which represent "don't cares." Thus, the asterisk $*$ matches either 0 or 1. For example, the schema $H = *0*$ contains the binary strings 000, 001, 100, and 101. Note that a bit string of length l belongs to 2^l different schemas. For example, the string 10 of length $l = 2$ belongs to the schemas $**$, $*0$, $1*$, and 10. Similarly, when $l = 3$, then any bit string of length $l = 3$ belongs to $2^3 = 8$ schemas. Elements of a set of N bit strings of length l are members of between 2^l and $N \times 2^l$ schemas. If all the strings are identical, then they belong to 2^l schemas; otherwise there may be up to $N \times 2^l$ schemas to which the N strings of length l can belong.

The parameters that characterize a given schema are its *order* and *defining length*. The order of a schema H, denoted $o(H)$, is the number of its fixed bits. For example, the order of $H = *01 * 1$ is $o(H) = 3$ because H has three bits fixed. The defining length of the schema H, denoted $d(H)$, is the distance between its first and the last fixed bits, which is the difference between the positions of the rightmost fixed symbol and the leftmost fixed symbol. The defining length of $H = *0 * 1$ is $d(H) = 4 - 2 = 2$. In further discussion, the term "schema" will be used to denote a set of binary strings represented by a template, as well as the template itself.

The capability of a chromosome to solve a given optimization problem—that is, the chromosome performance—is measured by its fitness. Similarly, we can characterize the performance of a schema by its average fitness. We now derive an expression for the average fitness of a given schema. Let $\boldsymbol{x}$ be a chromosome that matches the template H, that is, $\boldsymbol{x} \in H$. Then, we say that "$\boldsymbol{x}$ is an instance of H." Let H be a schema with at least one instant in the kth generation. Denote by $e(H, k)$ the number of instances of H in the kth generation, and let $\bar{u}(H, k)$ be the average fitness of H in the kth generation, that is,

$$\bar{u}(H, k) = \frac{\sum_{x \in H} f(\boldsymbol{x})}{e(H, k)}. \tag{10.9}$$

When the genetic algorithm is explicitly evaluating the fitness of N chromosomes from the current generation, it is implicitly evaluating the average fitness of up to $N \times 2^l$ schemas to which the N chromosomes belong.

We now find $\mathcal{E}(H, k+1)$, the expected number of instances of the schema H in the $(k+1)$st generation, given the kth generation. We first analyze the effect of the selection operator only. Assume that the roulette-wheel method is used in selecting parent chromosomes to the mating pool for reproduction. The probability that the chromosome $\boldsymbol{x}$ is going to be selected to the mating pool, $M(k)$, is

$$\frac{f(\boldsymbol{x})}{\sum_{i=1}^{N} f\left(\boldsymbol{x}_i^{(k)}\right)},$$

where $\boldsymbol{x}_i^{(k)}$ refers to the ith chromosome in the kth generation and $f(\boldsymbol{x})$ is the fitness of the chromosome $\boldsymbol{x}$. Therefore, the expected number of chromosomes equal to $\boldsymbol{x}$ that are selected to the mating pool is

$$N \frac{f(\boldsymbol{x})}{\sum_{i=1}^{N} f\left(\boldsymbol{x}_i^{(k)}\right)} = \frac{f(\boldsymbol{x})}{\bar{f}(k)},$$

where

$$\bar{f}(k) = \frac{1}{N} \sum_{i=1}^{N} f\left(\boldsymbol{x}_i^{(k)}\right)$$

is the average fitness of the kth generation. Let $m(k)$ be the number of chromosomes in $M(k)$ that are instances of H. The expected value, $\mathcal{M}(H, k)$, of the number of chromosomes in the mating pool that are instances of H is

$$\frac{\sum_{x \in H} f(\boldsymbol{x})}{\bar{f}(k)}. \tag{10.10}$$

Substituting (10.9) into (10.10) yields an alternative expression for the expected value of the number of chromosomes in the mating pool that are instances of H:

$$\boxed{\mathcal{M}(H, k) = \frac{\bar{u}(H, k)}{\bar{f}(k)} m(H, k)} \tag{10.11}$$

It follows from the above equation that a schema whose average fitness is greater than the population average will be found in an increasing number of chromosomes.

We will now quantify the effects of the one-point crossover on a given schema. Given two chromosomes, where at least one of them is an instance of the schema H, we say that the schema H survives crossover operation if at least one of the offspring is an instance of H. We will give a lower bound on the probability $S_c(H)$ that the schema H will survive the one-point crossover. A schema survives if the crossing site occurs outside the defining length. Crossovers that take place within the defining length can destroy the schema. For example, let $H = *00**$, and let the parent chromosomes be $\boldsymbol{p}_1 = 10011$ and $\boldsymbol{p}_2 = 11100$. Note that $\boldsymbol{p}_1 \in H$, while $\boldsymbol{p}_2 \notin H$. For the crossing site after the third position, there are two offspring chromosomes: $\boldsymbol{s}_1 = 10000$, which is an instance of H, and $\boldsymbol{s}_2 = 11111$, which is not an instance of H. Thus, the schema H survived this crossover. Suppose now that the crossing site is after the second position, which is within the defining length of the schema H. The offspring chromosomes are $\boldsymbol{s}_1 = 10100$ and $\boldsymbol{s}_2 = 11011$. None of the offspring chromosome is an instance of H, which means that H was destroyed by the crossover. If both parent chromosomes are instances of H, then the offspring chromosomes are also instances of H irrespective of the location of the crossing site. An upper bound on the probability $D_c(H)$ that the schema H will be destroyed by the one-point crossover is

$$D_c(H) \le p_c \frac{d(H)}{l-1}. \tag{10.12}$$

Therefore, a lower bound on the probability $S_c(H)$ that H survives is

$$\boxed{S_c(H) = 1 - D_c(H) \ge 1 - p_c \frac{d(H)}{l-1}} \tag{10.13}$$

Thus, the probability of survival under one-point crossover is higher for short schemas.

A schema H survives under mutation of its element, if all defined bits of this element of H remain unchanged. Let $S_m(H)$ denote the probability that the schema H survives under mutation of an instance of H. For each defined bit, the probability that it will not be mutated is $1 - p_m$. The probability that no defined bit of H will be mutated is obtained by multiplying $1 - p_m$ by itself $o(H)$ times. Therefore,

$$\boxed{S_m(H) = (1 - p_m)^{o(H)}} \tag{10.14}$$

We now combine (10.11), (10.13), and (10.14) to obtain a lower bound on the expected number of instances of H in the $(k+1)$st generation, given the k-generation:

$$\boxed{\mathcal{E}(H, k+1) \ge \frac{\bar{u}(H,k)}{\bar{f}(k)} m(H,k) \left(1 - p_c \frac{d(H)}{l-1}\right) (1 - p_m)^{o(H)}} \tag{10.15}$$

The above is known as the *schema theorem*. It follows from our discussion that the canonical genetic algorithm implicitly processes, in parallel, schemas. This implicit computation is performed with no operation being explicitly performed on schemas. To illustrate the last statement, consider the string 101. As the fitness of this string is evaluated, partial information is obtained about the expected fitness of all possible schemas to which the evaluated string belongs. In our example, computing the fitness of the string 101 gives partial information about the fitness of strings in the schemas $***$, $**1$, $*0*$, $1**$, $*01$, $1*1$, $\ldots$.

10.4 Simple Evolutionary Algorithm (EA)

Canonical genetic algorithms use binary representations of the elements of the parameter space. The reason behind this is the schema theory, discussed in the previous section, which allows one to analyze canonical genetic algorithms. When using a binary representation of the elements of the parameter space, canonical algorithms use encoding and decoding functions. A binary representation may have some disadvantages though. Indeed, let $h : \{0, 1\}^l \to \Omega$ denote a decoding function used in the canonical genetic algorithm implementation. Then the original problem of optimizing $f : \Omega \to \mathbb{R}$ is transformed into the problem of optimizing the combined objective function

$$f(h) : \{0, 1\}^l \to \mathbb{R}.$$

The combined objective function to be optimized may be more complex than the original objective function. For example, it may introduce extra nonlinearities or add extra optimizers, which can make the search process for a global optimizer more difficult.

In contrast to the canonical genetic algorithms, in the optimization procedure discussed next, the parameter space is no longer coded using the binary representation. The genetic operators work directly on real-valued vectors. A term used to describe an optimization algorithm using genetic-type operators working on the real-valued vectors is an *evolutionary algorithm*. In what follows, we present a very simple evolutionary algorithm for solving optimization problems of the form given by (10.1).

SIMPLE EVOLUTIONARY ALGORITHM (EA)

For a given optimization problem of the form (10.1), perform the following steps:

STEP 1 Initialize a population of candidate solutions (chromosomes) to the problem. Each chromosome is a real-valued vector of the same length as the vector $\boldsymbol{x}$ of decision variables.

STEP 2 Evaluate the fitness of each candidate solution in the population.

STEP 3 Replace the current population with a new population by applying selection, crossover, and mutation operations on the current population. The roulette-wheel selection method can be used for selection. A possible implementation of the crossover operation is to use the average crossover operator. This operator takes two parents, say $\boldsymbol{x}_j$ and $\boldsymbol{x}_k$, and returns one offspring

$$\boldsymbol{x} = \tfrac{1}{2}(\boldsymbol{x}_j + \boldsymbol{x}_k).$$

Another implementation of the crossover operator is a convex combination of two parents. Specifically, let $r \in (0, 1)$. Then, the offspring is

$$\boldsymbol{x} = r\boldsymbol{x}_j + (1 - r)\boldsymbol{x}_k.$$

As for the mutation operation, let $\boldsymbol{x}$ be a chromosome to be mutated. Then, an implementation of the mutation operation could have the form

$$\boldsymbol{x}' = \boldsymbol{x} + r\boldsymbol{d},$$

where $\boldsymbol{d}$ is a randomly generated feasible vector and $r \in (0, 1)$. Note that if the feasible set is convex, then the above genetic operators produce chromosomes that are elements of the feasible set. Otherwise, conditions should be imposed to ensure that candidate solutions are the elements of the feasible set.

STEP 4 If a stopping criterion is satisfied, terminate the algorithm and choose the fittest chromosome from the population as a solution, else go to Step 2.

In addition, both in the canonical genetic algorithm and in the evolutionary algorithm, the elitist strategy, introduced by K. DeJong in 1975, can be incorporated. This strategy copies the best chromosomes from the old population into the next population to prevent losing the best solutions in the succeeding iterations. As observed by Mitchell [206], among others, the elitist strategy may lead to the domination of a population by a "super chromosome." However, it appears that the elitist strategy improves genetic and evolutionary algorithms performance.

10.5 Evolutionary Fuzzy Logic Controllers

In this section we employ evolutionary algorithms to design tracking controllers. We describe three design techniques for evolutionary fuzzy logic controllers for the step-lane-change maneuver of a ground vehicle. The presentation of this section is based on the paper by Lee and Żak [179]. We mention that the step-lane-change maneuver is an important aspect in the design of the Intelligent Vehicle/Highway System (IVHS), the highway of the future.

An essential element in combining an evolutionary algorithm (EA) and a fuzzy logic controller (FLC) is the encoding method used to represent a fuzzy rule base as a chromosome on which the EA operates. Selecting an appropriate encoding procedure is a critical factor that affects the efficiency of the EA and the final solution. Another essential element is the fitness function. Each chromosome defines a fuzzy rule base. After conducting a simulation of the structure depicted in Figure 10.5 and measuring the tracking error, the fitness of the chromosome representing the fuzzy rule base of the simulated FLC is calculated. With all the chromosomes assigned a fitness value, genetic operators yield a new generation and the entire procedure is repeated until a stopping criterion, usually the number of iterations or the fitness value, is satisfied.

10.5.1 Vehicle Model and Control Objective

We use a "bicycle model" of a front-wheel-steering vehicle. Modeling of a ground vehicle is covered in detail in Subsection 1.7.2. The standard coordinate system of the Society of Automotive Engineers (SAE) is used. A schematic of the model used is shown in Figure 10.6. The modeling equations are

$$
\begin{aligned}
\dot{x}_1 &= v_x x_3 + \frac{1}{m}(F_{yf} + F_{yr}),\\
\dot{x}_2 &= x_3,\\
\dot{x}_3 &= \frac{2}{I}(l_1 F_{yf} - l_2 F_{yr}),\\
\dot{x}_4 &= -v_x x_2 - x_1,
\end{aligned}
$$

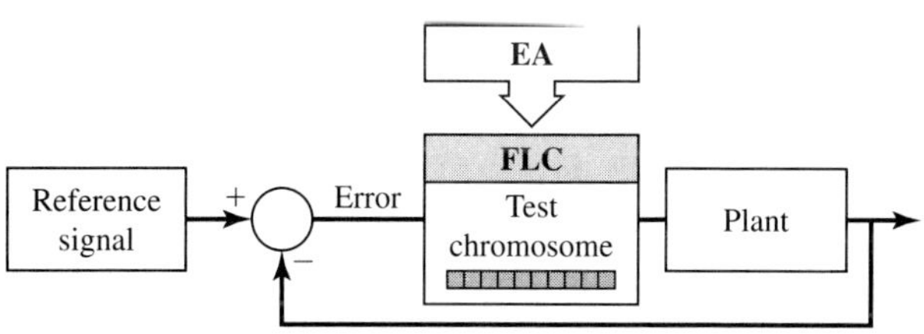

Figure 10.5 A configuration used to design a fuzzy logic controller using an evolutionary algorithm.

Table 10.1 Vehicle Data and Symbols Used in Modeling Equations

δ	Steering angle of the front wheel	
F_{yf}	Lateral force of the front tire	
F_{yr}	Lateral force of the rear tire	
m	Total mass of the vehicle	1280 kg
I	Moment of the inertia of the vehicle	2500 kg·m^2
l_1	Distance from center of gravity to front axle	1.2 m
l_2	Distance from center of gravity to rear axle	1.22 m
v_x	Longitudinal velocity of the vehicle	18.3 m/sec

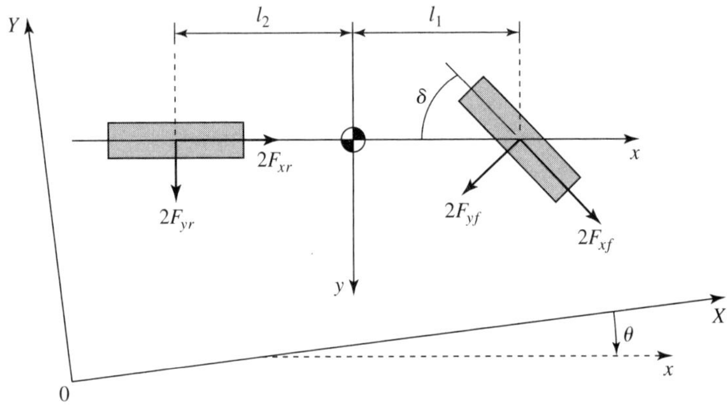

Figure 10.6 Schematic of a bicycle model of a front-wheel-steering vehicle.

where $x_1 = \dot{y}$ is the lateral velocity of the vehicle, $x_2 = \theta$ is the yaw angle, $x_3 = \dot{\theta}$ is the yaw velocity, and $x_4 = Y$ is the lateral position of the vehicle. Other parameters and their values are given in Table 10.1. The lateral forces $F_{yf} = F_{yf}(\alpha_f)$ and $F_{yr} = F_{yr}(\alpha_r)$ are nonlinear functions of respective slip angles with a typical plot of a lateral force versus tire slip angle shown in Figure 1.15. Recall that the slip angles for the front and rear tires are

$$\alpha_f = \delta - \frac{l_1 x_3 + x_1}{v_x} \quad \text{and} \quad \alpha_r = \frac{l_2 x_3 - x_1}{v_x}.$$

The input is the steering angle δ of the front wheel and the output is the lateral position of the vehicle, $x_4 = Y$.

The step-lane-change maneuver, along with cruising, traveling path selection, and highway exiting and entering, is one of the essential elements of automated vehicle control. However, automation of these tasks requires a roadway reference system that involves sensing for lane detection. In our simulation, we assume that the road reference system is already installed. Our objective is to design a fuzzy logic controller that forces the vehicle to track a given reference lateral position signal r. The reference signal is generated by a reference signal generator whose transfer function is $1/(s^3 + 1.75s^2 + 2.15s + 1)$. The tracking error e is the difference between the reference lateral position signal r and the actual lateral position Y of the vehicle. The role of the fuzzy controller is to produce the steering wheel angle δ using the information of the tracking error e and the change in the tracking error $\dot{e}$ so that the vehicle tracks the reference signal. The SIMULINK block diagram of the closed-loop vehicle system is depicted in Figure 10.7. In the following, we describe three different methods for an evolutionary algorithm (EA)-aided fuzzy logic controller (FLC) design.

10.5.2 Case 1: EA Tunes Fuzzy Rules

It is assumed in this case that the membership functions are fixed. In our simulation experiment, we use the membership functions shown in Figure 10.8. The labels of fuzzy sets LN, N, Z, P, and LP denote large negative, negative, zero, positive, and large positive. There are five fuzzy

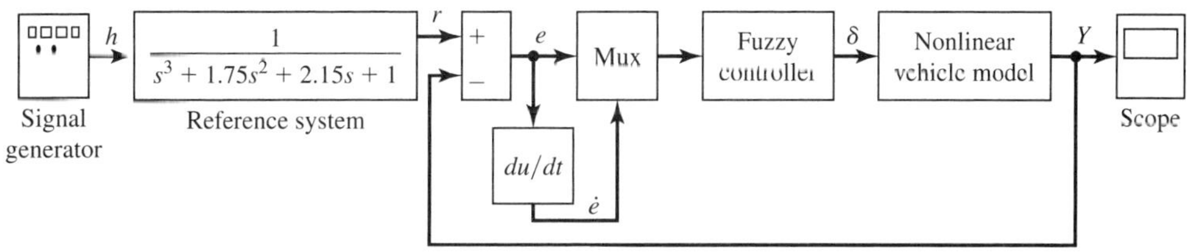

Figure 10.7 SIMULINK block diagram of the closed-loop vehicle system driven by a fuzzy logic controller.

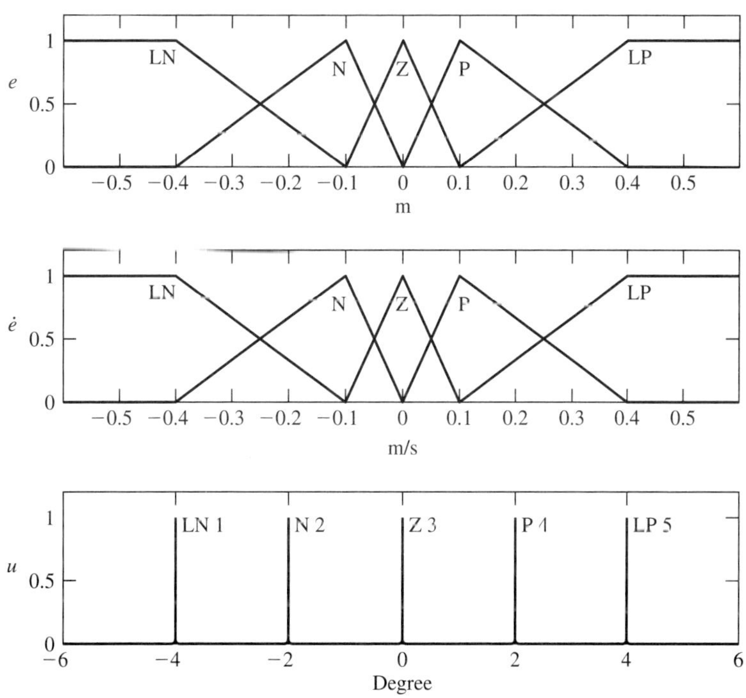

Figure 10.8 Fuzzy membership functions for e, $\dot{e}$, and u.

(a)

		$\dot{e}$				
		LN	N	Z	P	LP
e	LN	LP	LP	LP	P	Z
	N	LP	LP	P	Z	N
	Z	P	P	Z	N	N
	P	P	Z	N	LN	LN
	LP	Z	N	LN	LN	LN

(b)

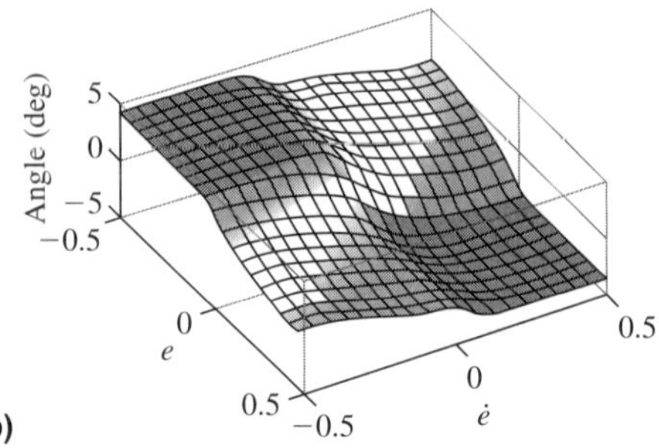

(c)

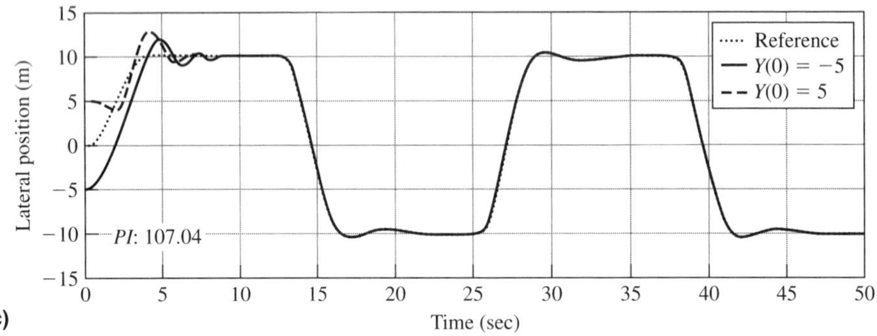

Figure 10.9 FLC constructed using a fuzzy rule base obtained heuristically. (a) Fuzzy rules table. (b) Control surface. (c) Tracking performance.

sets for e and five fuzzy sets for $\dot{e}$; hence we can have 25 different rules of the form

$$\text{IF } e \text{ is LN AND } \dot{e} \text{ is LN then } \delta \text{ is SP.}$$

$$\vdots$$

$$\text{IF } e \text{ is LP AND } \dot{e} \text{ is LP then } \delta \text{ is SN.}$$

These rules can be arranged into a rules table as shown in Figure 10.9(a). We mention here that the rules shown in Figure 10.9(a) were arrived at heuristically. Using the rules table, we construct an FLC using the center average defuzzifier given by (8.2). The control surface derived using the rules in Figure 10.9(a) is shown in Figure 10.9(b). Tracking performance for two different initial conditions $Y(0) = 5$ m and $Y(0) = -5$ m is illustrated in Figure 10.9(c). Tracking errors for each initial condition were calculated. We labeled the tracking error for the first initial condition as e_1 while the tracking error of the second initial condition was labeled as e_2. We then constructed the performance index (PI) as the sum of the integrals of the square of the error (ISE) over the simulation time:

$$PI = \int_0^{50} \left(e_1^2(t) + e_2^2(t)\right) dt.$$

The value of the PI in the above experiment is 107.04. The performance index, PI, was evaluated using the circuit shown in Figure 10.10.

Of course, we could include more error trajectories in the expression for the PI. However, for the sake of simplicity, we decided to use only two tracking error trajectories in our simulation experiments.

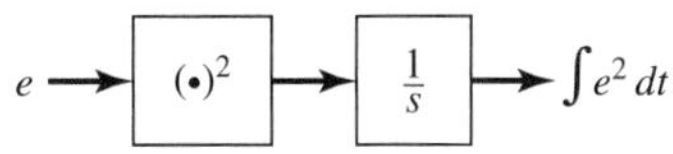

Figure 10.10 Circuit used to evaluate the performance index, *PI*.

We now employ an EA to generate a fuzzy rules table automatically. In order to represent the whole fuzzy rules table as a chromosome, we encode the labels of the fuzzy singletons corresponding to the controller outputs as integers, where LN is represented by 1, N by 2, Z by 3, P by 4, and LP by 5. This encoding of the fuzzy singletons of the control signal u is depicted in Figure 10.8. The actual control signal, however, is produced as the weighted sum of the fuzzy singletons using the center average defuzzifier. Because there are 25 rules, each chromosome is a string composed of 25 integers. For example, the rules table of Figure 10.9(a) can be represented as a chromosome of the form

5	5	5	4	3	5	5	4	3	2	4	4	3	2	2	4	3	2	1	1	3	2	1	1	1

In our experiment, we decided upon the population size of 30. The roulette-wheel method was used to select chromosomes to the mating pool. We employed the single-point crossover operator with the crossover rate $p_c = 0.9$. The role of the mutation operator was to replace each element of a chromosome randomly with an integer from the set $\{1, \ldots, 5\}$. We used the mutation rate $p_m = 0.05$.

We also employed an elitist strategy where two best individuals of the current generation were passed unchanged to the next generation. The fitness function employed was the inverse of the *PI* discussed above, that is,

$$\text{Fitness} = \frac{1}{PI} = \frac{1}{\int_0^{50} \left(e_1^2(t) + e_2^2(t)\right) dt}. \tag{10.16}$$

Recall that $e_1(t)$ is the tracking error with the initial lateral position at 5 m, and $e_2(t)$ is the tracking error with the initial lateral position at -5 m. The input to the reference system was a square wave with the amplitude of 10 m and the frequency of 0.04 Hz. The rules table of the best FLC, after 50 iterations of the EA, is shown in Figure 10.11(a). Figure 10.11(b) shows the fuzzy control surface based on the rules table of Figure 10.11(a). Two different initial conditions, $Y(0) = 5$ m and $Y(0) = -5$ m, were used to test the tracking performance of the fuzzy controller; Figure 10.11(c) depicts the results. The vehicle lateral position becomes almost indistinguishable from the reference position after about 7 seconds. The value of the *PI* is 97.10, which is an improvement over the performance of the FLC obtained using heuristically generated fuzzy rules.

10.5.3 Case 2: EA Tunes Fuzzy Membership Functions

In this case, fuzzy rules are assumed to be known and fixed. The role of an EA is to tune the input and output membership functions. We use the heuristically obtained rules table of Figure 10.9(a) as the given fuzzy rule base. The fuzzy sets for each input can be completely described using two real numbers if the corresponding fuzzy membership functions are symmetric with respect to the origin and, at each point of their universe of discourse, the sum of the values of the membership functions equals 1. The controller output fuzzy sets are also assumed to be symmetric with respect to the origin. Under the above assumptions, the controller output fuzzy singletons can be completely characterized using just two parameters. Hence, six real parameters can completely characterize all the input and output membership functions. So the chromosomes are strings of

(a)

e \ $\dot{e}$	LN	N	Z	P	LP
LN	LP	Z	P	LP	Z
N	LP	LP	LP	P	LN
Z	LP	P	Z	N	LN
P	LP	N	LN	LN	LN
LP	Z	LN	N	Z	LN

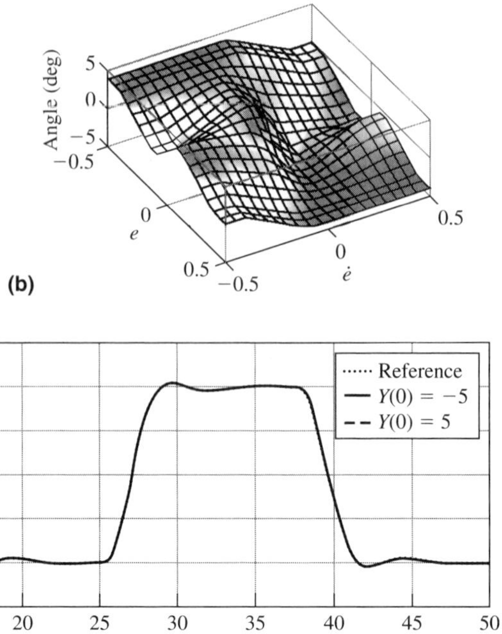

Figure 10.11 The best FLC of the 50th generation using a fuzzy rule base automatically generated by the EA. (a) Fuzzy rules table. (b) Control surface. (c) Tracking performance.

just six real parameters. For example, the fuzzy sets of Figure 10.8 can be represented as the string:

0.1	0.4	0.1	0.4	2	4

We now proceed to describe the EA that operates on the above-described fuzzy rule base. The population size, the selection method, the crossover operator and its rate, the mutation operator and its rate, and the fitness function, as well as other parameters, are the same as in the previous case. The resulting fuzzy sets, after 50 iterations of the EA, are shown in Figure 10.12. The control surface and the tracking performance of the closed-loop system driven by the FLC constructed using fuzzy sets of Figure 10.12 are depicted in Figure 10.13. The PI value is 43.51. The tracking error of the lateral position almost disappears in about 4 seconds.

10.5.4 Case 3: EA Tunes Fuzzy Rules and Membership Functions

We observed in the previous cases that changing fuzzy rules or modifying the shape of membership functions can improve the performance of the resulting FLC. We now analyze an EA that simultaneously operates on the fuzzy rules and fuzzy membership functions. This approach is a modification of the technique described in references 138 and 149.

In general, an ith fuzzy rule can have the form

$$\text{IF } x_1 \text{ IS } F_1^i \text{ AND } \cdots \text{ AND } x_n \text{ IS } F_n^i, \text{ THEN } u_1 \text{ IS } \theta_1^i \text{ AND } \cdots \text{ AND } u_m \text{ IS } \theta_m^i,$$

where $1 \le i \le r$, the x_j's are the state variables where $1 \le j \le n$, and the u_k's are the controller's

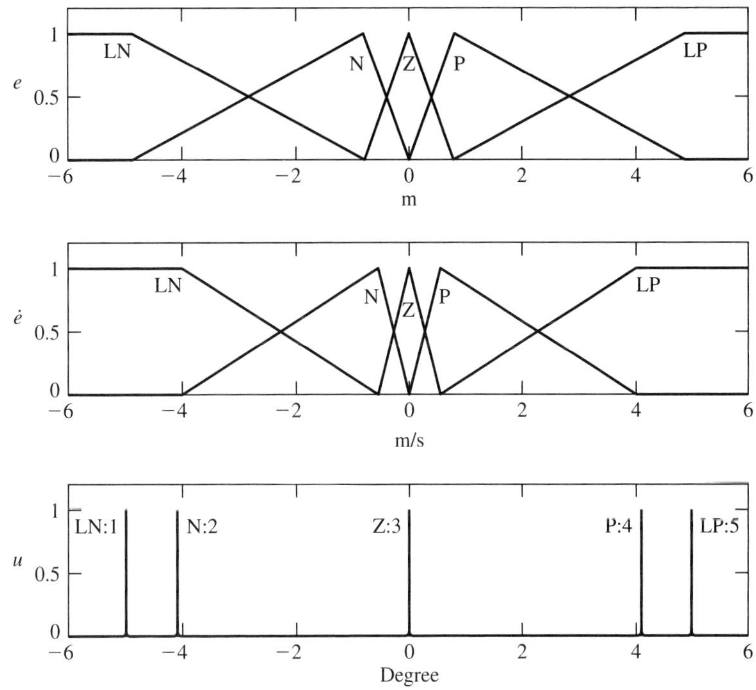

Figure 10.12 Input and output fuzzy membership functions of the best FLC of the 50th generation obtained by tuning fuzzy membership functions.

outputs, where $1 \leq k \leq m$. The F_j^i's are fuzzy sets with trapezoidal membership functions, and the θ_k^i's are the controller's output variables whose membership functions are singletons. A membership function, $\mu_{F_j^i}(x_j)$, of the fuzzy set, F_j^i, can be completely characterized using four parameters c_{ij}, d_{ij}, l_{ij}, and r_{ij}. It can be described as

$$\mu_{F_j^i}(x_j) = \begin{cases} \dfrac{x_j - l_{ij}}{c_{ij} - l_{ij} - d_{ij}} & \text{if } l_{ij} < x_j < c_{ij} - d_{ij}, \\ 1 & \text{if } c_{ij} - d_{ij} \leq x_j < c_{ij} + d_{ij}, \\ -\dfrac{x_j - r_{ij}}{r_{ij} - c_{ij} - d_{ij}} & \text{if } c_{ij} + d_{ij} < x_j < r_{ij}, \\ 0 & \text{otherwise,} \end{cases} \tag{10.17}$$

or, equivalently,

$$\mu_{F_j^i}(x_j) = \max\left(0, \min\left(1, \frac{x_j - l_{ij}}{c_{ij} - l_{ij} - d_{ij}}, -\frac{x_j - r_{ij}}{r_{ij} - c_{ij} - d_{ij}}\right)\right). \tag{10.18}$$

Figure 10.14 shows a generic fuzzy membership function of the form given by (10.17) or, equivalently, by (10.18). The fuzzy singletons, on the other hand, are described as

$$\mu_{\theta_k^i}(z) = \begin{cases} 1 & \text{if } z = \theta_k^i, \\ 0 & \text{otherwise.} \end{cases} \tag{10.19}$$

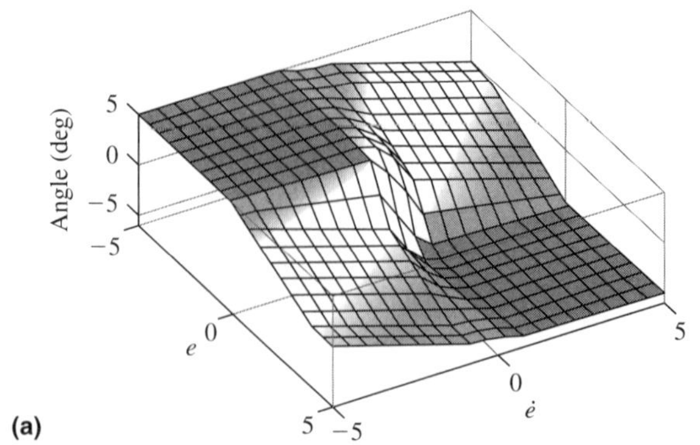

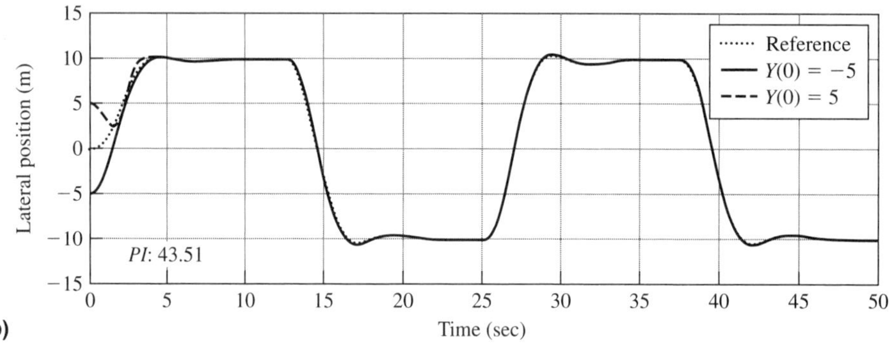

Figure 10.13 The best FLC of the 50th generation obtained by tuning fuzzy membership functions. (a) Control surface. (b) Tracking performance.

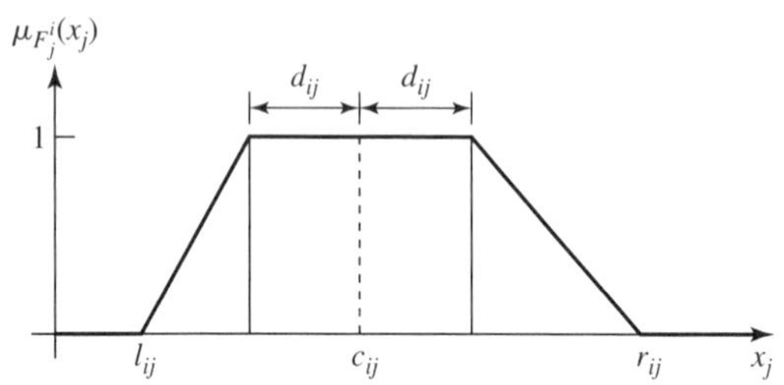

Figure 10.14 An example of a trapezoidal membership function.

The controller's signal u_k is calculated using the center average defuzzification formula,

$$u_k = \frac{\sum_{i=1}^{r} w_i \theta_k^i}{\sum_{i=1}^{r} w_i}, \tag{10.20}$$

where r is the number of fuzzy rules and $w_i = \mu_{F_1^i}\mu_{F_2^i} \times \cdots \times \mu_{F_n^i}$.

We now present a method of representing a given fuzzy rule base as a chromosome. In order to tune fuzzy rules and to alter the shape of membership functions, we require that a chromosome contains all the information pertaining to the structure of the fuzzy rules and the shape of membership functions explicitly. A chromosome is composed of two substructures: One substructure contains the fuzzy rules' structure matrix, whereas the other substructure contains the information concerning the parameters of the membership functions of the fuzzy rule base.

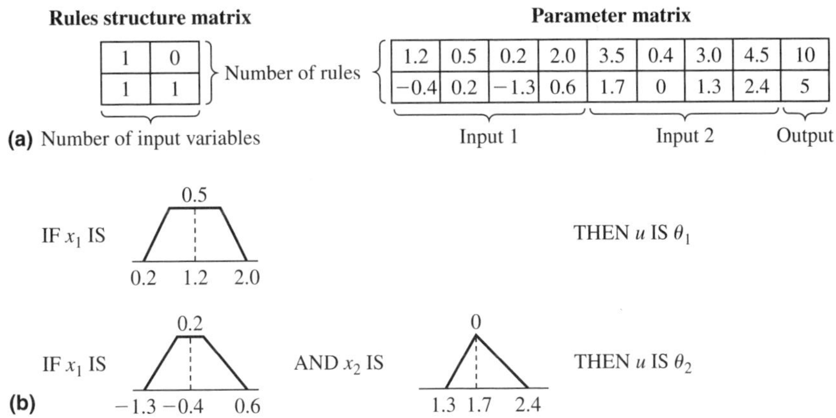

Figure 10.15 An example of a chromosome structure and its corresponding fuzzy rule base where the controller has two inputs, x_1 and x_2, and one output u.

The fuzzy rules' structure matrix contains the information related to the absence or presence of the premise part of the fuzzy rule corresponding to the particular controller input. Hence the elements of this matrix are from the set $\{0, 1\}$.

The parameter matrix contains the parameters that describe fuzzy membership functions. As previously mentioned, there are four parameters for each fuzzy trapezoidal membership function and one for each fuzzy singleton. The parameter matrix has the same number of rows as the number of rules, and the number of columns is equal to four times the number of the controller inputs plus the number of the controller's outputs. Figure 10.15(a) shows an example of the rules' structure matrix and the parameter matrix for the controller with two inputs and one output whose operation is described by two IF–THEN rules. The fuzzy rules and membership functions corresponding to the rules' structure matrix and the parameter matrix in Figure 10.15(a) are depicted in Figure 10.15(b). Note that a 0 in the second column of the first row of the rules' structure matrix means that the premise part of the first fuzzy rule corresponding to the second controller input is absent; that is, the first rule does not involve the controller's second input.

The fitness function has the form

$$\begin{aligned}\text{Fitness} &= \frac{1}{PI + \alpha \times (\text{number of rules})}\\ &= \frac{1}{\int_0^{50} \left(e_1^2(t) + e_2^2(t)\right) dt + \alpha \times (\text{number of rules})},\end{aligned} \tag{10.21}$$

where $e_1(t)$ is the tracking error with the initial lateral position at 5 m, and $e_2(t)$ is the tracking error with the initial lateral position at -5 m. The reference signal used has the amplitude of 10 m and the frequency of 0.04 Hz over 50 seconds. The design parameter α is employed to penalize for the excessive number of fuzzy rules. In our simulations, $\alpha = 1$.

We now describe genetic operators utilized by the EA. Three genetic operators are used with the probability p_g. We first discuss a genetic operator that modifies the number of fuzzy rules. Let r_{new} be the number of fuzzy rules after the genetic operator is applied, and let r_{old} be the number of fuzzy rules before the genetic operator is applied. Then,

$$r_{\text{new}} = r_{\text{old}} + \Delta r,$$

where Δr is the correction term that we analyze next. The number of fuzzy rules cannot exceed a prespecified bound $r_{\max}$. It is reasonable to assume that, on the average, r_{old} is about $r_{\max}/2$. Therefore, we wish to construct the correction term so that

$$\Delta r \in [-r_{\max}/2, r_{\max}/2].$$

Let Ψ be a random variable with $N(0, 1)$. The probability that Ψ will take its values in the interval $(-3, 3)$ is 0.99. Let integer$(\cdot)$ be the operator that produces the nearest integer value of its argument. Then, the random variable, integer$(\Psi r_{\max}/6)$, will take its value in the interval $(-\text{integer}(r_{\max}/2), \text{integer}(r_{\max}/2))$ with the probability 0.99. In view of the above, we use the following expression for updating the number of fuzzy rules:

$$r_{\text{new}} = r_{\text{old}} + \text{integer}(\Psi r_{\max}/6). \tag{10.22}$$

As follows from the discussion above, this particular form of the genetic operator will ensure that the resultant variation range of the rules' number is reasonable with respect to the maximum number $r_{\max}$ of the rules allowed. If the number of new rules, r_{new}, falls outside the allowed range, then we set $r_{\text{new}} = r_{\text{old}}$.

The second genetic operator acts upon the elements of the rules' structure matrix by taking the binary complement of each element—that is, $1 \rightarrow 0$ and $0 \rightarrow 1$ with the probability p_g.

The third genetic operator acts upon the parameter matrix as per the following expression:

$$p_{ij}^{\text{new}} = p_{ij}^{\text{old}} + \eta_{ij}\Psi e^{1-F_{\text{old}}/F_{\max}}, \tag{10.23}$$

where p_{ij}^{old} is the current value of the (ij)th element of the parameter matrix, p_{ij}^{new} is the updated value of the (ij)th element of the parameter matrix, Ψ is a random variable with $N(0, 1)$, F_{old} is the fitness value of the chromosome being operated upon, and $F_{\max}$ is the maximum fitness value of the chromosomes of the current population. Each design parameter η_{ij} was chosen according to the following recipe:

$$\eta_{ij} = (\text{maximum of the range of } p_{ij} - \text{minimum of the range of } p_{ij})/20.$$

The term $e^{1-F_{\text{old}}/F_{\max}}$ in equation (10.23) allows a chromosome with low fitness to vary the elements of the parameter matrix in a larger range than for the chromosomes with high fitness values.

In our implementation of the EA, we employ the elitist strategy to ensure that 20% of the fittest chromosomes in each generation are passed directly to the next generation without being acted upon by the genetic operators. The genetic operators, described above, are applied to the top 20% of the chromosomes in the current generation to produce the remaining 80% of the chromosomes of the new population.

After applying genetic operators to the chromosomes of the current generation, postprocessing may be required to ensure that the resulting chromosomes are feasible. If a zero row in the rules' structure matrix is created, then that row is deleted and the corresponding row in the parameter matrix is also deleted. The following conditions must be satisfied by the parameter matrix for the chromosome containing this matrix to be feasible:

$$\begin{aligned} d_{ij} &\geq 0, \\ l_{ij} &\leq c_{ij} - d_{ij}, \\ r_{ij} &\geq c_{ij} + d_{ij}. \end{aligned}$$

Whenever any one of the above conditions is violated by a parameter, then that parameter is assigned a boundary value. For example, if $d_{ij} < 0$, then d_{ij} is set to 0, and if $l_{ij} > c_{ij} - d_{ij}$, then l_{ij} is set to $c_{ij} - d_{ij}$, etc.

The above-described EA is used to construct a fuzzy logic controller for the front-wheel-steering ground vehicle model, whose dynamics are characterized by a certain type of symmetry that allows us to simplify computations. Specifically, suppose that in the rule base we have a fuzzy rule of the form

$$\text{IF } x_1 \text{ is } F_1 \text{ and } x_2 \text{ is } F_2, \text{ THEN } u \text{ is } \theta,$$

where the fuzzy sets F_1 and F_2 are specified by the membership functions $\mu_{F_1}(c_1, d_1, r_1, l_1)$ and $\mu_{F_2}(c_2, d_2, r_2, l_2)$, respectively. The parameters c_i, d_i, r_i, and l_i, $i = 1, 2$, characterize membership functions of the form shown in Figure 10.14. Because of the symmetry in the vehicle dynamics, the rule base also contains the rule

$$\text{IF } x_1 \text{ is } \tilde{F}_1 \text{ and } x_2 \text{ is } \tilde{F}_2, \text{ THEN } u \text{ is } -\theta,$$

where $\tilde{F}_1$ and $\tilde{F}_2$ have membership functions

$$\mu_{\tilde{F}_1} - \mu_{F_1}(-c_1, d_1, -l_1, -r_1) \quad \text{and} \quad \mu_{\tilde{F}_2} = \mu_{F_2}(-c_2, d_2, -l_2, -r_2),$$

respectively.

The size of the population in our simulations was 30, the maximum number of rules was $r_{\max} = 10$, and the probability p_g was equal to 0.1. A fuzzy rule base obtained after 100 iterations is shown in Figure 10.16. Only four rules were generated. The fuzzy rule base shown

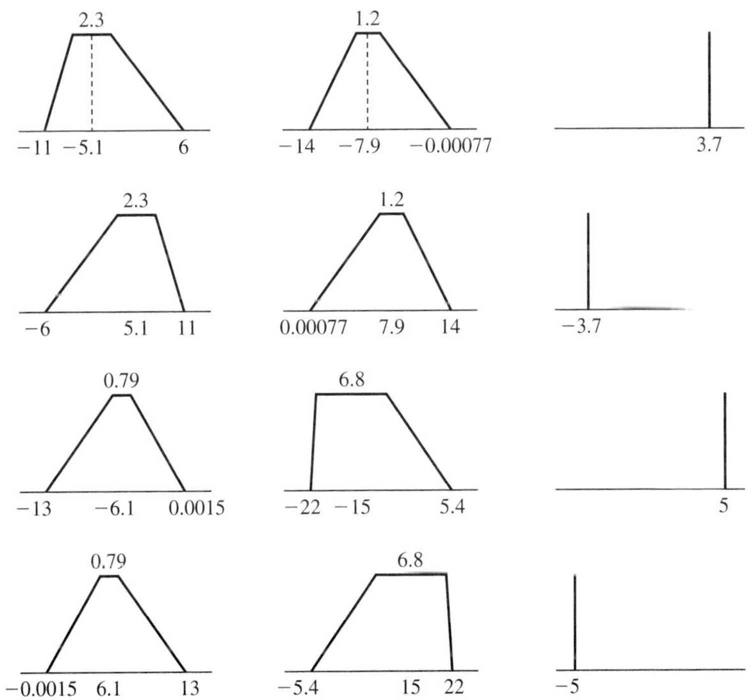

Figure 10.16 The best fuzzy rule base after 100 iterations of the EA operating on fuzzy membership functions and fuzzy rule base.

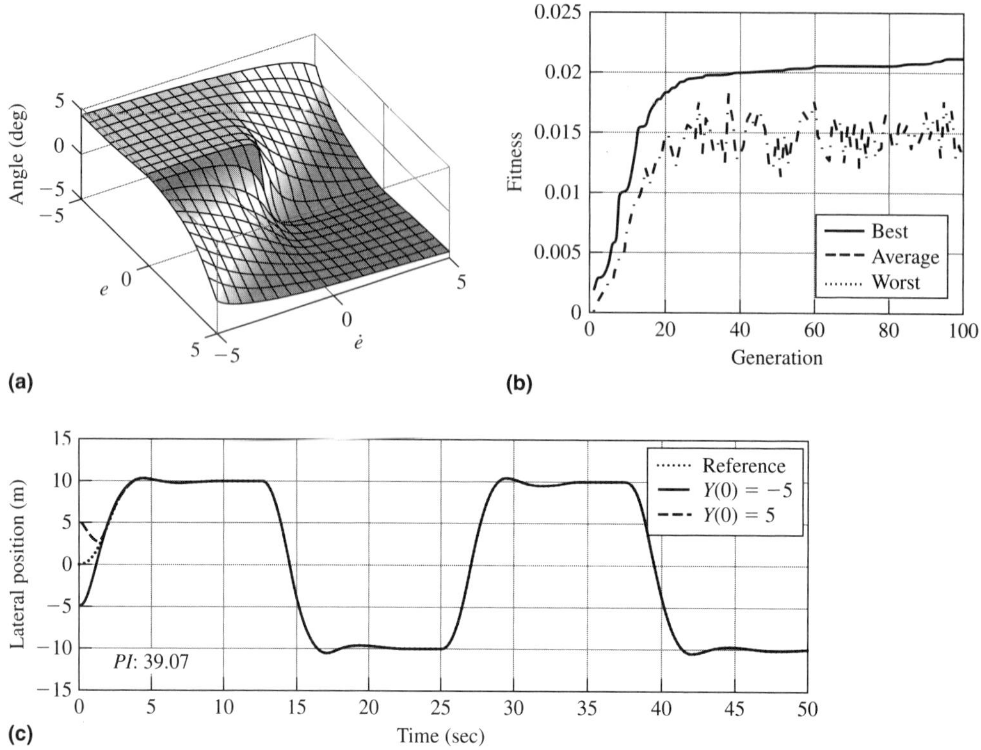

Figure 10.17 Summary of the simulation run of the closed-loop system driven by the fuzzy logic controller obtained from the fuzzy rule base shown in Figure 10.16. (a) Control surface. (b) Fitness values of the best, average, and worst chromosomes. (c) Tracking performance.

in Figure 10.16 was then used to construct a fuzzy logic controller that was connected to the vehicle model in the closed-loop configuration shown in Figure 10.7. The fuzzy control surface constructed using the obtained fuzzy rule base is shown in Figure 10.17(a). The results of a typical simulation run, along with the fitness values of the best, average, and worst chromosomes in each generation, are shown in Figure 10.17(b), while the tracking performance is illustrated in Figure 10.17(c). As in the previous cases, the simulations were performed for two different initial conditions, and excellent tracking was achieved in both cases. The value of the *PI* was reduced to 38.06, which is lower than in the two previous designs.

Notes

Basic principles of genetics have their roots in the experiments performed by a nineteenth-century monk, Gregor Mendel, in his monastery garden in Brno, now in the Czech Republic. Mendel worked with the garden pea (*Pisum sativum*) because it possesses traits that are easily recognizable and appear in widely contrasting forms. In addition, the garden pea lends itself well to controlled crosses. Mendel published his findings in 1866 in a paper entitled, "Experiments in Plant Hybridization." According to Elseth and Baumgardner [74, p. 3], "The publication of this paper marked the birth of genetics as a science. The principles stated in the paper still stand today as the cornerstone of modern genetics." The importance of Mendel's work went completely unrecognized for 34 years. "In 1900, three botanists, Hugo de Vries (Dutch), Carl Correns (German), and Erich von Tschermak-Seysenegg (Austrian), simultaneously discovered Mendel's paper on heredity, recognized its importance, and cited it in their own publications" [74, p. 7]. We add that the term *gene* was coined by W. L. Johannsen in 1903.

A seminal text on genetic algorithms by Holland [124] is definitely worth perusing. However, Holland's monograph is not intended as an introduction to genetic algorithms. A very well written introductory text to genetic algorithms is Mitchell's book [206]. A number of engineering design examples using genetic algorithms can be found in Gen and Cheng [96]. Bäck, Hammel, and Schwefel [15] survey the history as well as the current state of evolutionary computation, which is the term that encompasses evolution strategies, evolutionary programming, and genetic algorithms. Applications of genetic algorithms to automatic programming using evolving computer programs are described by Koza [166]. Automatic programming with genetic algorithms is called by Koza "genetic programming" (GP). Robust controller design with genetic algorithms are reported by Patton and Liu [224] and by Ge, Lee, and Zhu [94]. Kozek, Roska, and Chua [168], Pal and Bhandari [219], and Zamparelli [315] use genetic algorithms to train cellular neural networks. Kim and Myung [159] apply genetic algorithms to solve constrained optimization problems, while Koza et al. [167] employ genetic algorithms for the automatic design of electrical circuits. Yang, Hachino, and Tsuji [306] combine least squares techniques with a genetic algorithm to on-line identification of time-delay systems.

EXERCISES

10.1 Recall that the "elitism" is a genetic operator that forces the GA, or EA, to retain some number of the best individuals from the current population in the new population. Incorporate the elitism operator into your GA program and test it on the function (10.8). You may try, for example, to keep the best two individuals from the current population in the new population.

10.2 Implement the simple EA from Section 10.4 and test it on

(a) the function (10.8);

(b) the function $f(x) = x + |\sin(32x)|$, where $0 \le x < \pi$.

The function in (b) comes from Mitchell [206, p. 9].

10.3 Implement the simple EA from Section 10.4 that maximizes the MATLAB's peaks function

$$f(x, y) = 3(1-x)^2 \exp(-x^2 - (y+1)^2) - 10\left(\frac{x}{5} - x^3 - y^5\right)\exp(-x^2 - y^2) - \frac{1}{3}\exp(-(x+1)^2 - y^2),$$

where $-3 \le x, y \le 3$.

10.4 Implement the simple EA from Section 10.4 that maximizes the function

$$f(x, y) = 3 + y\sin(0.7x) + x\cos(1.5y),$$

where $-3.0 \leq x \leq 12.1$ and $4.1 \leq y \leq 5.8$.

10.5 Consider the classic traveling salesman problem (TSP): Suppose a salesman must visit clients in different cities and then return home. What is the shortest tour through those cities, visiting each one once and only once? Implement in MATLAB a simple variant of the genetic algorithm that optimizes the salesman route for 10 cities. How many different possible paths are there for 10 cities? Following Fogel [88, p. 29], use the following representation of any candidate solution: For a given map of the cities to be visited (including the salesman's home base), assign to each city a positive integer. Thus, for example, if there were six cities, one possible solution might be [1 2 3 4 5 6]. The above represents an order of progression of the salesman's trip. Note that because the problem requires a round trip, the first coordinate of the vector representing the first city of the trip is also the last city visited by the salesman. Because the trip is a closed loop, it does not matter from which city the salesman starts his trip. As far as a crossover operator implementation is concerned, we recommend using Fogel's idea [88, p. 29], where a single-parent operator produces an offspring by inverting the visiting order between two randomly chosen coordinates of the parent vector. As an example of this implementation of the single-parent crossover operator, suppose that the parent is as above and that the randomly chosen coordinates—that is, the inversion points—are the second and fifth ones. The offspring would then be [1 5 3 4 2 6].

Plot a map with the cities and mark the obtained salesman's route. The coordinates of the cities are given in Table 10.2.

Table 10.2 Locations of the Cities in Exercise 10.5

x	0.4306	3.7094	6.9330	9.3582	4.7758	1.2910	4.83831	9.4560	3.6774	3.2849
y	7.7288	2.9727	1.7785	6.9080	2.6394	4.5774	8.43692	8.8150	7.0002	7.5569

10.6 Consider a closed-loop system with a fuzzy logic controller (FLC) in the loop shown in Figure 10.18. Suppose that we use five membership functions for the FLC inputs and its output. A possible set of rules is given in Table 10.3, where LN denotes large negative, N means negative, ZE stands for zero, P means positive, and LP refers to large positive. Implement a variant of the simple EA from Section 10.4 that generates the set of rules to be used to synthesize the FLC. Assign to each fuzzy set a positive integer; for example, let LN be encoded as 1, N as 2, ZE as 3, P as 4, and LP as 5. Then,

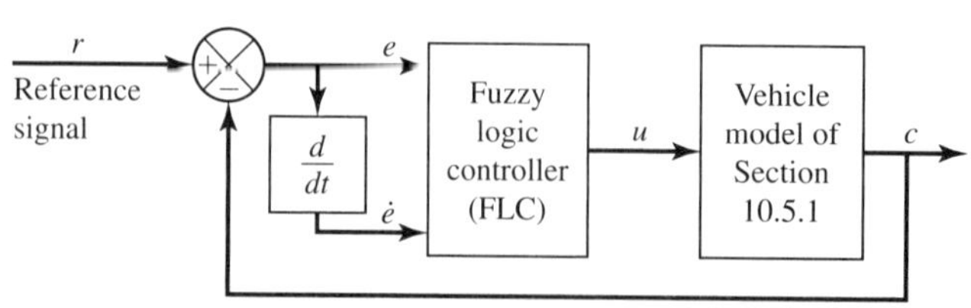

Figure 10.18 Closed-loop system with a fuzzy logic controller in the loop.

Table 10.3 Rules Matrix to Be Used to Construct a Fuzzy Logic Controller in Exercise 10.6

e \ $\dot{e}$	LN	N	ZE	P	LP
LN	LN	LN	LN	N	ZE
N	LN	LN	N	ZE	P
ZE	N	N	ZE	P	P
P	N	ZE	P	LP	LP
LP	ZE	P	LP	LP	LP

a candidate solution, which is shown in Table 10.3, can be represented in a vector form as

$$[1\ \ 1\ \ 2\ \ 2\ \ 3\ \ 1\ \ 1\ \ 2\ \ 3\ \ 4\ \ 1\ \ 2\ \ 3\ \ 4\ \ 5\ \ 2\ \ 3\ \ 4\ \ 5\ \ 5\ \ 3\ \ 4\ \ 4\ \ 5\ \ 5];$$

that is, a vector representation of the candidate solution given by the rules matrix is obtained using the stacking operation of the columns of the rules matrix. Use the one-point crossover operator similar to the one in the canonical GA. The mutation operation can be implemented by altering a randomly chosen coordinate in the given chromosome—that is, the vector representing a candidate solution. You can use the roulette-wheel method to determine the replacement. For example, if the seventh component was selected to be altered, then you can use the roulette-wheel method to determine an integer from the set $\{1, 2, 3, 4, 5\}$ that will replace the "1" that occupies the seventh coordinate of the above chromosome. Chan, Lee, and Leung [47] used a similar version of the EA to generate fuzzy rules to synthesize a fuzzy logic target tracking algorithm. You can use the fitness function given by (10.16).

CHAPTER 11 Chaotic Systems and Fractals

> An apparent paradox is that chaos is deterministic, generated by fixed rules which do not themselves involve any elements of change. We even speak of deterministic chaos. In principle, the future is completely determined by the past; but in practice small uncertainties, much like minute errors of measurement which enter into calculations, are amplified, with the effect that even though the behavior is predictable in the short term, it is unpredictable over the long run.
>
> —*Chaos and Fractals: New Frontiers of Science* [227, p. 11]

11.1 Chaotic Systems Are Dynamical Systems with Wild Behavior

The objective of this chapter is to show the reader "the wild things that simple nonlinear equations can do" [198, p. 459]. These simple nonlinear models with very complicated dynamics are examples of *chaotic systems,* which are deterministic nonlinear dynamical systems that exhibit a random-like behavior. A discrete-time model of population dynamics discussed in the next section, Newton's method for finding roots of certain polynomials, or a jetting balloon after its release are examples of dynamical chaotic systems. Trajectories of chaotic dynamical systems are sensitive to initial conditions in the sense that starting from slightly different initial conditions the trajectories diverge exponentially. Another property of a chaotic dynamical system is *loss of information* about initial conditions. We follow Moon [209, p. 4] to explain this property. Suppose that we are given a state-plane portrait of a chaotic dynamical system in the (x, y)-plane. Assume that we specify the x coordinate with accuracy Δx and the y coordinate with accuracy Δy. Then, we could form a grid in the (x, y)-plane, shown in Figure 11.1, where each area would be of size $\Delta x\ \Delta y$. Thus, if we select initial conditions to the stated accuracy, then we know that these initial conditions are located somewhere in the shaded box in the (x, y)-plane corresponding to the initial time t_0. For a chaotic system, this uncertainty grows in time to, say, N boxes at a later time t_1, as illustrated in Figure 11.1. The uncertainty growth, as mentioned above, is exponential and for continuous-time systems can be described by

$$N(t) \approx e^{\mu(t-t_0)} N(t_0), \tag{11.1}$$

where the constant μ is a measure of sensitivity with respect to initial conditions. Lyapunov exponents, which we discuss in a subsequent section, can be used to measure the initial condition

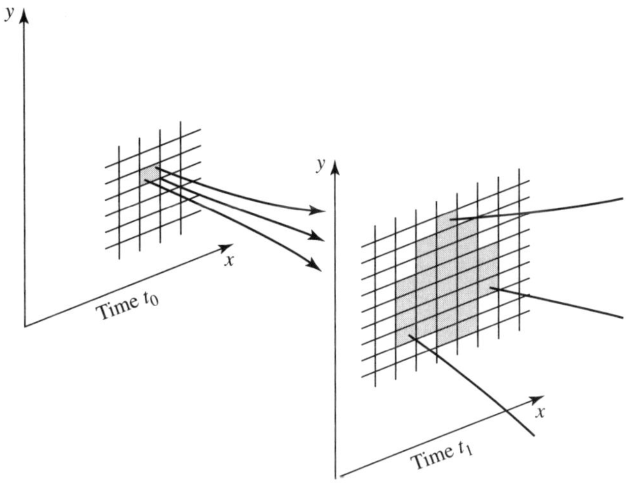

Figure 11.1 An illustration of the growth of uncertainty in a chaotic dynamical system.

sensitivity. Chaotic behavior can arise in low-order nonlinear dynamical systems. In the case of continuous-time systems, chaotic behavior can only occur for systems of dimension three or higher. However, in discrete-time systems, chaos can occur even in one-dimensional systems, as we will see in the following section.

11.2 Chaotic Behavior of the Logistic Equation

11.2.1 The Logistic Equation—An Example from Ecology

The so-called logistic equation, or logistic map, whose behavior we analyze in this section, arises in the analysis of the growth of a population, for example. Before proceeding, we define a few terms the we will use in our discussion.

Definition 11.1 A map is a function whose domain (input) space and range (output) space are the same. Let $\boldsymbol{x}(0)$ be a point in $\mathbb{R}^n$ and $\boldsymbol{f}$ be a map on $\mathbb{R}^n$. The orbit, or trajectory, of $\boldsymbol{x}(0)$ under $\boldsymbol{f}$ is the set of points

$$\{\boldsymbol{x}(0), \boldsymbol{f}(\boldsymbol{x}(0)), \boldsymbol{f}^2(\boldsymbol{x}(0)), \ldots, \boldsymbol{f}^k(\boldsymbol{x}(0)), \ldots\},$$

where $\boldsymbol{f}^k$ is the kth iterate of $\boldsymbol{f}$. The starting point, $\boldsymbol{x}(0)$, of the orbit is called the initial value of the orbit.

In what follows, we derive a discrete-time version of the logistic equation. The continuous-time logistic equation was analyzed in Subsection 1.7.5. We first analyze a situation where there are no predators and there is an unlimited supply of food. For the sake of argument, we consider a population of rabbits. Assume that, on the average, each rabbit gives birth to σ new rabbits in one unit of time and that no rabbit dies. Let $N(k)$ be the number of rabbits at time $t = k$; then the change in the population over a time period between times k and $k + 1$ is

$$N(k+1) - N(k) = \sigma N(k). \tag{11.2}$$

Thus,

$$N(k+1) = (1+\sigma)N(k). \tag{11.3}$$

Equation (11.3) is a very simple linear model of population growth. Given the number $N(0)$ of rabbits at time $k = 0$, the number of rabbits at time k is

$$N(k) = (1+\sigma)^k N(0). \tag{11.4}$$

One can see from (11.4) that the linear model given by (11.3) predicts the exponential grow of population. We now try to make the above model more realistic. We assume, as before, that the number of births is proportional to the size of the population. Let b be the birth rate; then the number of births at time k is $bN(k)$. We also assume that the number of deaths is proportional to the population size. Let d be the death rate; then the number of deaths at time k is $dN(k)$. The change in population in a time period between times k and $k+1$ is

$$N(k+1) - N(k) = bN(k) - dN(k). \tag{11.5}$$

We rewrite equation (11.5) as

$$N(k+1) = (1+b-d)N(k). \tag{11.6}$$

Let $\sigma = b - d$. Then, equation (11.6) has the same form as equation (11.3). In both cases, the population grows exponentially. We refer to the constant σ as the *unrestricted growth rate*. Over short periods of time, the above equation may be sufficient for purposes of an analysis of a population. Over long periods of time, the growth rates may change as the size of a population changes.

We now alter the above linear models by introducing the concept of saturation. We first assume that the environment of the population can only support a certain number of species. We denote this number by K, and we call it the *carrying capacity*. Thus, if $N(k) > K$, then there will not be enough food or space available, and more species will die than will be born. In such a situation the growth rate should be negative. If the population at time k is $N(k) < K$, then the growth rate should be positive. Finally, if $N(k)$ is much smaller than the carrying capacity, then the growth rate should be close to the unrestricted growth rate σ. A possible function that satisfies the above requirements has the form

$$\sigma\left(1 - \frac{N(k)}{K}\right). \tag{11.7}$$

The change in the size of population then can be described as

$$N(k+1) - N(k) = \sigma\left(1 - \frac{N(k)}{K}\right)N(k), \tag{11.8}$$

or, equivalently

$$N(k+1) = (1+\sigma)N(k) - \frac{\sigma}{K}N^2(k). \tag{11.9}$$

Equation (11.9) is called the *logistic equation*. The term $\frac{\sigma}{K}N^2(k)$ is referred to as the *damping term,* because it dampens the growth of the population. To simplify (11.9), we scale the variable $N(k)$ by introducing the new variable

$$x(k) = \frac{\sigma}{\rho K}N(k),$$

where

$$\rho = 1 + \sigma.$$

Then, equation (11.9) becomes

$$\boxed{x(k+1) = \rho x(k)(1 - x(k))} \tag{11.10}$$

We will now show that if $0 < \rho \leq 4$ and $0 \leq x(0) \leq 1$, then $0 \leq x(k) \leq 1$ for $k = 1, 2, \ldots$. Let

$$f(x) = \rho x(1 - x).$$

Then,

$$\frac{df(x)}{dx} = \rho - 2\rho x = 0$$

for $x = 1/2$ and $f_{\max}(x) = \rho/4$. Hence, for $k = 1, 2, \ldots,$

$$0 \leq x(k) \leq 1 \quad \text{if} \quad 0 < \rho \leq 4 \quad \text{and} \quad 0 \leq x(0) \leq 1.$$

11.2.2 Stability Analysis of the Logistic Map

We begin our investigation of the logistic equation (11.10) by seeking its equilibrium states (fixed points). These can be found by setting $x(k+1)$ equal to $x(k)$ in (11.10) to obtain

$$x(k)\,(\rho x(k) + (1 - \rho)) = 0.$$

It follows that the equilibrium solutions are

$$x^* = 0 \quad \text{and} \quad x^* = \frac{\rho - 1}{\rho}.$$

We will now study attracting properties of the fixed points as a function of the parameter ρ. Roughly speaking, a fixed point $x^* = f(x^*)$ of the system $x(k+1) = f(x(k))$ is attracting if, for a starting point close to x^*, the next iterate is still closer. Formally, a fixed point x^* is attracting if there is an $\varepsilon > 0$ such that when $|x(0) - x^*| < \varepsilon$, then $\lim_{k\to\infty} x(k) = x^*$.

A fixed point x^* is repelling if for any $\varepsilon > 0$ no matter how small, if $0 < |x(0) - x^*| < \varepsilon$, then $|x(k) - x^*| > \varepsilon$ for some k.

Theorem 11.1 Suppose $x^* = f(x^*)$ is a fixed point of $x(k+1) = f(x(k))$, where f is continuously differentiable. Then, x^* is attracting if

$$|f'(x^*)| < 1$$

and is repelling if

$$|f'(x^*)| > 1.$$

The case when $|f'(x^*)| = 1$ is inconclusive.

Proof We first show that if $x^* = f(x^*)$ is such that $|f'(x^*)| < 1$, then x^* is attracting. Because by assumption $|f'(x^*)| < 1$ and f' is continuous, there exists an interval $(x^* - \varepsilon, x^* + \varepsilon)$ about x^* such that for some $L > 0$,

$$|f'(x^*)| < L < 1 \quad \text{and} \quad |f'(x)| < L \qquad \text{for every } x \in (x^* - \varepsilon, x^* + \varepsilon).$$

Let $x(k) \in (x^* - \varepsilon, x^* + \varepsilon)$. We will show that $x(k+1)$ is closer to x^* than $x(k)$. We have

$$|x(k+1) - x^*| = |f(x(k)) - f(x^*)|$$

because $x^* = f(x^*)$ is a fixed point. Using the mean value theorem yields

$$|x(k+1) - x^*| = |f(x(k)) - f(x^*)| = |f'(\tilde{x})||x(k) - x^*|$$

for some $\tilde{x} \in (x(k), x^*)$ or $\tilde{x} \in (x^*, x(k))$ if $x^* < x(k)$. Since $x(k) \in (x^* - \varepsilon, x^* + \varepsilon)$, then $\tilde{x} \in (x^* - \varepsilon, x^* + \varepsilon)$ also. Thus, $|f'(\tilde{x})| < L < 1$, and therefore

$$|x(k+1) - x^*| < L|x(k) - x^*|.$$

From the above, we conclude that if $x(0) \in (x^* - \varepsilon, x^* + \varepsilon)$, then

$$|x(k) - x^*| < L^k|x(0) - x^*|.$$

Because $L < 1$, we have $\lim_{k\to\infty} |x(k) - x^*| = 0$, that is,

$$\lim_{k\to\infty} x(k) = x^*,$$

which means that x^* is an attracting fixed point.

Similarly, if $x^* = f(x^*)$ is such that $|f'(x^*)| > 1$, then it is a repelling fixed point. Indeed, if $|f'(x^*)| > 1$, then because f' is continuous, there is an interval $(x^* - \varepsilon, x^* + \varepsilon)$ about x^* and a constant $L > 1$ such that $|f'(x)| > L$ for all $x \in (x^* - \varepsilon, x^* + \varepsilon)$. We will now show that if $x(k) \in (x^* - \varepsilon, x^* + \varepsilon)$, then $x(k+1)$ will be further away from x^* than $x(k)$. Using the mean value theorem, we obtain

$$\begin{aligned}|x(k+1) - x^*| &= |f(x(k)) - f(x^*)| \\ &= |f'(\tilde{x})||x(k) - x^*| \\ &> L|x(k) - x^*|. \end{aligned} \tag{11.11}$$

If $x(k+1)$ is no longer in $(x^* - \varepsilon, x^* + \varepsilon)$, we are done. Otherwise, we repeat the process to get

$$\begin{aligned}|x(k+2) - x^*| &= |f(x(k+1)) - f(x^*)| \\ &> L|x(k+1) - x^*| \\ &> L^2|x(k) - x^*|. \end{aligned}$$

If $x(k+2)$ is no longer in the interval $(x^* - \varepsilon, x^* + \varepsilon)$, we are done. If not, we repeat the process to obtain

$$|x(k+m) - x^*| > L^m|x(k) - x^*|.$$

Since $(x^* - \varepsilon, x^* + \varepsilon)$ is a finite interval and $L^m \to \infty$ as $m \to \infty$, at some time, $x(k+m)$ must leave the interval, which means that x^* is a repelling fixed point. The proof of the theorem is complete.

Equation (11.11) tells us that the trajectory $\{x(k)\}$ diverges exponentially from x^* as $k \to \infty$. Thus, a trajectory with an initial condition close to x^* for the case when $|f'(x^*)| = L > 1$ will separate from x^* at a multiplicative rate of approximately $L > 1$ per iteration until it moves significantly away from x^*. In other words, the distance $|x(k) - x^*|$ for $x(0)$ close to x^* will grow exponentially; that is, this distance will be magnified approximately by L for each iteration. To quantify this multiplicative rate of separation of points close to x^*, the Lyapunov exponent term is introduced that we will discuss in Section 11.4.

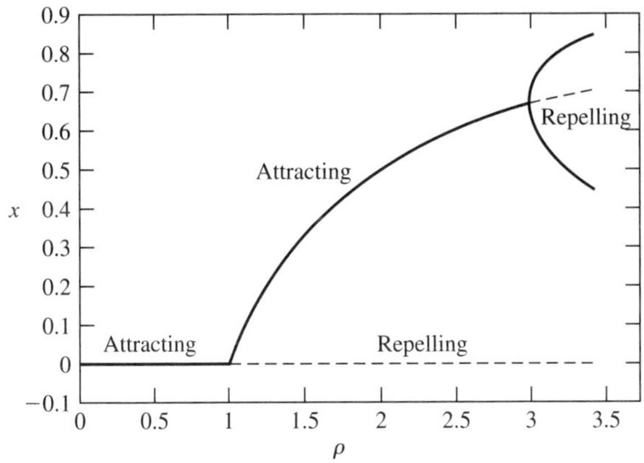

Figure 11.2 Exchange of attracting properties of the fixed points of the logistic equation for $0 < \rho \leq 3.44$.

We now consider the logistic equation (11.10). Recall that $f = \rho x(1 - x)$ denotes the right-hand side of the logistic equation. The derivative of f with respect to x is

$$f'(x) = \rho(1 - 2x).$$

Therefore, $f'(0) = \rho$. Since $|f'(0)| < 1$ when $\rho < 1$, by Theorem 11.1, the fixed point $x^* = 0$ is attracting for $\rho < 1$ and repelling for $\rho > 1$. The fixed point $x^* = (\rho - 1)/\rho$ is outside of the range of x for $\rho < 1$. Thus, if $0 < \rho < 1$, we have just one attractive fixed point. For $\rho > 1$, the fixed point $x^* = 0$ is repelling and $x^* = (\rho - 1)/\rho$ comes in within the range of the variable x. The derivative of f evaluated at $x^* = (\rho - 1)/\rho$ is

$$f'\left(\frac{\rho - 1}{\rho}\right) = 2 - \rho.$$

By Theorem 11.1, the fixed point $x^* = (\rho - 1)/\rho$ is attracting if $|2 - \rho| < 1$—that is, if

$$1 < \rho < 3.$$

Thus, for $\rho = 1$ there is an exchange of attracting properties from one fixed point to the other. This change in the qualitative behavior at a certain parameter value is called a *bifurcation*. In Figure 11.2, the fixed points $x^* = 0$ and $x^* = (\rho - 1)/\rho$ are shown and the ranges of ρ for which they are attracting. For $\rho > 3$, there are no attracting fixed points, and the solutions of the logistic equation become very complex. We discuss the logistic equation dynamics for $\rho > 3$ after displaying its solutions for $\rho = 0.7$, $\rho = 1.4$, $\rho = 2.8$, and $\rho = 3.6$ in Figure 11.3. The initial condition in all four cases was $x(0) = 0.9$.

Another way of displaying a solution of a first-order difference equation is a *cobweb*. To construct a cobweb for the given first-order discrete-time equation, $x(k + 1) = f(x(k))$, we first draw the function $y = f(x)$ representing the right hand side of the difference equation. Then, we draw the straight line $y = x$. The fixed points correspond to the points of intersection of these two graphs. The solution sequence, represented by the cobweb, starting at $x(0)$ is obtained as follows. Starting from $x(0)$, draw upward vertical line to intersect the graph of $f(x)$. This corresponds to the calculation of $x(1)$. The value of $x(1)$ is transferred to the straight line $y = x$. The x coordinate of the intersection is $x(1)$, which is then used to generate $x(2)$ using the above procedure. The process is repeated over and over again. In Figure 11.4 the cobwebs corresponding to the solutions displayed in Figure 11.3 are presented.

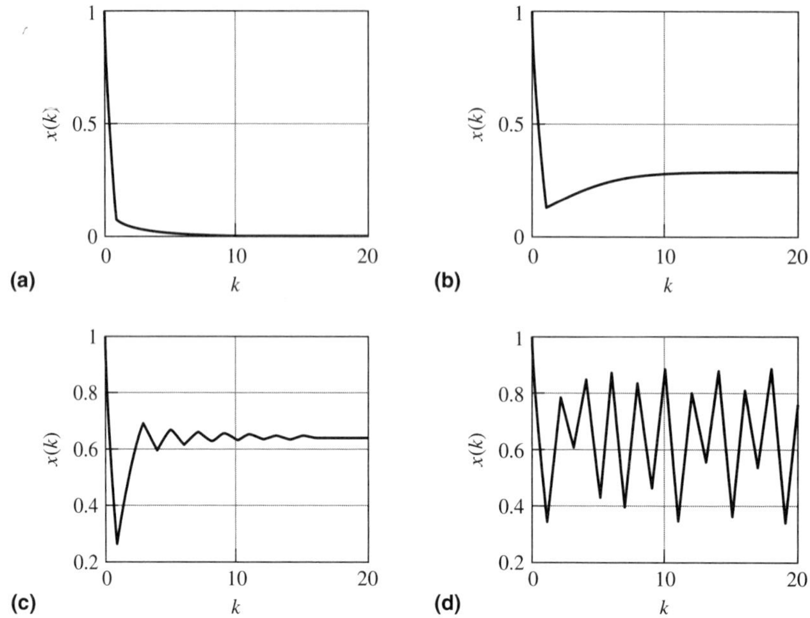

Figure 11.3 Solutions of the logistic equation for different values of the parameter ρ: (a) $\rho = 0.7$, (b) $\rho = 1.4$, (c) $\rho = 2.8$, and (d) $\rho = 3.6$.

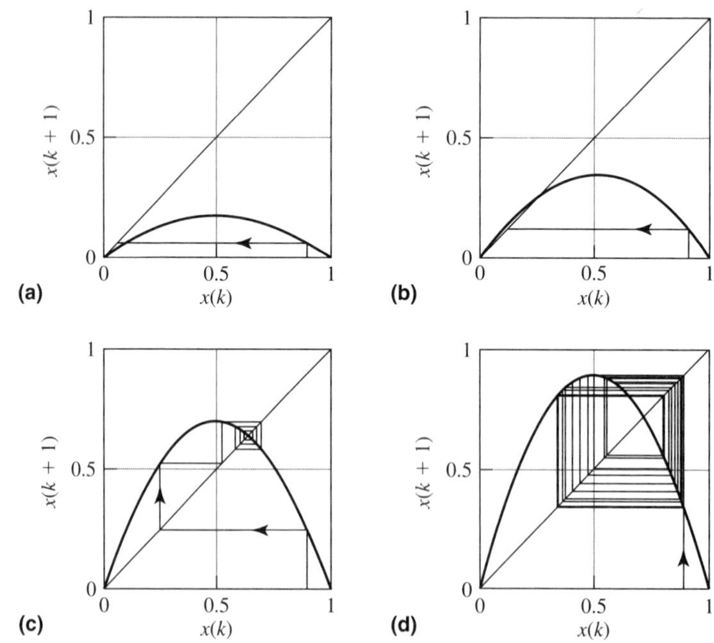

Figure 11.4 Cobwebs for the logistic equation: (a) $\rho = 0.7$, (b) $\rho = 1.4$, (c) $\rho = 2.8$, and (d) $\rho = 3.6$.

As we mentioned before, for $\rho > 3$ there are no attracting fixed points. As ρ increases, the solutions of the logistic equation exhibit increasing complexity. For ρ just slightly larger than 3, the solution sequence settles down into a steady oscillation of period 2 as shown in Figure 11.5. Then, as ρ is further increased, periodic solutions of period 4, 8, 16, ... appear. In Figure 11.6 a solution of period 4 is shown corresponding to $\rho = 3.5$. For $\rho = 3.9$ the solution is shown in Figure 11.7.

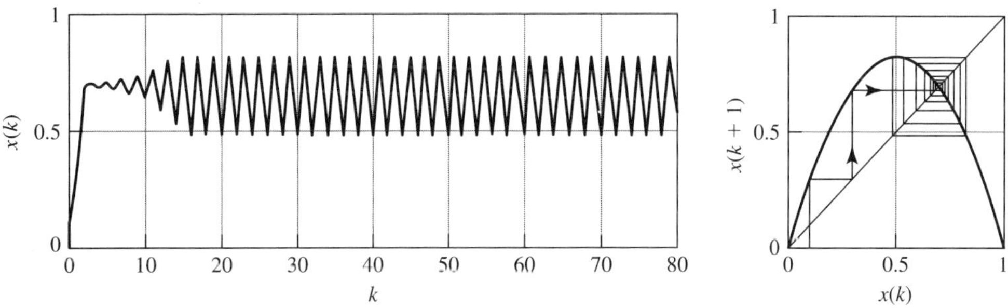

Figure 11.5 A two-cycle solution and its cobweb for $\rho = 3.3$ and $x(0) = 0.1$.

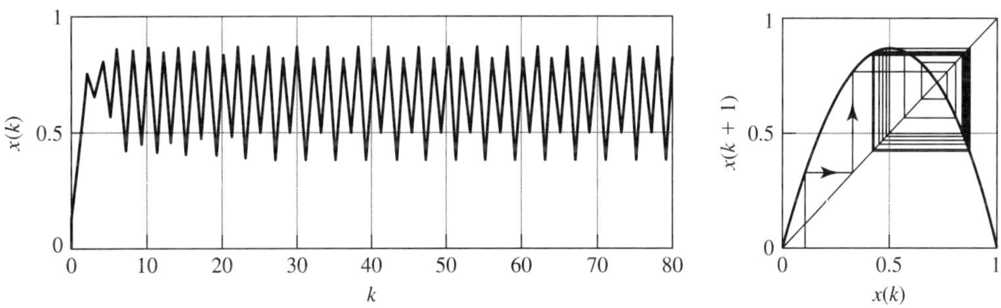

Figure 11.6 A period-four solution and its cobweb for $\rho = 3.5$ and $x(0) = 0.1$.

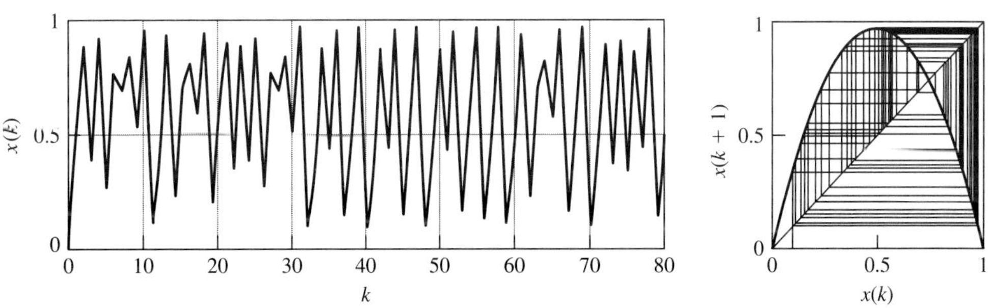

Figure 11.7 Aperiodic solution and its cobweb for $\rho = 3.9$ and $x(0) = 0.1$.

We now analyze the solutions for $\rho > 3$. To simplify the analysis, we first make the substitution

$$x(k) = z(k) + \frac{\rho - 1}{\rho}, \tag{11.12}$$

which is equivalent to shifting the fixed point $x^* = (\rho - 1)/\rho$ to the origin in the (ρ, z) coordinates. Note that

$$x(k+1) = z(k+1) + \frac{\rho - 1}{\rho}. \tag{11.13}$$

Substituting (11.12) and (11.13) into the logistic equation (11.10) yields

$$z(k+1) + \frac{\rho - 1}{\rho} = \rho\left(z(k) + \frac{\rho - 1}{\rho}\right)\left(1 - z(k) - \frac{\rho - 1}{\rho}\right). \tag{11.14}$$

After performing some manipulations, we represent (11.14) as

$$\boxed{z(k+1) = (2 - \rho)z(k) - \rho z^2(k)} \tag{11.15}$$

The fixed points of (11.15) are $z^* = 0$ corresponding to $x^* = (\rho - 1)/\rho$, and $z^* = (1 - \rho)/\rho$ corresponding to $x^* = 0$. We investigate the attracting properties of the fixed point $z^* = 0$ for $\rho > 3$. We see from Figure 11.5 that for ρ just greater than 3, the solution of the logistic equation goes to an attracting period-two cycle. Two numbers p and q form a period-2 cycle for a first-order discrete system if, when $x(k) = p$, then $x(k+1) = q$, $x(k+2) = p$, and so on. A period-2 cycle is attracting if there are intervals (p_1, p_2) and (q_1, q_2) about p and q, respectively, such that:

1. If $x(0) \in (p_1, p_2)$, then $\lim_{k\to\infty} x(2k) = p$ and $\lim_{k\to\infty} x(2k+1) = q$.
2. If $x(0) \in (q_1, q_2)$, then $\lim_{k\to\infty} x(2k) = q$ and $\lim_{k\to\infty} x(2k+1) = p$.

The fixed points p and q of the period-2 cycle of equation (11.15) can be found by solving the equation

$$z = g(g(z)), \tag{11.16}$$

where $g = (2 - \rho)z - \rho z^2$. We can use MATLAB's function `compose` to form (11.16), and then use the function `solve` to solve (11.16). We obtain four roots. Two of them are the fixed points that we analyzed above. The remaining two roots are the fixed points of the period-2 cycle:

$$p = -\frac{1}{2\rho}\left(\rho - 3 + \sqrt{\rho^2 - 2\rho - 3}\right) \tag{11.17}$$

and

$$q = -\frac{1}{2\rho}\left(\rho - 3 - \sqrt{\rho^2 - 2\rho - 3}\right). \tag{11.18}$$

Once we have found the 2-cycle, we next determine the range of ρ for which it is attracting. A 2-cycle p and q of the system $z(k+1) = g(z(k))$ is attracting if the fixed points p and q of the map $y(k+1) = h(y(k))$ are attracting, where $h(y) = g(g(y))$. By Theorem 11.1, the fixed points $p = h(p)$ and $q = h(q)$ are attracting if $|h'(p)| < 1$ and $|h'(q)| < 1$. Using the chain rule gives

$$h'(y) = \frac{d}{dy} g(g(y)) = g'(g(y))g'(y). \tag{11.19}$$

Because $g(p) = q$ and $g(q) = p$, we obtain

$$h'(p) = g'(g(p))g'(p) = g'(q)g'(p). \tag{11.20}$$

Similarly,

$$h'(q) = g'(p)g'(q). \tag{11.21}$$

Therefore, $|h'(p)| < 1$ and $|h'(q)| < 1$ when

$$|g'(p)g'(q)| < 1. \tag{11.22}$$

Using Theorem 11.1 and the arguments similar to the ones above, we conclude that the 2-cycle represented by p and q is repelling if

$$|g'(p)g'(q)| > 1. \tag{11.23}$$

We next substitute (11.17) and (11.18) into $g'(p)g'(q)$ and evaluate the resulting expression using MATLAB's function `subs`, to obtain

$$g'(p)g'(q) = (2 - \rho - 2\rho p)(2 - \rho - 2\rho q) = 4 - \rho^2 + 2\rho. \tag{11.24}$$

Thus, the 2-cycle represented by the pair (p, q) is attracting if

$$-1 < 4 - \rho^2 + 2\rho < 1, \tag{11.25}$$

where $\rho > 3$. Solving (11.25) yields

$$3 < \rho < 1 + \sqrt{6}. \tag{11.26}$$

When $\rho > 1 + \sqrt{6}$, an attracting 4-cycle appears. The bifurcation diagram in Figure 11.8 shows that as the value of ρ increases, the period doubling process takes place. Each of the branches of the bifurcation diagram produces a period-2 cycle of its own. The process continues as the parameter ρ approaches the value $\rho = 3.5699$, until no periodic cycles are created. For $\rho = 3.5699$ the behavior is not cyclic anymore. The behavior of the solutions of the logistic equation for $\rho \geq 3.5699$ is complex indeed, as can be seen, for example, by inspecting the bifurcation diagram shown in Figure 11.8. Various windows of ranges of $\rho > 3.5699$ correspond

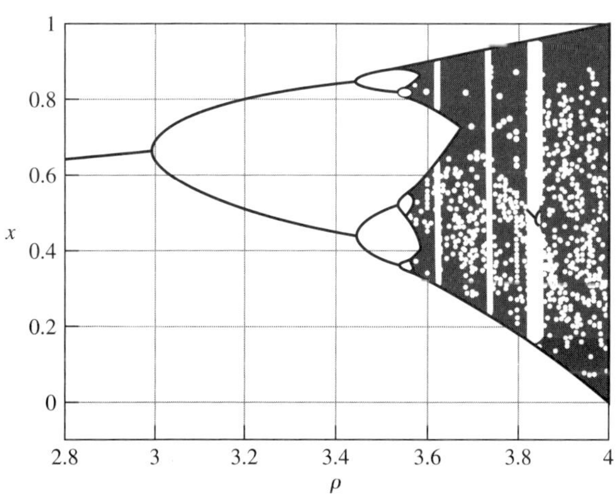

Figure 11.8 Bifurcation diagram of the logistic equation.

to different periodicities of solutions as well as to aperiodic solutions. These aperiodic solutions are called *chaotic solutions* or *chaotic motions*. Here, we quote a descriptive definition of chaotic motion due to Thompson and Stewart [281, p. 197]: "... one way to define chaos is in the negative—recurrent behavior that is not an equilibrium, a cycle or even a quasi-periodic motion—there is more to be said about chaos. Chaotic motion has some aspect that is provably as random as a coin toss. The randomness arises from sensitive dependence on imperfectly known initial conditions, resulting for example in broadband noise in the power spectra of observable time series. This seems remarkable because the dynamical system needs no stochastic input to achieve this." A precise definition of chaotic motion is given in Section 11.4.

11.2.3 Period Doubling to Chaos

We will now describe one of the ways in which chaotic motion can arise—by repeated flip bifurcations, that is, through period doubling. Such a behavior arises for $3 \leq \rho \leq 3.5699$. We will show that there is a sequence of numbers

$$c_1 < c_2 < \cdots < c_n < \cdots \tag{11.27}$$

such that when

$$c_n < \rho < c_{n+1},$$

then the logistic equation has an attracting 2^n-cycle. Thus, for example, if $c_8 < \rho < c_9$, then the solution will settle into an attracting 2^8-cycle. We already have $c_1 = 3$ and $c_2 = 1 + \sqrt{6} = 3.4495$. We are interested in finding the remaining elements of the sequence (11.27). The sequence (11.27) is convergent, where

$$\lim_{n\to\infty} c_n = c_\infty = 3.5699.$$

To find the elements of the sequence (11.27), we could, in principle, form composite maps $f^{2^n}(x)$ and use stability results from the previous subsection. We have observed that if the parameter ρ is slightly greater than $1 + \sqrt{6}$, then the 2-cycle is repelling and the solution goes to an attracting 4-cycle. A cobweb of an attracting 4-cycle is shown in Figure 11.9. We could find the four points

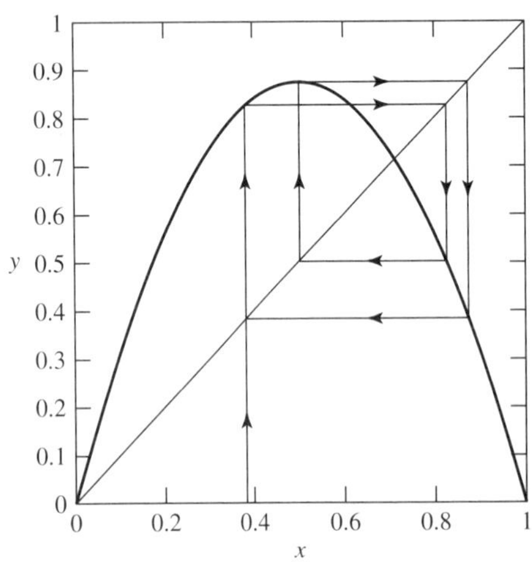

Figure 11.9 Cobweb of an attracting 4-cycle for $\rho = 3.5$.

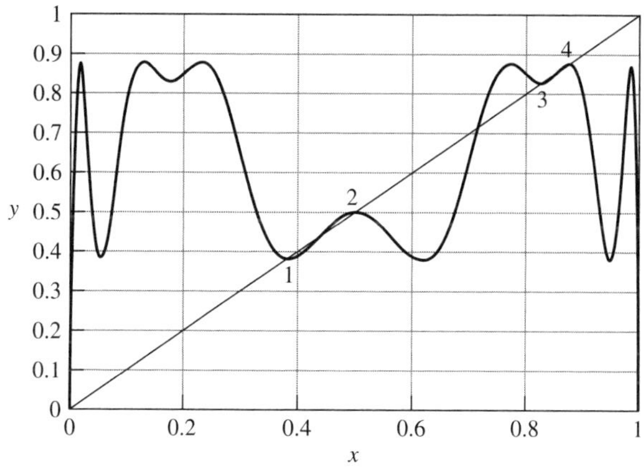

Figure 11.10 A graphical solution of $x = f^4(x)$, where $f(x) = 3.5x(1-x)$. Marked points, corresponding to slopes less than one, form an attracting 4-cycle.

corresponding to an attracting 4-cycle by solving the algebraic equation

$$x = f(f(f(f(x)))), \tag{11.28}$$

where $f(x) = \rho x(1-x)$. A graph of the function $y = f^4(x)$ is shown in Figure 11.10. The solution to (11.28) gives eight points. Four of these points, corresponding to $y = f^4(x)$ having negative slopes when intersecting the line $y = x$, form an attracting 4-cycle. Using results from the previous subsection, we could find the range of ρ resulting in attracting 4-cycle solutions. We could then proceed to find the range of ρ corresponding to solutions going to attracting 8-cycles, then 16-cycles, and so forth. However, this approach is only feasible for $n = 1$ and $n = 2$. We will have to take a different approach to compute the elements of the sequence (11.27). We will use an iteration technique consisting of four steps:

1. Finding the second iterate, $z(k+2) = g(g(z(k))$, of the logistic map (11.15).
2. Shifting one of the points of the 2-cycle of the second iterate to the origin.
3. Neglecting cubic and quartic terms in the shifted map.
4. Scaling the map obtained in step 3.

Performing iterations consisting of the above four steps will allow us to find the elements of the sequence (11.27).

The second iterate of (11.15) is

$$\begin{aligned} z(k+2) &= (2-\rho)z(k+1) - \rho z^2(k+1) \\ &= (2-\rho)((2-\rho)z(k) - \rho z^2(k)) - \rho((2-\rho)z(k) - \rho z^2(k))^2 \\ &= (2-\rho)^2 z(k) - ((2-\rho)\rho + \rho(2-\rho)^2)z^2(k) \\ &\quad + 2\rho^2(2-\rho)z^3(k) - \rho^3 z^4(k). \end{aligned} \tag{11.29}$$

The points p and q, given by (11.17) and (11.18), are fixed points of the map (11.29). We select one of the points, p or q, to work with. Suppose that we choose the point p. Consider the transformation

$$v(k) = z(k) - p. \tag{11.30}$$

Note that

$$v(k+2) = z(k+2) - p. \tag{11.31}$$

Substituting (11.30) and (11.31) into (11.29), performing manipulations, and neglecting terms of order three and four, we obtain

$$v(k+2) = (4 + 2\rho - \rho^2)v(k) - \rho\left(\rho^2 - 2\rho - 3 - 3\sqrt{\rho^2 - 2\rho - 3}\right)v^2(k). \tag{11.32}$$

One can easily verify the above result using MATLAB's function `subs` applied to (11.29)–(11.31). We will be studying fixed points of (11.32). We replace $k+2$, the argument of v on the left-hand side of (11.32), with $k+1$. Let

$$C_1(p) = \rho^2 - 2\rho - 3 - 3\sqrt{\rho^2 - 2\rho - 3}. \tag{11.33}$$

Then, (11.32) can be represented as

$$v(k+1) = (4 + 2\rho - \rho^2)v(k) - \rho C_1(p)v^2(k). \tag{11.34}$$

Note that we if we used q in (11.30) instead of p, we would get

$$v(k+1) = (4 + 2\rho - \rho^2)v(k) - \rho C_1(q)v^2(k), \tag{11.35}$$

where now $C_1(q) = \rho^2 - 2\rho - 3 + 3\sqrt{\rho^2 - 2\rho - 3}$. In the next step, we transform the v map given by (11.34) into the same format as the z-map given by (11.15). This will allow us to apply stability properties of the z-map to the v-map. We introduce an iterated parameter ρ_1 such that

$$2 - \rho_1 = 4 + 2\rho_0 - \rho_0^2, \tag{11.36}$$

and represent (11.34) as

$$v(k+1) = (2 - \rho_1)\, v(k) - \rho_0 C_1(p)v^2(k). \tag{11.37}$$

Finally, we scale (11.37) using the transformation

$$s(k) = \frac{\rho_0 C_1(p)}{\rho_1} v(k). \tag{11.38}$$

Substituting (11.38) into (11.37) gives

$$\frac{\rho_1}{\rho_0 C_1(p)} s(k+1) = \frac{\rho_1(2 - \rho_1)}{\rho_0 C_1(p)} s(k) - \frac{\rho_1^2 \rho_0 C_1(p)}{\rho_0^2 C_1^2(p)} s^2(k). \tag{11.39}$$

Simplifying (11.39) yields

$$\boxed{s(k+1) = (2 - \rho_1)\, s(k) - \rho_1 s^2(k)} \tag{11.40}$$

The above s-map has the same format as the z-map given by (11.15) and analyzed in the previous subsection. It follows from these results that the s-map given by (11.40) has an attracting 2-cycle for $3 < \rho_1 < 1 + \sqrt{6}$, which corresponds to an attracting 4-cycle of the z-map given by (11.15). Using (11.36), we conclude that $3 < \rho_1 < 1 + \sqrt{6}$ corresponds to

$$1 + \sqrt{6} < \rho_0 < 1 + \sqrt{4 + \sqrt{6}} = 3.5396.$$

Indeed, representing (11.36) as

$$\rho_0^2 - 2\rho_0 - (2 + \rho_1) = 0$$

and using the quadratic formula, we get

$$\rho_0 = 1 + \sqrt{3 + \rho_1},$$

where we took the solution that gives $\rho_0 > 0$. The above concludes the first iteration. The result of this iteration is the element $c_3 = 1 + \sqrt{4 + \sqrt{6}} = 3.5396$ of the sequence (11.27) and the scaling parameter $C_1(p)$ whose properties will be investigated in the following subsection.

We then proceed with the next iteration. In particular, executing the steps of the iteration method on the s-map yields the new map that we call the t-map. This map has the form

$$t(k+1) = \left(4 + 2\rho_1 - \rho_1^2\right) t(k) - \rho_1 C_2 t^2(k). \tag{11.41}$$

Introducing the iterated parameter, ρ_2, such that

$$\rho_2 = \rho_1^2 - 2\rho_1 - 2, \tag{11.42}$$

along with scaling the t-map using the transformation

$$u(k) = \frac{\rho_1 C_2}{\rho_2} t(k),$$

yields the u-map of the form

$$u(k+1) = (2 - \rho_2)\, u(k) - \rho_2 u^2(k). \tag{11.43}$$

The above map has an attracting 2-cycle for $3 < \rho_2 < 1 + \sqrt{6}$, which corresponds to an attracting 4-cycle of the s-map, which in turn corresponds to an attracting 8-cycle of the z-map. We calculate the range of the parameter ρ corresponding to attracting 8-cycles of the z-map by working backwards the recursive relation (11.42) to get

$$c_4 = 1 + \sqrt{3 + \rho_1} = 1 + \sqrt{3 + 1 + \sqrt{3 + \rho_2}} = 1 + \sqrt{4 + \sqrt{4 + \sqrt{6}}} = 3.5573.$$

In general, after $i + 1$ iterations we obtain

$$\rho_{i+1} = \rho_i^2 - 2\rho_i - 2 \tag{11.44}$$

and work backwards to find the term c_{i+3} of the sequence (11.27). We can also compute the elements of the sequence (11.27) using the recursive relation

$$\rho_i = \rho_{i+1}^2 - 2\rho_{i+1} - 2, \qquad i = 1, 2, \ldots. \tag{11.45}$$

Solving the above, using the quadratic formula, yields

$$\rho_{i+1} = 1 + \sqrt{3 + \rho_i}.$$

This means that the elements of the sequence (11.27) satisfy the difference equation

$$c_{i+1} = 1 + \sqrt{3 + c_i}, \qquad c(1) = 3. \tag{11.46}$$

The equation (11.46) has two equilibrium points that are obtained by solving the algebraic equation

$$c = 1 + \sqrt{3 + c}.$$

The equilibrium point

$$c_{eq} = \frac{3+\sqrt{17}}{2}$$

is attracting, and we have

$$\lim_{i\to\infty} c_i = c_\infty = c_{eq} = 3.5616.$$

An analysis without truncation when computing second iterates of the logistic equation yields $c_\infty = 3.5699$. Such an analysis can be found in the paper by Feigenbaum [81]. After the parameter ρ reaches the limiting value $\rho = 3.5699$, no stable solution patterns of the logistic equation emerge. We obtain a chaotic behavior.

11.2.4 The Feigenbaum Numbers

The sequence (11.27) and the scaling factors

$$\frac{\rho_i C_{i+1}}{\rho_{i+1}} \tag{11.47}$$

have certain properties that are universal for all quadratic maps. Here, we perform an approximate analysis of these properties following Kahn [145, Chapter 16].

We begin by rewriting equation (11.45) as

$$(\rho_{i+1} - 1)^2 = 3 + \rho_i. \tag{11.48}$$

Hence,

$$\rho_{i+1} - 1 = \sqrt{3+\rho_i} \tag{11.49}$$

and

$$\rho_i - 1 = \sqrt{3+\rho_{i-1}}. \tag{11.50}$$

Subtracting (11.50) from (11.49) gives

$$\rho_{i+1} - \rho_i = \sqrt{3+\rho_i} - \sqrt{3+\rho_{i-1}}. \tag{11.51}$$

Summing (11.50) and (11.49) yields

$$\rho_{i+1} + \rho_i - 2 = \sqrt{3+\rho_i} + \sqrt{3+\rho_{i-1}}. \tag{11.52}$$

Multiplying (11.51) and (11.52), we get

$$(\rho_{i+1} - \rho_i)(\rho_{i+1} + \rho_i - 2) = \rho_i - \rho_{i-1}. \tag{11.53}$$

We rewrite (11.53) as

$$\rho_{i+1} + \rho_i - 2 = \frac{\rho_i - \rho_{i-1}}{\rho_{i+1} - \rho_i}. \tag{11.54}$$

Let

$$\boxed{\delta_i = \frac{\rho_i - \rho_{i-1}}{\rho_{i+1} - \rho_i}} \tag{11.55}$$

The limiting value of δ_i as i tends to infinity is called the *Feigenbaum number* δ. To find its value, we use (11.55) to represent equation (11.54) as

$$\delta_i = \rho_{i+1} + \rho_i - 2. \tag{11.56}$$

We obtain an approximate value for the Feigenbaum number δ because of the truncation procedure that we used to obtain (11.47). We have

$$\delta_{\text{approx}} = \lim_{i\to\infty} (\rho_{i+1} + \rho_i - 2) = 2c_\infty - 2 = 5.1231 \tag{11.57}$$

The exact value of the Feigenbaum number δ is $\delta_{\text{exact}} = 4.669201\ldots.$

The other parameter that characterizes our iteration method is the scaling factor given by (11.47). The limiting value of (11.47) as i tends to infinity is called the *Feigenbaum number* α. Its approximate value is

$$\begin{aligned}\alpha_{\text{approx}} &= \lim_{i\to\infty} \frac{\rho_i C_{i+1}(p)}{\rho_{i+1}} \\ &= \frac{c_\infty C_\infty(p)}{c_\infty} \\ &= c_\infty^2 - 2c_\infty - 3 - 3\sqrt{c_\infty^2 - 2c_\infty - 3} \\ &= -2.24\end{aligned} \tag{11.58}$$

The exact value of the Feigenbaum number α is $\alpha_{\text{exact}} = -2.50\ldots.$

11.3 Fractals

This section is devoted to modeling of irregular shapes. These non-Euclidean objects are called *fractals*. The term *fractal* was coined by Mandelbrot [195], and it comes from the Latin adjective *fractus*, meaning "irregular." The corresponding Latin verb is *frangere,* which means "to break to create irregular fragments." An example of a simple fractal is the von Koch snowflake curve shown in Figure 11.11 along with an iterative procedure for constructing the curve. We start with dividing a line segment into thirds and then replacing the middle segment with two equal

Figure 11.11 The first four stages in constructing the von Koch snowflake curve.

segments that form two sides of an equilateral triangle. The outcome of this stage is a shape consisting of four segments. In the next stage, each of the four segments is divided into thirds, and the middle segments are replaced with two equal segments that form two sides of equilateral triangle. The process is repeated over and over again. The resulting curve is characterized by *self-similarity* or *scaling*.

One of the characteristics of a fractal is its dimension. A line segment is a one-dimensional Euclidean object, and we can divide this line segment into N identical parts. In such a case, the individual parts can be viewed as scaled down, by the factor $r = 1/N$, versions of the original object. Similarly, a square, a two-dimensional Euclidean object, can be divided into N self-similar parts. These parts can be viewed as scaled down copies by the factor $r = 1/\sqrt{N}$ of the original square. In general, a D-dimensional object can be divided into N copies of itself. Each copy can be viewed as a scaled down version by the factor

$$r = \frac{1}{N^{1/D}} \tag{11.59}$$

of the parent object. Equation (11.59) can be represented as

$$r^D N = 1. \tag{11.60}$$

We now generalize the above arguments to the objects that are not necessarily Euclidean. Taking logarithms of both sides of (11.60) yields the expression for the *fractal self-similarity dimension:*

$$\boxed{D = \frac{\log N}{\log(1/r)}} \tag{11.61}$$

We now use (11.61) to compute the dimension of the von Koch curve. Any segment of the curve is composed of four subsegments, that is, $N = 4$. Each subsegment is a scaled-down copy of its parent by the factor $r = 1/3$. Hence, the self-similarity dimension of the von Koch curve is

$$D = \frac{\log N}{\log(1/r)} = \frac{\ln 4}{\ln 3} = 1.26. \tag{11.62}$$

We can see from the above that the fractal dimension, unlike the dimension of a common Euclidean pattern, need not be an integer.

Another example of a fractal is the *Cantor set,* sometimes called *Cantor's ternary set*. The Cantor set consists of the points of the unit interval [0, 1] that remain after "middle third" intervals have been successively removed. This set is illustrated in Figure 11.12. It is an example of so-called *disconnected fractal* or *fractal dust*. For Cantor's set, we have $N = 2$ and $1/r = 3$.

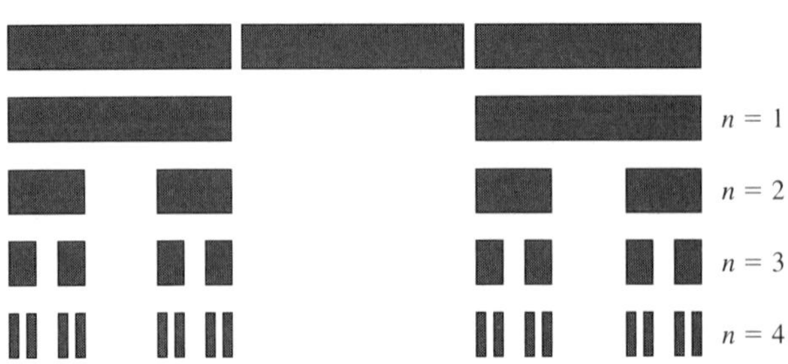

Figure 11.12 The first four iterations in constructing Cantor's ternary set.

Hence, the self-similarity dimension of the Cantor set is

$$D = \frac{\log N}{\log(1/r)} = \frac{\ln 2}{\ln 3} = 0.631. \tag{11.63}$$

Other types of fractal dimension are nicely explained in Peitgen, Jürgens, and Saupe [227].

11.3.1 The Mandelbrot Set

The Mandelbrot set, denoted $\mathcal{M}$, is the set of all complex $c \in \mathbb{C}$ for which the sequence

$$0, c, c^2 + c, (c^2 + c)^2 + c, \ldots \tag{11.64}$$

does not tend to ∞ as n tends to ∞. Thus, the Mandelbrot set is generated by the map in the complex plane,

$$z \mapsto z^2 + c, \qquad c \in \mathbb{C}, \tag{11.65}$$

or, equivalently,

$$z(k+1) = z^2(k) + c, \qquad c \in \mathbb{C}. \tag{11.66}$$

Let

$$z(k) = x(k) + jy(k) \quad \text{and} \quad c = a + jb.$$

Then,

$$\begin{aligned} z(k+1) &= x(k+1) + jy(k+1) \\ &= x^2(k) - y^2(k) + a + j(2x(k)y(k) + b) \\ &= x^2(k) + 2jx(k)y(k) + (jy(k))^2 + a + jb \\ &= z^2(k) + c. \end{aligned}$$

Thus, we can express the first-order quadratic complex dynamical system (11.66) as a second-order real dynamical system of the form

$$\begin{aligned} x(k+1) &= x^2(k) - y^2(k) + a, \\ y(k+1) &= 2x(k)y(k) + b. \end{aligned} \tag{11.67}$$

The representations (11.66) and (11.67) are indeed equivalent.

In terms of the dynamical system (11.67), the Mandelbrot set is the set of parameter values, a and b, for which the solutions of (11.67) starting at $x(0) = y(0) = 0$ remain bounded. The Mandelbrot set is shown in Figure 11.13.

11.3.2 Julia Sets

A filled Julia set corresponding to the given values of the parameters a and b of the dynamical system (11.67) is the set of initial conditions $(x(0), y(0))$ resulting in bounded system trajectories. For each value of $c = a + jb$ there is a different Julia set. In order to construct a Julia set, pick first a point (a, b), say in the Mandelbrot set. Then, pick an initial condition $(x(0), y(0))$ and compute the resulting trajectory of (11.67) starting from the chosen initial condition $(x(0), y(0))$. Plot only the initial conditions $(x(0), y(0))$ that give rise to bounded trajectories. The set of plotted points is the filled Julia set corresponding to the particular parameter values (a, b). The boundary of the filled Julia set is the Julia set. Two examples of filled Julia sets are shown in

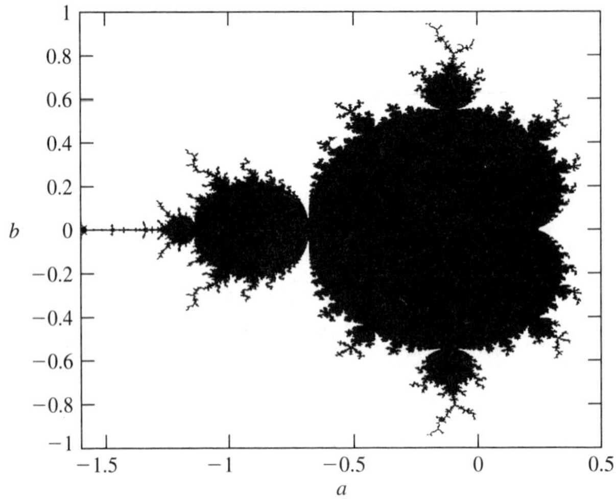

Figure 11.13 The Mandelbrot set for $x(0) = y(0) = 0$.

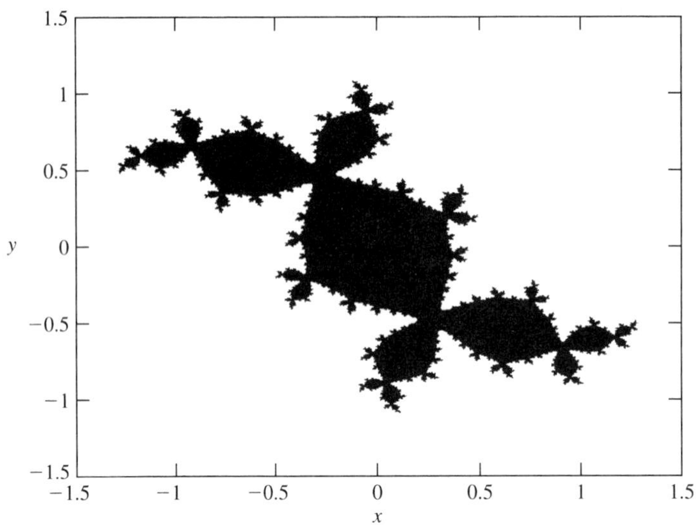

Figure 11.14 Filled Julia set for $c = -0.12256117 + j0.74486177$.

Figures 11.14 and 11.15. The filled Julia set shown in Figure 11.14 is called "Douady's Rabbit" by Devaney and Keen [63, p. 61]. Peitgen and Saupe [228, p. 199] observed that in a sense the Mandelbrot set can be considered as a one-page dictionary of all Julia sets. This means that the Mandelbrot set stores infinitely many different Julia sets.

11.3.3 Iterated Function Systems

An iterated function system (IFS) can be used to model natural objects such as clouds, smoke, leaves, or ferns. An IFS uses affine transformations that we discuss next. A two-dimensional affine transformation $\boldsymbol{w} : \mathbb{R}^2 \to \mathbb{R}^2$ from $\mathbb{R}^2$ into itself is defined by

$$\boldsymbol{w}(\boldsymbol{x}) = \boldsymbol{A}\boldsymbol{x} + \boldsymbol{b}, \tag{11.68}$$

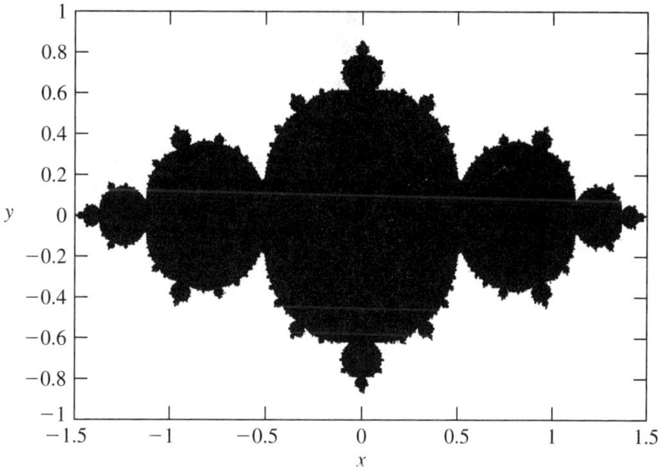

Figure 11.15 Filled Julia set for $c = -0.75$.

where $A \in \mathbb{R}^{2\times 2}$ and $b \in \mathbb{R}^2$. The affine transformation (11.68) satisfies the condition

$$\|w(x) - w(y)\| \leq L\|x - y\|, \tag{11.69}$$

where the Lipschitz constant L is

$$L = \|A\|.$$

The affine transformation w is contractive if $L < 1$. It is a symmetry if

$$\|w(x) - w(y)\| = \|x - y\|.$$

The affine transformation w is expansive if $L > 1$.

A two-dimensional IFS consists of a set of N affine transformations denoted

$$\{w_1, w_2, \ldots, w_N\},$$

each transformation mapping $\mathbb{R}^2$ into itself, together with a set of probabilities

$$\{p_1, p_2, \ldots, p_N\},$$

where each $p_i \geq 0$ and

$$p_1 + p_2 + \cdots + p_N = 1.$$

That is, an IFS is the set

$$\{w_n, p_n : n = 1, 2, \ldots, N\}.$$

Let L_n, $n = 1, 2, \ldots, N$, denote the Lipschitz constant for w_n. Then, we say that the IFS obeys the average contractivity condition if

$$L_1^{p_1} L_2^{p_2} \cdots L_N^{p_N} < 1. \tag{11.70}$$

Interesting mathematical properties of the IFS that satisfy the condition (11.70) are reported by Barnsley [21].

An IFS operates as follows. A set of N affine transformations is chosen and an initial point is picked. Then, the affine transformations are iteratively applied over and over again in a random order with probabilities p_n, for example, proportional to the determinants of $\boldsymbol{A}_n$. The result is a sequence of points from $\mathbb{R}^2$:

$$\boldsymbol{x}(0), \boldsymbol{x}(1), \ldots$$

As an example, suppose that $N = 4$, where

$$\boldsymbol{A}_1 = \begin{bmatrix} 0.02 & 0 \\ 0 & 0.21 \end{bmatrix}, \qquad \boldsymbol{b}_1 = \begin{bmatrix} -0.15 \\ -0.8 \end{bmatrix},$$

$$\boldsymbol{A}_2 = \begin{bmatrix} 0.8496 & 0.0255 \\ -0.0255 & 0.8496 \end{bmatrix}, \qquad \boldsymbol{b}_2 = \begin{bmatrix} 0 \\ 0.15 \end{bmatrix},$$

$$\boldsymbol{A}_3 = \begin{bmatrix} -0.1554 & 0.235 \\ 0.1958 & 0.1865 \end{bmatrix}, \qquad \boldsymbol{b}_3 = \begin{bmatrix} 0.05 \\ -0.55 \end{bmatrix},$$

$$\boldsymbol{A}_4 = \begin{bmatrix} 0.1554 & -0.235 \\ 0.1958 & 0.1865 \end{bmatrix}, \qquad \boldsymbol{b}_4 = \begin{bmatrix} -0.35 \\ -0.4 \end{bmatrix}.$$

The above data comes from Bunde and Havlin [38]. The elements of the set of probabilities

$$\{p_1, p_2, p_3, p_4\}$$

were constructed as

$$p_n = \frac{\det(\boldsymbol{A}_n)}{\sum_{n=1}^{N} \det(\boldsymbol{A}_n)}, \qquad n = 1, 2, 3, 4.$$

The initial point was chosen to be $\boldsymbol{x}(0) = [-0.1 \ \ -0.2]^T$. Then, the roulette-wheel method was used to select the affine transformation to operate on $\boldsymbol{x}(0)$. The roulette-wheel method operates as follows. Each affine transformation is assigned a slice of a wheel, where the slice is proportional to the probability p_n of the particular transformation $\boldsymbol{w}_n$. A random number from the interval [0, 1] is generated, which corresponds to spinning the wheel. The transformation under the wheel's marker is selected. Suppose that $\boldsymbol{w}_k$ was selected. Then $\boldsymbol{w}_k$ operates on $\boldsymbol{x}(0)$ and the result is

$$\boldsymbol{x}(1) = \boldsymbol{A}_k \boldsymbol{x}(0) + \boldsymbol{b}_k.$$

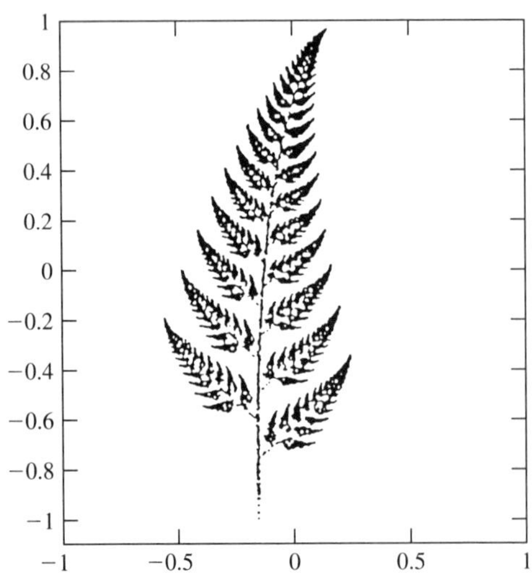

Figure 11.16 The Barnsley fern.

The roulette wheel is spun again to select the affine transformation to act upon $\boldsymbol{x}(1)$ in order to obtain $\boldsymbol{x}(2)$, and so on.

The result of 30,000 iterations is shown in Figure 11.16. The object is called the Barnsley fern because it was M. Barnsley [21] who first generated the fern fractal using an IFS.

11.4 Lyapunov Exponents

Trajectories of a chaotic dynamical system are characterized by a sensitive dependence upon initial conditions, which means that the distance between the adjacent trajectories grows exponentially with time as illustrated in Figure 11.17. For a discrete dynamical system, if x^* is a fixed point of a one-dimensional map f and if $|f'(x^*)| = a > 1$, then by Theorem 11.1, trajectories starting near x^* will separate from x^* at a multiplicative rate of approximately a per iteration until they move significantly far away from x^*. Let x_1 be a point that constitutes the k-cycle. Then, for a k-cycle, the trajectory starting from a point close to x_1, will separate from x_1 at a rate approximately equal to the product of the derivatives of the k points of the k-cycle. Let this product be equal to A and suppose that $A > 1$. Then, it takes k iterations to separate a trajectory starting close to x_1 by a distance A from x_1. We can say that in the case of a k-cycle the average rate of separation is $A^{1/k}$ per iteration. We now are ready to introduce the term *Lyapunov number* to describe the average multiplicative rate of separation of points close to x_1. For a k-cycle, the Lyapunov number $L(x_1)$ is

$$\begin{aligned} L(x_1) &= (|(f^k)'(x_1)|)^{1/k} \\ &= (|f'(x_1)||f'(x_2)|\cdots|f'(x_k)|)^{1/k}. \end{aligned} \tag{11.71}$$

The concept of the Lyapunov number is now extended to nonperiodic trajectories.

Definition 11.2 Let f be a differentiable map on the real line $\mathbb{R}$ and let $\{x(0), x(1), \ldots\}$ be a trajectory starting from $x(0)$. Then, the Lyapunov number $L(x(0))$ of the trajectory starting from $x(0)$ is

$$L(x(0)) = \lim_{N\to\infty} (|f'(x(0))|\cdots|f'(x(N-1))|)^{1/N}$$

if the limit exists.

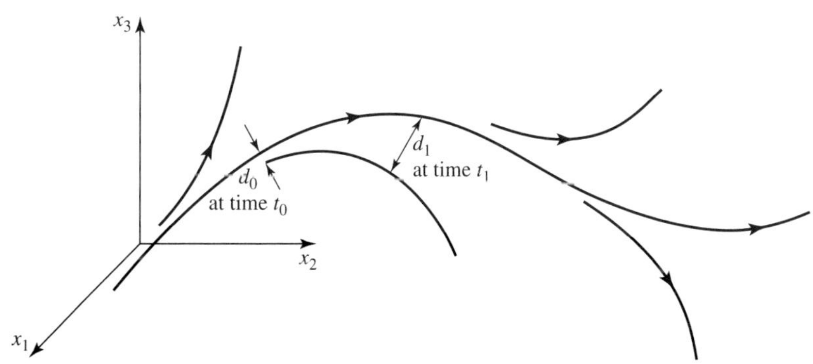

Figure 11.17 The change in distance between adjacent chaotic trajectories that is used to compute the largest Lyapunov exponent.

The Lyapunov exponent, $\mu(x(0))$, is then defined as the natural logarithm of the Lyapunov number; that is,

$$\mu(x(0)) = \ln L(x(0)) = \lim_{N\to\infty} \frac{\ln|f'(x(0))| + \cdots + \ln|f'(x(N-1))|}{N} \tag{11.72}$$

if the limit exists.

We now look at the Lyapunov exponent from another perspective. Let d_0 denote the distance between the two initial states of a continuous dynamical system. Then, for a chaotic motion, at a later time,

$$d(t) = e^{\mu t} d_0. \tag{11.73}$$

The measure of the trajectories' divergence is obtained by averaging the exponential growth at many points along a trajectory. This is because the chaotic trajectories' divergence can only be measured locally since if the trajectories are bounded, $d(t)$ in (11.73) cannot go to infinity. To define the measure of the divergence of chaotic trajectories, first a reference trajectory is obtained. Next a point on an adjacent trajectory is selected and $d(t)/d_0$ is measured. This is illustrated in Figure 11.17. Then, a new $d_0(t)$ is selected on new adjacent trajectory. Using the collected data, the Lyapunov exponent is computed as

$$\mu(x(0)) = \lim_{N\to\infty} \frac{1}{t_N - t_0} \sum_{k=1}^{N} \ln \frac{d(t_k)}{d_0(t_{k-1})}. \tag{11.74}$$

There are n Lyapunov exponents for an n-dimensional nonlinear dynamical system, just as there are n eigenvalues that govern behavior of an n-dimensional linear system. To define these Lyapunov exponents for an n-dimensional nonlinear dynamical system model, first choose an n-dimensional sphere centered at a point on a reference trajectory. Let the radius of this sphere be δ_0. We can view this sphere as an n-dimensional ellipsoid with n semiaxes of equal length $\delta_i(0) = \delta_0$. Next, at a later time t construct an n-dimensional ellipsoid with the property that all the trajectories emanating from the previously chosen sphere pass through this ellipsoid. Let the n semiaxes of the ellipsoid be denoted $\delta_i(t)$. Then, we can compute the Lyapunov exponents as

$$\boxed{\mu_i(\boldsymbol{x}(0)) = \lim_{t\to\infty} \left(\lim_{\delta_i(0)\to 0} \left\{ \frac{1}{t} \ln \frac{\delta_i(t)}{\delta_i(0)} \right\} \right), \qquad i = 1, 2, \ldots, n} \tag{11.75}$$

if the limits exist. It is common to order the Lyapunov exponents so that

$$\mu_1 \geq \mu_2 \geq \ldots \geq \mu_n.$$

Thus, the Lyapunov exponents are numbers that are used to measure the exponential attraction, or separation, in time of two neighboring trajectories of a dynamical system.

◆ Example 11.1

This example was adapted from Moon [209, p. 193]. Suppose that we are given a one-dimensional map

$$x(k+1) = f(x(k)). \tag{11.76}$$

We will demonstrate again that in the regions where f is differentiable, the Lyapunov exponent is determined by $|f'|$. Indeed, let $x(k)$ and $x(k) + \varepsilon$ be two initial conditions. Then, we have

$$d_0(k-1) = \varepsilon \quad \text{and} \quad d(k) = |f(x(k)+\varepsilon) - f(x(k))| \approx \left| f'|_{x=x(k)} \right| \varepsilon.$$

Using (11.74), we obtain

$$\mu(x(0)) = \lim_{N\to\infty} \frac{1}{t_N - t_0} \sum_{k=1}^{N} \ln \frac{d(t_k)}{d_0(t_{k-1})}$$

$$= \lim_{N\to\infty} \frac{1}{N} \sum_{k=1}^{N} \ln |f'(x(k))|,$$

which is the same as (11.72).

The notion of the Lyapunov exponent will be used to formally define chaotic motion. The other term that we will use in this definition is an asymptotically periodic trajectory that we define next. This definition is adapted from Alligood, Sauer, and Yorke [7, p. 108].

Definition 11.3 A trajectory $\{x(0), \ldots, x(N), \ldots\}$ of a smooth map f is called *asymptotically periodic* if it converges to a periodic trajectory as $N \to \infty$; that is, there exists a periodic trajectory $\{y_1, y_2, \ldots, y_k, y_1, y_2, \ldots\}$ such that

$$\lim_{N\to\infty} \|x(N) - y(N)\| = 0.$$

Using the definition of Lyapunov exponent and the above definition, we can define a chaotic trajectory. This definition is adapted from Alligood et al. [7, p. 110].

Definition 11.4 Let f be a map of the real line $\mathbb{R}$, and let $\{x(0), x(1), \ldots\}$ be a bounded trajectory of the map f. This trajectory is chaotic if:

1. The trajectory is not asymptotically periodic.
2. The Lyapunov exponent $\mu(x(0))$ is greater than zero.

For a map on $\mathbb{R}^n$, each trajectory has n Lyapunov numbers. The Lyapunov numbers, in this case, measure the rates of separation from the current point of the trajectory along n orthogonal directions. The first direction is the one where the separation between the nearby points is the greatest. The second direction is the one in which the separation is the greatest from all directions perpendicular to the first direction. The third direction will be the one where the separation is the greatest among directions perpendicular to the first two directions, and so on. We illustrate the above in two dimensions in Figure 11.18, where a unit disk is centered at $x(0)$. After N iterates

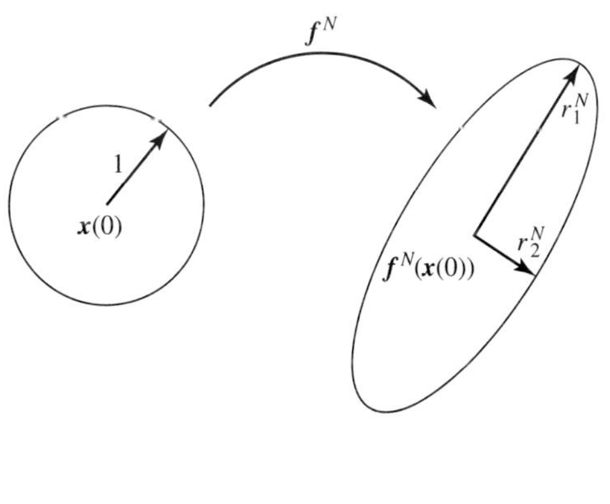

Figure 11.18 The unit disk is mapped into an ellipse by a two-dimensional map.

of the two-dimensional map f, the disk evolves into an ellipse. The per-iteration changes of the ellipse axes are the Lyapunov numbers of $x(0)$. Thus, the Lyapunov numbers measure the contraction or expansion near the trajectory of $x(0)$ during the first N iterations. The natural logarithm of each Lyapunov number is a Lyapunov exponent.

Let $Df(x(0))$ be the first derivative of f evaluated at $x(0)$. Let $J_N = Df^N(x(0))$ be the first derivative of the Nth iterate of f. We denote by B the unit sphere centered at $x(0)$. With the above notation, we are ready to give the definition of Lyapunov exponents for a map on $\mathbb{R}^n$. This definition is adapted from Alligood et al. [7, p. 195].

Definition 11.5 Let f be a smooth map on $\mathbb{R}^n$ and let r_k^N be the length of the kth longest orthogonal axis of the ellipsoid $J_N B$ for a trajectory with the initial state $x(0)$. The kth Lyapunov number is

$$L_k(x(0)) = \lim_{N\to\infty} \left(r_k^N\right)^{1/N}$$

if the limit exists. The kth Lyapunov exponent of $x(0)$ is

$$\mu_k(x(0)) = \ln L_k(x(0)).$$

Using the above definition, we can extend the definition of chaotic trajectories for maps on $\mathbb{R}^n$. This definition is adapted from Alligood et al. [7, p. 196].

Definition 11.6 Let f be a map on $\mathbb{R}^n$, $n \geq 1$, and let $\{x(0), x(1), x(2), \ldots\}$ be a bounded trajectory of the map f. The trajectory is chaotic if:

1. It is not asymptotically periodic.
2. No Lyapunov exponent is equal to zero.
3. $\mu_1(x(0)) > 0$.

For continuous systems, the above definition of a chaotic trajectory has to be modified—see Alligood et al. [7, pp. 385–386].

To numerically compute Lyapunov exponents, a reference trajectory of a given nonlinear system is constructed and the ellipsoid semiaxes lengths r_k's are found using linearized models. We now analyze this process in some detail. For this, suppose that we are given an nth-order continuous dynamical system model

$$\dot{x} = f(x), \tag{11.77}$$

where $x(t) \in \mathbb{R}^n$ is the system's state at time t, and $f : \mathbb{R}^n \to \mathbb{R}^n$ is called the vector field. The solution to (11.77) subject to initial condition $x(t_0) = x_0$ is called a trajectory and is denoted $\phi_t(x_0)$. For every t, the mapping

$$\phi_t : \mathbb{R}^n \to \mathbb{R}^n,$$

maps the state space $\mathbb{R}^n$ into itself. The mapping ϕ_t is called the *flow* of the system (11.77). For example, if (11.77) is linear—that is, $f(x) = Ax$, where $A \in \mathbb{R}^{n\times n}$—then

$$\phi_t(x_0) = e^{A(t-t_0)} x_0.$$

Consider now an equilibrium state x_e of (11.77). The local behavior of (11.77) in a neighborhood of x_e is determined by the linearized model of (11.77) about the equilibrium state x_e. The linearized model is

$$\frac{d}{dt}\Delta x = Df(x_e)\Delta x, \tag{11.78}$$

where $\Delta \boldsymbol{x} = \boldsymbol{x} - \boldsymbol{x}_e$. We can also linearize the system (11.77) about its trajectory $\boldsymbol{\phi}_t(\boldsymbol{x}_0)$. The linearized model about the trajectory $\boldsymbol{\phi}_t(\boldsymbol{x}_0)$ is

$$\frac{d}{dt}\Delta \boldsymbol{x} = D\boldsymbol{f}(\boldsymbol{\phi}_t(\boldsymbol{x}_0))\Delta \boldsymbol{x}, \qquad \Delta \boldsymbol{x}(t_0) = \Delta \boldsymbol{x}_0. \tag{11.79}$$

The solution to (11.79) is

$$\Delta \boldsymbol{x}(t) = \boldsymbol{\Phi}_t(\boldsymbol{x}_0)\Delta \boldsymbol{x}_0.$$

We call the $n \times n$ matrix $\boldsymbol{\Phi}_t(\boldsymbol{x}_0)$ the *transition matrix* of the system (11.79). The transition matrix satisfies the equation

$$\frac{d}{dt}\boldsymbol{\Phi} = D\boldsymbol{f}(\boldsymbol{\phi}_t(\boldsymbol{x}_0))\boldsymbol{\Phi}, \qquad \boldsymbol{\Phi}_0 = \boldsymbol{I}_n.$$

Let

$$\{m_i(t)\}_{i=1}^n$$

be the eigenvalues of the transition matrix $\boldsymbol{\Phi}_t(\boldsymbol{x}_0)$. Then, the Lyapunov exponents of $\boldsymbol{x}_0$ are

$$\boxed{\mu_i(\boldsymbol{x}_0) = \lim_{t\to\infty} \frac{1}{t}\ln|m_i(t)|, \qquad i = 1, 2, \ldots, n} \tag{11.80}$$

if the limits exist. The above is an alternative to (11.75) representation of the Lyapunov exponents.

◆ Example 11.2

The Lyapunov exponents of an equilibrium state $\boldsymbol{x}_e$ of the system (11.77), with $t_0 = 0$, can be found by first linearizing (11.77) about the equilibrium state to obtain a linearized model of the form (11.78). The transition matrix of (11.78) is

$$\boldsymbol{\Phi}_t(\boldsymbol{x}_e) = e^{D\boldsymbol{f}(\boldsymbol{x}_e)t}.$$

The ith eigenvalue of the transition matrix $\boldsymbol{\Phi}_t(\boldsymbol{x}_e)$ is $m_i(t) = e^{\lambda_i t}$, where λ_i is the ith eigenvalue of the matrix $D\boldsymbol{f}(\boldsymbol{x}_e)$. Therefore, the Lyapunov exponents of $\boldsymbol{x}_e$ are

$$\begin{aligned}
\mu_i(\boldsymbol{x}_e) &= \lim_{t\to\infty}\frac{1}{t}\ln|m_i(t)| \\
&= \lim_{t\to\infty}\frac{1}{t}\ln|e^{\Re\lambda_i t}e^{j\Im\lambda_i t}| \\
&= \lim_{t\to\infty}\frac{1}{t}\ln(|e^{\Re\lambda_i t}||\cos(\Im\lambda_i t) + j\sin(\Im\lambda_i t)|) \\
&= \lim_{t\to\infty}\frac{1}{t}t\,\Re\lambda_i \\
&= \Re\lambda_i.
\end{aligned} \tag{11.81}$$

Thus, the Lyapunov exponents in this case are the real parts of the eigenvalues of the linearized system about the equilibrium state. Note that if $\boldsymbol{x}_0$ is in the basin of attraction of $\boldsymbol{x}_e$, that is, if

$$\lim_{t\to\infty}\boldsymbol{\phi}_t(\boldsymbol{x}_0) = \boldsymbol{x}_e,$$

then the Lyapunov exponents of $\boldsymbol{x}_0$ are the same as the Lyapunov exponents of $\boldsymbol{x}_e$.

◆ Example 11.3

Suppose that we are given a discrete-time dynamical system model,

$$\boldsymbol{x}(k+1) = \boldsymbol{f}(\boldsymbol{x}(k)), \qquad \boldsymbol{x}(0) = \boldsymbol{x}_0. \tag{11.82}$$

Let $\boldsymbol{x}_e$ be the equilibrium state of the model (11.82), that is,

$$\boldsymbol{x}_e = \boldsymbol{f}(\boldsymbol{x}_e).$$

The linearized model is

$$\Delta\boldsymbol{x}(k+1) = D\boldsymbol{f}(\boldsymbol{x}_e)\Delta x(k).$$

The state transition matrix of the linearized model is

$$\boldsymbol{\Phi}_k(\boldsymbol{x}_e) = (D\boldsymbol{f}(\boldsymbol{x}_e))^k \tag{11.83}$$

If λ_i is the ith eigenvalue of the matrix $D\boldsymbol{f}(\boldsymbol{x}_e)$, then λ_i^k is the ith eigenvalue of the transition matrix (11.83). The Lyapunov exponents of $\boldsymbol{x}_e$, in this case, are

$$\begin{aligned}\mu_i(\boldsymbol{x}_e) &= \lim_{k\to\infty}\frac{1}{k}\ln|m_i(k)| \\ &= \lim_{k\to\infty}\frac{1}{k}\ln|\lambda_i|^k \\ &= \ln|\lambda_i|.\end{aligned} \tag{11.84}$$

In summary, Lyapunov exponents can be used to characterize a chaotic motion. Specifically, a necessary condition for a chaotic motion to occur in a dynamical system with bounded trajectories is that the largest Lyapunov exponent be positive. An efficient numerical method for evaluating the largest Lyapunov exponent can be found, for example, in the paper by Parker and Chua [221]; see also their book [222] on practical numerical algorithms for chaotic systems.

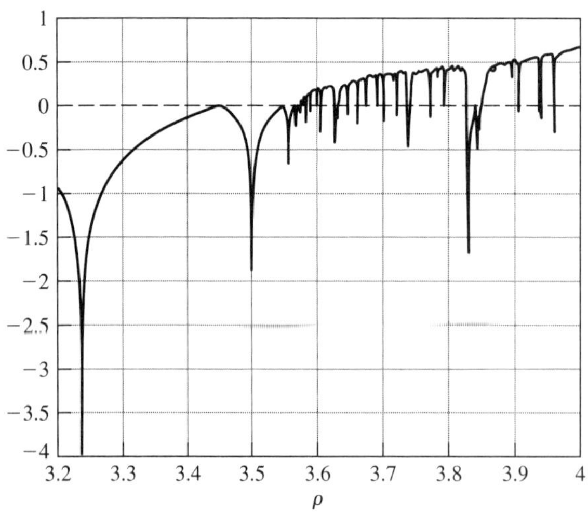

Figure 11.19 Lyapunov exponent versus parameter ρ for the logistic equation (11.10).

In Figure 11.19 we show a plot of the Lyapunov exponent versus the parameter ρ for the logistic equation (11.10).

11.5 Discretization Chaos

In this section we consider examples of chaotic systems that are discrete-time analogs of continuous-time systems, where the latter systems do not exhibit any chaotic behavior. We begin by considering a method of obtaining discrete models from continuous models. Suppose that we are given a scalar differential equation

$$\dot{x} = ax, \qquad x(t_0) = x_0. \tag{11.85}$$

The solution to the above differential equation is

$$x(t) = e^{a(t-t_0)}x(t_0). \tag{11.86}$$

Let $\Delta t = t - t_0$. Then, we can express (11.86) as

$$x(1) = e^{a\Delta t}x(0), \tag{11.87}$$

where $x(1) = x(t_0 + \Delta t)$ and $x(0) = x(t_0)$. More generally,

$$x(k+1) = e^{a\Delta t}x(k), \tag{11.88}$$

where $x(k+1) = x((k+1)\Delta t)$ and $x(k) = x(k\Delta t)$. The discrete model (11.88) of the continuous differential equation (11.85) is exact in the sense that the values of (11.86) and (11.88) are the same at the sampling times. We generalize the above method to a class of nonlinear differential equations. Consider a scalar differential equation

$$\dot{x}(t) = x(t)h(x(t)). \tag{11.89}$$

Dividing both sides of the above by x and then integrating yields

$$\int \frac{dx}{x} = \int h(x(t))\,dt. \tag{11.90}$$

Suppose now that we hold $h(x(t))$ constant on the interval $[k\Delta t,\ (k+1)\Delta t)$. Then, evaluating (11.90) over the above interval gives

$$\ln\left(\frac{x((k+1)\Delta t)}{x(k\Delta t)}\right) = h(x(k\Delta t))\Delta t.$$

Using the notation introduced after (11.88), we represent the above as

$$\boxed{x(k+1) = e^{h(x(k))\Delta t}x(k).} \tag{11.91}$$

◆ Example 11.4

Consider the system of differential equations

$$\begin{aligned} \dot{x}_1 &= \gamma x_1(1 - x_1 - 0.5x_2), \\ \dot{x}_2 &= 2\gamma x_2(1 - 0.5x_1 - x_2), \end{aligned} \tag{11.92}$$

where $\gamma = 2.5$. The above system models a pair of competing species. For a discussion of modeling of competing species population dynamics, the reader is

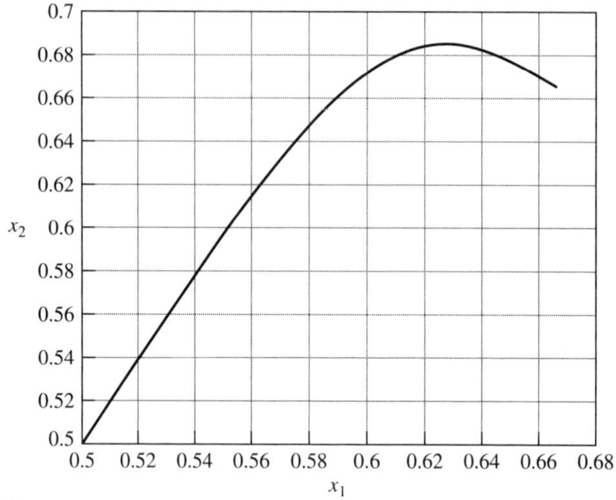

Figure 11.20 A trajectory of the system (11.92).

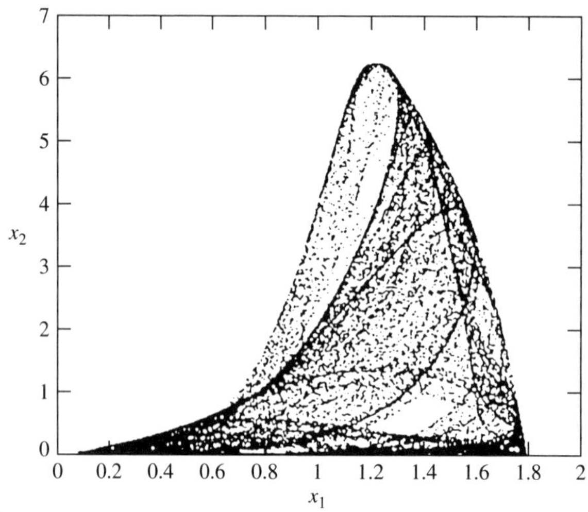

Figure 11.21 A trajectory of the system (11.93).

referred to Subsection 1.7.5. The system (11.92) was analyzed by Grantham and Athalye [106]. A trajectory of the system (11.92) for $x_1(0) = x_2(0) = 0.5$ is shown in Figure 11.20. We next discretize (11.92) using the formula (11.91). For $\Delta t = 1$, we obtain the following discrete-time analog:

$$\begin{aligned} x_1(k+1) &= x_1(k)\exp(\gamma(1 - x_1(k) - 0.5x_2(k))), \\ x_2(k+1) &= x_2(k)\exp(2\gamma(1 - 0.5x_1(k) - x_2(k))). \end{aligned} \tag{11.93}$$

The trajectory of the discrete model (11.93) starting from the same initial condition as its continuous-time analog is shown in Figure 11.21. The continuous-time model trajectory asymptotically converges to the equilibrium state $[2/3 \;\; 2/3]^T$, while the discrete-time trajectory wanders around chaotically.

11.6 Controlling Chaotic Systems

A trajectory of a chaotic system behaves in a random-like fashion. The set on which the chaotic motion is taking place attracts trajectories starting from a set of initial conditions. Roughly speaking, this attracting set is called a *chaotic attractor*. It is not possible to predict when, or if, a trajectory will pass through or near a given point in the chaotic attractor. However, with the probability equal to one, arbitrarily small neighborhood of any given point in the chaotic attractor will be visited by the chaotic system's trajectory in a finite time. This property of a chaotic system is referred to as *ergodicity*. Using the ergodicity property, one can propose a control algorithm whose task it is to bring the chaotic system's trajectory to a prespecified state in the system's state space. The idea behind the chaotic control algorithm, which we discuss next, is to set up "fishing nets" in the chaotic attractor in a neighborhood of an operating point to "catch" the system trajectory. Once the trajectory is "caught," a control law is used to bring the trajectory to the desired operating point. The controller is constructed using linear quadratic regulator (LQR) technique. The chaotic algorithm is illustrated with two examples: the Hénon map and the bouncing ball system.

11.6.1 Ingredients of Chaotic Control Algorithm

Chaotic motion is characterized by extreme sensitivity to initial conditions. This might suggest that it would be impossible, for example, to stabilize a chaotic system with respect to a desired operating point, because any perturbation resulting from applying a control action would lead to a chaotic motion's exponential grow in time. However, it turns out that precisely this property of a chaotic motion can be used to convert a chaotic attractor to a periodic motion. A chaotic attractor, as observed by Ott, Grebogi, and Yorke [217], typically has embedded within it an infinite number of unstable periodic orbits. These periodic orbits are dense in the attractor in the sense that they pass through any neighborhood, however small, of any point in the attractor. Ott et al. [217] noticed that these properties could be advantageous in achieving control objectives. They proposed a method, now called the OGY method, which involves converting, using small time-dependent perturbations of a system parameter, a chaotic attractor into a periodic orbit. The idea behind this method is to exploit the ergodicity property of a chaotic system. Specifically, all of the unstable periodic orbits of the chaotic attractor will eventually be visited by the system trajectory. Once the trajectory gets to a neighborhood of the desired unstable periodic orbit, a small perturbation is applied to stabilize the orbit of interest.

The chaotic control algorithm, which we discuss next, can be viewed as a variant of the OGY method. This algorithm was presented by Vincent [290] and is characterized by two ingredients: a chaotic attractor and a controllable target. These components of the chaotic algorithm are illustrated in Figure 11.22. We assume that either the system under consideration is chaotic or

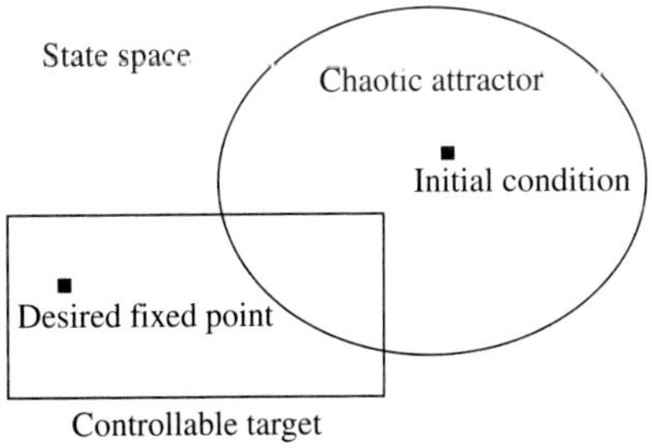

Figure 11.22 A pictorial representation of the chaotic algorithm.

that it can be made chaotic using an open-loop control. Next, we select a fixed point of interest and assume that a feedback law can be constructed such that the closed-loop system is nonchaotic and the chosen fixed point is asymptotically stable. The region of attraction of the fixed point should have a nonempty intersection with the chaotic attractor. A controllable target can be any subset of the region of attraction of the fixed point that has a nonempty intersection with the chaotic attractor. Note that the fixed point does not have to be located in the chaotic attractor. If we start the system from any initial state in the chaotic attractor, the resulting chaotic trajectory will at some time arrive at the intersection region of the chaotic attractor and the controllable target. The chaotic algorithm keeps track when the trajectory reaches the intersection region. When this happens, the open-loop control, which was used to create chaotic motion, is turned off and the feedback control is turned on to drive the system toward the fixed point. The region of attraction intersecting the chaotic attractor plays the role of a "fishing net" for the chaotic trajectory.

11.6.2 Chaotic Control Algorithm

We consider two classes of single-input dynamical systems. The first class is modeled by a nonlinear difference equation,

$$\boldsymbol{x}(k+1) = \boldsymbol{f}(\boldsymbol{x}(k), u(k)), \tag{11.94}$$

and the other class is modeled by a nonlinear differential equation,

$$\dot{\boldsymbol{x}}(t) = \boldsymbol{f}(\boldsymbol{x}(t), u(t)). \tag{11.95}$$

The control u, in both cases, is bounded. The control u is either a specified function of time or a specified function of the system state. In the first case we have an open-loop control, whereas in the latter case we have a closed-loop control. We assume that for a given system there exists an open-loop control such that the system has a chaotic attractor. In addition, we assume that for a specified constant control $u = u_e$ there is a corresponding equilibrium state located in a vicinity of the chaotic attractor. The equilibrium state satisfies the equation

$$\boldsymbol{x}_e = \boldsymbol{f}(\boldsymbol{x}_e, u_e)$$

for the discrete-time system, and it satisfies the relation

$$\boldsymbol{0} = \boldsymbol{f}(\boldsymbol{x}_e, u_e)$$

for the continuous-time system.

We now briefly describe the way we obtain a controllable target for the system of interest. We first linearize the system about the selected equilibrium state. The linearized discrete-time system has the form

$$\Delta\boldsymbol{x}(k+1) = \boldsymbol{A}\Delta\boldsymbol{x}(k) + \boldsymbol{b}\Delta u(k),$$

where $\Delta\boldsymbol{x}(k) = \boldsymbol{x}(k) - \boldsymbol{x}_e$ and $\Delta u(k) = u(k) - u_e$. The linearized continuous-time system is

$$\frac{d}{dt}\Delta\boldsymbol{x}(t) = \boldsymbol{A}\Delta\boldsymbol{x}(t) + \boldsymbol{b}\Delta u(t),$$

where $\Delta\boldsymbol{x}(t) = \boldsymbol{x}(t) - \boldsymbol{x}_e$ and $\Delta u(t) = u(t) - u_e$. Then, the LQR method, described in Chapter 5, is used to construct a state variable feedback control law, $\Delta u(\boldsymbol{\Delta x}) = -\boldsymbol{k}\Delta\boldsymbol{x}$, that guarantees the origin of the linearized system,

$$\Delta\boldsymbol{x}(k+1) = (\boldsymbol{A} - \boldsymbol{bk})\Delta\boldsymbol{x}(k), \tag{11.96}$$

or

$$\frac{d}{dt}\Delta\boldsymbol{x}(t) = (\boldsymbol{A} - \boldsymbol{b}\boldsymbol{k})\Delta\boldsymbol{x}(t), \tag{11.97}$$

to be asymptotically stable. This implies that for the corresponding nonlinear system driven by the state feedback controller, the equilibrium state about which the system was linearized, is asymptotically stable. We use a Lyapunov function obtained from the linearized system to estimate the region of asymptotic stability (RAS) of the closed-loop nonlinear system. The Lyapunov function has the form $V(\Delta\boldsymbol{x}) = \Delta\boldsymbol{x}^T \boldsymbol{P}_l \Delta\boldsymbol{x}$, where $\boldsymbol{P}_l$ is obtained by solving the corresponding Lyapunov equation for the optimal closed-loop linear system given by (11.96) or by (11.97). This Lyapunov function is then used to determine a controllable target for the nonlinear system driven by a bounded controller of the form

$$u = -\text{sat}(\boldsymbol{k}\Delta\boldsymbol{x}) + u_e, \tag{11.98}$$

where

$$\text{sat}(v) = \begin{cases} u_{\max} - u_e & \text{if} \quad v > u_{\max} - u_e, \\ v & \text{if} \quad u_{\min} - u_e \le v \le u_{\max} - u_e, \\ u_{\min} - u_e & \text{if} \quad v < u_{\min} - u_e, \end{cases}$$

and the bounds $u_{\min}$ and $u_{\max}$ are specified by a designer. Using the above control law, we then seek the largest level curve $V(\boldsymbol{x}) = V_{\max}$ such that for the points $\{\boldsymbol{x} : V(\boldsymbol{x}) \le V_{\max}\}$, for the discrete-time system we have

$$\begin{aligned} \Delta V(k) &= V(\boldsymbol{x}(k+1)) - V(\boldsymbol{x}(k)) \\ &= (\boldsymbol{x}(k+1) - \boldsymbol{x}_e)^T \boldsymbol{P}_l(\boldsymbol{x}(k+1) - \boldsymbol{x}_e) - (\boldsymbol{x}(k) - \boldsymbol{x}_e)^T(k)\boldsymbol{P}_l(\boldsymbol{x}(k) - \boldsymbol{x}_e) \\ &= (\boldsymbol{f}(\boldsymbol{x}(k), u(k)) - \boldsymbol{x}_e)^T \boldsymbol{P}_l(\boldsymbol{f}(\boldsymbol{x}(k), u(k)) - \boldsymbol{x}_e) - (\boldsymbol{x}(k) - \boldsymbol{x}_e)^T \boldsymbol{P}_l(\boldsymbol{x}(k) - \boldsymbol{x}_e) \\ &< 0, \end{aligned}$$

or for the continuous-time system we obtain

$$\begin{aligned} \dot{V}(\boldsymbol{x}) &= 2(\boldsymbol{x} - \boldsymbol{x}_e)^T \boldsymbol{P}_l \dot{\boldsymbol{x}} \\ &= 2(\boldsymbol{x} - \boldsymbol{x}_e)^T \boldsymbol{P}_l \boldsymbol{f}(\boldsymbol{x}, u) \\ &< 0. \end{aligned}$$

We next describe the chaotic algorithm. When the given nonlinear system trajectory is in the region $\{\boldsymbol{x} : V(\boldsymbol{x}) > V_{\max}\}$, an open-loop control is applied to keep the system in the chaotic attractor until the system reaches a point in the set $\{\boldsymbol{x} : V(\boldsymbol{x}) \le V_{\max}\}$, where the open-loop control is turned off and the feedback saturating control law is applied to force the system trajectory to asymptotically approach the equilibrium state. More formally, we state the chaotic algorithm in the form of the following algorithm:

IF $V(\boldsymbol{x}) > V_{\max}$

THEN $u =$ open-loop control to keep the system in the chaotic attractor

ELSE $u = -\text{sat}(\boldsymbol{k}\Delta\boldsymbol{x}) + u_e$

END

We illustrate the above algorithm when applied to the Hénon map.

◆ Example 11.5

The dynamical system known as the Hénon map is described by the difference equations

$$\begin{bmatrix} x_1(k+1) \\ x_2(k+1) \end{bmatrix} = \begin{bmatrix} -1.4x_1^2(k) + x_2(k) + 1 \\ 0.3x_1(k) \end{bmatrix}. \tag{11.99}$$

It was named after the French astronomer M. Hénon, who discovered chaotic behavior of the above system in 1974. The orbit of $\boldsymbol{x}(0) = [1 \;\; 0]^T$ of the Hénon map is shown in Figure 11.23. We modify the Hénon map by adding the control input into the first equation so that the map becomes

$$\begin{bmatrix} x_1(k+1) \\ x_2(k+1) \end{bmatrix} = \begin{bmatrix} -1.4x_1^2(k) + x_2(k) + 1 + u(k) \\ 0.3x_1(k) \end{bmatrix}. \tag{11.100}$$

There are two fixed states corresponding to $u(k) = u_e = 0$. They are obtained by solving two algebraic equations:

$$\begin{aligned} x_{1,eq} &= -1.4x_{1,eq}^2 + x_{2,eq} + 1, \\ x_{2,eq} &= 0.3x_{1,eq}. \end{aligned}$$

Solving the above gives

$$[0.6314 \quad 0.1894]^T \quad \text{and} \quad [-1.1314 \quad -0.3394]^T.$$

We select to work with the first of the above fixed states. We linearize the

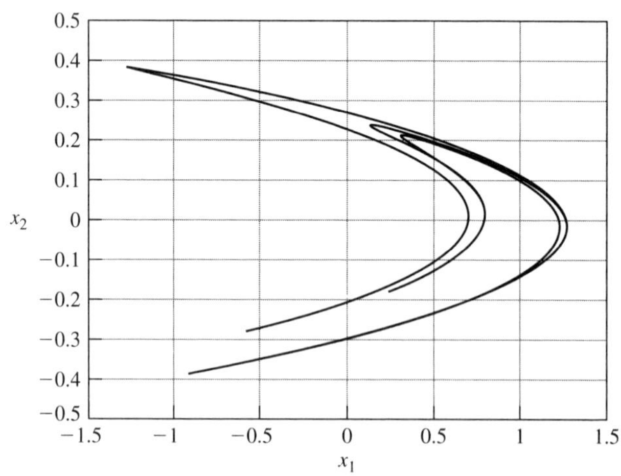

Figure 11.23 An orbit of the Hénon map.

system (11.100) about the first fixed state to obtain

$$\boldsymbol{A} = \begin{bmatrix} -2.8x_1 & 1 \\ 0.3 & 0 \end{bmatrix} \Bigg|_{x_1=0.6314} = \begin{bmatrix} -1.7678 & 1 \\ 0.3 & 0 \end{bmatrix}, \qquad \boldsymbol{b} = \begin{bmatrix} 1 \\ 0 \end{bmatrix}.$$

We then use the discrete LQR design method to find the state feedback. The weight matrices of the associated quadratic performance index are chosen to be $\boldsymbol{Q} = \boldsymbol{I}_2$ and $r = 1$. Solving the discrete algebraic Riccati equation yields

$$\boldsymbol{P} = \begin{bmatrix} 3.9351 & -1.5008 \\ -1.5008 & 1.7974 \end{bmatrix} \quad \text{and} \quad \boldsymbol{k} = [-1.5008 \quad 0.7974].$$

Hence,

$$\boldsymbol{A}_c = \boldsymbol{A} - \boldsymbol{b}\boldsymbol{k} = \begin{bmatrix} -0.2670 & 0.2026 \\ 0.3000 & 0 \end{bmatrix}.$$

Solving the discrete Lyapunov equation, $\boldsymbol{A}_c^T \boldsymbol{P}_l \boldsymbol{A}_c - \boldsymbol{P}_l = -\boldsymbol{I}_2$, gives

$$\boldsymbol{P}_l = \begin{bmatrix} 1.1902 & -0.0686 \\ -0.0686 & 1.0489 \end{bmatrix}.$$

The next step is to calculate V_{max} corresponding to the largest level curve such that the Lyapunov first difference $\Delta V < 0$ on this level curve for the nonlinear model. We obtain level curves by trial and error as follows. We first evaluate

$$\begin{aligned} V &= (\boldsymbol{x} - \boldsymbol{x}_e)^T \boldsymbol{P}_l (\boldsymbol{x} - \boldsymbol{x}_e) \\ &= [x_1 - 0.6314 \quad x_2 - 0.1894] \begin{bmatrix} 1.1902 & -0.0686 \\ -0.0686 & 1.0489 \end{bmatrix} \begin{bmatrix} x_1 - 0.6314 \\ x_2 - 0.1894 \end{bmatrix} \\ &= 1.1902(x_1 - 0.6314)^2 - 0.1372(x_1 - 0.6314)(x_2 - 0.1894) \\ &\quad + 1.0489(x_2 - 0.1894)^2. \end{aligned} \tag{11.101}$$

We next choose an initial state $\boldsymbol{x}(0)$ such that $V(\boldsymbol{x}(0)) = V_0$, where V_0 is a constant selected by the designer. Let $x_2(0) = 0.1894$, which is the value of the second component of the equilibrium state of interest. Substituting this value into (11.101) allows us to compute the first component of $\boldsymbol{x}(0)$ such that $V(\boldsymbol{x}(0)) = V_0$, where

$$\boldsymbol{x}(0) = \left[\sqrt{0.8402\, V_0} + 0.6314 \quad 0.1894\right]^T.$$

Then, a level curve is constructed by solving the differential equation

$$\nabla^T V \dot{\boldsymbol{x}} = 0,$$

subject to the initial condition $\boldsymbol{x}(0)$. To satisfy the above equation, we can select

$$\begin{aligned} \dot{x}_1 &= \frac{\partial V}{\partial x_2}, \\ \dot{x}_2 &= -\frac{\partial V}{\partial x_1}. \end{aligned} \tag{11.102}$$

After constructing a level curve, we evaluate ΔV at these points and check if

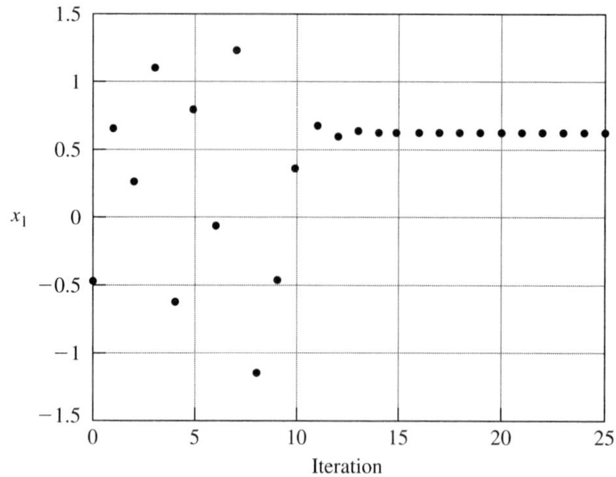

Figure 11.24 A plot of x_1 versus time for the controlled Hénon map of Example 11.5.

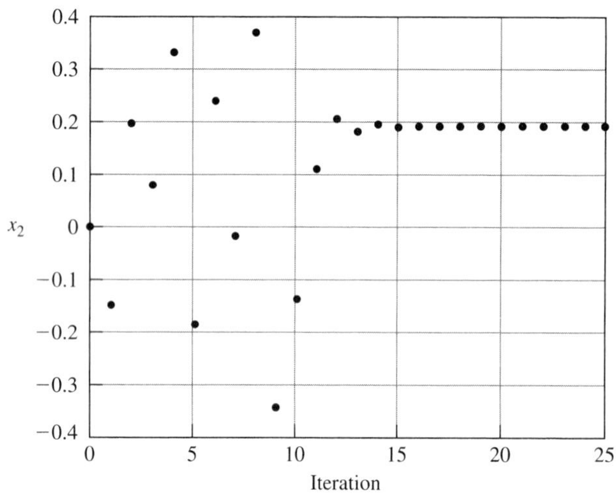

Figure 11.25 A plot of x_2 versus time for the controlled Hénon map of Example 11.5.

$\Delta V < 0$ at every point on the level curve. For the control law

$$u(k) = -0.3 \,\text{sat}(\boldsymbol{k}(\boldsymbol{x}(t) - \boldsymbol{x}_e)/0.3), \tag{11.103}$$

applied to the nonlinear system (11.100), we found $V_{\text{max}} = 0.0109$. Plots of x_1 and x_2 versus time for the closed-loop system (11.100), (11.103) are shown in Figures 11.24 and 11.25, respectively, for the initial condition $\boldsymbol{x}(0) = [-0.5 \;\; 0]^T$. The orbit of $\boldsymbol{x}(0) = [-0.5 \;\; 0]^T$ of the controlled Hénon map is shown in Figure 11.26, where the controllable target is also depicted as the set of points inside the ellipse $V(\boldsymbol{x} - \boldsymbol{x}_e) = 0.0109$.

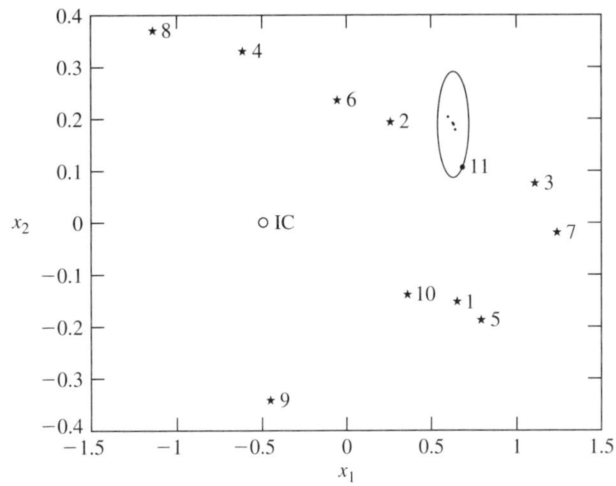

Figure 11.26 An orbit of the controlled Hénon map of Example 11.5, where the initial condition is denoted using "○" and labeled "IC". The subsequent iterations prior to hitting the controllable target are denoted as "∗". The iterations in the controllable target are denoted using "·".

◆ Example 11.6

Controlling the Bouncing Ball on a Vibrating Plate

The bouncing ball on a vibrating plate is another example of a chaotic system. In this example, we use the chaotic algorithm to design a control law for the vibrating plate, as shown in Figure 11.27, so that the ball, starting at rest on the plate, can be bounced up to and then maintained bouncing at a prescribed height. The amplitude of the vibrating plate is kept fixed. The control input is the frequency of the plate vibrations. To proceed, we need a mathematical model of the ball-plate system, which we derive next.

Modeling a Bouncing Ball on a Vibrating Plate

Let the displacement of the plate in the vertical direction be

$$y(t) = A\sin(\omega t),$$

where A is the amplitude of the plate displacement, ω is the angular frequency of vibrations and t is time. Then, the velocity of the plate is

$$W(t) = \frac{dy(t)}{dt} = A\omega\cos(\omega t). \tag{11.104}$$

Let U be the velocity of the ball just before impact and let V denote the velocity of the ball just after impact. In our modeling process, we use the principle of conservation of momentum, which states that momentum of the system before and after collisions is constant, that is,

$$\sum(\text{masses} \times \text{velocities})_{\text{before}} = \sum(\text{masses} \times \text{velocities})_{\text{after}}.$$

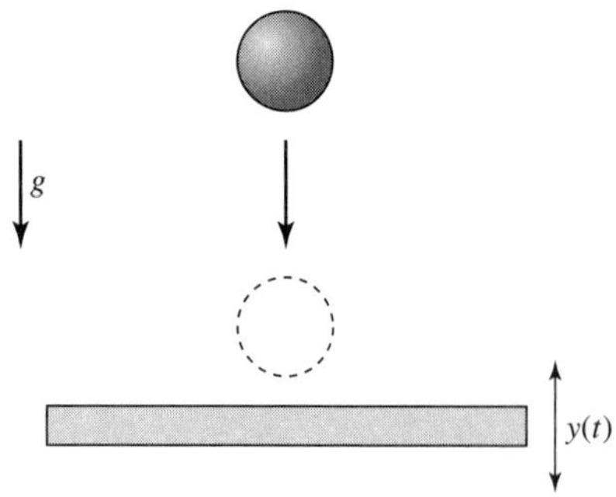

Figure 11.27 Bouncing ball on a vibrating plate.

Applying the conservation of momentum principle to our system consisting of the bouncing ball and the vibrating plate yields

$$m_{\text{ball}} U + m_{\text{plate}} W_{\text{before}} = m_{\text{ball}} V + m_{\text{plate}} W_{\text{after}}. \tag{11.105}$$

To proceed, we define the so-called *coefficient of restitution*

$$e = -\frac{V - W_{\text{after}}}{U - W_{\text{before}}}. \tag{11.106}$$

Performing simple manipulations on (11.106) yields

$$W_{\text{after}} = Ue - W_{\text{before}} e + V. \tag{11.107}$$

Substituting the above into (11.105) and rearranging gives

$$\begin{aligned} m_{\text{ball}} U &+ m_{\text{plate}} W_{\text{before}} \\ &= (m_{\text{ball}} + m_{\text{plate}}) V + e m_{\text{plate}} U - e m_{\text{plate}} W_{\text{before}}. \end{aligned} \tag{11.108}$$

Dividing both sides of the above by $(m_{\text{ball}} + m_{\text{plate}})$ and rearranging, we get

$$V = \frac{m_{\text{ball}} - e m_{\text{plate}}}{m_{\text{ball}} + m_{\text{plate}}} U + \frac{m_{\text{plate}} + e m_{\text{plate}}}{m_{\text{ball}} + m_{\text{plate}}} W_{\text{before}}. \tag{11.109}$$

Let

$$M = \frac{m_{\text{ball}}}{m_{\text{plate}}}.$$

Substituting the above into (11.109) gives

$$V = \frac{M - e}{M + 1} U + \frac{e + 1}{M + 1} W_{\text{before}}. \tag{11.110}$$

The ball is acted upon by a constant acceleration g and therefore $V(t) = gt$. Hence, from one bounce to the next, we have

$$t(k+1) = t(k) + \frac{2V(k)}{g}. \tag{11.111}$$

We assume that the amplitude, A, of the plate vibrations is negligible compared

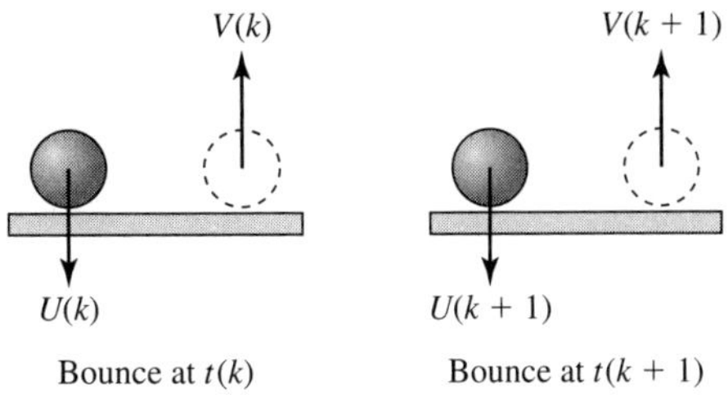

Figure 11.28 Velocities of the bouncing ball from one bounce to the next.

with the height of the bounce. Referring to Figure 11.28, we obtain

$$U(k+1) = -V(k). \tag{11.112}$$

Let

$$a_1 = \frac{e+1}{M+1} \quad \text{and} \quad a_2 = \frac{M-e}{M+1}.$$

Taking into account the above notation and (11.110), we obtain a relation between velocities of the bouncing ball and the vibrating plate at the bounce at a time $t(k+1)$:

$$V(k+1) = a_2 U(k+1) + a_1 W_{\text{before}}(k+1). \tag{11.113}$$

We assume that W_{before} is given by (11.104). Then, using this assumption and (11.112), we represent (11.113) as

$$V(k+1) = -a_2 V(k) + a_1 W(k+1). \tag{11.114}$$

Furthermore, by (11.104) and (11.111),

$$W(k+1) = A\omega \cos(\omega t(k+1)) = A\omega \cos\left(\omega\left(t(k) + \frac{2V(k)}{g}\right)\right).$$

Hence, we can represent (11.114) as

$$V(k+1) = -a_2 V(k) + a_1 A\omega \cos\left(\omega\left(t(k) + \frac{2V(k)}{g}\right)\right). \tag{11.115}$$

Let

$$\phi(k) = \omega t(k) \quad \text{and} \quad \psi(k) = 2\omega V(k)/g, \tag{11.116}$$

where $\phi(k)$ is the current phase angle of the plate and $\psi(k)$ is the phase angle change between the current bounce of the ball and the next. Then, we can write

$$\boxed{\begin{aligned} \phi(k+1) &= \phi(k) + \psi(k), \\ \psi(k+1) &= -a_2\psi(k) + \hat{a}_1\omega^2 \cos(\phi(k) + \psi(k)), \end{aligned}} \tag{11.117}$$

where $\hat{a}_1 = 2Aa_1/g$. The above dynamical system is known as *the ball map*. In our simulations, we use the same values of the ball map parameters as those used by Vincent [290], which are given in Table 11.1. Using these parameters, we compute $a_1 = 1.73333$, $\hat{a}_1 = 0.004594\ \text{sec}^2$, and $a_2 = -0.733333$.

Table 11.1 Simulation Parameters of the Ball Map

M	1/26
e	0.8
A	0.013 m

There are many possible periodic equilibrium solutions to the ball map (11.117). For instance, the ball can bounce to a fixed height at every n cycles of the plate, or it can bounce to m different heights at nm cycles of the plate. Here we are concerned with the case when $m = 1$. There are bands of frequencies of the vibrating plate that will not support periodic motion for $m = 1$. Examples of such frequency bands, as listed by Vincent [290], are

$$32.3966 < \omega < 33.0782,$$
$$36.4575 < \omega < 38.1954,$$
$$40.5278 < \omega < 42.7038,$$
$$44.4249 < \omega < 46.7796.$$

A frequency of the plate vibration from one of the above ranges will result in either a chaotic motion of the ball or a chaotic transient to a periodic motion with $m > 1$. In our example, we use $\omega = 45$ rad/sec to generate chaotic motion.

The ball map (11.117) may produce negative values for ψ corresponding to negative bounce heights, which is physically impossible. To overcome this problem, whenever the map produces a negative value for ψ, this value is multiplied by -1 to simulate the fact that the ball cannot go through the plate. In Figure 11.29, an orbit of the modified ball map for the initial conditions $\phi(0) = 0$ and $\psi(0) = 9.1743$ rads and $\omega = 45$ rad/sec is shown. Note that the values of ψ are divided by 2π before plotting. The chaotic attractor is located between the upper and lower bounds as

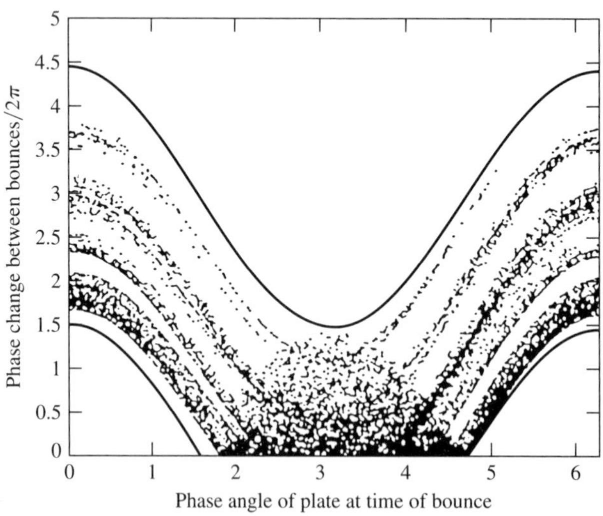

Figure 11.29 An orbit of the ball map.

depicted in the figure. The upper bound is given by

$$\psi_{\text{upper}} = 2\hat{a}_1\omega^2 + \hat{a}_1\omega^2\cos(\phi(k) + \psi(k)),$$

while the lower bound is

$$\psi_{\text{lower}} = \hat{a}_1\omega^2\cos(\phi(k) + \psi(k)).$$

When ψ_{lower} goes negative, it is set equal to zero.

Controller Construction for a Bouncing Ball on a Vibrating Plate

To proceed with the controller design, we first find fixed points of the ball map. They are found by solving the equations,

$$\begin{aligned}\bar{\phi} &= \bar{\phi} + \bar{\psi},\\ \bar{\psi} &= -a_2\bar{\psi} + \hat{a}_1\omega^2\cos(\bar{\phi} + \bar{\psi}).\end{aligned} \tag{11.118}$$

To obtain fixed points, we evaluate the right-hand side of the first of the above equations modulo $2n\pi$, where n is the number of cycles of the plate between bounces. The result of the above calculations is

$$\begin{aligned}\bar{\psi} &= 2n\pi,\\ \cos(\bar{\phi}) &= \frac{2n\pi(1 + a_2)}{\hat{a}_1\omega^2}.\end{aligned} \tag{11.119}$$

Linearizing the ball map about (11.119) gives

$$\begin{bmatrix} x_1(k+1) \\ x_2(k+1) \end{bmatrix} = \begin{bmatrix} 1 & 1 \\ a_{21} & a_{22} \end{bmatrix} \begin{bmatrix} x_1(k) \\ x_2(k) \end{bmatrix}, \tag{11.120}$$

where

$$\begin{aligned} x_1(k) &= \phi(k) - \bar{\phi},\\ x_2(k) &= \psi(k) - \bar{\psi},\\ a_{21} &= -\hat{a}_1\omega^2\sin(\bar{\phi}),\\ a_{22} &= -a_2 + a_{21}.\end{aligned}$$

We next construct a control law for selecting frequencies of the vibrating plate in such a way that the steady-state constant height of the bouncing ball is reached. Vincent [290] showed that for $n=1$ the range of frequencies for which there exists an asymptotically stable equilibrium is [19.0977 rad/sec, 28.9505 rad/sec]. The corresponding range of the bounce heights is [0.0578 m, 0.1327 m]. Thus, for example the height, h, of 0.1 m is in the above range and from the energy conservation law,

$$\tfrac{1}{2}m_{\text{ball}}V^2 = m_{\text{ball}}gh,$$

we obtain

$$V = 1.3994 \text{ m/sec}.$$

Using the above, (11.116), and (11.119), we get

$$\bar{\psi} = 2\pi = \frac{2\omega V}{g} \quad \text{and} \quad \cos(\bar{\phi}) = \frac{2\pi(1 + a_2)}{\hat{a}_1\omega^2},$$

from which we compute

$$\omega = 22 \text{ rad/sec} \quad \text{and} \quad \bar{\phi} = 41.1^\circ = 0.7173 \text{ rad}.$$

The linearized ball map about the above equilibrium state is

$$\boldsymbol{x}(k+1) = \begin{bmatrix} 1 & 1 \\ -1.4617 & -0.7283 \end{bmatrix} \boldsymbol{x}(k),$$

which has its poles in the open unit circle. Solving the discrete Lyapunov equation, $\boldsymbol{A}^T \boldsymbol{P} \boldsymbol{A} - \boldsymbol{P} = -\boldsymbol{I}_2$, gives

$$\boldsymbol{P} = \begin{bmatrix} 8.6414 & 4.6978 \\ 4.6978 & 5.9598 \end{bmatrix}.$$

Employing the same method as in the previous example, we obtain $V_{\max} = 5.006$. The controllable target is

$$\left\{ \begin{bmatrix} \phi \\ \psi \end{bmatrix} : V(\phi - \bar{\phi}, \psi - \bar{\psi}) = [\phi - 0.7173 \quad \psi - 2\pi]\, \boldsymbol{P} \begin{bmatrix} \phi - 0.7173 \\ \psi - 2\pi \end{bmatrix}^T \le V_{\max} \right\}.$$

The controllable target is depicted in Figure 11.30. We use $\omega = 45$ rad/sec when the system trajectory is outside of the controllable target, and we apply $\omega = 22$ rad/sec when the system trajectory hits the controllable target; that is, we use the following algorithm:

IF $V(\phi - \bar{\phi}, \psi - \bar{\psi}) > V_{\max}$

THEN $\omega = 45$ rad/sec

ELSE $\omega = 22$ rad/sec

END

An orbit of $\phi(0) = 0$, $\psi(0) = 9.1743$ of the controlled modified ball map using

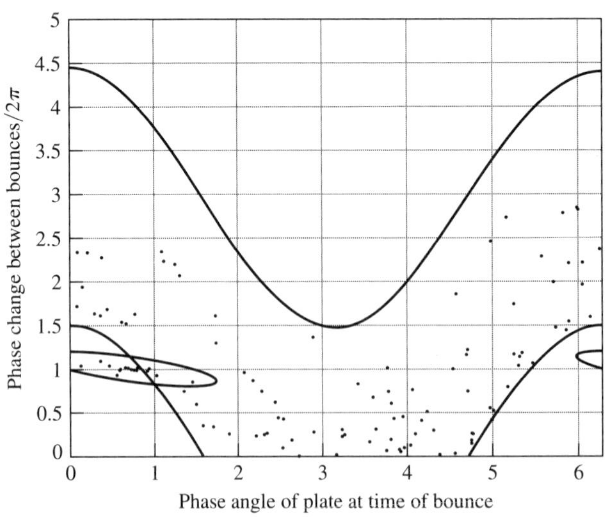

Figure 11.30 An orbit of the controlled modified ball map.

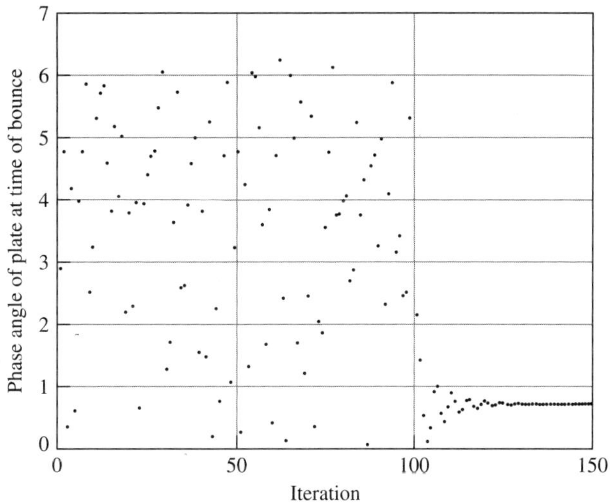

Figure 11.31 A plot of $\phi(k)$ versus time of the controlled modified ball map.

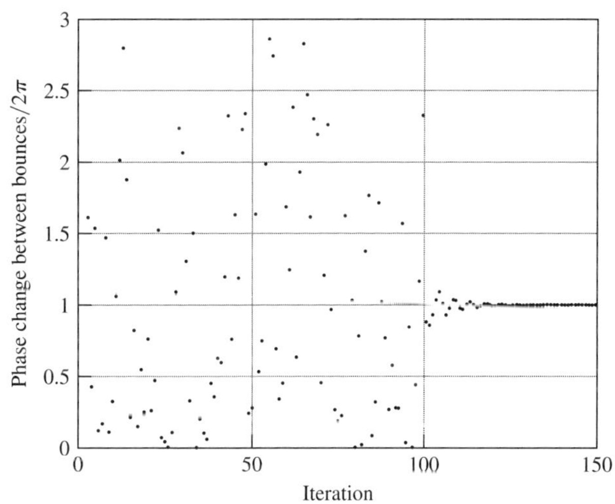

Figure 11.32 A plot of $\psi(k)$ versus time of the controlled modified ball map.

the above algorithm is shown in Figure 11.30. A plot of $\phi(k)$ versus time is shown in Figure 11.31. A plot of $\psi(k)$ versus time is shown in Figure 11.32.

Notes

Sandefur [253] is a lucid elementary introduction to chaotic discrete-time dynamical systems. Peitgen, Jürgens, and Saupe [227] and Alligood, Sauer, and Yorke [7] are superb introductions to chaos and dynamical systems. Papers by May [198] and Feigenbaum [81] are recommended for their originality and clarity. Introductory texts to nonlinear dynamics and chaos for scientists and engineers are the books by Moon [209] and by Thompson and Stewart [281]. Numerical algorithms for chaotic systems can be found in the book by Parker and Chua [222]. Fractals are covered in Mandelbrot [195], Peitgen and Saupe [228], Devaney and Keen [63], and Bunde and Havlin [38]. Barnsley [21] and Peitgen and Saupe [228] are recommended for their coverage of iterated function systems. A reader who is interested in a mathematical approach to chaotic behavior of dynamical systems should consult the book by Arrowsmith and Place [13]. Peterson [232] gives a remarkable account of the development of models of celestial mechanics leading to the discovery of chaos in the solar system. The use of fractals in education, art, music, fashion, chess, and medicine is described in an edited volume by Pickover [233]. Other, more recent papers on controlling chaotic systems, in addition to the ones already referenced in the previous section, are references 305, 308, and 317. Applications of chaos in information processing are discussed in the papers by Dmitriev, Panas, and Starkov [65] and by Andreyev et al. [10]. Aihara, Takabe, and Toyoda [5] and Adachi and Aihara [3] discuss the presence of chaos in biological neurons and propose mathematical models of chaotic neural networks. Applications of chaotic neural networks in the complex memory search for patterns with a sought feature are discussed by Nara, Davis, and Totsuji [214] and by Chan and Żak [46].

Edward Lorenz was a meteorologist at MIT. Lorenz's discovery of chaotic behavior of his mathematical model of weather is a fascinating one. As reported by Gleick [100, Chapter 1] and Alligood et al. [7, Chapter 9], to speed up his computer output which was simulating a weather model, Lorenz used just three significant digits to print out the results of simulations although the calculations were performed using several more digits of accuracy. One day in the winter of 1961, Lorenz came across a simulation that he decided to repeat. He typed in an initial condition for the new simulation from the printout using the three digit format. He then performed the simulation expecting the same behavior as before of the repeated simulation. Instead, "he saw something unexpected, something that planted a seed for a new science" [100, p. 16]. The resulting simulation was totally different from the previous one. The difference between the two solutions was the result of the Lorenz's weather model sensitive dependence on initial conditions. In the computer memory, according to Gleick [100], the initial condition for the first run was 0.506127, while to save space, Lorenz entered 0.506 for the second run. This seemingly inconsequential difference caused the two simulations to flow apart until there was no resemblance at all between them. Later, Lorenz greatly simplified his weather model to just three nonlinear equations that are given in Exercise 11.2. The phase portrait of these famous equations resemble butterfly's wings, which became a symbol for the early activity in chaotic systems.

EXERCISES

11.1 Design a 1-D map f with a stable limit cycle of period 5. Simulate the dynamical behavior of the map.

Hint A method of constructing 1-D maps with cycles of any period is described in Andreyev et al. [11].

11.2 Simulate the system of differential equations

$$\begin{bmatrix} \dot{x}_1 \\ \dot{x}_2 \\ \dot{x}_3 \end{bmatrix} = \begin{bmatrix} a(-x_1 + x_2) \\ bx_1 - x_2 - x_1x_3 \\ -cx_3 + x_1x_2 \end{bmatrix}, \qquad \boldsymbol{x}(0) = \begin{bmatrix} -8 \\ -8 \\ 24 \end{bmatrix}, \tag{11.121}$$

where $a = 10$, $b = 28$, and $c = 8/3$. The set of differential equations (11.121) is commonly referred to as the *Lorenz equations*—after the American meteorologist E. N. Lorenz, who first published these equations in 1963. They model fluid flows in the atmosphere. Plot x_2 versus x_1, x_3 versus x_1, and x_3 versus x_2. Use MATLAB's command `plot3` to obtain a 3-D plot of the solution of the Lorenz equations.

11.3 Simulate the system of the difference equations

$$\begin{bmatrix} x_1(k+1) \\ x_2(k+1) \end{bmatrix} = \begin{bmatrix} 1 - x_2(k) + |x_1(k)| \\ x_1(k) \end{bmatrix}, \qquad \boldsymbol{x}(0) = \begin{bmatrix} -0.1 \\ 0 \end{bmatrix}. \tag{11.122}$$

The solution "fills" a region that looks like a "gingerbread man." For this reason, Devaney [228, p. 149] refers to this example of a chaotic set in the plane as a *chaotic gingerbreadman*.

11.4 Simulate the system of differential equations

$$\begin{bmatrix} \dot{x}_1 \\ \dot{x}_2 \\ \dot{x}_3 \end{bmatrix} = \begin{bmatrix} -x_2 - x_3 \\ x_1 + a x_2 \\ b + x_3(x_1 - c) \end{bmatrix}, \tag{11.123}$$

where $a = 0.398$, $b = 2$, and $c = 4$. The set of differential equations (11.123) is referred to as the *Rössler equations*—after O. E. Rössler, who reported on their chaotic behavior in 1976. Rössler called the chaotic attractor of the above system the "spiral chaos." This type of chaotic attractor is also called a *folded band*. Select an initial condition, integrate the equations, then plot x_2 versus x_1, x_3 versus x_1, and x_3 versus x_2. Use MATLAB's command `plot3` to obtain a 3-D plot of the folded band attractor.

11.5 For the dynamical system

$$x(k+1) = x(k) + r(0.8 - \operatorname{atan}(x(k))), \tag{11.124}$$

plot a bifurcation diagram for the parameter r in the interval [3, 12]. The map (11.124) was analyzed in Hui and Żak [134].

11.6 For the dynamical system

$$x(k+1) = x(k) + r\left(0.8 - \frac{1 - e^{-x(k)}}{1 + e^{-x(k)}}\right), \tag{11.125}$$

plot a bifurcation diagram for the parameter r in the interval [10, 55].

11.7 For the dynamical system

$$x(k+1) = x(k) + r\left(0.8 - \frac{1}{1 + e^{-x(k)}}\right), \tag{11.126}$$

plot a bifurcation diagram for the parameter r in the interval [10, 45].

11.8 Write a MATLAB program of the following "chaos game": Select three points that are vertices of an equilateral triangle. Choose at random one of the vertices as a starting point. Then, choose at random another vertex and move halfway to it from the starting vertex. Mark this halfway point with a dot. This halfway point is your new starting point. Choose at random one of the vertices and move halfway to the chosen vertex from the new starting point. Mark the new halfway point with the dot. Repeat the process. The outcome of this chaos game is a fractal called the Sierpiński gasket. Generate a plot of the Sierpiński gasket. Other methods of constructing the Sierpiński gasket are described in Peitgen et al. [227] and in Exercise 11.9. The term "chaos game" was coined by M. F. Barnsley [21].

11.9 The result of the chaos game described in Exercise 11.8 is the Sierpiński gasket, which can also be constructed using the following procedure. Start with an equilateral triangle. Divide it into four equal subtriangles (quarters) and throw away the middle one—that is, the upside-down triangle. This step leaves three equilateral triangles. In the second step, divide each of the remaining three triangles into four equal quarter-triangles and throw away the middle, upside-down, triangle in each of the three triangles. This step leaves nine triangles. Repeat the process. The remaining area, as the number of iterations tends to infinity, is called the Sierpiński gasket.

(a) Show that the area of the Sierpiński gasket is zero.

(b) Determine the fractal dimension of the Sierpiński gasket.

11.10 In this exercise, we concern ourselves with another fractal, the Sierpiński carpet. Start with a square. Divide it into nine equal subsquares. Remove the central square. Repeat the procedure. The remaining area as the number of iterations goes to infinity is called the Sierpiński carpet.

(a) Show that the area of the Sierpiński carpet is zero.

(b) Determine the fractal dimension of the Sierpiński carpet.

11.11 For the double pendulum system shown in Figure 11.33, show that the Lagrangian has the form

$$\begin{aligned} L = {} & \tfrac{1}{2} m_1 l_1^2 \omega_1^2 + \tfrac{1}{2} m_2 \left(l_1^2 \omega_1^2 + l_2^2 \omega_2^2 + 2 l_1 l_2 \cos(\Delta\theta)\, \omega_1 \omega_2 \right) \\ & + m_1 g l_1 \cos(\theta_1) + m_2 g (l_1 \cos(\theta_1) + l_2 \cos(\theta_2), \end{aligned}$$

where $\omega_i = \dot{\theta}_i$, $i = 1, 2$, are the angular velocities of the pendulum arm, $\Delta\theta = \theta_1 - \theta_2$, and m_i and l_i are the masses and arm lengths, respectively. Assume that the input torques, T_1 and T_2 are applied at each joint. The parameter values are given in Table 11.2. Write the Lagrange equations of motion for the system and represent them in state-space format, where the state variables can be defined as $x_1 = \theta_1$, $x_2 = \dot{\theta}_1$, $x_3 = \theta_2$,

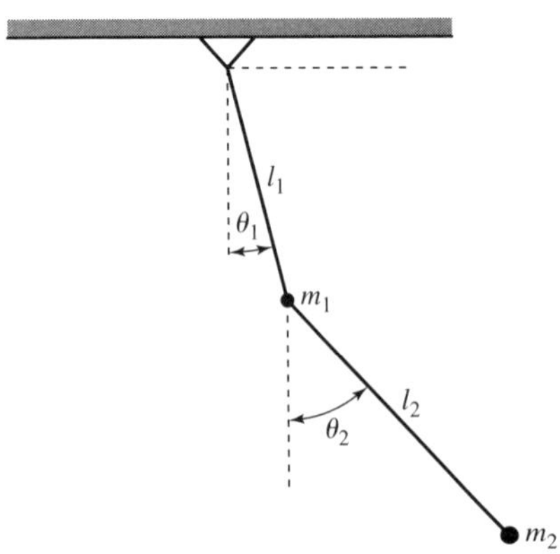

Figure 11.33 Double pendulum of Exercise 11.11.

Table 11.2 Parameter Numerical Values of the Double Pendulum System of Exercise 11.11

$m_1 = m_2$	1 kg
$l_1 = l_2$	0.5 m
g	9.8 m/sec^2

and $x_4 = \dot{\theta}_2$. Set the input torques to zero and simulate the system behavior starting from the initial state $\boldsymbol{x}(0) = [0\ \ 0\ \ \pi\ \ .001]^T$. Verify that one of the equilibrium states corresponding to zero input torques is $\boldsymbol{x}_e = [0\ \ 0\ \ \pi\ \ 0]^T$. Linearize the state-space model about this equilibrium point. Then, use the LQR approach to compute the state feedback matrix $\boldsymbol{K}$ for the weight matrices

$$\boldsymbol{Q} = \begin{bmatrix} 1 & 0 & 0 & 0 \\ 0 & 1 & 0 & 0 \\ 0 & 0 & 10 & 0 \\ 0 & 0 & 0 & 10 \end{bmatrix} \quad \text{and} \quad \boldsymbol{R} = \begin{bmatrix} 1 & 0 \\ 0 & 10 \end{bmatrix}.$$

Solve the Lyapunov equation

$$(\boldsymbol{A} - \boldsymbol{B}\boldsymbol{K})^T \boldsymbol{P}_l + \boldsymbol{P}_l(\boldsymbol{A} - \boldsymbol{B}\boldsymbol{K}) = -\boldsymbol{Q}.$$

Suppose that the bound on the input torques are $|T_1| \leq 5$ and $|T_2| \leq 2$. Use the method of Example 11.5 to find $V_{\max}$ and apply the chaotic control law to stabilize the closed-loop system with respect to $[0\ \ 0\ \ \pi\ \ 0]^T$. Plot x_1, x_2, and $V = (\boldsymbol{x} - \boldsymbol{x}_e)^T P_l(\boldsymbol{x} - \boldsymbol{x}_e)$ versus time.

Math Review

Do not worry about your difficulties in Mathematics. I can assure you mine are still greater.

—Albert Einstein

The last word, when all is heard: Fear God and keep his commandments, for this is man's all; because God will bring to judgment every work, with all its hidden qualities, whether good or bad.

—Ecclesiastes 12:13–14

The appendix reviews mathematical techniques that are used in the book. The appendix also serves as a glossary of symbols, notation, and concepts employed in the text. For more information on the topics covered here, the reader should consult references supplied at the end of the appendix.

In an attempt to keep the presentation as clear as possible, we employ the theorem-proof style whenever appropriate. Major results of independent interest are encapsulated as *theorems*. A *lemma* is used to denote an auxiliary result that is used as a stepping stone toward a theorem. We use the term *proposition* as a way to announce a minor theorem. A *corollary* is a direct consequence of a theorem or proposition. A *hypothesis* is a statement that is taken for further reasoning in a proof. For more information on the subject of writing for mathematical or technical sciences, we recommend a very clear and useful book by Higham [122].

A.1 Notation and Methods of Proof

We often use the term *set*. The great German mathematician G. Cantor (1845–1918), the creator of the theory of sets, defines a set as any collection of definite, distinguishable objects of our thought, to be conceived as a whole [193, p. 1]. In the definition the objects are the elements or members of the set. Usually, we use capital letters to denote sets and use lowercase letters to denote elements of the set. Thus, if X is a set, then we write $x \in X$ to mean that x is an element of X. When an object x is not an element of a set X, we write $x \notin X$. Often, we use the curly bracket notation for sets. For example, $\{x : x < 7\}$ denotes the set of all x less than 7.

The colon, :, following x is read "such that." The following sets are used in the book:

$$\mathbb{N} = \{1, 2, 3, \ldots\}, \text{ the set of positive integers,}$$
$$\mathbb{Z} = \{0, 1, -1, 2, -2, \ldots\}, \text{ the set of integers,}$$
$$\mathbb{R}, \text{ the set of real numbers,}$$
$$\mathbb{C}, \text{ the set of complex numbers.}$$

Sometimes, we take a fixed set and then carry out manipulations with reference to the fixed set. We call this fixed set the *universe of discourse,* or just the *universe,* for short.

Throughout the text, we work with statements that are combinations of individual statements. It is necessary to distinguish between sentences and the statements. A *sentence* is a part of a language. Sentences are used to make different statements. In classical logic, statements are either true or false. Consider the following two statements:

$$A : x > 7;$$
$$B : x^2 > 49.$$

We can combine the above two statements into one statement, called a *conditional,* that has the form "IF A, THEN B." Thus, a conditional is a compound statement obtained by placing the word "IF" before the first statement and inserting the word "THEN" before the second statement. We use the symbol "$\Rightarrow$" to represent the conditional "IF A, THEN B" as $A \Rightarrow B$. The statement $A \Rightarrow B$ also reads as "A implies B," or "A only if B," or "A is sufficient for B," or "B is necessary for A." Statements A and B may be either true or false. The relationship between the truth or falsity of A and B and the conditional $A \Rightarrow B$ can be illustrated by means of a diagram called a *truth table* shown in Table A.1. In the table, T stands for "true" and F stands for "false." It is intuitively clear that if A is true, then B must also be true for the statement $A \Rightarrow B$ to be true. If A is not true, then the sentence $A \Rightarrow B$ does not have an obvious meaning in everyday language. We can interpret $A \Rightarrow B$ to mean that we cannot have A true and B not true. One can check that the truth values of the statements $A \Rightarrow B$ and "not (A and not B)" are the same. We can also check that the truth values of $A \Rightarrow B$ and "not $B \Rightarrow$ not A" are the same. The above is the basis for three methods, inference processes, of proving statements of the form $A \Rightarrow B$. *Inference* is a process by which one statement is arrived at, using reasoning, on the basis of one or more other statements accepted as a starting point. *Reasoning* is a special kind of thinking in which inference takes place using the laws of thought. By a *proof* we mean the process of establishing the truth of a statement by arriving at it from other statements using principles of reasoning. Most of the time we are concerned with the statements of the form $A \Rightarrow B$ and use

Table A.1 Truth Table for Conditional

A	B	$A \Rightarrow B$
F	F	T
F	T	T
T	F	F
T	T	T

the following types of proofs of such statements:

1. The direct method
2. The proof by contraposition
3. The proof by contradiction

In the first case, the direct method, one starts with A and then tries to deduce B. In the proof by contraposition, one starts with "not B" and then deduces a chain of consequences to end with "not A." This method of proving is based on the fact that the truth values of $A \Rightarrow B$ and "not $B \Rightarrow$ not A" are the same. The proof by contradiction is based on the fact that the truth values of $A \Rightarrow B$ and "not (A and not B)" are the same. In this method, one starts with "A and not B" and tries to arrive at a contradiction, which is a statement that is always false.

It is possible to combine two conditionals to form a *biconditional,* which is a statement of the form "A if and only if B." Occasionally, we use the symbol "$\Leftrightarrow$" to write $A \Leftrightarrow B$ to mean "A if and only if B." Another way of expressing $A \Leftrightarrow B$ is "A is equivalent to B", or "A is necessary and sufficient for B."

A.2 Vectors

We define an n-dimensional column vector to be an n-tuple of numbers:

$$\begin{bmatrix} x_1 \\ x_2 \\ \vdots \\ x_n \end{bmatrix}, \tag{A.1}$$

where n is a positive integer. Thus a vector can be viewed as an ordered set of scalars. We use lowercase bold letters—for example, $\boldsymbol{x}$—to denote such n-tuples. We call the numbers $x_1, x_2, \ldots, x_n$ the *coordinates* or *components* of the vector $\boldsymbol{x}$. The number of components of a vector is its dimension. When referring to a particular component of a vector, we use its component—for example, x_i—or the subscripted name of the vector—for example $(\boldsymbol{x})_i$—to denote x_i. The set of all n-tuples, that is, n-vectors, with real coefficients is denoted $\mathbb{R}^n$ and called the real n-space. The transpose of the vector $\boldsymbol{x}$, denoted $\boldsymbol{x}^T$, is an n-dimensional row vector

$$\boldsymbol{x}^T = \begin{bmatrix} x_1 \\ x_2 \\ \vdots \\ x_n \end{bmatrix}^T = [x_1 \quad x_2 \quad \cdots \quad x_n]. \tag{A.2}$$

The sum of two n-vectors $\boldsymbol{x}$ and $\boldsymbol{y}$ is an n-vector, denoted $\boldsymbol{x} + \boldsymbol{y}$, whose ith component is $x_i + y_i$. The following rules are satisfied:

1. $(\boldsymbol{x} + \boldsymbol{y}) + \boldsymbol{z} = \boldsymbol{x} + (\boldsymbol{y} + \boldsymbol{z})$, that is, vector addition is associative.
2. $\boldsymbol{x} + \boldsymbol{y} = \boldsymbol{y} + \boldsymbol{x}$, which means that vector addition is commutative.
3. Let $\mathbf{0} = [0 \; 0 \; \cdots \; 0]^T$ denote the zero vector. Then,
$$\mathbf{0} + \boldsymbol{x} = \boldsymbol{x} + \mathbf{0} = \boldsymbol{x}.$$
4. Let $\boldsymbol{x} = [x_1 \; x_2 \; \cdots \; x_n]^T$ and let $-\boldsymbol{x} = [-x_1 \; -x_2 \; \cdots \; -x_n]^T$. Then,
$$\boldsymbol{x} + (-\boldsymbol{x}) = \mathbf{0}.$$

The product of an n-vector $\boldsymbol{x}$ by a scalar c is an nth vector, denoted $c\boldsymbol{x}$, whose ith component is cx_i.

For two vectors $\boldsymbol{x}$ and $\boldsymbol{y}$ from $\mathbb{R}^n$, we define their *scalar product,* also called the *inner product,* to be

$$\langle \boldsymbol{x}, \boldsymbol{y} \rangle = \boldsymbol{x}^T \boldsymbol{y} = x_1 y_1 + x_2 y_2 + \cdots + x_n y_n. \tag{A.3}$$

The scalar product of vectors from $\mathbb{R}^n$ is commutative, that is,

$$\langle \boldsymbol{x}, \boldsymbol{y} \rangle = \langle \boldsymbol{y}, \boldsymbol{x} \rangle,$$

and distributivity over vector addition holds:

$$\langle \boldsymbol{x}, \boldsymbol{y} + \boldsymbol{z} \rangle = \langle \boldsymbol{x}, \boldsymbol{y} \rangle + \langle \boldsymbol{x}, \boldsymbol{z} \rangle.$$

Two vectors $\boldsymbol{x}$ and $\boldsymbol{y}$ are said to be *orthogonal* if $\langle \boldsymbol{x}, \boldsymbol{y} \rangle = 0$.

A *vector norm,* denoted $\|\cdot\|$, satisfies the following properties:

VN 1. $\|\boldsymbol{x}\| > 0$ for any nonzero vector.

VN 2. For any scalar c, we have $\|c\boldsymbol{x}\| = |c|\|\boldsymbol{x}\|$.

VN 3. For any two vectors $\boldsymbol{x}$ and $\boldsymbol{y}$, the *triangle inequality* holds, that is, $\|\boldsymbol{x}+\boldsymbol{y}\| \le \|\boldsymbol{x}\|+\|\boldsymbol{y}\|$.

The *Euclidean norm,* or the *2-norm,* of a vector $\boldsymbol{x} \in \mathbb{R}^n$, denoted $\|\boldsymbol{x}\|_2$, is defined as

$$\|\boldsymbol{x}\|_2 = \sqrt{\langle \boldsymbol{x}, \boldsymbol{x} \rangle} = \sqrt{x_1^2 + x_2^2 + \cdots + x_n^2}. \tag{A.4}$$

Note that $\|\boldsymbol{x}\|_2^2 = \langle \boldsymbol{x}, \boldsymbol{x} \rangle = \boldsymbol{x}^T\boldsymbol{x}$. The Euclidean norm is a special class of norms called the *Hölder* or *p-norms* defined by

$$\|\boldsymbol{x}\|_p = (|x_1|^p + \cdots + |x_n|^p)^{1/p}, \qquad p \ge 1. \tag{A.5}$$

Other often-used norms that are special cases of the above class of norms are the *1-norm* and the *infinity norm,* defined respectively as

$$\|\boldsymbol{x}\|_1 = |x_1| + \cdots + |x_n| \tag{A.6}$$

and

$$\|\boldsymbol{x}\|_\infty = \max_i |x_i|. \tag{A.7}$$

Most of the time, we use the Euclidean norm for vectors. So for the sake of brevity, we write $\|\boldsymbol{x}\|$ to mean $\|\boldsymbol{x}\|_2$.

Theorem A.1 (Pythagoras) If $\boldsymbol{x}$ and $\boldsymbol{y}$ are orthogonal, then

$$\|\boldsymbol{x}+\boldsymbol{y}\|^2 = \|\boldsymbol{x}\|^2 + \|\boldsymbol{y}\|^2.$$

Proof We have

$$\begin{aligned}\|\boldsymbol{x}+\boldsymbol{y}\|^2 &= (\boldsymbol{x}+\boldsymbol{y})^T(\boldsymbol{x}+\boldsymbol{y}) \\ &= \|\boldsymbol{x}\|^2 + 2\langle \boldsymbol{x}, \boldsymbol{y} \rangle + \|\boldsymbol{y}\|^2 \\ &= \|\boldsymbol{x}\|^2 + \|\boldsymbol{y}\|^2,\end{aligned}$$

because by assumption $\langle \boldsymbol{x}, \boldsymbol{y} \rangle = 0$. The proof of the theorem is complete.

We now introduce the notion of *projection.* Let $\boldsymbol{x}$, $\boldsymbol{y}$ be two vectors, where $\boldsymbol{y} \ne \boldsymbol{0}$. Let $\boldsymbol{z} = c\boldsymbol{y}$, $c \in \mathbb{R}$, be the vector such that the vector $\boldsymbol{x} - \boldsymbol{z}$ is orthogonal to $\boldsymbol{y}$—see Figure A.1 for

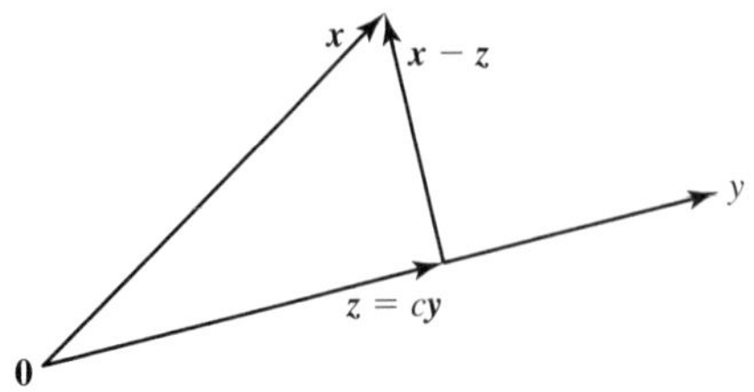

Figure A.1 Computing the projection of x along y.

an illustration of the above construction. Thus,

$$(x - z)^T y = 0. \tag{A.8}$$

Substituting $z = cy$ into (A.8) yields

$$(x - z)^T y = (x - cy)^T y = 0. \tag{A.9}$$

Hence,

$$c = \frac{\langle x, y\rangle}{\|y\|^2}. \tag{A.10}$$

The number c given by (A.10) is called the component of x along y. The projection of x along y is the vector

$$z = cy = \frac{\langle x, y\rangle}{\|y\|^2} y. \tag{A.11}$$

Theorem A.2 (Cauchy-Schwarz Inequality) Let $x, y \in \mathbb{R}^n$. Then,

$$|\langle x, y\rangle| \le \|x\|\|y\|. \tag{A.12}$$

Proof If $y = 0$, then both sides of (A.12) are equal to 0, and the assertion holds. Suppose that $y \neq 0$, and let c denote the component of x along y. Hence, c is given by (A.10). We then write

$$x = x - z + z = x - cy + cy. \tag{A.13}$$

For an illustration of (A.13), consult Figure A.1. Applying the Pythagoras theorem to (A.13), we obtain

$$\|x\|^2 = \|x - cy\|^2 + \|cy\|^2 = \|x - cy\|^2 + c^2\|y\|^2. \tag{A.14}$$

Hence,

$$c^2\|y\|^2 \le \|x\|^2. \tag{A.15}$$

On the other hand, by (A.10),

$$c^2\|y\|^2 = \frac{\langle x, y\rangle^2}{\|y\|^4}\|y\|^2 = \frac{\langle x, y\rangle^2}{\|y\|^2}. \tag{A.16}$$

Combining (A.15) and (A.16) yields

$$\frac{\langle x, y\rangle^2}{\|y\|^2} \le \|x\|^2. \tag{A.17}$$

Multiplying both sides of (A.17) by $\|y\|^2$ and taking the square root, concludes the proof.

Theorem A.3 (Triangle Inequality) Let $\boldsymbol{x}, \boldsymbol{y}$ be two vectors in $\mathbb{R}^n$. Then,

$$\|\boldsymbol{x}+\boldsymbol{y}\| \leq \|\boldsymbol{x}\|+\|\boldsymbol{y}\|. \tag{A.18}$$

Proof Both sides of (A.18) are nonnegative. Hence, to prove that (A.18) holds it suffices to show that the squares of the both sides satisfy the desired inequality,

$$(\boldsymbol{x}+\boldsymbol{y})^T(\boldsymbol{x}+\boldsymbol{y}) \leq (\|\boldsymbol{x}\|+\|\boldsymbol{y}\|)^2. \tag{A.19}$$

Expanding $(\boldsymbol{x}+\boldsymbol{y})^T(\boldsymbol{x}+\boldsymbol{y})$ and applying the Schwarz inequality yields

$$\begin{aligned}(\boldsymbol{x}+\boldsymbol{y})^T(\boldsymbol{x}+\boldsymbol{y}) &= \|\boldsymbol{x}\|^2+2\langle\boldsymbol{x},\boldsymbol{y}\rangle+\|\boldsymbol{y}\|^2\\ &\leq \|\boldsymbol{x}\|^2+2\|\boldsymbol{x}\|\|\boldsymbol{y}\|+\|\boldsymbol{y}\|^2\\ &= (\|\boldsymbol{x}\|+\|\boldsymbol{y}\|)^2,\end{aligned} \tag{A.20}$$

which completes the proof of the theorem.

Theorem A.4 Let $\boldsymbol{x}, \boldsymbol{y}$ be two vectors in $\mathbb{R}^n$. Then,

$$|\|\boldsymbol{x}\|-\|\boldsymbol{y}\|| \leq \|\boldsymbol{x}-\boldsymbol{y}\|.$$

Proof We have

$$\|\boldsymbol{x}\| = \|(\boldsymbol{x}-\boldsymbol{y})+\boldsymbol{y}\| \leq \|\boldsymbol{x}-\boldsymbol{y}\|+\|\boldsymbol{y}\|.$$

Hence,

$$\|\boldsymbol{x}\|-\|\boldsymbol{y}\| \leq \|\boldsymbol{x}-\boldsymbol{y}\|. \tag{A.21}$$

The above implies that

$$\|\boldsymbol{x}-\boldsymbol{y}\| = \|\boldsymbol{y}-\boldsymbol{x}\| \geq \|\boldsymbol{y}\|-\|\boldsymbol{x}\| = -(\|\boldsymbol{x}\|-\|\boldsymbol{y}\|). \tag{A.22}$$

Combining (A.21) and (A.22) yields

$$|\|\boldsymbol{x}\|-\|\boldsymbol{y}\|| \leq \|\boldsymbol{x}-\boldsymbol{y}\|,$$

thus completing the proof.

A.3 Matrices and Determinants

Let m, n be two integers such that $m, n \geq 1$. An array of numbers

$$\begin{bmatrix} a_{11} & a_{12} & \cdots & a_{1n}\\ a_{21} & a_{22} & \cdots & a_{2n}\\ \vdots & \vdots & \ddots & \vdots\\ a_{m1} & a_{m2} & \cdots & a_{mn}\end{bmatrix}$$

is called a *matrix*. Thus a matrix is a set of scalars subject to two orderings. We say that it is an m by n matrix, or $m \times n$ matrix. This matrix has m rows and n columns. If $m = n$, then we say that it is a square matrix. If the elements of an $m \times n$ matrix $\boldsymbol{A}$ are real numbers, then we write $\boldsymbol{A} \in \mathbb{R}^{m\times n}$. If the elements of the matrix $\boldsymbol{A}$ are complex numbers, then we write $\boldsymbol{A} \in \mathbb{C}^{m\times n}$. We define the zero matrix, denoted $\boldsymbol{O}$, to be the matrix such that $a_{ij} = 0$ for all i, j. We define the

$n \times n$ identity matrix, denoted $\boldsymbol{I}_n$, as the matrix having diagonal components all equal to 1 and all other components equal to 0.

Given an $m \times n$ matrix $\boldsymbol{A}$. The $n \times m$ matrix $\boldsymbol{B}$ such that $b_{ji} = a_{ij}$ is called the transpose of $\boldsymbol{A}$ and is denoted $\boldsymbol{A}^T$. Thus, the transpose of a matrix is obtained by changing rows into columns and vice versa. A matrix $\boldsymbol{A}$ that is equal to its transpose—that is, $\boldsymbol{A} = \boldsymbol{A}^T$—is called symmetric.

We now define the sum of two matrices. For two matrices $\boldsymbol{A}$ and $\boldsymbol{B}$ of the same size, we define $\boldsymbol{A} + \boldsymbol{B}$ to be the matrix whose elements are $a_{ij} + b_{ij}$; that is, we add matrices of the same size componentwise.

Let c be a number and let $\boldsymbol{A}$ be a matrix. We define $c\boldsymbol{A}$ to be the matrix whose components are ca_{ij}. Thus, to obtain $c\boldsymbol{A}$, we multiply each component of $\boldsymbol{A}$ by c.

We now define the product of two matrices. Let $\boldsymbol{A}$ be an $m \times n$ matrix and let $\boldsymbol{B}$ be an $n \times s$ matrix. Let $\boldsymbol{a}_i^T$ denote the ith row of $\boldsymbol{A}$, and let $\boldsymbol{b}_j$ denote the jth column of $\boldsymbol{B}$. Then, the product of $\boldsymbol{AB}$ is an $m \times s$ matrix $\boldsymbol{C}$ defined as

$$\boldsymbol{C} = \boldsymbol{AB} = \begin{bmatrix} \boldsymbol{a}_1^T\boldsymbol{b}_1 & \boldsymbol{a}_1^T\boldsymbol{b}_2 & \cdots & \boldsymbol{a}_1^T\boldsymbol{b}_s \\ \boldsymbol{a}_2^T\boldsymbol{b}_1 & \boldsymbol{a}_2^T\boldsymbol{b}_2 & \cdots & \boldsymbol{a}_2^T\boldsymbol{b}_s \\ \vdots & \vdots & \ddots & \vdots \\ \boldsymbol{a}_m^T\boldsymbol{b}_1 & \boldsymbol{a}_m^T\boldsymbol{b}_2 & \cdots & \boldsymbol{a}_m^T\boldsymbol{b}_s \end{bmatrix}. \tag{A.23}$$

It is not always true that $\boldsymbol{AB} = \boldsymbol{BA}$; that is, matrix multiplication is not necessarily commutative.

Let $\boldsymbol{A}$, $\boldsymbol{B}$, and $\boldsymbol{C}$ be matrices such that $\boldsymbol{A}, \boldsymbol{B}$ can be multiplied, $\boldsymbol{A}, \boldsymbol{C}$ can be multiplied, and $\boldsymbol{B}, \boldsymbol{C}$ can be added. Then,

$$\boldsymbol{A}(\boldsymbol{B} + \boldsymbol{C}) = \boldsymbol{AB} + \boldsymbol{AC}. \tag{A.24}$$

The above property is called the distributive law. If x is a number, then

$$\boldsymbol{A}(x\boldsymbol{B}) = x(\boldsymbol{AB}). \tag{A.25}$$

Furthermore, matrix multiplication is associative because

$$(\boldsymbol{AB})\boldsymbol{C} = \boldsymbol{A}(\boldsymbol{BC}). \tag{A.26}$$

Suppose now that $\boldsymbol{A}, \boldsymbol{B}$ can be multiplied. Then,

$$(\boldsymbol{AB})^T = \boldsymbol{B}^T\boldsymbol{A}^T. \tag{A.27}$$

Thus the transpose of a product is the product of transposed matrices in reverse order.

Let $\boldsymbol{A}$ be an $n \times n$ matrix. An inverse of $\boldsymbol{A}$, if it exists, is an $n \times n$ matrix $\boldsymbol{B}$ such that

$$\boldsymbol{AB} = \boldsymbol{BA} = \boldsymbol{I}_n. \tag{A.28}$$

If an inverse exists, then it is unique. A square matrix that is invertible is said to be *nonsingular*. We denote the inverse of $\boldsymbol{A}$ by $\boldsymbol{A}^{-1}$. Hence, we have

$$\boldsymbol{A}^{-1}\boldsymbol{A} = \boldsymbol{AA}^{-1} = \boldsymbol{I}_n.$$

The transpose of an inverse is the inverse of the transpose,

$$(\boldsymbol{A}^{-1})^T = (\boldsymbol{A}^T)^{-1}, \tag{A.29}$$

and we denote it as $\boldsymbol{A}^{-T}$.

We now present the Sherman–Morrison–Woodbury formula for inverting a special kind of matrices.

Lemma A.1 If A, D, and $D^{-1} + CA^{-1}B$ are invertible, then

$$(A + BDC)^{-1} = A^{-1} - A^{-1}B(D^{-1} + CA^{-1}B)^{-1}CA^{-1}.$$

Proof We prove the Sherman–Morrison–Woodbury formula by verification. We have

$$\begin{aligned}
&(A + BDC)(A^{-1} - A^{-1}B(D^{-1} + CA^{-1}B)^{-1}CA^{-1})\\
&= (A + BDC)A^{-1}\\
&\quad - (A + BDC)A^{-1}B(D^{-1} + CA^{-1}B)^{-1}CA^{-1}\\
&= I + BDCA^{-1}\\
&\quad - (B + BDCA^{-1}B)(D^{-1} + CA^{-1}B)^{-1}CA^{-1}\\
&= I + BDCA^{-1}\\
&\quad - BD(D^{-1} + CA^{-1}B)(D^{-1} + CA^{-1}B)^{-1}CA^{-1}\\
&= I + BDCA^{-1} - BDCA^{-1}\\
&= I,
\end{aligned}$$

as desired.

In the case when $D = I$, the Sherman–Morrison–Woodbury formula takes the form

$$(A + BC)^{-1} = A^{-1} - A^{-1}B(I + CA^{-1}B)^{-1}CA^{-1}. \tag{A.30}$$

Formula (A.30) is sometimes referred to as the *matrix inversion formula*.

A set of nonzero vectors $\{q_1, q_2, \ldots, q_r\}$ is said to be orthogonal if

$$\langle q_i, q_j \rangle = 0 \qquad \text{for } i \neq j.$$

If each vector in an orthogonal set is of unity length—that is, $\|q_i\| = 1$ for all $i = 1, 2 \ldots, r$—then this set of vectors is called *orthonormal*. A square matrix with orthonormal columns is called a *unitary matrix*. Sometimes in the literature, a unitary matrix is called an *orthogonal matrix*. If $Q \in \mathbb{R}^{n\times n}$ is a unitary matrix, then

$$Q^T Q = Q Q^T = I_n. \tag{A.31}$$

For a square, $n \times n$, matrix A, its trace, denoted trace A, is the sum of its diagonal elements,

$$\text{trace } A = \sum_{i=1}^{n} a_{ii}. \tag{A.32}$$

If B is also an $n \times n$ matrix, then

$$\text{trace}(AB) = \text{trace}(BA). \tag{A.33}$$

The *determinant* of a square matrix A, denoted det A, can be evaluated using the Laplace expansion as follows. Let c_{ij} denote the cofactor corresponding to the element a_{ij} of an $n \times n$ matrix A. The cofactor c_{ij} is $(-1)^{i+j}$ times the determinant of the $(n-1) \times (n-1)$ submatrix of A obtained by deleting the ith row and jth column from A. Then, for any fixed i, $1 \leq i \leq n$, we have

$$\det A = \sum_{j=1}^{n} a_{ij} c_{ij}. \tag{A.34}$$

This is the expansion of the determinant along the ith row. In a similar fashion, we can expand the determinant along a column. The determinant has the following properties:

D 1. $\det \boldsymbol{A} = \det \boldsymbol{A}^T$.

D 2. Let $\boldsymbol{A}$ and $\boldsymbol{B}$ be square matrices of the same size, then $\det(\boldsymbol{AB}) = \det \boldsymbol{A} \det \boldsymbol{B} = \det(\boldsymbol{BA})$.

D 3. $\det \boldsymbol{O} = 0$.

D 4. $\det \boldsymbol{I}_n = 1$.

We now list additional properties of determinants. Let $\boldsymbol{a}_i$ denote the ith column of $\boldsymbol{A}$. Then, we write

$$\det \boldsymbol{A} = \det(\boldsymbol{a}_1, \boldsymbol{a}_2, \ldots, \boldsymbol{a}_n).$$

D 5. Suppose that the ith column $\boldsymbol{a}_i$ is written as a sum

$$\boldsymbol{a}_i = \boldsymbol{b} + \boldsymbol{c}.$$

Then, we have

$$\begin{aligned}\det(\boldsymbol{a}_1, \ldots, \boldsymbol{a}_i, \ldots, \boldsymbol{a}_n) &= \det(\boldsymbol{a}_1, \ldots, \boldsymbol{b} + \boldsymbol{c}, \ldots, \boldsymbol{a}_n) \\ &= \det(\boldsymbol{a}_1, \ldots, \boldsymbol{b}, \ldots, \boldsymbol{a}_n) + \det(\boldsymbol{a}_1, \ldots, \boldsymbol{c}, \ldots, \boldsymbol{a}_n).\end{aligned}$$

D 6. If x is a number, then

$$\det(\boldsymbol{a}_1, \ldots, x\,\boldsymbol{a}_i, \ldots, \boldsymbol{a}_n) = x \det(\boldsymbol{a}_1, \ldots, \boldsymbol{a}_i, \ldots, \boldsymbol{a}_n).$$

D 7. If two columns of the matrix are equal, then the determinant is equal to zero.

D 8. If we add a multiple of one column to another, then the value of the determinant does not change. In other words, if x is a number, then

$$\det(\boldsymbol{a}_1, \ldots, \boldsymbol{a}_i, \ldots, \boldsymbol{a}_j, \ldots, \boldsymbol{a}_n) = \det(\boldsymbol{a}_1, \ldots, \boldsymbol{a}_i + x\boldsymbol{a}_j, \ldots, \boldsymbol{a}_j, \ldots, \boldsymbol{a}_n).$$

Properties D 5–D 8 hold also for rows instead of columns because $\det \boldsymbol{A} = \det \boldsymbol{A}^T$.

Using the cofactors of $\boldsymbol{A}$ and its determinant, we can give a formula for $\boldsymbol{A}^{-1}$. First, we define the adjugate of $\boldsymbol{A}$, denoted adj $\boldsymbol{A}$, to be the matrix whose (i, j)th entry is the cofactor c_{ji}; that is, adj $\boldsymbol{A}$ is the transpose of the cofactor matrix of $\boldsymbol{A}$. Then,

$$\boldsymbol{A}^{-1} = \left(\frac{1}{\det \boldsymbol{A}}\right) \operatorname{adj} \boldsymbol{A} = \frac{\operatorname{adj} \boldsymbol{A}}{\det \boldsymbol{A}}. \tag{A.35}$$

If $\boldsymbol{A}$ is $n \times n$ and $\boldsymbol{v}$ is a nonzero $n \times 1$ vector such that for some scalar λ,

$$\boldsymbol{Av} = \lambda \boldsymbol{v}, \tag{A.36}$$

then $\boldsymbol{v}$ is an *eigenvector* corresponding to the *eigenvalue* λ. In order that λ be an eigenvalue of $\boldsymbol{A}$, it is necessary and sufficient for the matrix $\lambda \boldsymbol{I}_n - \boldsymbol{A}$ to be singular; that is, $\det(\lambda \boldsymbol{I}_n - \boldsymbol{A}) = 0$. This leads to an nth-order polynomial equation in λ of the form

$$\det(\lambda \boldsymbol{I}_n - \boldsymbol{A}) = \lambda^n + a_{n-1}\lambda^{n-1} + \cdots + a_1\lambda + a_0 = 0. \tag{A.37}$$

This equation, called the *characteristic polynomial equation,* must have n, possibly nondistinct, complex roots that are the eigenvalues of $\boldsymbol{A}$. The *spectral radius* of $\boldsymbol{A}$, denoted $\rho(\boldsymbol{A})$, is defined as

$$\rho(\boldsymbol{A}) = \max_{1 \le i \le n} |\lambda_i(\boldsymbol{A})|,$$

where $\lambda_i(\boldsymbol{A})$ denotes the ith eigenvalue of $\boldsymbol{A}$.

Any $m \times n$ matrix $\boldsymbol{A}$ can be represented as

$$\boldsymbol{A} = \boldsymbol{U}[\boldsymbol{S}, \quad \boldsymbol{O}]\boldsymbol{V}^T \qquad \text{if } m \le n, \tag{A.38}$$

or

$$\boldsymbol{A} = \boldsymbol{U}\begin{bmatrix}\boldsymbol{S}\\ \boldsymbol{O}\end{bmatrix}\boldsymbol{V}^T \qquad \text{if } m > n, \tag{A.39}$$

where $\boldsymbol{U}$ is an $m \times m$ unitary matrix, $\boldsymbol{V}$ is an $n \times n$ unitary matrix, and $\boldsymbol{S}$ is a diagonal matrix of the form

$$\boldsymbol{S} = \operatorname{diag}(\sigma_1, \sigma_2, \ldots, \sigma_p), \qquad \text{where} \quad p = \min(m, n) \quad \text{and} \quad \sigma_i \ge 0, \quad i = 1, 2, \ldots, p.$$

The nonnegative numbers $\{\sigma_i\}$ are called the *singular values* of $\boldsymbol{A}$; and (A.38), or (A.39), is called the *singular value decomposition* (SVD) of $\boldsymbol{A}$. Following the convention, we write

$$\sigma_1 \ge \sigma_2 \ge \cdots \ge \sigma_p,$$

so that $\sigma_1 = \sigma_1(\boldsymbol{A})$ denotes the largest singular value of $\boldsymbol{A}$. The singular values of $\boldsymbol{A}$ are the square roots of the eigenvalues of $\boldsymbol{A}\boldsymbol{A}^T$ if $m \le n$ or of $\boldsymbol{A}^T\boldsymbol{A}$ if $m > n$. If $\boldsymbol{A}$ is symmetric, its singular values are the absolute values of its eigenvalues.

A.4 Quadratic Forms

A real *quadratic form* is the product $\boldsymbol{x}^T\boldsymbol{Q}\boldsymbol{x}$, where $\boldsymbol{Q}$ is a specified $n \times n$ matrix with real coefficients and $\boldsymbol{x}$ is an $n \times 1$ real, column vector. Without loss of generality, the matrix $\boldsymbol{Q}$ can be taken to be symmetric, that is, $\boldsymbol{Q} = \boldsymbol{Q}^T$. Indeed, if $\boldsymbol{Q}$ is not symmetric, we can always replace it with a symmetric matrix without changing the quadratic form values. Indeed, because $z = \boldsymbol{x}^T\boldsymbol{Q}\boldsymbol{x}$ is a scalar, hence $z^T = z$, and we have

$$(\boldsymbol{x}^T\boldsymbol{Q}\boldsymbol{x})^T = \boldsymbol{x}^T\boldsymbol{Q}^T\boldsymbol{x} = \boldsymbol{x}^T\boldsymbol{Q}\boldsymbol{x}.$$

Therefore, we can write

$$\boldsymbol{x}^T\boldsymbol{Q}\boldsymbol{x} = \boldsymbol{x}^T\left(\tfrac{1}{2}\boldsymbol{Q} + \tfrac{1}{2}\boldsymbol{Q}^T\right)\boldsymbol{x} = \boldsymbol{x}^T\left(\frac{\boldsymbol{Q} + \boldsymbol{Q}^T}{2}\right)\boldsymbol{x},$$

where $\frac{1}{2}(\boldsymbol{Q} + \boldsymbol{Q}^T)$ is a symmetric matrix. Thus, the quadratic form values are unchanged if $\boldsymbol{Q}$ is replaced with the symmetric matrix $\frac{1}{2}(\boldsymbol{Q} + \boldsymbol{Q}^T)$. From now on, in our study of quadratic forms, we assume that $\boldsymbol{Q}$ is symmetric.

A quadratic form $\boldsymbol{x}^T\boldsymbol{Q}\boldsymbol{x}$—or, equivalently, the matrix $\boldsymbol{Q}$—is called *positive semidefinite* if $\boldsymbol{x}^T\boldsymbol{Q}\boldsymbol{x} \ge 0$ for all $\boldsymbol{x}$. It is *positive definite* if $\boldsymbol{x}^T\boldsymbol{Q}\boldsymbol{x} > 0$ for all $\boldsymbol{x} \ne \boldsymbol{0}$. If $\boldsymbol{Q}$ is positive semidefinite, then we write $\boldsymbol{Q} \ge 0$; if it is positive definite, then we write $\boldsymbol{Q} > 0$. We write $\boldsymbol{Q}_1 \ge \boldsymbol{Q}_2$ to mean that $\boldsymbol{Q}_1 - \boldsymbol{Q}_2 \ge 0$. We say that $\boldsymbol{Q}$ is *negative semidefinite* if $\boldsymbol{x}^T\boldsymbol{Q}\boldsymbol{x} \le 0$ for all $\boldsymbol{x}$, and it is *negative definite* if $\boldsymbol{x}^T\boldsymbol{Q}\boldsymbol{x} < 0$ for all $\boldsymbol{x} \ne \boldsymbol{0}$. Note that $\boldsymbol{Q} \le 0$ if and only if $-\boldsymbol{Q} \ge 0$.

We now present tests for definiteness properties of symmetric matrices. We state these tests in terms of *principal minors* of a given real, symmetric, $n \times n$ matrix $\boldsymbol{Q}$ with elements q_{ij}. We also give tests for definiteness in terms of eigenvalues of $\boldsymbol{Q}$. The principal minors of $\boldsymbol{Q}$ are det $\boldsymbol{Q}$ itself and the determinants of submatrices of $\boldsymbol{Q}$ obtained by removing successively an ith row and ith column. Specifically, for $p = 1, 2, \ldots, n$, the principal minors are

$$\Delta_p\begin{pmatrix} i_1, & i_2, & \ldots, & i_p \\ i_1, & i_2, & \ldots, & i_p \end{pmatrix} = \det\begin{bmatrix} q_{i_1i_1} & q_{i_1i_2} & \cdots & q_{i_1i_p} \\ q_{i_2i_1} & q_{i_2i_2} & \cdots & q_{i_2i_p} \\ \vdots & \vdots & \ddots & \vdots \\ q_{i_pi_1} & q_{i_pi_2} & \cdots & q_{i_pi_p} \end{bmatrix}, \qquad 1 \le i_1 \le i_2 \le \cdots \le i_p.$$

The above are called the principal minors of order p.

In the following, we use the fact that if $\boldsymbol{Q} = \boldsymbol{Q}^T$, then its eigenvalues are all real.

Theorem A.5 The following are equivalent:

1. The symmetric matrix $\boldsymbol{Q}$ is positive semidefinite.
2. Principal minors of $\boldsymbol{Q}$ are nonnegative; that is, for $p = 1, 2, \ldots, n$,

$$\Delta_p\begin{pmatrix} i_1, & i_2, & \ldots, & i_p \\ i_1, & i_2, & \ldots, & i_p \end{pmatrix} \ge 0, \qquad 1 \le i_1 \le i_2 \le \cdots \le i_p.$$

3. Eigenvalues of $\boldsymbol{Q}$ are nonnegative, that is,

$$\lambda_i(\boldsymbol{Q}) \ge 0, \qquad i = 1, 2, \ldots, n.$$

We now give two tests for $\boldsymbol{Q} = \boldsymbol{Q}^T$ to be positive definite: one in terms of its leading principal minors, and the other in terms of its eigenvalues. The *leading principal minors* of $\boldsymbol{Q}$ are

$$\Delta_1\begin{pmatrix} 1 \\ 1 \end{pmatrix} = q_{11}, \qquad \Delta_2\begin{pmatrix} 1, & 2 \\ 1, & 2 \end{pmatrix} = \det\begin{bmatrix} q_{11} & q_{12} \\ q_{21} & q_{22} \end{bmatrix},$$

$$\Delta_3\begin{pmatrix} 1, & 2, & 3 \\ 1, & 2, & 3 \end{pmatrix} = \det\begin{bmatrix} q_{11} & q_{12} & q_{13} \\ q_{21} & q_{22} & q_{23} \\ q_{31} & q_{32} & q_{33} \end{bmatrix}, \ldots, \Delta_p\begin{pmatrix} 1, & 2, & \ldots, & p \\ 1, & 2, & \ldots, & p \end{pmatrix} = \det \boldsymbol{Q}.$$

Theorem A.6 The following are equivalent:

1. The symmetric matrix $\boldsymbol{Q}$ is positive definite.
2. Leading principal minors of $\boldsymbol{Q}$ are positive, that is,

$$\Delta_p\begin{pmatrix} 1, & 2, & \ldots, & p \\ 1, & 2, & \ldots, & p \end{pmatrix} > 0, \qquad p = 1, 2, \ldots, n.$$

3. Eigenvalues of $\boldsymbol{Q}$ are positive, that is,

$$\lambda_i(\boldsymbol{Q}) > 0, \qquad i = 1, 2, \ldots, n.$$

◆ Example A.1

We will find the range of the parameter γ for which the quadratic form

$$\boldsymbol{x}^T \begin{bmatrix} -5-\gamma & 6 \\ 0 & -2 \end{bmatrix} \boldsymbol{x}$$

is negative definite. We first symmetrize the above quadratic form to get

$$f = \boldsymbol{x}^T \begin{bmatrix} -5-\gamma & 6 \\ 0 & -2 \end{bmatrix} \boldsymbol{x} = \boldsymbol{x}^T \begin{bmatrix} -5-\gamma & 3 \\ 3 & -2 \end{bmatrix} \boldsymbol{x}.$$

A quadratic form is negative definite if and only if its negative is positive definite. So instead of working with f, we will find the range of the parameter γ for which $g = -f$ is positive definite, where

$$g = \boldsymbol{x}^T \begin{bmatrix} 5+\gamma & -3 \\ -3 & 2 \end{bmatrix} \boldsymbol{x}.$$

To finish to problem, we will investigate when the matrix

$$\begin{bmatrix} 5+\gamma & -3 \\ -3 & 2 \end{bmatrix}$$

is positive definite. The first-order leading principal minor of the above matrix is $\Delta_1 = 5+\gamma$, which is positive if and only if $\gamma > -5$. The second-order leading principal minor is

$$\Delta_2 = \det \begin{bmatrix} 5+\gamma & -3 \\ -3 & 2 \end{bmatrix} = 2\gamma + 1,$$

which positive for $\gamma > -1/2$. Thus the quadratic form f is negative definite if and only if $\gamma > -1/2$.

If a given matrix $\boldsymbol{Q} \in \mathbb{R}^{n\times n}$ is symmetric, then its eigenvalues are real and there exists a unitary matrix $\boldsymbol{U}$ such that

$$\boldsymbol{U}^T \boldsymbol{Q} \boldsymbol{U} = \begin{bmatrix} \lambda_1 & 0 & \cdots & 0 \\ 0 & \lambda_2 & \cdots & 0 \\ \vdots & & \ddots & \vdots \\ 0 & 0 & \cdots & \lambda_n \end{bmatrix} = \boldsymbol{\Lambda}. \tag{A.40}$$

Using the above, one can show that a symmetric positive semidefinite matrix $\boldsymbol{Q}$ has a positive semidefinite symmetric square root, denoted $\boldsymbol{Q}^{1/2}$ or $\sqrt{\boldsymbol{Q}}$, that satisfies

$$\boldsymbol{Q}^{1/2}\boldsymbol{Q}^{1/2} = \boldsymbol{Q}. \tag{A.41}$$

Indeed, it follows from Theorem A.5 that if $\boldsymbol{Q} = \boldsymbol{Q}^T \geq 0$ then its eigenvalues are all nonnegative. Hence,

$$\boldsymbol{\Lambda}^{1/2} = \begin{bmatrix} \lambda_1^{1/2} & 0 & \cdots & 0 \\ 0 & \lambda_2^{1/2} & \cdots & 0 \\ \vdots & & \ddots & \vdots \\ 0 & 0 & \cdots & \lambda_n^{1/2} \end{bmatrix} \tag{A.42}$$

is well-defined. Using the fact that $\boldsymbol{U}^T\boldsymbol{U} = \boldsymbol{I}_n$, we can define

$$\boldsymbol{Q}^{1/2} = \boldsymbol{U}\boldsymbol{\Lambda}^{1/2}\boldsymbol{U}^T, \tag{A.43}$$

which has the desired properties.

If $\boldsymbol{Q} = \boldsymbol{Q}^T$ is positive semidefinite and not positive definite, then some of its eigenvalues are equal to zero. Suppose that $\boldsymbol{Q}$ has r nonzero eigenvalues. Then, there is a unitary matrix, say $\boldsymbol{U}$, such that

$$\boldsymbol{U}^T\boldsymbol{Q}\boldsymbol{U} = \begin{bmatrix} \lambda_1 & 0 & \cdots & 0 & 0 & \cdots & 0 \\ 0 & \lambda_2 & \cdots & 0 & 0 & \cdots & 0 \\ \vdots & & \ddots & \vdots & & \ddots & \vdots \\ 0 & 0 & \cdots & \lambda_r & 0 & \cdots & 0 \\ 0 & 0 & \cdots & 0 & 0 & \cdots & 0 \\ \vdots & & & & & \ddots & \vdots \\ 0 & 0 & \cdots & 0 & 0 & \cdots & 0 \end{bmatrix}. \tag{A.44}$$

Let $\boldsymbol{V} \in \mathbb{R}^{n\times r}$ be a matrix that consists of the first r columns of the matrix $\boldsymbol{U}$, and let

$$\boldsymbol{C} = \begin{bmatrix} \lambda_1^{1/2} & 0 & \cdots & 0 \\ 0 & \lambda_2^{1/2} & \cdots & 0 \\ \vdots & & \ddots & \vdots \\ 0 & 0 & \cdots & \lambda_r^{1/2} \end{bmatrix} \boldsymbol{V}^T. \tag{A.45}$$

Note that the matrix $\boldsymbol{C}$ is of full rank, and

$$\boldsymbol{Q} = \boldsymbol{C}^T\boldsymbol{C}. \tag{A.46}$$

We refer to (A.46) as a full-rank factorization of $\boldsymbol{Q}$ because $\boldsymbol{C}$ is of full rank. We can generalize the above, using (A.38), to obtain a full-rank factorization of a nonsquare matrix. Specifically, given an $m \times n$ matrix $\boldsymbol{A}$ of rank r, we can represent it as

$$\boldsymbol{A} = \boldsymbol{B}\boldsymbol{C},$$

where $\boldsymbol{B}$ is an $m \times r$ matrix and $\boldsymbol{C}$ is a $r \times n$ matrix such that

$$\operatorname{rank}(\boldsymbol{B}) = \operatorname{rank}(\boldsymbol{C}) = \operatorname{rank}(\boldsymbol{A}) = r.$$

A.5 The Kronecker Product

Let $\boldsymbol{A}$ be an $m \times n$ and $\boldsymbol{B}$ be a $p \times q$ matrices. The *Kronecker product* of $\boldsymbol{A}$ and $\boldsymbol{B}$, denoted $\boldsymbol{A} \otimes \boldsymbol{B}$, is an $mp \times nq$ matrix defined as

$$\boldsymbol{A} \otimes \boldsymbol{B} = \begin{bmatrix} a_{11}\boldsymbol{B} & a_{12}\boldsymbol{B} & \cdots & a_{1n}\boldsymbol{B} \\ a_{21}\boldsymbol{B} & a_{22}\boldsymbol{B} & \cdots & a_{2n}\boldsymbol{B} \\ \vdots & \vdots & \ddots & \vdots \\ a_{m1}\boldsymbol{B} & a_{m2}\boldsymbol{B} & \cdots & a_{mn}\boldsymbol{B} \end{bmatrix}. \tag{A.47}$$

Thus, the matrix $\boldsymbol{A} \otimes \boldsymbol{B}$ consists of mn blocks. Using the definition of the Kronecker product, we can verify the following two properties:

$$(\boldsymbol{A} \otimes \boldsymbol{B})(\boldsymbol{C} \otimes \boldsymbol{D}) = \boldsymbol{AC} \otimes \boldsymbol{BD}, \tag{A.48}$$

$$(\boldsymbol{A} \otimes \boldsymbol{B})^T = \boldsymbol{A}^T \otimes \boldsymbol{B}^T. \tag{A.49}$$

Let

$$\boldsymbol{X} = [\boldsymbol{x}_1 \quad \boldsymbol{x}_2 \quad \cdots \quad \boldsymbol{x}_m]$$

be a $n \times m$ matrix, where $\boldsymbol{x}_i$, $i = 1, 2, \ldots, m$, are the columns of $\boldsymbol{X}$. Each $\boldsymbol{x}_i$ consists of n elements. Then, the *stacking operator* defined as

$$\begin{aligned} s(\boldsymbol{X}) &= \begin{bmatrix} \boldsymbol{x}_1^T & \boldsymbol{x}_2^T & \cdots & \boldsymbol{x}_m^T \end{bmatrix}^T \\ &= [x_{11} \quad x_{21} \quad \cdots \quad x_{n1} \quad x_{12} \quad x_{22} \quad \cdots \quad x_{n2} \quad \cdots \quad x_{1m} \quad x_{2m} \quad \cdots \quad x_{nm}]^T \end{aligned}$$

yields the column nm vector formed from the columns of $\boldsymbol{X}$ taken in order. Let now $\boldsymbol{A}$ be an $n \times n$, and let $\boldsymbol{C}$ and $\boldsymbol{X}$ be $n \times m$ matrices. Then, the matrix equation

$$\boldsymbol{AX} = \boldsymbol{C} \tag{A.50}$$

can be written as

$$(\boldsymbol{I}_m \otimes \boldsymbol{A})s(\boldsymbol{X}) = s(\boldsymbol{C}). \tag{A.51}$$

If $\boldsymbol{B}$ is an $m \times m$ matrix, then the matrix equation

$$\boldsymbol{XB} = \boldsymbol{C} \tag{A.52}$$

can be represented as

$$(\boldsymbol{B}^T \otimes \boldsymbol{I}_n)s(\boldsymbol{X}) = s(\boldsymbol{C}). \tag{A.53}$$

Using the above two facts, we can verify that the matrix equation

$$\boldsymbol{AX} + \boldsymbol{XB} = \boldsymbol{C}, \tag{A.54}$$

where $\boldsymbol{A}$ is $n \times n$, $\boldsymbol{B}$ is $m \times m$, and $\boldsymbol{C}$ is $n \times m$, can be written as

$$(\boldsymbol{I}_m \otimes \boldsymbol{A} + \boldsymbol{B}^T \otimes \boldsymbol{I}_n)s(\boldsymbol{X}) = s(\boldsymbol{C}). \tag{A.55}$$

◆ Example A.2

Consider the matrix equation (A.54), where

$$A = \begin{bmatrix} 0 & 1 \\ 0 & 1 \end{bmatrix}, \quad B = \begin{bmatrix} -2 & 0 \\ -3 & 1 \end{bmatrix}, \quad \text{and} \quad C = \begin{bmatrix} 1 & 0 \\ 1 & 1 \end{bmatrix}.$$

We will represent the above equation in the matrix–vector format as given by (A.55), that is,

$$(I_2 \otimes A - B^T \otimes I_2)s(X) = s(C).$$

Substituting into the above equation the given matrices yields

$$\left(\begin{bmatrix} 0 & 1 & 0 & 0 \\ 0 & 1 & 0 & 0 \\ 0 & 0 & 0 & 1 \\ 0 & 0 & 0 & 1 \end{bmatrix} + \begin{bmatrix} -2 & 0 & -3 & 0 \\ 0 & -2 & 0 & -3 \\ 0 & 0 & 1 & 0 \\ 0 & 0 & 0 & 1 \end{bmatrix} \right) \begin{bmatrix} x_{11} \\ x_{21} \\ x_{12} \\ x_{22} \end{bmatrix} = \begin{bmatrix} 1 \\ 1 \\ 0 \\ 1 \end{bmatrix}.$$

Performing simple manipulations yields

$$\begin{bmatrix} -2 & 1 & -3 & 0 \\ 0 & -1 & 0 & -3 \\ 0 & 0 & 1 & 1 \\ 0 & 0 & 0 & 2 \end{bmatrix} \begin{bmatrix} x_{11} \\ x_{21} \\ x_{12} \\ x_{22} \end{bmatrix} = \begin{bmatrix} 1 \\ 1 \\ 0 \\ 1 \end{bmatrix}.$$

The above system of linear equations has a unique solution because the matrix

$$\begin{bmatrix} -2 & 1 & -3 & 0 \\ 0 & -1 & 0 & -3 \\ 0 & 0 & 1 & 1 \\ 0 & 0 & 0 & 2 \end{bmatrix}$$

is nonsingular.

Let λ_i, $\boldsymbol{v}_i$ be the eigenvalues and eigenvectors, respectively of $\boldsymbol{A}$, and μ_j and $\boldsymbol{w}_j$ the eigenvalues and eigenvectors of $m \times m$ matrix $\boldsymbol{B}$. Then,

$$\begin{aligned}(\boldsymbol{A} \otimes \boldsymbol{B})(\boldsymbol{v}_i \otimes \boldsymbol{w}_j) &= \boldsymbol{A}\boldsymbol{v}_i \otimes \boldsymbol{B}\boldsymbol{w}_j \\ &= \lambda_i \boldsymbol{v}_i \otimes \mu_j \boldsymbol{w}_j \\ &= \lambda_i \mu_j (\boldsymbol{v}_i \otimes \boldsymbol{w}_j). \end{aligned} \tag{A.56}$$

Thus, the eigenvalues of $\boldsymbol{A} \otimes \boldsymbol{B}$ are $\lambda_i \mu_j$, and their respective eigenvectors are $\boldsymbol{v}_i \otimes \boldsymbol{w}_j$ for $i = 1, 2, \ldots, n$, $j = 1, 2, \ldots, m$.

◆ Example A.3

For the given two matrices

$$A = \begin{bmatrix} 0 & 0 \\ 2 & -2 \end{bmatrix} \quad \text{and} \quad B = \begin{bmatrix} 4 & 6 & -8 \\ 0 & -1 & 0 \\ 0 & 2 & 7 \end{bmatrix},$$

we will find the eigenvalues of the matrix $(I + A \otimes B^T)$, where I is the identity matrix of appropriate dimensions. We note that

$$\text{eig}(I + A \otimes B^T) = 1 + \text{eig}(A)\text{eig}(B),$$

because for any square matrix M, $\text{eig}(I+M) = 1+\text{eig}(M)$ and $\text{eig}(M^T) = \text{eig}(M)$. By inspection,

$$\text{eig}(A) = \{0, -2\} \quad \text{and} \quad \text{eig}(B) = \{4, -1, 7\}.$$

Hence,

$$\text{eig}(I + A \otimes B^T) = \{1, 1, 1, -7, 3, -13\}.$$

We will now use (A.56) to study the matrix equation (A.54), which we represent as

$$M\, s(X) = s(C), \tag{A.57}$$

where

$$M = I_m \otimes A + B^T \otimes I_n. \tag{A.58}$$

The solution to (A.57) is unique if, and only if, the $mn \times mn$ matrix M is nonsingular. To find the condition for this to hold, consider the matrix

$$(I_m + \varepsilon B^T) \otimes (I_n + \varepsilon A) = I_m \otimes I_n + \varepsilon M + \varepsilon^2 B^T \otimes A \tag{A.59}$$

whose eigenvalues are

$$(1 + \varepsilon\mu_j)(1 + \varepsilon\lambda_i) = 1 + \varepsilon(\mu_j + \lambda_i) + \varepsilon^2 \mu_j \lambda_i \tag{A.60}$$

because for a square matrix Q we have

$$\lambda_i(I_n + \varepsilon Q) = 1 + \varepsilon\lambda_i(Q).$$

Comparing the terms in ε in (A.59) and (A.60), we conclude that the eigenvalues of M are $\lambda_i + \mu_j$, $i = 1, 2, \ldots, n$, $j = 1, 2, \ldots, m$. Hence, M is nonsingular if and only if

$$\lambda_i + \mu_j \neq 0. \tag{A.61}$$

The above is a necessary and sufficient condition for the solution X of the matrix equation $AX + XB = C$ to be unique.

Another useful identity involving the Kronecker product is

$$s(ABC) = (C^T \otimes A)s(B). \tag{A.62}$$

For further information on the subject of the Kronecker product, we refer the reader to Brewer [33].

A.6 Upper and Lower Bounds

Consider a set $\mathcal{S}$ of real numbers. A number M is called an *upper bound* of the set $\mathcal{S}$ if

$$x \leq M \qquad \text{for all } x \in \mathcal{S}.$$

A set of real numbers that has an upper bound is said to be *bounded above.*

The Least Upper Bound Axiom Every nonempty set $\mathcal{S}$ of real numbers that has an upper bound has a *least upper bound,* also called *supremum,* and is denoted

$$\sup\{x : x \in \mathcal{S}\}.$$

Examples

1.
$$\sup\left\{\frac{1}{2}, \frac{2}{3}, \ldots, \frac{n}{n+1}, \ldots\right\} = 1;$$

2.
$$\sup\left\{-\frac{1}{2}, -\frac{1}{8}, -\frac{1}{27}, \ldots, -\frac{1}{n^3}, \ldots\right\} = 0;$$

3.
$$\sup\{x : x^2 < 3\} = \sup\left\{x : -\sqrt{3} < x < \sqrt{3}\right\} = \sqrt{3}.$$

Theorem A.7 If $M = \sup\{x : x \in \mathcal{S}\}$ and $\varepsilon > 0$, then there is at least one number x in $\mathcal{S}$ such that

$$M - \varepsilon < x \leq M.$$

Proof The condition $x \leq M$ is satisfied by all numbers x in $\mathcal{S}$ by virtue of the least upper bound axiom. We have to show that for any $\varepsilon > 0$ there is a number $x \in \mathcal{S}$ such that

$$M - \varepsilon < x.$$

We prove the statement by contradiction. We suppose that there is no such number in $\mathcal{S}$. Then,

$$x \leq M - \varepsilon \qquad \text{for all } x \in \mathcal{S}$$

and hence $M - \varepsilon$ would be an upper bound of $\mathcal{S}$ that is less than M, which is the least upper bound. This contradicts the assumption, and thus the proof is complete.

To illustrate the above theorem, we consider the following set of real numbers:

$$\mathcal{S} = \left\{\frac{1}{2}, \frac{2}{3}, \frac{3}{4}, \ldots, \frac{n}{n+1}, \ldots\right\}.$$

Note that

$$\sup\{x : x \in \mathcal{S}\} = 1.$$

Let $\varepsilon = 0.1$. Then, for example, the element of the set $\mathcal{S}$ that is equal to $99/100$ satisfies

$$1 - \varepsilon < \frac{99}{100} \leq 1.$$

A number m is called a *lower bound* of $\mathcal{S}$ if

$$m \le x \qquad \text{for all } x \in \mathcal{S}.$$

Sets that have lower bounds are said to be *bounded below*.

Theorem A.8 Every nonempty set of real numbers that has a lower bound has a *greatest lower bound,* also called *infimum,* and is denoted as

$$\inf\{x : x \in \mathcal{S}\}.$$

Proof By assumption, $\mathcal{S}$ is nonempty and has a lower bound that we call s. Thus

$$s \le x \qquad \text{for all } x \in \mathcal{S}.$$

Hence

$$-x \le -s \qquad \text{for all } x \in \mathcal{S};$$

that is, the set $\{-x : x \in \mathcal{S}\}$ has an upper bound $-s$. From the least upper bound axiom, we conclude that the set $\{-x : x \in \mathcal{S}\}$ has a least upper bound, supremum. We call it $-s_0$. Because

$$-x \le -s_0 \qquad \text{for all } x \in \mathcal{S},$$

we have

$$s_0 \le x \qquad \text{for all } x \in \mathcal{S},$$

and thus s_0 is a lower bound for $\mathcal{S}$. We now prove, by contradiction, that s_0 is the greatest lower bound, infimum, of the set $\mathcal{S}$. Indeed, if there was a number $\tilde{x}$ satisfying

$$s_0 < \tilde{x} \le x \qquad \text{for all } x \in \mathcal{S},$$

then we would have

$$-x \le -\tilde{x} < -s_0 \qquad \text{for all } x \subset \mathcal{S},$$

which contradicts the fact that $-s_0$ is the supremum of $\{-x : x \in \mathcal{S}\}$.

Using arguments similar to those we used in the proof of Theorem A.8, we can prove the following theorem:

Theorem A.9 If $m = \inf\{x : x \in \mathcal{S}\}$ and $\varepsilon > 0$, then there is at least one number x in $\mathcal{S}$ such that

$$m \le x < m + \varepsilon.$$

A.7 Sequences

A *sequence of real numbers* can be viewed as a function whose domain is the set of natural numbers $1, 2, \ldots, n, \ldots$ and whose range is contained in $\mathbb{R}$. Thus, a sequence of real numbers can be viewed as a set of ordered pairs $(1, a_1), (2, a_2), \ldots, (n, a_n), \ldots$. We denote a sequence

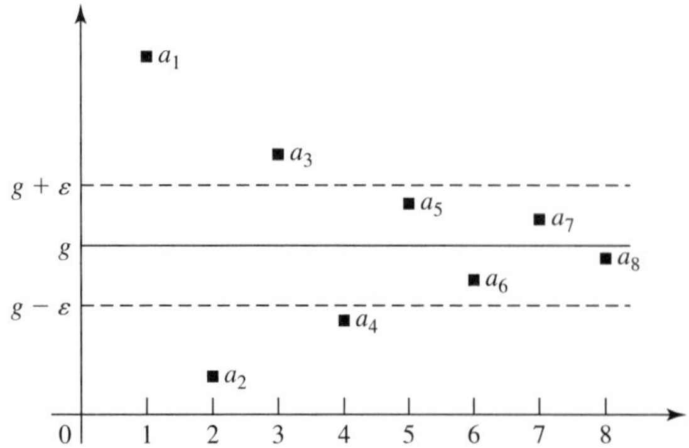

Figure A.2 An illustration of the notion of the limit of a sequence.

of real numbers

$$a_1, a_2, \dots, a_n, \dots$$

as $\{a_n\}$.

A sequence $\{a_n\}$ is increasing if $a_1 < a_2 < \cdots < a_n \cdots$. In general, a sequence is increasing if $a_n < a_{n+1}$. If $a_n \le a_{n+1}$, then we say that the sequence is nondecreasing. Similarly, we can define decreasing and nonincreasing sequences. Nonincreasing or nondecreasing sequences are called *monotone* sequences.

A number g is called the *limit* of the infinite sequence $a_1, a_2, \dots, a_n, \dots$ if for any positive ε there is a natural number $k = k(\varepsilon)$ such that for all $n > k$ we have

$$|a_n - g| < \varepsilon;$$

that is, a_n lies between $g - \varepsilon$ and $g + \varepsilon$ for all $n > k$. In other words, for any $\varepsilon > 0$ we have $|a_n - g| < \varepsilon$ for sufficiently large n's (see Figure A.2), and we write

$$g = \lim_{n\to\infty} a_n,$$

or

$$a_n \to g.$$

A sequence that has a limit is called a *convergent* sequence. A sequence that has no limit is called *divergent*.

The notion of the sequence can be extended to sequences with elements in $\mathbb{R}^p$. Specifically, a sequence in $\mathbb{R}^p$ is a function whose domain is the set of natural numbers $1, 2, \dots, n, \dots$ and whose range is contained in $\mathbb{R}^p$.

Suppose that $\boldsymbol{a}_1, \boldsymbol{a}_2, \dots, \boldsymbol{a}_n, \dots$, denoted $\{\boldsymbol{a}_n\}$, is a sequence in $\mathbb{R}^p$. Then, $\boldsymbol{g} \in \mathbb{R}^p$ is called the limit of this sequence if for each $\varepsilon > 0$ there exists a natural number $k = k(\varepsilon)$ such that for all $n > k$ we have

$$\|\boldsymbol{a}_n - \boldsymbol{g}\| < \varepsilon,$$

and we write

$$\boldsymbol{g} = \lim_{n\to\infty} \boldsymbol{a}_n,$$

or

$$\boldsymbol{a}_n \to \boldsymbol{g}.$$

Theorem A.10 If a limit of a sequence in $\mathbb{R}^p$ exists, then it is unique.

Proof We prove this theorem by contradiction. Let us assume that a sequence $\{\boldsymbol{a}_n\}$ has two different limits, say $\boldsymbol{g}_1$ and $\boldsymbol{g}_2$. Then, we have

$$\|\boldsymbol{g}_1 - \boldsymbol{g}_2\| > 0.$$

Let

$$\varepsilon = \tfrac{1}{2}\|\boldsymbol{g}_1 - \boldsymbol{g}_2\|.$$

From the definition of a limit, there exist $k_1 = k_1(\varepsilon)$ and $k_2 = k_2(\varepsilon)$ such that for $n > k_1$ we have $\|\boldsymbol{a}_n - \boldsymbol{g}_1\| < \varepsilon$, and for $n > k_2$ we have $\|\boldsymbol{a}_n - \boldsymbol{g}_2\| < \varepsilon$. Let $m = \max(k_1, k_2)$. If $n > m$, then $\|\boldsymbol{a}_n - \boldsymbol{g}_1\| < \varepsilon$ and $\|\boldsymbol{a}_n - \boldsymbol{g}_2\| < \varepsilon$. Adding $\|\boldsymbol{a}_n - \boldsymbol{g}_1\| < \varepsilon$ and $\|\boldsymbol{a}_n - \boldsymbol{g}_2\| < \varepsilon$ yields

$$\|\boldsymbol{a}_n - \boldsymbol{g}_1\| + \|\boldsymbol{a}_n - \boldsymbol{g}_2\| < 2\varepsilon. \tag{A.63}$$

It follows from the triangle inequality that

$$\|\boldsymbol{a} - \boldsymbol{b}\| \le \|\boldsymbol{a}\| + \|\boldsymbol{b}\|.$$

Using the above inequality, we obtain

$$\begin{aligned}\|-\boldsymbol{g}_1 + \boldsymbol{g}_2\| &= \|\boldsymbol{a}_n - \boldsymbol{g}_1 - \boldsymbol{a}_n + \boldsymbol{g}_2\| \\ &= \|(\boldsymbol{a}_n - \boldsymbol{g}_1) - (\boldsymbol{a}_n - \boldsymbol{g}_2)\| \\ &\le \|\boldsymbol{a}_n - \boldsymbol{g}_1\| + \|\boldsymbol{a}_n - \boldsymbol{g}_2\|. \end{aligned} \tag{A.64}$$

Applying (A.63) to (A.64) yields

$$\|-\boldsymbol{g}_1 + \boldsymbol{g}_2\| = \|\boldsymbol{g}_1 - \boldsymbol{g}_2\| < 2\varepsilon.$$

However, the above contradicts the assumption that $\|\boldsymbol{g}_1 - \boldsymbol{g}_2\| = 2\varepsilon$, which completes the proof.

A sequence in $\mathbb{R}^p$ is bounded if there exists a number $M \ge 0$ such that $\|\boldsymbol{a}_n\| \le M$ for all $n = 1, 2, \ldots$. The set of all bounded sequences is sometimes denoted as l_∞.

Theorem A.11 Every convergent sequence is bounded.

Proof Let

$$\boldsymbol{g} = \lim_{n\to\infty} \boldsymbol{a}_n.$$

Choose $\varepsilon = 1$. Then, there exists a natural number $k = k(\varepsilon)$ such that

$$\|\boldsymbol{a}_n - \boldsymbol{g}\| < 1 \qquad \text{for all } n > k.$$

By Theorem A.4,

$$\|\boldsymbol{a}\| - \|\boldsymbol{b}\| \le \|\boldsymbol{a} - \boldsymbol{b}\|.$$

Using the above inequality, we get

$$\|\boldsymbol{a}_n\| - \|\boldsymbol{g}\| \le \|\boldsymbol{a}_n - \boldsymbol{g}\| < 1 \qquad \text{for all } n > k.$$

Therefore,

$$\|\boldsymbol{a}_n\| < \|\boldsymbol{g}\| + 1 \qquad \text{for all } n > k.$$

Let

$$M = \max(\|\boldsymbol{a}_1\|, \|\boldsymbol{a}_2\|, \ldots, \|\boldsymbol{a}_k\|, \|\boldsymbol{g}\| + 1),$$

then we have

$$M \geq \|\boldsymbol{a}_n\| \qquad \text{for all } n,$$

which means that a convergent sequence is bounded, and the proof is complete.

Suppose we are given a sequence $\boldsymbol{a}_1, \boldsymbol{a}_2, \ldots$ in $\mathbb{R}^p$ and an increasing sequence of positive natural numbers

$$m_1, m_2, \ldots, m_n, \ldots.$$

The sequence

$$\boldsymbol{a}_{m_1}, \boldsymbol{a}_{m_2}, \ldots, \boldsymbol{a}_{m_n}, \ldots$$

is called a *subsequence* of the sequence $\boldsymbol{a}_1, \boldsymbol{a}_2, \ldots$ in $\mathbb{R}^p$, and we have

$$m_n \geq n.$$

Indeed, this is obvious for $n = 1$; that is, $m_1 \geq 1$ because m_1 is a positive integer. We now apply the principle of induction. We assume that $m_n \geq n$ for a given n. Then,

$$m_{n+1} > m_n \geq n.$$

Hence,

$$m_{n+1} \geq n + 1.$$

Thus, indeed $m_n \geq n$ for all n.

We can say that a subsequence is obtained from the given sequence by neglecting a number of elements in the sequence. We denote a subsequence $\boldsymbol{a}_{m_1}, \boldsymbol{a}_{m_2}, \ldots, \boldsymbol{a}_{m_n}, \ldots$ as

$$\{\boldsymbol{a}_{m_n}\}.$$

Theorem A.12 A subsequence of a convergent sequence in $\mathbb{R}^p$ is convergent to the same limit; that is, if

$$\lim_{n\to\infty} \boldsymbol{a}_n = \boldsymbol{g},$$

and if

$$m_1 < m_2 < \cdots < m_n < \cdots,$$

then

$$\lim_{n\to\infty} \boldsymbol{a}_{m_n} = \boldsymbol{g}.$$

Proof Let $\varepsilon > 0$ be given. Then, there exists a natural number $k = k(\varepsilon)$ such that $\|\boldsymbol{a}_n - \boldsymbol{g}\| < \varepsilon$ for any $n > k$. From the fact that $m_n \geq n$, it follows that

$$m_n > n > k,$$

and thus

$$\| a_{m_n} - g \| < \varepsilon$$

for any $m_n > n > k$, which means that

$$\lim_{n \to \infty} a_{m_n} = g.$$

The proof is complete.

We now state, without a proof, an important result due to Bolzano and Weierstrass.

Theorem A.13 (Bolzano–Weierstrass) Every bounded sequence in $\mathbb{R}^p$ contains a convergent subsequence.

Theorem A.14 (Cauchy) A sequence $\{a_n\}$ in $\mathbb{R}^p$ is convergent if and only if for every $\varepsilon > 0$ there exists an $r = r(\varepsilon)$ such that

$$\|a_n - a_r\| < \varepsilon$$

holds for all $n > r$.

Proof ($\Rightarrow$) Let

$$\lim_{n \to \infty} a_n = g,$$

and let $\varepsilon > 0$ be given. Then, for some $l = l(\varepsilon)$,

$$\|a_n - g\| < \tfrac{1}{2}\varepsilon \qquad \text{for all } n > l.$$

Let $r = l + 1$, then

$$\|a_r - g\| < \tfrac{1}{2}\varepsilon.$$

Adding the above inequalities, we obtain

$$\|a_n - g\| + \|a_r - g\| < \varepsilon \quad \text{for all } n > r.$$

On the other hand,

$$\|a_n - a_r\| \leq \|a_n - g\| + \|a_r - g\|.$$

Hence,

$$\|a_n - a_r\| < \varepsilon \qquad \text{for all } n > r.$$

($\Leftarrow$) We now assume that for every $\varepsilon > 0$ there exists an $r = r(\varepsilon)$ such that

$$\|a_n - a_r\| < \varepsilon \qquad \text{for all } n > r.$$

We then show that the above condition implies

$$\lim_{n \to \infty} a_n = g.$$

First, we show that $\{a_n\}$ is a bounded sequence. Let $\varepsilon = 1$. Then, an $r = r(\varepsilon)$ exists such that $\|a_n - a_r\| < 1$ for all $n > r$. Hence,

$$\|a_n\| - \|a_r\| \leq \|a_n - a_r\| < 1,$$

which means that

$$\|\boldsymbol{a}_n\| < \|\boldsymbol{a}_r\| + 1.$$

Let

$$M = \max\{\|\boldsymbol{a}_1\|, \|\boldsymbol{a}_2\|, \ldots, \|\boldsymbol{a}_{r-1}\|, \|\boldsymbol{a}_r\| + 1\},$$

then we have

$$\|\boldsymbol{a}_n\| \leq M \qquad \text{for all } n,$$

which implies that the sequence $\{\boldsymbol{a}_n\}$ is bounded. By the Bolzano–Weierstrass theorem the sequence $\{\boldsymbol{a}_n\}$ contains a convergent subsequence. Let

$$\lim_{n\to\infty} \boldsymbol{a}_{m_n} = \boldsymbol{g}, \qquad m_1 < m_2 < \cdots.$$

We will show that

$$\lim_{n\to\infty} \boldsymbol{a}_n = \boldsymbol{g}.$$

From the hypothesis of the theorem it follows that for some $r = r(\varepsilon/3)$ we have

$$\|\boldsymbol{a}_n - \boldsymbol{a}_r\| < \tfrac{1}{3}\varepsilon \qquad \text{for all } n > r. \tag{A.65}$$

On the other hand, the assumption that the subsequence $\{\boldsymbol{a}_{m_n}\}$ is convergent implies that for some $k = k(\varepsilon/3)$ we have

$$\|\boldsymbol{a}_{m_n} - \boldsymbol{g}\| < \tfrac{1}{3}\varepsilon \qquad \text{for all } n > k. \tag{A.66}$$

We can select k so that $k > r$. In such a case, (A.65) and (A.66) are satisfied for all $n > k$. Because $m_n \geq n > r$, we have for all $n > k$ the following:

$$\left\| \boldsymbol{a}_{m_n} - \boldsymbol{a}_r \right\| < \tfrac{1}{3}\varepsilon,$$

or, equivalently,

$$\left\| \boldsymbol{a}_r - \boldsymbol{a}_{m_n} \right\| < \tfrac{1}{3}\varepsilon. \tag{A.67}$$

Adding (A.65), (A.66), and (A.67) yields

$$\|\boldsymbol{a}_n - \boldsymbol{a}_r\| + \left\| \boldsymbol{a}_{m_n} - \boldsymbol{g} \right\| + \left\| \boldsymbol{a}_r - \boldsymbol{a}_{m_n} \right\| < \varepsilon.$$

Hence,

$$\begin{aligned}
\|\boldsymbol{a}_n - \boldsymbol{g}\| &= \left\| (\boldsymbol{a}_n - \boldsymbol{a}_r) + (\boldsymbol{a}_r - \boldsymbol{a}_{m_n}) + (\boldsymbol{a}_{m_n} - \boldsymbol{g}) \right\| \\
&\leq \|\boldsymbol{a}_n - \boldsymbol{a}_r\| + \left\| \boldsymbol{a}_{m_n} - \boldsymbol{g} \right\| + \left\| \boldsymbol{a}_r - \boldsymbol{a}_{m_n} \right\| \\
&< \varepsilon \qquad \text{for all } n > k,
\end{aligned}$$

which implies that

$$\lim_{n\to\infty} \boldsymbol{a}_n = \boldsymbol{g}.$$

The proof is complete.

Theorem A.15 Every monotone bounded sequence in $\mathbb{R}$ is convergent.

Proof We will prove the theorem for nondecreasing sequences. The proof for non-increasing sequences is analogous.

We assume that we are given a sequence $\{a_n\}$ in $\mathbb{R}$ that is bounded and nondecreasing. Let Z be the set of numbers belonging to the sequence, and let g be the least upper bound of Z. Hence

$$g \geq a_n, \qquad n = 1, 2, \ldots.$$

From the properties of the least upper bound (see Theorem A.7), it follows that for any $\varepsilon > 0$ there exists a k such that

$$g - \varepsilon < a_k.$$

Because the sequence $\{a_n\}$ is nondecreasing, the inequality $n > k$ implies that

$$a_k \leq a_n,$$

and thus

$$g - \varepsilon < a_n.$$

We then conclude that for any $\varepsilon > 0$, there exists a k such that for $n > k$,

$$g - \varepsilon < a_n < g + \varepsilon.$$

Hence,

$$|a_n - g| < \varepsilon,$$

which means that

$$\lim_{n \to \infty} a_n = g.$$

Thus, a monotone bounded sequence in $\mathbb{R}$ is convergent. The proof is complete.

Given a sequence $\{a_n\}$ of real numbers. Then the sequence $\{s_N\}$ defined by

$$s_N = \sum_{n=1}^{N} a_n = a_1 + a_2 + \cdots + a_N$$

is called the sequence of *partial sums* of the series

$$\sum_{n=1}^{\infty} a_n. \tag{A.68}$$

If the sequence of the partial sums has a limit—that is, $s_N \to s$ as $N \to \infty$—then we say that the series converges to the sum s, and write

$$s = \sum_{n=1}^{\infty} a_n.$$

The above formula can only make sense when the series (A.68) converges.

Theorem A.16 If the series $\sum_{n=1}^{\infty} a_n$ converges, then

$$a_n \to 0 \qquad \text{as } n \to \infty.$$

Proof Let s be the sum of the series $\sum_{n=1}^{\infty} a_n$. Then,

$$s_N = \sum_{n=1}^{N} a_n \to s \qquad \text{as } N \to \infty. \tag{A.69}$$

We also have

$$s_{N-1} \to s \qquad \text{as } N \to \infty, \tag{A.70}$$

and

$$\begin{aligned} a_N &= (a_1 + a_2 + \cdots + a_N) - (a_1 + a_2 + \cdots + a_{N-1}) \\ &= s_N - s_{N-1}. \end{aligned}$$

Therefore, by (A.69) and (A.70),

$$a_N \to (s - s) = 0 \qquad \text{as } N \to \infty,$$

and the proof is complete.

The converse of Theorem A.16 is false. The terms of the series

$$\sum_{n=1}^{\infty} \frac{1}{n}$$

tend to zero, but the series diverges.

A.8 Functions

We discuss functions of one and several variables and their basic properties. A function f from a set $\mathcal{A}$ to a set $\mathcal{B}$, denoted

$$f : \mathcal{A} \to \mathcal{B},$$

is a rule that assigns to each $\boldsymbol{x} \in \mathcal{A}$ a unique element $\boldsymbol{y} \in \mathcal{B}$. The set $\mathcal{A}$ is the domain of f and the range of f is a subset of $\mathcal{B}$, not necessarily the whole of $\mathcal{B}$.

Let $f : \mathcal{S} \to \mathbb{R}$, where $\mathcal{S}$ is a subset of $\mathbb{R}$. Then, f is said to be continuous at $z_0 \in \mathcal{S}$ if for every $\varepsilon > 0$, there exists $\delta = \delta(\varepsilon, z_0)$ such that if $z \in \mathcal{S}$, $|z - z_0| < \delta$, then $|f(z) - f(z_0)| < \varepsilon$. This is also written as $f(z) \to f(z_0)$ as $z \to z_0$. If f is continuous at every point of $\mathcal{S}$, then we say that f is continuous on $\mathcal{S}$.

We say that the function f is bounded on a set $\mathcal{S}$ if there is $M \geq 0$ such that for any $\boldsymbol{z} \in \mathcal{S}$ we have

$$|f(\boldsymbol{z})| \leq M.$$

Lemma A.2 If f is continuous on $[a, b] \subset \mathbb{R}$, then f is bounded on $[a, b]$.

Proof Consider the set

$$\{x : x \in [a, b], \text{ and } f \text{ is bounded on } [a, x]\}.$$

This set is nonempty and bounded above by b. Let

$$c = \sup\{x : f \text{ is bounded on } [a, x]\}.$$

We now show that $c = b$. Suppose that $c < b$. From the continuity of f at c, it follows that f is bounded on $[c - \delta, c + \delta]$ for some sufficiently small $\delta > 0$. Being bounded on $[a, c - \delta]$ and on $[c - \delta, c + \delta]$, it is bounded on $[a, c + \delta]$. This contradicts our choice of c. We therefore conclude that $c = b$. This means that f is bounded on $[a, x]$ for all $x < b$. From the continuity of f, we know that f is bounded on some interval $[b - \delta, b]$. Hence, f is bounded on $[a, b - \delta]$ and on $[b - \delta, b]$, which means that f is bounded on $[a, b]$.

With the help of the above lemma, we can now prove the maximum–minimum theorem of Weierstrass.

Theorem A.17 (The Maximum–Minimum Theorem) If f is continuous on an interval $[a, b]$, then it takes on both its supremum and infimum values on $[a, b]$.

Proof By Lemma A.2, the function f is bounded on $[a, b]$. Let

$$M = \sup\{f(x) : x \in [a, b]\}.$$

We now show that there is a $c \in [a, b]$ such that $f(c) = M$. Let

$$g(x) = \frac{1}{M - f(x)}.$$

If f does not take on the value of M, then g is continuous on $[a, b]$, and by Lemma A.2 the function g is bounded on $[a, b]$. But g cannot be bounded on $[a, b]$. Thus, the assumption that f does not take on the value M has led us to a contradiction. The other case that f takes on its infimum value can be proved in a similar manner.

A *sphere,* ball, around $\boldsymbol{x}$ in $\mathbb{R}^p$ is the set of the form $\{\boldsymbol{y} : \|\boldsymbol{x} - \boldsymbol{y}\| < \varepsilon\}$ for some $\varepsilon > 0$. A set Ω in $\mathbb{R}^p$ is said to be *open* if around every point $\boldsymbol{x}$ in Ω there is a sphere that is contained in Ω.

If $\boldsymbol{x} \in \mathbb{R}^p$, then any set that contains an open set containing $\boldsymbol{x}$ is called a *neighborhood* of $\boldsymbol{x}$. A sphere around $\boldsymbol{x}$ is a neighborhood of $\boldsymbol{x}$.

A point $\boldsymbol{x} \in \mathbb{R}^p$ is called a *boundary point* of a set $\Omega \subset \mathbb{R}^p$ if every neighborhood of $\boldsymbol{x}$ contains a point of Ω and a point that is not in Ω. The set of all boundary points of Ω is called the *boundary* of Ω. A set is said to be *closed* if it contains its boundary. Equivalently, a set Ω is closed if $\boldsymbol{x}_k \to \boldsymbol{g}$ with $\boldsymbol{x}_k \in \Omega$ implies $\boldsymbol{g} \in \Omega$. A set is *bounded* if it is contained within some sphere of finite radius. A set is *compact* if it is both closed and bounded. Any finite set $\Omega = \{\boldsymbol{x}_1, \boldsymbol{x}_2, \ldots, \boldsymbol{x}_k\}$ in $\mathbb{R}^p$ is compact.

We now generalize Lemma A.2 for real-valued functions of several variables.

Lemma A.3 Let $f : \mathbb{R}^p \to \mathbb{R}$ be a continuous function on a compact set $\Omega \subset \mathbb{R}^p$. Then, f is bounded on Ω.

Proof We prove the lemma by contradiction. We assume that f is continuous on a compact set Ω, but f is not bounded on Ω. Without loss of generality, suppose that f is unbounded above on Ω. Since f is unbounded above on Ω, we can find a sequence $\{\boldsymbol{x}_n\} \subset \Omega$ such that $f(\boldsymbol{x}_n) > r(n)$, where $r(n) \to \infty$ as $n \to \infty$, that is, $f(\boldsymbol{x}_n) \to \infty$ as $n \to \infty$. The sequence $\{\boldsymbol{x}_n\}$ must be bounded because, by assumption, the set Ω is compact. Therefore, by the Bolzano-Weierstrass theorem, on page 645, there is a convergent subsequence, say $\{\boldsymbol{x}_{n_k}\}$, that converges to a point $\boldsymbol{x}^* \in \Omega$. The function is continuous on Ω. Hence,

$$\lim_{k\to\infty} f(\boldsymbol{x}_{n_k}) = f\left(\lim_{k\to\infty} \boldsymbol{x}_{n_k}\right) = f(\boldsymbol{x}^*). \tag{A.71}$$

We arrived at a contradiction because, by assumption, the left-hand side of (A.71) is divergent, while $f(\boldsymbol{x}^*) < \infty$ because f is continuous on Ω. Therefore, a continuous function on a compact set in $\mathbb{R}^p$ must be bounded.

We next prove the maximum–minimum theorem of Weierstrass for real-valued functions of several variables.

Theorem A.18 (Weierstrass) If $f : \mathbb{R}^p \to \mathbb{R}$ is a continuous function on a compact set $\Omega \subset \mathbb{R}^p$, then f achieves its supremum and infimum.

Proof The proof follows that of Theorem A.17, with the interval $[a, b]$ being replaced with the compact set Ω and by referring to Lemma A.3 rather than to Lemma A.2.

Definition A.1 (Uniform Continuity) Let $f : \mathcal{X} \to \mathcal{Y}$. Then, f is called uniformly continuous on $\mathcal{X}$ if for every $\varepsilon > 0$ there exists $\delta = \delta(\varepsilon) > 0$, depending only on ε, such that $|x - \tilde{x}| < \delta$ implies $|f(x) - f(\tilde{x})| < \varepsilon$, where $x, \tilde{x}$ are in $\mathcal{X}$.

Every uniformly continuous function is continuous, but not necessarily conversely. Indeed, $f(x) = 1/x$ is continuous on $(0, 1)$ but not uniformly continuous.

Another type of continuous functions that we use in the book are absolutely continuous functions.

Definition A.2 (Absolutely Continuous Function) Let f be a function defined on the closed interval $[a, b]$. Then, the function f is said to be absolutely continuous on $[a, b]$ if for every $\varepsilon > 0$, there exists a $\delta > 0$ such that

$$\sum_{k=1}^{n} (b_k - a_k) < \delta \tag{A.72}$$

for all $a_1, b_1, \ldots, a_n, b_n$ such that $a_1 < b_1 \le a_2 < b_2 \le \cdots \le a_n < b_n$, we have

$$\sum_{k=1}^{n} |f(b_k) - f(a_k)| < \varepsilon. \tag{A.73}$$

Note that every absolutely continuous function is continuous, because the case $n = 1$ is not excluded in the above definition. A function f that satisfies the Lipschitz condition,

$$|f(x) - f(y)| \leq L|x - y|,$$

is absolutely continuous.

A.9 Linear Operators

A function

$$\boldsymbol{F} : \mathbb{R}^n \to \mathbb{R}^m$$

is said to be *linear* if it satisfies the following two properties:

LO 1. For any $\boldsymbol{x}$, $\boldsymbol{y}$ in $\mathbb{R}^n$, we have

$$\boldsymbol{F}(\boldsymbol{x} + \boldsymbol{y}) = \boldsymbol{F}(\boldsymbol{x}) + \boldsymbol{F}(\boldsymbol{y}).$$

LO 2. If c is a number, then

$$\boldsymbol{F}(c\boldsymbol{x}) = c\boldsymbol{F}(\boldsymbol{x}).$$

Linear functions are sometimes referred to as *linear maps* or *linear operators*.

Linear functions from $\mathbb{R}^n$ into $\mathbb{R}^m$ can be identified with the $m \times n$ real matrices as follows:

$$\boldsymbol{F} : \mathbb{R}^n \to \mathbb{R}^m \quad \text{is linear if and only if} \quad \boldsymbol{F}(\boldsymbol{x}) = \boldsymbol{A}\boldsymbol{x},$$

where $\boldsymbol{A} \in \mathbb{R}^{m \times n}$.

If $\boldsymbol{F} : \mathbb{R}^n \to \mathbb{R}^m$ is linear, then there exists a constant $L \geq 0$ such that

$$\|\boldsymbol{F}(\boldsymbol{x})\| \leq L\|\boldsymbol{x}\| \tag{A.74}$$

for all $\boldsymbol{x} \in \mathbb{R}^n$. The smallest such constant L for which this is true is denoted $\|\boldsymbol{F}\|$ and called the norm of the linear map $\boldsymbol{F}$. If $\boldsymbol{F}$ is the linear map induced by the matrix $\boldsymbol{A}$, then we define $\|\boldsymbol{A}\| = \|\boldsymbol{F}\|$. For any matrix $\boldsymbol{A}$ and vector norm $\|\cdot\|$, we can compute the matrix norm using the formula

$$\|\boldsymbol{A}\| = \sup_{\boldsymbol{x} \neq \boldsymbol{0}} \frac{\|\boldsymbol{A}\boldsymbol{x}\|}{\|\boldsymbol{x}\|}. \tag{A.75}$$

The matrix norm—that is, the norm of the linear map $\boldsymbol{F}$ given by (A.75)—is equivalent to

$$\boxed{\|\boldsymbol{A}\| = \max_{\|\boldsymbol{x}\|=1} \|\boldsymbol{A}\boldsymbol{x}\|} \tag{A.76}$$

A matrix norm produced using (A.75) or (A.76) is called the *induced norm*. Note that

$$\|\boldsymbol{A}\boldsymbol{x} - \boldsymbol{A}\boldsymbol{y}\| = \|\boldsymbol{A}(\boldsymbol{x} - \boldsymbol{y})\| \leq \|\boldsymbol{A}\|\|\boldsymbol{x} - \boldsymbol{y}\|;$$

that is, the linear map, $\boldsymbol{F}$, represented by the matrix $\boldsymbol{A}$ is uniformly continuous.

We now comment on the notation used in (A.75) and (A.76). First, the same symbol is used for the matrix norm as for the norm of a vector. Second, the norms appearing on the right-hand side of (A.75) and (A.76) are the norms of the vectors $\boldsymbol{x} \in \mathbb{R}^n$ and $\boldsymbol{A}\boldsymbol{x} \in \mathbb{R}^m$. Thus, if $\mathbb{R}^n$ is equipped with the vector norm $\|\cdot\|_s$ while $\mathbb{R}^m$ is equipped with the vector norm $\|\cdot\|_t$, then (A.76) should be interpreted as

$$\|A\| = \max_{\|x\|_s=1} \|Ax\|_t.$$

Vector norms and a matrix norm are said to be *compatible,* or *consistent,* if they satisfy

$$\|Ax\|_t \le \|A\|\|x\|_s. \tag{A.77}$$

We will show that (A.76) is well-defined. For this, we need the following lemma.

Lemma A.4 A vector norm $\|\cdot\| : \mathbb{R}^p \to \mathbb{R}$ is an uniformly continuous function.

Proof By Theorem A.4 on page 629, we have

$$|\|x\| - \|y\|| \le \|x - y\|.$$

Hence, for any $\varepsilon > 0$, if we take $\delta = \varepsilon$, then for any x, $y \in \mathbb{R}^p$, if $\|x - y\| < \delta$, then $|\|x - y\|| < \varepsilon$, as was to be shown.

By the continuity of the vector norm and the theorem of Weierstrass on page 650, a vector x_0 can be found such that $\|x_0\| = 1$ and $\|A\| = \|Ax_0\| = \max_{\|x\|=1} \|Ax\|$. Thus, the expression (A.76) is well-defined.

Next, we will show that (A.76) satisfies (A.77). Suppose that we are given a vector $y \neq 0$. Then,

$$x = \frac{1}{\|y\|} y$$

satisfies the condition $\|x\| = 1$. Hence,

$$\|Ay\| = \|A(\|y\|x)\| = \|y\|\|Ax\|.$$

Applying (A.76) to the right-hand side of the above yields

$$\|Ay\| \le \|y\|\|A\|.$$

Hence, any matrix norm produced by (A.76) is consistent.

We will now prove that a matrix norm obtained using (A.76) satisfies the following conditions:

MN 1. $\|A\| > 0$ if $A \neq O$ and $\|O\| = 0$.

MN 2. $\|cA\| = |c|\|A\|$, where c is a constant.

MN 3. $\|A + B\| \le \|A\| + \|B\|$.

MN 4. $\|AB\| \le \|A\|\|B\|$.

Proof of MN 1. Let $A \neq O$. Then a vector x, $\|x\| = 1$, can be found such that $Ax \neq 0$, and thus $\|Ax\| \neq 0$. Hence, $\|A\| = \max_{\|x\|=1} \|Ax\| \neq 0$. If, on the other hand, $A = O$, then $\|A\| = \max_{\|x\|=1} \|Ox\| = 0$.

Proof of MN 2. By definition, $\|cA\| = \max_{\|x\|=1} \|cAx\|$. Using a property of the vector norm, we can write $\|cAx\| = |c|\|Ax\|$, and hence

$$\|cA\| = \max_{\|x\|=1} |c|\|Ax\| = |c| \max_{\|x\|=1} \|Ax\| = |c|\|A\|.$$

Proof of MN 3. For the sum of two matrices, $A+B$, there exists a vector x_0, $\|x_0\| = 1$, such that

$$\|A+B\| = \|(A+B)x_0\|.$$

Then, using properties of the vector norm, we obtain

$$\|A+B\| = \|Ax_0 + Bx_0\| \le \|Ax_0\| + \|Bx_0\|.$$

Using now (A.76) yields

$$\|A+B\| \le \|A\|\|x_0\| + \|B\|\|x_0\| = \|A\| + \|B\|.$$

Proof of MN 4. For the product of two matrices AB, there exists a vector x_0, $\|x_0\| = 1$, such that $\|AB\| = \|ABx_0\|$. Then, using (A.76), we have

$$\|AB\| = \|A(Bx_0)\| \le \|A\|\|Bx_0\|.$$

Employing (A.76) again yields

$$\|AB\| \le \|A\|\|B\|\|x_0\| = \|A\|\|B\|.$$

The most frequently used matrix norms are the p-norms,

$$\|A\|_p = \max_{\|x\|_p=1} \|Ax\|_p.$$

The induced matrix norms corresponding to the one-, two-, and infinity-norm for vectors are:

- $\|A\|_1 = \max_j \|a_j\|_1$, that is, $\|A\|_1$ is the maximum of absolute column sum;
- $\|A\|_2 = \sigma_1(A)$, which is the largest singular value of A;
- $\|A\|_\infty = \max_i \|a_i^T\|_1$, which is the maximum of absolute row sum.

A matrix norm that is not induced by a vector norm is the *Frobenius norm,* defined for an $m \times n$ matrix A as

$$\|A\|_F = \left(\sum_{i=1}^{m} \sum_{j=1}^{n} a_{ij}^2 \right)^{1/2}. \tag{A.78}$$

The Frobenius norm satisfies the relation

$$\|A\|_F^2 = \operatorname{trace}(AA^T) = \operatorname{trace}(A^T A). \tag{A.79}$$

The Frobenius norm is compatible with the Euclidean vector norm, that is,

$$\|Ax\|_2 \le \|A\|_F \|x\|_2. \tag{A.80}$$

A.10 Vector Spaces

A given set of vectors of the same dimension $\{\boldsymbol{x}_1, \boldsymbol{x}_2, \ldots, \boldsymbol{x}_r\}$ is *linearly dependent* if the zero vector, $\mathbf{0}$, can be written as a nontrivial *linear combination* of the given vectors:

$$\alpha_1 \boldsymbol{x}_1 + \alpha_2 \boldsymbol{x}_2 + \cdots + \alpha_r \boldsymbol{x}_r = \mathbf{0}, \qquad \text{with } \alpha_i \neq 0 \text{ for at least one } i.$$

In other words, the set of vectors $\{\boldsymbol{x}_1, \boldsymbol{x}_2, \ldots, \boldsymbol{x}_r\}$ is *linearly independent* if

$$\alpha_1 \boldsymbol{x}_1 + \alpha_2 \boldsymbol{x}_2 + \cdots + \alpha_r \boldsymbol{x}_r = \mathbf{0}, \qquad \text{implies } \alpha_1 = \alpha_2 = \cdots = \alpha_r = 0.$$

A set $\mathcal{S}$ containing vectors of dimension p is a *subspace* of $\mathbb{R}^p$ if for any scalars α and β,

$$\boldsymbol{x}, \boldsymbol{y} \in \mathcal{S} \qquad \text{implies } \alpha \boldsymbol{x} + \beta \boldsymbol{y} \in \mathcal{S}. \tag{A.81}$$

It follows from the above that a subspace must contain the zero vector, $\mathbf{0}$. The set containing only the zero vector is a subspace. This set, $\{\mathbf{0}\}$, is called a *trivial* subspace.

A matrix–vector product

$$\boldsymbol{A}\boldsymbol{x} = [\boldsymbol{a}_1 \quad \boldsymbol{a}_2 \quad \cdots \quad \boldsymbol{a}_n] \begin{bmatrix} x_1 \\ x_2 \\ \vdots \\ x_n \end{bmatrix} = x_1 \boldsymbol{a}_1 + x_2 \boldsymbol{a}_2 + \cdots + x_n \boldsymbol{a}_n \tag{A.82}$$

is a linear combination of the columns of the matrix. Thus, linear dependence of the columns of $\boldsymbol{A}$ is equivalent to the condition

$$\boldsymbol{A}\boldsymbol{z} = \mathbf{0} \qquad \text{for some vector } \boldsymbol{z} \neq \mathbf{0}. \tag{A.83}$$

Linear independence of the columns of $\boldsymbol{A}$ is equivalent to the condition

$$\boldsymbol{A}\boldsymbol{z} = \mathbf{0} \qquad \text{implies } \boldsymbol{z} = \mathbf{0}. \tag{A.84}$$

The set of all vectors that are linear combinations of the columns of $\boldsymbol{A}$ is called the *range,* or *image,* of $\boldsymbol{A}$ and is denoted range$(\boldsymbol{A})$. Thus,

$$\text{range}(\boldsymbol{A}) = \{\boldsymbol{v} : \boldsymbol{v} = \boldsymbol{A}\boldsymbol{x} \quad \text{for some} \quad \boldsymbol{x}\}. \tag{A.85}$$

For a given $m \times n$ matrix $\boldsymbol{A}$ and an m-vector $\boldsymbol{b}$, we have $\boldsymbol{b} \in \text{range}(\boldsymbol{A})$ if and only if there exists an n-vector $\boldsymbol{x}$ such that $\boldsymbol{A}\boldsymbol{x} = \boldsymbol{b}$. We say that a system of linear equations represented by $\boldsymbol{A}\boldsymbol{x} = \boldsymbol{b}$ is compatible if $\boldsymbol{b} \in \text{range}(\boldsymbol{A})$.

The range of $\boldsymbol{A}$ is a subspace of $\mathbb{R}^m$. Indeed, for any $\boldsymbol{b}_1, \boldsymbol{b}_2 \in \text{range}(\boldsymbol{A})$, we have $\boldsymbol{b}_1 = \boldsymbol{A}\boldsymbol{x}_1$ and $\boldsymbol{b}_2 = \boldsymbol{A}\boldsymbol{x}_2$ for some $\boldsymbol{x}_1$ and $\boldsymbol{x}_2$. Using the fact that $\boldsymbol{A}$ represents a linear map, for any scalars α_1 and α_2, we obtain

$$\alpha_1 \boldsymbol{b}_1 + \alpha_2 \boldsymbol{b}_2 = \alpha_1 \boldsymbol{A}\boldsymbol{x}_1 + \alpha_2 \boldsymbol{A}\boldsymbol{x}_2 = \boldsymbol{A}(\alpha_1 \boldsymbol{x}_1 + \alpha_2 \boldsymbol{x}_2). \tag{A.86}$$

Hence, if $\boldsymbol{b}_1, \boldsymbol{b}_2 \in \text{range}(\boldsymbol{A})$, then $\alpha_1 \boldsymbol{b}_1 + \alpha_2 \boldsymbol{b}_2 \in \text{range}(\boldsymbol{A})$, which means that the range of $\boldsymbol{A}$ is a subspace of $\mathbb{R}^m$.

Similarly, we define the range of $\boldsymbol{A}^T$ to be the set

$$\text{range}(\boldsymbol{A}^T) = \{\boldsymbol{y} : \boldsymbol{y} = \boldsymbol{A}^T \boldsymbol{z} \quad \text{for some} \quad \boldsymbol{z}\}. \tag{A.87}$$

For a nontrivial subspace $\mathcal{S}$ there is a unique smallest positive integer called the *dimension* of the subspace $\mathcal{S}$ and denoted $\dim \mathcal{S}$. The dimension of the subspace is equal to the smallest number of linearly independent vectors of the subspace such that every vector of the subspace can be expressed as a linear combination of these vectors. Any such set of vectors is called a *basis* of the subspace. The vectors forming a basis are not unique.

The number of linearly independent columns of a matrix $\boldsymbol{A}$ is called the *rank* of $\boldsymbol{A}$ and is denoted $\operatorname{rank}(\boldsymbol{A})$. The rank of $\boldsymbol{A}$ is also equal to the number of linearly independent rows of $\boldsymbol{A}$.

For the product of an $m \times n$ matrix $\boldsymbol{A}$ and an $n \times p$ matrix $\boldsymbol{B}$, we have *Sylvester's inequalities:*

$$\operatorname{rank}(\boldsymbol{A}) + \operatorname{rank}(\boldsymbol{B}) - n \leq \operatorname{rank}(\boldsymbol{AB}) \leq \min\{\operatorname{rank}(\boldsymbol{A}), \operatorname{rank}(\boldsymbol{B})\}. \tag{A.88}$$

For an $m \times n$ matrix $\boldsymbol{A}$, the set of all n-vectors orthogonal to the rows of $\boldsymbol{A}$ is called the *kernel,* or *null space,* of $\boldsymbol{A}$ and is denoted $\ker(\boldsymbol{A})$, that is,

$$\ker(\boldsymbol{A}) = \{\boldsymbol{x} : \boldsymbol{Ax} = \boldsymbol{0}\}. \tag{A.89}$$

The kernel of $\boldsymbol{A}$ is a subspace of $\mathbb{R}^n$. Indeed, if $\boldsymbol{x}_1, \boldsymbol{x}_2 \in \ker(\boldsymbol{A})$, then $\boldsymbol{Ax}_1 = \boldsymbol{0}$ and $\boldsymbol{Ax}_2 = \boldsymbol{0}$. Hence, for any scalars α_1 and α_2, we have

$$\boldsymbol{A}(\alpha_1 \boldsymbol{x}_1 + \alpha_2 \boldsymbol{x}_2) = \alpha_1 \boldsymbol{Ax}_1 + \alpha_2 \boldsymbol{Ax}_2 = \boldsymbol{0}. \tag{A.90}$$

Thus, if $\boldsymbol{x}_1, \boldsymbol{x}_2 \in \ker(\boldsymbol{A})$, then $\alpha_1 \boldsymbol{x}_1 + \alpha_2 \boldsymbol{x}_2 \in \ker(\boldsymbol{A})$, which means that $\ker(\boldsymbol{A})$ is a subspace of $\mathbb{R}^n$. Similarly, we define the kernel of $\boldsymbol{A}^T$ to be the set

$$\ker(\boldsymbol{A}^T) = \{\boldsymbol{w} : \boldsymbol{A}^T \boldsymbol{w} = \boldsymbol{0}\}. \tag{A.91}$$

The intersection of $\operatorname{range}(\boldsymbol{A})$ and $\ker(\boldsymbol{A}^T)$ contains only the zero vector, that is,

$$\operatorname{range}(\boldsymbol{A}) \cap \ker(\boldsymbol{A}^T) = \{\boldsymbol{0}\}. \tag{A.92}$$

For a given subspace $\mathcal{S}$ of $\mathbb{R}^n$, the orthogonal complement $\mathcal{S}^\perp$ of $\mathcal{S}$ is defined as

$$\mathcal{S}^\perp = \{\boldsymbol{x} \in \mathbb{R}^n : \boldsymbol{x}^T \boldsymbol{y} = 0 \quad \text{for all} \quad \boldsymbol{y} \in \mathcal{S}\}. \tag{A.93}$$

The vectors in $\ker(\boldsymbol{A}^T)$ are orthogonal to the vectors in $\operatorname{range}(\boldsymbol{A})$, and vice versa. We say that $\operatorname{range}(\boldsymbol{A})$ and $\ker(\boldsymbol{A}^T)$ are *orthogonal complements* of one another and write

$$\operatorname{range}(\boldsymbol{A})^\perp = \ker(\boldsymbol{A}^T) \tag{A.94}$$

and

$$\ker(\boldsymbol{A}^T)^\perp = \operatorname{range}(\boldsymbol{A}). \tag{A.95}$$

Similarly,

$$\operatorname{range}(\boldsymbol{A}^T)^\perp = \ker(\boldsymbol{A}) \tag{A.96}$$

and

$$\ker(\boldsymbol{A})^\perp = \operatorname{range}(\boldsymbol{A}^T). \tag{A.97}$$

If $\boldsymbol{A}$ is an $m \times n$ matrix with linearly independent columns, which implies that $m \geq n$, then

$$\operatorname{rank}(\boldsymbol{A}) = n = \dim \operatorname{range}(\boldsymbol{A})$$

and

$$\dim \ker(\boldsymbol{A}^T) = m - n.$$

We now define a normed space.

Definition A.3 A vector space with a norm is called a *normed space*.

By the normed vector space $\mathbb{R}^n$, we usually mean the Euclidean space, which is $\mathbb{R}^n$ with the norm $\|\boldsymbol{x}\|_2 = \sqrt{x_1^2 + \cdots + x_n^2}$. The vector space $\mathbb{R}^n$ equipped with any of the p-norm forms a normed vector space. The absolute value is a norm in $\mathbb{R}$.

Let Ω be a compact subset of $\mathbb{R}^n$ and let $\mathcal{C}(\Omega)$ denote the set of continuous functions $f : \mathbb{R}^n \to \mathbb{R}$ on Ω. Thus, we view functions as vectors in the function space $\mathcal{C}(\Omega)$. Then, the function $\|\cdot\| : \mathcal{C}(\Omega) \to \mathbb{R}$ defined as

$$\|f\| = \max_{\boldsymbol{x}\in\Omega} |f(\boldsymbol{x})| \tag{A.98}$$

satisfies the norm axioms. The function space $\mathcal{C}(\Omega)$, with scalars from $\mathbb{R}$ and with the norm (A.98), is a linear normed space. In particular, the space of all real-valued functions, $f(x)$, continuous on some interval $[a, b]$ of the real line with the norm

$$\|f\| = \max_{x\in[a,b]} |f(x)| \tag{A.99}$$

is an example of a linear normed space. This space is denoted $\mathcal{C}([a, b])$.

We are also interested in sequences of elements of $\mathcal{C}([a, b])$. Let $f_1, f_2, \ldots$ be a sequence of elements of $\mathcal{C}([a, b])$. Denote this sequence as $\{f_k\}$. In the following, $f_k(x)$, refers to the value of the function f_k at x, while f_k denotes the function itself. We say that the sequence $\{f_k\}$ converges (pointwise) on the interval $[a, b]$ if there exists a function f such that for every $x \in [a, b]$ the sequence $\{f_k(x)\}$ converges to $f(x)$. In other words, the sequence $\{f_k\}$ converges (pointwise) if there exists a function f such that for a given $x \in [a, b]$ and a given $\varepsilon > 0$, we can find $N = N(x, \varepsilon)$ such that

$$|f(x) - f_k(x)| < \varepsilon \qquad \text{for } k > N. \tag{A.100}$$

The sequence $\{f_k\}$ converges uniformly, on the interval $[a, b]$, if there is a function f such that for any $\varepsilon > 0$ there exists $N = N(\varepsilon)$, depending on ε but not on x, such that (A.100) holds for every $x \in [a, b]$. The distinction between convergence and uniform convergence is that in the later case the same N can be used for any value of x to show the convergence of the sequence $\{f_k(x)\}$. One can show that the limit of an uniformly convergent sequence of continuous functions is continuous. It can be verified that the norm (A.99) defines the uniform convergence because $\{f_k\}$ converges uniformly to f if and only if $\max_{x\in[a,b]} |f(x) - f_k(x)| \to 0$. This is the reason the norm (A.99) is sometimes called the *norm of uniform convergence*. It can be shown that there is no norm on $\mathcal{C}([a, b])$ that defines pointwise convergence to an element in $\mathcal{C}([a, b])$—see Debnath and Mikusiński [58, p. 12].

Definition A.4 (Banach Space) A normed linear space is complete if every Cauchy sequence with elements form the space converges to an element in the space. A complete normed linear space is called a Banach space.

A sequence that satisfies conditions of Theorem A.14 is called a *Cauchy sequence*. The Euclidean space is an example of a Banach space, as is $\mathcal{C}([a, b])$ with the norm (A.99).

A.11 Least Squares

The theory of least squares was developed by Carl Friedrich Gauss (1777–1855) and published by him in 1809. Gauss first applied the least squares to devise a method of calculating orbits of celestial bodies. For more information on the history of the least squares development, we recommend an account by Campbell and Meyer [40, Section 2.5].

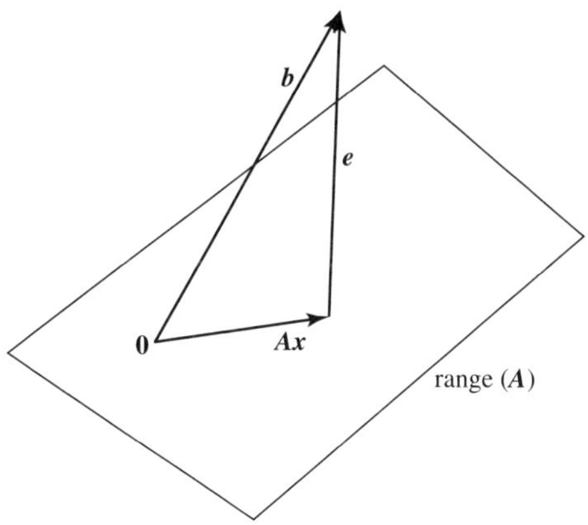

Figure A.3 Geometrical interpretation of the least squares solution to a system of linear equations.

Consider the system of equations

$$\boldsymbol{A}\boldsymbol{x} = \boldsymbol{b}, \tag{A.101}$$

where $\boldsymbol{A} \in \mathbb{R}^{m \times n}$, $\boldsymbol{x} \in \mathbb{R}^n$, and $\boldsymbol{b} \in \mathbb{R}^m$. A necessary and sufficient condition for the system (A.101) to have a solution is

$$\text{rank}(\boldsymbol{A}) = \text{rank}\left[\boldsymbol{A} \vdots \boldsymbol{b}\right]. \tag{A.102}$$

In particular, if the matrix $\boldsymbol{A}$ is invertible, then there is a unique solution to (A.101) given by

$$\boldsymbol{x}^* = \boldsymbol{A}^{-1}\boldsymbol{b}.$$

Here, we analyze the case when $m > n$ and $\text{rank}(\boldsymbol{A}) < \text{rank}[\boldsymbol{A} \vdots \boldsymbol{b}]$, that is, $\boldsymbol{b} \notin \text{range}(\boldsymbol{A})$, which means that the system (A.101) does not have an exact solution. We modify (A.101) by incorporating an error vector $\boldsymbol{e}$ to obtain

$$\boldsymbol{A}\boldsymbol{x} + \boldsymbol{e} = \boldsymbol{b}. \tag{A.103}$$

This is illustrated in Figure A.3. Because

$$\boldsymbol{e} = \boldsymbol{b} - \boldsymbol{A}\boldsymbol{x}, \tag{A.104}$$

we have

$$\begin{aligned} \|\boldsymbol{e}\|^2 &= \boldsymbol{e}^T\boldsymbol{e} \\ &= (\boldsymbol{b} - \boldsymbol{A}\boldsymbol{x})^T(\boldsymbol{b} - \boldsymbol{A}\boldsymbol{x}) \\ &= \boldsymbol{x}^T\boldsymbol{A}^T\boldsymbol{A}\boldsymbol{x} - 2\boldsymbol{x}^T\boldsymbol{A}^T\boldsymbol{b} + \boldsymbol{b}^T\boldsymbol{b}. \end{aligned} \tag{A.105}$$

The gradient of $\|\boldsymbol{e}\|^2$ with respect to $\boldsymbol{x}$ is

$$\nabla_{\boldsymbol{x}}\|\boldsymbol{e}\|^2 = 2\boldsymbol{A}^T\boldsymbol{A}\boldsymbol{x} - 2\boldsymbol{A}^T\boldsymbol{b}. \tag{A.106}$$

Solving

$$\nabla_{\boldsymbol{x}}\|\boldsymbol{e}\|^2 = \boldsymbol{0}$$

yields candidate minimizer points of (A.105). If the matrix $\boldsymbol{A}$ has full column rank, then the matrix $\boldsymbol{A}^T\boldsymbol{A}$ is invertible and the only critical point is

$$\boxed{\boldsymbol{x}^* = (\boldsymbol{A}^T\boldsymbol{A})^{-1}\boldsymbol{A}^T\boldsymbol{b}} \tag{A.107}$$

The above point satisfies the second-order sufficient condition to be the minimizer of (A.105) because the Hessian of (A.105) is

$$2A^T A,$$

which is positive definite. The solution given by (A.107) is called the least squares solution of the inconsistent linear system of equations, $\boldsymbol{A}\boldsymbol{x} = \boldsymbol{b}$, because it minimizes the sum of squares $\|\boldsymbol{e}\|^2 = \boldsymbol{e}^T\boldsymbol{e} = \sum_{i=1}^m e_i^2$. For a graphical interpretation of the least squares solution, we refer to Figure A.3.

Suppose now that we interpret the ith row of the matrix $\boldsymbol{A}$ and the ith component of the vector $\boldsymbol{b}$ as the ith data pair. Thus, the ith data pair has the form $(\boldsymbol{a}_i^T; b_i)$. Suppose that the k data pairs are represented by the pair $(\boldsymbol{A}; \boldsymbol{b})$. We know that the least squares solution to the system of equations represented by the data pair $(\boldsymbol{A}; \boldsymbol{b})$ is given by (A.107). If the new data pair, $(\boldsymbol{a}_{k+1}^T; b_{k+1})$, becomes available, then we could find the least squares solution to the system of equations

$$\begin{bmatrix} \boldsymbol{A} \\ \boldsymbol{a}_{k+1}^T \end{bmatrix} \boldsymbol{x} = \begin{bmatrix} \boldsymbol{b} \\ b_{k+1} \end{bmatrix} \tag{A.108}$$

by applying the formula (A.107) to (A.108). Let $\boldsymbol{x}_{k+1}$ denote the least squares solution to the above problem. Then,

$$\boldsymbol{x}_{k+1} = \left(\begin{bmatrix} \boldsymbol{A} \\ \boldsymbol{a}_{k+1}^T \end{bmatrix}^T \begin{bmatrix} \boldsymbol{A} \\ \boldsymbol{a}_{k+1}^T \end{bmatrix} \right)^{-1} \begin{bmatrix} \boldsymbol{A} \\ \boldsymbol{a}_{k+1}^T \end{bmatrix}^T \begin{bmatrix} \boldsymbol{b} \\ b_{k+1} \end{bmatrix}. \tag{A.109}$$

In the case of large k the above method of recomputing the new least squares solution from scratch may be time-consuming. We present a recursive method for computing least squares solutions as the data pairs become available. To proceed, let

$$\boldsymbol{P}_k = (\boldsymbol{A}^T\boldsymbol{A})^{-1}. \tag{A.110}$$

In view of the above, we can write

$$\begin{aligned} \boldsymbol{P}_{k+1}^{-1} &= \begin{bmatrix} \boldsymbol{A} \\ \boldsymbol{a}_{k+1}^T \end{bmatrix}^T \begin{bmatrix} \boldsymbol{A} \\ \boldsymbol{a}_{k+1}^T \end{bmatrix} \\ &= \begin{bmatrix} \boldsymbol{A}^T & \boldsymbol{a}_{k+1} \end{bmatrix} \begin{bmatrix} \boldsymbol{A} \\ \boldsymbol{a}_{k+1}^T \end{bmatrix} \\ &= \boldsymbol{A}^T\boldsymbol{A} + \boldsymbol{a}_{k+1}\boldsymbol{a}_{k+1}^T \\ &= \boldsymbol{P}_k^{-1} + \boldsymbol{a}_{k+1}\boldsymbol{a}_{k+1}^T. \end{aligned} \tag{A.111}$$

Hence,

$$\boldsymbol{P}_{k+1} = \left(\boldsymbol{P}_k^{-1} + \boldsymbol{a}_{k+1}\boldsymbol{a}_{k+1}^T \right)^{-1}.$$

Applying to the above the matrix inversion formula (A.30), we obtain

$$\boxed{\boldsymbol{P}_{k+1} = \boldsymbol{P}_k - \frac{\boldsymbol{P}_k\boldsymbol{a}_{k+1}\boldsymbol{a}_{k+1}^T\boldsymbol{P}_k}{1 + \boldsymbol{a}_{k+1}^T\boldsymbol{P}_k\boldsymbol{a}_{k+1}}.} \tag{A.112}$$

Using (A.110), we write

$$\begin{aligned} \boldsymbol{x}_k &= (\boldsymbol{A}^T\boldsymbol{A})^{-1}\boldsymbol{A}^T\boldsymbol{b} \\ &= \boldsymbol{P}_k\boldsymbol{A}^T\boldsymbol{b}. \end{aligned} \tag{A.113}$$

Employing (A.111), we obtain

$$\begin{aligned} \boldsymbol{x}_{k+1} &= \left(\begin{bmatrix} \boldsymbol{A} \\ \boldsymbol{a}_{k+1}^T \end{bmatrix}^T \begin{bmatrix} \boldsymbol{A} \\ \boldsymbol{a}_{k+1}^T \end{bmatrix}\right)^{-1} \begin{bmatrix} \boldsymbol{A} \\ \boldsymbol{a}_{k+1}^T \end{bmatrix}^T \begin{bmatrix} \boldsymbol{b} \\ b_{k+1} \end{bmatrix} \\ &= \boldsymbol{P}_{k+1}(\boldsymbol{A}^T\boldsymbol{b} + \boldsymbol{a}_{k+1}b_{k+1}). \end{aligned} \tag{A.114}$$

Computing $\boldsymbol{A}^T\boldsymbol{b}$ from (A.113), we get

$$\boldsymbol{A}^T\boldsymbol{b} = \boldsymbol{P}_k^{-1}\boldsymbol{x}_k. \tag{A.115}$$

Substituting the above into (A.114) yields

$$\begin{aligned} \boldsymbol{x}_{k+1} &= \boldsymbol{P}_{k+1}(\boldsymbol{A}^T\boldsymbol{b} + \boldsymbol{a}_{k+1}b_{k+1}) \\ &= \boldsymbol{P}_{k+1}\left(\boldsymbol{P}_k^{-1}\boldsymbol{x}_k + \boldsymbol{a}_{k+1}b_{k+1}\right). \end{aligned} \tag{A.116}$$

Computing $\boldsymbol{P}_k^{-1}$ from (A.111) gives

$$\boldsymbol{P}_k^{-1} = \boldsymbol{P}_{k+1}^{-1} - \boldsymbol{a}_{k+1}\boldsymbol{a}_{k+1}^T$$

Substituting the above into (A.116), we have

$$\boldsymbol{x}_{k+1} = \boldsymbol{P}_{k+1}\left(\left(\boldsymbol{P}_{k+1}^{-1} - \boldsymbol{a}_{k+1}\boldsymbol{a}_{k+1}^T\right)\boldsymbol{x}_k + \boldsymbol{a}_{k+1}b_{k+1}\right). \tag{A.117}$$

Performing simple manipulations yields

$$\boxed{\boldsymbol{x}_{k+1} = \boldsymbol{x}_k + \left(b_{k+1} - \boldsymbol{a}_{k+1}^T\boldsymbol{x}_k\right)\boldsymbol{P}_{k+1}\boldsymbol{a}_{k+1}.} \tag{A.118}$$

The two equations (A.112) and (A.118) constitute the *recursive least squares* (RLS) algorithm. To initialize the RLS, we can set

$$\boldsymbol{P}_0 = \left(\boldsymbol{A}_n^T\boldsymbol{A}_n\right)^{-1} \quad \text{and} \quad \boldsymbol{x}_0 = \boldsymbol{P}_n\boldsymbol{A}_n^T\boldsymbol{b}_n,$$

where the pair $(\boldsymbol{A}_n; \boldsymbol{b}_n)$ represents the first n data pairs.

A.12 Contraction Maps

We now discuss the contraction maps that found interesting applications in control and stability analyses of nonlinear systems. In particular, in 1964, Leibovic [180] presented an interesting idea of how to apply contraction maps in nonlinear and adaptive control. The proposed application required a significant computational power not available at that time. Leibovic's idea looks like a good challenge for today's computers. More recently, Kelly [152] used the contraction map theorem as well as Lyapunov functions to investigate stability properties of a class of neural networks.

An operator $\boldsymbol{F}$ that maps Euclidean space $\mathbb{R}^p$ into itself is called a *contraction map* if for any two elements $\boldsymbol{x}$ and $\boldsymbol{y}$ of $\mathbb{R}^p$,

$$\|\boldsymbol{F}(\boldsymbol{x}) - \boldsymbol{F}(\boldsymbol{y})\| \le L\|\boldsymbol{x} - \boldsymbol{y}\|, \qquad 0 < L < 1. \tag{A.119}$$

A point $\boldsymbol{x}^* \in \mathbb{R}^p$ is called a *fixed point* of $\boldsymbol{F}$ if

$$\boldsymbol{F}(\boldsymbol{x}^*) = \boldsymbol{x}^*. \tag{A.120}$$

The importance of the contraction operators lies in the fact that there is a unique fixed point. Furthermore, this point may be found by successive iterations as stated in the following theorem.

Theorem A.19 Let $\boldsymbol{F} : \mathbb{R}^p \to \mathbb{R}^p$ be a contraction map with the contraction constant $0 < L < 1$. Then $\boldsymbol{F}$ has a unique fixed point, and the sequence

$$\boldsymbol{x}^{(1)}, \boldsymbol{x}^{(2)}, \ldots$$

generated by $\boldsymbol{x}^{(n+1)} = \boldsymbol{F}(\boldsymbol{x}^{(n)})$, $n = 0, 1, 2, \ldots$, converges to $\boldsymbol{x}^*$ for any initial choice of $\boldsymbol{x}^{(0)} \in \mathbb{R}^p$.

Proof To show the uniqueness of the fixed point $\boldsymbol{x}^*$, suppose that there exist two distinct fixed points $\boldsymbol{x}^*$ and $\boldsymbol{y}^*$. Then,

$$\begin{aligned}\|\boldsymbol{x}^* - \boldsymbol{y}^*\| &= \|\boldsymbol{F}(\boldsymbol{x}^*) - \boldsymbol{F}(\boldsymbol{y}^*)\| \\ &\leq L\|\boldsymbol{x}^* - \boldsymbol{y}^*\|.\end{aligned}$$

Thus,

$$\|\boldsymbol{x}^* - \boldsymbol{y}^*\| \leq L\|\boldsymbol{x}^* - \boldsymbol{y}^*\|, \qquad 0 < L < 1,$$

which is impossible unless $\|\boldsymbol{x}^* - \boldsymbol{y}^*\| = 0$. Hence, $\boldsymbol{x}^* = \boldsymbol{y}^*$ which means that $\boldsymbol{F}$ has at most one fixed point.

We will now show that a fixed point exists and that the sequence

$$\boldsymbol{x}^{(1)}, \boldsymbol{x}^{(2)}, \ldots$$

generated by $\boldsymbol{x}^{(n+1)} = \boldsymbol{F}(\boldsymbol{x}^{(n)})$, $n = 0, 1, 2, \ldots$, converges to $\boldsymbol{x}^*$ for any initial choice of $\boldsymbol{x}^{(0)} \in \mathbb{R}^p$. If $\boldsymbol{F}(\boldsymbol{x}^{(0)}) = \boldsymbol{x}^{(0)}$, then the theorem is proved. In general, $\boldsymbol{F}(\boldsymbol{x}^{(0)}) \neq \boldsymbol{x}^{(0)}$. For any n, m with $m > n$, we have

$$\begin{aligned}\left\|\boldsymbol{x}^{(m)} - \boldsymbol{x}^{(n)}\right\| &= \left\|\boldsymbol{F}^m(\boldsymbol{x}^{(0)}) - \boldsymbol{F}^n(\boldsymbol{x}^{(0)})\right\| \\ &= \left\|\boldsymbol{F}(\boldsymbol{F}^{m-1}(\boldsymbol{x}^{(0)})) - \boldsymbol{F}(\boldsymbol{F}^{n-1}(\boldsymbol{x}^{(0)}))\right\| \\ &\leq L\left\|\boldsymbol{F}^{m-1}(\boldsymbol{x}^{(0)}) - \boldsymbol{F}^{n-1}(\boldsymbol{x}^{(0)})\right\| \\ &\;\;\vdots \\ &\leq L^n\left\|\boldsymbol{x}^{(m-n)} - \boldsymbol{x}^{(0)}\right\| \\ &= L^n\left\|\boldsymbol{x}^{(0)} - \boldsymbol{x}^{(m-n)}\right\|.\end{aligned} \tag{A.121}$$

We write

$$\begin{aligned}\left\|\boldsymbol{x}^{(0)} - \boldsymbol{x}^{(m-n)}\right\| = \Big\|&\boldsymbol{x}^{(0)} - \boldsymbol{x}^{(1)} + \boldsymbol{x}^{(1)} - \boldsymbol{x}^{(2)} + \boldsymbol{x}^{(2)} \\ &- \cdots - \boldsymbol{x}^{(m-n-1)} + \boldsymbol{x}^{(m-n-1)} - \boldsymbol{x}^{(m-n)}\Big\|.\end{aligned} \tag{A.122}$$

Substituting (A.122) into (A.121) and then applying the triangle inequality yields

$$\left\|\boldsymbol{x}^{(m)} - \boldsymbol{x}^{(n)}\right\| \leq L^n\left(\left\|\boldsymbol{x}^{(0)} - \boldsymbol{x}^{(1)}\right\| + \left\|\boldsymbol{x}^{(1)} - \boldsymbol{x}^{(2)}\right\| + \cdots + \left\|\boldsymbol{x}^{(m-n-1)} - \boldsymbol{x}^{(m-n)}\right\|\right). \tag{A.123}$$

Observe that

$$\begin{aligned}\left\| \boldsymbol{x}^{(j)} - \boldsymbol{x}^{(j+1)} \right\| &= \left\| \boldsymbol{F}\left(\boldsymbol{x}^{(j-1)}\right) - \boldsymbol{F}\left(\boldsymbol{x}^{(j)}\right) \right\| \\ &\leq L \left\| \boldsymbol{x}^{(j-1)} - \boldsymbol{x}^{(j)} \right\| \\ &\vdots \\ &\leq L^j \left\| \boldsymbol{x}^{(0)} - \boldsymbol{x}^{(1)} \right\|, \end{aligned} \tag{A.124}$$

We use the above to further evaluate the right-hand side of (A.123) to obtain

$$\begin{aligned}\left\| \boldsymbol{x}^{(m)} - \boldsymbol{x}^{(n)} \right\| &\leq L^n \left\| \boldsymbol{x}^{(0)} - \boldsymbol{x}^{(1)} \right\| (1 + L + \cdots + L^{m-n-1}) \\ &\leq L^n \left\| \boldsymbol{x}^{(0)} - \boldsymbol{x}^{(1)} \right\| (1 + L + \cdots + L^{m-n-1} + L^{m-n} + \cdots) \\ &= \frac{L^n}{1-L} \left\| \boldsymbol{x}^{(0)} - \boldsymbol{x}^{(1)} \right\| \end{aligned}$$

because $0 < L < 1$. Recall that a sequence $\boldsymbol{x}^{(0)}, \boldsymbol{x}^{(1)}, \ldots$ is a Cauchy sequence if for each $\varepsilon > 0$ there exists a positive integer $n = n(\varepsilon)$ such that $\|\boldsymbol{x}^{(m)} - \boldsymbol{x}^{(n)}\| < \varepsilon$ for all $m > n$. Given an $\varepsilon > 0$, let n be such that

$$L^n < \frac{\varepsilon(1-L)}{\left\| \boldsymbol{x}^{(0)} - \boldsymbol{x}^{(1)} \right\|}.$$

Then,

$$\left\| \boldsymbol{x}^{(m)} - \boldsymbol{x}^{(n)} \right\| \leq \frac{L^n}{1-L} \left\| \boldsymbol{x}^{(0)} - \boldsymbol{x}^{(1)} \right\| < \varepsilon \qquad \text{for all } m > n. \tag{A.125}$$

Therefore, the sequence $\boldsymbol{x}^{(0)}, \boldsymbol{x}^{(1)}, \ldots$ is a Cauchy sequence, and we will now show that the limit $\boldsymbol{g}$ of this sequence is the fixed point $\boldsymbol{x}^*$ of the map $\boldsymbol{F}$. Note that $\boldsymbol{F}$ is uniformly continuous. Indeed, for each $\varepsilon > 0$ there exists a $\delta = \delta(\varepsilon)$ such that if $\|\boldsymbol{x} - \boldsymbol{y}\| < \delta$, then $\|\boldsymbol{F}(\boldsymbol{x}) - \boldsymbol{F}(\boldsymbol{y})\| < \varepsilon$. Since $\|\boldsymbol{F}(\boldsymbol{x}) - \boldsymbol{F}(\boldsymbol{y})\| \leq L\|\boldsymbol{x} - \boldsymbol{y}\|$, given an $\varepsilon > 0$, we can choose $\delta = \varepsilon / L$. Because $\boldsymbol{F}$ is continuous, we have

$$\lim_{n\to\infty} \boldsymbol{F}\left(\boldsymbol{x}^{(n)}\right) = \boldsymbol{F}\left(\lim_{n\to\infty} \boldsymbol{x}^{(n)}\right) = \boldsymbol{F}(\boldsymbol{g}).$$

However, $\boldsymbol{x}^{(n+1)} = \boldsymbol{F}(\boldsymbol{x}^{(n)})$. Therefore,

$$\begin{aligned}\boldsymbol{g} &= \lim_{n\to\infty} \boldsymbol{x}^{(n+1)} \\ &= \lim_{n\to\infty} \boldsymbol{F}\left(\boldsymbol{x}^{(n)}\right) \\ &= \boldsymbol{F}\left(\lim_{n\to\infty} \boldsymbol{x}^{(n)}\right) \\ &= \boldsymbol{F}(\boldsymbol{g}).\end{aligned}$$

Hence, $\boldsymbol{g} = \boldsymbol{x}^*$ is a fixed point of $\boldsymbol{F}$, and the proof is complete.

Using the above proof we can give an estimate of the norm of $\|\boldsymbol{x}^{(n)} - \boldsymbol{x}^*\|$. Indeed, if we fix n in (A.125) and allow $m \to \infty$, then

$$\left\| \boldsymbol{x}^{(n)} - \boldsymbol{x}^* \right\| \leq \frac{L^n}{1-L} \left\| \boldsymbol{x}^{(0)} - \boldsymbol{F}\left(\boldsymbol{x}^{(0)}\right) \right\|. \tag{A.126}$$

A.13 First-Order Differential Equation

Let $\mathbb{R}$ denote the set of real numbers and let $D \subseteq \mathbb{R}^2$ be a domain, that is, an open connected nonempty subset of $\mathbb{R}^2$. Let f be a real-valued function defined and continuous on D. Let $\dot{x} = dx/dt$ denote the derivative of x with respect to t. We call

$$\dot{x} = f(t, x) \tag{A.127}$$

an *ordinary differential equation of the first order*. By a *solution* of the differential equation (A.127) on an open interval $I = \{t \in \mathbb{R} : t_1 < t < t_2\}$, we mean a real-valued continuously differentiable function ϕ defined on I such that $(t, \phi(t)) \in D$ for all $t \in I$ and

$$\dot{\phi}(t) = f(t, \phi(t)) \qquad \text{for all } t \in I.$$

Definition A.5 Given $(t_0, x_0) \in D$, the *initial value problem* for (A.127) is

$$\dot{x} = f(t, x), \qquad x(t_0) = x_0. \tag{A.128}$$

A function ϕ is a solution to (A.128) if ϕ is a solution of the differential equation (A.127) on some interval I containing t_0 and $x(t_0) = x_0$.

We can represent the initial value problem (A.128) equivalently by an integral equation,

$$\phi(t) = x_0 + \int_{t_0}^{t} f(s, \phi(s))\, ds. \tag{A.129}$$

To prove this equivalence, let ϕ be a solution of the initial value problem (A.128). Then $x(t_0) = x_0$ and

$$\dot{\phi}(t) = f(t, \phi(t)) \qquad \text{for all } t \in I.$$

Integrating the above from t_0 to t, we obtain

$$\int_{t_0}^{t} \dot{\phi}(s) ds = \int_{t_0}^{t} f(s, \phi(s))\, ds.$$

Hence,

$$\phi(t) - x_0 = \int_{t_0}^{t} f(s, \phi(s))\, ds.$$

Therefore, ϕ is a solution of the integral equation (A.129).

Conversely, let ϕ be a solution of the integral equation (A.129). Then $\phi(t_0) = x_0$, and differentiating both sides of (A.129) with respect to t gives $\dot{\phi}(t) = f(t, \phi(t))$. Therefore, ϕ is also a solution of the initial value problem (A.128).

The general *ordinary differential linear equation of the first order* has the form

$$\dot{x}(t) = a(t)x(t) + b(t), \qquad t \in (t_1, t_2), \tag{A.130}$$

where $a(t)$ and $b(t)$ are assumed to be continuous on the interval (t_1, t_2). We associate with equation (A.130) the initial condition

$$x(t_0) = x_0, \qquad x_0 \in \mathbb{R}.$$

To find the solution to (A.130), we write it in the form

$$\dot{x}(t) - a(t)x(t) = b(t). \tag{A.131}$$

We then multiply (A.131) by the "integrating factor"

$$e^{-\int_{t_0}^{t} a(s)\,ds} \tag{A.132}$$

to get

$$\dot{x}(t)e^{-\int_{t_0}^{t} a(s)\,ds} - a(t)x(t)e^{-\int_{t_0}^{t} a(s)\,ds} = e^{-\int_{t_0}^{t} a(s)\,ds}b(t). \tag{A.133}$$

The above equation can be represented in an equivalent form as

$$\frac{d}{dt}\left\{x(t)e^{-\int_{t_0}^{t} a(s)\,ds}\right\} = e^{-\int_{t_0}^{t} a(s)\,ds}b(t). \tag{A.134}$$

Integrating both sides of (A.134) yields

$$x(t)e^{-\int_{t_0}^{t} a(s)\,ds} - x_0 = \int_{t_0}^{t} e^{-\int_{t_0}^{\tau} a(s)\,ds}b(\tau)\,d\tau. \tag{A.135}$$

Hence,

$$x(t) = e^{\int_{t_0}^{t} a(s)\,ds}\left[x_0 + \int_{t_0}^{t} e^{-\int_{t_0}^{\tau} a(s)\,ds}b(\tau)\,d\tau\right], \tag{A.136}$$

or, in an equivalent form,

$$\boxed{x(t) = e^{\int_{t_0}^{t} a(s)\,ds}x_0 + \int_{t_0}^{t} e^{\int_{\tau}^{t} a(s)\,ds}b(\tau)\,d\tau} \tag{A.137}$$

A.14 Integral and Differential Inequalities

A.14.1 The Bellman-Gronwall Lemma

Lemma A.5 Let $v(t)$ and $w(t)$ be continuous real functions of t. Let $w(t) \geq 0$, and c be a real constant. If

$$v(t) \leq c + \int_0^t w(s)v(s)\,ds, \tag{A.138}$$

then

$$v(t) \leq ce^{\int_0^t w(s)\,ds}. \tag{A.139}$$

Proof Denote the right-hand side of (A.138) by $x(t)$; that is, let

$$x(t) = c + \int_0^t w(s)v(s)\,ds \tag{A.140}$$

and let

$$z(t) = x(t) - v(t). \tag{A.141}$$

Note that $z(t) \geq 0$, and $x(t)$ is differentiable. Differentiating both sides of (A.140) and using (A.141) yields

$$\begin{aligned}\dot{x}(t) &= w(t)v(t)\\ &= w(t)x(t) - w(t)z(t).\end{aligned} \tag{A.142}$$

We use (A.137) to represent (A.142) in the equivalent integral form as

$$x(t) = e^{\int_0^t w(s)\,ds} x_0 - \int_0^t e^{\int_\tau^t w(s)\,ds} w(\tau)z(\tau)\,d\tau. \tag{A.143}$$

Because $z(t) \geq 0$ and $w(t) \geq 0$, we have

$$x(t) \leq e^{\int_0^t w(s)\,ds} x_0. \tag{A.144}$$

From the definition of $x(t)$ given by (A.140), we have $x(0) = c$. On the other hand,

$$v(t) = x(t) - z(t) \leq x(t), \tag{A.145}$$

because $z(t) \geq 0$. Hence,

$$v(t) \leq c e^{\int_0^t w(s)\,ds},$$

and the proof is completed.

The Bellman–Gronwall lemma allows us to convert the investigation of integral equations into integral inequalities.

A.14.2 A Comparison Theorem

We consider the differential inequality

$$\dot{x}(t) \leq \omega(t, x(t)), \qquad x(t_0) = x_0,$$

where $\omega : \mathbb{R}^2 \to \mathbb{R}$ is a continuous function in a domain $\Delta \subset \mathbb{R}^2$. We assume that in Δ, the function ω satisfies the Lipschitz condition,

$$|\omega(t, x) - \omega(t, y)| \leq L|x - y|,$$

where $L \geq 0$ is a Lipschitz constant.
Consider now an associated differential equation

$$\dot{\tilde{x}}(t) = \omega(t, \tilde{x}(t)), \qquad \tilde{x}(t_0) = \tilde{x}_0.$$

We refer to the above equation as the *equation of comparison*. Denote the solution to the equation of comparison by

$$\tilde{x}(t) = \tilde{x}(t; t_0, \tilde{x}_0).$$

The following theorem, which can be found, for example, in Corduneanu [55, p. 49], relates the solutions of the differential inequality and the associated differential equation.

Theorem A.20 If $x(t)$ is a differentiable solution to the differential inequality and $\tilde{x}(t)$ is the solution to the associated differential equation with the corresponding initial conditions such that

$$x_0 \leq \tilde{x}_0,$$

then

$$x(t) \leq \tilde{x}(t)$$

for all $t \geq t_0$.

Proof We prove the theorem by contradiction. We suppose that $x_0 \leq \tilde{x}_0$ but that $x(t) \leq \tilde{x}(t)$ does not hold. Thus, there exists $t_1 > t_0$ such that

$$x(t_1) > \tilde{x}(t_1).$$

Let

$$y(t) = x(t) - \tilde{x}(t).$$

Note that $y(t_0) = x(t_0) - \tilde{x}(t_0) \leq 0$ and that $y(t_1) = x(t_1) - \tilde{x}(t_1) > 0$. Denote by τ the largest $t \in [t_0, t_1]$ such that $y(\tau) = 0$. Such a value of t exists by virtue of the continuity of y. Thus,

$$y(t) > 0 \qquad \text{for all } t \in (\tau, t_1].$$

On the other hand, for $t \in [\tau, t_1]$,

$$\begin{aligned}\dot{y}(t) &= \dot{x}(t) - \dot{\tilde{x}}(t)\\ &\leq \omega(t, x(t)) - \omega(t, \tilde{x}(t))\\ &\leq |\omega(t, x(t)) - \omega(t, \tilde{x}(t))|\\ &\leq L|x(t) - \tilde{x}(t)|.\end{aligned}$$

In the interval $[\tau, t_1]$,

$$y(t) = x(t) - \tilde{x}(t) \geq 0.$$

Hence,

$$\begin{aligned}\dot{y}(t) &\leq L|x(t) - \tilde{x}(t)|\\ &= L(x(t) - \tilde{x}(t))\\ &= L\,y(t).\end{aligned}$$

Therefore, in the interval $[\tau, t_1]$, the function y satisfies

$$y(t) \leq y(\tau)e^{L(t-\tau)} = 0$$

because $y(\tau) = 0$. But this is a contradiction to the previous conclusion that $y(t) > 0$ for all $t \in (\tau, t_1]$, which completes the proof.

A.15 Solving the State Equations

A.15.1 Solution of Uncontrolled System

Given a dynamical system's time-invariant model of the form

$$\dot{\boldsymbol{x}}(t) = \boldsymbol{A}\boldsymbol{x}(t) \tag{A.146}$$

subject to an initial condition

$$\boldsymbol{x}(0) = \boldsymbol{x}_0, \tag{A.147}$$

where $\boldsymbol{A} \in \mathbb{R}^{n \times n}$. The solution to (A.146) subject to (A.147) is

$$\boxed{\boldsymbol{x}(t) = e^{\boldsymbol{A}t}\boldsymbol{x}_0}$$

The above is a generalization of the situation when $\boldsymbol{A}$ is a scalar. The exponential matrix $e^{\boldsymbol{A}t}$ is defined as

$$e^{\boldsymbol{A}t} = \boldsymbol{I}_n + t\boldsymbol{A} + \frac{t^2}{2!}\boldsymbol{A}^2 + \frac{t^3}{3!}\boldsymbol{A}^3 + \cdots.$$

Observe that

$$\frac{d}{dt}e^{\boldsymbol{A}t} = \boldsymbol{A}e^{\boldsymbol{A}t} = e^{\boldsymbol{A}t}\boldsymbol{A}.$$

Suppose now that the initial condition $\boldsymbol{x}(0) = \boldsymbol{x}_0$ is replaced with a more general condition

$$\boldsymbol{x}(t_0) = \boldsymbol{x}_0,$$

where t_0 is a given initial time instant. Then, the solution to $\dot{\boldsymbol{x}}(t) = \boldsymbol{A}\boldsymbol{x}(t)$ subject to $\boldsymbol{x}(t_0) = \boldsymbol{x}_0$ is

$$\boxed{\boldsymbol{x}(t) = e^{\boldsymbol{A}(t-t_0)}\boldsymbol{x}_0}$$

The matrix $e^{\boldsymbol{A}(t-t_0)}$ is often written as

$$e^{\boldsymbol{A}(t-t_0)} = \boldsymbol{\Phi}(t, t_0)$$

and is called the *state transition matrix* because it relates the state at any instant of time t_0 to the state at any other time t. There are many methods for calculating the state transition matrix. For an overview of some of them, we refer the reader to the classic paper on the subject by Moler and Van Loan [208]. In the case when the matrix $\boldsymbol{A}$ is of low dimensions, we can use the formula that results from applying the Laplace transform to the time-invariant system $\dot{\boldsymbol{x}} = \boldsymbol{A}\boldsymbol{x},\ \ \boldsymbol{x}(0) = \boldsymbol{x}_0$. The Laplace transform of the above matrix differential equation is

$$s\boldsymbol{X}(s) - \boldsymbol{x}_0 = \boldsymbol{A}\boldsymbol{X}(s).$$

Hence

$$\boldsymbol{X}(s) = (s\boldsymbol{I}_n - \boldsymbol{A})^{-1}\boldsymbol{x}_0,$$

and the inverse transform yields

$$\boxed{e^{\boldsymbol{A}t} = \mathcal{L}^{-1}\{(s\boldsymbol{I}_n - \boldsymbol{A})^{-1}\}} \tag{A.148}$$

where $\mathcal{L}^{-1}$ denotes the inverse Laplace transform operator.

A.15.2 Solution of Controlled System

Consider a model of a controlled dynamical system,

$$\dot{\boldsymbol{x}}(t) = \boldsymbol{A}\boldsymbol{x}(t) + \boldsymbol{B}\boldsymbol{u}(t) \tag{A.149}$$

subject to a given initial condition $\boldsymbol{x}(0) = \boldsymbol{x}_0$, where $\boldsymbol{B}$ is an $n \times m$ matrix, and $m < n$ as is usually in practice. Premultiplying both sides of (A.149) by the integrating factor, e^{-At}, yields

$$e^{-At}\dot{\boldsymbol{x}}(t) = e^{-At}\boldsymbol{A}\boldsymbol{x}(t) + e^{-At}\boldsymbol{B}\boldsymbol{u}(t).$$

We then rearrange the above equation as

$$e^{-At}\dot{\boldsymbol{x}}(t) - e^{-At}\boldsymbol{A}\boldsymbol{x}(t) = e^{-At}\boldsymbol{B}\boldsymbol{u}(t). \tag{A.150}$$

Noting that

$$\frac{d}{dt}(e^{-At}\boldsymbol{x}(t)) = -\boldsymbol{A}e^{-At}\boldsymbol{x}(t) + e^{-At}\dot{\boldsymbol{x}}(t),$$

we can represent (A.150) in the form

$$\frac{d}{dt}(e^{-At}\boldsymbol{x}(t)) = e^{-At}\boldsymbol{B}\boldsymbol{u}(t).$$

Integrating both sides of the resulting equation gives

$$e^{-At}\boldsymbol{x}(t) - \boldsymbol{x}(0) = \int_0^t e^{-A\tau}\boldsymbol{B}\boldsymbol{u}(\tau)\,d\tau.$$

Hence,

$$\boldsymbol{x}(t) = e^{At}\boldsymbol{x}(0) + \int_0^t e^{A(t-\tau)}\boldsymbol{B}\boldsymbol{u}(\tau)\,d\tau.$$

If the initial condition is $\boldsymbol{x}(t_0) = \boldsymbol{x}_0$, then integration from t_0 to t yields

$$\boxed{\boldsymbol{x}(t) = e^{A(t-t_0)}\boldsymbol{x}(t_0) + \int_{t_0}^t e^{A(t-\tau)}\boldsymbol{B}\boldsymbol{u}(\tau)\,d\tau.} \tag{A.151}$$

As in the case of the uncontrolled system, we can also use the Laplace transform technique to obtain the solution to (A.149) subject to the initial condition $\boldsymbol{x}(0) = \boldsymbol{x}_0$. Indeed, taking the Laplace transform of (A.149) and performing some manipulations yields

$$\boldsymbol{X}(s) = (s\boldsymbol{I}_n - \boldsymbol{A})^{-1}\boldsymbol{x}(0) + (s\boldsymbol{I}_n - \boldsymbol{A})^{-1}\boldsymbol{B}\boldsymbol{U}(s).$$

Hence,

$$\boldsymbol{x}(t) = \mathcal{L}^{-1}\{(s\boldsymbol{I}_n - \boldsymbol{A})^{-1}\boldsymbol{x}(0)\} + \mathcal{L}^{-1}\{(s\boldsymbol{I}_n - \boldsymbol{A})^{-1}\boldsymbol{B}\boldsymbol{U}(s)\}.$$

If $\boldsymbol{x}_0 = \boldsymbol{x}(t_0)$, then

$$\boldsymbol{x}(t) = \mathcal{L}^{-1}\{e^{-t_0 s}(s\boldsymbol{I}_n - \boldsymbol{A})^{-1}\boldsymbol{x}(t_0)\} + \mathcal{L}^{-1}\{e^{-t_0 s}(s\boldsymbol{I}_n - \boldsymbol{A})^{-1}\boldsymbol{B}\boldsymbol{U}(s)\}.$$

A.16 Curves and Surfaces

Consider a point moving along a curve in n-dimensional space. The position of the point at time t can be described as

$$\boldsymbol{x}(t) = \begin{bmatrix} x_1(t) \\ x_2(t) \\ \vdots \\ x_n(t) \end{bmatrix}. \tag{A.152}$$

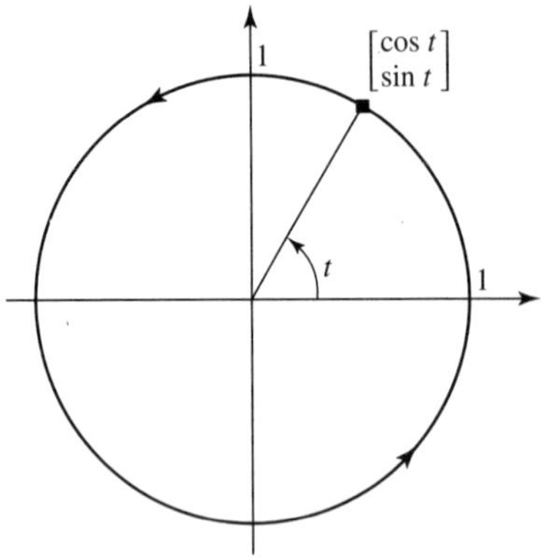

Figure A.4 An example of a curve in two-dimensional space.

For example, if the point is moving along a straight line passing through fixed points $\boldsymbol{y}$ and $\boldsymbol{z}$ in $\mathbb{R}^n$, then

$$\boldsymbol{x}(t) = \boldsymbol{y} + t\boldsymbol{z}. \tag{A.153}$$

As another example, consider

$$\boldsymbol{x}(t) = \begin{bmatrix} x_1(t) \\ x_2(t) \end{bmatrix} = \begin{bmatrix} \cos t \\ \sin t \end{bmatrix} \in \mathbb{R}^2. \tag{A.154}$$

Then, the point moves around a circle of radius 1 in counterclockwise direction as shown in Figure A.4. We used t as the variable corresponding to the angle corresponding to the position of the point on the circle. We are now ready for a formal definition of a curve.

Definition A.6 Let $I \in \mathbb{R}$ be an interval. A parameterized curve defined on the interval I is an association which to each point of I associates a vector, where $\boldsymbol{x}(t)$ denotes the vector associated to $t \in I$ by $\boldsymbol{x}$. We write the association $t \mapsto \boldsymbol{x}(t)$ as

$$\boldsymbol{x} : I \to \mathbb{R}^n.$$

We also call this association the parameterization of a curve. We call $\boldsymbol{x}(t)$ the position vector at time t.

The position vector can be represented in terms of its coordinates as in (A.152), where each component $x_i(t)$ is a function of time t. We say that this curve is differentiable if each function $x_i(t)$ is a differentiable function of time t. We define the derivative $d\boldsymbol{x}(t)/dt$ to be

$$\dot{\boldsymbol{x}}(t) = \frac{d\boldsymbol{x}(t)}{dt} = \begin{bmatrix} \dfrac{dx_1(t)}{dt} \\ \dfrac{dx_2(t)}{dt} \\ \vdots \\ \dfrac{dx_n(t)}{dt} \end{bmatrix}$$

and call it the velocity vector of the curve at time t. The velocity vector is located at the origin. However, when we translate it to the point $\boldsymbol{x}(t)$, as in Figure A.5, then we visualize it as tangent to the curve $\boldsymbol{x}(t)$.

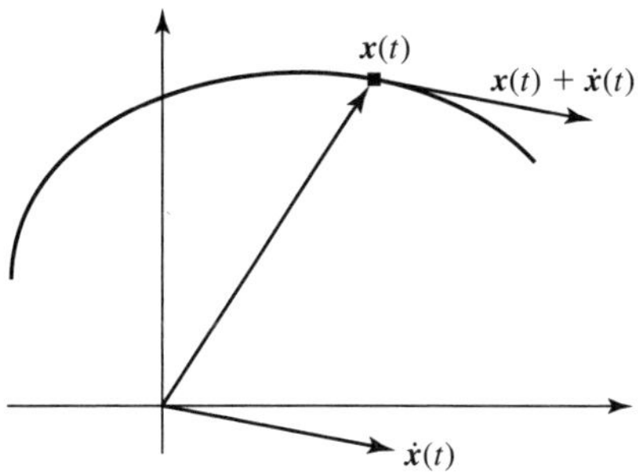

Figure A.5 Velocity vector of a curve at time t is parallel to a tangent vector at time t.

We define the tangent line to a curve $\boldsymbol{x}$ at time t to be the line passing through $\boldsymbol{x}(t)$ in the direction of $\dot{\boldsymbol{x}}(t)$, provided that $\dot{\boldsymbol{x}}(t) \neq \mathbf{0}$. If $\dot{\boldsymbol{x}}(t) = \mathbf{0}$, we do not define a tangent line. The speed of the curve $\boldsymbol{x}(t)$, denoted $v(t)$, is $v(t) = \|\dot{\boldsymbol{x}}(t)\|$. To illustrate the above, we consider the point moving on the circle. The point position at time t is

$$\boldsymbol{x}(t) = [\cos t \quad \sin t]^T.$$

The velocity vector of the curve at time t is

$$\dot{\boldsymbol{x}}(t) = [-\sin t \quad \cos t]^T,$$

and the speed of the curve $\boldsymbol{x}(t)$ at time t is

$$v(t) = \|\dot{\boldsymbol{x}}(t)\| = \sqrt{(-\sin t)^2 + \cos^2 t} = 1.$$

The set of points in $\mathbb{R}^3$ such that for some $F : \mathbb{R}^3 \to \mathbb{R}$ we have

$$F(\boldsymbol{x}) = F(x_1, x_2, x_3) = 0$$

and

$$DF(x_1, x_2, x_3) = \nabla F(x_1, x_2, x_3)^T = \begin{bmatrix} \frac{\partial F}{\partial x_1} & \frac{\partial F}{\partial x_2} & \frac{\partial F}{\partial x_3} \end{bmatrix} \neq \mathbf{0}^T$$

is called a surface in $\mathbb{R}^3$, where $\nabla F = \text{grad } F$. For example, let

$$F(x_1, x_2, x_3) = x_1^2 + x_2^2 + x_3^2 - 1;$$

then, the surface is the sphere of radius 1, centered at the origin of $\mathbb{R}^3$.

Now let

$$\boldsymbol{C}(t) = [x_1(t) \quad x_2(t) \quad \cdots \quad x_n(t)]^T$$

be a differentiable curve. We say that the curve lies on the surface described by

$$\begin{aligned} F_1(x_1, x_2, \ldots, x_n) &= 0, \\ &\vdots \\ F_m(x_1, x_2, \ldots, x_n) &= 0 \end{aligned}$$

if, for all t, we have

$$\begin{aligned} F_1(x_1(t), x_2(t), \ldots, x_n(t)) &= 0, \\ &\vdots \\ F_m(x_1(t), x_2(t), \ldots, x_n(t)) &= 0. \end{aligned}$$

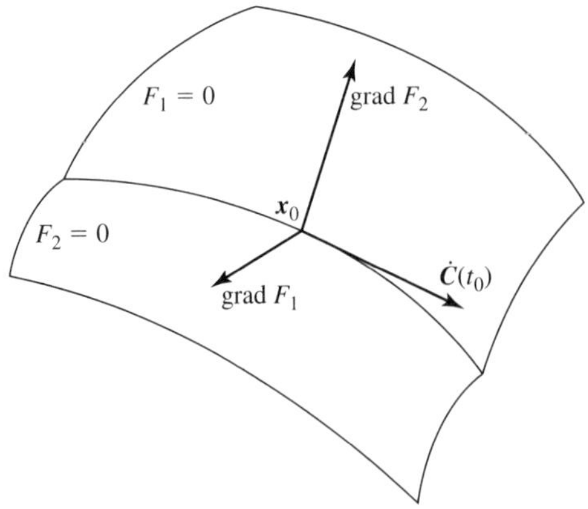

Figure A.6 Tangent space to a surface is the set of vectors orthogonal to the gradients of the functions defining the surface.

Differentiating the above equations, using the chain rule, we obtain

$$DF_i(\boldsymbol{C}(t))\dot{\boldsymbol{C}}(t) = \nabla F_i(\boldsymbol{C}(t))^T \dot{\boldsymbol{C}}(t) = 0, \qquad i = 1, 2, \dots, m.$$

Let $\boldsymbol{x}_0$ be a point of the surface, and let $\boldsymbol{C}(t)$ be a curve on the surface passing through $\boldsymbol{x}_0$; that is, for some t_0,

$$\boldsymbol{C}(t_0) = \boldsymbol{x}_0.$$

For this t_0 we have

$$\nabla F_i(\boldsymbol{x}_0)^T \dot{\boldsymbol{C}}(t_0) = 0, \qquad i = 1, 2, \dots, m,$$

where we assumed that $\dot{\boldsymbol{C}}(t_0) \neq \boldsymbol{0}$. Thus, the gradients ∇F_i at $\boldsymbol{x}_0$ are orthogonal to the tangent vector of the curve at the point $\boldsymbol{x}_0$. This is true for every differentiable curve on the surface passing through $\boldsymbol{x}_0$. For an illustration of the above properties in three-dimensional space, we refer to Figure A.6.

Definition A.7 The tangent space to the surface $\{\boldsymbol{x} : \boldsymbol{F}(\boldsymbol{x}) = \boldsymbol{0}\}$ at the point $\boldsymbol{x}_0$ is the set of vectors at $\boldsymbol{x}_0$ that are tangent to differentiable curves on the surface passing through $\boldsymbol{x}_0$.

A.17 Vector Fields and Curve Integrals

Definition A.8 Let $\mathcal{U}$ be an open subset of $\mathbb{R}^n$. By a vector field on $\mathcal{U}$ we mean an association

$$\boldsymbol{F} : \mathcal{U} \to \mathbb{R}^n,$$

which to every point of $\mathcal{U}$ associates a vector of the same dimension.

We can interpret a vector field as a field of forces. We can visualize a vector field as a field of arrows, which to each point associates an arrow as shown in Figure A.7. If $f : \mathcal{U} \to \mathbb{R}$ is a function, then

$$\boldsymbol{F} = \nabla f$$

is a vector field, and we have

$$\boldsymbol{F}(\boldsymbol{x}) = \nabla f(\boldsymbol{x})$$

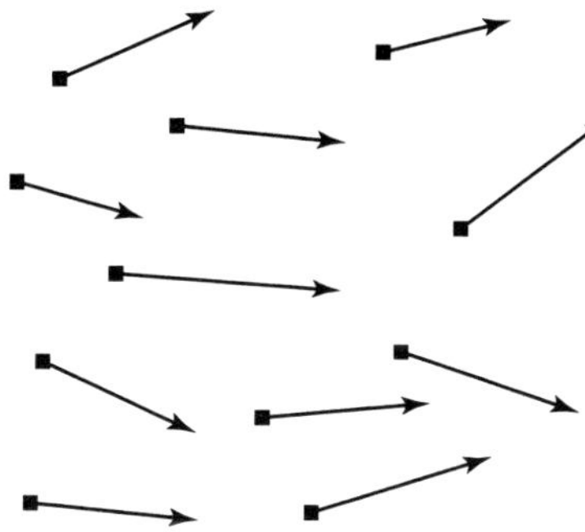

Figure A.7 A vector field.

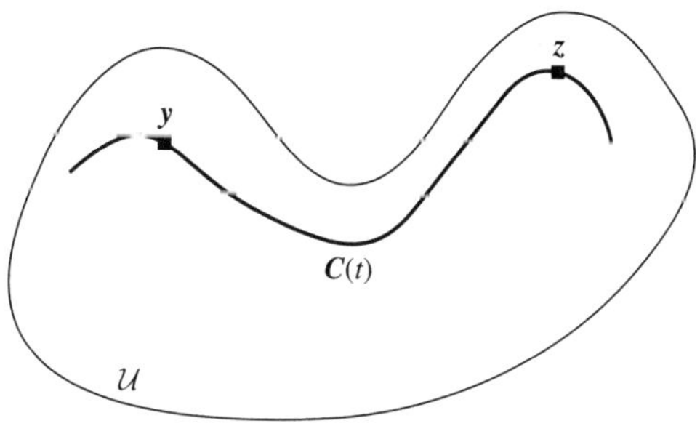

Figure A.8 A connected open set in which there are points that cannot be joined by a straight line.

for all $\boldsymbol{x} \in \mathcal{U}$. We now consider an inverse problem: When is a given vector field in $\mathbb{R}^n$ a gradient?

Definition A.9 Let $\boldsymbol{F}$ be a vector field on an open set $\mathcal{U}$. If f is a differentiable function on $\mathcal{U}$ such that

$$\boldsymbol{F} = \nabla f,$$

then we say that f is a potential function for $\boldsymbol{F}$.

We then can ask the question, Do the potential functions exist? To answer the question, we recast the problem into the one that involves solving n first-order partial differential equations:

$$\frac{\partial f}{\partial x_i} = F_i(\boldsymbol{x}), \qquad i = 1, 2, \ldots, n,$$

with the initial condition

$$\boldsymbol{F}(\boldsymbol{x}_0) = \boldsymbol{F}_0.$$

To answer the above question we need one more definition.

Definition A.10 An open set $\mathcal{U}$ is said to be connected if, given two points in $\mathcal{U}$, there is a differentiable curve in $\mathcal{U}$ that joins the two points.

If $\mathcal{U} = \mathbb{R}^n$, then any two points can be joined by a straight line. Of course, it is not always the case that two points of an open set can be joined by a straight line, as illustrated in Figure A.8. We say that the function $\boldsymbol{F} = [F_1 \ F_2 \ \cdots \ F_n]^T$ is a $\mathcal{C}^1$ function if all partial derivatives of all functions F_i exist and are continuous.

Theorem A.21 Let $\boldsymbol{F}(\boldsymbol{x})$ be $\mathcal{C}^1$ function defined on an open connected region $\mathcal{U}$ of $\mathbb{R}^n$. Then, there is a solution to the differential equation

$$\nabla f(\boldsymbol{x}) = \boldsymbol{F}(\boldsymbol{x}), \qquad \boldsymbol{F}(\boldsymbol{x}_0) = \boldsymbol{F}_0,$$

if and only if

$$\frac{\partial F_i}{\partial x_j} = \frac{\partial F_j}{\partial x_i};$$

that is, the Jacobian matrix of $\boldsymbol{F}$ is symmetric.

Proof We prove the theorem for $n = 2$, where the connected region $\mathcal{U}$ is a rectangle. We select a point $\boldsymbol{x}_0 = [x_{10} \;\; x_{20}]^T$ in the rectangle. We assume that $\boldsymbol{F} = [F_1 \;\; F_2]^T$ and that

$$D_2 F_1 = D_1 F_2; \qquad \text{that is,} \quad \frac{\partial F_1}{\partial x_2} = \frac{\partial F_2}{\partial x_1}.$$

Our goal is to find a potential function $f(x_1, x_2)$. We assume that the potential function has the form

$$f(x_1, x_2) = \int_{x_{10}}^{x_1} F_1(s, x_2)\, ds + u(x_2),$$

where $u(x_2)$ is a function of x_2 only and will be determined later in the proof. We illustrate the integration path in Figure A.9. Applying the fundamental theorem of calculus, we obtain

$$\begin{aligned}\frac{\partial f}{\partial x_1} &= \frac{\partial}{\partial x_1}\int_{x_{10}}^{x_1} F_1(s, x_2)\, ds + \frac{\partial u(x_2)}{\partial x_1} \\ &= F_1(x_1, x_2)\end{aligned}$$

because

$$\frac{\partial u(x_2)}{\partial x_1} = 0.$$

Hence, $D_1 f(x_1, x_2) = F_1(x_1, x_2)$ as required.

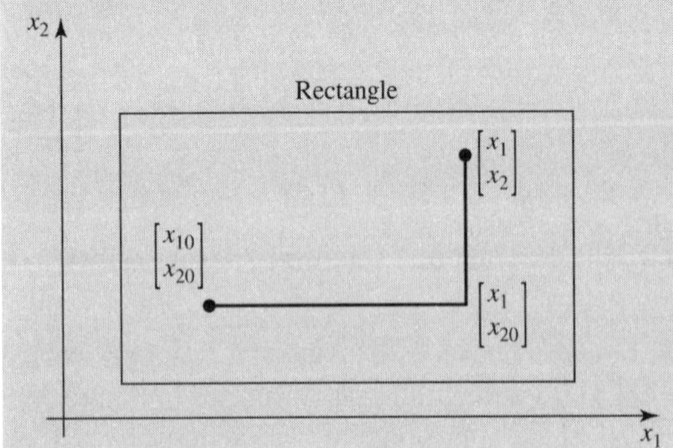

Figure A.9 Integration path used in the proof of Theorem A.21.

We now differentiate $f(x_1, x_2)$ with respect to x_2 to get

$$\frac{\partial f}{\partial x_2} = \int_{x_{10}}^{x_1} \frac{\partial F_1(s, x_2)}{\partial x_2} ds + \frac{\partial u(x_2)}{\partial x_2}$$

$$= \int_{x_{10}}^{x_1} \frac{\partial F_2(s, x_2)}{\partial x_1} ds + \frac{\partial u(x_2)}{\partial x_2}$$

because, by assumption, $D_2 F = D_1 F$. Therefore,

$$\frac{\partial f}{\partial x_2} = F_2(s, x_2)|_{x_{10}}^{x_1} + \frac{\partial u(x_2)}{\partial x_2}$$

$$= F_2(x_1, x_2) - F_2(x_{10}, x_2) + \frac{\partial u(x_2)}{\partial x_2}.$$

Let

$$-F_2(x_{10}, x_2) + \frac{\partial u(x_2)}{\partial x_2} = 0,$$

that is, $\frac{\partial u(x_2)}{\partial x_2} = F_2(x_{10}, x_2)$. Then, $D_2 f = F_2$ as required. To conclude the proof, we let

$$u(x_2) = \int_{x_{20}}^{x_2} F_2(x_{10}, x_2)\, dx_2,$$

which yields

$$\boxed{f(x_1, x_2) = \int_{x_{10}}^{x_1} F_1(s, x_2)\, ds + \int_{x_{20}}^{x_2} F_2(x_{10}, t)\, dt} \tag{A.155}$$

The proof is complete for $n = 2$.

We note that if the connected region $\mathcal{U}$ is an arbitrary open set, then the above proof may not work. This is because we needed to integrate over intervals that were contained in the region $\mathcal{U}$.

Suppose we are given a vector field $\boldsymbol{F}$ on an open set U in the plane. We can interpret $\boldsymbol{F}$ as a field of forces. We wish to calculate the work done when moving a particle from a point $\boldsymbol{C}(t_1)$ to a point $\boldsymbol{C}(t_2)$ along the curve $\boldsymbol{C}(t)$ in U, as illustrated in Figure A.10. Let

$$\boldsymbol{F}(\boldsymbol{x}) = [F_1(x_1, x_2) \quad F_2(x_1, x_2)]^T.$$

We assume that each component of $\boldsymbol{F}$ is differentiable; that is, $\boldsymbol{F}$ is a differentiable vector field. We next assume that the curve $\boldsymbol{C}(t)$ is defined on a closed interval $[t_1, t_2]$ with $t_1 < t_2$. For each $t \in [t_1, t_2]$ the value $\boldsymbol{C}(t)$ is a point in $\mathbb{R}^2$. The curve $\boldsymbol{C}$ lies in U if $\boldsymbol{C}(t)$ is a point in U for all $t \in [t_1, t_2]$. We further assume that the curve $\boldsymbol{C}$ is continuously differentiable and that its derivative $\dot{\boldsymbol{C}}(t) = d\boldsymbol{C}(t)/dt$ exists and is continuous; that is, we assume that $\boldsymbol{C} \in \mathcal{C}^1$.

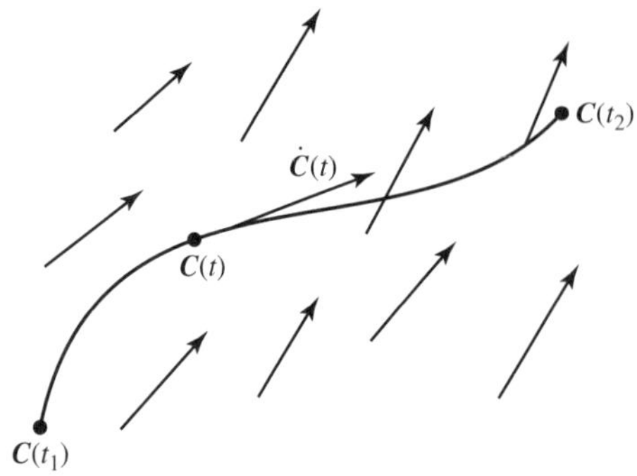

Figure A.10 A particle moving along a curve in vector field.

We define the integral of $\boldsymbol{F}$ along $\boldsymbol{C}$ to be

$$\int_C \boldsymbol{F} = \int_{t_1}^{t_2} \boldsymbol{F}(\boldsymbol{C}(t))^T \frac{d\boldsymbol{C}}{dt}\, dt$$
$$= \int_{t_1}^{t_2} \boldsymbol{F}(\boldsymbol{C}(t))^T d\boldsymbol{C},$$

which is the work done against the force field $\boldsymbol{F}$ along the curve $\boldsymbol{C}$. Let

$$\boldsymbol{C}(t) = [x_1(t) \quad x_2(t)]^T.$$

Then, we can write

$$\int_C \boldsymbol{F} = \int_C F_1\, dx_1 + F_2\, dx_2. \tag{A.156}$$

Theorem A.22 (Green) Let the vector field $\boldsymbol{F}$ be given on a region A that is the interior of a closed path $\boldsymbol{C}$ parameterized counterclockwise. Then,

$$\int_C F_1 dx_1 + F_2\, dx_2 = \iint_A \left(\frac{\partial F_2}{\partial x_1} - \frac{\partial F_1}{\partial x_2} \right) dx_1\, dx_2.$$

A.18 Matrix Calculus Formulas

Below, we list useful formulas from matrix calculus that involve computing the gradient (with respect to $\boldsymbol{x}$). Recall that in our convention the gradient is a column vector. We have

$$\nabla(\boldsymbol{x}^T \boldsymbol{y}) = \nabla(\boldsymbol{y}^T \boldsymbol{x}) = \boldsymbol{y}, \tag{A.157}$$

$$\nabla(\boldsymbol{x}^T \boldsymbol{x}) = 2\boldsymbol{x}, \tag{A.158}$$

$$\nabla(\boldsymbol{x}^T \boldsymbol{A} \boldsymbol{y}) = \boldsymbol{A}\boldsymbol{y}, \tag{A.159}$$

$$\nabla(\boldsymbol{y}^T \boldsymbol{A} \boldsymbol{x}) = \boldsymbol{A}^T \boldsymbol{y}, \tag{A.160}$$

$$\nabla(\boldsymbol{x}^T \boldsymbol{A} \boldsymbol{x}) = (\boldsymbol{A} + \boldsymbol{A}^T)\boldsymbol{x}. \tag{A.161}$$

Let $\boldsymbol{f} : \mathbb{R}^n \to \mathbb{R}^n$ and $\boldsymbol{g} : \mathbb{R}^n \to \mathbb{R}^n$ be two maps, and let $\partial \boldsymbol{f}/\partial \boldsymbol{x}$ and $\partial \boldsymbol{g}/\partial \boldsymbol{x}$ be the Jacobian matrices of $\boldsymbol{f}$ and $\boldsymbol{g}$, respectively. Then,

$$\nabla(\boldsymbol{f}(\boldsymbol{x})^T \boldsymbol{g}(\boldsymbol{x})) = \nabla(\boldsymbol{g}(\boldsymbol{x})^T \boldsymbol{f}(\boldsymbol{x})) = \left(\frac{\partial \boldsymbol{f}}{\partial \boldsymbol{x}}\right)^T \boldsymbol{g} + \left(\frac{\partial \boldsymbol{g}}{\partial \boldsymbol{x}}\right)^T \boldsymbol{f}. \tag{A.162}$$

For an $n \times n$ symmetric matrix $\boldsymbol{Q}$, we have

$$\nabla(\boldsymbol{f}(\boldsymbol{x})^T \boldsymbol{Q} \boldsymbol{f}(\boldsymbol{x})) = 2\left(\frac{\partial \boldsymbol{f}}{\partial \boldsymbol{x}}\right)^T \boldsymbol{Q} \boldsymbol{f}. \tag{A.163}$$

Notes

For more on the subject of proof techniques, we refer to Velleman [288]. For further reading on linear algebra, and in particular matrix theory, we recommend classical texts of Gel'fand [95] and Gantmacher [93]. Analyses of numerical methods for matrix computations can be found in Faddeeva [78], Householder [127], Golub and Van Loan [104], and Gill, Murray, and Wright [97]. Lang [175] and Seeley [255] are recommended for a review of calculus of several variables. A straightforward approach to mathematical analysis is presented in Binmore [28]. Principles of real analysis can be found in Natanson [215], Bartle [22], and Rudin [244]. Some facts from basic functional analysis that we touched upon in this appendix are discussed, for example, by Maddox [193] and by Debnath and Mikusiński [58]. Lucid expositions of differential equations are in Driver [69], Miller and Michel [205], and Boyce and DiPrima [32].

G. H. Hardy (1877–1947) argues in reference 114 (p. 113) that in truly great theorems and their proofs, "there is a very high degree of *unexpectedness,* combined with *inevitability* and *economy*. The arguments take so odd and surprising a form; the weapons used seem so childishly simple when compared with the far-reaching results; but there is no escape from the conclusions." He further states that "A mathematical proof should resemble a simple and clear-cut constellation, not a scattered cluster in the Milky Way."

EXERCISES

A.1 The open unit disc with respect to the norm $\|\cdot\|_p$ is the set

$$\Delta_p = \{\boldsymbol{x} \in \mathbb{R}^2 : \|\boldsymbol{x}\|_p < 1\}.$$

The unit discs are quite different for different choices of the norm $\|\cdot\|_p$. Sketch Δ_1, Δ_2, Δ_3, and Δ_∞.

A.2 Given a linear combination of matrices,

$$\boldsymbol{M} = \alpha_1 \boldsymbol{M}_1 + \cdots + \alpha_r \boldsymbol{M}_r, \tag{A.164}$$

where for $i = 1, \ldots, r$, $\alpha_i \in \mathbb{R}$, and $\boldsymbol{M}_i \in \mathbb{R}^{n \times n}$. Let $\boldsymbol{m}_i^j$ be the jth row of the matrix

M_i. Thus,

$$M_i = \begin{bmatrix} m_i^1 \\ m_i^2 \\ \vdots \\ m_i^n \end{bmatrix}.$$

Show that

$$\det M = \sum_{s_1=1}^{r} \cdots \sum_{s_n=1}^{r} \alpha_{s_1} \cdots \alpha_{s_n} \det \begin{bmatrix} m_{s_1}^1 \\ \vdots \\ m_{s_n}^n \end{bmatrix}. \qquad \text{(A.165)}$$

A.3 Let $X \in \mathbb{R}^{n\times m}$ and $Y \in \mathbb{R}^{m\times n}$. Show that

$$\det(I_n + XY) = \det(I_m + YX).$$

A.4 Let $\lambda_i(A)$ denote the ith eigenvalue of a matrix $A \in \mathbb{R}^{n\times n}$ and let α be a scalar. Show that

(a) $\lambda_i(\alpha A) = \alpha\lambda_i(A)$;

(b) $\lambda_i(I_n \pm \alpha A) = 1 \pm \alpha\lambda_i(A)$.

A.5 Show that the formulas (A.75) and (A.76) for matrix norms are equivalent.

A.6 Show that for any square matrix $A \in \mathbb{R}^{n\times n}$, we have

$$\|A\| \geq \max_{1\leq i\leq n} |\lambda_i(A)|. \qquad \text{(A.166)}$$

A.7 Show that if $P \in \mathbb{R}^{n\times n}$ is symmetric, then

$$\lambda_{\min}(P)\|x\|_2^2 \leq x^T P x \leq \lambda_{\max}(P)\|x\|_2^2.$$

A.8 Show that a sufficient condition for

$$\lim_{k\to\infty} A^k \to O$$

is

$$\|A\| < 1.$$

A.9 Let $A \in \mathbb{R}^{n\times n}$. Show that a necessary and sufficient condition for

$$\lim_{k\to\infty} A^k \to O.$$

is

$$|\lambda_i(A)| < 1, \qquad i = 1, 2, \ldots, n.$$

A.10 Let eig(M) denote the set of eigenvalues of the square matrix M and let Δ be the open unit disk in the complex plane. Prove that if eig$(M) \subset \Delta \cup \{1\}$, then $\lim_{k\to\infty} M^k$ exists if and only if the dimension of the eigenspace corresponding to the eigenvalue 1 is equal to the multiplicity of the eigenvalue 1; that is, the geometric multiplicity of the eigenvalue 1 is equal to its algebraic multiplicity.

A.11

(a) Show that the series

$$\sum_{n=1}^{\infty} \frac{1}{n}$$

diverges to $+\infty$.

(b) Show that if α is a rational number such that $\alpha > 1$, then the series

$$\sum_{n=1}^{\infty} \frac{1}{n^{\alpha}}$$

converges.

(c) Use MATLAB's Symbolic Math Toolbox to compute

$$\sum_{n=1}^{\infty} \frac{1}{n^2}.$$

A.12

(a) Let $0 < x < 1$. Show that

$$\ln(1 - x) \leq -x.$$

(b) Let $0 < x \leq 1/2$. Show that

$$\ln(1 - x) \geq -2x.$$

A.13 Let $0 < a_k < 1$ for $k \geq 0$. Use a proof by contraposition, Theorem A.16, and Exercise A.12 to show that

$$\prod_{k=0}^{\infty}(1 - a_k) = 0 \qquad \text{if and only if} \quad \sum_{k=0}^{\infty} a_k = \infty.$$

A.14 Prove Sylvester's inequalities (A.88).

A.15 Compute e^{At} for

(a)

$$A = \begin{bmatrix} 0 & -3 & 0 \\ 3 & 0 & 0 \\ 0 & 0 & -1 \end{bmatrix},$$

(b)

$$A = \begin{bmatrix} 0 & 1 & 0 \\ 0 & 0 & 1 \\ 0 & -2 & -3 \end{bmatrix}.$$

A.16 Find $\boldsymbol{x}(t)$ for the dynamical system model,

$$\dot{\boldsymbol{x}}(t) = \begin{bmatrix} 0 & 1 \\ 0 & 0 \end{bmatrix} \boldsymbol{x}(t) + \begin{bmatrix} 0 \\ 1 \end{bmatrix} u(t), \qquad \boldsymbol{x}(0) = \begin{bmatrix} 0 \\ 1 \end{bmatrix},$$

where $u(t)$ is the unit step function; that is, $u(t) = 1$ for $t \geq 0$ and $u(t) = 0$ for $t < 0$.

A.17 For

$$A = \begin{bmatrix} -2 & -6 \\ 2 & -10 \end{bmatrix},$$

find constants $M > 0$ and $\alpha > 0$ such that

$$\|e^{At}\|_2 \le Me^{-\alpha t}, \qquad t \ge 0.$$

A.18 Show that the function space $\mathcal{C}([a, b])$ of real-valued functions, $f : \mathbb{R} \to \mathbb{R}$, with the norm defined by (A.99) is a normed linear space. Show that this space is a Banach space.

A.19 Derive a sufficient condition for the convergence of the iteration process

$$\boldsymbol{x}^{(k+1)} = \boldsymbol{A}\boldsymbol{x}^{(k)} + \boldsymbol{b}, \quad k = 0, 1, \ldots,$$

for any initial $\boldsymbol{x}^{(0)}$, where $\boldsymbol{A} \in \mathbb{R}^{n \times n}$ and $\boldsymbol{b} \in \mathbb{R}^n$.

BIBLIOGRAPHY

1. P. P. Acarnley. *Stepping Motors: A Guide to Modern Theory and Practice*. IEE Control Engineering Series 19. Peter Peregrinus, London, 1982.
2. J. Ackermann. Robust control prevents car skidding. *IEEE Control Systems,* 17(3):23–31, June 1997.
3. M. Adachi and K. Aihara. Associative dynamics in a chaotic neural network. *Neural Networks,* 10(1):83–98, January 1997.
4. J. A. Adam and N. Bellomo, editors. *A Survey of Models for Tumor-Immune System Dynamics*. Birkhäuser, Boston, 1997.
5. K. Aihara, T. Takabe, and M. Toyoda. Chaotic neural networks. *Physics Letters A,* 144(6, 7):333–340, March 12, 1990.
6. A. W. Al-Khafaji and J. R. Tooley. *Numerical Methods in Engineering Practice*. Holt, Rinehart and Winston, New York, 1986.
7. K. T. Alligood, T. D. Sauer, and J. A. Yorke. *CHAOS An Introduction to Dynamical Systems*. Springer-Verlag, New York, 1997.
8. J. A. Anderson. *An Introduction to Neural Networks*. A Bradford Book, MIT Press, Cambridge, MA, 1995.
9. J. A. Anderson, J. W. Silverstein, S. A. Ritz, and R. S. Jones. Distinctive features, categorical perception, probability learning: Some applications of a neural model. In J. A. Anderson and E. Rosenfeld, editors, *Neurocomputing; Foundations of Research,* Chapter 22, pages 283–325. MIT Press, Cambridge, MA, 1989. Reprint from *Psychological Review,* 84:413–451, 1977.
10. Y. V. Andreyev, Y. L. Belsky, A. S. Dmitriev, and D. A. Kuminov. Information processing using dynamical chaos: Neural networks implementation. *IEEE Transactions on Neural Networks,* 7(2):290–299, March 1996.
11. Yu. V. Andreyev, A. S. Dmitriev, L. O. Chua, and C. W. Wu. Associative and random access memory using one-dimensional maps. *International Journal of Bifurcation and Chaos,* 2(3):483–504, September 1992.
12. P. J. Antsaklis and A. N. Michel. *Linear Systems*. McGraw-Hill, New York, 1997.
13. D. K. Arrowsmith and C. M. Place. *An Introduction to Dynamical Systems*. Cambridge University Press, Cambridge, Great Britain, 1990.
14. J.-P. Aubin and A. Cellina. *Differential Inclusions: Set-Valued Maps and Viability Theory*. Springer-Verlag, Berlin, 1984.
15. T. Bäck, U. Hammel, and H.-P. Schwefel. Evolutionary computation: Comments on the history and current state. *IEEE Transactions on Evolutionary Computation,* 1(1):3–17, April 1997.
16. F. N. Bailey. The application of Lyapunov's second method to interconnected systems. *Journal of SIAM on Control,* 3(3):443–462, 1966.
17. G. J. Bakker and H. Nijmeijer. Comments on 'Control design for multivariable nonlinear time-varying systems.' *International Journal of Control,* 41(6):1627–1629, 1985.
18. P. Ball, editor. *A Biography of Water: Life's Matrix*. Farra, Straus and Giroux, New York, 1999.

19. S. P. Banks. *Control Systems Engineering: Modelling and Simulation, Control Theory and Microprocessor Implementation*. Prentice-Hall International, Englewood Cliffs, NJ, 1986.
20. S. Barnett. *Introduction to Mathematical Control Theory*. Clarendon Press, Oxford, 1975.
21. M. Barnsley. *Fractals Everywhere*. Academic Press, San Diego, 1988.
22. R. G. Bartle. *The Elements of Real Analysis,* second edition. John Wiley & Sons, New York, 1976.
23. E. F. Beckenbach and R. Bellman. *Inequalities*. Springer-Verlag, New York, second revised printing, 1965.
24. R. Bellman. *Introduction to Matrix Analysis*. McGraw-Hill, New York, 1960.
25. R. Bellman. Vector Lyapunov functions. *Journal of SIAM on Control,* 1(1):32–34, 1962.
26. D. P. Bertsekas. Necessary and sufficient conditions for a penalty method to be exact. *Mathematical Programming,* 9(1):87–99, August 1975.
27. D. Bestle and M. Zeitz. Canonical form observer design for non-linear time-variable systems. *International Journal of Control,* 38(2):419–431, 1983.
28. K. G. Binmore. *Mathematical Analysis: A Straightforward Approach,* second edition. Cambridge University Press, Cambridge, Great Britain, 1982.
29. D. L. Boley. On Kalman's procedure for the computation of the controllable/observable canonical form. *SIAM Journal of Control and Optimization,* 18(6):624–626, November 1980.
30. V. G. Boltyanskii. *Mathematical Methods of Optimal Control*. Holt, Rinehart and Winston, New York, 1971.
31. T. L. Boullion and P. L. Odell. *Generalized Inverse Matrices*. Wiley-Interscience, New York, 1971.
32. W. E. Boyce and R. C. DiPrima. *Elementary Differential Equations and Boundary Value Problems,* sixth edition. John Wiley & Sons, New York, 1997.
33. J. W. Brewer. Kronecker products and matrix calculus in system theory. *IEEE Transactions on Circuits and Systems,* CAS-25(9):772–781, September 1978.
34. R. W. Brockett. *Finite Dimensional Linear Systems*. John Wiley & Sons, New York, 1970.
35. W. L. Brogan. *Modern Control Theory,* third edition. Prentice-Hall, Englewood Cliffs, NJ, 1991.
36. A. E. Bryson and Y. C. Ho. *Applied Optimal Control*. Blaisdell, Waltham, MA, 1975.
37. A. E. Bryson, Jr. *Dynamic Optimization*. Addison Wesley Longman, Menlo Park, CA, 1999.
38. A. Bunde and S. Havlin, editors. *Fractals in Science*. Springer-Verlag, Berlin, 1994.
39. J. B. Burl. *Linear Optimal Control; $\mathcal{H}_2$ and $\mathcal{H}_\infty$ Methods*. Addison Wesley Longman, Menlo Park, CA, 1999.
40. S. L. Campbell and C. D. Meyer, Jr. *Generalized Inverses of Linear Transformations*. Dover Publications, New York, 1991.
41. R. H. Cannon, Jr. *Dynamics of Physical Systems*. McGraw-Hill, New York, 1967.
42. S. G. Cao, N. W. Rees, and G. Feng. Quadratic stability analysis and design of continuous-time fuzzy control systems. *International Journal of System Science,* 27(2):193–203, 1996.
43. H. Y. Chan, S. Hui, and S. H. Żak. Synchronous and asynchronous Brain-State-in-a-Box information system neural models. *International Journal of Neural Systems,* 9(1):61–74, February 1999.
44. H. Y. Chan and S. H. Żak. On neural networks that design neural associative memories. *IEEE Transactions on Neural Networks,* 8(2):360–372, March 1997.
45. H. Y. Chan and S. H. Żak. Real-time synthesis of sparsely interconnected neural associative memories. *Neural Networks,* 11(4):749–759, June 1998.
46. H. Y. Chan and S. H. Żak. Chaotic neural fuzzy associative memory. *International Journal of Bifurcation and Chaos,* 9(8):1597–1617, August 1999.
47. K. C. C. Chan, V. Lee, and H. Leung. Generating fuzzy rules for target tracking using a steady-state genetic algorithm. *IEEE Transactions on Evolutionary Computation,* 1(3):189–200, September 1997.
48. C.-T. Chen. *Linear Systems Theory and Design,* third edition. Oxford University Press, New York, 1999.

49. E. K. P. Chong, S. Hui, and S. H. Żak. An analysis of a class of neural networks for solving linear programming problems. *IEEE Transactions on Automatic Control,* 44(11):1995–2006, November 1999.

50. E. K. P. Chong and S. H. Żak. *An Introduction to Optimization,* second edition. Wiley-Interscience, New York, 2001.

51. L. O. Chua and G.-N. Lin. Nonlinear programming without computation. *IEEE Transactions on Circuits and Systems,* CAS-31(2):182–188, February 1984.

52. A. Cichocki and R. Unbehauen. *Neural Networks for Optimization and Signal Processing*. John Wiley, Chichester, England, 1993.

53. S. D. Conte and C. de Boor. *Elementary Numerical Analysis: An Algorithmic Approach,* third edition. McGraw-Hill, New York, 1980.

54. I. M. Copi. *Introduction to Logic,* fifth edition. Macmillan, New York, 1978.

55. C. Corduneanu. *Principles of Differential and Integral Equations*. Allyn and Bacon, Boston, MA, 1971.

56. M. J. Corless and G. Leitmann. Continuous state feedback guaranteeing uniform ultimate boundedness for uncertain dynamic systems. *IEEE Transactions on Automatic Control,* AC-26(5):1139–1144, October 1981.

57. J. J. D'Azzo and C. H. Houpis. *Linear Control System Analysis and Design: Conventional and Modern,* fourth edition. McGraw-Hill, New York, 1995.

58. L. Debnath and P. Mikusiński. *Introduction to Hilbert Spaces with Applications*. Academic Press, San Diego, 1990.

59. R. A. DeCarlo. *Linear Systems: A State Variable Approach with Numerical Implementation*. Prentice-Hall, Englewood Cliffs, NJ, 1989.

60. R. A. DeCarlo, S. H. Żak, and G. P. Matthews. Variable structure control of nonlinear multivariable systems: A tutorial. *Proceedings of the IEEE,* 76(3):212–232, March 1988.

61. V. Del Toro. *Electric Machines and Power Systems*. Prentice-Hall, Englewood Cliffs, NJ, 1985.

62. M. J. Denham. Canonical forms for the identification of multivariable linear systems. *IEEE Transactions on Automatic Control,* AC-19(6):646–656, December 1974.

63. R. L. Devaney and L. Keen, editors. *Chaos and Fractals; The Mathematics Behind the Computer Graphics*. Proceedings of Symposia in Applied Mathematics, Volume 39. American Mathematical Society, Providence, RI, 1989.

64. B. M. Diong and J. V. Medanic. Dynamic output feedback variable structure control for system stabilization. *International Journal of Control,* 56(3):607–630, September 1992.

65. A. S. Dmitriev, A. I. Panas, and S. O. Starkov. Storing and recognizing information based on stable cycles of one-dimensional maps. *Physics Letters A,* 155(8, 9):494–499, May 27, 1991.

66. R. C. Dorf and R. H. Bishop. *Modern Control Systems,* eighth edition. Addison Wesley, Menlo Park, CA, 1998.

67. D. Driankov, H. Hellendoorn, and M. Reinfrank. *An Introduction to Fuzzy Control*. Springer-Verlag, Berlin, 1993.

68. M. Driels. *Linear Control Systems Engineering*. McGraw-Hill, New York, 1996.

69. R. D. Driver. *Ordinary and Delay Differential Equations*. Springer-Verlag, New York, 1977.

70. W. Dunham. *Journey Through Genius: The Great Theorems of Mathematics*. Wiley Science Editions, John Wiley & Sons, New York, 1990.

71. C. Edwards and S. K. Spurgeon. On the development of discontinuous observers. *International Journal of Control,* 59(5):1211–1229, May 1994.

72. C. Edwards and S. K. Spurgeon. Compensator based output feedback sliding mode controller design. *International Journal of Control,* 71(4):601–614, November 10, 1998.

73. C. Edwards and S. K. Spurgeon. *Sliding Mode Control: Theory and Applications*. Systems and Control Book Series. Taylor & Francis, London, England, 1998.

74. G. D. Elseth and K. D. Baumgardner. *Genetics*. Addison-Wesley, Reading, MA, 1984.

75. L. E. Elsgolc. *Calculus of Variations*. Pergamon Press, London, 1961.

76. K. D. Elworthy, W. N. Everitt, and E. B. Lee, editors. *Differential Equations, Dynamical Systems, and Control Science: A Festschrift in Honor of Lawrence Markus*. Lecture Notes in Pure and Applied Mathematics, Volume 152. Marcel Dekker, New York, 1994.

77. S. V. Emelyanov, editor. *Titles in Theory of Variable Structure Control Systems (1957–1989)*. International Research Institute for Mnagement Sciences, Moscow, 1989.

78. V. N. Faddeeva. *Computationl Methods of Linear Algebra*. Dover Publications, New York, 1959.

79. L. Fausett. *Fundamentals of Neural Networks; Architectures, Algorithms, and Applications*. Prentice-Hall, Englewood Cliffs, NJ, 1994.

80. L. Faybusovich. Dynamical systems that solve linear programming problems. In *Proceedings of the 31st Conference on Decision and Control, Tucson, Arizona,* pages 1626–1631, December 1992.

81. M. J. Feigenbaum. Universal behavior in nonlinear systems. *Physica D,* 7D(1–3):16–39, May 1983.

82. G. Feng, S. G. Cao, N. W. Rees, and C. K. Chak. Design of fuzzy control systems with guaranteed stability. *Fuzzy Sets and Systems,* 85(1):1–10, January 1, 1997.

83. A. F. Filippov. Differential equations with discontinuous right-hand side. In *American Mathematical Society Translations,* pages 199–231. American Mathematical Society, Providence, RI, 1964.

84. A. F. Filippov. *Differential Equations with Discontinuous Righthand Sides*. Mathematics and Its Applications (Soviet Series). Kluwer Academic Publishers, Dordrecht, The Netherlands, 1988.

85. K. Fischle and D. Schröder. An improved stable adaptive fuzzy control method. *IEEE Transactions on Fuzzy Systems,* 7(1):27–40, February 1999.

86. S. Flåm and J. Zowe. A primal-dual differential method for convex programming. In A. Ioffe, M. Marcus, and S. Reich, editors, *Optimization and Nonlinear Analysis,* pages 119–129. Longman Scientific & Technical, Essex, England, 1992.

87. S. D. Flåm. Solving convex programs by means of ordinary differential equations. *Mathematics of Operations Research,* 17(2):290–302, May 1992.

88. D. B. Fogel. What is evolutionary computation? *IEEE Spectrum,* 37(2):26–32, February 2000.

89. G. F. Franklin and J. D. Powell. *Digital Control of Dynamic Systems*. Addison-Wesley, Reading, MA, 1980.

90. G. F. Franklin, J. D. Powell, and A. Emami-Naeini. *Feedback Control of Dynamic Systems,* fourth edition. Prentice-Hall, Upper Saddle River, NJ, 2002.

91. B. Friedland. *Advanced Control System Design*. Prentice-Hall, Englewood Cliffs, NJ, 1996.

92. A. T. Fuller. Maxwell's treatment of governors fitted with differential gears. *International Journal of Control,* 65(3):385–408, October 1996.

93. F. R. Gantmacher. *Applications of the Theory of Matrices*. Interscience Publishers, New York, 1959.

94. S. S. Ge, T. H. Lee, and G. Zhu. Genetic algorithm tuning of Lyapunov-based controllers: An application to a single-link flexible robot system. *IEEE Transactions on Industrial Electronics,* 43(5):567–574, October 1996.

95. I. M. Gel'fand. *Lectures on Linear Algebra*. Interscience Publishers, New York, 1961.

96. M. Gen and R. Cheng. *Genetic Algorithms and Engineering Design*. Wiley-Interscience, New York, 1997.

97. P. E. Gill, W. Murray, and M. H. Wright. *Numerical Linear Algebra and Optimization,* Volume 1. Addison-Wesley, Redwood City, CA, 1991.

98. M. P. Glazos, S. Hui, and S. H. Żak. Sliding modes in solving convex programming problems. *SIAM Journal of Control and Optimization,* 36(2):680–697, March 1998.

99. M. P. Glazos and S. H. Żak. Practical stabilization of nonlinear/uncertain dynamic systems with bounded controllers. *International Journal of Control,* 62(1):153–171, July 1995.

100. J. Gleick. *Chaos: Making a New Science*. Penguin Books, New York, 1988.

101. I. Glynn, editor. *An Anatomy of Thought: The Origin and Machinery of Mind*. Oxford University Press, New York, 1999.

102. R. M. Golden. The "Brain-State-in-a-Box" neural model is a gradient descent algorithm. *Journal of Mathematical Psychology,* 30(1):73–80, 1986.

103. R. M. Golden. Stability and optimization analyses of the generalized Brain-State-in-a-Box neural network model. *Journal of Mathematical Psychology,* 37(2):282–298, 1993.

104. G. H. Golub and C. F. Van Loan. *Matrix Computations.* The Johns Hopkins University Press, Baltimore, MD, 1983.

105. D. Graham and D. McRuer. *Analysis of Nonlinear Control Systems.* John Wiley & Sons, New York, 1961.

106. W. J. Grantham and A. M. Athalye. Discretization chaos: Feedback control and transistion to chaos. In C. T. Leondes, editor, *Control and Dynamic Systems; Advances in Theory and Applications, Volume 34: Advances in Control Mechanics, Part 1 of 2,* pages 205–277. Academic Press, San Diego, CA, 1990.

107. H. J. Greenberg. Equilibria of the Brain-State-in-a-Box (BSB) neural model. *Neural Networks,* 1(4):323–324, 1988.

108. S. Grossberg. Nonlinear neural networks: Principles, mechanisms, and architectures. *Neural Networks,* 1(1):17–61, 1988.

109. V. B. Haas. EE 675 Class Notes. Purdue University, School of Electrical Engineering, West Lafayette, IN, Fall 1980.

110. M. T. Hagan, H. B. Demuth, and M. Beale. *Neural Network Design.* PWS Publishing Co., Boston, MA, 1996.

111. W. Hahn. Über die Anwendung der Methode von Ljapunov auf Differenzengleichungen. *Mathematische Annalen,* 136(1):430–441, 1958.

112. W. Hahn. *Theory and Application of Liapunov's Direct Method.* Prentice-Hall International Series in Applied Mathematics. Prentice-Hall, Englewood Cliffs, NJ, 1963.

113. F. Hakl. Recall strategy of B-S-B neural network. *Neural Network World,* (1):59–82, 1992.

114. G. H. Hardy. *A Mathematician's Apology.* Cambridge University Press, Cambridge, Great Britain, 1992. Canto edition with a foreword by C. P. Snow.

115. R. L. Harvey. *Neural Network Principles.* Prentice-Hall, Englewood Cliffs, NJ, 1994.

116. M. H. Hassoun. *Fundamentals of Artificial Neural Networks.* A Bradford Book, MIT Press, Cambridge, MA, 1995.

117. S. W. Hawking. *A Brief History of Time: From the Big Bang to Black Holes.* Bantam Books, New York, 1988.

118. S. Haykin. *Neural Networks: A Comprehensive Foundation,* second edition. Prentice-Hall, Upper Saddle River, NJ, 1999.

119. U. Helmke and J. B. Moore. *Optimization and Dynamical Systems.* Communications and Control Engineering Series. Springer-Verlag, London, 1994.

120. J. Hertz, A. Krogh, and R. G. Palmer. *Introduction to the Theory of Neural Computation.* Lecture Notes Volume I in the Santa Fe Institute Studies in the Sciences of Complexity. Addison-Wesley, Redwood City, CA, 1991.

121. M. R. Hestenes. *Calculus of Variations and Optimal Control Theory.* John Wiley & Sons, New York, 1966.

122. N. J. Higham. *Handbook of Writing for the Mathematical Sciences.* SIAM, Philadelphia, 1993.

123. H. Hochstadt. *Differential Equations: A Modern Approach.* Dover Publications, New York, 1975.

124. J. H. Holland. *Adaptation in Natural and Artificial Systems; An Introductory Analysis with Applications to Biology, Control, and Artificial Intelligence.* A Bradford Book, MIT Press, Cambridge, MA, 1992.

125. J. J. Hopfield. Neural networks and physical systems with emergent collective computational abilities. *Proceedings of the National Academy of Sciences of the USA, Biophysics,* 79:2554–2558, April 1982.

126. J. J. Hopfield. Neurons with graded response have collective computational properties like those of two-state neurons. *Proceedings of the National Academy of Sciences of the USA, Biophysics,* 81:3088–3092, May 1984.
127. A. S. Householder. *The Theory of Matrices in Numerical Analysis*. Dover Publications, New York, 1975.
128. J. C. Hsu and A. U. Meyer. *Modern Control Principles and Applications*. McGraw-Hill, New York, 1968.
129. S. Hui, W. E. Lillo, and S. H. Żak. Dynamics and stability analysis of the Brain-State-in-a-Box (BSB) neural models. In M. H. Hassoun, editor, *Associative Neural Memories; Theory and Implementation,* Chapter 11, pages 212–224. Oxford University Press, New York, 1993.
130. S. Hui and S. H. Żak. Dynamical analysis of the Brain-State-in-a-Box (BSB) neural models. *IEEE Transactions on Neural Networks,* 3(1):86–94, January 1992.
131. S. Hui and S. H. Żak. Robust control synthesis for uncertain/nonlinear dynamical systems. *Automatica,* 28(2):289–298, 1992.
132. S. Hui and S. H. Żak. Low-order state estimators and compensators for dynamical systems with unknown inputs. *Systems & Control Letters,* 21(6):493–502, December 1993.
133. S. Hui and S. H. Żak. Robust output feedback stabilization of uncertain dynamic systems with bounded controllers. *International Journal of Robust and Nonlinear Control,* 3(2):115–132, August 1993.
134. S. Hui and S. H. Żak. The Widrow–Hoff algorithm for McCulloch–Pitts type neurons. *IEEE Transactions on Neural Networks,* 5(6):924–929, November 1994.
135. J. Y. Hung, W. Gao, and J. C. Hung. Variable structure control: A survey. *IEEE Transactions on Industrial Electronics,* 40(1):2–22, February 1993.
136. L. R. Hunt, R. Su, and G. Meyer. Global transformations of nonlinear systems. *IEEE Transactions on Automatic Control,* AC-28(1):24–31, January 1983.
137. D. R. Hush and B. G. Horne. Progress in supervised neural networks: What's new since Lippmann? *IEEE Signal Processing Magazine,* 10(1):8–39, January 1993.
138. H. S. Hwang. Automatic design of fuzzy rule base for modelling and control using evolutionary programming. *IEE Proceedings—Control Theory Appl.,* 146(1):9–16, January 1999.
139. A. Isidori. *Nonlinear Control Systems,* third edition. Springer-Verlag, London, 1995.
140. U. Itkis. *Control Systems of Variable Structure*. A Halsted Press Book, a division of John Wiley & Sons, New York, 1976.
141. M. Jamshidi, M. Tarokh, and B. Shafai. *Computer-Aided Analysis and Design of Linear Control Systems*. Prentice-Hall, Englewood Cliffs, NJ, 1992.
142. J.-S. R. Jang, C.-T. Sun, and E. Mizutani. *Neuro-Fuzzy and Soft Engineering: A Computational Approach to Learning and Machine Intelligence*. MATLAB Curriculum Series. Prentice-Hall, Upper Saddle River, NJ, 1997.
143. T. Kaczorek. *Teoria Sterowania i Systemów*. Wydawnictwo Naukowe PWN—Polish Scientific Publishers, Warszawa, 1993.
144. T. Kaczorek. *Wektory i Macierze w Automatyce i Elektrotechnice,* second edition. Wydawnictwa Naukowe-Techniczne—Scientific-Technical Publishers, Warszawa, 1998.
145. P. B. Kahn. *Mathematical Methods for Scientists and Engineers: Linear and Nonlinear Systems*. Wiley-Interscience, New York, 1990.
146. T. Kailath. *Linear Systems*. Prentice-Hall, Englewood Cliffs, NJ, 1980.
147. R. E. Kalman. Mathematical description of linear dynamical systems. *SIAM Journal of Control, Series A,* 1(2):152–192, 1963.
148. R. E. Kalman. On the computation of the reachable/observable canonical form. *SIAM Journal of Control and Optimization,* 20(2):258–260, March 1982.
149. S.-J. Kang, C.-H. Woo, H.-S. Hwang, and K. B. Woo. Evolutionary design of fuzzy rule base for nonlinear system modeling and control. *IEEE Transactions on Fuzzy Systems,* 8(1):37–45, February 2000.

150. N. N. Karpinskaya. Method of "penalty" functions and the foundations of Pyne's method. *Automation and Remote Control,* 28(1):124–129, January 1967.
151. A. Kaufmann and M. M. Gupta. *Introduction to Fuzzy Arithmetic: Theory and Applications*. Van Nostrand Reinhold Electrical/Computer Science and Engineering Series. Van Nostrand Reinhold, New York, 1985.
152. D. G. Kelly. Stability in contractive nonlinear neural networks. *IEEE Transactions on Biomedical Engineering,* 37(3):231–242, March 1990.
153. T. Kenjo. *Stepping Motors and Their Microprocessor Controls*. Clarendon Press, Oxford, 1984.
154. T. Kenjo and S. Nagamori. *Permanent-Magnet and Brushless DC Motors*. Clarendon Press, Oxford, 1985.
155. M. P. Kennedy and L. O. Chua. Neural networks for nonlinear programming. *IEEE Transactions on Ciruits and Systems,* 35(5):554–562, May 1988.
156. H. K. Khalil. Adaptive output feedback control of nonlinear systems represented by input–output models. *IEEE Transactions on Automatic Control,* 41(2):177–188, February 1996.
157. H. K. Khalil. *Nonlinear Systems,* second edition. Prentice-Hall, Upper Saddle River, NJ, 1996.
158. T. Khanna. *Foundations of Neural Networks*. Addison-Wesley, Reading, MA, 1990.
159. J.-H. Kim and H. Myung. Evolutionary programming techniques for constrained optimization problems. *IEEE Transactions on Evolutionary Computation,* 1(2):129–140, July 1997.
160. D. E. Kirk. *Optimal Control Theory: An Introduction*. Prentice-Hall, Englewood Cliffs, NJ, 1970.
161. G. J. Klir, U. St. Clair, and B. Yuan. *Fuzzy Set Theory: Foundations and Applications*. Prentice-Hall PTR, Upper Saddle River, NJ, 1997.
162. D. E. Knuth. *The Art of Computer Programming,* Volume 1. Fundamental Algorithms, second edition. Addison-Wesley, Reading, MA, 1973.
163. C. Koch. Computation and the single neuron. *Nature,* 385(6613):207–210, January 16, 1997.
164. S. K. Korovin and V. I. Utkin. Using sliding modes in static optimization and nonlinear programming. *Automatica,* 10(5):525–532, September 1974.
165. B. Kosko. *Fuzzy Engineering*. Prentice-Hall, Upper Saddle River, NJ, 1997.
166. J. R. Koza. *Genetic Programming: On the Programming of Computers by Means of Natural Selection*. A Bradford Book, MIT Press, Cambridge, MA, 1992.
167. J. R. Koza, F. H. Bennett, III, D. Andre, M. A. Keane, and F. Dunlap. Automated synthesis of analog electrical circuits by means of genetic programming. *IEEE Transactions on Evolutionary Computation,* 1(2):109–128, July 1997.
168. T. Kozek, T. Roska, and L. O. Chua. Genetic algorithm for CNN template learning. *IEEE Transactions on Circuits and Systems—I: Fundamental Theory and Applications,* 40(6):392–402, June 1993.
169. P. C. Krause and O. Wasynczuk. *Electromechanical Motion Devices*. McGraw-Hill, New York, 1989.
170. A. J. Krener and W. Respondek. Nonlinear observers with linearizable error dynamics. *SIAM Journal of Control and Optimization,* 23(2):197–216, March 1985.
171. M. Krstić, I. Kanellakopoulos, and P. V. Kokotović. *Nonlinear and Adaptive Control Design*. John Wiley & Sons, New York, 1995.
172. V. Kučera. A contribution to matrix quadratic equations. *IEEE Transactions on Automatic Control,* AC-17(3):344–347, June 1972.
173. H. Kwakernaak and R. Sivan. *Linear Optimal Control Systems*. Wiley-Interscience, New York, 1972.
174. C. M. Kwan. On variable structure output feedback controllers. *IEEE Transactions on Automatic Control,* 41(11):1691–1693, November 1996.
175. S. Lang. *Calculus of Several Variables,* third edition. Springer-Verlag, New York, 1987.
176. J. P. LaSalle. *The Stability and Control of Discrete Processes*. Springer-Verlag, New York, 1986.
177. E. B. Lee and L. Markus. *Foundations of Optimal Control Theory*. Robert E. Krieger Publishing Company, Malabar, FL, 1986.

178. Y. Lee, J. Q. Gong, B. Yao, and S. H. Żak. Fuzzy adaptive robust tracking control of a class of nonlinear systems. In *Proceedings of 2001 American Control Conference,* pages 4040–4045. Arlington, VA, June 25–27, 2001.
179. Y. Lee and S. H. Żak. Genetic fuzzy tracking controllers for autonomous ground vehicles. In *Proceedings of 2002 American Control Conference,* pages 2144–2149. Anchorage, AK, May 8–10, 2002.
180. K. N. Leibovic. The principle of contraction mapping in nonlinear and adaptive control systems. *IEEE Transactions on Automatic Control,* AC-9(4):393–398, October 1964.
181. J. R. Leigh. *Modelling and Simulation.* IEE Topics in Control Series 1. Peter Peregrinus, London, 1983.
182. F. L. Lewis and V. L. Syrmos. *Optimal Control,* second edition. Wiley-Interscience, New York, 1995.
183. J.-H. Li, A. N. Michel, and W. Porod. Analysis and synthesis of a class of neural networks: Linear systems operating on a closed hypercube. *IEEE Transactions on Circuits and Systems,* 36(11):1405–1422, 1989.
184. W. E. Lillo, D. C. Miller, S. Hui, and S. H. Żak. Synthesis of Brain-State-in-a-Box (BSB) based associative memories. *IEEE Transactions on Neural Networks,* 5(5):730–737, 1994.
185. C.-T. Lin and C. S. G. Lee. *Neural Fuzzy Systems: A Neuro-Fuzzy Synergism to Intelligent Systems.* Prentice-Hall PTR, Upper Saddle River, NJ, 1996.
186. R. P. Lippmann. An introductuion to computing with neural nets. *IEEE ASSP Magazine,* 4(2):4–22, April 1987.
187. G. P. Liu, V. Kadirkamanathan, and S. A. Billings. Stable sequential identification of continuous nonlinear dynamical systems by growing radial basis function networks. *International Journal of Control,* 65(1):53–69, 10 September 1996.
188. D. G. Luenberger. Observers for multivariable systems. *IEEE Transactions on Automatic Control,* AC-11(2):190–197, April 1966.
189. D. G. Luenberger. An introduction to observers. *IEEE Transactions on Automatic Control,* AC-16(6):596–602, December 1971.
190. D. G. Luenberger. *Linear and Nonlinear Programming,* second edition. Addison-Wesley, Reading, MA, 1984.
191. X.-J. Ma, Z.-Q. Sun, and Y.-Y. He. Analysis and design of fuzzy controller and fuzzy observer. *IEEE Transactions on Fuzzy Systems,* 6(1):41–51, February 1998.
192. A. G. J. MacFarlane. The development of frequency-response methods in automatic control. *IEEE Transactions on Automatic Control,* AC-24(2):250–265, April 1979.
193. I. J. Maddox. *Elements of Functional Analysis,* second edition. Cambridge University Press, Cambridge, Great Britain, 1988.
194. E. H. Mamdani and S. Assilian. An experiment in linguistic synthesis with a fuzzy logic controller. *International Journal of Man–Machine Studies,* 7(1):1–13, January 1975.
195. B. B. Mandelbrot. *The Fractal Geometry of Nature.* W. H. Freeman, New York, 1983.
196. R. Marino and P. Tomei. *Nonlinear Control Design: Geometric, Adaptive and Robust.* Prentice-Hall Information and System Sciences Series. Prentice-Hall Europe, London, 1995.
197. J. C. Maxwell. On governors. *Proceedings of the Royal Society of London,* XVI:270–283, March 1868. Printed by Taylor and Francis, London, Red Lion Court, Fleet Street.
198. R. M. May. Simple mathematical models with very complicated dynamics. *Nature,* 261(5560):459–467, June 10, 1976.
199. O. Mayr. *The Origins of Feedback Control.* MIT Press, Cambridge, MA, 1970.
200. W. S. McCulloch and W. Pitts. A logical calculus of the ideas immanent in nervous activity. *Bulletin of Mathematical Biophysics,* 5:115–133, 1943.
201. J. M. Mendel. *Uncertain Rule-Based Fuzzy Logic Systems: Introduction and New Directions.* Prentice-Hall PTR, Upper Saddle River, NJ, 2001.
202. A. N. Michel and J. A. Farrell. Associative memories via artificial neural networks. *IEEE Control Systems Magazine,* 10(3):6–17, 1990.

203. A. N. Michel, J. A. Farrell, and W. Porod. Qualitative analysis of neural networks. *IEEE Transactions on Circuits and Systems,* 36(2):229–243, February 1989.

204. A. N. Michel, J. A. Farrell, and H.-F. Sun. Analysis and synthesis technique for Hopfield type synchronous discrete time neural networks with application to associative memory. *IEEE Transactions on Circuits and Systems,* 37(11):1356–1366, 1990.

205. R. K. Miller and A. N. Michel. *Ordinary Differential Equations.* Academic Press, New York, 1982.

206. M. Mitchell. *An Introduction to Genetic Algorithms.* A Bradford Book, MIT Press, Cambridge, MA, 1996.

207. R. R. Mohler. *Nonlinear Systems, Volume I: Dynamics and Control.* Prentice-Hall, Englewood Cliffs, NJ, 1991.

208. C. Moler and C. Van Loan. Nineteen dubious ways to compute the exponential of a matrix. *SIAM Review,* 20(4):801–836, October 1978.

209. F. C. Moon. *Chaotic Vibrations: An Introduction for Applied Scientists and Engineers.* Wiley-Interscience, John Wiley & Sons, New York, 1987.

210. Fr. M. Müller, C.S.S.R. *The Blessed Eucharist—Our Greatest Treasure.* TAN Books and Publishers, Rockford, IL, 1994. First published by Kelly & Piet of Baltimore, Maryland in 1868.

211. A. G. Nalecz. Sensitivity analysis of vehicle design attributes in frequency domain. *Vehicle System Dynamics,* 17:141–163, 1988.

212. A. G. Nalecz and A. C. Bindemann. Analysis of the dynamic response of four wheel steering vehicles at high speed. *International Journal of Vehicle Design,* 9(2):179–202, 1988.

213. A. G. Nalecz and A. C. Bindemann. Investigation into the stability of four wheel steering vehicles. *International Journal of Vehicle Design,* 9(2):159–178, 1988.

214. S. Nara, P. Davis, and H. Totsuji. Memory search using complex dynamics in a recurrent neural network model. *Neural Networks,* 6(7):963–973, 1993.

215. I. P. Natanson. *Theory of Functions of a Real Variable.* Frederick Ungar, New York, 1955.

216. K. Ogata. *Modern Control Engineering,* third edition. Prentice-Hall, Upper Saddle River, NJ, 1997.

217. E. Ott, C. Grebogi, and J. A. Yorke. Controlling chaos. *Physical Review Letters,* 64(11):1196–1199, March 12, 1990.

218. D. H. Owens. *Multivariable and Optimal Systems.* Academic Press, London, 1981.

219. S. K. Pal and D. Bhandari. Genetic algorithms with fuzzy fitness function for object extraction using cellular networks. *Fuzzy Sets and Systems,* 65(2/3):129–139, August 10, 1994.

220. J. Park, H. Cho, and D. Park. Design of GBSB neural associative memories using semidefinite programming. *IEEE Transactions on Neural Networks,* 10(4):946–950, July 1999.

221. T. S. Parker and L. O. Chua. Chaos: A tutorial for engineers. *Proceedings of the IEEE,* 75(8):982–1008, August 1987.

222. T. S. Parker and L. O. Chua. *Practical Numerical Algorithms for Chaotic Systems.* Springer-Verlag, New York, 1989.

223. R. V. Patel and M. Toda. Quantitative measures of robustness for multivariable systems. In *Proceedings of Joint Automatic Control Conference (JACC),* San Francisco, CA, page TP8-A, 1980.

224. R. J. Patton and G. P. Liu. Robust control design via eigenstructure assignment, genetic algorithms and gradient-based optimization. *IEE Proceedings—Control Theory Applications,* 141(3):202–208, May 1994.

225. W. Pedrycz. *Fuzzy Control and Fuzzy Sytems,* second, extended edition. Research Studies Press, Taunton, Somerset, England, 1993.

226. W. Pedrycz, editor. *Fuzzy Modeling: Paradigms and Practice.* Kluwer Academic Publishers, Boston, 1996.

227. H.-O. Peitgen, H. Jürgens, and D. Saupe, editors. *Chaos and Fractals: New Frontiers of Science.* Springer-Verlag, New York, 1992.

228. H.-O. Peitgen and D. Saupe, editors. *The Science of Fractal Images.* Springer-Verlag, New York, 1988.

229. R. Perfetti. A synthesis procedure for Brain-State-in-a-Box neural networks. *IEEE Transactions on Neural Networks,* 6(5):1071–1080, September 1995.

230. L. Personnaz, I. Guyon, and G. Dreyfus. Information storage and retrieval in spin-glass like neural networks. *Le Journal de Physique—Letters,* 46(8):L-359–L-365, 1985.

231. L. Personnaz, I. Guyon, and G. Dreyfus. Collective computational properties of neural networks: New learning mechnisms. *Physical Review A,* 34(5):4217–4228, 1986.

232. I. Peterson. *Newton's Clock: Chaos in the Solar System.* W. H. Freeman, New York, 1993.

233. C. A. Pickover, editor. *Fractal Horizons: The Future Use of Fractals.* St. Martin's Press, New York, 1996.

234. D. A. Pierre. *Optimization Theory with Applications.* Dover Publications, New York, 1986.

235. L. S. Pontryagin. *Ordinary Differential Equations.* Addison-Wesley, Reading, MA, 1962.

236. V. M. Popov. Invariant description of linear, time-invariant controllable systems. *SIAM Journal of Control,* 10(2):252–264, May 1972.

237. H. M. Power and R. J. Simpson. *Introduction to Dynamics and Control.* McGraw-Hill, London, 1978.

238. R. Pratap. *Getting Started with MATLAB 5: A Quick Introduction for Scientists and Engineers.* Oxford University Press, New York, 1999.

239. I. B. Pyne. Linear programming on an electronic analogue computer. *Transactions of the American Institute of Electrical Engineers,* 75:139–143, May 1956.

240. F. H. Raven. *Automatic Control Engineering,* fourth edition. McGraw-Hill, NY, 1987.

241. *Reader's Digest Practical Problem Solver. Substitutes, Shortcuts, and Ingenious Solutions for Making Life Easier.* The Reader's Digest Association, Pleasantville, New York, 1991.

242. A. Rodríguez-Vázquez, R. Domínguez-Castro, A. Rueda, J. L. Huertas, and E. Sánchez-Sinencio. Nonlinear switched-capacitor "neural" networks for optimization problems. *IEEE Transactions on Circuits and Systems,* 37(3):384–398, March 1990.

243. C. E. Rohrs, J. L. Melsa, and D. G. Schultz. *Linear Control Systems.* McGraw-Hill, New York, 1993.

244. W. Rudin. *Principles of Mathematical Analysis,* third edition. McGraw-Hill, New York, 1976.

245. W. J. Rugh. *Linear System Theory,* second edition. Prentice-Hall, Upper Saddle River, NJ, 1996.

246. W. J. Rugh and G. J. Murphy. A new approach to the solution of linear optimal control problems. *Transactions of the ASME: Journal of Basic Engineering,* 91(2):149–154, June 1969.

247. D. E. Rumelhart, J. L. McClelland, and the PDP Research Group. *Parallel Distributed Processing: Explorations in the Microstructure of Cognition,* Volume 1. A Bradford Book, MIT Press, Cambridge, MA, 1986.

248. M. V. Rybashov. Analog solution of algebraic and transcendental equations by the gradient method. *Automation and Remote Control,* 22(1):66–76, August 1961.

249. M. V. Rybashov. Some methods of solving systems of linear algebraic equations by electronic simulation. *Automation and Remote Control,* 23(2):223–231, July 1962.

250. M. V. Rybashov. The gradient method of solving convex programming problems on electronic analog computers. *Automation and Remote Control,* 26(11):1886–1898, November 1965.

251. M. V. Rybashov. Gradient method of solving linear and quadratic programming problems on electronic analog computers. *Automation and Remote Control,* 26(12):2079–2089, December 1965.

252. M. G. Salvadori and M. L. Baron. *Numerical Methods in Engineering,* second edition. Prentice-Hall, Englewood Cliffs, NJ, 1961.

253. J. T. Sandefur. *Discrete Dynamical Systems; Theory and Applications.* Clarendon Press, Oxford, 1990.

254. A. Schultz. Collective recall via the Brain-State-in-a-Box network. *IEEE Transactions on Neural Networks,* 4(4):580–587, July 1993.

255. R. T. Seeley. *Calculus of Several Variables: An Introduction.* Scott, Foresman and Company, Glenview, IL, 1970.

256. S. Seshagiri and H. K. Khalil. Output feedback control of nonlinear systems using RBF neural networks. *IEEE Transactions on Neural Networks,* 11(1):69–79, January 2000.

257. G. M. Shepherd, editor. *The Synaptic Organization of the Brain,* third edition. Oxford University Press, New York, 1990.

258. H. Sira-Ramírez and E. Colina-Morles. A sliding mode strategy for adaptive learning in adalines. *IEEE Transactions on Circuits and Systems—I: Fundamental Theory and Applications,* 42(12):1001–1012, December 1995.

259. H. Sira-Ramírez, E. Colina-Morles, and F. Rivas-Echeverría. Sliding mode-based adaptive learning in dynamical-filter-weights neurons. *International Journal of Control,* 73(8):678–685, May 20, 2000.

260. K.-Y. Siu, V. Roychowdhury, and T. Kailath. *Discrete Neural Computation; A Theoretical Foundation.* Prentice Hall PTR, Englewood Cliffs, NJ, 1995.

261. R. E. Skelton. *Dynamic Systems Control: Linear Systems Analysis and Synthesis.* John Wiley & Sons, New York, 1988.

262. J.-J. E. Slotine and W. Li. *Applied Nonlinear Control.* Prentice-Hall, Englewood Cliffs, NJ, 1991.

263. D. E. Smith and J. M. Starkey. Effects of model complexity on the performance of automated vehicle steering controllers: Controller development and evaluation. *Vehicle System Dynamics,* 23(8):627–645, November 1994.

264. D. E. Smith and J. M. Starkey. Effects of model complexity on the performance of automated vehicle steering controllers: Model development, validation and comparison. *Vehicle System Dynamics,* 24(2):163–181, March 1995.

265. M. C. Smith. Special issue on control education—The United Kingdom. *IEEE Control Systems,* 16(2):51–56, April 1996.

266. W. E. Snyder. *Industrial Robots: Computer Interfacing and Control.* Prentice-Hall, Englewood Cliffs, NJ, 1985.

267. O. A. Solheim. Design of optimal control systems with prescribed eigenvalues. *International Journal of Control,* 15(1):143–160, 1972.

268. R. Sommer. Control design for multivariable non-linear time-varying systems. *International Journal of Control,* 31(5):883–891, 1980.

269. E. D. Sontag. *Mathematical Control Theory: Deterministic Finite Dimensional Systems,* second edition. Springer-Verlag, New York, 1998.

270. R. Su. On the linear equivalents of nonlinear systems. *Systems & Control Letters,* 2(1):48–52, July 1982.

271. M. Sugeno and G. T. Kang. Structure identification of fuzzy model. *Fuzzy Sets and Systems,* 28(1):15–33, October 1988.

272. M. Sunwoo and K. C. Cheok. An application of explicit self-tuning controller to vehicle active suspension systems. In *Proceedings of the 29th Conference on Decision and Control, Honolulu, Hawaii,* pages 2251–2257, December 1990.

273. H. J. Sussmann and J. C. Willems. 300 years of optimal control: From the brachystochrone to the maximum principle. *IEEE Control Systems,* 17(3):32–44, June 1997.

274. K. Takaba. Robust servomechanism with preview action for polytopic uncertain systems. *International Journal of Robust and Nonlinear Control,* 10(2):101–111, February 2000.

275. T. Takagi and M. Sugeno. Fuzzy identification of systems and its applications to modeling and control. *IEEE Transactions on Systems, Man, and Cybernetics,* 15:116–132, January/February 1985.

276. K. Tanaka. Stability and stabilizability of fuzzy-neural-linear control systems. *IEEE Transactions on Fuzzy Systems,* 3(4):438–447, November 1995.

277. K. Tanaka and M. Sugeno. Stability analysis and design of fuzzy control systems. *Fuzzy Sets and Systems,* 45(2):135–156, January 24, 1992.

278. D. W. Tank and J. J. Hopfield. Simple 'neural' optimization networks: An A/D converter, signal decision circuit, and a linear programming circuit. *IEEE Transactions on Circuits and Systems,* CAS-36(5):533–541, 1986.

279. M. C. M. Teixeira and S. H. Żak. Stabilizing controller design for uncertain nonlinear systems using fuzzy models. *IEEE Transactions on Fuzzy Systems,* 7(2):133–142, April 1999.

280. T. Terano, K. Asai, and M. Sugeno. *Fuzzy Systems Theory and Its Applications*. Academic Press, San Diego, CA, 1992.

281. J. M. T. Thompson and H. B. Stewart. *Nonlinear Dynamics and Chaos*. John Wiley & Sons, Chichester, England, 1986.

282. Y. I. Toptscheev and A. P. Tsyplyakov. *Problem Solver in the Theory of Automatic Control.* Mashinostroyeniee, Moscow, 1977. In Russian.

283. J. G. Truxal. *Introductory System Engineering*. McGraw-Hill, New York, 1972.

284. V. I. Utkin. Variable structure systems with sliding modes. *IEEE Transactions on Automatic Control,* AC-22(2):212–222, April 1977.

285. V. I. Utkin. *Sliding Modes and Their Application in Variable Structure Systems*. Mir Publishers, Moscow, 1978.

286. V. I. Utkin. *Sliding Modes in Control and Optimization*. Springer-Verlag, Berlin, 1992.

287. I. Varga, G. Elek, and S. H. Żak. On the Brain-State-in-a-Convex-Domain neural models. *Neural Networks,* 9(7):1173–1184, October 1996.

288. D. J. Velleman. *How to Prove It: A Structured Approach*. Cambridge University Press, Cambridge, Great Britain, 1994.

289. M. Vidyasagar. *Nonlinear Systems Analysis,* second edition. Prentice-Hall, Englewood Cliffs, NJ, 1993.

290. T. L. Vincent. Control using chaos. *IEEE Control Systems,* 17(6):65–76, December 1997.

291. B. L. Walcott, M. J. Corless, and S. H. Żak. Comparative study of non-linear state-observation techniques. *International Journal of Control,* 45(6):2109–2132, 1987.

292. B. L. Walcott and S. H. Żak. State observation of nonlinear uncertain dynamical systems. *IEEE Transactions on Automatic Control,* AC-32(2):166–170, February 1987.

293. H. O. Wang, K. Tanaka, and M. F. Griffin. An approach to fuzzy control of nonlinear systems: Stability and design issues. *IEEE Transactions on Fuzzy Systems,* 4(1):14–23, February 1996.

294. L.-X. Wang. Stable adaptive fuzzy control of nonlinear systems. *IEEE Transactions on Fuzzy Systems,* 1(2):146–155, May 1993.

295. L.-X. Wang. *Adaptive Fuzzy Systems and Control; Design and Stability Analysis*. PTR Prentice-Hall, Englewood Cliffs, NJ, 1994.

296. L.-X. Wang. *A Course in Fuzzy Systems and Control*. Prentice-Hall PTR, Upper Saddle River, NJ, 1997.

297. R. Weinstock. *Calculus of Variations: With Applications to Physics and Engineering*. Dover Publications, New York, 1974.

298. J. R. Wertz, editor. *Spacecraft Attitude Determination and Control*. Astrophysics and Space Science Library, Volume 73. Kluwer Academic Publishers, Dordrecht, The Netherlands, 1997.

299. A. B. Will and S. H. Żak. Modeling and control of an automated vehicle. *Vehicle System Dynamics,* 27(3):131–155, March 1997.

300. J. L. Willems. *Stability Theory of Dynamical Systems*. John Wiley & Sons, New York, 1970.

301. W. A. Wolovich. *Linear Multivariable Systems*. Springer-Verlag, New York, 1974.

302. W. A. Wolovich. *Automatic Control Systems: Basic Analysis and Design*. Saunders College Publishing, Harcourt Brace College Publishers, Fort Worth, TX, 1994.

303. J. Y. Wong. *Theory of Ground Vehicles,* second edition. John Wiley & Sons, New York, 1993.

304. W. M. Wonham. On pole assignment in multi-input controllable linear systems. *IEEE Transactions on Automatic Control,* AC-12(6):660–665, December 1967.

305. T. Yang, C.-M. Yang, and L.-B. Yang. A detailed study of adaptive control of chaotic systems with unknown parameters. *Dynamics and Control,* 8(3):255–267, July 1998.

306. Z.-J. Yang, T. Hachino, and T. Tsuji. On-line identification of continuous time-delay systems combining least-squares techniques with a genetic algorithm. *International Journal of Control,* 66(1):23–42, January 10, 1997.

307. G. Yen and A. N. Michel. A learning and forgetting algorithm in associative memories: Results involving pseudo-inverses. *IEEE Transactions on Circuits and Systems,* 38(10):1193–1205, 1991.
308. X. Yu. Controlling chaos using input-output linearization approach. *International Journal of Bifurcation and Chaos,* 7(7):1659–1664, July 1997.
309. L. A. Zadeh. Fuzzy sets. *Information and Control,* 8(3):338–353, June 1965.
310. S. H. Żak and S. Hui. On variable structure output feedback controllers for uncertain dynamic systems. *IEEE Transactions on Automatic Control,* 38(10):1509–1512, October 1993.
311. S. H. Żak and S. Hui. Output feedback variable structure controllers and state estimators for uncertain/nonlinear dynamic systems. *IEE Proceedings—Control Theory Applications,* 140(1):41–50, January 1993.
312. S. H. Żak, W. E. Lillo, and S. Hui. Learning and forgetting in generalized Brain-State-in-a-Box (BSB) neural associative memories. *Neural Networks,* 9(5):845–854, July 1996.
313. S. H. Żak and C. A. MacCarley. State-feedback control of non-linear systems. *International Journal of Control,* 43(5):1497–1514, 1986.
314. S. H. Żak, V. Upatising, W. E. Lillo, and S. Hui. A dynamical systems approach to solving linear programming problems. In K. D. Elworthy, W. N. Everitt, and E. B. Lee, editors, *Differential Equations, Dynamical Systems, and Control Science: A Festschrift in Honor of Lawrence Markus,* Chapter 54, pages 913–946. Lecture Notes in Pure and Applied Mathematics, Volume 152. Marcel Dekker, New York, 1994.
315. M. Zamparelli. Genetically trained cellular neural networks. *Neural Networks,* 10(6):1143–1151, August 1997.
316. M. Zeitz. Comments on 'Comparative study of non-linear state-observation techniques.' *International Journal of Control,* 46(5):1861–1863, 1987.
317. Y. Zeng and S. N. Singh. Adaptive control of chaos in Lorenz system. *Dynamics and Control,* 7(2):143–154, April 1997.
318. A. S. I. Zinober, editor. *Deterministic Control of Uncertain Systems.* Peter Peregrinus, London, 1990.
319. M. Zribi and J. Chiasson. Position control of a PM stepper motor by exact linearization. *IEEE Transactions on Automatic Control,* 36(5):620–625, May 1991.
320. J. M. Zurada. *Introduction to Artificial Neural Systems.* West Publishing, St. Paul, MN, 1992.

INDEX